WITHDRAWN AND
SOLD BY
STOKE-ON-TRENT
CITY ARCHIVES
Reference Library

HORACE BARKS
Reference Library

INDUSTRIAL
CERAMICS

HORACE BARKS
Reference Library

INDUSTRIAL CERAMICS

FELIX SINGER
Dr. Ing., Dr. Phil.,
M.I.Chem.E., F.I. Ceram.

and

SONJA S. SINGER
M.A., B.Sc.

1963
CHAPMAN & HALL LTD.
LONDON

© SONJA S. SINGER 1963

First published 1963

Published by
Chapman & Hall Ltd.
37 Essex Street
London, W.C.2

Cat. No. 4/661

*Made and printed in Great Britain by
William Clowes and Sons, Limited, London and Beccles*

PREFACE

ALTHOUGH many new books have appeared in recent years dealing in detail with individual sections of ceramics, it was felt that there was a need to try to draw together the threads, and discuss as many aspects of ceramics as possible, in one volume. The word 'ceramics' from the Greek 'keramos', 'the Potter' has, however, come to mean a rather ill-defined number of subjects. The field considered in this book has therefore to be defined at the outset, and our definition is as follows:

Ceramics are products made from inorganic materials which are first shaped and subsequently hardened by heat.

This definition embraces the European use of the word, meaning ware made with clay, while allowing inclusion of non-clay new developments. It does, however, exclude the chemically related subjects of glass, enamels and cements which may be included in the term ceramics in the United States.

The word 'industrial' in the title is generally taken to mean any ceramic product made in a factory, and usually made in considerable numbers. The hand making of individual pieces by artists is not included, although some of the illustrations are perhaps borderline cases.

The text has been compiled largely from my father's large and well classified library, collected over fifty years, and also embodies his long practical experience as a ceramic consultant. It is very sad that after so many years of collecting data followed by several years of selecting from it he did not live to see the final completion of the manuscript, but I assure readers who remember him well that the choice of matter to be included was very largely his, and that he personally approved the bulk of the text. On the other hand I take responsibility for any errors that have not been eradicated, due to my desire to have the book published as quickly as possible in memory of him.

It would be impossible to list here the many people and firms who have helped us by allowing us to quote from their work, sent us information, or made illustrations available. Their names are given in the bibliography, which is arranged alphabetically, thus combining it with the author index.

Where machines, apparatus or products are described or illustrated these are not necessarily the best for the purpose, but it is hoped that they are representative of the types under consideration. Availability of information has often been a factor governing what has been included.

The units used throughout are the British ones with the metric units in brackets (except where Continental items are described when the metric unit may be given first). This applies also to pyrometric cone numbers, and where a foreign cone does not correspond exactly to a British cone the two neighbouring ones are given. A few tables that do not belong exclusively to any particular chapter are given at the end in the appendix.

Finally I want to thank most particularly my late mother and my husband for their encouragement, assistance and also their prolonged tolerance of the large amount of books and papers that have been amassed in their respective homes during the writing of this book.

South Croydon
March 1960

SONJA S. SINGER

CONTENTS

	Page
Preface	v
Introduction	xi

PART I

THE GENERAL PRINCIPLES OF CERAMIC MANUFACTURE

Chapter 1 **The Raw Materials**	3
Introduction to silicate chemistry	3
Plastic raw materials—Clays	9
Non-clay plastic raw materials—Talc	89
Non-plastic raw materials—Refractories—Fluxes—Colouring agents	96
Auxiliary raw materials	150
Chapter 2 **The Action of Heat on Ceramic Raw Materials**	167
Changes and reactions in the solid state	167
Melting, crystallisation and glass formation	198
Phase diagrams	215
Thermal stresses, fatigue, thermal shock	223
Colour	227
Chapter 3 **Winning and Purification of Clays**	236
China clay and kaolin	237
Sedimentary clays	253
Machinery used in clay winning	258
Treatment of clays	263

CONTENTS

	Page
Chapter 4 **The Ceramic Laboratory**	265
Chemical analysis	266
Physical tests	275
Investigation of:	
Clays	296
Fired properties	338
Glazes	369
Process control	380
Chapter 5 **Ceramic Bodies**	393
Chapter 6 **Glazes**	525
Glazes	525
Salt glazes	598
Ceramic colourants	605
Chapter 7 **The Mechanical Preparation of Ceramic Bodies and Glazes**	661
Crushing and grinding	661
Size grading	686
Mixing	696
Filtering	701
Wedging	708
Chapter 8 **Shaping**	716
Shaping of plastic bodies	717
Dry and semi-dry pressing	741
Casting	754
Chapter 9 **Glazing and Decorating**	777
Chapter 10 **Drying of Shaped Ware**	832
Dryers	834
Chapter 11 **Firing and Kilns**	854
Firing of ceramic ware	854
Kilns for firing ceramic ware	884
Choice of fuel	899

Instruments for observing, recording and controlling kiln and other works conditions	934
Saggars and kiln furniture	954
The setting of ware in kilns	956
Periodic or intermittent kilns	967
Continuous kilns (fire moving)	985
Tunnel kilns	1002
Frit kilns	1047

PART II
CERAMIC PRODUCTS

Chapter 12	**Ceramic Building Materials**	1065
Chapter 13	**Ceramics in the Home**	1089
Chapter 14	**Chemical and Technical Ceramics**	1103
Chapter 15	**Specialised Laboratory and Engineering Ware**	1135
Chapter 16	**Ceramics in the Electrical Industry**	1170
Chapter 17	**Constructional Refractories**	1212
Chapter 18	**Thermal Insulators**	1284

Appendixes

	Periodic Table	1300
I	Comparison of Standard Fine Test Sieves	1301
II	Weight solids, % water, % solids for given slip weight per pint	1302
III	Thermoscopes or heat work recorders	1304
IV	Ambiguous words	1307
V	British Standard Specifications	1308
	Normes Françaises	1311
	Deutsche Industrie Normen	1313
	American Society for Testing Materials, Standards	1317
VI	Health hazards and Factory Acts	1322

Bibliography and Authors Index   1327

Subject Index   1431

Introduction

THE word ceramics is taken to cover those articles that are made from inorganic substances first shaped and then hardened by fire. In ancient times this meant articles made from clay. In this century we have found out how to use ceramic production methods for a number of physically and chemically different substances, but the ceramic industry is still founded on a working knowledge of clay.

Chemically, the classical raw materials, clays, flint and feldspar, are compounds of silica. The properties which make them suitable for ceramics are precisely those by which they differ from other substances. It is therefore well worth considering the basic structures and properties of silicates before proceeding to the raw materials themselves.

PART I
The General Principles of Ceramic Manufacture

Chapter 1
The Raw Materials

INTRODUCTION TO SILICATE CHEMISTRY

ALTHOUGH silica is the commonest constituent of the earth's crust, the study of it and its compounds baffled investigators for many decades. There are several reasons for this. Firstly, silicates are almost all insoluble in anything except hydrofluoric acid, so that they cannot be separated or investigated by solution methods. Secondly, their thermal reactions of transition, inversion, melting or freezing are sluggish and ill-defined so that no information about compounds and their purity can be inferred from thermal curves. Thirdly, the distinctions between compounds and solid solutions and mixtures are not clear and phase diagrams are more difficult to plot than usual.

This resistance to the classical chemical methods of investigation in itself shows the silicates to be different types of compounds from the normal oxides, acids, bases and salts of inorganic chemistry.

Twentieth-century chemistry, physics and, in particular, X-ray crystallography, have now enabled us to elucidate the problem sufficiently to understand how it arose.

The Chemical Properties of the Silicon Atom

Silicon's place among the elements, determined by its atomic number and, therefore, chemical properties, is expressed in the Periodic Table (see Appendix 1300). It is a tetravalent element forming predominantly covalent bonds. These are directional and in the case of silicon are normally directed to the corners of a tetrahedron of which the silicon atom is the centre. Its atom is small so that only four other atoms can be packed round it; thus silicon has a coordination number of four.

Silicon atoms have no affinity for each other, and they do not form chains like carbon atoms. But silicon has a great affinity for oxygen. Oxygen is divalent so that the simplest theoretical silicon oxide molecule is

$$O=Si=O.$$

For the silicon molecule to take the above configuration, two bonds, normally some 109° apart, would have to be parallel. This is virtually impossible with silicon (although it occurs with carbon) so that silicon tends to combine with four half oxygen atoms instead of two whole ones. Four unsatisfied oxygen valencies are left, which can further combine with silicons.

The ability to form Si—O—Si—O—Si chains of undefined length is the basis of almost all the silicates, and its preponderance over all other combinations is unique amongst the elements.

The Building Up of Silicates

The fundamental unit of silicate chemistry then is the silicon–oxygen tetrahedron. This is found as such in the simplest natural silicates, the orthosilicates. These are crystalline ionic compounds of the tetravalent $[SiO_4]^{4-}$ anion with cations, and the commonest of these is olivine, containing about 85% magnesium and about 10% iron orthosilicates. This highly refractory mineral, softening point about 1700° C (3090° F), occurs but rarely near the earth's surface, but when it does it is found in a vast deposit. V. M. Goldschmidt,* however, believed that below the rocks comprising the accessible crust there is a great quantity of olivine, which must have crystallised out first, from which it can be inferred that the orthosilicate structure is very stable.

The SiO_4 tetrahedron is also the basic unit for building up the more complex silicates. Its representation presents some difficulty. The best method is to assume that atoms in combination are approximately spheres in close-packing. The average radii for these spheres have been found (**P20**); the easiest way to make use of them is, of course, to make solid models. Failing this, however, a comparatively simple arrangement can be represented by a perspective drawing with a portion cut away (Fig. 1.1(*a*)). Where

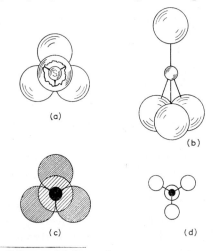

Fig. 1.1. *Representation of the $[SiO_4]^{4-}$ unit: (a) cutaway perspective view; (b) expanded perspective view; (c) top view of tetrahedron with top oxygen appearing transparent to show location of the silicon atom in the cavity formed by the four oxygens; (d) non-perspective expanded view (after Hauth, **H49**, and Hauser, **H43**)*

---

* Personal information.

confusion could arise from this an expanded view is sometimes given (Fig. 1.1(*b*)). In this and other diagrammatic presentations it must be remembered that the 'bond' separating the spherical atoms is there for visual convenience only and does not represent a physical fact.

Simpler non-perspective diagrammatic forms are shown in Figs. 1.1(*c*) and (*d*). The last (*d*) is the commonest schematic form.

The fundamental orthosilicate $[SiO_4]^{4-}$ unit is rarely found independently (olivines, chondrodite series, phenacite, garnet and the sillimanites). The unit readily joins up to form rings, chains, bands, sheets and three-dimensional networks. The oxygen atoms that are bonded to only one silicon bear a single negative charge. As the total charge is zero these negative charges must be balanced by cations, thus forming silicates. The double unit $[Si_2O_7]^{6-}$ is rare and does not occur in materials of ceramic interest. Of the various possible cyclic complexes only the $[Si_3O_9]^{6-}$ occurring in benitoite $BaTiSi_3O_9$ (**Z1**), wollastonite $CaSiO_3$ (**B12**), and catapleite $Na_2ZrSi_3O_9.2H_2O$ (**B128**), and the $[Si_6O_{18}]^{12-}$ in beryl $Be_3Al_2Si_6O_{18}$ (**B106**) are known to exist as discrete ions.

The next degree of complexity is found in the chain ions. The pyroxenes have twisted single chains of SiO_4 units giving the composition $(SiO_3)_n$, they include enstatite $MgSiO_3$ and diopside $CaMg(SiO_3)_2$ (**W9**), jadeite $NaAl(SiO_3)_2$ and spodumene, $LiAl(SiO_3)_2$ (Fig. 1.2).

The amphiboles have double chains and hence the composition $(Si_4O_{11})_n$ (Fig. 1.3).

By sharing three of the four oxygen atoms of the SiO_4 unit with other units, silicon–oxygen sheets can be built up. Two kinds of these are known, one containing rings of six silicons, and the other with alternate four-rings and eight-rings. The former is of major importance in the structures of micas and clay minerals (the latter is found in the mineral apophyllite) (Fig. 1.4).

The three-dimensional network $(SiO_2)_n$ is found in pure silica, feldspars and zeolites.

An essential feature of silicon–oxygen structures is that the oxygen atoms are bound by the silicons and take up positions leaving spaces of various sizes. This is in contrast with ionic oxides as will be seen. The different metal silicates are formed by balancing the negative charge of the silica skeleton with positively charged metal ions. These fit into the holes in the network.

The three-dimensional network can be so arranged as to have no surplus charge (except at the edge of a crystal); it is then pure silica. Here the actual arrangement of the SiO_4 tetrahedra has three possibilities, giving quartz, tridymite and cristobalite, and the Si—O—Si angle can vary, giving the low- and high-temperature forms of each isomorph. The changes between quartz, tridymite and cristobalite involve breaking and remaking of Si—O bonds; these changes are known to be very sluggish. The changes between the α- and β-forms of each variety involve only a small rotation of the atoms and are quick (Fig. 1.5).

FIG. 1.3. *Silicate band* $[(Si_4O_{11})^{5-}]n$ *(Hauth,* **H49**)

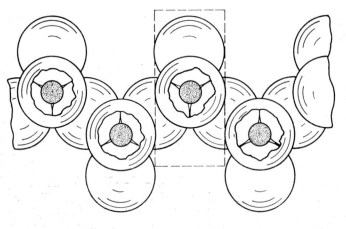

FIG. 1.2. *Silicate chain* $[(SiO_3)^{2-}]n$, *perspective cutaway view (Hauth,* **H49**)

THE RAW MATERIALS

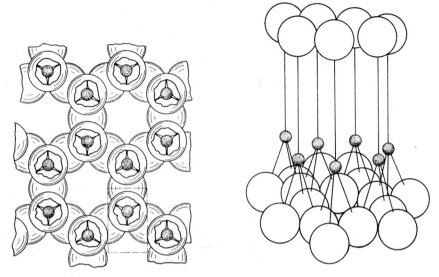

(a) With the unit $[(Si_2O_5)^{2-}]n$, cutaway and expanded views (Hauth, **H49**)

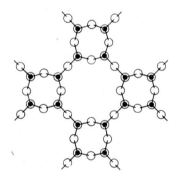

(b) With 4- and 8-rings (Wells, **W40**)

FIG. 1.4. *Silicate layers*

There is, strictly speaking, therefore, no independent molecule of any silicate (other than the orthosilicates). The ordered structure of a crystalline silicate continues to the edge of the crystal. The chemical formulae ascribed to such compounds are merely expressions of the ratios of different atoms present.

These structures amply illuminate the observed properties of the silicates, their resistance to chemical attack and particularly their sluggish melting and freezing. The melting point of a silicate may be so undefined that the term has been replaced by the pyrometric cone equivalent P.C.E., the temperature at which it attains a certain viscosity. In fact the solid merely softens and gradually becomes less and less viscous. Because of the unsaturated nature

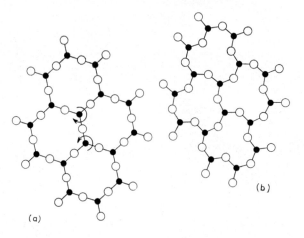

Fig. 1.5. *Temperature inversion between α- and β-forms of a silica isomorph (Hauth, **H49**)*

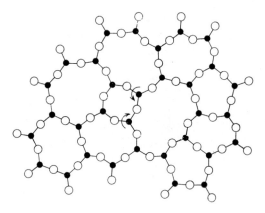

Fig. 1.6. *The semi-random arrangement of a vitreous structure*

of the SiO_4 basic units, bonds are re-formed and re-broken constantly in the melt. When it cools it becomes more and more viscous so that the units have increasing difficulty in arranging themselves in the ordered crystalline lattice network and easily join up with neighbouring units in a random network. This action produces a glass (Fig. 1.6).

Strangely enough, although impurities often prevent proper crystallisation, the presence of a trace of certain substances, termed 'mineralisers', in a silicate melt will cause it to crystallise instead of forming a glass. It is such impurities that caused the crystallisation of the mineral silicates of the primary rocks on the earth's crust.

THE RAW MATERIALS

Other Major Structural Constituents of Silicates

ALUMINIUM

The aluminium atom has a dual role. It can only become a positively charged trivalent ion with difficulty, and so quite easily forms three covalent bonds like the four bonds of silicon. Its size is such that although the aluminium atom should be surrounded by six oxygen atoms, it can fit into the hole left in the middle of a tetrahedral oxygen arrangement.

The mineral gibbsite $Al(OH)_3$ has a layer structure that recurs in several of the clay minerals (Fig. 1.7).

MAGNESIUM

Magnesium is divalent. It is larger than aluminium and is always in octahedral coordination (surrounded by six oxygen atoms). The mineral brucite, $Mg(OH)_2$, has a layer structure (Fig. 1.8). Such layer structures occur combined with silica layers in a number of minerals of major importance in ceramics, as will be seen.

Isomorphous Substitution

The large structures of suitably combined $[SiO_4]^{4-}$ tetrahedra, $[AlO_6]^{9-}$ and $[MgO_6]^{10-}$ octahedra may also be considered as oxygen atoms so stacked that the holes between them are either tetrahedrally or octahedrally surrounded. These are then filled with atoms, the small ones in the small holes and the large ones in the large holes until the total charge is zero. Thus in an oxygen network of charge $12-$, one could insert $4Al^{3+}$ or $3Si^{4+}$. A crystalline structure is obtained when the cations are in regularly repeated positions.

Bearing this theoretical concept in mind it is easy to see that one atom can be substituted by another of the same size. This is isomorphous substitution. The substituting ion may or may not be of the same valency. If it is not, some other adjustment such as loss of a hydrogen or addition of a sodium occurs.

Thus it is often found that an aluminium atom has replaced a silicon atom in a tetrahedral hole. As aluminium is trivalent, whereas silicon is tetravalent, the charge is neutralised by an additional univalent ion, *e.g.* sodium, which fits in one of the larger holes (Fig. 1.9).

PLASTIC RAW MATERIALS

Introduction

In primitive and early times the only raw materials for pottery were the natural plastic clays. In modern ceramics many other raw materials play an important role, but that of clay is still a major one. The term 'clay' is

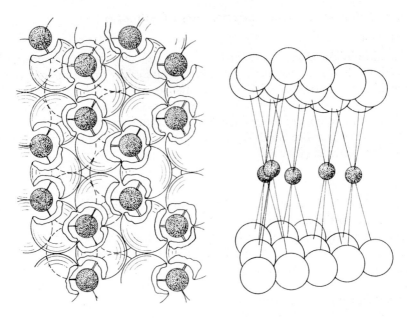

Fig. 1.7. *Gibbsite structure* [Al(OH)$_3$]; left, *cutaway view*; right, *expanded perspective view. Large spheres are hydroxyls* (OH) *and small spheres, aluminiums* (Hauth, **H49**)

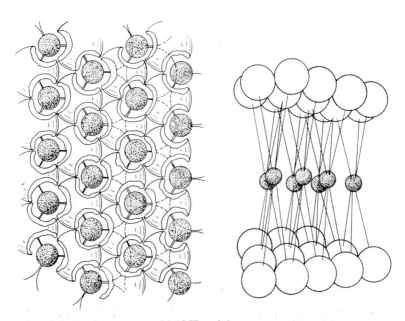

Fig. 1.8. *Brucite structure,* Mg(OH)$_2$; left, *cutaway view*; right, *expanded perspective view. Large spheres are hydroxyls* (OH) *and small spheres are magnesiums* (Hauth, **H49**)

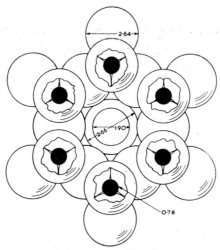

Fig. 1.9. *Cutaway view of hexagonal arrangement of silicon–oxygen tetrahedra. The two circles in the centre space represent potassium and sodium ions. The dimensions marked are in Ångström units (Hauser, H43)*

applied to those natural earthy deposits which possess the singular property of plasticity. This property, so easily detected and yet so hard to define, will be discussed later.

Clays occur in deposits of greatly varying nature in many parts of the world. No two deposits have exactly the same 'clay' and frequently different samples of clay from the same deposit differ. It is therefore worth while to give brief consideration to the origin and mineralogy of clay. Firstly, clay is a secondary 'rock', that is, it has been formed by weathering of certain other rocks. Secondly, clay is a mixture. (**P14**)

Geology

Primary igneous rocks that gave clays on weathering were the granites, gneisses, feldspars, pegmatites, etc. The weathering of these primary rocks was achieved by the mechanical action of water, wind, glaciers and earth movements working together with the chemical action of water, carbon dioxide, humic acids and, more rarely, sulphurous and fluorous gases, assisted by elevated temperatures.

The weathered rocks have in some cases remained in their original position. These are the residual clays. More frequently the weathering agents and other influences have transported the small particles and deposited them elsewhere. During the transportation sorting by size takes place, also mixing in of weathering products from other sources occurs. The deposits from water are always layered. Deposits from wind transportation are known as 'loess'; they are not stratified and have a much more porous and crumbly structure.

Mineralogy

The basic rocks from which clays are formed are complex aluminosilicates. During the weathering these become hydrolysed, the alkali and alkaline earth ions form soluble salts and are leached out, the remainder consists of hydrated aluminosilicates of varying composition and structure, and free silica. This remainder is therefore more refractory than the original igneous rock. Unchanged rock particles, *e.g.* feldspar, mica and quartz remain in the clay, too.

This process can be represented by chemical equations, *e.g.*

(1) $K_2O.Al_2O_3.6SiO_2 + 2H_2O \rightarrow Al_2O_3.6SiO_2.H_2O + 2KOH$ hydrolysis
    feldspar
(2) $Al_2O_3.6SiO_2.H_2O \rightarrow Al_2O_3.4SiO_2.H_2O + 2SiO_2$ desilication
    pyrophyllite
(3) $Al_2O_3.6SiO_2.H_2O \rightarrow Al_2O_3.2SiO_2H_2O + 4SiO_2$ desilication
(4) $Al_2O_3.2SiO_2.H_2O + H_2O \rightarrow Al_2O_3.2SiO_2.2H_2O$ hydration
    kaolinite
(5) $Al_2O_3.2SiO_2.H_2O \rightarrow Al_2O_3.H_2O + 2SiO_2$ desilication
    diaspore
(6) $Al_2O_3.H_2O + 2H_2O \rightarrow Al_2O_3.3H_2O$ hydration
    gibbsite

The liberated silica is probably hydrated (**N36**).

This is a convenient way of obtaining a picture of the processes. It must, however, be remembered that these chemical formulae have little physical significance as most of the substances under consideration exist as giant molecules. The actual weathering process must be highly complicated. (In general, equations will not be given for any but the most straightforward reactions.)

The hydrated silicates of aluminium are the 'clay substance' which gives the clays their main defined characteristics. One of the predominant properties of these substances is the extreme fineness of their particles. This factor, so vital to their physicochemical nature, was for a long time a major stumbling block to investigation. With the aid of the microscope, the electron microscope, X-ray diffraction and differential thermal analysis it has now been established that clay particles are extremely small flake-like particles of *crystalline* minerals.*

A number of these clay minerals have now been investigated and the knowledge obtained makes it possible to divide them into groups. The actual structures of the clay minerals have been very clearly put forward by Schofield (**S27**) and Jasmund (**J41**).

Basically the structures are dominated by the distribution of the

---

* It will be remembered that at first all 'clay substance' was considered to be amorphous, later only those fine particles small enough to be colloidal were so classified. Modern concepts of molecular structure make it difficult to conceive an amorphous nature for giant molecules and the X-ray powder technique has proved the crystalline nature of all clay particles.

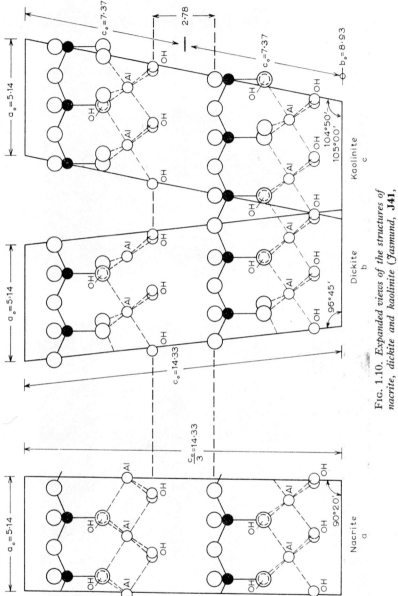

Fig. 1.10. *Expanded views of the structures of nacrite, dickite and kaolinite (Jasmund, **J41**, after Gruner, **G93**)*

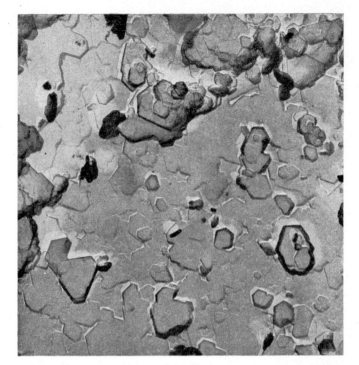

1.11 (a)

1.11 (b)

1.11 (c)

FIG. 1.11

(a) Kaolinite from Langley, South Carolina. Fresh fracture. Electron micrograph of preshadowed carbon replica. × 19 300 (Bates and Comer, **B20**)

(b) Kaolinite from Langley, South Carolina. Fresh fracture, showing edge view of kaolinite book. Electron micrograph of preshadowed carbon replica. × 30 900 (Bates and Comer, **B20**)

(c) Dickite, Schuykill County, Penn. Electron micrograph of preshadowed carbon replica. × 16 400 (Bates and Comer, **B20**)

commonest and largest atoms, namely oxygen. In between them are tetrahedral, octohedral and polyhedral spaces. The way in which these are filled determines how they are attracted and held together. The five main clay mineral groups are outlined below.

The simplest clay mineral group is the *kaolinite* group. This includes: kaolinite, dickite, nacrite; anauxite; halloysite, high- and low-temperature forms; livesite. Their basic structure consists of oxygen atoms arranged to give alternate layers of tetrahedral holes and octahedral holes. Where these layers are filled with silicon in the tetrahedral holes and aluminium in two-thirds of the octahedral ones we get the common mineral kaolinite, and the more perfect and rarer minerals dickite and nacrite (Figs. 1.10 and 1.11). Where silicons replace aluminiums in octrahedral spaces (hydrogens being expelled to keep the charge right) a continuous series is obtained with anauxite, $Al_2O_3.3SiO_2$, as the end member. These crystals are all thin hexagonal plates. The basal spacing is 7·2 Å. They tend to become stacked on top of each other and loosely cemented into aggregates.

Halloysite crystals are elongated. The adjacent oxygen layers are able to take up a unimolecular water layer, making the basal spacing 10 Å. A continuous series from kaolinite to halloysite exists, intermediate members are termed livesite (Figs. 1.12, 1.13). Inside the crystals there is no excess charge. On the surface distortions occur and excess negative charges may be set up. Positive ions are then adsorbed to neutralise it.

The livesite of English fireclays is described as a kaolin type of mineral randomly oriented along the b-axis, and with ultimate particles much smaller

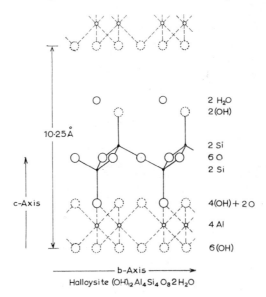

Fig. 1.12. *Schematic representation of the crystal structure of halloysite*
(Grim **G79** after Edelman and Favejee, **E3**)

Fig. 1.13. *Electron micrograph of surface replica of halloysite from Wendover, Utah, showing tubular morphology of crystals.* × 84 300 (*Bates and Comer*, **B20**)

than those of kaolinite from china clays, a proportion of them being less than 0·1 μ. The structure of the crystals is further disordered by the substitution of Fe^{3+} and Mg^{2+} ions for Al^{3+}, leaving an inherent negative in the structure. The mineral therefore readily attracts cations, mostly Ca^{2+}, to balance this charge (**W88**).

The *montmorillonite* group of clay minerals present a very different picture:

Montmorillonite (Fig. 1.14),
$Al_2O_3.4SiO_2.H_2O + Aq$, or $(MgCa)O.Al_2O_3.5SiO_2.nH_2O$.
Beidellite, $(Mg, Ca)O.Al_2O_3.4SiO_2.4H_2O$, or $Al_2O_3.3SiO_2,nH_2O$.
Saponite, $2MgO.3SiO_2.nH_2O$.
Stevensite (perhaps a talc-saponite interlayered mineral (**B116**).
Nontronite, $(Al.Fe)O_3.3SiO_2.nH_2O$.
(Sauconite), $2ZnO.3SiO_2.nH_2O$.

Fig. 1.14. (a) *Montmorillonite from Montmorillon, France. Electron micrograph of preshadowed carbon replica.* × 20 225 (Bates and Comer, **B20**)

There are two layers of tetrahedral holes to every one of octahedral ones. These minerals have the common property of absorbing large quantities of water between adjacent layers, changing the basal spacing from 10 Å to 20 Å. The oxygen lattice is such that its spaces may be filled by different atoms; the octahedral spaces may have aluminium, magnesium, ferric ion or zinc, the tetrahedral may have silicon or aluminium. Larger spaces that will accommodate alkali cations are also present. The charges are often unbalanced and large numbers of cations are adsorbed and may be easily exchanged (Fig. 1.15).

These structures make it easy to split the particles into very fine charged fragments ideally suited to go into colloidal sols.

The next group of clay minerals is the *illite*, or hydromica group (Fig. 1.16). These resemble the micas and they have larger spaces which contain cations to keep the charge neutral. Their finely divided state, however, makes many of these cations accessible for exchange. Unlike the montmorillonites water does not enter into the lattice itself and expand it, as adjacent layers are held together by potassium ions (Fig. 1.17).

The general formula is:

$$(OH)_4 K_y (Al_4 . Fe_4 . Mg_4 . Mg_6) Si_{8-y} . Al_y O_{20}$$

with y varying from 1 to 1·5.

 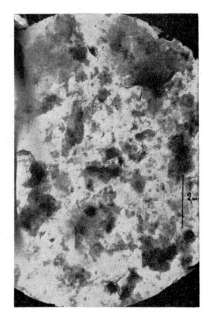

FIG. 1.14. (b) *Montmorillonite from Gaura, Siebengebirge, Germany*: left, coarse fraction $<2\mu$; right, fine fraction $<0.5\mu$. × 7000 (*Jasmund*, **J41**)

An unusual clay mineral is *attapulgite* which is fibrous. It is found only in Florida and Georgia (U.S.A.), fuller's earths and clays at Mormoiron, France (**G79**) (Figs. 1.18, 1.19). Suggested formulae are:

$(OH_2)_4(OH)_2Mg_5Si_8O_{20}.4H_2O$, some Mg replaced by Al (**B103**)
$Si_3O_{12}(Al_{\frac{4}{3}}.Mg_2)H_8$ (**D23**)

the former giving rise to the structural diagram shown in Fig. 1.20.

The term '*allophane*' is used to cover non-crystalline mutual solutions of silica, alumina and water. Although a constituent of clays this is not a true mineral. Future investigation will throw more light on this group.

These main groups divide the minerals according to their general properties as well as their chemical composition. It may be useful for general reference to include a table of theoretical composition. Unfortunately nomenclature of minerals is not universal and so a mineral may occur under two names. Some of the named minerals have already been proved to be mixtures, others may prove to be so. Table 1, however, is a guide to names one may come across.

The possible variations of chemical composition of a given mineral have been summarised by Engelhardt (**E19**), and are presented here in Table 2. Most natural clays are dominated by one clay mineral but also contain smaller quantities of a few others.

Clays also have a number of other constituents which are not in themselves plastic. The chief one is quartz, which, together with feldspar and mica,

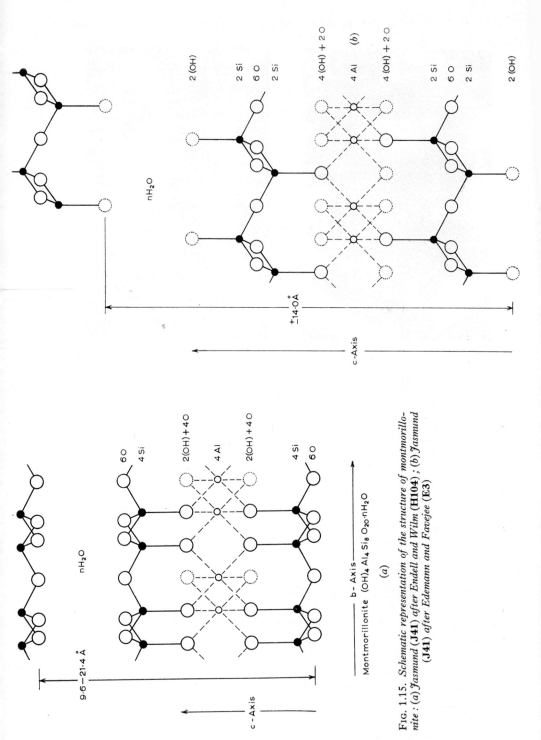

FIG. 1.15. *Schematic representation of the structure of montmorillonite:* (a) *Jasmund* (**J41**) *after Endell and Wiltm* (**H104**); (b) *Jasmund* (**J41**) *after Edemann and Favejee* (**E3**)

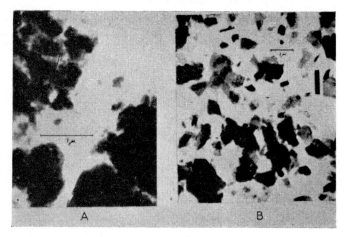

Fig. 1.16. *Illites*. A, from Fithian, Illinois. B, from N.E. Pennsylvania (*Jasmund*, **J41**, after Bates, **B19**)

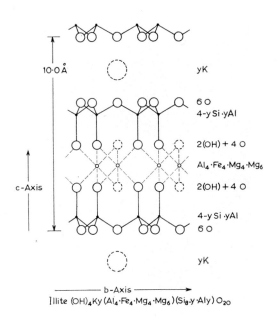

Illite $(OH)_4 Ky(Al_4 \cdot Fe_4 \cdot Mg_4 \cdot Mg_6)(Si_{8-y} \cdot Al_y)O_{20}$

Fig. 1.17. *Schematic representation of the crystal structure of illite* (Grim, Bray and Bradley, **G78**)

TABLE 1

Clay Mineral Nomenclature
(after Greaves-Walker (**G73**))

Mineral	Formula			$Al_2O_3:SiO_2$
Schroetterite	$8Al_2O_3$	$3SiO_2$	$30H_2O$	1:0.38
Collyrite	2	1	9	1:0.5
Allophane	1	1	5	1:1
Kochite	2	3	5	1:1.5
Kaolinite	1	2	2	1:2
Clayite	1	2	2	1:2
Nacrite	1	2	2	1:2
Dickite	1	2	2	1:2
Halloysite	1	2	2 + Aq	1:2
Newtonite	1	2	5	1:2
Anauxite	1	3	2	1:3
Leverrierite	1	3	3	1:3
Pholerite	1	3	4	1:3
Beidellite	1	3	5	1:3
Montmorillonite	1	4	1 + Aq	1:4
Pyrophyllite	1	4	1	1:4
Cimolite	2	9	6	1:4.5

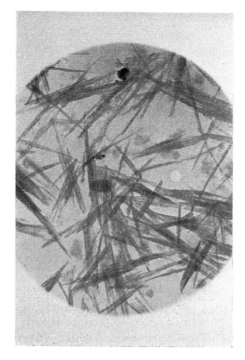

FIG. 1.18. *Attapulgite.* ×15 000 (*Jasmund*, **J41**)

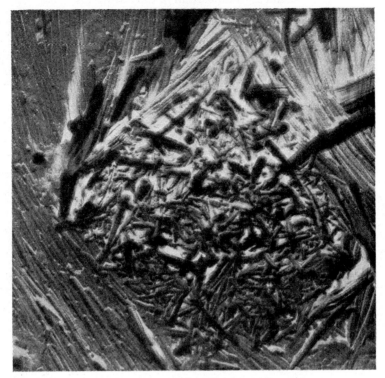

Fig. 1.19. *Attapulgite, Attapulgus, Georgia. Electron micrograph of pre-shadowed carbon replica, parallel fracture surface. Some end sections are seen.* × 41 600 (*Bates and Comer*, **B20**)

are unaltered remainders of the parent rock. The hydromicas are partially altered fragments. Under certain high-temperature conditions the weathering has proceeded beyond the clay stage to give free aluminium hydrates, namely gibbsite and diaspore; these are, however, unusual.

Iron compounds, often the oxides, are frequently present, and constitute the main colouring agents in clays. Many other minerals occur in clays, e.g. calcite $CaCO_3$, aragonite $CaCO_3$, dolomite $CaCO_3.MgCO_3$, gypsum $CaSO_4.2H_2O$, rutile TiO_2, tourmaline (a complex aluminium borosilicate), glauconite (a variable hydrous silicate of iron and potassium), hornblende (silicate of calcium and magnesium also containing iron, manganese, sodium and potassium), garnet (silicate minerals of special structure), vanadates, wavellite $Al_6(OH)_6(PO_4)_4.9H_2O$, manganese oxides, and vivianite $Fe_3(PO_4)_2.8H_2O$. Organic inclusions also occur and may play a very important role. Many of these are introduced during the transportation.

It is therefore hardly surprising that every clay is different. Classification of clays is a big task and leads to a different result according to the viewpoint taken, whether geological, mineralogical, with regard to properties or according to use. A geological classification, as made by Ries, gives some idea about

TABLE 2
Chemical Composition of Clay Minerals
(after Engelhardt (E19))

	Kaolinite	Halloysite	Montmorillonite
SiO_2	43·6 –54·7	40·0–45·8	47·9–51·2
Al_2O_3	30·0 –40·2	33·8–39·2	20·0–27·1
Fe_2O_3	0·3 – 2·0	0 – 0·4	0·2– 1·4
MgO	0 – 1·0	0·3	2·1– 6·6
CaO	0·03– 1·5	0·1– 0·8	1·0– 3·7
K_2O	0 – 1·5	0·3	0·2– 0·6
Na_2O	0 – 1·2	0·1– 0·2	0·3– 0·8
TiO_2	0 – 1·4	—	—
H_2O	11·0 –14·3	13·4–23·7	17·1–23·7

	Beidellite	Nontronite	Illite
SiO_2	45·3–47·3	31·1–47·6	50·1–51·7
Al_2O_3	12·2–27·8	0·4–22·7	21·7–32·8
Fe_2O_3	0·8–18·5	15·2–40·8	0 – 6·2
MgO	0·2– 3·0	0·1– 4·0	2·0– 4·5
CaO	0·5– 2·8	0·6– 4·5	0 – 0·6
K_2O	0·1	0·1– 0·4	6·1– 6·9
Na_2O	0·1– 1·0	0 – 0·2	0·1– 0·5
TiO_2	0·8	0 – 0·1	0·5
H_2O	17·3–22·6	5·1–13·0	6·4– 7·0

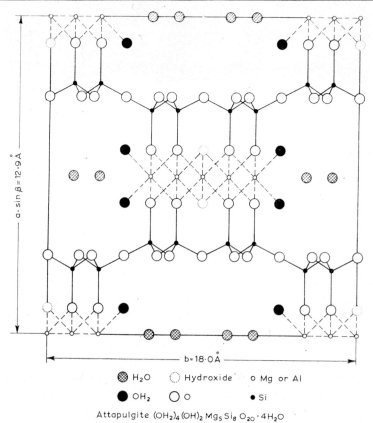

Attapulgite $(OH_2)_4 (OH)_2 Mg_5 Si_8 O_{20} \cdot 4H_2O$

FIG. 1.20. *Schematic representation of the crystal structure of attapulgite* (*Jasmund*, **J41**, *after Bradley*, **B103**)

THE RAW MATERIALS

the position as well as the nature of clays, and is a useful preliminary guide to the ceramic industry.

Ries' Classification (R28)

A. Residual clays. Formed in place by rock alteration due to various agents, of either surface or deep seated origin.
    I. Those formed by surface weathering, the processes involving solution, disintegration, or decomposition of silicates.
       (a) Kaolins, white in colour and usually white burning.

Parent Rock	Shape
Granite, pegmatite, rhyolite, limestone, shale, feldspathic quartzite, gneiss, schist, etc.	Blankets; tabular steeply dipping masses; pockets or lenses

       (b) Ferruginous clays, derived from different kinds of rocks.
    II. White residual clays formed by the action of ascending waters, possibly of igneous origin:
       (a) formed by rising carbonated waters;
       (b) formed by sulphate solutions;
    III. Residual clays formed by the action of downward percolating sulphate solutions.
    IV. White residual clays formed by replacement, due to action of waters, supposedly of meteoric origin (indianite).

B. Colluvial clays, representing deposits formed by wash from the foregoing and of either refractory or non-refractory character.

C. Transported clays.
    I. Deposited in water.
       (a) Marine clays or shales, deposits often of great extent;
          ball-clays, white burning clays;
          fireclays or shales, buff burning;
          impure clays or shales $\begin{cases} \text{calcareous.} \\ \text{non-calcareous.} \end{cases}$
       (b) Lacustrine clays (deposited in lakes or swamps);
          fireclays or shales;
          impure clays or shales, red-burning;
          calcareous clays, usually of surface character.
       (c) Floodplain clays, usually impure and sandy.
       (d) Estuarine clays (deposited in estuaries), mostly impure and finely laminated.
       (e) Delta clays.
    II. Glacial clays, found in the drift, and often stony. May be either red- or cream-burning.
    III. Wind-formed deposits (some loess).
    IV. Chemical deposits (some flint-clays).

Ernst, Forkel and Gehlen (**E22**) have put forward a method of shorthand description of clays on a mineralogical basis, as they consider this to be a scientific foundation from which properties and uses can be derived. For instance, although mineral content cannot be derived from chemical analysis, the reverse can give a fair value. Also they have shown that firing behaviour, *i.e.* P.C.E. of mixtures of minerals can be plotted on diagrams and then compared with phase rule data with which it does not coincide, the former being of greater use to the ceramist.

Their proposed system is to use initial capital letters to denote mineral groups, *e.g.* K, kaolinite, Q, Quartz, F, Feldspar, with index letters when the mineral is known more accurately. The letters are used in the order showing the relative quantities, and/or percentages are given. Added to these are values for M the 'half weight particle size' (*i.e.* such that the particles making up half the weight of material are smaller than this size and the other half bigger) and the surface factor.

Of greater industrial use is a classification according to the properties and therefore uses of the clays themselves as given by Norton (**N36**):

A. White-burning clays (used in whiteware).
   (1) Kaolins:
      (*a*) residual;
      (*b*) sedimentary.
   (2) Ball clays.

B. Refractory clays (having a fusion point above 1600° C but not necessarily white burning).
   (1) Kaolins (sedimentary).
   (2) Fire clays:
      (*a*) flint;
      (*b*) plastic.
   (3) High alumina clays:
      (*a*) gibbsite;
      (*b*) diaspore.

C. Heavy clay-products clays (of low plasticity but containing fluxes).
   (1) Paving brick clays and shales.
   (2) Sewer-pipe clays and shales.
   (3) Brick and hollow tile clays and shales.

D. Stoneware clays (plastic, containing fluxes).

E. Brick clays (plastic, containing iron oxide).
   (1) Terra-cotta clays.
   (2) Face and common brick.

F. Slip clays (containing more iron oxide).

We will now consider the different important clays.

Kaolin

The name is a corruption of the Chinese 'kao-liang', meaning 'high ridge', a local designation for the area where a white china clay was found. The kaolins or china clays, the latter expression often being reserved for the Cornish product, are white-burning clays, generally of low plasticity, and high refractoriness (cone 34 to 35) (1750–1770° C, 3182–3218° F). When mined they are rather siliceous, *e.g.* of the Cornish clay rock only about 13% is extracted as china clay. But after washing the chemical composition of the clay approximates to that of kaolinite; this mineral does in fact predominate in kaolins but others are present.

Many theories have been advanced about the conversion of feldspar to kaolin. The three most satisfactory being: (1) the igneous emanation theory; (2) the surface weathering theory; and (3) the bog or moor water theory.

According to the first theory igneous gases, originating in the centre of the earth and containing superheated steam, boron and fluorine compounds, carbon dioxide, etc., are the active agents. This theory accounts for deep deposits like the world-famous ones in Cornwall and West Devon, the depth of which is not known.

Kaolinisation due to surface weathering, *i.e.* downward percolating water containing carbon dioxide, is necessarily of limited depth, and there is a graduation from rock base to the fully weathered product. The china-clay deposits of Auvergne (France) and Passau (Germany) are supposed to have been formed in this way.

The fact that many of the German kaolin deposits are near beds of lignite points to the fact that the drainage water from bogs containing ammoniacal salts and organic acids may have been active kaolinisating agents.

The most famous European deposits are the Cornish ones, followed by those at Zettlitz near Karlsbad in Czechoslovakia, Kemmlitz, Börtewitz and Amberg in Germany.

In the United States the main deposits of residual kaolin lie in a band from Vermont to Georgia and up the Mississippi valley with a few scattered deposits in the West, the chief ones being near Spruce Pine in North Carolina. Sedimentary kaolins occur in the United States in South Carolina, Georgia and Florida.

Less than half of the kaolin produced is used for ceramics, the rest being employed as a filler in the paper, rubber, textiles and numerous other industries. Of that used in ceramics some goes to white-burning pottery and some to refractories.

Ball Clays

These are sedimentary plastic refractory clays which are dark in the unfired state because of organic impurities but burn white or cream coloured as long as they are not vitrified fully. They have a large proportion of kaolinite but also contain a variety of impurities and probably have some montmorillonite attached to the edges of the kaolinite platelets (**B83**). The name is derived from the English mining method of cutting the clay out in cubes or 'balls'.

There are three important English deposits of ball clays in Dorset, North Devon and South Devon. Notable European sources are those at Sezanne in France, those near Meissen in Germany, and those near Wildstein in Czechoslovakia. In the United States they occur in Florida, Tennessee, Kentucky, Alabama and New Jersey.

Ball clays are used in whitewares (earthenware, porcelain, etc.) to make the body more plastic and workable.

Stoneware Clays

Stoneware clays are refractory or semi-refractory but contain enough flux to fire to a dense body at comparatively low temperatures (*ca.* 1100° C). They are comparatively plastic without showing too much air- and fire-shrinkage. Stoneware clays include those clays that resemble ball clays in every respect except that they do not burn to a white product.

Fireclays

The use of this term has unfortunately become increasingly wide and thereby loose in its application. Strictly it should be applied only to refractory clays and shales which occur in hard masses that do not in their natural state take up water and become plastic, but on fine grinding will do so. True fireclays occur in Great Britain, Czechoslovakia and in the U.S.A.

Unfortunately the term 'fireclay' has been used to cover all types of clay deposited in swamps or coal basins, *i.e.* associated with coal measures, without regard to its fusibility or firing behaviour.

This has necessitated classification of fireclays. The primary classification is according to physical character into:

(1) Plastic fire clays.

(2) Semi-flint fire clays (**G73**). These resemble the plastic fireclays but develop plasticity only after working and are somewhat more refractory.

(3) Flint fire clays (the true fire clays). These hard clays break with a conchoidal fracture and are refractory. They find great use in the refractories industry.

(4) Nodular flint fire clays. Deposits of these clays are rare. They are flint clays containing nodules of gibbsite or other hydrous aluminium oxides and are therefore the most refractory.

Classification according to fusibility is also of importance, the Ries (**R28**) system being usually used.

(1) Highly refractory clays fusing above cone 33.

(2) Refractory clays fusing from cone 31–33 inclusive.

(3) Semi-refractory clays fusing between cones 27 and 30.

(4) Clays of low refractoriness, 'low heat duty clays', fusing between cones 20 and 26.

This classification deals only with the upper limit of use, the fusion point. It gives no guidance to the temperatures of incipient or complete vitrification which one needs to know for the manufacture of certain products.

Purdy and Moore (**P77**) and subsequently Purdy (**P79**) take this need into consideration in classifying clays according to changes in porosity and specific gravity curves for different temperatures.

No. 1 fireclays show a small regular decrease in porosity from the beginning of firing to above cone 11.

No. 2 fireclays show an increase in loss of porosity with consequent early vitrification, the change beginning about cone O2 and becoming marked with increasing temperature.

No. 3 fireclays show a marked change with increasing porosity at cone 4 and seldom have a fusion point above cone 16. (No. 3 fireclays would not be termed fireclays in Great Britain.)

Another practical classification on similar lines is given by Matthews (**M35**) who considers the temperatures at which the following occur:

(1) Some of the constituents fuse to a glass and cement the more refractory grains together, termed 'incipient vitrification'.

(2) Sufficient fusion occurs to close all the pores, the maximum shrinkage taking place, termed 'complete vitrification'.

(3) So much fusion takes place that the body is just no longer self-supporting, termed 'viscosity'.

He groups the clays in the following way:

(*a*) Clays for which the interval between incipient and complete vitrification is more than that between complete vitrification and viscosity.

(*b*) Those for which the range is approximately half of the total range from incipient fusion to viscosity.

(*c*) Those for which the range from incipient to complete vitrification is less than that from vitrification to viscosity.

Investigation of the constitution of British refractory clays was undertaken by Roberts and his collaborators (**C12, C13, G84, G85, G86, G87, R39**). The major clay mineral component was at first thought to be halloysite but later proved to be another mineral of the kaolinite group, namely livesite. A secondary clay mineral component is illite. The main non-clay component is quartz. The third important mineral in British fireclays is hydrous mica, $0{\cdot}3K_2O.3Al_2O_3.6SiO_2.4{\cdot}5H_2O$, which is probably an intermediate product in the breakdown of mica to livesite, or it may be a mixture of muscovite and livesite. Livesite, hydrous mica and quartz make up 90–95% of fireclays. Fireclays may also contain some iron compounds, but in general they are very pure and particularly free from soluble salts.

The clay mineral constituents of United States clays are principally kaolinite and hydromica, the hydromica probably being what has elsewhere been referred to as illite (**G73**). Some Ohio deposits contain a large quantity of dickite (**R57**).

The respective roles of the three main minerals in British fireclays have been outlined by Grimshaw (**G87**). Livesite confers refractoriness and a high potential plasticity although this may require weathering to make it active. Livesite gives a green strength considerably greater than a corresponding amount of kaolinite but less than halloysite and montmorillonite. Hydrous mica may assist in developing the plasticity of livesite by opening up the structure of the mined clay. Its chief function is its fluxing action which is greater in high-silica than in high-alumina clays and also depends on particle size. Fine-grained hydrous mica gives a large firing shrinkage up to 1200° C (2192° F) but a small after contraction on reheating to 1400° C (2552° F) whereas coarse-grained material gives a smaller firing shrinkage but a larger after contraction. Quartz acts as grog up to about 1350° C (2462° F) but thereafter acts as an active flux and can lead to rapid contraction of the body.

Fireclays are used chiefly for refractories, *e.g.* firebricks, retorts, furnace linings, and also for sanitary ware and certain tiles.

High-alumina Clays

The hydrated alumina minerals diaspore and gibbsite (see p. 114 alumina) frequently occur together with kaolinite and may be used for making refractories in the mixture in which they occur. The mixtures containing diaspore 'diaspore clays' are preferred to those with gibbsite because they have more favourable shrinkage properties. They usually contain over 60% alumina. The gibbsite-containing mixtures are termed bauxites, 'bauxitic clay' having less than 50% gibbsite and 'argillaceous bauxite' more than 50%.

Bentonite

This clay is derived from volcanic ash. It is widely distributed, occurring in beds from a few inches to ten feet deep.

The main clay mineral of bentonite is montmorillonite. This makes the clay take up water readily and swell to four or five times its dry volume. It is extremely plastic, has a low fusion point, and gives a coloured product.

The chief use of bentonite is as a plasticiser. Addition of 1% bentonite may improve plasticity more than 10% ball clay would, making it particularly useful for moulding sands.

Brick Clays

Large clay deposits have become such a mixture of various minerals that they will fire to coloured bodies at a relatively low temperature. Complete vitrification to a stoneware is not possible, a porous product being obtained.

Correns (**C56**) describes the origin and nature of the minerals present. Plate-like minerals are predominant, this including not only the clay minerals but also micas and chlorites. It is possible that the mica is formed after the sedimentation of the clay. Certainly the sulphides are formed *in situ* with the aid of organic matter, as are also nitrates, the former being of more concern to the brickmaker than the latter. There is also considerable migration of carbonates in the clay leading to local concentrations. It is therefore not surprising that brick clays are localised and give rise to ware called by place names.

The very important British Fletton brick clays have the great advantage of containing about 5% carbonaceous matter which reduces the fuel requirements for firing the clay by about $\frac{5}{6}$.

Loess

This clay, also called brick-earth, cave-earth, adobe, gumbo or till, unlike other clays, is a *windborne* sedimentary rock. It occurs as a loose or fragmentary yellowish rock which readily crumbles when it is dry. The particles are uniformly small, and may contain quartz, unweathered feldspar, iron minerals and calcium carbonate; this last cements the porous mass together.

Loess occurs in the central part of North America, Northern Europe, Southern Russia and right across Asia to Northern China. None occurs in Great Britain (**S16**).

Loess usually matures at low temperatures to products of varying colours. It is largely used for bricks.

A Glossary of Trade Names compiled by Robertson (**R42**) was issued in 1954 by the clay minerals group of the Mineralogical Society of Great Britain and Ireland and is most useful.

Table 3 gives a small selection of analyses of various types of clay, a few details other than the analyses being included. So far as possible the data have been obtained from the firms that mine and/or supply the clays. Some of these firms furnish considerably more details as regards grain size distribution and drying and firing shrinkage.

Nos. 1–14, china clays and kaolins.

Nos. 15–40, ball clays and similar clays.

Nos. 41–43, off-white fat clays.

Nos. 44–48, siliceous clays.

Nos. 49, 50, ferruginous clays.

Nos. 51–54, red clays.

Nos. 55–59, stoneware clays.

Nos. 60–85, fireclays and refractory clays.

Nos. 86–89, bentonites.

Nos. 90–91, special clays.

TABLE

Selection of Analyses of

[Cols. 1 to 9: China

	(1)	(2)	(3)
Name / District	J.M. Cornish clay	M Cornish china clay	Kernick or No. 10 Cornish china clay
Mining firm			E.C.L.P.
Ref.	R42	R42	R42
Country	Brit.	Brit.	Brit.
Analysis: SiO_2	47·17	45·45	48·1
TiO_2	0·01	0·34	0·03
Al_2O_3	38·90	38·33	37·0
Fe_2O_3	0·25	0·55	0·55
MgO	0·15	0·12	0·26
CaO	0·17	0·06	0·14
Na_2O	0·04	0·43	0·04
K_2O	1·09	2·22	1·6
MnO			0·02
SO_3	—		
Loss on ignition	12·37	12·44	12·2
Vitrification			
Initial deformation refractoriness cone			
Special properties			high exchange capacity
Modulus of rupture (lb/in²)			
(kg/cm²)			
Rational analysis: Method			
Clay substance			
Potash mica			
Soda mica			
Quartz			
Feldspar			
TiO_2			
Fe_2O_3			
MgO			
CaO			
Organic matter			
Fired colour Temp. (°C)			
Dry-to-fired contraction Temp. (°C) (%)			
Uses			

THE RAW MATERIALS

3

Various Types of Clay

Clays and Kaolins]

(4)	(5)	(6)	(7)	(8)	(9)
L.M.B. or 24 or Lee Moor Best Cornish china clay	S.C. Devon china clay	C.C. china clay Devon	Kemmlitz kaolin	Zettlitz kaolin	Börtewitz raw kaolin
E.C.L.P.	Watts, Blake	Watts, Blake			Erbslöh
R42	**R42**	**R42**	**S83**	**S83**	**E21**
Brit.	Brit.	Brit.	Germ.	Czechoslovakia	Germ.
47·0	46·96	46·72	58·30	46·09	75·6
0·06	0·13	0·13			
38·2	38·78	37·52	29·31	39·28	17·6
0·5	0·86	0·50	0·87	0·76	0·2
0·19	0·04	0·19	trace		0·2
0·22	0·09	0·18	0·51	0·15	
0·2	0·34	0·20	} 1·26	0·03	} 0·1
1·0	0·78	2·10		0·12	
0·02					
	12·51	12·00	10·62	13·58	6·2
	32	32			34
low-exchange capacity					
	85	35·4			
	6	2·5			
					Kallauner-Steinbrecher
			54·9	98·3	43·9
			43·8	1·2	54·5
			1·3	0·4	1·6
				white	
	bone china	earthenware			
	porcelain	white wall tiles			
	earthenware	bone china			
	white wall tiles	porcelain			

TABLE 3

Selection of Analyses of

[Cols. 10 to 14: China Clays and Kaolins

		(10)	(11)	(12)
Name } District }		Börtewitz washed kaolin	Hirschau washed kaolin I	Hirschau washed kaolin IV
Mining firm		Ersblöh	Dörfner	Dörfner
Ref.		E21	D53	D53
Country		Germ.	Germ.	Germ.
Analysis: SiO_2		50·5	47·42	53·10
TiO_2		—	0·20	0·13
Al_2O_3		35·6	38·30	33·68
Fe_2O_3		0·6	0·48	0·40
MgO		0·1	0·13	trace
CaO		0·1	0·20	0·34
Na_2O		} 0·3	} 0·48	} 0·95
K_2O				
MnO		—	—	—
SO_3				
Loss on ignition		12·7	12·79	11·40
Vitrification				
Initial deformation refractoriness cone		35	35	35
Modulus of rupture (lb/in²)			50	21
(kg/cm²)			3·5	1·5
Rational analysis:	Method	Kall-Steinbrecher	Kallauner	Kallauner
	Clay substance	90·0	93·0	82·0
	Potash mica			
	Soda mica			
	Quartz	8·5	4·0	14·0
	Feldspar	1·5	3·0	4·0
	TiO_2			
	Fe_2O_3			
	MgO			
	CaO			
	Organic matter			
Fired colour			white	white
Temp. (°C)			1410	1410
Dry-to-fired contraction	Temp. (°C)		cone 14	cone 14
(%)			10·5	6·8
Uses			porcelain glazes	refractory saggars

THE RAW MATERIALS

—continued

Various Types of Clay
Cols. 15 to 18: Ball Clays and Similar Clays]

(13)	(14)	(15)	(16)	(17)	(18)
Sergeant china clay	N. Carolina kaolin	No. 1 Dorset blue ball clay	No. 71 C.W. Dorset ball clay	A.B.M. Dorset carbonaceous ball clay	G.R. Dorset ball clay
Bell		Pike Bros.	Pike Bros.	Pike Bros.	Pike Bros.
B31	S110	R42	R42	R42	R42
U.S.A.	U.S.A.	Brit.	Brit.	Brit.	Brit.
44·90	48·46	51·18	51·76	49·00	55·60
1·32	0·06	1·75	1·75	1·20	1·45
38·90	36·53	32·54	32·45	30·24	29·05
0·40	0·15	1·03	0·83	1·75	1·38
0·10	0·03	0·39	0·68	0·22	0·36
0·06	0·13	0·56	0·95	1·13	0·48
0·22	0·21	0·30	0·67	1·32	0·96
0·20	1·34	2·56	1·89	2·18	3·08
14·20	13·03	9·42	9·27	13·00	8·02
		32	32/33	31	30/31
75		585	1072	1187	573
5		41	75	83	40
white					
		off-white dust-pressed tiles, ivory earthenware, bonding, refractories	earthenware sanitary ware, bonding	earthenware	ivory earthenware, electrical porcelain, bonding

36 INDUSTRIAL CERAMICS

TABLE 3

Selection of Analyses of

[Cols. 19 to 27: Ball

		(19)	(20)	(21)
Name } District }		E.W.O.A. S. Devon black ball clay	T.W.V.C. S. Devon ball clay	B.W.S. S. Devon ball clay
Mining firm		Watts, Blake	Watts, Blake	Devon & Courteney
Ref.		**R42**	**R42**	**D35**
Country		Brit.	Brit.	Brit.
Analysis: SiO_2		45·53	50·50	47·96
TiO_2		0·60	1·15	0·93
Al_2O_3		37·36	33·70	35·52
Fe_2O_3		0·94	0·88	0·92
MgO		0·16	0·52	0·02
CaO		0·46	0·24	0·29
Na_2O		0·28	0·28	0·25
K_2O		0·70	2·48	2·08
MnO				
SO_3				
Loss on ignition		14·46	10·60	12·19
Vitrification				1200° C
Initial deformation refractoriness cone		33+	33+	33
Modulus of rupture (lb/in²)		316	571	540
(kg/cm²)		22	40	38
Rational analysis:	Method			calculated
	Clay substance			69·7
	Potash mica			17·6
	Soda mica			3·1
	Quartz			6·1
	Feldspar			
	TiO_2			0·9
	Fe_2O_3			0·9
	MgO			0·0
	CaO			0·3
	Organic matter			1·5
Fired colour				white
Temp. (°C)				1120° C
Dry-to-fired contraction	Temp. (°C)			1120, 1200, 1250
	(%)			10·0, 10·5, 11·0
Uses		earthenware, sanitary earthenware, white wall tiles	white wall tiles, earthenware, sanitary earthenware, enamelling, abrasives	earthenware, sanitary earthenware, vitrified ware, white and coloured tiles, electrical insulators

Various Types of Clay
Clays and Similar Clays]

(22)	(23)	(24)	(25)	(26)	(27)
O.B.F. S. Devon ball clay Whiteway	C.K. S. Devon ball clay Whiteway	Wildstein ball clay	Meissen ball clay	Tambach clay Ortenburg	Vallendar whiteburning semi-fat clay Capitan
D35	D35	S83	S83	O9	C6
Brit.	Brit.	Germ.	Germ.	Germ.	Germ.
59·06	47·86	49·00	60·09	58·94	70·09
1·10	1·20			0·70	
27·45	36·29	36·18	28·15	28·30	22·98
0·76	0·76	2·58	1·24	1·27	0·69
0·77		0·13		0·27	0·23
0·68	0·33	0·32	0·22	0·56	5·01
0·63					0·55
2·58	1·65	} 0·87			0·41
					0·09
7·30	11·76	11·18	10·63	9·98	
	>1250				
	31	34+	33–34	30	
486	395				
34	28				
calculated	calculated			calculated	
40·4	78·3	95·6	75·3	72·70	50·41
21·8	14·0				
7·8	3·0				
26·7	5·0	2·5	24·7	23·18	46·38
		1·9		4·12	3·21
1·1	1·2				
0·8	0·8				
0·8					
0·7	0·3				
0·3	0·2				
deep ivory	white		white	off-white, yellow	white
1120	1120			1000, 1410	
1120, 1200, 1250	1120, 1200				
6·5, 7·0, 7·0	11·6, 13·7				
fireplace and floor tiles, with fireclay in sanitary ware, lower grade earthenware, low-tension porcelain, electrical refractories, stoneware, kiln furniture	earthenware sanitary earthenware, white and coloured glazed tiles, floor tiles, vitreous china, electro-ceramics		earthenware		

38 INDUSTRIAL CERAMICS

TABLE 3

Selection of Analyses of

[Cols. 28 to 36: Ball

	(28)	(29)	(30)
Name / District	Vallendar white burning fat clay	Goldhausen extra fat white clay	Lämmersbach light, fat clay
Mining firm	Capitain	Jäger	Jäger
Ref.	**C6**	**J9**	**J9**
Country	Germ.	Germ.	Germ.
Analysis: SiO_2	62·52	58·7	60·9
TiO_2			
Al_2O_3	34·68	35·1	36·6
Fe_2O_3	0·56	1·1	0·6
MgO	1·43		
CaO	0·09		
Na_2O	0·20		1·2
K_2O	0·56		1·6
MnO			
SO_3			
Loss on ignition		9·8	7·9
Vitrification		ca. 1130	ca. 1180
Initial deformation refractoriness cone			
Modulus of rupture (lb/in²)			
(kg/cm²)			
Rational analysis: Method			
Clay substance	78·8		
Potash mica			
Soda mica			
Quartz	16·6		
Feldspar	4·6		
TiO_2			
Fe_2O_3			
MgO			
CaO			
Organic matter			
Fired colour	white	white	very light
Temp. (°C)			
Dry-to-fired contraction Temp. (°C)			
(%)			
Uses		earthenware, fine stoneware, sanitary ware, insulators, electrical pressed ware, porcelain (binder), wall tiles.	fine stoneware, oven ware, sanitary ware, pressed electrical ware, wall tiles.
Mineralogical composition			

THE RAW MATERIALS

—*continued*

Various Types of Clay
Clays and Similar Clays]

(31)	(32)	(33)	(34)	(35)	(36)
'Stoss' clay Langenaubach	Grossalmerode Osmose clay	Westerwald casting clay	Westerwald casting clay	Westerwald fat clay	Westerwald white burning clay
Stephan	Grossalmerode	Weston	Hassia	Hassia	Fuchs
S186	G90	W44	H39	H39	F34
Germ.	Germ.	Germ.	Germ.	Germ.	Germ.
64·95	68·8	74·42	74·42	50·19	64·76
1·76	2·0	0·96	0·96	0·71	} 25·09
23·58	19·6	15·17	15·17	35·72	
0·38	1·3	0·64	0·64	0·87	0·75
0·09	0·2	0·17	0·17	0·09	0·09
0·44	0·2	0·21	0·21	0·18	0·07
0·59	} 1·1	2·08	2·08	1·51	1·48
1·83		1·39	1·39	0·99	1·10
trace					
0·09					
6·26	6·8	4·72	4·72	9·76	6·85
1150/1180				cone 7/8	
30/31		28/30	28/30	34	30+
			107	311	
			7·50	21·90	

	Kall-				
Berdel	auner	Mica		Berdel	Kallauner- Mateyka	Berdel
	Ma- teyka					
63·90	38·12	34·50		52·90	92·21	65·83
		28·80				
32·33	47·15	32·50		43·83	5·73	31·31
3·77	14·13	3·80		3·27	2·06	2·86
		0·40				

			yellowish grey cone 14	off-white cone 14	white
				cone cone cone	
			cone 14	0·5 1 6	
white			4·69	0·3 1·8 2·0	earthenware,
1400			artware	bond in fine ceramic industry *e.g.* hotel ware	fine stoneware, stoneware, tiles.
casting, electrical porcelain.					

overall:
  quartz 65%
  kaolinite
  +serizite 35%
under 1 μ:
  quartz 15%
  kaolinite 70%
  serizite 15%

TABLE 3

Selection of Analyses of

[Cols. 37 to 40: Ball Clays and Similar Clays] [Cols. 41 to

		(37)	(38)	(39)
Name } District }		Tennessee special ball clay	Mayfield, Ky ball clay	Tennessee ball clay
Mining firm		Bell		
Ref.		**B31**	**N35**	**N35**
Country		U.S.A.	U.S.A.	U.S.A.
Analysis: SiO_2		55·00	56·4	46·9
TiO_2		1·60		
Al_2O_3		29·79	30·0	33·2
Fe_2O_3		0·98		2·1
MgO		0·30	trace	} 0·8
CaO		0·30	0·4	
Na_2O		} 0·90	2·0	} 0·7
K_2O			3·3	
MnO				
SO_3				
Loss on ignition		11·27	7·9	16·5
Vitrification				
Initial deformation refractoriness cone		31/32		
Modulus of rupture (lb/in²)		570		
(kg/cm²)		40		
Rational analysis:	Method			
	Clay substance			
	Potash mica			
	Soda mica			
	Quartz			
	Feldspar			
	TiO_2			
	Fe_2O_3			
	MgO			
	CaO			
	Organic matter			
Fired colour		light cream		
Temp. (°C)		cone 9/10		
Dry-to-fired contraction	Temp. (°C)	cone 9/10		
	(%)	1·58		
Uses				

THE RAW MATERIALS

—*continued*

Various Types of Clay
43: Off White Fat Clays]   [Cols. 44, 45: Siliceous Clays]

(40)	(41)	(42)	(43)	(44)	(45)
H.P.I. Canadian ball clay	Westerwald light, fat clay Jäger	Westerwald Wirges half fat, light clay Jäger	Westerwald coloured clay Fuchs	H.V.A. S. Devon siliceous clay Watts, Blake	St. A. S. Devon siliceous clay Watts, Blake
S110	**J9**	**J9**	**F34**	**R42**	**R42**
Canada	Germ.	Germ.	Germ.	Brit.	Brit.
57·62	58·2	71·1	55·50	60·42	71·14
0·91			} 31·27	1·50	1·65
28·55	29·1	19·0		26·37	19·37
0·58	1·1	0·7	1·25	0·81	0·44
0·63		0·1	0·04	0·33	0·33
0·23		0·2	0·04	0·61	0·18
	} 2·1	1·6	1·70	} 3·26	0·23
		1·2	1·22		1·78
11·47	8·8	4·9	9·19	7·10	4·92
	ca. 1150	*ca.* 1250	32/33	30/31	26
				344	382
				24	27
			Berdel Kall-Mat.		
			80·84 51·87		
			16·52 26·54		
			2·64 21·59		
	very light	off-white	off-white		
	fine stoneware	fine stoneware, sanitary ware, floor tiles.	stoneware, tiles, faience, majolika.	faience, stilts, buff tiles, ivory earthenware, enamelling, bonding, core bonding, abrasives.	stoneware, ivory earthenware, buff tiles.

TABLE 3

Selection of Analyses of

[Cols. 46 to 48: Siliceous Clays]   [Cols. 49 and 50:

		(46)	(47)	(48)
Name District		F.P. S. Devon siliceous clay	C.V.4 S. Devon siliceous clay	Colbond Clayne Village, Co. Cork, Ireland
Mining firm		Devon and Courtenay	Devon and Courtenay	Clayne Minerals
Ref.		D35	D35	R42
Country		Brit.	Brit.	Ireland
Analysis: SiO_2		66·10	66·41	61·57
TiO_2		1·56	1·07	0·96
Al_2O_3		20·60	20·13	18·56
Fe_2O_3		0·88	3·13	3·08
MgO		0·51	0·20	0·62
CaO		0·64	0·10	1·07
Na_2O		0·48	1·28	0·41
K_2O		1·84	1·90	0·24
MnO				
SO_3				
Loss on ignition		7·10	6·07	13·41
Vitrification			<1150	
Initial deformation refractoriness cone		19/20	14	
Modulus of rupture (lb/in^2)		543	615	
(kg/cm^2)		38	43	
Rational analysis:	Method	calculated	calculated	
	Clay substance	31·0	19·4	
	Potash mica	15·6	16·1	
	Soda mica	5·9	15·8	
	Quartz	41·8	42·7	
	Feldspar			
	TiO_2	1·6	1·1	
	Fe_2O_3	0·9	3·1	
	MgO	0·5	0·2	
	CaO	0·6	0·1	
	Organic matter	1·8	1·9	
Fired colour		cane or buff	deep cane	
Temp. (°C)		1120	1120	
Dry-to-fired contraction	Temp. (°C)	1120, 1200, 1250	1120, 1200, 1250	
	(%)	5·5, 6·0, 6·0	6·0, 6·0, 6·5	
Uses		floor tiles, stoneware, low-tension porcelain, electrical refractories, kiln furniture.	stoneware pipes, stoneware, bonding.	bonding

THE RAW MATERIALS

—continued

Various Types of Clay

Ferruginous Clays]   [Cols: 51 to 54: Red Clays]

(49)	(50)	(51)	(52)	(53)	(54)
B.S. Dorset ferringinous clay	C.Y. Dorset ferringinous clay	Tambach red clay	Odenwald red slip clay	Manganese clay	Pfalz red engobe clay
Pike Bros.	Pike Bros.	Ortenburg	Jäger	Sieg-Lahn	Sonnenberg
R42	**R42**	**O9**	**J9**	**S72**	**S150**
Brit.	Brit.	Germ.	Germ.	Germ.	Germ.
48·19	48·16	58·85	58·8	32·90	50·45
0·95	0·95	0·86			0·46
31·94	31·75	21·17	18·7	19·00	18·96
5·02	5·14	7·86	8·1	16·80	12·34
0·02	0·56	1·43	0·9		2·26
0·92	0·66	0·69	0·2		0·75
0·27	0·06	} 1·47	1·8		1·07
2·59	1·74		7·9		4·06
				8·60	0·25
	0·36				0·32
9·90	10·18	7·78	4·5	15–20	8·97
			ca. 1000		S.K. 08/07
32	32	12			
506	1005				
36	71				
		55·0			
		33·0			
		3·0			
		9·0			
		brick red cone O9	sealing wax red		
saggars, bonding, floor tiles, faience.	refractories, kiln furniture, bonding.		fine stoneware, floor tiles, architectural ware, binder for abrasives.	colouring of bricks	

TABLE 3

Selection of Analyses of

[Cols. 55 to 59: Stoneware Clays]   [Cols.

		(55)	(56)	(57)
Name ⎱ District ⎰		Best stoneware clay, S. Devon	Briesen stoneware clay	Zinzendorf best stoneware clay
Mining firm		Devon and Courtenay		
Ref.		**D35**	**S90**	**S90**
Country		Brit.	Germ.	Germ.
Analysis: SiO_2		67·65	42·10	47·64
TiO_2		0·52		
Al_2O_3		22·21	39·48	36·82
Fe_2O_3		0·41	2·65	1·84
MgO		0·46	0·15	trace
CaO		0·83	0·27	0·03
Na_2O		1·23	⎫ 0·18	0·10
K_2O		2·11	⎭	0·19
MnO				
SO_3				
Loss on ignition		4·58	14·90	13·44
Vitrification		1180° C		
Initial deformation refractoriness cone		18		
Modulus of rupture (lb/in²)		461		
(kg/cm²)		32		
Calculated rational analysis:	Clay substance	23·5	M.F.	M.F.
	Potash mica	17·8	MgO, 0·0844	Na_2O, 0·0517
	Soda mica	15·2	CaO, 0·1100	K_2O, 0·0735
	Quartz	41·5	NaKO, 0·0526	CaO, 0·0292
	Feldspar		FeO, 0·7530	FeO, 0·8456
	TiO_2	0·5	Al_2O_3, 8·8202	Al_2O_3, 13·28
	Fe_2O_3	0·4	SiO_2, 15·98	SiO_2, 29·2
	MgO	0·5		
	CaO	0·8		
	Organic matter	0·0		
Fired colour		light cane or buff		
Temp. (°C)		1120		
Dry-to-fired contraction	Temp. (°C)	1120, 1200, 1250		
	(%)	4·5, 5·0, 6·5		
Uses		stoneware, fireplace and coloured wall tiles, floor tiles, low-tension porcelain, electrical refractories, fireclay sanitary ware.	stoneware	stoneware

—*continued*

Various Types of Clay
60 to 63 Fire Clays and Refractory Clays]

	(58)	(59)	(60)	(61)	(62)	(63)
	Haydenville Ohio, Mingo-clay	Ohio clay	Glenboig fireclay	Derbyshire fireclay	Garnkirk fireclay	Stourbridge glasshouse pot fireclay
	S90	S90	S57	S57	S57	S57
	U.S.A.	U.S.A.	Brit.	Brit.	Brit.	Brit.
	72·24	57·45	56·42	62·35	44·37	65·10
			1·15	1·10		
	16·87	21·06	26·35	18·47	38·59	22·22
	0·16	7·54	1·33	4·77	1·82	1·92
	trace	1·22	0·55	1·36	0·30	0·18
	4·20	0·29	0·60	trace	0·51	0·14
	} 1·09	0·39				
		3·27	0·48	2·47		0·18
	5·44	5·90	13·75	9·37	14·47	9·86
	M.F.	M.F.				
	NaKO, 0·152	Na_2O, 0·0388				
	CaO, 0·826	K_2O, 0·2029				
	FeO, 0·022	MgO, 0·1771				
	Al_2O_3, 1·822	CaO, 0·0302				
	SiO_2, 13·26	FeO, 0·5510				
		Al_2O_3, 1·206				
		SiO_2, 5·6				
	stoneware	stoneware				

TABLE 3

Selection of Analyses of

[Cols. 64 to 72: Fireclays

	(64)	(65)	(66)
Name / District	Stannington fireclay	Normandy fireclay	Stourbridge Old Mine Fireclay A
Mining firm			
Ref.	**S57**	**S57**	**P8**
Country	Brit.	French	Brit.
Analysis: SiO_2	48·04	58·00	68·1
TiO_2			1·1
Al_2O_3	34·47	30·85	27·2
Fe_2O_3	3·05	1·55	1·95
MgO	0·45		0·72
CaO	0·66	0·80	0·35
Na_2O			0·68
K_2O	1·94		0·6
MnO			
SO_3			
Loss on ignition	11·15	9·70	9·46
Vitrification			
Initial deformation refractoriness cone			
Modulus of rupture (lb/in²)			
(kg/cm²)			
Rational analysis: Clay substance			56·7
Potash mica			
Soda mica			
Quartz			28·2
Feldspar			
TiO_2			orthoclase 3·2
Fe_2O_3			albite 5·2
MgO			anorthite 1·6
CaO			serpentine 1·4
Organic matter			limonite 1·9
Fired colour			
Temp. (°C)			
Dry-to-fired contraction Temp. (°C)			
(%)			
Uses			

THE RAW MATERIALS

—*continued*

Various Types of Clay
and Refractory Clays]

(67)	(68)	(69)	(70)	(71)	(72)
Yorkshire siliceous fireclay	Meissen aluminous clay	Ayrshire bauxitic clay	Coalbrookdale aluminous clay	Leutendorf saggar clay Schmidt	Grossalmerode glasshouse-pot clay Grossalmerode
P8	P8	P8	P8	S74	G90
Brit.	Germ.	Brit.	Brit.	Germ.	Germ.
75·5	62·9	42·36	53·79	73·20	71·8
1·06	1·02	3·31	1·64		1·8
19·1	33·9	52·48	38·86	20·50	17·6
1·4	1·41	1·16	2·80	0·90	1·3
0·42	0·8	trace	0·28	0·40	0·2
0·45	0·36	trace	0·47	0·50	0·2
0·30	0·27	0·35	0·09	} 0·20	} 1·0
		0·34	2·07		
7·2	12·3	14·13	14·43	4·30	6·1
				31/32	28/29
45·0	72·6	72·3	77·4	49·80	
46·1	18·5	nil	1·9	49·60	
				0·60	
		1·7	10·5		
2·4	2·0	2·5	0·5		
2·1	1·6		2·0		
0·9	1·6		0·6		
2·2	1·4	1·1	2·7		
		free alumina			
		15·7			
				saggars	

TABLE 3
Selection of Analyses of
[Cols. 73 to 81: Fireclays

		(73)	(74)	(75)
Name		Grossalmerade	Westerwald	Westerwald
District		binding clay	refractory clay	'fireclay'
Mining firm		Grossalmerade	Westerwälder	Westerwälder
Ref.		**G90**	**W45**	**W45**
Country		Germ.	Germ.	Germ.
Analysis: SiO_2		49·0	56·24	69·37
TiO_2		1·1		
Al_2O_3		34·2	41·66	28·60
Fe_2O_3		2·1	1·35	1·42
MgO		0·2	trace	trace
CaO		0·3	trace	trace
Na_2O		} 1·4	trace	trace
K_2O				
MnO				
SO_3				
Loss on ignition		11·7		
Vitrification				
Initial deformation refractoriness cone		33	34	29
Modulus of rupture (lb/in^2)				
(kg/cm^2)				
Rational analysis:	Method			
	Clay substance		96·5	57·2
	Potash mica			
	Soda mica			
	Quartz		2·54	41·3
	Feldspar		0·84	1·26
	TiO_2			
	Fe_2O_3			
	MgO			
	CaO			
	Organic matter			
Fired colour				
Temp. (°C)				
Dry-to-fired contraction	Temp. (°C)			
	(%)			
Uses			refractories	fireclay sanitaryware

—continued
Various Types of Clay
and Refractory Clays]

	(76)	(77)	(78)	(79)	(80)	(81)
	Pfalz clay Pfälzer Chamotte	Pfalz clay Pfälzer Chamotte	Pfalz glass-house pot clay Sonnenberg	Limburg bond clay Hintermeilingen	Oberpfalz refractory clay Austria	Westerwald refractory clay Fuchs
	P28 Germ.	P28 Germ.	S150 Germ.	H99 Germ.	A200 Germ.	F34 Germ.
	74·34	45·05	49·84	46·36	52·04	48·36
	0·96	1·12	0·84	0·70		
	16·88	36·27	33·50	34·99	44·51	} 34·72
	0·96	2·21	1·71	1·11	2·36	1·36
	0·05	0·08	0·07	0·11	0·16	0·10
	0·04	0·11	0·10	0·28	0·12	0·32
	1·75	1·60	2·08	} 0·81	} 0·81	1·46
	0·42	0·65	1·08			1·47
	trace	trace	trace			
	trace	trace	trace			
	4·56	13·01	10·90	15·64		12·08
				2		
	cone 3/4	cone 03/02	33	34/35	34/35	32/33
	28	33/34		20		
	24	23		15·46		
	17	16		Kal-		
				launer Berdel		
	calculated	calculated	Berdel	84·0 95·15		Berdel
	34·27	73·24	68·46			98·27
	46·62	2·19	7·62	8·75 3·33		0·80
	17·27	17·36	23·92	7·25 1·52		0·93
	0·96	1·12				
	0·96	3·21				
	0·05	0·08				
	0·04	0·11				
		rest 2·69				
	off-white	pink		greyish		
	1100	1000		1120		
	1000|1100|1200	900|1000|1100		1120		
	1·57|1·98|6·08	1·83|3·46|12·81		19		
				bond for refractories	refractories	refractories

50 INDUSTRIAL CERAMICS

TABLE 3

Selection of Analyses of

[Cols. 82 to 85: Fireclays and Refractory Clays. Cols.

		(82)	(83)	(84)
Name District		Lawrence, Ohio, plastic fireclay	Carter, Ky, flint fireclay	Clearfield, Pa. semiflint fireclay
Mining firm				
Ref.		N35	N35	N35
Country		U.S.A.	U.S.A.	U.S.A.
Analysis: SiO_2		58·10	44·78	43·04
TiO_2		1·40	2·22	1·79
Al_2O_3		23·11	35·11	36·49
Fe_2O_3		1·73	1·18	1·37
FeO		0·68	0·74	0·83
MgO		1·01	0·55	0·54
CaO		0·79	0·77	0·74
Na_2O		0·34	0·29	0·46
K_2O		1·90	0·44	1·10
MnO		0·01	0·02	0·01
ZrO_2		0·01	0·01	0·01
FeS_2		0·55	0·14	0·24
P_2O_5		0·17	0·02	0·10
SO_3		0·03	0·01	0·01
Loss on ignition		10·52	14·09	13·56
Vitrification				
Initial deformation refractoriness cone				
Modulus of rupture (lb/in^2) (kg/cm^2)				
Rational analysis:	Method			
	Clay substance			
	Potash mica			
	Soda mica			
	Quartz			
	Feldspar			
	TiO_2			
	Fe_2O_3			
	MgO			
	CaO			
	Organic matter			
Fired colour Temp. (°C)				
Dry-to-fired contraction	Temp. (°C) (%)			
Uses				

THE RAW MATERIALS

—continued

Various Types of Clay
86 to 89: Bentonites.   Cols. 90 and 91: Special Clays]

(85)	(86)	(87)	(88)	(89)	(90)	(91)
Schippach refractory clay	Clay spur Bentonite Wyoming and Nevada, U.S.A.	Geisenheimer calcium-bentonit	Geisenheimer keram-bentonit	Volclay swelling bentonite	Attapulgus clay, Ga. U.S.A.	Klingenberg pencil clay
Blaschek	Silica Products	Erbslöh	Erbslöh	American Colloid Co.	Attapulgus Clay Co.	Klingenberg
B78	**R42**	**R42**	**R42**	**R42**	**R42**	**K36**
Germ.	U.S.A.	Germ.	Germ.	U.S.A.	U.S.A.	Germ.
49·86	62·67	57·30	59·10	64·0	53·96	52·12
0·95			0·10		0·24	0·76
33·37	22·45	22·68	19·60	21·0	8·56	28·73
1·80	3·89	4·68	3·80	3·5	3·10	5·68
					0·19	
0·02	2·46	4·08	3·10	2·3	10·07	0·09
0·52	0·75	2·34	1·90	0·5	2·01	0·75
} 1·34	0·80	0·74	} 4·80	2·6	0·03	} 0·40
		0·45		0·4	0·39	
	Cl 0·20					
	0·30					
12·16	6·18	8·26	7·60	5·2		11·42
cone 1						
33						
Berdel 83·2						
15·9						
0·9						
cone cone cone						
0·5 6 14						
5·07 13·01 12·27						
refractory grog, crucibles, glasstank blocks.				bonding		pencils

PROPERTIES OF CLAYS

Harman and Parmelee (**H31**) show the relationship of the external and observable properties of clays to the basic nature of the constituents in the following grouping of properties:

Group I: Compositions:
 (a) Mineralogical composition: types and relative amounts of mineral species.
 (b) Physical compositions: particle sizes, distributions, and shapes of each mineral.
 (c) Chemical composition.

Group II: Properties or factors derived from Group I:
 (a) Adsorption phenomena.
 (b) The packing and textural characteristics.

Group III: Those physical properties controlled by (a), or (a) and (b) of Group II:
 (a) Dry strength.
 (b) Drying shrinkage.
 (c) Plasticity.
 (d) Flow characteristics.
 (e) Stability of suspensions.
 (f) Drying behaviour.
 (g) Permeability.
  and others.

The previous section dealt with the compositions of clays. Here the properties in Groups II and III will be considered.

The Properties of Clay Particles

ADSORPTION PHENOMENA OF CLAY COLLOIDS

The extremely small particles of clay minerals exhibit 'colloidal' properties derived from the charged nature of their surfaces. Very many of the ceramic properties of clays are related to the surface phenomena arising from these charges on the particles.

Charges arise on clay particles from at least two causes:

(1) Broken bonds, due to subdivision of the giant crystal.
(2) Residual charges in the lattice due to disordered structure containing ions of incorrect valency.

The first cause is bound to occur in all clay minerals, although it is often not the more important quantitatively. The continuous nature of silicate molecules has been discussed (pp. 4–7). The clay minerals have been shown to have layered structures with silicate sheets. Cleavage between sheets does not break chemical bonds and is therefore easier than cleavage of the layers themselves. The first predominates and leads to plate-like particles. The second involves breaking bonds and leaves these unsaturated (Fig. 1.21).

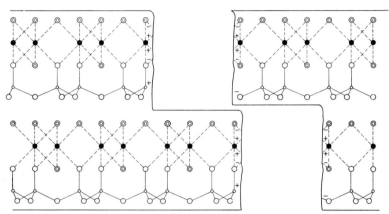

Fig. 1.21. *Fractured crystal of kaolinite showing unsaturated areas due to broken valence bonds* (Johnson and Norton, **J49**)

Where there are unsaturated bonds or electrical charges, there counterions can be adsorbed. This effect would be dependent on particle size, and the fact that this has not been conclusively proved shows that its contribution is relatively small (**W88**).

The second cause of charged clay particles is the structural disorder found in montmorillonites and also more recently in livesite (**W88**) where isomorphous substitution of Al^{3+} for Si^{4+} and Mg^{2+} for Al^{3+} occurs, giving rise to inherent negative charges. This is independent of particle size.

These surface effects exist throughout the clay, but as the size of a particle decreases so the relative importance of the surface properties increases, until a size is reached at which the surface forces predominate and are unaffected by any forces due to the mass of the particle. *These particles exhibit colloidal properties* and it is on them that the properties of the clay as a whole depend. In the case of kaolinite an average equivalent spherical diameter of less than 4μ appears to be necessary for plastic and other characteristic properties to be shown (**W63**).

Kaolinite particles are known to be hexagonal plates, so that this equivalent diameter is composed of one very small and two larger figures. Only the small one is within the normal size range for colloidal particles, namely $0{\cdot}1$–$0{\cdot}001\mu$. A halloyisite particle is lath-like, two of its dimensions are small and one large. The two smaller ones lie in the colloidal range (**H44**). Similarly a spherical clay particle would have all three dimensions less than $0{\cdot}1\mu$. The nature of the clay depends very largely on the quantity of the colloidal particles present.

TABLE 4

The Relation of Surface Area to Particle Size in a Given Quantity of Material shown by subdividing a Unit Cube

(adapted from Ostwald (**S111**))

Length of edge	Number of cubes	Total surface
1 cm = 0·3937 in.	1	6 cm^2 = 0·93 in^2
1 mm = 0·0394 in.	1 000	60 cm^2 = 9·3 in^2
0·1 mm = 0·0039 in.	1 000 000	600 cm^2 = 93 in^2
0·01 mm = 0·0004 in.	1 000 000 000	6000 cm^2 = 6·46 ft^2
1·0 μ = 0·001 mm	1 000 000 000 000	6 m^2 = 64·58 ft^2
0·1 μ = 0·0001 mm	1 000 000 000 000 000	60 m^2 = 645·83 ft^2

Any charged particle will tend to adsorb other charged particles, usually ions, in order to neutralise the charge. Charged clay particles are always considered in contact with water, with or without dissolved ionised matter.

Where the surface charge is positive (the rarer occurrence) it is thought that hydroxyl ions are adsorbed from the water to form a fixed inner layer. As the excess charge on the particle occurs in fractional electron units the adsorption of hydroxyls leads to an over-all negative charge which is then counterbalanced by an outer layer of cations (Fig. 1.22). These cations are

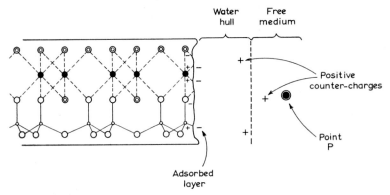

FIG. 1.22. *Micelle of a kaolinite fragment showing position of adsorbed cations* (*Johnson and Norton*, **J49**)

exchangeable, the hydroxyl is not usually exchangeable for other anions.

More frequently the clay particle is negatively charged and adsorbs cations directly. The arrangement of the cations in the water sheath surrounding clay particles depends on their size, charge and water of hydration and this alters the electric field surrounding the particles (Fig. 1.23). The distance from the particle at which its electrical field of influence becomes zero at a given concentration of cations is governed by the size and charge of these cations. Small ions can pack in closer than large ones and so

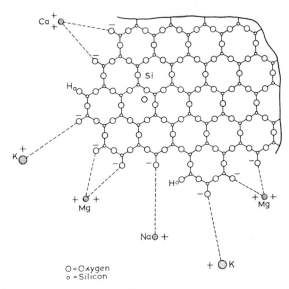

FIG. 1.23. *Unsaturated oxygen atoms in Si—O tetrahedra showing equilibrium distance of cations (redrawn according to Hofman and Bilke, **H105**)*

neutralise the field nearer to the surface. Similarly cations of large charge are required in smaller numbers than those of small charge and can pack into a smaller space. The type of cation can therefore considerably influence the relative importance of the two opposing forces acting on clay particles.

(1) A mutual force of attraction akin to gravity but because of the very small size acting only when the particles are extremely close together.

(2) A force of repulsion arising from the negative charge on each particle.

According to these, if two particles in a clay suspension approach one another they will be repelled unless they can come sufficiently close together for the attractive force to predominate. If the counterions are large, *e.g.* K^+ and Na^+, repulsion occurs and a stable suspension can be made (deflocculation). If the counterions are small and/or highly charged, *e.g.* Al^{3+}, Mg^{2+}, Ca^{2+} agglomeration occurs (flocculation) (**G88**). The relative effects of cations are given.

Li^+ Na^+	K^+ Rb^+ NH_4^+ Cs^+	Ca^{2+} Mg^{2+} Zn^{2+} Cu^{2+} Fe^{3+} Al^{3+} H^+
deflocculants	intermediate	flocculants

The nature of the surface charges on clay particles and therefore their state of flocculation or deflocculation are also affected by anions in solution and/or relative acidity or alkalinity. Deflocculation occurs with hydroxyl and in the presence of those sodium and lithium salts of weak acids which will give an alkaline reaction, *i.e.* carbonate, silicate, pyrophosphate. For instance, if the counterions around the clay particles are sodium, as is frequently the case in natural clays, and the clay is dispersed in pure water, the sodium ions can enter into the water hull sufficiently closely for the slight attraction to occur. In this condition the clay is highly plastic. If, however, the water medium contains from 0·1 to 0·2% sodium hydroxide, giving a pH of 9–10, the clay becomes deflocculated and liquid.

Flocculation is brought about by sulphates and acetates.

Bloor (**B83**) states that clay particles have only negative surface charges when in an alkaline medium whereas both positive and negative charges occur at different points on the surface in acid media. The former condition therefore requires more cations to achieve electrical neutrality. In the acid medium, moreover, flocculation can occur by attraction of oppositely charged edges and faces of the particles giving a bulky and water-retentive condition. Flocculation of a clay in alkaline conditions, on the other hand, occurs by face-to-face stacking of the particles, giving a dense product.

The actual thickness of the water film adsorbed by clay particles is therefore greatly influenced by the counterions and ionic medium. This is demonstrated when a clay of fixed water content can be made either plastic or liquid. An idea of the thickness of the water film is given by the drying shrinkage.

Drying tests show that montmorillonite adsorbs a much thicker water layer than kaolinite, and X-ray diffraction has proved that the water penetrates between the layers in the crystal itself so that considerable swelling occurs. It is therefore not surprising that whereas deflocculated kaolinite with its plate-like particles will flow readily, the swollen montmorillonite particles of less definite shape and surface will not. A clay containing any appreciable quantity of montmorillonite cannot normally be made into a satisfactory casting slip. This ability to absorb large quantities of water can be hindered by replacing the relatively small inorganic cations by the much larger and irregularly shaped organic ones. Hendricks (**H79**) points out that this is due to two factors, namely the steric hindrance by the organic ions and the destruction of the hexagonal arrangement of the water molecules on the remainder of the clay surface because of the shape of the organic ions.

CATION EXCHANGE

The cationic counterions surrounding colloidal clay particles, often termed 'bases', can under suitable circumstances be interchanged. That is, the total charge of the counterions present must be constant but the actual ions can be exchanged.

The cation-exchange capacity (c.e.c.) of a clay is the sum of the exchangeable metallic, hydrogen and ammonium cations. It is expressed in milliequivalents (m.e.) per 100 g material (1 m.e. Na_2O is 0·030997 g).

The relative cation-exchange capacities of clays is shown by the following figures which are further compared with the particular zeolites known as permutites, which have a very high capacity and are used for water softening (**S216**):

>Kaolin, under 5 m.e.
>Kaolinite, 6 m.e.
>Flint clays, *ca.* 5–7 m.e.
>Plastic clays, *ca.* 10–30 m.e. (**G88**).
>Bentonites, *ca.* 100 m.e.
>Permutites, 400–500 m.e.

The cation-exchange capacity is typical for any particular clay mineral, especially if the charge on its particles is predominantly due to disorder in the structure, so that it is little affected by particle size, *e.g.* livesite and montmorillonite. It is possible that it can be used to assist in identifying clay minerals (**W88**).

If all the exchangeable cations are replaced by hydrogen the clay is termed 'hydrogen clay' or in the 'hydrogen form'. Similarly if all the exchangeable ions are sodium we get the 'sodium form'.

Washing a clay with water tends to convert it to the hydrogen form but may take a long time (months or even years). The hydrogen form can be obtained rapidly by leaching with excess acid, *e.g.* HCl, and then washing with distilled water.

Any other form of the clay can be obtained in a pure state from the hydrogen form by adding the calculated quantity of the oxide or hydroxide of the required cation. The product is then free of soluble salts.

Exchange of one base by another is of course also achieved by washing with a soluble salt of the second base, and this is the procedure employed commercially. The ease with which the reaction will go in the required direction is governed by the relative strength of adsorption of the cations to the clay particle, as seen in the Hofmeister series, the small highly charged ions being most strongly adsorbed:

$$H^+ > Al^{3+} > Ba^{2+} > Ca^{2+} > Mg^{2+} > NH_4^+ > Na^+ > Li^+$$

In a simple equilibrium reaction using soluble salts of the type:

$$\text{clay-}A^+ + B^+ \rightleftharpoons \text{clay-}B^+ + A^+$$

the equilibrium lies on the side of the cation that is higher in the Hofmeister series, so that an alkali clay could only be obtained from one containing another cation by using excess alkali. If, however, the strongly absorbed cation can be precipitated as an insoluble salt during the exchange reaction then the sodium clay is readily made, *e.g.*

$$\text{Clay-}Ca^{2+} + Na_2CO_3 \rightarrow \text{clay-}Na^+ + CaCO_3 \downarrow$$

carbonates, silicates and phosphates are used.

To change from a deflocculated sodium clay to a flocculated hydrogen or alkaline earth clay almost any soluble salts may be used (**G88**).

THE PACKING OF CLAY PARTICLES

Clay particles are not spherical; many of them are plate-like while some are rod-like. From geometric considerations alone, therefore, there is a tendency for them to settle, pack, slide over one another, etc., in certain preferred ways. The platelets tend to 'stack' parallel with each other. Clays with oriented particles show different properties in the direction up the 'stack' to that perpendicular to it, as is seen in differential permeability, drying shrinkage and green strength, firing shrinkage, and fired strength.

Orientation of clay particles occurs most readily in the ceramic process of casting; it is also found in filter-pressed and in extruded clay. Very thorough plastic working eliminates orientation. It has also been shown that there is a greater tendency to parallel orientation in alkaline than in acid aqueous media (p. 56).

The Bulk Properties of Clays

THE FLOW PROPERTIES OF CLAYS

Clay–water mixtures show a number of types of flow, depending on the nature of the clay and also on soluble additions made to the water.

Although some clay suspensions may appear to flow like liquids, they rarely behave exactly like homogeneous liquids which would show Newtonian viscous flow. The various types of flow are shown schematically on the graph in Fig. 1.24. A true Newtonian liquid gives a straight line through the

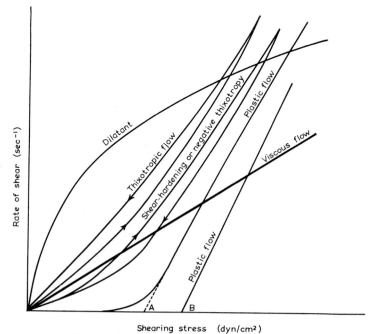

FIG. 1.24. *Types of flow. A, extrapolated yield point. B, yield point.*

origin, showing that the viscosity, or its reciprocal the fluidity, is independent of the rate of shear at which it is measured. Most clays when mixed with water have an infinitely high viscosity at low shearing stress which decreases at higher rates of shear. This is termed 'plastic flow' and the graph obtained is either a curve bending away from the stress axis and not passing through the origin, or a line cutting the stress axis. In the latter case the point of intersection is termed the 'yield point'. In the case of the curve, this often becomes linear and can be extrapolated back to give a so-called yield point. The two should not be confused, however, as one is practical and the other not.

When deflocculants are added a curve passing through the origin, nearer that of a viscous liquid, is obtained, although the behaviour is almost never linear. Certain suspensions give a curve bending away from the rate of shear axis, showing that their apparent viscosity is lowest at low rates of shear; they are termed 'dilatent'.

Any of these types of flow curves may have hysteresis loops superimposed on them. Certain suspensions have the property of 'thixotropy'. That is when a fluid mixture is left standing, it stiffens and becomes a gel, without separating into two layers. On stirring the original state is regained. Thus the viscosity decreases with increase of both the rate and time of stirring (as opposed to the viscosity of a plastic mixture which decreases with increasing rate of shear but is independent of time). If the shearing stress curve is obtained at both increasing and decreasing rates of shear an anticlockwise hysteresis loop is obtained, showing that the viscosity of the stirred liquid is lower than that of the rested one, at any given rate of shear.

In certain more uncommon instances a clockwise hysteris loop is obtained showing shear-hardening or negative thixotropy (**N34**).

THIXOTROPY

The property of certain clay–water–chemical mixtures to have flow characteristics dependent on previous periods of rest or motion is of considerable importance in ceramic works.

The mechanism of thixotropy has been investigated by Hofmann and his co-workers (**H106, F2**). They used a number of experimental methods including centrifuging thixotropic gels, allowing them to stand in contact with water, freezing them and subsequently sublimating off the ice, and electron microscopic investigations. The conclusions arrived at were that when mechanical agitation of a thixotropic clay–water slip is stopped the plate- or rod-like particles are at first kept in motion by Brownian movement. However, as the charged plate edges or rod ends happen to make contacts, they tend to attract each other and so remain together. In this way a random lattice is gradually built up throughout the slip. This entraps both the liquid and the non-colloidal solids present so that the whole appears solid. The forces holding the lattice together are weak, and easily broken down by

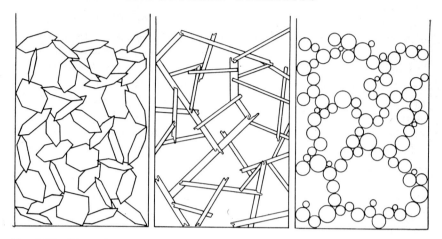

Fig. 1.25. *Schematic representation of the lattice construction in thixotropic gels, with the different shapes of basic colloidal particles: (a) plates, e.g. kaolinite, bentonite, graphite; (b) rods or tubes, e.g. halloysite; (c) spheres, e.g. lampblack, gamboge (Hofmann,* **H106**)

agitation (Fig. 1.25). This theory of the random lattice explains the following facts:

(1) Thixotropic slips will form gels over a wide range of dilution, the more dilute taking longer to reach the same degree of stiffness.

(2) A gel once formed is not easily dewatered by centrifuging, but once this is achieved it does not take up the supernatant water again.

(3) Only very dilute slips separate into a gel and liquid.

(4) Very concentrated gels do not take up water put in contact with them.

These facts all lead to the conclusion that there is no fixed optimum structure with a minimum potential energy. It has also been found that gentle rhythmical motion accelerates the setting of a thixotropic gel (**H42**). This 'rheopexy' probably assists the orientation.

Experiments with viscometers that can be run at a number of different, but in each case constant, rates of shear show the time factor and reversibility of thixotropic behaviour.

The schematic graph (Fig. 1.26) shows how the shearing stress drops as the period of stirring increases, when starting with the rested gel. The descending portion of the curve does not show steps although obtained in the same way. It is found that for a given system and a given rate of shear there is a tendency towards equilibrium values for the shearing stress and apparent viscosity which are independent of whether the previous rate of stirring was faster or slower.

As thixotropy depends on the surface properties of colloidal clay particles, it is perforce affected by any electrolytes present in the water, which change the particle interactions. Hofmann (**H106**) has investigated the effect of different cations on kaolinite and montmorillonite. In order to avoid confusion of the issue by base exchange, carefully prepared clay minerals

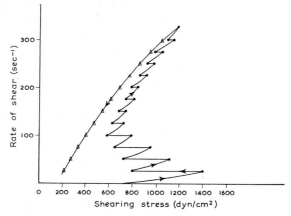

Fig. 1.26. *The shearing stress of a thixotropic liquid if the rate of shear is changed in steps of 25 r.p.m. and held at each speed for 20 sec*

were used, containing only the one exchangeable cation that was to be tried in the experiment. The structure of montmorillonite, with its tendency to take up water and cations in between its lattice layers, makes it necessary to consider it apart from kaolinite, as regards actual thixotropic behaviour. However, interesting parallels can be drawn about the interactions of these two clay minerals with water in the presence of salts. The distance between the montmorillonite layers under different circumstances were measured and it was found that in weak solutions ($<2 \cdot 0\ n$) of the alkali chlorides the uptake of water between the layers was so large that they could be considered to constitute separate platelets. In the $2 \cdot 0\ n$ solutions of LiCl, NaCl, KCl and in solutions of $MgCl_2$, $CaCl_2$ and $BaCl_2$ of $2 \cdot 0\ n$ to 0 concentration the interlayer distance was measurable, and in the latter cases increased slightly with greater dilution of the salt. In general then, montmorillonite has the greatest affinity for water when it has absorbed alkali cations and is surrounded by dilute solutions of alkali salts, and in these circumstances, therefore, the attraction between montmorillonite particles is least.

In the parallel experiments on kaolinite the volume of solution per 3 g of kaolin was found that would form a thixotropic gel under defined conditions. The results showed the highest values for the $2 \cdot 0\ n$ solutions of alkali chlorides, the values for the $0 \cdot 2\ n$ and more dilute solutions of the same salts were less than half as high, while those for the Group II metal chlorides varied little with concentration and were intermediate. In other words the tendency for kaolinite to form thixotropic gels is greatest under the conditions when montmorillonite has the least distance between its layers and vice versa.

The effect of the cations on the attraction between clay particles applies equally to the montmorillonite, but those cations that discourage adhesion at the edges to form the lattice structure of the thixotropic gel also encourage cleavage giving more particles to build up the structure. Hence a change in the soluble environment that may prevent thixotropy of kaolinite may increase that of montmorillonite. For instance, kaolinite is at its most thixotropic in concentrated common salt solutions whereas montmorillonite is at its least so.

Thixotropy is met with in various fields of ceramic work: in apparently plastic bodies that become liquid when jarred; in casting slips that solidify in narrow parts of the mould and do not pour out properly; and in the prevention of thickening of glazes by sedimentation (conversely glazes that sediment too easily are deliberately made more thixotropic).

PLASTICITY

The ceramic industry could be said to be founded on the fact that clays have the property of 'plasticity', and yet this all-important feature cannot be properly defined or measured. When applied to clay the term 'plastic' is accepted as meaning that a clay will take up water and, with a given quantity, achieve a condition when applied pressure can deform it without rupture, yet when the pressure is released the new shape is retained. If it is then dried the ability to be deformed is gradually lost and the clay becomes relatively hard and brittle. Such a description of the term, however, has no defined physical meaning, the plasticity of clay having a much wider meaning than 'plastic flow' as considered under flow properties (p. 59).

Plasticity is a highly complex property. In order to deform a clay there must be some flow of the particles over each other. In order to retain the shape there must be resistance to flow. In other words there is initially a resistance to applied pressure, the clay behaving elastically, until the pressure reaches a certain minimum, the yield point; after this flow occurs. The flow then becomes proportional to the excess of rate of shear. The straight portion of the curve of rate of flow against stress does not pass through the origin, and therefore the coefficient of viscosity is not constant but varies with the rate of shear. This type of flow is termed elasticoplastic flow, as differentiated from viscous flow exhibited by liquids.

Plasticity is inherently bound up with the physicochemical relations between colloidal clay particles and water, and between each other, and is therefore effected by the following four factors:

(1) Mineralogical composition.
(2) Particle size and particle-size distribution.
(3) Cation-exchange capacity, cations and pH.
(4) Surface tension of the water.

The plasticity of natural clays depends on the nature of the very fine fraction, which often differs from the remainder of the clay. In this fraction are most of the clay minerals and the finest part has sometimes been found to contain clay minerals not present in the coarser fractions, *e.g.* montmorillonites in kaolinite, which have a very large influence on the plasticity of the whole.

The relative plasticity of the clay minerals themselves has been listed as follows:

Low: dickite < flint < illite < nontronite < hectorite < kaolinite < montmorillonite. *High* (**B82**).

This is related to the particle fineness in which these minerals occur and very much more to their charge or their cation exchange capacity, from which are derived their relative attraction and repulsion forces. Minerals of high cation exchange have high plasticity but their plasticity is more susceptible to changes of cationic environment.

Sullivan (**S216**) investigated the effect of exchangeable cations on the working properties of clays of equal water content by twisting bars to breaking point:

Increasing yield strength: Li, Na, Ca, Ba, Mg, Al, K, Fe, NH_4, H.
Increasing maximum torque: Li, Na, Ca, Ba, Mg, Al, Fe, K, H, NH_4.
Decreasing angle at maximum torque: Li, Na, Ba, Mg, K, Ca, NH_4, Al, H, Fe.

The fact that the order is not always the same illustrates the impossibility of directly relating laboratory techniques to 'plasticity'.

In more general terms small cations of high charge lead to the clay colloid holding water on to its surface more strongly and in greater quantity. The water films around individual grains which then merge to give a single plastic piece. The force required to rupture the piece depends on the degree of bonding of the water to the clay surface.

When alkali counterions are adsorbed the water sheath is less strongly held than when H^+ or Ca^{2+} are present and these clays are more deformable but more easily ruptured. More water is required to produce a plastic mass of the same strength with a H-clay or a Ca-clay than with a Na-clay, but the green body is stronger and less likely to lose its shape (**G88**).

It is seen that the hydrogen ion is at or near the end of each series; it is also known to be the strongest flocculant, whereas the hydroxyl ion is a strong deflocculant. It follows that changes in pH have a marked influence on plasticity. Many natural red brick, ball and other plastic clays show acidity (*i.e.* their pH < 7·0) and many pottery china clays show alkalinity (pH > 7·0).

By addition of sodium carbonate to acid clays and acetic acid (vinegar) to alkaline clays, their pH can be adjusted to an optimum value when the workability is at its best. Barker and Truog (**B8**) found that the optimum value for acid clays lies in the range pH 6·0 to pH 8·5 whereas for alkaline clays it is in the range pH 7·3 to pH 10·5. The actual value must be found independently for individual clays, as the best workability is obtained only over a small pH range. For instance, for a given porcelain body using several Continental clays the best working condition was found to be at pH 7·6–7·7, but for a similar body using English clays it was at pH 8.0.

Unfortunately it has only recently become practicable to measure the pH of a clay when in the plastic state. It is more usually done with a clay–water mixture that is liquid enough to be stirred. Barker and Truog (**B9**) have shown that the pH of a clay varies with the clay–water ratio. The pH increases with increasing water dilution but the change is not regular nor can it be predicted for one clay from data on another one. Thus values obtained by different observers are not necessarily comparable. Dal (**D2**) has worked out more repeatable methods of finding the pH of liquid slips by very prolonged stirring, in his work on finding the optimum additions of soda to clay.

The best pH for a given clay is found by testing the original clay, and then gradually adding acid (acetic acid) or alkali (sodium carbonate, often termed 'soda ash') to bring the clay towards neutrality, testing the workability at regular small intervals.

Constant check must be kept on the pH of an established body, as the pH of tap water may vary from day to day.

The improvements obtained on treating a clay to bring it to its optimum pH are summarised as follows (**B8**):

(1) The treated clay is more plastic than the non-treated clay.

(2) Less mixing water is usually required to give the desired plasticity.

(3) Less power is required to pass the clay through a die because it is more plastic.

(4) The physical structure of the clay column is improved and with clays which have a tendency to laminate, proper treatment will often eliminate the laminations.

(5) The clayware is more perfectly formed and has less cracking and corner defects.

(6) Dried, treated ware is less pervious to water and often can be placed in water for a period of ten minutes without disintegrating.

(7) The firing temperature is usually lowered because the clay has a closer body.

(8) The compressive and transverse strengths of the fired ware are greater.

(9) Moisture absorption is lower, especially when laminations are reduced.

(10) The colour of the clayware is usually improved. The reds are more brilliant, and some light cream high-calcium clays will fire red if there is some iron in the clay.

Each clay requires a different quantity of water to bring it to its maximum plasticity. Ries (**R28**) has summarised published data about various clay types:

Water Plasticity of Different Kinds of Clay (R28)

Crude kaolin	36·69–44·78
Washed kaolin	44·48–47·50
White sedimentary kaolin	28·60–56·25
Ball clays	25·00–53·30
Crucible clays	26·84–50·85
Refractory bond clays	32·50–37·90
Glass-pot clays	19·64–36·50
Plastic fire clays	12·90–37·40
Flint fire clays	8·89–19·04
Sagger clays	18·40–28·56
Stoneware clays	19·16–34·80
Face brick clays	14·85–37·50
Sewer pipe clays	11·60–36·20
Paving brick clays	11·80–19·60
Brick clays	13·20–40·70

In the plastic clay the water sheaths round the particles and in the pores play a vital role. Surface tension forces form them into skins which tend to hold the mass together. Lowering the surface tension of water with certain solutes therefore diminishes this skin effect and decreases the yield point, maximum strength and workability of the plastic clay. Fortunately most additions usually made to the water increase the surface tension (**S43, P82**.)

The water between the particles also acts as a lubricant for their movement, which is assisted by their flaky structure, softness, rough surface and good cleavage (**S6**). It is found that the internal forces opposing the tearing apart of clay particles are 10 times greater than those opposing their relative displacement or shear (**P48**).

The Working Range of a Plastic Clay. Given a clay containing the right minerals at a suitable grain size distribution, and with the right metallic counterions its plasticity depends on the water content. Starting with a dry mobile powder, addition of water makes it crumbly, until a water content is reached when the powder will hold together, at which point the lower limit of the working range has been attained. The upper limit occurs when the plastic mass becomes sticky.

Memory of Clay. Although most considerations of plasticity assume that deformed clay retains its shape this is not always so. Where deformation has been rapid signs of strain may appear only after the force has been removed. This is seen in hand-thrown pieces that subsequently develop a twisted appearance. Considerable evidence is also seen in certain auger extruded columns.

Moore (**M89**) has also described conditions where an initial deformation by torsion appears normal and permanent, but if a second deformation is undertaken it is found to be much easier to return the piece to its original shape than to continue the deformation.

The Effect of Pressure on Plasticity and Water Content. It follows from what has been said about the function of the water between mutually attracted charged particles that if higher pressure is applied the same plasticity can be achieved using less water (**M53**).

The curve plotted in Fig. 1.27 is obtained by measuring the pressure required to make a given body flow at different water contents. It will be noted that below 30% water the curve rises very steeply (**S90**). Curves

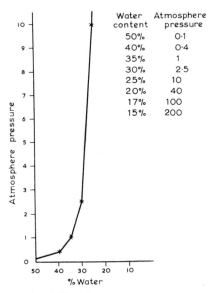

Water content	Atmosphere pressure
50%	0·1
40%	0·4
35%	1
30%	2·5
25%	10
20%	40
17%	100
15%	200

FIG. 1.27. *The effect of pressure on the water content of a plastic clay*

obtained with different clays are similar but not identical in shape. Bodies containing less water can therefore be shaped by pressure; they have less drying shrinkage, and hence less strain is set up inside the body during drying and so the green strength is greater (**W49**).

High-pressure shaping of articles from bodies of very low water content is a direct corollary of these facts. It should be remembered that high pressure does not exert a beneficial effect on bodies of fairly high water content as water is merely pressed out (**N19**).

THE PERMEABILITY OF CLAY

The ability of water to pass through a layer of clay depends on the way the particles are packed together, and on the water films adsorbed around and in between the particles.

The water film depends on the thoroughness of the wetting and dispersing of the clay, and on the cations which determine its thickness. Harman and Parmelee (**H31**) used three different methods for preparing the clay before measuring its permeability. In the first the clay–water mixture was allowed to stand for twelve hours, in the second it was agitated for seventy-two hours, and in the third it was deflocculated, agitated for seventy-two hours, and then reflocculated. The water films therefore were allowed to form more completely as they passed from method (1) to (3), and it was found that the permeability increased accordingly. In other words, if there is a continuous water film, addition of water to one side merely induces a shift of the water film through the clay and it appears permeable. Where the particles are inadequately wetted the continuous water membrane is not present and dry spots resist the passage of water. The effect of cations on the permeability as found by Lutz (**L50**) is on the same lines, the order of increasing permeability being the same as the order of the cations that attract thicker water films, *i.e.* least permeable $Li < Na < K \ll Ca < Ba < H$ most permeable.

Harman and Parmelee's (**H31**) work also showed that although the permeability of a clay is affected by its base-exchange capacity there is no direct correlation. They conclude that the packing characteristics of the clay play a considerable role.

The permeability of a clay is a major factor in the drying characteristics of a clay.

THE AFFINITY OF DRY CLAY FOR WATER; THE HEAT OF WETTING

The affinity of clay for water is shown by the fact that heat is given out when clay is wetted. Montmorillonite, which absorbs a great deal of water has a very much larger heat of wetting than kaolinite which adsorbs only a little water. Illite has an intermediate value. Fired ware generally has a low heat of wetting.

The graphs in Fig. 1.28 show the heat of wetting of three typical clays after heating to different temperatures. Parmelee and Fréchette (**P19**) deduce from their work that the heat of wetting may be an indication of the susceptibility of the specimen to destruction by contact with water, and that a high heat of wetting indicates high capillary forces and great strains within the ware as a result of varying humidity. This factor is rarely determined or considered in the industry. It may, however, prove to be an answer to drying difficulties experienced with some clays and ceramic bodies. Care in making up bodies, bearing the heat of wetting in mind, could eliminate such losses.

DRYING

In air, moist clay gradually gives up water, at first at a constant rate and subsequently at a decreasing rate, until it contains no free water. During the constant-rate period the surface is always equally moist and shrinkage of the whole piece occurs. Thereafter the surface begins to appear dry and evaporation takes place within the piece, little or no shrinkage occurring.

In Fig. 1.29 it is seen how a piece of plastic clay of initial volume and water content P at first shrinks by an amount equal to the volume of water given off, until the point L. At this point the clay is in the leather-hard

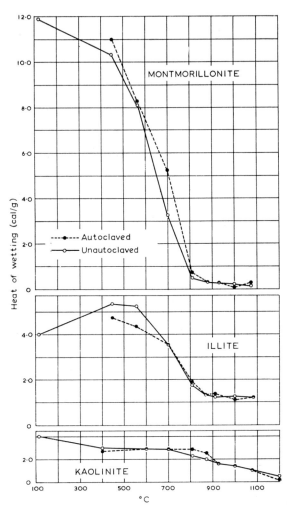

Fig. 1.28. *The heat of wetting of dried and fired clays* (Parmelee and Fréchette, **P19**)

condition. Further reduction in the moisture content produces little or no volume change right down to dryness, point *D*. The generally accepted explanation is that while the water actually separating particles evaporates shrinkage occurs until they are in contact. Thereafter water is lost from the pores and no further shrinkage occurs.

The more water a clay has absorbed the higher is the shrinkage, so that high plasticity is offset by the danger of cracking due to irregular shrinkage.

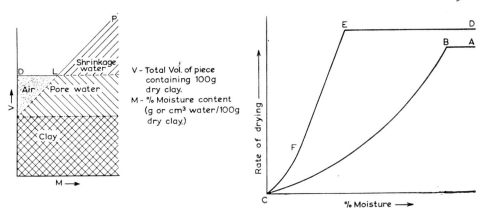

FIG. 1.29. *Volume changes of a plastic clay–water mix during drying* (Newitt and Coleman, **N12**)

FIG. 1.30. *Typical drying rate curve* (Pearse, Oliver and Newitt, **P21**)

Linear and Volume Air Shrinkage of Different Clays (Ries (**R28**))

Clay type	% linear air shrinkage	% of volume air shrinkage
Crude kaolin	5·0 – 7·60	14·11–20·92
Washed kaolin	3·3 –10·80	20·20–28·91
White sedimentary kaolin	4·5 –12·50	7·53–36·46
Ball clay	5·25–12·0	21·9 –31·92
Crucible clay	6·7 –15·0	18·98–55·05
Refractory band clay	4·25–11·0	30·48–45·00
Glass pot clay	3·31– 9·5	20·57–48·80
Plastic fire clays	1·7 – 9·4	6·67–28·80
Flint fire clays	0·78– 6·59	2·36–21·12
Sagger clays	2·8 –10·8	9·10–25·0
Stoneware clay	4·8 – 9·3	15·11–41·2
Face brick clay	2·4 – 5·66	7·60–17·95
Sewer pipe clay	3·5 –10·5	16·12–22·9
Paving brick clay	0·9 – 5·82	3·54–18·29

Harman and Parmelee (**H31**) found that if clays are made up to the same plasticity, the drying shrinkage is proportional to the base-exchange capacity and is little affected by the nature of the adsorbed cation. This is perhaps due to the compensating effects of more water and higher porosity in the hydrogen clay and less water and lower porosity in the sodium clay.

Figure 1.30 shows typical drying curves for clay. During the constant-rate period *AB* and *DE*, drying takes place at the exposed surface only, by diffusion of the vapour through an adhering stationary layer of air. The water from the interior of the piece moves towards the surface, constantly replacing that which has evaporated. The rate of drying is dependent only on external conditions, unless it becomes so great that the surface dries before being replenished from the interior. In fact Gillihand (**G49**) has shown that a wide variety of materials dry at the same rate, and the empirical relationships found for the evaporation of water from free surfaces hold.

TABLE 5

Evaporation Rates for Various Materials under Constant Conditions (Gillihand (**G49**))

Material	Rate of evaporation (g hr^{-1} cm^{-2})
Water	0·27
Whiting pigment	0·21
Brass filings or turnings	0·24
Sand (fine)	0·20–0·24
Clays	0·23–0·27

According to Powell and Griffiths (**P54**) the total rate of evaporation W, in g/sec, from a horizontal plane of length L cm, and breadth B cm, maintained at a uniform temperature T^0, to a tangential air stream of velocity, V cm/sec is given by:

(1) for values of V between 100 and 300 cm/sec,
$$W = 2{\cdot}12 \times 10^{-7} L^{0{\cdot}77} B (P_w - P_a)(1 + 0{\cdot}121 V^{0{\cdot}85})$$
(2) for values of V below 100 cm/sec,
$$W = 4{\cdot}3 \times 10^{-7} L^{0{\cdot}73} B^{0{\cdot}8} (P_w - P_a)(1 + 0{\cdot}121 V^{0{\cdot}85})$$

where P_w is the saturation pressure at T^0 and P_a is the vapour pressure in the air stream, both in mm Hg.

The falling-rate period is sometimes curved throughout, being concave towards the rate of drying axis, or with other materials it can be subdivided into a linear *EF* and a curved portion *FC* (**P21**).

During the linear first falling-rate period the water still evaporates from the surface, but as this is beginning to dry out the rate falls. This stage, like the first, is greatly influenced by environmental conditions. The second falling-rate period involves evaporation within the pores of the body followed by movement of the vapour to the surface. It is not very susceptible to external conditions.

The question immediately arises as to the cause of the liquid movement during drying, particularly the forces that keep the surface continually moist during the constant-rate period. The only forces that are likely to affect the movement of free water in a solid bed are those due to gravity, friction, convection and capillarity (or surface tension). Their relative magnitudes are dependent on such factors as the structure of the bed and its component parts and on those (*e.g.* temperature) which may cause changes in the physical properties of water. Furthermore, water movements may be induced by temperature gradients and uneven pressures across the solid bed (**P21**).

One of the first workers to investigate the mechanism of drying was Pukall (**P69, P70**). In 1893 (**P61**) he had made the observation that a fine-pored ceramic filter vessel is impervious to gases when wet, *i.e.* that although water drives air out of the pores easily the reverse is not the case. If such a vessel is filled with water, the neck fitted with a stopper and a long glass tube and the whole inverted so that the tube dips into a container of water and mercury, a drying process can be observed. The glass tube is at first kept dipping in the water. The surface of the porous vessel then appears moist all the time, and the water level in the beaker gradually falls. The evaporation of water therefore occurs at the surface, which, however, is kept continually moist by water moving from the interior to the surface. The force motivating this must be of considerable size in order to draw water up a long tube against gravity. This is further proved by adjusting the tube to dip in the mercury. This too is gradually drawn up the tube to a height only a small amount below the barometric height, remaining there as long as the filter's surface is moist. As water evaporation proceeds further, dry places begin to appear on the surface, and the mercury column drops as air is able to enter.

Similar experiments were in due course made with a number of clay and ceramic body samples (**P70**). These were prepared very carefully, by maturing, kneading, etc., to give a truly uniform piece, to minimise cracking. The piece was then made up into a sphere and a glass tube inserted and sealed in, by a carefully worked-out technique. The tube was first evacuated, then allowed to fill with air-free water, and finally dipped in mercury. In every case the mercury rose about 700 mm (27·5 in.) but the course of the experiment varied somewhat with the clay or body used (Fig. 1.31).

Pukall summarises his work as follows. The experiments prove:

(1) The drying of clay and clay bodies and other fine-pored bodies at ordinary temperatures occurs at the surface, without this, however, getting ahead of the uniform drying of the internal layers of the piece.

(2) It is independent of atmospheric pressure.

(3) As water evaporates at the surface, cooling occurs in the centre of the piece.

(4) During the drying a reduction of pressure occurs in the interior of the piece.

(5) This can be measured and is somewhat greater than 700 mm height of mercury (equivalent to 135 lb/in^2 pressure).

(6) It occurs in a similar manner and magnitude in both fat and lean clays and clay bodies, which differ only in the time factor.

(7) The strains set up by the reduced pressure are independent of the strength of the piece.

(8) They usually reach their maximum before the drying is complete.

(9) Where deformation of the piece can occur, as in clay bodies, the

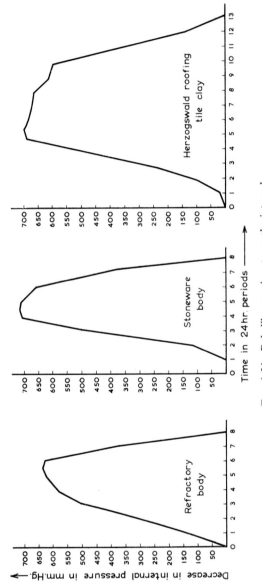

Fig. 1.31. Pukall's experiments on the internal tension set up during the drying of clays (**P70**)

internal strain causes shrinkage in direct proportion to its magnitude. Drying and shrinkage are therefore intimately bound up with each other.

(10) The evaporation process is dependent on the ability of surrounding air to take up water vapour, *i.e.* its saturation, temperature and speed.

(11) Better conditions for removing water vapour must not be so applied that the surface dries before the interior, or that the boiling point of the internal water at its considerably reduced pressure is reached.

Returning to Pearse, Oliver and Newitt's (**P21**) list of possible motivating forces for the water movement (p. 70) towards the surface, Pukall's experiments with the spherical sample eliminate gravity, friction and convection. Capillarity is clearly a major cause and the actual force has been termed 'capillary suction'.

Further work was, however, required to determine what actually happened in the inside of a clay piece when under reduced pressure, whether a vapour bubble tended to form there or not. Westman (**W46**) describes a more complex apparatus for finding the capillary suction of clays and bodies. His results, given in Table 6, show that Pukall's experiments did not estimate this force to anything like its fullest extent. This, and subsequent work, led to the belief that capillary action alone could not be responsible for the large suction forces involved. Newitt and Coleman (**N12**) suggest the use of the term 'suction potential' as a general description of the force promoting liquid flow in particulate solids.

TABLE 6

Capillary Suction of Various Specimens
(Westman (**W46**))

Composition of specimens				No. of tests	Average		Probable error (lb/in^2)
Flint	Feldspar	Ball clay	Kaolin		(lb/in^2)	cm of Hg	
100	—	—	—	—	5	26	—
—	100	—	—	3	10	52	—
—	—	100	—	3	880*	4550	—
—	—	—	100	3	263	1360	±5·5
—	—	40	60	3	400	2068	±6·4
50	—	20	30	6	182	941	±6·1
—	50	20	30	6	198	1024	±5·4
20	30	20	30	4	192	993	±5·8
10	15	30	45	6	263	1360	±6·4
30	45	10	15	5	33·9	175	±2·0

* This may not be the maximum pressure. The values obtained were 800, 760 and 880 lb/in^2.

The 'capillary suction' theory of drying of solid granular beds, when applied to clays, totally ignores the electrochemical nature of the surface of the clay particles. As this is the origin of the clay properties of plasticity, base-exchange, deflocculation, etc., it must also be involved in the drying process, and would seem more likely to set up water movements by osmotic rather than by capillary forces. Newitt and Coleman (**N12**) showed that whereas the drying of a silica bed was influenced when the surface tension of the water was altered by adding Teepol, that of clay was hardly affected, *i.e.* that capillary suction moves the water in the silica bed but does not predominantly do so in clay. On the other hand, the addition of electrolytes does affect the drying of clay, which leads to the conclusion that here osmotic forces play a major part.

They continue:

'While being very different in nature, both types of suction promote the flow of liquid to regions of higher suction potential. Since these regions in a homogeneous body are always of lower moisture content, whether the suction potentials be capillary or osmotic, the available water will always tend to distribute itself evenly. This tendency will be modified by the hydrostatic head through the depth of the body, if this is comparable to the suction potentials present. The equilibrium distribution may be calculated by the method applied to capillary suction potentials by Ceaglske and Hougen (**C28**) and probably applicable to osmotic suction potentials.

'Frictional or viscous resistance to flow is another and more important factor modifying the moisture distribution. Where this is small, in the coarser non-shrinking materials and in clays of high moisture content, the distribution will be unchanged from the equilibrium relation. Where it is appreciable, considerable moisture distribution gradients may be set up, of steepness according to the rate-of-flow and viscosity of the liquid, the permeability of the solid structure and the suction potential difference across the body.'

Macey (**M7**) investigated not only the suction potentials of clays but also the moisture gradients and aqueous conductivities. He too, considers capillary suction incapable of exerting the forces found to be present. He postulates that as water evaporates from between particles they are drawn together. As, however, the particles are electrically charged they repel each other, and a region of lower hydrostatic pressure arises, to which water from moister regions will flow.

The idea that water movements in clay are largely governed by the electrochemical nature of the clay particle surfaces also accounts for the anomalous viscosity of the water.

The shrinkage following on drying is the cause of the cracking, so often found in dried clay pieces. As the drying process sets up moisture gradients, so unequal shrinkage occurs and strains are set up. Therefore some internal strain will always occur, although it is only when it exceeds the natural strength of the clay that cracking will take place. Luikov (**L46**) has shown that cracking due to moisture gradients can occur even when no drying is taking place, *i.e.* when a piece is heated in a water-saturated atmosphere.

GREEN STRENGTH

The property of the dried clay to hold together, known as the 'green strength', is essential to the ceramic industry. It depends on the electrostatic attractions between clay colloid particles which increase as the separating water film is removed. It is therefore essential that the fine particles should be evenly distributed over the coarser ones and particularly over the non-plastic ingredients, quartz and feldspar, in order to hold the whole together.

As the green strength depends on the colloidal particles present in the clay, it follows that it will be affected by some of the same factors as plasticity, *e.g.* weathering of the raw clay, maturing of the clay body, presence of electrolytes. Kohl (**K62**) gives the following examples:

A well prepared earthenware body had a transverse strength when air-dried of 143·6 lb/in^2 ±2·5% (10·1 kg/cm^2), whereas if it was matured a week or two the transverse strength rose to 160·7 lb/in^2 ±1·9% (11·3 kg/cm^2).

Replacement of hydrogen counterions by sodium greatly increased the dry strength. This is due to difference in packing characteristics and pore structure. Porosity–strength curves show that the pore structures of sodium and hydrogen clays are of different types. It is concluded that the dry strength is a function of the porosity for a given type of pore structure (**H31**). Figures given by Grimshaw (**G88**) give similar results (*see* Table 7).

TABLE 7

The Effect of Different Counterions on the Behaviour of a Pennsylvania Shale Clay

(c.e.c. 12·6 m.e./100 g (**G88**))

	Forming water (%)	Drying shrinkage (%)	Porosity after drying (%)	Dried transverse strength (lb/in^2)
Raw clay	21·2	14·0	26·1	1275
H-clay	22·3	15·5	28·2	1150
Ca-clay	19·7	12·4	26·3	1250
Na-clay	18·6	11·0	24·6	1410

The green strength of a body is very largely dependent on the method of shaping it, as on this depends the orientation of the particles and the thickness and completeness of the original water hull holding them together. These determine the homogeneity of the body, and any strains present. Experience has shown that extruded bodies have the highest strength and that this drops for turned and moulded bodies, and is least for cast ones.

The green strength of a clay body increases as the water content decreases. Drying at room temperature lowers the water content to about 2–3%. Further drying in ovens at increasing temperatures increases the green strength very considerably as shown in the curves by Kohl (**K62**) (*see* Fig. 1.32). In pieces dried at temperatures up to 70° C, if they are remoistened

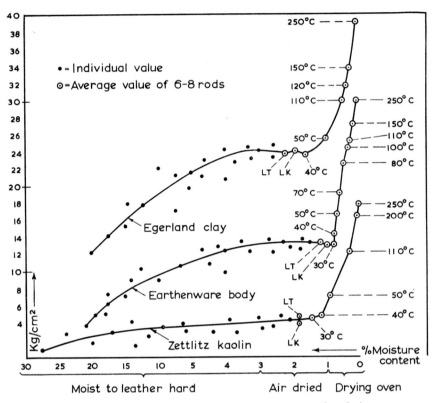

Fig. 1.32. *The dependence of bending strength on moisture content and on drying temperature* (Kohl, **K62**)

and then air-dried the original strength is attained, but if they are dried above 80° C this is not found to be so. An irreversible change that can be considered as part of the firing process begins at about 80° C when alteration of the colloid particles sets in.

The strength of dry clay is an important property with direct practical bearing on problems of moulding, handling and drying of ware, since a high strength enables the clay to withstand the shocks and strains of handling. Through it also the clay is able to 'carry' non-plastic material, *e.g.* flint, feldspar, grog, etc.

The green strength of clay is generally determined by finding its transverse strength and expressed as the modulus of rupture which has values ranging from about 20 to 1500 lb/in² (1·4 to 105 kg/cm²) (**R28**).

The Firing of Clays

The irreversible hardening reaction that occurs when a clay is strongly heated is another of its features that has been fundamental to the ceramic industry from earliest times. Briefly, on further heating of dried clay more water is given off and in due course a hard but porous piece forms. Thereafter vitrification occurs leading to a strong dense piece but is eventually followed by softening and fusion. Release of gases may also give rise to a swollen appearance known as 'bloating'. During these reactions some expansion may occur but there is almost always over-all shrinkage.

The whole series of reactions is governed by the mineralogical, chemical and grain-size composition of the clay and is therefore different for each clay.

The actual reactions on firing of clay have been investigated most thoroughly for kaolins freed as far as possible of matter other than clay minerals. Combined study by differential thermal analysis (see p. 313), optical observations, electron microscopy, chemical analysis and X-ray diffraction have still not furnished an entirely satisfactory picture, but sufficient is known about the reactions to be of great help.

When the dry clay is heated to a given temperature in the range 450–550° C (842–1022° F) it begins to lose combined water and to absorb heat. This continues progressively over a temperature rise of about 130° C (266° F). Long heating at one of the lower temperatures will not expel the water that a short heating at a specific higher temperature will remove (**I6**). The kinetics of the dehydroxylation of clays have recently been investigated and described by Evans and White (**E25**).

The product of this dehydration is known as 'metakaolin' and was long considered to possess no crystal structure although the hexagonal shape of the kaolinite particles remains. The alumina can be dissolved out with hot hydrochloric acid (**C48**). Work on the thermal decomposition of kaolinite and halloysite (**R64**) and nacrite (**B117**) has now (1955) shown that there is a residual structure at this stage most probably associated with the Si—O hexagonal layers.

At about 950° C (1742° F) there is a further thermal effect when heat is evolved. X-rays show that small quantities of γ-alumina and/or mullite crystals begin to appear at this temperature. After the exothermic effect is over mullite formation becomes fairly rapid. McVay (**M17**), who did not find that γ-Al_2O_3 was always formed before mullite, was not able to account for the exothermic effect, as he considered the amount of mullite at that temperature too small. Insley and Elwell (**I6**), however, found that γ-Al_2O_3 always formed first. Pure amorphous alumina begins to change to γ-Al_2O_3 at 600° C (1112° F) and is largely crystalline by 900° C (1652° F), whereas when amorphous silica is present it does not begin to change until 925° C (1697° F). Silica retards the crystallisation, probably by steric hindrance, and once the energy is acquired for it to start it proceeds much more rapidly than the unretarded crystallisation at lower temperatures. Prolonged heating at temperatures below 925° C (1697° F) enables some γ-Al_2O_3 to form with corresponding decrease of the exothermic effect. A few degrees above the peak of this effect mullite begins to form. Roy, Roy and Francis (**R64**) assign the exothermic peak to the formation of mullite.

Grim (**G82**), too, considers that there is a gradual change from metakaolin to mullite with some structural carry-over. Observations of the X-ray diffraction patterns of samples were made during the time that they were being heated. He concludes that the metakaolin retains kaolin structure at first, and then there is a shift in the bonding, particularly the octahedral portion of the structure. It may or may not require additional temperature and time to cause this to become sufficiently crystalline to show an X-ray diffraction pattern. A well crystallised kaolinite changes to well developed mullite with less energy input than a poorly crystallised one.

Conversion of kaolinite $Al_2O_3.2SiO_2.2H_2O$ to mullite $Al_2O_3.SiO_2$ leaves free silica. The precise temperature at which this crystallises is not clear; tridymite may form at about 1000° C (1832° F) (**K75, R36**) as an intermediate, and cristobalite almost certainly appears above 1200° C (2192° F) (**M17**) if the silica has not meanwhile combined with other materials. Thus the result of firing pure kaolin to 1000° C (1832° F) is the formation of mullite and amorphous silica or tridymite, and on firing to 1200° C (2192° F) the silica appears as cristobalite. The product fuses at 1770° C (3218° F).

Grim (**G82**) summarises his work on the heating of montmorillonite as follows:

'The structure of the montmorillonite remains the same, approximately, or expands a little bit when the hydroxyl water goes out of it, that is, the structure is retained if the temperature is raised up to about 800° C (1472° F) and then there is an interval of 50–100° C, depending on the particular kind of montmorillonite, when there is no crystalline phase. At higher temperatures the first development is a quartz phase followed by development of crystobalite which increases in abundance and crystallinity as the quartz phase disappears. Later cordierite, magnesium aluminium silicate, and mullite appear in varying abundance depending on the character of the montmorillonite.'

Of considerable interest is the fact that although there is a gap when no crystallinity can be detected, use of oriented particles shows that there is a relationship between the structures of the low- and high-temperature phases. The whole sequence is also very sensitive to the exchangeable bases in the clay, 2 or 3 m.e. of sodium in place of calcium changes the sequence of the appearance of high temperature phases.

Presence of other substances also alters the temperatures of all these changes very considerably. Many oxides will act as mineralisers in mullite formation, increasing the speed at which this occurs (*see* p. 190). Presence of a liquid phase due to the fusion of fluxes accelerates and alters reactions.

Fluxing action is noticeable in clays of large cation exchange capacity, if the counterion is sodium. The nature of the counterions also determines the packing and texture of the dried clay and leads to more or less ready vitrification in the firing. Here again the sodium clays give dense well-packed green pieces with maximum contact surface, which have higher firing shrinkage and greater fired strength and denseness than calcium or hydrogen clays (Table 8, **G88**).

THE RAW MATERIALS

TABLE 8
Effect of the Counterion on Firing Behaviour of Clays (G88)

Shale clay	Firing temperature							
	800° C		900° C		1000° C		1100° C	
	Porosity (%)	Transverse strength (lb/in²)	Porosity (%)	Transverse strength (lb/in²)	Porosity (%)	Transverse strength (lb/in²)	Porosity (%)	Transverse strength (lb/in²)
Raw clay	25·6	148	20·3	935	10·3	3279	4·8	6540
H-clay	26·1	126	22·3	622	15·2	1850	7·9	4800
Ca-clay	25·2	136	19·2	1020	8·1	4210	bloated	—
Na-clay	23·9	170	16·1	1400	5·2	6379	bloated	—

FIRING SHRINKAGE

Owing to the chemical changes that occur during firing there are considerable changes in volume of the clay. These changes are irregular, rapid expansion or contraction, being associated with the rapid chemical changes which are also evident as exothermic or endothermic reactions in the differential thermal analysis (Fig. 1.33). In general a fired clay piece is

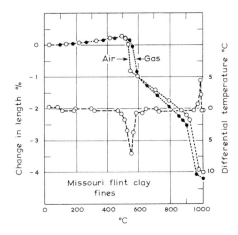

FIG. 1.33. The changes in length (upper curves) and thermal effects (lower curve) of a Missouri flint clay, when heated at 2·5° C/min, in air and in the products of gas combustion (Heindl and Mong, **H75**)

smaller than a raw one; there are, however, exceptions where gases are given off and then trapped in the clay. Just as the different clay minerals have different thermal reactions so their expansion curves are different (Fig. 1.34).

Hyslop and McMurdo (**H132**) have investigated the expansion and contraction of different clay minerals, and their results show how individual they are. They found that the curves for halloysite and china clay are similar but for the flattening of the latter between 700° C and 900° C and the cessation of contraction in the halloysite at 1000° C. The beidellite curve shows progressive contraction from 200° C to 900° C with accelerations at

420° C and 800° C. The contraction ceases at 900° C due to a gas reaction and the test piece heated to 1100° C was found to be bloated. The Marquoketa shale and the Fithian underclay are similar except for the sharp expansion of the former between 950° C and 1000° C caused by the gas reaction due to carbonates.

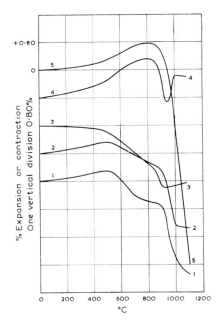

FIG. 1.34. *The thermal expansion of some clay minerals heated at 10° C/min*
(1) *China clay (English)* $Al_2O_3.2SiO_2.2H_2O$
(2) *Halloysite (Missouri)* $Al_2O_3.2SiO_2.4H_2O$
(3) *Beidellite (Putnam clay, colloidal fraction)* $R_2O_3.3SiO_2.xH_2O \pm MO$
(4) *Sericitic mineral (Maquoketa shale)* $2K_2O.3MO.8R_2O_3.24SiO_2.12H_2O$
(5) *Sericitic clay (Fithian underclay)*
(Hyslop and McMurdo, **H132**)

Although these curves show primarily the shrinkage of different clay minerals the effects of non-clay matter do become apparent. Silica has a marked effect and Bigot (**B50**) concludes that only clays containing no free silica do not expand somewhat below 1000° C.

Lépingle (**L30**) has done some approximate measurements of some commercial clays. His graphs are schematic and should be taken qualitatively rather than quantitatively. The rate of heating was 10° C in twelve minutes (Fig. 1.35).

Westman (**W48**) shows schematically the volume changes on drying and firing three typical clays and includes the relationship between the clay matter itself and the volume it takes up when packed dry, or with water in the plastic state. Thus for each clay he shows in the top line the heat treatment of a theoretical 100 cm³ cube of solid crystalline material of zero porosity. On heating to 400° C this loses its water of crystallisation but does not contract. Thereafter the change to mullite sets in and contraction occurs without further loss in weight. In the second line he shows first the

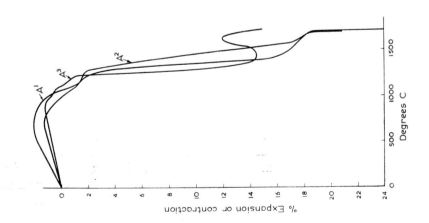
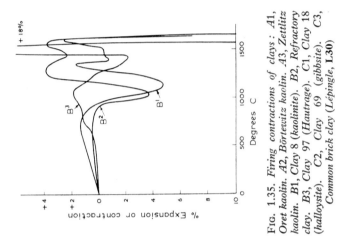
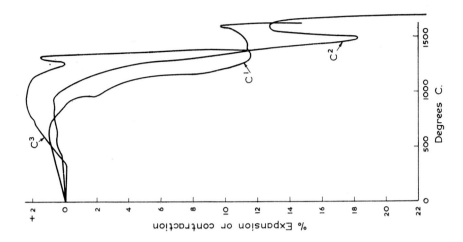

FIG. 1.35. *Firing contractions of clays*: A1, Oret kaolin. A2, Börtewitz kaolin. A3, Zettlitz kaolin. B1, Clay 8 (kaolinite). B2, Refractory clay. B3, Clay 97 (Hautrage). C1, Clay 18 (halloysite). C2, Clay 69 (gibbsite). C3, Common brick clay (Lépingle, L30)

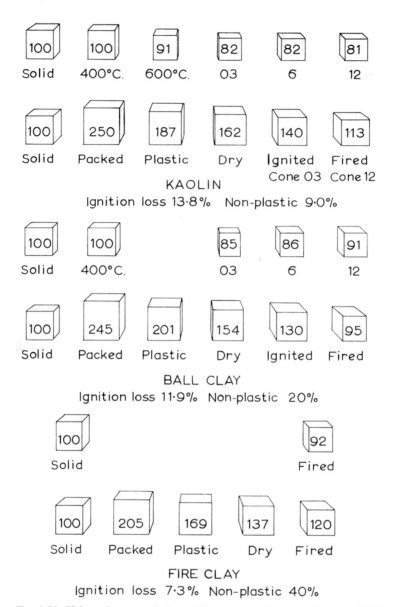

FIG. 1.36. *Volume changes on drying and firing three typical clays (Westman, W48)*

theoretical solid, then the same amount of clay made up of particles of the normal commercial size packed as a tight cake (packed); it is then made plastic with water, dried, fired to cone 03 and then cone 12. Comparison of the two sets shows whether contraction is due to loss of water, shrinkage of clay particles or shrinkage of pores (Fig. 1.36). Any orientation of clay mineral particles during preparation of pieces gives rise to differential shrinkages in different planes.

The importance of knowing the shrinkage curves of clays and clay bodies lies not only in determining the fired size of ware but also in adjusting the firing schedule to reduce the speed of volume changes.

In this connection it can be of greater value to have curves of volume changes occurring during firing of a clay under load, conditions simulating those of pieces at the bottom of the kiln setting being used. Freeman (**F30**) obtained such curves for brick clays, and found that these could be divided up into four types which proved to be related to the composition of the clays (Fig. 1.37). Type A curves were given by clays having no $CaCO_3$ or quan-

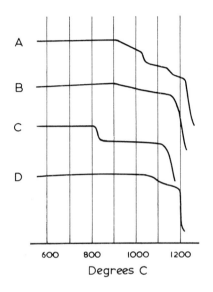

FIG. 1.37. *Firing-under-load curve types for the four categories of brick clays* (*Freeman*, **F30**)

tities of it less than 9% (5% CaO). Type B is given by clays of $CaCO_3$ content between 9 and 25%. Type C are high lime clays of from 25 to 30% $CaCO_3$ content. Type D curves are associated with clays of high natural grog content, normally sand.

In order to obtain ware of the same size throughout a kiln setting it is important that those parts of it last to be heated to the right temperature for the reactions causing volume changes are kept at this temperature for long enough for the reaction to be completed. Keeping some of the ware at this temperature for longer does not increase the volume change but even prolonged heating at a lower temperature will not bring it about. Where the volume changes are very rapid slow heating is very important to prevent cracking, etc., of the ware.

Other factors that affect the shrinkage of a given body to a lesser extent are: (1) atmospheric conditions during firing; (2) grain size; (3) deairing (**H126**).

The firing shrinkage of a raw clay can be diminished without changing the composition by mixing grog made by burning the same clay until the main shrinkage has occurred. This does not shrink in its second firing.

The shrinkage can also be increased without changing the composition by adding to the raw clay a material that will burn out completely, *e.g.* certain coal dusts. Certain secondary clays do in fact contain finely divided coal particles of colloidal dimensions which burn out during the firing causing high shrinkage.

The firing shrinkage of ceramic bodies is made use of in the Buller's ring system of measuring heat treatment (p. 938).

The Correlation between the Mineral Content, the Grain Size and the Ceramic Properties of Clays

At the beginning of the discussion on the properties of clays a classification was given dividing them into three groups. Having considered the different properties it is now possible to go into their correlation. Linseis (**L34**) has gone into this very carefully in his recent (1950) series of articles. Linseis investigated four kaolins, four pottery clays and three clays for heavy clay products, namely:

(1) Amberg kaolin KO.
(2) Hirschau kaolin DIIa.
(3) Konnersreuth kaolin.
(4) Schnaittenbach kaolin.
(5) Aussernzell clay.
(6) Axtham clay.
(7) Grossalmerode clay.
(8) Hebertsfelden clay.
(9) Neuötting clay.
(10) Schwarzenfeld clay.
(11) Tambach clay.

The mineralogical constitution of the clays was determined by: (*a*) chemical and the Berdel (**B39**) rational analysis; (*b*) X-ray analysis; (*c*) thermal and differential thermal analysis.

The properties derived from the mineral composition determined were: (*a*) grain size distribution; (*b*) base-exchange capacity; (*c*) pH value.

The ceramic properties investigated were: (*a*) green strength; (*b*) shrinkage; (*c*) viscosity; (*d*) plasticity.

It is generally accepted that the ceramic properties are chiefly dependent on the mineralogical composition and the grain sizes of clays. The question of how best to represent these properties numerically in order to demonstrate correlation is not so easy.

In 1933 Vieweg (**V18**) investigated the possibility of correlation of average particle size with drying shrinkage, firing shrinkage and apparent porosity on firing. He found that there is no apparent relationship between average particle size and linear drying shrinkage, a property probably more dependent on the very fine particles and the grading; there is some relationship between particle size and firing shrinkage, the coarser clays shrinking less; and a fairly smooth curve is obtained by plotting particle size against porosity of the fired clay.

The cohesion of particles in a system with varying particle sizes must

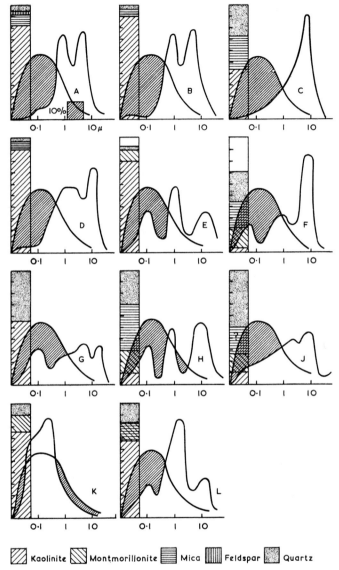

FIG. 1.38. *Grain size distribution, distribution factor and mineral composition of eleven clays* : A, Amberger kaolin. B, Hirschauer kaolin. C, Konnersreuther kaolin. D, Schnaittenbacher kaolin. E, Ausssernzeller clay. F, Axthamer clay. G, Grossalmeroder clay. H, Hebertsfeldener clay. J, Neuöttinger clay. K, Schwarzenfelder clay. L, Tambacher clay (Linseis, **L34**)

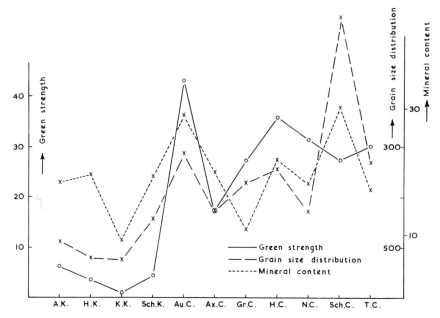

(a) *Relation of green strength to mineral content and grain size distribution*

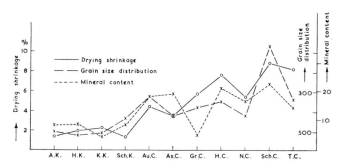

(b) *Relation of drying shrinkage to mineral content and grain size distribution*

Fig. 1.39

depend very largely on the possibility of packing them together satisfactorily and therefore on the grading.

Linseis (**L34**) plotted the particle size distribution and then superimposed a Maxwell distribution curve chosen to correspond with the best ceramic clays. He then measured the area (shaded) where the two curves do not overlap and used this factor of divergence from ideality as a grain size factor (Fig. 1.38). His mineral composition factor was derived from the montmorillonite content plus one-fifth of the content of other clay minerals.

It can be seen from the graphs in Fig. 1.39 that in general the ceramic properties are related to grain size distribution and mineral content. The

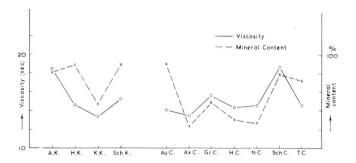

(c) *Relation of viscosity to mineral content (the viscosity of the kaolins is measured at a different concentration to that of the other clays, so the curves must be separated)*

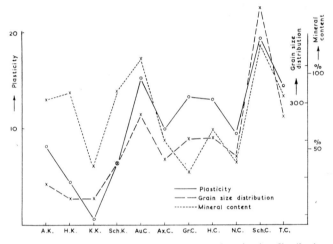

(d) *Relation of plasticity to mineral content and grain size distribution (Linseis, L34)*

FIG. 1.39 (*continued*)

general conclusions drawn by Linseis (**L34**) from these investigations are that plasticity, drying shrinkage and dry strength increase as the grain-size distribution approaches the ideal, and with increase of clay mineral content. In the case of plasticity the relationship is proportional to:

kaolinite content + 3 × montmorillonite content − $\frac{1}{3}$ quartz content

The drying shrinkage is proportional to:

kaolinite content + 10 × montmorillonite content

and the green strength is proportional to:

kaolinite + 5 × montmorillonite content

It will be noted that illite has been omitted from these considerations. Linseis amongst others experienced difficulty in estimating this mineral.

Moisture Expansion of Fired Clay Bodies

Compared with raw clay, fired clay is almost completely unaffected by water, but long exposure to moisture (or autoclaving) can bring about changes that are noticeable under certain conditions. Moisture is taken up by ceramic bodies in three ways:

(1) Absorption into the larger pores, whence it can be expelled by heating to 105° C.
(2) Adsorption on the body surface, reducing the surface energy of the body, which behaves elastically and expands.
(3) Chemical combination with the body, water entering the crystal lattices by diffusion and enlarging or changing them.

Both adsorption and chemical combination give rise to expansion of the body. If the piece is glazed on one side, expansion of the body puts the glaze in tension and can cause it to craze. This is the main cause of delayed crazing in wall tiles.

Moisture expansion is affected by firing temperature and composition. Both the mineralogical composition and the pore size distribution of a fired piece are affected by the firing temperature. For bodies normally matured at about 1000° C (1832° F) porosity decreases continually once vitrification has started but the moisture expansion gradually increases to a maximum at about 1000° C (1832° F), thereafter decreasing with the porosity. Adsorption of water therefore depends not only on the surface area but also on the nature of the surface. It has been established that crystalline body constituents shrink when treated with steam whereas glassy ones expand. The glass phase is therefore the chief cause of moisture expansion. Certainly the moisture expansion of pure kaolinite is increased if glassy material is formed at lower temperatures by adding sodium or potassium carbonates, whereas moisture expansion is decreased if calcium carbonate is added (**M81**).

As well as the slow moisture expansion observable on porous pieces in use, a faster expansion lasting only a number of hours is found to occur in pieces cooled in a dry kiln atmosphere and subsequently exposed to normal moist air (**N25**).

EXPANSION OF FIRED CLAY BODIES, DUE TO SOLUBLE SALTS

Norris, Vaughan, Harrison and Seabridge (**N25**) have also found that permanent expansion of porous bodies is brought about by small quantities of soluble salts (~ 1 mg/g). These salts can only be dissolved out of the pores with great difficulty. The actual expansion depends on the salt used, carbonates being more effective than chlorides or sulphates, and also on the way it is introduced, repeated small additions being more effective than one large one. The effect of firing temperature on expansion due to soluble salts is the reverse of its effect on moisture expansion, so that high fired pieces may have a salt-caused expansion exceeding the moisture one.

NON-CLAY PLASTIC RAW MATERIALS

There are two further minerals with structures akin to those of the clay minerals which are somewhat plastic. These are the hydrated magnesium silicates talc and steatite, and the hydrated aluminium silicate pyrophyllite. Their structures are layered, leading to cleavage into flakes, and they can be shaped by pressure when moist, particularly if finely ground; inclusion of these substances in a ceramic body does not lower the plasticity of the clay so much as a truly non-plastic ingredient does.

Talc and Steatite

Talc and steatite are different forms of hydrated magnesium silicate which has a composition varying between the limits:

$$3MgO. \; 4SiO_2. \; H_2O \quad \text{and} \quad 4MgO. \; 5SiO_2. \; H_2O$$
$$31.8\% \quad 63.5\% \quad 4.7\% \quad\quad\quad 33.5\% \quad 62.7\% \quad 3.8\%$$

associated with impurities introducing alumina, iron, lime, alkalies and more water.

Talc and steatite are secondary rocks formed by various interactions of water together with magnesium salts on the primary rocks. Although there are a large number of deposits, workable sources of a pure product are not nearly so frequent as those of clay. In Great Britain magnesium silicate is mined chiefly in Argyllshire and other parts of Scotland; in the United States important deposits occur in California and on the eastern seaboard; talc resources are also found in France, India, Egypt, China, Australia,

TABLE 9

Analyses and Properties of some Talcs

(after Stone and Gunzenhauser (**H48**))

Type or Origin	China	France	Germany	Italy	India	Manchuria	U.S.A. California	U.S.A. N.Y.	U.S.A. high fired steatite	U.S.A. low fired bodies	U.S.A. high alumina bodies
SiO_2	62.35	60.0	60.3	59.4	61.24	61.4	61.9	55.9	60.20	57.00	32.10
MgO	32.1	31.8	31.0	31.6	32.42	32.5	31.3	30.7	30.37	26.95	30.35
Al_2O_3	0.17	⎫	⎫	⎫	⎰ 1.42	0.56	0.45	⎫	⎰ 1.87	1.86	22.06
Fe_2O_3	0.05	⎬ 2.3	⎬ 3.1	⎬ 2.2	⎱ 0.02	0.45	0.85	⎬ 1.3	⎱ 0.88	0.26	2.99
CaO	—	⎭ 0.3	⎭ —	⎭ —		0.14		⎭ 6.2	0.76	8.58	0.65
K_2O	0.77	—	—	—			⎱ 0.20		⎰ 0.24	0.28	0.15
Na_2O	0.19	—	—	—							
Loss on ignition	4.37	5.6	5.6	6.8	4.90	4.95	5.30	5.90	5.41	4.42	11.48
Acid solubles as CaO									0.72	2.15	1.55
pH value									9.3	9.5	9.1
S.G.									2.74	2.85	2.80
Apparent density (lb/ft³) (g/cm³)									26.70 0.43	32.05 0.51	27.58 0.44

Austria, Bavaria, Roumania, Manchuria, Norway and Czechslovakia. The mineral is white to light green, extremely soft (Moh's scale 1) and has a greasy feel. Where the mineral is laminated and very soft it is termed talc; impure varieties are known as soapstone. The more massive and relatively pure varieties are known as steatite.

Hausner and Gunzenhauser (**H48**) summarise the requirements for talc for steatite insulators (of which it is the major constituent). It should be soft, substantially free from foliated, flaky or fibrous material and from gangue which makes dark spots when fired. The colour of the raw talc is of no importance, but should be white after heating to 1350° C (2462° F) in an oxidising or neutral atmosphere. The desirable chemical composition as compared with the theoretical is:

Theoretical		Desired limits	
$3MgO.4SiO_2.H_2O$			
SiO_2	63·4%	SiO_2 not less than	60%
MgO	31·9%	MgO not less than	30%
H_2O	4·7%		
		Al_2O_3 not more than	2·5%
		CaO not more than	1·0%
		Fe_2O_3 not more than	1·5%
		$Na_2O + K_2O$ not more than	0·4%
		Loss on ignition not more than	6·0%
		Acid soluble lime not more than	1·0%

PRODUCTION

Talc is very difficult to purify by gravity separation but lends itself admirably to froth flotation methods. It floats readily with pine oil through the pH range from 5 to 12, and also with cresylic acid or alcohol (**N11**).

DATA AND PROPERTIES

Vitrification range, 2320° to 2460° F ($\sim$ 1270° to 1350° C) (**N11**).
Melting point, 2714° F (1490° C) (**A89**).
Specific gravity, 2·6–2·8.
Hardness, 1–2 Moh.
Thermal conductivity (2·3–2·8)/360 c.g.s.
Specific heat, 0·2–0·3 c.g.s.
Coefficient of cubical expansion, *ca.* 56×10^{-7}.
Coefficient of linear expansion, $4·5 \times 10^{-6}$.

REACTIONS OCCURRING IN TALC ON HEATING

Talc undergoes decomposition followed by several transitions on heating. The temperatures at which the changes occur vary not only with the nature and amounts of impurities but also on the fineness of grinding and hence cannot be given exactly.

On heating a talc, of total ignition loss of 5·5% (slightly above average), water is lost, and this occurs in three stages. Some 0·4% which is held as H_2O molecules comes off in the ranges from 120° to 200° C (250 to 390° F) and 350° to 500° C (660 to 930° F). The remaining 5·1% is present as chemically bound hydroxyl groups and comes off at 600° to 1050° C (1100 to 1920° F) (**A202**). At the same time free silica splits off, leaving magnesium metasilicate. Talc does not split into free magnesium and silicon oxides. The intermediate phases now formed are variously designated; for example according to Avgustinik and Vignerganz (**A202**) β-magnesium metasilicate is formed at about 600° C (1100° F) in a hydrated orthogonal amphibole form, and loses further water at 1000° C (1830° F) to give α-magnesium metasilicate in an anhydrous monoclinic amphibole form, and the silica dissolves in these. At about 1200° C (2190° F) there is a further recrystallisation to protoenstatite. Kedesdy (**K9**) gives the dehydration temperature as 800° C (1470° F) a single intermediate product termed protoenstatite (more correctly enstatite) and at 1400° C (2550° F) formation of clinoenstatite (protoenstatite, see below), whereas Ewell, Bunting and Geller (**E29**) state that the talc decomposed to enstatite, amorphous silica and water vapour at 800 to 840° C (1470 to 1540° F) the enstatite gradually changing to clinoenstatite (now called protoenstatite) around 1200° C (2190° F) and the amorphous silica to cristobalite around 1300° C (2370° F). They found two endothermic heat effects, a small one at 530 to 572° C (986 to 1062° F) and a larger one at 860 to 953° C (1580 to 1747° F).

According to recent work by Foster (**F20**) the relationship of the enstatite phases is:

enstatite, stable low-temperature form obtained by formation or crystallisation below about 1260° C (2300° F) (perhaps as low as 1140° C (2084° F)) and converting on heating to:
protenstatite, the stable high-temperature form which does not normally reinvert to enstatite on cooling, instead forming:
clinoenstatite.

The salient feature in all these phase relationships during the heating of talc is that a magnesium silicate is present throughout and this sets up a strong network (Fig. 1.40).

Magnesium silicate readily reacts either by solution, formation of eutectics, etc., or chemically, with alumina, iron oxides, lime, etc. In the system MgO–Al_2O_3–SiO_2 alone the melting points vary from about 1200 to 2500° C (2190 to 4530° F) and this system includes cordierite $2MgO.2Al_2O_3.5SiO_2$ which has exceptionally low thermal expansion. The results obtained by introducing varying amounts of talc with different impurities into different ceramic bodies are therefore very varied. A large range of effects can be achieved by using talc, but often only within carefully defined limits of composition and firing temperature. If this fact is not appreciated results obtained may be entirely different to claims made by others.

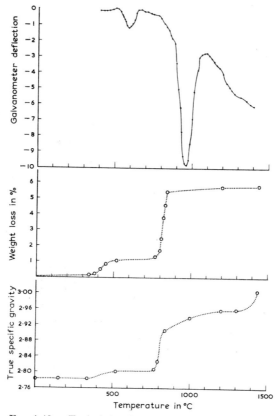

FIG. 1.40. *Typical heating curve, and curves showing weight-loss/temperature and true specific gravity/temperature relations for a nearly pure talc (Ewell, Bunting and Geller,* **E29**)

USES

Suitable quantities of talc or steatite introduced into certain ceramic bodies impart a number of advantageous properties.

As a major constituent:

(1) Where talc is the main constituent (70 to 90%) bodies highly suitable as electrical insulators are obtained and the body type has become known as 'steatite'.
(2) Where the body composition approximates to that of cordierite $2MgO.2Al_2O_3.5SiO_2$, bodies of low thermal expansion, and therefore high thermal shock resistance, together with good electrical properties are obtained.

In smaller quantities:

(3) The fluxing action of the magnesia lowers maturing temperature, reduces porosity and increases strength of semivitreous bodies but shortens the firing range of vitreous bodies.
(4) Bodies of high specific heat.
(5) Bodies of high resistance to acid attack.
(6) Bodies of decreased moisture expansion and hence less delayed crazing.

There is frequently, however, an undesirable colour due to the iron content, the colour of which is augmented on the surface by the magnesia.

The main uses of talc and steatites are therefore those taking full advantage of the good electrical properties and thermal shock resistance, that is for electrical porcelain, and saggers and kiln furniture for use up to 1250° C (2280° F).

In electrical porcelain the term 'steatite' denotes the final product where the body contained from 70 to 90% ground talc or steatite with small additions of clay, feldspar or alkaline earth compounds. The product after firing to cone 12 to 14 has a uniform structure of crystalline magnesium silicate. The modulus of rupture of such a body fired to cone 12 is given as greater than 17 000 lb/in^2 (1195 kg/cm^2). It has a good dielectric strength (**A89**).

Some 5% of the mined steatite is directly machined to shape and fired. Solid steatite contains no hygroscopic water as does some talc and the loss of the combined water causes a firing shrinkage of only 1%, so that articles can be made to accurate dimensions. Such production is, however, very limited (**A89**).

Talc is used in amounts up to 50% in earthenware wall tiles, when delayed crazing is greatly reduced. It is used in quantities up to 10% in some porous dinnerware bodies to improve the strength and resistance to shock, and decrease moisture crazing. It also partially replaces feldspar for fluxing.

Pyrophyllite (Agalmatolith)

Pyrophyllite is often confused with talc because of the remarkable similarity of their physical properties. It is, however, a hydrous aluminium silicate, $Al_2O_3.4SiO_2.H_2O$, and hence chemically more akin to the clay minerals. Its structure is similar to the ideal structure for montmorillonite, it is perfectly crystalline and hence does not absorb ions and water.

Important deposits of pyrophyllite occur in North Carolina, Pennsylvania, California and Newfoundland, Brazil, China, and South Africa where it occurs as the main mineral in the 'Union Stone' or 'Wonder Stone'. There are three different varieties of pyrophyllite in the North Carolina deposit alone. These are descriptively termed 'foliated', 'radiated granular' and 'massive compact', and on grinding give, respectively, flaky, needle and normal particles (**G72**).

The main impurity in pyrophyllite is quartz. This makes the mineral abrasive, in spite of the softness of pyrophyllite itself. As the specific gravities are almost the same sedimentation separation is useless. Some purification can be achieved by differential grinding and air flotation. The most promise is shown by froth flotation methods where a 97% pure pyrophyllite was produced in test runs from a crude product containing 56% (**L12**).

DATA AND PROPERTIES

Chemical composition:

$$Al_2O_3.4SiO_2.H_2O$$
$$28.3\% \quad 66.7\% \quad 5.0\%$$

Typical analysis (**S110**):

SiO_2	75·30
Al_2O_3	20·74
Fe_2O_3	0·08
CaO	0·08
K_2O	0·21
loss on ignition	3·67
total	100·08

Melting point, 1700° C (3092° F) (**N17**).
Vitrification point, variously given as cone 8, 1250° C (2282° F) or cone 15, 1435° C (2615° F) (**A121**).
Specific gravity, 2·8–2·9.
Hardness, Moh 1–2.
The shrinkage on firing is small, and the product is fairly resistant to thermal shock by virtue of its high thermal conductivity and low coefficient of thermal expansion (**N17**).

PYROCHEMICAL PROPERTIES

On heating pyrophyllite loses its water of constitution between 400° and 700° C (750 and 1290° F) and the alumino-silicate layers separate somewhat, giving a permanent linear expansion. Between 1000° and 1100° C (1830° and 2010° F) the pyrophyllite structure breaks down to mullite and free silica. During both reactions no marked thermal effects are shown, as is seen by the curve obtained, compared with that given by kaolinite (**P15**) (*see* Fig. 1.41). Pyrophyllite hardly shrinks at all on firing.

USES

Pyrophyllite may be used to replace a part or all of the flint or feldspar or even kaolin in certain ceramic bodies. The product is more resistant to

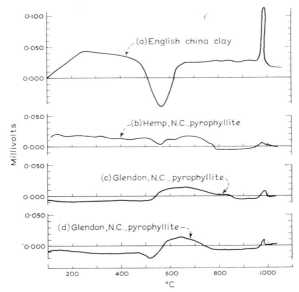

Fig. 1.41. *Differential thermal curves for pyrophyllite compared with kaolinite (Parmelee and Barrett, P15)*

sudden temperature changes because of its lower thermal expansion. It reduces the shrinkage of the body considerably, even more than might be expected from the percentage of pyrophyllite present, and it increases the firing range and the fired strength (**N11**).

It is used extensively in wall tiles and to a lesser extent in dinnerware. Because it lowers the plasticity it cannot be added in quantities greater than 40% to plastic bodies.

Pyrophyllite has also found use for electrical insulators containing 94% to 96% of it, and for special refractories (**A89**). It is also used in unfired refractory bricks, bonded with refractory clay and dry-pressed at 5000 lb/in² (350 kg/cm²). These bricks withstand temperatures up to cone 31 (3074° F, 1690° C) (**A36**).

Sericite Pyrophyllite

A pyrophyllite containing up to 15% of the micaceous-type mineral sericite has been used successfully in certain pottery bodies to replace flint, thereby increasing the fired strength (**A94**). Sericite is:

$$K_2O. \; 3Al_2O_3. \; 6SiO_2. \; 2H_2O$$
$$11\cdot83\% \quad 38\cdot40\% \quad 45\cdot25\% \quad 4\cdot52\%$$

It has been claimed that sericite has been used in place of other fluxes to make a body with only 6 to 8% firing shrinkage (**R54**).

THE NON-PLASTIC RAW MATERIALS

There are several major and numerous minor non-plastic raw materials used in the ceramics industry. Used in clay bodies they reduce the plasticity and with it the drying time and shrinkage; increase the green and sometimes the fired strength; alter the maturing range and temperature and the properties of the burnt product. The three most important materials for the fine ceramic industry, silica, feldspar and bone ash are discussed first. There are also numerous non-plastic materials which, when bonded with a plastic one, constitute the major component of certain ceramic bodies.

Silica

After oxygen, silicon is the most plentiful element on the earth's crust. It occurs as its oxide either free or combined with metallic oxides as silicates. Free silica is abundant and amongst the purest of naturally occurring minerals. Silica crystallises in different forms at different temperatures but as the changes are slow the unstable forms occur naturally as well as the far more common stable form which is α-quartz.

QUARTZ

Quartz crystals occur in the primary rocks, granite, gneiss, etc., sometimes by themselves as individual veins and often as grains interspersed between the other minerals present. Quartz is almost unaffected by weathering, and the veins may remain intact; the quartz amongst rock that weathers and disintegrates becomes washed away with it. It then may either be deposited separately or remain intermixed with the weathering product as is found in crude china clay.

SANDSTONE, GANISTER, QUARTZITE, SAND

Sedimentary quartz makes up the sandstones, quartzites, ganisters, sands, etc. The shape of sand grains in free sand or sandstone and quartzite rocks depends on the amount of erosion they have undergone and any other matter that may have coated them. In the sandstones the sand grains may be cemented together by silica, lime, clay, iron oxide, mica, etc., and their usefulness for different purposes depends on the impurities thus introduced; ganister is a fine-grained sandstone containing some clay; quartzite is a metamorphosed silica-bonded sandstone with the deposition of the secondary silica cementing the quartz grains together so strongly that breakage will occur across the grains (**F4**). Sandstones are widespread, ganisters occur in Great Britain and quartzites are fairly well distributed, but the North German Findlings quartzite is particularly fine grained and specially well suited for making silica bricks (**C26**).

FLINT

In England, France and Denmark there are large quantities of flint pebbles. These consist of cryptocrystalline quartz with a small quantity of water and carbonaceous matter. They are generally coated with calcareous matter. Flint breaks with a characteristic conchoidal fracture and appears black.

DIATOMITE

Amorphous silica is the main chemical constituent of diatomaceous earth, also called diatomaceous silica, diatomite and Kieselguhr, which is made up of the skeletons of diatoms. The mineral has a high porosity which gives it a very low thermal conductivity. It is used for special insulation bodies. In Denmark there is an especially useful deposit of diatomaceous clay containing siliceous diatom skeletons and plastic clay making it suitable for direct manufacture of insulating bricks. It is termed Moler.

The metastable crystalline forms of silica, cristobalite and tridymite occasionally occur in nature but are not of commercial importance as raw materials (Tables 10 and 11).

TABLE 10

Silica Raw Materials

Material	S.G.	Hardness	M.P.
α-Quartz	2·65 (2·5–2·8)	7 (**N17**)	< 1470° C (2678° F)
α-Tridymite	(2·27–2·35) 2·32	6·5–7 (**N17**)	1670° C (3038° F)
α-Cristobalite	2·32–(2·33)	6–7 (**N17**)	1710° C (3110° F)
Quartz sand	2·62–2·67 Apparent 0·19		
Sandstone	2·4 Apparent 0·22 (**N17**)		
Flint (flint A)	2·62 (**H109**)	6–7 (**S83**)	
Flint A′ calcined at 450° C	2·52 (**H109**)		
Flint A′ calcined at 1100° C	2·48 (**H109**)	easily crushable	
Flint B calcined at 1250° C	2·23 (**H109**)		
Opal	1·9–2·5 (**N17**)	5–6	
Diatomaceous earth	2·2–2·36 (**N17**) Apparent 0·16–0·72 (**N17**)		

PRODUCTION

Various forms of crude quartz are mined by normal quarrying methods. As the resources of the pure raw material become used up so the less pure ones must be used and methods evolved for purifying them. The mined

TABLE 11

Properties of Welsh, Sharon, Findlings and South African Quartzites (C26) and German Quartz Sands (Q1)

	(1)	(2)	(3)	(4)	(5)	(6)	(7)
	Coarse-grained quartzite, low titania content		Chalcedonic silica matrix, high titania content		Dörentrup crystal quartz sand	Duing crystal quartz sand	Bonn-hausen crystal quartz sand
	Normal quartzite (washed) Bwlchgwyn, N. Wales	Sharon conglomerate U.S.A.	Findlings quartzite Germany	Opaline silcrete Albertinia, S. Africa			
Micro-examination	Compact tough grey rock composed of interlocked strained quartz grains with some interstitial sericite and kaolinite	Compact translucent pebbles composed of closely sintered quartz grains, often much strained and containing abundant inclusions. Grain size generally 0·1–0·5 mm	Compact fawn rock breaking with conchoidal fracture. Composed of scattered quartz grains set in an extremely fine-grained chalcedonic matrix	Hard translucent fawn to pale grey rocks breaking with conchoidal fracture: 'ring' under hammer and are rather brittle. Composed of a few scattered quartz grains set in a compact matrix of extremely fine-grained transparent chalcedonic silica			
Chemical analysis:							
loss on ignition (%)					0·091	0·040	0·131
SiO_2	97·39	98·0	97·8	96·1– 97·8	99·816	99·700	99·659
TiO_2	0·11	0·1	—	1·50– 1·93		0·030	0·048
Al_2O_3	0·73	0·3	0·4	0·38– 0·47	0·060	0·070	0·140
Fe_2O_3	0·78	0·5	0·3	0·15– 0·56	0·015	0·015	0·015
CaO		$Al_2O_3+TiO_2+$ alkalies 0·5			0·010	trace	0·021
MgO					0·004	trace	trace
NaKO					0·011	trace	
Physical properties:							
As received:							
Specific gravity	2·66	2·64	2·64	2·62– 2·67			
Porosity (%)	1·6	0·6	0·3	1·1– 7·4*			
After being fired:	2 hr at 1450 °C	1 hr at 1450° C	1 hr at 1450° C	2 hr at 1500° C			
Specific gravity	2·37	2·37	2·41	2·27– 2·30			
Porosity (%)	16·8	15·8	3·0	3·9 – 9·9*			

* High values due to interstitial cavities and large pores.

THE RAW MATERIALS

rock is crushed to free the grains but not break them and then thoroughly washed and magneted. Impurities may then be removed by froth flotation or other methods (**D10, S12**). It is then thoroughly ground to uniform particle size.

European flint is too hard to grind. It is therefore first calcined to 300 to 900° C (570 to 1650° F) when the structure opens up. The specific gravity of the product depends on the temperature and time of calcining. The coating on flint pebbles is of variable composition, a fact which is usually not taken into account when the whole is ground up, giving a somewhat variable product.

In the United States the term 'potter's flint' is applied to all ground silica used in ceramics. This is inaccurate.

The different specific gravities affect volume measurement of the raw material, and the hardness affects the choice of material for crushing and grinding so that only sands, sandstones, calcined flint A' and burnt diatomaceous earth are crushed.

The Action of Heat on Silica. The stable form of silica at room temperature is α-quartz. On heating to 573° C (1063° F) this changes rapidly and reversibly to β-quartz with a volume increase of 2%. If this is heated rapidly it will melt somewhat below 1470° C (2680° F), but on further slow heating β-quartz changes to $β_2$-tridymite at 870° C (1600° F) with a volume increase of 12%, and then at 1470° C (2678° F) to β-cristobalite with a volume increase of 5%; this in turn fuses at 1710° C (3110° F). These changes are all reversible under suitable conditions of slow cooling and the presence or absence of other substances, but under different conditions of cooling the silica may be arrested in one of the higher temperature forms, *i.e.* vitreous silica, cristobalite, or tridymite. On cooling the high-temperature forms of tridymite and cristobalite change rapidly to the lower-temperature forms at 163 and 117° C (325 and 263° F) and 220 to 275° C (428 to 527° F), respectively, the cristobalite contracting by 5%, so that by introducing the raw material in the form of quartz the product after firing may be either α-quartz, α-tridymite, α-cristobalite or vitreous silica, each with its own specific gravity, the piece undergoing strains whenever sudden volume changes occur. The different crystalline forms and the conditions that affect their formation will be discussed in greater detail in Chapter 2.

The diagram represents the different forms of silica. The continuous line shows the form that is stable at the given temperature, the broken line metastable forms. The specific gravity of each is given and the percentage volume increase when it changes to another form either directly (continuous arrow) or indirectly (broken arrow) (Fig. 1.42).

When heated in the presence of feldspar, which fuses to a glass at a much lower temperature, silica tends to dissolve in this latter. The speed at which it does this is determined not only by the particle size but by the fine structure of the particles themselves. Thus a quartz whose structure is interrupted either by

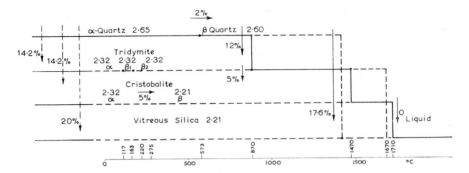

Fig. 1.42. *Silica, its forms, inversion temperatures, S.G., and expansions (adapted from Niederleuther, **N17**, and Norton, **N36**)*

inclusion of impurities, especially mica, or by the conditions of its crystallisation, will dissolve and react more readily than a pure quartz sand. So that if a translucent porcelain is desired, the more soluble slightly impure forms of silica give a better result than the hard (and therefore often the less finely ground) pure quartz sands (**D41**).

USES

Finely ground silica is a major component (about 20 to 50%) of normal ceramic bodies for making earthenware, porcelain and sanitary ware. As such it may be introduced either as ground calcined flint, or as ground quartz sand, sandstone, etc., in Great Britain and Europe but normally only in the latter form in the United States (where it is nevertheless termed 'flint'). Normal surface factors of this ground silica have been given as follows: for earthenware, 230 ± 10; for tiles, 210 ± 10.

The effects of the two raw materials are somewhat different. Flint is more readily converted to cristobalite than quartz, and the contraction on cooling β-cristobalite to the α-form is somewhat less sudden although larger than the one for the $\beta \rightarrow \alpha$-quartz change; hence there is less danger from cracking, and the glaze is compressed, thus preventing crazing. Alternatively if ground quartz is used it is less susceptible to thermal shock at 220–275° C, although more so at 573° C than the body containing flint.

A major snag in the use of ground flint, the traditional material for pottery in Great Britain, is the variable quantity of impurities, which are rarely adequately removed, and generally not sufficiently checked. The chief offenders are iron, which affects the colour, and lime, which alters the pH of the moist body, and the temperature and type of reactions on firing. Such variations from batch to batch may cause large losses. Washed ground sand usually has a more reliable composition, but because of its more perfect and compact structure enters into inversion reactions and solution less readily; its fewer impurities therefore do not become distributed and the fired product may be more discoloured than one containing more iron impurities differently distributed and combined.

A large quantity of coarser alkali and lime-free silica or siliceous material is used for refractory silica and semi-silica bricks. The British raw material for these is crushed ganister containing some clay which acts as a bond in the fired product. Elsewhere quartzites (especially the North German Findling variety), sandstones and sands are used. The raw material is bonded with an organic binder, generally sulphite lye, and quicklime.

Feldspar

Feldspar is the most important flux used in ceramic bodies and glazes. The term feldspar covers a number of alkali or alkaline earth aluminium silicates. It is an igneous mineral, one of the commonest in primary rocks, where it occurs chiefly mixed with quartz and often with mica.

Natural feldspars are usually a mixture in varying proportions of the aluminium silicates of sodium, potassium, calcium, lithium and occasionally barium and caesium. A small quantity of rubidium is also always associated with the potassium in these primary crystalline rocks. V. M. Goldschmidt's work on the conditions for simultaneous crystallisation makes clear that this is because of the chemical and size similarity between potassium and rubidium; it also follows that rubidium does not naturally occur together with sodium.

The feldspars all have a three-dimensional silicon–aluminium–oxygen framework with the exception of spodumene which has the chain $(SiO_3)n$ ion (**W40**).

It is worth noting that the ratio of basic oxide to alumina to silica tends to be 1:1:6 for the alkali metals and 1:1:2 for the alkaline earths (Table 12). (Artificial rubidium and strontium compounds also fall into line.)

For ceramic bodies the potash spars are the most important, the lower fusing soda spars find more use in glazes, and lithium spars are gaining considerably in importance now because of their greater fluxing action.

Although feldspar is widespread in igneous rocks its occurrence in sufficient concentration to be worth mining is not very frequent. The chief European deposits are in Sweden, Norway, Finland, France, Italy, and Czechoslovakia (some is mined in the islands off Scotland). Canada and the United States have important sources, the chief one being in North Carolina. Most British feldspar is mined and used in the form of Cornish or china stone (p. 103).

As the purer deposits become worked out so methods of purifying lower quality raw material are developing. The most efficient method is that of froth flotation (**K51**).

Froth flotation of feldspar must be done in an acid medium of sulphuric and hydrofluoric acids in order to depress the silica. It is therefore necessary to use flotation machines with wooden tanks, hoods and standpipes, and

TABLE 12

Feldspar Minerals

Mineral		Formula and % composition	S.G.	Hardness	Crystalline form and colour
Orthoclase	potash spar	$K_2O.Al_2O_3.6SiO_2$ 16·9 18·3 64·8	2.56	6	monoclinic. colourless, white, pale yellow, flesh red to grey.
Microcline	potash spar	same	2·54–2·57	6·0–6·5	triclinic. white, yellowish, grey, green or red.
Albite	soda spar	$Na_2O.Al_2O_3.6SiO_2$ 11·8 19·4 68·8	2·61–2·64	6·0–6·5	triclinic. grey or rarely coloured.
Anorthite	lime spar	$CaO.Al_2O_3.2SiO_2$ 20·1 36·6 43·3	2·70–2·76	6·0–6·5	triclinic. colourless, white or greyish.
Plagioclase	potash-soda spar	$[NaKO].Al_2O_3.6SiO_2$ variable			
Oligoclase	lime-soda spar	$Na_2O.Al_2O_3.6SiO_2$ $+ CaO.Al_2O_3.2SiO_2$ variable	2·62–2·67	6–7	triclinic. white, grey, greenish, reddish.
Labradorite	Labrador feldspar intermediate between albite and anorthite	same ratio 1:2 to 1:6	2·70–2·72	5·0–6·0	triclinic. greyish, brown or green.
Celsian	baryta spar (barium anorthite)	$BaO.Al_2O_3.2SiO_2$ 40·85 27·15 32·00	3·37	> 6	triclinic. **D49**
Hyalophane	potash (orthoclase)– baryta spar	$K_2O.Al_2O_3.6SiO_2$ $+ BaO.Al_2O_3.2SiO_2$ variable	2·835 **(D49)**	6–6·5 **(N6)**	monoclinic. colourless.
Pollucite		$2Cs_2O.2Al_2O_3.$ $9SiO_2.H_2O$ 42·50 15·38 40·76 1·36 **(N36)** or $Cs_2O.2Al_2O_3.$ $4SiO_2$ or $(Cs,Na)_2O.$ $Al_2O_3.5SiO_2.H_2O$ **(F4)** variable	2·87–2·90	6·5	cubic. colourless.
Spodumene Hiddenite Kunzite		$Li_2O.Al_2O_3.4SiO_2$ 8·03 27·40 64·57 $Li_2O.Al_2O_3.6SiO_2$ 6·07 20·71 73·22 **(N36)**	2·64–2·65	5·5–6·0	monoclinic. white, grey, green, pink or purple.

moulded rubber impellers and wearing plates. The collectors and frothers found to be suitable are kerosene, fuel oil, amine acetates, methyl isobutyl carbinol and pine oil.

After separation the feldspar must be freed from the water-repelling film on it. This is done by the addition to every ton of concentrate of 2 to 5 lbs of kaolin, bentonite or both, to absorb the reagent (**D28**).

PROPERTIES

The different feldspars, even when free from quartz and other impurities, naturally have a range of values for their physical properties.

> Melting point, 1110° to 1532° C (2030° to 2790° F)
> Specific gravity, 2·56 to 2·63.
> Hardness, 6·0 to 6·5.

Feldspars have a vitreous to pearly lustre, the colours ranging through white, cream, pink, buff brown, red, grey, green and bluish; they may be clear or milky. The fracture is uneven.

Feldspar is the most convenient form for introducing nearly insoluble alkalies. The alkalies in massive feldspar are not water soluble but fine grinding does nevertheless release some of the alkali. The alkali content of the feldspar will be somewhat lowered by wet grinding followed by filter pressing and drying. A dry ground feldspar will, however, on addition to a wet batch release free alkali and alter the pH.

The presence of these alkalies gives feldspars a comparatively low melting point, so that in the normal mixture clay–feldspar–quartz the feldspar softens and becomes glassy or even liquid while the clay and quartz remain as solid particles. The molten feldspar wets the solid particles and the surface tension gradually pulls them together, while the feldspar becomes distributed through the pores. The molten feldspar also dissolves some of the solids and reacts chemically so that the different phases of the resulting fired product differ from those of the raw materials.

The actual temperature at which feldspars soften and melt depends on the nature and quantity of the different individual feldspathic minerals making them up. This complex matter will be taken up again in Chapter 2. The temperature is also affected by contamination and the other materials in the ceramic body, as well as the grain sizes (normal surface factor 240 ± 10) so that it has proved almost impossible to generalise about the effects to expect.

Feldspars have to be tested both for their chemical composition, particularly harmful iron impurities, and for their P.C.E., as this is not directly related to the over-all chemical composition as was previously thought.

Cornish Stone

In Great Britain, Cornish stone (also called china stone and Cornwall stone) is commonly used where feldspar is employed in other countries. It is a partly kaolinised potash and soda feldspathic rock containing feldspar, quartz, kaolin, mica and a small amount of fluorspar. There are four main varieties in degrees of kaolinisation of the feldspar which grade into each other.

(1) Hard purple. The most valuable kind with a high feldspar content. The purple colour is due to fluorspar.

(2) Mild purple.

104   INDUSTRIAL CERAMICS

TABLE

Analyses of

Name Type Origin Ref.	Potash Spar B102	Soda Spar B102	Spodu- mene B102	Soda Spar C58	LL. Feldspar C58	Forshammer Feldspar Sweden H98
SiO_2	66·72	66·22	63·84	67·84	67·17	75·93
Al_2O_3	18·52	21·76	28·61	20·53	18·00	14·29
Fe_2O_3	0·08	0·04	0·19	0·09	0·15	0·11
CaO	0·30	2·06	0·04	0·19	0·42	0·23
MgO	trace	trace	trace	0·11	trace	0·20
K_2O	12·10	0·41	—	0·83	11·27	4·76
Na_2O	1·82	9·37	—	9·69	2·46	4·06
Li_2O	—	—	7·26			
TiO_2						0·03
Loss on ignition	0·36	0·16	0·16	0·42	0·41	0·46
Rational analysis:						
K_2O feldspar						
Na_2O+CaO feldspar						
Total feldspar						
Quartz						
Clay substance						
Molecular formula						
Molecular weight						

(3) Dry white; fluorspar is now nearly absent.

(4) Buff stone; this is contaminated by iron.

The kaolin content increases at the expense of the feldspar on passing from (1) to (4), with it the melting point rises and the physical hardness of the stone falls.

Typical analyses of the purple and buff stones are given as:

	Kaolin	Feldspar	Quartz
Purple	6·7	77·2	16·1
Buff	14·6	55·5	30·9

It will be noted that the fluorspar content is omitted.

A rock of similar composition to Cornish stone is found in Carolina (U.S.A.) and marketed as Carolina stone.

13
Some Feldspars

No. 42 U.S.A.	No. 54 U.S.A.	Body Spa U.S.A.	N.H. Potash Spa U.S.A. S110	Bavarian Feldspar K30/11 Germ M24	Scandinavian Feldspar M24	Scandinavian Feldspar M24	Spodumene A79
70·50	64·85	66·84	68·30	70·32	65·68	65·54	62·91
16·50	20·76	18·97	17·90	17·94	19·71	18·58	28·42
0·06	0·21	0·05	0·07	0·22	0·19	0·17	0·53
0·40	1·18	0·03	0·31	1·40	0·94	0·47	0·11
—	0·17	0·01	—	0·16	0·15	0·58	0·13
10·00	4·64	11·28	2·87	1·52	8·09	12·30	0·69
2·70	7·84	2·66	10·14	7·89	4·56	2·10	0·46
							6·78
0·10	0·20	0·42	0·26	0·66	0·72	0·31	0·28
				9·01	47·81	73·01	
				73·90	43·25	20·07	
86·0	100·0	89·6		82·91	91·06	93·08	
12·8		4·8		15·26	6·21	5·28	
1·2		5·6		1·83	2·05	1·38	
0·277 Na_2O	0·621 Na_2O	0·261 Na_2O					
0·673 K_2O	0·242 K_2O	0·730 K_2O					
0·045 CaO	0·103 CaO	0·002 MgO					
0·005 FeO	0·021 MgO	0·003 CaO					
1·03 Al_2O_3	0·013 FeO	0·004 FeO					
7·415 SiO_2	1·000 Al_2O_3	1·133 Al_2O_3					
	5·281 SiO_2	6·754 SiO_2					
548·2	489·1	550·2					

TABLE 14
Comparative Analyses of Cornish and Carolina Stone (A89)

	Cornish Stone (%)	Carolina Stone (%)
SiO_2	71·10	72·30
Al_2O_3	16·82	16·23
Fe_2O_3	0·16	0·07
CaO	1·60	0·62
MgO	0·05	trace
K_2O	6·57	4·42
Na_2O	2·29	4·14
CaF_2	0·50	1·23
Ignition loss	1·25	1·06
Total	100·34	100·07

Nepheline Syenite

A mineral used in place of feldspar (especially in the United States and Canada) in certain bodies is nepheline syenite. This is an igneous rock resembling granite but containing no free quartz, its chief constituents being nepheline, microcline (potash feldspar) and albite (soda feldspar). Nepheline is a comparatively rare mineral of formula and composition:

$$K_2O \cdot 3Na_2O \cdot 4Al_2O_3 \cdot 9SiO_2$$
$$7.67\% \quad 15.14\% \quad 33.19\% \quad 44.00\%$$

The chief source of nepheline syenite is Ontario, Canada. It is also found in New England, India and Russia. The Canadian product is ground and magnetic iron removed. The percentage content of the major constituents is given in Table 15.

TABLE 15

Analyses of Some Nepheline Syenites

Ref.		(S110)	(S110)	(S110)
SiO_2	60.24	59.30	48.92	50.20
TiO_2		0.002	—	<0.01
Al_2O_3	24.05	24.70	31.69	30.40
Fe_2O_3		0.06	0.10	0.13
Na_2O	10.03	9.91	13.70	12.88
K_2O	5.01	5.10	4.06	3.40
MgO		0.02	0.06	—
CaO	ca. 0.15	0.27	0.86	0.71
Loss on ignition		0.44	0.40	0.41
	99.48	99.802	99.79	98.14

PROPERTIES

Specific gravity, 2.54 when crystalline; 2.28 in the glassy state.

Hardness, about 6.

Sintering range, cone 02 to 7 (**N11**).

Nepheline itself has hexagonal crystals which are colourless, white, yellow, dark green or brownish and have a vitreous lustre; the hardness is 5.5 to 6; specific gravity 2.5 to 2.6 (**A89**).

USES

Nepheline syenite is used to replace part or all of the feldspar where it is desirable to have a body maturing at lower temperatures. It begins to sinter at cone 02 and has a P.C.E. of about cone 7. This long sintering range is partly due to the eutectics which form between nepheline and soda feldspar.

Raw nepheline syenite behaves like raw feldspar. Its greater fluxing action, however, lowers the firing temperature, saving fuel and time. It finds use in sanitary bodies, floor and wall tiles, electrical porcelain, earthenware (semi-vitreous) and vitreous bodies. These last may be produced from clay–nepheline syenite mixtures at cone 4 to 5 (**N11**).

Pumice or Volcanic Ash

Pumice has recently come into use to replace feldspar in both bodies and glazes, although its high iron content makes it unsuitable for white bodies, and transparent or white glazes. Analyses are given as follows:

	European (**V9**)	United States (**A89**)
SiO_2	55·28	72·51
Al_2O_3	21·90	11·55
Fe_2O_3	2·66	1·21
K_2O	6·21	7·84
Na_2O	5·10	1·79
TiO_2	0·28	0·54
CaO	1·88	0·68
MgO	0·37	0·07
H_2O	5·64 {drying at 98° C 1·42; ignition loss 4·22}	Ignition loss 3·81
	99·32	100·00

The molecular formulae calculated from these analyses are:

European

$$\left.\begin{array}{l} Na_2O\ 0{\cdot}367 \\ K_2O\ 0{\cdot}294 \\ MgO\ 0{\cdot}041 \\ CaO\ 0{\cdot}150 \\ FeO\ 0{\cdot}148 \\ \hline 1{\cdot}000 \end{array}\right\} Al_2O_3\ 0{\cdot}959 \left\{\begin{array}{l} SiO_2\ 4{\cdot}109 \\ TiO_2\ 0{\cdot}016 \end{array}\right\} H_2O\ 1{\cdot}399$$

United States

$$\left.\begin{array}{l} Na_2O\ 0{\cdot}208 \\ K_2O\ 0{\cdot}584 \\ MgO\ 0{\cdot}012 \\ CaO\ 0{\cdot}087 \\ FeO\ 0{\cdot}109 \\ \hline 1{\cdot}000 \end{array}\right\} Al_2O_3\ 0{\cdot}815 \left\{\begin{array}{l} SiO_2\ 8{\cdot}686 \\ TiO_2\ 0{\cdot}005 \end{array}\right\} H_2O\ 1{\cdot}153$$

Low melting material of similar composition is to be found in most localities and could be useful if exploited as auxiliary fluxes for bodies or glazes.

Perlite

A siliceous volcanic glass that has recently found a number of applications is called 'perlite'. On quick heating it yields an expanded product. It is also a flux and can replace feldspar where colour does not matter.

USES

Perlite can be used as a lightweight aggregate for thermal and sound insulation with ceramic (or non-ceramic) bonds, and as a flux in lightweight brick, sewer pipes and building blocks.

Bone Ash, Apatite and Tricalcium Phosphate

The use of bone ash as a major constituent in a ceramic body gives rise to the specialised ware known as bone china. Attempts have been made to replace bone ash by the mineral apatite or the pure chemical calcium phosphate.

Bone ash is produced from thoroughly calcined bones, finely ground. The process involves cleaning of the bones followed by steam treatment to remove the fat and alter the bone glue to a soluble form. After thorough washing the bones are then calcined in an air stream sufficient to ensure complete oxidation of remaining organic matter. To produce white bone china, bones of low iron content must be used, the most suitable are those of cattle; horse and pig bones have a higher iron content which gives rise to yellowish ware.

The calcined bone ash is then ground, usually in wet pans. When fine enough it is magneted and allowed to age for three or four weeks, during which time physical and chemical reactions occur which give the bone ash a certain plasticity. Afterwards the clear liquid is decanted and the remainder dried in flat kilns heated through the floor for three to four days to about 12% moisture (**G24**). Bone ash contains 67 to 85% calcium phosphate $Ca_3(PO_4)_2$, 3 to 10% calcium carbonate, 2 to 3% magnesium phosphate and a little caustic lime and calcium fluoride. It is sometimes given the approximate formula (**A89**):

$$4Ca_3(PO_4)_2.CaCO_3.$$

The normal surface factor for ground bone ash used in Great Britain is about 270 with about 75 to 80% less than 0·01 mm.

Apatite is a natural calcium phosphate mineral containing a little fluorine or chlorine:

$$Ca_4(CaF)(PO_4)_3 \quad \text{or} \quad Ca_4(CaCl)(PO_4)_3.$$

It is fairly widespread in granular limestones, igneous rock and metalliferous ores in North America, North Africa and Europe.

A typical chemical composition of the material now being introduced in the United States is (**A89**):

THE RAW MATERIALS

Apatite

Calcia (lime)	54·0%
Phosphorus Pentoxide	40·5%
Magnesia	0·14%
Alumina	0·27%
Ferric oxide	0·15%
Titania	0·04%
Silica	0·85%
Fluorine	2·25%

PROPERTIES OF APATITE

Specific gravity, 3·197 crystalline, 2·972 glass.

Melting point, if all the halogen is chlorine, 1530° C, and if fluorine, 1650° C. Where mixed crystals occur, the m.p. is around 1300° C.

The change of specific gravity on converting the crystalline to the glassy state by the action of heat denotes a volume increase of 7·57% (**N17**).

Tricalcium phosphate is prepared by precipitation on mixing of solutions of sodium phosphate and calcium chloride:

$$Ca_3(PO_4)_2, \text{ M.W. } 310·28. \quad CaO \ 54·24\%, \ P_2O_5 \ 45·76\%.$$

PROPERTIES OF TRICALCIUM PHOSPHATE

Specific gravity, 3·14.

Melting point, 1670° C (3038° F).

Solubility in water at 20° C (68° F), 0·023–3 g/100 ml. It is decomposed by hot water.

THE ACTION OF HEAT

No thermal dissociation is initiated by heat on calcium phosphate. The crystalline form is, however, converted to a glass with an over-all expansion.

USES OF BONE ASH

The chief use of bone ash is as a major constituent of bone china. The standard recipe is:

china clay	25%
bone	50%
stone	25%

This is often modified to:

china clay	23%
ball clay	2%
stone	25 or 30%
bone	50 or 45%

The second body is more plastic and relies less on the plasticity of properly-prepared bone ash, but the introduction of ball clay diminishes the translucency and whiteness.

The function of the bone ash in the china is complex and has not been clearly elucidated. A certain amount of it (about one-third) seems to act as a flux, but not before cone 8, whereas the remainder is refractory. The portion that fuses gives a phosphate glass. The remainder seems to become incorporated in the glassy matrix, building up a rigid framework.

True bone china is made only in Great Britain and one factory in Sweden, and is very specialised. The body is not very plastic, shows large shrinkage and easily warps (**G24**).

In U.S.A. bone ash is also suggested for use as a flux in other dinnerware, by the addition of small quantities, when the calcium carbonate present increases the fluxing action of the feldspar.

Other Alumina- and Silica-containing Raw Materials

There are numerous materials which, although not widespread in the industry, have found specific uses in producing bodies and finished ware of particular composition and properties.

The Sillimanite Group $Al_2O_3.SiO_2$

There are three minerals with the same chemical composition:

(1) Sillimanite, also known as fibrolite.
(2) Kyanite, also known as cyanite and disthene.
(3) Andalusite, special varieties are known as chiastolite and viridine.

Sillimanite, andalusite and kyanite are orthosilicates. That is, they have discrete $(SiO_4)^{4-}$ ions. Their formula does not directly indicate this but it is not necessary for all the oxygen atoms to be linked to silicons. In each of the three forms of Al_2SiO_5 half of the Al atoms are six-co-ordinated, the other half are four-, five- and six-co-ordinated in sillimanite, andalusite and kyanite, respectively. The last is accordingly the most compact structure with the oxygen atoms in cubic close packing. These variations on a basic structure account for the differences found when these minerals are heated.

Sillimanite, the mineral giving the name to the group (and also erroneously to the other minerals and to products) is the least common. It is found in quantity only in India, though deposits also occur in South Dakota, U.S.A., and South Africa.

Kyanite is more plentiful, important deposits occur in the United States, India, Africa and Russia. In the United States the ore is crushed to pass on 35-mesh sieve and then separated by froth flotation. Kyanite is sometimes heated to above 1350° C before introduction into a body because of its disadvantageous irreversible volume increase at that temperature.

Andalusite, the most useful of the trio, is unfortunately uncommon. It is mined commercially in California, and other promising reserves have been found in Nevada, New England, Brazil, Australia, Sweden, France, Russia, Germany and Spain. The mined ore generally contains 75–85% andalusite, together with pyrophyllite, mica, rutile and corundum.

TABLE 17

Properties of Alumino-Silicates

(compiled from C26, K60, A149)

	Sillimanite	Kyanite	Andalusite	Mullite
Name	Sillimanite	Kyanite	Andalusite	Mullite
Formula	$Al_2O_3 . SiO_2$	$Al_2O_3 . SiO_2$	$Al_2O_3 . SiO_2$	$3Al_2O_3 . 2SiO_2$
Crystal system	Rhombic	Triclinic	Rhombic	Rhombic
Form	Slender prismatic crystals—felted mass of fibres	Broad elongated plates—tabular parallel to 100	Euhedral crystals or coarse columnar aggregates. Cross-section nearly square	Long prismatic habit with nearly square cross-section
Refractive indices: alpha	1·680	1·728	1·643	1·654/1·673
beta	1·660	1·720	1·639	1·644 {1·640
gamma	1·659	1·712	1·634	1·652
2V (Axial angle)	20°	82°	84°	45–50°
Optic sign	+	—	—	+
Colour	Colourless	Colourless to pale blue	Usually colourless	Colourless
Birefringence	0·021	0·016	0·019	0·012 to 19
Extinction	Rather strong and up to second-order blue Parallel	Moderate, up to first-order red On 100 is about 30° with length of crystals	Moderate, first-order yellow Parallel in most sections	Moderate, first-order yellow Parallel
Specific gravity	3·23	3·60	3·15–3·16	3·16
Hardness	6–7	4–5 parallel, 6–7 across the blades	7·5	
Temperature of decomposition:				
(1) to give mullite and vitreous silica	1545° C	1350–1380° C	1380–1400° C	—
(2) to give corundum and liquid silica	1810° C	1810° C	1810° C	1810° C

TABLE 16

Some Typical Analyses of Andalusite and Kyanite

	Theoretical	Andalusite				Kyanite	
		California	Massachusetts	Brazil	N. Carolina	Kenya	Virginia, U.S.A.
SiO_2	37·07	33·78	39·78	37·24	37·70	By difference 58–60	37·65
Al_2O_3	62·93	56·89	58·56	62·07	61·40		58·72
Fe_2O_3	—	—	0·61	2·30	0·15	not exceeding 0·75	1·17
Ignition loss		3·67			0·10	TiO_2 not exceeding 1·5	Cr_2O_3 0·20
Others		5·74			0·28	fluxes not exceeding 0·5	TiO_2 1·25 MgO Trace KNaO 0·59 Ignition loss 0·38 P.C.E. cone 36–37
	100·00	100·08	98·26	101·61	99·78 (A89)	(K17)	(K93)

ACTION OF HEAT

The sillimanite minerals are all mullite-forming materials, that is, on heating they decompose to mullite and vitreous silica. Kyanite does this most readily, starting slowly at about 1310° C (cone 10) and becoming rapid with disrupting 17% expansion at about 1350–1380° C (2462–2516° F), cones 12 to 13. It therefore normally has to be calcined before use. The less common andalusite changes at 1380–1400° C (2516–2552° F) cone 13 to 14 with a gradual volume increase of only 5 to 6%. It can therefore be used directly in a body, taking into account the anticipated expansion. Sillimanite does not change to mullite until 1545° C (2813° F), with an expansion of from 5 to 6%.

The mullite formed in these reactions develops as interlocking crystals which impart great strength.

On heating to 1810° C (3290° F) mullite decomposes to corundum and liquid silica, and is usually considered to have an incongruent melting point. Barsakowski (**B14**) claims to have melted mullite congruently in a vacuum at 1870° C (3398° F).

USES

Sillimanite minerals are used predominantly for refractories, and technical porcelains. Formerly almost all the andalusite was taken up for sparking plugs, but such silicate sparking plugs are losing favour because they are attacked by the lead tetraethyl in the petrol.

Calcined kyanite is being introduced into very many ceramic bodies thereby imparting mechanical strength.

Topaz

A further mullite-forming raw material is topaz, found in South Carolina and California (U.S.A.), Brazil and South Africa.

Topaz is $Al_2(F, OH)_2SiO_4$, the fluoride and hydroxyl radicals being interchangeable in the structure so that their ratio varies. An analysis of South Carolina topaz is given:

	%
SiO_2	37·2
Al_2O_3	53·2
Fe_2O_3	0·4
F	14·1
Ignition loss	0·9
	105·8
less O and F	5·9

Hardness, 8.
Specific gravity, 3·4 to 3·6.

Topaz is normally calcined before introduction into a body. On heating, the crystals begin to change at 815° C (1500° F); the inversion to mullite and silica glass occurs from 1090° to 1260° C (2000° to 2300° F) the mullite crystals continuing to grow until 1650° C (3000° F). During the change to mullite fluorine, silicon tetrafluoride, and water are lost. The product has a specific gravity of 3·1 and composition of about 70% Al_2O_3 to 30% SiO_2.

This high aluminous calcined topaz is used in refractories particularly where an alumina content higher than that of kyanite is desired (**A89**).

Dumortierite

This is a low-temperature mullite-forming mineral containing the flux boric oxide. It is mined in Nevada (U.S.A.).

$$8Al_2O_3.\ 6SiO_2.\ B_2O_3.\ H_2O$$
$$64\cdot54\%\ \ 28\cdot52\%\ \ 5\cdot51\%\ \ 1\cdot43\%$$

Specific gravity, 3·2 to 3·3.
Hardness, 7.

It breaks down to mullite and silica at 1230° C (2246 °F), cone 6. Dumortierite is used to introduce alumina in a number of ceramic bodies, *e.g.* laboratory porcelain.

Synthetic Mullite $3Al_2O_3.2SiO_2$

Mullite itself is a rare mineral found on the Isle of Mull but not occurring in workable deposits. However, its refractory properties are so desirable that synthetic mullite is prepared from pure silica and alumina, by heating to about 1800° C (3272° F), just below the decomposition temperature of mullite. This material can be very carefully controlled as regards chemical composition, crystal growth, grain size, etc.

Other Means of Introducing Alumina

Aluminium oxide either free or combined with other oxides has already been seen to be an important constituent of most raw materials and hence most ceramic bodies and glazes. By itself alumina is highly refractory and the sintered product has an electrical resistance second only to sintered beryllia.

Alumina occurs in crystalline form as corundum and the less pure emery, these are used raw as abrasives. It also occurs hydrated as bauxite, diaspore and gibbsite. The different minerals differ in crystalline form as well as composition. Alumina itself has two forms, α and γ. The naturally occurring hydrated minerals correspond to these forms. α-Alumina is strictly the high-temperature form but it does not revert to the low-temperature γ-form on cooling. γ-Alumina is produced by controlled low-temperature dehydration of hydrates at temperatures up to 500° C (930° F). On strong heating it is converted irreversibly to the α-form at 1150–1200° C (2100–2190° F) **(G57)**. β-Alumina, previously thought to be a modification of the α-form, is now known to be a sodium aluminate of composition $Na_2O.11Al_2O_3$ **(B27)**, or $Na_2O.12Al_2O_3$; it forms in the slightly contaminated commercial alumina which has been electrically fused or calcined at high temperature **(A164)**.

TABLE 18

Crystallographic Relationship between Alumina and its Hydrates

γ-Series (low temperature)	Composition	α-Series (high temperature)
Gibbsite, Hydrargillite	$Al_2O_3.3H_2O$	(Bayerite?)
Bauxite	$Al_2O_3.H_2O$	Diaspore
γ-Alumina	Al_2O_3	Corundum

TABLE 19 (E8)

Chemical formula	Corresponding Mineral Name	British Symbol	American Symbol
Al_2O_3	Corundum	α	α
Al_2O_3	—	γ	γ
$Al_2O_3.H_2O$	Boehmite	γ	α
$Al_2O_3.H_2O$	Diaspore	α	β
$Al_2O_3.3H_2O$	Gibbsite	γ	α
$Al_2O_3.3H_2O$	(Bayerite?)	α	β

The use of the symbols α, β, and γ is not the same in Britain and the United States. The two systems compare as shown above.

Of the hydrated minerals the commonest is bauxite. It is prepared by calcination or fusion and then ground coarsely for refractory and heavy clay products or finely for special pottery, filters, grinding wheels, etc. (Fig. 1.43).

Fig. 1.43. *Bauxite mine near Mackensie, British Guiana (Demerara Bauxite, D24)*

Many aluminous raw materials are treated chemically to extract pure alumina, the main process (the Bayer) involving the solution of alumina in strong aqueous caustic soda in an autoclave at a pressure of 4–6 atm, followed by its precipitation when the solution is diluted and seeded with freshly precipitated alumina. Pure alumina finds innumerable uses in fine ceramics, both by itself as sintered alumina electrical products, as a refractory, in composite bodies and glazes, and as a bedding material in saggers.

PROPERTIES

The main properties of Al_2O_3 are shown in tabular form below:

	γ-alumina	α-alumina
S.G.	3·5–3·9	4·00
Crystalline form	hexagonal	trigonal
M.W.	101·94	
M.P.	2050° C (3722° F)	
B.P.	2250° C (4082° F)	
Solubility at 29° C	0·000098	

Calcium and Magnesium Oxide Sources

There are a number of widespread minerals consisting chiefly of calcium and/or magnesium carbonates. Both the separate and the mixed carbonates and the oxides derived from them are used in the ceramic industry.

CALCIUM CARBONATE $CaCO_3$

CaO 56·03%, CO_2 43·97%, M.W. 100·09.
CO_2 is lost at about 900° C (1652° F).

Calcium carbonate occurs pure in the mineral calcite (also known as calcspar), and in the limestone rocks which may be more or less pure and include marble, marls, oolitic limestones, chalk and argillaceous limestones. There is a continuous series from the argillaceous limestones to the lime-rich clays.

Most calcium carbonate is used in the finer ceramic bodies and glazes. For this it needs to be finely ground, and in Britain the commonest form is finely ground English chalk known as 'whiting'. In the United States certain forms of ground limestone are sold as whiting.

TABLE 20

Physical Properties of Calcium and Magnesium Carbonates

Material	Formula	S.G.	Hardness	M.P. (°C)	B.P. (°C)	Solubility	
Calcium carbonate (calcite) pure	$CaCO_3$	2·7	2–3	decomp. 900		at 25° 0·0014	at 75° 0·0018
Calcium oxide	CaO	3·4		2572	2850	at 10° 0·131	at 80° 0·07
Chalk	$CaCO_3$	app. 1·85					
Marble	$CaCO_3$	2·7	3–4				
Dense limestone	$CaCO_3$	2·6	3–4				
Lime marl	Chiefly $CaCO_3$	2·3–2·5					
Dolomite	$CaCO_3 \cdot MgCO_3$	2·8–2·9	3·5–4	decomp. 500–700 and 800–1000			
(Magnesite) Magnesium carbonate	$MgCO_3$	3	3·5–4·5	decomp. 350		at 20° 0·0106	
Magnesium oxide	MgO	3·7		2500–2800		at 20° 0·00062	at 30° 0·0086

DOLOMITE

Calcium and magnesium carbonates occur together in dolomite, in almost equal molecular quantities, $CaCO_3 \cdot MgCO_3$ (56% $CaCO_3$ and 44% $MgCO_3$), and also in dolomitic limestones which have 25 to 44% $MgCO_3$ and magnesian limestones with from 5 to 25% $MgCO_3$. Dolomite is found in primary deposits, having been formed in solution, and also in secondary rocks formed by alteration of limestones by the percolation of magnesium salt solutions. Primary dolomites tend to be purer, more porous and softer than those of secondary origin.

'Doloma' is the product of calcination of dolomite and is a mixture of lime and magnesia. Although magnesium carbonate decomposes at about 700° C (1292° F) and calcium carbonate at about 900° C (1652° F) the dolomite has to be heated to 1700° C (3092° F) in order to obtain a less porous product that can be stored. Most British doloma is produced in cupolas or vertical shaft kilns. In the U.S.A. rotary kilns are normally used (**C26**).

TABLE 21

Composition and Properties of some British Dolomites
(after Chesters (C26))

Origin and formation	Chemical Analysis						Physical Properties			Remarks
	SiO_2	Mix'd R_2O_3	Lime CaO	Magnesia MgO	Loss on ignition	Total	S.G.	Bulk density (g/ml)	Porosity (%)	
Theoretical			30·41	21·87	47·72	100				
S. Yorks, Derby and Notts: Permian lower magnesium limestone	0·33 0·87	0·52 0·60	30·63 30·32	21·50 21·23	47·37 47·13	100·35 100·15	2·84 2·85	2·47 2·39	13·0 16·3	Soft
Durham: Permian system	0·89	0·96	30·6	20·6	46·95	100·00	2·85	2·53	11·2	Medium to soft
Leicestershire: Carboniferous limestone	0·46	1·80	31·08	20·42	46·65	100·41	2·81	2·54	9·6	Medium to hard
N. Wales: Carboniferous limestone	2·06	0·83	31·78	19·35	45·94	99·96	2·87	2·68	6·2	Hard
S. Wales: Carboniferous limestone	1·28	0·81	32·48	19·41	45·15	99·13	2·82	2·77	1·8	Very hard
Scotland: Appin limestone formation	3·70	1·66	28·9	20·85	44·80	99·91	2·87	2·85	0·5	Friable

MAGNESITE $MgCO_3$

Magnesite, magnesium carbonate (MgO 47·80%, CO_2 52·20%), occurs in two different forms, the cryptocrystalline or compact magnesite and the coarsely crystalline magnesite. The compact form is commoner and originates from the decomposition of magnesium silicates such as serpentine, talcum, olivine, etc., by carbonated water. It occurs in quantities worth exploiting in Czechoslovakia (Bohemia), Austria, Yugoslavia, Italy (Piedmont and Tuscany), India (Salem and Mysore), Manchuria (between Mukden and Dairen), California, British Columbia (associated with hydromagnesite), Queensland, New South Wales, Southern Australia, Tasmania, New Caledonia and the Transvaal. The crystalline form is a metamorphic formation of limestone coming in contact with magnesium-containing thermal springs. It contains varying quantities of iron, since there is a continuous transition from magnesite ($MgCO_3$) to siderite ($FeCO_3$); the iron-rich magnesites are termed 'breunnerite'. The main deposits of this occur in Austria (Veitsch), Slovakia, Sweden, Norway (Snarum), Spain, Urals (Satka), Siberia (Birobidjan), Manchuria, U.S.A. (Washington, California and Vermont), Canada (Quebec), Sudan.

TABLE 22

Natural Magnesites (S105)

	MgO	CO_2	CaO	SiO_2	FeO	Al_2O_3	Total
Amorphous:							
Kraubath (Styria)	48·4	50·9	—	0·2	—	—	99·5
Euboea (Greece)	43·5	48·2	1·0	5·0	0·8	0·85	99·35
Crystalline:							
Veitsch (Styria)	42·2	50·4	1·7	—	3·5	0·03	97·8
Snarum (Norway)	47·3	51·4	—	—	0·8	—	99·5
Slovakia	44·5	48·6	0·50	1·31	4·62	0·51	100·04

Magnesite begins to dissociate, giving off carbon dioxide, at 500° C (932° F). 'Caustic magnesite' is produced by burning at 800° C (1472° F) and can be completely hydrated by contact with water or steam. 'Dead burned magnesite', burned at 1500° C (2732° F) cannot be hydrated. Magnesite required for making refractories is further heated to 1600–1800° C (2912–3272° F) to sinter it, during which process a shrinkage of about 25% occurs. After-shrinkage is then relatively small. Iron-containing magnesites sinter much more readily than the iron-free ones, apparently because of the formation of magnesium ferrite above 1200° C (2192° F) (**S105**).

A further useful magnesia mineral is brucite, magnesium hydroxide, and occurs in Canada (Ontario and Quebec) and Nevada (U.S.A.).

The source of magnesia to be most recently exploited is the sea, and almost the whole of Great Britain's supply of refractory magnesia now comes from the sea-water dolomite process, which in 1936 was not yet under way. The process depends on the fact that magnesium hydroxide is less soluble than calcium hydroxide. Dolomite is calcined to give a mixture of calcium and magnesium oxides. These hydrate when added to sea-water and the calcium hydroxide reacts with the soluble magnesium salts there present (0·2% MgO in sea-water) to precipitate magnesium hydroxide. A number of precautionary measures must be taken to eliminate the impurities that contaminate the magnesium hydroxide either by precipitation from the sea-water, or from the dolomite, and these complicate the plant considerably. The magnesium hydroxide obtained is 'dead burned' in the same way as rock magnesite (**C26**), by firing in a rotary kiln with a maximum temperature of 1650° C (3002° F) (**W55**).

For certain specialised purposes magnesia is electrically fused, allowed to recrystallise and then ground and graded before marketing (**N27**).

USES

Although both lime and magnesia are in themselves refractory they react with silica and soda to give low-melting products and therefore act as fluxes in bodies and glazes (magnesia is generally not used in glazes in England). In general they react with clay, flint and feldspar to give a glassy bond, this increases strength and decreases porosity but also increases shrinkage. Lime and dolomite may be used in earthenware, sanitary ware, floor and wall tiles, such bodies containing from 3% to 20%.

Magnesite and dolomite are used extensively for making basic refractory bricks for the steel and cement industries. Developments in the last ten to fifteen years have produced refractory bricks that stand up to loads at high temperatures and are water resisting.

Wollastonite

This fibrous calcium silicate mineral became commercially available in the United States during 1952 when an extensive and very pure deposit was opened up near Willsboro, New York.

PROPERTIES

Theoretical composition:

	SiO_2	51·75
	CaO	48·25
	total	100·00

Actual composition:

	SiO_2	51·76
	CaO	47·56
	FeO	0·46
	MnO	0·04
	H_2O	0·09
	total	99·91

Specific gravity, 2·9.
Hardness, 4·5–5·0.
Solubility in water, 0·0095 g/100 ml.
pH of 10% slurry, 10·1–10·3.
Action of heat, transition from β- to α-form at 1200° C (2192° F). Melting point, 1540° C (2804° F).
Wollastonite is built up of $(Si_3O_9)^{6-}$ rings and calcium ions (**B12**).

USES

The applications of wollastonite are not yet fully worked out. It has a high fluxing action bringing down the maturing point of bodies. Added to this is the fact that, unlike calcium carbonate, wollastonite does not evolve a gas on heating so that it can successfully be used in low-temperature once-fired products, where the glaze melts at a temperature below that at which the carbonate decomposes. Its incorporation in wall tiles can replace double by single firing. When used to replace other fluxes the firing shrinkage may be reduced by nine-tenths and the strength increased up to 50% because of the fibrous nature of the wollastonite particles. Wollastonite bodies have good resistance to thermal shock.

Wollastonite is also used for low-loss electrical bodies fired at lower temperatures (cone 4) than otherwise possible (**C1 a, b, c**).

Sierralite

Sierralite is a recently introduced raw material. It is a hydrous magnesium aluminium silicate and has been shown to be a chlorite.

Composition of Sierralite

SiO_2	36·24%
Al_2O_3	23·56%
Fe_2O_3	1·19%
CaO	1·47%
MgO	23·39%
Na_2O	0·10%
K_2O	0·25%
CO_2	0·86%
Moisture loss at 105° C	0·18%
Chemically combined water	12·19%
	99·43%

It fires to a greenish-grey body of zero absorption at about 1290° C (2354° F) with a diameter shrinkage of 9·80% and volume shrinkage of 34·7%. Porous buff bodies are obtained at lower temperatures.

USES

The composition of sierralite is such that the cordierite composition can be obtained by using it together with clay only, in 50:50 mixtures, whereas other magnesia raw materials required the addition of alumina which is abrasive to the dies, etc. (**L10**).

Natural Cordierite

Although it is the only three-component compound in the system $MgO–Al_2O_3–SiO_2$ the mineral cordierite $2MgO.2Al_2O_3.5SiO_2$ is very rare. Two deposits are reported but ceramic cordierite is always synthetic.

TABLE 23
Analyses of Wyoming and Greenland Cordierites (B122)

	Theoretical	Wyoming	Greenland
SiO_2	51·4	50·10	49·17
Al_2O_3	34·9	29·70	31·71
MgO	13·7	6·82	10·16
MnO			0·33
Fe_2O_3		12·12	
FeO			8·32
CaO		1·77	

Other Fluxes

There are numerous low-melting reactive materials that promote the maturing of ceramic bodies and glazes at lower temperatures. They are compounds of the alkali metals, lithium, sodium and potassium; magnesium and the alkaline earths, calcium, strontium and barium, boron, zinc, lead and bismuth. They will be considered in this order derived from the Periodic Table which therefore pays no regard to the relative quantities used.

Lithium Compounds

Lithium minerals and chemicals have a very strong fluxing action both by themselves and because of the eutectics they form with other fluxes. Details of the minerals are given in Tables 24 and 25 and of some of the chemicals in Table 26. The lithium content is usually considered in terms of the oxide lithia, Li_2O, which cannot, however, be used as such.

DATA FOR LITHIA Li_2O

Molecular weight, 28·88.
Specific gravity, 2·013.
Melting point, > 1700° C.
Solubility in water, at 0° C with decomposition 6·67 g/100 ml; at 100° C 10·02 g/100 ml.

CHIEF USES

(1) To lower vitrification temperature and decrease final porosity of hard porcelains, especially electrical porcelain and acid resisting porcelain. Feldspar is replaced by lepidolite, spodumene or petalite, in such quantity that the Li_2O content of the body is not more than 2% (greater quantities of lithia alter the expansion of the body and necessitate changing the glaze). The maturing temperature may be lowered by from 3 to 6 cones.

(2) To improve frit porcelains, giving lower expansion and good acid resistance using 2·5 to 6% lithia preferably in the form of lepidolite, or if spodumene or petalite is used a little lithium fluoride should be added.

TABLE

Lithium

(compiled from

Name	Average lithia content (%)	Structural formula	Ceramic formula and theoretical % composition
Phosphates: Amblygonite	6–9	(Li,Na)Al[PO$_4$]F	2LiF . Al$_2$O$_3$. P$_2$O$_5$ 17·54% 34·46% 48·00%
Triphylite (with Fe) (Lithiophilite, with Mn)	3–8	Li(Mn,Fe)[PO$_4$]	(Mn, Fe)$_2$O$_3$.P$_2$O$_5$.Li$_2$O
Dilithium sodium phosphate		Li$_2$Na[PO$_4$]	2Li$_2$O.Na$_2$O.P$_2$O$_5$
Silicates: Petalite	3–5	(Li,Na)Al[Si$_4$O$_{10}$]	(Li,Na)$_2$O . Al$_2$O$_3$. 8SiO$_2$ 4·88% 16·65% 78·47%
Eucryptite	3–5	LiAl[SiO$_4$]	Li$_2$O.Al$_2$O$_3$.2SiO$_2$
Spodumene	5–7	LiAl[Si$_2$O$_6$]	Li$_2$O . Al$_2$O$_3$. 4SiO$_2$ 8·03% 27·40% 64·57%
Micas: Lepidolite	3·5–5	(Li,Al)$_3$[(Si,Al)$_4$O$_{10}$] K(OH,F)$_2$	2Li$_2$O.K$_2$O.2Al$_2$O$_3$.6SiO$_2$. 2F$_2$.H$_2$O or LiF.KF.Al$_2$O$_3$.3SiO$_2$ 7·08, 15·87, 27·84, 49·21%
Zinnwaldite	3·5–5	(Li,Al,Fe,Mg)$_3$ [(Si,Al)$_4$O$_{10}$]K(OH,F)$_2$	2Li$_2$O.K$_2$O.2(Fe,Mg)O. 2Al$_2$O$_3$.6SiO$_2$.F$_2$.H$_2$O
Double salt: Cryolithionite	to 8	Li$_3$Na$_3$Al$_2$F$_{12}$	

Bodies of good transparency and surface gloss can be obtained.

(3) To lower sintering temperature of titanate and zirconate bodies, lithium titanate or zirconate being used accordingly. Expansion is also lowered.

(4) In lithium–alumino–silicate bodies of carefully regulated zero, small negative or small positive thermal expansion.

(5) As flux and regulator of expansion and interaction with the body, in glazes (**M44**).

Sodium Compounds

NATURAL MINERALS

The sodium feldspars albite, plagioclase, oligoclase and labradorite have already been described.

24

Minerals
V10, M44, A89)

Crystal System	Hardness	Density	Action of heat	Occurrence
triclinic	6	3·0–3·1	P.C.E. ~ cone 5	S.W. Africa, S. Rhodesia, Belgian Congo, Brazil, N.S. Wales, Spain, Madagascar, Sweden, France, Russia, Black Hills, U.S.A. S.W. Africa
rhombic	4–5	3·4–3·6		California
monoclinic	6·5	2·4	900° C conversion to β-spodumene and amorphous silica, softening point ~1350° C	S.W. Africa, S. Rhodesia W. Australia
hexagonal		α, 2·67 β, 2·35	970° C α- to β-inversion 1400° C incongruent melting point	
monoclinic	6·5–7	α, 3·1–3·2 β, 2·406	700–900° C α- to β-inversion 1421° C congruent melting point	S.W. Africa, Belgian Congo, N. Carolina, S. Dakota, U.S.A., Manitoba, Canada, Brazil, W. Australia
monoclinic	3	2·8–2·9		S.W. Africa, S. Rhodesia, W. Australia, U.S.A.
monoclinic		2·9–3·1		
cubic	2·5	2·78		

Cryolite, sodium aluminium fluoride, 3 NaF. AlF$_3$, occurs primarily in Greenland, and also in Colorado, U.S.A., and in the Urals, U.S.S.R. Three grades of natural cryolite are marketed, the best containing a minimum of 98% sodium aluminium fluoride with a maximum of 1·5% silica, 0·25% iron oxide, 1% lime. The other two grades have 93 to 94% cryolite, and iron oxide not exceeding 0·75%.

Artificial cryolite is made in several European countries, Canada and U.S.A.

PROPERTIES

Composition, Na, 32·8%, Al, 12·9%, F, 54·3%.
Molecular weight, 210.

TABLE 25

Analyses of Some Lithium Minerals
(for Spodumene, see Feldspars, Table 13)

Name Ref.	Amblygo- nite A89	Amblygo- nite A79	Lepidolite A89	Lepidolite A79	Lepidolite V10	Lepidolite California, U.S.A. V10	Petalite A89
Li_2O	8·48	8·48	4·30	4·65	4·87	4·92	4·2
Na_2O	1·63	1·63	0·94	0·13	1·05	—	
K_2O	0·30	0·30	10·33	10·33	11·29	17·48	} 0·5
Rb_2O						0·97	
Cs_2O						0·30	
MgO	0·41	trace	0·31	0·31	0·18	0·16	
CaO	trace	0·15		0·92	0·35	0·18	
FeO					0·33	—	
Al_2O_3	34·42	22·96	26·77	26·77	28·69	22·81	17·8
Fe_2O_3	0·30	0·02	0·06	0·19			
Mn_2O_3					0·75	0·41	
SiO_2	0·36	5·16	52·89	52·89	47·63	50·18	77·7
MnO_2			0·15	0·59			
P_2O_5	46·75	54·42					
F	2·67	2·67	5·60	3·68	3·94	1·20	
H_2O					1·59	2·10	
Ignition loss	4·80	4·80	0·66	0·66			
	100·12	100·59	102·01	101·12	100·67	100·71	100·2

Specific gravity, 2·98.
Hardness, 2·5.
Solubility in water at 25° C, 0·04g/100ml.
Action of heat, transition from monoclinic to isometric or possibly cubic, 565° C (1049° F). Melting point, 1020° C (1868° F).

USES

Cryolite finds use as a flux in lead- and boron-free glazes, in special glazes for crucibles, and as an auxiliary flux in certain pottery bodies. It should be fritted to avoid excessive blistering owing to the volatile fluorides produced in the reactions.

SODIUM CHEMICALS

Sodium oxide, Na_2O. Na_2O is not used as such but is the form of expressing the sodium content of raw materials and batches. Molecular weight, 61·994. Sodium oxide sublimes at 1275° C (2327° F) and may therefore be lost at high temperatures. Sodium in glazes produces a high coefficient of expansion which may have detrimental effects, causing crazing.

Sodium carbonate, Na_2CO_3.
Composition, Na_2O, 58·48%, CO_2 (loss on ignition), 41·52%.
Molecular weight, 106·00.

TABLE 26

Lithium Chemicals

Compound	Formula	M.W.	Composition	S.G.	M.P. (°C)	Solubility g/100 ml	Use
Lithium carbonate	Li_2CO_3	73·89	Li_2O Loss on ignition 40·44% 59·56%	2·111	618	1·54 1·33 0·72 @ 0°C @ 20°C @ 100°C	Frit component of leadless glazes
Lithium silicate	Li_2SiO_3	89·94	Li_2O . SiO_2 33·22% 66·78%	2·52	1201	insol. cold, decomp. hot	Frit component of leadless glazes 0·5 to 1·0%
Lithium aluminate	$LiAlO_2$	65·91	Li_2O . Al_2O_3 22·67% 77·33%		~1625		Limited use in ceramic colours
Lithium titanate	Li_2TiO_3	109·78	Li_2O . TiO_2 27·22% 72·78%			insoluble	Flux in titanate bodies Flux in leadless glazes producing opacity 0·5 to 1·5%
Lithium zirconate	Li_2ZrO_3	153·10	Li_2O . ZrO_2 19·52% 80·48%			insoluble	Flux in zirconium bodies Flux in glazes to improve gloss and prevent pinholing
Lithium zirconium-silicate	$2Li_2O.ZrO_2.SiO_2$	243·04	Li_2O . ZrO_2 . SiO_2 24·59% 50·70% 24·71%				Mill addition to tile and sanitary glazes eliminating eggshelling and pinholing, 0·5 to 1·5%

Specific gravity, 2·509.
Melting point, 851° C (1564° F).
Solubility at 0° C, 7·1 g/100 ml; at 100° C, 45·5 g/100 ml.

Sodium carbonate is occasionally added as an auxiliary flux in a fritted glaze. Together with other sodium chemicals it finds numerous other uses in the ceramics industry, which are discussed elsewhere.

Potassium Compounds

NATURAL MINERALS

The potassium feldspars orthoclase, microcline, plagioclase, which have been discussed separately are the principal form of introducing potassium in ceramic bodies and glazes.

POTASSIUM CHEMICALS

Potassium oxide, K_2O. K_2O is not used as such but is the form of expressing the potassium content of raw materials, bodies and glazes. Molecular weight, 94·192. The inclusion of potassium produces a stronger glaze than sodium (**R58**).

Potassium carbonate, K_2CO_3.
Composition, K_2O, 68·15%, CO_2 (loss on ignition), 31·85%.
Molecular weight, 138·20.
Specific gravity, 2·4.
Melting point, 891° C (1635° F).
Solubility in water at 20° C, 112 g/100 ml; at 100° C, 156 g/100 ml.

Potassium carbonate is used as a flux in glazes; it also has a marked effect on some colouring agents.

Potassium nitrate, KNO_3.
Composition, K_2O, 46·58%, N_2O_3 (loss on ignition) 53·42%.
Molecular weight, 101·10.
Specific gravity, 2·1.
Transition at 129° C (264° F); melting point, 334° C (633° F); decomposes at 400° C (752° F).

Solubility in water at 0° C, 13·3 g/100 ml; at 20° C, 31·6 g/100 ml; at 100° C, 247·0 g/100 ml. This high solubility makes it necessary to frit potassium nitrate. It is a strong oxidising agent up to about 400° C (752° F) when its decomposition is complete.

Magnesium Compounds

Although magnesia, MgO, is itself highly refractory it forms low-melting eutectics and compounds and may be used as a flux. This is more commonly done in bodies than glazes.

THE RAW MATERIALS 127

The magnesium minerals have all been discussed. They are summarised together with some common chemicals in Table 27.

TABLE 27

Magnesium Minerals and Chemicals

Name	Formula and percentage composition	M.W.	S.G.	Hardness Moh	Vitrification (°C)	M.P. (°C)
Talc, p. 89	$3MgO \cdot 4SiO_2 \cdot H_2O$ 31·8% 63·5% 4·7% to $4MgO \cdot 5SiO_2 \cdot H_2O$ 33·5% 62·7% 3·8%	~379	2·6–2·8	1–2	1270–1350	1490
Dolomite, p. 116	$CaO \cdot MgO \cdot (CO_2)_2$ 30·41% 21·87% 47·72%	184	2·8–2·9	3·5–4		decomp. 500–700 800–1000 m.p. 2500
Magnesite, p. 117 Magnesium carbonate	$MgO \cdot CO_2$ 47·80% 52·20%	84·32	3	3·5–4·5		d. 350
Sierralite, p. 120	Hydrous magnesium aluminium silicate					
Magnesium oxide	MgO	40·32	3·22–3·654			2800
Magnesium chloride	$MgCl_2 \cdot 6H_2O$	203·3	1.56			d. 186
Magnesium hydroxide	$Mg(OH)_2$	58·34	2·34			d. 100
Magnesium sulphate	$MgSO_4 \cdot 7H_2O$	246.5	1.68			d. 150

Calcium Compounds

The calcium feldspars anorthite and oligoclase; the carbonate in limestones, chalk, etc.; the oxide, lime; and the silicate, wollastonite have been described and are summarised in Table 28.

FLUORSPAR (FLUORITE)

CaF_2, M.W. 78·08, Ca, 51·33%, equivalent to CaO, 71·82%, F_2, 48·67%. S.G., 3·2. M.P., 1360° C (2480° F), decrepitates when heated. Hardness, 4. Solubility in water at 18° C, 0·0016 g/100 ml; at 26° C, 0·0017 g/100 ml.

The use of fluorspar as a flux in bodies and glazes is not widespread. It is an excellent flux but the reactions with the other glaze constituents produce volatile silicon, aluminium and sodium fluorides. Steger (**S182**) investigated this and found that the losses of these compounds from a glaze are rather unpredictable, being easily affected by small differences of batch composition, firing schedule and kiln atmosphere, so that the final product is not uniform. Also the escaping gases produce pinholes and an altogether unsatisfactory glaze texture.

TABLE 28
Calcium Minerals and Chemicals

Name	Formula and percentage composition	M.W.	S.G.	M.P. (°C)
Anorthite or lime spar, p. 102	$CaO \cdot Al_2O_3 \cdot 2SiO_2$ 20·1% 36·6% 43·3%	278	2·70–2·76	~1550
Oligoclase, lime-soda spar, p. 102	$Na_2O.Al_2O_3.6SiO_2 + CaO.Al_2O_3.2SiO_2$, variable		2·62–2·67	
Bone ash, p. 108	$\sim 4Ca_3(PO_4)_2.CaCO_3$			
Apatite, p. 108	$Ca_4(CaF)(PO_4)_3$ or $Ca_4(CaCl)(PO_4)_3$			
Tricalcium phosphate, p. 109	$Ca_3(PO_4)_2$. CaO, 54·24%, P_2O_5, 45·76%	310·28	3·14	1670
Calcium carbonate Calcite or calcspar Limestone Whiting, p. 116	$CaCO_3$ CaO, 56·03%, CO_2, 43·97%	100·09		d. 900
Dolomite, p. 116	$CaCO_3.MgCO_3$, CaO, 30·41%, MgO, 21·87%, CO_2, 47·72%	184	2·8–2·9	d. 500–1000 m.p. 2500
Lime, calcium oxide	CaO	56·08	3·4	2572
Wollastonite, p. 119	$CaO.SiO_2$, CaO, 48·25%, SiO_2, 51·75%	116·14	2·9	1540
Fluorspar	CaF_2 ~CaO, 71·82%	78·08	3·2	1360
Calcium hydroxide	$Ca(OH)_2$ CaO, 75·6%	74·1	2·3	$-H_2O$, 580
Calcium borate	$Ca(BO_2)_2$ CaO, 44·6%, B_2O_3, 55·4%	125·7		1154

The full effects of fluorine in ceramic bodies and glazes have not been worked out systematically and whereas some manufacturers use it successfully it is not generally advisable.

Fluorine and its compounds, especially the acidic one, are toxic so that any gases given off must be removed before they can be inhaled by workers. Nor is discharge from a tall chimney stack sufficient precaution as the surrounding vegetation suffers badly. Adequate scrubbing of fluorine fumes is necessary (**B127**).

Strontium Compounds

CELESTINE (CELESTITE)

This mineral, found in England, Germany, Sicily and Mexico and less pure in Texas, is essentially strontium sulphate, $SrSO_4$, with some calcium and barium impurities. Most of it is converted to strontium carbonate.

STRONTIANITE AND CHEMICALLY PREPARED STRONTIUM CARBONATE $SrCO_3$

Strontianite is unfortunately a rather rare mineral found as a commercial deposit only in Germany. Most of the strontium carbonate required therefore has to be prepared from celestine.

$SrCO_3$, M.W. 147·64, SrO, 70·19%, CO_2 loss on ignition, 29·81%. S.G., 3·7. Decomposes without melting at about 1340° C (2440° F). M.P.

of resulting SrO, 2430° C (4406° F). Solubility in water at 18° C, 0·0011 g/100 ml, at 100° C, 0·065 g/100 ml.

The introduction of strontia in glazes can be very beneficial although its limited supply and the fact that it is still somewhat more expensive than calcium compounds limit its use to special purposes, or for the elimination of faults that cannot otherwise be dealt with. Strontia is also used in special titanium bodies.

In glazes replacement of some of the calcium or barium by strontium increases the complexity and so decreases the fusion point. Substitution of strontium for lead in certain coloured glazes is more successful than using calcium, as the latter reduces the firing range. Satisfactory leadless glazes are made with strontia.

Barium Compounds

There are barium feldspars, celsian and hyalophane.

BARYTES (BARITE OR HEAVY SPAR) $BaSO_4$

This is the most readily available barium mineral. Little of it is used directly, the bulk being converted to barium carbonate.

WITHERITE AND BARIUM CARBONATE $BaCO_3$

The mineral witherite is mined in England and California. Most of the commercial barium carbonate is obtained from barytes.

$BaCO_3$, M.W. 197·37. BaO, 77·70%, CO_2 loss on ignition, 22·30%. S.G., 4·43. Action of heat on witherite: transition to α-$BaCO_3$ at 811° C (1492° F); transition of α-$BaCO_3$ to β-$BaCO_3$ at 982° C ($\sim$1800° F); decomposition of $BaCO_3$ at 1450° C (2640° F); M.P. of resultant BaO, 1923° C (3490° F). Solubility in water of all three varieties at 18 to 20° C, 0·002g/100 ml; at 100° C, 0·006 g/100 ml.

Although rather refractory in itself and fairly slow to combine with silicates, once it has done so, barium carbonate is nearly as active a glaze flux as lead oxide. It therefore finds more use in glazes, where the interaction can proceed towards completion, than in bodies. It makes a good flux for glazes particularly when used in quantities less than 0·1 equivalents; it promotes a matt structure.

In pottery bodies barium carbonate seems to impart better translucency but reduces the vitrification range; it increases the strength, but causes excessive shrinkage and blistering. In low-loss steatite bodies barium carbonate plays a very important part. Replacement of half the feldspar flux by barium carbonate reduces the loss factor from 3·5% to 1·5% and complete replacement reduces it to 0·4%. A low-alkali clay has to be added with the barium carbonate to the talc.

Boron Compounds

Borax and other borates are used as fluxes in glazes. Boron also plays an important structural function in the glaze akin to that of silicon and unlike the bond-breaking function of the alkalies and alkaline earths. This will be described in Chapter 2.

BORIC OXIDE B_2O_3

Boric oxide is not found or used free. It is the form of expressing the boron content of a raw material and a composite body.

B_2O_3, M.W. 69·64. S.G., 1·8. M.P., 294° C (561° F). Solubility in water where it hydrates at 0° C, 1·1 g/100 ml; at 100° C, 15·7 g/100 ml.

BORIC ACID H_3BO_3

This occurs in the Tuscan 'soffioni' or 'fumaroles'. It is also prepared from borocalcite.

H_3BO_3, M.W. 61·84. B_2O_3, 56·3%. S.G., 1·4. M.P. with decomposition, 185° C (365° F). Solubility in water at 0° C, 1·95 g/100 ml; at 21° C, 5·15 g/100 ml; at 100° C, 39·1 g/100 ml.

Boric acid is occasionally used for introducing boron into a glaze if for some reason no alkali may be added.

BORAX OR TINCAL $Na_2O.2B_2O_3.10H_2O$

KERNITE OR RASORITE $Na_2O.2B_2O_3.4H_2O$

These minerals are purified to give the readily available pure borax, sodium diborate decahydrate, or the products with less water, sodium diborate pentahydrate and anhydrous sodium diborate (Table 29).

TABLE 29

The Composition of Hydrated and Anhydrous Sodium Diborate

	Anhydrous $Na_2O.2B_2O_3$	Pentahydrate $Na_2O.2B_2O_3.5H_2O$	Decahydrate $Na_2O.2B_2O_3.10H_2O$
Boric oxide, B_2O_3	69·27	47·80	36·52
Sodium oxide, Na_2O	30·73	21·28	16·25
Water, H_2O	—	30·92	47·23
	100·00	100·00	100·00
Molecular weight	201·27	291·35	381·43
Ratio of weights containing the same amount of B_2O_3	53	76	100 (S127)
S.G.	2·37	1·82	1·73 (S127)
M.P.	741° C	120° C	62° C
B.P.	1575° C decomp.	120° C	62° C

Solubility of Borax $Na_2O.2B_2O_3.10H_2O$ in H_2O

at   0° C,   2·29 g/100 g H_2O
at   20° C,   5·14 g/100 g H_2O
at 100° C, 191·1 g/100 g H_2O.   (**B93**).

On heating the penta- and deca-hydrates fuse and boil while the water of crystallisation escapes, which may have a beneficial stirring effect. Unless this property is desired, the anhydrous salt is much more economical in space, transport, handling, and fuel.

Borax, sodium diborate pentahydrate and anhydrous sodium diborate are the most usual means of introducing boric oxide in glazes (**S127**).

PANDERMITE OR BOROCALCITE $4CaO.5B_2O_3.7H_2O$

This hydrated calcium borate is found near Panderma in Asia Minor, resembles a fine-grained marble and is practically insoluble in water. The formula is approximate. The average composition is: B_2O_3, 46%, CaO, 31%, H_2O, etc., 23% (**B93**).

Most of this mineral is converted to boric acid but some is directly used in glazes; being almost insoluble it does not require fritting.

COLEMANITE $2CaO.3B_2O_3.5H_2O$

This mineral is found in California and Nevada and was for a time the chief domestic source of boron in the United States.

B_2O_3, 50·81%, CaO, 27·28%, H_2O, 21·91%.

Being almost insoluble colemanite finds direct use in certain raw glazes, especially lead ones (**A89**).

USES

Boric oxide is an admirable flux, giving easily fusible frits with good fluidity and glazes maturing at low temperatures. It can be used satisfactorily to replace part or all of the lead in a glaze, thus eliminating this poisonous substance.

Boric oxide lowers the coefficient of thermal expansion thus increasing the resistance to thermal shock. It increases the tensile strength and elasticity, thereby reducing crazing. It increases the hardness of a glaze and exerts a strong solvent action on the body, bringing about a more intimate interaction between body and glaze, which in itself also reduces crazing. It improves the chemical resistance of glazes to steam, water, weathering and alkaline chemicals. It imparts lustre and brilliance to glossy glazes and is also beneficial in matt glazes.

Being acidic boric oxide sometimes has a marked effect on glaze colourants which if taken into account can be used to great advantage.

Certain applications of boron compounds in ceramic bodies have also been described. Cool (**C51**) shows that whereas borax has useful properties in increasing plasticity of a clay body and acting as a flux, its tendency to migrate and then form large crystals leads to fused spots on the surface. Calcium borate, on the other hand, is insoluble and a useful flux, but cannot be beneficial to the raw clay. Furthermore it is more expensive. He claims that very useful results are obtained by using mixtures of borax with hydrated lime in equivalent proportions of 2 to 1. In contact with water colloidal ulexite together with sodium monoborate are slowly formed:

$$4Na_2B_4O_7.10H_2O + 2Ca(OH)_2$$
$$= Na_2O.2CaO.5B_2O_3.16H_2O + 6NaBO_2.4H_2O + 2H_2O.$$
borax ulexite

This reaction may be carried out in a separate vessel and the ulexite filtered off and washed free of sodium monoborate, or the dry ingredients are added to the body or glaze mixture allowing the reaction to occur *in situ*.

Zinc Compounds

Zinc is a common element occurring in all five continents. It is found throughout Europe; the British deposits of zinc ores consisting chiefly of the sulphide sphalerite ZnS, are in S. Scotland, N.W. England, N. and W. Wales and Cornwall. There are numerous deposits in the U.S.A., also in Mexico, Canada and Australia.

The ores are generally not used directly in ceramics.

ZINC OXIDE

This may be prepared by the 'French process' by the oxidation of zinc vapour obtained by boiling the metal, or by the 'American process' combining reduction of the zinc ore with coal which gives off zinc vapour and oxidation of this before it condenses (**A89**).

The product has a very fine particle size, averaging at 0.21 to 0.26 μ, the finest particles being 0.12 to 0.18 μ, incorporation of which into a glaze may give it a tendency to crawl, so it is often specially calcined before use. Whereas the pure zinc oxide remains fluffy after calcination at cone 4 to 5 (1165° C, 2129° F) a product of uniform particle size is obtained if the zinc oxide is contaminated with 0.2 to 0.5% lead oxide (introduced as lead sulphate) and then heated to from cone 4 (1160° C, 2120° F) to cone 10 (1300° C, 2372° F) (**T24**).

ZnO, M.W. 81·38. S.G., 5·5. M.P., 1975° C (3263° F) (**B142**). Solubility in water at 29° C 0.00016 g/100 ml.

Zinc oxide is amphoteric, acting either as a base or as an acid according to its environment. Zinc can therefore undertake two different functions in a glaze (see p. 213).

In normal glazes not more than 16% zinc oxide should be used (the average is 10%), when according to circumstances it can be used in brilliant or matt, transparent or opaque, glossy or crystalline, white or coloured glazes.

Introduction of a large quantity of zinc gives beautiful highly crystalline glazes.

Zinc can replace lead in the glaze structure, and its addition to a lead glaze decreases the solubility of the lead and makes it whiter. Zinc decreases the coefficient of expansion and hence the tendency to craze, and increases the resistance to scratching. It has a marked effect on some glaze colourants making them more brilliant, so that it is often used as a carrier for the colouring agent (**S120**).

The refractory zinc oxide is sometimes replaced by its lower melting compounds.

	Formula	ZnO %	M.P. (° C)
Zinc Borates	$ZnO.B_2O_3$	53·89	1000(**I3**)
	$3ZnO.2B_2O_3$	63·67	980(**P4**)
	$5ZnO.2B_2O_3$	74·50	1080(**I3**)
Zinc Orthophosphate	$3ZnO.P_2O_5$	63·23	900

Lead Compounds

Lead compounds are among the chief traditional glaze fluxes. They give a glaze of low fusion point and viscosity which results in a product of high brilliance, lustre and smoothness. They are also said to be less susceptible to errors of composition and processing during manufacture. Thus lead compounds are still in widespread use, although it is well known that any that are soluble are a danger to the health of both maker and consumer. Much work, thought and legislation have gone into the effects of lead on health, resulting in increased fritting of lead compounds to give the much less dangerous lead silicates, and in development of special low-solubility glazes (**S114**) (Chapter 6).

LEAD OXIDES

Litharge, PbO, M.W. 223·21. S.G., 9·53. M.P., 888° C (1630° F). Solubility in water at 20° C, 0·0017 g/100 ml.

Red lead, Pb_3O_4, M.W. 685·63. PbO, 97·67%, O_2 2·33%. S.G., 9·1. Decomposes to PbO and oxygen at about 500° C (930° F).

Red lead is sometimes preferred to litharge. It is somewhat less difficult to keep in suspension. It decomposes, giving off oxygen at 500° C, but unfortunately this cannot entirely prevent reduction to free lead which may well occur at higher temperatures. The lead oxides are used as glaze fluxes and fritted before use to convert them to silicates.

LEAD CARBONATE (WHITE LEAD. BASIC LEAD CARBONATE)

$2PbCO_3.Pb(OH_2)$, M.W. 775·67. PbO, 86·33%, loss on ignition, 13·67%. S.G., 4. Decomposes at from 320° C to 400° C (600° F to 750° F).

Commercial basic lead carbonate is produced in a chemically pure form with particle sizes suitable for suspension and dispersion in water. Its decomposition at 320–400° C with the evolution of gases leaves the resultant oxide in a finely divided state that readily reacts with the other glaze constituents. Thus fusion of the glaze is quicker than when the oxides are used.

The finer state of division of white lead, however, makes it a greater hazard to workers, who might inhale the dust, and although insoluble in water it is soluble in the gastric juices, so that it should always be fritted. White lead is a more expensive frit component than litharge or red lead so the latter are generally used.

LEAD SILICATES

The fritted forms of lead are much less harmful than the dusty carbonate and oxides. The composition of a relatively insoluble lead silicate may vary and a number of products are on the market. The solubility of the lead drops markedly if a small quantity of alumina is added to the mixture. The lead–alumina–silica eutectic mixture is:

PbO, 0·254, Al_2O_3, 1·91 SiO_2. PbO, 61·2%, Al_2O_3, 7·1%, SiO_2, 31·7%, M.P., 650° C (1200° F) (**S108**).

There is also the commercial 'lead bisilicate': PbO, 65%, SiO_2, 34%, Al_2O_3, 1%. M.P. about 815° C (1499° F) (**A89**).

Bismuth Oxide

Bismuth oxide is sometimes used as a flux for applying glaze colours which are hardened on at about 800° C and also in conjunction with certain coloured oxides to produce different colours.

Bi_2O_3, M.W. 466·00. S.G., 8·2–8·9. M.P., 820–860° C (1508–1580° F) depending on the crystalline form. Insoluble in water.

FURTHER REFRACTORY AND SPECIAL RAW MATERIALS

Zirconium Compounds

ZIRCON $ZrSiO_4$

This mineral originally widely distributed in acid igneous rocks has been concentrated by weathering agents when the rocks have disintegrated. Zircon itself was unattacked but being heavy it has concentrated on beaches. The chief sources are Australia, India and Florida.

S.G., 4·2 to 4·6. Hardness, 7 to 8. Thermal dissociation into zirconia and silica glass at from 1500 to 1800° C (2730 to 3270° F).

Zircon is used for refractories, in electrical and chemical porcelains, *e.g.* in sparking plugs, and in glazes as an opacifier or to control texture and crazing; it is also used to stabilise colours.

THE RAW MATERIALS

TABLE 30
Analysis of a Commercial Granular Zircon

	Commercial		Theoretical
ZrO_2	66·5	to 66·9	67·23
SiO_2	32·5	33·0	32·77
Al_2O_3	0·05	0·12	
TiO_2	0·05	0·07	
P_2O_5	0·04	0·05	
Fe_2O_3	0·01	0·12	
Rare earths	0·01	0·01	
		(A89)	

ZIRCONIUM OXIDE (ZIRCONIA), BADDELEYITE, ZIRKITE AND FAVAS ZrO_2

Zirconium oxide occurs in Brazil in the mineral baddeleyite together with hydrated zirconium oxide and zirconium silicate.

Different grades give analyses in the following ranges: ZrO_2, 65 to 75%. SiO_2, 10 to 14. Fe_2O_3, 3 to 5. S.G., 5·5–6·0, Hardness, 6·5. The natural mineral can be used directly for refractories.

Other crude zirconium oxides on the market are termed: zirkite, 70–80% ZrO_2, Favas, 80–85% ZrO_2, and there is also the chemically purified and sometimes fused white zirconium oxide, 98% ZrO_2.

ZrO_2, M.W. 123·22. S.G., 5·5–5·7. M.P. of baddeleyite, 2700° C (4890° F); of zirconia, 2950–3000° C (5340–5430° F). Insoluble in water.

On heating, the pure zirconia undergoes one or more rapid and reversible inversions with volume changes that can shatter a solid piece of the material (Chapter 2, p. 170). However, by heating to about 1700° C (3092° F) with additions of lime and magnesia totalling about 5%, a stable cubic structure is obtained. Such stabilised zirconia is specially made and marketed, and is used for fine refractories, crucibles, etc. (Fig. 1.44).

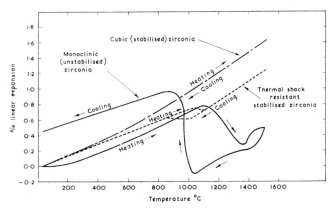

FIG. 1.44. *Volume changes of pure monoclinic zirconia and of stabilised cubic zirconia* (N26)

Zirconia and several synthetic zirconium silicates of very fine grain size less than 5 μ are used as opacifiers (p. 144) and zirconia is also used to develop certain colours.

Silicon Carbide

Silicon carbide, SiC, is an entirely synthetic material produced in electric resistance furnaces from mixtures of pure sand, petroleum coke, sawdust and salt. When heated to temperatures of up to 2500° C ($\sim$4500° F) the silica is reduced and the free silicon combines with the coke, the sawdust burns out and keeps the mass porous for the escape of gases and the salt assists in removing impurities at high temperatures.

SiC, M.W. 40·10. Oxidised above 800° C (1472° F); dissociates at 2300° C (4172° C); maximum operating temperature, 1540° C (2804° F). S.G., 3·20. Hardness (Knoop) 2500. Electrical resistance, low for a non-metal (**N26**).

Silicon carbide is incorporated in abrasive and refractory bodies, and as a major constituent of electric resistance heaters, but does not enter into reactions with the other body components. It is very resistant to thermal shock, making it very useful for kiln furniture.

Carbon Raw Materials

The chief sources of free carbon are coal and mineral oils. Graphite also occurs naturally in some places, notably Ceylon, Madagascar and for pencil graphite, Bavaria. The materials used in the ceramic industry as major body constituents are:

Foundry coke, having a relatively low ash content of up to 10 to 12%, combined with good strength and low reactivity. (Gas coke and blast furnace coke are unsuitable.)

Anthracite.
Bituminous coking coal.
Natural graphite.
Synthetic graphite.
Charcoal.

The following carbon materials may be used as bonds:

Tars, *e.g.* dehydrated coke oven tar, water-free steel works tar, tar containing 70% pitch and 30% anthrazenol.

Mixture of tar, anthracene oil and bitumen plus a little naphthalene.
High-melting asphalt.
Pitches.
Resins may be used as fillers.

Carbon differs from most other ceramic materials in its ease of oxidation. It is thus necessary to fire it in almost sealed containers and the application of the products is restricted. Under reducing conditions, however, carbon is highly refractory. It is also an electric conductor and graphite is an excellent thermal conductor.

TABLE 31
Properties of Carbon (C, At. Wt. 12·00)

	Charcoal	Gas carbon	Graphite
S.G.			2·3
Hardness			1–2
M.P.		3500° C (6332° F)	3700° C (6690° F)
Thermal conductivity (c.g.s.)	$0 \cdot 13 \times 10^{-3}$	10×10^{-3}	300×10^{-3}
Specific electrical resistivity		0·004–0·007	0·003

USES

The chief uses of carbon minerals are in refractories, refractory holloware, as heat exchangers and in chemical plant.

Chrome-Iron Ore

Chrome–iron ores are a series of complex spinelloids:

$$\begin{matrix} FeO & & Cr_2O_3 \\ & \times & \\ MgO & & Al_2O_3 \end{matrix}$$

the end member, iron chromite $FeO.CrO_3$, is only found in meteorites. It has the composition Cr_2O_3, 68%, FeO, 32%. Chrome–iron ore has a chrome content of from 35 to 50%, and also contains a certain amount of silica in the form of magnesium silicates.

About 40% of the world production of chrome–iron ore is used for refractories. The chief producing countries are U.S.S.R., South Africa and Southern Rhodesia, Turkey, Cuba, New Caledonia and Pakistan.

The melting point of pure chromite is 2180° C (3956° F) but that of chrome–iron ores lies in the range 1700 to 2100° C (3090 to 3810° F). Ores useful for refractories have melting points above 1900° C (3450° F). The melting point is not directly related to the chrome content of the ore, and the suitability of an ore to make a useful refractory seems to depend on its mineralogical character and grain size distribution. The silicates present appear to combine with the iron oxide and form a matrix which holds the chrome particles together.

Neutral refractory chromite bricks are made directly from ground chrome–iron ore on the addition of a small quantity of binder. Their chief use is to make a neutral zone in furnace linings between a basic and an acidic section to prevent interaction of the two active linings (**R46**). Chromite bricks are also being used increasingly for other parts of furnaces because they are somewhat less temperature sensitive than magnesite bricks and are cheaper to produce.

INDUSTRIAL CERAMICS

Chrome ore is also used with magnesite to make chrome–magnesite refractories. Although chrome–magnesite bricks are more expensive to make than chromite, magnesite or silica bricks, their durability is much greater. Thus chrome–magnesite bricks are generally replacing other types in open hearth furnaces (**R47**).

Olivine

The mineral olivine, found in quantity only in Norway, North Carolina and Washington, U.S.A., and as a small deposit on the Ise of Harris, Scotland, consists of mixed crystals of magnesium and iron orthosilicates (*see* silicate structures).

Magnesium orthosilicate, Mg_2SiO_4, is known as the rather rare mineral 'forsterite'; iron orthosilicate, Fe_2SiO_4, occurs as the mineral 'fayalite', often found as a constituent of metallurgical slags. Their compositions are:

Forsterite Mg_2SiO_4 or $2MgO.SiO_2$	Fayalite Fe_2SiO_4 or $2FeO.SiO_2$
MgO, 57·2%	FeO, 70·4%
SiO_2, 42·8%	SiO_2, 29·6%
M.P., 1910° C (3470° F)	M.P., 1205° C (2012° F)

The melting-point of olivine itself lies between the two, decreasing as the iron content increases.

High magnesian olivine is used chiefly for forsterite refractories. The raw olivine is mixed with magnesium oxide before shaping so that the fayalite,

TABLE 32

Forsterite (Theoretical) and Natural Olivines (S122)

	$2MgO$ SiO_2	Norway	North Carolina	Pacific & W. Coast	Scotland	
SiO_2 (%)	42·8	41·81	41·10	44·87	39·25	39·10
MgO (%)	57·2	50·31	49·27	42·99	45·15	44·76
FeO (%)	—	5·83	5·79	⎱ 7·68	7·69	7·82
Fe_2O_3 (%)	—	0·25	0·18	⎰	2·67	2·67
MnO (%)	—	0·12	0·18	—	0·80	0·57
Cr_2O_3 (%)	—	0·37	0·16	2·20	1·81	1·75
Al_2O_3 (%)	—	0·22	0·16	0·81	0·60	0·32
CaO (%)	—	—	0·30	—	0·05	0·07
NiO (%)	—	0·38	n.d.	—	—	—
CoO (%)	—	0·02	—	—	—	—
$K_2O + Na_2O$ (%)	—	0·01	—	—	—	—
Loss on ignition (%)	—	0·49	0·50	0·33	1·80	2·72
Totals	100·0	99·81	98·67	99·88	98·82	99·78

which would otherwise reduce the refractoriness, is converted to forsterite and magnesium ferrite ($MgFe_2O_4$) during the burning.

Forsterite refractories are particularly useful for basic high-temperature furnaces, glass tanks, checkers, etc., containing alkali (**A89**).

Beryllium Compounds

BERYL

This mineral, occurring in Brazil, India, North America, Argentina, South Africa and Australia, is the only commercial source of beryllium oxide. It is a beryllium aluminium silicate:

$$3BeO \,.\, Al_2O_3 \,.\, 6SiO_2$$
$$13\cdot97\% \quad 18\cdot97\% \quad 67\cdot06\%$$

The commercial product usually has from 10 to 12% BeO. A typical analysis is (**A89**):

SiO_2	62·00%
Al_2O_3	22·50
BeO	11·85
Na_2O	1·13
K_2O	0·22
Fe_2O_3	0·41
CaO	1·40
MgO	0·30
P	0·11
S	0·02
	99·94

S.G., 2·5–2·9. Hardness, 7·5 to 8. M.P., 1410 to 1430° C (2570 to 2606° F).

This silicate contains the unusual cyclic ion $(Si_6O_{18})^{12-}$ (**B106**).

Beryl is used as a flux in some high-temperature electrical porcelains. Under certain conditions it will bring about recrystallisation of alumina. It is an excellent catalyst for the formation of mullite and hence strong bodies, but also assists in the formation of cristobalite. Beryl is used for glazes for low-expansion, thermal shock-resisting bodies containing talc.

BERYLLIUM OXIDE

The oxide, produced from beryl, is used when it is undesirable to add the quantities of alumina, silica and alkalies associated with the beryllium oxide in the mineral.

BeO, M.W. 25·02. S.G., 3·025. M.P., 2570° C (4660° F). B.P., ~3000° C (5430° F). Solubility in water at 20° C 0·00002 g/100 ml.

Beryllium oxide is used pure to make crucibles that have a high thermal shock resistance and chemical inertness. It is also used together with oxides of the rare earths and other oxides to make special refractory bodies, *e.g.* one of the composition (**A110**):

$$48\text{BeO} \cdot \text{Al}_2\text{O}_3 \cdot \text{ZrO}_2 + \text{CaO}$$
$$82 \cdot 6\% \quad 7 \cdot 0\% \quad 8 \cdot 4\% \quad 2\%$$

Incorporation of beryllium oxide in electrical porcelain greatly increases the thermal conductivity.

Care must be taken with beryllium compounds, which are highly toxic, the oxide less so than other compounds (**A89**).

Titanium Compounds

Titanium oxide occurs in three forms, rutile, anatase and brookite (*see* Chapter 2, p. 170) which seem analogous to the forms of silica and zirconia (**N35**):

SiO_2	Quartz	Cristobalite	Tridymite
ZrO_2	Baddeleyite	β-ZrO_2	α-ZrO_2
TiO_2	Brookite	Anatase	Rutile

Although very widespread in small quantities, when it frequently appears as a disturbing impurity in ceramic raw materials, there are few workable deposits, so that it is often classed as 'rare'. The titanium minerals are listed in Table 33 and some analyses given (Table 34). These are treated to give pure titania.

TABLE 33

Titanium Minerals

Name	Formula	Percentage composition	M.W.	Density	Hardness
Rutile	TiO_2		79·90	4·23	6–6½
Anatase	TiO_2		79·90	3·8–3·9	6
Brookite	TiO_2		79·90	3·9–4·2	6
Ilmenite	$FeTiO_3$	TiO_2, 52·67, FeO, 47·33	151·75	4·75	5–6
Perowskite	$CaTiO_3$	TiO_2, 58·90, CaO, 41·10	135·98	4·00	5½
Titanite	$OCaTi(SiO_4)$	TiO_2, 40·82, CaO, 28·57, SiO_2, 30·61	196·07	3·51	5–5½
Ilmenorutile		Rutile with niobium and iron			
Mossite		Rutile with niobium			
Tapiolith		Rutile with tantalum			
Knopite		Perowskite with cerium replacing calcium			
Loparite		Perowskite with niobium			

TITANIUM OXIDE, TITANIA, RUTILE TiO_2

TiO_2, M.W., 79·90. S.G., 4·17. Hardness, 5·5–6·5. M.P., 1830° C (3330° F).

In the presence of carbon monoxide or hydrogen, titania is easily reduced to titanium suboxide, TiO (**N35**).

USES

Pure titania is used for high permittivity, electrical ceramics and also as an opacifier in glazes.

TABLE 34
Analysis of Australian Titanium Minerals (V11)

Rutile		Ilmenite		
TiO_2	96·02	TiO_2	46·90	
ZrO_2	0·44	FeO	26·60	} 42·07
SiO_2	0·42	Fe_2O_3	15·47	
Al_2O_3	0·28	Cr_2O_3	2·40	
Fe_2O_3	0·22	MgO	1·70	
S	0·02	MnO	1·40	
P_2O_5	0·01	Al_2O_3	1·30	
		V_2O_5	0·15	
	97·41	P_2O_5	0·03	
			95·95	

Thorium Oxide ThO_2

Thorium oxide is obtained from the rare earth phosphate, monazite, the most important deposit of which is in India.

ThO_2, M.W. 264·12. S.G., 9·7. M.P., $\sim$3000° C (5430° C); the highest melting oxide. B.P., 4400° C (7950° F).

Crucibles of electrically fused thoria cast and fired to 1800° C (3270° F) are used for high-melting metals, which do not wet it. Thoria is also used to increase the density of zirconium ware.

Vermiculite

This hydrated biotite mica that has undergone alteration by hydrothermal agencies is given the formula:

$$(OH)_2(MgFe)_3(SiAlFe)_4O_{10}.4H_2O.$$

It possesses the remarkable property of exfoliating from sixteen to twenty times its original volume on heating. This makes it a very good insulator and it is used in acoustical tile and thermal insulators.

Vermiculite insulating bricks are stable up to 1100° C (2010° F), the melting point being about 1350° C (2460° F).

Rare Elements

The development of various new industries in the world, particularly those connected with nuclear energy, is making available as by-products a number of materials formerly too rare to be considered for application in the ceramic industry.

Hans Vetter (**V11**) tabulates and discusses the occurrence and mineralogy

TABLE 35

	No.	Name	Symbol	At. Wt.
Group Ia	37	Rubidium	Rb	85·48
	55	Caesium	Cs	132·91
Group IIIa	21	Scandium	Sc	45·10
	39	Yttrium	Y	88·92
	57	Lanthanum	La	138·92
	58	*Cerium*	Ce	140·13
	59	*Praesodymium*	Pr	140·92
	60	*Neodymium*	Nd	144·27
	61	Illium*	Il	147
	62	*Samarium*	Sm	150·35
	63	Europium	Eu	152·00
	64	Gadolinium	Gd	157·26
	65	Terbium	Tb	158·93
	66	*Dysprosium*	Dy	162·51
	67	*Holmium*	Ho	164·94
	68	*Erbium*	Er	167·27
	69	*Thulium*	Tm	168·94
	70	*Ytterbium*	Yb	173·04
	71	Lutetium	Lu	174·99
Group IIIb	31	Gallium	Ga	69·72
	49	*Indium*	In	114·76
	81	*Thallium*	Tl	204·39
Group IVa	72	Hafnium	Hf	178·6
Group IVb	32	Germanium	Ge	72·60
Group Va	23	*Vanadium*	V	50·95
	41	*Niobium*†	Nb	92·91
	73	*Tantalum*	Ta	180·95
Group VIa	42	*Molybdenum*	Mb	95·95
	74	*Tungsten*	W	183·92
	92	*Uranium*	U	238·07
Group VIIa	75	*Rhenium*	Re	186·31
Group VIII	44	*Ruthenium*	Ru	101·1
	45	*Rhodium*	Rh	102·91
	46	*Palladium*	Pd	106·4
	76	*Osmium*	Os	190·2
	77	*Iridium*	Ir	192·2
	78	*Platinum*	Pt	195·09

* also called Promethium Pm
† also called Columbium Cb

of these. Some of them are already used as ceramic raw materials and have been discussed, the rest are listed below although some are used as opacifiers (p. 144) and colourants (p. 146) (set in italics in Table 35).

OPACIFIERS

Normal glazes are transparent, they offer a homogeneous medium to light rays which may be reflected at the surface, or refracted on entering and leaving it. In opaque glazes the medium is not homogeneous, particles are suspended in the glassy matrix each of which reflects, refracts and diffracts light. The refractive index of the suspended particle must differ from that of the glass phase, and irregularly shaped particles are the most effective. The particles may be:

(1) A finely divided raw material that has not reacted and dissolved.

(2) A compound formed in the reactions that is immiscible with the rest of the glass.

(3) Crystallites formed on cooling.

(4) Minute gas bubbles (this makes the glaze mechanically weak and is inapplicable).

The choice of opacifier therefore depends on the nature of the glaze. Materials used under appropriate conditions are tabulated in Table 36 (**A89**).

COLOURING AGENTS

The cause of visible colour will be discussed later on (Chapter 2). Here it will suffice to say that certain atoms are capable of appearing coloured, their actual colour depending on the other atoms around them and the way in which they are bound. Thus copper compounds are almost all coloured but the colours range from green, turquoise and light blue to deep brown, and the metal itself can impart a brilliant red. There are also certain specific combinations of colourless elements that are coloured.

It is therefore not surprising that the raw material used to give a certain colour in a ceramic product is in itself quite a different colour, and that the colour produced may vary with circumstances, for instance, composition, temperature and kiln atmosphere.

The large number of colouring agents are presented in Table 37. The order taken has been alphabetical order of the colouring oxides or metal, each of which are immediately followed by other compounds containing them (thus lithium cobaltite will be found under cobalt not lithium).

There are in addition many commercial ceramic colours and frits. These contain true colouring agents fused together with sand, clay, etc., and are more or less finely ground. Instructions for their use generally accompany them. It must, however, be emphasised that only very finely ground 'colours' can be used.

TABLE 36

Opacifiers (compiled from A89, W34)

Name	Formula	M.W.	S.G.	M.P. (°C)	M.P. (°F)	Oxides entering fusion	M.W.	Conversion Factor	Remarks
Antimony oxide	Sb_2O_3	291.5	5.2–5.7	656	1213	Sb_2O_3 or Sb_2O_5	291.5 323.5	1 1.10	Only remains white in leadless glazes. Poisonous unless oxidised.
Sodium Antimonate	$Na_2O.Sb_2O_5.\frac{1}{2}H_2O$	394.5		d. 1427	d. 2600	Sb_2O_5 Na_2O	323.52 61.99	0.820 0.157	Non-poisonous.
Arsenic oxide	As_2O_3	197.8	3.7–3.8	subl. 193	379	As_2O_3	197.8	1	Good opacifier at cones 05 to 1 (1000–1100° C) only; poisonous.
Cerium oxide	CeO_2	172.1	6.74	1950	3540	CeO_2	172.1	1	Good if pure and only below cone 03 to 02 (1050° C).
Stannic oxide (tin oxide)	SnO_2	150.7	6.6–6.9	1127 d.	2060 d.	SnO_2	150.7	1	Traditionally the best opacifier for all temperatures.
Titanium oxide (rutile)	TiO_2	79.9	4.26	1640	2980	TiO_2	79.9	1	Tends to produce colour if cooled slowly.
Zinc oxide	ZnO	81.38	5.5	>1800	>3270	ZnO	81.4	1	Excellent under certain conditions.
Zinc spinel	$ZnO.Al_2O_3$	183.3				ZnO Al_2O_3	81.4 101.9	0.444 0.556	
Zirconium oxide	ZrO_2	123.2	5.5–5.7	2950–3000	5340–5430	ZrO_2	123.2	1	Very good over wide range of temperature and composition if the grain size is smaller than 5 μ. The silica as well as the zirconia in the silicate acts as an opacifier.
Zirconium silicate (zircon)	$ZrSiO_4$	183.3	4.56	2550	4620	ZrO_2 SiO_2	123.2 60.1	0.673 0.327	

Name	Formula					Oxides			Notes
Barium zirconium silicate	approx. BaO.ZrO$_2$.SiO$_2$*	336.6	4.65			BaO ZrO$_2$ SiO$_2$	153.36 123.22 60.06	0.411 to 0.455 0.355 to 0.361 0.175 to 0.212	Should only be used in small quantities.
Calcium zirconium silicate	approx. CaO.ZrO$_2$.SiO$_2$	239.4	4.3	1248	2280	CaO ZrO$_2$ SiO$_2$	56.08 123.22 60.06	0.234 0.515 0.251	Synthetic; generally used in conjunction with each other and zirconium oxide and silicate.
Magnesium zirconium silicate	MgO.ZrO$_2$.SiO$_2$	223.6	4.35	1790	3260	MgO ZrO$_2$ SiO$_2$	40.32 123.22 60.06	0.180 0.551 0.267	
Zinc zirconium silicate	ZnO.ZrO$_2$.SiO$_2$	264.7	4.80	2076	3773	ZnO ZrO$_2$ SiO$_2$	81.38 123.22 60.06	0.307 0.466 0.227	Decomposes with loss of ZnO.
Zirconium spinel	1.26ZnO.Al$_2$O$_3$. 1.7ZrO$_2$.2.2SiO$_2$*		4.70	1766	3115	ZnO Al$_2$O$_3$ ZrO$_2$ SiO$_2$	81.38 101.94 123.22 60.06	0.195 0.194 0.399 0.253	
Sodium phosphates:									
sodium dihydrogen phosphate	NaH$_2$PO$_4$.H$_2$O	138.01	2.04	−H$_2$O 100 d. 204	−H$_2$O 212 d. 399	Na$_2$O P$_2$O$_5$	61.99 141.96	0.224 0.514	
sodium monohydrogen phosphate	Na$_2$HPO$_4$.2H$_2$O	179.01	2.07	−H$_2$O 92.5	−H$_2$O 198.5	Na$_2$O P$_2$O$_5$	61.99 141.96	0.346 0.396	
sodium ortho phosphate	Na$_3$PO$_4$.10H$_2$O	344.03	2.54	100	212	Na$_2$O P$_2$O$_5$	61.99 141.96	0.270 0.207	
sodium pyro phosphate	Na$_4$P$_2$O$_7$	265.94	2.53	880	1616	Na$_2$O P$_2$O$_5$	61.99 141.96	0.466 0.534	
Calcium phosphate (tri)	Ca$_3$(PO$_4$)$_2$	310.29	3.2	1670	3040	CaO P$_2$O$_5$	56.08 141.96	0.542 0.458	Only above cone 8.
Magnesium phosphate	Mg$_3$(PO$_4$)$_2$	262.92	2.6	1383	2520	MgO P$_2$O$_5$	40.32 141.96	0.460 0.539	Only above cone 8.
Bone ash	See p. 108								
Cryolite	3NaF.AlF$_3$	209.96	2.9	1000	1830	Na$_2$O Al$_2$O$_3$ F$_2$	61.99 101.94 38	0.443 0.243 0.543	Good if cooled quickly but dangerous fluorine fumes.

* The formula has been calculated from the percentage composition, the figures for which are given in the conversion factor column.

TABLE
Ceramic

No.	Name	Formula	M.W.	S.G.	M.P. (°C)	M.P. (°F)
1	Antimony oxide	Sb_2O_3	291·52	5·2–5·7	656	1213
2	Lead pyroantimonate	$Pb_2Sb_2O_7$	769·94	6·72	—	—
3	Lead ortho antimonate	$Pb_3(SbO_4)_2$	993·15	– –	—	—
4	Potassium pyroantimonate	$K_2H_2Sb_2O_7.4H_2O$	507·79	—	—	—
5	Potassium meta antimonate	$KSbO_3xH_2O$	—	—	—	—
6	Cadmium sulphide	CdS	144·47	3·9–4·8	1750 at 100 atm subl. in N_2 980	3180 at 100 atm subl. in N_2 1796
7	Cadmium oxide	CdO	128·41	6·95–8·15	d. 900	d. 1652
8	Cadmium carbonate	$CdCO_3$	172·42	4·258	d. < 500	d. < 932
9	Cadmium sulphate	$CdSO_4$	208·47	4·691	1000	1832
10	Cerium oxide pure commercial	CeO_2 CeO_2xH_2O	172·13	7·3	1950	3542
11	Chromium oxide	Cr_2O_3	152·02	5·2	1990	3610
12	Potassium dichromate	$K_2Cr_2O_7$	294·21	2·7	398 d. 500	748 932
13	Potassium chromate	K_2CrO_4	194·20	2·7	968	1774
14	Sodium dichromate	$Na_2Cr_2O_7.2H_2O$	298·05	2·5	$-2H_2O$ 100 320 d. 400	$-2H_2O$ 212 608 750
15	Sodium chromate	Na_2CrO_4	162·00	2·7		
16	Ammonium dichromate	$(NH_4)_2Cr_2O_7$	252·10	2·2	d.	d.
17	Barium chromate	$BaCrO_4$	253·37	4·5	d. ~1000	d. ~1830
18	Lead chromate	$PbCrO_4$	323·22	6·1–6·3	844	1551
19	Strontium chromate	$SrCrO_4$	203·64	3·9		
20	Zinc chromate	$ZnCrO_4$	181·39	—	—	—
21	Chromium phosphate (ortho)	$Cr(PO_4)2H_2O$	183·06	2·42		
22	Cobalt oxide (cobalt black)	$Co_2O_3.CoO$	240·82	6	d. 900	1650
23	Cobalt oxide (grey)	CoO	75			
24	Cobalt carbonate	$CoCO_3$	118·95	4	d.	
25	Cobalt chloride	$CoCl_2.6H_2O$	238			
26	Cobalt chromate	$CoCrO_4$	174·95		d.	
27	Cobalt nitrate	$Co(NO_3)_2.6H_2O$	291			
28	Cobalt sulphate anhydr.	$CoSO_4.7H_2O$ $CoSO_4$	281·11 155·00	2 3·7	97 981	207 1798
29	Lithium cobaltite	$LiCoO_2$	97·88			
30	Copper oxide (cupric)	CuO	79·57	6·4	d. 1026	d. 1879
31	Copper carbonate	$CuCO_3.Cu(OH)_2$	221·17	4	d. 200	d. 390

Colouring Agents

No.	Oxides entering fusion	M.W.	Conversion factor	Colours obtainable
1	Sb_2O_3	291.52	1	With lead gives Naples yellow. Gives yellow body stain with rutile or TiO_2. Greys and buffs with tin.
2	Sb_2O_3 PbO	291.52 223.21	0.381 0.580	⎫ ⎬ Naples yellow.
3	Sb_2O_3 PbO	291.52 233.21	0.294 0.674	⎭
4	Sb_2O_3 K_2O	291.52 94.2	0.574 0.186	⎫ ⎬ With lead for Naples yellow. With tin—greys and buffs.
5	Sb_2O_3 K_2O	291.52 94.2	— —	⎭
6	CdS	144	1	Yellow, only up to cone 010. With selenium—red.
7	CdO	128.41	1	⎫
8	CdO	128.41	0.745	⎬ With S—yellow.
9	CdO	128.41	0.616	⎭ With Se—red.
10	CeO_2	172.13	1	Ivory.
11	Cr_2O_3	152	1	⎫
12	K_2O Cr_2O_3	94.2 152	0.32 0.517	⎬ In absence of zinc, magnesium, tin and with only small amount of lead—green.
13	K_2O Cr_2O_3	94.2 152	0.485 0.39	⎭
14	Na_2O Cr_2O_3	62 152	0.208 0.510	With tin, calcium, and silica—pink. With zinc—brown, or under certain circumstances—red. High lead content—yellow.
15	Na_2O Cr_2O_3	62 152	0.383 0.469	
16	Cr_2O_3	152	0.603	
17	BaO Cr_2O_3	153.4 152	0.605 0.3	Chrome yellow, generally over-glaze, also coloured glazes.
18	PbO Cr_2O_3	223 152	0.69 0.235	⎧ Chrome yellow if in acid flux. ⎨ Red opaque with basic flux. ⎩ With tin gives pink.
19	SrO Cr_2O_3	103.6 152	0.509 0.373	Chrome yellow.
20	ZnO Cr_2O_3	81.38 152.02	0.449 0.419	
21	Cr_2O_3 P_2O_5	152.02 141.95	0.415 0.388	
22	CoO	75	0.934	⎫ Blues in bodies, under-glazes and coloured glazes. Very little is required and the colour is unchanged over a variety of firing temperatures and conditions so long as the atmosphere is oxidising.
23	CoO	75	1	⎬
24	CoO	75	0.630	
25	CoO	75	0.316	⎭
26	CoO Cr_2O_3	75 152	0.429 0.434	With aluminium and zinc oxides for special light blue and light green.
27	CoO	75	0.257	
28	CoO	75	0.268	Water soluble blue body stain. Sometimes for sprayed decoration.
29	Li_2O CoO	30 75	0.484 0.153	Incorporates fluxing action with blue stain.
30	CuO	80	0.765 1	⎧ Oxidising conditions. Turquoise to green. Susceptible to quantity and nature of alkali and alkaline earth components and aluminium and boron content.
31	CuO	80	0.725	⎩ Reducing conditions. Reds and purples.

TABLE

No.	Name	Formula	M.W.	S.G.	M.P. (°C)	M.P. (°F)
32	Copper sulphate	$CuSO_4.5H_2O$	249·71	2·3	d. 110	d. 230
	anhydr.	$CuSO_4$	159·63	3·6	200	390
					d. 650	d. 1200
33	Gold	Au	197·20	19·3	1063	1945
34	Gold chloride	$AuCl_3$	303·57	3·9	d. 254	d. 489
35	Iron oxides, ferrous	FeO	71·84	5·7	1420	2588
	ferric	Fe_2O_3	159·68	5·2	1565	2849
	ferroso-ferric	Fe_3O_4	231·52	5·2	d. 1538	d. 2800
36	Iron dichromate (ferric)	$Fe_2(Cr_2O_7)_3$	759·74			
37	Manganese oxide (pyrolusite)	MnO_2	86·93	5·0	d. 535	d. 995
38	Potassium permanganate	$KMnO_4$	158·03	2·7	d. <240	d. <464
39	Manganous monohydrogen phosphate	$MnHPO_4.3H_2O$	205·01	—	—	—
40	Manganic orthophosphate	$MnPO_4.H_2O$	167·97	—	—	—
41	Manganous sulphate	$MnSO_4.4H_2O$	223·05	2·107	—	—
42	Neodymium phosphate	$NdPO_4$	239·24	—	—	—
43	Nickel oxides					
	nickelous oxide	NiO	74·69	6·6–7·5	2090	3790
	nickelic oxide	Ni_2O_3	165·38	4·8	d. 600	d. 1112
44	Nickel sulphate	$NiSO_4$	154·75	3·4–3·7	d. 840	d. 1544
		$NiSO_4 6H_2O$	262·85	2·0	d.	d.
		$NiSO_4 7H_2O$	280·86	1·9	d.	d.
45	Ochres	Variable mixtures of ferric oxide, manganese oxide, limonite with clay and sand	—	—	—	—
46	Palladium	Pd	106·7		1551·5	2825
47	Platinum	Pt	195·23	21·5	1773	3223
48	Praseodymium phosphate	$PrPO_4$	235·89	—	—	—
49	Selenium	Se	78·96	4·2–4·8	217	423
50	Sodium selenite	$Na_2O.SeO_2.5H_2O$	263·04			
51	Sodium selenate	$Na_2O.SeO_3$	188·95	3·0		
52	Barium selenite	$BaO.SeO_2$	280·32	4·7	d.	d.
53	Silver (powdered, fluxed or paste)	Ag	107·88	10·5	960	1760
54	Silver oxide	Ag_2O	231·76	7·1	d. 300	d. 570
55	Silver carbonate	Ag_2CO_3	275·77	6·1	218 d.	424 d.
56	Silver chloride	AgCl	143·34	5·6	455	851
57	Sulphur	S	32·06	2·0	~120	~250
58	Tin oxide (stannic)	SnO_2	150·70	6·6–6·9	1127 d.	2060 d.
59	Titanium oxide (processed mineral rutile)	TiO_2 with impurities	79·90	3·8–4·2	1640 d.	2984 d.

37—(contd.)

No.	Oxides entering fusion	M.W.	Conversion factor	Colours obtainable
32	CuO	80	0·320	⎱ Water soluble.
	CuO	80	0·500	⎰
33	—	—	—	Metallic gold on-glaze decoration. It is suitably suspended for application.
34	—	—	—	Means of applying metallic gold in precipitated form. With stannous and stannic chlorides gives purple of cassius.
35	Fe_2O_3	160	1·11	Generally undesirable colouring impurity.
	Fe_2O_3	160	1·0	Cause of reds, browns, yellows in bricks, etc.
	Fe_2O_3	160	0·035	
36	Fe_2O_3	160	0·210	Brown under-glaze colours, alone or with manganese or zinc oxides.
	Cr_2O_3	152	0·600	Black under-glaze colour with cobalt oxide.
37	MnO	71	0·817	Insoluble ⎱ Red, yellow, brown, purple or black body
38	K_2O	94·2	0·298	⎰ and glaze colours. Temperature stable up
	MnO	71	0·45	Water soluble ⎰ to 1300° C (2372° F) cone 10.
39	MnO	71	0·346	
	P_2O_5	142	0·346	
40	MnO	71	0·423	
	P_2O_5	142	0·423	
41	MnO	71	0·3185	
42	Nd_2O_3	336·54	0·703	
	P_2O_5	141·95	0·297	
43	NiO	74·7	1·0	⎱ Blue, green, grey, brown, yellow depending on composition of the glaze.
	NiO	74·7	0·903	⎰
44	NiO	74·7	0·483	⎱
	NiO	74·7		⎰ Water soluble.
	NiO	74·7		
45	—	—	—	Yellows, browns and reds for engobe slips, under-glaze and on-glaze decoration.
46	—	—	—	Used for metallic decoration.
47	—	—	—	Used for metallic decoration and in-glazes to give certain lustres.
48	Pr_2O_3	329·84	0·699	
	P_2O_5	141·95	0·301	⎱
49				
50	Na_2O	62	0·236	
	Se	79	0·300	
51	Na_2O	62	0·328	⎰ Red with cadmium sulphide.
	Se	79	0·418	
52	BaO	153	0·546	
	Se	79	0·282	⎰
53	—	—	—	Metallic decoration; tarnishes.
54	Ag	108	0·931	Yellow. With lime or zinc—brownish. With boric oxide—grey. Reduced—metallic lustre.
55	Ag	108	0·783	Iridescent stains or sheens and lustres.
56	Ag	108	0·755	Yellows, purple of cassius, lustres.
57				With cadmium sulphide and selenium in selenium red. Readily causes defects and often an undesirable impurity.
58	SnO_2	151	1	With chromates and lime—pinks and maroons. With vanadium compounds—yellows. With gold chloride—purple of cassius.
59	TiO_2	80	1	Glaze and body stains for light cream to dark ivory in oxidising conditions only. Reducing—greys. Rare special conditions—blue.

TABLE

No.	Name	Formula	M.W.	S.G.	M.P. (° C)	M.P. (° F)
60	Tungsten oxide	WO_3	231·92	7·2	1473	2683
61	Tungstic acid	H_2WO_4	249·94	5·5	$-\tfrac{1}{2}H_2O.$ 100	$-\tfrac{1}{2}H_2O.$ 212
62	Uranium oxides					
	uranous	UO_2	270·07	10·9	2176	3949
	uranous-uranic	$UO_2.2UO_3$	842·21	7·3	d.	d.
	uranyl	UO_3	286·07	7·3	d.	d.
63	Sodium uranate (commercially known as yellow uranium oxide)	$Na_2U_2O_7$	634·122	—	—	—
64	Vanadium pentoxide	V_2O_5	181·90	3·4	690	1274

AUXILIARY RAW MATERIALS

There are a number of materials which do not form a part of finished ceramic products but are nevertheless vital to the preparation of the ware. The most important of these is water; there are the water-soluble salts that modify the interaction of the clay colloids with the water; there are also the organic binders which assist in shaping and in holding the particles of green ware together and chemicals to aid its drying. Further important auxiliary materials worked in the ceramic factories are plaster of paris for moulds for ware pressed or jiggered in plastic clay, or slip cast, and iron and steel moulds and dies for dry pressing and extrusion.

Water

Water is of such prime importance to life that settlements rarely rise up in its absence; there is therefore no need to discuss its distribution.

Normal sources of water in approximate order of their purity are: underground wells, pumps and fresh springs, lakes, rivers and the sea. The impurities are organic and suspended matter, soluble magnesium, calcium, sodium and potassium carbonates, chlorides and sulphates. This water is filtered and freed from bacteria and sometimes softened before being supplied in normal town mains. The soluble salts that are harmless in drinking water are left in and their nature and quantity may vary.

The physical properties of water are now known very accurately and frequently form the unit for different measuring systems.

H_2O, M.W. 18·016. S.G., 1·0 at 4° C. Sp. ht., 1. M.P., 0° C (32° F) under one atmosphere pressure. The change of boiling point with pressure is frequently made use of to produce high temperatures. A short table of corresponding values is therefore given (Table 38). Normal boiling point under 1 atm pressure (760 mm of Hg), 100° C (212° F).

37—(contd.)

No.	Oxides entering fusion	M.W.	Conversion factor	Colours obtainable
60	WO_3	232	1	Yellow or blue, susceptible to conditions.
61	WO_3		0·928	
62				
	UO_2	270	1·0	Black, brown.
	UO_2	270	0·962	Black, browns, greys or yellows, reds.
	UO_2	270	0·944	Greens or yellows.
63	Na_2O	62	0·098	Tomato red.
	UO_2	270	0·852	Present restrictions rule out use of uranium.
64	V_2O_5	182	1	With tin—yellow. Must be calcined together before use. Cones 04 to 14. With copper to give stable greens and bluey-greens up to cone 9. With zirconium and silica—blues and greens up to cone 12.

TABLE 38

Change of Boiling Point of Water with Pressure

Temp. (° C)	Temp. (° F)	Pressure		
		(mm of Hg)	(lb/in²)	(atm)
100	212	760	14·696	1
110	230	1074·6	20·779	
120	248	1489·1	28·795	~2
130	266	2026·2	39·180	~3
140	284	2710·9	52·421	~4
150	302	3570·5	69·042	~5
160	320	4636·0	89·646	~6
170	338	5940·9	114·879	~8
180	356	7520·2	145·417	~10
190	374	9413·3	182·025	~12
200	392	11659·2	225·451	~14

In the ceramic industry water is used as a suspension and washing medium in a number of the preliminary processes, *e.g.* the washing out of china clay from the parent rock and numerous purification methods by sedimentation, etc. It is then used in most preparations of bodies for shaping, increasing amounts being required as we pass from 'dry' pressing, through plastic pressing, moulding, jiggering and throwing, on to slip casting.

In discussing the properties of clay, the physicochemical interactions of clay and water and the part played by ions present in the water were described. Ions already present in the water supply therefore play an important part in the preparation of clay bodies.

It has been shown that plasticity of clay is affected by the pH of the water, being highest when the pH has a specific value within the range 6·0 to 8·5.

Mains water has a variable pH value, London water fluctuates between pH values of 7·6 and 9·6 from day to day or even hour to hour. If such water is used in constant quantity to make a clay plastic, the result is variable and the throwers and jiggermen experience difficulties in working it. It is therefore most desirable to test the pH of the water in use and make suitable adjustments. Numerous pH-meters are available which will give a result in a matter of minutes; there is also a continuous-recording instrument which can be fitted in the water pipe. (Both types are illustrated in the section on testing.) In most cases the tap water is too alkaline and this can be remedied by adding a small quantity of acid, *e.g.* vinegar. It may, however, be found more reliable to demineralise (*see* below) the water and then add acid or alkali to bring it to the right pH.

When making the clay body into a casting slip the nature of all the impurities in the water is of much greater importance. A casting slip is a deflocculated clay suspension, generally, the less water required to produce a slip of given viscosity, the better. Flocculants prevent the formation of a good casting slip and, amongst the ions frequently present in water, calcium, magnesium and the sulphate radicle are all flocculants and can be very disturbing. Ideally the water for making casting slips should have no dissolved foreign matter at all. It is not, however, usually possible to use distilled water, but in one factory in the United States where tap water was replaced by distilled water the necessary quantity of deflocculant dropped from 0·3–0·5% to 0·1%. Normal water softening by base exchange is of some assistance as it removes calcium and magnesium but replaces them by sodium, and it does not remove the sulphate ion. Demineralisation of water does, however, achieve a very satisfactory end product of higher purity than distillation and at much less cost. This double process involves first, the replacement of all the cations by hydrogen, followed by replacement of the anions by hydroxyl or absorption of the acids formed in the first process and then removal of carbon dioxide gas. Control is automatic by measurement of the conductivity (**A161**). The final decision, however, of what treatment, if any, is applied to the water supply must be peculiar to each individual factory.

Deflocculants

There are a number of chemicals used for the deflocculation of clay for the preparation of casting slips. They are sodium and lithium salts of weak acids which therefore give alkaline reactions. The maximum beneficial result obtainable with sodium compounds occurs at a specific concentration. Addition of too much reagent will produce flocculation. The beneficial effect of lithium compounds can be obtained over a much wider range of concentrations.

There are also a number of organic deflocculants. These have the advantage of not attacking plaster moulds and of burning out completely; but their unpleasant and often irritating smells have frequently made their use impossible.

TABLE 39
Inorganic Deflocculants

Name	Formula	M.W.	S.G.	Solubility in water (g/100 ml)	Remarks
Sodium hydroxide (caustic soda)	NaOH	40·01	2·1	42 at 0° C, 347 at 100° C	
Sodium silicate (water glass)	Variable from $Na_2O.1·6SiO_2$ to $Na_2O.4SiO_2$	158·09 to 302·23		from 156 to 43	The best deflocculant in widespread use.
Best composition of sodium silicate for deflocculation	$Na_2O.3·3SiO_2$	260·19			
Sodium carbonate (soda ash) (washing soda)	Na_2CO_3 $Na_2CO_3.10H_2O$	106·00 286·17	2·5 1·4	7·1 at 0° C, 45·5 at 100° C 21·52 at 0° C, 421 at 100° C	Less effective per equivalent of Na_2O than the silicate but used in conjunction with it.
Sodium phosphates: (pyrophosphate)	$Na_4P_2O_7$	266·03	2·5	3·16 at 0° C, 40·26 at 100° C	Very efficient deflocculant. Not in widespread commercial use.
(tetraphosphate) (hexametaphosphate)	$Na_6P_4O_{13}$	469·90			
Sodium aluminate	$Na_2O.Al_2O_3$	163·94		Soluble in cold, very soluble in hot	For porcelain. Can be used instead of the silicate and carbonate with better results.
Sodium oxalate	$Na_2C_2O_4$	134·01	2·34	3·7 at 20° C, 6·33 at 100° C	See ammonium oxalate.
Sodium gallate					Very good deflocculant.
Sodium tannate					
Ammonium oxalate	$(NH_4)_2C_2O_4.H_2O$	142·12	1·5	2·54 at 0° C, 4·0 at 16·8° C, 11·8 at 50° C	Essentially to precipitate soluble calcium, in conjunction with other deflocculants.
Lithium hydroxide	LiOH	23·95	1·4	12·7 at 0° C, 14·9 at 100° C	
Lithium carbonate	Li_2CO_3	73·89	2·1	1·54 at 0° C, 1·33 at 20° C, 0·72 at 100° C	
Lithium aluminate	$LiAlO_2$	65·91			
Lithium citrate	$Li_3C_6H_5O_7.4H_2O$	281·99		74·5 at 25° C, 66·7 at 100° C	

The actual quantity of deflocculant required depends on the nature of the clay, the nature of any adsorbed ions, impurities in the water and the particular deflocculant.

The deflocculating agents are tabulated in Tables 39 and 40.

TABLE 40
Organic Deflocculants (W64)

Diethylamine
Di-*n*-Propylamine
Monoamylamine
Monoethylamine
Mono-*iso*-butylamine
Mono-*n*-butylamine
Mono-*n*-propylamine
Mono-sec-butylamine
Triethylamine
Pyridine
Piperidine
Ethylamine
Tetramethylammonium hydroxide
Polyvinylamine
'Clay Deflocculant'
'Pantarin'

PROTECTIVE COLLOIDS

In the preparation of ceramic bodies either in the plastic condition or by deflocculation for making casting slips it is frequently advantageous to add certain, generally organic, materials which act as protective colloids. Some materials play the dual role of deflocculant and protective colloid. The same substance may be helpful for deflocculating one clay while making another more plastic.

The commonest protective colloids, some of which may be present in the natural clay are: humic acid (Kassler Braun), tannic acid, lignin, peat, colloidal silicic acid, and sulphite lye.

BARIUM CARBONATE (PRECIPITATED)

This is used as an auxiliary to the deflocculating agents to remove flocculants and is added to brick clays to prevent efflorescence. It precipitates sulphates as barium sulphate and calcium and magnesium as the carbonates.

$BaCO_3$, M.W. 197·37. S.G., 4·4. Solubility, 0·0022 g/100 ml at 18° C; 0·0065 g/100 ml at 100° C.

Flocculants

MAGNESIUM SULPHATE (EPSOM SALTS)

$MgSO_4.7H_2O$, M.W. 246·49. S.G., 1·6–1·7. Solubility, 71 g/100 ml at 20° C; 91 g/100 ml at 40° C.

Magnesium sulphate is a flocculant and is used in specific instances to stiffen casting slips.

CALCIUM CHLORIDE

$CaCl_2$, M.W. 110·99, S.G., 2·5. Solubility, 59 g/100 ml at 0° C; 159 g/100ml at 100° C.

Calcium chloride is used as a flocculant and also as an 'anti-settle' (often incorrectly termed 'anti-set') to prevent settling out of solids during and after milling.

Organic Binders

There are numerous occasions when a body or glaze does not contain sufficient plastic clay to make it possible to shape it, or to give it sufficient green strength. Such bodies and glazes are supplemented by organic binders which hold the particles together when they are raw but burn out at about 400–500° C (750–930° F) so that the resultant ware is unaffected by them. Organic binders therefore make an ordinary clay body easier to handle, and in a non-clay body they may be the only means of shaping and handling it at all. There are numerous products available; different circumstances make one more suitable than another. Table 41 has therefore been compiled to embrace as many products as possible with but brief comment on each. It should be noted, however, that water-soluble binders only act in the presence of water and are not suitable for dry-pressed bodies.

Lubricants and Anti-sticking Agents

Lubricants are used both in ceramic bodies to assist forming and at die faces to prevent sticking. Water is the natural lubricant for clay and is used in the traditional shaping methods of throwing and modelling. Water or steam are used successfully to lubricate the extrusion mouth pieces for bricks, tiles and other ware of fairly large and simple cross-section. Aqueous solutions of stearates are also used.

For extrusion of tubes or of complicated shapes, and for pressing of ware, oily lubricants are usually required on the dies. Where a dry body is being pressed it may also require an internal lubricant. It is essential that these lubricants should not lower the qualities of the body they are used on. They should burn out without charring or fluxing the surface, and without leaving holes where the lubricant has formed a large pore to migrate to the surface. Pottery oils are all basically mixtures of petroleum oils and fatty material, a mixture of lard oil and kerosene being very popular.

Van-Horn (**V4**) suggests that less expensive oils for surface lubricating can be compounded from one part of low-grade fatty materials to ten parts of low-viscosity petroleum oils, the whole further diluted with kerosene if desired. Fatty oils have the property of forming a tenacious film on metal surfaces. This is thought to be due to the affinity of the carboxyl groups

TABLE 41

Organic Binders (W64, A89)

Name	Comment	Water Solubility
Alginates, sodium ammonium		
Bitumen emulsions		
Casein		
Cellulose acetate	Thermoplastic dry-press-body binder.	
Corn flour	Plastic, extrusion and dry-press bodies.	
Dextrin	Bodies and glazes especially talc low clay bodies.	S
Gelatin	Glazes	
Glucose	Automatic dry pressing of intricate pieces.	
Glue (impure gelatin)	Hardening of glazes	
Glutrine	Bodies, with Gulac.	
Glycerin	Bodies and glazes.	
Gulac (Goulac)	With glutrine. Wet and dry-pressed bodies.	
Gum arabic	General electrical porcelains.	S
Gum ghatti	Good bond.	S
Gum tracacanth	Spraying glazes.	S
Lignin extract (same as sulphite lye)	General. The most economical binder.	S
Lignin sulphonate	Calcium or sodium salts, general.	S
Methyl cellulose (methocel)	General.	
(tylose SL25) (**K3**)	Joining, steatite and electrical porcelain, dry pressing, refractories and pyrometric cone bond.	S
Methyl ethyl cellulose (cellofas A)	Joining, steatite, electrical porcelain, dry pressing, refractories and pyrometric cone bond.	
Sodium carboxy—methyl cellulose(**C60**) (courlose) (cellofas B)	Glazes, engobes, bodies. Excellent for spraying.	
(tylose MGC 25 and 2000) (**K3**)	Glazes, sprayed, dipped and brushed. Engobes.	S
Molasses	Glazes.	
Peptone	General.	
Plaskon thermalplastic	General.	
Polyvinyl alcohol	Glazes.	
Rosin emulsion	Mixes with water, dries to tough bond.	
Sevco gum	Glazes.	
Squeegee oil	For application of colours by silk screen method the composition being adjusted to the particular circumstances.	

THE RAW MATERIALS

TABLE 41—(*contd.*)

Name	Comment	Water Solubility
Starch	Glazes that settle out and crawl; bodies and dry pressing (sours on standing).	S
Stearine	Low melting; dry pressing (**L20**).	
Sulphite lye (sulphite paper waste) (same as lignin extract)	General.	
Tannic acid	General; glaze anti-settle; susceptible to pH changes.	
Urea	Low melting; dry pressing (**L20**).	
Veegum	Glazes, improving suspension and dry hardness.	
Wax emulsions		
carnauba wax	with paraffin;	
carbowax	non-plastic and dry-pressed bodies;	
ceremul C wax	dry pressing;	
zophar C49 wax		
ceresin wax	bodies or glazes of low clay content.	

Certain of these organic materials tend to ferment on standing. This can be prevented by adding a suitable odourless phenol derivative.

---

for metal, resulting in the hydrocarbon chains standing away from the surface parallel to one another and forming a base for a second molecular layer. The second layer may be either fatty or hydrocarbon but the latter is preferable because of its lower affinity for ceramic bodies.

For internal lubrication it is important that the agent should be easily dispersable in water to ensure an even distribution. The so-called 'soluble' oils and wax emulsions are therefore very suitable. These consist of petroleum oils, paraffin and microcrystalline waxes treated with emulsifying agents to make them easily dispersable in water. By correct choice of the emulsifier an emulsion can be made anionic, cationic or non-ionic. The wax particles range in size from 1 to 3 μ. These water emulsions can be uniformly distributed through wet or moist bodies. They have the advantage over water-soluble agents of not migrating to the surface during drying (**V4**).

There are also non-oily lubricants that are used in the powdered solid form, namely the stearates of zinc, magnesium, barium, and aluminium and stearic acid itself (Table 42).

Drying Aids

AMMONIUM CARBONATE AND BICARBONATE

These substances are used to assist the drying of ware containing water-soluble substances, particularly organic binders which clog the pores. The ammonium carbonates decompose at low temperatures into volatile components which break through skins of deposited matter in the surface pores. Quantities used are 0·1 to 0·5% of the wet body (**S121**).

TABLE 42
Lubricants (W64)

Badilla (a hydrous magnesium silicate gel).
Baking flour.
Carbowax.
Cellulose acetate (requires heat as well as pressure).
Diglycol stearate.
Stearic acid–water emulsions.
Magnesium stearate.
Zinc stearate.
Barium stearate.
Aluminium stearate.
Graphite.
Oils (often deleterious to binders) mixture of petroleum oils and fat, *e.g.* kerosene and lard oil.
Paraffin emulsion.
Plaskon thermal plastic (heat as well as pressure).
Wax emulsions.

Ammonium carbonate, $(NH_4)_2CO_3.H_2O$, M.W. 114·11. Decomposes at 58° C (136° F).

Ammonium bicarbonate, NH_4HCO_3, M.W. 79·06. Decomposes at 36–60° C (97–140° F) or 80° C (176° F) (**K12**).

Both compounds start to decompose at room temperature but do so less readily when dissolved in water.

PLASTER OF PARIS

Plaster of Paris is essentially calcium sulphate hemihydrate, $CaSO_4.\frac{1}{2}H_2O$. It is made from the mineral gypsum, $CaSO_4.2H_2O$, which is of fairly widespread occurrence, both almost pure and with various impurities which colour it and modify the properties of the plaster made from it.

Gypsum, $CaSO_4 2H_2O$, M.W. 172·18, $CaSO_4$, 79·1%, H_2O, 20·9%.
Plaster of Paris, $CaSO_4\frac{1}{2}H_2O$, M.W. 154·16. $CaSO_4$, 93·8%, H_2O, 6·2%.
Anhydrite, $CaSO_4$, M.W. 136·15. $CaSO_4$, 100%.

The production of satisfactory plaster from gypsum is complicated by the number of possible dehydration products. There are two hemihydrates, α and β; the α-form makes much stronger and generally more satisfactory plasters and is therefore the desired dehydration product. There are also four anhydrous calcium sulphates, obtained by stronger heating of the gypsum. The α-hemihydrate forms by recrystallisation of gypsum from water above 115° C (239° F), its formation is therefore favoured by heating gypsum in a sufficiently damp atmosphere for there to be a thin adsorbed water layer on the particles. β-hemihydrate is formed when gypsum is heated rapidly in a dry atmosphere above 100° C (212° F).

On heating ground gypsum the temperature rises until 128° C (262° F) when violent 'boiling' occurs. The temperature does not rise again until this has ceased and the plaster enters 'the first settle'. On further heating a second, shorter, period of boiling begins at 163° C (325° F) after which the plaster enters the 'second settle'. On continued further heating the hemihydrate begins to decompose, giving off water and being converted into the anhydrous salt until at 800–1000° C (1470–1830° F) the dead-burned gypsum, or Keene's cement is attained.

First-settle plaster is more plastic and easier to work than second-settle material; the latter, however, gives a denser and stronger set.

The old method of producing plaster from gypsum in periodic pans or in 'kettles' is still in general use, although some continuous rotary calciners have been introduced. The English heating pans are made of brick 12 ft (3·7 m) in diameter and 8–9 in. (20–23 cm) deep. The charge of about 3 tons is spread out 6 in. (15 cm) deep and stirred by rotating arms dragging chains. It is heated to 120° C (250° F) for 3 hr and then raked into a second pan where heating is completed at 160° C (320° F). When the escape of steam slackens and the plaster enters the first settle it is discharged (**A46**). Other figures are given for the temperature of the calcination, *e.g.* 150–155° C (302–311° F) or 120–170° C (248-338° F) or about 170° C (340° F); this arises from the fact that the product is not only dependent on the temperature but also on the efficiency of stirring and the water vapour pressure.

In the United States and Canada the kettle system is used. The kettles are large steel cylinders 8 or 10 ft high (2·4–3·0 m) with arched bottoms holding from 8 to 15 tons. The raw gypsum is crushed, magneted, ground and then filled into the calciner. This must be stirred continuously during heating to prevent local overheating. The size of the charge and the ventilation system of the calciner keeps the humidity of the air surrounding the material sufficiently high to produce enough α-hemihydrate to give a good plaster. The temperature is kept at about 120° C (250° F) until most of the water has escaped. Most of it is discharged in the first settle at 150° C (300° F) (**A46**), although some is heated until the second boil has started and removed at 176° C (350° F) and some is heated to the second settle and discharged at 190° C (370° F). Second-settle plaster is also known as 'soluble anhydrite', as its water content is often lower than that of the hemihydrate yet it will take up water and set.

A product consisting almost entirely of α-hemihydrate, termed 'alpha-plaster', which therefore gives a very strong plaster, is made by calcining the gypsum under pressure so that the atmosphere is saturated with water vapour.

Calcined gypsum or plaster of Paris readily takes up water and rehydrates, it recrystallises as gypsum in interlocking crystals which then set as a hard mass. The actual quantity of water required for this chemical combination is 18·6% of the dry weight of plaster. The dry plaster must, however, be mixed with more water than this in order to distribute the water properly and form a plastic mass which can be shaped. The amount of water required to bring 100 parts of plaster to a standard fluidity is termed the 'normal consistency'. A plaster paste is said to be highly plastic if 'when forced through an orifice, it presents a continuous, clean surface. A plastic plaster is neither sticky nor sandy. It feels soft and velvety and a trowel passes over it as over an oiled surface' (**W67**). When the plaster has set the uncombined water evaporates, leaving open pores. Therefore the more water added in mixing the plaster the more porous the set and dried product, and the faster it absorbs water. It will be shown that absorption of water is not the only role of plaster moulds, but it is nevertheless often desirable for moulds to be used with casting slips to have a somewhat higher porosity than those used with a plastic clay body. When little water is used the set plaster is dense and hard, it has much higher compressive strength and resistance to abrasion but is not so absorbing. Alpha-plaster requires less water to bring it to the same consistency as ordinary plaster and gives a very dense set (Fig. 1.45).

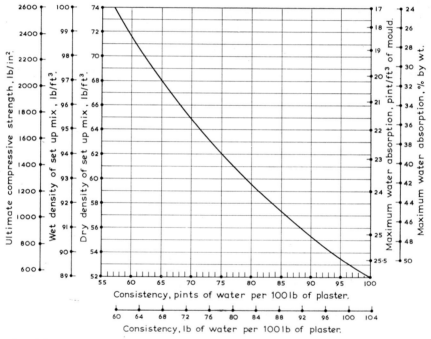

FIG. 1.45. *Effect of consistency of plaster of Paris on the wet and dry density, dry compressive strength and water absorption of the dried cast* (**C30**)

The setting time of plaster is about 15 to 30 min. As the hemihydrate combines with the water and sets there is noticeable evolution of heat. On setting, plaster expands about 0·1 to 0·2%, the amount being inversely proportional to the water content (**A165**) and also dependent on the pressure exerted by the mould, etc. The setting time can be retarded or accelerated by addition of small quantities of other substances. Clay is a most effective retarder and is already present in some less pure plasters. Other retarders are various organic glues, etc. Temperature also affects the setting time, warmth shortening it. Gypsum crystals accelerate the set of plaster. This is generally undesirable and due to insufficient cleaning of the mixing vessels before starting on a new batch. Where acceleration of the set is really desired, ground calcined gypsum gives more reliable results. On the whole it is better if neither retarder nor accelerator is used in ceramic plasters, because of the difficulty of their truly uniform distribution.

Plaster is used in the ceramic industry for three related purposes:

(1) Model making. This requires a dense uniform plaster which can be readily carved.
(2) Making moulds for pressing and jiggering shapes in plastic clay.
(3) Making moulds for slip casting ware.

In the shaping of clay in plaster moulds the plaster plays a dual role. Not only does it absorb water and thereby make the clay stiffer but the slight solubility of the calcium sulphate in the water accompanying the clay furnishes flocculating ions, both the calcium and the sulphate ions being efficient flocculants. Therefore a deflocculated casting slip poured into a plaster mould is flocculated on the mould surface to give a stiff cast. This highly important physicochemical action of the plaster must not be overlooked when working out the desired porosity of a mould. It is for this reason that a mould for casting slips does not necessarily need to be more porous than one for jiggering plastic clay; and also the reason why a casting slip will set even in a damp mould that cannot withdraw water from it.

Typical plaster mixes for jiggering and casting are:

jiggering moulds, 78 parts by weight water to 100 parts plaster;
casting moulds, 90 parts by weight water to 100 parts plaster (**N36**).

But the optimum plaster to water ratio for any particular plaster for any particular job of work should be determined in each case by experiment.

Once the optimum mixture has been determined to give a plaster of the correct porosity and hardness it is essential that a carefully worked out procedure schedule is followed correctly every time a batch is mixed. (As temperature affects the reaction, warmth accelerating the set, it may be necessary to have a summer and a winter mixture.)

In order to abide by the weight ratios of plaster and water it is often necessary to check the moisture content of the raw plaster as this sometimes absorbs water even when kept in sealed multi-wall paper bags. There are several speedy moisture testers available (Chapter 4). The materials must then be weighed out, making allowance for absorbed water. Then follows mixing, and it is of extreme importance that this is thorough and uniform so that moulds of uniform porosity can be made. A mould that absorbs water at different rates in different places will give slip cast ware of uneven thickness and shrinkage, which will later distort and crack. Although plaster mixing is still frequently done by hand the mixing is much best done mechanically.

A non-rusting metal bucket with a propeller shaft stirrer is used. The bucket dimensions should be such that the base diameter is two-thirds of the height and the top diameter equal to the height. The batch should three-quarters fill it. The propeller should clear the bottom by 1 to 2 in., and be so directed as to force the liquid downwards. The shaft is often set at 15° to 20° to the vertical.

The plaster is added to the unstirred water and allowed to soak for 2 to 5 min. It is then stirred until it reaches a creamy air-free consistency. The calcining of the gypsum entraps a large number of tiny air bubbles round and in each particle, and it is most important that the mixing process should release and remove all of the air. Proper stirring takes from 2 to 5 min. Its completion can be tested by pressing half a teaspoonful of the mixture between two glass plates until it is $\frac{1}{64}$ in. thick and holding it up to the light. Air bubbles show up as specks, the number of which should not exceed ten. Failure to 'cream' the plaster mix may result in particles settling out before it has set, making the lower part of the mould denser, and results in pinholes and 'nodules' (dense spots). There is now also a vacuum plaster blender, which should remove all entrapped air (**E6**). Plaster mixed in vacuum is considerably stronger than the ordinary type (**L38**). The plaster mixture is then cast into moulds. If the moulds are made of plaster they must be suitably coated to prevent the fresh plaster from sticking to them. Ceramists find soap the best partant as it does not affect the freshly cast surface. Numerous layers are applied and the lather removed until a hard, smooth waxy surface is obtained. Setting takes about a quarter to half an hour. There are numerous well-known methods for determining the actual setting time for any particular plaster mixture; it can readily be observed by noting when a sharp temperature rise occurs.

Litzow (**L38**) describes how mechanisation of the plaster shop can reduce the space required by 45% and increase the output per man-hour by 90%. The plaster is delivered loose in special rail trucks instead of in bags. It is transferred to storage silos by compressed air conduits. From here the correct quantities are sieved and automatically tipped into the mixing vessel holding the right quantity of water, and then mixed in a vacuum. The mixed plaster is delivered by flexible pipe to the caster. The models move along a conveyor belt to the casting position and continue thereafter to the opening and finishing.

After setting follows the drying of the moulds. The dry plaster has up to twice the strength of damp plaster. However, if the drying is carried out at too high a temperature in dry air, dehydration may result and the mould may crack and become useless. The best conditions are moderate temperatures not above 45° C (113° F) in turbulent moist, but not saturated, air. 'Burning' of the mould does not actually occur until 70° C (158° F) but mould life is greatly prolonged by never heating above 45° C (113° F).

The threefold action of plaster moulds in shaping ceramic ware by containing, flocculating and dewatering it has been described. The moulds therefore should be redried each time they have been used. Well cared for moulds may last for up to 200 to 300 castings and in exceptional cases up to 1000 plastic mouldings. Moulds that are used wet deteriorate rapidly.

The soluble chemicals used in deflocculating clay slips unfortunately have an undesirable effect on plaster. Sodium carbonate and silicate gradually react with calcium sulphate, giving calcium carbonate and silicate and sodium sulphate. The soluble substances effloresce at the surface when the plaster dries and make it unserviceable. Drying the mould from the non-casting side reduces this trouble on the casting surface. The minimum of deflocculants should be used. Organic deflocculants do not attack plaster (**W67, S103, K90, N36, A89**), but can only rarely be used because of their smell and price.

Attempts have been made to protect moulds from the action of the sodium salts. It has been found that potassium polyacrylate forms an insoluble thin film on the surface of plaster moulds which achieves this end. It is not stated in the data available whether the action of the mould is impaired thereby (**K1**).

POLYVINYL CHLORIDE AND OTHER SYNTHETIC FLEXIBLE MATERIALS

There are two major difficulties in preparing plaster moulds for casting or plastic pressing from plaster master dies. The first is the fact that the master has to be covered with a partant to prevent the new cast sticking to it. This inevitably leads to a certain loss of surface detail, as the soap builds up. The second is the rigid character of both master and cast coupled with the fact that the cast expands. This calls for multiple masters which give rise to seams on the cast, or the cast must have more pieces than are really necessary.

Wiss, Wagner and Beaver (**W78**) (1949) have worked out methods for using flexible synthetic materials, in particular polyvinyl chloride, for the master die. This does not need a partant when plaster is cast on it, so that surface detail can be reproduced more faithfully, and its flexibility withstands the expansion of the cast and makes it more easily removable.

Litzow (**L38**) (1958) refers to four years of successful industrial application of plastics as 'masters', especially polyesters and epoxyresins.

Steels for Machinery Parts

Various types of iron and, more particularly, steel, are used in the construction of ceramic machinery. The compositions and properties of some are given in Tables 43–45.

The 13% manganese steel has the remarkable property of work hardening on its wearing face under the effects of pressure to a greater extent than any other metal, *i.e.* to about $2\frac{1}{2}$ times its initial value. It is also very tenacious and ductile. It is only machined with difficulty although not essentially a hard material. Because of its ductility it may spread or burr under any excessive hammering or squeezing action. The work hardening of manganese steel only takes place when the crushing is accompanied by a certain intensity of hammering or pressure. If the pressures developed are insufficient, rapid wear may occur under apparently easy conditions. Manganese steel is therefore used for parts that receive very hard wear, such as the teeth on the front of excavator buckets, surfaces of jaw crushers, gyratory crushers and hammer mills.

For parts that have to be machined into alignment periodically, chromium or high-carbon steels are used. Chromium steels are also used for the liners and grinding plates of steel ball and tube mills as the ductility of manganese steel makes it unsuitable. Where wear is less and replacement easy, such as the knives of a clay cutter, carbon steels are used (**M76**).

TABLE 43

Steels for Heavy Clayworking Machinery—Chemical Composition (M76)

Steel type	Composition							
	C	Si	S	P	Mn	Cr	Mo	V
Manganese	1·1 –1·2	0·2 –0·4	0·01–0·02	0·05–0·08	12·0–13·0	—	—	—
Chromium	0·5 –0·55	0·15–0·25	0·06 max	0·06 max	0·5 –0·8	0·8–1·0	—	—
Carbon 'B'	0·28–0·32	0·2 –0·4	0·02–0·03	0·04–0·06	0·7 –1·0	—	—	—
Carbon 'C'	0·3 –0·35	0·2 –0·3	0·03–0·04	0·04–0·06	0·7 –1·0	—	—	—
High-carbon high-chromium die	1·8 –2·0	0·2 –0·4	—	—	0·2 –0·4	12·0–13·0	0·7 –0·9	0·2 –0·3

TABLE 44

Uses of Steels and Heat-treatments (M76)

Steel type	Application	Heat-treatment	Brinell hardness
Manganese	Wearing surfaces	1050/1100° C Water quench	180–220
Carbon 'B'	Jaw crusher body	Annealed 850/900° C	150–170
Carbon 'C'	Pitman, clay cutters, constructional parts		180–200
Chromium	Toggle seatings, tube mill liners, grids claws	850° C—oil quench Temper 620° C	241–269
Chromium	Ball mill stepped grinding plates	750° C—oil quench Temper 560° C	269–302
Carbon–chromium die	Brick mould liner plates	1020° C—air cool Temper 200° C	630
Chromium	Brick mould liner plates	850° C—oil quench No temper	600

TABLE 45

Alloy Cast Irons for Heavy Clayworking Machinery (E28)

Iron Type	C	Si	Mn	Cr	Mo	Ni	Brinell hardness
White cast iron 'A'	2·45	1·00	0·45				321
Alloy cast iron 'H'	2·53	0·66	0·74	28·14			460
Alloy cast iron 'X'	3·40	0·95	0·70	1·50		4·62	700
Alloy cast iron '63'	2·96	1·41	0·50	1·36			461
Alloy cast iron 'R16'	3·89	0·99	1·09		1·00		321
Alloy cast iron 'R20'	3·13	0·83	1·63	1·06	1·05		477
Alloy cast iron 'TP12'	2·00	1·25	0·50	2·95	4·95		650+
Malleable iron 'J'	2·25	1·00	0·45				217

Everhart (**E28**) tested the relative wear of various cast irons, alloy cast irons, malleable irons and steels when used for pugmill knives, muller tyres and die liners in the heavy clay industry. He found that the alloy cast irons were generally best. For pugmill knives, low-chromium and molybdenum irons offer considerable improvement over unalloyed iron. For die liners, high-chromium cast iron was best. For dry-pan screen plates malleable irons were the best. Brinell hardness cannot be taken as a measure of wearing property.

New abrasion-resisting alloys are reported to outlast chilled iron by from three to four times and manganese steel twice. They require no work hardening (**F14**).

Very much prolonged life can also be obtained from heavy clay machinery by hard-surfacing the parts that get most wear, either as a repair measure or as preventive maintenance (**K68**).

EXTRUSION AND PRESSING DIES

The dies for extruding clay columns, and those for pressing individual shapes from plastic or almost dry bodies may be made from a number of materials. Probably the most usual are various irons and steels, although brass, rubber or sillimanite may be used, the last so far only for extrusion dies (**A112, D1**).

The choice of iron or steel alloy is governed by the amount of machining required to achieve the desired shape, the size of the die, the pressure to which it will be submitted and the abrasive properties of the material to be pressed in it. Thus chilled cast iron may be used for thicker sections and lower qualities. Carbon steel is very much favoured because of its ability to be forged, and because a skilled workman can restore to its original dimensions a hole enlarged by use. Of the alloy steels chromium, nickel, tungsten and molybdenum steels all have their advocates. They are harder and more abrasion-resisting so that the higher initial cost is repaid where large quantities of identical products are required. Most of them can be forged and machined

166 INDUSTRIAL CERAMICS

and are thereafter hardened. Some alloys which are particularly resistant to abrasion have to be finished by grinding.

A few examples of irons and steels suitable for ceramic dies are given. The range is, however, much wider.

TABLE 46

	High-carbon high-chrome steels			Graphite steel U.S.A.	Nickel–chromium iron Brit. (I9)
	Brit. (E5)	Brit. (E5)	U.S.A.		
Carbon	1·8– 2·0	1·2– 1·4	1·5– 1·75	1·5 av.	3·0–3·4
Manganese	0·2– 0·4	0·2– 0·4	0·5– 0·6	0·5 max.	0·6–0·9
Silicon	0·2– 0·4	0·2– 0·4	0·4– 0·5	0·8 av.	1·0–1·5
Phosphorus	—	—	under 0·02	0·025 max.	—
Chromium	12·0–13·0	12·0–13·0	12·0–13·0	—	0·6–1·0
Nickel	—	—	—	—	2·5–3·0
Vanadium	0·2– 0·3	—	0·5– 0·6	—	—
Cobalt	—	2·5– 3·0	1·1– 1·2	—	—
Molybdenum	0·7– 0·9	1·4– 1·6	1·1– 1·2	0·25 av.	—
Sulphur	—	—	under 0·02	0·025 max.	—

Steels are usually delivered by the steel works ready annealed for machining. Machining to shape should be undertaken in gradual stages reannealing the part after each major dimensional change. When the final dimensions are achieved the die is hardened. The actual temperatures and times for these processes vary from steel to steel, and are supplied by the steel manufacturer. As an example the recommended treatment for the first steel listed above will be given:

'Annealing for machining should be carried out between 780 and 800° C (1440 and 1470° F). This temperature should be maintained for a period varying from two to four hours. The steel should then be allowed to cool in the furnace.

'The steel should be pre-heated slowly between 750 and 800° C (1380 and 1470° F) then brought up to 1050° C (1920° F) and air cooled. Temper at 200–250° C (390–480° F). If required it can be oil hardened from 950° C (1740° F) and tempered at 200–250° C (390–480° F) ' (**E5**).

Proper pre-treatment of a steel die before putting it into use prolongs its life very considerably.

A method of producing a hard and abrasion-free die surface, without having to machine a very hard steel, is to chromium-plate the die. This produces a very satisfactory surface.

The lifetime of a die depends on the steel used, the annealing and hardening treatment it has been given, the nature of the material to be extruded or pressed, the pressure, etc. Good dies last for from 1 250 000 to 4 000 000 pressings.

Chapter 2

The Action of Heat on Ceramic Raw Materials

WHEN the traditional, heterogeneous clay body is fired it shrinks and becomes hard. Optical examination shows the most apparent reaction to be that part of the body softens and distributes itself round the more refractory grains, partly dissolving them and holding them together by capillary and surface tension forces. On cooling this portion generally forms a glassy matrix, part of which may, however, crystallise, giving crystals interlocking with the more refractory particles. This then is the main action of the fluxes in a ceramic body.

However, closer investigation of the portions that did not fuse shows that they are not identical with the starting material, *i.e.* they have undergone changes of shape, size, structure or composition. Furthermore, pure oxides or mixtures containing no fusible component undergo considerable changes, on heating to temperatures well below their melting point.

These reactions in the solid state are one of the most important features in modern ceramics.

CHANGES AND REACTIONS IN THE SOLID STATE

I. *Changes not Altering the Chemical Composition*

POLYMORPHISM

Silica

(*a*) *Fast low–high inversions*

The three silica polymorphs, quartz, tridymite and cristobalite, each have a low (α) and a high (β) temperature form. The change from one to the other is reversible and quite sharp, which indicates that it is not a radical change involving bond breaking and rearrangement of the atoms.

These high–low transformations in fact are merely a change of bond angles and slight adjustment of inter-atomic distances. The high-temperature forms are more symmetrical than the low ones. A diagrammatic representation of such a change is shown in Fig. 2.1.

The volume change brought about by these rapid inversions is very important in the ceramic industry, particularly the α-quartz⇌β-quartz change at 573° C (1063° F) which has a 2% cubical expansion or contraction. The rapidity of the change may cause failure in silica bricks.

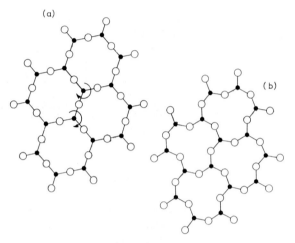

Fig. 2.1. *Crystalline forms, interconversion of which involves no bond breaking*

(b) Slower changes from one polymorph to another

There are three crystalline forms of silica, namely quartz, tridymite and cristobalite. Each is stable only in a given temperature range, but when at a temperature outside this often continues to exist. The fields of these stable and metastable states have been shown diagrammatically (Fig. 1.42, p. 100), and they are also tabulated in Table 47.

The reason for forms existing at temperatures at which they are not stable is that the transformation is difficult; chemical bonds must be broken and atoms rearranged. Such changes do not occur readily, and when they do they are slow, starting at the surface of a crystal and gradually working towards its centre. The volume changes, even if large, as in the case of the quartz–tridymite inversion, do not occur suddenly and so do not disrupt a body.

The quartz structure seems to have linked spiral chains of SiO_4 units more densely packed than the other two structures of hexagons. Tridymite and cristobalite only differ in the relative orientation of neighbouring tetrahedra.

Table 48 shows the specific gravities of the various forms of silica and the expansion when one changes to another, whether directly (when the temperature is given) or indirectly.

TABLE 47
Modifications of Silica (H49)

Form	Properties
Low quartz (α)	Stable at atmospheric temperature and up to 573° C (1063° F).
High quartz (β)	Stable from 573° to 870° C (1063° to 1598° F); capable of existence above 870° C but is then not the stable form.
Low tridymite (α)	Capable of existence at atmospheric temperature and up to 117° C (243° F) but is not the stable form in this range.
Lower–high tridymite (β_1)	Capable of existence between 117° C (243° F) and 163° C (325° F) but is not the stable form in this range.
Upper–high tridymite (β_2)	Capable of existence above 163° C and is the stable form from 870° to 1470° C (1598° to 2678° F); above 1470° C is again unstable; melts at 1670° C (3038° F).
Low cristobalite (α)	Capable of existence above 163° C (325° F) and up to from 200° to 275° C (392° to 527° F) but is not the stable form in this range.
High cristobalite (β)	Capable of existence above 200° to 275° C (392° to 527° F), and is stable from 1470° to 1710° C (2678° to 3110° F), its melting point.
Vitreous silica	Capable of existence at atmospheric temperature and up to 1000° C (1832° F) and above, where it begins to crystallise with measurable rapidity; but it is an unstable, undercooled liquid at all temperatures below 1710° C (3110° F).
Non-crystalline transition phase	Occurs as transition phase when quartz is converted into cristobalite (C21).

TABLE 48
Inversions of Forms of Silica (S83, A89, N17)

From	S.G.	Temp. (° C)	Temp. (° F)	To	S.G.	% cubical expansion
α-Quartz	2·65	573	1063	β-Quartz	2·60	2·4
α-Tridymite	2·32	117	243	β_1-Tridymite	2·32	0
β_1-Tridymite	2·32	163	325	β_2-Tridymite	2·32	0
α-Cristobalite	2·32	220–275	428–527	β-Cristobalite	~2·21	5·6
β-Quartz	2·60	870	1598	β_2-Tridymite	2·32	12·7
β_2-Tridymite	2·32	1470	2678	β-Cristobalite	2·21	5
β-Cristobalite	2·21	1128	2062	Vitreous silica	2·21	0
α-Quartz	2·65	indirect		Vitreous silica	2·21	20
β-Quartz	2·60	indirect		Vitreous silica	2·21	17·6
β-Tridymite	2·32	indirect		Vitreous silica	2·21	5
Quartz		<1470	<2678	Liquid silica		
Tridymite		1670	3038	Liquid silica		
Cristobalite		1710	3110	Liquid silica		
α-Quartz	2·65			α-Tridymite	2·32	14·2
α-Quartz	2·65			α-Cristobalite	2·32	14·2

Alumina

In the case of alumina although the active γ-form is converted quite readily at from 1150° to 1200° C (2100° to 2190° F) to the α-form, this does not change back on cooling.

Zirconia

The natural mineral baddeleyite, ZrO_2, is monoclinic. Pure monoclinic zirconia is often termed the 'C-form'; it is stable up to 1000° C (1832° F), when it changes reversibly to the tetragonal β-form, with considerable volume decrease. The tetragonal β-ZrO_2 is transformed irreversibly at 1900° C ($\sim$3450° F) to trigonal α_2-ZrO_2. On cooling α_2-ZrO_2 changes quickly and reversibly to α_1-ZrO_2 at 625° C (1157° F). This is trigonal too.

These inversions and the temperatures at which they occur are considerably affected by the presence of impurities. For instance, if zirconium nitrate, oxychlorate, oxalate or hydroxides are heated to 600° C ($\sim$1110° F) a metastable tetragonal zirconia forms which changes to the stable monoclinic one on further heating. If silica is present, however, the tetragonal ZrO_2 remains unchanged until 1460° C ($\sim$2660° F), where it is transformed to zircon, $ZrSiO_4$. Magnesia, lime and certain other oxides in quantities from 3 to 28% force the zirconia into a cubic structure, on heating to about 1700° C ($\sim$3090° F); it remains stable on cooling (**N35**). This stable cubic zirconia is of great industrial importance as a refractory, as the sharp volume change at 1000° C (1832° F) makes the natural monoclinic–tetragonal form useless. Stabilised zirconia is produced from zircon sand $ZrSiO_4$ by fusing it with coke, iron borings and lime until the silica is reduced to silicon which alloys with the iron. On cooling the zirconia crystallises in the cubic form (**W65**).

Titania

Titania occurs as the minerals brookite, anatase and rutile. Anatase is tetragonal and has two forms with a rapid inversion (**S33**):

$$\text{anatase I} \underset{(1188° \text{F})}{\overset{(642° \text{C})}{\longleftrightarrow}} \text{anatase II}$$

The higher temperature form changes to the tetragonal rutile at 915° C (1679° F) (**S33**), which in its turn changes to the rhombic brookite at 1300° C (2372° F) (**J59**). These changes are slow. Rutile may not change to brookite before melting at 1825° C (3317° F) (**K71**). The melting point of brookite is 1830° C (3326° F) (**B143**).

Others

Certain silicates also have high–low rapid inversions. Some of the systems that have been investigated are:

	Inversion Temperature	
$Li_2O.2SiO_2$	936° C	1717° F
$Na_2O.2SiO_2$	678° C	1252° F
and	707° C	1305° F
$K_2O.2SiO_2$	594° C	1101° F
$K_2O.4SiO_2$	592° C	1098° F
$2CaO.SiO_2$	675° C	1247° F (10% volume decrease)
	1420° C	2588° F
$K_2O.Al_2O_3.2SiO_2$ Kaliophite–kalsilite	1540° C	2804° F

Certain silicates have slow inversions (**K72**):

	Inversion Temperature	
$CaSiO_3$ Wollastonite–pseudowollastonite	1125° C	2057° F
$MgSiO_3$ Orthoenstatite–clinoenstatite	1140° C	2084° F
$NaAlSiO_4$ Nepheline-carnegieite	1254° C	2289° F

SINTERING

Sintering is the reaction between individual solid particles of a substance to give a hard, less porous and somewhat smaller product. Sintering is brought about by heat but occurs at a temperature below the melting point. The process is of great importance in making ceramic articles from pure refractory oxides.

The sintering of pure compounds has made it possible to investigate interactions between solid particles; the results also apply largely to reactions between different substances in the solid state, although heterogeneous systems cannot often be studied satisfactorily.

Investigation of the sintering reaction in a body shows that the many very small crystalline particles of the raw body are gradually replaced by large interlocking crystals in the product. It appears that adjacent crystallites can interact so as to become one crystal. But as the process starts in numerous places, some junctions arise where relative orientations are not suitable for the two now adjacent crystals to become one. The product is therefore not one single crystal.

The chief reason why sintering occurs lies in the basic structure of the crystals. It has been shown how in each structure each atom has to be surrounded by certain others in a specific way to neutralise the charges and valency forces. This occurs in the inside of a perfectly arranged crystal; on the surface, however, the environment of the atoms is disorganised, making it unsaturated and reactive. Decrease of the surface area therefore represents a decrease of energy and so gives rise to a more stable state.

The actual mechanism of sintering is not completely elucidated but depends on the following factors. Each particular case is of course different and so different factors may predominate.

(1) All atoms in solids oscillate about a mean position. The amplitude increases with temperature and may eventually become large enough for an atom to move into the field of another and not return to its own mean position.

(2) The actual pressure at point contacts may be sufficient to form a chemical bond which will be followed up by atom movements. Such reaction depends on contact between dissimilar atoms. Certain cations of low polarisability and high field strength do not appear at the surface of the crystals of their oxides, *e.g.* SiO_2, MgO, Al_2O_3, BeO; they absorb extra oxygen ions which are backed by a layer of cations thus setting up an electrical double layer causing repulsion between particles. This is only broken down at high temperatures or by the addition of certain impurities, 'mineralisers'. Other cations like zinc and ferric iron can be so deformed as to appear in the surface of their oxide crystals. With these contact pressure can start sintering.

(3) When a lattice has some defect causing a vacant position an atom may move into it, leaving a free position in its wake; this in turn may be filled, so that a vacant position moves through a crystal while atoms move in the opposite direction.

Defect lattices occur fairly readily where the cation can either have more than one valency, *e.g.* iron, or where it may attract electrons, *e.g.* Zn^{2+} forms Zn^+ and Zn. Zinc oxide sinters at 1000° C (1832° F) to a relatively dense body in a few hours.

Other oxides containing noble gas type ions can be induced to form defect structures by introducing other ions of the same size but different charge, *e.g.* Fe^{3+} for Mg^{2+} or F^- and OH^- for O^{2-}, when zero charge can only be achieved by leaving points vacant.

(4) Atoms move into interstitial positions temporarily, that is they force their way in between others whose mean positions are on the normal lattice points. This involves like atoms passing each other and requires considerable energy, especially for those of high field strength and low polarisability, thus such migration is more difficult for magnesium than zinc and it only occurs at high temperatures. Interstitial movement is probably the chief method of material transport at high temperatures, being possibly the only one accounting for the rapid movement involved in sintering.

The contraction that occurs when a dry-pressed body of low porosity is sintered is due in part to the filling up of some of the remaining pores, but also to the elimination of the surface electrical double layers that repelled adjacent particles (**W52**).

Most of the principles of sintering or recrystallisation apply also to reactions in the solid state between dissimilar materials.

In practice sintering is greatly affected by the purity of the material, its grain sizes, shapes and their relative orientation, and to a certain extent by its previous thermal history. The method of shaping and the pressure applied has considerable bearing.

II. *Changes altering the Chemical Composition of Individual Phases*

SOLID SOLUTIONS

Adjacent crystallites of two different materials of similar structures may react together in a manner resembling sintering by going into solid solution with each other. (Solid solutions are also obtained when certain liquid mixtures solidify.) There are two types of solid solution. The first and more usual involves two substances of similar lattice type and whose ionic radii do not differ by more than 10%. When lattice defects give rise to vacant positions either component can move into them, and thus gradually intermingle. This process generally does not go to completion to form a homogeneous product. The second type of solid solution is the interstitial one, where small atoms tend to occupy the spaces between large ones, and may force them apart slightly. The solution of carbon in iron is of this type.

CHEMICAL REACTION

Many chemical reactions take place on heating of solid reactants without fusing them. Migration of the atoms occurs as in sintering or in the formation of solid solutions, but the product, if the reaction goes to completion, has a definite composition and structure differing from the starting materials.

Such reactions may be of direct addition, combination by elimination of another compound, decomposition to several compounds, etc. An important feature of these reactions is that they are irreversible, the compound formed remaining intact on cooling (high–low inversions may occur).

One of the first to realise the importance of such reactions in the solid state as a method of synthesis was Pukall (**P65**), who made a large number of compounds that it had not been possible to prepare before. His results show the scope and importance of solid phase reactions, which it can be assumed occur on heating of most inorganic, inhomogeneous bodies, and are given in Table 49.

Pukall was unsuccessful in preparing the following compounds by the methods that otherwise proved so fruitful: sodium and potassium orthosilicates and aluminates; ferric compounds; copper and chromium ortho- and metastannates; iron, cobalt, nickel, copper and cadmium orthozirconates; manganese, iron, cobalt, nickel metazirconates.

In his recent work Hedvall (**H72**) has tabulated as many of the known solid phase reactions as possible.

Study of this table yields some very interesting data. It is well known that a basic metal in contact with a solution of a salt of a less basic metal will

TABLE 49

Pukall's Inorganic Syntheses by Reactions in the Solid State (P65)

The raw materials mentioned were very finely ground and heated for one to four periods of $6\frac{1}{4}$ hr at 950° C (1742° F) and if necessary further periods of 5 hr at 1270° C (2318° F) and 35 hr at 1370° C (2498° F).

Starting Material	Heating Process °C	Product	Formula	Comments
Orthosilicates from hydrated silicic acid ($SiO_2.7H_2O$) *and the following*:				
Lithium hydroxide	Twice at 950	Lithium orthosilicate	Li_4SiO_4	Largely decomposed by water.
Beryllium hydroxide	Twice 1270	Beryllium orthosilicate	Be_2SiO_4	Fairly resistant to acid and alkali.
Magnesium oxide	1370	Magnesium orthosilicate	Mg_2SiO_4	Virtually resistant to water, slow solution in hot HCl.
Lime	Once 950	Calcium orthosilicate	Ca_2SiO_4	Takes up water and swells, dissolves slightly; soluble in HCl.
	1370			Disintegrates on cooling, no reaction with water.
Strontium oxide	Twice 950	Strontium orthosilicate	Sr_2SiO_4	White; slight reaction with water.
Barium hydroxide	Twice 950	Barium orthosilicate	Ba_2SiO_4	Bluish-green; slightly soluble in water; readily soluble in HCl.
Aluminium hydroxide	Several 1370	Aluminium orthosilicate	$Al_4(SiO_4)_3$?	Only slight reaction, this compound is questionable.
Manganese carbonate	Twice 950	Manganous orthosilicate	Mn_2SiO_4	Reddish colour.
Ferrous oxide (air excluded)	950	Ferrous orthosilicate	Fe_2SiO_4	Dark brown; insoluble in water; soluble in acid; oxidised by heating in air.
Cobalt hydroxide	Once 1270	Cobaltous orthosilicate	Co_2SiO_4	Violet crystals; insoluble in water; soluble in HCl.
Nickel hydroxide	Several 1270	Nickel orthosilicate	Ni_2SiO_4	Grey; insoluble in water; slightly soluble in HCl.
Zinc oxide	Several 1370	Zinc orthosilicate	Zn_2SiO_4	Loose powder ⎫ readily soluble in HCl. Sinters and crystallises ⎭
Cadmium oxide	Once 950	Cadmium orthosilicate	Cd_2SiO_4	Light yellowish brown; readily soluble in HCl.
Lead oxide	Less than 950	Lead orthosilicate	Pb_2SiO_4	Yellow; insoluble in water; readily decomposed by acids.
Bismuth oxide	Less than 950	Bismuth orthosilicate	$Bi_4(SiO_4)_3$	Insoluble in water; soluble in HCl.
Metasilicates from hydrated silica and the following:				
Lithiumh ydroxide	Twice at 950	Lithium metasilicate	Li_2SiO_3	Pinkish; slightly soluble in water; soluble in HCl.
Beryllium hydroxide	Twice 1270	Beryllium metasilicate	$BeSiO_3$	Hardly affected by HCl.
	and once 1370			
Magnesium oxide	1370	Magnesium metasilicate	$MgSiO_3$	Hardly affected by HCl.

TABLE 49—(contd.)

Starting Material	Heating Process		Product	Formula	Comments
		°C			
Metasilicates from hydrated silica and the following—(contd.)					
Calcium carbonate		950	Calcium metasilicate	$CaSiO_3$	Formation incomplete; strong hydration reaction with water.
		1370			Formation almost complete; less hydration reaction with water.
Strontium oxide	Twice at	950	Strontium metasilicate	$SrSiO_3$	Slightly soluble in water; soluble in HCl.
Barium hydroxide	Twice	950	Barium metasilicate	$BaSiO_3$	Slightly soluble in water; soluble in HCl.
Aluminium hydroxide	Several	1370	Aluminium metasilicate?	$Al_2(SiO_3)_3$	Compound formation not proved.
Manganous carbonate	Twice	950	Manganous metasilicate	$MnSiO_3$	Insoluble in water; slightly soluble in HCl; flesh coloured.
Ferrous oxide (air excluded)	Twice	1270	Ferrous metasilicate	$FeSiO_3$	Easily oxidised.
Cobalt oxide	Occurs at Completed	950 1270	Cobaltous metasilicate	$CoSiO_3$	Violet; insoluble in water; soluble in hot HCl.
Nickel oxide	Twice	950	Nickel metasilicate	$NiSiO_3$	Grey; insoluble in water; slightly soluble in hot HCl.
Silver oxide			Silver metasilicate	Ag_2SiO_3	Dark yellow; easily reduced when hot; partially soluble in HNO_3.
Zinc oxide	Twice	950	Zinc metasilicate	$ZnSiO_3$	Insoluble in water; soluble in HCl.
Cadmium oxide	Twice	950	Cadmium metasilicate	$CdSiO_3$	Insoluble in water; almost insoluble in HCl.
Red lead	Less than	950	Lead metasilicate	$PbSiO_3$	Insoluble in water; almost insoluble in HCl.
Orthostannates from stannic acid ($SnO_2.1\cdot54H_2O$) *and the following*:					
Lithium carbonate	Once at	1270	Lithium orthostannate	Li_4SnO_4	Flesh coloured; slow solution in hot HCl.
Beryllium carbonate	Several	1270	Beryllium orthostannate	Be_2SnO_4	Pinkish; insoluble in hot HCl.
Magnesium oxide	Twice	1270	Magnesium orthostannate	Mg_2SnO_4	Insoluble in water; slowly soluble in hot HCl.
Calcium carbonate	Twice	1270	Calcium orthostannate	Ca_2SnO_4	Insoluble in water; soluble in warm HCl.
Strontium carbonate	Twice	1270	Strontium orthostannate	Sr_2SnO_4	Swells in water; soluble in cold HCl.
Barium oxide	Twice	1270	Barium orthostannate	Ba_2SnO_4	Slowly decomposes in water to $BaSnO_4$; soluble in HCl.
Zinc oxide	Several	1270	Zinc orthostannate	Zn_2SnO_4	Flesh coloured; slightly attacked but not dissolved by HCl.
Cadmium oxide	Several	1270	Cadmium orthostannate	Cd_2SnO_4	Clear yellow; slowly dissolved by hot HCl.
Red lead	Four times	950	Lead orthostannate	Pb_2SnO_4	Light yellow; soluble in HCl to give double salt.

TABLE 49—(contd.)

Starting Material	Heating Process	Product	Formula	Comments
Metastannates from stannic acid ($SnO_2 \cdot 1 \cdot 54H_2O$) *and the following:*	°C			
Lithium carbonate	Once at 950	Lithium metastannate	Li_2SnO_3	White; soluble in HCl.
Beryllium oxide (hydrate)	Several 1270	Beryllium metastannate	$BeSnO_3$	Insoluble in HCl.
Magnesium oxide	Three times 1270	Magnesium metastannate	$MgSnO_3$	Only slightly soluble in hot HCl.
Calcium carbonate	Three times 1270	Calcium metastannate	$CaSnO_3$	Slowly soluble in hot HCl.
Strontium carbonate	Twice 1270	Strontium metastannate	$SrSnO_3$	Slow hydration in water; soluble in HCl.
Barium hydroxide	Twice 1270	Barium metastannate	$BaSnO_3$	Slow hydration in water; soluble in HCl.
Zinc oxide	Several 1270	Zinc metastannate	$ZnSnO_3$	White; slightly attacked by hot HCl.
Cadmium oxide	Several 950	Cadmium metastannate?	$CdSnO_3$?	Yellowish grey; slightly attacked by HCl.
Red lead	Four times 950	Lead metastannate	$PbSnO_3$	Insoluble; very slow solution in hot HCl.
Ferrites from hydrated ferric oxide ($Fe_2O_3 \cdot 2 \cdot 05H_2O$)	°C			
Calcium oxide	Twice at 950	Calcium ferrite	$CaFe_2O_4$	Deep violet red; insoluble in water; slightly soluble in hot HCl.
Calcium oxide	Once at 1270	Dicalcium ferrite	$Ca_2Fe_2O_5$	Bronze; insoluble in water; soluble in hot HCl.
Calcium oxide	Once 1270	Tricalcium ferrite	$Ca_3Fe_2O_6$	Dark grey; takes up water; readily soluble in HCl.
Calcium oxide	Several 950	Tetracalcium ferrite	$Ca_4Fe_2O_7$	Light grey; takes up water readily, swelling and hardening.
Orthotitanates from hydrated titania ($TiO_2 \cdot 0 \cdot 18H_2O$)				
Lithium carbonate	Several at 950	Lithium orthotitanate	Li_4TiO_4	Reddish grey; soluble in hot HCl with difficulty.
Beryllium oxide (hydrated)	Once 1270	Beryllium orthotitanate	Be_2TiO_4	Light yellow; insoluble in HCl.
Magnesium oxide	Twice 1270	Magnesium orthotitanate	Mg_2TiO_4	Very light pink; very slow solution in hot HCl.
Calcium oxide	Twice 1270	Calcium orthotitanate	Ca_2TiO_4	Brownish-red; somewhat soluble in hot HCl.
Strontium carbonate	Once 1270	Strontium orthotitanate	Sr_2TiO_4	Light reddish grey; readily soluble in warm HCl.
Barium carbonate	Once 950	Barium orthotitanate	Ba_2TiO_4	Reddish grey; readily soluble in warm HCl.
Manganous carbonate	Several 1270	Manganous orthotitanate	Mn_2TiO_4	Dark brown; very slow solution in hot HCl.
Ferrous oxide (no air)	Several 1270	Ferrous orthotitanate	Fe_2TiO_4	Black; very slow solution in hot HCl.
Cobalt oxide (no air)	Several 950	Cobalt orthotitanate	Co_2TiO_4	Greenish black; very slow solution in hot HCl.
Nickel hydroxide	Several 950	Nickel orthotitanate	Ni_2TiO_4	Greenish yellow; little attacked by hot HCl.
Copper oxide (hydrated)	Once 950	Copper orthotitanate?	Cu_2TiO_4?	Compound questionable; CuO dissolves in HCl.
Zinc oxide	Once 1270	Zinc orthotitanate	Zn_2TiO_4	Yellowish white; soluble in hot HCl.
Red lead	Twice 950	Lead orthotitanate	Pb_2TiO_4	Bright yellow; slow solution in HCl.

TABLE 49—(contd.)

Metatitanates from hydrated titania ($TiO_2, 0.18H_2O$) *and*:

Starting Material	Heating Process °C	Product	Formula	Comments
Lithium carbonate	Once at 950	Lithium metatitanate	Li_2TiO_3	Reddish grey; soluble in hot HCl.
Sodium bicarbonate	Melted at 1060	Sodium metatitanate	Na_2TiO_3	Cream; slowly soluble in hot HCl.
Potassium bicarbonate	Melted at 1060	Potassium metatitanate	K_2TiO_3	Cream; easily soluble in hot HCl.
Beryllium oxide (hydrated)	Once 1270	Beryllium metatitanate	$BeTiO_3$	Light yellow; hardly attacked by hot HCl.
Magnesium oxide	Twice 1270	Magnesium metatitanate	$MgTiO_3$	Hardly attacked by hot HCl.
Calcium carbonate	Several 1270	Calcium metatitanate	$CaTiO_3$	Cream; hardly attacked by hot HCl.
Strontium carbonate	Twice 1270	Strontium metatitanate	$SrTiO_3$	Somewhat attacked by HCl.
Barium carbonate	Twice 1270	Barium metatitanate	$BaTiO_3$	Yellow; soluble in hot HCl.
Manganous carbonate	Twice 1270	Manganous metatitanate	$MnTiO_3$	Dark brown; somewhat attacked by hot HCl.
Ferrous oxide	Once 1270	Ferrous metatitanate	$FeTiO_3$	Black; somewhat attacked by hot HCl.
Cobalt carbonate	Twice 1270	Cobalt metatitanate	$CoTiO_3$	Dark green; somewhat attacked by hot HCl.
Nickel oxide (hydrated)	Twice 1270	Nickel metatitanate	$NiTiO_3$	Yellow; not attacked by hot HCl.
Zinc oxide	Once 1270	Zinc metatitanate	$ZnTiO_3$	Off-white; slow solution in hot HCl.
Cadmium oxide	Once 950	Cadmium metatitanate	$CdTiO_3$	Light yellow; slight reaction with hot HCl.
Red lead	Twice 950	Lead metatitanate	$PbTiO_3$	Light yellow; slow reaction with hot HCl.

Orthozirconates from zirconia ($ZrO_2, 0.04H_2O$) *and*:

Starting Material	Heating Process °C	Product	Formula	Comments
Lithium carbonate	Once at 950	Lithium orthozirconate	Li_4ZrO_4	White; almost insoluble in hot HCl.
Beryllium oxide (hydrated)	Several 1270	Beryllium orthozirconate	Be_2ZrO_4	Yellowish white; almost insoluble in hot HCl.
Magnesium oxide	Several 1270	Magnesium orthozirconate	Mg_2ZrO_4	White; almost insoluble in hot HCl.
Calcium carbonate	Twice 1270	Calcium orthozirconate	Ca_2ZrO_4	White; readily soluble in warm HCl.
Strontium carbonate		Strontium orthozirconate	Sr_2ZrO_4	Light brown; readily soluble in warm HCl.
Barium carbonate	Twice 1270	Barium orthozirconate	Ba_2ZrO_4	Yellowish white; readily soluble in warm HCl.
Zinc oxide	Several 1270	Zinc orthozirconate	Zn_2ZrO_4	Yellowish; attacked but not dissolved by hot HCl.
Red lead	Once 950	Lead orthozirconate	Pb_2ZrO_4	Light yellow; soluble in warm HCl.

Metazirconates from zirconia ($ZrO_2, 0.04H_2O$) *and*:

Starting Material	Heating Process °C	Product	Formula	Comments
Lithium carbonate	Several at 1270	Lithium metazirconate	Li_2ZrO_3	White; very slow solution in hot HCl.
Sodium bicarbonate	Once 1270	Sodium metazirconate	Na_2ZrO_3	Greyish; slightly hygroscopic; slow solution in hot HCl.

177

TABLE 49—(contd.)

Starting Material	Heating Process		Product	Formula	Comments
Metazirconates from zirconia ($ZrO_2.0.04H_2O$) *and* —(contd.)		°C			
Potassium carbonate	Once at	1270	Potassium metazirconate	K_2ZrO_3	White; hygroscopic but only slightly soluble; almost insoluble in HCl.
Beryllium oxide (hydrated)	Several	1270	Beryllium metazirconate	$BeZrO_3$	Slightly attacked by hot HCl.
Magnesium oxide	Several	1270	Magnesium metazirconate	$MgZrO_3$	Pinkish; slightly attacked by hot HCl.
Calcium carbonate	Several	1270	Calcium metazirconate	$CaZrO_3$	White; slowly dissolved by hot HCl.
Strontium carbonate	Twice	1270	Strontium metazirconate	$SrZrO_3$	Pinkish; readily dissolved by warm HCl.
Barium carbonate	Twice	1270	Barium metazirconate	$BaZrO_3$	Readily soluble in warm HCl.
Zinc oxide	Several	1270	Zinc metazirconate	$ZnZrO_3$	Yellowish; slightly attacked by hot HCl.
Cadmium oxide	Difficult to bring to completion		Cadmium metazirconate	$CdZrO_3$	White.
Red lead	Once	950	Lead metazirconate	$PbZrO_3$	Yellow; readily soluble in warm HCl.
Aluminates from alumina (Al_2O_3) *and the following*:					
Magnesium oxide	Starts at 950, completed 1270		Magnesium aluminate	$MgAl_2O_4$ or $MgO.Al_2O_3$	No reaction with water.
Calcium carbonate		1370	Calcium aluminate	$CaAl_2O_4$ or $CaO.Al_2O_3$	Sets with water to hard cement.
Strontium carbonate	Twice	1270	Strontium aluminate	$SrAl_2O_4$ or $SrO.Al_2O_3$	Hydrates with water; a certain amount dissolves; soluble in HCl.
Barium carbonate	Twice	1270	Barium aluminate	$BaAl_2O_4$ or $BaO.Al_2O_3$	Slowly fairly soluble in water; soluble in HCl.
Ferrous oxide	Four times	950	Ferrous aluminate	$FeAl_2O_4$ or $FeO.Al_2O_3$	Insoluble in water; oxidised by heating in air.
Magnesium oxide	Several	1270	Dimagnesium aluminate?	$Mg_2Al_2O_5$ or $2MgO.Al_2O_3$	Compound questionable.
Calcium oxide	Several	950	Dicalcium aluminate	$Ca_2Al_2O_5$ or $2CaO.Al_2O_3$	Sets with water; soluble in warm HCl.
Strontium carbonate	Once	1270	Distrontium aluminate	$Sr_2Al_2O_5$ or $2SrO.Al_2O_3$	Fairly soluble in water; hydrates to interlocking needles.
Barium carbonate	Once	1270	Dibarium aluminate	$Ba_2Al_2O_5$ or $2BaO.Al_2O_3$	Fairly soluble in water; readily soluble in warm HCl.
Manganese carbonate	Several	950	Dimanganous aluminate	$Mn_2Al_2O_5$ or $2MnO.Al_2O_3$	Insoluble in water; not oxidised by heating in air.

TABLE 49—(contd.)

Starting Material	Heating Process		Product	Formula	Comments
		°C			

Aluminates from alumina (Al_2O_3) and the following—(contd.)

Starting Material	Heating Process		Product	Formula	Comments
Ferrous oxide (no air)	Several at	950	Diferrous aluminate	$Fe_2Al_2O_5$ or $2FeO.Al_2O_3$	Insoluble in water; oxidised by heating in air.
Calcium oxide	Once	1370	Tricalcium aluminate	$Ca_3Al_2O_6$ or $3CaO.Al_2O_3$	Hydrates, swells and dissolves slightly in water; soluble in cold HCl.
Strontium carbonate	Once	950	Tristrontium aluminate	$Sr_3Al_2O_6$ or $3SrO.Al_2O_3$	Hydrates, swells and dissolves slightly in water; soluble in cold HCl.
Barium carbonate	Once	950	Tribarium aluminate	$Ba_3Al_2O_6$ or $3BaO.Al_2O_3$	Ditto, slightly more soluble.
Manganous carbonate	Several	950	Trimanganous aluminate	$Mn_3Al_2O_6$ or $3MnO.Al_2O_3$	After heating in air is soluble in hot HCl.
Ferrous oxide (no air)	Once	950	Triferrous aluminate	$Fe_3Al_2O_6$ or $3FeO.Al_2O_3$	Soluble in hot HCl.
Calcium oxide			Tetra calcium aluminate Quinqua calcium aluminate, etc.	$4CaO.Al_2O_3$ $5CaO.Al_2O_3$	

Manganites from manganous carbonate $MnCO_3$:

Starting Material	Heating Process		Product	Formula	Comments
Calcium oxide	Three times at	950	Tetracalcium manganite	$Ca_4Mn_2O_7$ or $4CaO.Mn_2O_3$	Reddish-brown to black; soluble in warm HCl.

Chromites from chromic hydroxide $Cr(OH)_3$: °C

Starting Material	Heating Process		Product	Formula	Comments
Calcium oxide	Several at	950	Calcium chromite	$CaCr_2O_4$	Faintly coloured; partially soluble in HCl.
	Several	950	Dicalcium chromite	$Ca_2Cr_2O_5$ or $2CaO.Cr_2O_3$	Light green; almost completely soluble in HCl.
	Several	950	Tricalcium chromite	$Ca_3Cr_2O_6$ or $3CaO.Cr_2O_3$	Greyish green; almost completely soluble in HCl.
	Several	950	Tetracalcium chromite	$Ca_4Cr_2O_7$ or $4CaO.Cr_2O_3$	Dark green; hydrates in water; completely soluble in cold HCl.

TABLE

Reactions in

(Columns 1, 2, 3, 4, 5, 7, 9

No.	Reactants		Products		Reaction starts at	
	(1)	(2)	(3)	(4)	(5)	(6)
					(°C)	(°F)
1	Li_2CO_3	Al_2O_3	$Li_2O.Al_2O_3$	CO_2	450	842
2	Li_2O	SiO_2	$Li_2O.SiO_2$			
3	Li_2O	$2SiO_2$	$Li_2O.2SiO_2$			
4	$2Li_2O$	SiO_2	$2Li_2O.SiO_2$			
5	Li_2CO_3	SiO_2	$Li_2O.SiO_2$	CO_2	585	1085
6	Na_2O	SiO_2	$Na_2O.SiO_2$		140	284
7	Na_2CO_3	SiO_2	$Na_2O.SiO_2$	CO_2	300	572
8	Na_2CO_3	SiO_2	$Na_2O.SiO_2$	CO_2	600	1112
9	Na_2CO_3	$2SiO_2$	$Na_2Si_2O_5$	CO_2	600	1112
10	Na_2SiO_3	SiO_2	$Na_2Si_2O_5$		720	1328
11	Na_2CO_3	$4SiO_2$	$Na_2O.4SiO_2$	CO_2	395	743
12	Na_2CO_3	SiO_2	$Na_2O.SiO_2$	CO_2	348	1378
13	$2Na_2CO_3$	SiO_2	$2Na_2O.SiO_2$	CO_2	375	707
14	Na_2CO_3	Al_2O_3	$Na_2O.Al_2O_3$	CO_2		
15	$2K_2O$	SiO_2	$2K_2O.SiO_2$			
16	K_2O	SiO_2	$K_2O.SiO_2$			
17	K_2O	$2SiO_2$	$K_2O.2SiO_2$			
18	K_2O	$4SiO_2$	$K_2O.4SiO_2$			
19	K_2CO_3	SiO_2	$K_2O.SiO_2$	CO_2		
20	$3BeO$	$2Ag_3PO_4$	$Be(PO_4)_2$	$3Ag + 3/2O_2$	~500	~932
21	BeO	MoO_3	$BeMoO_4$		400	752
22	BeO	WO_3	$BeWO_4$			
23	BeO	Fe_2O_3			600	1112
24	CuO	ZnS	CuS	ZnO	400	752
25	CuO	Al_2O_3	$CuO.Al_2O_3$		700	1292
26	$2CuO$	SiO_2	Silicate		650	1202
27	$3CuO$	V_2O_5	$Cu_3(VO_4)_2$		450	842
28	CuO	Nb_2O_5	Niobate		650	1202
29	CuO	Sb_2O_3	Antimonate		415	779
30	CuO	MoO_3	$CuMoO_4$		615	1139
31	CuO	WO_3	$CuWO_4$		{ 480 600	896 1112
32	CuO	UO_3	$CuUO_4$		340	644
33	CuO	Fe_2O_3	$CuO.Fe_2O_3$		650	1202
34	MgO	$Mg_2P_2O_7$	$Mg_3P_2O_8$			
35	MgO	$ZnMoO_4$	$MgMoO_4$	ZnO	550	1022
36	MgO	$ZnWO_4$	$MgWO_4$	ZnO	550	1022
37	MgO	$CdWO_4$	$MgWO_4$	CdO	750	1382
38	MgO	Al_2O_3	$MgO.Al_2O_3$		{ 800 900	1472 1652
39	$2MgO$	SiO_2	$MgO.SiO_2$, $2MgO.SiO_2$		1170	2138
40	$MgCO_3$	SiO_2	$MgO.SiO_2$, $2MgO.SiO_2$		1200	2192
41	$2MgO$	TiO_2	{ $MgTiO_4$ $2MgO.TiO_2$		725 { 700 1090	1337 1292 1994
42	MgO	TiO_2	$MgOTiO_2$		850 { 700 800	1562 1292 1472
43	MgO	$2TiO_2$	$MgO.2TiO_2$		1090	1994

50

the Solid State

from Hedvall (**H72**)

No.	Reaction completed at		Ref. (9)	Replacement Reaction (*see* text) (10) p. 188	Solubility of Products			
					Water		Acid	Alkali
	(7)	(8)			cold	hot		
	(°C)	(°F)						
1	600	1112	K85					
2			K83, K89		i.	s.d.	s.dil.HCl	
3			K83, K89					
4			K83, K89		i.	d.		
5	680	1256	K83, 85, 87		i.	s.d.	s.dil.HCl	
6	765	1409	H119		s.	s.d.		
7	600	1112	K83, K89		s.	s.d.		
8			O2		s.	s.d.		
9			S165		s.	s.		
10			J10		s.	s.		
11	800	1472	H114		s.	s.		
12	800	1472	H114		s.	s.d.		
13	800	1472	H114					
14			M11		s.	v.s.		
15			K83, 89					
16			K83, 89		s.	s.		
17			} K83, 84					
18					s.	s.		
19			} 86, 89		s.	s.		
20	780	1436	J13		s.	s.		
21	650	1202	T7					
22			T7					
23	800	1472	K30, 31					
24			T7		i.	i.	s.	s.
25	850	1562	H56					
26	1000	1832	S132, H104					
27	600	1112	T7					
28			J14					
29			T7					
30	720	1328	T7					
31	560	1040	J36				d.	
	820	1112	T7					
32	600	1112	T8					
33	850	1562	K31, J21					
34			J14					
35			J23	✓				
36			J23	✓	i.		d	
37	750	1382	J23	✓				
38			T11				v.sl.sol.dil.HCl	
	1000	1832	J33, H118					
39	1170	2138	T7, B130, J29					
40			T16					
41			T7					
	850	1562	J34					
	1090	1994	J32					
	1090	1994	J32, B144					
42			J34					
			J34					
43	1090	1994	J32					

TABLE

No.	Reactants		Products		Reaction starts at	
	(1)	(2)	(3)	(4)	(5)	(6)
					(°C)	(°F)
44	$2MgO$	SnO_2	$2MgO.SnO_2$			
45	MgO	V_2O_5	$MgO.V_2O_5$		530	986
46	$3MgO$	Sb_2O_3	$Mg_3(SbO_3)_2$		500	932
47	$3MgO$	Nb_2O_5	$Mg_3(NbO_4)_2$		500	932
48	$3MgO$	Ta_2O_5	$Mg_3(TaO_4)_2$		485	905
49	MgO	Cr_2O_3	$MgCr_2O_4$		600	1112
50	MgO	MoO_3	$MgMoO_4$		425	797
51	MgO	WO_4	$MgWO_4$		300	572
52	MgO	Fe_2O_3	$MgFe_2O_4$		600	1112
53	MgO	$FeO.Cr_2O_3$			{ 670 570	1238 1058
54	$3MgO$	$Co_3(PO_4)_2$	$Mg_2(PO_4)_2$	$3CoO$	540	1004
55	$2MgO$	$NiSiO_4$	Mg_2SiO_4	$2NiO$		
56	CaO	$MgSiO_3$	$CaSiO_3$	MgO	562	1044
57	CaO	$Mg_2P_2O_7$	$[Ca_3(PO_4)_2]$		502	936
58	$2CaO$	$Mg_2P_2O_7$	$[Ca_3(PO_4)_2]$		505	941
59	$3CaO$	$Mg_2P_2O_7$	$Ca_3(PO_4)_2$	$2MgO$	506	943
60	CaO	Cu_2S+2O_2	$CaSO_4$	Cu_2O	375	707
61	CaO	$CuO.Al_2O_3$	$CaO.Al_2O_3$	CuO	760	1400
62	CaO	$ZnS+2O_2$	$CaSO_4$	ZnO	467	873
63	CaO	$ZnSO_4$	$CaSO_4$	ZnO	520	968
64	CaO	$ZnO.Al_2O_3$	$CaO.Al_2O_3$	ZnO	475	887
65	CaO	$Zn_2P_2O_7$	$Ca_3(PO_4)_2$	ZnO	510	950
66	$2CaO$	$Zn_2P_2O_7$	$Ca_3(PO_4)_2$		511	952
67	$3CaO$	$Zn_2P_2O_7$	$Ca_3(PO_4)_2.2ZnO$		509	948
68	CaO	Al_2O_3	$CaO.Al_2O_3$		800	1472
69	$2CaO$	Al_2O_3	Calcium aluminate			
70	CaO	$Al_2O_3.SiO_2$	$CaO.Al_2O_3.SiO_2$		532	990
71	$CaO(CO_3)$	SiO_2	$mCaO.nSiO_2$		{ 1000	1832
72	1, 2, or 3 CaO	1 or $2SiO_2$	$CaO.SiO_2$ $2CaO.SiO_2$ $3CaO.SiO_2$ $3CaO.2SiO_2$ Calcium silicate		700 400	1292 752
73	CaO	TiO_2	$CaO.TiO_2$		675	1247
74	CaO	SnO_2	$CaO.SnO_2$		900	1652
75	CaO	ZrO_2	$CaO.ZrO_2$		1000	1832
76	CaO	$PbSiO_3$	$CaPbSiO_4$		500	932
77	$3CaO$	$Pb_3(PO_4)_2$	$Ca_3(PO_4)_2$	$3PbO$	524	975
78	$3CaO$	$Pb_2P_2O_7$	$Ca_3(PO_4)_2$	$2PbO$	529	984
79	$2CaO$	$PbSiO_4$	Ca_2SiO_4	$2PbO$	550	1022
80	CaO	SbO_3			490	914
81	CaO	V_2O_5			400	752
82	$3CaO$	V_2O_5	$Ca_3(VO_4)_2$		375	707
83	$3CaO$	$2CrPO_4$	$Ca_3(PO_4)_2$	Cr_2O_3	517	963
84	CaO	Fe_2O_3	$CaO.Fe_2O_3$		{ 300–500 550 800	572–932 1022 1472
85	CaO	$FeO.Fe_2O_3$	$CaO.Fe_2O_3$	FeO	525	977
86	$CaO(CaCO_3)$	MoO_3	$CaMoO_4$		425	797
87	$CaO(CaCO_3)$	WO_3	$CaWO_4$		580	1076
88	CaO	$MnSiO_3$	$CaMnSiO_4$		563	1045
89	CaO	$CoSO_4$	$CaSO_4$	CoO	533	991
90	CaO	$CoO.Al_2O_3$	$CaO.Al_2O_3$	CoO	548	1018

50—(contd.)

No.	Reaction complete at		Ref. (9)	Replacement Reaction (see text) (10) p. 188	Solubility of Products		
					Water	Acid	Alkali
	(7)	(8)			cold hot		
	(°C)	(°F)					
44			T10				
45	650	1202	J38				
46			T7				
47			J25				
48			J25				
49	1000	1832	K30, H118				
50	650	1202	T7				
51	600	1112	T7		i.	d.	
52	1000	1832	L44, H118			s.conc.HCl	
53	1000	1832	—				
			H67				
54	580	1076	J13	✓	i.	s.	
55			J27	✓			
56			H59	✓	sl.s.	s.	
57			H58	✓	sl.s. d.	s.	
58			H58	✓	sl.s. d.	s.	
59			H58	✓	sl.s. d.	s.	
60			H60	✓	sl.s.	d.	
61			H58	✓	d.	s.	
62	560	1040	H60	✓	sl.s.	s.	s.
63	680	1246	H57, T7	✓	sl.s.	s.	s.
64			H60	✓	d.	s.	s.
65			H58	✓	sl.s. d.	s.	s.
66			H58	✓	sl.s. d.	s.	s.
67			H58	✓			
68			Z5		d.		
69			A176		d.		
70			H59				
71			H55, M25, D6			s.HCl	
	600	1112	T7				
72			K88		sl.s	s.HCl	
			K88, H69				
			K88				
			K88, 89				
	1200	2192	A176				
73			T9				
	820	1508	T7, H71				
74			T11, 3, 5				
75			T7				
76			H70				
77			H58	✓	sl.s.	s.	s.
78			H58	✓	sl.s.	s.	s.
79			H70	✓	sl.s.	s.	s.
80			T7				
81			T7				
82			T7				
83			H58	✓	sl.s.		
	1000	1832	H68, K30				
84	900	1652	C34				
			S131				
85	800	1472	H59	✓		s.	
86	680	1256	B6, J33		i.	s.	
87	950	1742	B6, T7		s.		sNH$_4$Cl
88	800	1472	H70				
89			H57, J13	✓	sl.s.	s.	
90			H59	✓	d.	s.	

TABLE

No.	Reactants (1)	Reactants (2)	Products (3)	Products (4)	Reaction starts at (5) (°C)	Reaction starts at (6) (°F)
91	$3CaO$	$Co_3(PO_4)_2$	$Ca_3(PO_4)_2$	$3CoO$	520–532	968–990
92	CaO	$CoO.Cr_2O_3$	$CaO.Cr_2O_3$	CoO	512	954
93	CaO	UO_3	$CaUO_4$		160	320
94	CaO	$PbSO_4$	$CaSO_4$	PbO	510	950
95	$CaCO_3$	Al_2O_3	$CaO.Al_2O_3$	CO_2	600 / 900 / 700	1112 / 1652 / 1292
96	$CaCO_3$	SiO_2	Calcium silicates	CO_2	900 / 600	1652 / 1112
97	$CaCO_3$	Fe_2O_3	$CaO.Fe_2O_3$	CO_2	600	1112
98	$CaSO_4$	SiO_2	$CaSiO_3$	SO_2		
99	SrO	$MgCO_3$	$SrCO_3$	MgO	456	853
100	SrO	$MgSiO_3$	$SrSiO_3$	MgO	453	847
101	SrO	$MgSO_4$	$SrSO_4$	MgO	441	826
102	SrO	$CaSiO_3$	$SrSiO_3$	CaO	454	849
103	$3SrO$	$Ca_3(PO_4)_2$	$Sr_3(PO_4)_2$	$3CaO$	450	842
104	$3SrO$	$Ca_2P_2O_7$	$Sr_3(PO_4)_2$	$2CaO$	468	874
105	SrO	$CuO.Al_2O_3$	$SrO.Al_2O_3$	CuO	420	788
106	SrO	UO_3	$SrUO_4$		125	257
107	$SrO(CO_3)$	Nb_2O_5	Niobate		450 (420)	842 (788)
108	$SrO(CO_3)$	Ta_2O_5	Tantalate		400 (410)	752 (770)
109	$SrCO_3$	MoO_3	$SrMoO_4$	CO_2	700	1292
110	$SrCO_3$	WO_3	$SrWO_4$	CO_2		
111	SrO	$ZnO.Al_2O_3$	$SrO.Al_2O_3$	ZnO	427	801
112	SrO	$Al_2O_3.SiO_2$	$SrSiO_3$	Al_2O_3	429	804
113	$SrO(CO_3)$	SiO_2	Strontium silicate			
114	$3SrO$	$Pb_3(PO_4)_2$	$Sr_3(PO_4)_2$	$3PbO$	453	847
115	$3SrO$	$2CrPO_4$	$Sr_3(PO_4)_2$	Cr_2O_3	464	867
116	SrO	$Cr_2O_3.FeO$	$SrO.Cr_2O_3$	FeO	403	757
117	SrO	$CoO.Cr_2O_3$	$SrO.Cr_2O_3$	CoO	403	757
118	$3SrO$	$Co_3(PO_4)_2$	$Sr_3(PO_4)_2$	$3CoO$	466	871
119	SrO	$CoO.Al_2O_3$	$SrO.Al_2O_3$	CoO	435	815
120	$SrCO_3$	Al_2O_3	$SrO.Al_2O_3$	CO_2	900	1652
121	BaO	$MgSiO_3$	$BaSiO_3$	MgO	354	669
122	$2BaO$	Mg_2SiO_4	Ba_2SiO_4	$2MgO$	345	653
123	BaO	$CaSiO_3$	$BaSiO_3$	CaO	350	662
124	$3BaO$	$Ca_3(PO_4)_2$	$Ba_3(PO_4)_2$	$3CaO$	340	664
125	$3BaO$	$Sr_3(PO_4)_2$	$Ba_3(PO_4)_2$	$3SrO$	350?	662?
126	$3BaO$	$Ca_2P_2O_7$	$Ba_3(PO_4)_2$	$2CaO$	341	646
127	$3BaO$	$Sr_2P_2O_7$	$Ba_3(PO_4)_2$	$2SrO$	336	637
128	BaO	$2AgCl$	$BaCl_2$	Ag_2O	324	615
129	BaO	$ZnS+2O_2$	$BaSO_4$	ZnO	321	610
130	BaO	$ZnSO_4$	$BaSO_4$	ZnO	340–346	644–655
131	BaO	$ZnO.Al_2O_3$	$BaO.Al_2O_3$	ZnO	345	653
132	$1-3BaO$	ZnP_2O_7	$Ba_3(PO_4)_2$		380	716
133	BaO	$Al_2O_3.SiO_2$	$BaSiO_3$	Al_2O_3	357	675
134	$BaO(CO_3)$	SiO_2	Barium silicates		375 / 420 / 735 / 310	707 / 788 / 1355 / 590
135	$3BaO$	$Pb_3(PO_4)_2$	$Ba_3(PO_4)_2$	$3PbO$	335	635
136	BaO	U_2O_5				
137	$3BaO$	$2CrPO_4$	$Ba_3(PO_4)_2$	Cr_2O_3	342	648
138	BaO	$Cr_4(P_2O_7)_3$		Cr_2O_3	347	657
139	BaO	$FeO.Cr_2O_3$	$BaO.Cr_2O_3$	FeO	347	657
140	BaO	MoO_3	$BaMoO_4$		290	554

50—(contd.)

No.	Reaction complete at		Ref. (9)	Replacement Reaction (see text) (10) p. 188	Solubility of Products			
					Water		Acid	Alkali
	(7)	(8)			cold	hot		
	(°C)	(°F)						
91			H58, J13	√	sl.s.	d.	s.	
92			H59	√			s.	
93	600	1112	T8					
94	700	1292	T7	√	sl.s.		s.	
95	1200	2192	J11		d.			
	1400	2552	D66					
	1400	2552	K88, 89		sl.s.		s.HCl	
96			H55					
97	1200	2192	J28					
98			J11					
			B65		sl.s.		s.HCl.	
99			H71	√			s.	
100			H59	√	i.	i.	s.	
101			H57	√	sl.s.		s.	
102			H59	√	i.	i.	s.	
103			H58	√			s.	
104			H58	√			s.	
105			H59	√			s.	
106	600	1112	T8				s.	
107			J25					
108			J25					
109			T7		sl.s.		s.	
110			T7		sl.s.		d.	
111			H59	√			s.	s.
112			H59	√	i.	i.	s.	s.
113	920	1688	J28		i.	i.		
114			H58	√			s.	
115			H58	√				
116			H59	√			s.	
117			H59	√			s.	
118			H58	√			s.	
119			H59	√			s.	
120	1200	2192	J30					
121			H59	√	s.	d.	s.	
122	640	1184	T6	√			s.	
123	634	1173	T6	√	s.	d.	s.	
124			H58	√	i.	i.	s.	
125			H58	√	i.	i.	s.	
126			H58	√	i.	i.	s.	
127			H58	√	i.	i.	s.	
128	360	680	H62	√	s.			
129			H60	√	i.	i.	s.	
130	550	1022	T7, H57	√	i.	i.	s.	
131			H59	√			s.	
132			H57	√	i.	i.	s.	
133			H59	√	s.	d.	s.	s.
134	590	1094	H55		s.	d.	s.	
	600	1112	T6					
	600	1112	J19, 21, 22 T7					
135			H58	√	i	i.	s.	
136			T7					
137			H58	√	i.	i.	s.	
138			H58	√				
139			H59	√			s.	
140	580	1076	T7		sl.s.		sl.s.	

TABLE

No.	Reactants		Products		Reaction starts at	
	(1)	(2)	(3)	(4)	(5)	(6)
					(C°)	(°F)
141	BaO	WO_3	$BaWO_4$		300	572
142	BaO	UO_3	$BaUO_4$		240	464
143	BaO	$MnSiO_3$	$BaSiO_3$	MnO	355	671
144	BaO	$2Fe_2O_3$	$BaO.2Fe_2O_3$		550	1022
145	BaO	$6Fe_2O_3$	$BaO.6Fe_2O_3$		750	1382
146	8BaO	$4Fe_2O_3$	$Ba_8Fe_8O_{21}$		750	1382
147	4BaO	$2FeSi_2 + 5\frac{1}{2}O_2$	$4BaSiO_3$	Fe_2O_3	329	624
148	BaO	$CoO.Cr_2O_3$	$BaO.Cr_2O_3$	CoO	331	628
149	3BaO	$Co_3(PO_4)_2$	$Ba_3(PO_4)_2$	3CoO	354	669
150	BaO	$CoO.Al_2O_3$	$BaO.Al_2O_3$	CoO	350	662
151	BaO_2	Cu_2O	BaO	2CuO	130	266
152	$BaO(CO_3)$	Nb_2O_5	Niobate		325 (450)	617 (842)
153	$BaO(CO_3)$	Ta_2O_5	Tantalate		310 (480)	590 (896)
154	$BaCO_3$	SnS	SnS.nBaO	CO_2	860	1580
155	$BaCO_3$	MoO_3	$BaMoO_4$	CO_2	660	1220
156	$BaCO_3$	WO_3	$BaWO_4$	CO_2	300	572
157	$BaCO_3$	Fe_2O_3	$BaO.Fe_2O_3$ $2BaO.Fe_2O_3$	CO_2	650	1202
158	ZnO	Al_2O_3	$ZnO.Al_2O_3$		700 (−5)	1292 (1301)
159	1·2ZnO	SiO_2	Zinc silicate		775 (580)	1427 (1076)
160	ZnO	TiO_2	$ZnTiO_3$		700	1292
161	ZnO	ZrO_2	$ZnZrO_3$		980	1796
162	ZnO	V_2O_5			500	932
163	ZnO	As_2O_3	$Zn(AsO_2)_2$		250	482
164	3ZnO	Nb_2O_5	Zinc niobate		520	968
165	ZnO	Sb_2O_3	$Zn(SbO_2)_2$		485	905
166	3ZnO	Ta_2O_5	Zinc tantalate		500	932
167	ZnO	Cr_2O_3	$ZnO.Cr_2O_3$		400	752
168	ZnO	UO_3	$ZnUO_4$		200	392
169	ZnO	Fe_2O_3	$ZnO.FeO_3$		500–600	932–1112
170	ZnO	CoO	Rinmann's green			
171	ZnO	Co_3O_4	$ZnOCo_2O_3$	CoO		
172	CdO	ZnS	CdS	ZnO	400	752
173	CdO	PbS	CdS	PbO	440	824
174	CdO	$ZnMoO_4$	$CdMoO_4$	ZnO		
175	CdO	$ZnWO_4$	$CdWO_4$	ZnO		
176	CdO	UO_3	$CdUO_4$		425	797
177	CdO	Fe_2O_3	$CdFe_2O_4$		800	1472
178	Al_2O_3	$BaSO_4$	$Ba(AlO_2)_2$	$SO_2 + \frac{1}{2}O_2$	1200	2192
179	$2AlF_3$	$2SiO_2$	$Al_2SiO_4F_2$	SiF_4	750	1382

Most of the reactions of alumina are shown in the other sections

No.	(1)	(2)	(3)	(4)	(5)	(6)
180	$4CeO_2$	$6SiO_2$	$2Ce(SiO_3)_3$	O_2	290	554
181	$2CeO_2$	$3TiO_2$	$Ce_2(TiO_3)_3$	$\frac{1}{2}O_2$	700	1292
182	CeO_2	Sb_2O_3			300	572
183	$2CeO_2$	V_2O_5	$2CeVO_4$	$\frac{1}{2}O_2$	350	662
184	$4CeO_2$	$6MoO_3$	$2Ce_2(MoO_4)_3$	O_2	200	392
185	$4CeO_2$	$6WO_3$	$2Ce_2(WO_4)_3$	O_2	240	464
186	2PbO	SiO_2	Pb_2SiO_4			
187	PbO	SiO_2	Lead silicate		580 / 450	1076 / 842
188	1·2PbO	TiO_2	$PbO.TiO_2$		470 (570)	878 (1058)

50—(contd.)

No.	Reaction complete at		Ref. (9)	Replacement Reaction (see text) (10) p. 188	Solubility of Products			
					Water		Acid	Alkali
	(7)	(8)			cold	hot		
	(°C)	(°F)						
141	550	1022	T7					
142	600	1112	T8					
143			H59	√	s.	d.	s.	
144	800	1472	—					
145	1100	2012	—					
146								
147			H60		s.	d.	s.	
148			H59	√			s.	
149			H58	√	i.	i.	s.	
150			H59	√			s.	
151			H55		s.	s.	s.	
152			J25, 26					
153			J25					
154	860	1580	T4					
155			T7, J20, 22		sl.s.		sl.s.	
156			T7, J22, 24		sl.s.	sl.s.	d.	
157								
158	900 (5)	1652(-61)	J31		i.	i.	i.	sl.s.
159			J37, T7		i.	i.		
160			T7					
161			T7					
162			T7					
163			T7, D60					
164			J25					
165			T7					
166			J25					
167			K30					
168			T8					
169			H117, S34, K29, 30, 31				s.conc.HCl.	i.
170			H54					
171	750	1382	H63	√			s.	
172	600	1112	T7	√	sl.s.		s.	s.
173	600	1112	T7	√	sl.s.		s.	
174			J23	√			s.	s.
175			J23	√	sl.s.		s.	s.NH$_4$OH
176	600	1112	T8					
177	800	1472	F31					
178	1400	2552	A201					
179	950	1742	S24					
180	800	1472	T7					
181	1000	1832	T7					
182			T7					
183			T7					
184	650	1202	T7					
185	600	1112	T7					
186			J22		sl.s.		d.	
187	600	1112	T7		sl.s.		d.	
	570	1058	H65, V1					
188	650	1202	T7, P2		i.	i.		

TABLE

No.	Reactants (1)	Reactants (2)	Products (3)	Products (4)	Reaction starts at (5) °C	Reaction starts at (6) °F
189	PbO	ZrO_2	$PbO.ZrO_2$		700	1292
190	3PbO	V_2O_5	Lead vanadate		325	617
191	PbO	As_2O_3	Lead arsenite		250	482
192	PbO	Sb_2O_3	Lead orthoantimonite		400	752
193	PbO	MoO_3	$PbMoO_4$		460	860
194	PbO	WO_3	$PbWO_4$		480	896
195	PbO	UO_3	$PbUO_4$		375	707
196	PbO	Fe_2O_3	$PbO.Fe_2O_3$			
197	Cr_2O_3	$3UO_3$			230	446
198	2FeO	TiO_2			700	1292
199	FeO	MoO_3	$FeMoO_4$		320	608
200	FeO	WO_3	$FeWO_4$		260	500
201	Fe_2O_3	SiO_2			575	1067
202	Fe_2O_3	TiO_2				
203	Fe_2O_3	P_2O_5	Phosphate		300	572
204	Fe_2O_3	UO_3				
205	CoO	Al_2O_3	$CoO.Al_2O_3$		550	1022
206	2CoO	SiO_2	Co_2SiO_4		1100	2012
207	Co_3O_4	SiO_2			550	1022
208	CoO	Ta_2O_5	$CoO.Ta_2O_5$			
209	CoO	Cr_2O_3	$CoO.Cr_2O_3$			
210	CoO	WO_3	$CoWO_4$			
211	CoO	UO_3	$CoUO_3$		230	446
212	NiO	Al_2O_3	$NiO.Al_2O_3$		1000 (1005)	1832 (1841)
213	NiO	SiO_2	Nickel silicate		570	1058
214	2NiO	SiO_2	Nickel silicate		800	1472
215	NiO	U_2O_5			515	959
216	NiO	UO_3	$NiUO_4$		340	644
217	$NiSiO_3$	CaS	$CaSiO_3$	NiS	400	752
218	$NiSiO_5$	FeS			400	752

replace it in solution. This has given rise to the determining of the electrochemical series. The same occurs in solid phase reactions so that in a mixture containing the free oxide of a base, say CaO, and the salt of a weaker base, say $ZnO.Al_2O_3$, the result of the reaction which starts at 520° C (968° F) and ends at 680° C (1256° F) is ZnO and $CaO.Al_2O_3$. This feature of replacement reactions is emphasised in Table 50, items ticked in column (10), as are also those reactions that give water-soluble products.

Note: The temperatures in the columns 'Reaction Complete' denote that further increase of temperature does not increase the quantity of the product. The percentage of reactants converted to the products may nevertheless be small.

Solid-phase reactions are probably of even greater importance when considering the action of heat on the more complex materials that make up

50—(contd.)

No.	Reaction complete at		Ref. (9)	Replacement Reaction (see text) (10) p. 188	Solubility of Products			
					Water		Acid	Alkali
	(7)	(8)			cold	hot		
	(°C)	(°F)						
189	750	1382	T7					
190	720	1328	T7		sl.s.		d.	
191			D60		i.		s.HNO$_3$	
192	500	932	T7		i.			
193	700	1292	T7, J22		i.		s.	s.
194	700	1292	T7, J22					
195	600	1112	T8					
196			L44					
197			T8					
198	800	1472	T7					
199	580	1076	T7					
200	800	1472	T7					
201	700–900	1292–1652	H61					
202			T7, P51, C31					
203	900	1652	B109		sl.s.		s.	
204			T8					
205	1100	2012	H66		i.	i.	i.	
206			H66		i.		s.dil.	
207	950	1742	H64					
208	1100	2012	H54					
209	1100	2012	H54					
210			H54		i.		s.hot.conc.	
211			H54, T8					
212	1100 (1105)	2012 (2021)	J25					
213	1000	1832	H64					
214			S132					
215	605	1121	T7					
216			T16					
217	900	1652	S204		sl.s.	d.	s.	sl.s.
218	800	1472	S204					

the bulk of ceramic raw materials and products. Steger (**H72**) discusses some of these in connection with production of refractories, electrical bodies, porcelain and colours. In the refractory field there is the conversion of olivine containing magnesium and ferrous orthosilicates to forsterite and magnesium ferrite by the reaction with magnesite; the production of silica bricks involves the quartz–cristobalite inversion and the combination of lime and quartz to give first, calcium orthosilicate, and then the metasilicate which form the bond for the brick; chromite and chrome–magnesite refractories involve reactions to form the highly resistant spinels.

In the electrical field there are the reactions producing enstatite (magnesium metasilicate $MgSiO_3$) the main constituent of steatite bodies from talc, and those giving cordierite ($\sim 2MgO.2Al_2O_3.5SiO_2$) from kaolin, talc, alumina, magnesite and silica in suitable proportions; the preparation of suitably bonded titania bodies with magnesium orthotitanate and barium metatitanate; solid-phase reactions are also often used to prepare ceramic colouring agents, particularly the refractory spinels.

Table 51 gives a summary of some of the data available on the temperatures of solid-state reactions of ceramic raw materials. Variations of temperatures given are probably due to particle size differences and the unknown presence of impurities.

REVERSIBLE REACTIONS

Certain reactions of importance in ceramics are reversible, their direction depending on the temperature. Thus certain coloured oxides, even when in combination as silicates, decompose on heating, losing oxygen and changing their colour. On cooling they are able to take up oxygen and return to their original state and colour. This can, of course, only occur on the surface and only so long as that is not impervious to oxygen.

Two such oxides are those of iron and manganese.

Ferric oxide, Fe_2O_3, is a common mineral and occurs to a greater or lesser extent in all ceramic raw materials. It is coloured a reddish brown which it imparts to any body containing it unless calcium compounds are present, when a yellow is produced. Heat decomposes it, however, with the loss of oxygen to give the ferrosoferric oxide Fe_3O_4 which is grey and a less powerful colourant then ferric oxide. On cooling oxygen is again taken up to give ferric oxide.

The temperature at which this decomposition takes place is given variously within the range 1200° C–1300° C (2192° F–2372° F); however, in the presence of the other constituents of a ceramic body it occurs below 1200° C (2192° F) and in some cases as low as 1100° C (2012° F).

Similar reversible reduction and oxidation of ferric oxide through ferrosoferric oxide to ferrous oxide can be brought about below the decomposition temperature by a reducing or oxidising atmosphere (**D29**).

Manganese dioxide MnO_2 is dark brown or black. On heating above 300–500° C (572–932° F) it gradually loses oxygen becoming successively Mn_2O_3, Mn_3O_4, which are dark brown, and finally well above 1000° C (1832° F) manganous oxide which furnishes the pink manganous ion. On cooling the reverse occurs if oxygen can enter.

MINERALISERS

It has already been shown why silicates react slowly and have difficulty in crystallising from a melt or in recrystallising during some solid-state changes. In numerous instances these changes are accelerated, or even made possible, by the presence of a small amount of certain other substances. It is assumed that it was due to such impurities that the igneous minerals are crystalline instead of glassy solids and so they are called 'mineralisers'.

Mineralisers are vital to the ceramic industry to catalyse changes that

TABLE 51
Ceramic Reactions in the Solid State

Starting Materials				Temperature of noticeable reaction		Product		Ref.
Name	Formula	Name	Formula	(°C)	(°F)	Name	Formula	
Quartz	SiO_2	Lime	CaO	600	1112	Calcium silicates	$2CaO.SiO_2$, then $2CaO.SiO_2$ — $3CaO.2SiO_2$ — $3CaO.SiO_2$	B133 J28 J28
Quartz	SiO_2	Sodium Carbonate	Na_2CO_3	1300	2372	Sodium metasilicate Sodium orthosilicate	$CaO.SiO_2$ Na_2SiO_3 Na_4SiO_4	G45 N4 N4
Quartz	SiO_2	Magnesia	MgO	726–805	1339–1481	Forsterite	Mg_2SiO_4	N4
Quartz	SiO_2	Forsterite	Mg_2SiO_4	1200–1400	2192–2552	Clinoenstatite	$MgO.SiO_2$	N4
Quartz	SiO_2	α-Alumina	Al_2O_3	1400–1500	2552–2732	Mullite	$3Al_2O_3.2SiO_2$	N4
Quartz	SiO_2			1500–1700	2732–3092	Mullite	$3Al_2O_3.2SiO_2$	C19
Alumina	α-Al_2O_3	Magnesia	MgO	1300–1500	2372–2732	Spinel	$MgO.Al_2O_3$	N4
				1000	1832	(very dependent on particle size)		C27
				1000–1200	1832–2192			
				1450–1500	2642–2732			
Alumina	α-Al_2O_3	Nickel oxide	NiO	1000–1100	1832–2012	Nickel spinel	$NiO.Al_2O_3$	J35
Alumina Magnesia	Al_2O_3 MgO	Silica	SiO_2	1400	2552	Very little cordierite	$2MgO.2Al_2O_3.5SiO_2$	N4
Alumina Talc	Al_2O_3 $3MgO.4SiO_2.H_2O$	Silica	SiO_2	1400	2552	Small amount cordierite	$2MgO.2Al_2O_3.5SiO_2$	N4
Alumina Talc	Al_2O_3 $3MgO.4SiO_2.H_2O$	Silica Kaolin	SiO_2 $Al_2O_3.2SiO_2.2H_2O$	1400	2552	Small amount of Cordierite	$2MgO.2Al_2O_3.5SiO_2$	N4
Magnesite 23% Talc	$MgCO_3$ $3MgO.4SiO_2.H_2O$	Kaolin 77% Kaolin	$Al_2O_3.2SiO_2.2H_2O$ $Al_2O_3.2SiO_2.2H_2O$	1300–1400 above 1000	2372–2552 above 1832	Cordierite Cordierite	$2MgO.2Al_2O_3.5SiO_2$	N4 K76
Fayalite (in olivine)	Fe_2SiO_4	Magnesite	$MgCO_3$	1000–1200	1832–2192	Forsterite and magnesium ferrite	Mg_2SiO_4 + $MgFe_2O_4$	J29

TABLE 51—(contd.)

Starting Material		Temperature of noticeable reaction		Products				Ref.
Name	Formula	(°C)	(°F)	Name	Formula	Name	Formula	
Sillimanite	$Al_2O_3.SiO_2$	1545	2813	Mullite	$3Al_2O_3.2SiO_2$	Vitreous silica	SiO_2	A89
Andalusite	$Al_2O_3.SiO_2$	1380–1400	2516–2552	Mullite	$3Al_2O_3.2SiO_2$	Vitreous silica	SiO_2	
Kyanite (cyanite)	$Al_2O_3.SiO_2$	1310–1380	2390–2516	Mullite	$3Al_2O_3.2SiO_2$	Vitreous silica	SiO_2	
Mullite	$3Al_2O_3.2SiO_2$	1810	3290	Corundum	$\alpha\text{-}Al_2O_3$	Liquid silica	SiO_2	{M17, C48
Kaolin	$Al_2O_3.2SiO_2.2H_2O$	650	1202	Non-crystalline product				{I6, M17, H10, N4, M17
		950	1832	Mullite	$3Al_2O_3.2SiO_2$	Vitreous silica	SiO_2	
		1000	1832	Mullite	$3Al_2O_3.2SiO_2$	Vitreous silica	SiO_2	
		1200	2192	Mullite	$3Al_2O_3.2SiO_2$	Cristobalite	SiO_2	
		1225–1275	2237–2327	Mullite	$3Al_2O_3.2SiO_2$	Amorphous silica		
		800–840	1472–1544	Enstatite	$MgO.SiO_2$			
Talc	$3MgO.4SiO_2.H_2O$ to $4MgO.5SiO_2.H_2O$	1145	2093	Clinoenstatite (protoenstatite)	$MgO.SiO_2$	Amorphous silica	SiO_2	E29
		1200–1300	2192–2372	Clinoenstate (protoenstatite)	$MgO.SiO_2$	Cristobalite	SiO_2	
Nomenclature of products according to Foster (F20) in brackets		600	1112	Hydrated β-magnesium metasilicate	$MgO.SiO_2$	Cristobalite		
		1000	1832	α-magnesium Metasilicate (enstatite)	$MgO.SiO_2$	Cristobalite		{A 202
		1200	2192	Clinoenstatite (protoenstatite)	$MgO.SiO_2$	Cristobalite		
		800	1472	Protoenstatite (enstatite)	$Mg.SiO_2$	Cristobalite		
		1400	2552	Clinoenstatite (protoenstatite)	$Mg.SiO_2$	Cristobalite		K9
Magnesium carbonate magnesite	$MgCO_3$	350	662	Magnesia	MgO	Carbon dioxide	CO_2	F32
		570	1058					

Calcium carbonate Calcite and aragonite	$CaCO_3$	825 895	1517 1643	Lime	CaO	Carbon dioxide	CO_2	F32
Strontium carbonate	$SrCO_3$	1340	2440	Strontia	SrO	Carbon dioxide	CO_2	
Barium carbonate	$BaCO_3$	1450	2642	Baryta	BaO	Carbon dioxide	CO_2	F32
Cerussite	$PbCO_3$	330	626	Lead monoxide	PbO	Carbon dioxide	CO_2	F32
Basic lead carbonate		430	806	Lead monoxide	PbO	Carbon dioxide and water	$CO_2 + H_2O$	
Zinc spar	$ZnCO_3$	400	752	Zinc oxide	ZnO	Carbon dioxide	CO_2	F32
Siderite	$FeCO_3$	450	842	Ferrous oxide	FeO	Carbon dioxide	CO_2	F32
Rhodochrosite	$MnCO_3$	540	1004	Manganous oxide	MnO	Carbon dioxide	CO_2	F32

otherwise take too long to be of economic value. They probably serve several functions. In crystalline solids an impurity distorts the lattice and upsets the regularity of atomic arrangement. It therefore puts some of the atoms in a slightly different environment to the rest, so that their energy is not the same and they are more able to start a change. Once this has been initiated the rest can follow. In a liquid, atoms or groups tend to surround themselves regularly. If an impurity has stronger orientating forces than the other constituents of an irregularly bonded melt it may set up a sufficiently large independent nucleus that is regularly arranged to induce crystallisation. That a mineraliser breaks some of the bonds in a silicate melt is seen by the lowered viscosity.

Foreign matter that distorts a crystal lattice excessively, or is of an incompatible chemical nature, prevents crystallisation and has the exact reverse effect to a mineraliser. Mineralisers are specific.

In many instances a flux acts as a very active mineraliser by fusing to a thin liquid layer between particles and making material transport much easier. The reactions are then not strictly solid-phase ones but approach them closely.

Ideally a mineraliser would be one that can be removed completely after it has produced the desired result. Water fulfils these conditions and it is well known that traces of water or water vapour have a surprising catalytic effect in many instances. Other gases, especially when they are freshly evolved during the reaction also have accelerating action, notable amongst these are carbon dioxide, CO_2, sulphur dioxide, SO_2, hydrogen, H_2, and oxygen, O_2 (**S11**).

Certain salts such as the tungstates, fluorides, vanadates, borates, or phosphates are effective in bringing about recrystallisation of polymorphic forms (**K72**). The most well-known example of this is in the classic work of Fenner (**F6**) on the silica inversions, where sodium tungstate made a very efficient mineraliser for the quartz–tridymite and tridymite–cristobalite inversions, the temperature for which could not otherwise be determined. The crystalline form produced in the presence of mineralisers is not, however, always the stable one for that temperature. Silica glass can be induced to crystallise in the presence of alkalies, the product being cristobalite even when the temperature is below 1470° C (2678° F), and in the presence of sodium tungstate the conversion can occur at lower temperatures. In the latter conversion tridymite forms, but here again even below 870° C (1598° F) when, however, it does very slowly change to quartz (**F6**). A lithium–sodium silicate melt in contact with silica also acts as an accelerator for the quartz–tridymite inversion (**K72**), and ferrous and ferric oxides catalyse the quartz–cristobalite change. It is due to the presence of suitable mineralisers that the North German Findlings quartzite is more easily converted to tridymite and cristobalite and hence better suited for making silica bricks than any other (**H72**).

Another solid-state reaction of technical importance is the sintering of alumina, where it is most desirable to reduce the firing temperature without incorporating sufficient flux to alter the properties of the product. Here, treatment of the alumina with hydrochloric acid assists, presumably because it attacks the surface slightly, causing irregularities that can react more easily, and leaving some volatile aluminium chloride. The silicates and titanates of bivalent metals are useful accelerators, the iron and manganese compounds being the best, although they are changed by the atmospheric conditions. The alkalies were said to retard the sintering (**Y2**).

Clear instances of foreign atoms present in solid solution distorting the lattice are seen in the inversions of some calcium and magnesium silicates. The slow change from pseudowollastonite, $CaSiO_3$, to wollastonite at 1125° C (2057° F) is accelerated by the presence of magnesium in the lattice, added as $CaMgSi_2O_6$. The orthoenstatite to clinoenstatite, $MgSiO_3$, change at 1140°C (2084° F) is accelerated by $FeSiO_3$ which also lowers the inversion temperature. But the otherwise rapid high–low inversion of $CaSiO_3$ is retarded by the foreign ions Na, Mg, Mn which go into solution and block the new structure, and also by B_2O_3 and P_2O_5 (**K72**).

Probably the most important solid-phase change in the ceramic industry is that of kaolinite to mullite. As the kaolinite breaks down to a very poorly crystalline state some time before the mullite forms, mineralisers are required here to assist in crystallisation, rather than for breaking down a crystal. As a residual structure is considered to be present, once nuclei of mullite have been formed the rest can crystallise readily. Kyanite changes to mullite more easily than kaolinite and if present in small quantities furnishes small nuclei and accelerates the kaolinite change. Introduction of mullite crystals in the form of calcined kyanite does not have the same effect as they are well formed and inert (**T15**).

Parmelee and Rodriquez (**P17**) give a review of literature on mullite catalysis by addition of foreign matter. The salient suggestions are:

A soda-lime glass consisting of 70 parts by weight of silica, 15 parts lime and 15 parts of soda (**N28**).

2·5% cupric oxide (**N4**).

Sodium tungstate or superheated steam (**A210**).

Aqueous solutions of aluminium salts preferably containing sulphate and chlorate ions (**D34**).

Alkali silicates and magnesium oxide (**K60**).

Parmelee and Rodriquez (**P17**) made a systematic study of the effects of different metallic oxides on mullite formation at different temperatures. Their results are summarised in Table 52, columns (1)–(5).

Budnikov and Shmukler (**B134**) investigated the effects of oxides whose metals had different values of r/e (r = radius, e = charge of the ion). They found that mineralisers may lower the temperature at which mullite formation begins by 100° to 200° C. Their results are shown in Table 52, columns (6)–(8).

TABLE 52

Effectiveness of Mineralisers in Mullite Formation at Different Temperatures *

(1) 1200° C, 1 hr		(2) 1300° C		(3) 1400° C		(4) 1400° C		(5) 1500° C	(6) 1400° C	(7) 1500° C	(8) 1600° C
	Yield (%)		Yield (%)		Yield (%)		Yield (%)	Yield (%)			
1% MgO	29	1% MgO	79	1% ZnO	100	4% Fe_2O_3	100	None 100	Decreasing effectiveness →	2% MnO	2% MnO
1% Li_2O	24	1% ZnO	72	1% MgO	97	2% Fe_2O_3	96		2% MnO	2% TiO_2	2% TiO_2
1% ZnO	11	1% Li_2O	70	1% MnO_2	95				1% $MgCl_2$	4% Cr_2O_3	4% $MgCl_2$
1% MnO_2	10	1% Ce_2O_3	67	1% Fe_2O_3	94	2% B_2O_3	91		2% TiO_2	4% $MgCl_2$	4% LiCl
1% TiO_2	8·3	1% MoO_3	67	1% Li_2O	91	4% Ce_2O_3	90		4% CaF_2	4% $Fe(OH)_3$	
1% CaO	8·1	1% Fe_2O_3	65	1% Ce_2O_3	91	2% Ce_2O_3	88			2% $CuCl_2$	
1% K_2O	7·2	1% B_2O_3	59	1% MoO_3	91	4% B_2O_3	81				
1% ZrO_2	6·1	1% MnO_2	58	1% CaO	86						
1% Fe_2O_3	5·9	1% CaO	50	1% B_2O_3	84						
None	4·7	1% TiO_2	44	1% ZrO_2	80						
1% B_2O_3	4·1	1% ZrO_2	39	1% Na_2O	79						
		1% K_2O	36	1% SnO_2	75	4% Li_2O	65				
		None	36	1% K_2O	72	4% TiO_2	61				
				1% TiO_2	58						
				None							

* Columns (1)–(5) from Parmelee and Rodriquez (**P17**); columns (6)–(8) from Budnikov and Shmukler (**B134**).

INCOMPLETE AND COMPLETE REACTIONS

As early as 1907 Mellor (**M48**) made the following statement: 'The reaction between the different constituents of the body, in firing, is arrested before the system is in a state of equilibrium. The chemistry of pottery is therefore largely a chemistry of incomplete reactions, and many erroneous inferences have been made from the failure to appreciate this fact. Thus, two bodies may have the same ultimate composition, and yet behave very differently on firing, because the velocity of the reaction, under constant conditions, is different when the bodies are compounded with different materials so as to furnish the same ultimate composition, or with the same materials in a different state of subdivision.'

Numerous workers have subsequently expressed similar thoughts about the incompleteness of reactions in ceramic bodies, which have been summarised by Foster (**F19**). Examination of thin sections of fired earthenware and some refractories shows it to be undoubtedly true for the common ceramic bodies where the raw materials are not very finely ground.

Many of the properties of these bodies depend on the incompleteness of the reactions initiated in the kiln. The fact that the product depends not only on the over-all chemical composition but also on the individual compositions, crystalline structures and particle sizes of the raw materials, and is often considerably influenced by minor impurities, makes it very difficult to make quantitative predictions about the optimum composition, firing schedule, etc., for a desired product.

However, modern methods of production of special bodies from very finely ground and relatively purer raw materials, formed under pressure and fired at high temperatures, have been shown by modern techniques in identification of fine-grained crystalline phases to approach by solid-phase reactions the state of equilibrium fairly closely (**G23**). For such bodies careful interpretation of phase diagrams may assist development considerably (**F19**). In either case the results of experiments on solid-phase reactions are a valuable aid to the interpretation of practical results obtained.

On turning to glazes quite different matters come under consideration. Glazes are formed by cooling a molten mixture in a thin layer on the surface of a ceramic body. The reactions between the components occur in the liquid state and are therefore generally complete or nearly so, and the result is reasonably uniform in composition.

Ceramic bodies are produced by incomplete or almost complete solid–solid or solid–liquid phase reactions; whereas glazes are produced by complete liquid phase reactions.

MELTING, CRYSTALLISATION AND GLASS FORMATION

It has already been mentioned that a characteristic property of silicates is their extreme sluggishness in melting. At all temperatures above the absolute zero, atoms oscillate and vibrate. In solids the movement is about a mean position which remains stationary in relation to the mean position of other neighbouring atoms. With increasing temperature the amplitude of the oscillation increases until it is so large that atoms break away from the surface. At this point, in normal crystalline substances, heat is absorbed without raising the temperature and the crystal melts. But with few exceptions, e.g. lithium metasilicate (**K69**), silicates do not fuse promptly at their melting points. The classic example of this is the work of Day and Allen (**D17**) on the feldspar albite. Although the true melting point of this is now known to be 1118° C (2044° F) (**G77**) they found that the mineral did not melt completely even at 1247° C (2277° F) in spite of thirty minute exposures at various constant temperatures. They also found no arrest in the thermal curve, as is caused in normal melting during the absorption of latent heat. By holding an initially crystalline substance at a constant elevated temperature for a long time and then quenching it (rapidly cooling to room temperature) and examining a thin section under a petrographic microscope they were able to see how slowly a substance melts when held at a temperature only a little above its melting point.

The very slow melting of silicates to a very viscous liquid is of extreme importance in the firing of ceramic ware and will be considered in detail again.

The liquid formed on melting may or may not have the same composition as the solid. Solids that give a liquid of the same composition are said to have a 'congruent melting point'; those that give a liquid of different composition together with another different solid are said to have an 'incongruent melting point'. A number of crystalline silicates fall into the latter group.

Some examples of silicates that melt incongruently are:

zircon $(ZrO_2.SiO_2) \rightleftharpoons ZrO_2 +$ liquid (**G22**).
protoenstatite $(MgO.SiO_2) \rightleftharpoons$ forsterite $(2MgO.SiO_2) +$ liquid (**B97**).
orthoclase $(K_2O.Al_2O_3.6SiO_2) \rightleftharpoons$ leucite $(K_2O.Al_2O_3.4SiO_2) +$ liquid
(**M92, S15**).
cordierite $(2MgO.2Al_2O_3.5SiO_2) \rightleftharpoons$ Mullite $(3Al_2O_3.2SiO_2) +$ liquid (**R2**).

Mullite is generally supposed to have an incongruent melting point at 1810° C (3290° F) as follows:

mullite $(3Al_2O_3.2SiO_2) \rightleftharpoons$ corundum $(Al_2O_3) +$ liquid (**B98**)

Recent investigation has thrown some doubt on this. If the surface of the mullite is covered with tungsten in order to prevent vaporisation of silica it has a congruent melting point at about 1860° C (3380° F) (**T35**).

Solid solutions (excluding eutectic mixtures) all melt incongruently. This matter will be considered with the phase diagrams.

Molten silicates are very viscous at temperatures little above their melting points. The SiO_4 tetrahedra link up with each other in a random way, the bonds being broken and remade with another group frequently. It is therefore very difficult for the SiO_4 units to rearrange themselves to give an ordered crystalline structure when cooled below their melting point, and in fact this generally does not occur. On cooling the silicate melt becomes solid without becoming crystalline and is known as a glass. Structurally a glass resembles the liquid in the random arrangement of SiO_4 tetrahedra but like a crystal the bonds to neighbouring atoms are permanent. Glasses can appropriately be termed 'supercooled liquids' or 'amorphous' solids (Fig. 2.2).

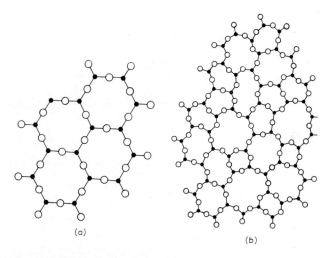

FIG. 2.2. *Schematic representation in two dimensions of the difference between:* (a) *crystalline silica;* (b) *glassy silica (Zachariasen,* **Z2**)

In a traditional ceramic body a part of the mixture fuses on heating and distributes itself round the more refractory particles. On cooling this part generally becomes a glass and acts as a permanent solid bond. The amount of glass depends on the refractoriness of the constituents relative to the firing temperature. The quantity that can be permitted without the body deforming at a given temperature depends on the viscosity of the liquid form. Thus the partial fusion of a high-grade clay gives a very viscous liquid which increases the ability of the body to bear a load by holding the solid particles together. Even addition of the flux feldspar, which increases the amount of glass present, does not weaken the body until very high temperatures. An example of the viscosity of feldspar above its melting point is given by Morey (**M93**). A mixture of the composition $K_2O.Al_2O_3.6SiO_2$ which has an incongruent melting point at 1150° C (2102° F) was heated for about a week at

1400° C (2552° F) with intermediate cooling and grinding, when a spongy sintered mass resulted. Heating was continued until the platinum crucible containing it melted (m.p. platinum 1773° C, 3123·4° F) but no molten homogeneous glass could be obtained. Thus 600° C (1112° F) above the temperature at which melting began and 200° C (392° F), above the temperature of complete fusion the viscosity was still so high that the mass would not flow together.

In the case of the lime–silica bond in silica bricks, however, the glassy bond, although contributing both as a bond, and also by assisting the quartz–tridymite–cristobalite inversions and accompanying recrystallisations, is also

FIG. 2.3. *Mullite crystallisation on a piece of fused Corhart. Slightly reduced* (*Halm*, **H10**)

the source of weakness of the bricks at high temperatures. As can be seen from the phase diagram and explained in connection with glass structure the eutectic temperature in the $CaO-SiO_2$ system is high, namely 1425° C (2577° F). But the liquid that forms has a low viscosity and can no longer assist structurally (**M93**).

In bone china containing a fair quantity of calcium phosphate a large proportion of the body is allowed to fuse in order to give a translucent product on cooling. It is, however, difficult for the ware to bear its own weight during the firing and each piece has to be supported individually.

The glaze differs essentially from the body by being so made up that it fuses entirely, becomes homogeneous, and then spreads evenly over the body. The majority of glazes solidify to clear, hard, glasses on cooling. Some, however, when suitably made up and treated, produce a mass of small crystals imbedded in a glassy matrix becoming velvet matt, while others form beautiful large crystals in the glass making 'crystal glazes', which are very attractive on individual pieces of artware.

Crystallisation from a silicate melt occurs in certain ceramic bodies too (Fig. 2.3). Although mullite has an incongruous melting point, careful heat treatment of a melt of the mullite composition $3Al_2O_3.2SiO_2$ produces beautiful long needles of square section with practically no glass. On cooling the liquid first deposits corundum, Al_2O_3, between 1925° and 1810° C (3497–3250° F) and only then does mullite begin to separate and the silica-rich liquid tend to react with the corundum to give more mullite (**M93**).

THE STRUCTURE OF GLASSES AND GLAZES

Glazes are specialised glasses forming a thin adherent film on the body surface. As such they are difficult to investigate so that our primary knowledge on them is obtained from work on glasses.

The basic concepts of the modern idea of glass structure were first put forward by Zachariasen (**Z2**) in 1932. His views, although they were questioned and criticised at first, proved correct and have since been tested, defined and amplified.

A number of elements, certain oxides and a few other compounds are known to form glasses more or less readily on cooling of their liquid forms. We will concern ourselves only with the oxides. In his classical work Zachariasen (**Z2**) predicted theoretically that oxides of the type A_nO_m could form glasses if the m/n ratio is 1·5–2·5, the A—O bonds are substantially covalent and therefore directional, and the following four rules are obeyed:

(1) an oxygen atom is linked to not more than two atoms A;
(2) the number of oxygen atoms surrounding atoms A must be small (now known to have a maximum of 4);
(3) the oxygen polyhedra share corners with each other, not edges or faces;
(4) at least three corners in each oxygen polyhedron must be shared.

The arrangement of the polyhedra, which are either triangles or tetrahedra, is irregular, thereby accounting for the isotropic properties of glass and also for its gradual melting because of irregular energy distribution.

Smekal (**S130**) gives the following definition: 'glasses and vitreous products in general are regarded as *polymers* characterised by *rigid irregular* frameworks.' On Zachariasen's theory the oxides B_2O_3, SiO_2, GeO_2, P_2O_5 and As_2O_5 should form glasses readily; As_2O_3 and Sb_2O_3 with some difficulty and P_2O_3, V_2O_5, Sb_2O_5, Nb_2O_5 and Ta_2O_5 might do so. There is still considerable doubt about the last group (**S167**).

The common glass-forming oxides, B_2O_3, SiO_2 and P_2O_5, frequently become vitreous simply by rapid cooling of their melts. This is considered to be due to the fact that the bonding and orientation conditions between close neighbours required for glass formation are normally present in the crystalline form, and are assumed to tend to occur in the liquid at low temperatures. Thus in silica the normal crystalline structure is of linked SiO_4 units; these probably tend to be present in the liquid where temporary polymerisation gives irregular groups of SiO_4 units making it very viscous. On cooling these often do not rearrange to the regular crystalline periodic structure but instead remain as irregular polymers, *i.e.* glass. These oxides are termed *network formers*.

Other oxides, namely those of Pb^2, Tl^1, Bi^3, Zn^2, Cd^2, Al^3 and possibly Ti^4, Zr^4 and Ta^5, are able to form four-coordinated partially covalent bonds although this may not necessarily be their normal way of combining. They are able to take part in a glass network initiated by Si, B or P, taking up positions equivalent to the network formers (where there is a valency change neutrality is achieved by the addition of cations). Thus Al^3 will replace a Si^4 in tetrahedral coordination although in crystalline Al_2O_3 the Al is six-coordinated. This group of elements is termed *network co-formers*, they sometimes also include Mg^2 and Be^2.

Zachariasen's postulate that the glass-forming oxides were able to form glasses, because of the similarity in the basic structures of their crystalline and glassy forms, was also frequently taken to mean that where the crystalline form did not have the characteristics suitable for a similar glass the oxide could not be prepared as a glass. Certainly it does not seem easy to do so, but it is of interest that Al_2O_3 and TiO_2 have now been prepared as glasses by condensation of the oxidation products of the vaporised metals on well-cooled surfaces (**H87**). This is a considerable achievement bearing in mind that the metals are both six-coordinated in their crystalline oxides, four-coordinated in composite glasses and that Al_2O_3 must be three-coordinated in its pure oxide glass (**S130**).

Glasses also contain a number of larger cations, *e.g.* Na^+, K^+, Rb^+, Cs^+, Ca^{2+}, Sr^{2+}, Ba^{2+}. These do not form directional covalent bonds with oxygen and their coordination number is generally greater than four. On the incorporation of, say, sodium oxide in a silica glass some of the Si—O—Si bonds are broken, the oxygen ion from the sodium oxide taken up to form Si—O O—Si which is then negatively charged and the free Na^+ fits in an interstice in the network. This weakens the structure and lowers the melting point. These cations are termed *network modifiers*.

The structure of glasses is investigated by taking photographs of the X-ray diffraction pattern formed by passing monochromatic X-rays through a small specimen of the glass. The diffraction pattern is scanned by a photometer and the intensities used to plot the X-ray scattering curve. Warren and his cooperators (**W12**) worked out a method for applying the mathematical Fourier analysis to the scattering curve to give the radial distribution curve

ACTION OF HEAT ON RAW MATERIALS

which represents interatomic distances and gives the number of neighbours. Their systematic application of these principles gives a clear picture of glasses.

Two years after the publication of Zachariasen's theory on the atomic arrangement in glass (**Z2**), Warren was able to prove it experimentally by the X-ray diffraction method (**W10**). Warren's work on the pure glass-forming oxides showed the following results:

SILICA

'In SiO_2, for example, each silicon is tetrahedrally surrounded by four oxygens at 1·60 Å and each oxygen is shared by two such tetrahedral groups. The mutual orientation of the two groups about their common direction of bonding is random, and hence it is a random three-dimensional network which is built up. The configuration does not repeat at regular intervals and hence is non-crystalline; it is this feature which distinguishes the glassy state from the crystalline.' Germanium oxide glass has a similar structure (**W10**).

Further work on the application of Fourier analysis to the problem, the importance of which lies in the fact that it involves no assumptions about the structure being investigated, gives the silicon–oxygen distance as 1·62 Å and the oxygen–oxygen distance as 2·65 Å. This is in good agreement with the values found in crystalline silicates (**W12, W13**).

BORIC OXIDE

The tendency for boric oxide to become vitreous is so great that its crystalline form was unknown until 1933 when it was obtained by prolonged heating in a vacuum slightly below its melting point. In this way it greatly resembles the sluggish crystallisation properties of the silicates.

X-ray diffraction determination of boric oxide glass affords a clear explanation of this. In this glass the boron is in triangular coordination lying between three oxygen atoms, the boron–oxygen distance being 1·39 Å and the oxygen–oxygen distance 2·42 Å. Each oxygen is shared between two borons, the three generally not being in a straight line so that a random network is set up. The basic unit of B_2O_3 is a planar BO_3 unit (**W12**). Irregular linking of these in the liquid makes it virtually impossible for them to become crystalline on cooling.

Compared with the SiO_4 unit with its external four links, the BO_3 unit with only three is less tightly held. It has accordingly a lower melting point, higher coefficient of thermal expansion and is water soluble. B_2O_3 glass is, however, very tough.

SILICA AND BORIC OXIDE

Incorporation of B_2O_3 in a SiO_2 glass is accompanied by a lowering of the melting point and a stabilisation of the vitreous state. The change in properties on the addition of B_2O_3 to SiO_2 is continuous. The mixed glass like both of its components has covalent directional bonds (Fig. 2.4).

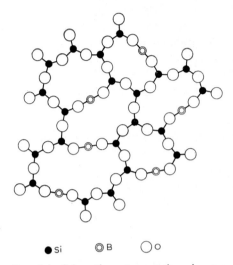

● Si   ◎ B   ○ O

Fig. 2.4. *Schematic representation in two dimensions of the structure of boric oxide–silica glass. To correspond to two dimensions, the silicons are shown bonded to three oxygens rather than four, and the borons to two rather than three (Biscoe, Robinson and Warren,* **B67**)

PHOSPHORUS PENTOXIDE

Like boric oxide, phosphorus pentoxide forms a glass easily and is very resistant to devitrification. The structural unit is the tetrahedrally arranged PO_4 group, but in this case only three of the four oxygens are linked to a second phosphorus. P_2O_5 glass is therefore structurally much weaker than SiO_2 glass, having a low melting temperature and high thermal expansion. It is also very chemically reactive and readily forms glasses with basic oxides and alumina.

Introduction of Network Modifiers

The oxides of the network modifiers (*see* p. 202) are ionic, the bond between the cation and the oxygen is not directional and permanent. In crystalline ionic compounds the ions are held together by electrostatic forces, each cation is surrounded by a given number of anions and vice versa.

Introduction of ionic oxides into a glass therefore furnishes free oxygen ions. These are taken up by the silicons which always tend to be bonded to four oxygens each but thereby break away from each other. The ensemble becomes negatively charged. The positively charged metal ion finds a place in a suitable hole in the network where such surplus negative charge is situated. The over-all structure is weakened and the melting point lowered. The first example to this to be fully worked out was that of the soda–silica system.

SODA–SILICA

In this system (as elsewhere) the immediate surroundings of the silicon atoms remain a tetrahedral arrangement of four oxygen atoms at 1·62 Å (Fig. 2.5). Part of the oxygens are shared between two silicons and thus build up a silicon–oxygen framework; the others are bonded to only one silicon. The sodium atoms are located at random in the different holes in the framework surrounded by about six oxygens at 2·35 Å (**W11, W14**). ['The existence in the glass of simple discrete molecules such as SiO_2, Na_2O, $Na_2Si_2O_5$ and Na_2SiO_3 is uniquely ruled out by the distribution curves' (**W14**).]

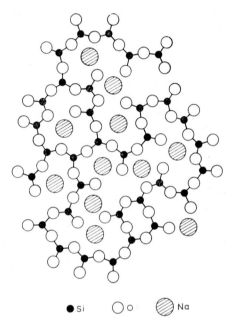

FIG. 2.5. *Schematic representation in two dimensions of the structure of soda–silica glass* (*Warren and Biscoe*, **W14**)

Each Na_2O molecule converts one Si—O—Si unit to

$$\text{Si—O}^- \quad \overset{Na^+}{\underset{Na^+}{}} \quad {}^-\text{O—Si}$$

When the proportion of soda to silica is increased to $Na_2O:2SiO_2$ each SiO_4 tetrahedron has only three Si—O—Si links to neighbouring tetrahedra. At the composition $Na_2O:SiO_2$ only a chain structure could exist, although in practice devitrification occurs a little before this composition is reached (**M90**).

The fact that the sodium atoms are present as ions that are not directionally bonded makes movement through the network possible and the glass therefore becomes an electrical conductor to a small degree.

POTASH–SILICA

The general arrangement is similar to that of soda–silica glass but as the potassium atom is larger than the sodium one it is surrounded by ten oxygens whereas the sodium had only six (**B68**).

Potassium opens up the network more than sodium, thereby lowering the density, the viscosity and the index of refraction, and increasing the co-efficient of expansion. The larger potassium atom is, however, less able to move through the silicon–oxygen network so that potash glass has a lower electrical conductivity than soda glass.

DIVALENT CATIONS

When divalent cations (Mg^{2+}, Ca^{2+}, Sr^{2+}, Ba^{2+}) are used instead of the monovalent ones (Li^+, Na^+, K^+, Rb^+, Cs^+) it is found that at low concentrations two immiscible liquids form instead of one homogeneous one. One layer contains almost all the network modifier. At higher concentrations of

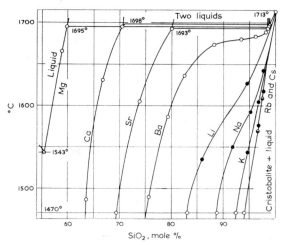

Fig. 2.6. *Equilibrium diagrams for the alkali and alkaline earth oxide–silica systems (Warren and Pincus, W15)*

the cation a homogeneous glass is formed. This tendency to immiscible liquids in a glass system is stronger the higher the charge and the smaller the size of the cation, as is seen in the diagram showing the effect of additions of different cations to silica (Fig. 2.6).

Warren and Pincus (**W15**) attempt to explain this by application of the ideas of crystal chemistry already found to be useful in the glass sphere.

The network formers silicon, boron and phosphorus tend to bond all available oxygen atoms and thereby favour miscibility. But the network modifiers need to be properly surrounded by singly-bound oxygens. When they are present in small quantity and the singly-bound oxygens are distributed throughout the glass there are no holes in the network that give adequate surroundings to a cation so that there is an opposing tendency for the cation to form a separate layer with only a small quantity of the network former.

In the glasses with the alkali metals the tendency for miscibility dominates at all concentrations. In those with the alkaline earths and magnesium with their higher charge the tendency for immiscibility rules at low cation concentrations and that for miscibility at higher ones. Calculations based

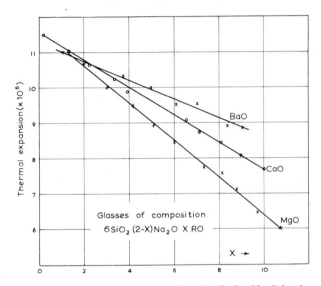

FIG. 2.7. *Effect on thermal expansion of replacing Na_2O by the indicated oxides (plotted on molal basis) (Biscoe, **B70**)*

on this assumption and on bond lengths give a value of 33% for the minimum CaO content for miscibility in a lime–silica glass. The observed value is 28% which the authors consider to be sufficiently close to make their theory plausible.

The addition of about 1% of Na_2O or of Al_2O_3 makes lime soluble in silica in all proportions. This is thought to be due to the furnishing of some extra singly-bound oxygens (**W15**).

X-ray diffraction investigation of soda–lime–silica glass showed that the positions taken up by the calcium atoms are very similar to that of the sodium atoms but calciums generally are surrounded by eight oxygens whereas the usual number for sodiums is six.

The greater affinity of the divalent cations for oxygen has an effect on the physical properties of the glass when they replace sodium (Fig. 2.7). They reduce the coefficient of expansion and in the case of the two smallest, magnesium and calcium, even increase the density (**B70**).

In general, replacement of a network-modifying cation by another of the same size but higher valency increases the density, the viscosity and the refractive index and lowers the electrical conductivity and the coefficient of expansion.

The difference in energy between an ordered crystalline environment for the calcium ion and the disordered vitreous state is greater than that for the sodium ion, so that crystallisation during cooling of a calcium glass is more likely to occur than with a sodium glass (**B69**) (*see* Zachariasen's rules p. 201).

Similar effects occur with barium and strontium as with calcium except that these larger ions require larger numbers of oxygen atoms to surround them.

Divalent ions do not appear to migrate through the silicate structure.

R_2O—RO—SiO_2 GLASSES

The number of breaks in the silicate structure should correspond to the total number of R_2O and RO molecules introduced. But it is found in practice that this total can exceed the proportion that would give a chain structure leading to devitrification. It therefore appears that the divalent ions can act as bridges between SiO_4 units in these conditions (**M90**).

LIME-PHOSPHORIC OXIDE

X-ray diffraction investigation of two lime–phosphoric oxide glasses containing 23% and 28% CaO, respectively, showed that here the phosphorus is tetrahedrally surrounded by oxygens, like silicon in silicates. The P—O bond is 1·57 Å. Each oxygen is bound to one or two phosphorus atoms. The calcium ions are situated in suitable holes in the phosphorus–oxygen network usually surrounded by seven oxygens (**B69**). As a quarter of the oxygens in the pure P_2O_5 glass are bound to only one phophorus there should be much less tendency for immiscibility on the introduction of small quantities of CaO to P_2O_5 glass.

NETWORK MODIFIERS AND BORIC OXIDE

Continuous addition of network modifiers to silica or phosphorus pentoxide glasses produces a continuous change of properties, but similar action with boric oxide glass produces a continuous change in the other direction up to a certain concentration (about 16% Na_2O and 22% K_2O), and another different continuous change similar to that in silica glass thereafter. This is seen in Fig. 2.8 (a)–(f). This became known as the 'boric oxide anomaly', and was not satisfactorily explained until the X-ray diffraction technique was applied to the problem. Biscoe and Warren (**B66**) investigated soda–boric oxide glasses of the following molal compositions: 0·114 Na_2O, 0·225 Na_2O, and 0·333 Na_2O. The B—O distances were found to be 1·37, 1·42 and

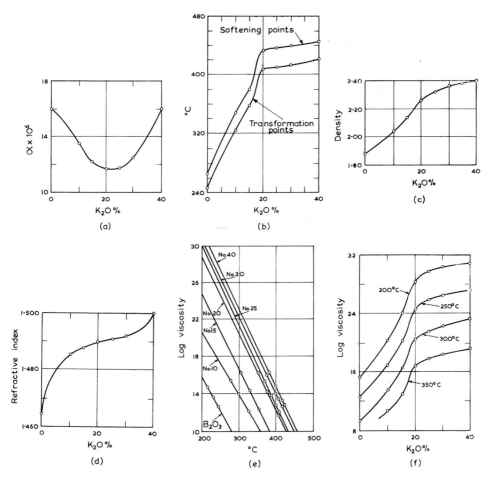

Fig. 2.8. *Boric oxide–potash glasses :*

(a) *coefficient of expansion of potassium borate glasses ;*
(b) *transformation and softening points of potassium borate glasses ;*
(c) *density of potassium borate glasses ;*
(d) *index of refraction of potassium borate glasses ;*
(e) *viscosity of potassium borate glasses ;*
(f) *log viscosity of potassium borate glasses* (Greene, **G74**)

1·48 Å, respectively, and the number of oxygens about a boron in the three cases was 3·2, 3·7 and 3·9. This points to a change in the coordination number of the boron. As the soda furnishing free oxygens is added the boric oxide network takes them up by some of the boron atoms becoming four-coordinated. This tightens up the network, whereas the bond-breaking effect of adding soda to silica glass loosens the network. The sodium ions take up suitable positions in the network surrounded by six oxygens at 2·4 Å (Fig. 2.9).

A maximum in this change of coordination of the boron atoms is, however, reached when 16% Na_2O, *i.e.* when the molal ratio is $R_2O.5B_2O_3$ showing that BO_4 groups must be separated by at least two BO_3 triangles. Thereafter B—O—B bonds break and become B—O O—B and the network loosens and the properties change in the same direction as in a soda–silica glass. The relation between physical properties and coordination number of the network former is seen in Fig. 2.10.

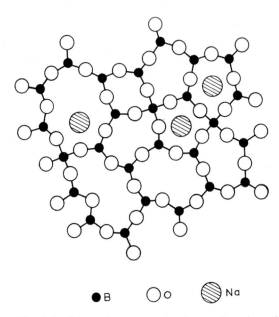

Fig. 2.9. *Schematic representation in two dimensions of the structure of soda–boric oxide glass of low soda content. All B—O bonds are shown in one plane* (Biscoe and Warren, **B66**)

Similarly Greene (**G74**) found that with potash–boric oxide glasses ranging from 0 to 40% K_2O the boron changes from three- to four-fold coordination, reaching a maximum at 22% K_2O, *i.e.* $K_2O.5B_2O_3$. This is in complete agreement with the experimental curves (Fig. 2.8).

The same is found in lime–boric oxide glasses (**B69**) but here there is the further complication of the tendency for mixtures of low lime content to separate into two layers. Calculations similar to those for lime–silica glass (p. 207) can be made assuming that either the boron stays in triangular coordination when a value of 34% CaO minimum for miscibility is obtained, or that all the oxygen being supplied changes the borons to tetrahedral coordination when the value is 19%. The observed value is 23%, which shows that the change of coordination dominates over the tendency for immiscibility (**W15**).

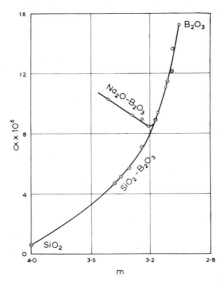

Fig. 2.10. Thermal expansion coefficients against average coordination number m (Warren, **W16**)

BORO-SILICATE GLASSES

Glasses can be made of silica and boric oxide alone. In these the boric oxide must occur as BO_3 triangles. They are readily attacked by water.

In glasses containing SiO_2 and B_2O_3 and one of the alkalies boron can become four-coordinated. Again every two BO_4 tetrahedra must be separated by two other units but these may be SiO_4's instead of BO_3's so that in a mixture containing $B_2O_3:8SiO_2:2Na_2O$ the whole of the boron could be four-coordinated.

However, boric oxide has a strong tendency to combine with alkali and make a separate phase of $R_2O.5B_2O_3$ which complicates the considerations. It now seems that whether the maxima or minima of the 'boric oxide anomaly' occur at $R_2O.B_2O_3$ or $R_2O.5B_2O_3$ depends on the SiO_2 contents and probably also on the rate of cooling of the molten glass down to the annealing temperature (**M90**).

Parts played by Other Glass Constituents

It has been seen how the chemical nature, size, charge, valency, etc., determines what elements are network formers and which others network modifiers. There are, however, numerous other elements that are used in glasses and glazes that do not fall strictly into either class. They differ in one or more ways from the characteristic properties of the groups and hence their structural role in glasses is intermediate. They may be subdivided approximately into several groups.

CONTRACTORS (LITHIUM, BERYLLIUM AND MAGNESIUM)

True network modifiers both loosen the network by breaking its bonds and lower the density because of their relatively large size. In limited circumstances the density–composition relationship can be calculated for them (**S166**). In the case of lithium, beryllium and magnesium the density decrease is markedly less than anticipated and they are therefore termed 'contractors'. These three elements have small highly charged ions which have a strong affinity for oxygen (**S188**).

It is found that the lithium ion is small enough to fit in the space between four close-packed oxygen atoms. Beryllium even forms covalent BeO_4 structural tetrahedra, but then requires the inclusion of one Li_2O, Na_2O or K_2O molecule for every BeO. Magnesia plays an intermediate part. There is some evidence that under certain conditions MgO_4 tetrahedra form with two alkali ions in close association with each tetrahedron but most of the Mg^{2+} ions enter interstices like the alkaline earths (**M90**).

NETWORK COFORMERS

Aluminium. The best known coformer is aluminium, imparting very useful properties on glasses and glazes. The trivalent aluminium ion is six-coordinate, but it is not very stable alone and easily goes over to covalent binding when it becomes smaller and can be four-coordinate. Alumina itself can only be made as a glass in very special circumstances (*see* p. 202) and by comparing its data with the Zachariasen rules this difficulty can be readily appreciated. However, if enough singly-bound oxygens, furnished by network modifiers, are present for it to be surrounded by four oxygens it will take up a position in a silica network analogous to that of the silicon itself, *i.e.* as a network former. This is seen conclusively in the continuous three-dimensional crystalline silica–alumina networks found in the feldspars and in the following isomorphous trio:

$SiSiO_4$	$NaAlSiO_4$	$Na_2Al_2O_4$
Cristobalite (high temperature form)	Carnegieite	Sodium aluminate

The crystal structure of calcium aluminate shows the presence of AlO_4 units, some linked to each other at all four corners, the others at three (**B138**). Comparison with Zachariasen's rules shows this to be a potential glass-forming system and indeed Stanworth (**S167**) has succeeded in obtaining small quantities of such glass. Much larger quantities of a quick-setting glass were obtained when about 6% silica was added to the calcium aluminate.

In general then alumina is a coformer. Its addition to an alkali-containing glass or glaze imparts strength, durability, higher chemical resistivity and higher resistance to devitrification. It raises the softening and melting points.

It must be remembered that these very good network-forming properties of aluminium in silica glass depend on an adequate supply of oxygen from network modifiers, *i.e.* every Al_2O_3 molecule requires at least one R_2O or RO and the free cations resulting must remain closely associated with the two AlO_4 tetrahedra formed to preserve local electroneutrality. The AlO_4 groups must also be separated by at least one SiO_4 or BO_3 group (**M90**). Therefore continued addition of alumina to a given mixture to improve a given property will do so only until a certain point is reached, thereafter a decay of this improvement sets in. At this point the mixture is saturated with four-coordinate network-forming aluminium and further additions introduce six-coordinated aluminium which acts as a network modifier and loosens the structure. This has occasionally been termed the 'alumina anomaly'; the composition at which it occurs is, however, perforce less predictable than with the boric oxide anomaly (p. 208) (**S115**).

It is very interesting to note that whereas alumina will hardly react with boric oxide, and forms a completely miscible series of liquids with silica which, however, crystallise on cooling, it readily combines with phosphorus pentoxide to form a glass even at the orthophosphate composition $AlPO_4$. In its crystalline form $AlPO_4$ shows polymorphic changes remarkably like those of SiO_2 and exhibits glass-forming properties like it, both as a unitary glass and as the chief ingredient in a whole range of complex glasses. This shows how the trivalent aluminium alternating with the quinquevalent phosphorus and four oxygens between them can act very like tetravalent silicon (**P42**).

Titanium, Zirconium and possibly Tantalum. Data from physical measurements, particularly density, suggest that these elements are network coformers similar to aluminium. As they are, however, normally larger, their coordination number in glasses cannot be directly predicted and has not yet been determined (**S166, 167**).

Zinc and Cadmium. It is well known that it is possible to obtain glasses that do not devitrify easily with a far higher percentage of some cations than of others. Amongst these are zinc and cadmium. Thus zinc orthosilicate, Zn_2SiO_4, forms a glass and borate glasses are possible with 61 mol and 57 mol per cent, respectively, of ZnO and CdO.

Zinc and cadmium are in group IIB and differ from group IIA elements calcium and strontium by the ions having an outer complement of eighteen electrons instead of eight. Their bonds to oxygen have a more covalent character. In fact measurements of the bond lengths and coordination number of zinc in different compounds point clearly to its sometimes being an ion of radius 2·04 Å and coordination number six, and at other times a predominantly covalently-linked atom of radius 1·92 Å and coordination number four, the latter being the case in zinc orthosilicate.

In the zinc orthosilicate glass a continuous network of SiO_4 units cannot exist, as each is separated by a zinc atom so the zinc must almost certainly take on network-forming work in this instance (**K79**), and as much other evidence points to its forming ZnO_4 tetrahedra in conjunction with one RO_2 in other glasses it should be termed a coformer (**S120, M90**). Under the circumstances where this condition prevails the addition of zinc to glazes is most beneficial, increasing the strength, chemical resistance and resistance to weathering. But on addition of zinc a maximum amount for the required improvement is found after which zinc acts like the standard network modifiers. Here again, an adequate supply of singly-bound oxygens will enable it to become an ion which has a larger radius than the covalent zinc (**S120**).

The individual role of cadmium is somewhat more difficult to interpret, as even when covalent it is probably too large for tetrahedral coordination (**S167**).

Lead Pb^2, *Bismuth* Bi^3 *and Thallium* Tl^1. Lead is a classical glass constituent for imparting fusibility, transparency, brilliancy and high refractive power. It too can form a glass with extraordinarily small quantities of a glass former, *e.g.* the orthosilicate Pb_2SiO_4. It has now been established that the lead takes part in the network, being joined to two oxygens which are corners of two different SiO_4 tetrahedra (**S167**). Even $2\cdot5$ $PbO.SiO_2$ can be a glass (**M90**).

The proportion of PbO that can be introduced into glassy lead borates is also very high. Whereas alkali borates are at their stablest at a ratio of $RO:5B_2O_3$ lead borates are so at $2PbO:5B_2O_3$. This is seen especially in the relative solubility of the lead and the boric oxide in lead borates. Where the ratio is less than $2:5$ boric oxide is more readily extracted, whereas when it is greater than $2:5$ it is the lead oxide that is more easily removed. This ratio of maximum stability is, of course, affected by other constituents in a complex mixture with silica, alkalies, etc., but it is probable that by varying constituents progressively an optimum can be found (**M90, B3**).

The anomalous behaviour of lead when compared with the divalent alkaline earths is due to its atomic structure. Like trivalent bismuth and monovalent

TABLE 53

The Groups of Glass Constituents (**P42, S167, 188, K79**)

Network formers: Si^4, B^3, P^5; Ge^4, As^3, As^5, Sb^3, (V^5)
Network co-formers:
  (a) replacing Si^4 in network:

$$Al^3, Ti^4, Zr^4, Ta^5,$$

  (b) forming covalent bonds and allowed in high %:

$$Pb^2, Tl^1, Bi^3,$$
$$Zn^2, Cd^2$$

Network contractors: Li^+, Be^{2+}, Mg^{2+}.
Network modifiers: Na^+, K^+, Rb^+, Cs^+, Ca^{2+}, Sr^{2+}, Ba^{2+}.

thallium it has a large ion with an outer electron shell of 18+2 electrons and it is therefore very easily deformed. All three behave in glasses as if they make covalent network-forming bonds (**S167**). Recent work on the infra-red absorption spectra of glasses containing lead in widely varying proportions has, however, failed to show bonds which can be attributed to definite lead–oxygen groups (**M90**).

PHASE DIAGRAMS

It is not expedient to give here a full explanation of the phase rule and its applications. Its importance in ceramics is shown by the fact that Hall and Insley under the auspices of the American Ceramic Society have brought out a supplement to the *Journal* containing eighteen pages of general discussion of phase diagrams and 507 such diagrams that are relevant to ceramics (**H9**). (Several hundred more diagrams have been published subsequently.)

The phase rule itself is expressed by the equation:

$$P+F = C+2$$

where $P=$ number of phases present at equilibrium.
 $F =$ number of degrees of freedom.
 $C =$ number of components of the system.

It describes systems at equilibrium and phase diagrams are schematic representations of the conditions of a one-, two- or three-component system when the temperature, pressure or composition are altered.

The sluggish physical and chemical reactions of the materials of interest to ceramists make impossible the otherwise generally used method of plotting phase diagrams from data obtained from thermal curves. Most of them are plotted from data obtained by the quenching technique. A given mixture is held at a sufficiently elevated temperature for it to become partly liquid, until equilibrium is believed to have been reached. It is then cooled quickly, 'freezing the equilibrium', when the liquid part becomes glassy. The product can be examined at leisure by the petrographic microscope. The method requires time and expensive equipment and is usually only undertaken by specialists.

Foster (**F21**) suggests that a certain amount of very useful information can be obtained by working at temperatures and compositions when only solid-state reactions occur; this requires less specialised equipment and time.

Thermodynamic data, *e.g.* free-energy data, can also be used to construct phase diagrams (**K41**).

There is no doubt about the fact that phase diagrams are not easily or quickly interpreted. The theoretical side of this will not be discussed here but some of the aspects of it that are of specific interest to ceramists will be mentioned.

Use for Ceramic Bodies

The application of phase-diagram data to interpretation of reactions and products in bodies must be undertaken with some care. As Mellor (**M48**) pointed out, a large number of bodies containing relatively coarse grains, and as they are often not fired at very high temperatures, do not reach equilibrium. This applies particularly to earthenware (**W53**) and certain refractories. Knowledge of the equilibrium state cannot therefore be of more than qualitative assistance.

However, numerous modern bodies made up of very finely ground raw materials fired at high temperatures, where the main reaction occurs in the solid state, are now believed to attain a condition very close to that of equilibrium. In such circumstances a great deal of valuable information can be obtained from phase diagrams (**F19, A158**).

PRIMARY FIELDS

Triangular phase diagrams are usually divided clearly into 'primary fields' or 'fields of stability of crystalline phases in contact with liquid'. This means that if a mixture of composition falling within a particular field is heated, the last crystalline phase to melt is the one of that field; and when liquid of that composition is cooled that phase is the first to appear. These facts are not of very much help in considering ceramic bodies where only a small proportion melts, and we must look further into the diagrams to obtain more useful information.

COMPATIBILITY (COMPOSITION, OR SOLIDUS) TRIANGLES

Many phase diagrams for three-component systems are further subdivided into compatibility triangles (for more than three components these become polyhedra) and where this is not so they can be readily inserted from a study of the primary fields.

A phase diagram of considerable interest in certain ceramic fields will be used as an example (Fig. 2.11). Here the primary fields are marked in thick lines and it is seen that there are always three adjoining fields that meet at a point, *i.e.* three crystalline phases that can exist simultaneously. The compatibility triangle is constructed by joining the points representing the compositions of these three phases (lighter lines).

With the use of these triangles the phases present after a solid-phase reaction has gone to completion can be predicted both qualitatively and quantitatively. The over-all composition of the mixture will fall either in one of the compatability triangles, on one of its sides, or at an apex. In the last case only the one phase will occur, in the second the two phases joined by the line will be present, and in the first all three phases joined by the triangle. The relative quantities are given by the position of the point *up* the line passing from the base of the triangle through the point to the apex representing the phase under consideration.

Naturally both the primary fields and the compatibility triangles show what phases cannot coexist, *e.g.* periclase and corundum, which will therefore tend to react to give spinel until one of them is used up. Spinel can exist together with either periclase or corundum (**F19**).

It is of interest to note that the highly desirable product, cordierite, although having a very small primary field, is compatible at all compositions except those in the triangles at the MgO and Al_2O_3 apexes, *i.e.* in well over half the possible compositions.

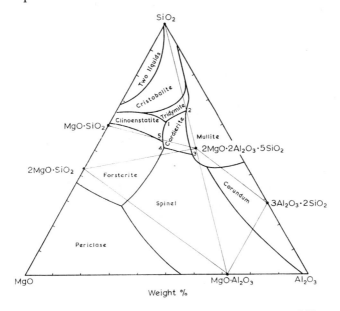

Fig. 2.11. *Equilibrium diagram of the system $MgO-Al_2O_3-SiO_2$ (after Rankin and Merwin, **R2**, modified by Bowen and Greig, **B98**, and Greig, **G76**)*

INVARIANT POINTS

These include eutectic points and reaction (peritectic) points or solid solution minima. They do not differ greatly but can be distinguished according to whether or not they fall inside or outside the pertinent compatibility line or triangle. In all systems, except where a continuous series of solid solutions forms, addition of either component to the other lowers the melting point until a certain fixed minimum is reached. At this, the eutectic point, not only is the lowest melting point of the system reached, but, since the melting or solidifying are faster, and the solid and liquid are of the same composition, the solid is homogeneous.

For systems at equilibrium with only compatible phases present the invariant point for these phases gives the temperature at which melting starts for all compositions within the compatibility triangle (line or polyhedron) concerned. Its location shows the composition of the liquid first formed. This is of importance in determining a suitable firing temperature for the body and in ascertaining the nature of the glassy bond formed.

In systems where equilibrium is not reached by solid-phase reactions before melting commences and hence incompatible phases are present some liquid will appear at the lowest invariant point within the triangles, one apex of which represents a phase that is present. Thus in a mixture of silica, clay and forsterite whose total composition does not fall in the silica–clinoenstatite–cordierite triangle the first liquid will nevertheless appear at 1347° C (2457° F), the invariant point of that triangle. But this melting is transient. The liquid formed reacts more readily with the solids to bring the mixture towards the equilibrium conditions (**F19**).

MELTING INTERVAL

With the exception of pure congruently melting compounds and eutectics, all mixtures of substances melt gradually over a 'melting interval'. The invariant point gives the temperature at which a system at equilibrium will begin to melt. The amount of liquid present at any subsequent temperature can be calculated from the phase diagram (*see* **H9**, p. 5, top of second column). This is of considerable use in connection with the firing range of a body. Vitrification occurs when a certain proportion of liquid is present, distortion occurs when there is too high a liquid content. (The actual percentages of liquid required for vitrification and subsequently for distortion depend on its viscosity, being lower the more freely it flows.) For satisfactory commercial production the firing range must be sufficient for the product to be unaffected by the normal fluctuations of the firing conditions. Use of the calculated proportions of liquid present at different temperatures, together with their compositions, which give an idea of their viscosities, should lead to an estimate of the possibility of using a certain mixture as a ceramic body.

THE COURSE OF MELTING

It has been shown that phase diagrams can give data about:

(1) The crystalline phases present in an equilibrium mixture before melting starts.

(2) The temperature of initial melting and the composition of the first liquid formed.

(3) The last crystalline phase to remain before complete melting.

There is, therefore, still the missing link of data on the order of disappearance of phases during melting, *i.e.* the course of melting. Unfortunately there has been little work done on this so far. The theoretical course of melting can be calculated from phase-diagram data and the usefulness of such work is seen in Fig. 2.12 (*a*) and (*b*). Here the course of melting is plotted for four mixtures of about the same $Al_2O_3:SiO_2$ ratio and MgO content ranging over only 8·5%, and the considerable differences are most interesting. It is not surprising then that phase-rule data misinterpreted or applied to a mixture not quite at equilibrium can easily differ radically from practical experience.

ACTION OF HEAT ON RAW MATERIALS 219

The direct applications of phase-rule data to ceramic bodies will be considered in the relevant sections when it will also be seen that the practical test can never be entirely replaced by theoretical considerations.

DEFORMATION-EUTECTIC MIXTURES

From the practical point of view ceramists are often more interested in the lowest temperature at which a mixture becomes sufficiently fluid to deform

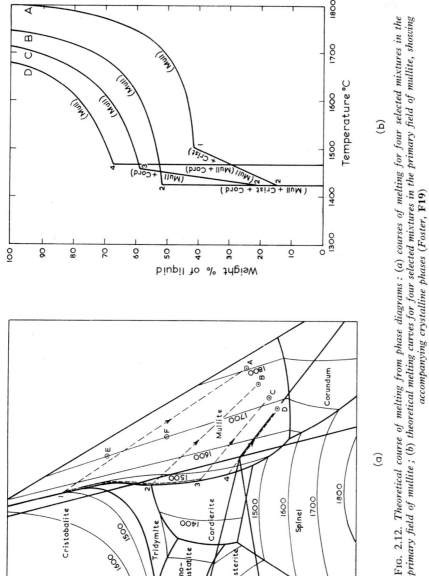

FIG. 2.12. Theoretical course of melting from phase diagrams: (a) courses of melting for four selected mixtures in the primary field of mullite; (b) theoretical melting curves for four selected mixtures in the primary field of mullite, showing accompanying crystalline phases (Foster, **F19**)

under its own weight, than the equilibrium data on melting. They therefore differentiate between the 'deformation-eutectic' and the 'equilibrium-eutectic' as defined by McCauhey (**M1**):

Deformation-eutectic: that composition within a system that develops a sufficient amount of liquid to cause deformation at the lowest temperature.

Equilibrium-eutectic: the composition and temperature of a melt in a system of n components, in equilibrium with n crystal phases, and on cooling the system there is a positive crystallization in n crystal phases without changing the composition and temperature of any of these phases.

Deformation-eutectic mixtures will probably be found to be very useful as auxiliary fluxes in certain bodies. Watts and his collaborators (**W29**) have therefore determined some deformation eutectics and then investigated their reactions with feldspars, the more usual body flux.

Table 54 gives the compositions and temperatures of a number of deformation eutectics determined by the bending of pyrometric cones.

TABLE 54
Deformation Eutectics

(1) Data obtained with test cones made up of pulverised material 100% through 200-mesh sieve and 85% through 325-mesh sieve. Heating rate was 200° C/hr up to 100° C below critical temperature, *i.e.* about 800° C, thereafter 25 to 30° C/hr. The deformation temperature range is from the first appearance of deformation until the cone is down.

(2) The deformation-eutectic mixtures have been abbreviated after their first appearance in the table, *e.g.* the CaO–Al_2O_3–SiO_2 deformation eutectic of composition CaO 23·0; Al_2O_3 14·7; SiO_2 62·3% is termed CAS-D.

(3) Although different references are given the work was carried out in the same laboratory and is therefore comparable.

Ref.	Composition	%	Deformation temperature or range	
			(°C)	(°F)
T17	(1) Microcline $K_2O.Al_2O_3.6SiO_2$ $CaCO_3$	98·0 2·0	1205–1225	2201–2237
T17	(2) Microcline $K_2O.Al_2O_3.6SiO_2$ $MgCO_3$	97·0 3·0	1190–1215	2174–2219
T17	(3) Microcline $K_2O.Al_2O_3.6SiO_2$ $BaCO_3$	56·0 44·0	1140–1150	2084–2102
T17	(4) Microcline $K_2O.Al_2O_3.6SiO_2$ ZnO	98·0 2·0	1200–1215	2192–2219
T17	(5) Microcline $K_2O.Al_2O_3.6SiO_2$ $CaCO_3$ $MgCO_3$	97·5 1·0 1·5	1195–1215	2183–2219
T17	(6) Microcline $K_2O.Al_2O_3.6SiO_2$ $CaCO_3$ $BaCO_3$	57·05 0·05 42·90	1140–1145	2084–2093
P40	(7) K feldspar (prefused)	100	1177–1205	2151–2201
P40	(8) Na feldspar (prefused)	100	1090–1118	1994–2044
T13	(9) K feldspar } (raw) Na feldspar (termed KNa feldspar-D)	36 64	1200	2192

TABLE 54—(contd.)

Ref.	Composition		%	Deformation temperature or range	
				(°C)	(°F)
P40	(10) KNa feldspar-D (prefused)			1070–1090	1958–1994
P40	(11) CaO		23·0	1200–1210	2192–2210
	Al₂O₃		14·7		
	SiO₂ (raw)		62·3		
	(termed CAS-D)				
P40	(12) CAS-D (prefused)			1168–1175	2134–2147
P40	(13) MgO		20·3	1343–1363	2449–2485
	Al₂O₃		18·3		
	SiO₂ (raw)		61·4		
	(termed MAS-D)				
P40	(14) MAS-D (prefused)			1320–1337	2408–2439
P40	(15) BaO		36·25	1185–1191	2165–2176
	Al₂O₃		11·25		
	SiO₂ (raw)		42·50		
	(termed BAS-D)				
P40	(16) BAS-D (prefused)			977–1005	1791–1841
T13	(17) CaO		20·9	1185	2165
	MgO		2·1		
	Al₂O₃		14·0		
	SiO₂ (raw)		63·0		
	(termed CMAS-D)				
T13	(18) CMAS-D (prefused)			1145	2093
P40	(19) CAS-D	(raw)	6·25	1170–1182	2138–2160
	K feldspar	(raw)	93·75		
P40	(20) CAS-D	(prefused)	90·0	1160–1168	2120–2134
	K feldspar	(raw)	10·0		
P40	(21) CAS-D	(prefused)	20·0	1121–1144	2050–2091
	K feldspar	(prefused)	80·0		
P40	(22) CAS-D	(raw)	30·0	1167–1184	2133–2163
	Na feldspar	(raw)	70·0		
P40	(23) CAS-D	(prefused)	85·0	1159–1166	2118–2131
	Na feldspar	(raw)	15·0		
P40	(24) CAS-D	(prefused)	3·75	1038–1071	1900–1960
	Na feldspar	(prefused)	96·25		
P40	(25) CAS-D	(raw)	15·0	1125–1132	2057–2070
	NaK feldspar-D	(raw)	85·0		
P40	(26) CAS-D	(prefused)	77·5	1160–1167	2120–2133
	NaK feldspar-D	(raw)	22·5		
P40	(27) CAS-D	(prefused)	10·0	1095–1110	2013–2030
	NaK feldspar-D	(prefused)	90·0		
P40	(28) MAS-D	(raw)	22·5	1190–1202	2174–2196
	K feldspar	(raw)	77·5		
P40	(29) MAS-D	(prefused)	25·0	1174–1178	2145–2152
	K feldspar	(raw)	75·0		
P40	(30) MAS-D	(prefused)	20·0	1155–1168	2111–2134
	K feldspar	(prefused)	80·0		
P40	(31) MAS-D	(raw)	25·0	1155–1163	2111–2125
	Na feldspar	(raw)	75·0		
P40	(32) MAS-D	(prefused)	22·5	1152–1161	2106–2122
	Na feldspar	(raw)	77·5		
P40	(33) MAS-D	(prefused)	5·0	1080–1110	1976–2030
	Na feldspar	(prefused)	95·0		

TABLE 54—(contd.)

Ref.	Composition		%	Deformation temperature or range (°C)	(°F)
P40	(34) MAS-D	(raw)	19·0	1130–1143	2066–2089
	NaK feldspar-D	(prefused)	81·0		
P40	(35) MAS-D (prefused)		20·0	1130–1138	2066–2080
	NaK feldspar-D	(raw)	80·0		
P40	(36) MAS-D	(prefused)	10·0	1065–1086	1949–1987
	NaK feldspar-D	(prefused)	90·0		
P40	(37) BAS-D	(raw)	95	1176–1182	2149–2159
	K feldspar	(raw)	5		
P40	(38) BAS-D	(prefused)	Devitrifies, No eutectic		
	K feldspar (raw or prefused)				
P40	(39) BAS-D	(raw)	73·75	1176–1180	2149–2156
	Na feldspar	(raw)	26·25		
P40	(40) BAS-D	(prefused)	Devitrifies, No eutectic		
	Na feldspar	(raw)			
P40	(41) BAS-D	(prefused)	87·5	971–988	1780–1810
	Na feldspar	(prefused)	12·5		
P40	(42) BAS-D	(raw)	85·0	1171–1175	2140–2147
	NaK feldspar-D	(raw)	15·0		
P40	(43) BAS-D	(prefused)	Devitrifies, No eutectic		
	NaK feldspar-D (raw) or (prefused)				
T13	(44) CMAS-D	(raw)	19	1135	2075
	NaK feldspar-D	(raw)	81		
T13	(45) CMAS-D	(prefused)	No eutectic		
	NaK feldspar-D	(raw)			
P40	(46) BAS-D	(raw)	81·25	1118–1130	2044–2066
	CAS-D	(raw)	18·75		
P40	(47) BAS-D	(prefused)	80·0	971–1002	1780–1836
	CAS-D	(prefused)	20·0		
P40	(48) CAS-D	(raw)	90·0	1165–1184	2129–2163
	MAS-D	(raw)	10·0		
P40	(49) CAS-D	(prefused)	90·0	1163–1168	2125–2134
	MAS-D	(prefused)	10·0		
P40	(50) MAS-D	(raw)	4·5	1167–1176	2133–2149
	BAS-D	(raw)	95·5		
P40	(51) MAS-D	(prefused)	No deformation eutectic found		
	BAS-D	(prefused)			
P40	(52) BAS-D	(raw)	78·0	1113–1118	2035–2044
	CAS-D	(raw)	16·0		
	MAS-D	(raw)	6·0		
P40	(53) (BAS-D, CAS-D) D (raw) or (prefused)		No deformation eutectic found		
	NaK feldspar-D	(raw)			
P40	(54) (CAS-D, MAS-D) D (raw)		19	1130–1145	2066–2093
	NaK feldspar-D	(raw)	81		
P40	(55) (CAS-D, MAS-D) D (prefused)		85	1145–1154	2093–2109
	NaK feldspar-D	(raw)	15		
P40	(56) (MAS-D, BAS-D) D (raw)		90	1165–1171	2129–2140
	NaK feldspar-D	(raw)	10		
P40	(57) (MAS-D, BAS-D) D (prefused)		90	1120–1127	2048–2061
	NaK feldspar D	(raw)	10		

Glazes

On the whole glazes are much more likely than bodies are to be in equilibrium conditions when in the kiln. Application of phase-rule data on eutectics and liquid compositions can prove extremely useful in determining suitable glaze compositions (**S108**).

THERMAL STRESSES, FATIGUE, THERMAL SHOCK

Many ceramic materials and bodies are highly refractory, withstanding temperatures far exceeding the softening and melting points of metals, but their use is restricted by their brittle character and inability to withstand changes of temperature. This becomes apparent in two slightly different forms.

(1) *Fatigue*

This was first discussed at length by the senior writer (**S91**), in 1930, in connection with hard porcelain laboratory ware, which was found to withstand heat cycling a considerable number of times, and then to fail suddenly under easier conditions than it had repeatedly withstood. Some forms of mechanical fatigue also occur.

(2) *Thermal Shock*

Most materials can withstand sudden temperature change only to a certain extent, failure occurring when the temperature gradient that has been set up exceeds a given value. Such failure often occurs in a complicated way, it is governed not only by the nature of the material but also by shape, etc. It has become very much more important recently when the use of ceramic refractories in gas turbines, rocket motors and nuclear reactors is under consideration.

There is no firm dividing line between thermal fatigue and thermal shock, and both arise from similar internal features of ceramic bodies.

Thermal Stresses

Heating or cooling of ceramic bodies can give rise to internal stresses from a number of causes, irrespective of thermal gradients. These have been discussed by Singer (**S91**) and White (**W56**).

(1) Differences in thermal expansion and contraction between various phases in polycrystalline bodies.
(2) Volume changes due to polymorphic transitions.
(3) Volume changes due to devitrification processes.
(4) Non-isotropic expansion or contraction of crystal grains.
(5) Volume changes due to reversible oxidation and reduction processes.
(6) Glaze fit.

(1) The first bodies to show serious thermal fatigue and, therefore, to be studied, were chemical porcelain and chemical stoneware. Both have a number of crystalline phases bedded in a glassy matrix. Each phase has its own expansion coefficient. For example, a typical porcelain may have the following composition:

mullite	30%
quartz	27%
glass	43%

the composition of the glass is estimated to be:

K_2O	20·7%
Al_2O_3	17·0%
SiO_2	62·3%

which is calculated to give an expansion coefficient of $10·3 \times 10^{-6}$. The expansion coefficients for quartz are $13·37 \times 10^{-6}$ parallel to the main axis and $7·97 \times 10^{-6}$ perpendicular to it, and for mullite aggregate it is $5·3 \times 10^{-6}$. During the firing a stress-free condition can be attained, but once the glassy phase has solidified during cooling each phase will tend to contract according to its own expansion coefficient. If the body structure is strong, stresses arise, if it is weaker, cracks may occur. The expansion coefficient of the finished porcelain is 3 to 4×10^{-6}. During any subsequent heating and cooling each component again tends to expand and contract according to its own expansion coefficient, but is restrained from doing so by the packing arrangement in the body; however, some relative and permanent movement may occur, so that the condition of the body at the end of a cycle is not identical with that at the beginning. Repeated cycling produces a progressive change leading to mechanical weakening of the body.

The presence of strains in a two-phase system has been demonstrated by Newkirk and Sisler (**N13**) in nickel–titanium carbide cermet. Owing to differences in thermal expansion of the two phases, the nickel is in triaxial tension and the titanium carbide particles are in triaxial compression. The brittleness and low impact resistance of the cermet is attributed to this.

Kingery (**K26**) has derived a formula to estimate the order of the stresses set up in two-phase materials in which the expansion coefficients differ. Using it, he estimates that in a silica glass containing 10% cristobalite, when cooled from 1200° C (2192° F), where it is assumed there is no stress, there would be stresses of the order of 1 000 000 lb/in² (70 307 kg/cm²).

(2) The abrupt volume changes associated with the α–β changes in quartz tridymite and cristobalite have long been a limiting factor in the heating and cooling rates permitted in silica refractories. The fastest safe rate being 5° C/min up to 600° C. Changes of modulus of elasticity, and of compressive strength on heating and cooling pieces containing silica, or consisting

largely of it, show sharp changes of gradient at temperatures corresponding to the α–β inversions. For instance, a body known to contain cristobalite has a relatively low modulus of elasticity. On heating through the α–β cristobalite inversion temperature there is an abrupt increase of the modulus, with a corresponding fall, at a slightly lower temperature on cooling. This is explained as being due to the sudden contraction of the cristobalite during cooling after firing, leading to stresses sufficient to produce a system of microcracks, that lowers the modulus of elasticity. On heating through the inversion temperature, expansion occurs, sealing the cracks and so increasing the modulus.

(3) Bodies of fairly high glass content are subject to a tendency for further devitrification to occur gradually at elevated temperatures. This is accompanied by a volume change and, if the temperature is not high enough for softening to relieve them, stresses will occur.

(4) Even in single-phase polycrystalline materials thermal stresses may be set up where the thermal expansion of individual crystal grains is anisotropic. (This will of course also contribute to the causes of stress in polyphase systems.) Certain titanates, e.g. aluminium titanate, and also rutile, show marked anisotropy of expansion coefficients and readily become severely cracked along grain boundaries during cooling of sintered pieces. The ratio between minimum and maximum expansion coefficients for rutile is 1·5. Even in alumina, where this ratio is only 1·08, sintered pieces of coarse crystal size may show such cracking.

The thermal expansion and contraction curves of materials of anisotropic expansion characteristics also show hysteresis. This can be explained qualitatively in terms of simplified models as shown in Fig. 2.13.

(5) Iron changes its state of oxidation with temperature, tending towards the ferrous state at high temperatures and the ferric at lower ones. Volume changes may accompany this and are not always completely reversible. In some cases the ferric–ferrous change is accompanied by other chemical changes which may have volume effects. Certain refractories containing iron show a small increase in size after each heating cycle, a process referred to as thermal ratcheting. The growth eventually makes the piece friable due to internal cracks.

(6) In glazed pieces the relative expansion coefficients of body and glaze play a very important role in the strength of the body. Seriously dissimilar expansions lead to the glaze defects of crazing and peeling (Chap. 6, p. 551) but even where such obvious incompatibility is not present, the glaze may weaken or strengthen the body, according to whether it has a larger or a smaller expansion. Reheating of glazed pieces that do not show glaze faults, but may have stresses, may also gradually give rise to crazing which weakens the piece and allows complete failure to occur.

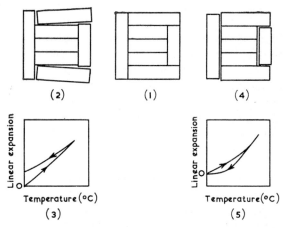

Fig. 2.13. *Model to illustrate mechanism of thermal expansion hysteresis in a single-phase polycrystalline body. Crystal grains are represented as rectangular prisms and are assumed to have large coefficient of thermal expansion in length direction and small in transverse direction. (1) Close-packed initial structure. (2) Fissuring caused by heating structure* (1). *(3) Thermal expansion hysteresis due to non-recovery of thermal expansion on cooling structure* (2). *(4) Fissuring caused by cooling structure* (1). *(5) Thermal expansion and contraction during heating and cooling of structure* (4). *In initial stage of heating large longitudinal expansions are taken up in fissures. As fissures close, expansion coefficient of body increases. At maximum temperature reached fissures are assumed to have welded together and on cooling the large longitudinal contractions dominate the contraction of the body until stress relief by cracking with reopening of fissures occurs*
(White, **W56**)

Thermal Gradients

The factor that most usually leads to failure because of thermal stresses is a thermal gradient. Heating or cooling of any piece can inevitably never be completely uniform. It must occur at the surface and be transmitted through the body to the centre, and it frequently occurs at only part of the surface. A thermal gradient therefore arises. During heating the surface expands more than the centre of the body and is therefore put in compression. During cooling the surface contracts first and so is placed in tension. Ceramic bodies are much weaker in tension than in compression and failure is found almost always to occur during cooling.

Calculations of the stresses that arise from thermal gradients are made using values of K, the thermal conductivity, α, the coefficient of expansion, E, the modulus of elasticity, μ, the value of Poisson's ratio, etc. But they always involve assumptions or approximations regarding the values of these properties, as they vary with temperature.

Mechanical Failure of Brittle Solids

As ceramic bodies have very much higher compressive strengths than tensile strengths, failure due to thermal stresses occurs when the stress at any point exceeds the tensile strength of the material. Failure takes the form of irregular cracking which probably originates at microcracks (Griffiths cracks) already present in the surface, and flaws present in the main body. Cracking is therefore dependent on the probability of the presence of flaws and has to be considered statistically, which accounts for the scatter of results usually obtained with ceramic bodies. Furthermore the probability of failure depends on the volume of material under stress, so that strength will vary with the method of test.

COLOUR

A large section of the ceramic industry produces ware whose appearance and hence colour is important. In Chapter 1 the choice of raw materials to produce ware that is white is discussed, and the materials for achieving coloured products are listed in Table 37, p. 146. Some aspects of the nature of colour, as relevant to ceramic ware, are therefore included here.

Colour is a visual perception. Light is that portion of the electromagnetic wave series that can be detected by the human eye. It is a range of wavelengths, each of which appears a different colour. If the eye receives all of the wavelengths together as in sunlight the sensation is termed 'white light'. An object is seen when light falling on it is reflected to the eye. If the white light falling on the object is in some way split up so that not all the wavelengths are transmitted to the eye the object appears coloured.

The main ways in which white light is split up so that the resultant light appears coloured are shown below. In every case the resultant colour depends on the spectral composition of the original light. As sunlight and the different forms of artificial light are not made up of the same wavelengths, in the same proportions, removal of certain wavelengths leaves a different residue and so the colour seen is not the same.

I. Absorption. Certain given wavelengths have exactly the right energy to cause a change of energy level of electrons in the substance; they are therefore absorbed and the rest transmitted. The observed colour depends on the length of the path of the light beam through the absorbing substance, *i.e.* if it is transparent its thickness, if in small particles their size, etc.

II. Selective reflection occurs at certain opaque surfaces depending on the refractive index and the extinction coefficient.

III. Scattering and diffraction occurs where very small particles are present if their radius is of the same order as that of the wavelength of light. Colloidal particles are of such a size. The particles scatter light of different wavelengths to varying degrees so that the longer red waves are transmitted and the shorter blue ones scattered.

IV. *Interference.* Light reflected by two surfaces only very slightly separated is out of phase and shows interference colours (**B96**).

The most usual cause of colour is absorption, but the others play their part too, particularly in certain special glazes (lustres, copper red, etc.). The actual definition and measurement of colour is complex and difficult. To the ceramist one of the main aspects is the appearance. Perception of colour involves three quantities:

(1) Brightness, as compared with a complete series of greys ranging from black to white.

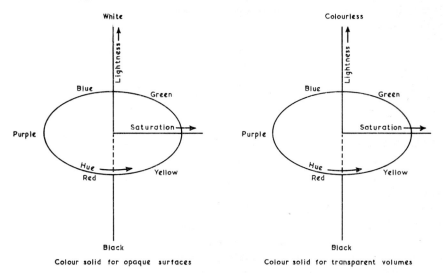

FIG. 2.14. *Dimensions of solids representing the colour perception of objects: lightness is represented by distance above the base line; hue by angle about the central black–grey–white axis; and saturation by distance from the axis (Judd, **J58**)*

(2) Hue. The normal eye can detect 150 hues in the spectrum and a further twenty hues can be made by mixing the two end members, namely red and violet.

(3) Saturation or tint, the extent to which the hue is diluted with white (**J58**).

The brightness is determined by the amount of light received by the eye, the hue by the wavelength or combinations of wavelengths, the saturation by the amount of alteration that white light has undergone.

The three attributes of colour perception can be represented by a solid cylinder as shown in Fig. 2.14.

The visual perception and matching of colours is made difficult by several factors. Firstly, the colour seen is relative and partly dependent on the area of colour and on its surroundings. For instance a grey object will look light grey when placed on a black background but dark grey when on a white one. Secondly most common colours are mixtures of pure spectral colours. In a given light the same coloured appearance may be achieved by different spectral mixtures but when viewed under a different light the mixtures will no longer match. The limitations of any colour measuring or matching device must therefore be carefully considered.

Colour produced by Absorption

The absorption of certain electromagnetic waves by atoms or groups of atoms is of major importance in colour phenomena.

The atom is made up of a small, heavy, positively charged nucleus, surrounded by a cloud of light, negatively charged electrons. These latter are distributed in a series of energy levels, and by absorption of energy they can move to a higher vacant energy level. Light energy is in quanta, which are the product of h, Planck's constant and ν, the frequency. In order to change its energy level an electron absorbs one whole quantum. As the energy difference between levels is fixed for any one atom so the wavelength of the light quantum absorbed is determined. Most of these electronic transitions require a lot of energy and can only be achieved with the high-frequency waves of the ultra-violet region. Such substances therefore appear white or colourless. But where the energy levels are close together absorption occurs in the visible region. This is so for the transition elements and for numerous organic compounds.

In ceramics the chief sources of colour are inorganic compounds. Elements whose compounds may appear coloured are shown in Table 55.

An atom forms a bond with a neighbouring atom by altering the course and energy levels of its electrons to a more stable mutual arrangement. This of necessity changes the difference between certain levels and so often changes the colour. *The colour of an atom in combination therefore depends on its environment.*

In *ceramic bodies* where the reactions initiated by the firing do not go to completion (see p. 197) the interaction of a colouring agent with other body constituents cannot be successfully predetermined. It is therefore desirable to introduce the colouring agent in a form that resists attack and it has been found that crystals possessing the same structure as the mineral spinel, $MgO.Al_2O_3$, are more resistant to silicates, borates, etc., than other compounds.

The spinel structure is unlike that of silicates (*see* p. 4) and borates (*see* p. 203). Although it is a complex oxide the bonds between the oxygen atoms and the cations present are essentially ionic so that the oxygens cannot be regarded as being attached to one sort of neighbouring cation rather than the others, and no complex ion can be distinguished. Oxygen ions are much larger than the majority of cations and they form a close-packed arrangement with the smaller positive ions in the interstices between them.

In the spinel structure, of empirical formula AB_2O_4 the oxygen atoms are in cubic close packing. In the true spinels the A atoms occupy tetrahedral holes and the B atoms octahedral ones. The crystallographic unit cell of the

TABLE 55

Elements frequently showing Colour

Group: IA	IIA	IIIA	IVA	VA	VIA	VIIA	VIII		
			Titanium	Vanadium	Chromium	Manganese	Iron	Cobalt	Nickel
				Niobium	Molybdenum		Ruthenium	Rhodium	Palladium
				Tantalum	Tungsten	Rhenium	Osmium	Iridium	Platinum
		Certain of the rare earths, *e.g.* cerium			Uranium				

Group: Ib	IIb	IIIb	IVb	Vb	VIb	VIIb
					Sulphur	
Copper				Arsenic	Selenium	Bromine
Silver	Cadmium	Indium		Antimony	Tellurium	Iodine
Gold	Mercury	Thallium	Lead	Bismuth		

spinel structure contains thirty-two oxygen atoms which surround eight tetrahedral and sixteen octahedral holes. In certain spinels the eight tetrahedral holes are occupied by half the B atoms and the sixteen octahedral holes by the A atoms and the second half of the B atoms distributed at random. These are termed 'inversed' spinels and should be represented $B(AB)O_4$.

The spinel structure is found for compounds AB_2O_4 of three types: (1) A^{2+}, B^{3+} (2) A^{4+}, B^{2+} (3) A^{6+}, B^+ referred to as the 2:3, 4:2 and 6:1 spinels (**W40**). The sizes of A and B are important factors, calcium, strontium and barium being too large to enter the structure, and beryllium too small (**S112**).

Some examples of spinels are:

	Normal structure	Inversed structure
2:3	$MgAl_2O_4$ $Fe^2Al_2O_4$ $CoAl_2O_4$ $NiAl_2O_4$ $ZnAl_2O_4$ $MgCr_2O_4$ $NiCr_2O_4$ $CdCr_2O_4$ $ZnFe_2O_4$ $CdFe_2O_4$	 $Fe(CuFe)O_4$ $Fe(MgFe)O_4$ $In(MgIn)O_4$ $Fe^3(Fe^2Fe^3)O_4$ (**W40**)
	Also other combinations of MgO, MnO, FeO, NiO, CoO, CuO, ZnO and CdO with Al_2O_3, Ga_2O_3, Cr_2O_3, Mn_2O_3 Fe_2O_3, V_2O_3, Ti_2O_3, Rh_2O_3 (**S112**).	
4:2		$Zn(ZnTi)O_4$ $Zn(SnZn)O_4$ $Co(SnCo)O_4$ (**W40**) $Mg(MgTi)O_4$ $Mg(SnMg)O_4$ $Fe(FeTi)O_4$ $Ni(NiGe)O_4$ $Ni(NiTi)O_4$ $Ni(NiSn)O_4$
6:1	$MoAg_2O_4$	(**S112**)

The reason for the stability of the spinels to attack by silicates, borates, etc., may lie in the close-packed oxygen framework which remains stable even if some of its cations are removed or replaced, so long as it remains electrically neutral. This probably prevents an attack from penetrating.

The spinels can theoretically be prepared by one of three methods:

(1) By fusing the correct quantities of the two oxides together.
(2) By completely sintering the well-mixed raw materials to a dense mass.
(3) By allowing the raw materials to react in the solid state at elevated but not excessive temperatures.

But as very high temperatures are required to fuse or completely sinter spinels the first two methods are impracticable because of the use of fuel, the tendency of the atmosphere to become reducing at such high temperatures and the reactions with the crucibles. The interactions in the solid state have been discussed (*see* pp. 173 & 190) and a porous product is obtained.

The colour of each spinel depends not only on the element that absorbs visible light but also on the other constituents because they alter the strength of bonds and the distribution of electrons, so that $NiO.Al_2O_3$ is not the same colour as $NiO.Ga_2O_3$, and $MgO.V_2O_3$ is different from $ZnO.V_2O_3$.

The spinels used to colour bodies are finely ground and then intimately mixed with the body. Many of them are resistant to further attack, others, for instance $CoO.Al_2O_3$, are attacked by silica but the result is nevertheless more uniform than if free oxides had been used (**S112**).

In *glazes* where interaction of the constituents is more or less complete (p. 197) the colouring ion is affected by the rest of the glaze make up.

In the consideration of silicate and glass and glaze structures it was shown that the main bulk consists of oxygen interspersed with the usually smaller cations and other atoms. Thus any particular atom is surrounded predominantly by oxygens. The colour of an atom is therefore directly affected by its interaction with neighbouring oxygens which in its turn is regulated by the other atoms adjacent to the oxygens concerned.

A number of methods have been proposed for calculating the effect of a cation on oxygen or for calculating the bond strengths between a cation and oxygen. Satisfactory qualitative results are obtainable but quantitative prediction of colour changes is not yet possible. Blair and Moore (**B76**) have evaluated the effect of a silicate melt on a colour centre using the calculations of colour factors suggested by other authors. They made up a series of stains containing a given colouring oxide in fixed molar ratio to colourless oxides whose cation was varied. They measured the reflectance in the visible region and from the area under the curve found the percentage reflectance. This was plotted against the colour factor, and the following curves obtained (Fig. 2.15).

The curves are all of the same type. The flat portions occur where the cations are either the network modifiers, Li Na, etc. (top portion), which relinquish oxygen, or the network formers, Si, P, B, which hold oxygen firmly; the steep central portion occurs where the cations are of the coformer type in the glass structure. (*See* p. 201 for glass structure.)

In the case of copper colours it is well known that the colour is noticeably different if alkali is present. Alkali increases the frequency of the light

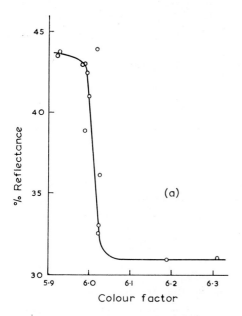
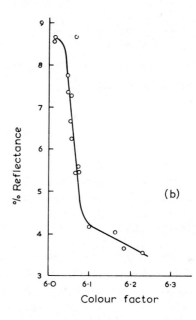
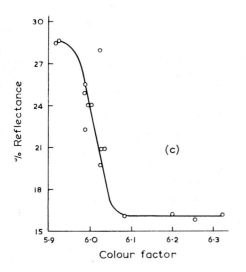
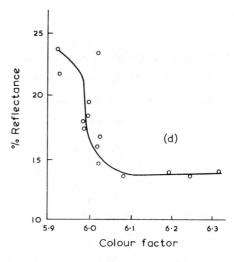

FIG. 2.15. *Relation between ionic potentials (proposed by Pincus, **P41**) and colours : (a) vanadium colours ; (b) cobalt colours ; (c) copper colours; (d) chromium colours (Blair and Moore, **B76**)*

absorbed from red to orange so that the resulting coloured glaze is bluish instead of green, thus:

$$1{\cdot}0PbO.0{\cdot}2Al_2O_3.2{\cdot}0SiO_2 + 1\text{--}2\%\ CuO$$

is emerald green,

and
$$\left.\begin{array}{l} 0{\cdot}1Li_2O \\ 0{\cdot}3Na_2O \\ 0{\cdot}3K_2O \\ 0{\cdot}3CaO \end{array}\right\} \left.\begin{array}{l} 0{\cdot}3B_2O_3 \\ 0{\cdot}1Al_2O_3 \end{array}\right\} 1{\cdot}8SiO_2 + 1\text{--}2\%\ CuO$$

is turquoise.

However, a turquoise glaze is little affected by changes of alkali content, and a green glaze is little affected by change of silicon or boron content; yet variation of the content of zinc, lead or aluminium leads to marked colour changes (**S126**).

Colour Effects by Scattering of Light

Particles of colloidal dimensions scatter light, they scatter shorter waves (blue) in preference to longer ones (red) so that the light transmitted is yellowish red and that scattered is blue. The amount of scattering is dependent on the particle size which therefore determines the apparent colour.

If, in addition to scattering, the material of the colloid absorbs selectively, as do metal colloids, a wide range of brilliant colours occur, varying with the relative contributions from the absorption and the scattering. For instance, gold has an absorption maximum in the green, a reflection maximum in the yellow and its colloidal particles scatter light. Sols containing very small particles of radius about 200 Å do not scatter very much light and they are ruby-red by transmitted light due to the absorption of the green. When the particles are larger, scattering plays a part and the sol appears blue or violet.

In the case of copper sols the very fine particles appear yellow, the intermediate ones red and the coarser ones blue. These effects are made use of in copper red (**M58**) and gold 'famille rose' glazes.

Colour effects produced by Selective Reflection and Interference

Finely divided metals, whether as free colloidal particles or in a very thin layer, absorb light of certain wavelengths and reflect the rest. This gives a colour effect. Also very thin transparent layers lying on a reflecting surface reflect light from both their upper and their lower surface, producing interference and thus splitting up light to its spectral colours (as is so often seen in a patch of oil on a wet road). Such iridescence is produced in lustres on glazes obtained by suitable application of finely divided metals, gold, silver, platinum, palladium, cobalt, bismuth, iron, aluminium, cadmium, etc.

Colouring Agents

The use of colouring agents to give a desired colour depends on the environment but in Table 56 a summary made by Wolf (**W81**) shows what materials will give certain colours under given conditions.

TABLE 56

Colouring Agents

Compiled by Wolf (**W81**)

Colour	Oxidising Conditions	Reducing Conditions
White	Magnesia, magnesite, alumina, calcium borate, titania, zinc oxide, arsenic, zirconia, tin oxide, antimony oxide, cerium compounds, metallic silver.	
Grey	Soluble salts of platinum, iridium, rhodium, palladium, ruthenium and osmium, iridium sesquioxide, antimony grey.	Carbon and organic compounds, nickel compounds, stannous oxide, vanadium salts, molybdenum compounds, metallic antimony with antimony oxides and stannic oxide, mixtures of the oxides of iron, chromium, cobalt, nickel, uranium, manganese and copper.
Black	Pyrolusite, iridium sesquioxide, mixtures of the oxides of iron, chromium, cobalt, manganese, nickel, uranium and copper; mixtures of chrome iron ore, pyrolusite and cobalt oxide.	Uranium oxide, iridium sesquioxide, nickel oxides, molybdenum compounds, bismuth salts, lead salts, carbon, carbides, sulphides.
Yellow	Titania, rutile, manganese oxide, pyrolusite, ceric oxide, vanadium stannate, chromates, lead chromate, barium chromate, zinc chromate, iron oxide, nickel oxide, molybdenum salts, silver salts, Naples yellow (lead–antimony oxides), litharge, praseodymium salts, uranium yellow, cadmium sulphide, gold sulphide, bismuth salts, metallic gold.	Vanadium stannate, praseodymium compounds.
Orange	Rutile, iron titanate, basic lead chromate, manganese oxides and titanates, iron oxide and chromate, chrome iron ore, manganese tungstate, uranium yellow, lead uranate, bismuth uranate, uranium titanate, cadmium sulphide and selenide.	
Red	Basic lead chromate, uranium yellow and uranium protoxide in lead glazes, lead uranate, bismuth uranate, manganese oxide, manganese pink, iron oxide and and salts, cadmium–selenium red, purple of Cassius, neodymium salts.	Copper red (colloidal copper).
Pink	Chrome–tin combinations.	
Violet	Pyrolusite, purple of Cassius, nickel oxide.	Rutile, metallic colloidal copper.
Deep blue	Cobalt and neodymium compounds.	Rutile, vanadous compounds.
Ice blue	Nickel–zinc oxides, copper oxide and compounds.	Rutile.
Sea green	Nickel–zinc oxides, copper oxide and compounds in lead glazes, cobalt antimonate.	Cobalt titanate, chromic oxide.
Leaf green	Chromic oxide, cobalt titanate, nickel–zinc oxides, copper antimonate and other compounds, praseodymium salts, copper–vanadium compounds.	Cobalt titanate, praseodymium salts.
Gold	Metallic gold.	
Silver	Metallic silver, platinum and palladium.	

Since completion of Chaps. 1 and 2, the third edition has been published of *The Chemistry and Physics of Clays and Other Ceramic Materials* by Alfred B. Searle and Rex W. Grimshaw in which much of the matter treated briefly here is set out in greater detail.

Chapter 3

Winning and Purification of Clays

PROSPECTING

Economic mining of clay depends on thorough prospecting and charting of the deposit by core drilling followed by sinking of shafts to recover sufficient clay for testing (1–2 tons, 1000–2000 kg). Where the clay is known to be near the surface small pits may be dug to obtain samples of the clay and to ascertain the area of the deposit.

The more usual method of prospecting is by core drilling (Fig. 3.1). The distance between drill holes is fairly large (800 ft, 240 m) in preliminary prospecting. The cores removed are investigated and any clay present tested. Once clay is found the holes are drilled closer and closer together (100 ft, 30 m) to obtain a complete chart of the clay bed (**B104**) (Fig. 3.2).

THE WINNING OF CLAY

The diversity of clay deposits demands a number of different ways of winning it. The primary kaolins of Cornwall, which are still intimately mixed with the parent rock, are washed out with a strong water jet. Other clays are quarried or mined in the solid state. The ball clays and other soft clays can be cut with picks and spades, pneumatic spades, and mechanical excavators. Hard fireclays have to be blasted. Many clays are sufficiently near the surface to enable them to be worked from open-cast pits, others are mined underground.

The extent to which the working of clay deposits can be mechanised depends largely on the regularity and depth of the deposit. Where a number of different materials lie in narrow beds on top of each other manual digging, assisted by pneumatic spades, and sorting of the pieces from different layers is essential. Where the exposed clay face is uniform a universal excavating machine with shovel, dragline, dragshovel, skimmer, etc., attachments or a multi-bucket, endless-chain excavator, or, for hard clays, a shaleplaner are used.

Fig. 3.1. Core drilling a test hole at Langley, U.S.A. (Huber, **H116**)

CHINA CLAY AND KAOLIN

It is proposed to describe the older method of mining and purifying Cornish china clay in some detail, because although the mining method is not used elsewhere the method of purification is fundamental and shows clearly the basis for newer methods. More modern methods will be described more briefly thereafter.

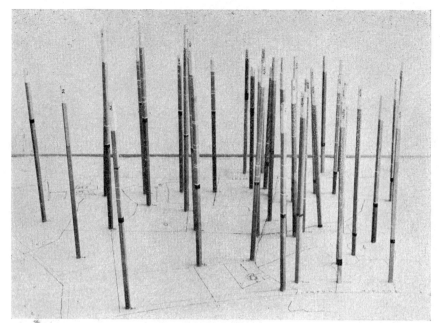

FIG. 3.2. *Stick model showing section of a prospected pit (London Brick,* L41)

Cornish China Clay

The Cornish china clay deposits are unique as regards their extent, depth and purity. The kaolin is still in the position in which it was formed, and therefore in intimate contact with unchanged granite, quartz, feldspar and mica. The tremendous depth of these deposits—one pit is being worked 360 ft (110 m) (1948) below the surface—gives reason to suppose that the disintegrating agents were gases emanating from below rather than descending surface waters.

The average size of the majority of pits is 8–10 acres (3–4 hectare) in extent and 150–200 ft (45–60 m) deep. (Fig. 3.3) The overburden, which has an average depth of 12–20 ft (4–6 m), is removed mechanically. If it is shallow, bulldozers and scrapers are used, and if deeper, excavators and dumpers are needed. It consists of surface soil and stained clay. Beneath this the material has a greyish-white crumbly appearance. On the pit sides veins of undecomposed granite or quartz from a few inches to several feet thick may be seen. The clay may need initial loosening by blasting, but once a steep-sided pit has been formed most of the work is done by powerful water jets. Water is supplied under pressure through steel pipes to a ball and joint mounted nozzle, known as a 'monitor' and directed against the pit sides (Fig. 3.4). The slurry produced is directed to sand pits situated towards the bottom of the clay pit. Here the coarser impurities and sand settle out. The sand pits are worked intermittently, and the accumulated sand removed and sent by truck to the top of the conical dumps. There are

WINNING AND PURIFICATION OF CLAYS

FIG. 3.3. *General view of the Lee Moor china clay pit near Plymouth (English Clays, Lovering and Pochin,* E20)

from 5 to 8 tons of waste for every ton of clay. The clay slurry then flows through shallow channels, called 'drags', where more of the coarse sand grains settle out.

The clay slurry is pumped to the surface for further refining. The old Cornish pump, previously used in all the pits and restricting their expansion, has (in 1948) been superseded in all but two pits by the much more mobile centrifugal pump. The slurry is delivered to a semicircular pit about 30 in. (76 cm) in diameter and 2–3 ft (0·6–0·9 m) deep in which some settling may occur, it then flows into the 'micas'. In these long shallow concrete troughs, some 150 ft (45 m) long and 30 ft (9 m) wide, subdivided into 20 in. (50 cm) wide channels with transverse baffles every 25 ft (7·5 m), the rate of flow is so reduced that virtually all the fine sand and mica flakes settle out. The settled product may be run to waste, or sold as 'mica-clay' or 'mica'. The purified clay slurry, now containing about 2% solids, next goes to the settling pits. These are circular tanks 20–40 ft (6–12 m) in diameter, 15–20 ft (4·5–6 m) deep and cone-shaped for about 4–5 ft (1·2–1·5 m) from the bottom. As settling occurs water is drawn off from the top until the slurry has about 10–12% solids. It is then piped to a convenient rail or sea dispatch point before final drying. This involves a journey under gravity of up to 10 miles (16 km). (Fig. 3.5)

The old method of treatment of this slurry which is still in use at some works is to allow it to settle further in large tanks for eight weeks. The resulting mass of buttery consistency (55–65% moisture) is then shovelled into the drying kiln. This is a long building, some 200–350 ft (60–110 m)

Fig. 3.4. *Hydraulic mining of china clay. Washing down the clay from the face by high pressure water jets* (*Watts, Blake, Bearne*, **W23**)

Fig. 3.5. *Mica drags and settling pits* (*English clays, Lovering and Pochin*, **E20**)

long and 15–50 ft (4–15 m) wide with porous fireclay floor tiles. It is heated by flues running under the floor. The clay takes about twenty-four hours to dry at the hot end and from three to four days at the cooler one. As it dries it is scored across into squares by a cutter. This process involves very heavy and messy work and a lot of fuel. It is largely being superseded by filter presses (see p. 702) working at about 100 lb/in² (7 kg/cm²). The clay from the settling tank is efficiently dewatered to 30–35% moisture and then requires less time and fuel to complete the drying in the kiln.

Fig. 3.6. *Throwing shredded clay to stock heaps, a method that ensures uniformity* (*Watts, Blake, Bearne*, **W**23)

The dried clay is then shovelled, manually or mechanically into the store or 'linhay' which generally lies alongside and a little below the kiln floor. Clay destined for export is usually crumbled or ground and bagged (**C70, A49, T38, A122**) (Fig. 3.6).

These fundamental steps in the purifying of crude china clay are the basis for modernisation.

The Board of Trade Working Party Report (1948) (**T38**) on the china-clay industry strongly advocates mechanisation to reduce labour and fuel costs and speed up production. Intensive research has been going on and has been applied to Lee Moor, the largest clay pit in the world, and the adjacent Whitehill Yeo pit. The manual scraping and shovelling of sand, mica and wet clay has been eliminated and production stepped up from 400 tons to 2000 tons a week using the same labour force. The new process is as follows:

The clay is washed down from the pit face to the sand pits as before. Here a mechanised scraper removes the coarse sand and quartz directly into self-dumping skips which carry it to the top of the surface dump.

FIG. 3.7. *Rotary dryer for china clay* (*Watts, Blake, Bearne,* **W23**)

The clay slurry is run underground straight to the first of three new refining tanks lying on the hillside below the clay pit. In these tanks the liquid is kept revolving slowly by means of a rotor arm near the bottom. The increasingly pure clay slurry slowly passes through the second and third tanks and then through fine screens. It is next piped to the works from which it can be dispatched. Here it passes through highly-vibrating screens which are fine enough to remove all remaining impurities (the test screens for the final product have 90 000 holes/in^2 (13 950 holes/cm^2). It is then passed into the hydraulic filter presses which deliver rectangular slabs of clay weighing 1 cwt each.

These slabs are mechanically broken up and fed by overhead conveyor belts into the three giant rotary kilns (Fig. 3.7). Each kiln, heated by steam from the new power-generating plant, contains twenty-six revolving trays, through which the clay passes gradually (for an hour) until it emerges ready to be packed in self-sealing bags, and marketed (Fig. 3.8) (**A148**).

Kaolin

MINING

The winning of crude kaolin from all the other deposits in the world is by more conventional methods of removing the solid material. It is mined in both open pits and from underground shafts.

The famous and extensive kaolin deposit at Zettlitz in Bohemia is one of a

WINNING AND PURIFICATION OF CLAYS 243

FIG. 3.8. *Bagging and storage of finely ground airfloated clays* (*Watts, Blake, Bearne*, **W23**)

series that lie along the WSW to ENE fault that gives rise to thermal springs. It is therefore not surprising that the kaolinised rock extends downwards to an as yet unfathomed depth, and that the disintegration of the primary granite was presumably achieved by uprising gases or waters. Zettlitz hill lying near Carlsbad is apparently entirely kaolinised and is surrounded by a containing wall of granite. The overburden consists of Tertiary rocks containing brown coal, followed by a layer of coarse sand, and then a layer of cemented sand.

Mining of the Zettlitz kaolin is affected by the proximity to the fashionable watering resort of Carlsbad. Since open-cast mining is unsightly, shafts have to be sunk. These, however, may not be sunk lower than the level of the adjacent thermal springs because of the danger of tapping them.

The mined kaolin contains quartz, feldspar and mica, of which the particles are mostly large enough to be separated out by screens. The rest is removed in settling tanks. After this washing the kaolin contains only about 4% fine sand. It is then filter pressed (**H115**). The better qualities of mined product contain 40–60% of useful kaolin (**R51**). The average is about 30%.

The Zettlitz kaolin is remarkably plastic for a primary kaolin.

In the United States there are important kaolin deposits in the pegmatite dikes of the Southern Appalachians in North Carolina and Georgia (**W25**); other residual kaolin deposits occur in California, Delaware, Idaho and Vermont; Georgia and South Carolina have very important sedimentary kaolin deposits (Fig. 3.9).

FIG. 3.9. *Mining crude kaolin at Huber, Georgia (Huber,* **H116**)

Mining methods of course depend on the nature of the deposit. In Chodau near Carlsbad (Czechoslovakia) where the clay lies in an almost horizontal bed 60 ft (18 m) thick beneath an overburden of 30 ft (9 m) (**I2**) and in Georgia and South Carolina where the clay bed is 5–40 ft (1·5–12 m) thick and the overburden 10–60 ft (3–18 m) (**H116**) mechanical stripping of first the overburden and then the clay bed is possible. The mined clay is transported to the processing plant by continuous rope railway (Chodau) or trucks and trailers.

In the pegmatite dikes the situation is different. This kaolinised pegmatite occurs as inclusions of irregular shape in the massive gneiss of the mountains. Each deposit must therefore be prospected and worked individually. Both open-cast and shaft mining are used, the most economical method is frequently a combination of the two using open-cast methods for the first 20 to 30 ft (6 to 9 m) and shafts thereafter.

Although the dikes dip at an angle from 70 to 80 degrees, the greater ease of sinking vertical shafts has made it the practice followed. The shaft is usually started in the hanging wall adjacent to the dike so that it cuts into the dike from 2 to 3 ft (0·6–0·9 m) below the surface and strikes the footwall from 50 to 75 ft (15–23 m) below the surface. The shafts are circular from 14 to 20 ft (4–6 m) in diameter, and from 40 to 110 ft (12–33 m) in depth. The dike material and the disintegrated gneiss that forms the walls are both friable, so timbering has to be placed as rapidly as the shaft is sunk.

When the shaft has been sunk robbing is commenced from the bottom upwards. Short tunnels are dug from the bottom of the shaft as far as is safe without timbering, the pillars between them are then carefully removed and the robbed level filled with waste material. Some of the timbering of the shaft is then removed to expose another level.

PURIFICATION

Most of these crude kaolins are processed by first bringing them into aqueous suspension by suitable blunging, 'washing', disintegrating and dispersing operations. The purification may then proceed on lines similar to those used in Cornwall, with progressive sedimentation of the impurities assisted by a sand wheel (**W25**) followed by settling, or filter pressing and drying.

Consideration of the old methods of allowing impurities to settle out of clay slip in various troughs and pits shows that not only are they time, space and labour consuming but also wasteful of clay which adheres to the settled impurities. Newer methods attempt to overcome these faults.

Dorr Bowl Classifier. A method much used in western North Carolina, Georgia and Florida, involves the use of the Dorr bowl classifier (**I2, A189**). This classifier consists of two main parts, the bowl and the washing compartment. The clay slip is led into the bowl centrally; there, settling of the sand occurs and the purified slip slowly overflows as a thin film over a strip which extends completely around the circumference of the bowl. The size of bowl required depends on the volume of clay slip to be handled and the necessary slowing down of its flow for the finest sand to settle out. It can be worked out accurately.

The sand deposited on the bottom of the bowl is moved towards the centre by plough blades attached to a set of slowly revolving arms. It then passes through to the washing compartment against a counter-stream of clean water which removes adhering clay. The process is completed in the washing compartment where the sand is raked through a spray of water. The Dorr bowl classifier therefore produces not only sand-free clay but also clay-free sand and no clay is wasted (**A190**). The clay slip is subsequently filter pressed and dried. (Fig. 3.10)

Continuous Centrifuge. Time, space and labour can all be saved by

246   INDUSTRIAL CERAMICS

accelerating sedimentation by centrifuging, particularly if a continuous centrifuge is used (**H116**).

The Bird continuous centrifuge consists essentially of a conical rotating vessel fitted internally with a screw conveyor rotating at a slightly different speed. The slurry to be separated into fine and coarse-grained portions, or solid and clear liquid, is fed in centrally. The coarser particles are thrown to

(a)

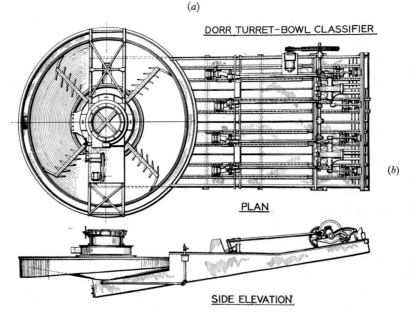

(b)

FIG. 3.10. *Dorr bowl classifier, with turret type drive for bowl rakes:* (a) *photograph;* (b) *diagram* (Dorr-Oliver, **D54**)

WINNING AND PURIFICATION OF CLAYS 247

the walls of the vessel and brought to the neck by the screw conveyor, being washed by the draining liquid on the way. The liquid or fine suspension discharges at the wider end (Fig. 3.11).

The machines are clean and compact, as shown in Fig. 3.12, and handle large quantities of slurry quickly. The large machine taking up 150 ft² (14 m²) handles 100 tons of material per hour (**V12**).

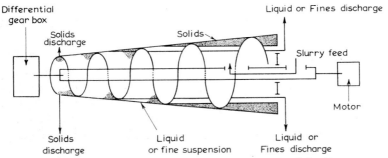

FIG. 3.11. *Schematic diagram of continuous centrifugal classifier*

FIG. 3.12. *Bird continuous centrifugal classifiers, in the foreground a* 24×38 *in. machine, and two* 18×28 *in. machines in the background* (*Vickery*, **V12**)

Electro-osmosis. A method of purifying and dewatering fine kaolin that was introduced and fairly often undertaken in Continental Europe in the second and third decades of this century is the electro-osmosis method. The method, evolved by Count Schwerin in about 1910, proved to be very expensive and in 1931 it was estimated that at ½d. per unit of electricity the power cost per ton of purified clay produced was 3s. 8d. (**I2**). In 1953, there was apparently only one European firm using the method. As it has, however, certain advantages it is worth describing for possible use in regions where cheap power is available.

The mined clay is brought into aqueous suspension and passed through sand troughs where the sand and coarse mica are removed, the clay slip passes on to settling channels some 525 ft (160 m) long, 1 ft 7 in. (0·5 m) deep and 4 ft 3 in. (1·3 m) wide. Here it is allowed about 70 min to settle from a thick thoroughly dispersed slip containing about 31% solids. The settled clay is regarded as second grade, it is dried and used chiefly as paper filler.

The overflow from the settling trough carrying the high-grade clay passes on to the electro-osmosis machines. These consist of a semicircular trough containing a half-cylinder cathode grid and a central rotating anode 4 ft 8½ in. (1·5 m) long by 2 ft (0·6 m) diameter. It is fed with direct current at 170 A and 100 V. Sodium silicate is used as electrolyte. Electropositive impurities such as free iron and titanium oxides and sulphides should collect on the cathode grid, while the electronegative clay attaches itself to the surface of the drum from which it is scraped in a sheet $\frac{1}{8}$ in. (3 mm) thick. It is essentially a dewatering process. In 1924 it was found that only 9% of the suspended clay material is so removed (**H93**). The overflow is partly returned to the blungers for reprocessing, the remainder being filter pressed. Both the electro-osmosis and the filter-pressed products are thereafter pugged, dried and bagged (**I2**).

The electro-osmosis method was originally developed to purify clay because it was erroneously thought that whereas the clay particles are negatively charged the quartz, mica and iron ore particles are positive and so would separate on the cathode. Unfortunately this is not so, and has resulted in the process only being useful for dewatering. Its use is therefore not general (**C72**).

The European firm still employing an osmosis machine from the Count Schwerin Company to produce a special high-grade clay gives the following figures: the power consumption for the preparation of the clay suspension and its dewatering by osmosis is 110 kWh/ton, and the rate of production 2·5 tons in 8 hr (**G90**).

In Great Britain the method was investigated in about 1912–14 (**O8**) but not taken up.

In the United States Curtis (**C71**) (1938) has shown that in very fine-particled clays, where filter pressing is difficult, the electrical dewatering is successful and suggests that where power is cheap it should be considered.

Another electrolytic method in use at Chodau (Czechoslovakia) is the electro-osmosis filter press, which combines the normal filter press with electrical dewatering by placing an anode in each filter chamber and a cathode behind the filter cloth. The former assists in forming the filter cake and the latter attracts the water (**C72**).

Rotary Vacuum Filter. An economical dewatering process for clays is with rotary vacuum filters. This method has been used successfully in the United States and manufacture of machines began in England towards the end of 1952. The water content is reduced from about 70% to 25 to 30% during the passage round an evacuated drum dipping in the slip. The filtering rate is about 8 lb of dry solids per square foot of filtering area (39 kg/m^2) per hour (*see* p. 705).

Soluble Salts. A matter that must be considered carefully in connection with all methods of purifying clay suspended in water is whether or not it is desirable to alter the soluble salt content of the clay during the purification. The processes described all involve two stages, first preferential sedimentation of the impurities leaving the clay in suspension, then separation of the clay from the water.

The first is greatly assisted by complete deflocculation of the clay so that the particles to be held in suspension are as small as possible and the slip mobile. In this condition, however, the purified clay will not readily settle out. Subsequent flocculation will accelerate the settling. Both deflocculation and flocculation involve adding chemicals to the clay. By careful selection this can be used to beneficiate the clay but a wrong choice may be harmful.

The most usual deflocculant is sodium silicate. This is sometimes enhanced by an acid. Flocculants include acids, calcium, aluminium and ferric iron salts and organic hydrophilic colloids, *e.g.* starch. If the final product is required to have a definite alkaline pH value but the salt content is immaterial, the process can be achieved by deflocculating with sodium silicate and acid, and then flocculating with acid which need not be added in sufficient quantity to neutralise all the alkali present (**P5**). Use of calcium, aluminium and iron salts, and organic colloids is precluded if the clay is to be used for casting slips thereafter; and iron salts may colour the clay. The advantage of starch, etc., is that it burns out in the firing so that no soluble salts remain in the product.

Plants often have cause to require their clay in a clean condition but without added chemicals. Such production has been developed in the United States. The lump clay from the mine is blunged with synthetically purified water, from which the hardness-producing salts have been removed. Care is taken to blunge until all the clay particles have fallen away from the impurities and separated thoroughly. The slurry is then screened and passed through magnetic separators. It is dewatered by evaporation, as settling or filter pressing removes soluble salts with the water. The purified slip is sprayed into a drying chamber against a flow of dry hot air and the dry clay particles collected (**A75**).

250   INDUSTRIAL CERAMICS

Where it is undesirable to add flocculants but loss of salts in water that is removed does not matter, the settling may be accelerated by use of hot water (up to 80° C); this also improves the plasticity of the product. If, however, pyrites are present the hot water may convert them to iron sulphates which discolour the body (**A114**).

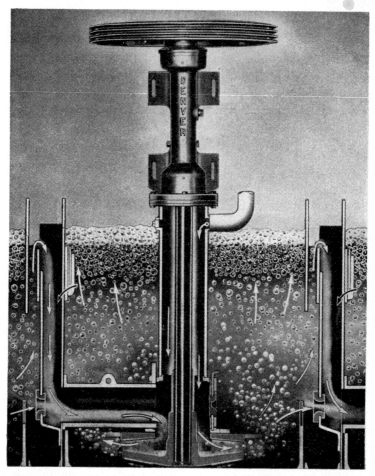

FIG. 3.13. *Section of a froth flotation cell showing the three zones:* Bottom zone: *intense agitation and aeration. Pulp is intimately mixed with air.* Central zone: *separation; here the bubbles attach themselves to mineral particles and rise to the top while worthless gangue particles sink to the bottom.* Top zone: *concentration takes place as millions of tiny bubbles accumulate on the surface. Bubbles and their valuable load of mineral are raked off the flotation cell. They are then dewatered, dried and are ready for the mineral market* (Denver, **D27**)

Froth Flotation. A method of purifying kaolin based on a quite different principle is that using froth flotation. It was introduced in the 1930s. Froth flotation depends on the different surface activities of minerals towards given reagents. These, known as collectors, are generally organic reagents with a polar group which adheres to the mineral, while the rest of the molecule is water repelling. A particle thus coated will tend to collect at an air–water interface. Addition of a frothing agent followed by passage of air will therefore collect the mineral most affected by the 'collector'.

One example of the use of froth flotation is in the separation of kaolin from quartz that is so fine that it will not separate effectively by sedimentation. Here the clay slurry of 10% solids is brought to pH3 and a dispersant, sodium lignin sulphonate, added and agitated; the collector, lauryl amine hydrochloride, is then added and after agitation the kaolin-bearing froth removed (**K13**).

The method can also be used to facilitate the removal of fine clay and mica from the coarser impurities (**K5**), or to separate the mica in a pure form (**S209**). It is, however, more generally used for non-clay minerals.

Froth flotation is developing rapidly and involves a large number of chemicals, most of which have trade names. Suppliers of these can give better information on their use than a work of this general nature. (Fig. 3.13)

Dry Methods. There are, too, a number of processes that dispense with wetting the clay to purify it. They employ some mechanical means of pulverising the crude clay combined with drying and air separation.

Crude clays containing some 20 to 25% moisture may be dried to 1 to 2% in a rotary kiln before being pulverised. Alternatively the clay may be dried during the pulverising process by introducing hot air into the mill. This is known as 'flash drying'. The turbulence of the air and powder and the fact that dried surfaces are immediately ground off, exposing the underlying moist parts, make it a very efficient heat-exchange system and the drying is almost instantaneous. Pulverisation of a very sticky clay can be assisted by introducing fine dry clay into the mill. The dry or partially dry clay is pulverised in ball mills, roller mills, imp mills, etc.

The pulverised material then passes to an air separator, which is essentially a system for separation by sedimentation from air instead of water, the fine product being removed. One such machine is illustrated in Fig. 3.14. Description of its operation will readily show why such airborne separation is speedier than sedimentation from slow-moving water. The material to be separated, fed in through the hollow shaft, is spread out by the rotating disk. Here it meets an upward rotary current of air, circulated by means of a fan and whizzer blades which force air down the outside of the inner cone whence it enters the cone through the deflector vanes. The largest particles drop down under gravity, the other oversize particles are thrown to the cone sides by the centrifugal action and drop towards the tailings spout. The fines are carried upwards and then down between the cones to their discharge tube. The desired fineness of the product is obtained by varying either or both of the number of fan blades and whizzers and also the speed of the separator (**I7**).

The disadvantage of dry grinding followed by air separation, when compared with disintegration of a crude clay by blunging with water, is that although the mill will grind soft clay material in preference to hard impurities, the latter will be somewhat affected and will contaminate the clay. There is also the danger of introduction of foreign, metallic matter produced by wear of the mill.

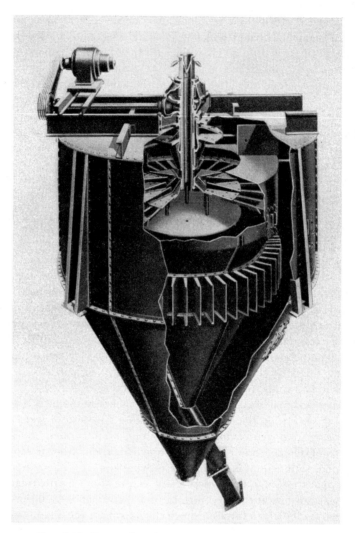

FIG. 3.14. *Raymond mechanical air separator* (*International Combustion*, **17**)

In practice many machines are built with mills and air separators in closed circuit so that the ground product has the particles of the correct fineness removed and the oversize returned to the mill for regrinding. The process thereby ceases to be one of purifying and becomes merely a method of obtaining a well-mixed product of uniform particle size. Where, however,

WINNING AND PURIFICATION OF CLAYS

the impurities are both larger and harder than the desired product, *e.g.* sand and quartz in kaolin or clays, the circuit can be fitted with a 'throw-out attachment'. This device skims the oversize material being returned from the separator to the mill and removes the largest particles. This greatly increases the usefulness of dry grinding.

SEDIMENTARY CLAYS

All clays other than the primary or residual kaolins are secondary or sedimentary. They therefore occur in layers. The depth of the layers and their variation of composition from one to another depend on the conditions prevalent when the beds were laid down and are different in every pit.

In some cases it may be most desirable to keep the clay from each layer separate and use it for different purposes. More supervision and care are therefore needed in the excavation and it is often necessary to do it by hand. In other cases a uniform mixture of the layers is required. Such pits benefit greatly by mechanical devices that scrape the whole exposed surface, *e.g.* the dragline and the shaleplaner.

British Ball Clay

Ball clays occur only in Dorset, North and South Devon and Cornwall. They include clays for a number of purposes and are chiefly subdivided into the white burning clays suitable for earthenware and the dense burning off-white stoneware clays. Here it should be noted that many clays have a white fired colour when porous, but if fired to a higher temperature so that the product is dense it is coloured. The stoneware clays, then, are those ball clays with more fluxes present. The clays therefore have to be worked and sorted carefully to obtain the best benefit from the layers of pure clays. Both open-cast and shaft mining are employed. Mechanisation has only been introduced of very recent years.

In open-cast workings the overburden is stripped off and the bed laid bare. Where manual digging is still used this is cut across into 9 in. (23·4 cm) squares with a spade and the blocks or 'balls' dug out about 8 in. (20 cm) deep with a tool known as a 'tubill' (**S46**). They are then passed to the surface on spiked tools, 'poges', or hauled up in tubs, etc.

In the more modern pits the overburden is broken up by a ripper and removed with a scraper drawn by tractors (Fig. 3.15). Pockets of gravel are removed with an excavator. The clay is dug with pneumatic spades, and then removed to stock heaps by dumpers (Fig. 3.16). Different superimposed layers can be worked independently if the pit face is shelved as shown in Fig. 3.16. The pits must be drained and even then winter working is not possible everywhere

Shaft mining for white burning clays is done differently in North Devon and Dorset than in South Devon. In South Devon a vertical rectangular

Fig. 3.15. *Open-cast working showing veins of clay over 30 ft thick, and the mechanical removal of overburden by modern methods* (*Watts, Blake, Bearne*, **W23**)

Fig. 3.16. *Hand selection of clay by using pneumatic spades* (*Watts, Blake, Bearne*, **W23**)

WINNING AND PURIFICATION OF CLAYS 255

shaft is sunk to the required depth, the average being 120 ft (40 m). The size of the shaft depends on the haulage system employed. In the old pits it is 4×6 ft ($1 \cdot 2 \times 1 \cdot 8$ m) and is subdivided into a square portion 4×4 ft ($1 \cdot 2 \times 1 \cdot 2$ m) for bucket haulage, and 4×2 ft ($1 \cdot 2 \times 0 \cdot 6$ m) portion with ladders for the miners. Where modern cage gear raises the whole wagon of clay, the shaft size must be adjusted accordingly. The sides are heavily timbered.

FIG. 3.17. *Underground mining. Work at the clay face showing the careful method of winning the clay (Watts, Blake, Bearne, W23)*

When the clay bed is reached, levels are driven along it, if possible slightly upwards to facilitate drainage. The clay is removed to give a passage 6 ft (1·8 m) high and 7 ft (2 m) wide which is timbered or supported by steel hoops. In older works they seldom extend more than 90 ft (27 m) from the shaft, in the newer ones they may be 300 ft (90 m) long. In these latter transport is by wagons running on narrow-gauge rails.

The clay is won by hand or mechanical picks and spades, and then loaded in wheelbarrows or wagons for removal to the shaft (**A124, S46**) (Fig. 3.17). Only some shafts require water pumping. Ventilation may be achieved with an intermittent fan or by allowing some of the water pumped up to return down the shaft.

In North Devon and Dorset the shaft commences practically at the outcrop and follows the dip of the clay bed, the inclination being about 75° at the surface but getting less as the depth increases. The vertical depth of the deepest shaft in 1929 was about 350 ft (110 m) while its length was about 600 ft (180 m). The working levels are driven horizontally into the clay bed and the clay brought to the surface on a tramway running in the shaft.

Shaft mining requires skilled labour, and also it is generally impossible to remove all the clay from a level as can be done with open-cast methods. However, it has to be used for many of the valuable ball clays required for white ware (**T37**).

The clay produced by either method is hand sorted into grades on reaching the surface. A certain amount of it is then weathered (**S46**). Much ball clay is dispatched in the 'balls' as dug, and is treated by blunging and sieving at the pottery works. In mechanised mining and processing works, however, the clay is also dried, shredded and then pulverised. The fine particles are separated by air flotation and then bagged under pressure without allowing dust to get into the atmosphere. They are then automatically weighed ready for dispatch.

The dried, ground and bagged clay has found more favour on the export market where the higher cost is offset by the saving in freight charges. The dried clay is, of course, in any case easier to handle; being of known moisture content and arriving in weighed quantities, it is very suitable for dry mixing and making up of special bodies. It is considerably used in the United States. It does, however, remove the necessity for a slip house and is therefore of less interest to those manufacturers who have recently modernised their slip houses and making shops (**A123**).

A new German clay digger has recently been described (**F33**). It is designed for the careful digging of good quality clays to replace manual digging and pneumatic spades. The cutter consists of an electrically driven motor mounted above two blades set at right angles to each other. The blades are driven downwards alternately. The machine weighs 35 lb and consumes 1 kWh of power to cut 10 tons of clay; it is cheaper, quicker, quieter and gives less shock than other machines for the same work.

Fireclays

It is said that a great deal of Great Britain has a substratum of fireclay, it is also fairly widespread in the United States and Canada, but occurs less frequently in Continental Europe. The methods of working are again varied, both open-cast and underground methods being employed. The fireclays are usually too hard to be dug and have to be blasted, or scraped with a shaleplaner.

In open-cast mining the overburden is removed by power shovels, bulldozers, scrapers, etc., and may be finished off by hand. The clay face is then drilled and blasted. Where sufficiently large areas yield the same clay it can be loaded mechanically (**B104**), but where the different layers are narrower, hand sorting and loading ensures a more uniform product.

Underground mining of clay is undertaken when the first three or the last of the following conditions exist:

(1) When the material is of sufficient value to warrant such method, for instance, fireclays which are to be used in the production of comparatively high-priced clay products; sewer-pipe clays, the quality of which is superior to those occurring on the surface, or for building brick when the material will produce brick of higher quality than can be had from surface materials in close proximity and can, therefore, command comparatively higher price.

(2) When the ratio of the thickness of desired material to overburden is less than 1 to 3.

(3) When the beds of the material are at least 4 ft (1·2 m) in thickness.

(4) When the clay beds occur in association with coal (**M12**).

Another factor is the relative land value for the district where the clay is found (**K80**).

Taking point (4) first, much of the British fireclay occurs above and/or below coal measures and is mined either simultaneously with the coal or by the same methods. Where the clay occurs by itself there are two main methods of mining—the advancing, and the retreating. There are also advocates of a combined method.

Both systems involve main entries, room entries and pillars. In the advancing method the material is retrieved as the work progresses, side rooms being worked as the main entry is cut. This makes clay immediately available but means leaving larger pillars and putting in more timbering to keep the working safe.

In the retreating system a main entry is cut and timbered from the outcrop or the shaft to the limit of the deposit. The clay is then removed from rooms beginning at the furthest end. Pillars are left until the room is finished but can then often be removed. More of the clay can therefore be won from the deposit although it takes longer to get a new working into production.

In order to overcome this initial wait for material a combined system is used. The advancing method is used while the main entry is driven through to the limit of the deposit. Thereafter the retreating method is used (**M12**).

Most underground clay is hard and has to be blasted. It is then loaded into trucks; this may be done mechanically, except where careful sorting is required.

The question of purification of sedimentary clays is naturally very different from that of residual clays. Each layer contains those components that have settled out under identical conditions, so that if the mining is carefully done a uniform product that does not contain particles of rock, etc., can be produced directly. The question of separating the minerals therefore does not often arise. The most pressing problems are impurities containing iron compounds, calcium carbonate, certain soluble salts, etc. These will be considered later on.

Certain clays, however, have a high content of fine quartz, and are sometimes worth separating. Sedimentation methods are clearly inapplicable but froth flotation is suitable.

Other Secondary Clays, Brick Clays, Shales, etc.

Clays for cheaper ware are common and widespread. Their winning must be as simple and inexpensive as possible, and is therefore usually by open-cast mining and frequently in small pits. The hardness of such clays varies considerably, so that while some may be dug others have to be blasted or scraped. For cheaper products sorting of the layers is often unnecessary, so that mechanical excavating is suited to the task.

Bentonite

Bentonite does not occur in Great Britain. In the United States there are widespread occurrences, the chief being in Wyoming (Fig. 3.18).

It usually occurs in stratified deposits of from a few inches to 4 ft (1·2 m) thick sufficiently near the surface to be mined by open-cast methods. The overburden is removed mechanically and finally swept off to give a clean surface. The bentonite is fairly hard if its water content is below 30% and is broken away with picks. The bentonite is loaded into trucks and removed to the plant where it is crushed and dried (**S77**).

The Californian shoshonite and amargosite consists essentially of montmorillonite and is therefore a bentonite. Its characteristics, however, differ from those of the Wyoming bentonite. It occurs in beds from 4 to 12 ft (1·2–3·7 m) thick under 50–100 ft (15–30 m) sandstone, clay and marl. This is mined from drifts. On exposure to the arid air the bentonite crumbles. In order to prevent this the workings are provided with airtight doors. After removal from the mine the bentonite is exposed to the air for drying (**D13**).

MACHINERY USED IN CLAY WINNING

The use of mechanical digging, scraping, planing, etc., devices has been mentioned. They may be generally classified as follows.

(1) Power shovels: steam; electric; diesel.

(2) Scrapers—taking fairly deep slices of from 6 to 8 in. (15–20 cm) from the horizontal or slope: wheel scrapers, animal or tractor drawn; cable-drawn scraper, dragline; scraper working from boom, skimmer.

(3) Planers—taking a thin slice from the horizontal, slope, or vertical: shale planer; clay planer.

(4) Others: dredges, etc. (**M12**).

The choice of machine depends largely on the hardness of the material, the slope of the deposit, whether there is water in the pit or not and the condition in which the excavated material must be delivered to the plant (**A26**). The commonest is the power shovel which will dig soft clays or gather up harder ones that have been shot down by blasting. It is strong and the capacity is from $\frac{3}{8}$ yd³ (0·3 m³) to 3 yd³ (2 m³) the majority being about $\frac{3}{4}$ yd³ (0·6 m³).

A recent invention is the front end loader (Fig. 3.19) which is able to deliver its clay to trucks or cars without turning, thereby saving time and space. It is mobile and easily moved from one bank to another. Its speedy action is particularly useful in stripping operations. Its digging action is through the power-driven shovel and does not require movement of the wheels. The charge is released at the full height of the shovel by opening its base.

There are a number of tractor-drawn scrapers in general use now both for stripping overburden and for removing a thin layer of clay, giving a well-mixed product of the area being worked. Their main advantage is that they both dig the clay and transport it to the storage dump on the plant (**A25**). The dragline also scrapes over a large radius. It is particularly useful for a sloping cut, especially when the machine must be situated above the pit because of water or other reasons. The skimmer takes a horizontal slice by movement of the bucket along the boom. It is well adapted to removal of overburden on level sites. A universal excavator is readily convertible for use with different forms of equipment (Fig. 3.20).

The shale planer was specifically designed for the digging of shales for the structural clay products industry (Fig. 3.21). It consists of a steel structure carrying an electrically driven forged steel chain that is forced against the wall of shale. It works either upwards or downwards, scraping away from $\frac{1}{2}$ to 1 in. (1·27 to 2·54 cm) of the shale, which drops to the bottom of the face. It is best applied to a face that has already been developed. It moves automatically along the face, and develops it at an angle of from 68° to the horizontal to almost vertical. An endless belt catches the fallen material and carries it to hoppers or trucks. The shaleplaner gives a uniform cut over the whole face, which may be up to 80 ft (25 m) high, and delivers in small pieces thereby reducing crushing and grinding operations.

The ditcher is a specialised machine which develops a vertical face of maximum height 10 ft (3 m). It is a combination digger and loader. It cuts away from 1 to 12 in. (2·5–30·5 cm) at a time and is useful where bands of undesirable material must be carefully removed.

The multi-bucket endless-chain clay planer (Fig. 3.22), which will dig any material that can be worked by hand without blasting, is a versatile excavator, and can be adapted to the shape of the clay face. The chassis may be situated above or below the face to be worked so that the machine is also useful in wet pits. Where the clay is under water some sort of dredge must be used (Fig. 3.23).

The cost of using mechanical plant has been studied by the Building

Fig. 3.18. *Outcrops of Wyoming bentonite. An unusual deposit where the bentonite occurs as an outcrop up to 100 ft (30 m) above the level ground. It is dry enough to grind without artificial drying* (Big Horn Bentonite, **B38**)

Fig. 3.19. *Front end loader* (Dempster, **D25**)

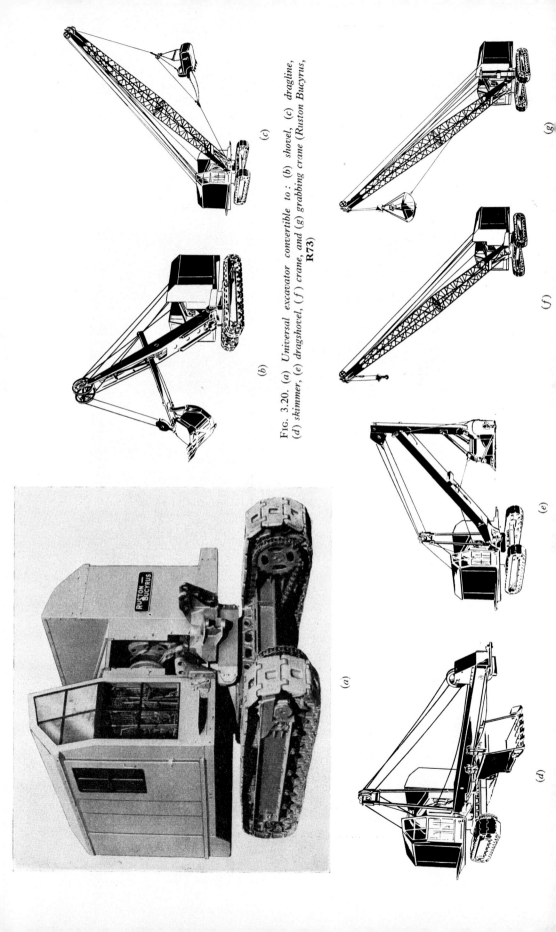

Fig. 3.20. (a) Universal excavator convertible to: (b) shovel, (c) dragline, (d) skimmer, (e) dragshovel, (f) crane, and (g) grabbing crane (Ruston Bucyrus, **R73**)

Fig. 3.21. *Shaleplaner. Model on four crawlers, two powered. Clay is delivered to plant by diesel locomotive and dump car. Texas, U.S.A. (Eagle Iron Works, E1)*

Fig. 3.22. *Multi-bucket excavator (Stothert and Pitt, S201)*

WINNING AND PURIFICATION OF CLAYS

Fig. 3.23. *Bucket chain excavator working under water; when working clay emptying knives are fitted (Lübecker Maschinenbau, L45)*

Research Station, and a digest (No. 113) issued in which the various costs of replacement, interest, maintenance, licences, insurance, administration, fuel and labour are discussed and tabulated. An extract of those matters relevant to clay working has also appeared (**A42**). Riedig (**R21**), considering mechanical plant available in Germany, gives tables showing useful technical details of the different makes.

TREATMENT OF CLAYS

WEATHERING

Weathering consists of storage of the raw clay in the open and submitting it to rain, frost, air and sun. It is generally done for a few weeks or months either in the summer, 'summering', or the winter, 'wintering', but may also be done for a few years. Freshly dug clay is generally in tough lumps which are very difficult to work. Weathering produces an improvement in workability not readily achieved by mechanical means (**S94, 116**).

Medium-sized lumps are piled from 18 in. to 2 ft 6 in. (46 to 76 cm) high and if necessary watered, but they must not become waterlogged. During wintering the water in the pores freezes and expands, thereby breaking the lumps apart. As soon as a layer has frozen another can be added on top. Wintering is particularly beneficial to very firm fireclays and also very plastic clays.

During summering, which is more favourable for less plastic clays, thin layers are used; the clay dries, shrinks and cracks, gradually breaking up.

Weathering benefits plasticity by breaking up the clay into small particles and by enabling these to take up water. It is this latter reaction in particular that is slow and cannot be simulated by mechanical breaking up of the raw clay.

Other properties of the clay may be improved by weathering, namely the rotting and gradual washing out of organic matter, perhaps assisted by bacteria, and the leaching out of carbonates of iron and calcium, with humic and atmospheric acids, and oxidation of sulphides followed by solution of the sulphates, etc. Where sulphates form intermediates, weathering must be continued for long enough for these to be washed out, as their presence may be more harmful than the original sulphide, causing efflorescence, discoloration and preventing deflocculation. If properly weathered, improvement of colour will result.

Weathering may be accelerated by raising the temperature. This is found in hot climates. The higher temperature not only increases the rate of chemical reaction but by decreasing the surface tension and viscosity, and increasing the vapour pressure of the water enables it to penetrate more rapidly.

British examples of weathering are given by the Scottish fireclays which are weathered for from one to three years, and the ball clays of Devon and Dorset of which a portion is weathered for several months (**S46**).

Chapter 4
The Ceramic Laboratory

SCIENTIFIC methods of testing and research are finding ever-increasing application in the ceramic industry in a number of different forms. There are three main fields of application, which, however, overlap considerably.

(1) *Pure Research*, the investigation of the nature of ceramic materials and their reactions and the nature and cause of their properties. The results of such work lead to the descriptions given in Chapters 1 and 2 and are an important aid to development.

(2) *Development*, the systematic improvement of ceramic products towards a desired end.

(3) *Control*, the checking of properties of materials, processes and products to achieve uniform results of the kind indicated by development projects, and the tracing of sources of trouble.

The methods used may be applicable to one or more of these fields. Some are simple and may be described in detail here; others require costly apparatus and/or skilled technicians, and for these only the principle will be described.

Throughout the work in ceramic laboratories there is frequently the need to make up samples on lines similar to those used in the works. Small-scale equipment for preparing these is therefore required, *e.g.* crushing rolls, rotary crushers, jaw crushers, edge runner mills, jar mills; mixers, blungers, sieves, magnets, filter presses, pugmills; extruding augers, hand presses; drying ovens, small kilns; masonry saw and grinders for preparing samples for tests. The manufacturers of full-size equipment often also produce a laboratory model. German (**G27**) gives an outline of the type of small-scale equipment required in a works laboratory.

RECORDING OF LABORATORY TESTS

The need for adequate recording of methods, samples and results cannot be over-emphasised. Whereas research tests may be adapted to the sample, control tests, which are very frequently purely comparative *must* be done by standard methods which are laid down in writing.

It is also essential that accurate record be kept of every sample entering the laboratory with its date, its origin, any treatment it has already undergone in the works and in which machines, followed by the test results. It is only in this way that any systematic error will be detected (**W62**).

WIDELY APPLICABLE TESTS
CHEMICAL ANALYSIS

Chemical analysis has very wide applications in research, development and control, even though the nature of ceramic raw materials and their reactions necessitates the use of many other methods besides analysis. Knowledge of the chemical composition of raw materials, bodies, glazes and ceramic products is frequently very useful, but because complete analyses of silicate materials by 'classical' methods take a week to perform, they were not formerly used as routine control measures. Sometimes only one or two components must be checked, and quick chemical methods were evolved for them. Instruments now greatly speed up chemical analysis so that it can be used increasingly for control purposes.

Methods of carrying out complete or 'ultimate' chemical analyses of silicate materials and any other raw materials required in the ceramic industry have been described in the detail which is their due in books devoted entirely to the subject, *e.g.* Hillebrand, *Analysis of Silicates* (1919) (**H91**), Groves, *Silicate Analysis* (1937) (**G92**), Jakob, *Chemische Analyse der Gesteine und Silikatschen Mineralien* (1952) (**J16**). We will therefore not enter into the subject here. The classical method of complete analysis is mainly gravimetric, occasionally assisted with colorimetric measurements. It relies on previous knowledge of the elements present and takes from six to ten days to complete.

For speedier work microchemical techniques are frequently very suitable. The field has become too wide to be dealt with here. There are numerous works on the subject, *e.g.* Emich, *Microchemical Laboratory Manual* (**E16**).

There are also, many specific tests for individual components. These generally involve specific chemical reactions followed by gravimetric, volumetric, or colorimetric estimation. This last has been developed to give quick and accurate results by the use of physical instruments instead of the human eye.

Bennett, Hawley and Eardley (**B37**) describe in detail a method of carrying out a complete analysis of silicate materials by rapid methods for the individual constituents, using normal chemical laboratory equipment plus a flame photometer. The following oxides can be determined: SiO_2, Al_2O_3, Fe_2O_3, TiO_2, CaO, MgO and the alkalies. Interference to the methods is found if phosphates (*e.g.* bone, or bone china body) and vanadates are present, or the magnesia content is high, *e.g.* in magnesite. A complete analysis can be carried out in one day and individual determinations in four hours.

Colorimetric Analysis

Colour tests in qualitative analysis are widespread. The use of colour for quantitative work is possible when there is only one coloured substance present. A previous qualitative analysis is therefore necessary. If no disturbing substances are present the test colour is usually obtained by adding suitable reagents and then compared with the depth of colour of the same compound in a solution of known concentration. (It is essential that the two solutions are the same colour.)

A notable application of this in ceramics was evolved by Pukall (**P62**) as far back as 1906 to determine the solubility of lead in lead glazes. The solutions obtained from the glazes are very dilute and difficult to assess gravimetrically, but accurate results were obtained from the depth of colour of the lead sulphide suspension obtained by passing H_2S.

The unaided human eye is not, however, really suitable for making colour comparisons; it can tell when two shades are of the same depth and when they are different but cannot tell the degree of difference. Modern colorimeters for visual observation have optical systems that bring the test sample and the comparison standard into the two halves of the same field of vision. The depth of the one can then be adjusted until the field is uniform (Dubosq colorimeter).

It is, however, more accurate to replace the human eye by the photoelectric cell as in the 'Spekker' and similar instruments (Fig. 4.1). In the 'Spekker' light from a central lamp is divided through adjustable slits, passes through heat filters and then takes two different paths before being registered by two photoelectric cells connected in opposition to a galvanometer. On the testing side the sample and comparison liquids are placed in pairs of cells of 1, 4 or 20 cm^3 on a sliding shelf, so that either can easily be brought into the light beam. The aperture on this side has a drum so graduated that its readings are proportional to the concentration of the liquids. On the compensating side the aperture adjustment is not graduated. Comparisons are effected by inserting the darker liquid and adjusting the compensation until a null-reading is obtained. The lighter liquid is then inserted, the compensating aperture left as it is, and the calibrated drum altered to give a null reading. The method is both speedy and accurate.

The choice of suitable coloured compounds for colorimetric analysis is a vast subject dealt with in specialised works, *e.g.* E. B. Sandell, *Colorimetric Determination of Traces of Metals* (**S10**), Noel L. Allport, *Colorimetric Analysis* (**A188**), Bruno Lange, *Kolorimetrische Analyse* (**L8**). Nearly every element and some radicals can now be estimated colorimetrically and the method is particularly useful for minor constituents of samples being analysed. Adaptation for use with silicate materials is described by Richter (**R18**).

Spectrophotometric Analysis

Spectrographic methods for qualitative and quantitative analysis have considerable advantages over ordinary wet methods. Firstly the analysis is completely objective, determining the unexpected as well as the expected elements. Secondly it will determine smaller quantities of rare constituents.

(a)

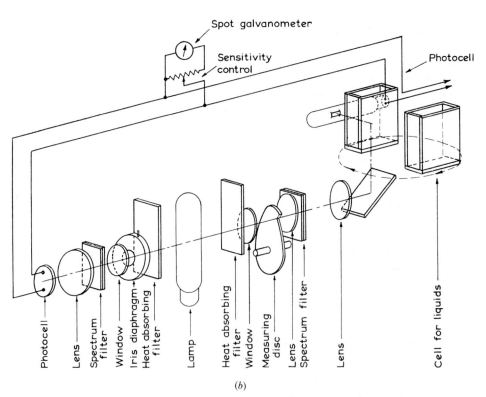

(b)

FIG. 4.1. *Spekker photoelectric absorptiometer*: (a) *external view*; (b) *simplified diagram of optical system and photocell circuit* (Hilger and Watts, **H88**)

Thirdly it is quick, a complete silicate analysis takes only from three to four hours (**H74**), and requires only small samples.

The apparatus consists of a spectrograph (Fig. 4.2), an excitation source, means of photographic recording, viewing and comparison. For emission spectra the sample has to be so prepared that it will volatilise completely in the arc, and all its constituents will register their presence.

For qualitative work the spectrogram obtained is compared with a pure spectrum which has numerous lines, *e.g.* that of iron, and thus enables the exact position of the lines given by the test specimen to be identified. For quantitative work the intensities of carefully chosen lines (very strong lines show

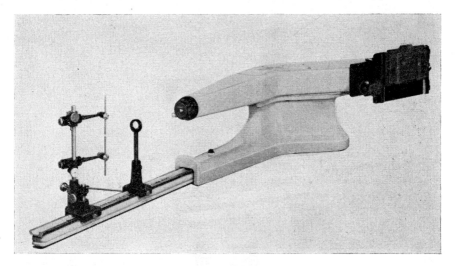

FIG. 4.2. *The medium quartz spectrograph* (*Hilger and Watts*, **H88**)

reversal on the photograph) are measured comparative to standards photographed under identical conditions (Fig. 4.3). Because of the sensitivity of the method, standards for comparison must be made up from spectroscopically pure substances (**J47**).

Application of spectroscopy to ceramic materials is not entirely straightforward because some constituents are very much more readily volatilised than others. Zoellner (**Z8**) and Smith and Hoagbin (**S136**) discuss some of the difficulties that arise initially.

The methods worked out for complete analyses all involve more than one determination. Hegemann and Zoellner (**H74**) in 1952 used three different electrode set-ups as shown in Fig. 4.4, together with a quartz spectroscope which makes the ultra-violet region available. The three determinations are: (1) Complete qualitative analysis plus quantitative determination of minor constituents (less than 1%) other than the alkalies. Minor constituents determined to $\pm 10\%$ of their content. (2) Determination of major constituents (over 1%) to from ± 2 to 3%. The sample mounted on filter paper is rotated between the pure carbon electrodes. (3) Determination of the alkalies. Na_2O can be determined down to 0·01% and K_2O down to 0·05% with an error of ±2% of the content. This may also be done with a flame photometer.

Brown (**B123**) in 1958 describes a method for analysis of ceramic raw materials and also frits, glazes and fired products. The equipment used is a medium quartz spectrograph (*see* Fig. 4.2), a non-recording microphotometer (*see* Fig. 4.3), spectrum projector and B.N.F. source unit, with electromagnetic timing unit. Again three main methods for introducing the sample

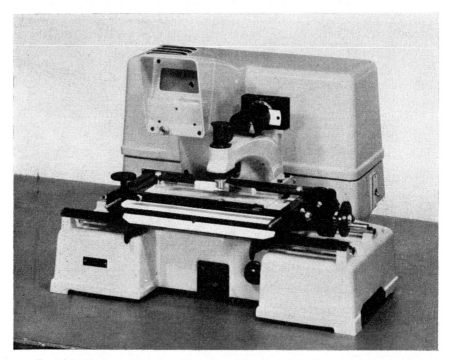

Fig. 4.3. *Non-recording microphotometer with galvoscale* (*Hilger and Watts*, **H88**)

into the discharge are described: (1) The finely ground material, or material mixed with another substance which acts as an internal standard and/or spectrographic buffer, is placed in a cavity made in the lower of two pure graphite or carbon electrodes, usually the anode, and is then volatilised in a d.c. arc. (2) The finely ground sample is mixed with briquetting graphite, an internal standard and a flux, and pressed into a briquette. A controlled spark discharge is used for excitation. (3) The material is brought into solution and the solution is sparked either directly or after being allowed to soak into a graphite rod. The accuracy obtainable in raw material control analyses using the d.c. arc excitation is about ±10–15% of the amount of impurity present. In the solution spark method it is ±4% except for SiO_2, for which the deviation is approximately ±6%.

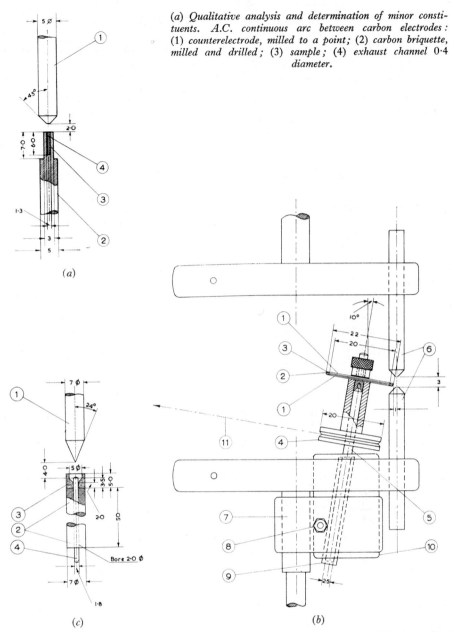

(a) Qualitative analysis and determination of minor constituents. A.C. continuous arc between carbon electrodes: (1) counterelectrode, milled to a point; (2) carbon briquette, milled and drilled; (3) sample; (4) exhaust channel 0·4 diameter.

(b) Determination of major constituents. A.C. intermittent arc between carbon electrodes: (1) two plates, diameter 20, thickness 1·15 (electrolytic copper); (2) disk, diameter 22 (quantitative filterpaper); (3) sample powder; (4) pulley (brass); (5) axle (silver steel); (6) electrodes (special carbon); (7) two jaws (brass); (8) screw for jaw (brass); (9) insulating tube (rubber); (10) two insulating leaves (Pertinax); (11) belt drive (linen thread)

(c) Determination of the alkalies. Copper electrodes: (1) upper Cu electrode; (2) lower electrode; (3) sample; (4) Cu wire diameter 1·8

FIG. 4.4. *Electrodes for a complete spectroscopic silicate analysis by the Hegemann and Zoellner method* (**H74**). *Dimensions in mm*

Innes (**I4**) describes spectrographic methods for determining the minor constituents in refractories, namely Fe_2O_3, TiO_2, ZrO_2, Na_2O and K_2O which occur in quantities of less than 6%, circumstances when spectrographic analysis is at least as accurate as chemical analysis and the latter is particularly tedious.

In the new absorption spectrophotometer the sample is brought into solution and inserted in a glass cell (Fig. 4.5). The wavelength of incident light can be varied gradually by the turn of a knob and the absorption observed and recorded for each wavelength. In qualitative work the absorption in the ultra-violet and visible regions gives the elements present and that in the infra-red gives evidence about the compounds present. Suitable lines for comparison are then chosen and compared with those given by a standard solution contained in an identical cell which can be easily interchanged with the test solution.

THE FLAME PHOTOMETER

A special adaptation of spectroscopic analysis for quantitative determination of certain particular elements is the flame photometer. In ceramics it is used predominantly for determining the alkalies (lithium, sodium and potassium) and calcium.

The intensity of the spectral lines emitted by an excited atom is proportional to its concentration. In the flame photometer a suitable solution (**B49**) of the component to be determined is sprayed into a flame, so that emission occurs. Lines characteristic of the element are selected by inserting filters in the beam of light produced by a lens system focused on the flame. The intensity can then be determined with a photoelectric cell and a galvanometer. There are various types of the instrument available, Evans Electroselenium Ltd.'s model (**E24**), the Zeiss model (**Z9**), and the Riehm-Lange model (**L28**). The Hilger absorption spectrophotometer also has a flame photometer attachment (**H88**).

Lange (**L8**) has tabulated the most suitable wavelength for each element and the sensitivity of its determination (Table 57).

The Evans Electroselenium flame photometer is briefly described by Collins and Polkinhorne (**C45**) in connection with the diagrams (Fig. 4.6). 'Compressed air is supplied to a small annular-type atomiser (1), through a control valve (2), at a pressure of 10 lb/in² as indicated on the gauge (3), mounted on the front of the instrument. The air supply can be derived either from a cylinder with a reducing valve or from a suitable compressor. The flow of air through the atomiser draws the sample from the beaker (4) up the stainless steel capillary (5) and sprays it as a fine mist through the ebonite plug (6) into the mixing chamber (7). Here the larger droplets fall out and flow to waste through the drain tube (8). Gas is introduced into the mixing chamber through the inlet tube (9) from the internal gas pressure stabiliser (10) and control valve (11). The gas–air mixture passes to a multi-jet burner mounted above the mixing chamber where it burns as a broad flat

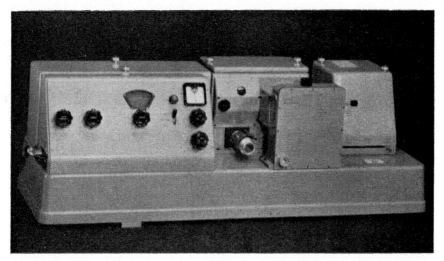

FIG. 4.5. *Ultra-violet and visible photoelectric spectrophotometer (Hilger and Watts, H88)*

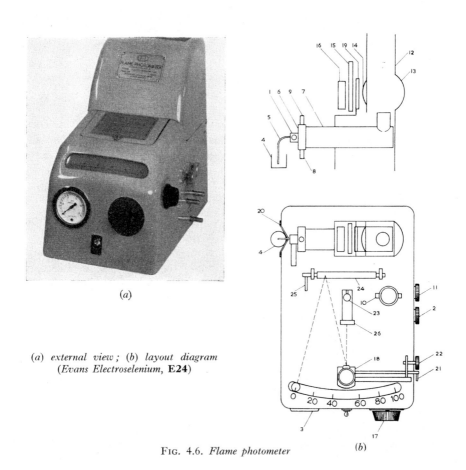

(*a*)

(*a*) *external view;* (*b*) *layout diagram*
(*Evans Electroselenium,* **E24**)

FIG. 4.6. *Flame photometer*

flame, and the hot gases pass up a well-ventilated chimney (12). Coal gas, butane, propane or any of the proprietary bottled gases, such as Calor gas or Botto-gas, can be used without any modification of the instrument.

TABLE 57

Element	Wavelength (mμ)			Sensitivity (mg/ml)
B	548	521 (1/2)		0·7
Ba	554	516 and 527 (3/4)		0·4
Ca	616	556 (1/3)		0·01
Co	535			1
Cr	521	387 (1/5)		0·7
Cs	852			0·04
Cu	521	515 (1/1)		0·4
Fe	536	526 (1/1)	643 (1/2)	0·5
K	767			0·0006
Li	671	610 (1/6)		0·003
Mg	553	383 (1/2)	518 (1/2)	0·3
Mn	601	561 (1/2)		0·2
Mo	557	642 (1/2)		0·7
Na	589			0·0003
Ni	548	618 (3/4)	472 (1/2)	0·6
Pb	561	406 (1/5)		5
Rb	780	795 (1/3)		0·0005
Sn	607	452 (1/2)		2
Sr	616	707 (1/1)		0·05
Tl	535	377 (1/2)		3

'The light emitted by the flame is collected by a reflector (13) and focused by a lens (14) through the interchangeable optical filters (15) on to an "EEL" barrier-layer photocell (16). The current generated by this cell is taken through a potentiometer (17) to a Tinsley taut suspension galvanometer unit (18). A glass window (19) is interposed between the lens and filter for cooling purposes.

'The sensitivity of the instrument is such that full-scale deflection is obtained with 5 p.p.m. of sodium and 10 p.p.m. of potassium.'

The instrument must be calibrated with solutions as nearly the same as the test solution as possible, *i.e.* containing the same anions and cations. For instance, for measuring potassium content a calibration on potassium chloride does not apply to solutions of potassium sulphate (**C45**). Similarly the curve given by potassium alone is different from that given by potassium in the presence of sodium and/or calcium (**L28, Z10**). The method is therefore of most use where large numbers of samples containing different quantities of the same constituents are to be determined.

The advantages of the method are the speed, without any sacrifice of accuracy, and the small amount of material required (**O14**). A worker who was able to estimate potassium in twelve solutions a day by gravimetric-colorimetric methods was able to determine thirty samples an hour with a flame photometer.

The disadvantage mentioned above, of having to calibrate the flame photometer with complex solutions as like as possible to the test one, has been overcome by a method described by Debras and Voinovitch (**D20**). Interference in the determination of lithium, sodium and potassium is found to arise from the presence of Mg, Al, Fe, Ti and Ca. These are therefore first precipitated from the test solution before it is placed in the flame photometer. Reproducible results can then be quickly obtained.

Electrolytic Analysis with the Polarograph

By means of the polarograph dilute (10^{-3} to 10^{-4} moles) solutions containing several cations can be analysed qualitatively and quantitatively in a short time. The apparatus consists of an electrolytic cell with one small polarisable electrode, namely a dropping mercury electrode, and the other large and non-polarisable, namely a pool of mercury. This cell is connected to a battery, resistance, galvanometer and a means of varying and measuring the voltage input.

The test solution is freed from oxygen (by passing N_2 or H_2) and placed in the cell. The voltage is gradually increased and the current passing either observed or automatically recorded. The resulting curve has a number of steps. The position of these on the voltage scale gives the nature of the ion and their height gives the quantity present. The instrument has to be calibrated under identical conditions with a standard solution. The instrument is highly sensitive and gives reproducible results; it is quick and uses very little material. The makers particularly recommend it for measuring the alkali content of ceramic materials.

Chromatography

The relatively new technique of chromatography, first evolved by biologists and biochemists, may also find application in ceramic chemical analysis. It depends on the fact that solutions of different substances move through an absorbing medium, *e.g.* filter paper, columns of inert gels, etc., at different rates, so that mixtures become separated into individual materials at different levels and can then be identified and estimated (see *The Elements of Chromatography* by T. I. Williams, Blackwell, Oxford, 1954).

WIDELY APPLICABLE TESTS
PHYSICAL TESTS

Moisture Content

It is important to know the moisture content of raw materials, both to make allowance for it when weighing and, in the case of clays used as nearly as possible in their natural condition, in order to make the necessary adjustment to the optimum water content. It is also very important to control the moisture content of ceramic materials prepared for shaping, especially the almost dry ones.

Samples may be weighed, dried at 110° C, cooled in a desiccator and re-weighed. It is more convenient to give the moisture content as a percentage of the dry weight, but as this is not always done the method adopted should be stated.

Adequate routine control may require frequent taking and testing of samples, perhaps as often as every ten minutes.

Commercially developed moisture meters can be divided into two classes:
(1) Direct determination of moisture.
(2) Determination of a property related to the moisture content.

Apparatus for weighing and drying a number of samples simultaneously is available, but the collection of samples and measurement of weight loss of each still requires a full-time operator.

Of the derivative methods, the best known is the measurement of the pressure of acetylene gas freed by the moisture in the sample, in the 'Speedy' moisture tester. An operator is required and the results vary slightly with different operators.

Other instruments measure electrical resistance, but only give repeatable results when the material is compressed to a standard density and are therefore not suitable for continuous measurements.

Considerable advance has been made recently in instruments which depend on the change in capacitance at the measuring head caused by the moisture of the sample. Errors can, however, arise from changes of dielectric constant due to other causes.

With these inadequacies of instrumentation in mind an attempt has now been made to develop an automatic moisture indicator suited to the conditions of the ceramic industry. The prototype is described by Stedham (**S172**). In it a sample of approximately the right size is collected mechanically, measuring by volume. It is transferred to a container which can be heated. The increase in weight of the container is measured and recorded temporarily, after which heat is applied while the container is rotated to promote the escape of steam. At the end of the heating period the loss in weight is subtracted from the initial weight to give a final result on the 'dry basis'. The prototype works on a six-minute cycle.

Specific Gravity

Specific gravity, being the ratio of the density of the material to that of water at 4° C, is a pure number and requires no conversion to the various units. It is numerically equal to the density when expressed in grammes per cubic centimetre (g/cm^3). The specific gravity of both powders and pieces can be measured by normal physical methods. Very fine powders tend to entrap air and need to be evacuated before introducing the water.

It is not normally necessary to check the specific gravity of raw materials with the exception of calcined flint. The S.G. of this varies with the degree of calcination so that the dry content of a slip of given pint weight will vary.

Regular checking is therefore necessary. S.G. is also a measure of the dead burning of sintered magnesite and is used as a check. S.G. measurements are used more extensively for control of fired products.
(B.S.S. 2701:1956.)

Melting Behaviour

The melting behaviour of ceramic raw materials, mixtures and products is of considerable interest in a number of fields varying from research into phase relationships to the practical need for refractory materials to remain solid at high temperatures and the requirements for low-melting fluxes.

Many ceramic materials have no definite melting point, they show progressive sintering, then softening, leading to fusion. Many empirical tests can be made by heating the material at a firing schedule similar to that for which it is required and withdrawing samples at intervals. For many ceramic materials also the 'heat–work' test of the pyrometric cone equivalent (P.C.E.) is very useful.

For comparative work on feldspars, with their complex melting behaviour Zwetsch (**Z12**) has developed a method using a heatable plate microscope and standard photographs. Aldred and White (**A180**) describe several types of apparatus: tungsten-strip furnace for observing melting points up to 2400° C (4352° F) of materials stable in a vacuum; hot-stage attachment for the microscope for melting points and phase changes up to 1820° C (3308° F); arc-image furnace for very high temperatures, *e.g.* melting of thoria 3200° C (5800° F).

PYROMETRIC CONE EQUIVALENT

The refractoriness of ceramic raw materials, mixtures, or products is estimated by comparison with mixtures of known properties, *i.e.* pyrometric cones. The pyrometric cones are 'heat–work' recorders (*see* p. 934), that is they do not register temperature directly, but a combination of temperature and rate of heating. To give reproducible results in P.C.E. tests, therefore, a standard rate of heating must be adopted.

The material to be tested is shaped in a metal mould to give a cone identical with the standard commercial cones. If the test material is not plastic, an alkali-free organic binder is used with it. The cone may then be pre-fired to give it stability. For products, a cone of the right shape is cut out. A number of cones of the test material are then set alternately with a series of standard cones in a refractory plaque or a single sample cone is set in the middle, surrounded by standard cones. Correct setting is assisted by making a plastic plaque in a mould, marking equidistantly spaced positions with a prepared stamp, and then setting the cones at the correct angle with the aid of a jig. The prepared plaque is then dried before setting in the test furnace. These are of various kinds, *e.g.* gas-fired, carbon-resistance, electrically heated, oxyacetylene. Here heating up is taken at the prescribed rate and the cones observed by means of a spy-hole through coloured glass. A neutral to slightly oxidising atmosphere should be maintained. A cone is said to be 'down' when its tip touches the plaque. The position of the standard cones is noted when the test cones are down. The P.C.E. of the test material is given as the cone number or numbers to which it most closely approximates.

Petrie (**P27**) indicates two sources of error in P.C.E. determinations of refractory materials. The first is interaction between the cone and the plaque with lower melting constituents migrating from one to the other and a false result being obtained. This can be avoided by choosing a plaque material that is as similar as possible to the cone to be tested. The second is bloating of raw clays which is sometimes aggravated by certain organic binders. The calcining of the raw ground clay to 950°–1000° C (1742°–1832° F) for about half an hour before making up into cones, is recommended.

Grain-size Determination

In investigating raw clays the separation of grain-size fractions followed by their determination assists considerably in separating the main mineral types. Such fractions may be termed:

> Stones, > 20 mm
> Gravel, 20–2 mm
> Coarse sand, 2–0·2 mm
> Fine sand, 0·2–0·02 mm
> Silt, 0·02–0·002 mm
> Clay, < 0·002 mm

Knowledge of the content of the finest fraction therefore gives information on the active colloidal clay present.

Grain size has a strong influence on solid-state reactions when the greater the surface the faster the reaction, and similarly on solid–liquid reactions. For such purposes knowledge of the surface area is the most useful.

TABLE 58

Normal Lower Limits for Various Methods for Particle Size Analysis (H84)

Method of Analysis	Lower Limit (μ)
Sieving	50 Brit., 40 U.S.A.
Elutriation	10
Sedimentation, gravitational	2
	1 with temperature control
centrifugal	0·1

Size grading influences packing density, mechanical strength and porosity. There are therefore three aspects to grain-size determinations:

(1) Diameters.
(2) Surface area.
(3) Grading.

As the range of sizes is large a number of different methods have to be used.

SIEVING

All the grading of coarser particles can be achieved by passing the mixture through a series of sieves. Sieves may be made of wire, silk or nylon. The sieve number is the number of strands per linear unit. As different types of strand are of different thickness the actual aperture will vary. It is therefore necessary to state the exact sieve specification (p. 1301).

Sieving is a statistical process, that is, there is always an element of chance as to whether a particle will pass through a given aperture or not. Reproducibility can only be achieved by standardising the time and method of sieving.

In order to eliminate the human element in sieving there are a number of mechanical shaking or vibrating devices for performing screen analyses. These will hold a nest of graded test sieves together with a receiver and lid and vibrate for a given time in a reproducible manner. They are operated by a fractional horse power motor (Fig. 4.7).

Wet sieving is preferable to dry sieving, especially for fine particles. In dry sieving great care must be taken to use really dry material to prevent any 'balling up', in which case dust may be lost.

(B.S.S. 410:1943; B.S.S. 1796:1952.)

STOKES' LAW

Elutriation and sedimentation depend on the fact that particles of the same specific gravity but different size fall through water at different rates. This is expressed by Stokes' Law:

$$v = \frac{2gr^2}{9} \cdot \frac{(D_1 - D_2)}{\eta} \qquad (1)$$

where v is the velocity of the falling particle (cm/sec);

r is the radius of the particle (cm);

D_1 and D_2 are the specific gravity (g/cm^3) of the particle and the liquid, respectively;

η is the viscosity of the liquid (g/cm sec);

g is acceleration due to gravity (981 cm/sec^2).

Stokes' Law applies to smooth, spherical bodies. Clay particles are flat disks. The results obtained by applying Stokes' Law to the sedimentation of clays are therefore empirical 'equivalent radii', i.e. the radius of a sphere that would fall at the same speed as the clay particle.

Fig. 4.7. *Sieve analysis with a testing sieve shaker* (*International Combustion*, 17)

For a given solid in suspension, a given liquid medium, at a given temperature $(D_1-D_2)/\eta$ is constant and can be evaluated. The formula then becomes

$$v = C.r^2$$

e.g. for $D_1 = 2\cdot60$ (an average for clays and soils) and using water at 20° C, $C = 34\,720$.

The velocity of falling of a particle is also given by the equation

$$v = \frac{h}{t} \qquad (2)$$

where h is the height (cm) the particle has fallen in time t sec. By substituting into (1), the relationship between equivalent radii and time is obtained, h being fixed, i.e.

$$r = \left\{ \frac{9}{2g} \times \frac{h}{t} \times \frac{\eta}{(D_1-D_2)} \right\}^{1/2} \qquad (3)$$

These equations are used in evaluating results obtained in the apparatuses to be described.

PARTICLE SIZE DETERMINATION BY ELUTRIATION

As particles of different size fall through still water at different velocities, application of an upward velocity of water makes it possible to reverse the motion of the slower and therefore smaller particles, and carry them out with the water stream. Careful selection of the velocity of the water should make it possible to remove all particles smaller than a given known size.

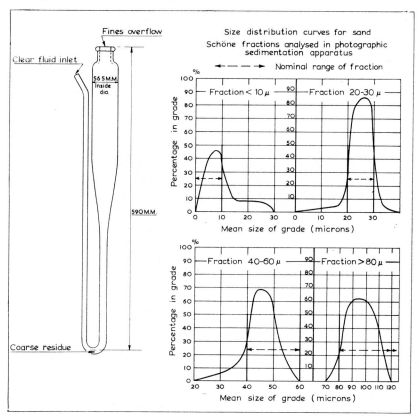

FIG. 4.8. *Schöne elutriator* (*Stairmand*, **S160**)

Construction of an apparatus suitable for elutriation involves designing a vessel where the flow is constant across the whole diameter and where there is no turbulence, and a means of supplying a water flow at a known constant rate.

The first successful apparatus was that of Schöne (**S25**) in 1866 (Fig. 4.8). It was subsequently improved on. Fig. 4.8 shows the general shape and dimensions of the elutriating vessel. The Schöne apparatus has the disadvantage of being fragile and that a large volume of liquid has to be evaporated to extract the elutriated particles for determination by weight.

The apparatus first designed by Schulze and improved by Harkort (**H22**) has a wider conical vessel tapped at the bottom with a controlled water stream running through a vertical tube which enters the cone from the top, with its mouth near the bottom of the cone. This apparatus gained considerable popularity (Fig. 4.9).

Straightforward elutriation does not, however, give clear size cuts. Any accidental turbulence will allow larger particles to be carried over with the small ones, and fines remain in the coarse sediment due to insufficient 'scrubbing'.

The Andrews' kinetic elutriator (Fig. 4.10) overcomes the difficulties of breaking up aggregates and removing fines, by the waterborne impact of the initial mixture against a stationery cone. With this commercially

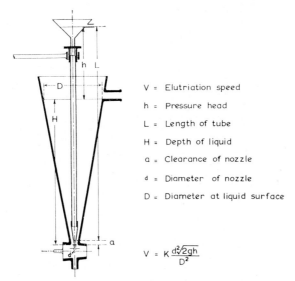

V = Elutriation speed
h = Pressure head
L = Length of tube
H = Depth of liquid
a = Clearance of nozzle
d = Diameter of nozzle
D = Diameter at liquid surface

$$V = K\frac{d^2\sqrt{2gh}}{D^2}$$

FIG. 4.9. *Schulze–Harkort elutriation apparatus (Harkort, H22)*

available and well standardised apparatus any sample of fine powder can be divided into three grades. The smallest fraction is of particles $< 20\ \mu$.

Elutriation finds considerable application in large-scale preparation of raw materials, as has been seen in Chapter 3.

AIR ELUTRIATION

It is sometimes necessary to carry out particle-size analysis on a material without wetting it. Even where this is not essential air elutriation of a ground material has several advantages over that carried out with water. In principle it is the same as water elutriation in that the dispersed particles are allowed to settle against an air stream, the rate of settling being governed by Stokes' Law.

Mechanical difficulties have now been overcome and Roller (**R53**) describes an apparatus that gives results comparable with those obtained by other direct methods (Fig. 4.11). The chief advantage is speed; the rate of fall of particles in air is some 100 times greater than in water, no chemical deflocculating agent is required, and tedious separation from large quantities of water is eliminated. Also, as the viscosity of air varies little with temperature, elaborate temperature control is not required.

Deflocculation of the sample is achieved by direction of an air jet against it in the U-bend at the bottom of the system. This is so arranged that the sample always moves against the jet. The size of the nozzle on the end of the air inlet tube is so chosen that the pressure is from 1·0 to 1·5 in. of mercury. This was found to be sufficient to effect complete deflocculation without disrupting the particles. The air velocity is regulated strictly in accordance with Stokes' Law so as to float particles of maximum desired size. The fine particles removed by the jet are carried into a settling chamber whence only the desired fraction is carried over into the sample thimble.

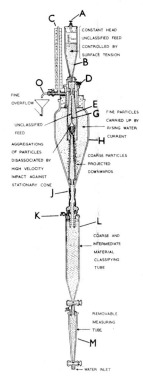

FIG. 4.10. *Andrews kinetic elutriator*
(*International Combustion,* **17**)

PARTICLE SIZE DETERMINATION BY SEDIMENTATION

The progress of sedimentation is followed either by removing sample portions at a known height below the surface, or by following the density changes. The simplest method of sampling is with a tall cylinder with a side arm and tap as suggested by Atterberg. This has now been superseded by withdrawal with a pipette. The original apparatus was designed by Andreasen. A slightly modified form is commercially available so it will be described in some detail before considering more modern modifications.

The Andreasen Pipette

The 'pipette' (Fig. 4.12) itself consists of a graduated cylinder (W) with a long tube fitted through the centre of the stopper. Above the stopper a three-way tap regulates the tube inside the cylinder, an accurate 10 ml bulb (P), and a draining tube. Auxiliary apparatus comprises an aspirator attached to the measuring bulb and a thermostat in which to immerse the pipette. The stopper holding the pipette itself can be replaced by a plain one.

The analysis is carried out with a suspension of the clay containing from 1 to 3% by volume of solids. The cylinder is filled to a given mark with the well-shaken suspension at thermostat temperature. The pipette-holding stopper is inserted and a stopclock started. Samples (10 ml) are withdrawn at given times and transferred to silver evaporating dishes (S), and the solid content found by evaporation and drying at 110° C.

Assuming that Stokes' Law is applicable the size of the largest particle that can be present at the depth below the surface at the time of withdrawal

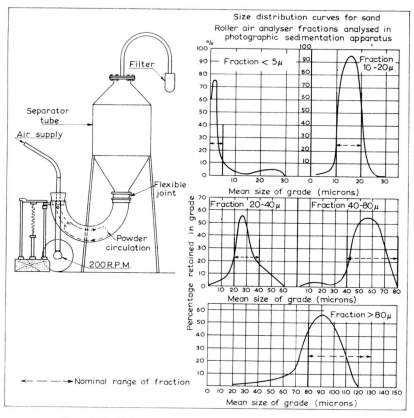

FIG. 4.11. *Roller particle size air analyser* (*Stairmand,* **S160**)

can be calculated (it must be remembered that as material is withdrawn the upper surface drops). The sample therefore contains all sizes smaller than this. The experimental results can be plotted as % by weight of particles under a given size against size of particles, giving the undersize cumulative curve. From this curve the % by weight of particles in each size range (usually 5 μ) can be read off. A further curve of particle-size distribution can then be plotted (**A193**).

The main advantage of the Andreasen pipette is its ease of operation which requires no adjustments of height, etc. The depth of the sample is a fixed constant of the apparatus. Its chief disadvantage is the length of time required for estimating the smaller particles, some 204 hours (8½ days) for particles of 0·5μ, and also the fact that the suspension used is too concentrated for true free falling of the particles.

Another factor that has aroused some investigation is the shape of the pipette tip and the position of the hole(s) through which the sample is withdrawn. If a normal drawn-out end is used, and suction is sufficiently fast, the liquid is withdrawn from a sphere whose centre is the tip of the pipette. The withdrawn sample is therefore not strictly from the one depth. This can be overcome by sealing the tip and piercing from one to four holes slightly above the end. Zimmermann (**Z6**) does, however, show that the error introduced with the straightforward tip is about 0·04% when the pipette is immersed to 10 cm and withdraws 10 ml, this is within the experimental error; and Vinther and Lasson (**V19**) show that if the applied suction is slow the withdrawn liquid comes from an almost ideally horizontal layer. There have been numerous suggestions for overcoming the inherent disadvantages of the Andreasen pipette. In 1933 Vinther and Lasson (**V19**) put forward an adaptation to shorten the time required for determining the smaller particles. By reducing the depth of immersion of the stem of the pipette from 20 cm to 20 mm the following time savings are made:

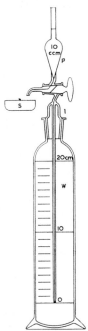

FIG. 4.12.
Andreasen pipette
(A203)

Maximum grain size present	0·36 μ	0·22 μ	0·13 μ
200 mm depth of immersion	9 days, 14 hr	23 days, 19 hr	69 days, 20 hr
20 mm depth of immersion	1 day (24 hr)	2 days	4 days

Greater precautions are necessary in determining the exact depth of immersion of the tip and in slow and steady withdrawal of the sample.

Zimmermann (**Z6**) describes a simple apparatus with a movable pipette. The stem of the pipette itself is graduated so that the depth of immersion can be read off directly. This system has the tremendous advantage of being able to determine different particle sizes at convenient times by adjusting the depth of immersion. It has the disadvantage of disturbing the suspension when height adjustments are made.

Schmied (**S22**) (1958) has developed this idea further by constructing an apparatus more like the Andreasen pipette but with five pipettes fixed in the ground-glass stopper of the sedimentation vessel. Each pipette has its tip at a different height. The withdrawal of the samples is carefully regulated through the attached respirator.

None of these modifications of the Andreasen pipette overcome its inherent defect of using suspension concentrations too high to allow true

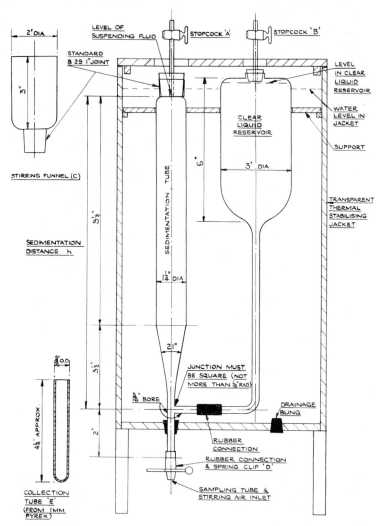

Fig. 4.13. *I.C.I. sedimentation apparatus* (*Stairmand*, **S161**)

free falling. In 1947 Stairmand (**S161**) described an apparatus designed for easy operation but using concentrations of less than 0·1% (Fig. 4.13). The essential method here is to allow settling to occur into a clear medium and to wash out the accumulated sediment at given times, of up to 4–5 hr. It seems possible that this type of apparatus will soon be more generally used.

The second main method of following sedimentation is by either observing the fall in the density of the suspension at a given height, with a Wiegner pipe, or following the sinking of a level of given density, with a hydrometer.

Wiegner Pipe. The Wiegner pipe (Fig. 4.14) uses the method of balancing columns in a U-tube (**W66**). The suspension is allowed to settle in the wider tube (*A*). A narrow side arm (*B*) is attached part of the way up the tube and contains the suspension medium alone. As sedimentation proceeds the density of the suspension decreases so that the balancing arm of clear liquid gradually drops. The method is suitable for particles of from 0·1

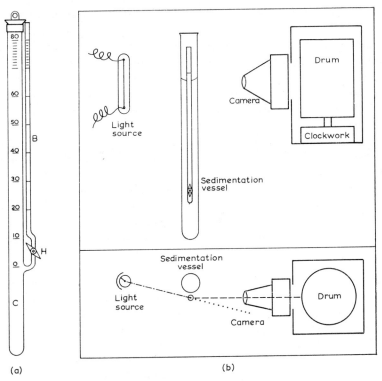

Fig. 4.14. *Wiegner–Gessner sedimentation apparatus:* (*a*) *Wiegner pipe;* (*b*) *Gessner photographic recorder* (*Gessner,* **G30**)

to 0·005 mm diameter. It has been made considerably more valuable by recording the level of the clear column photographically (**G29, 30, 54**).

Gollow (**G54**) points out the particular usefulness of the photographic recording apparatus in routine checking where curves can be compared visually, thus eliminating calculation, but for accurate work the Wiegner pipe has been superseded.

Hydrometer

Another simple method which can be so accurately calibrated as to compare favourably with the Andreasen pipette is that of the hydrometer.

Essentially it consists of measuring the density of the suspension at the level of the centre of buoyance of the hydrometer. As the density of the suspension decreases the hydrometer sinks in further, so the calibration curve obtained from Stokes' Law must be adjusted accordingly. Constant temperature and initial density are essential. Ratcliff and Webb go into the calculations in detail (**R4**). Given good calibration this method of grain-size determination is quick, simple and fairly accurate. It is the most commonly used method for grain-size determination.

The main objection to the hydrometer method is the size of the instrument. It does not give the density of a point but rather the mean density over a considerable depth. Also sedimentation may occur on the shoulders of the bulb causing totally erroneous results, removal of the hydrometer to wipe it causes turbulence, upsetting the steady sedimentation.

These difficulties are overcome by using small 'divers' which are completely immersed. The construction and use are described in detail by Berg (**B40**) who first introduced them in 1940. They are small sealed glass bodies suitably ballasted so that they come to rest at a point in the suspension, thus giving the density at that point. Being very much less bulky than a hydrometer they give a more accurate location of the density they measure. Once at equilibrium the diver moves downwards with the velocity of the largest particles present. There is therefore little likelihood of particles settling on it. Berg compared the accuracy of the determination with divers with that with the Andreasen pipette and found it to be equal. It is to be hoped that such divers will soon become commercially available.

Centrifugal Sedimentation

The lower limit of size of particle which can be determined by gravitational sedimentation is controlled not only by the practical consideration of available time, but also by inaccuracies introduced by temperature variation causing circulatory currents, by gradual flocculation of the particles, and diffusion or Brownian movement of the smallest particles.

The sedimentation time can, however, be greatly reduced by using centrifugal instead of gravitational acceleration. Stokes' formula requires adaptation, because the velocity of a particle is not constant but increases as the radius of rotation increases with settlement. It becomes (**H84**)

$$v = \frac{d(h+r)}{dt} = \frac{d^2(D_1 - D_2)}{18\eta} \omega^2(h+r)$$

$$d = \left\{ \frac{18\eta}{(D_1 - D_2)\omega^2 t} \cdot \log_e \frac{(h+r)}{r} \right\}^{1/2}$$

where d is the equivalent diameter;
h the depth below the liquid surface;
r the radius of rotation of the liquid surface;
ω the angular speed of rotation in radians per sec;
The other symbols remain as before (*see* Stokes' Law, p. 279).

'The settling times for particles below 1 μ diameter may be reduced to practical values by quite moderate speeds of rotation, and Table 59 shows the gravitational and centrifugal settling time for particles of quartz, density 2·65, settling a depth of 5 cm in water at 15° C, the radius of rotation of the liquid surface being also 5 cm.'

TABLE 59

Settling Times in Minutes for Quartz Particles Falling 5 cm in Water at 15° C

Particle diameter (μ)	Gravitational Sedimentation	Centrifugal sedimentation $(h+r)/r = 2$			
		500 r.p.m.	1000 r.p.m.	2000 r.p.m.	4000 r.p.m.
20	2·6	—	—	—	—
10	10·6	—	—	—	—
5	42·3	2·1	—	—	—
2	264	13·1	3·3	—	—
1	1057	52·4	13·1	3·3	—
0·5	70 hours	209·6	52·4	13·1	3·3
0·2	18 days	—	328	82·0	20·5
0·1	10 weeks	—	—	328	82·0
0·05	—	—	—	—	328
0·02	—	—	—	—	2050

'The figures in the above table show that there is no difficulty in effecting the sedimentation of particles 0·1 μ diameter, and that even smaller particles could be sedimented with specially constructed centrifuges. Measurement during rotation of the concentration or density of the suspension is, however, a problem requiring continued investigation.' (**H84**)

The small divers devised by Berg (**B40**) can also be used to measure the density of suspensions being centrifuged. This is probably one of their most useful applications. There are numerous models of laboratory centrifuges on the market, varying in size and speed (**B4**). The change of acceleration due to increase of the radius of rotation during the sedimentation is minimised by using a long arm centrifuge (**N32**).

Optical Methods of Following Sedimentation

Particles of colloidal dimensions scatter light, so that the path of a beam of light passed through a colloidal suspension is clearly visible, each particle appearing as a point of light. The intensity of the emerging light beam is of course different from that of the one entering. This principle is used in various instruments.

In the turbidimeter the optical density of a suspension is found by measuring the light extinction. A very dilute suspension is used, when conditions for free falling and settling are approached. A narrow parallel beam of light is passed through the suspension at a known depth below the surface and the intensity of the emerging beam measured with a photoelectric cell.

This gives the optical density, which represents the surface area of the particles. It is converted to the weight per cent of particles present. It is this conversion that has not been fully elucidated yet and an empirical method must be used. Schweyer (**S45**) has shown that it is applicable only to ground sand, silica and certain cements. It is to be hoped that further work will lead to standard instruments calibrated to measure clay particle sizes as the method would then be simple and quicker than any involving evaporation and weighing.

Photographing of a suitably illuminated and magnified particle during its fall through a liquid gives an accurate value of its size. This photo-sedimentation method evolved by Carey and Stairmand (**C11**) is of particular use for checking other quicker and simpler methods (**S160**).

THE MICROSCOPE

Actual measurement and counts of particles can be achieved by observation under a microscope. The method has the tremendous advantage that the particles are observed individually, so that their shape can be determined as well. It is assumed that the particles come to rest on the microscope slide in their most stable position so that the dimensions observed are the two largest; the third must be estimated. Particle sizes are usually expressed as the diameter of a circle having the same area as the projected image of the particle when viewed in its most stable position.

The weight of material investigated under a microscope is inevitably small and if there is much variation in size a large number of particles must be measured and counted to give a representative value. Fairs (**F1**) in his description of the practical technique of the method states that for difficult samples the time taken for preparation, actual sizing, calculation and plotting of the results is not more than 3–4 hr.

Resolution by a microscope using visible light is limited by the wavelength of light to $0\cdot2\,\mu$. This is a theoretical limit that cannot be passed, however powerful the microscope. For higher resolution electromagnetic waves of shorter wavelength must be used, *e.g.* ultra-violet, electron beams. The resolution limits are summarised in Table 60.

The Electron Microscope. This instrument which has a resolving power down to 50–100 Å is no longer a rare novelty. (Until the importation of

TABLE 60 (H84)

Method	Lower Limit (μ)
Microscopical measurement	
Visible light	0·2 ⎫
Ultra-violet light in air	0·1 ⎬ particles resolved
Ultra-violet light in a vacuum or N_2	0·03 ⎬
Electron microscope	0·01 ⎭
Ultramicroscope	0·005 particles detected

six of them in 1942 under lease-lend there was only one in this country.) It is now being produced commercially and in a form relatively simple to operate, giving images either on a fluorescent screen or on a photographic plate. Its main problem is that of suitable preparation of specimens for observation.

Walton (**W7**) has summarised the main points to be borne in mind when using an electron microscope for particle size measurement as:

'(i) The specimen must not evaporate or deform when placed in a hard vacuum, and it must be resistant to fairly intense electron bombardment.
'(ii) The specimen has to be mounted on or in a delicate membrane over a minute hole in a small metal disk.
'(iii) The field of view covers a very small specimen area, approximately $30 \mu^2$.
'(iv) Accurate visual assessment of the electron image is difficult, for quantitative work each field of view must be recorded.
'(v) The total specimen area accessible for examination consists either of a single circle 50–100 μ diameter, or a number of such apertures lying within a $\frac{1}{2}$ mm diameter circle.
'(vi) Smooth, regular traversing of a specimen is difficult, although any desired area can usually be manipulated into the field of view. The position of any particular field with respect to the specimen screen is difficult to specify.
'(vii) The depth of focus is large and stereoscopic micrographs of deep structures can be taken with satisfactory sharpness.'

It is with the aid of the electron microscope that the actual size and shape of the minute clay particles has been determined, *e.g.* that kaolinite and montmorillonite are plate-like whereas haloysite is lath- or rod-like.

The Ultramicroscope. Microscopic observation of a colloidal suspension perpendicular to a transmitted beam of light is possible with this instrument. The colloidal particles appear as points of light and can be counted. By altering the plane of focus of the microscope the particles present in a volume of given depth can be counted. By evaporation the weight present is found and together with data on the density the particle size can be calculated. The ultramicroscope detects particles down to 0.005μ.

X-rays. The use of X-ray diffraction for determination of very small particles, 1μ to 0.01μ, is more satisfactory than the ultramicroscope.

The principle of the Debye–Scherrer powder X-ray photographs is that if small crystals are randomly orientated one photograph of the diffraction of monochromatic X-rays gives all the lines that would be obtained if a single crystal were rotated. If the crystals are large too few act as diffraction gratings and only part of the pattern is produced. If the crystals are very small diffraction is incomplete and broadening of the lines appears. This happens progressively for crystals from about 1 μ to 0·001 μ so that the size can be determined from the sharpness of the lines.

SURFACE FACTOR

The interpretation of the results of grain size analyses on a comparative basis is not easy. It is therefore sometimes expedient to convert the results to a single numerical value. As it is the surface of a particle that is involved in physical and chemical interactions, and as this varies with the grain size, a factor based on the surface area has been evolved.

First suggested in different forms by Jackson (**J5**) and Purdy (**P74**), the 'standard surface factor' now generally used is that evolved by Mellor (**M50**). It is a numerical way of expressing the texture (fineness) of clays and other ground materials which makes comparisons easy, but it is based on assumptions which we know do not necessarily hold and so must not be considered an absolute value.

The formula for the surface factor is derived as follows. A fraction of evenly distributed spherical particles of sizes varying between two known limits has total weight W. The specific gravity of the material is S. The 'mean' diameter of the particles is d (measured in millimetres). The average volume of the particles is

$$\frac{1}{6}\pi d^3$$

and the average weight

$$\frac{1}{6} S\pi d^3$$

The total number of particles, N, is

$$\frac{W}{\frac{1}{6}S\pi d^3}$$

and the total surface area is $N.\pi.d^2$, or

$$\frac{6W}{S\pi d^3} \times \pi d^2 = \frac{6W}{Sd}$$

If we have four fractions of weights W_1, W_2, W_3, W_4, and mean diameters d_1, d_2, d_3, d_4, respectively, the total surface area is:

$$\frac{6}{S}\left(\frac{W_1}{d_1}+\frac{W_2}{d_2}+\frac{W_3}{d_3}+\frac{W_4}{d_4}\right)$$

When $W_1+W_2+W_3+W_4=1$ this expression gives the value of the 'standard surface factor'.

The question then arises as to the best way of deriving d, the average diameter of a given fraction, from the known values of the two extreme sizes, d_a and d_b. It can easily be shown that the arithmetic mean,

$(d_a+d_b)/2$, is totally unrepresentative. Mellor suggests that the best 'average diameter' is deduced from the average volume found by integrating the complete range of possible volumes.

$$\text{Average volume} = \frac{\dfrac{\pi}{6}\displaystyle\int_{d_b}^{d_a} x^3\,dx}{\displaystyle\int_{d_b}^{d_a} dx}$$

where x is the diameter of the actual particles.

$$\text{Average volume} = \frac{\pi}{24}\frac{d_a^4-d_b^4}{d_a-d_b}$$

$$\text{Therefore average diameter} = \sqrt[3]{\frac{d_a^4-d_b^4}{4(d_a-d_b)}}$$

or, expressed in a more easily evaluated form:

$$\text{Average diameter} = \sqrt[3]{\frac{(d_a+d_b)(d_a^2+d_b^2)}{4}}$$

This is the formula generally used.

It will have been seen from the course of the argument what assumptions have to be made if the formulae are to be applied to real fractions of graded material. These are:

(1) That the particles are spherical.
(2) That the particles are evenly graded and that each size has the same number of particles.
(3) That the density is constant throughout.

It is the fact that they are not true for most materials actually used that limits the reliability of the 'standard surface factor'.

In industrial practice the 'standard surface factor' is found by dividing the ground material under examination into a specific number of fractions. One method employs sieving followed by elutriation to give four fractions (**H52**):

(1) Residue on IMM 100 lawn.
(2) Residue on IMM 200 lawn.
(3) The fraction remaining in the elutriation apparatus after fraction (4) has been removed.
(4) The fraction washed out with a stream of water at constant velocity.

Fraction	Limiting diameters of spherical particles (mm)	Average diameter (mm)
(1)	>0·127	—
(2)	0·127–0·063	0·098
(3)	0·063–0·010	0·042
(4)	<0·010	0·0063

Where the hydrometer has replaced the elutriator either two or four fractions are taken as follows:

Fraction	Limiting diameters (mm)
(1)	0·063 (200-mesh)–0·01
(2)	0·01–zero
(1)	0·063 (200-mesh)–20 μ
(2)	20 μ–10 μ
(3)	10 μ– 5 μ
(4)	5 μ– zero

The second method gives a higher value for the surface factor because the minus 0·01 mm fraction usually contains more fine material than is allowed for in the calculation of the average diameter for the 0·01 mm to zero range.

To recapitulate:

$$\text{'standard surface factor'} = \frac{6}{S}\left(\frac{W_1}{d_1} + \frac{W_2}{d_2} \cdots \frac{W_n}{d_n}\right)$$

where $W_n \times 100$ is the percentage weight present in a fraction of given extreme sizes d_a and d_b, and d_n, the average diameter of that fraction, is given by

$$d_n = \sqrt[3]{\frac{(d_a+d_b)(d_a^2+d_b^2)}{4}}$$

Normal Values for the Surface Factor

Earthenware: flint, 230 ± 10; stone 230 ± 10.
Tiles: flint, 210 ± 10; stone, 230 ± 10.
Bone china: stone 245 ± 5; bone 270 with 75 to 80% finer than 0·01 mm.
General ceramic use: feldspar 240 ± 10; whiting 200–250.

MEASUREMENT OF RELATIVE SPECIFIC SURFACE BY THE AIR PERMEABILITY METHOD

It is often sufficient for control purposes to know the specific surface (surface area per unit weight) of a sample rather than the particle size and distribution. This has gained favour because of the simplicity and ease of operation of the apparatus required to determine the permeability of the powder to air.

In the apparatus devised by Lea and Nurse (**L18, 19**) the dry test powder is compressed to a definite porosity in a special permeability cell. Air is then made to flow through, the pressure and rate of flow being measured

(Fig. 4.15). If the particles are spherical it is possible to calculate an absolute value for the specific surface. For non-spherical particles like those used in the ceramic industry the value obtained is not absolute but gives a figure useful in comparisons.

Lea and Nurse derive the equation:

$$S = \frac{14}{\sigma(1-\epsilon)} \sqrt{\frac{\epsilon^3 A h_1}{CL h_2}}$$

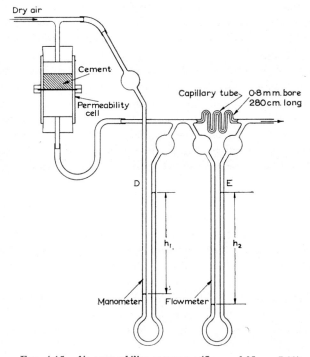

FIG. 4.15. *Air permeability apparatus (Lea and Nurse, L19)*

where h_1 and h_2 are manometer and flowmeter readings as indicated on the diagram;

ϵ is the porosity (volume of pore space in 1 cm³ bed);
A is area of bed (cm²);
L is depth of bed (cm);
C is a constant for a given capillary;
σ is the density of the solid (g/cm³) (**L19**).

A simplified method used industrially is to find the resistance of the powder to a current of air. A cylinder is filled with a given weight of the powder with a certain degree of packing and the time taken for the passage of a certain volume of air through it is noted. The values are proportional to approximate specific surface and therefore directly comparable.

INVESTIGATION OF CLAYS. BASIC DATA

Preliminary Tests on 'Natural Clay' Deposits

When a deposit of clayey material is found, preliminary tests must be made to determine its possible uses (**K35**).

(a) VISUAL AND TEXTURAL INSPECTION

Unfortunately the colour of the raw clay does not always give a fair indication of that of the fired clay. Thus most but not all clays that are white or cream when raw remain so on burning. A very dark clay may also burn white. Most yellow clays burn yellow, other coloured clays may change their colour on firing.

Colour does not directly define the use of a clay, for although fine ware is predominantly white many beautiful art products have a deep colour; similarly, although heavy clay products are generally made from red- or yellow-burning clays, the refractories are often quite light coloured.

The texture of the raw clay will be helpful only if it is soft when the relative fineness can be felt by hand. But many fine-grained clays occur in hard masses and have to be ground before their properties can be determined.

(b) SCREENING

It is useful to break the clay up without grinding the individual particles and screen it. If most of it remains on the 18-mesh sieve the material is unlikely to contain much clay substance and is only of specialised use in ceramics. If there is a small residue on the 18-mesh sieve and fair residues on 60-, 100- and 200-mesh sieves, the material is probably suitable for coarser ceramic products and on grinding may be good for fine ones. A material with less than 3% residue on the 200-mesh and $\frac{1}{2}$% on the 325- (U.S.A.) mesh is very fine-grained and probably suitable for fine ware.

(c) PLASTICITY

This may be followed by a crude plasticity test, for instance gradually working water into the powdered clay by hand and making an estimate of plasticity, *i.e.* none, poor, fair, good, exceptional.

(d) CHEMICAL TESTS

(1) For carbonates. If effervescence occurs with dilute cold HCl (1:1) calcite is present, on warming dolomite will effervesce too. Presence of carbonates makes a clay troublesome in ceramics and such clays should be avoided if possible.

(2) Iron. Treatment of the clay with hot conc. HCl and an oxidising agent ($KMnO_4$, $K_2Cr_2O_7$) will generally dissolve at least part of the iron. This is then precipitated with NH_4OH and the volume and colour give an estimate of the iron.

(3) Free alumina. This is soluble in hot caustic soda solution and can be reprecipitated and estimated. A very high alumina content indicates a refractory clay (*see* silicate chemistry).

(4) Phosphates. The sample is dissolved in aqua regia, suitably diluted and then tested with ammonium molybdate. Phosphates are of importance for certain refractories, and certain glazes.

(5) Water-soluble salts. The dried weighed sample is agitated with water, then allowed to settle, or centrifuged and the clear liquid evaporated to dryness. It is essential to know how much water-soluble salt is present in a ceramic clay. For instance the high magnesium sulphate content of clay in the Middle East makes special treatment necessary before it can be deflocculated.

(*e*) DRYING AND FIRING

If these preliminary tests show that a clay may be suitable for the ceramic industry it is well worth making drying and firing tests before going into a detailed investigation. Small bars are shaped and observed during drying and firing for excessive shrinkage, cracking and warping, formation of scums, bloating and fired colour. It must then be decided whether a clay showing defects is worth improving by certain treatments or suitable mixing with other clays.

Sampling

The composition and nature of a clay deposit frequently varies considerably when passing down or along a single seam. The method of sampling the clay for investigation must therefore be suited to the type of result that is required and must form part of the record made. Tests carried out on samples taken directly from the clay beds, in fact, refer only to the samples themselves unless suitable precautions have been taken to make them represent the bulk. For instance if the variations in a seam are to be investigated small samples are taken from widely but regularly spaced positions and tested individually. If changes in average properties are to be ascertained small samples are taken at regular intervals during the day and mixed to give a bulk sample whose properties are compared with similar bulk samples taken at, say, weekly intervals.

For most circumstances samples of raw clays should weigh about 100 lb (45·3 kg) so that several tests can be performed with up to twelve test pieces for each.

Grain-size Determination

Many of the properties of clays are related to the particle size and size distribution of their constituents. Particle size determines the surface area and therefore the relationship with water which causes plasticity and viscosity, and shrinkage. Adsorption of ions occurs at the surface, the particle becomes charged and affects other particles.

Particle-size distribution determines the packing of the particles and is related to plasticity, dry strength, and porosity. The densest packing of spheres of uniform size which is independent of the actual size has a porosity of 25·95%. By filling in the holes with four more decreasing sizes the porosity is reduced to 14·9%, and with further very fine sizes it can be brought to 3·9%. Even where the particles are not spheres a good range of sizes gives closer packing, and an ideal distribution has been worked out (**L34**).

Although many properties of clay are related to particle size they are unfortunately not linear functions. Harman and Fraulini (**H25**) have investigated the properties of size fractions of a pure kaolin. They found that base-exchange capacity, heat of wetting and drying shrinkage, although they vary with particle size, are not simple functions of it, but they are linear functions of each other. From their results they are able to deduce that the larger particles are aggregates, and therefore have a corrugated surface of larger area than a sphere of the same radius.

This non-linearity of clay properties with particle sizes makes it essential to control the sizes and distribution in an industrial product, as variations will give unpredictable results.

Another important factor greatly influenced by particle size is the firing temperature. The interactions in heated ceramic ware occur primarily between solids; they are therefore greatly assisted by the intimate contact obtained with fine particles and will occur at lower temperatures. Fired density is therefore also related to the original particle size.

The general methods of determining grain size are given above. Clays must, however, first be brought into an aqueous suspension in which particles are properly separated from each other without additional cleavage occurring.

Dispersion of the Raw Clay

The individual particles in the clay must be separated without disrupting them. They must, therefore, be dispersed in an aqueous medium which prevents coagulation.

In order to eliminate stray coagulating ions distilled water is used. A dispersing agent which influences the pH is added, *e.g.* sodium carbonate, caustic soda, ammonia, sodium silicate (water glass), sodium oxalate, sodium pyrophosphate, potassium citrate. Individual clays give different results with the various dispersing agents, the actual quantities not being unimportant. In some cases proper dispersion only occurs in a narrowly confined range. It is best to have a few trial runs to find the best overall dispersing agent. In general sodium oxalate and sodium pyrophosphate seem to be the most favoured.

The dispersion is achieved by shaking for a long period (Zimmermann (**Z6**) recommends seventeen hours), or shaking and warming, or boiling (half an hour). He has shown that organic matter, which can be removed by treatment with hydrogen peroxide, and lime, which can be removed with dilute hydrochloric acid, do not affect the grain-size determination and therefore evidently do not permanently stick the particles together.

Chemical Analysis

Chemical analysis of known clays acts as a check on uniformity. On new clay deposits it can give qualitative ideas on what to expect.

For instance the sum of the Fe_2O_3 and TiO_2 contents affects the fired colour; if it is less than 2% the clay will usually burn white or light coloured. This may have been masked by dark organic matter in the raw clay which burns out. Consideration of the silica and alumina content is also worth while. Remembering that the 'ideal' clay substance, namely $Al_2O_3.2SiO_2.2H_2O$, contains 46·51% SiO_2, 39·53% Al_2O_3 and 13·95% water (loss on ignition), a silica content of above 46·5% may indicate free silica or quartz. A high alkali content generally indicates a vitreous fired product. A loss on ignition greater than 14% indicates carbonaceous matter, such clay may be plastic and often deflocculates readily for casting purposes unless soluble salts interfere (**A50**).

The complete chemical analysis is also required to derive calculated rational analyses.

Rational Analysis

The old 'rational analysis', first introduced as a great advance by Seger (**S52a**) in 1876 and subsequently frequently altered and improved on, is now known to be totally unsuitable for a true mineralogical analysis of clays. It does, however, form an empirical method for comparison of similar material, and as such is still much used in many of its different forms.

The complete chemical analysis of clays shows the exact quantities of the elements present, but it gives no data on the way in which these are combined. It will be clear from the description of clays, their make-up and their properties, that their characteristics are dependent on the mineral composition rather than the total chemical composition.

Seger (**S52a**), realising this in 1876, evolved the rational analysis, which divided a clay into parts according to its solubility in concentrated sulphuric acid, the soluble part being termed 'clay substance' and the insoluble part being further subdivided into quartz and feldspar by the reaction with hydrofluoric acid. The procedure was improved on by a number of workers, *e.g.* Berdel (**B39**), Bollenbach (**B89**) and Koerner (**K61**).

In 1897 Vogt (**V20**) pointed out that certain clays contain micaceous matter that dissolves in sulphuric acid but does not act as a clay mineral. This was taken up by Kallauner and Matejka (**K2**) in 1914, whose method of decomposing the 'clay substance' by heat at 750° C and determining the liberated aluminium oxide by solution in hydrochloric acid (1:1) does not affect the mica.

Keppeler and Gotthardt (**K18, 19**) combined the two methods, thus obtaining a value for the 'mica' from the difference between the results given by the sulphuric acid and the calcination methods. They also realised the importance of determining the water-soluble salts and the organic matter present in clays.

The Hirsch–Dawihl (**H103**) method estimates the free quartz by dissolving the clay substance, mica and feldspar in phosphoric acid.

Each method is useful for one of the major components and can be summarised as follows (**D33**):

Berdel (H_2SO_4): feldspar (unweathered).
Kallauner–Matejka (calcination): clay substance (without intermediate weathering products).
Keppeler (combined method): mica.
Hirsch–Dawihl (H_3PO_4): quartz.

The presentation of the results obtained from these methods of 'rational analysis' depend on several assumptions. Firstly that a clay consists of the minerals 'clay substance', 'mica', 'feldspar' and 'quartz', each of which has a definite chemical formula; and secondly that the selective treatment acts on the whole of one mineral leaving another totally unchanged. It is now known that the first three so-called minerals are groups of minerals with varying chemical compositions, and that the chemical reactivity is affected by particle size so that the very fine particles of the less reactive mineral may be attacked before the larger grains of the more reactive ones have dissolved. The rational analysis must therefore be considered with care and is generally only applicable to the purer clays.

The difficulty of finding a reagent specific for a certain mineral is overcome by calculating the rational analysis from the ultimate analysis. Here again the assumption of a definite chemical formula for a given mineral is very misleading. For instance, in the older methods, originally worked out for kaolins, for which they give reasonable results, the alkalies and lime are allocated to feldspar ($K_2O.Al_2O_3.6SiO_2$) the remaining alumina to theoretical kaolinite ($Al_2O_3.2SiO_2.2H_2O$) and the remaining silica is considered to be quartz. Application of this to ball clays which rarely contain feldspar, the alkalies being present in illites, montmorillonites, mica, tourmaline, etc., is entirely incorrect. The use of the ideal kaolinite formula for all the clay minerals is also erroneous, as can be seen from Table 61, which lists the commoner ones.

Accordingly McVay (**M19**) suggests a different method of computing ultimate analysis results for ball clays to give a basis for comparisons. All the alumina present is allocated to kaolinite, any remaining silica is termed 'excess silica', leaving open the possibility of its being combined in a mineral other than quartz; the total alkalies, total alkaline earths, and the iron are reported as such and the titanium is considered to be rutile.

This calculation of a rational analysis from the ultimate analysis by application of a qualitative knowledge of minerals usually found in clays is taken further by Shipley (**S67**). He suggests making a series of assumptions

TABLE 61

Clay Minerals

(After Grim (**M19**))

Name	Chemical Composition	Remarks
Kaolinite	$Al_2O_3.2SiO_2.2H_2O$	Anauxite and kaolinite form an isomorphous series.
Anauxite	$Al_2O_3.3SiO_2.2H_2O$	
Halloysite	$Al_2O_3.2SiO_2.xH_2O$	
Beidellite	$Al_2O_3.3SiO_2.xH_2O$	Beidellite and nontronite form an isomorphous series.
Nontronite	$Fe_2O_3.3SiO_2.xH_2O$	
Montmorillonite	$Al_2O_3.4SiO_2.xH_2O$	Montmorillonite, beidellite and nontronite probably contain essential alkalies or alkaline earths.
Illite	$(OH)_4K_y(Al_4.Fe_4.Mg_4.Mg_6)(Si_{8-y}.Al_y)O_{20}$	

about the minerals present, allotting the elements found in the complete analysis to complex formulae. The quantity of the clay mineral deduced by this method gives a theoretical loss on ignition very close to the experimental value. The method is nevertheless empirical, moreover rather complicated and not in general industrial use.

Klug (**K38**), although still adhering to the old calculated rational analysis which allocates all the alkalies to feldspars, suggests that more help can be derived from consideration of the ratios:

$(SiO_2 + TiO_2):(K_2O + Na_2O)$, or the total flux content
$(SiO_2 + Al_2O_3)$:total flux content.

This idea clearly originates from consideration of glazes and it is essential to remember that whereas glasses and glazes are the products of a completed reaction and are homogeneous, ceramic bodies are not so. The usefulness of these ratios is therefore empirical, and questionable.

The problem of interpreting the chemical analysis of a raw material containing several minerals arises in cases other than the clays.

The question of the mineralogical composition of feldspar has been tackled by Koenig (**K50**) and subsequently by Coffeen (**C44**). The former's method of calculation is as follows. Let

$A = \% H_2O$
$B = \% SiO_2$
$C = \% Al_2O_3$
$D = \% CaO$
$E = \% Na_2O$
$F = \% K_2O$

from the chemical analysis.

and

$U = \%$ albite, $Na_2O.Al_2O_3.6SiO_2$
$V = \%$ microcline, $K_2O.Al_2O_3.6SiO_2$
$W = \%$ anorthite, $CaO.Al_2O_3.2SiO_2$ } calculated mineral composition.
$X = \%$ muscovite, $K_2O.3Al_2O_3.6SiO_2.2H_2O$
$Y = \%$ quartz, SiO_2
$Z = \%$ kaolinite, $Al_2O_3.2SiO_2.2H_2O$

The second section being calculated from the first by means of the following equations:

$U = 8.458E$
$V = 15.442A - 5.459C + 9.923D + 8.977E + 11.815F$
$W = 4.960D$
$X = 7.813C - 22.097A - 14.201D - 12.847E - 8.455F$
$Y = B + 1.178C - 6.666A - 4.284D - 7.751E - 5.101F$
$Z = 2.740F + 14.323A - 2.532C + 4.602D + 4.163E$

Small errors in the analysis and presence of minerals, other than the six anticipated, readily upset this calculation, in particular a small error in the water content makes a large error in the muscovite (X) content and often gives a negative answer. Coffeen (**C44**) therefore proposes to adapt the method. The value of X is found first, and if it proves to be negative he suggests taking it as zero and making the subsequent calculations as follows:

$U = 8.458E$

$V = \dfrac{F - 0.118X}{0.169}$

$W = 4.960D$

$Z = \dfrac{C - 0.195U - 0.183V - 0.365W - 0.385X}{0.395}$

$Y = B - 0.687U - 0.648V - 0.430W - 0.452X - 0.465Z$

Even the improved method has the same disadvantages as well as advantages as similar calculations for clays already described.

It is now generally agreed that direct chemical means are of little use in determining the mineralogical make-up of clays. For satisfactory data physicochemical means must be used, *i.e.* one or more of the following: differential thermal analysis (D.T.A.), X-ray analysis, microscopic and electromicroscopic investigation.

Various forms of chemical rational analysis either with definite reagents or by calculation from the complete analysis still find their uses in process control. They will detect changes in proportion of the same mineral but they may easily miss a radical change of minerals and cause considerable losses.

Methods of Determining Mineral Content

We have mentioned the principal minerals that occur in clays and that their individual characteristics become part of the properties of the clay itself. Study of the minerals present in a clay together with knowledge of their grain size will therefore give valuable information. Unfortunately such data are difficult to obtain. Grimshaw and Roberts (**G84**) and Linseis (**L36**), give summaries of the methods they have used to produce as complete as possible a picture of a clay's constitution. Although these methods are, in general, supplementary to one another, certain individual ones may prove very useful by themselves.

INVESTIGATION OF THE CLAY SOURCE

This most obvious source of information should not be ignored. Visual examination of the clay in bulk, if possible *in situ* may reveal stratification and the presence of mica along the bedding planes. Other larger mineral fragments may be visible. (Not only does such examination identify some of the clay constituents but it is obvious that the mixed product will be irregular.) The presence of organic matter may be inferred from the colour, the remains of roots and plants and sometimes shale or coal particles.

Knowledge of the geological age and nature of the occurrence of a clay will give information about the speed of sedimentation and therefore the particle sizes. The geological environment of a clay influences the impurities, *e.g.* clays in sandstone areas will contain quartz, those in limestone areas, calcium carbonate.

SEPARATION OF THE MINERALS

This is a difficult task and must be carried out in most cases by physical methods. Separation into grain sizes is the first step. Many minerals are concentrated in the larger or smaller grained fractions.

Several of the methods used for determining the grain sizes of clays are also suitable for separating the mixture into grain-size fractions. The clay is first thoroughly dispersed, and then passed through sieves followed by elutriation and/or sedimentation and centrifuging. The last two do not normally produce clean fractions and have to be repeated several times. The benefit of grain sizing towards separating the minerals present is shown by these two examples (Tables 62 and 63).

Table 62 shows clearly how all the 'clay mineral' which in this case is the very fine-grained montmorillonite concentrates in the smallest-sized fraction.

In the next example a primary kaolin was divided into three portions centrifugally. These were analysed chemically and the results are shown in Table 63.

It is seen that the alumina content is greater in the finer fractions and the silica content decreases correspondingly. From this it is deduced that there is more quartz in the coarser fractions. The decreasing alkali content, with fineness, is interpreted as denoting decrease of sericite mica.

As the different methods overlap in size range some may be omitted. The scheme adopted by Grimshaw and Roberts is outlined (**G86**). The

TABLE 62

Grain Size Distribution and Mineral Content of the Diluvial Clay from Papendorf

(Schlünz (S18))

	<2 μ	2–11 μ	11–24 μ	24–60 μ	>60 μ
Grain size distribution	34·8	56·3	5·7	2·3	0·9
Quartz	30–50	18·7	23·4	25·6	28·7
Feldspar	—	7·3	11·2	16·1	17·6
Calcite	—	13·5	12·9	15·0	13·6
Hornblende	—	6·6	7·5	6·7	7·0
Biotite	—	2·7	5·4	5·8	7·0
Muscovite	30–50	26·6	18·6	15·4	11·5
Chlorite and serpentine	—	5·1	5·5	3·6	5·7
Montmorillonite	10–30	—	—	—	—
Minerals of high refractive index	—	5·6	4·7	3·7	2·4
Non-transparent minerals	—	3·3	3·1	1·6	1·1
Undetermined minerals	—	10·6	7·7	6·5	5·4

TABLE 63 (L53)

	Coarse (%)	Intermediate (%)	Fine (%)
H_2O	0·53	0·66	0·58
Ignition loss	6·54	8·86	12·35
SiO_2	63·92	54·42	47·40
TiO_2	1·33	1·29	1·01
Fe_2O_3	0·98	0·89	0·49*
Al_2O_3	22·79	30·44	36·50
Alkalies	3·60	3·34	1·62
Total	99·69	99·90	99·65

* Iron content of fines reduced slightly by chemical bleaching.

clay is first thoroughly dispersed. The suspension is then poured through graded sieves, ranging from 60 B.S.S. to 240 B.S.S. The largest retains a high proportion of ironstone nodules and a few of the largest mica flakes. The smaller sieves retain mica, quartz and most of the heavy minerals.

The suspension passing through the finest sieve is allowed to settle for twenty-four hours. The sediment contains all the remaining quartz and heavy minerals and can be washed free of any retained clay or other fine-particled mineral.

The remaining suspension now contains all the clay minerals originally present and any other minerals of comparable particle size.

It is divided up into grain-size zones by *centrifuging*. Separation by centrifuging is governed by the same laws (Stokes' Law) as sedimentation, except that the force of gravity is greatly increased. A supercentrifuge is capable of producing a g of 50 000. The speed of separation makes the relative starting position of different sized particles an important factor, and fractions contain a fair amount of material of the wrong size. By re-centrifuging, however, the particles can be separated into closely mono-dispersed fractions.

Marshall (M31) describes a method for making clean separations of grain sizes. If a thin layer of a clay suspension is placed above a column of aqueous solution of higher density (*e.g.* with sugar or glycerol) the particles will move downwards with their characteristic speeds, unimpeded by lower particles of different size. If this is carried out in a centrifuge disturbing counter-effects are eliminated. Marshall uses a centrifuge with six 50 ml tubes which can be run at from 1000 to 6000 r.p.m. For quantitative work 0·5% suspensions, and for qualitative work 1% suspensions are used. Some results obtained with clays dispersed by bringing to pH9 with NaOH are tabulated in Table 64.

TABLE 64

The Distribution of Particle Sizes in Four Clay Types (M31)

	Concentration %	2 μ–1 μ	1–0·5 μ	500–200 mμ	200–100 mμ	100–50 mμ	<50 mμ
Kaolin (English)	0·2	66·0	21·0	7·0	6·0	—	—
Bentonite (U.S.A.)	1·0	2·0	31·0	16·0	12·0	39·0	
Putnam (Missouri U.S.A.)	0·5	7·8	6·6	11·8	11·6	21·3	40·9
Rothamsted Exp. Stn.	0·5	15·2	12·1	18·7	14·3	10·3	29·4

Grimshaw and Roberts (G86) also suggest further separation of centrifuged fractions by elutriation with very slow liquid flow.

FURTHER SEPARATION OF THE MINERALS BY SPECIAL METHODS

Where several different minerals occur in the same grain size fraction further specific methods of separation must be applied. *Physical methods* include: (1) heavy liquid separation; (ii) froth flotation; (iii) electrostatic separation; (iv) magnetic separation.

(i) *Heavy Liquid Separation.* Where minerals of different density are present, and the particles are not too fine, separation can be achieved with a liquid of intermediate density. Small particles tend to flocculate and restrict free sedimentation. The choice of liquid also makes a difference. For instance, specific gravity separation of clays in organic liquids proves to be successful in bromoform (S.G.2.89), but coagulation occurs in tetrabromethane (S.G.2.96) (D59).

(ii) *Froth Flotation.* If a mineral is covered by a unimolecular layer of a water-repelling substance it will tend to collect at an air–water interface in preference to other minerals present that are wettable. If air is bubbled through the mixture it will collect in the froth, particularly if this is stabilised. The water-repelling cover is known as a 'collector'; it is an organic compound with a polar group which forms a loose chemical attachment with the mineral. It is therefore theoretically possible to separate any two chemically different minerals by froth flotation.

The method is used a great deal for concentrating minerals of metals, but is also applicable in the ceramic field since the recognition that some minerals are collected successfully with 'anionic reagents' whereas others are collected by 'cationic reagents'. This is because most mineral particles have a surface charge and will react with molecules having a polar group of opposite charge. Quartz, mica and the clay minerals are negatively charged and so require collectors whose surface-active constituent is a positive ion. Thus short-chain amines, such as di-n-butyl amine, are effective collectors of minerals containing 5% water, *e.g.* talc, pyrophyllite, sericite, clays and weathered mica; whilst anhydrous minerals require amines with long chains such as lauryl or stearyl amine hydrochloride, stearyl or cetyl trimethyl ammonium bromide. Quartz and mica can now be separated. For example, pyrophyllite can be separated from a mixture of 56% pyrophyllite and 43% quartz with small amounts of iron oxide, mica and ilmenite by using quarternary ammonium compounds or amines for floating the required mineral and preventing the flotation of quartz with hydrofluoric acid (**L12**).

(iii) *Electrostatic Separation.* As most mineral particles carry an electric charge, and as different minerals generally have different charges, it should be possible to attract them differentially to charged plates. In electrostatic separation the ground material is allowed to fall freely between two oppositely charged plates. Muscovite mica is deflected appreciably by an applied voltage of only 3000 V; feldspar is readily separated from quartz. Unfortunately many of the other minerals associated with clays behave similarly in an electric field, so very high voltages have to be applied to get a separation.

(iv) *Magnetic Separation.* This can be applied to larger particles of haematite or other magnetic minerals. Its application is perforce limited.

Chemical methods for the separation of minerals can be applied in two ways: (*a*) If one constituent is unattacked by a reagent that removes all the others it can be determined directly; (*b*) one constituent can be removed selectively.

(*a*) For instance the quartz content of a clay can be separated from the rest and determined by removing all the other constituents (**T41**). The sample is strongly fluxed with potassium pyrosulphate in a silica glass crucible when the minerals are decomposed liberating free silicic acid. This is dissolved out with hot concentrated caustic soda solution. The quartz remains as residue. Such a determination takes only eight hours.

(*b*) For instance dilute acid dissolves carbonates. Free Fe_2O_3 is almost completely dissolved out by shaking for three days with 0·1N sodium oxalate solution. Talc is dissolved almost instantaneously by 2·5% HF whereas muscovite and tremolite remain unaffected for some time (**D59**).

IDENTIFICATION AND ESTIMATION OF MINERALS IN CLAYS

The more selective methods of separating the minerals also identify and estimate them. The other methods, which do not necessarily depend on previous separation, can be divided into four groups:

(1) Methods dependent on the chemical nature of the minerals.
(2) Methods dependent on the external crystallographic features.
(3) Methods dependent on the internal atomic, ionic and crystal lattice arrangements.
(4) Methods dependent on some physical or chemical change which can be measured under controlled conditions.

Chemical Analysis. The main subject of chemical analysis is discussed at some length elsewhere. In connection with the identifying and estimating of clay minerals it can be useful for determining minerals containing elements not present in any other mineral. For instance in many clays that do not readily absorb ions the alkali present is all part of feldspars, micas or hydromicas. Determination of the nature of the mineral followed by the determination of the alkali content gives the quantity of the mineral.

Colour reactions are given by aromatic amines and phenols with certain clay minerals. Benzidine (benzidine sulphate) may be used to distinguish kaolinite from montmorillonite, the latter giving a blue colour (**H78**). Interference may be caused by soluble compounds, like those of iron and manganese, which oxidise benzidine to purple or purplish-black colours, or others which have reducing properties. Endell (**E18**) found that some montmorillonites do not give the characteristic blue, and Siegl (**S71**) suggests the addition of a few drops of ferric chloride solution when all montmorillonites react. It is clearly necessary to elucidate this reaction further before saying that it is infallible. Another colour test developed in the United States uses safranin. The sample is heated with strong hydrochloric acid, distilled and filtered. It is then stained with a few drops of safranin and investigated under the microscope; some clays turn blue, bluish purple, reddish purple, or violet, and others absorb the dye without changing colour (**A90**).

Moisture Content. Keeling (**K10**) describes a method dependent on the physicochemical relationship between clay mineral particles and water. The ratio of the loss on ignition to the moisture absorption is found to be characteristic of the individual clay minerals, *e.g.*

China clays and primary kaolins	7–16
Fireclays	2·1–3·9
Ball clays (Dorset and N. Devon)	1·9–3·1

Ball clay (S. Devon) (high value, probably due to china clay)	2·1–6·7
Etruria marls	1·6–2·8
Keuper marls	0·8–1·4
Sepiolithic marl	0·6–0·7
Weald clays	0·1–1·8
London clays	0·8–1·0

In making the measurements allowance must be made for minerals that have a loss on ignition but no moisture absorption, such as the micas.

Optical Methods. Microscopic investigation of both the original clay and the fractions can give valuable information. If the clay has been fractionated and is in aqueous suspension a suitable dilute drop of this can be mounted on the microscope slide. Where the original clay is sufficiently hard it is impregnated with bakelite and a thin section is cut. As this method of examination is also used for finished ceramic products it will be described here somewhat more fully.

The material to be investigated is impregnated with bakelite varnish, sometimes with a dye added. It is slit with a diamond cutting disk to $\frac{1}{8}$ to $\frac{1}{16}$ in., and then ground down to $\frac{1}{16}$ in., mounted on a slide and further ground until it transmits sufficient light. Adequate transmission occurs when the specimen has the thickness of only very few (one to three) crystals. As some of the crystals are of the order of $1\,\mu$ it is a good plan to grind a sloping upper surface which tapers out to the smallest possible thinness. The specimen is covered with a cover slip when the thin section becomes ready for microscopic examination.

In favourable circumstances a trained observer can identify the minerals present from examination of the thin section or aqueous suspension alone. However, the data obtained are frequently used in collaboration with the refractive indices found by examining the powdered material, and are further assisted by knowledge of the chemical composition. Examination of the mounted specimen also makes possible a quantitative estimate of the minerals present, the grain size and grading, the texture, and the pore size of a solid (**R32**).

In order to identify a mineral certain properties must be ascertained. Two of the most important of these can be obtained by examination of the thin section; namely the crystal group (whether the mineral is isotropic, uniaxial positive, uniaxial negative, biaxial positive, biaxial negative) and the birefringence (the difference between the maximum and minimum refractive indices in the crystal). These together with the refractive index, which is obtained independently, are usually sufficient to identify a mineral. Additional confirmation is given by data on the optical orientation, the colour and pleochromism and the axial angle and dispersion.

The refractive index of a mineral can be determined to ± 0.002 by immersion in liquids of known refractive index. Rigby suggests a series of liquids in ascending steps of 0·01. Correns (**C57**) evolved a system to enable the same specimen to be used throughout by using a mixture of liquids whose refractive index gradually changes on evaporation. Rigby (**R32**) gives details of the optical properties of minerals.

As is well known the use of the optical microscope is limited by the wavelength of visible light to 0·2 μ. Further resolution can be obtained by using ultraviolet light, and still further by using ultraviolet light in a vacuum or a nitrogen atmosphere, the limit being 0·1 μ and 0·03 μ, respectively. The optical system must be specially constructed of material transparent to ultraviolet light, e.g. quartz, and photographic recording is necessary.

The Electron Microscope. The use of very much shorter waves in this instrument makes it possible to resolve particles down to 0·01 μ. Specimens have to be specially mounted on a delicate cellulose membrane, and the field of view is very small, 30 μ^2. However, as visual observation on a fluorescent screen is possible it is easier to use than the microscope using ultra-violet light.

It is with the electron microscope that it has been possible to determine the actual sizes and shapes of clay particles (*see* illustrations of the clay minerals, pp. 14–23) and to account for variation of properties in different specimens that other instruments show to be of the same mineral, e.g. certain kaolins show higher viscosity in a suspension and greater green strength of a shaped body than others. They cannot be differentiated by X-rays or differential thermal analysis, but the electron microscope showed that the less viscous samples have thick simple crystals (type A) whereas those of medium viscosity have thin complex crystals (type B) (**C35**) (Fig. 4.16).

Improvements in replica technique have now made possible the high magnification study of textural characteristics and surface features of clay aggregates, as found in nature or in the laboratory. Features of the clay minerals have been observed that had not beforehand been seen in electron micrographs of samples that had been 'prepared' (and disturbed) (**B20**) (*cf*. Figs. 1.14 (a) with (b), 1.18 with 1.19, 1.11 (a) with 4.16).

X-rays. The wavelength of X-rays is so short that they can penetrate in between atoms in a solid. The regular atomic arrangements in a crystal then act as a diffraction grating so that the emerging rays occur at definite angles to the incident beam. From these the structure of the crystal can be found. A large crystal can be examined by observing the diffraction of a monochromatic beam of X-rays impinging on it at various angles in turn. Where only a powder is available all the orientations are observed at once, this is the Debye–Scherrer powder technique.

X-ray diffraction is the means by which we know the internal structure of the minerals and therefore the surest way of identifying them. Where it has been possible to separate the minerals such identification is straightforward.

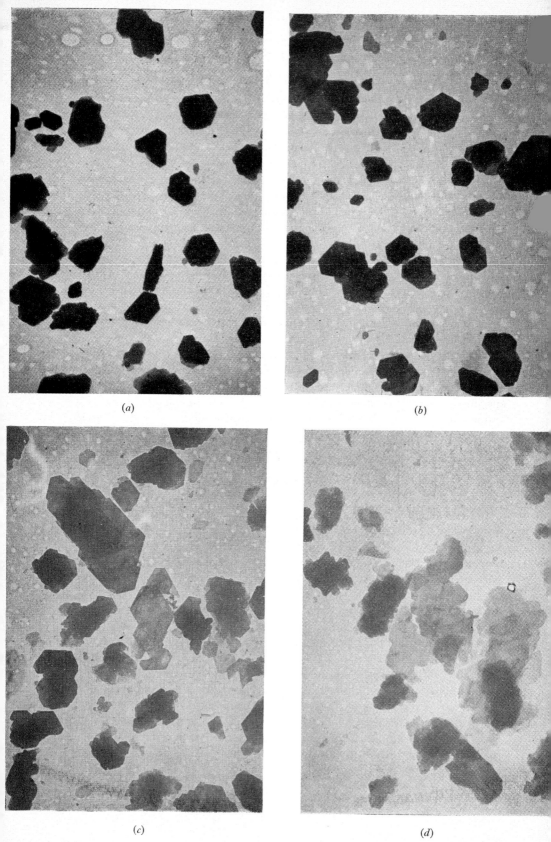

Fig. 4.16. *Electron microscope pictures of two types of kaolinite:* Type A: (a) *Georgia kaolin;* (b) *English kaolin A*
Type B: (c) *English kaolin B;* (d) *Zettlitz kaolin (Clark,* **C35**)

Unfortunately, however, where several clay minerals are present, the lines given by one may dominate the others in a straightforward powder photograph. For instance Keppeler (**K20**) has shown that small quantities of very fine-grained montmorillonite present with kaolinite are not detected by X-rays although their influence on the clay–water relationships may be sufficient to detect them in that way.

Rivière (**R38**) describes a method by which the number of lines in the X-ray photograph is reduced. He takes advantage of the plate-like shape of the individual particles. On slow evaporation of a suspension the flakes settle on top of each other with all the flat sides parallel. They can then be investigated like a single crystal giving a few characteristic lines. It is also easy to measure the height of the unit layer, which is characteristic for the clay mineral groups (7·1 Å, kaolinite; 7·3 Å, haloysite; 10 Å and more, montmorillonite).

Grim (**G82**) describes how the changes occurring in clay minerals on heating can be followed by X-ray diffraction data obtained during the heating.

The apparatus required for X-ray diffraction investigations consists essentially of a source of monochromatic X-rays, a suitable method of mounting the specimen and a camera for receiving and recording the diffracted rays. The wavelength is partly determined by the target in the X-ray tube so that a series of these must be available to give a full range. Suitable filters are also necessary for further selection. The voltage and current must also be adjustable.

X-ray diffractions of powders are done in a circular camera. This is the one most used in mineral investigation (Fig. 4.17).

There are numerous instruments available for X-ray diffraction work. Fig. 4.18 shows the Philips Norelco diffraction unit (**P37**). This has a central X-ray tube with interchangeable targets, and four windows fitted with rotating filter disks. It is possible to mount a number of test and observation devices on the four camera platforms. Those shown are two circular powder diffraction cameras of different diameter hence different separation of the lines, and a special camera for the rays scattered only a little from the incident beam.

Serwatzky (**S56**) describes the greater detail obtainable by recording the X-ray diffraction pattern with an X-ray goniometer or Geiger counter diffractometer. In this, the film of a Debye–Scherrer powder camera is replaced by a Geiger–Müller counter, feeding its impulses via an integrator into an automatic recorder. Greater resolution is obtainable in mineral mixtures. Legrand and Nicholas (**L24**) use the same apparatus to develop a method of determining the quartz content of clays.

Electron Diffraction. As electrons can be considered as a wave function of very small wavelength, they too can be used to determine crystal structure by diffraction. Reflections corresponding to very small planar spacings can be detected, so that the information obtainable by X-ray diffraction can be

Fig. 4.17. *Plan view of Hilger powder camera (Hilger and Watts,* **H88***)*

Fig. 4.18. *X-ray diffraction unit. Front view showing large and small type powder camera and special low-angle scatter camera (Phillips,* **P37***)*

supplemented, especially for complex crystals. Most modern electron microscopes can be adapted for electron diffraction work.

Infrared and Microwave Spectroscopy. The development of radar instruments has made possible the accurate investigation of the spectrum in the microwave region which lies between the infrared and wireless waves, *i.e.* about 30 μ to 2000 μ. In this region of the electromagnetic wave band, absorption and emission are due to vibration of the molecules. Each compound will therefore give a characteristic spectrum which can be used for qualitative and quantitative analysis and sometimes for structural diagnosis.

The application of this method to identification of clay minerals was brought to general notice at a conference on clay technology in 1952 by Nahin (**N2**) and it is to be hoped that development will make possible its general use in the near future.

Differential Thermal Analysis. Thermal analysis is based on the observation or measurement of heat evolution or absorption when a material undergoes a physical or chemical change. Some seventy years ago it was realised that clays have distinctive thermal characteristics. It is now known that every clay mineral has its own thermal properties, minerals of the same group being similar but not identical. Testing is done by the differential thermal method in which the temperature of the test material is measured relative to an adjacent inert material.

The advantages of the method are that it does not require separation of the constituents of the clay, it is quick and sensitive, it requires relatively little apparatus.

Grimshaw, Heaton and Roberts (**G84**) amongst others, *e.g.* Norton (**N33**) and Grim (**G80**), describe the apparatus they use in some detail. We will describe the apparatus developed by Roberts and his collaborators, as it is the more recent, and will refer to its main differences from that of other workers.

A fixed quantity (0·8 g) of the sample under test is fitted into one 1 cm³ compartment of a thin walled refractory container. The adjacent compartment is filled with an inert material, generally a calcined oxide. Norton placed the two materials in a metal block which minimises thermal gradients but allows heat transfer from the test sample and so lowers the temperature differences during thermal reactions. The two materials are then fitted with chromel–alumel thermocouples connected to recording instruments in such a way that the temperature difference is amplified (**K4**) and recorded continuously together with the temperature of the inert material at fixed time or temperature intervals. The whole is then enclosed in a refractory block which is inserted in an electric furnace. This is heated as steadily as possible at a rate of from 10 to 20° C per min. The constancy of the heating rate is less important than its reproducibility.

The apparatus must be standardised with samples of pure minerals, when distinctive curves are obtained for minerals of different groups (Fig. 4.19) and

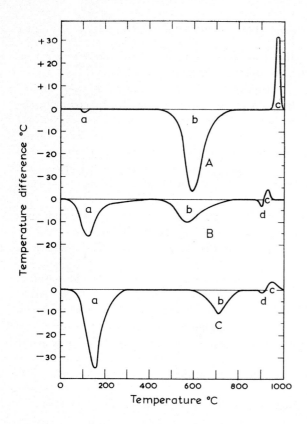

FIG. 4.19. *Differential thermal curves for clay minerals of the main types: A, kaolinite; B, illite; C, montmorillonite* (Grimshaw, **G84**)

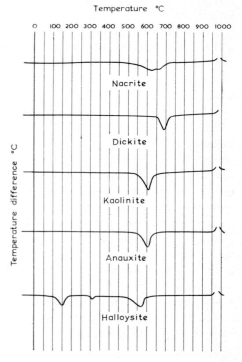

FIG. 4.20. *Differential thermal analysis curves for clay minerals of the same group (the exothermic peak at 980° C is not completely shown)* (Norton, **N33**)

only slightly differing curves are obtained for minerals of the same group (Fig. 4.20). In this group kaolinite and anauxite are not distinguishable.

In mixtures each component records its own curve independently so that its presence can be spotted. The area of the characteristic peak relative to the area of the peak in the curve given by the pure mineral gives its quantity (Fig. 4.21).

Organic matter gives a large and indefinite exothermic peak at low temperatures obliterating other effects; it must therefore be removed with hydrogen peroxide or alkaline permanganate beforehand (**G81**).

Alkali-bearing clay constituents such as micas (except the hydromicas or illites) and feldspars are thermally inert and cannot be detected by this method (*Bibliography of Differential Thermal Analysis*, **S145**).

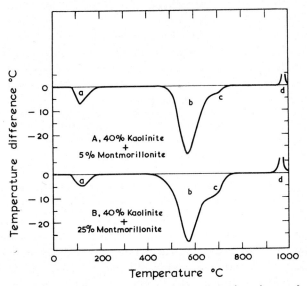

FIG. 4.21. *Differential thermal analysis curves for mixture of kaolinite and montmorillonite* (Grimshaw, **G84**)

Differential Thermogravimetric Analysis. Keyser summarises the work done on following weight changes during the heating of minerals. This supplements differential thermal analysis by distinguishing between reactions involving weight loss and those that do not. In cases of well defined and identifiable reactions it can also be used for quantitative analyses for constituents that decompose during heating, *e.g.* carbonates, various hydrates.

However, direct weighing of samples throughout a heating cycle does not prove entirely satisfactory, largely because of the wide range of changes, so that a small weight loss following close on a large one will not be distinguished. Differentiation of the weight loss with respect to temperature to give a plot of dw/dt against t gives a more useful curve with sharp peaks and troughs where w against t was an indeterminate curve. Keyser therefore constructed an apparatus to plot this curve automatically.

The differential thermobalance consists of a balance arm from which two identical samples of the test material are suspended in two identical electric furnaces. The two furnaces can be heated at exactly the same rate but with the one a small but fixed amount cooler than the other. The condition of the balance arm is observed by an illuminated mirror which reflects a light spot on to a photosensitive rotating chart. With this apparatus the change in weight of the sample when heated from the temperature in the cooler furnace to that in the hotter one is automatically registered.

Complex Thermal Analysis. Koehler (**K46**) describes a Russian apparatus that will automatically perform and record in one operation the D.T.A., continuous measurement of shrinkage and of weight loss, for temperatures up to 1450° C (2642° F). Photosensitive paper records curves for rate of temperature rise, D.T.A., shrinkage and weight loss with temperature references at every 100° C.

Checking against the Complete Chemical Analysis. It is as well to compare the total chemical content present in the identified and estimated minerals with the complete chemical analysis of the clay. If there is a discrepancy it points to a mineral that has evaded detection. The nature of the difference gives valuable evidence on the missing mineral.

Although a complete mineralogical analysis may be very lengthy and difficult, considerable useful data can be obtained from the X-ray, differential thermal and particle-size analyses taken together (**L35**).

COMPARISON OF RESULTS OBTAINED BY DIFFERENT METHODS IN ORDER TO DERIVE THE BEST VALUES FOR THE MINERAL CONTENT

Linseis (**L34**) investigated clays by chemical and the Berdel (**B39**) rational analysis, X-rays and differential thermal analysis. He concludes that the rational analysis is of little ultimate value; that X-ray analysis although involving special apparatus and skilled handling gives good results, particularly for the constituents that cannot be determined by differential

TABLE 65

Linseis' Comparison of Methods of Mineralogical Analysis (L34)

Amberg Kaolin

Chemical Analysis:

SiO_2	48·3%	CaO		0·15%
Al_2O_3	35·1	Alkalies		2·6
Fe_2O_3	0·48	Loss on Ignition		11·06
TiO_2	0·17			

Mineral Content:

	Kaolinite	Montmor.	Mica	Feldspar	Quartz
X-ray analysis	80–85%	—	10–15%	3%	3%
D.T.A.	86				
Rat. anal. (Berdel)	87 clay min.*	—	—	9	4
Chem. anal.	83	—	8	3	3.8
Total	84	—	10	3	3

* Clay minerals

TABLE 65—(contd.)

Hirshau Kaolin

Chemical Analysis:

SiO$_2$	47·8%	CaO	0·65%
Al$_2$O$_3$	38·17	Alkalies	0·66
Fe$_2$O$_3$	0·7	Loss on Ignition	12·2
TiO$_2$	0·16		

Mineral Content:

	Kaolinite	Montmor.	Mica	Feldspar	Quartz
X-ray	90%	—	10–15%	—	3%
D.T.A.	90	—	—	—	—
Rat. anal. (Berdel)	95·2 clay min.*	—	—	1·75%	3
Chem. anal.	90	—	7·2	—	3
Total	90	—	7	—	3

Konnersreuth Kaolin

Chemical Analysis:

SiO$_2$	61·74%	CaO	1·37%
Al$_2$O$_3$	28·4	MgO	1.07
Fe$_2$O$_3$	1·07	Loss on Ignition	7·4
TiO$_2$	0·2		

Mineral Content:

	Kaolinite	Montmor.	Mica	Feldspar	Quartz
X-ray	45%	—	ca. 30%	—	25%
D.T.A.	45	—	—	—	—
Rat. anal. (Berdel)	68 clay min.*	—	—	1·1%	30·9
Chem. anal.	43·2	—	30	—	28
or	46	—	26	—	28
Total	45	—	30	—	25

Aussernzell Clay

Chemical Analysis:

SiO$_2$	58·9%	MgO	0·1%
Al$_2$O$_3$	29·6	Alkalies	2·0
Fe$_2$O$_3$	0·7	Loss on Ignition	7·3
CaO	0·3		

Mineral Content:

	Kaolinite	Montmor.	Mica	Feldspar	Quartz
X-ray	80%	15%	—	—	4%
D.T.A.	75	10–15	—	—	—
Rat. anal. (Berdel)	76·5 clay min.*	—	—	3·85%	19.1
Chem. anal.	70	12	—	—	15
Total	80	10	—	—	4

Hebertsfelden Clay

Chemical Analysis:

SiO$_2$	65·4%	CaO	0·72%
Al$_2$O$_3$	15·95	Alkalies	3·4
Fe$_2$O$_3$	7·05	Loss on Ignition	4·8

Mineral Content:

	Kaolinite	Montmor.	Mica	Feldspar	Quartz
X-ray	7%	20%	45%	—	28%
D.T.A.	10	10–20	—	—	—
Rat. anal. (Berdel)	42·5 clay min.*	—	—	16·8%	40.6
Chem. anal.	5	20	40	—	22·4
or	8	20	35	—	25
Total	10	20	40	—	28

* Clay minerals.

thermal analysis, namely quartz, feldspar and mica, but does not show sufficient distinction between individual clay minerals especially minor constituents; differential thermal analysis is particularly useful for determining the clay minerals, kaolinite to $\pm 3\text{--}4\%$ and montmorillonite to $\pm 10\%$.

Linseis' (**L34**) tabulated results are well worth consideration (Table 65). Each case has been considered individually to determine which values are the most accurate if a choice occurs.

Determination of Cation-exchange Capacity of Clays (C.E.C.)

There are two distinct types of reaction that are termed base-exchange reactions. The one under consideration is the exchange of cations adsorbed in the *outer* layer round colloid particles. The second type is the replacement of cations *in* the crystal lattice.

Cation-exchange determinations on clays include measurement of the total exchange capacity (C.E.C.) and/or of the individual cations (Na, K, Ca and Mg, exchangeable Fe and Al being rare). Total exchange capacity is determined by electrodialysis or by replacement leaching. Graham and Sullivan (**G65**) give a detailed survey of the available methods.

ELECTRODIALYSIS

This is carried out in a three-compartment cell, the compartments being separated by a semipermeable membrane, *e.g.* parchment paper. The anode of carbon plate, and the cathode of sheet nickel are immersed in distilled water in the two end compartments. The clay slip is placed in the centre cell and kept in suspension by a mechanical glass stirrer. A direct current at 200 V is applied for 48 hr for most clays and at least 72 hr at low current density for ball clays and bentonite. The anolyte and catholyte are periodically replaced with fresh water. This electrodialysis converts the clay to its hydrogen form.

The clay is then dried at 105° C and broken down to pass a 100-mesh Tyler sieve. Weighed samples are transferred to a bottle and treated with standard NaOH, agitated for 72 hr and then centrifuged. The pH of the supernatant liquid is then determined with a glass electrode or other suitable pH-meter. The plot of pH against amount of alkali added shows an inflection which gives the exchange capacity (Fig. 4.22).

REPLACEMENT LEACHING

In this method all the adsorbed cations are replaced by the ammonium ion by treatment with excess ammonium acetate solution. The removed metallic ions can then be determined individually, and the adsorbed ammonium is analysed by the Kjeldahl method.

Replacement of the metallic cations by allowing the ammonium acetate solution to percolate through the clay is very slow and tedious. Speedy replacement can, however, be achieved by agitating a sample of the clay with 1N ammonium acetate solution for from 30 to 40 min and then centrifuging the suspension until the liquid is clear, *i.e.* 10 to 60 min. The liquid is decanted and kept for analysis of the replaced ions and the leaching of the clay sediment repeated four or five times. The decanted liquid is then analysed by normal chemical procedure for the metallic cations Na, K, Ca and Mg.

The clay is washed with 80% neutral ethyl alcohol to remove all excess ammonium acetate, by the same agitation and centrifuging method as was used for the leaching. Its ammonium content is then determined by the Kjeldahl method.

Worrall, Grimshaw and Roberts (**W88**) consider this to be the most reliable method provided that sufficient sample, say, 5–10 g for a kaolinite, is available.

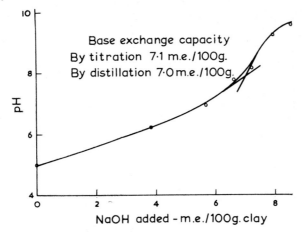

Fig. 4.22. *Determination of cation-exchange capacity by dialysis followed by titration (Graham and Sullivan, G65)*

Bower and Truog amended the method by replacing the ammonium salt by salts which lend themselves to a rapid determination by colorimetric or photometric methods. For instance, if the clay is treated with alcoholic manganous chloride the amount retained by the clay can be determined colorimetrically by oxidation to permanganate. This method is useful for small samples (**W88**).

CERAMIC PROPERTIES OF CLAYS AND BODIES

In this section methods of measuring the properties mentioned in the definition of ceramics, are discussed. Namely, workability, castability, plasticity and other matters concerned with shaping, drying and firing.

The Measurement of pH

For clear, colourless liquids the pH can be measured by using one of the numerous colour indicators that are available. The choice of indicator must be carefully made, and Fig. 4.23 shows over what ranges the different indicators change, with special reference to the Universal Indicator that shows a number of changes between pH3 and 11. Colour indicators always suffer from the personal factor and are probably better for showing change of pH during a titration than for giving absolute values.

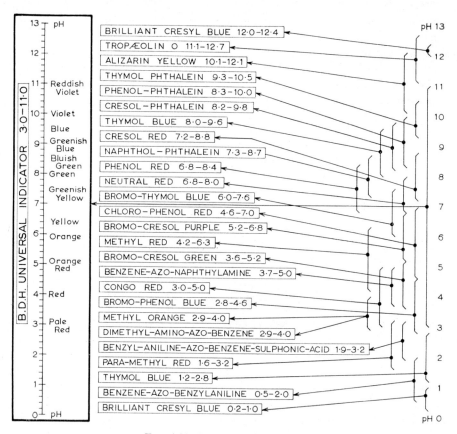

FIG. 4.23. *B.D.H. indicators* (**C43**)

Where the solution is coloured or contains colloidal suspended matter it is usual to use the potentiometric method, which gives more accurate results in all cases. There are a number of commercial models of pH-meters with compact cases containing the potentiometer, etc., some are run off the mains, others from batteries. They are usually fitted with glass electrodes, but in some the electrodes can be changed. Fig. 4.24 shows one such instrument. There are also continuous, recording pH-meters.

When measuring the pH of clays considerable attention must be paid to the preparation of samples. Barker and Truog (**B9**) have shown that the pH of clay varies with the clay-to-water ratio and the length of time the clay has been in contact with water together with the speed of stirring the mixture.

So that in order to obtain comparable results a standard method of preparation must be adopted. Two such methods which will give reproducible results are given:

'Method No. 1: Place 20 g of clay screened to pass a 20-mesh sieve on a smooth glass plate. Carefully add 20 ml of water. Work this water thoroughly into the clay with a spatula until a bright shiny surface is obtained. Add additional water to the clay and thoroughly work in with a spatula until the mass becomes too soft to stay on the glass plate. Transfer the clay to a beaker and add the remaining water necessary to constitute a total

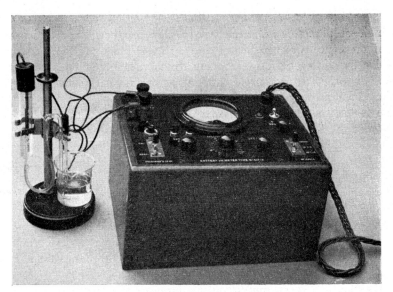

FIG. 4.24. *Electrode assembly and pH-meter* (Muirhead, **M99**, note that this firm at present only make electrodes and not meters)

of 50 ml. Stir the mixture with a glass rod until it is evenly mixed and then determine the pH with a glass electrode. A clay sample prepared thus will give a maximum pH which can be checked accurately by repeating the performance.

'Method No. 2: Add 20 g of clay which has passed a 20-mesh sieve to 50 ml of water in a beaker. Stir the mixture violently with an electric propeller (such as is used in a malted-milk mixer) for 15 min, and then determine the pH of the clay with the glass electrode. This method is accurate and is the method recommended for use in our laboratories.' (**B9**)

The pH of the water used is also of importance, and, as even distilled water does not have a constant pH, must be checked.

As these methods use dry clay they are not easily applicable for checking the pH of plastic or suspended clay during processing, but a comparable method can be evolved using a given amount of clay in a known state and diluting it with water of known and constant pH.

More recently a pH-meter with a pointed electrode bulb has been put on the market. This is said to be suitable for measuring the pH of gels, viscous clay mixtures and clays in the plastic state. In the latter case the makers consider it advisable to make a conical impression with a glass rod before inserting the spear-point electrode (**M27**). Measurement of the pH of plastic clay should make checking of prepared bodies very much more straightforward.

(B.S.S. 1647:1950).

The Determination of Soluble Sulphates in Clay

Soluble sulphates are detrimental to clay products causing scumming and efflorescence during drying and firing. They are normally made insoluble by adding barium carbonate which then precipitates barium sulphate. In order to reckon the amount of barium carbonate required the sulphate content of the clay must be determined.

Clay is shaken, stirred, boiled or otherwise extracted with water to give a solution of the soluble sulphates present in a given weight of dry clay. In order to avoid the time-consuming task of completely filtering the suspension obtained it is allowed to settle, and a fixed amount of clear solution decanted, or the almost clear upper liquid filtered.

The sulphates present may then be determined in a number of ways, depending largely on the accuracy of the results required, and the time, equipment and personnel available.

In the analytical laboratory accurate methods such as the gravimetric determination of precipitated barium sulphate, or titration against barium chloride, using adsorption indicators, are used.

In the plant, however, more rapid methods are required if each new batch of clay is to be tested. The commonest quick methods are turbidity measurements of barium sulphate suspensions. These consist of having some standard of transmission of the suspension to which test suspensions are adjusted by altering their depth or dilution. For instance in Jackson's method a tube is supported over a candle and the suspension is poured in until the flame disappears. In the Parr instrument a sighting tube is lowered into the suspension until a given object at the bottom is visible. With the Hagan instrument the suspension is diluted until the light source is visible. All these methods require calibration with known solutions.

A quick method of this type requiring inexpensive equipment is one developed and recommended by an American company (**T42**). A graduated tube open at both ends is fixed perpendicular to a disk with a suitable mark on it for sighting purposes. Sulphate solution (300 ml) made in the prescribed way is put in a 400 ml beaker and 30 ml HCl solution (47 ml conc. HCl per litre) and 30 ml $BaCl_2$ solution (120 g hydrated barium chloride crystals per litre) added. This is stirred for two minutes and the turbidimeter then lowered into the suspension until the lettering on the disk is unidentifiable. The finger is placed on the top of the tube to seal it and it can then be raised and the liquid level easily read off on the graduations. Charts are supplied with the turbidimeter from which sulphate concentration, or pounds barium carbonate required per ton of clay, can be read off. However, as the personal factor plays a fairly large part greater accuracy can be achieved if each operator makes his own chart. The makers claim that readings made by the same operator on the same solution should be reproducible to within 2 mm on the tube, that is 0·001% SO_3 or 0·05 lb barium carbonate per ton of clay (**T42**).

Measurement of Viscosity

The viscosity of a liquid is the internal friction opposing its flow. The coefficient of viscosity (η) can be defined as follows:

When two planes, 1 cm apart, are immersed in a liquid, and are moving over each other with a relative velocity of 1 cm per sec, the viscosity of the liquid is equal to the force in dynes exerted on each square centimetre of the planes.

The unit of viscosity is the poise.

The equation for the determination of viscosity as expressed by the Frenchman Poiseuille is:

$$V = \frac{\pi p r^4}{8 l \eta}$$

where V = volume of liquid flowing per sec;
p = pressure of liquid (g/cm^2);
r = radius of tube (cm);
l = length of tube (cm);
η = viscosity of liquid.

This can be rearranged to:

$$\eta = \frac{p \pi r^4}{8 l V}$$

There are three basic methods of measuring viscosity: (*a*) time of flow through an orifice under a fixed or a varying pressure, or head; (*b*) force required to shear layers of liquid past one another; (*c*) rate of fall of a sphere or plunger through a tube of the liquid. There are also a number of specialised testing methods.

Method (*a*) is simple, requiring little apparatus or skill when used for checking purposes. The essential requirements are a vertical tube of constant cross-section, fitted with an orifice of known diameter at the bottom. The top may have an overflow to give a constant head of liquid, or may not have one, when a variable head is obtained. A stop-watch, a receiving vessel and a thermometer are further essentials. The time is taken for a given quantity of liquid to flow through the orifice. The viscosity is obtained from the Poiseuille equation given above.

Manufactured viscometers of this type are made of brass, or glass and brass, and may have a set of orifices of different sizes. (German model complies with DIN 53211 (**A101**).)

A newer viscometer operating on the same principle forces a column of the liquid under test along a horizontal glass capillary tube, which is subjected to a constant air pressure on a reservoir of the liquid. It is then not necessary to have a specific volume of the liquid or to know its density. Absolute values can be obtained for true viscous Newtonian liquids and accurate comparative values for anomalous non-Newtonian ones. The shear stress can also be varied so that anomalous liquids can be investigated over a range, and the results represented graphically (**A45**).

Method (*b*) requires more elaborate instruments but is capable of a high order of accuracy. The liquid is sheared between concentric cylinders, a cup and a bob or spindle, one of which is rotated, and the torque on the suspension of the bob is recorded. There are a number of models of such viscometers, which can give accurate absolute values for true viscous liquids. They also give comparative values for anomalous liquids if a standard procedure is adopted.

One model, considerably used in the British ceramic industry, consists essentially of a metal cylinder suspended in the sample liquid by means of a torsion wire. This is attached at its upper end to an adjustable torsion head at the top of the instrument. At the lower end of the wire is a flywheel disk with a pointer fixed on the edge reading on a circular scale graduated 0–360°. The cylinder is supported from the lower side of the flywheel. The zero point on the scale is fitted with a stop and release pin which engages the pointer on the flywheel. The sample is placed in a cup or beaker on a support below the cylinder so that the latter is immersed about $\frac{1}{2}$ in. below the surface.

The viscosity of a sample is determined by its damping effect on the suspension. The flywheel disk is rotated a complete turn and held by a release pin. When released it swings back the complete turn plus a further distance depending on the viscosity of the sample. The greater the viscosity the less the rotation beyond zero. The instrument can be modified for estimating thixotropy (**G1**).

For obtaining absolute values for liquids whose viscosity varies with the rate of shear, a number of special points have to be borne in mind.

A fixed known rate of shear must be used and measurements made for a number of rates. This is best achieved on instruments using a rotating cup, which can be geared to an accurate variable-speed motor. The measurements must be made at each speed until a steady state has been reached and also a check made on whether the same viscosity is found when the rotation speed is attained from a higher one as from a lower one. With these liquids it is difficult to calculate a correction for the viscous drag on the bottom of the bob or spindle (the main shear being on the curved surface). This can be reduced by using a small gap between the curved surfaces and a large one underneath the bob, or by having a bob of small radius, or by having a hollow bob in a double cup with a solid centre. There are then two narrow calculable annular spaces and the minimum bottom. The viscosity of anomalous liquids is best expressed graphically.

Method (c) is used largely for viscous oils, etc., although almost any liquid could be so tested. A glass tube of uniform bore with two marks near the top and bottom, respectively, is set at a fixed angle (say, 80° to horizontal) and filled with the liquid. It may be surrounded by a jacket through which thermostat water can be run. A polished sphere with its centre of gravity accurately central, and whose diameter and weight bear the correct relationship to the bore of the tube, is then rolled down it without causing turbulence. The time taken for it to fall from one mark to the other is taken. Stokes' Law corrected for 'sidewall' and 'end' effects is then applied.

Testing the viscosity of slips containing clay together with grog, whose particles are relatively much coarser is very unsatisfactory in any of the viscometers described. Williams (**W72**) describes a reproducible method using an 'Irwin slump tester'. After determining S.G. and pH the test slip is placed in a bottomless cup resting on the centre of a metal plate, marked in concentric circles. The cup is suspended from a pulley system with a weight at the other end. When the latter is released the cup rises vertically, allowing the slip to flow out. The diameter of the circle covered can be read off and the area found, both are a measure of flow characteristics.

Woodward (**W85**) describes a vibrating-plate viscometer suitable for following the drying and setting process of slips in plaster moulds. The instrument consists basically of an electromagnetic vibrator motivating a thin circular (diameter 0·200 in. (5·08 mm)) blade at about 800 c/s. The principle on which the measurements are based is that the amplitude of vibration of the blade varies with the viscosity, η, and density, ρ, of the liquid in which it is immersed. The amplitude is measured by means of an associated electronic circuit which gives a meter reading of $\eta\rho$, and which can be further coupled to an automatic graphical recorder. The change in viscosity of a casting slip can thus be continuously recorded while a test cast is being made.

(B.S.S. 188:1937.)

Castability

There are three stages in the testing of a clay, or body, for castability.

(1) Deflocculation, the determination of the optimum water and deflocculant additions to give a slip that holds the nonplastic materials in suspension without dependence on thixotropy.

(2) Flow characteristics, especially checking that the slip is not so thioxtropic that blockage will occur in pipelines.

(3) Setting rate, ease of removal of the cast and its strength.

These tests are usually of an empirical nature derived directly from the factory conditions anticipated and are discussed under *casting* pp. 755–764.

Although casting rate is usually tried in plaster moulds Weintritt and Perricone (**W37**) describe the use of a laboratory pressure filter which gives fast and reproducible results. The measure of casting rate is the volume of water pressed through the filter in a given time at standard pressure, starting with a standard volume of slip.

Measurement of Plasticity

This most important property of clays virtually cannot be measured directly by any method other than that of human touch. The practised worker can tell very accurately whether a clay is 'right' or not but that is not very much help in plant control. It is therefore necessary to measure other and simpler properties of the clay and correlate them to the plasticity. Having considered the complexity of the property termed 'plasticity' it is clearly apparent that measurement of only one property is totally inadequate.

Plasticity is influenced by: (1) past history of the clay; (2) type of clay; (3) amount of working the mixture has received; (4) time of contact of clay and water; (5) temperature; (6) amount of water; (7) presence of electrolytes (**H7, D47**).

Some measurements are made on clays while one of these factors, usually the water content, is being altered. Graphs can then be plotted and some optimum value obtained. For other measurements the clay is first brought to its maximum workability by altering the various factors (3) to (7), careful records being kept of the treatment given. 'Maximum workability' itself is ill-defined, for instance 'for practical purposes a body may be said to have attained maximum plasticity when on squeezing in the hand it retains an impression of the fine lines in the hand without being sticky and causing parts to remain on the hand' (**A145**).

Properties that are measured are outlined below.

A. RELATED TO THE DRY CLAY

(1) Water of plasticity (expressed as a percentage).

(2) Range of water of plasticity between the flow limit, when neighbouring clay pastes will unite by jarring, and the rolling limit, when rolling out a rod leads to crumbling. A larger difference is found in the more plastic clays which are also those showing the finest grain sizes (Atterberg plasticity number).

(3) Water absorption of clay by 'Enslin's' method. An exact weight of dry clay is placed in a glass crucible with a porous base. This is connected by a U-tube, stopcocks and filling funnel to a graduated, accurately horizontal glass tube which is at the same height as the sample.

The Enslin value (E) is then given.

$$E = \frac{V \times 100}{p}$$

where V is volume absorbed (ml);
   p is weight used (g).
The Enslin value gives a measure of the affinity of a clay to water as related to its grain size and its mineralogy and puts clays in relative order that agrees with measurements of workability (**L33**).

B. CLAY SLIPS

(1) *Viscosity.* Various viscosity measurements in concentric cylinder viscometers are used to assist in determining general flow characteristics. Wilson and Hall (**W75**) used a Bingham apparatus to evaluate two results for each sample, namely (*a*) the water content required to give a certain mobility, and (*b*) the yield value at this mobility. This is much more satisfactory than the use of the Bingham plastometer (**B51**) or the MacMichael torsional viscometer (**H7**) alone.

C. PLASTIC BODY

(1) *The deformability or amount of possible flow before rupture.* This includes very simple empirical devices such as finding the smallest circle that a clay column can be bent round into, or finding the smallest thickness that a disk of given size can be rolled out to, or compressing a sphere.

Several methods measure the deformability of a clay by compressing it in the form of a given cylinder or sphere until cracks occur, when the deformation prior to cracking, together with the weight required determine the plasticity. 'Plasticity indices' can be derived from the results in one of the following forms:

$$\text{plasticity index} = (\text{load to produce first crack}) \times (\text{deformation}) \ (\mathbf{Z4})$$

$$\text{plasticity index} = \frac{\text{total deformation at the point of failure}}{\text{average stress beyond the proportional limit}} \ (\mathbf{T2})$$

A similar example of an empirical standardised test is that for the 'workability index' of fireclay and plastic refractories. 'Workability index has been defined as the measure of the ability of plastic refractories to flow without cracking when rammed, and not deform during subsequent drying. The present A.S.T.M. test uses a sand rammer consisting of a cylindrical mould of 2 in. (50·8 mm) inside diameter and $4\frac{3}{4}$ in. (120·65 mm) long, open at both ends, with a plunger, and a 14 lb (6·35 kg) weight which slides on the shaft. Test specimens are formed by putting 300 g (10·6 oz) of plastic material in the mould and subjecting it to ten impacts of the weight at each end. The percentage deformation, workability index, is calculated on the original length. Acceptable limits are 15–35%, and the specimen must not crumble under three impacts. Steel disks used in the ends of the cylinder while forming the test specimens are recommended.

'Different mountings of the rammer, lubrication or lack of lubrication of the mould, the method of putting the material in the mould, and variability of water content in different parts of the specimen are factors found to affect results appreciably. In some cases, twenty impacts in forming did not give a sample of maximum bulk density.' (**A108**)

More precise information is obtained by compressing cylinders and plotting the decrease in height against load and converting this to a stress/strain diagram. Comparative data on similar clays or bodies can be obtained but clays of widely differing natures should not be compared on this basis (**B82**).

(2) Deformation without rupture by indentation tests. One involves a Vicat needle of given cross-section (7 mm^2) (0·045 in^2) and the weight required to force it 4 cm (1·6 in.) into the clay in half a minute is measured (**G91**); in another, the indentation made by a freely falling steel ball or cone is observed (**B41**). Flat-ended and hemispherical plungers have also been used. All of these are suitable only for comparisons.

Where a mathematical treatment is required cone indenters are now used and the deformation can give values for 'hardness' by a number of equations (**B82**).

(3) The resistance to flow or deformation, as seen when extruding clay from an orifice.

(4) The rates of flow at different pressures. These are used in combination by Cassan and Jourdain (**C18**). They use three variables:

$$x = \frac{\text{wt. of water}}{\text{wt. of dry clay}},$$
$$y = \text{function of deforming pressure},$$
$$z = \text{function of deforming rate}.$$

Using dried sieved clay they add known quantities of water and find the pressure required to press the plastic mass through a nozzle at a given fixed rate. Then, having kept z constant, x can be plotted against y. They obtain hyperbolic curves which differ in shape for even quite similar clays. The sequence of plasticity of the clays is given by the y values for a given x value or the x values for a given y value.

If x, the weight ratio, is replaced by X the volume ratio and $\log X$ is plotted against P (the deforming pressure) linear graphs are obtained. The equation to these lines is $c = P.X^b$ where c and b are constants for the clays concerned, and are related to the inherent plasticity and the mineralogical content that attracts water, respectively.

(5) Cohesion between particles or tensile strength, e.g. the length of an extruded column that can support its own weight.

Tensile strength of plastic clay is a complicated function varying with the manner in which the stress is applied, whether intermittent or continuous, slow or fast. Bloor (**B82**) summarises results obtained by others.

Tensile strength tests often form part of the composite empirical tests discussed below.

(6) Substitution of torsional forces for compressive forces gives reliable

values of the plasticity or the 'workability' (an acknowledgement of the fact that practical workability may not be identical with plasticity). Suitable apparatuses are described by Wilson (**W73**), Norton (**N31**), and Graham and Sullivan (**G66, S216**). The tests comprise the extrusion of a clay bar or tube in the stiff-mud consistency range, its formation for use in the torsion machine where failure must occur at the centre of the bar and not at the chucks, followed by its distortion in a motor-driven machine, the chuck of which rotates at 3 r.p.m. A spring-loading device resists and measures the torque transmitted through the bar. A continuous curve of torque against angular deformation is automatically recorded. These torsion tests give data on the yield strength, maximum torque, angle at which maximum torque occurs and the angle at ultimate failure. They therefore give valuable information on the moulding of a clay and on how it will stand handling.

Moore's (**M89**) work with torsion tests also gave evidence for the 'memory of clay' and delayed recovery from distortion.

(7) Shear strength. Shearing properties are involved in any deformation of clays but cannot normally be measured thereby.

Using a shear box, unidirectional shearing can be approached. In this, a rectangular box is divided horizontally into two halves, and filled with a rectangular prism of clay. This is subjected to a constant vertical compression while an increasing horizontal force is applied to the upper half of the box. Tests are made on a number of identical specimens at different vertical compressions.

The concentric cylinder type of viscometer can also be adapted to work with clay pastes. Slippage at the cylinder walls is prevented by using toothed surfaces and tearing is prevented by closing the ends and perhaps keeping them under pressure (**B82, G75**).

(8) Hybrid methods. Many tests measure at least two properties which are used in a calculation to give some 'plasticity index', etc.

Deformability and Tensile Strength

Zschokke (**Z11**) measured these together. The thoroughly worked clay was moulded into cylinders 60 mm (2·4 in.) high and 30 mm (1·2 in.) in diameter and then pulled in two in a specially designed machine. The amount of expansion showed the degree of deformability and the force required to pull the cylinder in two showed the tensile strength. The product of the two being the plasticity coefficient. It was found, however, that higher figures are obtained if the stretching is done more rapidly or in a succession of rapid jerks.

This method has been successively improved on.

Tensile Strength and Resistance to Flow

Linseis (**L36**) gives details of his apparatus. He has separated the two measurements. The two instruments used are represented diagrammatically

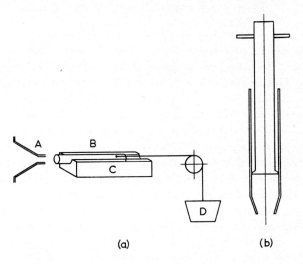

Fig. 4.25. *Linseis' plasticity apparatus: (a) tensile strength; (b) resistance to shear* (**L36**)

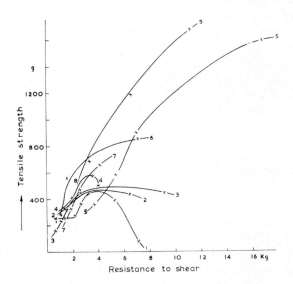

Fig. 4.26. *Linseis' plasticity measurements* (**L36**)

(1) *Amberg kaolin, 08*
(2) *Kemmlitz kaolin 'Meka'*
(3) *Schnaittenbach kaolin, R*
(4) *Zettlitz kaolin*
(5) *Aussernzell clay, AIX*
(6) *Klardorf clay*
(7) *Schwarzenfeld clay*
(8) *Tambach clay*

(Fig. 4.25). The first gives the tensile strength which is a measure of the cohesion between particles. The clay is forced through the opening A with an auger. The cylinder B is brought into contact with A until 10–20 mm of the clay column is in B. A and B have the same internal diameter of 8·5 mm (0·33 in.). B is then disconnected from A and allowed to slide along the holder C while the clay column is further extruded until B is exactly 100 mm (3·9 in.) from A. Weights are then added to the pan D until the column breaks. The weight, in grammes, used is taken as a measure of the tensile strength.

The second instrument is used to give a measure of the resistance to shear. The force (in grammes) required to press the clay from a cylinder of 9 mm (0·35 in.) diameter through a nozzle of 6 mm (0·24 in.) diameter is measured.

Linseis makes these measurements on a clay with gradually increasing water content. The results are plotted with tensile strength as ordinates and resistance to shear as abssicae with the corresponding values for each water content plotted. The curve is therefore independent of the mixing water (Fig. 4.26).

The curves show maxima or tendencies towards a maximum. The measure of plasticity is taken as the distance of the maximum from the origin. From the point of the maximum the water content can be read off from the original result tables.

THE STRENGTH OF DRIED CLAYS AS A MEASURE OF PLASTICITY

It has become general practice to test the strength of the dried clay instead of the plastic clay. It is assumed thereby that a measure of the bonding strength of a clay is more or less proportional to the plasticity. As both are due to the colloidal clay mineral particles the assumption of qualitative relationship is justified but a quantitative relationship does not necessarily exist. This was pointed out by Mellor in 1922 (**M53**). Kohl (**K62**) found a remarkable instance of such a lack of direct correlation between plasticity and green strength. He measured the transverse strength of a number of different clays and found that although normal values were in the region of 200 lb/in^2 (14 kg/cm^2) the highest value was given by a certain Sommerfeld brick clay which was not very plastic to touch. His values are:

Raw clay: air dried, 459·3 lb/in^2 (32·3 kg/cm^2) $\pm 2·7\%$; dried at 110° C, 846·1 lb/in^2 (59·5 kg/cm^2) $\pm 6·1\%$.

After wintering followed by steam treatment: air-dried, 676·9 lb/in^2 (47·6 kg/cm^2) $\pm 4·2\%$; dried at 110° C, 974·3 lb/in^2 (68·5 kg/cm^2) $\pm 5·7\%$.

Green strength increases as the water content decreases and particularly as the temperature of drying is raised. In work on the transverse strength of dried clay bars Kohl (**K62**) found that if they had been heated above 80° C the strength increased considerably but the change caused by the greater drying was irreversible. Hence measurement of this strength cannot be representative of the plasticity of the wet clay.

Measurement of Permeability of Clays

The permeability is measured by finding the time taken for a given quantity of water to pass through a given layer of clay. In the methods employed by Harman and Parmelee (**H31**) 10 g of the clay sample in 50 ml distilled water is used. The manner in which the clay is dispersed in the water affects the result and so must be standardised, *e.g.* (i) allow to stand for a given time, or (ii) shake for a given time, or (iii) deflocculate, shake and reflocculate. The suspension is then transferred to a Buchner funnel of given diameter (*e.g.* 9 cm) and a good vacuum applied. When a firm layer has formed a graduated cylinder is placed inside the Buchner flask and the time noted for a given volume of water to be sucked through the clay layer.

Measurement of the Heat of Wetting

The heat of wetting can be measured directly in a suitable calorimeter. A simple form is a large (about ½ to 1 litre) Dewar flask in reasonably stable thermal surroundings. It is fitted with a stirrer, a Beckmann thermometer and an electric heater (**J39, H31, P19**).

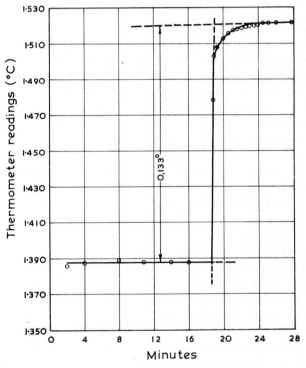

Fig. 4.27. *Heat of wetting of English china clay; weight of dried sample 39·847g; heat of wetting 1·04 cal/g (Harman and Parmelee,* **H31**)

The pulverised clay sample is dried at from 105 to 110° C (221–230° F), weighed and sealed into a bottle. This is immersed in the water in the calorimeter and allowed to come to thermal equilibrium over a considerable period of time, preferably overnight. With the stirrer going, readings of time and temperature are taken over a period of time slightly longer than the anticipated wetting reaction (15 to 20 min). This constitutes the 'fore drift' and together with a similar 'after drift' makes possible the evaluation of the correction to the experimental heat of wetting necessitated by heat interchange with the surroundings. The dry clay sample is then quickly tipped into the calorimeter and temperature readings taken at suitable time intervals (about every ½ min) and continued on to obtain the 'after drift'.

An example of a time–temperature curve obtained by Harman and Parmelee (**H31**) showing how the actual temperature change can be obtained by producing the linear parts of the curve is shown (Fig. 4.27).

The water equivalent of the apparatus is found for each experiment by repeating it and introducing a measured amount of electrical energy in place of the dry clay. The result is expressed in calories per gramme of dry clay. Its value may vary from about 2 to 12 cal/g.

The Measurement of Shrinkage

Shrinkage is generally expressed as a percentage and may concern the linear, area or volume shrinkage. The choice of measure depends on practical considerations, so that with a rod the linear shrinkage is important whereas with a tile the area may be more useful. The three are related mathematically by formulae using a the percentage linear shrinkage:

$$\% \text{ square shrinkage} = 2a - \frac{a^2}{100}$$

$$\% \text{ cubic shrinkage} = 3a - \frac{3a^2}{100} + \frac{a^3}{100^2}$$

For low values of a this approximates to $2a$ and $3a$, respectively. Some values are given in Table 66.

It is necessary when discussing shrinkage always to state whether linear, square or cubic shrinkage is under consideration; unfortunately this is often not done. It is also essential to state clearly the exact conditions of drying and firing and, in the case of firing, if the rate of heating is not uniform, a time–temperature curve is the best means of doing this (**R22**).

The most usual methods of measuring shrinkage employ callipers, travelling microscopes, etc., on the sample before and after given treatment. Volume shrinkage may be found by immersing the moist and dry pieces in mercury and measuring the displacement. Samples drying in air can be measured at any convenient time intervals throughout the process. When drying ovens are used samples must be withdrawn periodically. There are

TABLE 66
Percentage Shrinkages (R22)

Linear	Square	Cubic
1	1·99	2·97
2	3·96	5·88
3	5·91	8·73
4	7·84	11·53
5	9·75	14·26
6	11·64	16·94
7	13·51	19·56
8	15·36	22·08
9	17·19	24·64
10	19·00	27·10
11	20·79	29·50
12	22·56	31·85
13	24·31	34·15
14	26·04	36·39
15	27·75	38·59
16	29·44	40·71
17	31·11	42·82
18	32·76	44·86
19	34·39	46·86
20	36·00	48·80

now various devices for the continuous measurement of shrinkage fitted with pens and recording drums.

A laboratory apparatus for the continuous recording of loss of water and linear shrinkage of a test bar has been described. The test bar is hung from the top of a suitable framework suspended on one arm of a balance. It is counterpoised at the beginning of the experiment and its gradual rise as water is lost is recorded by a pen fixed to the frame. Another pen attached to the bottom of the bar records the shrinkage. The pens mark graduated paper on a drum revolving at fixed speed. Once the graph of moisture content against shrinkage has been obtained for a given clay or clay body the measure of one gives a value for the other so long as no distortion or cracking has occurred. Thus works control need consist only of checking either weight or shrinkage.

A continuous and recording apparatus for measuring the linear shrinkage of ware while it passes through a tunnel dryer is described by Frank (**F26**). It consists of a 200 mm plate with two blades at the corners of one end which fix it to the ware and a roller at the other end. The movement of the roller as the surface under it moves during shrinkage is recorded by a pen on paper graduated in percentage linear shrinkage, mounted on a revolving drum. The clockwork and the ink have been developed to withstand the temperature and humidity of the dryer. The whole instrument is small enough to cause very little displacement when put on one piece in a normal

THE CERAMIC LABORATORY

setting. This is important as rearrangement of the setting to incorporate a large instrument changes the drying conditions so that the measurements obtained are worthless.

SUMMARY OF USEFUL EQUATIONS FOR CALCULATING SHRINKAGES AND VOLUMES

Symbols

  a, % linear shrinkage;
  b, % square shrinkage;
  c, % cubic shrinkage.
  V_u, over-all volume (including pores) dry.
  V, over-all volume (including pores) fired.

$$b = 2a - \frac{a^2}{100}$$

$$c = 3a - \frac{3a^2}{100} + \frac{a^3}{100^2}$$

$$a = 100 - 10\sqrt[3]{(1000 - 10c)}$$

$$a = 100 - \left[1 - \sqrt[3]{\left(\frac{V_u}{V}\right)}\right]$$

$$c = 100\left(1 - \frac{V}{V_u}\right)$$

$$V_u = \frac{100V}{100 - 3a - (3a^2/100) + (a^3/100)}$$

Measurement of the Green Strength of Clays

The strength of dry clay may be determined by transverse, tension or compression test, the first being the most usual (**R28**).

TRANSVERSE STRENGTH

Kohl (**K62**) describes a reproducible method of measuring the transverse strength of a specially shaped bar of a given body dried to a given extent. Hans Zoellner (**Z7**) has improved on the apparatus to give a neat set-up useful in both research and industrial laboratories.

The test bar, specially shaped to ensure that failure occurs at the centre, is supported on two knife edges. A third one lies above its centre and is connected to a weight. This is just counterbalanced by the funnel of water. As water is run out into the measuring cylinder the pressure on the test bar increases. When failure occurs the fall of the weight releases a device to divert the water from the measuring cylinder to a neighbouring vessel. The water collected in the measuring cylinder gives an accurate value for the applied force. The area of the broken test bar is measured.

The choice of the shape of test bar is debatable. A bar under stress will break at the weakest point near the application of the load. Kohl and Hans Zoellner chose to make a particularly weak spot by narrowing the bar at the centre to ensure its breaking there. This assumes that the strength of the body itself is absolutely uniform throughout, and affected only by its geometrical shape. But a clay body is known to be inhomogeneous and the

FIG. 4.28. *Modulus of rupture apparatus designed by B.C.R.A.* (*Malkin*, **M23**)

strength may vary from point to point, a particularly weak spot occurring where there is an occluded air bubble. If the test bar is of uniform cross-section it will break at the nearest weak spot to the applied load and give a more realistic value for the strength. In the writers' experience the most strain-free and comparable test bars are extruded cylinders.

A modulus of rupture apparatus using uniform rectangular or circular section bars has been designed by the British Ceramic Research Association and can be used to test both unfired and fired clays and bodies (Fig. 4.28). The specimens are made by extrusion or by plastic pressing. With an adaptor the apparatus can be used to estimate the adhesion of cup handles to the body of a cup (**M23**).

It is essential that the test be carried out under the same conditions of preparation of the clay and drying temperature as the process which the test is desired to control; because green strength is altered by many of the factors affecting plasticity, *e.g.* weathering of the raw clay, maturing of the clay body, presence of electrolytes; also the method of shaping, whether extruded or cast, and by the temperature of drying.

The *modulus of rupture* (*R*) is deduced from the results by the formula:

$$R = \frac{3wl}{2bh^2}, \quad \text{or} \quad \frac{8wl}{\pi d^3}$$

where w = weight required to break bar;
l = distance between supports;
b = width of bar ⎫
h = height of bar ⎬ rectangular specimen (**R28**);
d = diameter of a round specimen.

The result is expressed in weight per unit area, *i.e.* lb/in² or kg/cm².

TENSILE STRENGTH

Although the tensile strength of the green clay is of interest its measurement has been abandoned because it cannot be carried out sufficiently accurately to give reliable and comparable results.

Firing Behaviour

Samples are fired to different temperatures or withdrawn at intervals from a test kiln to follow the development of a number of features:

(1) Colour.
(2) Shrinkage.
(3) Porosity.
(4) Sintering.
(5) Vitrification.
(6) Softening.
(7) Strength.
(8) Relation to glazes.
(9) P.C.E. test.

Most of the tests employed are either straightforward physical tests or they are described elsewhere.

Clays containing a noticeable amount of combustible matter that has to be allowed for in the firing programme are checked in a bomb calorimeter.

FIRING SHRINKAGE UNDER LOAD

Bricks, tiles and similar heavy goods are stacked on top of each other in the kiln so that the load on the bottom layer may become considerable. If it becomes too great, there is a possibility that not only greater shrinkage will occur, but that deformations leading to collapse of the stack might arise. The behaviour of such materials during firing under load must therefore be ascertained.

One apparatus for this test has been devised by Macey, Moore and Davenport (**M9**). Six briquettes $3 \times 1\frac{1}{2} \times 1$ in. ($7 \cdot 6 \times 3 \cdot 8 \times 2 \cdot 5$ cm) are set on edge, in pairs, three courses high, two stretchers, on two headers, on two stretchers. A load of 5 lb/in^2 ($0 \cdot 35$ kg/cm^2), which is of the same order as that on the bottom course of a kiln brick setting, is applied through a silica tube on to a sillimanite bar, on the top two briquettes. The briquettes are placed in a small electric furnace $6\frac{1}{2} \times 6\frac{1}{2} \times 6$ in. ($16 \cdot 5 \times 16 \cdot 5 \times 15 \cdot 2$ cm) and heated at a rate of 1° C/min, governed by a programme controller. During the heating the downward movement of the load beam which bears on the silica tube is recorded by a clock-driven drum with a magnification of 5:1. The expansion of the fused silica tube during the heating is negligible compared with the subsidence of the samples. As the temperature increase is at a constant rate the graph traced out on the drum is subsidence against temperature. For building bricks the apparatus automatically switches off when 1200° C (2192° F) or 16% linear shrinkage have been obtained, whichever occurs first.

Apparatuses for creep tests and furnaces for the preparation of samples and physical testing are described by Aldred and White (**A180**):

(1) Gas-fired furnaces for temperatures up to 1800–1900° C (3272–3452° F), the atmosphere always containing the products of combustion.
(2) Wire-wound electric furnaces: Pt–10% Rh up to 1600° C (2912° F); Pt–40% Rh up to 1700° C (3092° F); molybdenum wound on alumina up to 1900° C (3452° F).
(3) Graphite resistor furnace up to 3000° C (5432° F) but the atmosphere usually contains CO.
(4) Graphite resistor alumina tube furnace up to 1700° C (3092° F) but with practical advantages over the wire-wound furnaces.
(5) Silicon carbide furnaces up to 1500° C (2732° F).

FIRED PROPERTIES

The Investigation, Testing and Control of Fired Bodies

It is important to differentiate between investigations of the properties of ceramic materials and of the properties of ceramic products. The first are scientific tests of use in research and development, the second are practical applications of the first for use as guarantees of the ware. This applies particularly to their physical properties.

Ceramic pieces are difficult subjects for accurate scientific research. Owing to their heterogeneous nature more than one physical property is frequently involved in any particular test, so that results are often only true of test pieces of specified size and shape, and there is a wide scatter of results due to faults and flaws, etc. In fact absolute values can rarely be obtained. Nevertheless the aim has been to evolve tests for clearly defined properties, with the minimum interference from forces not under test, which will give reproducible results with any one particular material, and results that can be compared for different materials. For this it is essential to set out standard practice using test pieces of agreed size and shape and carrying out the test in a precise way. Figures thus obtained for different materials can be used as a basis for selecting the most suitable one for a given application. Further tests can then be carried out to find the variation of the properties with size and shape.

The reproducibility of results is in itself a useful measure of the suitability of the test method and/or of the homogeneity of the test pieces. Thus, given reliable apparatus, much can be learnt from the number and size of the deviations from the average value. A well made set of ten test pieces of suitable size and shape should give results that do not vary from the average by more than $\pm 5\%$. If one or two do, they should be discarded and replaced by similar pieces. If it proves impossible to obtain ten results falling within the desired range then it can be concluded that, either the test method is unsuitable (which can be verified by using another material) or the material under test cannot be shaped, dried and fired to give a reliable result in the test under discussion, and cannot therefore be compared with other materials in this way.

Testing of finished products is an essential step in production control for the output of reliable ware, possibly with various guarantees. Such tests are empirical, being adapted for each shape and type of piece and involving complexes of forces peculiar to the expected conditions of use. There are two main types of such tests. One is the testing to destruction of samples, the other is a simulation of conditions in use applied to samples or to all the pieces. Engineering ware is usually sold in various grades according to whether each piece or only samples have been tested.

Mineralogical Investigation

Thin sections of fired ceramic bodies or glazed pieces can yield very valuable information when viewed under the following instruments:

Microscope (transmitted light for silicates; reflected light for opaque minerals).
Polarising microscope.
Thermal microscope.
Electron microscope, for three-dimensional data.
Microradiography, in which an X-ray shadow is cast by the thin section on to a very fine-grained photographic plate with which it is in contact.

X-ray diffraction powder photographs using both the Debye–Scherrer technique for major constituents and the Guinier multiple focusing camera for minor ones are also valuable. Chemical analysis helps in apportioning the phases observed.

Such mineralogical data are exceptionally useful for the study of reactions, both those that occur in firing and those that happen in use, *e.g.* of refractories. Similar materials can be studied comparatively and the causes of their differences may become apparent, *e.g.* in porcelains, differences in the size of the quartz particles and/or absence of quartz, orientation or segregation of one mineral, relative quantities of glass (**M43, C42, G70**).

Colour

Many ceramic bodies and glazes are white, presence of colour being a sign of impurity or wrong firing. They may require visual inspection to remove disfigured pieces. In coloured pieces too, certain divergences from the desired colour are pointers to wrong firing, such as 'black core' in red bricks.

It is in the nature of ceramic manufacture that it is difficult to produce large quantities of pieces of identical colour, and more or less exact checking and sorting may be required according to the application. This may range from the hand selection of multi-colour facing bricks to give a regular all-over effect, to the exact matching required for wall tiles and sanitary ware.

Although 'colour' means little except as perceived by the human eye (p. 227), it cannot be measured or even accurately matched by eye. The colour seen is affected by the spectral composition of the illuminant and by the background. Colours that appear to match in daylight may not do so in artificial light, and a grey looks much lighter on a black background than on a white one. Nevertheless much colour matching and checking are done by comparison with standards. There are numerous colour charts, including some of international standing.

(1) The Munsell System. *Munsell Book of Colour.* Munsell Color Co., Baltimore, Md. (1929). Contains about 400 panels.
(2) *The Maerz and Paul Dictionary of Color.* McGraw-Hill, New York (1930). Contains about 7000 panels.
(3) *The Ostwald Colour Album.* Winsor and Newton, London (1931). Contains 600 panels.
(4) *Ridgway Color Standards and Color Nomenclature.* Hoen, Baltimore, Md. (1912). Contains 100 panels.

In conjunction with such charts colours are designated by numbers, numbers and letters, names, etc.

Where higher precision, independent of the observer and the illuminant are required physical instruments must be used, *e.g.* automatic recording spectrometer which plots the reflectance when the sample is illuminated by all of the pure spectrum colours in turn.

With the help of instruments true colour matches can be differentiated from 'drifting' colour matches, the latter being those whose differences change with the spectral composition of the illuminant. Data can be used to compute numerical figures on various systems, *e.g.* the tristimulus specification. It can be used to determine the significance of differences (**W51**).

Size

Every ceramic product must be checked for size either individually or by samples. There are causes for variations in the size throughout the production:

(1) Variation of chemical composition or grain size of raw materials.
(2) Errors in body preparation, especially water content.
(3) Inaccuracies and wear of dies and tools.
(4) Variations in firing shrinkage due to different temperatures attained, different soaking periods, different atmospheres.
(5) Warping, bloating.

Consumers should therefore always be asked to allow the maximum possible tolerance in order to keep down the cost which perforce arises if the tolerance is very small.

Size testing may be done with gauges, callipers, etc., varying in accuracy according to requirements. For instance, for checking bricks, six random samples are placed in a wooden box-like gauge and the deviation of their total lengths, widths and depths, respectively, read off on a scale in $\frac{3}{32}$ in., that is $\frac{1}{64}$ in. per brick. Wall tiles must be measured individually and sorted into sizes varying by $\frac{1}{32}$ in. This can be done on electric machines which show lights of particular colours according to the size of tile being tested. One operator can size and box fifty 6 in. by 6 in. tiles per minute (**P50**).

The required method of testing and tolerance allowed is often given in Standard Specifications.

X-ray Inspection of Green and Fired Goods

By passing an X-ray beam through a solid piece faults can be detected without its destruction. Voids, laminations and inclusions are shown up. This is of considerable importance in precision engineering materials, *e.g.* sparking plugs. Irregular glaze application can also be quickly detected (**A83**).

There are a number of X-ray cameras available for such inspection. A mobile unit has been developed by Philips (**P37**). It consists of a high-tension transformer mounted on the base of a trolley. The shock-proof X-ray tube is mounted on the vertical column with universal movement. The portable control table and its angle-iron frame fit over the transformer on the trolley.

Density

TRUE SPECIFIC GRAVITY

'True specific gravity' is also termed 'powder specific gravity', 'absolute specific gravity' or 'specific gravity'.

Samples, free of glaze, are crushed to pass a given standard size, this depending on the type of body, and the specific gravity found in a specific gravity bottle or a Rees-Hugill flask.

APPARENT SOLID DENSITY

This is the ratio of the mass of a material to its apparent solid volume, *i.e.* the volume of the material plus the volume of sealed pores.

BULK DENSITY

This is the ratio of the mass of a material to its bulk volume, *i.e.* the volume of the material plus all its pores. It is also known as 'apparent specific gravity'.

Both bulk density and apparent solid density can be obtained from the same test procedure. A cut or broken sample piece is dried, weighed and then immersed in a liquid in such a way that all the open pores become filled, *e.g.* by evacuating before immersion or by boiling. It is then weighed, first immersed in the liquid and then in air. If

W_a = wt. of dry test piece;
W_b = wt. of test piece soaked and suspended in the immersion liquid;
W_c = wt. of test piece soaked and suspended in air;
D = density of immersion liquid at temperature of test;

then

$$\text{bulk volume} = \frac{W_c - W_b}{D}$$

$$\text{bulk density} = \frac{W_a}{W_c - W_b} \times D$$

$$\text{apparent solid volume} = \frac{W_a - W_b}{D}$$

$$\text{apparent solid density} = \frac{W_a}{W_a - W_b} \times D$$

Porosity

Apparent porosity is the ratio of the volume of the open pores to the bulk volume of a material. True porosity is the ratio of the volume of the open and the sealed pores to the bulk volume. Sealed porosity is the ratio of the volume of the sealed pores to the bulk volume, also called closed porosity.

Porosity is dimensionless and usually expressed as a percentage; it is related to density as follows:

$$\text{apparent porosity} = 100\left(1 - \frac{\text{bulk density}}{\text{apparent solid density}}\right)$$

$$\text{true porosity} = 100\left(1 - \frac{\text{bulk density}}{\text{true density}}\right)$$

$$\text{sealed porosity} = 100 \times \text{bulk density}\left(\frac{1}{\text{app. solid density}} - \frac{1}{\text{true density}}\right)$$

Porosity can therefore be deduced from the absorption measurements used to obtain bulk and apparent solid densities together with the true specific gravity. Owing to experimental variations porosity is often given in the form of 'water absorption'.

Where the true specific gravity of a material is known and can be assumed to be constant more rapid evaluation of porosity data can be obtained direct from bulk volume and bulk density measurements without the time and work involved in water absorption data. Where the shape is regular simple measurement and weighing give the answer. Where the shape is irregular mercury volume vessels are used. These have to be sealed and brought under pressure in order to immerse the sample. A number of models exist. Mercury is also used in rapid bulk density measurements of open-pored pieces, in special scales incorporating needles to which known forces are applied in order to sink the specimen (**C74**).

WATER ABSORPTION TESTS

A dry (heated to 110–115° C, 230–239° F) ceramic piece with any open pores will gain weight by immersion in water. The amount of water absorbed, and therefore the measure of porosity obtained, however, varies considerably with the method of immersion, *i.e.* time and temperature. Well-known methods include:

(1) Weigh, immerse in water, bring to boil, boil for one hour, allow to stand in same water for a further twenty-three hours, wipe off surface water, weigh.

(2) Weigh, place in vacuum desiccator, evacuate to low pressure, slowly admit water until specimen completely submerged, restore pressure, weigh first suspended in water, and then remove and wipe surface and reweigh.

(3) Weigh, soak in water at 15·5–30° C (59·7–86° F) for twenty-four hours, wipe surface, weigh, place on grid in water tank and heat to boiling, boil for five hours, allow to cool for from sixteen to nineteen hours to 15·5–30° C (59·7–86° F), weigh immersed, remove, wipe surface, weigh. This method is used for bricks. From this the following data are calculated:

If a = wt. of dry brick;
  b = wt. of brick after 24 hr immersion in cold water;
  B = wt. of brick after 5 hr immersion in boiling water;
  c = wt. of brick when suspended in cold water after 5 hr boiling;

then

$$\text{absorption by weight after 24 hr cold immersion} = \frac{100(b-a)}{a}\%$$

$$\text{absorption by volume after 24 hr cold immersion} = \frac{100(b-a)}{B-c}\%$$

$$\text{absorption by weight after 5 hr boiling immersion} = \frac{100(B-a)}{a}\%$$

$$\text{absorption by volume after 5 hr boiling immersion} = \frac{100(B-a)}{B-c}\%$$

$$\text{saturation coefficient} = \frac{b-a}{B-a}$$

$$\text{apparent porosity} = \frac{100(B-a)}{B-c}$$

$$\text{bulk density} = \frac{a}{B-c}$$

DYE ABSORPTION TESTS

Water absorption tests cannot differentiate sharply between completely vitrified pieces and those with a very few open pores. In order to detect the smallest absorption such pieces are immersed in dye solutions. Evacuation and/or boiling are used to ensure penetration of the dye into any pores present. Any dye absorbed can be readily observed. The depth of penetration can also be seen if the piece is broken open.

MOISTURE MOVEMENT

Most porous ceramic ware expands slightly when it becomes wet, although the process may be slow to reach completion. Moisture expansions are not always equalled by drying contractions. Measurements are made of the distance between two carefully located points either during expansion or during contraction. The method chosen should be specified.

MAXIMUM PORE DIAMETER

Permeable ceramics used for filters, diffusers, diaphragms, etc., are graded by their maximum pore sizes.

The maximum pore diameter is measured by finding the minimum pressure required to force a bubble through the filter. The pores in the test piece are first filled with the test liquid (usually water, for very fine pores carbon tetrachloride or amyl acetate) the piece is then submerged in the liquid and air pressure is gradually applied to one side until the first bubble breaks away. The calculation is based on the passage of an air bubble through a capillary tube of circular cross-section:

$$D = \frac{4 \times 10^5 \alpha}{981(P - g\rho h)}$$

where D = maximum pore diameter (μ);
P = bubble pressure recorded (mm H_2O);
h = depth (mm) below surface at which bubble appeared (can usually be neglected);
α = surface tension of liquid (dyn/cm);
ρ = density of liquid.

Some measure of the uniformity of pore diameters can be demonstrated by increasing the air pressure a little beyond the point where the first bubble appeared (**S123**).

PERMEABILITY

The permeability of a material to a fluid is the rate at which the fluid in question will pass through the material under definite difference of pressure; permeability, unlike porosity, is a directional property.

The permeability of materials of low permeability, *e.g.* refractories, is measured by timing the passage of a measured volume of air through the sample at known constant pressure (B.S.S. 1902:1952):

$$\text{permeability} = \frac{\text{volume of air passed through (ml)} \times \text{height of test piece (cm)}}{\text{time of flow (sec)} \times \text{area of cross-section of test piece (cm}^2\text{)} \times \text{pressure head (cm } H_2O\text{)}}$$

The permeability of filters and other highly permeable materials is more difficult to measure because of the high flow rate. It is done with water. (B.S.S. 1969:1953).

PORE DISTRIBUTION AND SURFACE TEXTURE

Bouvier and Kaltner (**B95**) give these two investigations as examples of the uses of radioactive isotopes in ceramic research. For the pore distribution the sample is impregnated *in vacuo* with a self-hardening resin bearing radioactive sulphur. A polished section is then prepared and the pore distribution can be found from autoradiographs.

Mechanical Properties

MODULUS OF ELASTICITY

The modulus of elasticity of a material is the ratio of stress to strain. The strain may be a change in length, Young's modulus; a twist or shear, rigidity modulus; or a change in volume, bulk modulus. The stress required to produce unit strain being in each case expressed in dynes per square centimetre (dyn/cm^2) or pounds per square inch (lb/in^2).

The modulus of elasticity is closely related to important properties of engineering materials, such as strength, thermal shock resistance, behaviour in service under stress. Its measurement is therefore increasingly necessary as the use of ceramics as engineering materials spreads.

As ceramics are mostly brittle, very little strain occurs before failure. Special arrangements must therefore be made to detect very small changes of shape if a direct method of measuring stress and strain is to be used. Vassiliou and Baker (**V6**) have, however, developed a quick direct method of sufficient accuracy. The specimen, in the form of a bar or a whole brick if necessary, of uniform cross-section throughout its length, is placed on knife-edge supports. A lever is brought into contact with its underside by screw adjustment. At the other end of the lever is a mirror of focal length 1 m (3·28 ft) which reflects an incident beam on to a scale one metre away. The spot light can be adjusted to a standard starting point for each test. This arrangement can detect deflections of the sample of one-millionth of an inch, while a load is applied to the specimen by 3- or 4-point loading.

A more usual method for finding the modulus of elasticity is the sonic one. The method involves the determination of the resonant oscillation frequency of a regular specimen when excited in longitudinal, flexural or torsional vibration. The resonant frequency depends on the dimensions, density and appropriate elastic modulus of the specimen. The equations involved are set out by Lakin (**L3**). Apparatus necessary for determining the resonant frequency consists of:

(1) An audio-frequency oscillator for generating frequencies continuously variable from 15 to 50 000 c/s with an accuracy of ± 1%.

(2) A single-stage power amplifier fed from the oscillator and driving (3).

(3) A vibration generator which is placed in contact with the specimen (4).

(4) The test specimen supported on foam rubber supports.

(5) A gramophone crystal pick-up with the stylus placed in light contact with the specimen.

(6) A double-beam oscillograph. One trace receives the output from the pick-up, suitably amplified, while the other trace is actuated direct from the oscillator; the latter signal is also used to synchronise the time base (**L3**).

This apparatus can be adapted to make measurements at high temperatures (**L4**).

Another commonly used instrument for measurement is a strain gauge. It involves less expensive equipment than the sonic method. In it, a metal alloy in the form of a grid is fixed to the surface of the material under test. The change of resistance of the grid when stress is altered is a measure of the strain on the surface of the material. Although accurate, the method is very slow in dealing with a large number of samples as long rest periods are necessary after fixing the grid before readings can be taken (**V6**).

STRENGTH AND ALLIED PROPERTIES

'Strength' is the ability of a material to retain shape when subjected to various forces. There are various types of failure when strength is exceeded:

(1) A material is 'brittle' if it fails suddenly when subjected to a sharp blow or other shock.

(2) It is 'friable' if it is easily and gradually crushed.

(3) It is 'malleable' if it yields slowly under a succession of sharp blows to form a coherent mass of different shape.

(4) It is 'ductile' if its shape can be altered by pulling or extrusion; the maximum increase in length without fracture is its 'extensibility'.

A material that resists forces that tend to pull it out of shape is 'tough', one that regains its shape after the deforming force has been removed is 'elastic', one whose shape is altered by an applied force is 'deformable'.

Strength is measured in a number of ways according to the nature of the applied force:

(1) Compressive strength, or crushing strength.

(2) Tensile strength.

(3) Transverse strength, or bending strength, or modulus of rupture.

(4) Torsional strength.

(5) Shear strength.

(6) Impact strength.

Compressive Strength or Cold Crushing Strength. This is high for most ceramics and strong mechanical or hydraulic machines are required to test it. The platens must be designed to give axial loading.

Dry test pieces of rectangular, cubical, or cylindrical shape are used, care being taken that the bearing faces are plane and perpendicular to the direction of pressure application. Where whole bricks are used there are standard methods of filling the frogs and general irregularities with suitable mortar (Britain and Germany), cement paste, shellac and plaster of Paris (U.S.A.). Packings of fibreboard, asbestos mill board, tarred roofing felt, etc., are also used.

Cylindrical objects, *e.g.* drainage pipes may be placed between bearers

Fig. 4.29. *Testing the compressive strength of a brick pier (London Brick,* **L41**)

Fig. 4.30. *Hydraulic tensile test machines for electrical porcelain insulators (Doulton,* **D56**)

and packed out with rubber, for the test. It may also be desirable to test structures using a number of ceramic pieces suitably bonded (Fig. 4.29). The results are given in pounds per square inch or kilograms per square centimetre. It may also be necessary to specify the direction of application of the load.

Ball Compression Test. This measures the compressive strength under point loading axial with a point support.

Tensile Strength. Ceramic bodies are at their weakest when in tension, furthermore this property is difficult to measure, often giving a wide scatter of results. Failure occurs at the weakest point and some workers like to ensure that this is at the centre of the piece by constricting this. Others use straight rods. It is also important that the tensile stress be applied along the geometrical axis of the piece.

The tensile strength may then be measured on a special machine (Fig. 4.30) or on an adapted compressive strength machine. The result, expressed in pounds per square inch or kilograms per square centimetre, is given by the force required divided by the cross-sectional area at the break. (B.S.S. 1610:1952).

Transverse Strength, Cross-bending Strength, Modulus of Rupture. It is very difficult to get a true value for the modulus of rupture of ceramics. It is therefore necessary to use standard methods of test as regards sample size, distance between and dimensions of the supports and rate of application of load, if results are to be comparable.

The test consists essentially of supporting the test piece on two bearers and applying a load to a third one half-way between the supports. Whole bricks, refractory bricks, wall tiles, etc., may be so tested, or rods cut from materials. Roofing tiles are previously soaked in water at 25° C (77° F) for twenty-four hours (**C41**).

Torsional Strength. This is not frequently measured. Rods of square cross-section are clamped into a torsion machine. The force required for failure divided by the cross-sectional area gives the result in pounds per square inch or kilograms per square centimetre.

Impact Strength. Most ceramic products have at some time to withstand dynamic forces, it is therefore necessary to make dynamic tests. Theoretically the resistance to impact in compression, tension, shear and cross-bending could be found, but impact tensile and shear tests are impracticable. The senior writer was instrumental in developing a reliable impact bending strength test (**S82**).

Impact bending strength tests are carried out on a pendulum machine, usually on the Charpy principle. In this, a weight is allowed to fall through a definite arc, more than sufficient for it to break the test piece, and its residual energy is measured by its rise on the other side. The test pieces are in the shape of rods or bars and are supported in an adjustable angle at the base of the pendulum's path. The radius of the striking surface of the weight is specified (Fig. 4.31).

Another test method is to find the height from which the pendulum must be released so that it just breaks the test piece without swinging on. If this is done by gradually increasing the height of release until a given test piece fails, it raises the problem of how to calculate the cumulative effect of impacts that did not cause failure. Alternatively, fresh samples must be used for each successive trial. Pendulum machines can also be adapted for testing whole plates both centrally and at the edges.

Fig. 4.31. *Pendulum impact cross-bending strength machine*
(*Wolpert*, **W83**)

Impact compressive strength tests suffer from the same difficulty of the cumulative effect. For these tests small cubes or briquettes are placed on a firm bedding plate and a weight dropped on to them from increasing heights. A personal factor also enters here regarding what constitutes 'failure', which is not so clear cut as the breaking of a rod (**S31**).

Impact strength data should be given in terms of momentum, inch-pounds or gramme-centimetres.

HARDNESS BY MOHS' SCALE

Mohs' scale of hardness is as follows:

Talc	1
Gypsum	2
Calcspar	3
Fluorspar	4
Apatite	5
Feldspar	6
Quartz	7
Topaz	8
Corundum	9
Diamond	10

The material to be examined and the minerals of the hardness scale should if possible be in the form of sharp-edges or sharp-cornered pieces and, at least in part, of fresh plane surfaces. If the test material is not scratched by one of the scale minerals and does not in turn scratch it, it is said to have the same Mohs hardness. If it scratches one scale mineral and is scratched by the next highest then its hardness lies between the two. (B.S.S. 860:1939; B.S.S. 891:1940).

HYDRAULIC TESTS

Vessels and pipes are tested for leaks and strength by hydraulic tests. All air must be expelled before starting the test, pressure is then applied at a given maximum rate and the maximum pressure required must be held for a given number of seconds.

In a pipe under internal pressure the force tending to burst it is the pressure per unit area (p) multiplied by the area of a section across the diameter (d). The resisting power is the thickness of wall (w) multiplied by the length of the pipe (l) and by the ultimate strength (u):

$$p \times d \times l = u \times 2w \times l$$

$$w = \frac{p \times d}{2u}$$

w must then be multiplied by a safety factor in order to give the required wall thickness.

ABRASION RESISTANCE

Drum Test. Cubes of given side length are either made specially or cut from a finished piece. They are dried and weighed, placed in an irregularly shaped drum and rotated for a given period. They are then reweighed and the wear expressed as a percentage of the initial weight. This test involves both impact and abrasion.

Sand Blast Test. A weighed piece has a circular patch exposed to a sand blast of given intensity for a standard period of time. The weight loss is found, divided by the density and then expressed per unit area of exposed surface.

Rubbing Test. A weighed flat tile is made to oscillate under a stationary cast-iron pan with a slot in it through which silicon carbide powder of specified grit size is fed. The loss in weight over a period of hours is found. (B.S.S. 784:1953).

Thermal Properties

COEFFICIENT OF REVERSIBLE THERMAL EXPANSION

The coefficient of linear expansion is the increase in length per unit length caused by a rise in temperature of 1° C. It has no units. The temperature

FIG. 4.32. *Thermal expansion apparatus* (*Malkin*, **M23**)

range of the measurement should be stated. Expansion data may also be expressed as the percentage increase in length when heated through a stated range of temperature.

Expansion data may be obtained by heating suitably marked pieces in a furnace with spy holes and following the movement of the marks with optical apparatus. It is more convenient to use mechanical measurement with a dial gauge for the temperature range for which this is available, up to 1200° C (2192° F). Both a horizontal and a vertical type of apparatus are described for B.S. methods of test (*e.g.* B.S.S. 1902:1952; 784:1953). In these a test rod butts on to a fixed disk at one end and a movable fused silica distance rod at the other. The whole is enclosed in an electrically heated furnace, fitted with thermocouples. The movement of the silica rod during heating and cooling is shown on a micrometer dial gauge. The return to the original reading after cooling should be checked (Fig. 4.32).

Such apparatus can also be used: (1) to plot the curve of size change with firing temperature for raw clays and for bodies; (2) to estimate the degree of calcination of flint; (3) to estimate the proportion of free silica in certain clays; (4) measure the coefficient of expansion of bulk glaze samples up to their softening point (**M23**).

For higher temperatures, up to 1450° C (2642° F) or even 1550° C (2822° F), an electric furnace with spy holes is necessary and the measurements are made with a telescope with a vernier scale. Tubes corresponding to the viewing tubes that allow light to enter the furnace opposite the viewing arrangement are also essential (B.S.S. 1902:1952). Errors due to uneven refraction of light in air of changing temperature can occur but are much more marked in a vertical furnace than a horizontal one. The latter is therefore recommended. In this case, however, the specimen should be on rollers to allow free movement during the expansion.

In order to eliminate the effect of the unequal refraction of light rays passing through the furnace Czedik-Eysenberg (**C74**) has evolved an apparatus using X-rays and a photographic plate. Greater accuracy can be achieved at high temperatures.

SPECIFIC HEAT

The specific heat of a substance is the quantity of heat required to raise the temperature of unit weight of it through one degree. It is usually expressed in calories per gramme per degree centigrade (cal/g per °C). Specific heat is normally measured by well-known calorimetric methods, *i.e.* method of mixtures.

THERMAL CONDUCTIVITY

Thermal conductivity, or heat conductivity, is the rate of transfer of heat along a body by conduction. It is defined as the quantity of heat which will flow through unit area of material of unit thickness with unit difference of temperature between faces, in unit time.

The units are:

(1) B.t.u. per ft^2 per hr for 1 in. thickness and 1° F difference in temperature (B.t.u.-in./ft^2-hr-°F), often denoted 'B.t.u'.
(2) Calories per cm^2 per sec for 1 cm thickness and 1° C difference in temperature (cal-cm/cm^2-sec-°C).
(3) Kilocalories per m^2 per hr for 1 m thickness and 1° C difference in temperature (kcal-m/m^2-hr-°C).

It is frequently not specified which of the metric units are being used and their conversion involves a factor of 360, instead of the more usual decimal point. To convert cal-cm/cm^2-sec-°C to kcal-m/m^2-hr-°C, multiply by 360.

Ceramic materials are relatively poor thermal conductors, so that all measurements are made using the principle of the disk method, with a sample of large cross-section and small thickness. Heat input is electrical, including silicon carbide elements for very high temperatures. The heat flow can be measured in a number of ways, Nicholls (**N15**) recommends a

water flow calorimeter. Norton (**N29**) has shown that it is very difficult to obtain reproducible results on different, but similar, forms of apparatus. He describes an apparatus (Fig. 4.33) that is satisfactory, but emphasises the need to standardise construction and operation.
(B.S.S. 874:1956.)

Diffusivity, temperature diffusivity, thermal diffusivity or *temperature conductivity*, are terms which sometimes occur. Diffusivity governs the flow of the temperature wave through a body, it is a measure of the heat

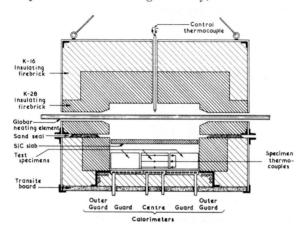

Fig. 4.33. *Diagrammatic section through thermal conductivity testing apparatus (Norton, **N29**).*

stored by a body through which heat is being conducted. It is related to the thermal conductivity as follows:

$$\text{diffusivity} = \frac{\text{thermal conductivity}}{\text{bulk density} \times \text{specific heat}}$$

Himsworth (**H92**) describes a relatively fast and simple method of measuring thermal diffusivity to give values of thermal conductivity for insulating bricks.

THERMAL TRANSMITTANCE

As many ceramic materials are used in building construction, their value as thermal insulators when so used is of greater importance than the laboratory measured thermal conductivity.

The thermal transmittance, or U value, of a wall is the quantity of heat which will flow from air on one side to air on the other side, per unit area, in unit time, for unit difference in air temperature. The unit is British thermal units per square foot per hour for one Fahrenheit degree difference in temperature, B.t.u./ft^2-hr-°F (kcal/m^2-hr-°C, conversion factor 4·882).

Thermal resistance, R, is the resistance to heat flow either of a material of any given thickness or of a combination of materials. It may be defined as the number of degrees Fahrenheit difference between the surfaces when unit heat flows through unit area in unit time. The unit used is Fahrenheit degrees difference in temperature per British thermal unit for one square foot in one hour, °F/B.t.u.-ft^2-hr (°C/kcal-m^2-hr, conversion factor 0·2048).

Thermal conductance, C, is the rate of heat flow either through a material of any given thickness or through a combination of materials. It is the reciprocal of thermal resistance. The unit used is British thermal units per square foot per hour for one Fahrenheit degree difference in temperature, B.t.u./ft^2-hr-°F (kcal/m^2-hr-°C, conversion factor 4·882).

Heat flux is the quantity of heat flowing through unit area in unit time. The unit used is British thermal units per square foot per hour, B.t.u./ft^2-hr.

Thermal transmittance is frequently calculated from values of thermal conductivity. This involves assumptions regarding moisture content, and about the transfer of heat from one medium to another. The result obtained has now been shown to be appreciably different from that found in prolonged practical measurement (Table 67). As acceptable U values for structures are included in certain forms of legislation, its proper measurement is important.

TABLE 67

Differences between Published Calculated Data and Measured Data for Thermal Properties of Constructions (D43)

Description	Published value	Measured value
U value of 11-in. cavity Fletton brick wall	0·30	0·28
U value of 11-in. external brick/cavity/cellular concrete block wall	0·18	0·22
U value of 11-in. external brick/cavity/4-in. clinker block wall	0·22	0·26
k value of 4½-in. Fletton outer leaf	8·0	6·9
k value of 4½-in. Fletton brick inner leaf	8·0	4·4
k value of 4-in. cellular concrete block inner leaf	1·4	1·85
k value of glass wool as cavity fill (density 3–4 lb/ft^3)	0·25–0·28	0·50

Two main methods of measuring thermal transmittance of external walls have been described. In the one a test panel forms the north wall of a room which is maintained at constant temperature and the necessary electrical input is measured (**D63**). In the other a test panel, forming part of a north wall of a building, is heated by a guarded hot plate whose heat output is kept constant, and the temperature differences across the whole structure and each section of it measured with thermocouples. Measurement is carried out continuously throughout the six winter months (**D43**).

REFRACTORINESS

Refractoriness is usually measured by the pyrometric cone equivalent (P.C.E.) test (p. 277). The required cones are cut from a fired piece, no external face being included.

REFRACTORINESS UNDER LOAD

For many purposes the results of refractoriness under load tests are more useful than refractoriness itself. They give an indication of the probable

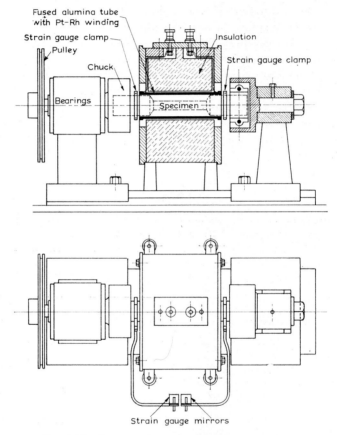

Fig. 4.34. *Apparatus for high-temperature torsional creep measurement (Greaves and Mackensie, G71)*

behaviour of refractory materials when in a construction, *e.g.* a furnace arch, and are also a better measure of the general quality of a product. There are two main types of test:

(1) Rising temperature test, in which a loaded specimen is heated at a prescribed rate and the temperatures at which deformation starts, reaches a certain value, and leads to collapse are noted.

(2) Maintained temperature test, in which the loaded specimen is heated

at a prescribed rate to a predetermined temperature, which is maintained either for a specified time, or until a certain deformation occurs, or until collapse.

Gas, or oil-fired or carbon granule resistance furnaces are used. No flame may impinge on the specimen. The rate of heating must be controllable and a zone of uniform temperature should extend a short distance above and below the specimen. There must be a suitable means of applying the load and of detecting and recording the deformation. The loads used for refractories in B.S.S. 1902:1952 are 50 lb/in² (3·5 kg/cm²) for silica bricks, and 28 lb/in² (2 kg/cm²) for other materials.

Chesters (**C26**) points out the relevance of the loads chosen; the calculated stress in a typical open hearth furnace roof is about 25 lb/in² although uneven pressures may give rise to local loadings of 100 lb/in², in vertical walls the pressure rarely reaches 25 lb/in² unless insufficient expansion allowance has been made.

Until recently the rising temperature test was preferred but is now being superseded by the maintained temperature test because it gives much more realistic data about materials to be used, for instance in arch construction. There is even a trend towards making prolonged creep tests. An international standard test has been proposed (**A102**).

Other tests of strength at high temperature, using slow torsional and tensional loadings, or finding shear strength, give useful results that relate to the relative service performance of various types of refractory bricks (**G71**) (Fig. 4.34).

PERMANENT LINEAR CHANGE ON REHEATING

This is also known as 'after-contraction' or 'after-expansion'. This test reveals facts about the hardness of the previous firing of the material and the presence of excessive flux. Soft-fired pieces show the most after-contraction. In the case of stabilised dolomite this is also an indirect measure of the probable stability in storage. It is, however, difficult to get reproducible results, especially with fireclay bricks (**C26**).

The test is carried out in a carefully regulated furnace in an oxidising atmosphere. The rate of increase of temperature is laid down and the specimen is kept at the test temperature for a given number of hours. In Great Britain this is two hours, in U.S.A. twenty-four hours.

There is some difficulty in applying results obtained from this test to the way a brick heated from one end only will act over very prolonged times of service.

THERMAL SHOCK RESISTANCE

The resistance to thermal shock of ceramic materials is a highly complicated property, and cannot be measured directly to give values that are immediately applicable to the behaviour the material will show in use. Numerous thermal shock tests that can give comparative results are, however, in use, and some are given in standard specifications. These tests involve careful heating of samples of standardised dimensions followed by sudden cooling in water or still air, either at room temperature or at 0°C (32° F). The heating is then repeated either at the same temperature, the number of cycles until failure being counted, or at progressively higher temperatures, the size of the temperature step and the temperature of failure being noted. In vitreous bodies, failure is usually obvious when it occurs, for a sizeable crack shows. In porous bodies more progressive, small cracks are formed, followed by larger ones, spalling off of fragments, and finally the piece falls apart or may be pulled apart by hand. Each stage should be noted.

The effect of thermal cycling before cracking occurs can be measured by finding the transverse strength of suitably shaped pieces removed from the bulk of pieces under test after each cycle, *e.g.* for chemical stoneware (B.S.S. 784:1953) thirty-five test pieces of the shape required for the transverse strength test are placed in a basket in the test furnace and quenched simultaneously. Five pieces are removed and their transverse strength found after heating the batch to 50, 100, 175, 200, 225 and 250° C (all $\pm 5°$ C) in turn. Refractories may also be tested for compressive strength after cycles that have not produced cracks.

It is important to choose temperature changes that are related to those that the bodies can withstand. For instance porcelain intended for sparking plugs and heating devices may be heated to 900° C (1652° F), cooled to room temperature and its loss of transverse strength determined. Chemical stoneware so tested would be destroyed, and as it is not intended for such treatment the test cannot give useful data (**R48**).

In the case of refractories attempts are made to simulate the conditions of use in ways more realistic than the $3 \times 2 \times 2$ in. prism. For instance, most refractory bricks are heated from one face only and are under a constant thermal gradient. This is recognised in tests that heat a brick from one end by means of an electric hot plate, and the rate of heating to produce cracking is found.

The American standard test uses a panel of fourteen $9 \times 4\frac{1}{2} \times 2\frac{1}{2}$ in. bricks set in refractory kaolin and surrounded by further suitable refractory 'dummy' bricks (A.S.T.M. C38–36, C107–36 and C122–37). The completed panel is first preheated for twenty-four hours to induce any vitrification, shrinkage, etc., that might develop in service and lead to structural spalling (1600° C, 2912° F for high heat-duty fireclay bricks, and 1650° C, 3002° F, for super-duty fire-clay bricks). It is allowed to cool naturally and then inspected. It is then transferred to the spalling furnace and subjected to twelve cycles of alternate ten-minute periods of heating at 1400° C (2552° F) and cooling with an air/water-mist blast. At the end of this the bricks are freed from the mortar and any detached pieces and reweighed. The percentage loss is recorded. The method represents severe furnace conditions but is lengthy and requires special equipment.

Banerjee and Nandi (**B7**) have compared it with the British standard method (B.S.S. 1902:1952) and believe that although the latter is much more empirical with a little modification it is quite representative and can be carried out easily at individual works.

Visual methods of determining whether cracking has occurred combined with the manual method of pulling a specimen apart prove to be particularly

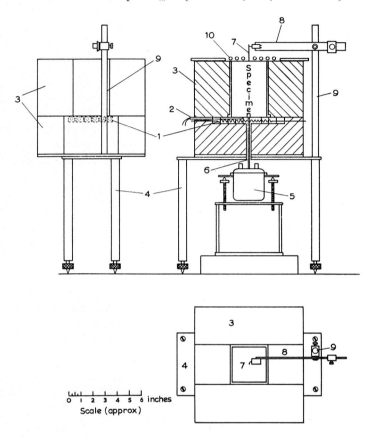

FIG. 4.35. *Apparatus for sonic spalling test*
(1) *Heater brick, cut and grooved refractory brick bearing six heating coils*
(2) *Thermocouple* (3) *Diatomite insulation* (4) *Metal table*
(5) *Vibrator unit* (6) *Fused alumina rod* (7) *Contact wire*
(8) *Pickup arm* (9) *Upright* (10) *Cooling coil*
(*Clements and Davis,* **C38**)

unreliable when the piece is very porous as in insulating bricks. Clements and Davis (**C38**) describe a test in which the first crack can be detected. The test piece is made to vibrate at its resonant frequency while it is being heated. Cracking is indicated by sudden changes of frequency. With this apparatus accurate and reproducible measurements can be made of the rate of heating that will crack the test piece, and the temperature at which the crack appears (Fig. 4.35).

For modern engineering ceramics a better analysis of thermal shock data is necessary if predictions about specific pieces under definite conditions are to be made. Buessem (**B135**) discusses this aspect: 'The prediction of the performance of a device requires:

'(1) recognition and definition of all the important factors which influence the performance;
'(2) measurement of these factors;
'(3) knowledge of possible variations of these factors;
'(4) determination of the dependence of the performance on these factors.'

From the mathematical point of view the 'performance' is the dependent variable and the 'factors' are independent variables. The relation between the two is the function which must be determined. The variables must be measurable in physical units. The function may be expressed either in an analytical formula or in tables or graphs. The performance function:

$$P = \frac{\sigma_{max}}{s} = F(f_1, f_2, f_3 \ldots)$$

where P = performance index;
σ_{max} = maximum stress;
s = strength of the material;
f_1, f_2, f_3 = factors affecting the performance.

If P is always below 1 no failure occurs and the difference between the maximum value of P and 1 constitutes the safety margin.

The factors affecting the performance can be listed in three groups:

(1) Factors defining the thermal shock conditions.

(a) Δt = temperature difference between solid body and cooling or heating medium.
(b) h = heat transfer coefficient (measured in cal-cm^{-1}-sec^{-1}-°C^{-1}).

(2) Factors defining the dimensions of the solid body.
r_m = half thickness or radius or any representative dimension $\bar{a}$ of the body or a group of such dimensions.

(3) Factors defining the material properties.
(a) R = first thermal stress resistance factor (measured in °C):

$$R = \frac{s(1-\mu)}{E.\alpha}$$

(b) R' = the second thermal stress resistance factor (measured in cal-cm^{-1}-sec^{-1}):

$$R' = \frac{s(1-\mu)}{E.\alpha}k$$

where s = strength;
E = modulus of elasticity;
μ = Poisson's ratio;
α = thermal expansion coefficient;
k = thermal conductivity.

From 3(*a*) and (*b*):

$$\frac{R'}{R} = k$$

The performance function may then be written:

$$P = \frac{\Delta t}{R}\left(\bar{h}, \bar{a}, \frac{R'}{R}\right) = \frac{\Delta t}{R}\left(\bar{h}, \bar{a}, k\right)$$

Buessem and Bush (**B137**) describe an apparatus for making 'ring tests' where the thermal stress resistance factors R and R' can be determined. In this, a pile of rings of the material are heated from the inside by a heating element and cooled from the outside by a calorimetric chamber (Fig. 4.36). The centre rings of the pile have radial heat flow. The radial temperature gradient in them can be measured and also the number of calories flowing through their unit surface area. Under equilibrium conditions these measurements give the thermal conductivity of the ring material. The radial temperature gradient, which produces thermal stresses, is then gradually increased until failure occurs ($P=1$).

Astarita (**A195**) also makes both a theoretical and practical study of the parameters involved in the study of thermal shock.

Chemical Properties

SOLUBLE SALTS

Suitable samples of ware are crushed and ground. A weighed portion is stirred repeatedly with distilled water. The solution obtained is filtered off into a graduated flask. The residue and the filter paper are stirred with a further quantity of distilled water and then filtered and washed into the flask, which is then made up to the mark. Some of the solution so obtained is evaporated to dryness to give the total soluble content, while the rest is used for quantitative determinations of possible constituents, *e.g.* SiO_2, $Al_2O_3 + Fe_2O_3$, CaO, MgO, Na_2O, K_2O, SO_3 (**C41**).

EFFLORESCENCE

The appearance of salt deposits on porous ware after repeated wetting and drying is known as 'efflorescence'. Several pieces (usually bricks) are stood on end in a flat dish containing distilled water and left at room temperature until they have absorbed the water and appear dry again. More water is added and the process repeated. The surface of the pieces

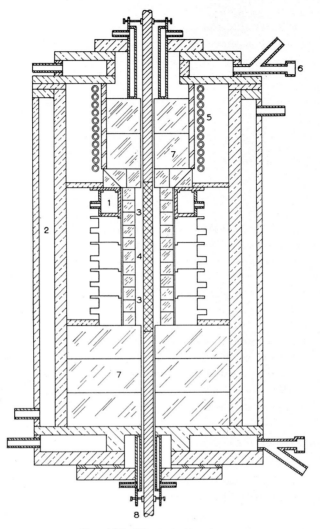

Fig. 4.36. *Ring test apparatus*

(1) *One of five interlocking calorimetric cells*
(2) *Outer calorimeter jacket*
(3) *Guard ring specimens*
(4) *Test ring specimens*
(5) *Cooling coil of copper tubing*
(6) *Thermoresistor socket*
(7) *Insulating refractory brick*
(8) *Heating element (Buessem and Bush,* **B137**)

is then examined for deposits of salts, powdering, and flaking of the surfaces (**C41**).

HYDRATION RESISTANCE

In the field of refractories materials may be used that are not normally stable in water. Where these are specially stabilised, *e.g.* stabilised dolomite, the hydration resistance must be checked. Two methods are described by Chesters (**C26**). In the first, the finely ground product is subjected to the action of steam for five hours and the percentage hydration found by the loss on ignition at 1000° C, after drying. In the second, samples of known grading are subjected to atmospheric conditions for periods of some weeks.

ACID RESISTANCE

Ware to be tested for acid resistance is crushed, ground, and sieved, the portion passing B.S. 18-mesh sieve and retained on B.S. 25-mesh is used for the test. This is first boiled with distilled water, washed and dried. A weighed portion is then treated with a mixture of sulphuric and nitric acids and heated until all the nitric acid has evaporated and the sulphuric begins to fume strongly. The mixture is cooled, more nitric acid and distilled water added and the heating repeated. Repeated washings follow until they are free of sulphuric acid and the sample then dried to constant weight. The result is given as a percentage (B.S.S. 784:1953).

Acid Soluble Iron. In chemical ware it is important that certain acidic liquids should not become contaminated by iron. The test for acid-soluble iron consists of taking a sample of the body free from glaze and crushing and grinding it to specified size grading. It is then boiled in hydrochloric acid for a number of hours, filtered, washed and the iron content of the filtrate determined colorimetrically (B.S.S. 784:1953).

RESISTANCE OF REFRACTORIES TO SLAG ATTACK

The attack of refractories by slags, metals, gases, etc., is always a complicated process involving physical as well as chemical reactions. There is therefore no single test for measuring such attack and in fact so many have been described that there have been at least three surveys of the literature: by Ferguson (**F7**) in 1928, Simpson (**S79**) in 1932 and Hurst and Read (**H128**) in 1942. The last arranged the test methods in a number of classes.

Type I. This class is characterised by the use of relatively small amounts of slag in contact with large amounts of refractory. The slag is placed on the refractory or in holes and hollows in it at room temperature, and the materials are then heated together to the test temperature. The resultant penetration of slag, corrosion of the refractory, sagging of the refractory, etc., are used as measures of the resistance to corrosion. The argument against this method of test is that once the slag has penetrated the refractory it is not replenished by more molten slag and no simulation of erosion occurs.

Type II. Here the tests involve immersion of the refractory in the corrosive fluid and hence relatively large amounts of slag to small amounts of refractory. Both corrosive liquids and slags and destructive gases and vapours have to be considered. The refractories under test may be incorporated in the walls of pot furnaces filled with molten slag, etc., or they may be suspended in the slag. Actual service tests placing the refractories below the slag level also fall in this category.

Type III. Impingement of slag against hot refractory test specimens is claimed to simulate service conditions for materials to be used above the slag level. The chief criticism is in the methods adopted for producing the accelerated reactions necessary for laboratory test work. Furnaces of various shapes with panels of the test specimens or entirely lined with them are used together with a variety of methods of admitting the slag. Measurements are made of the depth of surface layer removed by erosion, area of erosion, channelling of the refractory caused by slag flow, widening of any cracks that result from solution of the refractory, destruction of joints, weight changes, volume changes and slag penetration.

Type IV. Cone fusion and similar tests, which do not simulate service conditions but instead they investigate the possible reactions between slag and refractory under highly accelerated conditions. Slag and refractory are ground and then mixed in varying proportions and softening and fusion data obtained by P.C.E. or other tests. Hyslop, Stewart and Burns (**H133**) have further refined the method by using superimposed cubes of 80% slag:20% refractory and 20% slag:80% refractory standing on the same refractory.

All these test types, except P.C.E. tests, require some means of measuring the slag attack that takes place and again a variety of methods have been listed.

(1) Chemical analysis, either complete, or tracing the movement of particular components, *e.g.* alkalies.
(2) Optical methods ranging from visual observation of the surface, through planimetric measurement of the attacked area as seen in sections, to petrographic analysis.
(3) Changes of physical properties such as volume, weight, porosity and modulus of rupture.

RESISTANCE TO CARBON MONOXIDE ATTACK

Refractories used in the upper parts of blast furnace linings are subjected to carbon monoxide gas which may have a disintegrating effect. To test for the resistance to this B.S.S. 1902:1952 describes the heating of cylindrical specimens to 450–500° C (842–932° F) in a stream of carbon monoxide for 200 hr or until the test pieces have disintegrated, whichever is the shorter. Any general discoloration or carbon deposition at separate nuclei or cracking during the test is also noted.

Electrical Properties

Standardised grading of electrical ceramics is of great importance so that standard tests have necessarily been laid down in most countries, *e.g.*

B.S.S. 16:1949 for telegraphic materials.
B.S.S. 137:1941 for insulators for overhead power lines.

B.S.S. 223:1956 for high-voltage bushings.
B.S.S. 1540:1949 for electrical insulating materials for use at radio frequencies.
B.S.S. 1598:1949 for materials for telecommunication and allied purposes.
D.I.N. 40685
D.I.N. 40686
D.I.N. 41341

These specifications lay down a number of tests regarding porosity and mechanical strength, followed by electrical tests. Tables assigning rating numbers are also given. Many of these tests are made on completed articles rather than on specimens of material.

Measurements of dielectric properties of materials are affected by a number of factors such as temperature and relative humidity. At high temperatures the nature of the atmosphere whether reducing or oxidising plays a role. Hausner (**H45**) has further shown that the relation between humidity and dielectric properties is affected by almost the same factors that determine the dielectric properties themselves, namely composition of the material, compacting pressure, firing temperature and time, and surface conditions, to which are added the frequency of measurements, temperature and air pressure. This explains why measurements made at the same relative humidity on the same type of ceramic can differ so widely.

FIG. 4.37. *Wet flash over test on high-voltage insulators* (*Doulton*, **D56**)

TESTS FOR HIGH-VOLTAGE INSULATORS SIMULATING WORKING CONDITIONS

The voltages are found for:

(*a*) minimum dry spark over;
(*b*) minimum wet spark over;
(*c*) dry spark over maintained for one minute;
(*d*) wet spark over maintained for one minute (rain test);
(*e*) puncture.

Alternatively, at the voltage given in the rating table, puncture, etc., may not occur. (Fig. 4.37.)

TABLE
Schedule for the Testing

Name	Building bricks	Roofing tiles	Salt glazed drainpipes	Floor tiles
B.S.S.	1257:1945 657:1950	402:1945 1424:1948	65:1952 540:1952 539:1951 1143:1955	1286:1945
Size	√	√	√	√
Colour	√	√	√	√ + light reflection
True specific gravity	—	—	—	—
Bulk specific gravity	—	—	—	—
Water absorption	√	√	√	√ + frost resis.
Moisture movement	√	—	—	—
True porosity	—	—	—	—
Permeability	—	—	—	—
Hardness	—	—	—	—
Abrasion resistance	—	—	—	√
Strength,				
compressive	√ single, piers and walls	—	√	√
tensile	—	—	hydraulic internal pressure	—
bending	—	√	—	√
impact	—	—	—	—
Modulus of elasticity	—	—	—	—
Torsional strength	—	—	—	—
Shear strength	—	—	—	—
Coefficient of expansion	—	—	—	—
Crazing tests	—	—	—	—
Thermal conductivity	√	—	—	—
Diffusivity	—	—	—	—
Specific heat	—	—	—	—
Thermal shock resistance	—	—	—	—
Pyrometric cone equivalent	—	—	—	—
Maximum temperature of use	—	—	—	—
Refractoriness under load	—	—	—	—
Linear change on refiring	—	—	—	—
Rising temperature test	—	—	—	—
Dielectric strength	—	—	—	—
Resistivity	—	—	—	—
T_e value	—	—	—	—
Dielectric constant	—	—	—	—
Power factor	—	—	—	—
Loss factor	—	—	—	—
Temperature coefficient of capacity	—	—	—	—
Chemical analysis	√	√	√	√
Soluble salts	√ + efflorescence	√	—	—
Resistance to:				
Acids, inorganic	—	—	√	√
Hydrofluoric acids	—	—	—	—
Organic liquids	—	—	—	—
Corrosive liquids	—	—	—	—
Corrosive gases	—	—	—	—
Alkalies	—	—	—	—
Hot conc. caustics	—	—	—	—
Organic solvents	—	—	—	—
Oxidising agents	—	—	—	—
Phosphoric acid	—	—	—	—
Resistance to:				
Molten acid slags	—	—	—	—
Basic slags	—	—	—	—
Metals	—	—	—	—
Millscale bursting	—	—	—	—
Resistance to CO	—	—	—	—

of Ceramic Products

Wall tiles	Sanitary ware	Table-ware	Oven-ware	Chemical ware	Engineer-ware	Elec-trical	Refractories	Thermal insulators
	MOE 1–7:1947			784:1953 974:1953			1902:1952	
√ √	√ √	√ √	√ √	√ √	— —	— —	√ —	√ —
— — √ — —	— √ — — —	— √ — — —	— √ — — —	√ √ √ — √ filters	√ √ √ √ —	√ √ √ — —	√ √ — √ √	— √ — √ √
— — —	— — —	— — —	— — —	√ √ √	√ √ √	√ — √	— √ √	— √ —
√ — — —	— — — —	√ — — —	√ — — —	√ √ √ —	√ √ — —	√ √ √ —	√ — — √	√ — √ —
√ — — —	√ √ — —	√ — — —	√ — — —	√ — √ √	√ — √ √	√ — √ √	√ — √ √	— — √ √
— — — —	— — — —	— — — —	√ — — —	√ √ — —	√ √ √ —	√ √ √ —	√ √ √ √	√ √ √ √
— — — — — —	— — — — — —	— — — — — —	— — — — — —	— — — — — —	— — √ — — —	√ √ √ √ √ √	— — — — — —	— — — — — —
√ —	√ —	√ —	√ —	√ —	√ —	√ —	√ —	— —
— — — — — — —	— — — — — — —	√ — — √ — — —	√ — — √ — — —	√ √ √ √ √ √ √	— — — — — — —	— — — — — — —	— — — — — — —	— — — — — — —
— — — —	— — — —	— — — —	— — — —	— — — —	— — — —	— — — —	√ √ √ √	— — — —

DIELECTRIC STRENGTH OR BREAKDOWN STRENGTH

This is the minimum stress in volts per mil (inch/1000), inch, centimetre or millimetre, required to puncture a dielectric. It is a measure of its resistance to high voltages.

The value for dielectric strength obtained is very dependent on test conditions particularly specimen thickness, frequency and wave shape of the applied voltage, electrostatic field distribution, and temperature of the test piece; d.c. also gives much higher results than the normal 50-cycle a.c. Very thin specimens give much higher values than thicker ones.

The test requires disk-shaped specimens with smooth, plane parallel faces free from irregularity. Opposite faces are either metallised or have flat electrodes placed on them. Voltage across these is then gradually increased until puncture occurs, and the thickness at this point measured accurately with a micrometer.

DIELECTRIC CONSTANT ϵ (ALSO PERMITTIVITY)

This is the ratio of the capacitance of a capacitor with dielectric between the electrodes to the capacitance when air is between the electrodes. Dielectric constant is used to classify ceramic insulators.

Tests are made on disks of standard thickness and diameters varying according to the constant. The flat surfaces are completely metallised. For low and medium frequencies, capacity is measured by a bridge method, *e.g.* the Schering bridge; for high frequencies a resonance method is used. The dependence of dielectric constant on frequency and temperature is also determined.

POWER AND LOSS FACTORS

The power factor is the sine of the loss angle δ, the phase difference angle between current and voltage when they leave the insulator. It is a measure of the electrical energy lost as heat. Where the loss angle is less than 5°, as is practically always the case with ceramic dielectrics, sine $\delta \sim \tan \delta$.

The loss factor is the product of the power factor and the dielectric constant.

Testing methods are all based on a balanced electrical bridge, *e.g.* the conjugate Schering bridge. A thin test disk metallised on both sides is connected as a capacitor in parallel with a variable precision air capacitor, and a reading taken. The specimen is removed and the circuit again balanced by means of the variable capacitor so that the equivalent parallel capacitance of the test specimen can be calculated. The Q factor which is the reciprocal of the power factor can then be derived. Readings are taken at at least three frequencies and at different temperatures.

VOLUME RESISTIVITY

Volume resistivity is the resistance in ohms per cubic centimetre (Ω/cm^3). Insulators must have a minimum value of $10^6\ \Omega/cm^3$.

The measurement is made on circular disks with electrodeposited terminals set in a guard ring to eliminate surface currents. The voltage drop across the specimen and the current flowing are accurately measured, when the resistance of the specimen can be found by Ohm's law:

$$\text{volume resistivity} = \frac{\text{resistance of sample} \times \text{area (cm}^2\text{)}}{\text{depth (cm)}}$$

Volume resistivity varies with temperature and is measured at a number of temperatures. T_e value is the temperature in °C at which the resistivity drops to $10^6\ \Omega/cm^3$.

GLAZES

Testing glazes

The basic conditions and properties of glazes as thin layers bonded on to body surfaces make them particularly difficult to investigate. Many investigations are therefore carried out on pieces of glass made from the glaze batch. However, the information so obtained is not completely applicable to true glazes, for two main reasons. Firstly the composition of a matured glaze does not correspond exactly to the one calculated from the batch because of losses of constituents due to volatilization, and also interaction with the body surface which may add or remove material. Secondly, the properties of a thin sheet of glass are not quite the same as that of bulk glass, for instance tensile strength increases when one or more dimensions are considerably reduced. To this must be added the physical constraint imposed by adherence to the body surface. It is therefore very useful to obtain information by testing actual glazes, either *in situ* or by removing samples after firing.

Testing and control of raw glaze batches are also necessary if they are to be applied properly.

Raw, Glaze Slips

Tests similar to those used on raw body slips are applicable to glaze slips for S.G. or pint weight, grain size, etc. Glaze consistency may be checked with efflux, or ball or plunger viscometers. Many empirical tests are made regarding the 'feel' of glazes. Experienced workers can differentiate very well between a correctly made-up slip and a wrong one; the remedy for the latter is not always obvious.

HARMAN'S METHOD FOR GLAZE-SLIP CONTROL BY MEASURING 'COHERENCE', 'RECEPTIVITY' AND 'PICK-UP'

Coherence Test. The coherence value C of a glaze is defined as that weight of it adhering to unit area of a glass plate after immersion in it, and withdrawal at a constant rate of 1 cm/sec. Its units are grammes per square centimetre (g/cm^2).

A 3 in. square piece of glass is cleaned with fine scouring powder and sodium oleate, followed by rinsing with distilled water and then dried. A hook is attached to the centre of one side. By this it can be suspended first from the scales for weighing and then from a chain wound round a pulley driven by a small motor through a suitable speed reducer. By this it is lowered and then withdrawn from a beaker of slip. It is then reweighed.

Receptivity Test. Test samples of raw or fired bodies have all except the required test area covered in paraffin, and are weighed. They are then attached to the end of a pivoted arm. On release this arm is pulled through an arc by means of a string attached to a pulley and a motor. In so doing the sample is dipped in a trough of water for a fixed amount of time. The samples are then reweighed.

The receptivity R is the weight in grammes of water absorbed per square centimetre of body surface multiplied by 10^4.

Glaze Pick-up Test. This test is conducted in the same way as that for receptivity with the exception that glaze slip is placed in the dipping trough instead of water. The glaze pick-up P of a body is the weight in grammes of dry glaze deposited per square centimetre of body surface.

For both R and P value tests the body should be dry but at room temperature. The desiccator used for cooling the dried specimens should contain calcium chloride and not sulphuric acid (**H26, 27, 28**).

Results. Plots can be made of C against clay content of glaze, they may or may not be linear or have abrupt maxima. Linear graphs are obtained by plotting R against P for constant C, P against C for constant R. These follow the equation (**H28**):

$$P = K\left(\frac{R}{100}\right)(C - 0.001)$$

Bulk Glazes

Samples made from melted batches of the glaze are used to find physical properties in the normal way. For instance: coefficient of thermal expansion, modulus of elasticity, surface tension, tensile strength, hardness, softening and melting points, viscosity. More specialised methods for finding the last items are also described below.

VISCOSITY OF GLAZES AT LOW TEMPERATURES, 500–800°C

James and Norris (**J17**) describe the method they used for measuring the viscosity of numerous pottery glazes over the range 500–800° C (932–1472° F). It is a bending beam method. Rectangular bars of glass $2 \times \frac{1}{2} \times \frac{1}{4}$ in. ($5.1 \times 1.2 \times 0.6$ cm) are used for viscosities in the range $10^{6.5}$–10^{11}, but for higher viscosities fibres must be used, although these give slightly less accurate results. The specimen is mounted horizontally in a rotatable chuck. It is heated in an electric furnace to the required temperature and held there with a variation of $\pm \frac{1}{4}°$ C. During this heating deformation is prevented by slowly rotating the chuck. When the glass has reached a stable condition at the temperature of the experiment, rotation is stopped, and the rate of sag observed with a cathetometer, through a window of heat-resisting glass. At high viscosities the movement can be timed with a stop-watch, but at lower ones a recording drum is needed.

Detailed investigation showed that this formula is sound and reliable:

$$v = \frac{g\rho l^4}{2d^2\eta}$$

where v = rate of displacement;
g = acceleration due to gravity;
ρ = density;
l = length of overhang;
d = depth of specimen;
η = viscosity.

The method is claimed to have the following advantages:

'(1) It uses a horizontal furnace which enables more uniform temperature to be attained than in a vertical furnace.

'(2) The temperature is constant throughout any one viscosity determination, thus enabling the glass to be stabilised at the temperature of measuring.

'(3) The deformations being reversible, the bar can assume its original form, permitting several determinations at different temperatures on the same specimen' (**J17**).

VISCOSITY OF GLAZES AT HIGH TEMPERATURES 800–1300°C (1472–2372°F)

James and Norris (**J17**) describe their apparatus for measuring viscosities in the range $10^{2 \cdot 5}$–$10^{7 \cdot 0}$ poises by means of the movement of a long, thin, loaded rod down a cylinder containing the liquid. In it a platinum rod 6 in. long by $\frac{1}{10}$ in. diameter (15 cm long by 2·54 mm) is suspended by means of a platinum chain from a balance arm. The driving load of the rod is adjusted by weights on the balance pan. Its movement is observed by means of a cathetometer sighted on a mark on the suspension. The rod penetrates axially a cylindrical crucible of glaze 1·7 in. (4·3 cm) internal diameter. These crucibles had to be made of cast zircon, as both sillimanite and recrystallised alumina are attacked by leadless glazes under the conditions of the experiment. The crucible is heated in an accurate vertical electrical furnace wound in eight sections with independent controls. Measurements of displacement and time are taken, and used in the formula:

$$\frac{dx}{dt} = \frac{(m - \pi a^2 \sigma x \lambda)gk}{2\pi\eta\lambda x}$$

(a term involving d^2x/dt^2 may be ignored)

where x = displacement of the rod;
t = time;
a = radius of the rod;
σ = density of the fluid;
$\lambda = \dfrac{r_0^2}{(r_0^2 - a^2)}$ where r_0 = inner radius of the crucible;
g = acceleration due to gravity;
$k = \log_e \dfrac{r_0}{a} - \dfrac{(r_0^2 - a^2)}{(r_0^2 + a^2)}$
η = viscosity;
[m presumably load].

The advantages of this method are:

'(1) Simplicity of design, ease of construction, and simple experimental procedure.

'(2) Glazes can be studied at very low rates of shear that approximate to the conditions under which glazes flow in practice.

'(3) The method is absolute' (**J17**).

PLANT CONTROL TESTS FOR GLAZE FLUIDITY

A number of simple tests have been devised to ascertain or check glaze or frit fluidity by allowing it to flow down an incline under standard heat and time conditions. These are known variously as the 'button test' in which a standard-sized pellet is preheated on a horizontal plate and then placed at a slope for a fixed period of time (**M26**), the channel viscometer where the glaze flows down an inclined channel, etc. (**L27**). The distance of flow is measured and used for comparisons.

THE GLOSS POINT

Useful information can be obtained by observing the reflecting ability of a glass as it is heated. A normal glaze undergoes a fairly sudden transition from dull to bright after a definite amount of heat-work has been done on it. The gloss point can be determined by observing the glaze, applied to a suitable flat body specimen, during heating in a furnace, through a viewing-aperture. At temperatures above 700° C (1292° F) sufficient light is radiated from the furnace walls to show when the glaze becomes reflective, below that temperature an incandescent lamp must be used as the light source. However, this visual method of determining the gloss point sometimes gives an unaccountably wide spread of values.

De Vries (**V24**) describes a small furnace fitted with light source, mirrors, photoelectric cells, and suitable automatic recording arrangements to plot graphs of reflected light against temperature. With this it is found that although some glazes have simple curves others show a number of peaks and troughs. Rapid freezing of samples heated to various points on such curves, followed by microscopic and X-ray diffraction analysis, showed that alternate fusion and crystallisation may occur several times before final fusion is achieved.

THE SOLUBILITY OF LEAD FRITS AND GLAZES

There are numerous ways described in the literature of measuring the solubility of lead combinations (**S114**).

The measured solubility of a lead frit or glaze depends not only on the amount of lead in it and the other constituents present, but also on the method of testing, that is the particle size of the sample, the nature and strength of the acid reagent, and the temperature and duration of the test.

In Great Britain the Factory Acts lay down the general method of approach in the Ministry of Labour and National Service regulations governing the use of frits and glazes in the pottery industry in 'Pottery (Health and Welfare) Special Regulations 1950':

'A weighed quantity of the material which has been dried at 100° C and thoroughly mixed is to be continuously shaken for one hour, at the common temperature, with 1000 times its weight of an aqueous solution of hydrochloric acid containing 0·25% of hydrochloric acid. This solution is thereafter to be allowed to stand for one hour, and to be passed through a filter. The lead salt contained in an aliquot portion of the clear filtrate is then to be precipitated as lead sulphide, and weighed as lead sulphate.'

This regulation, however, does not lay down the particle size of the sample or the temperature of the test, both of which are very important. It was therefore necessary to define the test more rigidly. In working out a reproducible method Norris (**N23**) found that for solubilities up to 10% (or with small error 15%) the shaking period can be dispensed with, when thermostating becomes feasible.

The interim report from the British Ceramic Research Association is as follows:

'The "intrinsic solubility" of a lead glaze shall not exceed 5%, that of a lead frit for high lead glazes shall not exceed 1%, that of a lead frit for low lead glazes may be up to 2·5% under certain circumstances.'

The 'intrinsic solubility' is measured as follows:

'28 lb of frit shall be taken in a random manner from the batch to be tested. This quantity shall be crushed and passed through 4- and 20-mesh sieves. Only that material which has passed the coarser sieve and been retained on the finer sieve shall be used for the test. A suitable sized test sample shall be obtained from this material by a process of repeated quartering.

'The whole test sample shall then be dry ground by hand, using an agate pestle and mortar, to pass a 170-mesh B.S. sieve, the final material used for the solubility determination being that which will pass a 170-mesh B.S. sieve and be retained on a 240-mesh B.S. sieve.

'The determination of the solubility of this sample shall be carried out by the method described in the next section of this interim specification and shall be performed in duplicate. If the two percentage solubilities obtained differ by more than 0·3, they shall be rejected and the test repeated. The mean value obtained shall be known as the "intrinsic solubility" of the frit.'

Measurement of Lead Solubility. The basic requirements of the lead solubility test are laid down in the Factory Regulations. Recent work at the British Ceramic Research Association and elsewhere has shown that in some ways a rather more closely defined set of requirements is desirable. The following method is suggested:

'A half-gramme sample of the frit to be tested is weighed on a watch glass. The sample is then carefully washed into a 500 ml Stohman flask, with about 480 ml of a solution containing 0·25% HCl. The exact time of addition is noted.

'Stopper the flask and shake vigorously by hand for a few seconds. The remaining acid solution is introduced to bring the level up to the 500 ml mark, care being taken to wash all exposed particles down into the acid. The flask is then allowed to stand for a total period of 2 hr from the time of the first acid addition, during which period it is maintained at a temperature of $20 \pm 1°$ C. (This may conveniently be done in a water bath.)

'During this period a Buchner funnel is prepared in the following fashion:

'A filter paper of Whatman No. 42 type is placed on the grid of a Buchner funnel and then covered with a $\frac{1}{4}$ to $\frac{3}{8}$ in. layer of paper pulp (prepared by pulping filter paper or filter paper clippings in water). A second No. 42 filter is placed over the pulp, and the prepared pad is dried under suction.

'At the end of the 2 hr period the contents of the Stohman flask are filtered through the Buchner funnel, care being taken to ensure that the total time taken does not exceed 5 min, and that filtration is adequate. As a further precaution the resultant filtrate may be filtered through a No. 42 paper in an ordinary funnel.

'450 ml of filtrate are accurately measured out and the lead content is determined by one of the several methods available.

'Note: This method is considered adequate for solubilities of 10% or less. The error is likely to be small for solubilities up to 15% but above this value may be of the order of 2–3%. It is necessary for the test to be carried out in duplicate' (**N23**).

Glazes in situ

Finished glazes are inspected visually or with a hand lens for colour, gloss, crazing, bubbles, orange peeling, etc.

MICROSCOPIC EXAMINATION

Microscopic examination may be used both for control purposes (**M43**) and for research. Edwards and Norris (**E7**) summarise seven microscopic methods of examining the state of a glaze and describe a new method. They emphasise that the information obtained depends to a considerable extent on the method used.

Non-Destructive Methods

(1) *Oblique Illumination.* Surface detail can be seen if it involves a sharp change of slope, *i.e.* scratches but not undulations. If the glaze is clear some inclusions are seen but contrast is poor, *i.e.* bubbles can be seen.

(2) *Vertical Illumination.* Sharp changes of surface slope are seen in good contrast (scratches) but gentler changes (orange peel) are not seen. Very little is visible beneath the surface (fewer bubbles seen than in (1)).

(3) *De-centred Vertical Illumination.* Slight changes of slope (orange peel) are shown in strong relief but only on fairly flat surfaces and at low magnification. Large changes of slope are in too great contrast. None of the interior of the glaze is seen.

(4) *Catoptric Illumination.* The surface details are eliminated and inclusions at various depths can be focused on at will.

(5) *Replication.* Collodion replicas of the surface can be examined at very high magnifications including the electron microscope, 50 000 times.

Destructive Methods

(6) *Cross-section.* A cross-section, prepared as a thin section can be examined by transmitted or by reflected light and is useful for studying body–glaze reactions. It is of little use for studying the glaze surface.

(7) *'Vitroderm.'* A specimen is prepared by grinding away the body. It can then be examined by ordinary or polarised transmitted light. Scratches, devitrification, bubbles show up well. 'Orange peel' is not seen. Crystalline matter not seen by other methods is revealed.

(8) *'Fill-in' Method.* New method for quick assessment of the 'state of maturity of a glaze'. A fluid paste made by dispersing carbon black in a 5% solution of gum arabic is smeared on the specimen and the surface 0·001 to 0·002 in. (0·025–0·051 mm) is ground off with a fine carborundum stone. The bubbles in the glaze are thereby opened up and filled with carbon. The specimen, smeared with the mixture, is baked for a short time at 40–50° C (104–122° F) and the surplus carbon then carefully wiped off. The prepared specimen is then examined visually or under a low-power microscope fitted with a graticule.

GLAZE FIT

Where pieces show no crazing when they come out of the kiln accelerated thermal cycling may be applied to check the glaze fit. Depending on the use of the piece the requirements vary, *e.g.* for domestic pottery a minimum number of quenchings from boiling to cold water are required, whereas for engineering ware the upper temperature is increased in 10° C steps, perhaps to 200° C (392° F). Autoclaving or boiling in glycerine at 170° C (338° F) are used.

A measure of the tension or compression of the glaze is given by the improved ring method. Extruded, hollow, cylindrical rings, 4 in. (10·2 cm) outer diameter, are made of the body and when dry they are sprayed on the outside only with the glaze. They are then fired. Afterwards reference points are marked and accurately measured. The ring is cut between the reference marks and the distance between them again measured. The change found indicates the magnitude of the tension or compression of the glaze (**N11**).

INVESTIGATION OF THE PHYSICAL PROPERTIES OF TRUE GLAZES

Investigating the physical properties (thermal expansion, modulus of elasticity, viscosity and surface tension) of actual glazes that have been matured on the body surface is very much more complicated than using samples of bulk glass of the same initial batch composition. In some cases a sample of the glaze can be removed from the ware and then tested like a bulk glass sample, in other instances tests on the glazed piece must be devised.

Glaze Removal. With practice and skill it is possible to remove specimens of glaze several inches long and about $\frac{1}{2}$ in. wide from flat tiles with thick glazes. It is easier if glaze and body have different colours. The specimen can then be tested as for bulk glass, and Table 69 compares results for a blue glaze removed from a tile and a glass fired to the same schedule. It will be noted that although there is no difference in the mean thermal expansion and transformation temperature the Young's modulus and viscosities are appreciably different, although no detectable interaction between body and glaze could be found in a microscopic investigation of a thin section.

TABLE 69

Comparison of Physical Properties of Glass and Glaze of same Composition (N24)

Property	Bulk Glass	Glaze specimen removed from Tile
Total thermal expansion (20–500° C) (68–932° F)	0·31%	0·31%
Transformation temperature	*ca.* 550° C (1022° F)	*ca.* 550° C (1022° F)
Young's modulus measured with resistance-strain gauges	11×10^6 lb/in² $7·7 \times 10^5$ kg/cm²	8×10^6 lb/in² $5·6 \times 10^5$ kg/cm²
$\log_{10}$ viscosity by bending beam method at 680° C (1256° F)	9·0	9·8
at 700° C (1292° F)	8·5	9·3
at 720° C (1328° F)	8·1	8·9

The advantages of this technique are:

'(1) Once the specimen has been obtained its characteristics can be determined by the standard methods used for bulk samples.

'(2) Because the specimen is of reasonable size, the results will not be unduly influenced by chance local variations.

'(3) It is unaffected by glaze opacity and there are no restrictions on composition.'

The disadvantages are:

'(1) The result is a mean value and yields no information about the variation through the depth (which may be more important than the change of mean value).

'(2) The technique cannot be used for the very thin layers of glaze on much pottery.

'(3) The technique cannot be used if the original piece is curved.

'(4) The viscosity at high temperatures cannot be measured.' (**N24**)

Measurements on Glazes in situ. These are always complicated by the very fact that the glaze is *in situ* and largely influenced by its body. However, there is partial compensation for this in not having to remove a glaze specimen intact. Two methods have been devised for estimating the Young's modulus of a glaze on one side of a flat specimen. In the first, two small resistance strain-gauges are attached, one at the centre of each face. The strains developed when a load is applied, combined with the ratio of the glaze and body thicknesses, enable the ratio of elastic moduli to be determined. By grinding off the glaze and measuring Young's modulus of the body, that of the glaze can be calculated. In the second method a single gauge is used on the unglazed face and the strain measured for a known load. The glaze is then ground off and the new strain for the same load found. The accuracy of the method falls off if the glaze is very thin and also if the ratio of the moduli is less than about 3. It is therefore useful for earthenware but not good for vitreous china, bone china and porcelain, where the moduli of body and glaze are similar. The result is, of course, a mean one giving no information about variation with depth unless the glaze is ground off in stages.

Where a particular property of a fired glaze cannot be measured directly an indirect method can be adopted. For instance, to investigate the change in composition of a glaze with depth, due to interaction with the body, a series of bulk samples are made of the glaze alone and with different percentages of body in addition. These are all given the same firing treatment as the ware. Any specimens in which the body has completely dissolved can be used in the subsequent tests. Required properties are measured on the bulk specimens and any two properties that change monotonically with composition are interrelated and the one can be used to deduce the other. Thus the refractive index which can be readily measured on a glossy glaze *in situ* can be used to follow the change of composition due to body glaze interaction by progressively grinding off the glaze, polishing the new surface and measuring its refractive index (**S133**).

DIAMOND PYRAMID HARDNESS TEST

Standard equipment is available for this test, and consists of a four-sided diamond pyramid indenter mounted in such a way that a known load can be applied to bring it in contact with a preselected spot on the test piece at known speed. The angle between opposite faces of the pyramid is 136°. The speed of the indenter at contact should not exceed 1 μ/sec, suitable loads for testing glazes are from 30 to 100 g, time of contact should be 15 sec.

The indenter is mounted in the centre of the front face of a 6 mm objective lens so that the precise spot to be indented can be preselected by focusing on the surface. This selected spot is then tested by bringing the indenter in contact with it. The diagonal lengths of the indentations obtained are measured either by using a graticule in the eyepiece of the microscope or by taking a photograph of the impressions and making measurements on an enlarged print. The latter method is more accurate.

The diamond pyramid hardness number is given by the applied load divided by the area of the sides of the impression and in the case of the 136° pyramid this is:

$$\text{D.P.H. No.} = \frac{1854P}{D^2} \text{ kg/mm}^2$$

where P is load (g);
D is diameter (μ).

If twenty-four measurements are made, six at each of four loads, an accuracy of about 1% is obtained.

The significance of these determinations as a measure of bond strength and therefore of composition is discussed on p. 557. Taking measurements at different loads and therefore different depths of penetration can also give information on any changes of composition through the depth of glaze tested, as these will be apparent as changes of hardness (**A177**). (B.S.S. 427:1931.)

TESTS FOR THE DURABILITY OF ON-GLAZE DECORATION

The chief concern regarding on-glaze decoration durability is for tableware which undergoes acid attack from foods, alkali attack during washing up and abrasion from cutlery and stacking. A number of laboratory tests have been evolved and in some cases compared with wear and tear in service conditions. Some are based on determining quantitatively the amount of one or more of the materials which go into solution when either a known or a fixed weight of controlled grain size are submitted to treatment by hot or cold acid or alkali of known strength and temperature. Dale and Francis (**D5**) point out that these methods, although indicating any health hazards from soluble lead, do not directly measure the durability of the design.

Instead, qualitative tests comparing treated and untreated samples are more helpful. Such tests are usually arbitrary in a number of ways, such as the duration, temperature, nature and concentration of the solvent, degree of agitation, convective movement of the solvent, and in the methods of reporting the results. A number of workers have adopted a scale of visible alterations.

TABLE 70

Scale for Reporting Results of On-glaze Durability Tests

Condition	Numbers by Koenig and Watts (K47)	Numbers by Sharratt and Francis (S58)
No detectable attack	1	0
Less glossy appearance	2	1
Slight fading of colour	3	2
Pronounced fading of colour	4	3
Colour destroyed	5	—
Colour removed to a considerable extent	—	4
Colour changes in hue	—	c

Endell (**E17**) tested samples with $\frac{1}{4}\%$ solutions of eight different detergents at both 40–50° C (104–122° F) and 80–90° C (176–194° F) for 14 hr under reflux condensers. Similar samples were passed through washing-up machines using the same detergents at 60° C (140° F) 500 times. The first laboratory test was found to correspond well with the service test and all the decorated porcelain tested withstood these adequately. The high-temperature test proved detrimental to a number of decorations.

Kohl (**K63**), however, found that some violets that resist boiling in 1% soda solution for an hour are nevertheless gradually removed in washing-up machines. This he believes to be due to food acids going into solution during the washing-up and not being immediately removed. He therefore recommends that decoration types must be tested with 3% hydrochloric acid at room temperature for 5 hr and 1% boiling soda solution for $\frac{1}{2}$ hr.

Koenig and Watts (**K47**) used the following tests on tableware broken into pieces, sufficient samples being laid aside for comparison (Table 71).

TABLE 71

Test No.	Agent	Temperature	Time (hr)
I	$\frac{1}{3}\%$ trisodium phosphate	65 ± 2° C (149° F)	3 × 5
II	$\frac{1}{3}\%$ sodium carbonate	65 ± 2° C (149° F)	3 × 5
III	$\frac{1}{3}\%$ commercial cleanser	65 ± 2° C (149° F)	3 × 5
IV	4% acetic acid	22° C (71·6° F)	24
V	5% (or 3%) HCl	22° C (71·6° F)	5
VI	1% sodium carbonate	100° C (212° F)	$\frac{1}{2}$

They estimate that tests I, II, and III simulate service conditions for ware washed for 1 min, six times a day, for 150 days.

Table 72 shows tests described in a French paper (**A173**).

TABLE 72

Test No.	Agent	Temperature	Time (hr)
(1)	1% cleanser No. 1 (type not stated)	80° C (176° F)	3
(2)	2% cleanser No. 2 (type not stated)	80° C (176° F)	3
(3)	4% solution Na_2CO_3	80° C (176° F)	3
(4)	4% acetic acid; stir	cold	20
(5)	Slice of lemon placed on the colour	cold	20
(6)	3% HCl; shake	cold	5

Decorations resisting tests (1) to (5) are suitable for hotel use and those resisting test (6) can be used with all assurance. Decorations failing in tests (4) and (5) should only be used for ornamental ware. For luxury ware tests (1) and (3) are of less importance.

Sharratt and Dale (**S58**) found that acetic acid tests gave a good indication of the effects of the food acids such as citric, malic, etc., and that sodium carbonate was always slightly more vigorous than sodium phosphate, as used in many detergents. They therefore used only two tests (Table 73).

TABLE 73

Test No.	Agent	Temperature	Time (hr)
(1)	Vinegar (5% acetic acid)	cold	overnight
(2)	% sodium carbonate	90–95° C	½ hr

PROCESS CONTROL

Raw Materials

It has been pointed out that by nature of their geological origin no two clays are alike and single deposits frequently lack uniformity. Similarly many of the non-clay raw materials are heterogeneous and variable. The consumer, however, requires a uniform reliable product of fixed chemical and physical properties, of definite shape and finish. Immense losses are still incurred from time to time in the ceramic industry due to deviations from standard qualities such as incorrect porosity, inadequate strength, crazing or peeling of the glaze, changing shrinkage, non-matching colour, lack of gloss of the glaze, incorrect electrical characteristics, etc.

Many such losses can be avoided by adequate process control, starting with the raw materials. The Portland cement industry has been able to establish such control. Their chemists worked out the correct composition of the final product, and instituted rigid control of the variable raw materials and their consequent accurate compounding to a mixture of constant analysis (**H96**).

The ceramic industry is faced with a far harder problem and has lagged behind. Any control of raw materials by testing them in any way involves segregating each batch, taking proper samples from it, and then not using it until the results of the tests have been evaluated. It is therefore not surprising that in the past it has been normal practice to try to stabilise the raw materials by other methods and to take laboratory action only when something goes wrong.

The chief method of obtaining a raw material of reasonably constant composition is to blend a number of similar clays from different sources, or to blend different batches from the same source; this is done by placing each batch in a different bunker or hopper and making up the material used with some from each. This method is very successful and will probably not be superseded by quicker analysis in most smaller works.

Where space permits and the value of the product demands it, proper testing of raw materials does now take place. It must furthermore be preceded by correct sampling techniques if useful data are to be obtained.

Binns (**B59**) describes the nature of such tests, referring generally to the manufacture of chemical stoneware and electrical porcelain. He divides tests into indicative tests, such as moisture, particle size and loss on ignition, and decisive tests, on which the material is accepted or rejected, and which he considers should be based on the properties of the final product. For the latter therefore the raw material should be tested in the mixings in which it is to be used, and processed and fired under conditions as like to those in the works as possible. Stocks of raw materials liable to variation, *e.g.* ball clay, should be kept in the laboratory so that comparative mixes using the old and the new materials can be made up.

Really complete control of raw materials would, however, involve more extensive tests including complete chemical analysis, petrographic and mineralogical analysis and a number of other tests depending on the type of raw material, whether clay, grog, clinker, etc., as shown in Tables 74–75.

Complete chemical analysis by classical methods takes at least a week and costs about £20 (1959). The quicker methods of partial and rational analysis have been discussed (pp. 266 and 298) and are only of use where they are measuring variable quantities of the same substance. Chemical analysis of raw materials has therefore only become a possibility since the introduction of physical methods of rapid analysis.

Spectrographic Analysis. In 1949 Hind (**H96**) made the following statement in this connection: 'In view of recent advances in spectroscopic analysis, I do not think that these obstacles to full chemical knowledge of our raw materials and products need persist much longer. The prospect is that very soon it will be possible to obtain fully adequate control data by this method in a few hours per sample and at a rate of many samples per day from a single skilled operative and one set of equipment. This seems, on examination of capital outlay (say £2000) and running expenses of £1000 to £2000 p.a., very expensive by comparison with the cost of a works laboratory employing one skilled analyst. However, when one views the probability that such spectroscopic outlay will cope with work otherwise costing up to thirty times as much and, above all, will be able to deliver its results within twenty-four hours at the most, there can be no doubt as to which procedure will be the more practical and economical.'

Koenig and Smoke (**K58**) point out that spectrographic analysis of raw materials is not only infinitely faster than the old wet methods but also more accurate. It determines all unexpected impurities which would not have been determined by normal wet chemical methods. It is therefore exceptionally useful for testing those materials where impurities take a large part in their behaviour.

It is therefore to be hoped that full working out of spectrographic methods will shortly make them more generally available (p. 268).

Polarographic Analysis. Another means of rapid qualitative and quantitative analysis well suited to control work is the polarograph (p. 274).

Flame Photometer. Rapid determination of the all-important alkali content is best achieved with the flame photometer (p. 270).

X-ray Spectrographic Analysis This method is exceptionally good for elements of prime importance in the ceramic industry such as Si, P and S. The procedure is rather difficult but the identification is extremely simple. The substance under test is made the target of the X-ray tube and this must then be highly evacuated. Only a few spectral lines result, though, so that identification is very easy. A relatively new method is to irradiate the specimen with an ordinary X-ray tube, when it generates a secondary characteristic fluorescent X-ray spectrum (**K58**).

Accurate knowledge of the chemical analysis of the raw materials will not always in itself make possible compensation for variations, although it makes the vital initial step of detecting them. It is necessary to know how the mineral content is varying. The extent to which this must be investigated depends on the anticipated variables.

Loss on Ignition. For instance in a clay whose clay mineral is always the same containing a fixed percentage of water and where no other hydrated mineral is present and the carbonate and carbon content is known the loss on ignition will give the clay mineral content after making allowance for losses of carbon dioxide.

X-ray and Differential Thermal Analyses. If, however, more than one type of clay mineral is present and their relative quantities vary a D.T.A.

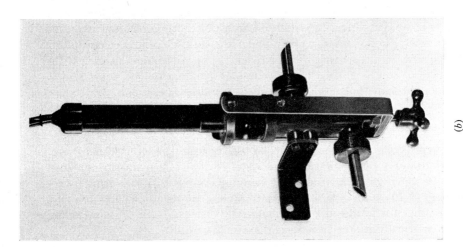

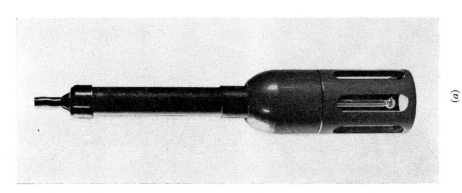

(a)   (b)

(c)

FIG. 4.38. *Continuous electrometric pH-recorder*: (a) *tank-type electrode, miniature version*; (b) *flow-type electrode, miniature version*; (c) *electrometer withdrawn from case* (Kent, **K15**).

is called for. X-ray analyses are better for the quartz and feldspar contents and combination of the two methods and the complete chemical analysis will give data on the elusive illites and micas.

Petrographic Investigation. Mehmel (**M43**) gives examples of the usefulness of the polarising microscope in works control: *e.g.* contamination of quartz by clay and feldspar; particle size of grinding mill product; checking condition of templates and plaster moulds; checking casting.

PARTICLE-SIZE DISTRIBUTION OF RAW MATERIALS, BODIES AND GLAZES

Here again it is essential that the control tests are speedy so that large quantities of material do not have to be held back awaiting the result of the test.

Suggested methods are the long-arm centrifuge (p. 287) which achieves an analysis in 6 hr that would take 96 hr with the Andreason pipette, the broadening of X-ray lines (p. 291) for individual crystalline particles of less than 1000 Å which is very useful for control of grinding operations, and the development of colloidal particles in reactions and inversions.

Closer control on ball milling has been made possible through the use of an electric ear (Hardinge Co. Inc.). The sound during grinding is picked up by a microphone placed usually under the mill, amplified electronically and recorded on a recording milliammeter. The continuous record on the differences of sound emitted in the course of grinding can be duplicated to reproduce the same fineness in subsequent grinding operations.

CONTROL OF pH

A recent development that will make the control of pH very much easier is the continuous electrometric pH recorder (**K58**). This consists of a suitably constructed glass electrode used in conjunction with a calomel half-cell. The Kent (**K15**) model offers two types of cell, one for immersion in a tank and the other for fitting in a flow channel. The e.m.f. set up is measured potentiometrically and recorded graphically (Fig. 4.38).

Works Control in the Fine Ceramic Industry

TABLE 74

Routine Control and Laboratory Investigations which have to be repeated at Regular Intervals in the Fine Ceramic Industries

RAW MATERIALS

Ball and China Clays:
  (1) Water content in the state of delivery.
  (2) Particle size:
      (*a*) by sieve analysis;
      (*b*) by sedimentation.
  (3) Test pieces for shrinkage:
      (*a*) water content;
      (*b*) drying shrinkage;
      (*c*) burning shrinkage and sag;
      (*b*)+(*c*) = total shrinkage;
      (*d*) burning colour;
      (*e*) porosity.
  (4) pH measurement.
  (5) Rational analysis including determination of the mica content.
  (6) Determination of the fluorine content.

Feldspar:
  (1) Sieve analysis of the material after leaving the dry edge runner mill.
  (2) Fusion test at the highest temperature.
  (3) Cone melting point in the laboratory test kiln.
  (4) Determination of the fluorine content.

Quartz:
  (1) Sieve analysis of the material, after crushing in the dry edge runner mill.
  (2) Burning test at the highest temperature.
  (3) Sedimentation of the powder.

Talc:
  (1) Burning test at the highest temperature.
  (2) Sedimentation.
  (3) Shrinkage.
  (4) Porosity.

Coal:
  (1) Water content.
  (2) Ash content.
  (3) Melting point of the ash.
  (4) Content of gases.

Plaster:
  (1) Water content after drying at 150° C.

Grog:
  (1) Sieve analysis every three months.

Dry Crushed Pitchers for Glaze Purposes:
  (1) Sieve analysis.
  (2) Manufacture of a trial glaze.

TABLE 74—(contd.)

CLAY PREPARATION ROOM

Manufacture of the Body:

(1) Sedimentation (body for dust pressing twice a month; other bodies once a month).
(2) Specific gravity of the slurry.
(3) Homogeneity of the pressed cakes (four tests per day).
(4) pH measurement.
(5) Magnetic content.

GLAZE MANUFACTURE

(1) Sedimentation (each glaze each month).
(2) pH measurement
(3) Viscosity at higher temperatures.
(4) Tests on glazed pieces.
   (a) cross-breaking;
   (b) ring test for glaze fit;
   (c) abrasion resistance;
   (d) brittleness;
   (e) thermal expansion;
   (f) electrical resistivity;
   (g) colour.

SAGGAR BODY

(1) Water content.
(2) Sieve analysis.
(3) Warping test.
(4) Porosity determination of the test pieces used for warping (each three months).

INVESTIGATION OF THE WATER

(1) pH measurement.
(2) Content of sulphates.
(3) Temporary hardness.
(4) Total hardness.

BODY FOR JIGGERING

(1) Mechanical strength of test pieces unglazed and glazed (each month).
(2) Tensile strength of different shapes.
(3) pH measurement.
(4) Plasticity (determination by the bending strength of the dried body).

CASTING SLIP

(1) Specific gravity (degree Beaumé).
(2) pH measurement.
(3) Viscosity:
   (a) with the degree Bé in use;
   (b) with 60° Bé.

BODIES FOR DUST PRESSING

(1) Water content of the dry body.
(2) Water content of the prepared body.
(3) Measurement of the shrinkage of test pieces (twice a week).
(4) Porosity tests with the Fuchsin dye, investigated not with special test pieces but with pieces from the kiln (twice each month).

TABLE 74—(contd.)

Bodies for Dust Pressing—(contd.)
    (5) Pressing of test pieces and thereby the measurement of:
        (a) extent of pressure;
        (b) rate of extension;
        (c) differences in porosity in the biscuited state;
        (d) differences in porosity in the glost state;
        (e) mechanical strength in the biscuited state;
        (f) mechanical strength in the glost state;
        (g) shrinkage in a vertical direction;
        (h) shrinkage in a horizontal direction;

Glazing
    (1) thickness or weight per unit area of glaze applied.

Drying Room
    (1) Loss of weight during drying of pieces about fifteen measurements per cycle (3–4 days).
    (2) Ten thermometers to be checked fifteen times per cycle, this being the basis for the regulation of:
        (a) increased temperature;
        (b) ventilation.

Control of Burning
    (1) Draught meter.
    (2) CO_2 and CO meter.
    (3) Heat work meters.
    (4) Thermocouples.

Finished Ware
    Inspection and testing according to specification.

Works Control on Refractory Making

Gilbert (**G46**) gives the following plan for complete control of refractory production adding that a factory making 50 000 tons of mixed production would require a control staff of six. He gives a number of general criteria:

(1) All tests must be simple—only essential tests so work is kept at a minimum.
(2) Reasonably cheap in materials and personnel.
(3) Results must be available to guide production depts., *i.e.* quick results.
(4) Records must be in such a form to show up variations.
(5) All concerned must be kept in the picture and on no account must technical control be considered an extra mural or 'oncost' activity.
(6) Keep trials for at least six months for visual comparison.
(7) All plant experiments should be given experimental order numbers and costed.
(8) Never relax controls.
(9) Keep standard of all phases of production for rapid comparison.

TABLE 75 (G46)

Raw Material

(1) Establish criteria of acceptance for *all* materials used.
(2) Complete survey and mining data to key in with ceramic tests.

    (i) Borehole samples.
    (ii) Trial trenches along outcrop.
    (iii) Study of associated deleterious impurities.

    (*a*) Ironstones.
    (*b*) Sandstones.
    (*c*) Calcareous materials.
    (*d*) Soluble salts.
    (*e*) Surface earth.
    (*f*) Low refractory shales.
    (*g*) Carbonaceous matter.

(3) Complete picture of technical properties well in advance of manufacture.
(4) Seasonal variations precautions and getting difficulties.
(5) If received in wagon or lorry loads ensure that correct samples are taken.
(6) Separation into clay sizes—silt sizes—sand sizes.
(7) Strength of clay green and fired (modulus of rupture and tensile strength).

Clays and Grogs

(1) Daily.
    (i) Bulk density.
    (ii) Moisture.
    (iii) Kilning test.
(2) Weekly.
    (i) Shrinkage.
    (ii) Porosity.
    (iii) Density.
    (iv) Slaking property.
    (v) Fines.
(3) Monthly.
    (i) Chemical analysis.
    (ii) Fired semi-technical trials.
    (iii) Full size brick.
    (iv) Underload test.
    (v) Refractoriness.
    (vi) After contraction 1450° C.
    (vii) Spalling test.
    (viii) Slag test.
    (ix) Reversible thermal expansion.
    (x) Ad hoc tests for specific applications for special orders.

Milling

(1) Moisture content.
(2) Grading test

    +10s   wet and
    −10+60s   dry sieve
    −60s   tests.

(3) Tamping tests to determine packing density.
(4) Daily check of:
    (i) circulating load of pan mills;
    (ii) pan speed;
    (iii) ploughs.
(5) Check mill house for excessive dust.

TABLE 75—(contd.)

SCREENING

(1) Grading test:
    (i) delivery to screens;
    (ii) undersize, oversize, tailings.
(2) Check on screening efficiency.
(3) Draw cumulative log; plot gradings.

STORAGE

(1) Test for alterations in moisture.
(2) Test for segregation.
(3) Effects of souring or maturing on processing.

MIXING

(1) Check delivery of measuring devices.
(2) Check for segregation.
(3) Grading test of mixed fractions.
(4) Moisture variations.
(5) Packing density of each batch.

TEMPERING

(1) Moisture content each batch or at definite intervals in continuous processes.
(2) Check for temp. rise (joule effect).
(3) Correlate 'feel' with ceramic tests.
(4) Ramming test for semi-dry processes.
(5) Workability test for plastic bodies.
(6) Check bulk density.
(7) Screening test.
(8) Plasticity test.
(9) Test binding power in high grog bodies.

FORMING

(1) Details of dies and moulds used. Shrinkage allowance checked by pilot manufacture.
(2) Special dimensions emphasised and marked on working drawing.
(3) Dry press.
    (i) Moisture.
    (ii) Pressure cracks or exfoliation.
    (iii) Laminations.
    (iv) Weight of brick.
    (v) Segregation defects.
(4) Extrusion and repress.
    (i) Check clot size.
    (ii) Check press size.
    (iii) Moisture content.
    (iv) Dry and section to examine flow lines.
(5) Test for uniformity, die wear.
(6) Pipe machine.
    (i) Moisture content.
    (ii) Temperature of body (*a*) after pugging; (*b*) before extruding; (*c*) after extruding.
    (iii) Check die lubrication.

TABLE 75—(contd.)

FORMING—(contd.)

(7) Hand moulding. Pneumatic ramming, check for:
    (i) correct filling;
    (ii) folds or laminations;
    (iii) excessive oil on surface;
    (iv) level working surface.

DRYING

(1) Initial moisture.
(2) Safe drying shrinkage.
(3) Critical points during drying schedule.
(4) Bulk density before and after drying.
(5) Determine rate of moisture removal inside and outside shape.
(6) Measure air circulation.

GREEN INSPECTION

(1) Analyse rejections into types—if faults are diagnosed take action.
(2) Test for hair cracks by painting with kerosene oil at possible locations.

SETTING

(1) Draw scale drawings of solid matter and kiln spaces.
(2) Calculate setting density (i) green; (ii) after firing.
(3) Check setting level (plumbline, spirit level).
(4) Check bedding of grog or sand between courses.
(5) Alter settings if there is evidence of soft or overburnt zones.

FIRING

(1) Check firing schedule (thermocouples, optical pyrometers, cones, bullers rings, etc., draw trials at fixed time intervals over 24 hr day and night).
    (i) Water smoking period.
    (ii) Oxidation period.
    (iii) Vitrification.
(2) Fuel economy. Routine observations of recuperation or regeneration.
(3) Work out fuel consumption per standard unit of production.
(4) Record chimney draughts and temperatures.

DRAWING

Examine setting for:
    (i) Overfired zones.
    (ii) Underfired zones.
    (iii) Moisture patches.
    (iv) Excessive fly ash.
    (v) Excessive movement or cracking from improper tying of courses and setting benches.

INSPECTING

(1) Systematic layout.
(2) Must be to customer's specification.
(3) If product fails to pass (i) down grade; (ii) inform customer.
(4) Use 'go' and 'not go' gauges and jigs where possible.
(5) Test 'ring' on at least 10% of shapes.
(6) For standards take overall measurements for ten bricks: (i) end to end; (ii) edge to edge.
(7) Section bricks on saw: (i) examine internal structure for discoloured hearts, folds or other forming defects; (ii) examine hardness.

TABLE 75—(contd.)

INSPECTING—(contd.)
(8) Sort into various faults and classify broadly.
    (i) Dimensional.
    (ii) Body.
    (iii) Forming.
    (iv) Drying.
    (v) Setting.
    (vi) Firing.

ASSEMBLY
    When there are a number of shapes check against assembly drawings. Fire adjacent shape in same kiln or on same tunnel kiln car.

Testing of Fuels

Although complete chemical analyses of fuels are rarely necessary routine checks for the suitability and uniformity are essential.

GENERAL

    B.S.S. 526:1933. Definitions of gross and net calorific value.

COAL AND COKE

    B.S.S. 735:1944. Sampling and analysis of coal and coke for performance efficiency tests on industrial plant.
    B.S.S. 1016:1942. Methods for the analysis and testing of coal and coke.
    B.S.S. 1017:1942. Methods for the sampling of coal and coke.

SAMPLING

    Scoopfuls of coal or coke should be taken from each barrowful during unloading of a consignment, and placed in a metal bin with a lid to prevent evaporation. This gives the gross sample.

    To obtain smaller samples from this it is poured on to a metal plate to form a cone and then flattened out. From evenly distributed points on the flattened heap ten equal small portions are taken. A smaller sample is obtained from this by repeating the coning, flattening, etc.

    Large coal is crushed between the first and second cone sampling to $\frac{1}{2}$ in. (1·27 cm) size and then used for the moisture determination. Samples for analysis are ground to pass 36-mesh sieve between successive cone samplings.

    Moisture Content. This is determined of the crushed original sample and the ground sample prepared for other analyses. The original sample is weighed and dried in air at 100°–110° C (212–230° F) until the weight is constant. The weighed 36-mesh sample is dried in a stream of nitrogen at $108° \pm 2°$ C (226° F).

Ash Content. From 1 to 2 g air-dried 36-mesh sample are weighed into an open shallow porcelain or silica dish and heated to $775 \pm 25°$ C ($1427 \pm 45°$ F) in a muffle furnace in an oxidising atmosphere until constant in weight. The dish is covered and cooled in a desiccator before each weighing. The residual ash is expressed as a percentage of the coal used.

Volatile Matter Content. Air-dried 36-mesh coal is ground to 72-mesh. 1 g is weighed into a translucent silica crucible of specified dimensions and placed in a muffle furnace preheated in standard manner to 925° C (1697° F). It is left here for 7 min, removed, cooled and weighed. The weight loss is expressed as a percentage, and after subtraction of the moisture content, gives the volatile matter.

Sulphur Content. A 1 g, 72-mesh sample is heated with 3 g Eshka mixture and the sulphur content converted to soluble sulphate which is determined gravimetrically as barium sulphate (**C41**).

Other tests include: fusion temperature of ash, chlorine content, calorific value, specific gravity, swelling of coal, shatter, true and apparent specific gravity, trommel test, abrasion of coke.

LIQUID FUELS

B.S.S. 742:1947. Fuel oils for burners. Tests include: specific gravity; viscosity; flash-point; calorific value (**S57**).

GASES

These can only be estimated by a proper gas analysis, unless only the carbon dioxide content is required when a single apparatus can be used, and even fitted with automatic recording instruments, etc. (B.S.S. 526:1933).

Chapter 5
Ceramic Bodies

THE term 'body' covers both the mixture of raw materials prepared for making any one product, and also the main part of that product, as opposed to the glaze.

The nomenclature of ceramic bodies is very confusing. In general the traditional bodies have distinctive names which have come to mean particular features of colour, texture and shapability. These bodies are largely based on clay, or on the triaxial diagram of clay–feldspar–flint. Modern bodies based on better knowledge of the composition, structure and reactions of materials are described either by the specialised use to which they can be put, or by the chief raw material or by the chief compound formed in the body. Ideally one or the other of these last two should be adopted but except where the same term has been applied to more than one body the most general practice has been followed here. A brief definition of the terms set out in the table will therefore be given before describing body compositions.

(1) *Brickware* (this term has been coined specially) has a coarse coloured body made by the most direct methods from natural clays, water absorption is about 5 to 20% and P.C.E. up to cone 26 (1580° C, 2876° F). Its main uses are for bricks and tiles.

(2) *Refractories* are essentially bodies of P.C.E. of cone 26 or above, up to cone 42 (1580° C–2000° C, 2876° F–3632° F). Constructional refractory bodies are of any colour and medium to coarse texture. They are sub-divided according to their refractoriness and by their chemical nature.

(3) *Thermal insulation bodies* are made of brickware or refractories made especially porous.

(4) *Stoneware* is basically a dense but inexpensive body. It is of any colour, impervious but opaque, breaking with a conchoidal or stony fracture. Traditionally made of fine-grained plastic clays it can be shaped into very large pieces.

(5) *Fine stoneware* made from more carefully selected, prepared and blended raw materials is used for tableware and art ware.

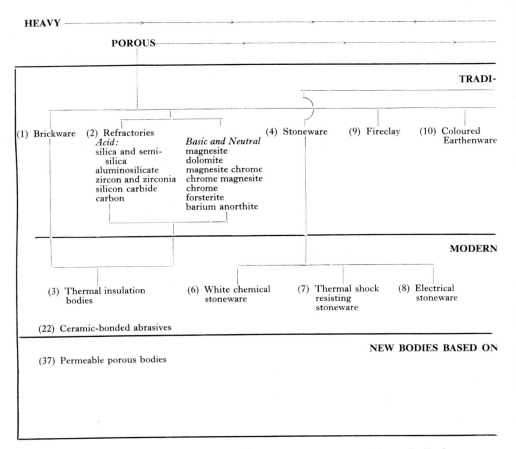

TABLE
Ceramic

(6) *White chemical stoneware* requires purer raw materials and eliminates any possibility of contamination.

(7) *Thermal shock resisting stoneware* has special additions to make it more suitable to this stress.

(8) *Electrical stoneware* has special additions.

(9) *Fireclay* bodies are moderately fine, porous, off-white bodies of the triaxial type using natural fireclays in place of the china clays, ball clays or stoneware clays used elsewhere. The open structure allows the making of large strong pieces.

(10) *Coloured earthenware*, common pottery, majolica, terracotta are some of the terms applied to fairly fine or even fine porous but coloured bodies. (The term earthenware strictly covers all articles made of clay 'earth'.)

(11) *White earthenware* is a fine, porous and white body, readily manufactured into a wide range of products. Water absorption 10–15%.

76 Bodies

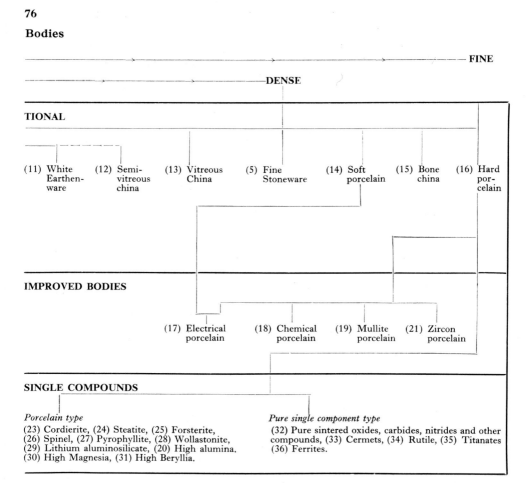

(12) *Semi-vitreous china* is the American development of earthenware. It has a smaller water absorption, 4–10%, and higher strength (**N11**). Its body type is, however, similar to earthenware and has been treated as such in the following tables. It is used for tableware.

Semi-vitreous porcelain is once-fired American earthenware with low water absorption, 0·3–4%, high strength and a little translucency (**N11**). It is used for tableware.

(13) *Vitreous china* has a white, opaque, vitrified body with a water absorption of from 0 to 1·0%.

American hotel china is a nontranslucent dense and very strong white body of the vitreous china type.

(14) *Soft porcelains* are essentially fine, white or ivory and more or less translucent bodies originally made in Europe to simulate imported Chinese porcelain. A number of different types of composition are included. Firing temperatures are cones 7 to 11.

(15) *Bone china* can also be classed as a soft porcelain. It is a highly translucent ivory to white body.

(16) *Hard porcelain* is pure white, completely vitrified, translucent, strong and hard body. It is the most demanding of the triaxial bodies, requiring very pure raw materials, skilled making and high firing, cone 12 to 15.

(17) *Electrical porcelain*, whiteness, translucency and refractoriness are slightly relaxed in order to use fluxes that give better electrical properties.

(18) *Chemical porcelain*, appearance is sacrificed for mechanical and chemical strength.

(19) *Mullite porcelain* has composition and procedural adjustments to develop more mullite in the fired body.

(20) *High alumina porcelain* has a high alumina content in the mixture.

(21) *Zircon porcelain* has zircon additions.

(22) *Ceramic bonded abrasives* are composite bodies.

(23) *Cordierite* bodies are compounded to give the maximum of the phase cordierite in the fired body. This has very low thermal expansion.

(24) *Steatite* bodies are made from steatite or talc.

(25) *Forsterite* bodies contain forsterite when fired.

(26) *Spinel* bodies contain spinel when fired.

(27) *Pyrophyllite* bodies are made from pyrophyllite.

(28) *Wollastonite* bodies are made with wollastonite.

(29) *Lithium aluminosilicate* bodies are made from compositions in particular areas of the triaxial diagram $Li_2O–Al_2O_3–SiO_2$. They have exceptional thermal expansion characteristics.

(30) *High magnesia.* (31) *High beryllia.* (32) *Pure sintered compounds.* (34) *Rutile.* (35) *Titanates.* (36) *Ferrites.* These are all made largely from the pure materials mentioned in their names.

(33) *Cermets* are materials containing ceramic and metallic ingredients.

(37) *Permeable porous bodies* are based on a number of the above technical bodies but have a specially open texture.

More general divisions in the ceramic industry are:

(1) 'Heavy clay' covering brickware and some stoneware bodies.

(2) 'Refractories.'

(3) 'Pottery' (Britain) or 'Whitewares' (U.S.A.) covering earthenware, fireclay, vitreous china, porcelains and all the specialised bodies.

Each type of body has numerous variations so that there are bodies of borderline properties that are hard to place in a particular section. In fact, ceramic bodies can be produced with properties varying continuously from brickware to hard porcelain.

The nature of the fired ceramic body is affected by each stage of its manufacture:

(1) Batch composition, the nature and amounts of its raw materials.

(2) Total composition.

(3) Physical condition of the raw materials, in particular, the grain size.
(4) Method of preparation of the raw body.
(5) Method of shaping.
(6) Firing.
(7) Surface treatment by glazing, grinding or polishing.

The varied nature of the raw materials and the peculiar flow properties of plastic bodies always make it difficult to produce homogeneous bodies. All ceramic preparation, making, drying and firing processes are developed to make uniform products but nevertheless study of finished pieces, *e.g.* by soaking in 2% potassium permanganate solution, shows inhomogeneity, as has been described by Salmang (**S9**): casting skin, greater density of the surface due to surface tension of water, compression of the surface while still moist, smoothing, etc., irregular distribution of water in the raw body as used for shaping, irregular porosity of plaster moulds, drying out and migration of salts, clay colloids, etc., to the surface where greater vitrification therefore occurs when fired.

The Composition of Ceramic Bodies

The compositions of the traditional ceramic bodies lie within certain limits. They are governed by a number of factors. Firstly, the shaping of the ware, which necessitates the inclusion of sufficient plastic material. After shaping the clay also holds the ware together as it dries. Secondly, the drying and firing of the ware, without cracking, which calls for the incorporation of non-clay materials that do not shrink during the drying and firing. These are flint, quartz, etc., and for large bodies grog, which is essentially a ground-up fired clay, or ceramic body. The third factor is the fluxing of the body so that sufficient glassy material forms to hold it together when fired and yet not so much that it warps. The traditional flux is feldspar, or minerals containing it, such as Cornish stone.

So the traditional bodies are essentially made up of clay, flint or quartz, and feldspar, and can be represented by a triaxial diagram (Fig. 5.1).

Now, it has become possible to make almost any body sufficiently plastic for shaping, either by using bentonite in small quantity instead of a large amount of ordinary clay, or by using organic binders. It is therefore theoretically possible to use almost any combination of inorganic materials to make ceramic products of properties very similar, or, if desired, very different from the traditional ones.

In the following sections traditional ceramic bodies will be considered first followed by modifications possible by the use of different raw materials. Lastly the new fields will be discussed.

The presentation of the composition of ceramic bodies in a comparative way is very difficult. The most natural method is by means of the batch composition. But it was shown in the section on raw materials that the traditional ones, namely clays and feldspars, are of variable compositions

398 INDUSTRIAL CERAMICS

and properties. Thus the batch composition for a particular sort of ware from given works depends on the source of the raw materials and cannot be directly used with different clay and feldspar, etc.

One simple way out of this difficulty was thought to be the application of rational analysis, dividing up a composite clay into 'clay substance, quartz and feldspar'. The fields of usefulness and uselessness of this have been discussed (p. 299), and it was concluded that the simple, chemical, rational analysis of the turn of the century does not give a true idea of the make-up of a clay but can be helpful for checking the variations in a particular clay from a particular pit.

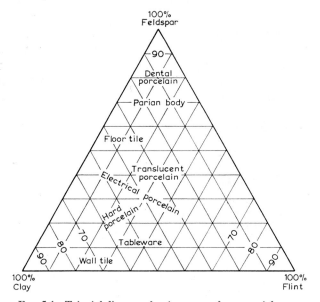

FIG. 5.1. *Triaxial diagram showing areas of commercial wares*

The next step is to attempt comparison by a complete chemical analysis. Here one touches the question of the completeness of the reaction (p. 197). This depends not only on the chemical compositions of the raw materials but also on their grain size, surface condition, method of grinding and shaping, thermal history and finally the temperature and time of firing. In general, a coarse body fired at relatively low temperature does not reach chemical equilibrium, whereas a fine body fired at a high temperature for a considerable time may almost do so. In the latter case the chemical analysis of the body is related to the product. In the former the analysis does give the relative amount of fluxes, but it must be used with caution.

The actual figures used for batch compositions are usually parts by weight or percentage weights. For comparisons the second must be used. The chemical analysis is given in terms of oxides, either as percentage weights or, as moles or percentage moles. All three have their merits, but that with most is the one using moles, with either the mono- and divalent fluxing oxides (RO) summed to unity, or the Al_2O_3 as unity. The oxides are best divided into columns, as in the glaze molecular formulae, according to valency. The first column for mono- and divalent oxides then contains the fluxes, and the second and third columns for tri- and tetravalent elements have the more refractory materials. Column four with pentavalent phosphorus is again fluxing and glass-forming. For bodies where the firing does not bring about completed reactions probably the most useful way of setting out the molecular formula is with the sum of the fluxing oxides (other than P_2O_5) as one. For high-temperature bodies whose properties depend largely on the mullite and silica contents the method putting the Al_2O_3 content as unity may be more helpful. Some fired bodies contain unchanged insoluble sulphates. This should not be ignored when working out the molecular formula.

The calculation of molecular formulae plays a much more important part in the consideration of glazes and will be discussed in greater detail in that section (p. 526).

These points should be borne in mind when considering the examples of various body compositions which have perforce been taken from a number of uncorrelated sources.

1. BRICKWARE

This term has been coined to signify coarse, porous, coloured ware of unspecified refractoriness, which finds application in the building trade. (It differs from the term 'heavy clay product' by excluding dense ware such as blue bricks, sewage pipes and floor tiles.) The necessity to keep down the price of bricks and tiles by minimising both production and transport costs has led to brickworks springing up as near to the chief centres of population as possible. For this purpose the most readily available clays, shales and marls that find no higher grade application are therefore brought into use. They are won and processed on the spot and whenever possible no specially transported additions are made.

Brickware bodies are therefore individual local products and are usually known by the name of their place of origin together with such terms as 'multi', common, facing, semi-engineering, engineering, etc. Clays used come from most geological periods and geographical locations as described in greater detail by Clews (**C41**). In the United Kingdom the clay type used for the highest percentage of bricks is the Carboniferous one, associated with coal seams. 33% of the 1938 output was made of clays of this origin. These Carboniferous clays are the only deep-lying brick clays that can be used economically as they are worked together with coal seams, or may have to be removed in order to get at the coal seams.

Of recent years colliery waste has also become of interest as brickmaking material. Robinson (**R45**) (1958) states that of the 7000 million building bricks manufactured per annum, 500 to 600 million, *i.e.* 7 to 8½% are made of colliery waste. There are three types of colliery waste: (1) the old tips; (2) newly wrought materials; (3) coal washery tailings. This last is highly plastic and additions of grog or power station fly ash are made (**G47**).

A close second to the Carboniferous clays are the Jurassic clays particularly the Lower Oxford clay, which was used for 32% of 1938's brick production. In the area between Bletchley and Peterborough this clay occurs in thick uniform beds with little overburden, which are suitable for very large-scale open-cast working. Furthermore the clay contains about 5% of carbonaceous matter which greatly reduces fuel requirements. This clay is used primarily for the distinctive common brick known as 'Fletton'.

The compounding of brickware bodies is done by one or more of the following methods:

(1) Use of clay as dug with or without hand picking of pockets of undesirable material.

(2) Mixing of several superimposed seams of different clays in the proportion in which they occur, usually during the winning operation.

(3) Mixing of several superimposed seams of different clays in proportions other than those occurring naturally. The seams must then be dug individually by suitable terracing.

(4) Admixing sand, sifted town ashes or refuse, dense or porous grog, fly ash, colliery shale or other waste products (**H101**).

Whitaker (**W54**) describes bricks made from soft plastic limestone of over 90% calcium carbonate. About 10% red clay is added to improve strength and give a slight yellow colour.

The surface texture and colour of bricks and tiles may be altered and improved by applying an engobe to the dry but raw pieces. This may consist of a quite different clay to the base body or it may contain the same base body more highly refined by washing and grinding, with or without additions. Engobes may also be artificially coloured. When applying engobes certain precautions with regard to the relative shrinkage of body and engobe must be taken. The matter is discussed more fully under 'decoration'.

Terracotta

Brickware bodies of a finer type which have been prepared by grinding and mixing are used for decorative constructional work and are then called terra-cotta. The lighter pieces that are glazed may also be termed 'terra-cotta'. The eighteenth-century 'Coade artificial stone', used for much London statuary, is a form of terra-cotta.

2. REFRACTORIES

The term 'refractory' strictly covers all materials of P.C.E. greater than cone 26. However, these materials can be divided fairly easily into the heavy and/or coarse grained products used for construction of furnaces, kilns, ovens, retorts, melting pots, etc., and the small, fine pieces required for specialised work. In general the former only are to be discussed here although some overlap will occur. Nevertheless a very wide range of products falls into the group.

Refractories may be classified in three ways:

(1) Chemicomineralogical.
(2) Refractoriness.
(3) Method of manufacture.

Their performance in service depends on all three, so that a very large number of products are made.

The chemicomineralogical classification varies a little but uses the three groups: acid, basic and neutral. These run into each other (Table 77).

TABLE 77

Acid	Zirconia	
	Zircon	
	Silica	⎫
	Semi-silica	⎪
	Fireclay	⎪
	Kaolin	⎬ aluminosilicate range
	Aluminous fireclay	⎪
	Sillimanite	⎪
↓	Mullite	⎪
Neutral	Bauxite	⎪
	Alumina	⎭
	Carbon	
	Silicon carbide	
	Chrome	
	Chrome–alumina	
↓	Chrome–magnesite	
Basic	Magnesite–chrome	
	Magnesite	
	Periclase	
	Dolomite	
	Forsterite	

The classification according to refractoriness can be made as follows:

Low heat duty	P.C.E. 19–28.
Intermediate heat duty	P.C.E. 28–30.
High heat duty	P.C.E. 30–33.
Super duty	P.C.E. above 33.

Silica Refractories

This body consisting of 93–98% SiO_2 endeavours to utilise the remarkable properties of silica (*see* p. 99) in particular its refractoriness, for use in the steel industry.

The chief raw materials for silica brick are ganisters and quartzites containing at least 97% SiO_2, Al_2O_3 less than 1%, alkalies less than 0·3% (*see* p. 96). It is necessary to add a suitable bonding agent that will hold the nonplastic, non-fusing silica grains in both the raw and fired state. This is hydrated lime 1 to $2\frac{1}{2}$% being used and also sulphite lye to add green strength (**C26**).

The conversion of quartz via the transition phase to cristobalite and then to tridymite is perforce slow (*see* p. 168) but can be accelerated by the presence of mineralisers, *e.g.* 1·5% Na_2O and 1·5% Fe_2O_3. Acid oxides, *e.g.* Al_2O_3, B_2O_3, P_2O_5 hinder the conversion (**S7**).

Although in use at high temperatures cristobalite will eventually become the predominant form of silica in the brick it is still debated whether the conversion should be completed in manufacture or not (**K70**). In both Great Britain and the United States the general view of the steelman is that the brickmaker should fire silica bricks sufficiently to convert most of the raw quartz to cristobalite or tridymite, preferably the latter, so that he will not have to allow for a serious after-expansion in service. In Germany the steelman prefers the higher thermal shock resistance of a medium-fired brick. But with these there is always the danger of soft-fired bricks occurring and shattering a structure by their greater expansion when heated.

Apart from the useful mineralisers it is highly desirable to exclude impurities, and as silica bodies appear to reach equilibrium rapidly, study of phase diagrams shows which of these are the most deleterious to refractoriness.

The phase diagram of the system SiO_2–CaO clearly shows that additions of up to 30% lime to silica reduce the melting point by only very little because the liquids formed by interaction are immiscible (magnesia acts similarly). The presence of certain other oxides, notably alumina, destroys this immiscibility and leads to a marked drop in refractoriness.

The SiO_2–FeO diagram shows that up to 40% FeO results in only a small drop of melting point, again because of immiscible liquids, and is the reason for this refractory's amazing resistance to attack in steel furnaces.

The SiO_2–Al_2O_3 diagram shows that the eutectic occurs at only 5% Al_2O_3 and its melting point is as low as 1545° C (2813° F). As clays are frequent impurities in silica rocks, this is of great importance, 12% clay impurity corresponding approximately to the eutectic mixture. Clays are also often used in siliceous cements and patching materials and can lead to softening and lubrication of joints between high-grade silica bricks.

The SiO_2–TiO_2 eutectic occurs with 10% TiO_2, but low alumina bricks with 2% natural titania give outstandingly good performances.

Study of phase diagrams has led to the development of super-duty silica bricks which can be used under load up to 1705° C (3101° F) as compared with ordinary silica which fails at about 1650° C (3002° F). Harvey and Birch (**H37**) classify the impurities that may occur in silica bricks according to their effect on refractoriness. Group A: fluxing oxides, *i.e.* Al_2O_3, TiO_2, Na_2O, K_2O. They recommend that the sum of their percentage contents should not exceed 0·5 and preferably be less than 0·4. (But work with South African 'Silcrete' has emphasised that titania is not as harmful as alumina and need not necessarily be so severely restricted (**W55**).) Group B: oxides forming immiscible liquids with silica, *i.e.* CaO, MgO, SrO, MnO, ZnO, FeO, NiO, CoO. The total percentage of this group should lie between 2·5 and 5·0%.

As the composition of these refractories is intimately bound up with their properties they are given together in Chapter 17.

Dale (**D3**) describes the processes occurring during the firing of silica bricks as follows. 'At approximately 750° C (1382° F), the unfired brick consists of granules of quartz separated by a groundmass or matrix containing essentially lime, quartz flour, and the smaller grades of quartz particles.

'Above 800° C (1472° F), the interaction between the lime and the free quartz surfaces commences. For simplicity, this reaction was regarded as one of solution, resulting in the formation of a lime–silica glass, the silica content of which is continually increasing as time proceeds and temperature rises. This glass can, however, devitrify and precipitate cristobalite below 1250° C to 1300° C (2282–2372° F) in local regions of high silica content.

'At a temperature above 1200 to 1300° C (2192–2372° F) the surface of the quartz particles commences to convert directly to cristobalite, and the velocity at which this conversion proceeds inwards increases with further rise of temperature. Thenceforward, with rise of temperature, cristobalite dissolves in the lime–silica glass of the groundmass.

'As the temperature rises towards the soaking temperature, the conversion of the larger quartz granules is proceeding towards the cores. Small particles of quartz in the groundmass are completely converted, and the silica content of the lime–silica glass is still increasing rapidly by solution of minute cristobalite granules and the neighbouring surfaces of the larger grain.

'During the soaking period, the lime–silica glass of the matrix becomes saturated at a rate dependent, amongst other factors, on the actual temperature, which is critical. At some degree of saturation, the glass commences to devitrify, and precipitates silica in that crystalline form called 'tridymite'. This phenomenon of devitrification naturally lowers the degree of saturation of the glass; more cristobalite dissolves and the reactions cristobalite → saturated glass → tridymite, proceed continuously.'

Chaklader and Roberts (**C21**) have subsequently shown that even pure quartz is converted to cristobalite only indirectly, namely via a non-crystalline phase. This transition phase occurs in fired silica bricks to the extent of 20–25% and is converted to cristobalite when the bricks are reheated, the change beginning slowly at around 1500° C (2732° F) and being rapid at 1600–1650° C (2912–3002° F).

Florke (**F13**), on the other hand, shows that at room temperature little or no amorphous or glassy phase exists in silica bricks. In the past the relative proportion of cristobalite and tridymite in a specimen has been estimated by comparing its X-ray diagrams with those obtained from pure and perfectly ordered specimens. Any material not accounted for in this way being assigned to 'glass'. Both cristobalite and tridymite frequently show disorder in one dimension of their lattices in varying degrees. This is seen, and its degree can be measured, from thermal expansion curves, the disordered lattices showing inversion at lower temperatures than the ordered ones. These disordered specimens show different X-ray diagrams, and a series of sample curves for the various degrees of disorder are necessary to compare with the test curves. When this is done there is no 'remainder' to place in the glass category. Florke considers that even at high temperatures there is less glass present than is calculated from the content of fluxes in the mixture, as a proportion of the alkali and alkaline earth ions diffuse into the silica grains, inducing the change to tridymite. The remainder, he calculates, can easily devitrify completely during cooling.

Throughout the production of silica bricks the degree to which conversion into the various forms becomes completed depends on the thermal treatment given. Small variations of time and temperature may have noticeable results. As the inversions involve volume changes, changes of porosity, permeability and bonding can occur. This has been investigated by Eaton, Glinn and Higgins (**E2**) both in raw lump quartzite and in silica compacts. Lahr and Hardy (**L2**) have shown that variations of porosity, where these are independent of changes in Al_2O_3 content, have a small but significant effect on performance in open-hearth roofs. In the range 15 to 30% porosity bricks of the lower values are superior.

Refractories are not only subjected to high temperatures but also to changing temperatures which may cause failure by spalling. Performance in this respect can be very variable in silica bricks. Hargreaves (**H21**) tested fifteen brands and found maximum safe heating rates ranging from 1° to 5° C/min. He attributes this chiefly to differences in Al_2O_3 content, the form in which Al_2O_3 occurs and the type of rock from which the brick was made.

Semi-silica Refractories

These bodies have 88 to 93% (in U.S.A. 76 to 78%) silica as against the 93 to 98% in true silica bricks. The remainder is made up of fireclay either from a natural sand–clay mixture or as an addition in the form of clay slip. The composition of some examples together with their properties is given in Chapter 17.

Aluminosilicate Refractories (including Fireclay)

Natural fireclays were the raw materials for the first refractories. Their usefulness in this respect is founded on the high proportion of clay minerals and other aluminosilicates they contain and which are contaminated by only small amounts of alkalies, iron compounds and other fluxes. Their properties vary in respect of chemical composition, refractoriness, plasticity, drying and firing shrinkages, and after contraction, so that it is normal practice to use several clays mixed together for any particular product.

As regards composition the B.S.S. 1902:1952 defines the nomenclature as follows: a 'firebrick' is a brick that in the fired state consists essentially of aluminosilicates and silica and shows on analysis less than 78% of silica and less than 38% of alumina. An 'aluminous firebrick' contains between 38 and 45% alumina the remaining 62 to 55% being essentially silica. A 'siliceous firebrick' contains from 78 to 92% silica. In the U.S.A. where the nomenclature is according to heat duty rather than composition (*see* p. 401) high-alumina clays containing diaspore extend the range of natural clay refractories towards the alumina end.

The principal impurities in refractory fireclays are silica, as quartz, and iron oxide. A few per cent of the latter can lower the melting point by 20 to 30° C. Alkalies, if present in greater amounts than 1 to 2%, can cause serious vitrification and lead to shrinkage and spalling in service. Other impurities such as titania, lime and magnesia have less serious effects.

The make-up of the fireclay body is largely governed by the plasticity and shrinkage properties of the clays. In Great Britain these are usually such that a proportion must be prefired and crushed to form grog to give stability to the body. The grog may be made from wasters or from rotary kiln fired clay. The quantity used varies from 0 to 90%, but is generally between 15 and 20%. The amount of grog and also its size and grading considerably affect the properties of a brick, even when the composition is the same. In the U.S.A. flint fireclays, which show little or no plasticity and have a conchoidal fracture, are used in place of grog.

The aluminosilicate range of bodies is extended beyond the range of compositions of natural fireclay by the addition of non-clay high-alumina raw materials such as sillimanite, kyanite and andalusite (all three forms of $Al_2O_3.SiO_2$) and corundum, Al_2O_3, etc. In this way the fullest use can be made of data available from phase diagrams. In the Al_2O_3–SiO_2 binary system there are several points of interest. Firstly both silica and alumina are highly refractory, having melting points of 1710° C (3110 °F) (cristobalite) and 2050° C (3722° F), respectively, but they form a eutectic mixture melting at 1545° C (2813° F). Secondly, the composition of the eutectic lies very near the silica end, namely, at 94·5% SiO_2, 5·5% Al_2O_3, thereby making the range of 'silica' refractories very small but leaving a wide range for aluminosilicates. Thirdly, a refractory compound occurs in the system, this is mullite, $3Al_2O_3.2SiO_2$, but it has an incongruent melting point at 1810° C (3290° F) forming corundum and siliceous liquid. All compositions having a silica content in excess of that required for the mullite composition begin to melt at 1545° C (2813 °F), the amount of liquid formed being proportional to the excess of silica present. Compositions richer in alumina than $3Al_2O_3.2SiO_2$ only begin to melt at 1810° C (3290° F).

Bowen and Grieg (**B99**) further point out that the siliceous liquid associated with mullite in compositions on the silica side of the mullite ratio makes a suitable bond for refractories, as it is highly viscous, and on cooling forms a glass with high thermal shock resistance. This is readily spoilt by small amounts of alkalies, lime and magnesia, which not only increase the amount of liquid but also lower its viscosity. This is very important as the service duty of a refractory depends less on the temperature at which melting occurs than on the conditions which give a critical quantity of low viscosity liquid.

Broadly speaking the usefulness as a refractory of alumino–silica mixtures from the eutectic mixture to pure alumina increases with the alumina content.

The nomenclature of aluminosilicates is confusing, some being given names derived from their raw materials, *e.g.* firebrick, sillimanite, while others have names descriptive of the product, *e.g.* mullite (*see* Table 79). Although the phase diagrams would lead one to expect that the dominating compound and mineral formed in a well-fired aluminosilicate brick is mullite, this is in practice very dependent on the raw material and the manufacturing process, as seen in the Table 78 and in greater detail in composition and property tables, p. 1243.

A high-alumina brick, although it does not contain a large amount of mullite, frequently develops mullite during high-temperature service.

TABLE 78

Comparison of two Refractories made from Bauxite by Differing Processes (K43)

	Mullite type	High-alumina type
Analysis:		
SiO_2	26·87%	23·10%
Al_2O_3	68·21	70·83
Fe_2O_3	1·41	1·60
TiO_2	2·92	3·25
CaO	0·03	0·07
MgO	0·01	0·35
Alkalies	0·59	0·80
Mineralogical analysis:		
Mullite	88·6	34·4
α-Alumina	4·1	53·0
Glass	7·3	12·6
Physical properties:		
P.C.E.	39	37
Shrinkage at 1592°C (2900° F) under load of 25 lb/in² (1·8 kg/cm²)	−2·8	−14·2
Reheat shrinkage 5 hr 1703°C (3100° F)	−0 to 0·5	−2·0 to 2·5

TABLE 79
Aluminosilicate Refractory Bodies

Ref.	Composition				Nomenclature		Raw materials
	Al_2O_3	SiO_2	Others	P.C.E.	British	U.S.A.	
				41/42		High alumina	Fused alumina base.
				40		99% Al_2O_3	
				39		90% Al_2O_3	
				36		80% Al_2O_3	
				35		70% Al_2O_3	Clay and calcined alumina from diaspore or bauxite.
				34		60% Al_2O_3	
						50% Al_2O_3	
N27	52–72			38	Sillimanite	Mullite or sillimanite	One of sillimanite, andalusite, calcined kyanite, dumortierite or topaz together with bonding clay; or pure Bayer alumina, and silica; or kaolin and alumina, e.g. 68–70% kaolin and 32–30% alumina.
S93	62	38	Fe_2O_3	over 40			
S93	80	18·4	1·56	39			
S93	72·14	26·3	1·56	33–39	Kaolin or molochite	Kaolin	China clay, or precalcined molochite.
S93	63·20	34·0	1·66	31–34	Firebrick:		
S93	44–45	51–53			Aluminous		
C26	38–45	62–55		34–39		Fireclay:	Diaspore or bauxitic clays.
N27	43–44	51–53		31–33	Firebrick	Super-duty	
N27	35–42	52–60				High-duty	
C26	<38	55–78		29–31		Intermediate-duty	Natural fireclays suitably blended.
S93	31–33	65·42	2·34	19–30		Low-duty	
S93	26–34	70·50	2·28		Siliceous		
C26	22–8	78–92		27	Semi-silica		
		88–93		31/32	Silica		Silica rocks.
		93–96		33	Super-duty silica		
		96–98					

Mullite refractories may be defined as: 'a refractory product consisting predominantly of mullite crystals ($3Al_2O_3.2SiO_2$) formed either by conversion of one or more of the sillimanite group of minerals or by synthesis from appropriate materials employing either melting or sintering processes' (**A1**). A good mullite refractory should contain from 82 to 93% mullite, averaging at 85%. The glass content should not exceed 15%, 9% being the average, and its composition should be highly siliceous so that its viscosity is high. There is then also a content of α-alumina of about 3 to 6%.

Fortified mullite refractories contain additional pure alumina and have improved load bearing capacity, spalling resistance, and resistance to slag attack. The mineralogical constitution of the brick as manufactured is about 70% mullite, 20% α-alumina and 7% glass. More mullite forms during service at high temperatures (**K43**).

Bodies for Saggars and Kiln Furniture

The ceramic industry itself requires specialised refractory ware for supporting and protecting finer ceramic products during firing. Much of this is made from aluminosilicate refractory types of body. Cordierite and silicon carbide are also used.

Saggars themselves are flat-bottomed, vertical sided vessels, which are filled with ware to be fired and then stacked on top of each other. Other kiln furniture consists of bats, props and smaller pieces (Chapter 11). Basic requirements are as follows:

(1) *Refractoriness.* They must withstand repeated heating not only to the soaking temperature of the ware but also to the extra heat found immediately opposite fireboxes, burners, etc.

(2) *Thermal shock resistance.* As the saggars, etc., are nearly always about $\frac{1}{2}$–1 in. (1·27–2·54 cm) thick they are thicker than the ware so that their inherent thermal shock resistance must be greater if they are to withstand the firing cycle repeatedly. The shorter firing cycles possible in tunnel kilns, as compared with intermittent ones, make this a more serious problem.

(3) *Hot strength and dimensional stability.* The method of stacking the filled saggars means that the cumulative weight bearing on the lower ones is very considerable. They must be able to withstand this without softening or contraction, even if a flame is playing on them. In tunnel kilns the stacks are much smaller than in large intermittent kilns, it is therefore possible to use thinner saggars. Similar considerations apply to other kiln furniture.

(4) There may not be any tendency to 'dust', *i.e.* for particles of the saggar to split off it and so foul the ware.

(5) The kiln auxiliaries must withstand prolonged use if the cost of their replacement is not to be uneconomical.

It can be seen that the first three requirements are quantitatively dependent on the firing cycle (maximum temperature and time) and on the height of the setting. There can therefore be no universally good saggar and both saggars and kiln furniture should be purposely made for the particular circumstances for which they are required. This entails choosing different raw materials, grain sizes and firing temperatures accordingly.

To fulfil these stringent requirements it has been found that the fired body should have the following properties:

(1) A very high proportion of the finished material should be crystalline, held together by the minimum of glassy material which should have the highest possible viscosity at the temperature of use (**W1**).

(2) The water absorption should be kept between 18 and 28% (except for small props, etc., which may be impervious).

The desirability of a high proportion of crystalline matter is shown by the following figures by Walker (**W1**) (Table 80).

TABLE 80

Creep Rate of Four Fine-textured Materials at $1000°$ C under Stress of 7000 lb/in^2

Body	Chemical Composition	Structure	Creep Rate 10^{-6} (in./in. per hr)
A	Alumina	Crystalline	0·27
B	Mullite	Crystalline	0·34
C	Aluminosilicate	Largely crystalline	16·6
D	Aluminosilicate	Mullite with appreciable glass	1060·0

Walker also studied photomicrographs of two mixings of the type used successfully for impervious tubing for posts and props. 'A' vitrifies within the range 1400–$1450°$ C (2552–$2640°$ F), and was seen to have a high proportion of glassy phase. 'B' contains more alumina and has been so fired as to produce the maximum of crystalline mullite. Rods made of both mixings were held horizontally at $1500°$ C ($2732°$ F) for the same amount of time. 'A' sagged noticeably, whereas 'B' remained straight.

COMPOSITION

Saggar bodies are usually composed of two parts grog to one part of plastic material. The raw materials must be chosen with care. The plastic material should be a fireclay or ball clay with the minimum of alkalies and other fluxes. The actual refractoriness required depends on the temperature of use. Clays suitable for earthenware saggars may be useless for porcelain ones, but need nevertheless not be rejected for the former use. Refractoriness itself is, of course, not the only property that must be investigated when choosing a clay.

Scottish fireclays are often very suitable both as binding clays and for grog for saggars. They occur with as little as 0·4% potassium and sodium which is less than any ball clay. By fine wet grinding they can be made highly plastic. Kaolin, either raw or washed, also has the advantage of low alkali content and rather coarser grain size than many plastic clays. Nevertheless stoneware clays are frequently used because of their high plasticity.

Clay binders can be supplemented by organic binders to make the body more plastic or suitable for pressing. Bentonite should not be used because of its high flux content of about 10%, although in some instances where only 2% is required the over-all flux introduction is only 0·2% and it has been used.

Adjusting the pH of the body to an optimum, generally about pH 8·0, and maturing are both very beneficial.

The grog used may be simply fired batches of the clay used as the binder, or when production is under way it will consist largely of ground broken saggars. The best constituents to keep the body highly crystalline are alumina and mullite; furthermore the glassy binder, which is an aluminosilicate, has the highest viscosity when the over-all alumina content is high, and the fluxes are at a minimum. Alumina is therefore frequently added as calcined alumina or high-grade bauxite. Mullite may be added directly in the form of calcined or fused sillimanite, kyanite or artificial mullite, or be allowed to form *in situ* by the reaction between alumina and clay, etc. (**W1**). Free silica makes a crystalline body but its volume changes give low thermal shock resistance and it is therefore to be avoided.

A number of raw materials which are harmless at cone 9 to 10 (1280° C, 2336° F to 1300° C, 2372° F) for earthenware saggars form eutectics which melt before cone 14 (1410° C, 2570° F), and so must be excluded from saggars for porcelain, *e.g.* small quantities of alkalies and alkaline earths in the presence of alumina and silica and magnesia. Magnesia, which forms cordierite in the body, is very beneficial to the thermal shock resistance of earthenware saggars, but it melts to a mobile liquid before cone 14.

The bats used for pusher-type tunnel kilns and in the structures used on kiln cars for open settings are frequently found to have the best properties if the 'grog' used is silicon carbide. Unfortunately its price prevents its widespread use.

Making saggars and bats porous assists considerably in lowering the proportion of glassy phase required to hold the crystalline phases together and also in increasing thermal shock resistance. The limits of 18–28% water absorption have been found by practical experience for the usual bodies. Talc bodies where cordierite develops are satisfactory with a lower porosity, provided their temperature limit is not exceeded. Bodies containing a quantity of iron compounds which readily flux alumina are often unsatisfactory even when the water absorption is over 18%. Saggars used for cone 14 are often found to vitrify gradually in the successive firings and thereby to decrease their porosity. Once this becomes less than 15% deterioration is very rapid. They should have more calcined alumina added to the body, making them initially more porous than necessary.

The question of grain size of the non-plastic material is not quite straightforward. Coarse-grained powders require little binder and make an open porous body, but there can be little solid-phase interaction and recrystallisation, which in itself produces strength. Fine-grained material requires more binder, but if it is of the correct composition, solid-phase reactions giving interlocking crystals can take place. In particular, finely ground alumina undergoes favourable reactions with fine grog.

In general some coarse-grained material must always be used, although this should be finer, *e.g.* sieve No. 16 for thin saggars, than for thick ones where a grain size up to about 5 mm is used.

In the older descriptions of saggar preparation a complicated system of size grading is recommended. For instance Searle (**S50**) suggests the following procedure:

(1) Remove 'dust', less than $\frac{1}{32}$ in. (0·8 mm, sieve 18) diameter, by sifting.

(2) Determine maximum size. For thick saggars this may be $\frac{3}{8}$ in. (9·5 mm); for thin ones not more than three-eighths of the wall thickness.

(3) Divide the permitted remainder into three grades, by sifting:

(*a*) coarse grog, $\frac{1}{8}$ in. to $\frac{3}{8}$ in. or permitted maximum (3·2 mm to 9·5 mm);

(*b*) medium grog, $\frac{1}{16}$ in. to $\frac{1}{8}$ in. (1·6 mm to 3·2 mm);

(*c*) fine grog, $\frac{1}{32}$ in. to $\frac{1}{16}$ in. (0·8 mm to 1·6 mm).

This is then mixed in definite proportions according to the properties required of the saggar.

It is obvious that this requires a good deal of equipment and time, and will always result in a remainder of grog dust and perhaps even of one of the grades (depending on whether the proportions produced and required are the same), which is useless.

It is now known that this inconvenience is quite unnecessary. If a saggar body is well compounded with the minimum of flux and sufficient mullite-forming material the whole of the crushed grog, including fines, can be used. For instance, in example (1) in Table 81 the alumina is bought 'finely ground', *i.e.* about sieve No. 150, crushed broken saggars contain about 50% fines passing sieve No. 16 these are mixed together with a suitable clay, brought to the optimum pH, shaped and then fired to cone 14, to give a strong and thermal shock resistant body with water absorption lying between 18 and 28%.

Silicon Carbide

Although not stable in oxidising atmospheres at very high temperatures the high thermal conductivity, good thermal shock resistance, excellent hot

TABLE

Saggar

No.	Ref.	Description	Burning Cone No.	Plastic Clay		Lean Clay		Kaolin
(1)	—	for cone 14 water absorption 18–28%		40				
(2)	D31			Wildstein F. 21	}	Bennstedt 32		
(3)	D31			Wildstein B. 8	}			
				Wildstein F. 22	}			Kamig 15
(4)	D31			Wildstein B. 8	}			
(5)	D31			Ponholz 20		Schönhaider 40		
						Bennstedt 35	}	
						Luckenauer 12	}	
						Schönhaider 12	}	
(6)	D31			Wildstein 12		30		Halle 12
(7)	H120	Australian cordieritic, $\alpha = 4 \cdot 0 \times 10^{-6}$ porosity 40%	1390–1400° C	sillimanitic clay 30 refractory shale 50			}	
(8)	H120	Australian cordierite, $\alpha = 3 \cdot 2 \times 10^{-6}$ porosity 20%	1260–1280° C	sillimanitic clay 55 ball clay 25			}	
(9)	—	cordieritic for earthenware, water absorption 10·8%						
(10)		for cone 14 hard porcelain		10				
(11)		for cone 15 hard porcelain		Wildstein B. 10		Bornholm ChCl. 35		
(12)		for cone 15 hard porcelain		Wildstein B. 10		Bornholm ChCl. 35		

strength and dimensional stability make silicon carbide a useful refractory for special purposes, either as the non-plastic material in a fireclay body, or alone.

Of the clay-bonded bodies the highest grade have 85% silicon carbide, others have 40–78%. The influence of the quantity on physical properties is seen in Table 82. Such bodies are fired at 1450–1500° C (2642–2732° F). Alumina may be included, e.g. SiC, 40%; Al_2O_3, 40%; clay, 20%.

A specialised silicon carbide body for extremely severe conditions of thermal shock is bonded with silicon nitride (**C9**).

Silicon carbide is also used alone and 'sintered', or 'recrystallised', the latter at 2200 to 2350° C (3992–4262° F) (**S14**). Another method of producing a very stable product is to subject a skeletal carbon structure to the action of superheated molten silicon. The silicon combines with the carbon *in situ*, any excess silicon filling the spaces between the newly formed SiC

81
Bodies

No.	Organic binder	Grog	Silicon Carbide	Sillimanite	Feldspar	Talc	Alumina	Analysis
(1)	tannin 0·2	Saggar 40					20	
(2)		39						
(3)		55						
(4)		40						
(5)		41						
(6)		46						
(7)						20		
(8)						20		
(9)								SiO_2, 64·95% $Al_2O_3 + Fe_2O_3$, 30·80% CaO, 1·20% MgO, 2·60% NaKO, not determined total 99·55%
(10)			45				45	
(11)				25	10		20	
(12)				35	—		20	

crystals. On heating in use to 1400° C (2552° F) no change occurs, above this silicon exudes from the body but it does not change shape or undergo disintegration (**K57**).

TABLE 82
Some German Silicon Carbide Refractories (A6)

Silicon Carbide %	Clay %	Crushing Strength (tons/in²)	Bending Strength (tons/in²)	Young's Modulus (lb/in² × 10⁶)
0	100	21·5	1·35	3·1
60	40	17·4	0·76	4·9
80	20	10·8	0·48	4·7
90	10	4·3	0·35	3·5

Carbon Bodies

Ceramic bodies based on carbon have been developed to serve varied purposes in the refractories, chemical engineering and electrical fields.

Normal carbon bodies are made from coke bonded with tars or pitches, moulded when the latter are hot and then allowed to set before firing to carbonise the bond. The product is strong, porous and largely composed of amorphous carbon.

Graphite bodies are obtained by heating a carbon body to the maximum temperature of the electric furnace, when crystallisation occurs. The changes occurring are shown in Table 83.

Both carbon and graphite bodies can be made impervious by impregnation with resins, the product is known as 'Karbate'.

TABLE 83
Typical Changes in Characteristics of Carbon due to Graphitisation (W42)

Properties	Carbon	Graphite
Resistivity (Ω-in.)	0·0014–0·0018	0·0003–0·0005
Thermal conductivity (B.t.u.)	3–4	70–80
Threshold oxidation temperature in air (°C)	660	750
(°F)	1220	1382
Oxidation rate % in 24 hr on $2 \times 2 \times 4$ in. samples in:		
(1) 98% excess steam at 800° C (1472° F)	2·7	0·2
(2) 98% excess CO_2 at 800° C (1472° F)	1·0	0·1

Bodies of higher than the normal porosity can also be obtained by suitable selection and grading of the raw materials and they make very good filters and diffusing media, combining the desirable chemical and physical properties of carbon and graphite with the unique advantage of being able to have a material of uniform and controlled pore size. The materials are known as 'Carbocell' and 'Graphicell'.

A carbon body that can be made with a porosity as low as 2% or with controlled higher porosities is made direct from bituminous coal by the special 'Delanium' process. In this, a coking coal is finely ground and graded from 200-mesh to 10 μ. It is blended with substances to prevent swelling and then shaped. It is fired in a reducing atmosphere at a carefully controlled rate, when shrinkage of as much as 45% occurs. A very fine-grained material results. This has a high proportion of α-carbon which is characterised by its hardness, high compressive and tensile strength and an optical reflectivity which is high compared with that of coal or amorphous carbon.

The formation of α-carbon during carbonisation appears to be a process of successive polymerisation reactions involving the formation of cross-linkages between neighbouring molecular units to form a hard coherent structure. The process is accompanied by the liberation of volatile constituents and a shrinkage of the order of 45%. Whether the shrinkage results in a porous piece of the same size as the original, a swollen piece full of gas bubbles as seen in coking, or a dense piece, depends at which stage the material exhibits plasticity. In the 'Delanium' process adjustment of the chemical composition together with careful regulation of the rate of heating during carbonisation controls the plasticity within a narrow range so that reorientation and cross-linkage of the larger molecular units can occur without destroying the macrostructure. Thus the fine pores in the coal structure permit gas to escape without swelling and subsequent shrinkage gradually closes up the pore spaces left (**N22**).

Heating 'Delanium' carbon above 3000° C (5432° F) converts it to 'Delanium' graphite of high thermal conductivity.

CARBONACEOUS REFRACTORIES

Refractory hollow-ware of particularly useful properties in the metallurgical industries is made from bodies containing graphite (or sometimes amorphous carbon in the form of coke or charcoal). The graphite is bonded with a ball clay or plastic fireclay which should vitrify relatively easily, but the finished body shows a refractoriness far above that of the bond clay, even though no chemical interaction takes place between the graphite and the clay. Graphite in fairly large flakes is far more suitable for making refractory crucibles than granular material, which leads to the supposition that it forms some sort of rigid framework in the fired piece.

Graphite crucible bodies have high heat conductivity which is helpful in reducing heating time, good thermal shock resistance, toughness and small tendency to oxidise the contents.

As well as graphite and clay the following raw materials may be included in these bodies:

(1) Feldspar, as an additional flux to obtain good vitrification of the bond. This should be quartz-free and very finely ground.

(2) Quartz was formerly added to reduce plasticity and increase refractoriness. Its unsatisfactory thermal expansion has led to its replacement by other materials wherever possible.

(3) Silicon carbide is a very useful non-plastic addition to graphite bodies, increasing strength, refractoriness and heat stability.

(4) Grog from scrapped crucibles may also be included so long as the graphite and silicon carbide have not been too severely oxidised.

The proportions of these raw materials are regulated both by the nature of local materials available and by the metal to be heated in the finished crucible (*see* Table 84).

Basic and Neutral Refractory Bodies

The various types can be summarised as follows.

(1) Burned magnesite brick containing 85 to 95% of magnesia. (Chemically bonded and unburned magnesite bricks which have greater spalling resistance are not true ceramics under the definition used in this book.)

TABLE 84

Graphite or Plumbago Bodies

Ref.	M97	S50	S50	S50	S50	S50	S50	S50	S50	S50
Uses	General body type		Hard steel German	Mild steel German	Razor steel	Very pure steel	Copper alloys	Copper alloys	Cast iron	Cast iron	Sheffield steels
Ceylon graphite	20										
Madagascar graphite	20										
Graphite	20–50 / 30–40										
Fireclay		40–50	54	40	12	3	8	12	53	50	4–7
Ball clay		30–40	36	38	40	87	67	50	43	40	73–86
China clay	32				40	10		13			10–20
Bentonite	0–2										
Silicon carbide	10–20	10–20									
Sand	17		10	22	8		25	25	4	10	
Graphite body grog	4	0–5									
Feldspar	3										

(2) Magnesite–chrome brick, made from blends of dead-burned magnesite and chrome-ore, in which the amount of magnesite predominates. Two modifications are made:
  (a) Burned brick, having excellent strength and resistance to abrasion and erosion at high temperatures.
  (b) Chemically bonded brick, having outstanding resistance to spalling.
(3) Chrome–magnesite brick made from mixtures of chrome-ore and magnesite in which the amount of chrome-ore predominates.
  (a) Bricks of extremely hard burn, which have greater strength and load-carrying capacity at high temperature than the chemically bonded brick.
  (b) Chemically bonded bricks, which have greater resistance to spalling than burned brick.
(4) Burned chrome brick.
(5) Forsterite brick, made from the mineral olivine with added magnesia. Outstanding properties are stability of volume and strength at high temperatures.
(6) Dolomite. Stabilised dolomite rock having very high refractoriness.

Their main distinguishing characteristics are:
  (a) Great resistance to chemical attack by basic slags and by basic oxides, such as iron oxide, lime and alkaline oxides.
  (b) Extremely high melting point.
  (c) High rate of thermal expansion—two to three times that of fireclay brick.
  (d) High weights.
  (e) High thermal conductivity, especially in compositions consisting chiefly of magnesite (**M3**).

Magnesite Refractories

This body consists primarily of dead burned magnesium oxide, obtained by calcining natural magnesite (magnesium carbonate) rocks, or magnesium hydroxide obtained from sea water. The magnesia content is normally between 85% and 95%. The most important minor constituent is iron oxide which may vary from 8% down to 0·5% or less.

The traditional fired magnesite refractory consists of relatively small grains of periclase (magnesia) which do not interlock, the whole being bonded by a glassy matrix. The glass is a silicate mixture which has low viscosity. The strength of the brick depends almost entirely on this glass and it may fail under load when as little as 10% liquid is present, whereas silica bricks only fail when 30% liquid has been formed (**B62**).

Natural magnesites sinter at varying rates, and the ones from the Austrian deposit at Veitsch sinter very easily. These magnesites have a high iron content.

As higher iron contents make sintering easier but the finished brick somewhat less refractory it was formerly assumed that the matrix was an iron-rich glass. Birch (**B62**), however, shows that ferrous oxide together with manganese and some other minor divalent oxides and also magnesio-ferrite $MgO.Fe_2O_3$ go into solid solution in the periclase grain, whilst the glass is based on CaO–Al_2O_3–SiO_2 together with magnesia, alkalies, etc. The softening of a magnesite refractory is therefore governed largely by the melting behaviour of the silicates which he assumes consist largely of CaO, Al_2O_3 and SiO_2. The equilibrium data for the ternary system for these three oxides is of assistance in determining the softening behaviour of a magnesite refractory.

Nadachowski's (**N1**) experiments also show that magnesium ferrite enters into solution in the crystalline periclase instead of going into equilibrium with the silicates, when a eutectic mixture would form at 1350° C (2462° F). When magnesium ferrite is replaced by magnesium aluminate in a refractory magnesite body this does not dissolve in the periclase but does form a eutectic with the liquid silicate phase as expected.

Kriek, Ford and White (**K82**) have investigated the effect of other additions on the sintering of magnesia. CaO is found to inhibit sintering. But TiO_2, Al_2O_3 and SiO_2 promote it to some extent, although there is an optimum addition for each, after which inhibition occurs. They suggest that these three oxides promote sintering in the same way as ferric oxide by providing cations that enter the magnesia lattice and then create cation vacancies.

Sintering of pure magnesia to a hard product is much more difficult than the iron-rich magnesites and the refractory so produced is sometimes termed 'artificial periclase' to differentiate it. Where it is unnecessary to obtain such a high-grade product the sintering of pure magnesia is promoted by additions of iron oxide, chromite ore, etc.

The addition of 0·25% or more of certain lithium compounds, particularly the halides, facilitates the pressing and sintering of reactive forms of high purity magnesium oxide (**A197**).

Normal magnesite bricks have a high coefficient of thermal expansion ($13·5 \times 10^{-6}$ at 20–800° C, $13·2 \times 10^{-6}$ at 20–1400° C) which leads to poor thermal shock resistance. This is avoided in special brands by one or both of the following methods. (*a*) Careful grading, in particular use of one-size granules bonded with fine powder. (*b*) Addition of from 2 to 6% Al_2O_3, *e.g.*

MgO	82·8%
Fe_2O_3	6·8%
Al_2O_3	4·7%
CaO	2·6%
SiO_2	2·1%

The bonding of calcined magnesia for shaping of the bricks may be achieved without additions other than water which hydrates any caustic magnesia or lime that may be present, or it may require the addition of one of the following: alkalies, caustic lime, boric acid, coal ashes, clay, cement, magnesium salts, tar, molasses, etc.

Dolomite Refractories

Although dolomite is a readily available and highly refractory material, its use for making bricks has only been made possible very recently (during World War II), by careful research followed by tight control in manufacture.

Dolomite rock is calcium magnesium carbonate $CaMg(CO_3)_2$. When it is fully calcined carbon dioxide is given off and a mixture of lime and magnesia remains and the lowest melting point of this mixture is 2300° C (4172° F). But although magnesia loses its tendency to hydrate by dead burning, lime never does, so that bricks made of pure calcined dolomite (doloma) whilst super refractories, will always be attacked and disintegrated by atmospheric moisture, a process known as 'perishing'. Since most dolomite bricks are intended for use in basic open-hearth furnaces, where moisture is absent, such bricks can be coated with pitch to make them waterproof during transit and a limited storage period, so long as they are not damaged. They are termed 'semi-stable dolomite bricks'.

A brick that is stable throughout is, of course, very much more desirable. Stabilisation involves making additions of silica or iron oxides to combine with the free lime and so render it inactive towards water. This, however, brings a secondary trouble with it. Only two of the lime silicates are sufficiently refractory for this purpose, namely, tricalcium silicate ($3CaO.SiO_2$) and dicalcium silicate ($2CaO.SiO_2$) which do not melt below 2000° C (3632° F), although the tricalcium silicate breaks down at 1900° C (3452° F) to dicalcium silicate and free lime. Dicalcium silicate, however, exists in three allotropic forms of which the low-temperature or γ-form develops from the β-form spontaneously and, as it is accompanied by a 10% volume increase, almost explosively. This inversion will cause even a strong brick to disintegrate into a material as fine as face powder and the result is referred to as 'dusting'. 'Perishing' and 'dusting' are readily confused, but unless differentiated they cannot be prevented.

The β to γ inversion of dicalcium silicate can be prevented by the addition of small amounts of Cr_2O_3, B_2O_3, P_2O_5 or other stabilisers. To produce a 'stable dolomite brick' it is therefore necessary to combine the free lime and simultaneously to make any β-dicalcium silicate present stable.

The stabilisation effected by various silicate minerals has been investigated, *e.g.* flint, steatite, olivine, ball and china clays, slags, etc., and serpentine, the last having proved the most successful and is being used in the commercial bricks.

The resultant phases, and therefore quantity of dicalcium silicate can be predicted fairly accurately from the $CaO–Al_2O_3–SiO_2$ phase diagram (**B62**). The refractoriness of dolomite–serpentine mixtures can be seen from the diagrams recently worked out by Braniski (**B108**). These show two minima at 40% dolomite–60% serpentine, 1520° C (2768° F), and 3% dolomite–97% serpentine, 1740° C (3164° F), and a maximum at 25% dolomite–75% serpentine, 1825° C (3317° F). In the whole range only a small portion is non-refractory (below cone 26), namely 38–47% dolomite–62–53% serpentine. The addition of serpentine to stabilise the free lime in dolomite does not therefore have undue ill effects on the refractoriness.

It is recommended to add the serpentine in such quantities as to form the maximum tricalcium silicate, according to the following equation:

$$6(CaCO_3 \cdot MgCO_3) + 3MgO \cdot 2SiO_2 \cdot 2H_2O \rightarrow 2(3CaO \cdot SiO_2) + 9MgO + 2H_2O + 12CO_2$$
\quad dolomite \qquad serpentine

As this would require very tight control it is usual to operate slightly on the silica-rich side, and add small amounts of boric oxide or phosphates to stabilise any dicalcium silicate that may form.

TABLE 85

Typical Analyses of British Semi-stable and Stable Dolomite Bricks (C26)

	Semi-stable %	Stable %
SiO_2	4·3	14·4
Fe_2O_3	2·5	3·4
Al_2O_3	2·2	1·5
CaO	51·7	40·0
MgO	38·2	40·4
Loss on ignition	1·1	0·3

Chrome Refractories

Neutral refractories were originally made directly from selected chrome–iron ores with the sole addition of binders, even though the chemical composition of these ores may vary considerably (H18):

Cr_2O_3 \quad 30–50% ⎱ sum to be not
Al_2O_3 \quad 13–30% ⎰ less than 60%.
FeO \quad 12–16%.
MgO \quad 14–20%.
SiO_2 \quad 3–6%.
CaO \quad up to 1%.

Certain chrome ores, however, are associated with serpentine which melts readily and forms a very fluid liquid. By careful addition of magnesia the serpentine can be converted to the more refractory and viscous forsterite. Most modern chrome brick mixtures contain 10% magnesia in the form of caustic or sinter magnesia (C26).

Other additions that may be made to increase the refractoriness of the bond of chrome bricks include: brick clay, loam, fireclay, china clay, bauxite, aluminium hydroxide, grog, hydrargillite, lime, limestone, dolomite, artificial spinel, chrome salts, barytes, etc. (**R46**).

ACTION OF HEAT ON CHROME ORE ALONE OR WITH MAGNESIA ADDITIONS

Chrome ore is theoretically the spinel ferrochromite $FeO.Cr_2O_3$ but always also contains other RO and R_2O_3 oxides, largely MgO, Al_2O_3, Fe_2O_3. In general the molecular proportions of $RO:R_2O_3$ are 1:1, *i.e.* the spinel is balanced. The firing behaviour of natural and synthetic spinels has been investigated by Rigby, Lovell and Green, and Rait (**R31**), and Konopicky (**K64**), and has been summarised (**H18**). Raw chrome ores are unaffected by heating in a reducing atmosphere but when heated in an oxidising one most of the ferrous oxide is oxidised to ferric oxide. No volume change occurs in the spinel grains but two solid phases result: (a) spinel crystals consisting mainly of $MgO.R_2O_3$; and (b) a solid solution of the excess R_2O_3 constituents (Fe_2O_3, Cr_2O_3 and Al_2O_3). Most of this solid solution is believed to dissolve in the spinel crystals but some of it probably migrates into the ground mass or matrix. Where the ratio $Al_2O_3:FeO$ is high, the tendency to oxidise below 1450° C (2640° F) is lessened.

Heat treatment in a reducing atmosphere of such an oxidised chrome ore reduces the ferric oxide to the ferrous which probably re-enters the spinel crystal structure. This is usually accompanied by expansion, and increase in porosity and sometimes cracking. Some chrome ores become friable by this process, if repeated, others do not.

When a body mixture containing chrome ore and magnesia is fired under reducing conditions the ferrous oxide of the spinel is replaced, at least in part, by magnesia. The FeO diffuses outwards from the grain into the matrix as the MgO diffuses into the grain and the spinel crystals outwardly appear unchanged. When such a chrome ore–magnesia mixture is fired under oxidising conditions and the FeO is oxidised to Fe_2O_3 the MgO not only replaces the FeO in the original spinel structure but also combines with the newly formed Fe_2O_3, thus preserving the spinel structure. While the Fe_2O_3 in the spinel $MgO.Fe_2O_3$ is easily reduced, solid solutions of $MgO.Fe_2O_3$ with $MgO.Al_2O_3$ and $MgO.Cr_2O_3$ are more stable. Oxidation of FeO in a chrome spinel with simultaneous absorption of MgO may be accompanied by a volume increase of 5–10%. The effect is, however, sometimes masked by the shrinkage of silicates, etc., in forming a matrix.

Addition of magnesia to chrome ore also increases the refractoriness of the silicate groundmass by converting it into forsterite $2MgO.SiO_2$. The increase of volume of the spinel grains together with this lowering of the binding power of the matrix may result in a chrome–magnesite brick having smaller cold strength than a chrome brick, but the reverse is true at high temperatures.

MAGNESITE–CHROME AND CHROME–MAGNESITE REFRACTORIES

A range of refractories has been developed with magnesite and chrome refractories as the two end points. In those compositions where magnesite predominates the product is termed magnesite–chrome, whereas in those where chrome predominates (70:30 or 60:40) the refractory is called chrome–magnesite. These intermediate compositions have very favourable properties which both end members lack, particularly a very good resistance to thermal spalling.

In Great Britain a tentative specification has been drawn up for the chrome ore to be used in all-basic furnace roof bricks, it suggests:

$$SiO_2 \quad 3\text{–}6\%.$$
$$FeO \quad <18\%.$$
$$CaO \quad <1\%.$$
$$Cr_2O_3 \quad 40\text{–}45\%.$$

It is satisfactory but probably not exclusive.

Olivine and Forsterite

The natural rock olivine consists substantially of mixed crystals of forsterite, $2MgO.SiO_2$, and fayalite $2FeO.SiO_2$, together with small quantities of serpentine $3MgO.2SiO_2.2H_2O$, and talc $3MgO.4SiO_2.2H_2O$, the relative quantities of the different minerals varying with the source of the rock.

Forsterite has desirable refractory properties while the other minerals present with it in natural olivine rock do not, in particular serpentine and talc give large amounts of very fluid liquid at 1559° C (2838° F) and 1544° C (2811° F), respectively. The separate reactions can be summarised:

Forsterite, $2MgO.SiO_2$, M.P. 1910° C (3470° F)

Fayalite, $2FeO.SiO_2$, M.P. 1205° C (2201° F)

Serpentine, $3MgO.2SiO_2.2H_2O$, $\xrightarrow{1559°C \; (2838°F)}$ 60·5% forsterite
$\qquad\qquad\qquad\qquad\qquad\qquad\qquad\qquad\qquad + 39·5\%$ siliceous liquid

Talc, $3MgO.4SiO_2.2H_2O$, $\xrightarrow{1544°C \; 2811°F}$ 4·4% cristobalite + 95·6% liquid

Fayalite $\xrightarrow{\text{oxidising heating}}$ $Fe_2O_3 + SiO_2$

The application of olivine rock in the refractories field was initiated by V. M. Goldschmidt in 1925 (**G52, G53**), and his work has been summarised by the senior writer (**S122**).

Natural olivine of high forsterite content is a very useful refractory in itself. As it has no firing shrinkage or expansion ($\pm 0\text{–}1\%$ compared with magnesite, -60%; fireclay, -15%; quartzite, $+15\%$), and no crystal change or gas evolution on firing, the natural impervious rock can be hewn into shape and used directly for furnace bottoms, reheating furnaces and other applications where a non-porous refractory is particularly useful. Crushed olivine rock can also be shaped and fired as a normal refractory brick. The resultant body still contains a mixed crystal of forsterite and fayalite, and so by Goldschmidt's terminology should be called an 'olivine refractory'. However, the term 'forsterite refractory' is frequently applied.

The true 'forsterite refractory' is one in which the silicate is largely present as magnesium orthosilicate and the iron as non-siliceous compounds, e.g. magnesio–ferrite. This is achieved by the addition of the correct amount of magnesia to olivine, when most of the magnesia and silica present can be converted to forsterite, together with magnesio–ferrite ($MgO.Fe_2O_3$) and periclase according to the following reactions:

Mixed crystals of forsterite and fayalite $\xrightarrow[heat]{oxid.}$ free forsterite + $Fe_2O_3 + SiO_2$

$Fe_2O_3 + MgO \rightarrow MgO.Fe_2O_3 \rightarrow$ solid solution in periclase

$SiO_2 + 2MgO \rightarrow$ forsterite

2 serpentine, $3MgO.2SiO_2.2H_2O \rightarrow 3$ forsterite + $SiO_2 \xrightarrow{2MgO} 2MgO.SiO_2$

2 talc, $3MgO.4SiO_2.2H_2O \rightarrow 3$ forsterite + $5SiO_2 \xrightarrow{10MgO} 5(2MgO.SiO_2)$

(H18)

Such a mixture has a melting point of 1800–1900° C (3272–3452° F). It must, however, be emphasised that the olivine used must have an iron content of less than 6–7% if a good refractory is to be achieved. The main raw materials of a forsterite refractory body are olivine and dead-burnt magnesite. To these must be added a suitable binder, one of the best being a very finely ground mixture of one part dead-burned magnesite to two parts low alumina (maximum 12% Al_2O_3) chrome ore. A binder recommended by Goldschmidt is: 48% chromite, 32% sinter magnesite, 15% caustic magnesite, 5% china clay.

A forsterite body can also be prepared from raw materials not predominantly made up of forsterite, e.g. by heating talc or serpentine with magnesium oxide, when a solid-state reaction begins at 1000° C (1832° F) and become quantitative at 1200° C (2192° F). This is not done if a natural olivine is available.

Olivine and forsterite refractories do not react with magnesite, chrome-magnesite and chromite bricks at any temperature. There is no reaction with silica up to 1450° C (2642° F), above which they must be separated by chromite. However, serious reactions occur with aluminous materials above 1250° C (2282° F) as follows:

$$2MgO.SiO_2 + 4SiO_2 + 2Al_2O_3 = 2MgO.2Al_2O_3.5SiO_2$$
M.P.   1910°C   1710°C   2050°C   1460°C
      (3470°F)   (3110°F) (3722°F)   (2660°F)

Lime or very basic compounds of lime found in slags react at temperatures exceeding 1400° C (2552° F).

$$2MgO.SiO_2 + CaO = MgO + CaO.MgO.SiO_2$$
M.P.   1910°C   3000°C   1500°C
      (3470°F)   (5432°F)   (2732°F)

With excess lime dicalcium silicate forms, whose volume change at the inversion temperature leads to dusting.

The reaction with silica occurs above 1450° C (2642° F):

$$2MgO.SiO_2 + SiO_2 = 2(MgO.SiO_2)$$
$$\text{M.P.} \quad 1910°C \quad\quad 1710°C \quad\quad\quad 1560°C$$
$$(3470°F) \quad (3110°F) \quad\quad (2840°F)$$

Olivine and forsterite are reduced by carbon at temperatures above 1630° C (2966° F), also by high-grade ferro–silicon, calcium carbide, etc., when magnesium vapour and volatile silicon monoxide are formed. Iron vapour will also substitute for magnesium in forsterite crystals at sufficiently high temperatures. Molten iron does not affect forsterite up to 1800° C (3272° F).

Barium Aluminium Silicate Bodies

The equilibrium curve for the system barium oxide to clay substance shows two eutectics and between them a distectic of considerable height. The melting point of the intermediate compound is 1780° C (3236° F) and its composition is that of the mineral celsian, or barium anorthite, $BaO.Al_2O_3.2SiO_2$ (BaO 40·8%, Al_2O_3 27·1%, SiO_2 32·1%).

Bodies of the following range of composition are refractory and have exceptional resistance to certain corrosive agents, *e.g.* molten aluminium and salt glazing vapours, vapours and condensates containing lead and boron, which destroy other refractories. Range:

$$BaO \quad 31–51\%.$$
$$Al_2O_3 \quad 17–37\%.$$
$$SiO_2 \quad 22–42\%.$$

The optimum being the true celsian composition. Such bodies can be made into both refractory bricks for kiln and furnace linings and porcelain-like bodies of high density and impervious to X-rays.

The raw materials used are barium carbonate and a pure clay whose composition approaches the ratio $Al_2O_3:2SiO_2$ and which has little or no free quartz.

Müller-Hesse and Planz (**M98**) describe modern investigations of the $BaO–Al_2O_3–SiO_2$ refractories which are being used increasingly in the construction of electric tunnel kilns. With the aid of chemical analysis, X-ray photography, dynamic differential calorimetry, optical and heated-stage microscopy, dilatometry, etc., it is now possible to work out the optimum raw materials, preparation, particle-size distribution, manufacturing and firing conditions. It was also found that during firing an unstable form of celsian is formed first which converts to the stable form after continued heating.

3. THERMAL INSULATION BODIES

There are a number of applications where it is desirable to make a structure a good thermal insulator, ranging from domestic buildings to high-temperature kilns and furnaces.

The principle of thermal insulation by the incorporation of air spaces in solid materials was known long before it was understood. Heat transfer occurs by conduction, convection and radiation. Air being a much poorer heat conductor than solid materials, the larger the proportion of air in a brick the less heat conducted through it. Convection requires air movement and is prevented if the air spaces are small. Radiation depends on the temperature difference between opposite solid faces. The smaller the air gap, the smaller the temperature difference and hence the less radiation. The three factors have different relative importance at different temperatures, as shown by Halm and Lapoujade (**H11**) in their curves of heat transmission against temperature. Fine-pored materials show a linear increase of transmission with temperature, whereas for coarse pored ones the curve starts at lower transmission and is linear at first but thereafter rises much more steeply. These show that at low temperatures transmission is due to conduction and the lighter the material the better the insulation. At high temperatures radiation is the chief method of transmission and fine-pored materials are better insulators, *e.g.* at 1027° C (1881° F) a pore of 0·1 mm diameter transmits only half as much by radiation as by conduction whereas one of 5 mm diameter transmits thirty times as much. The pore structure of an insulating brick must therefore be adapted to the anticipated temperature of application.

Insulating bricks are made from the materials used in normal building bricks, firebricks, and refractory bricks together with materials which will make them porous. The nature of the brick-making material largely determines the refractoriness of the finished brick but this is also influenced by the method of making the pores. Insulating materials are divided into three main groups, the ceramic insulating bricks falling into the intermediate and high temperature groups.

(i) Low temperature, cold face down to −15° C (5·0° F); hot face up to 40° C (104° F).
(ii) Intermediate temperature, hot face up to 900° C (1652° F).
(iii) High temperature, hot face up to 1550° C (2822° F).

The main methods of producing porous pieces are as follows:

(1) Incorporation of diatomaceous earth (moler), vermicullite or other natural or manufactured air-filled inorganic cellular material.
(2) Use of raw materials that remain porous if not fired too high.
(3) Addition of organic matter which burns out during firing, *e.g.* coal,

anthracite, coke, lignite, pitch, petroleum residues, sawdust (green, weathered or roasted), cork, charcoal, paper, straw, chaff, husks of cornflower seeds, etc.

(4) Addition of material that can be volatilised, *e.g.* naphthalene.

(5) Introduction of gas bubbles by: (a) chemical evolution; (b) by foaming with surface-active or wetting agents followed by setting with plaster of Paris.

(6) Fire bloating. A number of the more impure clays and shales used for brickmaking give off gases during firing. If the firing is rapid this may occur after some fusion and glass formation has occurred. If the viscosity of the glass is sufficiently high gases become trapped and the whole body expands.

(7) 'Sintering'. By various combinations of fire bloating and inclusion of combustible matter highly porous bodies can be made (**S124**).

METHOD (1)

Diatomaceous earth (the hollow siliceous remains of diatoms), silex (unicellular, microporous siliceous shells remaining after the chaff, etc., of rice waste has been burnt), exfoliated vermiculite and expanded perlite are excellent insulators, although each has a maximum temperature of use above which collapse of the cells occurs. By bonding with plastic clays excellent intermediate temperature insulators are made.

Porous grog of almost any desired refractoriness made by one of the methods described below is used to increase the porosity and assist in shaping and shrinkage characteristics of insulators made by other methods.

METHOD (2)

All ceramic bodies that contain water during their processing are porous at some stage during their firing before vitrification sets in. Suitably compounded bodies are sufficiently rigid at this stage not to require further firing. Moderately light bricks can therefore be made by this direct method.

The decomposition of carbonates, *e.g.* in raw magnesite and dolomite also leaves a porous mass, and this has been used for certain basic refractories although they do not show great strength (**C26**).

The firing expansion of raw kyanite also gives a porous body (**O5**).

METHOD (3)

The addition of combustible material has found the most widespread use, primarily because the same plant and general method of mixing, shaping, drying and firing can be applied as are used for normal bricks and refractories. Three factors determine the nature and quality of the product:

(a) bonding clay and general body composition;
(b) nature and size grading of the combustible;
(c) shaping method.

The general nature of the body composition is of course determined by the type of insulator required, *e.g.* building block, silica, basic, zircon, fireclay, etc., but the bonding clay can be further selected with regard to the need for high plasticity to hold the additional non-plastic material. Improvement of the clay by fine grinding or by chemicals is sometimes necessary. Deflocculation of clays before mixing with the combustible followed by flocculation with calcium chloride (0·25%) is recommended. Additions of bentonite, sulphite lye, or tannin–lignin mixtures are advantageous.

The most important feature of combustible materials is the nature of the inevitable ash that remains. Coal, coke, anthracite, etc., leave a residue of iron oxide which both discolours a clean-burning brick and acts as a flux, particularly as the reducing conditions prevailing while the combustible burns out will enable ferrous silicates to form. Sawdust, wood flour, straw and chaff generally contain alkalies which, although they do not discolour the product, have a serious fluxing action. The refractoriness of a refractory insulator made porous by the addition of combustible material must always therefore be less than that of the original body and this lowering of refractoriness may be as much as 2 cones if material containing unsuitable ash is used.

The size grading of the combustible is of importance not only with regard to the exclusion of oversize material, but also because the inclusion of fines leads to excessive firing shrinkage. Coal, coke, anthracite, etc., have a greater tendency to contain fine dust than sawdust, etc. Fibrous materials such as straw, peat and softwood sawdust are difficult to size grade, but hardwood sawdust and cork can be carefully graded to give a regularly pored product. The surface texture of the combustible grains influences the amount of bonding clay required to cover them, irregular surfaces requiring more than the smooth ones. The absorptive nature of sawdust and cork suggest the advisability of soaking these before incorporation in the body mixture as they can otherwise take moisture from the body and swell up, leading to cracking of the piece.

These mixtures can be shaped by: (*a*) slop moulding, when large drying facilities are required but the clay can carry the largest amount of non-plastics; (*b*) extrusion; (*c*) semi-dry or dry pressing; (*d*) tamping, when porous grog can be included as well as the combustible.

Firing of bodies containing combustibles has to be supervised very carefully as local overheating can easily occur at the time of ignition (**O5, S107**).

METHOD (4)

The incorporation of relatively pure materials that can be removed by sublimation (and thereby recovered for re-use) overcomes some of the disadvantages of method (3). Firstly, no fluxing ash is left in the body so that insulators of the same refractoriness as dense bodies can be made. Secondly, as the pore-forming agent is removed at low temperatures no danger of local overheating leading to distortion can occur in the kiln. Volatile materials that have been experimented with are sulphur, ammonium chloride, paradichlorobenzene; but the only one that has found widespread application is naphthalene.

Naphthalene in the form of press cakes (bought) and flakes (from the recovery plant) is mixed with the other raw materials and then shaped by similar methods to those used for bodies incorporating insulators. The bricks are then placed in special chambers similar to intermittent hot air dryers and hot air passed through, and then led into condensation chambers. The dry porous bricks are rather fragile and cannot be set in the kiln to heights of more than about 3 ft. 6 in. (1 m) but they can safely be fired very fast.

TABLE 86

Successful Batch Mixtures used by the Senior Writer (S107)

Raw ball clay (water content 15%)	60	60
Dense refractory grog	40	40
Raw pressed naphthalene: 90% pure	50	
95% pure		15
75–85% pure	15	
Refined naphthalene, moist, 90% pure		20

METHOD 5 (*a*)

The development of gas bubbles by chemical action. A porous structure with fine and regular pores can be achieved by the evolution of gas within the mixed body followed by stabilisation of the bubble structure. The gas is evolved by the reaction between two evenly and finely dispersed substances and must be controllable and possess a sufficient period of induction to allow the mixing to be completed before the reaction is properly under way, otherwise the mixing would destroy the bubble structure. Materials that can be used are carbonates with acids, carbides with water, and powdered metals with acids or alkalies. The first is used most, the second is accompanied by the unpleasant smell of acetylene and some danger from its combustion, the third deposits alkalies in the body, lowering the refractoriness. A fourth method involving no permanent additions is the decomposition of hydrogen peroxide by heat or catalytically in the presence of small quantities of manganese salts. Ground dolomite and sulphuric acid are the commonest reactants.

The stabilisation of the mixture at the stage of maximum bubble development is best achieved with plaster of Paris, using at least 6% of this. Where the addition of plaster of Paris has adverse effects on the refractory body some foaming agents, soaps, saponin, or organic preparations may be used.

The pore size is determined more by the grain size and structure of the body than by the grain size of the dolomite or other chemical.

CERAMIC BODIES

The body raw materials, the dolomite and the plaster of Paris are mixed together dry, and then blunged with the dilute acid to make a casting slip in which the effervescence occurs. The fullest advantage is thereby made of the binding properties of any plastic material which can therefore be reduced to a small quantity (2–5%). The method is then applicable to making porous pieces of silica, high-alumina and sillimanite refractory bodies.

METHOD 5 (b)

The body, which may have a minimum of plastic content, is made into a slip and then mixed with a foam made by whipping up a solution of a surface-active agent, *e.g.* saponin, resin, soaps, sodium resinate. Stabilising agents, such as glue, gum arabic or plaster of Paris are also required. Bodies that can be made porous are highly grogged fireclay, quartzite, chromite, magnesite, corundum.

METHOD 6

Fire bloating will occur in some natural clays under any conditions but it occurs more readily if they are fired very rapidly. The method is usually used to make porous aggregate for light-weight concrete, but expanded clay bricks can be made directly by restraining the clay in the kiln.

Fire bloating can be induced in clays by small additions and such synthetic fire-bloating mixtures can be better controlled to give absorbing or impermeable products at will. Thus soda ash and magnesite additions give open porous structures while soda ash with dolomite can give products with closed cells and very small absorption.

METHOD 7

'Sintering' has been applied to the method of making light-weight aggregate by mixing clays or shales, including colliery shale, with coal dust or other combustible, and running it through a swift firing furnace where air is drawn through it to foster combustion. The aggregate is then bonded with raw clay and made into bricks or hollow bricks. Bell and McGinnis (**B29**) noted that the conditions that must be controlled are: nature and quantity of fuel; tempering water; size of pellets made from mixture; rate of travel through sintering furnace; air flow. The first sinter-hearth machine to produce light-weight aggregate in Great Britain was described by Catchpole (**C 20**) in 1957.

Insulating bricks are made by one or more of these methods from the raw materials normally used in brick manufacture and in the production of the following refractories. The most normal method of production is indicated.

Brick clays, shales, etc.: methods (1), (2), (3), (4), (5), (6), (7).
Vermicullite: method (1).
Fireclays: methods (3), (4), (5*a*) and (*b*).

Kaolinite bodies: methods (3), (4), (5a) and (5b).
Silica: methods (3), (5a) and (5b).
High-alumina: methods (5a) and (5b).
Sillimanite: methods (3), (5a) and (2).
Dolomite: method (2).
Chromite: method (5b).
Chrome–magnesite: methods (5b), (2).
Magnesite serpentine: method (2).
Zircon and zirconia: method (3) (only experimental).

4. STONEWARE

Stoneware is defined as a vitreous opaque body of optional colour. It is therefore in a central position amongst ceramic bodies, bordering on every main type. Porous bodies when 'improved' are made more vitreous and eventually a stoneware body results; fine vitrified bodies when required for technical purposes can relax specifications regarding colour and translucency and again a stoneware results (*see* Table 87).

TABLE 87

Relationship of Stoneware to Other Bodies

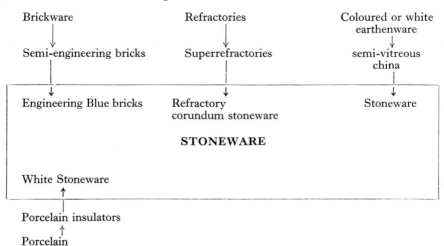

The range of possible compositions is very large, but the properties of the finished body are also considerably dependent on factors other than the composition. Preparation and size grading, and firing being the main ones.

Composition

The stoneware composition is distinguished from other bodies by the fact that it uses up to 70% of secondary clays, *i.e.* ball clays, stoneware clays, etc., which exhibit high plasticity and high green strength. These clays also vitrify well without excessive temperatures, because of their natural flux content, potash, soda and iron being good fluxes whereas lime and magnesia should usually be kept below 2%. By comparison hard porcelain, which has to be pure white and translucent contains only kaolin, which has much lower plasticity and green strength, requires much higher firing, contracts more and deforms more easily during the firing. The individual nature of these secondary clays has led to the regional character of traditional stoneware bodies. Originally these bodies were made directly from natural clays. However, many stoneware clays have large drying and firing shrinkages and it is better to introduce non-plastic material, *e.g.* grog or sand.

The chemical composition of general stoneware bodies lies in the range given by the following molecular formula:

$$RO.\ 0{\cdot}33{-}7{\cdot}0Al_2O_3.\ 4{\cdot}0{-}44SiO_2$$

where RO varies between $0{\cdot}7(CaO+FeO)+0{\cdot}3K_2O$ and $0{\cdot}3(CaO+FeO)+0{\cdot}7K_2O$.

TABLE 88

Stoneware Bodies' Chemical Composition (S90)

Description	SiO_2	Al_2O_3	Fe_2O_3	CaO	MgO	K_2O	Na_2O
Altröm	65·62	27·94	1·60	1·25	1·33	0·39	1·42
Vauxhall	74·00	22·04	2·00	0·60	0·17	1·06	
Helsingborg	74·60	19·00	4·25	0·62	trace	1·30	
Voisonlieu	74·30	19·50	3·90	0·50	0·80	0·50	
Baltimore	67·40	29·00	2·00	0·60	trace	0·60	
Wedgwood	66·49	26·00	6·12	1·04	0·15	0·20	
China	62·00	22·00	14·00	0·50	trace	1·00	
China	62·04	20·30	15·58	1·08	trace	trace	
Japan, grey	71·29	21·07	1·25	2·82	1·98	1·03	0·44
Japan, brown	73·68	19·20	4·37	0·70	0·32	1·41	0·32
Bitterfeld	71·24	25·25	2·11	0·11	0·21	0·64	
Krauschwitz	53·77	41·34	3·34	0·03	0·01	1·40	0·10
Muskau	68·05	29·22	1·31	0·13	0·08	0·91	0·24
Rhine	62·6	34·2	1·7	0·3	0·1	0·9	0·4

By substituting other oxides in one or more of the valency groups special physical properties are developed.

RO group
  MgO: (*a*) increased electrical resistance;
  (*b*) increased resistance to bases and molten metals;
  (*c*) if the general composition leads to cordierite formation, low thermal expansion, therefore better thermal shock resistance.
  SrO: properties between CaO and BaO stonewares.
  BaO: (*a*) increased resistance to bases and molten metals;

BeO: (a) increased bending strength;
(b) increased electrical resistance;
(c) lower expansion coefficient.

ZnO: (a) increased resistance to molten metal without loss of acid resistance.

MnO.
CoO.
NiO.
PbO.

R_2O_3 group

Al_2O_3: (a) increased mechanical strength;
(b) increased refractoriness.

Cr_2O_3: (a) increased resistance to bases without loss of acid resistance;
(b) increased refractoriness.

B_2O_3.

RO_2 group

ZrO_2: (a) lower electrical conductivity;
(b) increased electrical puncture strength;
(c) exceptional resistance to acids and bases.

SnO_2.
CeO_2.
TiO_2.

Other additions

P_2O_5: (a) lowering of refractoriness;
(b) considerably increased resistance to HF.

SiC
Ferrosilicon } increased thermal conductivity, therefore increased thermal shock resistance.
Silicon

Fused quartz. Lowers expansion coefficient, therefore increases thermal shock resistance.

WO_3.
V_2O_5.
C
(S90, 117).

Investigation of the triaxial systems $RO-Al_2O_3-SiO_2$ with RO as one or more of MgO, CaO, SrO, BaO, ZnO and FeO has led to some interesting stoneware bodies with specialised properties which form links with other ceramic bodies, *e.g.* with MgO to steatite, and to basic stoneware, with BaO to highly alkali-resistant bodies and also to refractory bodies. When the following three phases are present (a) mullite alone or as mixed crystals with RO-silicate or aluminosilicate, (b) RO-orthosilicate or RO-metasilicate, (c) glass, then exceptionally low expansion coefficient values are found (**S90**).

Preparation and Size Grading

The properties of stoneware are considerably influenced by the nature of the sintering, fusion, recrystallisation and glass-forming reactions which occur during firing, which depend in part on the grain size and the size grading of the raw materials. Washing and grinding of the clays and fluxes to give the finest particle sizes makes solid state reactions easier. On the other hand angular-shaped grog particles give better skeletal support than round ones.

Firing

The properties of stoneware bodies are highly dependent on, and can therefore be regulated by, the firing. The atmosphere, whether oxidising, neutral or reducing, is important. Rate of heating, cooling and time of soaking greatly affect the crystal structure.

Prolonged soaking and slow cooling assist in approaching equilibrium, which frequently entails greater crystallisation from the glassy phase. The resultant piece has smaller expansion coefficient and therefore greater thermal shock resistance. Such crystallisation is encouraged by the presence of small quantities of those phases expected to arise from the reaction, *e.g.* mullite, or one of the sillimanite minerals, metasilicates of magnesium and the alkaline earths (enstatite, forsterite, wollastonite), and aluminosilicates (cordierite, triclinic plagioclase, hexagonal nepheline, celsian, etc.). The following materials also act as mineralisers to accelerate crystallisation of the glassy phase: compounds of cerium, zirconium, chromium, manganese, tungsten and vanadium.

Any overheating or too lengthy soaking can, however, lead to excessive fusion, giving a melt from which quite different and possibly unsuitable phases will crystallise on cooling (**S90**).

The bending strength of certain crystalline stoneware bodies has been shown to be improved by quenching from 260° C (500° F), when it is thought that a stress relieving action occurs (**R48**).

5. FINE STONEWARE

Better-defined modern bodies used for chemical stoneware or tableware are composite bodies based on:

Clay	30–70%.
Feldspar	5–25%.
Quartz	30–60%.

The proportion of the body which is prefired and introduced as grog depends partly on the size of the pieces to be made. Larger pieces require more grog to prevent distortion and cracking but the resultant fired piece is slightly more porous, 2·0–2·5% as compared with 0·5%. Higher grog content also lowers resultant tensile strength.

TABLE

Stoneware

No.	Ref.	Description	Burning cone No.	China clay	Stoneware clay	Other clay	Feldspar	Flint
1	S83	General						
2		General						
3	S96	General						
4	S90	General			Bunzlau 78·26		10·99	
5		General			Lämmersbach fat 48·0 Lämmersbach lean 40·0		12·0	
6		General			Lämmersbach 81·8		13·9	
7		General			Goldhausen 55		15·0	
8		General			Halle 60		22	
9		General			Löthain 55		15	
10		General			Löthain 59·2		11·8	
11		Cooking ware	8–9		40		35	
12		Cooking ware	8–9		40		20	
13		Cooking ware	7		30		20	
14		Cooking ware	8		30		20	
15		Cooking ware	8		40		30	
16		Cooking ware	9	36	36		10	
17		Cooking ware	9		32			
18		Cooking ware	9	14	26			10
19		Cooking ware	9	14	26			10
20		Grinding cylinders			33·3 calcined (900° C) 4·1		12·5 200/–*	
21		Grinding cylinders	8–9		32	Bentonite 2	30	
22	A23	Chemical		Amberg 6·82 Geisenheim 2·27	Palatinate 35·20 Waldhilsbach 6·82 Oberpleis 14·79		11·35	
23	A11	Chemical		9·09	56·81		11·35	
24		Chemical			60–70	Sand clay 3–5	8–10	
25		White chemical		Amberg	Palatinate 27 Löthain 6 Kaschka 3 Lieskau 5		Wunsiedel 10 Stroebel 8	

6. WHITE STONEWARE

Careful selection of clays, feldspars and quartzes together with high-grade preparation and optimum firing schedules can give highly improved stoneware bodies which are white or are truly completely vitrified with zero water absorption.

7. IMPROVING THE THERMAL SHOCK RESISTANCE OF STONEWARE

One of the main adverse properties of stoneware, which replaces metals on grounds of corrosion resistance in so many applications, is its relative sensitivity to thermal shock. This is aggravated by the fact that it can be made into large, and perforce thick, pieces where thermal stresses can easily be set up. All stoneware must therefore be heated slowly, and cooled even more slowly, the latter process setting up tensile stresses even more easily.

89

Bodies

No.	Quartz	Calcined alumina	Calcined bauxite	Miscellaneous	Porcelain grog	Grog	Water absorption	Rational analysis		
								Clay subs.	Quartz	Feldspar
1								50	38	12
2								30–40	30–60	5–25
3	Hohenbaka							30–70	30–60	5–25
4	sand 10·75									
5										
6	Quartz sand 3·3			Magnesite 1·0						
7	Quartz 30									
8	Quartz 18									
9	Quartz 30									
10	Quartz sand 29·5			Magnesite 0·5						
11		20		Talc 5						
12			40							
13		20		Talc 5		25				
14		20		Talc 5		5				
15		20		Talc 5						
16					0–2 mm 15		3·6			
17				{ Rutile 60 Magnesite 8						
18				BaSO₄ 48			0			
19				{ BaSO₄ 48			0			
20				Plaster 2						
21					72/100* 18·3 100/200 13·3 200/– 18·3					
					{ 20/30* 9 40/80 6 80/– 12 160/– 9					
22						22·75				
23						22·75				
24						0–3 mm 17–21				
25					54* 7 54/900 11 }	Firebrick 5				

* Sieve grading

Ordinary stoneware will not withstand rapid temperature changes between room temperature and 100° C (212° F) and repeated slow changes between these temperatures also weaken it.

The usual methods of increasing thermal shock resistance in any body are:

(1) Increasing porosity.
(2) Increasing thermal conductivity.
(3) Decreasing thermal expansion.

Increased porosity allows some of the thermal expansion to be taken up within the piece by reducing the size of the pores. This method must, however, be ruled out for any type of chemical stoneware and also for frost-resisting structural stoneware, both of whose major properties depend on having as small a water absorption as possible.

Increasing thermal conductivity is the best method of improving thermal shock resistance of chemical stoneware. This is usually done by incorporating grog of high thermal conductivity such as corundum, silicon

TABLE

Stoneware Floor

No.	Ref.	Description	Burning cone No.	China clay	Ball clay	Stoneware clay	Flint and quartz
1	A89	American	8–9	19	14		5
2	A67	American	7	18	15		25
3	A67	American	5	31	20		20
4	A67	American	10	40	7		25
5		American	9	E.P.K. Florida 15 Pecatess Socar 10	Ky Tennessee 7		
6		American		Kamec 9 English 10 Edgar plastic 14			
7		American	12	Edgar plastic 26	Ceramic No. Cowl 12		
8	A67	Once fired	4	18	16		15
9		Once fired	4	20	24		10
10	B17	German	8			46	Quartz
11	A17	German. Red	8			Eppenrod 54	[305
12		German. Buff	6–8			Bechlinghof 22 Albach 22 Kaerlich 22	
13		German. Vitreous	9–11	5		Westerwald 75	
	V5	German. For wet preparation method:					
14		White	8			White 45	
15		Yellow	8			White 18 Yellow 28	
16		Red	8			Red 40	Sand 12
17		Black	8			White 20 Red 10 Manganese clay 26	Sand 9
		For semi-dry or dry preparation:					
18		White	10	10		White 60	Sand 2
19		Yellow	10			White 15 Yellow 64 Red 1	
20		Red	8			Red 60 Yellow 25	
21		Black	8	10		White 20 Red 10 Manganese clay 25	
22		Green		White body + 3·5% chrome ore			

carbide, ferrosilicon, silicon 65–75% or graphite. This should be accompanied by a maximum reduction of porosity in order to eliminate the insulating effect of air spaces. In this way thermal conductivity can be increased from 0·01 c.g.s. for normal stoneware, to 0·03–0·05 c.g.s. units. Alteration of the over-all composition so that the glassy matrix of the stoneware has a higher thermal conductivity is also beneficial.

Decreasing thermal expansion of stoneware materially assists thermal shock resistance. Two constituents are likely trouble makers in the fired body. One is crystalline silica with its inversions and large expansion

CERAMIC BODIES

Tile Bodies

No.	Feldspar	Whiting	Talc	Pyrophyllite	Magnesite	Misc.	Grog	Water absorption
1						Nepheline 62		
2	35		7		—			
3	4		10		15			
4	3		10		15			
5	Derry 55			Pyrax Pyr. 13				
6	Potash 62							
7	60		2					
8	3	2	5	40	1			
9	3	2	5	35	1			
10	20					Flux 4		
11	Birkenfeld 14						Saggar 30	1–1·5
12	Birkenfeld 28						Saggar 6	3
13	Birkenfeld 20							0–0·2
14	Mixed 55							
15	Mixed 54							
16	Mixed 43						Red tile 5	
17	Mixed 18						Saggar 17	
18	Mixed 28							
19	Mixed 20							
20	Mixed 5						Saggar 10	
21	Mixed 15						Saggar 20	
22								

coefficient. Higher alumina content of the body leads to mullite formation. Use of sillimanite minerals instead of free quartz or flint assists considerably. The second constituent that should be reduced in quantity is the glassy matrix and it should be as free from alkaline fluxes as possible (**R48, W8**).

8. ELECTRICAL STONEWARE

The need for large single piece electrical insulators points to the use of stoneware bodies. Unfortunately ordinary stoneware has insufficient puncture strength. This property is, however, dependent not only on the denseness of the piece but also on its chemical composition and on the physical condition of its constituents. Improvement is achieved by introducing titanium dioxide and/or titanates, better values are obtained when ferric and titanic compounds are present together. Small quantities of cerium, zirconium, chromium, manganese, phosphorus, tungsten or vanadium compounds act as mineralisers in these titanate stonewares, bringing about crystallisation from the glassy phase (**S90**).

9. FIRECLAY

Bodies suitable for sanitary ware and large tiles are based on fireclay. They are used where earthenware has insufficient strength both during making and when finished, such as thick-walled, complicated pieces of sanitary ware. The body has up to 50% of the clay prefired, and ground down to grog. This is bound with the same or similar clays prepared by grinding, washing, etc. China clays, quartz, feldspar, etc., may be added similarly to earthenware bodies, but fireclay bodies usually have a very localised character related to the clay pit giving the main ingredient. The body is usually shaped by slipcasting. Only a few published examples of fireclay sanitary ware bodies are available (Table 91).

The fireclay body is normally covered with an engobe of fine-textured white firing body. This is applied on the green piece (Table 92).

Opaque glazes are also applied on top of the engobe when green as these large pieces cannot withstand firing twice.

10. COLOURED EARTHENWARE

Many studio or craft potteries use a coloured earthenware body. Frequently the clay is of local origin and its treatment and use is a local tradition. The principles for compounding such coloured bodies are the same as those for white earthenware except for the precautions in choosing

TABLE
Fireclay Sanitary

No.	Ref.	Description	Burning cone No.	China clay or kaolin	Clay	Quartz	Grog
1	A17	German		Bavarian 17	Westerwald 29	Sand 8	46
2	A17	German		Kriegsheim 31·1	Römer 3·5 Hohewiese 13·2		fine 26·1 med. 23·4 coarse 2·6
3	A8	German sanitary ware	10	} Amberg 20	Westerwald 20	5	50
4		German stove tiles	07				

CERAMIC BODIES 439

only white-burning ingredients. Also the coloured clays tend to vitrify at lower temperatures than the white-burning ones and so need less flux.

A more specialised form of coloured earthenware is the deliberate addition of colourants to a known white body. This is discussed under 'Decoration', Chapter 9.

11. WHITE EARTHENWARE
AND
12. SEMI-VITREOUS CHINA

The white-burning porous body makes a general purpose body in three main fields of application: tableware, wall tiles and sanitary ware. The compositions and raw materials vary a little for these three classes, but the rules for any fine ceramic regarding care in the choice of clean raw materials of known composition and grain size apply.

The traditional general purpose earthenware has three composition types of the average compositions shown:

Calcareous earthenware, low fired: clay 50%—flint 35%—whiting 15%.
Mixed earthenware, medium fired: clay 50%—flint 40%—whiting 7·5% and feldspar 2·5%.
Feldspathic earthenware, higher fired: clay 50%—flint 45%—feldspar or stone 5% (**S83**).

The general limits of composition as defined by various authors are shown in Table 94.

A number of materials are used in each category. In the clay group, china clays or kaolins and a variety of ball clays are used. Ball clays tend to vary from one delivery to another and so it is normal practice to use a number of different ones, or to stock each delivery separately and use a mixture.

Of the siliceous materials, calcined flint is the commonest in Britain,

91
Ware Bodies

Feldspar	Name	Chemical Analysis of Materials							
		SiO_2	Al_2O_3	Fe_2O_3	CaO	MgO	K_2O	Na_2O	Loss
	Bavarian kaolin	50·00	35·50	0·75	0·77	0·16	0·49	—	12·50
	Westerwald clays	65·86	23·64	1·44	0·19	0·14	—	0·99	7·60
	Grog (Bavarian porcelain saggars)	67·64	27·64	1·44	1·20	—	—	—	0·13
5									

TABLE
Engobes for Fireclay

No.	Ref.	Description	Burning cone No.	China clay	Ball clay	Quartz
1	A17	for fireclay body No. 1			Saxony 26	18
2	A17	for fireclay body No. 2		26·5 Calcined 22·0		15·0
3		Vitreous	7	48	5	
4		Vitrified		35	5	
5		Porous		59·2		
6		—		16·7		10·3
7		—		25·8	2·5	
8		Dense				

elsewhere quartz and sand are used. These benefit by being wet ground before use, when a water film forms around each particle, but overfine grinding of the flint (below 200-mesh) necessitates a larger quantity of flux. Pitchers, *i.e.* ground, burnt earthenware body, make an additional non-plastic.

Of the calcareous fluxes whiting, chalk and limestone are the commonest. Dolomite finds application here, and for low-fired ware, cone 03–cone 1 (1040° C, 1904° F–1100° C, 2012° F) reduces the tendency to craze and show moisture movement (**S200**).

The traditional feldspathic fluxes are the feldspars, Cornish stone and Carolina stone, to which can be added nepheline syenite which has a long sintering range, beginning to soften at cone 2 (1120° C, 2048° F) but with a P.C.E. of cone 8 (1250° C, 2282° F).

Other fluxes that have found application especially in U.S.A. and frequently more readily for wall tiles are:

> Lepidolite (**K49**).
> Spodumene (lithium feldspar).
> Talc (**A52**).
> Pyrophyllite (**S155**).
> Glass cullet.

Their application is shown in Table 94.

Collins (**C47**) investigated the possibility of replacing part of the feldspar of earthenware and vitreous bodies by fusible glasses. Of the twenty-five mixtures selected from the literature six proved promising. Most of them had an objectionably high water solubility and so they were fused with some high potash feldspar. The compositions and molecular formulae of these frits is given in Table 93.

Sanitary Ware Bodies

No.	Feldspar	Cornish stone	Flint	Miscellaneous	Molecular formula
1	29			Crushed porcelain 27, gelatine or glue	
2	22·0			Zinc oxide 5·7, Whiting 8·5, tylose	
3		30	9	plaster 3, talc 5, zircosil 10	
4	20		31	tin oxide 4, Whiting 5	
5		35·5		plaster 5·3	
6	54·2			magnesite 10·9, zinc oxide 7·9	
7		54·8	12·0	plaster 1·0, tin oxide 3·9	
8					K_2O 0·3, MgO 0·5, ZnO 0·2 } Al_2O_3 0·5 . SiO_2 4·0

TABLE 93

Glassy Frits for use as Auxiliary Fluxes in Ceramic Bodies (C47)

No.	Composition	%	Fusion temperature (°C)	% Potash feldspar added	Final formula
1	$CaSiO_3$ / Na_2SiO_3	20·0 / 80·0	932	25	0·7573 Na_2O, 0·0483 K_2O, 0·1943 CaO } 0·069 Al_2O_3. 1·342 SiO_2
2	$CaSiO_3$ / Na_2SiO_3	69·0 / 31·0	1130	0	0·3021 Na_2O, 0·6978 CaO } 1·000 SiO_2
3	$BaSiO_3$ / Na_2SiO_3	40·0 / 60·0	908	25	0·6644 Na_2O, 0·0560 K_2O, 0·0040 CaO, 0·2756 BaO } 0·805 Al_2O_3. 1·3954 SiO_2
4	$MgSiO_3$ / Na_2SiO_3	10·0 / 90·0	921	35	0·7915 Na_2O, 0·0624 K_2O, 0·1416 MgO, 0·0047 CaO } 0·898 Al_2O_3. 1·4404 SiO_2
5	Na_2O / SiO_2	18·4 / 81·6	860	10	0·9571 Na_2O, 0·0401 K_2O, 0·0028 CaO } 0·0578 Al_2O_3. 4·6516 SiO_2
6	Na_2O / B_2O_3 / SiO_2	24·21 / 35·26 / 40·53	570	25	0·9227 Na_2O, 0·0721 K_2O, 0·0051 CaO } 1·1488 B_2O_3, 0·1037 Al_2O_3 } 2·1602 SiO_2

Approx. formula of Canadian feldspar used

0·3097 Na_2O, 0·6951 K_2O, 0·0488 CaO } 1·000 Al_2O_3. 5·972 SiO_2. 0·0855 H_2O

Equivalent weight 549·13

TABLE

Earthenware

No.	Ref.	Body type	Burning cone No.	China clay or kaolin	Ball clay	Flint	Feldspar	Stone	Dolomite	Nepheline
1		English general		20–30	30–20	30–40		15–20		
2		U.S.A. dinnerware, semi-vitreous china		19	25	25	21			
3		U.S.A. general		25	25	38	12			
4				15	22	—	5			
5				18	18	10	—			
6	N11	U.S.A. dinnerware, earthenware		31	20	20	4			
7	N11	U.S.A. semi-vitreous porcelain	one fire	32	20	36	12			
8		Continental general		20–30	30–20	30–35	12–20			
9		Continental Feldspathic	7–10							
10	F36	Mixed	2–6							
11		Calcareous	07–6							
12			1–4							
13	K24		5–10 (7–9)							
14	A3		6							
15	P68	Calcareous	4							
16		Feldspathic								
17		General	08		30					
18		General	08		80	10				
19		General	08		70	20				
20		General	08		60	30				
21		General	08		50	40				
22		General	08		40	50				
23		General	08		30	60				
24		General	08		70	10				
25		General	08		60	20				
26		General	08		50	30				
27		General	08		40	40				
28		General	08		30	50				
29		General	08–1		30					
30		General	08–1		50	40				
31		General	08–1		30	60				
32	S17	Talc body	05–6		fat 40–30 / lean 30–35	15–10				
33		German general	8	50		30	11			
34		German general	3a	50		42	3			
35		German general	03a	50		25	—			
36		General	02	mixed 13	mixed 28	42		mixed 10		
37		General	3–4	20	12	39	9			
38		General	3–4	11	16	39	9			
39	H2	General	2–3	Florida 16	Ky No. 4 18	14	Conn. 5			
40	H2		2–3	26·1	Ky 18·9	15	5			
41	H2		2–3		Ky 20	12	8			
42	H2		2–3	Florida 15	Bell 10 / Cooley 12		5			
43	H2		6–7	Florida 12·5	Ky 11·5 / Cooley 11·0	16	5·5			
44		General		24·3	21·5	39·4		14·8		
45		General		15·8	25·7	38·5		20·0		
46		General	2–6	mixed 20	mixed 32	34·4	9·2	—		
47		General	2–6	mixed 14	mixed 28	41	—	9		
48		General	2–6	13	28	42		10		
49		General	5½–6	10	30	44	12			
50		General	5½–6	13	30	44	9			
51		General	5½–6	12	30	44	6			4
52		General	5½–6	20	30	40	5		5	
53		General	5½–6	18·0	30	40	6		2	4
54		General	5½–6	13·0	30	45	10		2	
55		General	6	mixed 30	mixed 22	33		15		
56		General	4	20	12	41	7			
57				12	16	39	9			
58				12	16	41	7			
59				4	32	40	4			
60				2	32	46	6			
61				—	32	41	8			

Bodies

No.	Whiting	Talc	Pyro-phyllite	Magnesite	Pitcher	CaCO₃	Clay subs.	Quartz	Feldspar	Silica content	Water absorption
						\multicolumn{4}{c}{Rational analysis}		
1	0–5										4–10
2											
3	—	—	—								
4	—	40	18								
5	2	12	40								10–15
6		10	15								
7											0·3–4
8	0–5										
9						—	40–55	55–42	5–3		
10						5–7·5	45–50	48–42	1–3		
11						20–5	40–55	~40	—		
12						—	45–50	40–45	5–10		
13						2–5	45–50	35–40	8–15		
14						5	50	42	3		
15						15	50	35			
16						—	50	45	5		
17		70									19·9
18		10									18·6
19		10									18·0
20		10									14·3
21		10									17·4
22		10									19·7
23		10									20·6
24		20									23·7
25		20									23·4
26		20									23·2
27		20									22·2
28		20									20·7
29		70									
30		10									
31		10									
32		35–15									
33				1							
34	5			—							
35	20			5							
36	6			ZnO 1·0							17·4
37		20									8·0
38		27									10·1
39	3	40	· 9								
40		40									
41		40	20								
42		40	23								
43		40	9								
44											
45											34·0–7·1
46	4·4										15·0
47	8										
48	7										7·5–20
49	4										13–18
50	4										11·5–15·0
51	2	2									18·6–20·0
52											13·6–15·0
53											13·7–15·4
54										68–70	
55											8·6
56		20									6·0
57		24									8·6
58		24									3·4
59		20									1·3
60		20									1·1
61		20									

TABLE

No.	Ref.	Body type	Burning cone No.	China clay or kaolin	Ball clay	Flint	Feldspar	Stone	Dolomite	Nepheline
62	D50	General (Amer.)	5	56		30		Lepidolite 7		6
63		English general	5	30	22	33		15		
64	M77	General	5	25	20	37		18		
65	K59	General	5	a 12·0 b 9·0 c 6·0	a 7·0 b 5·0 c 7·0 d 7·0	33·5	13·5			
66		Dinnerware	5	10	30	35	5			
67			5	9	30	34	4			
68			4	20	20	40	5			
69		Steatite	6	20	12	41				
70			6	12	16	41				
71			6	12	16	39				
72		General	6	14	30	41		11	4	
73		Ivory	6	20	20	33				
74		Ivory	6	20	20	33				
75		Ivory	6	10	30	32				
76		General	6	mixed 14	mixed 30	Belgian 44	8		4	
77		General	6	50		42	3			
78		General	10	40		55	5			
79			6	48		50	Pegmatite 8		4	
80		General	6	20·5	31·9	35·6		11·9		
81		Lime	6	13	30	40		13		
82		New body	6	13	30	44	9			
83	H2	Dinner ware body	7–8	Lundy Kaolin 12 Florida 8	Kent No. 4 20	37	Conn. 6			
84	H2	Dinnerware	7–8	Lundy 12 Florida 5	20	35	6			
85	H2	Dinnerware	7–8	12	20	37	6			
86		General	9½	32	20	35	13			
87		General	9½	32	20	35	13			
88		General	9½	31·3	21·5	33·5	13·7			
89		American	9½	China clay 5·84 Georgia K. 16·254 Florida K. 8·127 N. Carolina K. 8·127	Old Mine No. 4 13·40	34·52	No. 10 13·69			
90		American	9½	Georgia K. 16·0 Florida K. 8·0 N. Carolina K. 8·0	Old Mine No. 4 20	35·0	No. 42 13·0			
91		American	9½	China clay 10·75 Georgia K. 23·50 Florida K. 7·80	Old Mine No. 4 10·75	33·50	No. 42 13·70			
92		American	9½	32·508	19·240	34·520	13·690			
93		American	9	35	20					
94		Tableware		16·5	24·2	30·6		25		
95		Ivory tableware		11·25	31·0	32·0		18·25		
96		Lime–feldspar		13·00	28·0	45·0	7·0			
97		Low firing	3	12	18	Quartz 50		—		
98		contraction	3	13	28	42		10		
99		bodies	5	42	—	41		9		
100			5	—	30	30		13		
101	A3	German A		Unwashed K. 19 Pilsen OA K. 8	Meissen K. 1 7 Saxon OR II 10 Römer 6 Fundgrube 6	Dörentrup sand 30	Amberg Pegmatite 12			

94—(contd.)

No.	Whiting	Talc	Pyro-phyllite	Magnesite	Pitcher					Silica content	Water absorption
62		1									
63										68	8–12
64											
65											
66		20									
67	3	20									
68		20									
69		20									
70		24									
71		24									
72											
73	7	20									14·3
74	7	20									11·9
75	7	20									9·5
76											Average 12–13
77	5									83·8	
78										80	
79											
80											
81	4										12·7
82	4										12·7
83	2	Loamis 15									
84	2	20									
85	2	15	8								
86											
87											
88											

Rational analysis

	Clay Subs.	Quartz	Feldspar	
89	45·081	36·346	18·531	
90	44·040	37·784	18·176	
91	46·060	36·576	17·364	
92	45·081	36·346	18·531	8·5

No.	Whiting	Talc	Pyro-phyllite		Pitcher						Water absorption
93		20	25								18
94					3·4						
95					4·0						

	% drying shrink.		% firing shrink.	
96	7·0	5·0	0·8	19·2
97	20	5·4	1·1	16·3
98	7	—	1·0	24·8
99	8	2·7	1·6	11·8
100	—	23		

Rational analysis

					CaCO₃	Clay subs.	Quartz	Feldspar	
101	chalk 2				3	2	42·8	46·4	8·8

TABLE

No.	Ref.	Body type	Burning cone No.	China clay or kaolin	Ball clay	Flint	Feldspar	Stone	Dolomite	Nepheline
102	A3	German B		20	Lantersheim 10 Stoss 20 Klingenberg 8	Sand 20	Pegmatite 13			
103		White Bone-ash	4	10	30	40	Bone ash 20			
104		Earthenware	4	10	30	30	Bone ash 30			
105		Wall tile	08		90					
106		Wall tile	08		80					
107		Wall tile	08		70					
108		Wall tile	08		60					
109		Wall tile	08		50					
110		Wall tile	08		40					
111	A67	Once-fired tiles	4	18	16	15	3			
112		Once-fired tiles	4	20	24	10	3			
113		Wall tile	2–5	18	18	10				
114		Wall tile	1–5	18	18	10				
115				35	20					
116			Biscuit 6 Glost 02	26·32		15·79	7·89			
117			2	mixed 13	mixed 28	42		12		
118				mixed 15	mixed 25	44		12		
119			6	50		43	3			
120				18·89	19·80	43·74		17·57		
121				43		50	7			
122		Tile	2–5	18	18	20				
123		Once-fired wall tile	5	18	16	15	3			
124		Once-fired wall tile	5	20	24	10	3			
125		Tile	5½–6	10	30	30	12			
126		Tile	8	30	30	30	14			
127		Tile		35	20					
128		Tile	6–8	21	32	20	6			
129		Tile	6–8	21	32	20	6			
130	S155	Amer. wall tile	7	13 15	27	33	12			
131	S155	Amer. wall tile	7	13 15	27	10	3			
132	S155	Amer. wall tile	7	15 15	27	33	10			
133	S155	Amer. wall tile	7	15 7	27	10	4			
134	S155	Amer. wall tile	7	17 17	20	37	9			
135	S155	Amer. wall tile	7	26	30	32	15			
136	S155	Amer. wall tile	7	18 18		15	3	12		
137	S155	Amer. wall tile	7	18 18		15	3			
138	—	Amer. wall tile	7	Florida 6 N. Carolina 6	Ky. No. 4 15	26	Conn. 13			
139	R69		8	8 20	25	35	12			
140	S155	Amer. wall tile	9	15	18	15	3			
141	S155	Amer. wall tile	9	15	18	15	3			
142	S155	Amer. wall tile	9	15	18	15	3			
143	S155	Amer. wall tile	9	15	18	15	3			
144	S155	Amer. wall tile	9	15	18	15	3			
145	S155	Amer. wall tile	9	15	18	15	3			
146	S155	Amer. wall tile	9	15	18	15	3			
147	S155	Amer. wall tile	9	15	18	15	3			
148	S155	Amer. wall tile	9	15	18	15	3			
149	H1	Amer. wall tile	8–10		25	30	5			
150	H2	Amer. wall tile	10	Florida 15·5 N. Carolina 15·0	Ky. No. 4 24·0	33	Conn. 6·0			
151	S155	Amer. wall tile	12	26	30	32		11		
152	S155	Amer. wall tile	12	17 16	21	33	2			
153	S155	Amer. wall tile	12	20	30	10	2			
154	S155	Amer. wall tile	12	20	30	10	2			
155		Wall tile		25	25	40		10		
156				23	26	40		11		
157				14·3	21					

94—(cont.)

No.	Whiting	Talc	Pyro-phyllite	Magnesite	Pitcher	Rational analysis				Silica content	Water absorption
						CaCO$_3$	Clay subs.	Quartz	Feldspar		
102	chalk 4				5	1·3	55·1	38·3	5·4		
103											19
104											16
105		10									24·4
106		20									26·1
107		30									20·6
108		40									21·9
109		50									20·0
110		60									
111	2	5	40	1							
112	2	5	35	1							
113	2	12	40								14·8–15·6
114		14	40								
115		20	25								
116	7·89	31·58	10·53								17·2
117	5										
118	4										
119	5										
120											
121											
122	2	12	40								
123	2	5	40	1							
124	2	5	35	1							
125			16								
126	2		16								
127		20	25								
128	—	6	15								
129	2	4	15								
130											13·2
131			32								17·4
132											15·4
133			37								16·6
134											15·9
135											16·5
136	1	4	37	4							15·5
137	3	4	38	1							15·7
138		Loan's 2·5	6	9							6
139											
140	2		45	2							14·2
141	2	5	40	2							14·8
142	2	10	35	2							14·8
143	2	15	30	2							14·5
144	2	20	25	2							15·0
145	2	25	20	2							13·3
146	2	30	15	2							7·5
147	2	95	10	2							—
148	2	40	5	2							—
149		40									
150		6·5									
151											12·4
152											10·5
153			38								15·1
154		4	34								15·7
155										73	
156										73	
157	47	38·1									

TABLE

No.	Ref.	Body type	Burning cone No.	China clay or kaolin (%)	Ball clay	Flint	Feldspar	Stone	Dolomite	Nepheline
158		Wall tile		H.N. 9·45 No. 6 9·44	A 9·90 W.M. 4·95 K 4·95	43·74		17·57		
159	S26	Wall tile		25	29	32	6			
160	A12	Industrial wall tiles (German) (a)		40	4 sorts 55					
161		(b)			7 sorts 50·6					
162	A12	German white wall tiles (a)		Grosslohner 10·0 Kemmlitz 12·5 Amberg 7·0 Börtewitz 9·0 Hohburg 10·0	Westerwald 8·0 Priem 9·5	Amberg Spat Sand 7·0 Sand 25·0			2	
163		(b)		Kemmlitz 13·5 English 12·5 Brechmitz 9·0 Amberg 5·0	English 9·0 Priem 9·0 Goldhausen 9·0	Sand 26·0			2	
164		(c)		Kemmlitz 12·0 Raw F.K.I. 9·0 Amberg 6·0	Priem 8·0 Goldhausen 8·0 Wildstein 6·0 Marl Sennewitz 15·0	Quartz Frechem 27·0	Amberg 3·0			
165		(d)		Saxony 12·7 Doberan 6·7 Pilsen 10·0 Geisenheim 16·8	Lieskau 5·1 Wildstein B 4·2 Muskau B 5·9 Muskau C 5·9 Wildstein C 2·5 Westerwald 1·7 Michelob 3·4 Goldiltz C 5·9 Jessen 1·7	Sand, Hanover 3·3				
166	A12	German cream wall tiles (a)		Kemmlitz 11·8	6 sorts 63·7				1·0	
167		(b)		3 sorts 33·6	2 sorts 28·3 Marl 16·1	Quartz 2·2	Limespar 2·9			
168		(c)		3 sorts 10·6	12 sorts 49·2 Marl 3·9	Sand 3·9				

94—(cont.)

No.	Whiting	Talc	Silica content	Pyro-phyllite	Magnesite	Pitcher	Chemical composition				Water absorption
158							SiO_2 TiO_2 Al_2O_3 Fe_2O_3	78.7% 0.3 17.2 0.52	MgO CaO K_2O Na_2O	0.2 0.9 1.6 0.6	
159		4									
160											
161						5.0 { Grog 22.5 Pitcher 26.9					
162											
163						Biscuit 5.0					
164											
165	chalk 1.6					12.6					
166						{ Biscuit 7.8 Grog 15.7					
167						{ Saggar 6.6 Biscuit cream 5.5 Biscuit white 4.8					
168	chalk 1.5					{ Grog 11.2 Pitcher 19.5					

450   INDUSTRIAL CERAMICS

TABL[E]

No.	Ref.	Body type	Burning cone No.	China clay or kaolin	Ball clay	Flint	Feldspar	Stone	Dolomite	Nephelin
169	R66	French research bodies with talc	05–1	20	30	Sand 10			—	
170				20	30	Sand 10			—	
171				20	30	Sand 10			—	
172				20	30	Sand 10			10	
173				20	20	Sand 20			—	
174				20	20	Sand 20			—	
175				20	20	Sand 20			—	
176				20	20	Sand 20			10	
177				10	20	Sand 30			—	
178				10	20	Sand 30			—	
179				10	20	Sand 30			—	
180				10	20	Sand 30			10	
181		Sanitary ware	6	14	30	44	8		4	
182		Sanitary ware	5–6	25	3 sorts 30	36	4			2
183		Sanitary ware	5–6	23	3 sorts 27	40	4			2
184		Sanitary	7–9	mixed 23	mixed 27	quartz 46	3			
185		Sanitary	5a	45		40	10 {Feldspar 6·6	6·3		
186	A12	Sanitary (German)		2 sorts 24·5	4 sorts 26·2	Quartz sand 26·4	{Peg-matite 5·8			
187		Sanitary (German)		2 sorts 29·0	2 sorts 33·0	17·0	Peg-matite 21·0			

Use of 2% of these auxiliary fluxes in typical earthenware bodies made it possible to reduce the feldspar content to 8% (increasing the flint) and fire at cone 8 (1250° C, 2282° F) to any desired degree of vitrification between 0·50 and 7·0%.

Watts (**W29**) has investigated a number of auxiliary fluxes and found their deformation-eutectics and the reactions of these deformation-eutectic mixtures with feldspars (p. 220).

13. VITREOUS CHINA

This body has been developed largely for utilitarian reasons to give a strong and relatively impermeable body that can be easily manufactured into sanitary ware and crockery. It has frequently been found expedient to sacrifice complete vitrification to a zero water absorption for the sake of greater mechanical strength and less warping during firing. The reason for maximum strength being reached shortly before complete densification has been treated at length by the senior writer (**S110**). Briefly, the mechanical strength of a ceramic body depends largely on the interlocked crystalline components. If complete densification involves solution of the crystalline phases into the glassy phases, strength is lost. Where densification is brought about by solid phase reaction strength can be increased. Many commercial vitreous china bodies therefore have a water absorption of 0–1·0%. The U.S.A. Bureau of Standards Specification (**A95**) states that a fractured

CERAMIC BODIES

4—(cont.)

No.	Whiting	Talc	Water Absorption Fired at		Chemical composition of raw materials							Loss on ignition
			1000° C	1080° C		SiO₂	Al₂O₃	TiO₂	CaO	MgO	K₂O	
169	chalk —	40	16·9	16·6	Ball clay Saint Loup.	65·9	23·6	1·8	0·4	0·2	0·4	7·9
170	chalk 5	35	19·0	18·3	Kaolin d'Arvor.	46·5	38·8	—	—	—	0·8	13·9
171	chalk 10	30	19·6	19·1								
172	—	30	20·2	19·2	Crushed Nemours sand	99·7	—	—	—	—	—	0·3
173	—	40	17·4	17·2	Talc Luzenac	54·1	8·9	—	—	31·5	—	5·5
174	chalk 5	35	18·9	17·7	Chalk	—	—	—	56·0	—	—	44·0
175	chalk 10	30	22·1	21·6	Dolomite	—	—	—	30·0	22·0	—	48·0
176	—	30	19·7	20·1								
177	—	40	17·5	18·0								
178	chalk 5	35	17·7	20·5								
179	chalk 10	30	20·2	20·9								
180	—	30	19·2	20·9								
181												
182	1	2	Av. 12–13									
183	1	3	Av. 12–13									
184	1											
185	5		Av. 12–13									
186	Chalk 4·0											
187												

piece of material taken from any part of a vitreous china sanitary fitting, after being immersed in red aniline ink for one hour and then broken, shall not show absorption to a depth of more than ⅛ in. (3·175 mm) below the surface of the fracture at any point.

Vitreous china bodies are based on mixtures of china clays, ball clays, quartz or flint and feldspar. A number of examples use nepheline syenite instead of feldspar, others include lithium minerals such as lepidolite and spodumene. Whiting is fairly frequently used and occasionally talc. Examples are given in Table 95. The firing temperatures vary from cone 5 to 12, 8 to 10 being the most usual.

14. SOFT PORCELAIN

Soft porcelain is the term covering a number of different bodies that are all vitreous, white and translucent. They vitrify below cone 12 and most are vitrified in the biscuit firing and subsequently glost fired at a lower temperature. They are subdivided according to the raw materials.

Seger Porcelain, American Household China, British Electrical Porcelain

These are all bodies made up of china clay, some ball clay, flint or quartz and feldspar or Cornish stone or nepheline syenite (an excellent flux is

TABLE
Vitreous China

No.	Ref.	Body type	Burning cone No.	China clay or kaolin	Ball clay	Flint	Quartz	Feldspar	Pegmatite
1		General							
2	K49	General	02	27	18	5			
3	A89	General	6–7	24	16	6			
4	G18			ChCl 29 / Kaolin 8	7·5	36		18	
5	G18	General	5	30	20		30	20	
6	G18	General	5	10	30		10	30	20
7	G18	General	5	20	35	10		35	
8		General	5	20	35	13		22	10
9	A86	General	13	45		33			
10			7	45		27			
11	—	Quartz free	8	28	24			10	
12	—	Quartz free	8	28	24			5	
13		All American	9	Florida 10	Tennessee No. 9 14 / Kentucky No. 4 6	27		36	
14			11	English 16 / Florida 4	Tennessee No. 5 12 / Old Mine No. 4 5	27		36	
15	N11	Tableware American hotel china		28	15	39		15	
16a	N11	Hotel china	10–12	33·0	9·5	35·0		21·0	
b	N11	Hotel china	10–12	37·0	7·5	36·0		18·0	
c	N11	Hotel china	10–12	28·3	15·0	38·9		14·8	
17	N36	Hotel china	10/11	34·8	7	35		22	
18	A89	Dinnerware	3–4	11	29	8		8	
19		Wall tile	5	20	25	30		25	
20		Wall tile	5	20	20			25	30
21		Wall tile	5	40	5		10	15	30
22	N36	Wall tile	10/11	27	29	11		33	
23		Sanitary	7	English 25	Jernigan 15 / Champion & Challenger 10	15			
24			6	mixed 15	mixed 20	41		9	
25	C61	Sanitary	6	58·5		9·0		32·5	
26	C61	Sanitary	6–7	58·5		9·0		22·75	
27	C61	Sanitary	5	46·3		19·7			
28	C61	Sanitary	9	58·3		13·5			
29	C61	Sanitary	5–6	46·3		13·7			
30	—	Sanitary	8	28	24		33	15	
31		Sanitary	8	23	29	14		24	
32	A89	Sanitary ware	11	26	26	26			
33	—	Sanitary	5	40	5			25	30
34	—	Sanitary	5	20	32			20	15
35	N36	Sanitary	11/12	28	20	20		32	
36	S110	Sanitary I	9 / 9 / 9 / 9						
37		II	Biscuit 01 / Glost 8–9	11–14	30–33 / Fire clay 0–12		18–22	20–22	
38		III	Biscuit 3–6 / Glost 02	Halle 24	Meissen 28		Hohenbock 35		Weihersham 13
39		IV	Biscuit 09 / Glost 9 / or Biscuit 9 / Glost 6 / or once-fired 9	Salzmünde 28	Löthain–Meissen 18		21	Ströbel 33	
40		V	Biscuit 8–10 / Glost 1	25 / 55 / Kemmlitz 56	30 / 6 / Bergarten 15		20 / 40	20 / 5 / 26	
41		VI	Biscuit 9–10	Kemmlitz 48 / Geppersdorf 37	Bergarten 24 / Bergarten 15 / Bennstedt 10 / Siersham 10			27 / 28	

Compositions

No.	Whiting	Miscellaneous fluxes	Talc	Nepheline syenite	Pitcher	Rational analysis					
						Clay substance	Quartz	Feldspar	Marble	Magnesite	Dolomite
1						33–50	20–48	10–30			
2		Lepidolite 15	5	30							
3	1·5			54							
4											
5											
6	5										
7											
8											
9											
10		Lepidolite 14	3	11							
11		BaSO₄ 10	28								
12		BaSO₄ 10	33								
13											
14											
15											
16a		Dolomite 1·5									
b	1·5	Dolomite 3·0									
c	1·2										
17	7										
18			37								
19											
20	5										
21											
22											
23				25							
24			15								
25											
26		Spodumene 9·75									
27		6·8	6·8	20·4							
28		14·0		14·0							
29		12·0		28·0							
30											
31					10						
32				22							
33											
34											
35											
36						50	30	20			
						50	30	12–18	8–2		
						50	30	10·6–17·6		9·4–2·4	
						50	30	11·8–17·9			8·2–2·1
37					Scrap 10–18	%: Na₂O 0·6–0·7, K₂O 3·4–4·2, MgO 0·1–0·2, CaO 0·2–0·4, Al₂O₃ 21·4–22·6, Fe₂O₃ 0·3–0·6, SiO₂ 66·2–66·9, TiO₂ 0·6–1·0.					
38						M.F.: Na₂O 0·16–0·20, K₂O 0·65–0·70, MgO 0·05–0·08, CaO 0·06–0·08, Al₂O₃ 3·51–4·00, Fe₂O₃ 0·04–0·06, SiO₂ 17·5–20·17, TiO₂ 0·16–0·19.					
39						40	40	20			
40	2				Scrap 10						
	4				Scrap 5						
	2				China grog 10						
41	2				10						
	—				20						

%: Na₂O 0·6–0·7, K₂O 3·4–4·2, MgO 0·1–0·2, CaO 0·2–0·4, Al₂O₃ 21·4–22·6, Fe₂O₃ 0·3–0·6, SiO₂ 66·2–66·9, TiO₂ 0·6–1·0.

M.F.: Na₂O 0·16–0·20, K₂O 0·65–0·70, MgO 0·05–0·08, CaO 0·06–0·08, Al₂O₃ 3·51–4·00, Fe₂O₃ 0·04–0·06, SiO₂ 17·5–20·17, TiO₂ 0·16–0·19.

No.	Ref.	Body type	Burning cone No.	China clay or kaolin	Ball clay	Flint	Quartz	Feldspar	Pegmatite
42	S110	Sanitary VII	Biscuit 07 Glost 9–11		45		28	Norwegian 27	
43		VIII	Biscuit 8 Glost 6	38 37	20 17 Meissen 34·8		22 16	Scandinavian 20 Ströbel 30	
44		IX	Bis. 010–07 Glost 9–10 or once-fired 9/10				16	Ströbel 49·2	
45		X	Biscuit 8–10 Glost 1–4	25 55	30 6		20 40	20 5	
46		XI	Biscuit 10 Glost 1						
47		XII	Biscuit 12 Glost 12 or 10	Florida 8·0 English 29·0	Tennessee No. 5 7·5	36·0		18·0	
48		XIII	Biscuit 2 Glost 12 or 10	Florida E.P.K. 6·0 English Al. 22·0 English M.G.R.2 11·0	English 8·0		34·5	16·0	
49		XIV		17–32	Canadian H.P.l. 6–23	38–40		15–20	
50		XV			Canadian H.P.l. 20–25 Blending clay 20–25	30–40		15–20	
51		XVI		Georgia 5 N. Carolina 5 China clay 8	Kentucky 5 Canadian H.P.l. 22	36		15	
52		XVII–XXV	7–12	English V Cl. 25–27	Jernigan 15 Champion & Challenger 10	25–29			
53	A12	Sanitary (German) (a)		(1) 9·7 (2) 6·3 (3) 10·5	(1) 9·0 (2) 3·7 (3) 4·1 (4) 4·5		Quartz sand 21·0	(1) 12·3 (2) 15·9	
54		(b)		Spergau 6·0 Hirschau 6·0	(1) 14·0 (2) 9·7 (3) 13·3		24·0	Schmidt– Retsch 27·0	
55		(c)		(1) 12·4 (2) 13·7	(1) 5·4 (2) 6·1 (3) 9·0		18·2	35·0	
56		(d)		Spergau 10·0 Kemmlitz 8·0	(1) 15·0 (2) 8·0 (3) 10·0		21·0	24·0	

85% nepheline and 15% talc (**L55**)). They form a continuous series with the lower fired hard porcelains and also vitreous china. Some compositions are therefore included in those sections.

Seger porcelain differs from most soft porcelains in being fired like hard porcelain, biscuit to cone 010 and glost at cone 8–10.

Frit Porcelain, Belleek China, American Fine China

These are all low-fired bodies of high translucency depending for this on a large proportion of glass frit used together with smaller quantities of clay, quartz and possibly whiting.

Dental Porcelain

Largely feldspar with a few per cent flint and clay. It is self-glazing.

95—(cont.)

No.	Whiting	Miscellaneous	Talc	Nepheline syenite	Pitcher	Clay substance	Quartz	Feldspar	Whiting	Magnesite	Dolomite
42						33	40	27			
43		Bentonite 4				45	35	20			
44		Bentonite 4				30	48	30			
45	4				Grog 10	40–50	20–30	15–30	0·5–1·0		
46					5	40	40	20			
47	1·5										
48	2·5										
49	1–6										
50	1–6										
51	4										
52				or feldspar 18–25							
53					3·0						
54											
55											
56											

Parian Ware

A self-glazing porcelain used for unglazed figurines, etc., it has a high feldspar content or perhaps some frit.

Jaspar Ware

Composed of more than half barium sulphate together with clay and flint. It takes body stains extremely well and is used mainly for biscuit ware.

Basalt Ware

A jet black biscuit body (**S83, N36**).

TABLE
Soft

No.	Ref.	Body type	Burning cone No.	China clay or kaolin	Ball clay	Flint	Feldspar	Frit
1	S81	Japanese porcelain						
2		and European seger porcelain						
3			8					
4	S81	Seger porcelain						
5		Soft porcelain	7	Zettlitz 50		Quartz 15	30	
6			9	Zettlitz 52		Quartz 15	30	
7			11	Zettlitz 54		Quartz 15	30	
8	G68	Soft porcelain	4		25	Quartz 45	30	
9	S14	German soft porcelain	Biscuit 010	30	Earthenware clay 5	Quartz 30	35	
10				20	Earthenware clay 20	Quartz 30	30	
11			Glost 8/9	40	Earthenware clay 10 Stoneware clay 3	Quartz 25	20	
12	S83	Soft porcelain						
13		Intermediate porcelain						
14		Hard porcelain						
15	N11	American household	10–11	32	9	23	36	
16	N36	American hotel china	8–9	34·8	7	35	22	
17	S81	Frit porcelain						
18								
19	S83	Frit porcelain			Lime marl 8			78
20	N11	American belleek china	9	32	4	—	63	
21	C46	Belleek	8	35	15	20		30
22		Belleek china (all American)	8	Edgar Florida 31	Old Mine No. 4 4		62	
23	N36	Dental porcelain	10/11	5		14	81	
24	N11	Dental porcelain		5		Up to 15	Nepheline 80	
25	S83	Dental porcelain				Quartz 24·6	73	
26	N11	Copeland statuary Parian (1851)		36			40	24
27	N36	Parian	8	30	10		60	
28	N36	Jasper	10/11		30	7		
29	A141	Jasper (as made by Josiah Wedgwood)		18	26	11		
30	N36	Basalt	10/11	30	15			

TABLE 97

Frits for Belleek China (C46)

No.	Canadian feldspar	Soda feldspar	Flint	Boric acid	Bone ash	Dolomite	Whiting
1	35	35		30			
2	25	25	20		20	10	
3	25	25	20	20			10

Porcelains

No.	Frit composition	Miscellaneous	Chemical composition or glaze	Rational analysis			
				Clay subs.	Quartz	Feldspar	CaO
1			Range	25–35	41–45	20–35	
2			Av.	30	43	27	
3				40–55	15–25	30–40	1–2
4			SiO_2 77.5, $Al_2O_3 + Fe_2O_3$ 17.5, CaO 0.3, MgO 0.2, K_2O 3.8, Na_2O 0.7.				
5		Dolomite 5					
6		Dolomite 3					
7		Dolomite 1					
8							
9							
10			Glaze, average composition				
11		$CaCO_3$ 2	K_2O 0.30, MgO 0.10, CaO 0.60 } Al_2O_3 0.4 . SiO_2 2.5–3.0				
12				40	30	30	
13				50	25	25	
14				55	22.5	22.5	
15							
16		$CaCO_3$ 1.2					
17			SiO_2 76.75, $Al_2O_3 + Fe_2O_3$ 2.23, CaO 13.40, MgO —, K_2O 5.00, Na_2O 2.50				
18			SiO_2 74.52, $Al_2O_3 + Fe_2O_3$ 2.70, CaO 16.10, MgO 0.61, K_2O 3.45, Na_2O 2.63				
19	KNO_3 22.0, Salt 7.2, Alum. 3.6, Soda 3.6, Plaster 3.6, Quartz 60.0	Calcspar 17	Glaze; Litharge 33, Quartz 38, Potash 15, Soda 9				
20							
21	See below, Table 97						
22		Grd pitcher 3					
23							
24		Chalk 0–5					
25		Marble 2.3					
26	White sand 57, Cornish stone 11, Potash 8						
27							
28		$BaSO_4$ 63					
29		$BaSO_4$ 45					
30		Fe_2O_3 36, MnO_2 19					

15. BONE CHINA

Bone china was first made in England by the younger Spode in 1794; it is still used for most of the fine ware made in England and is little made elsewhere. Its best-known characteristic is its translucency (Fig. 5.2). It offers opportunities for a much larger pallette of under-glaze colours than hard porcelain, and on-glaze colours also fuse into its softer glaze better than on the hard porcelain one. It has the best impact and chipping resistance of the fine European tablewares. The traditional body composition is:

> bone ash   50%,
> china clay   25%,
> Cornish stone 25%,

458   INDUSTRIAL CERAMICS

although it will be seen in the tables that there are a number of variants incorporating small quantities of ball clay and flint, replacing Cornish stone by feldspar, replacing bone ash by apatite, etc.

There are three inherent manufacturing difficulties in the traditional bone china body. One is the lack of plasticity. In order to produce a good white body the purest Cornish china clay is normally used, but this lacks natural plasticity. Addition of ball clay to overcome this can have adverse effects

FIG. 5.2. *The characteristic translucency of bone china (Doulton,* **D56**)

on colour and translucency. Small quantities of bentonite are now sometimes used, as are also organic binders. Mellor (**M52**) points out how essential proper sliphouse practice is to the development of the maximum available plasticity of the body. Both feldspar, and therefore Cornish stone, and bone ash react gradually with water when in prolonged contact with it. Feldspar releases alkali silicates which make the water alkaline and viscous. Bone ash reacts more quickly giving soluble acid calcium phosphates or even phosphoric acid, which definitely increase the plasticity of the body. 'The longer the materials have been in association with water during their preparation, the more "buttery" the body in the workshops.' On the other hand a bone china body cannot be matured, as the organic matter remaining in the bone ash after calcination putrefies and gases are formed. The use of antiseptics such as formaldehyde is not recommended because of the possible ill effects on the plasticity.

The second cause of heavy losses is the short firing range and large firing shrinkage. Warped pieces occur very easily. St. Pierre's (**S3**) phase equilibria work has shown that the composition of bone china is near the eutectic of the phases present in the fired body, namely tricalcium phosphate, anorthite and silica, so that a large quantity of the liquid phase forms rather suddenly. In itself this contributes to the vitrification of the body at such a low temperature and to its translucency, but it also leads to a very short firing range and to warping. In order to obtain a longer firing range without losing the bone china characteristics it would be necessary to use proportionately more bone ash, at the expense of the china clay, which would make shaping even more difficult. Mellor (**M52**) has also investigated the cause of oversize overfired plates. These contain gas bubbles. A temperature excess of $20°$ C above the optimum maturing temperature can lead to this fault. The gas appears to be phosphorus and is more readily formed in slightly reducing than in oxidising conditions.

The third cause of considerable losses is the tendency for the body to go blue or brown and to blister. Mellor (**M52**) investigated this fault and traced the trouble-maker to the $1\frac{1}{2}\%$ iron present in the body. If the composition and firing conditions favour the decomposition of the calcium phosphate to give any free phosphate or phosphorus this rapidly unites with the iron present. Iron phosphate can be white, but is oxidised by air to blue, blueish-green and finally brown compounds. In general then, there is a tendency for the bone china body to go blue if the kiln atmosphere is reducing, and if steam or carbonaceous matter are given off by the body or saggars. Furthermore some bone-ash contains organic carbon but the bad colour of ware is not directly related to the carbon content of the ash but rather to the ease with which it will burn out. The rate of heating during the calcination of the bone ash seems to determine this, slow heating leaves a carbon that burns out easily before the china body vitrifies, whereas fast heating leaves a slow-burning dangerous form of carbon. Work on the triaxial diagram clay–bone–stone showed that discoloration is very dependent on composition, there being a very small margin of safety around the traditional body composition. Outside this, bluish green china easily forms even in an oxidising atmosphere and with no carbon present. In general the more clay and the less stone, or the 'drier' the stone, the greater the tendency to blue ware. Higher alkali content seems to hinder the transfer of phosphate from the bone-ash to the iron. On the other hand if the optimum amount of stone is exceeded by only a little blistering and bloating occur. So that the change from a soft to a hard stone can make a good white body into a blue one, and the change from hard to soft stone can cause blistering. Weighing out of stone or china clay whose moisture content has changed from that expected can also lead to these faults. The margin

TABLE

Bone China

No.	Ref.	Description	Burning cone No.	China clay	Ball clay	Feldspar	Cornish stone	Bone ash	Biscuit grog
1	S81	Very strong English							
2	S48	General bone china		30			18	47	5
3				28			30	42	
4				23			31	46	
5					23		23	40	
6				23			27	46	
7				30	4		7	50	
8	M87	General		26				44	
9	S83	General		25			25	50	
10	B55	China body		26			26	45	
11	B55	General	8	33–35		15–19		42–32	
12	B55	Fenton china		24			27	46	
13	B55	Common china		14	14		32	35	
14	B55	Medium china		25				44	5
15	B55	Special china		24	2			39	
16	B55	Longport china		31			18	48	
17	B55	Mason's china		29			24	42	
18	B55	Desert china		24	3		29	44	
19	B55	Alcock's figure		24			32	44	
20	B55	Spode's china		12			25	38	
21	B55	Nantgarw china		15				50	
22	B55	French china		28	3		28	34	
23	B55	Old Worcester china		27	4		42	27	
24	B55	Lime china		33			25	33	
25	B55	Apatite china		24	3		27	—	
26	D15	Apatite		33		15			
27	D15	Apatite		34		16			
28	D15	Apatite		34		17			
29	D15	Apatite		35		19			
30		General		30		10		47	
31	S3, 4, 5	General	9	26			26	45	
32			9	24			27	46	
33			9	31			18	48	
34			9	29			24	42	
35			9	24			32	44	
36			9	12		25	25	38	
37	A82	All American	9	Florida 30				45	
38	M77		9	25	Sometimes	0–10	25	50	
39	J6	English research		0–90			7·5–50	0–50	Al$_2$O$_3$ 0–6·37
40	M52	English research	9				20	80	
41			9				40	60	
42			9				60	40	
43			9				80	20	
44	J1	English research	11	45			30	45	
45			11	25			31	47	
46			11	Florida 25			31	47	
47	W24	Range by American research	8–12						
48	W24	Examples: English							
49	W24	Amer., very fine							

Bodies

No.	Flint	Whiting	Apatite	Frit	Frit composition		% Composition
1							Na_2O, 1·35; K_2O, 1·48; MgO, 0·50; CaO, 25·63; Fe_2O_3, 0·19; Al_2O_3, 17·46; SiO_2, 32·27; P_2O_5, 21·21.
2							
3							
4							
5	14						
6				3	Cornish stone	56–60 parts	
7	3			6	Quartz	20–30	
					Soda	8	
					Borax	8–10	
					Zinc oxide	8	
8	30						
9							
10	3						
11	10–14						
12	3						
13	5						
14	26						
15	35						
16	3						
17	5						
18							
19							
20	25						
21				35	Lynn sand	100 parts	
					Pearl ash	14 parts	
22				7	Lynn sand	100 parts	
					Pearl ash	8 parts	
23							
24		9					
25			46				
26	10		42				
27	11		39				
28	14		35				
29	14		32				
30	13						
31	3						
32	3						
33	3						
34	5						
35	—						
36	25						
37							
38							
39	0–90						
40							
41							
42							
43							
44							
45							
46							

Molecular Formula

No.					
47	0·2–0·4 K_2O 1·8–2·8 CaO	} 1·0 Al_2O_3	{ 2·8–4·00 SiO_2 0·6–0·933 P_2O_5
48	0·080 K_2O 1·756 CaO ——— 1·836	} 1·0 Al_2O_3	{ 3·200 SiO_2 0·585 P_2O_5
49	0·220 K_2O 2·170 CaO ——— 2·390	} 1·0 Al_2O_3	{ 4·45 SiO_2 0·72 P_2O_5

TABL
Bone Chin

No.	Ref.	Description	Burning cone No.	China clay	Ball clay	Feldspar	Cornish stone	Bone ash	Alumina
50	W24	Ernest Mayer							
51	W24	English							
52	W24	Ernest Mayer							
53	W24	English							
54		German research		25			25	50	
55				23	2		25	50	
56				23	2		30	45	
57	E14	German research	9	25		Limoges Pegmatite } 24.1 {		50.9	
58				22	3			50.9	
59			10	24			26	47	

of safety with regard to the quantity of bone ash used, its moisture content, etc., is much larger than for clay and stone.

Where the unstable white ferrous phosphate has formed in biscuit firing but has not become oxidised the body appears white but it may turn blue or brown in the glost or enamel firings or after a period of time in a show case or in use.

German, Ratcliffe and others (G25) have investigated the relationship between translucency and composition in the tertiary system. China clay–stone–bone. The trials were all fired in an industrial bone china kiln and so were not necessarily fired under optimum conditions for each individual composition. The results and conclusions were as follows:

Results

(1) Binary mixtures of china clay and bone exhibit very little or no translucency. Mixtures containing stone and bone, and stone and china clay, show translucencies which increase with increasing stone content.

(2) In the ternary system bone–china clay–stone the following results were obtained:

Relative translucency increases with

(*a*) increasing stone content, the china clay being kept constant;
(*b*) increasing stone content, the bone being kept constant,
(*c*) increasing bone content, the stone being kept constant (except in some bodies low in stone);
(*d*) increasing stone content, the ratio of china clay and bone being kept constant.

Bodies

No.	Flint	Whiting	Apatite	Frit	Frit composition	Molecular formula
50					0.192 K_2O 2.487 CaO ———— 2.679	$\left.\begin{array}{l}\\ \\\end{array}\right\}1{\cdot}0\ Al_2O_3\ \left\{\begin{array}{l}2{\cdot}770\ SiO_2\\0{\cdot}829\ P_2O_5\end{array}\right.$
51					0.137 K_2O 2.737 CaO ———— 2.874	$\left.\begin{array}{l}\\ \\\end{array}\right\}1{\cdot}0\ Al_2O_3\ \left\{\begin{array}{l}4{\cdot}25\ SiO_2\\1{\cdot}09\ P_2O_5\end{array}\right.$
52					0.1054 K_2O 3.1690 CaO ———— 3.2744	$\left.\begin{array}{l}\\ \\\end{array}\right\}1{\cdot}0\ Al_2O_3\ \left\{\begin{array}{l}3{\cdot}580\ SiO_2\\1{\cdot}056\ P_2O_5\end{array}\right.$
53					0.230 K_2O 5.200 CaO ———— 5.430	$\left.\begin{array}{l}\\ \\\end{array}\right\}1{\cdot}0\ Al_2O_3\ \left\{\begin{array}{l}7{\cdot}12\ SiO_2\\1{\cdot}50\ P_2O_4\end{array}\right.$
54						
55						
56						
57						
58						
59						

Relative translucency decreases with

(e) increasing china clay, the ratio of bone and stone being kept constant.

Conclusions

(1) Translucency is related to thickness by an exponential law.

(2) Mixtures of bone ash, china clay and stone seem to increase in relative translucency as the fluxing materials, stone and bone, are increased at the expense of china clay.

(3) Additions of ball clay to bone china body lower translucency, though the effect is small with the additions used commercially.

(4) Grinding the body in a ball mill increases translucency.

(5) On what little is known of the composition of bone china, it seems that a glassy phase with a refractive index higher than that found in porcelain is present. Anything which aids the formation of this is likely to increase the translucency. The practical limits to increasing translucency by, for example, additions of fluxing material, are determined by the narrow range of safety in firing, and by the colour effects which develop in the body, as reported by Dr. Mellor.

German (G28) has also reported on experiments on replacing the bone ash by chemically prepared phosphates, namely tricalcium phosphate $Ca_3(PO_4)_2$, dicalcium phosphate $CaHPO_4$, and hydroxyapatite $3Ca_3P_2O_8.Ca(OH)_2$. Although translucent bodies could be produced and in some cases at lower temperatures than with bone ash the raw bodies were so short that making was very difficult and drying and firing shrinkages were so high that high losses occurred at both stages. It was not considered practicable to use these phosphates in commercial production.

Odelberg (**O1**) reports a satisfactory bone china made in Sweden from both European and British raw materials, namely Zettlitz kaolin, the most plastic of all kaolins, English ball and china clays, Swedish orthoclase and bone ash.

16. HARD PORCELAIN

The word porcelain has come to cover a wide range of bodies. Primarily it implies a high-quality, dense body, the direction in which the quality is considered sometimes being shown by adjectives. True Continental hard porcelain is an intensely white, translucent, fully vitrified body, consisting largely of potash aluminosilicates and was developed as a copy of Chinese ware for tableware and artware. Once its technical qualities of strength, relative refractoriness, electrical insulation, etc., were discovered these were developed to give specialised bodies for ovenware, chemical and laboratory apparatus, and electrical insulation. For these translucency is not of importance, nor is whiteness essential. Already the drift towards other vitrified ceramic bodies has started, but these bodies retain the name 'porcelain'.

Further development has led to the inclusion of materials quite foreign to traditional porcelain, especially pure refractory oxides. Such bodies, which may contain quite large amounts of one such material together with some clay and flux, are frequently also called 'porcelains'. They bridge the gap between 'hard porcelain' and 'sintered oxides'.

Hard Porcelain

European hard porcelain is composed chiefly of kaolin, potash feldspar and quartz:

	Normal composition%	Range, %	Average, %
Kaolin	50	42–66	54
Potash feldspar	25	17–37	27
Quartz	25	12–30	21

giving a rational analysis:

	Range, %	Average, %	
Clay substance	25–35	30	
Quartz	41–45	43	
Feldspar	20–35	27	(**S81**)

and chemical composition silica   58–73
                         alumina 18–36
                         potash   1– 8
                         lime   0– 4   (**D31**)

During the firing of such bodies the feldspar melts first and attacks the clay, the ensuing solution then gradually dissolves the quartz. Kaolin also decomposes to mullite which crystallises, together with cristobalite which is dissolved preferentially by the glass. The most translucent porcelains, when investigated petrographically show no free quartz.

In bodies with a very high quartz content the quartz is not entirely dissolved during firing and remains recognisable. They have much-reduced translucency. Prolonged high firing of such bodies may convert some of this quartz to cristobalite, which with its higher expansion coefficient leads to reduced thermal shock resistance.

Firing temperatures for hard porcelain range between cone 9, 1280° C (2336° F) and cone 16, 1460° C (2660° F) (**D31**).

Hard porcelain is more seriously affected by variations at any stage of its production than the less vitrified bodies. Factors to be watched are: (*a*) composition of raw materials (iron impurities); (*b*) moisture content; (*c*) particle size, *i.e.* the whole grinding process must be thoroughly controlled; (*d*) water-soluble salts and pH value; (*e*) water content of the prepared body. The composition and general nature of the various raw materials vary and with them the properties of the porcelain manufactured. The impurities especially can have very important and undesirable influences.

Iron impurities are a major cause of discoloured ware. Raw materials of low iron content must be selected. Sieving and magneting during the body preparation help to remove larger particles and any remaining must be very finely ground and evenly distributed throughout the body, when reduction during firing will convert the iron to the almost colourless ferrous form.

Unfortunately, if titania is present as well as iron, reduction during firing produces worse black discoloration than oxidation as ferrous titanate spinel forms (**K77**).

Mica, which is frequently present in kaolin, has a marked effect on the firing and physical properties of porcelain. During firing it expands when the mica-free body would contract and so retards vitrification and the attainment of the final specific gravity, etc. It therefore leaves the biscuit-fired porous porcelain body more porous and weaker (**R24**).

Wetzel (**W50**) investigated the effect of 3% of various materials when added to a porcelain body of normal composition, namely Zettlitz kaolin 50%; Hohenbock sand 25%; Norwegian feldspar 25%; fired at cone 15 and with a P.C.E. of 29/30. He summarises his results as follows.

(1) CaO. CaO lowers the P.C.E. by six cones. There is considerable deformation at cone 10. The modulus of elasticity is fairly small, impact bending strength high, as is also the coefficient of thermal expansion and the thermal shock resistance therefore correspondingly low. Microscope shows much undissolved quartz. Transparency at cone 10 is poor.

(2) MgO. MgO lowers the P.C.E. the most (down to cone 17). The interval between softening and melting is the smallest ($\sim 280°$ C). The

elasticity modulus is similar to the normal body. In other respects like the CaO body.

(3) Talc. Interval between softening and melting about 420° C. Modulus of elasticity greater than for standard body. Highest value for the impact bending strength. Coefficient of expansion and thermal shock resistance similar to bodies 1 and 2.

(4) Dolomite. Very similar to the MgO body, except that transparency increases more slowly with increasing temperature.

(5) CaF_2. Similar to the talc body with regard to softening and melting, but with a lower modulus of elasticity, the impact bending strength is the same as the CaO body. The coefficient of expansion is fairly high, and the thermal shock resistance very low.

(6) Fe_2O_3. Very early vitrification (1150°C). Interval between softening and melting very large ($\sim 470°C$). The elasticity modulus is similar to the standard body. Impact bending strength and thermal shock resistance are relatively low. Transparency poor.

(7) ZnO. P.C.E. higher than the standard body at cone 31; the interval between softening and melting is very large ($\sim 470°$ C). Impact bending strength small. Smallest coefficient of expansion and correspondingly good thermal shock resistance. Microscope shows almost complete solution of the quartz.

(8) Al_2O_3, (9) TiO_2, (10) ZrO_2. Late sintering, above 1200° C. P.C.E. slightly lowered by TiO_2 or ZrO_2 to 27/28, but raised by Al_2O_3 to cone 32. Modulus of elasticity is increased somewhat, especially with Al_2O_3, as is also the impact bending strength. With ZrO_2 the transparency is very small, and with TiO_2 there is none. In all cases there is much mullite formation.

In comparing the experimental bodies with each other it should be noted that numbers 1–6 were fired at cone 10, whereas 7–10 and the standard body were fired at cone 15. If higher firing without deformation were possible with bodies 1–6 some of their properties might approach those of bodies 7–10.

Royer (**R65**) claims a lowering of 100° C in the firing temperature of porcelain by the inclusion of 3–4·5% talc previously fritted with the feldspar.

Not only the chemical composition of raw materials is of importance in making satisfactory hard porcelain. Hirsch (**H102**) has investigated the effect of using different minerals for introducing the quartz into the body.

>   Body: clay   50;
>   quartz   30;
>   feldspar   20.

Several series were undertaken using different pure or mixed clays together with: four types of rock quartz; three sands; flint, both raw and calcined; the sand by-product from washing kaolins. Firing temperature, shrinkage, fired colour and water absorption were little affected throughout. Expansion coefficient was irregular, being partly affected by particle size. Properties considerably affected were:

(*a*) transparency, here the true quartz undoubtedly gives the best result;

CERAMIC BODIES

(b) ball compressive strength, best values are given by calcined flint and one sand;
(c) impact bending strength is best with the kaolin washing by-product sand;
(d) tensile strength, the washings sand is best followed by calcined flint;
(e) puncture strength is greatest with calcined flint and sand, least with raw flint.

The plasticity, drying and firing shrinkage of a given body can be regulated by using clays of different plasticity, *e.g.* including up to 10% white-firing ball clay, and by using both biscuit and glost porcelain pitchers.

The *particle size* of feldspar has considerable influence on the reactions occurring during firing. If it is too coarse, insufficient homogeneity leads to centres of pure feldspar glass. Overground feldspar reacts too fast with the kaolinite before the resultant liquid attacks the quartz and leads eventually to clusters of mullite crystals unevenly dispersed in the body. Medium grinds lead to an even network of mullite crystals.

The quartz particle size is also important. If too coarse it does not dissolve in the fluxes and lowers transparency of the finished piece, it also generally raises the finishing temperature. Fine-ground quartz assists in the fluxing action of the feldspar. Overfine quartz leads to crazing.

Shrinkage of porcelain bodies during processing varies with the shaping method and can be directional.

TABLE 99
Shrinkage of Hard Porcelain (D31)

	Shaping method		
	Throwing	Casting	Pressing
General—			
Green to biscuit	7	6	2·5
Biscuit to glost	10	9	13·5
Green to glost	17	15	16
	lengthwise	breadthwise	
Turned insulators—			
leather hard to bone hard	3·5	2·5	
Bone hard to fired	15	10·5	
Leather hard to fired	18·5	13·0	

17. ELECTRICAL PORCELAIN

The requirements for porcelain to be used for both low- and high-tension electrical insulation differ from those for tableware.

Absolutely vitreous bodies are essential for any pieces liable to absorb damp from the atmosphere. Low-tension insulators may have a water absorption up to 0·5%, for high tension it must be zero. Strength is important. Many parts have metal pressed on to them or are bolted to metal or wood, etc. Others have considerable tensile stress. Dimensional accuracy is required for assembly. Ease of shaping into intricate and accurate shapes is necessary, and dry pressing is frequently used. On the other hand, colour and transparency are not important.

It is therefore possible to take advantage of the easier shaping obtained by using ball clays and white stoneware clays, and the better strength obtained with calcined flint or fine quartz sands.

Mechanical strength is improved in compositions nearer the high clay corner of clay–feldspar–quartz triangle. Undissolved quartz is detrimental to strength particularly to tensile strength and improvements are obtained either by grinding the quartz so finely that it dissolves or by replacing it by glost pitchers (**M34**), calcined ball clay, alumina or zirconia. The use of some material that does not dissolve during firing is favoured in large high-tension insulators.

The expansion coefficient is high for high feldspar and high free quartz, but low for high clay content. The thermal shock resistance is therefore best in the last instance. This property is considerably affected by firing treatment which affects crystallisation. Abrasion resistance is best in high-quartz bodies and lowest in high feldspar low-quartz ones. The electrical conductivity is lowered by complete exclusion of sodium, replacing it by potassium. The T_e value is raised if all alkalies are excluded.

The power factor also rises with the alkali content and with any porosity that could absorb humidity. Sodium feldspars can give rise to porosity by their premature decomposition with vaporisation of the soda. Carbonaceous matter can also lead to pore formation.

The dielectric constant rises slowly with increased feldspar content.

18. CHEMICAL PORCELAIN

The emphasis on the properties required for porcelain to be used in laboratories or chemical industry tends towards mechanical strength and thermal shock resistance although whiteness is still of importance.

There is therefore need to eliminate any free quartz in the fired body. If it has a high silica content this can be introduced in the form of silica glass with its very low expansion coefficient.

Higher mechanical strength is obtained by increasing the alumina content beyond that obtainable through using kaolin and feldspar; aluminium hydroxide, alumina or a sillimanite mineral may be added. This leads to the development of high-alumina bodies (**S87**).

Thermal Shock-resistant Porcelains for Ovenware and Laboratory Ware

The Société Française de Céramique has published a literature research into work on thermal shock resistant porcelains based on traditional kaolin–quartz–feldspar bodies (**A106**).

'(1) High temperature firing combined with a long soaking period is necessary to allow the maximum solution of quartz and conversion to mullite. Porcelains fired at cone 13–14 show noticeable smaller expansion coefficients than those fired at cone 9.

'(2) Compositions with at least 50–55% kaolin and less than 25–27% quartz are the most thermal shock resistant.

'Silica with its large expansion coefficient is the least thermal shock resistant component in porcelain and should be replaced as much as possible by alumina, *e.g.* by high kaolin content. Different forms of silica raw materials have different effects.

'The main fluxes should be feldspar and whiting.

'(3) Inclusion of pitchers improves tensile strength and thermal shock resistance.

'(4) Substitution of CaO, BaO, MgO for the feldspar may increase or decrease the expansion according to the amounts used. Use of talc decreases it to a minimum and subsequently increases it.

'In a traditional formula such as:

$$\left.\begin{array}{l} 0{\cdot}66\ CaO \\ 0{\cdot}34\ K_2O \end{array}\right\}, \quad 2{\cdot}8Al_2O_3, \quad 8SiO_2$$

replacement of CaO by MgO gives an optimum at 0·4–0·5 MgO.

'Bodies based on lithia have exceptionally small expansion coefficients.

'Soda feldspar is capable of dissolving eight times more quartz than potash feldspar, at high temperatures.

'(5) The development of mullite crystals is assisted by the addition of zircon, calcium fluoride or calcined kyanite.

'(6) A feldspathic glaze maturing at cone 13–14 may be used.

'(7) Proper adjustment of the glaze to the body is essential, especially when one face only of the piece is glazed.

'(8) Shape and thickness of the pieces are important, thin sections and spherical or elliptical shapes being best. Homogeneity of the body and even thickness of the piece are important.

'(9) Heat treatment by three times heating to 900° C in an electric kiln followed by air cooling reduces the expansion coefficient and increases thermal shock resistance.'

A number of examples of hard porcelain batches and compositions are given in Table 100. Hegemann-Dettmer (**D31**) gives 234 such examples and indicates what glazes should be applied and the suitability for six classes of ware: tableware; ovenware; electrical goods; luxury goods; biscuit; porcelain flowers.

TABL[E]

Hard Porcelai[n]

No.	Ref.	Description	China clay or kaolin	Ball clay	Flint	Quartz	Feldspar	Pegmatite	Miscellaneous
1	S96	General European hard porcelain; Range	42–66			20–30	17–27		
2	S81	Range	42–66			12–30	17–37		
3		Average	54			21	27		
4		Normal	50			25	25		
5		Seger porcelain							
6	A171	Summary; Chemical	65.5			9.7	24.8		
7		Electrical low clay	43.3			37.3	19.4		
8		Electrical high clay	50.6			27.7	21.7		
9		Low tension	49.0			27.5	23.5		
10		Domestic ware	48.7			27.1	24.2		
11		Hotel ware	51.8			24.6	23.6		
12		Cooking ware	60.1			17.6	22.3		
13		Art ware	44.4			28.2	27.4		
14		Soft paste porcelain	36.3–47			23.3–34.8	24.2–34		
15	D31	Tableware No. 73	47			35.5	17.5		
16	D31	Tableware No. 107	50			25	20		Limestone 5
17		Dinnerware	Zettlitz 48.17				Bavarian 8.17	Tirschenreuth 41.18	
18		Dinnerware	Zettlitz 48				Swedish 18	Weiden 34	
19		Dinnerware	Zettlitz 43.2			Czechoslovak 31.4	Swedish 3.4 / Czechoslovak 22.0		
20		Dinnerware	Zettlitz 28.8 / Kemmlitz 24.4			Swedish 19.2	Swedish 24.0		
21		Dinnerware	Zettlitz 30.0 / Kemmlitz 22.0			Swedish 20.0	Swedish 24.0		
22		Dinnerware	Zettlitz 40.0 / Kemmlitz 12.0			Swedish 20	Swedish 24		
23		Dinnerware	Zettlitz 48			Swedish 20	Swedish 25		
24		Translucent dinnerware	Kemmlitz 10 / Zettlitz 30 / English 15						
25		Translucent tableware	Zettlitz 14.9 / Osmose 16.6 / Kemmlitz 22.9			Swedish 20.6	Swedish 25		
26		Translucent tableware	Zettlitz 30 / Kemmlitz 25			Swedish 20	Swedish 25		
27		Very translucent, white tableware with few losses in glost firing	Zettlitz 46			Swedish 26	Swedish 18		
28		Very translucent tableware	Zettlitz 36.9 / Hirschau 3.8			Bavarian sand 30.8	Swedish 25.7		
29		Tableware	Zettlitz 37.0 / Kemmlitz 6.5 / Kaaden 5.6			Bohemian 18.5	Bohemian 23.1 / G.N. 5.6		
30	A3	Tableware	Zettlitz 45.3 / English 3.8			Sand 28.8	Potash F. 21.2		ZnO 1.0
31		Hotelware	Zettlitz 24.6 / Kemmlitz 25.9			Kronach Sand 34.1	Bavarian 12.2		
32		Hotelware	Zettlitz 26.9 / Kemmlitz 23.6			Kronach sand 34.1	Bavarian 12.2		
33		Hotelware	Zettlitz 48.17				Bavarian 8.17	Tirschenreuth 41.18	
34		Hotelware	Zettlitz 48				Swedish 18	Weiden 34	
35		Hotelware	Zettlitz 43			Bavarian 7	Swedish 27	Tirschenreuth 3.0	
36		Hotelware	Zettlitz 45.7			Bavarian sand 7.6	Swedish 2.9	Tirschenreuth 39.0	
37		Hotelware. Blue for dining cars	Zettlitz 26.9 / Kemmlitz 23.6			Kronach sand 34.1	Bavarian 12.2		
38		Hotelware	English 24.1 / Zettlitz 28.9			Kronach sand 34.1	Bavarian 9.9		

Bodies

No.	Glost pitchers	Biscuit pitchers	Burning cone No.		Rational Analysis		
					Clay	Quartz	Feldspar
1							
2							
3					25–35	41–45	20–35
4							
5							
6							
7							
8							
9							
10							
11							
12							
13							
14							
15							
16							
17	1·24	1·24					
18							
19							
20							
21							
22							
23							
24							
25							
26							
27	Unglazed but high fired 6·5	Biscuit 3·5	14		52·1	22·9	25·0
28		2·8			47·88	27·17	24·95
29	3·7		14				
30							
31		3·2					
32		3·2			53·9	16·8	29·3
33		2·48					
34							
35	2	18					
36	1·9	2·9					
37		3·2		0·1% cobalt oxide, wet ground for 100 hr, dried and then added to ball mill batch.			
38		3·0					

No.	Ref.	Description	China clay or kaolin	Ball clay	Flint	Quartz	Feldspar	Pegmatite	Miscellaneous
39	D31	Porcelain flowers 224	Zettlitz 15 / Halle 25			35	20		limestone 5
40	D31	Luxury goods 114	40			30	25		limestone 2
41	D31	Biscuit 209	Zettlitz 32 / English 25			3	40		
42	D31	Biscuit body	Zettlitz 45			Swedish 25 21·0	Swedish 30 Keystone 30		
43	H2	Amer. press body	Cherokee 11·0 / Florida 11·0	Bell Dresden 12·5 / KCM 12·5					Loomis Talc 2·0
44	S99	General	Zettlitz 40			Swedish 42	Swedish 18 15		
45	A3	General	Mixed 50					Bavarian 35	
46	A8	General (Rosenthal)	Zettlitz 14·5 / Chodau 14·5 / Kemmlitz 14·1				Haidhaus 4·4 / Krokenham 4·9	Edelweiss 47·1	
47	H127	Porcelain vacuum tube	English 40	English 15	20		25		
48	A57	French fireproof							
49		Ovenware 32	40			35	25		
50		Thermal shock resistant ovenware	Zettlitz 26 / Kemmlitz 31			Hohenbock sand 12	16		
51	A106	Thermal shock I resistant,	d'Avor 50	Pomaride 14		Decize sand 16	Norwegian 15		Chalk 5
52		ovenware II	d'Avor 50	Pomaride 15		Decize sand 20	Norwegian 10		Chalk 5
53		III	d'Avor 46	Pomaride 14		Decize sand 28	Norwegian 9·7		Chalk 0·6 Calcine MgO
54		Cooking ware	10	Pajot 29				Nepheline 40	calc. Al_2O_3 15 Steatite
55			10	Pajot 30			20	Nepheline 15	calc. Al_2O_3 20 Steatite
56				Wildstein 30			20		calc. Al_2O_3 20 Steatite
57	D31	Laboratory ware (Berlin)	Halle 55	Halle 27·5			Swedish 17·5		
58	D31	Chemical. All Amer.	N. Carolina 33·6 / Florida 22·4				18·0		Whiting 1·0
59	D31	Chemical. All Amer.	N. Carolina 33 / Georgia 11 / Florida 11		14		Potash 22 Soda 6		Whiting 1·0
60	D31	Chemical. All Amer.	N. Carolina 34 / Edgar Florida 24	Old Mine No. 4 5	14		23		
61		Electrical insulator bodies, English	Eng. ch.cls. mixed 10·0	Eng. ball clays mixed 30·0	35		Scandinavian 17·5		Talc 7·5
62			Eng. ch. cls. mixed 19·0	Eng. ball clays mixed 35·0	15		Scandinavian 31·0		
63		Insulator	China cl. 25 / Börtewitz 9	Siershahn 7		Limbourg sand 28	Scandinavian 22		
64	M88	Brit. res. body 11/2	Ch.cl. 28	5		Aylesbury sand 12			Dry cornish stone 55
65	M88	Brit. res. body 9/1	32	10		Aylesbury sand 18			Hard purple Cornish stone 40
66	M88	Brit. res. body 1/3	40	10					Westwood stone 30 / Dry white or purple hard Cornish stone 20
67	A8	Germ. Low tension (Rosenthal)	Kemmlitz 18·75 / Börtewitz 18·75	Halle 23·75		Freihung feldspar sand 14·58	Bavarian 9·58	Freihung 14·58	
68		High tension (Rosenthal)	Poschezan 30·0	Wildstein 10·3		Freihung feldspar sand 5·4	Norwegian 10·7	Freihung 43·7	

0—(contd.)

No.	Glost pitchers	Biscuit pitchers	Burning cone No.		Rational analysis		
					Clay	Quartz	Feldspar
49							
50							
51							
52				Grind for 100 hr.			
53			10				
54			14		49·5	27·75	22·75
55							
56					52	26	22
57			10				
58				SiO_2, 61·61: Al_2O_3, 30·01: Fe_2O_3, 1·56: CaO, 3·56: K_2O, 3·26.			
59							
60	15				60	20	20
61			13/14	SiO_2 Al_2O_3 Fe_2O_3 CaO MgO NaKO 61·62 30·00 1·56 3·56 — 3·26	0·605 CaO 0·395 NaKO } 2·913 Al_2O_3 9·874 SiO_2		
62				60·93 32·86 3·32 — 2·89	0·663 CaO 0·337 NaKO } 3·617 Al_2O_3 11·403 SiO_2		
63				64·09 30·53 0·52 1·87 2·99	0·100 CaO 0·400 NaKO } 3·185 Al_2O_3 11·366 SiO_2 0·500 MgO		
				Glaze; 0·253 CaO 0·222 K_2O } 1·120 Al_2O_3 9·29 SiO_2 0·525 Na_2O			
54	chalk 1		8				
55			7				
56		Grog 25	14				
57				SiO_2, 68·05: Al_2O_3, 28·66: Fe_2O_3, 0·26: CaO, 0·62: MgO, trace: K_2O, 0·35: Na_2O, 2·15: loss on ign. 0·15.	55·0		24·3
58			12				
59			12				
60			14				
61			7				
62			8				
63	8						
64–66							

Raw materials used, materials dried at 109° C

	Cornish stone		Aylesbury sand	J.M. china clay	Ball clay	Westwood stone
	Purple	Dry white				
Silica (SiO_2)	72·10	72·88	96·12	47·56	47·50	89·22
Titanic Ox. (TiO_2)	—	—	—	—	1·52	0·58
Alumina (Al_2O_3)	15·96	17·33	0·96	38·50	34·54	6·14
Ferric Ox. (Fe_2O_3)	0·24	0·25	0·11	0·62	1·11	0·17
Magnesia (MgO)	0·13	0·04	0·32	0·03	0·38	0·02
Lime (CaO)	2·54	0·76	0·44	0·29	0·42	—
Potash (K_2O)	3·34	5·16	0·92	1·32	1·80	2·06
Soda (Na_2O)	2·60	1·30	1·32	0·10	0·65	0·38
Loss on ignition	1·96	2·40	0·10	12·02	11·20	1·31

TABLE

No.	Ref.	Description	China clay or kaolin	Ball clay	Flint	Quartz	Feldspar	Pegmatite	Miscellaneous
69	A8	General. Berlin S.P.M.	Halle 50–53			(in clay) 22–25	Norway 24		
70	A72	Vitreous electrical body. Engl.	10–20	20–35		25–40	20–30		
71	A139	Engl. electrical	10–20	25–35	Flint or Quartz 20–30		25–40		
72		Electrical 227	Mügeln 41	Halle 21		18	20		
73	R8	Insulator	58·7			21·8	19·5		
74		Insulator		50			Scandinavian 20		Talc 5 Al$_2$O$_3$ 20
75		Insulator	40				Soda feldspar 25		Talc 5 Al$_2$O$_3$ 30
76	A139	Electrical	10–20	25–35		20–30	25–40		
77		Insulator	Poschezan 9·6 30·78	Wildstein 9·6			Bavarian 4·8	Steinfels 48·36	
78		Insulator	Osmose 32	Wildstein 10·0			Bavarian 4·8	Steinfels 48·0	
79		Insulator	Kemmlitz 38	Halle 24			Bavarian 16	Steinfels 18	
80		Insulator	Hirschau 20·80 Zettlitz 15·85	Wildstein 10·90			Norwegian 12·85	Weiden 35·70	Dolomite 0·80
81		Insulator or cooking ware	Zettlitz 40				Swedish 25		Steatite 5 Al$_2$O$_3$ 30
82		Insulator	Ch. cl. 24·1 Zettlitz 28·9			Kronach sand 34·1	Bavarian 9·9		
83		Insulator	Zettlitz 24·6 Kemmlitz 25·9			Kronach sand 34·1	Bavarian 12·2		
84		Insulator	Zettlitz 26·9 Kemmlitz 23·6			Kronach sand 34·1	Bavarian 12·2		
85		Insulator or cooking	Zettlitz 42	Wildstein 8			Norwegian 29		Steatite 1 Al$_2$O$_3$ 20
86		Electro porcelain							
87		Electropress body							

19. MULLITE PORCELAIN

Porcelains of better refractoriness, better thermal shock resistance and higher thermal conductivity are obtained under the following conditions:

(1) The fired body should contain no free silica whether in the form of quartz, cristobalite or tridymite.

(2) The fired body should have the minimum glassy matrix and the maximum newly-formed crystals (as opposed to introducing ready-made crystals).

These conditions make necessary the use of low-quartz raw materials which are so finely ground that all the quartz will dissolve or combine during firing and the addition of alumina to combine with the free silica and make the maximum of crystallised mullite. A low feldspar content is required.

Some examples of such bodies are given in Table 101. They are often called 'high alumina', 'sillimanite' or 'mullite' bodies. The first two names refer to the raw materials used while the third gives an idea of the

—(contd.)

Glost pitchers	Biscuit pitchers	Burning cone No.		Rational analysis		
				Clay	Quartz	Feldspar
		15	Glaze 0·3 K$_2$O, 0·6 CaO, 0·1 MgO } 1·0 Al$_2$O$_3$, 10·0 SiO$_2$			
		6	Glaze; Stone 80, Whiting 15, Zinc Oxide 5			
1 unglazed but vitreous 1	2·5					
		11	Drying shrinkage 2%, Firing shrinkage at 900° C 1%, Firing shrinkage at 1300° C 14% } Total 17%	43·2	30·0	26·8
	Saggar grog 5	14/15				
		14/15				
	4·0					
	2·2					
3·10			Linear exp. coeffd. = 3·79 × 10^{-6}	40 56·2	40 21·9 dolomite 0·8	20 21·8
		14				
	3·0					
	3·2			53·9	29·3	16·8
	3·2					
				43·92 38·2	34·76 29·8	21·32 32·0

nature of the fired body. As 'high alumina' is also used for bodies that are largely sinter alumina the term 'mullite body' will be used here.

A number of practical conditions are very important in the successful production of mullite bodies. The quartz content must not only be below 10–15% but must be ground to pass sieve No. 200 or even No. 250. The alumina content is raised by using either high-alumina china or ball clays, calcined as necessary to reduce plasticity, or calcined sillimanite or calcined alumina. The two last materials must be very finely ground (120 hours), but even so free alumina reacts faster than sillimanite. Alumina should be irreversibly converted to corundum before grinding in order to get the best results. The grinding should be in rubber-lined mills with sinter alumina or mullite porcelain balls as the wear on flints is so great that allowance would have to be made for the extra silica in the body calculations. The very fine grinding of these materials assists in the sintering of the body, so that even if the feldspar content is reduced, vitrification is not necessarily at a very much higher temperature. Small additions of steatite, preferably lime-free, help flux the body and magnesia is one of the best mineralisors for the crystallisation of mullite.

20. HIGH ALUMINA

As the alumina content of the porcelain type of body is increased it eventually exceeds the proportion that can combine to give mullite ($3Al_2O_3.2SiO_2$) and free corundum becomes part of the fired body. With increasing alumina content and decreasing silica the body approaches sintered alumina. Many bodies termed 'high alumina' are modifications of the pure sintered alumina body, made to obtain sintering at lower temperatures. Thus alumina is mixed with small quantities, totalling 8–16%, of: talc; dolomite; whiting; barium carbonate; fluorspar; chromium oxide; beryllium silicate; clay.

Bodies of over 92% alumina are called 'sinter alumina' and discussed under 'pure oxide bodies'.

Glazing of alumina bodies presents a number of difficulties summarised by Fisher and Twells (**F11**):

'(1) High porosity of the body with high capacity for absorption at the glaze fusion temperature.

TABLE

Mullite and High

No.	Ref.	Description	Firing cone No.	Alumina or sillimanite	China clay	Ball clay	Grog	Feldspar
1		Mullite		calc. A 20		Aluminous 30	Saggar 25	20
2		Mullite		calc. A 30	Aluminous 40			Soda 25
3		Mullite		calc. A 25		35 Bentonite 10		Norwegian 25
4		Mullite	10	Sillimanite 30		55		15
5	A9	Mullite	18		Zettlitz or Halle 50		calc. A 60% Feldspar 40% } 50	
6	A9	Mullite	31/32	calc. A 50	Zettlitz 50			
7	A9	Mullite	31–36	calc. A 70	Zettlitz 30			
8	A9	Mullite	18	calc. A 46·5	Mixed 52			1·5
9	A9	Mullite	18	fused A 50	Zettlitz 50			
10	A9	Mullite	18	fused A 60	Zettlitz 40			
11	A9	Mullite	26/27	Kyanite 50		17	33	
12	F11	Mullite	18	calc. A 60	E.P.K.	30		
13	F11	Mullite	18	calc. A 72		20		
14	F11	Mullite	18	calc. A 84		10		
15	F11	High alumina	18	calc. A 84		0		
16	F5	High alumina		92·25	3·50			
17	F5	High alumina		91·68	3·50			
18	F5	High alumina		92·04	3·50			
19	A5	High alumina 'Pyronit 2'		fused A 82·13	Zettlitz 8·51 Seilitz calcined 4·25			
20	G34	High alumina improved sparking plug		95				

CERAMIC BODIES 477

'(2) Difference in thermal expansion between body and glaze.
'(3) Chemical activity of body and glaze materials.'

They found that it is preferable to use feldspathic glazes with the RO group composed of alkalies from high-potash feldspars augmented if necessary by small amounts of potash frits. Their summary:

'(1) Keep the firing temperatures as low as possible (preferably not above cone 20).
'(2) Fire as rapidly as practical to avoid glaze absorption by the body.
'(3) Use a simple glaze formula with alkalies, preferably potash, for the RO group; keep the glaze solubility low.
'(4) For one-fire cone 18 glazing, keep the alumina content of the glaze near 1·50 and silica near 15·0 equivalents.
'(5) For two-fire glazing at cone 18, keep the alumina between 1·50 and 1·75 and silica near 15·0 equivalents.
'(6) Alumina bodies with between 10 and 20% clay are simpler to glaze at cone 18 than bodies of lower and higher clay content.'

Alumina Bodies

No.	Steatite or talc	Miscellaneous						Properties
1	Egyptian 5	suitable glaze	0·75 MgO	0·9 Al_2O_3 . 8·0 SiO_2				
2	Indochinese 5		0·25 K_2O					
3	Indochinese 5							
4								
5								Use up to 1700° C, M.P. 1800° C
								Impermeable to gases, even unglazed.
6								Impermeable to gases, even unglazed.
7								Impermeable to gases, even unglazed.
8								Use up to 1700° C, M.P. 1820° C, requires feldspathic glaze.
9								Requires feldspathic glaze.
10								Use up to 1700° C, M.P. 1850° C, requires feldspathic glaze.
11								
12	Talc 5	Dolomite 5	one-fire glaze ~ 1·0 RO (mostly K_2O), 1·5 Al_2O_3, 15·0 SiO_2					
13	4	4						
14	3	3	two-fire glaze ~ 1·0 RO (mostly K_2O), 1·75 Al_2O_3, 15·0 SiO_2					
15	8	8						
16	Talc 1·25	Whiting	Barium	Fluorspar 2·00	Chromium 1·0			
17	1·25	2·57	carbonate		oxide 1·0			
18	1·25	0·44	0·89		1·0	Beryllium silicate 0·88		
19		Fine chalk 1·45	Magnesia 1·54	Fine quartz sand 2·13				
20	1·5	Zircon 3·5						stronger than 100% alumina.

TABLE 102
Zircon Porcelains

Ref.	Description	Burning cone No.	China clay	Ball clay	Zircon.	Zirconium compounds	Feldspar	Talc	Aluminium hydrate	Magnesia	Properties
N11		8–10		Ky No. 5 18·52	59·25	Calcium zirconium silicate 7·41 Magnesium zirconium silicate 7·41 Barium zirconium silicate 7·41					Dielectric constant 9·8 Power factor 0·05–0·06% Loss factor 0·5–0·6% Resistivity > $1·0 \times 10^{14}$ ohm-cm
N11		10–12		Ky No. 5 11·1	66·7	Calcium zirconium silicate 22·2					
N11		10–12		8·9	68·5	Calcium zirconium silicate 22·6					
N11		10–12		11·2	59·2	Calcium zirconium silicate 29·6					
N11		10–12		25	35	Calcium zirconium silicate 20 Magnesium zirconium silicate 10 Barium zirconium silicate 10					
B46	Thermal shock resis. vitreous	12		30	300-mesh 55		13	2			Absb. 1·40%, flexural strength 9600 lb/in² After thermal shock, flexural strength 3460 lb/in²
	Thermal shock resis. porous	11	20	30	200-mesh 30			15	5		Absb. 14·80%, flexural strength 3400 lb/in² After thermal shock, flexural strength 2130 lb/in²
	'Unbreakable' porcelains			Bentonite 4 Bentonite 4		Zirconia 55–90 Zirconia 55–90 Zirconia 55–90 Zirconia 55–90		45–10 41–6		41–6 45–10	

21. ZIRCON BODIES

Zircon can be substituted for all or part of the clay, flint or feldspar of a hard porcelain body. As it is very dense almost twice the weight of zircon can replace an equivalent volume of flint or feldspar without altering the working qualities. When fired at high temperatures, cone 16 or above, very high mechanical strengths are obtained. Zircon has very low thermal expansion (4·1 compared with flint at 13·4) and its use in bodies confers very high thermal shock resistance. It also increases thermal conductivity. Chemical resistance is very good. These properties are good over quite a wide range of compositions.

Zircon bodies, especially if alkali-free have high dielectric strength even at elevated temperatures, and also low dielectric loss (**B46**).

Zircon decomposes to zirconia and silica between 1600° C (2912° F) and 1700° C (3092° F) and bodies containing it should therefore not be fired above 1600° C (2912° F). The firing range is, however, larger than for many specialised bodies, namely 30–50° C.

Russell and Mohr (**R71**) give the general composition of zircon 'porcelains' as: zircon, not less than 50%; fluxes, often double zirconium silicates of alkaline earth oxides, up to 30%; clay, up to 30%; bentonite, optional.

Individual compositions are given in Table 102.

Such bodies fire to a white or cream colour. If iron and titania are present in the zircon the fired pieces are red-brown externally and grey internally but these two impurities together with phosphates narrow the firing range. Alkalies, iron, chromium, manganese, cobalt, nickel, vanadium and copper all have an adverse effect on the power factor although ferrous iron has a smaller effect than ferric (**B80**).

22. CERAMIC-BONDED ABRASIVE GRINDING WHEELS

A large number of abrasive grinding wheels consist of abrasive grain bonded by a ceramic body and fired. There are two main types of abrasive grain, (1) natural corundum, fused alumina or artificial corundum, and (2) silicon, carbide, each in several grades and a wide range of grit sizes. Other abrasive grains of less importance in the field of ceramic-bonded wheels are: diamond, boron carbide, emery, natural forms of silica.

The ceramic bond may be the traditional type of mixture of clay with feldspar, calcspar, magnesite or glassy frit, fired to cone 10 to 13, or it may have a more glassy nature, fired to, say, cone 1.

The properties of the finished grinding wheel are dependent on both the type, size and hardness of the abrasive and on the proportion and composition of the binder. The binder is required to show strength and toughness and in general this was considered to be better achieved with bodies fired to high temperatures (Table 103). Rieke and Haeberle (**R25**) have, however, shown that low-fired bodies can serve equally well (Table 104).

TABLE

Traditional Grinding

German, for Alumina

	I	II	III	IV	V
Feldspar	48	55·5	66	74	35
Clay	32	35·5	44	26	59
Frit	20				5
Quartz					1
Powdered sodium silicate		9·0			
Water glass					
Iron oxide					
Calcium carbonate					
Zinc oxide					
Silica					
% of dry bond used for various grades	8·8–15·4	8–24	30		

TABLE

Cone 02–1 Fritted Grinding

Batch in mols.

	(1)	(2)	(3)	(4)	(5)	(6)	(7)	(8)	(9)	(10)
Feldspar	0·25	0·25	0·25	0·25	0·85	0·25	0·25	0·25	0·25	0·25
Kaolin	0·75	0·75	0·75	0·585	0·50	0·50	0·75	0·75	0·75	0·75
Borax	0·25	0·25	0·25	0·25	0·25	0·25	0·25	0·25	0·25	0·25
Magnesite	0·25	0·25	—	—	0·25	—	0·25	—	—	—
Calcspar	0·25	—	—	—	0·25	—	0·25	—	—	—
$Ca_3(PO_4)_2$		0·08	0·08	0·08		0·08	—	0·08	0·08	0·08
$Mg_3(PO_4)_2$			0·08	0·08		0·08		0·08		—
$Zn_3(PO_4)_2$				—						0·08
$AlPO_4$				0·33						
ZnO				—					0·25	
Sand				0·33					—	
Alumina							0·25	0·25	—	

Sodium silicate bond, baked at 205–245° C (401–473° F)
sodium silicate
zinc oxide   } $ZnO \cdot Na_2O \cdot 2SiO_2$
clay   (N11)

Wheel Binders

cone 10–12. A15

		for Silicon carbide					American, cone 8–9. N11	
							for Silicon carbide	for Alumina
VI	VII	I	II	III	IV	V		
72·5	51·5	38	74	33·9	76	58·6	45	25
19·3	34·7	19	26	33·4	24	21·8	35	75
4·8	5·0			20·6				
				12·1		19·6	20	
	4·0							
1·2	1·0							
	4·0	17						
2·4								
		26						
13·2–16·0						14–21		

Wheel Binders (R25)

Molecular formulae and percentage composition

No.		Na$_2$O	K$_2$O	MgO	CaO	ZnO	B$_2$O$_3$	Al$_2$O$_3$	SiO$_2$	P$_2$O$_5$
1	M.F.	0·25	0·25	0·25	0·25	—	0·50	1·00	3·0	—
	%	4·1	6·2	2·6	3·7		9·2	26·9	47·4	
2	M.F.	0·25	0·25	0·25	0·25	—	0·50	1·00	3·0	0·082
	%	4·0	6·0	2·5	3·6		8·9	26·0	45·9	3·1
3	M.F.	0·25	0·25	0·25	0·25	—	0·50	1·00	3·0	0·165
	%	3·8	5·8	2·5	3·5		8·7	25·2	44·6	5·9
4	M.F.	0·25	0·25	0·25	0·25	—	0·50	1·00	3·0	0·33
	%	3·6	5·5	2·4	3·3		8·2	23·9	42·2	10·9
5	M.F.	0·25	0·25	0·25	0·25	—	0·50	0·75	3·0	—
	%	4·4	6·6	2·9	3·9		9·8	21·6	51·0	
6	M.F.	0·25	0·25	0·25	0·25	—	0·50	0·75	3·0	0·165
	%	4·1	6·2	2·7	3·7		9·3	20·2	48·3	6·2
7	M.F.	0·25	0·25	0·25	0·25	—	0·50	1·25	3·0	—
	%	3·8	5·8	2·5	3·5		8·6	31·5	44·4	
8	M.F.	0·25	0·25	0·25	0·25	—	0·50	1·25	3·0	0·165
	%	3·6	5·5	2·3	3·3		8·1	29·6	41·9	5·5
9	M.F.	0·25	0·25	—	0·25	0·25	0·50	1·00	3·0	0·082
	%	3·9	5·9		3·5	5·0	8·7	25·4	44·8	2·9
10	M.F.	0·25	0·25	—	0·25	0·25	0·50	1·00	3·0	0·165
	%	3·7	5·7		3·4	4·9	8·5	24·7	43·6	5·7

23. CORDIERITE BODIES

During his systematic work on stoneware bodies in the triaxial system $MgO-Al_2O_3-SiO_2$, the senior writer discovered bodies of exceptionally low expansion coefficient and was able to trace this desirable property to the phase 'cordierite'. Cordierite occurs in Nature as a rather rare mineral, it constitutes the phase found in the centre of the triaxial diagram of $MgO-Al_2O_3-SiO_2$ bounded by the lines joining five eutectic points (Fig. 5.3).

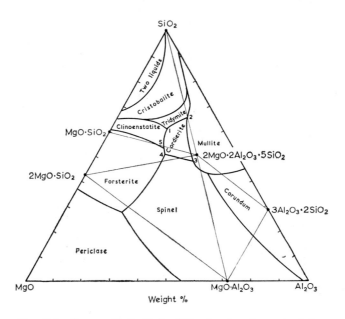

Fig. 5.3. System $MgO-Al_2O_3-SiO_2$ showing the area of the phase cordierite with its five surrounding eutectic points numbered. These are detailed in Table 105 (Rankin and Merwin, **R2**, Bowen and Greig, **B98**, and Greig, **G76**)

TABLE 105

The Eutectic Points Surrounding the Cordierite Area as shown in Fig. 5.3

Pt.	Crystalline phases	Percentage comp.			Temp. (°C)
		MgO	Al_2O_3	SiO_2	
1	$MgO-SiO_2$, SiO_2, $2MgO-2Al_2O_3-5SiO_2$	20·3	18·2	61·5	1345
2	SiO_2, $3Al_2O_3-2SiO_2$, $2MgO-2Al_2O_3-5SiO_2$	10·0	23·5	66·5	1425
3	$3Al_2O_3-2SiO_2$, $MgO-Al_2O_3$, $2MgO-2Al_2O_3-5SiO_2$	16·1	34·8	49·1	1460
4	$MgO-Al_2O_3$, $2MgO-SiO_2$, $2MgO-2Al_2O_3-5SiO_2$	25·7	22·8	51·5	1370
5	$2MgO-SiO_2$, $MgO-SiO_2$, $2MgO-2Al_2O_3-5SiO_2$	25·0	21·0	54·0	1360

CERAMIC BODIES

TABLE 106

Molecular Formula of the Eutectic Points Surrounding the Cordierite Area

Molecular formula	Percentage comp.			Melting point (°C)
	MgO	Al_2O_3	SiO_2	
$MgO-0·33\ Al_2O_3-1·45\ SiO_2$	25·0	21·0	54·0	1360
$MgO-0·35\ Al_2O_3-1·35\ SiO_2$	25·7	22·8	51·5	1370
$MgO-0·36\ Al_2O_3-1·98\ SiO_2$	20·3	18·3	61·4	1345
$MgO-0·85\ Al_2O_3-2·05\ SiO_2$	16·1	34·8	49·1	1460
$MgO-0·93\ Al_2O_3-4·46\ SiO_2$	10·0	23·5	66·5	1425

Cordierite shows polymorphism, three forms having been recognised. Of these the α-form is the stable high-temperature form and the only one normally found in Nature or obtained in ceramic bodies. β- and μ-cordierite can only be formed under special conditions (**K6**).

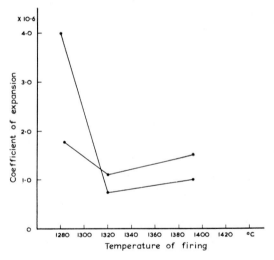

FIG. 5.4. Variation of expansion coefficient of cordierite with the firing range

The development of cordierite in a ceramic body is not achieved merely by using raw materials whose chemical composition sums up to that of cordierite. Two conflicting factors affect the result. One is the fact that equilibrium is rarely reached during a ceramic firing. The other is the proximity of the various eutectic points so that if equilibrium is approached, only small deviations from the correct composition will produce melting and/or unwanted phases. Bodies of high cordierite content inevitably have a short firing range, under-firing (below 1350° C, 2462° F) fails to develop the cordierite and over-firing (above 1450° C, 2642° F) results in its deterioration into forsterite and mullite, both with much higher expansion coefficients (**B24**) (Fig. 5.4). It is generally easier to produce a porous cordierite body than a vitreous one, both find a number of uses.

The range of body composition to give useful cordierite bodies within the system $MgO-Al_2O_3-SiO_2$ is:

MgO 2·60–13·8%;
Al_2O_3 25·50–38·8%;
SiO_2 51·40–64·9%;

the most favoured is:

MgO 13·8%;
Al_2O_3 34·8%;
SiO_2 51·4%;

which gives a molecular ratio of $2MgO.2Al_2O_3.5SiO_2$ (**S125**).

The most usual raw materials are clay, talc and alumina. In order that no quartz is present in the fired body a quartz-free clay should be used. As talc frees silica on heating, it is best replaced by magnesite for the lowest expansion bodies. Another way of introducing the magnesia is with sierralite (**L10**).

Much research has gone into investigating methods of increasing the firing range of cordierite bodies sufficiently to obtain commercially producible vitreous bodies. Gebler and Wisely (**G13**) found that this could be achieved by calcining magnesite, clay and talc and then blending this with raw clay to make up the body (*see* Table 108). Most other work involves making additions of fluxes or even refractories to the body.

Thiess (**T22**) describes prolonged experimental work on vitrified cordierite bodies, and as a few examples of his compositions given in Table 108 show, he included small quantities of various fluxes.

More recently Lamar and Warner (**L11**) made an extensive study of cordierite bodies for the three following reasons:

(1) To gain a better understanding of some of the fundamentals involved in the firing of cordierite ceramics.
(2) To develop both porous and non-porous types of cordierite which could have practical values for refractories, dielectrics, or other commercial applications.
(3) To find methods of broadening the firing range and reducing the maturing temperature of cordierite bodies.

Throughout this series of trials the ratio of $MgO:Al_2O_3:SiO_2$ was kept as close to 2:2:5 as possible, as this has been shown to give the lowest expansion coefficient. This composition was, however, arrived at by using several combinations of raw materials.

Series A. $\begin{cases} \text{kaolin–Yellowstone talc–MgCO}_3 \\ \text{kaolin–talc} \end{cases}$

Series B and D. $\begin{cases} \text{kaolin–sierralite (prochlorite)} \\ \text{kaolin–sierralite–talc} \end{cases}$

Series C. kaolin–talc–α-alumina

(The sierralite was used in two different particle size distributions.)

To the basic cordierite body were added varying quantities of $BaCO_3$ and $PbSiO_3$ either separately or together. Test pieces were mixed and pressed dry and fired in an electric laboratory kiln. Differential thermal analyses were made of each raw material and of each unfired body. Each fired body was investigated by X-ray diffraction methods, its thermal expansion and dielectric properties found. By these methods the temperature of different relevant reactions and the relative proportions of minerals in the products can be found.

D.T.A. shows that cordierite crystallises at 1300° C (2372° F) except in the presence of $PbSiO_3$ when it does so at 1260° C (2300° F). In the presence of $BaCO_3$ a new phase is found, first identified by Wisely and Gebler (**W76**) and for which the formulae $2MgO.BaO.3Al_2O_3.9SiO_2$ and $3MgO.BaO.4Al_2O_3.9SiO_2$ were suggested. (This phase is represented as Ba–C.) This phase crystallises at 1250° C (2282° F) and the reaction is sufficiently rapid for it to do so in preference to cordierite until all the Ba^{2+} are used up. $PbSiO_3$ lowers this crystallisation temperature by some 60° C. Cordierite was found to form more rapidly in series B than series A, probably because of the type of combination in which the Al_2O_3 is present in the raw materials.

X-ray diffraction showed which bodies were made up almost entirely of cordierite and/or Ba–C and which also contained appreciable quantities of mullite and spinel. The values for the thermal expansion of the bodies that are almost completely cordierite is extremely low. Ba–C expands almost twice as much as cordierite but even this is small enough to confer high thermal shock resistance. Mullite has a high expansion coefficient, and it is probably its presence in very small quantities that has upset previous measurements on cordierite. $PbSiO_3$ leads to a certain amount of lead glass, this increases the expansion either in itself or because of the decrease of cordierite content.

In general the presence of $BaCO_3$ and $PbSiO_3$ lowers the maturing temperature and widens the firing range sufficiently to make production of dense low expansion bodies much simpler, *e.g.* A–11, B–2, B–6 and C–1. Some bodies with low expansion and good refractoriness although porous show possibilities for kiln furniture, etc., *e.g.* A–A, A and 823–D mix.

Table 107 has been compiled from Lamar and Warner's numerous tables to give an idea of the results they obtained. It is a selection only.

Hedvall, Bjökengren and Rähs (**H73**) have made a systematic study of the fluxing action of BaO on the system $MgO–Al_2O_3–SiO_2$.

TABLE 107

Composition and Properties of Cordierite Bodies
(after Lamar and Warner) (L11)

Body	Composition (parts by weight)							Absorption % after maturing at °C	Mineral composition from intensities of principal X-ray peaks				Expansion coefficient × 10⁻⁶				
	E.P.K. kaolin	Yellowstone talc	Sierralite (pro-chlorite)	Al₂O₃	MgCO₃	BaCO₃	PbSiO₃		Cordierite	Ba-C	Mullite	Spinel	100 °C	300 °C	500 °C	700 °C	900 °C
A–A	67·9	14·7			17·4			13·4 at 1288°C	Very high	—	very low	very low	0·13	0·47	0·99	1·39	1·60
A	72·3	21·9			5·5			2·2 1343	high	—	low	—	1·07	0·83	0·95	1·08	1·25
A–0	75·0	25·0						1·5 1316	high	—	appreciable	low	2·00	2·18	1·83	1·91	1·92
A–1	67·9	14·7			17·4	10·0		7·8 1288	lower	high	—	—	0·67	1·38	1·87	2·22	2·52
A–2	67·9	14·7			17·4		10·0	8·9 1316	high	—	—	low	0·27	0·76	1·26	1·78	1·99
A–11	75·0	25·0					10·0	0·2 1288					0·67	0·84	1·41	1·85	2·09
B–B	50·0		50·0					0·0 1204	high	—	very low	fairly low	1·33	2·29	2·76	3·14	3·56
B	50·0		40·8					0·5 1232	high	—	very low	fairly low	0·67	1·05	1·50	1·94	2·28
B–0	50·0		50·0			10·0		0·7 1288	lower	high	—	fairly low	1·20	2·00	2·36	2·76	3·14
B–1	50·0		50·0				10·0	0·1 1149	high	—	—	fairly low	0·93	1·67	2·15	2·70	2·86
B–2	50·0	9·2	50·0				10·0	0·4 1260		high	—	fairly low	1·33	2·26	2·72	3·32	3·45
B–6	50·0	9·2	40·8				10·0	2·6 1232					1·07	2·00	2·54	3·05	2·96
C–C	41·0	44·0		15·0				9·1 1316	high	—	very low	fairly low	0·53	1·13	1·58	1·93	2·28
C	41·0	44·0		15·0		10·0		1·5 1288	—	high	—	—	1·33	2·29	2·61	3·05	3·32
C–0	41·0	44·0		15·0				0·0 1288	medium	—	—	higher	1·60	2·44	2·92	3·42	3·75
C–1	41·0	44·0		15·0		10·0	10·0	7·8 1260	—	high	—	—	1·87	2·58	3·12	3·63	3·77
C–2	41·0	44·0		15·0		5·0	5·0	0·4 1288	low	low	—	very low	1·87	2·54	3·10	3·57	3·94
823–D mix	50·0		50·0*					1·8 1316					0·16	0·44	0·82	1·15	1·49

* Sierralite of different composition.

Bushan and Roy (**B149**) describe experiments on the addition of lithium carbonate to cordierite type bodies and achieved a more easily vitrified body of low expansion coefficient.

Lamar (**L10**) found that the firing range could be considerably lengthened without increasing the expansion coefficient by adding 5 to 20% of zircon.

Owing to their low expansion coefficient cordierite bodies are difficult to glaze successfully, especially as it is frequently desirable to fire them only once. Thiess (**T22**) found that some of his experimental bodies can be glazed with slips made from the same body mixed with coloured oxides; namely Nos. 5, 23, 28, which are shown in Table 108. A characteristic of these bodies is the high feldspar content, which is probably the source of the self-glazing property. The senior writer found that the best self-glazing is achieved only if the feldspar used has a high sodium content and if firing is above cone 13.

Twice-fired cordierite can be glazed with ordinary low-solubility glazes as used for bone china or earthenware at 900–1100° C (1652–2012° F) although the appearance is often less glossy (**S125**).

Since the general interest in low-expansion cordierite has increased, natural deposits of the mineral have been found. Brown (**B122**) reports on a preliminary investigation of a large deposit in Wyoming containing 70–80% of the mineral cordierite. This shows an Fe_2O_3 content of 12% on analysis, and so fires to a dark colour. When bonded with bentonite, or bentonite with sufficient magnesia to convert it to theoretical cordierite, low-expansion bodies comparable with synthetic cordierite were obtained. These had very high thermal shock resistance but lacked transverse strength and had poor dielectric properties. The firing range was narrow.

24. STEATITE

In the U.S.A. this body is called 'clinoenstatite', but see below. Amongst the non-clay plastic raw materials is talc or steatite. This forms the main constituent of a body which when fired consists largely of magnesium metasilicate and is used for electrical insulators.

Steatite bodies are made up of talc, 40–90% but generally around 80%, together with clay to plasticise the body, and a flux. The nature of the flux greatly influences the electrical properties and immediately divides steatite bodies into: (*a*) bodies with feldspar or other fluxes containing alkali metals suitable for general low-tension work; (*b*) bodies excluding the alkalies and using the alkaline earths instead, and suitable for low-loss high-frequency work.

The specialised nature of the applications of steatite bodies requires adjustments to give particular properties, a range of which can be achieved by careful choice and pretreatment of the main raw materials together with specific minor additions.

TABL
Cordieri

No.	Ref.	Description	Burning cone No.	China clay	Ball clay	Quartz	Miscellaneous refractory	Feldspar	Miscellaneous fluxes
1	S84	1st Industrial cordierite body. F. Singer, 1926			Zinzendorf 27				
2	A23 A22b	German, body Ost		Kemmlitz 43	Leskau 18 Luckenau 14 Egar 20		Grog 225-mesh 10		
3	A23 A22a	German, Sipa H	9–12		Egar stoneware clay 8·5 Bennstedt clay 8·5		Sillimanite 30 Cordierite pitcher 3	5	
4	—	Saggar body							
5	A186	U.S.A. Alsimag 72 and 202							
6		U.S.A. saggar body							
7	A22c	U.S.A. pyrostat			44		Saggar grog 30		Iron oxide 0·
8			7	25	Pikes 30	4		11	
9		'Ardostan' German		44	Bentonite 4		Sillimanite 18	10	Witherite
10		French		25	30		Zircon 4	11	
11		French		20	30		Zircon 3	11	
12	B80, A22d	'Ardostan'			41		Sillimanite 20	14	
13	A22e	Stemag heater plate body							
14	T22 and N11	Experimental. Recommended for flame resistant cooking ware	12 Self glazing		22			12·8	Zinc oxide 3·2
15	S197	Commercial body	13/14	35·3			Zirconia 25·0		
16	B80	Scheing and Lutz			50		Grog 25		
17	B80	Turbine blades			48		Sillimanite 18	10	BaCO₃ 3
18	B47	Cordierite with lithia B₄	7 (1220°C)		61·5				Li₂CO₃ 14·8
19		D₄	6/7 (1210°C)		41·5				Li₂CO₃ 14·8
20	—	For small intricate parts of electrical heating apparatus		60	15		Saggar grog 10		
21		For large parts of electrical apparatus		31			Saggar grog 55		Iron oxide 1·
22	—	German	14		23				
23			14		27				
24			14		29				
25			14		31				
26			14		35				
27	D31	Italian low expansion electrical porcelains	10		Fat, lime, low melting 9	28	Porcelain pitchers 28	7	magnesite 28
28	D31		10	Schio 41·7	White Meissen 5			5·0	
29	D31		10	Low Al₂O₃ High CaO 33·6	10·9			6·7	
30	D31		10	31·0	10·6			7·1	
31	D31		10	28·4	19·7			6·3	
32	D31		10	27·6	11·7			7·4	
33	D31		10	27·0	10·8			9·8	
34	D31		10	26·3	11·2			11·9	
35	D31		10	—	27·4			9·0	
36	D31	Magnesia stoneware	12/13	25·0	12·5	20·0	Pitchers 5·0	20·0	
37	T22	Thiess Exp. No. 1	14		Tennessee 35				
38	T22	No. 5	13		29			Canadian feldspar 16	
39	T22	No. 11	13		29			11	Spodumene 5

No.	Steatite	Talc	Alumina	Molecular formula or other data	% Composition	Water absorption	Exp. coeff. ×10⁶
1		Göpfergrün 40	Hydrated 33	MgO.Al$_2$O$_3$.2SiO$_2$	MgO, 15·3 ; Al$_2$O$_3$, 38·3 ; SiO$_2$, 46·4		0·15
2		15		MgO.2·4Al$_2$O$_3$.7·4SiO$_2$	MgO, 4·8 ; Al$_2$O$_3$, 30·1 ; SiO$_2$, 54·1		
3		25		MgO.1·25Al$_2$O$_3$.5SiO$_2$	MgO, 8·5 ; Al$_2$O$_4$, 25·5 ; SiO$_2$, 58·7		
4					MgO, 2·60 ; Al$_2$O$_3$ + Fe$_2$O$_3$, 30·80 ; SiO$_2$, 64·95 ; CaO, 1·20 ;	10·8	
5				2MgO.2Al$_2$O$_3$.5SiO$_2$	MgO, 13·8 ; Al$_2$O$_3$, 34·8 ; SiO$_2$, 51·4		
6				MgO.4Al$_2$O$_3$.15SiO$_2$	MgO, 2·60 ; Al$_2$O$_3$, 38·80 ; SiO$_2$, 64·95		
7		25		MgO.1·7Al$_2$O$_3$.3·8SiO$_2$	MgO, 8·0 ; Al$_2$O$_3$, 34·1 ; SiO$_2$, 56·2		
8	Calcined 30						
9	21						1·1
10	de Luzenac 30						
11	de Luzenac 30						
12		20		MgO.2Al$_2$O$_3$.6SiO$_2$	MgO, 6·9 : Al$_2$O$_3$, 32·7 ; SiO$_2$, 56·25 : Fe$_2$O$_3$, 0·6 ; TiO$_2$, 0·28		
13				MgO.3·2Al$_2$O$_3$.7·6SiO$_2$	MgO, 4·85 ; Al$_2$O$_3$, 39·4 ; SiO$_2$, 56·0		
14		46	16			0·0	1·6
15	29·7		10·0				
16		25					
17		21					
18		34	5·5			0·0	2·4
19		44·5	14·0			0·0	2·0
20		Göpfersgrün 15·0					
21		13·0					
22		42	Hydroxide 35			20	1·85
23		40	33			19·6	1·5
24		39	32			9·7	1·6
25		38	31				1·4
26		36	29			14·5	2·05
27							
28		48·3					
29		48·8					
30		51·3					
31		45·6					
32		53·3					
33		52·4					
34		50·6					
35		63·6					
36		17·5		Glaze; quartz 30, feldspar 40, kaolin 1, dolomite 1, pitchers 28			
37		California 43	Calcined 22	MgO.1·01Al$_2$O$_3$.2·47SiO$_2$		9·9	
38		36	19	0·911 MgO 0·065 K$_2$O } 1·16 Al$_2$O$_3$. 0·029 Na$_2$O } 3·65 SiO$_2$		0·0	
39		36	19	0·911 MgO 0·057 K$_2$O } 1·12 Al$_2$O$_3$. 0·019 Na$_2$O } 2·85 SiO$_2$		0·0	

No.	Ref.	Description	Burning cone No.	China clay	Ball clay	Miscellaneous refractory	Feldspar	Miscellaneous fluxes
40	T22	No. 16	14		32		Canadian 4·8	Beryl
41	T22	No. 23	13		29		Canadian 12·8	Zinc ox
42	T22	No. 27	13		30		Nepheline syenite 12	
43	T22	No. 28	13		29		16	
44	R23	Experimental 3	11/2	80				
45	R23	(Rieke and Thurnauer) 4	10	70				
46	R23	5	10	60				
47	R23	6	9	50				
48		Experimental bodies	9, 11, 14		Stoneware clay 23			
49			9, 11, 14		27			
50			9, 11, 14		29			
51			9, 11, 14		31			
52			9, 11, 14		35			
53	L10	Experimental bodies with wider firing range		50		Zircon 5–20		
54	—		7	25		Pikes 30, Sillimanitic clay 30	Quartz 4	11
55	H120	Australian saggar body	13/14			Warren refractory shale 50		
56	—		12	Silica free 80				Magne
57	B24	Exp. low ex. 8C	14	39·27				
58	B24	5C	14	59·1				
59	B24	7C	14	29·95				
60	G13	Calcine No. 23	6	35·7		—		MgCO
61	G13	Calcine No. 29	6	—		19·2		,,
62	G13	Calcine No. 4	6	26·5		9·4		,,
63	G13	Body F-5	10	58·0			Calcine 23,50	Weights
64	G13	Body F-25	10			68·0	Calc. 29,42·0	to 100
65	G13	Body F-39	10	41·1		35·45	Calc. 23, 34·25	units fired
66	G13	Body F-42	10	38·1		32·8	Calc. 4, 39·1	weight

Talcs

These vary in quality, having variable $MgO:SiO_2$ ratio and can contain combined alumina, calcium, iron and trace elements. These impurities influence the firing-properties and, if not allowed for, lead to over-firing. Klug (**K39**) suggests that the necessary adjustment of the body composition is more readily done by using the molecular formulae as is done for glazes. The impurities also increase the dielectric loss factor.

CERAMIC BODIES

8—(contd.)

Steatite	Talc	Alumina	Molecular formula or other data	% Composition			Shrinkage	Water absorption	Exp. coeff. ×10⁶	
				MgO	Al_2O_3	SiO_2			20–150°C	20–900°C
	42	20	$0.940MgO$ $0.015Na_2O$ $0.030K_2O$ $0.015BeO$ $\Big\} Al_2O_3 \cdot 2.45SiO_2$					0.0		
	36	19	$0.809MgO$ $0.017Na_2O$ $0.050K_2O$ $0.124ZnO$ $\Big\} 1.09Al_2O_3 \cdot 3.55SiO_2$					0.0		
	38	20	$0.896MgO$ $0.028Na_2O$ $0.076K_2O$ $\Big\} 0.978Al_2O_3 \cdot 2.46SiO_2$					0.0		
	36	19	$0.877MgO$ $0.029Na_2O$ $0.093K_2O$ $\Big\} 0.982Al_2O_3 \cdot 2.5SiO_2$					0.0		
								Difference between sintering temp. and P.C.E. in °C.		
	20		$MgO.2.08Al_2O_3.5.62SiO_2$	6.8	36.0	57.2		170	0.03	0.25
	30		$MgO.1.23Al_2O_3.3.88SiO_2$	10.1	31.4	58.5		150	0.03	0.23
	40		$MgO.0.79Al_2O_3.3.00SiO_2$	13.3	26.7	60.1		100	0.04	0.27
	50		$MgO.0.54Al_2O_3.2.48SiO_2$	16.4	22.5	61.1		10	0.06	0.30
	Gopfersgrün 42	$Al(OH)_3$ 35					10	17.0		4.55
								20		2.1 1.85
							10	19.2 20 19.6		5.2 1.4 1.5
	40	33					10	12.4 12 9.7		3.9 1.65 1.6
	39	32					13	4.7		4.45 1.25 1.4
	38	31					11	13.0 14.7 14.5		5.05 1.9 2.05
	36	29								
(Sierralite 50) Calcined 30 Tumby bay 20								40		4.0
										0.15
	47.91	12.82	96% crystalline cordierite							0.98
	36.7	4.2	97% crystalline cordierite							1.02
	46.75	23.30	95% crystalline cordierite							1.22
	18.5	—								
	18.5	$Al(OH)_3$ 16.3								
	17.3	$Al(OH)_3$ 3.7								
								0.03 0.00 0.02 0.03		0.98 0.87

The flaky 'soapy' nature of talc which gives it the desirable property of plasticity also easily gives rise to orientation of the particles leading to differential shrinkage and laminations. Plasticity is increased by long milling, about 60 hours, but the tendency for orientation can only be destroyed by calcination followed by milling. When calcining the talc, magnesium carbonate may be added in the required quantity to combine with any excess silica released from the talc. Synthetic magnesium metasilicate can also be added to the body, both to break up the talc structures and to assist in obtaining the desired composition.

Calcining destroys the plasticity of talc and so limits the proportion of it that can be so treated. Little or no calcined talc is used in dry-press bodies, for wet pressing half to two-thirds of the talc may be calcined and for casting, plastic work and extrusion two-thirds is calcined.

The heating of talc decomposes it to magnesium metasilicate and silica according to the equation:

$$3MgO.4SiO_2.H_2O = 3(MgO.SiO_2) + SiO_2 + H_2O$$
$$84\cdot6\% 15\cdot4\%$$

Magnesium metasilicate exists in three polymorphic forms, enstatite which is orthorhombic and found in igneous rocks, proto-enstatite, and clino-enstatite which is monoclinic and found in some meteorites. The relationship between the three is difficult to establish due to very sluggish inversion but on the basis of Foster's (**F20**) work is believed to be as follows:

Enstatite is the stable low-temperature form when the compound is formed or crystallised below about 1260° C (2300° F) (perhaps as low as 1140° C (2084° F)).

Proto-enstatite is the stable high-temperature form arising from the heating of enstatite, the inversion being very slow but possibly starting at 1140° C (2084° F). Foster found it noticeable at 1260° C (2300° F). Cooling does not normally lead to reinversion of proto-enstatite to enstatite, although it is possible that this might occur in the presence of mineralisers. Normal cooling of proto-enstatite either leaves it in a metastable unchanged condition or gives rise to clino-enstatite. Proto-enstatite crystals are more stable if small and in the presence of certain glassy phases.

Clino-enstatite is a low-temperature form exclusively obtained by cooling proto-enstatite and not obtainable from enstatite although it is possible that under suitable conditions clino-enstatite could be converted to enstatite.

This polymorphism is of importance in steatite bodies because of the disintegration or 'dusting' that is associated with the inversion of proto-enstatite to clino-enstatite. It is now fairly well established that 'dusting' occurs after prolonged or repeated heating, when crystal growth of the proto-enstatite would occur and lead to readier inversion, and also to the absence of an adequate glassy bond.

Clay

From 5 to 15% clay is added to steatite bodies, largely to give them plasticity. The clays chosen should have the maximum plasticity consistent with a minimum alkali, titania, and iron content. Bentonites, although very helpful in augmenting plasticity, must be kept to a small percentage because of the alkalis and iron introduced. In general those bodies to be shaped by plastic processes must contain more clay than the dry-press ones but have higher power factors. During firing the clay lowers the sintering temperature of pure steatite and by supplying alumina to the glassy phase prevents its devitrification and thereby the inversion of proto-enstatite to clino-enstatite. However, if more than 10% clay is present a eutectic in the system $MgO-Al_2O_3-SiO_2$ may be approached and the firing range seriously shortened.

Fluxes

Low-tension steatite bodies can be fluxed with from 5 to 10% feldspar (potash spar is preferred) which gives a longer firing range and a glassy bond that does not easily devitrify, although dielectric losses are higher than in alkali-free bodies. The power factor also increases much more rapidly with temperature.

In the low-loss bodies, alkali is eliminated. The most usual flux is then barium carbonate. This frees barium oxide which forms a eutectic with silica and alumina at $1175 \pm 20°$ C ($2147°$ F) of composition BaO, 28; Al_2O_3, 11; SiO_2, 61%. Further reactions with magnesium metasilicate lead to other eutectics. Considerable reduction of sintering temperature is therefore achieved by using barium carbonate, although with quantities above 5% the firing range becomes shortened. Strontia acts similarly to baria. Calcium, magnesium, and iron compounds either as additions or from the impurities contribute to the glass. There is considerable devitrification of this alkaline earth glass on cooling. Variations in the amounts, proportions and kinds of these alkaline earth fluxes can cause large variations in the dielectric properties of the body (**R34**).

Smoke (**S137**) has shown that steatite bodies can be given better thermal shock resistance by replacing the barium carbonate by barium zirconium silicate which markedly reduced the thermal expansion.

The substitution of lead silicate for all or part of barium carbonate gives lower firing temperatures and better dielectric properties.

Other Body Constituents

Magnesium carbonate may be added to combine with excess silica freed by the decomposition of the talc. It raises the sintering temperature. It improves the power factor and specific resistance.

Kaolin and alumina are sometimes added, they reduce firing temperature and assist in stabilising the glassy phase.

If the composition of the body is expressed as a molecular formula:

$$1 \cdot 0RO.xAl_2O_3.ySiO_2$$

and RO contains no alkalies then the power factor decreases with the alumina content.

Beryllia improves power factor and specific resistance (**B80, A2**).

The structure of a vitrified steatite body differs from hard porcelain by its much smaller content of glass. Under the microscope it appears as a largely crystalline body.

TABLE

Steatite

No.	Ref.	Description	Burning cone No.	China clay	Ball clay	Bentonite	Steatite talc	Feldspar
1	S137	Thermal shock resistant bodies No. 2		E.P.K. 10			70	
2		No. 11		E.P.K. 20			40	
3	A19	Calite			Wildstein 16·2		65·8	
4		—	8	25	Pikes 30		Calcined 30	Stone 15
5	S83	Average composition						
6	—	German Holenbrun	14		Wildstein 6·5		raw 62·5 / calc. 24·5	Ströbel 6·5
7		German			8–9		raw 82–84	8–9
8		German			5		90	
9	—	Italian	3					
10		Italian	6					
11		Italian	7		24		58	18
12		Italian			7		63	30
13		Italian			11		63	25
14		Italian	8		8		72	
15	—	English	7		24		58	18
16			7	8			72	
17			7	15·9			63·2	
18			7	15·1			71·4	
19			7	14·7			69·7	
20			7		5		73	
21			7	10	5		63	
22			7	15	5		55	
23	T30	U.S.A.	14				Calif. 89	11
24			14				Calif. 89	
25			14	10			70	
26	S198	U.S.A.	14		5	0·5	85·3	
27			14		15		79·3	
28			14				90	10
29	N11	Normal	14	Plastic kaolin 7			87	Potash 6
30	N11	Normal	13	Plastic kaolin 5			90	Potash 5
31	N11	High-frequency	12–13	Plastic kaolin 15·0			60·0	
32	N11	High-frequency	13	Plastic kaolin 5			88	
	B80	Hermsdorf Schomberg Calit bodies:						
33		286 Pressing, casting, plastic		2 types 15		0·063	Raw 19 / Calcined 50	
34		610 Non HF work			9	1	Raw 30 / Calcined 56	
35		617			11		Raw 29 / Calcined 45	
36		618 Dry press			10		Raw 46 / Calcined 18	
37		623 Plastic extrusion			13·5		Raw 18 / Calcined 37	
38	A20	Sicalit			Wildstein 9·7		39·5	

Bodies

No.	Barium car-bonate	Mag-nesium car-bonate	Calcium car-bonate	Miscellaneous grog	Misc. fluxes	Molecular formula or glaze
1					Barium zirconium Silicate ⎰20 ⎱40	
2						
3	10·8	7·2				
4						0·876–0·962 MgO
5						0·0–0·223 CaO }0·058–0·281 Al_2O_3
						0·0–0·017 Na_2O 1·476–2·202 SiO_2
						0·007–0·046 K_2O
						0·026–0·038 FeO
6						Glaze: Body mix 34·2%, quartz 12·8%, feldspar 38·5%, Zettlitz kaolin 6·8%, MnO_2 5·1%, chrome oxide 2·6%.
7						
8		5				
9						0·007 NaKO
						0·721 MgO
						0·111 CaO } 0·066 Al_2O_3 . 1·00 SiO_2
						0·057 BaO
						0·004 FeO
10						0·003 NaKO
						0·847 MgO
						0·005 CaO } 0·089 Al_2O_3 . 1·14 SiO_2
						0·105 BaO
						0·006 FeO
11						
12						
13						
14	20					
15						
16	20					
17		calc. MgO 3·9			$BaSO_4$ 17·0	
18	13·5					
19					$BaSO_4$ 15·6	
20					$BaSO_4$ 22	
21					$BaSO_4$ 22	
22					$BaSO_4$ 25	
23						
24	8			flint 3		
25	5			flint 15		
26	7		2·2			
27	4		1·7			
28						
29						
30						
31	17·5	7·5				
32	6		1			
33	10	6				
34		5				
35				Fired scrap 15		
36	6	7		⎰ Fired scrap 8 ⎱ Sand 5		
37	9	6		Fired scrap 16·5		
38	6·5	4·3		SiC 40		

496 INDUSTRIAL CERAMICS

TABLE

No.	Ref.	Description	Burning cone No.	China clay	Ball clay	Bentonite	Steatite talc	Feldspar
39	B80	Steatit Magnesia: Low tension dry press	13	6·9			Raw 87	6·1
40		Low tension wet press	13	6·9			Raw 40 Calcined 47	6·1
41		Frequenta AS 42a	14	4			Raw 87	
42		Frequenta for pressing		6		0·5	Raw 87	
43		Frequenta casting				2·0	Frequenta pressing body 75	
44	B80	Rosenthal: Plastic			14	2	Raw 32 Calcined 35	
45		Dry press			3·3	2·5	Raw 69	
46	B80	Lutz group			6		Raw 84	
47		Frequenta	14		3–4	0·5	87·9	
48		Low loss	14	2·0	7·0		82	
49		Low tension	6–10		6·9		87	6·1
50	—	General low tension			7–9		82–84	7–9
51		General HF			5		90	
52	C69				10		29–58	
53	A2	Low loss 1		Zettlitz 6	Wildstein 12		Raw 15	
54		2		Zettlitz 13			Raw 70	
55		3			Wildstein 6	Ca 1	Raw 84	
56		4			Wildstein 9		Raw 49 Calcined 35	
67	S198	U.S.A.	14		5		90	5
58			14		5		90	
59			14		5		90	
60			14		5		90	
61		Best low-loss body in experiments with yellowstone talc M	8/9	E.P.K. 3·6			Yellowstone 82·1	
62		Best in further c	3/4				Yellowstone 80	
63		exps. for (1)	5	E.P.K. 10			Yellowstone 70	
64		ultra low loss (7)	8/9	E.P.K. 5			Yellowstone 80	
65	A4	German radio valve	12–13	Zettlitz 11·7			82·6	Potash 1·9

Bichowsky and Gingold (**B47**) investigated the relationships between composition and physical properties in three clay–talc series. They used a kaolin, a white-burning fairly plastic clay and a dark-firing fat clay each with Göpfersgrün talc, varying the compositions in each series in 10% steps. Samples were fired at cones 11 and 14. The best firing results were obtained in the kaolin series, much deformation or melting occurred in the coloured clay series. However, the kaolin series had the highest porosity and the dark clay the lowest. The porosity curves showed maxima in the kaolin and white clay series and a minimum in the dark clay one. Impact bending strength increases with talc content. The coefficient of expansion reached a

109—(contd).

No.	Barium carbonate	Magnesium carbonate	Calcium carbonate	Miscellaneous grog	Misc. fluxes	No.	Properties
39							
40							
41	8		1				
42	8			Aluminium hydroxide 2			
43				Fired frequenta 25			
44	4			Calcined body 7, Fired scrap 6			
45	5	2·5	1·0	Calcined body 12·3, Fired scrap 5			
46	4	—	—	Sand 6			
47	6–8		1·0	—			
48			3·5	Sand 5·5			
49							
50							
51	5						
52	24–35		6		Boric acid 1–10, Zinc oxide 1–10, $MgSiO_3$ 53		
53	6	9					
54	9	8		Aluminium hydroxide 2			
55	7						
56	7						
57							
58	5						
59			5				
60				5			
61	9·1	5·2					Power factor 0·0004, dielectric constant 6·11. Loss factor 0·0024, Q_x 2851.
62					$PbSiO_3$ 20		
63	10				$PbSiO_3$ 10		
64	5	5			$PbSiO_3$ 5		
65				ZrO_2 3·9			

No.	Q_x	Power factor	Dielectric constant	Loss factor
62	3360	0·000296	5·96	0·00176
63	20 160	0·0000493	6·56	0·000324
64	5220	0·000192	6·04	0·00116

minimum in each series but at different talc contents. It was also always reduced by higher firing and sometimes by longer firing.

Rigterink, Grisdale and Morgan (**R34**) show that if the principal constituent of a low-loss steatite body is talc variations in the ratio of clay to talc have little effect on the dielectric constant.

Differential Shrinkage of Talc in Steatite Bodies

Different talcs have different forms when crushed and ground, some being granular, some fibrous, some micaceous and some irregularly shaped. This influences the occurrence of differential shrinkage in extruded and dry-pressed pieces made from bodies containing them. Uniform granular talcs show the least differential shrinkage, the unsymmetrical particled talcs showing more of it in extruded than dry-pressed pieces, and more at higher pressing pressures. Differential shrinkage therefore arises from orientation.

Lamar (**L9**) describes work on such talc, using a numerical method of comparison by 'specific shape factor', and reaching the following conclusions.

(1) Particle shape controls differential shrinkage.

(2) Particle size and particle size distribution affect total shrinkage but not differential shrinkage.

(3) Generally the more irregular the particles, the higher the 'specific shape factor' and the greater the differential shrinkage of the particular substance.

(4) The various values for linear and diametrical shrinkage suggest that differential shrinkage can be largely controlled by blending talcs of different shapes.

(5) Tendency towards micaceous form caused more differential shrinkage than tendency towards fibrous form.

Hausner and Naporan (**H46**) investigated the peculiar tendency for either the outside or the inside of an extruded steatite tube to shrink more than the other. This seems to be related in a rather complicated way to the relative values of outside and inside diameters. It is also affected by the pressure distribution during extrusion and it is possible that it could be regulated thereby.

25. VITRIFIED FORSTERITE BODIES

Braniski (**B108**) has shown that in the system serpentine–dolomite there is a small section where the mixtures are not refractory (P.C.E. less than cone 26) but that these make excellent electrical insulator bodies. The composition range is:

$$\text{natural dolomite} \quad 40\text{--}47\%;$$
$$\text{natural serpentine} \quad 60\text{--}53\%;$$

or

$$\text{pure dolomite} \quad 40\text{--}46\%;$$
$$\text{pure serpentine} \quad 60\text{--}54\%.$$

With the natural materials the product is brown.

Vitrified forsterite bodies can also be usefully made from olivine using the fines removed during the manufacture of forsterite refractories. Here again the product is brown (**S122**).

A forsterite body derived from talc and magnesite is given as follows:

calcined talc	15%
raw talc	18%
kaolin	16%
bentonite	1%
magnesite	50%
	100% (D31)

Another:

steatite	40–60%
magnesium hydroxide	45–20%
clays	5–10%
alkaline earth compounds	about 10% (B18)

26. SPINEL BODIES

Smoke (S144) investigated the possible usefulness of bodies consisting largely of spinel $MgO.Al_2O_3$, and found that low-loss ceramic insulating materials could be achieved (see Table 110).

TABLE 110

Compositions and Properties of Spinel Bodies

	R1*	R2	R6	R3	R7	R8
Composition						
MgO	28.4	26.0	24.0	23.0	20.0	17.0
Al_2O_3	71.6	64.0	62.0	57.0	50.0	43.0
SiO_2	—	10.0	14.0	20.0	30.0	40.0
Batch Formula						
Magnesia	28.4	25.5	23.0	21.7	18.3	14.9
Alumina	7.16	53.5	48.0	37.8	22.5	10.2
Clay (E.P.K.)	—	21.0	29.0	40.5	59.2	74.9
Maturing temp. (°F)	3540	2800	2750	2650	2600	2600
Maturing temp. (°C)	1947	1536	1508	1453	1425	1425
Physical Properties						
Modulus of rupture (lb/in^2)	—	12,700	—	18,500	—	—
Coef. of expansion ($\times 10^{-6}/°C$)	8.0	6.98	—	6.61	—	—
Thermal conductivity (150° F tempilstik)	—	3.06"	—	2.75"	—	—
Electrical Properties (1Mc)						
Power Factor ($\times 10^{-2}$)	0.37†	0.0375	0.0445	0.0468	0.650	0.858
Dielectric constant	7.2†	7.12	7.38	6.62	6.25	5.82
Loss factor ($\times 10^{-2}$)	2.66	0.278	0.328	0.310	4.06	4.98
Crystalline Phases	Spinel	Spinel	Spinel	Spinel	Spinel and Cordierite	Spinel and Cordierite
Thermal Shock (°F cycle)	—	400	—	400	—	—

* All data for R1 from literature.   † Tested at 100 Kc/s.

27. PYROPHYLLITE BODIES

Stevens (**S190**) describes some preliminary experiments using pyrophyllite as a major body constituent. An almost vitreous, white and glossy body can be made. This compares favourably with porcelain in mechanical and electrical characteristics, lacking only zero porosity and high puncture strength. In high-frequency applications pyrophyllite is superior to porcelain but inferior to steatite.

28. WOLLASTONITE BODIES

Bodies based on the mineral wollastonite, calcium silicate $CaO.SiO_2$, are found to vitrify at 1200–1250° C (2192–2282° F) and give a body of low dielectric loss and good mechanical strength. These bodies contain between 70 and 85% wollastonite, 10–20% ball clay and a flux. Jackson (**J7**) describes experiments with various fluxes including barium carbonate, barium zirconium silicate, boron phosphate, and found that boron phosphate gave the best dielectric properties especially together with small additions of alumina and zircon (*see* Table 111).

TABLE 111
Wollastonite Bodies

Ref.	Description	Burning cone No.	Clay	Wollastonite	$BaCO_3$	PbO	BPO_4	Alumina	Zircon	
C2	Experimental	4	Old Mine No. 4 : 30	60	5	5				loss factor 0·371%
J7	Low-loss experimental	4	Old Mine No. 4 : 8	80			5	4	3	loss factor 0·360% dry 0·392% wet

Wall tile bodies can also be made from wollastonite with a clay or glass binder. They have very small firing shrinkage so that larger and thinner pieces can be made without warping. The product is light in weight and can be cut, drilled and shaped with ordinary tools (**M84**).

29. LITHIA PORCELAINS AND BODIES

The discovery that certain bodies in the field Li_2O–Al_2O_3–SiO_2 have respectively, very low, zero or negative expansion coefficients has been a tremendous step forward in producing pieces of high thermal shock resistance and any desired expansion coefficient.

Roy and Osborn (**R62**) first suggested that since the lithium ion is monovalent, like sodium and potassium, but is similar to the magnesium ion in size, it might be expected that the Li_2O–Al_2O_3–SiO_2 system might show resemblances to both the alkali–Al_2O_3–SiO_2 and the MgO–Al_2O_3–SiO_2 systems and might find application in ceramics. The Li_2O–Al_2O_3–SiO_2

system is represented in Nature by the minerals eucryptite (ratio 1:1:2), spodumene (1:1:4) and petalite (approx. 1:1:8). The properties and amongst them the expansion coefficients of these minerals were therefore investigated. The early experiments showed that these lithium minerals have very low thermal expansion, comparable to or lower than that of fused silica. Further work in this field therefore seemed very worth while (S139).

Smoke (S139) and Hummel (H121) each followed the matter up and made the remarkable discovery that samples of petalite and spodumene actually initially contracted on heating, regaining their initial lengths at 730° C (1346° F) and 400° C (752° F) respectively and thereafter having very low expansion coefficients. As the minerals themselves are difficult to handle they then turned to synthetic mixtures to give compositions in the Li_2O–Al_2O_3–SiO_2 systems. Smoke found two areas in which the bodies obtained had negative expansion coefficients, one surrounding the composition of eucryptite and the second larger one embracing the compositions of petalite and spodumene (see Fig. 5.5).

The compositions are given in Table 112 and shown on the phase diagram. Some of the expansion curves are given in the graph (Fig. 5.6).

TABLE 112
Lithia Bodies
Compositions (Wt. %) and Linear Thermal Expansions of QA Series (S139)

QA Series No.	Oxide compositions			Batch compositions					Linear thermal expansion	Zero expansion temperature (°C)*
	Lithia	Alumina	Silica	Spodumene	Lithium carbonate	Clay (E.P.K.)	Flint	Alumina		
1				100					N	400
4	4·6	17·6	77·8		10·0	41·8	48·2		N	N
5	6·0	22·9	71·1		13·0	54·5	32·5		N	N
6	8·5	31·5	60·0		16·6	67·7	15·8		N	N
7	3·0	12·0	85·0		6·7	29·8	63·5		P	P
8	10·0	20·0	70·0		20·0	44·0	36·0		O	450
9	4·0	26·0	60·0		8·4	60·3	31·3		P	P
10	2·4	43·0	54·6		5·0	95·0			P	P
11	5·2	41·6	53·2		10·0	90·0			P	P
12	10·0	12·0	78·0		20·6	27·3	52·1		P	P
13	14·0	16·0	70·0		27·1	34·1	38·8		P	P
14	20·0	15·0	65·0		36·2	30·0	33·8		P	P
15	25·0	25·0	50·0		41·8	46·0	12·2		P	P
16	14·0	29·0	57·0		25·9	59·1	15·0		O	460
17	4·0	16·0	80·0		8·7	39·4	51·9		P	P
18	9·0	34·0	57·0		17·3	72·1	10·6		N	425
19	10·0	36·0	54·0		18·3	75·8	5·9		P	P
20	15·0	32·0	53·0		27·2	63·8	9·0		P	P
21	5·0	15·0	80·0		10·8	35·8	53·4		O	350
22	5·0	12·0	83·0		10·9	29·0	60·1		N	N
23	5·0	9·0	86·0		11·0	22·1	66·9		P	P
24†	11·9	40·4	47·7		22·6	75·1	2·3		N	N
25	13·0	45·0	42·0		24·0	67·0	9·0		P	P
26	11·0	38·0	51·0		20·4	77·0	1·9		N	N
27	15·0	37·0	48·0		26·8	72·7	0·5		N	N
28	8·0	41·0	51·0		15·2	84·0		0·8	N	125

* Temperature at which curve crosses zero line.
† Eucryptite (Li_2O–Al_2O_3–$2SiO_2$).
N—Negative expansion.
P—Positive expansion.
O—Zero expansion.

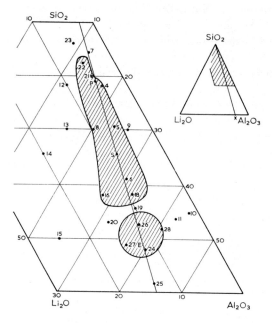

Fig. 5.5. Areas of negative linear thermal expansion in the system Li_2O–Al_2O_3–SiO_2 (Q series) (Smoke, **S139**)

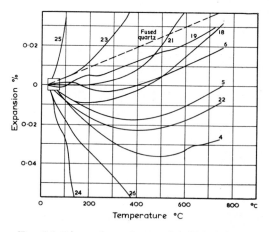

Fig. 5.6. Linear thermal expansion (QA series), *see* Table 112 for compositions (Smoke, **S139**)

Smoke (**S143**) followed this with a more detailed study of the properties of all the compositions within the area of negative thermal expansion. Sixty compositions were chosen at $2\frac{1}{2}\%$ lithia and 5% silica increments. The raw materials lithium carbonate, clay, flint and alumina were dry mixed, then water was added and the batch extruded. The specimens were dried and fired.

These mixtures were found to have a very short firing range so that a firm body could not be obtained with less than 1% water absorption. However, by preparing part of the body as a frit and then combining this with clay before firing to $1020-1120°$ C ($1868-2048°$ F) a body with 0.06% water absorption was obtained. Its transverse strength was 6000 lb/in^2; dielectric constant 4.06; power factor 5.23%; and loss factor 26%.

White and Rigby (**W58**) undertook similar investigations simultaneously with Smoke, but using different raw materials. They found that the same ultimate composition might have a negative expansion coefficient when made from one set of raw materials but a small but positive one when a different group of raw materials is used. It seems, therefore, that the mineralogical, chemical and probably physical properties of raw materials used and the nature of the preparation and firing will determine whether the product has positive or negative thermal expansion properties. They suggest that this may be connected with the proportion of glassy phase present, as glasses of the composition of eucryptite and spodumene have higher expansions than the corresponding crystalline phases (**S125**).

Both porous and vitreous bodies are now made with small positive, zero or negative expansion coefficients. Their main application is for thermal shock-resistant pieces, the vitreous ones being used in this connection in the electrical and low-loss fields.

Phase rule data on the $Li_2O-Al_2O_3-SiO_2$ and adjacent systems have recently appeared:

$Li_2O-MgO-Al_2O_3-SiO_2$, Prokopowicz and Hummel (**P58**).
Compositional and stability relationships among eucryptite, spodumene and petalite, Roy, Roy and Osborn (**R63**).
Lithium metasilicate–forsterite–silica, Murthy and Hummel (**M103**).
Lithium metasilicate–eucryptite, Murthy and Hummel (**M102**).
$Li_2O-BaO-SiO_2$, Dietzel, Wickert, Köppen (**D42**).
$Li_2O-ZrO_2-SiO_2-Al_2O_3$, Bonlett and Thomas (**B90**).

30. HIGH-MAGNESIA BODIES

Snyder and Ruh (**S147**) describe bodies containing over 90% magnesia, together with ball clay or talc, which together with good low-loss electrical properties have thermal conductivities ten times that of steatite and about twice that of high alumina porcelain.

TABLE 113
High-magnesia Bodies

Ref.	Description	Firing temp. °C	MgO	Ball clay	
S147	No. 184	1540	95	5	
	183	1540	92·5	7·5	
	181	1540	97·5	2·5	
	180	1540	95·0	5·0	
	179	1540	92·5	7·5	
G36	refractory electric insulator		>65	the rest	Borosilicate glass ~5

31. BERYLLIA BODIES

Beryllia can be introduced into ceramic bodies either as the mineral beryl, a beryllium aluminosilicate, or as the pure oxide BeO.

Beryl can be substituted for part of the feldspar and flint in a hard porcelain type of body, when the following properties are altered:

thermal expansion: lowered.
thermal conductivity: increased.
thermal shock resistance: greatly improved, by 132%.
firing shrinkage: decreased.
impact resistance: increased 44·0%.
transverse strength: increased 22·5%.
dielectric strength: +16·5%.
electrical resistance: increased.

Beryllia acts as a very strong mineraliser for the formation of mullite and thereby assists in achieving great mechanical strength. However, if all the feldspar in a normal porcelain is replaced by beryl much free cristobalite is released with all its attendant disadvantages. Proper adjustments to dissolve or combine the free silica must be made.

A number of triaxial oxide systems with one component being BeO have been investigated in order to develop bodies of special physical properties. These are shown in Table 114. Many of these bodies have high thermal conductivity. Investigation of such bodies has shown that they are almost entirely crystalline, having vitrified by sintering with no intermediate fusion.

In the systems $BeO-Al_2O_3-\genfrac{}{}{0pt}{}{TiO_2}{ZrO_2}$ addition of magnesia lowers firing
$\qquad\qquad\qquad\qquad\qquad\quad ThO_2$
temperature but also the thermal conductivity.

[There is a considerable health hazard in the use of beryllia; Breslin (**B113**) describes safe techniques.]

32. SINGLE-COMPONENT REFRACTORY BODIES

Although all traditional ceramic bodies have several solid phases bound together with a glassy matrix, it is now possible to take full advantage of the refractoriness of certain pure metals, oxides, borides, nitrides and carbides, by sintering them with the minimum of impurities.

The composition of these bodies is described almost entirely by the name given them, although a small quantity of additions may be necessary to induce the required type of crystal growth.

Refractory Oxides

Alumina, M.P. 2050° C, finds considerable use.
Beryllia, M.P. 2530° C, very high refractoriness.
Chrome oxide, M.P. 2435° C, little used because easily reduced and high vapour pressure.
Magnesia, M.P. 2800° C, very high refractoriness.
Titania, M.P. 1830° C, little used.
Thoria, M.P. 3050° C, application restricted because it is fissionable.
Zinc oxide, M.P. >1800° C, experimental.
Zirconia, M.P. 2690° C, in the stabilised form finds applications.
Lime, M.P. 2572° C, baria, M.P. 1923° C, strontia, M.P. 2430° C, subject to hydration.
Rare earth oxides: CeO_2, M.P. 2600° C, La_2O_3, M.P. 2320° C, Y_2O_3, M.P. 2410° C, HfO_2, M.P. 2810° C, are not sufficiently available.

ALUMINA

This was the first pure oxide body to be exploited commercially and finds widespread application as sparking plugs. There is a tendency for undesirable crystal growth to occur during sintering, which weakens the product. This is prevented by additions of: Cr_2O_3, MgF_2 or $AlPO_4$. Sintering is at 1800–1950° C (3272–3542° F). Sintering at lower temperatures, 1300–1400° C (2372–2552° F), can be induced by using very fine grained alumina with 3–4% additions of $MnO+TiO_2$ or $CuO+TiO_2$ (**C73**).

BERYLLIA

Interest in sintered beryllia has chiefly grown because it is one of the few efficient moderators for nuclear reactors available for use at high temperatures in oxidising conditions. For this use it is essential that it be of the highest chemical purity, no additions to facilitate sintering being permissible. This pure material is difficult to sinter to a completely pore-free piece, which is also essential. Hyde, Quirk and Duckworth (**H130**) describe preliminary

TABLE
Beryllia

No.	Ref.	Description	Burning cone No.	China clay or kaolin	Ball clay	Quartz	Feldspar	Beryl
1	L47	Hard porcelain F	15	45	10	30	15	0
2	L47	Beryl porcelain B	17	45	10	30	0	15
3	L47	Beryl porcelain B-2	15	45	10	0	0	45
4	L47	Beryl porcelain	10/11	48·5		5·8	24·0	21·6
5	T44	Beryl insulator porcelain	9–11	48·54		0·0–5·83	13·30–23·98	21·65–37·63
			Maturing range					
6	L6	High beryllia body	1500–1600° C					
7	B114							
8	S29							
9	S29							
10	S29							
11	S29							
12	S29							
13	S29							
14	S29							
15	S29							
16	S29							
17	L5							
18								
19	L7							
20	R41					5		
21	A113							
22								

		Moles							
	G23	Al_2O_3	MgO	BeO					
23		1	—	1					
24		—	1	10					
25		1	1	—					
26		1	1	24	1600–1725				
27		1	4	24	1600–1630				
28		1	1	8	1625–1725				
29		1	4	8					
30		1	1	4	1680–1725				
31		1	4	96	1500–1635				
32		1	4	4					

work on finding the best ways to obtain maximum density. One of the governing factors is the previous thermal treatment of the beryllia used. Beryllia is obtained from pure beryllium sulphate or hydroxide, and it was found that the most active powder for subsequent sintering was obtained if the heating is to not more than 1000° C (1832° F). (Above this temperature well-defined crystals which are visible under a microscope form and these do not sinter so well.) Hydrostatically pressed pieces can then be

114
Bodies

No.	Beryllia	Alumina	Titania	Magnesia	Thoria	Zirconia	Calcia	Properties	
1								Modulus of rupture (lb/in²)	Thermal conductivity
1								10 420	0·0032 c.g.s.—100° C
2								11 630	0·0042
3								13 850	0·0054
4									
5								Very good electrical properties.	
6	76	4				20	2	High strength at room temperature and 1000° C.	
7	84	7				7	2	Bulk density 3·0. Developed to have high strength and low creep, e.g. for gas turbine blades.	
8	80	10		0	10				
9	80	10		4	10				
10	80	10		8	10				
11	80	10		15	10				
12	90	5		0, 4, 8, 15	5				
13	95	2·5		0, 4, 8, 15	2·5				
14	80	10		0, 4, 8, 15		10			
15	90	5		0, 4, 8, 15		5			
16	95	2·5		0, 4, 8, 15		2·5			
17	√			√		√			
18	√			√	√				
19	√	√	√						
20	95							Creep per ton per in² per hr 1700×10^{-8} at 1000° C.	
21	84					8	8		
22	10			10		80			
								Crushing strength (tons/in²)	Relative thermal conductivity
23	19·7	80·3						53	
24	86·1			13·9				70	
25		71·7		28·3				44	
26	80·9	13·7		5·4				75	10
27	69·5	11·8		18·7				59	12·5
28	58·4	29·8		11·8				68	
29	43·2	22·0		34·8				61	7
30	41·3	42·1		16·6				63	
31	90·1	3·8		6·1				55	11
32	27·5	28·1		44·4					3

sintered to almost theoretical (3·025) density in a hydrogen atmosphere at 1370° C (2498° F) upwards.

Strength of a piece consisting solely of interlocking brittle crystals with no glassy matrix depends on the crystal size, the smaller the better. Denton and Murray (**D26**) found that the crystal size distribution in a sintered beryllia piece may be very irregular with sections of large crystals set in a fine-grained background. They attribute this to the use of too large a range of particle sizes in the raw material. Both hot-pressed pieces and ones slip-cast followed by sintering were investigated and it was found that the application of pressure at high temperatures increased the general grain size.

THORIA

It normally densifies only very slowly when sintered below 2000° C (3632° F) but it can be successfully densified at 1500° C (2732° F) if small amounts of CaF_2, V_2O_5, SrF_2, SnO_2, Bi_2O_3, $CaBr_2$, $CaCl_2$ or NH_4I are added. A 5 wt. % addition of -325-mesh stabilised zirconia is also beneficial (**A212**).

ZINC OXIDE

Zinc oxide is extremely refractory, and does not melt when heated to 1800° C (3272° F), although if it is in powder form it may sublime below that temperature. If it is pressed in moist condition, 20% water, it can be sintered at 1400° C (2552° F) to a very hard product. Libman (**L32**) found that during the sintering the shrinkage increases along a smooth curve but the porosity although decreasing at first in the normal way reached a minimum at 1325° C (2417° F) and then increased again. He attributed this to recrystallisation producing denser single crystals with widening fissures between them. This might make a pure zinc oxide body gradually disintegrate at high temperatures.

ZIRCONIA

It is essential to stabilise the low-temperature cubic lattice to prevent disintegration due to the rapid inversion. The nature of this stabilisation is discussed by Weber and Schwartz (**W33**). It is achieved by one of the following additions:

(a) 8–12% CaO;
(b) 6·7% CaO + 3·7% MgO;
(c) 2·3% CaO + 6·4% MgO (**N35**);

or for special crucibles in which titanium can be heated without contamination titanium metal is added to the zirconia, 15 atomic % being used, *i.e.* 93·55 wt % ZrO_2 plus 6·45 wt % Ti metal powder (**W32**).

Sintering is considerably assisted by the addition of 4% ferric oxide, this body sintered at 1500° C (2732° F) is as dense as untreated zirconia fired at 1700° C (3092° F) (**H129**).

Refractory Spinels

Barium spinel, $BaO.Al_2O_3$, M.P. 2000° C (3632° F).
Chrysoberyl, $BeO.Al_2O_3$, M.P. 1870° C (3398° F).
Magnesia spinel, $MgO.Al_2O_3$, M.P. 2135° C (3875° F).

CERAMIC BODIES

These can all be made from their constituent oxides, in some cases assisted by boric acid as a fugitive flux. These materials can be shaped and sintered by the methods used for other single-component materials.

Refractory Silicates

Beryllium orthosilicate, $2BeO.SiO_2$, M.P. 2000° C (3632° F).
Calcium orthosilicate, $2CaO.SiO_2$, M.P. 2130° C (3866° F).
Forsterite, $2MgO.SiO_2$, M.P. 1890° C (3434° F).
Zircon, $ZrO_2.SiO_2$, M.P. 2500° C (4532° F).

Zircon and forsterite are major constituents of the refractory bricks named after them. All four materials can also be used pure to make sintered ware.

The sintering of zircon is assisted by adding 4% Fe_2O_3, when fired at 1400° C (2552° F) it gives as dense a product as the pure material at 1700° C (3092° F) (**H129**).

Refractory Zirconates

Barium zirconate, $BaO.ZrO_2$, M.P. 2620° C (4748° F).
Calcium zirconate, $CaO.ZrO_2$, M.P. 2350° C (4262° F).
Magnesium zirconate, $MgO.ZrO_2$, M.P. 2150° C (3902° F).
Strontium zirconate, $SrO.ZrO_2$, M.P. > 2800° C (> 5072° F).
Thorium zirconate, $ThO_2.ZrO_2$, M.P. 2800° C (5072° F).

Little is known about these materials.

Refractory Carbides

The carbides are amongst the most stable of solids, some having softening points above 3500° C (6332° F). Several are exceptionally hard and therefore find application as cutting tools. They have fair thermal and electrical conductivity, but only limited stability in air at high temperatures.

Beryllium carbide Be_2C, M.P. 2100° C (3812° F)d.
Boron carbide, B_4C, M.P. 2350° C (4262° F).
Calcium carbide CaC_2, M.P. 2300° C (4172° F).
Hafnium carbide HfC, M.P. 4160° C (7520° F).
Molybdenum carbide, Mo_2C, M.P. 2380° C (4316° F).
Neodymium carbide, NdC_2, d.
Silicon carbide, SiC, M.P. 2700° C (4892° F).
Tantalum carbide, TaC, M.P. 3700–3800° C (6692–6872° F).
Thorium carbide, ThC_2, M.P. 2773° C (4523° F).
Titanium carbide, TiC, M.P. 3140 ± 90° C (5684° F).
Tungsten carbide, WC, M.P. 2777° C (5031° F).
Uranium carbide, UC_2, M.P. 2400° C (4352° F).
Vanadium carbide, VC, M.P. 2830° C (5126° F).
Zirconium carbide, ZrC, M.P. 3540° C (6404° F).

Silicon carbide is made by heating coke, sand, sawdust and salt in a resistance furnace with a coke heating element to 2200° C (3992° F). It is usually bonded with clay varying from 8 to 60% of the batch.

Tungsten and molybdenum carbides are prepared by grinding the powdered metal with carbon black and heating to 1600° C (2912° F) for 2 hr in a graphite container, and cooling in hydrogen atmosphere. The material is wet ball milled in carbide mills to the desired size.

Titanium, vanadium and neodymium carbides are made by heating the oxides with carbon to 1900° C (3452° F).

Boron carbide is made from a mixture of boric oxide and petroleum coke with 10% excess boric oxide to allow for volatilisation. The mixture is melted in an arc furnace at 2600° C (4712° F). The product is ground in steel mills, acetone washed, and dried (*see also* metal-bonded carbides).

Refractory Nitrides

There are a number of nitrides of high refractoriness and sufficient stability when heated in vacuum. They are mostly electrical conductors

Aluminium nitrides, AlN, M.P. 2200° C (3992° F)d.
Barium nitride, Ba_3N_2, M.P. 2200° C (3992° F).
Beryllium nitride, Be_3N_2, M.P. 2200° C (3992° F).
Boron nitride, BN, M.P. 2730° C (4946° F).
Hafnium nitride, HfN, M.P. 3300° C (5972° F).
Silicon nitride, Si_3N_4, M.P. 2170° C (3938° F) Subl.
Scandium nitride, ScN, M.P. 2650° C (4802° F).
Tantalum nitride, TaN, M.P. 3360° C (6080° F).
Thorium nitride, Th_3N_4, M.P. 2100° C (3812° F).
Titanium nitride, TiN, M.P. 2900° C (5252° F).
Vanadium nitride, VN, M.P. 2050° C (3722° F).
Zirconium nitride, ZrN, M.P. 2950° C (5342° F).

The preparation of some of these materials is outlined below.

TITANIUM NITRIDE

$$TiO_2 + 2CaH_2 \rightarrow TiH_2 + 2CaO + H_2$$
$$TiH_2 \rightarrow Ti \text{ (finely divided)} + H_2$$
$$Ti \text{ (finely divided)} + NH_3 \rightarrow TiN + \tfrac{3}{2}H_2$$

or

$$TiCl_4 + xNH_3 \rightarrow TiCl_4 \cdot xNH_3$$
$$TiCl_4 \cdot xNH_3 \rightarrow TiN + 4NH_4Cl + H_2 + \tfrac{1}{2}N_2 + (x-6)NH_3$$

Zirconium, thorium, beryllium, tantalum and vanadium nitrides are made similarly, although percentage conversion varies.

BORON NITRIDE

$$Na_2B_4O_7 + 2N_2 + 7C \xrightarrow{1400°C} 2Na + 7CO + 4BN$$
$$\text{borax} + \qquad\quad \text{lampblack}$$

Boron nitride has good chemical stability so long as fair sized crystals (min. 3 μ) have developed, great refractoriness, very high electrical resistivity, low thermal conductivity and light weight. Moulded and self-bonded pieces can be produced to densities up to 2·10 g/cm³. It should find nuclear applications (**N27**).

Refractory Silicides

Some of the silicides of Groups III and VII are refractory, *e.g.* tantalum silicides; molybdenum silicides, $MoSi_2$ density 6·24, M.P. 1850° C (3362° F); tungsten silicide (**K57, M84**).

Refractory Sulphides

Some of the sulphides of barium, cerium, thorium and uranium show possibilities as refractories for use in high vacua as they have low vapour pressures (**N35**). Crucibles have been made from individual sulphides or their mixtures and are useful in the casting of molten metals (**K57**).

Refractory Borides

The borides of the transition metals of Groups IV, V and VI in the Periodic Table show considerable refractoriness, hardness and high conductivity at all temperatures. The following having been tried as ceramic materials:

Borides

Boride	M.P. (°C)	M.P. (°F)	% boron by wt.	Density
Calcium boride, CaB_6			61·8	2·45
Titanium boride, TiB_2	2900	5252	31·1	4·50
Zirconium diboride, ZrB_2	3060	5540	19·2	6·08
Zirconium dodecaboride, ZrB_{12}			58·8	
Uranium dodecaboride, UB_{12}			35·3	(**N27**)
Chromium boride, CrB	2760	5000		6·17
Cerium boride				
Niobium boride	2000	3632		
Tantalum boride	2000	3632		
Molybdenum borides				
Tungsten borides, W_2B and WB				
Thorium boride				(**K57**)

33. CERMET BODIES

Characteristic properties of selected ceramic bodies combine high refractoriness, resistance to oxidation, high compressive strength not easily lost on heating, and electrical insulation with brittleness and poor thermal shock resistance. Metals on the other hand are ductile and have excellent

TABLE 115
Cermet Bodies

Ref.	Composite type	Non-metal	Metal	Characteristics
W43	Impregnated	Preformed porous carbon	Molten silicon	Refractory for furnace parts and heat exchangers with superior thermal shock resistance, moderate strength, oxidation resistance and abrasion resistance.
W43	Laminated	Powdered graphite plus silicon carbide forms.	Flake silver or copper	Current collector brushes of low friction, high wear resistance, high electrical and thermal conductivity, and lower electrical resistance parallel to direction of current flow than perpendicular.
B72	Bonded particulate	Alumina 70	Chromium 30	Very suitable for high temperature service, good impact resistance and thermal shock resistance, high modulus of rupture and slow oxidation during service.
F29	Bonded particulates	Silicon carbide / Titanium carbide / Zirconium carbide / Boron carbide / Niobium carbide / Tantalum carbide / Tungsten carbide	Cobalt / Nickel / Chromium / Iron / Ferrosilicon / Niobium / Silicon (various combinations)	
F29	Self glazing / Bonded particulates	Titanium carbide / Tungsten carbide / Zirconium carbide / Titanium carbide / Titanium carbide	Nickel / Cobalt / Niobium / Chromium / Iron	High temperature scale resistant material. Applicable to stainless steels, high duty alloy steels, inconel, etc., adding 82° C (150° F) to the operating range of high-duty alloys.
A92 / A68	Flame-sprayed coating	plus small amount Chromium boride Magnesia	molybdenum and silicon Nickel-chromium Nickel	
A70	Cold slip coating	Ceramic frit 9·5 Enamelling clay 4·8	Cr-B-Ni 85·7	Applied as aqueous slip and then fired to fuse the frit, the coating gives protection against oxidation to low-alloy steels at 815° C (1500° F).
C66	Bonded particulate	Thoria	Molybdenum	Thermionic cathodes that can be heated by impressing a high voltage directly across the terminals, whose resistivity can be varied with composition, low vapour pressure, high refractoriness, adequate thermionic and secondary electron emission, high hot strength, smooth surface.
H47	Bonded particulate	Alumina 40-70	Iron 60-30	Turbine blades.

H47	Bonded particulate	ZrO$_2$	Fe$_2$O$_3$	Cu	Electrical resistivity (ohm-cm)	Temp. coeff. of resistance (room temp.)
		20	80	—	3×10^3	Neg.
		20	60	20	6×10^3	Neg.
		40	60	—	7×10^3	Neg.
		40	40	20	3×10^4	Neg.
		60	40	—	4×10^6	Pos.
		60	20	20	9×10^4	Neg.
		80	20	—	8×10^6	Pos.
		80	—	20	8×10^5	Neg.

TABLE 115—(contd.)

Ref.	Composite type	Non-metal	Metal	Characteristics
S219d	Bonded particulate	Molybdenum boride $Mo_2B+Mo_2NiB_2$	Nickel	Cutting tool, Rockwell A hardness 75 to 87. Somewhat better for machining titanium than tungsten carbide of same hardness.
D37	Bonded particulate	Titanium carbide	Various alloys including Molybdenum Aluminium Chromium	Service in aircraft engines where refractoriness, thermal shock resistance and resistance to oxidation required.
H35	Refractory coating	3 coats, fusible base coat and glaze (same composition) and refractory cover coat	Molybdenum	Protection against oxidation of molybdenum. Long term at 900°C (1652°F), short term at 1650°C (3002°F).
		details →		(see sub-table below)
M84	Bonded particulate	Alumina silicate	71.4–91.0% stainless steel	
M84	Refractory coating	Molybdenum disilicide	Graphite	Graphite nozzles for hot gases.
M84	Refractory coating	Alumina		
M84	Refractory coating	Zirconia		
M84	Refractory coating	Silicon carbide		
M84	Refractory coating	Nickel magnesia cermet	Stainless steel, Inconel	
M84	Refractory coating	High barium frit plus phosphate, beryllia, lime, ZnO, titania, CeO_2, Cr_2O_3/CeO_2		High temperature parts in nuclear reactors.

Details for H35:

Coat type	Frit	ZrO_2 −325	Fla. kaolin	Calc. kaolin	ZrO_2 −40+80	Na_2NO_3	Water	Milling time (hr)	Firing temp. °C	Exp. coeff. $\times 10^{-6}$
Base	80	20	5			0.1	42.5	4	1180	5.14
Base	80		5			0.1	42.5	4	1180	5.26
Cover		95	5				50	$\frac{1}{2}$	1120	6
Cover			5	94	20		35	$2\frac{1}{2}$	1120	6.22
Cover			6		95		46	$2\frac{1}{2}$	1120	4.25

Frit:
SiO_2 73.3%
B_2O_3 9.1
Al_2O_3 6.6
Na_2O 6.7
BaO 2.1
CaO 0.8
K_2O 0.7

thermal shock resistance while having poor refractoriness, hot strength and oxidation resistance. Nuclear, jet and electronic engineers, however, require materials of high refractoriness, hot strength, good thermal shock resistance and particular electrical conductivities. Attempts were therefore made to combine ceramic oxides, carbides, borides, nitrides, etc., with metals to fulfil these special needs. The products are known as 'cermets' or sometimes 'ceramals'. Most of the research in this field is unfortunately 'classified', so that very little has been published. A brief outline only can be given here.

Composite bodies may be classified according to physical form of the combination:

(1) Coated bodies.
(2) Laminates.
(3) Impregnated bodies.
(4) Bonded particulate mixtures.
(5) Fibre composites.

Westbrook (**W43**) summarises factors that must be considered when trying new metal–ceramic combinations.

(1) The possibility of forming a bond between the two materials. Although little positive knowledge is available about bonding mechanisms a qualitative understanding has been arrived at which describes four types of bond.

(*a*) Mutual solubility, resulting either from the limited solubility of one component in the other as in cobalt-cemented carbides, or to solubility in the ceramic component of a surface phase on the metallic one as in chromium-bonded alumina where chromium oxide forms on the metal surface.

(*b*) Chemical reaction, as is believed to occur in iron–alumina combinations to give iron spinel $FeO—Al_2O_3$.

(*c*) Mechanical bonding due to good wetting of one phase by the other as in silver–graphite bodies.

(*d*) Addition of a small amount of a third phase to give a bond by solution or reaction with the major components.

(2) Relative thermal expansion coefficients over the range of fabrication and application.

(3) Possibility of undesirable chemical reactions or formation of eutectics.

(4) Mechanical design of components to allow for the difference of physical properties between metals and cermets, in particular the negligible or limited ductility of the latter. Cermets also have much smaller tensile strength than compressive strength.

TABLE 115

Metal-bonded Carbides or 'Hard-metal' or Cemented Carbides

The refractory carbides are difficult to sinter in the pure state, so that where their hardness is of greater importance than their refractoriness they are bonded with a small percentage of metal, usually cobalt. 3–20% of metal are normal quantities. The powdered carbide and powdered metal are mixed, dry-pressed and then sintered. The product therefore although containing free metal has been made by ceramic methods.

The greatest application of this type of body is in making tungsten carbide cutting tools. Small quantities of tantalum and titanium carbide may be added to the tungsten carbide. The metal bond is always cobalt (for compositions, see Chapter 15, p. 1154).

Investigations on metal-bonded boron and titanium carbides have also been reported (**A73**). The two carbides were used both separately and mixed together in varying proportions. The metals used were 20–40% of nickel, iron, cobalt, titanium and chromium. It was found that the two carbides tend to interact at 1650° C (3000° F) to give titanium boride, TiB_2, which is highly refractory and will not readily sinter. The metals all react with boron carbide to give more or less refractory borides which also inhibit sintering, as does the free graphite which is liberated in the reaction. In general, then, it is necessary to study the high temperature reactions of these materials more thoroughly.

34. RUTILE BODIES

The mineral rutile (TiO_2) has a high dielectric constant of 173 parallel to the crystal axis, and 89 perpendicular to it, giving a mean of 114 for random arrangement of the crystals. It also has the unusual feature of a negative thermal coefficient of dielectric constant. It therefore finds application as a major constituent of specialised electrical ceramic bodies.

Very pure and fine titania is used for these bodies, but as this has high shrinkage it is wholly or partly calcined before use. Calcination temperatures are 950–1150° C (1742–2102° F) which therefore also convert any anatase (density 3·84) to rutile (density 4·26), the transition temperature being $915 \pm 15°$ C (1679°F). Anatase is more plastic than rutile and is therefore advantageous when using raw titania.

Pure titania cannot readily be vitrified so that fluxes must be added. Even small quantities of these reduce the dielectric constant and increase the power factor. Fluxes that have found application are:

Titanates of Ba, Sr, Ca, Mg, Be and Pb.
Calcium titanium silicate $CaO.TiO_2.SiO_2$.
Glasses containing TiO_2, SiO_2, B_2O_3, PbO, Al_2O_3, CaO including the
  eutectic SiO_2, 43·3; Al_2O_3, 13·8; TiO_2, 10·2; CaO, 32·8.
MgO, BeO, $BaCO_3$.
Talc, feldspar, clay.

Zirconium and lanthanum oxides lengthen the firing range, and may prevent reduction during firing, and also the development of coarse crystallinity.

Tungstic oxide, bismuth oxide and antimony pentoxide are said to act similarly, but may cause semi-conductivity. However, Zr, Sn, rare earths, Al, Zn, Cd and B should be used in only small amounts if high dielectric constants and low power factors are required.

The use of clay or bentonite is necessary for shaping large pieces but should normally be kept to a minimum. In some cases clay is deliberately used to give a body of zero capacitance change (**N11**, **B80**).

The change of permittivity and temperature coefficient with clay content is given approximately in Table 116.

TABLE 116

Rutile	Clay	Dielectric constant	Temperature coefficient parts per 10^6 per 1° C
100	—	114	−800
90	10	80	
80	20	50	
70	30	40	
60	40	30	
50	50	20	
40	60	18	
30	70	12	—
20	80	10	0
10	90		+

Rutile bodies have a low power factor at high frequencies but it increases at low frequencies. This can be overcome by incorporating about 10% zirconia in the body, as is seen in many of the commercial bodies.

For insulating bodies it is essential that titania be fired in an oxidising atmosphere. Reduction leads to loss of oxygen and the development of darker colour, higher power factor and eventually semi-conduction. This fact is used in the manufacture of semi-conducting thread guides which can dissipate static electricity (**B80**).

Rutile bodies have also been developed having dielectric constants considerably greater than pure rutile, *i.e.* from 100 to 2000. These bodies usually have a higher power factor than normal rutile bodies and a positive temperature coefficient of the order of +1800 parts per 10^6 per °C.

In some cases titanium peroxide plays a part in the composition and assists in stabilising the various properties. Some of the bodies are fired in a reducing atmosphere containing carbon dioxide or nitrogen and their properties are very greatly dependent on the firing conditions (**A18**).

35. TITANATE BODIES

Titania combines with magnesia, zirconia, baria, lead oxide or beryllia to form titanates. Such bodies have dielectric constants lower than pure rutile bodies, but have very low power factors and also their temperature coefficient of permittivity is very small, which gives them preference for oscillatory circuits for radio-equipment.

Magnesium titanate, $MgO.TiO_2$, finds use as the main constituent of such a body (dielectric constant ~ 18). Like pure rutile it has better dielectric properties, especially breakdown strength, at high frequencies than at lower ones, but this can be corrected by the addition of zirconia or zinc titanate. The body is made from calcined mixes of magnesite or magnesia with titania, clay and fluxes and fired at 1500–1600° C (2732–2912° F).

Calcium and strontium titanates have higher dielectric constant and power factor than magnesium titanate and temperature coefficients more negative than that of titania. Calcium titanate is sometimes used in titania bodies to reduce power factors due to impurities, and in magnesium titanate bodies to give a zero temperature coefficient of capacity.

Barium titanate shows characteristics not found in other materials in the ceramic field. The body is made basically from barium carbonate and titania in equal molecular proportions by mixing, calcining, grinding and firing, but the pure body has a firing range of only 10° C. Even with traces of impurity or slight excess of either component the maturing temperature varies between 1250 and 1450° C (2282–2642° F). Attempts have been made to obtain a wider firing range by mixing dry-milled high-maturing body with fine wet milled low-maturing mixture. From 0·05 to 5% of magnesium titanate, calcium fluoride or magnesium fluoride improve firing properties. Fluorides improve thermal shock resistance and magnesium can improve dielectric properties (**B80**).

Barium titanate exists in several modifications:

below $-70°$ C trigonal, ferroelectric;
-70 to $+5°$ C orthorhombic, ferroelectric;
$+50°$ C to $+120°$ C tetragonal, ferroelectric;
above 120° C cubic, non-ferroelectric.

Ferroelectric properties are characterised by high dielectric permeability, ferroelectric hysteresis, polarisation saturation, piezo effect and a Curie temperature.

The high dielectric constant but not the other properties make the material suitable for capacitor use. By suitable adjustments and additions to the body composition the undesired properties can be sufficiently suppressed to give useful capacitor bodies. However, the fact that barium titanate bodies show an ageing effect in their dielectric constant and power factor must be allowed for.

It is possible that the ferroelectric properties of barium titanate will find their uses as capacitors in non-linear circuit elements. Ideally a body should show small capacity change with temperature, but large capacity change with variations in applied field strength is required.

Barium titanate possesses true piezoelectric properties, showing a strong

TABLE

Rutile and

No.	Ref.	Description		Burning cone No.	Raw titania	Calcined titania		Clay	Bentonite
1	N11	Rutile				90·00			
2	A18	Lutz & Co.	C267		9·5	@1000° C 80·7			
3		Rutile body	C260 or 987			@1000° C 80 / @950° C 5			
4		Rutile body	C721 or 964			@1000° C 69 / @950° C 10			
5		Rutile body	C722			@1050° C 69 / @950° C 10			
6		Rutile body	336		97				
7		Rutile body	C250		TiO₃ 2	@1000° C 75 / @950° C 5			
8		Lutz & Co. high permittivity rutile bodies:	164		TiO₃ 13	@950° C 70 / @1000° C 10			
9			56		94				
10			75		78	@950° C 14			
11	T31	Steatit Magnesia High voltage	U	7–8	53	cone 11	20	Schippach 2	6
12		Kerafar bodies:	W	7–8	40	cone 11	15	Schippach 2	6
13			X	10	42			Luckenau 3 Kemmlitz kaolin 4	2

response to treatment with an electric polarising field for a short period of time. The material to be polarised is heated above its Curie temperature, an electric polarising field is applied, and the material is cooled under applied voltage. The piezoelectric response of prepolarised barium titanate drops to about 80% of its initial value after a short period but remains practically constant thereafter. The strength of the piezoelectric effect as measured by the coupling coefficient is 50% which makes it very useful in transducers.

Even better piezoelectric properties, higher operating temperatures and better stability are obtained by admixing some lead titanate with the barium titanate (**H100**). Other additions, however, markedly reduce the piezo-

117

Titanate Bodies

No.	RO	Titanates	RO$_2$	Miscellaneous	Temp. coeff. parts/10^6	Diel. Const.	Diel. Strength	Power Factor
1	MgO 3·33 BeO 3·33	Calcium titanium silicate 3·33				85	144	0·00036
2	BaCO$_3$ 4·5		Zirconium hydrate 5·0		−700	97		0·00004
3	Lanthanum oxide hydrate 10		Zirconium hydrate 5·0		−600	85		0·0001
4	Lanthanum oxide hydrate 10 BeCO$_3$ 1		Zirconium hydrate 10		−700	87		0·0001
5	Lanthanum oxide hydrate 10 BeCO$_3$ 1		Zirconium hydrate 10		−700	110		0·0001
6	Lanthanum oxide hydrate 1		Zirconium hydrate 2		−800	105		0·0001
7	Lanthanum oxide hydrate 10		Zirconium hydrate 8		—	105		0·00015
8	Lanthanum oxide hydrate 7				+1800 to +2000	400 to 1400		0·018
9	Li$_2$CO$_3$ 4		Cerium carbonate 2			840		0·144
10	Lanthanum oxide hydrate 2		Cerium carbonate 6			910		0·255
11	BaCO$_3$ 6		ZrO$_2$ calc. cone 14 8	Chinese talc 5	−650 to −750	64		0·0003 to 0·0010
12	BaCO$_3$ 6 CoO 0·1		ZrO$_2$ calc. cone 14 8	Chinese talc 23	−350 to −450	32 20		0·0003 to 0·0010
13	BaCO$_3$ 8		ZrO$_2$ calc. cone 14 38	Feldspar 3	−50			0·0003 to 0·0008

electric coupling of barium titanate, namely one mole per cent additions of silica, magnesia, iron oxide or excess titania (**B107**).

In the processing of piezoelectric and high-K type of dielectric bodies, of which barium titanate is a major component, firing conditions have a marked effect on the resulting ceramic and electrical properties. It is indicated that gas-fired kilns are more conducive to producing the optimum properties than electric-fired kilns when normal operating conditions are used. However, when humid air, oxygen or carbon dioxide atmospheres are introduced into the electric-fired kiln, optimum ceramic and electrical properties are obtained which are comparable to those attained with the gas-fired kiln;

atmospheres of argon and nitrogen produce no improvement over normal firing conditions. Results also indicate that fired density rather than moisture absorption control the optimum electric properties of this type of ceramic (**B44**). Brajer (**B107**) describes barium titanate ceramics fired in oxygen instead of air that have densities of about 99% of the single crystal density.

Much research into the relationships in the titanate systems has occurred recently. Phase rule data on the systems $BaO-TiO_2$ (**R3, M14**), $BaTiO_3-CaTiO_3$ (**D36**), $BaTiO_3-SrTiO_3$ (**B16**), $CaO-TiO_2-SiO_2$ (**D37**), (Ba, Ca, Sr)TiO_3 (**M14**), (Ba, Ca, Pb) TiO_3 (**M16**) being of considerable assistance in the development of mixed bodies such as $BaTiO_3 + SrTiO_3$, $BaTiO_3 + PbTiO_3$, $PbTiO_3 + SrTiO_3$, $BaTiO_3 + SrTiO_3 + PbTiO_3$, and also additions of stannates and zirconates (**B80**).

36. FERRITES

The earliest known magnetic substance, magnetite, Fe_3O_4, is ferrous ferrite $FeO.Fe_2O_3$. It has a structure similar to the mineral spinel. Other ferrites of divalent metals also have spinel structures, some having the normal spinel structure and some the inverse spinel structure, also found in the spinel series. Those ferrites which crystallise in the inverse spinel structure are ferromagnetic, *e.g.* nickel, cobalt, manganese, magnesium and copper ferrites, whereas zinc and cadmium ferrites which have the normal spinel structure are non-magnetic (**M15**).

Ferrites are non-metallic and can be sintered into hard, dense pieces by ceramic methods. As they also have high electrical resistance compared with metallic magnets their obvious advantages have led to intensive research and development. The first commercial magnetic ceramics were made in 1947 by Snoek (**S146**), eleven years later Dean (**D19**) reviewed progress.

The simple magnetic spinels have low permeabilities and high losses, the latter being largely due to magnetic anisotropy. This is temperature dependent and becomes small near the Curie temperature. The Curie temperature is reduced to within the range 100–200° C (212–392° F) by forming a solid solution of zinc ferrite in a magnetic ferrite, the best being Mn–Zn ferrite and Ni–Zn ferrite. Furthermore all single ferrites have negative magneto-restriction with the one exception of magnetite, which has a very large positive magneto-restriction. By mixing manganese zinc ferrite of magneto-restriction -4×10^{-6} with ferrous ferrite with a value of $+30/40 \times 10^{-6}$ a very low value is obtained with a considerable increase in the permeability.

Simple ferrites other than ferrous ferrite have high electrical resistivity, up to $10^8 \ \Omega/cm^3$, which makes eddy current losses small. Inclusion of ferrous ferrite to give better permeability lowers resistivity to 100–500 Ω/cm^3 but good properties up to 1 Mc can be obtained. Nickel zinc ferrites have high resistivity but low permeability but at high frequency freedom from eddy current loss is more important than high permeability so that this material is useful up to 20 Mc.

Manganese–zinc ferrites containing ferrous ions must be very carefully fired with controlled atmosphere both during firing and cooling as they contain two elements capable of two valency states, and slight changes in their state of oxidation lead to marked changes of magnetic properties. The resistivity of this material can be increased by small additions of calcium which forms calcium ferrite at the grain boundaries.

Hysteresis losses are affected by grain size, grain size distribution and grain shape of the ferrite, the best being uniform-sized small grains of minimum surface area. This requires most careful control of the raw materials and their preparation.

The temperature coefficient of permeability is controlled by the amount of ferrous ion present over small ranges and over a wider range by replacing some of the ferric ions by other trivalent ions, *e.g.* aluminium.

Rectangular hysteresis loops are obtained with magnesium manganese ferrites and copper-manganese ferrite, their coercive force being greatly reduced by incorporating zinc ferrite. They find extensive use for information storage in high speed computers.

Ferrites of hexagonal instead of cubic structure are formed by ions of larger size than the ferrous ion. Barium ferrite $BaO.6Fe_2O_3$ has a very large crystal anisotropy and has a very high coercive force, it can be used as a permanent magnet material. Various additions modify this type of structure (**D19**).

Blackman (**B74**) describes detailed work on magnesium–manganese ferrite.

37. PERMEABLE POROUS BODIES

A number of specialised porous bodies have been developed of recent years, whose pores are connected and of carefully regulated sizes. The solid part of these bodies is derived from one of the bodies already described for other technical purposes, but the processing is different. There are three ways of achieving porosity:

(1) Cessation of firing before vitrification sets in.
(2) Inclusion of combustible or volatile matter.
(3) Bonding grog of a single size.

1. Unvitrified Ware

Body mixtures resembling those used for technical porcelain but adjusted to contain less flux can be fired to a fine-grained porous texture especially suitable for electrolytic diaphragms. In general the method is not suitable for highly porous or large pored bodies as these are very fragile and easily become irregular.

2. Inclusion of Combustible or Volatile Matter

This method, already described under 'Thermal Insulation Bodies' leads to irregular-sized pores and poor strength. It may be used for inexpensive ware.

3. Bonding of Grog

Carefully prepared and graded natural or ceramic granules are mixed with the minimum of bonding material and dry-pressed into shape. Maximum pore volume with the most regular pore size is obtained by using granules of as small a size range as possible.

TABLE 118
Some of the Granule Types for Permeable Ceramics

Granules	Bond	Characteristics
Diatomite	Low-melting soda–lime glass	
Quartz sand	Ground glass	
	Silicates	
Stoneware	Clay and water glass	
Porcelain		
Alumina	Aluminous glass	Refractoriness up to 1000° C
Activated alumina	Finer activated alumina	(D18)
Fireclay grog	Clay and water glass	Refractoriness
Silicon carbide		
Carbon	Tar	Alkali proof
(Gas coke)		Thermal shock resistant
Graphite	Tar	Alkali proof
		Thermal shock resistant
		Electrical conductor

General characteristics of permeable porous ceramics when compared with other permeable materials are their rigidity, mechanical strength, chemical resistance and wide available temperature range, very high accuracy and uniformity of pore size combined with good permeability.

PHOSPHATE BONDS

The ability of phosphoric oxide and phosphates to form glasses of structures not dissimilar to silicates leads to the supposition that phosphates might make useful silica-free bonds for ceramic materials. In fact aluminium phosphate, $AlPO_4$, resembles silica, $SiSiO_4$, very closely in structure and some properties. Kingery (**K25**) made a literature review on phosphate bonding, and followed this by experimental work. The former showed that both siliceous material and certain oxides would form a bond with phosphoric acid on heat treatment. Certain acid phosphates had been used as refractory bonds, but the only data available were on dental cements which are cold setting and use zinc phosphates. The experimental work showed that cold-setting cements may be formed by many cations by oxide–phosphoric acid, phosphate–phosphoric acid reactions or direct use of liquid phosphates. For oxide–phosphoric acid reactions to give suitable bonds the oxide should be weakly basic or amphoteric, since excess basicity causes violent reactions and acidic or inert materials do not react. In all cases the bond is an acid phosphate. For optimum bond strength a weakly basic cation of small ionic radius is required so that a disordered structure results. This leads to the further investigation of aluminium phosphate.

Heating mortars of aluminium phosphate showed no weak zone, indicating that the water loss and crystallisation of the bond are gradual. Aluminium phosphate is adsorbed on the surface of clays, especially when these are ground finely. In this condition it does not migrate to the surface during drying as does a soluble salt or the phosphate mixed with an unground clay.

TABLE 119

Typical Phosphate Bonded Alumina Composition

	% wt.
Tabular alumina, 14–28-mesh	38
Tabular alumina, 28–65-mesh	13
Tabular alumina, 65–150-mesh	9
Tabular alumina, minus 150-mesh	13
Tabular alumina, minus 325-mesh	4
Aluminium hydrate	9
Ammonium fluoride	1
85% phosphoric acid with 1% Rodine 78 added	9
Water	4

Experimental mortars were then made with fireclay grog, kaolin and aluminium phosphate $Al(H_2PO_4)_3$, or magnesium phosphate $Mg(H_2PO_4)_2$. The monomagnesium phosphate mortars are satisfactory up to 1500° C (2732° F), the monoaluminium phosphate ones above that temperature. A mortar using fused alumina grog and aluminium phosphate bond appeared good up to at least 1850° C (3362° F).

Buckner (**B132**) claims a satisfactorily bonded diamond abrasive using aluminium phosphate.

Recently Sheets, Bulloff and Duckworth (**S63**) have shown that alumina, zirconia, mullite, beryllia or silicon carbide can all be phosphate bonded with cold-setting mixtures that are as refractory as the main refractory material. In the case of alumina it was found more satisfactory to use phosphoric acid to react with alumina *in situ* than aluminium phosphate. However, the acid then reacts with any metallic impurities present releasing hydrogen and causing bloating. This side reaction was successfully inhibited with 'Rodine

78', a complex amine. Setting without drying has to be encouraged by additives, *e.g.* 1% ammonium fluoride in the presence of the Rodine.

Phosphate bonding of talc is reported to be very satisfactory by Comeforo, Breedlove and Thurnauer (**C49**). Here orthophosphoric acid, aluminium phosphate $Al(H_2PO_4)_3$ and magnesium phosphate $Mg(H_2PO_4)_2$ were tried and the first found to be the most satisfactory. Ground talc of suitable grain size grading was mixed with 3·6 to 12·6% phosphoric acid, dry-pressed or hydrostatically pressed and cured at 850° C (1562° F). Hot-pressing can also be used. Such bonded talc is then similar to natural block talc, it can be machined with normal steel tools and has the advantage of not having veins or faults as in the mineral.

Chapter 6

Glazes

FUNCTIONS

THE thin, hard, shiny and usually transparent layer covering a great many ceramic wares has a very important role to play.

Glazes are thin layers of glass fused on to the surface of the body. They are applied to bodies to make them impervious, mechanically stronger and resistant to scratching, chemically more inert and more pleasing to the touch and eye.

Glazes are required to 'fit' bodies of varying chemical and physical natures, they must mature at a variety of temperatures and exhibit specific but various properties when finished, so it is not surprising that there are numberless different glaze compositions, making it very difficult to classify them systematically.

GOVERNING FACTORS FOR GLAZE COMPOSITION

The method of preparation and application of glazes is one of their determining factors. The constituents are finely ground in aqueous suspension which is then applied to the dry raw or biscuited body. It is therefore necessary that the raw materials put in the mill are water insoluble and this may entail their previous treatment by fritting. The glazed body is next dried, when the glaze must adhere regularly, otherwise crawling may occur. It is then fired, when the glaze mixture must fuse and become homogeneous without becoming so fluid that it begins to flow off the vertical or inclined portions of the article. Where the body has not been previously fired the glaze composition must be such that it matures under the same conditions as the body. Where the body has already been fired to completion it is naturally convenient if the glaze matures at as low a temperature as is compatible with its subsequent use. Unlike a glass, which is made in large tanks which can be stirred and where scum can be removed, the glassy layer that is to become a glaze must become homogeneous without any mechanical aid. It is therefore very important that the raw materials should be of known composition and should not introduce any matter that will not become part of the vitreous phase. *The first essential then is to make up a mixture that will fuse to a homogeneous, viscous glass at a desired temperature.*

During and after fusion the glaze components react with the body surface to form a bonding intermediate layer. Proper interaction is very important and depends not only on the total composition of the glaze but also on the individual compounds used to introduce the constituent oxides. *The glaze batch must be made so that the right amount of interaction with the body occurs.*

On cooling the fired glazed body the whole contracts. If the coefficients of expansion of the glaze and body are not sufficiently close together stresses and strains will be set up resulting in 'peeling', 'crazing', or even 'dunting'. *The coefficients of expansion of body and glaze must be mutually compatible.*

The finished glaze must be hard, smooth and glossy (except matt, etc., glazes). This is not only for the visual effect; a smooth surface is more resistant to chemical and physical attack, it is less likely to fracture. By applying a glaze of slightly lower coefficient of expansion than that of the body the cooled glaze is brought into slight compression and the mechanical strength of the piece improved.

Visually a transparent, colourless, brilliant glaze may not always be desirable. A discoloured body can be made to appear white by application of a white opaque glaze. Further, matt glazes with a wax-like surface, crystalline glazes with a few large crystals, and numberless coloured glazes can be made. *The glaze composition must be adjusted to give individual chemical, mechanical and optical properties.*

The finished glaze may differ in composition from the glass that could have been made from the same raw materials due to differential volatilisation and interaction with the body. The extent of such losses of the constituents depends in part on the batch composition of the glaze and the order in which the various materials in it react and fuse.

THE FORMULATION OF GLAZES

Unlike the glasses described on p. 201, which were developed for research into their structure, most commercial glazes have a large number of constituents. It is therefore desirable to have a system of representing the composition that permits easy comparisons. The first step is to convert the 'batch composition' or recipe to relative quantities of the constituent oxides. In the glass industry it is customary to use the percentage composition as a general means of comparison, but in ceramics it is usual to use some form of molecular ratio. The two are of course readily interconvertible.

The actual presentation of a molecular formula can be done in several ways.

Zachariasen (**Z2**) suggested the form $A_m B_n O$ where A are network modifiers and B network formers, m and n being calculated for O as unity. Others make Si unity and calculate the values of A and O. Difficulty arises in comparing formulae containing network formers of a different valency to that of silicon, e.g. B^3, or when network modifiers of varying valencies are used, and also when elements of intermediate properties are present it is hard to know whether to place them under A or B. Such methods therefore although used in research work are not in general industrial ceramic use.

Mellor suggested using percentages of molecules but as the Seger method had already taken root Mellor's idea was never put into practice (**M63**).

TABLE 120

Modified Seger Method of Presenting Molecular Formulae of Glazes

Mono- and divalent (summing to one)		Trivalent		Tetravalent		Pentavalent		Other anions
IA	Li_2O							
	Na_2O							
	K_2O							
	Rb_2O							
	Cs_2O	III	B_2O_3	IV	SiO_2	V	P_2O_5	F_2
IIA	BeO		Al_2O_3		GeO_2	VA	V_2O_5	
	MgO	VIA	Cr_2O_3	IIIA	CeO_2	VB	As_2O_5	
	CaO	VIII	Fe_2O_3	IVA	TiO_2		Sb_2O_5	
	SrO	VB	As_2O_3		ZrO_2			
	BaO		Sb_2O_3	IVB	SnO_2			
VIII	FeO		Bi_2O_3					
	CoO							
	NiO							
IIB	ZnO							
	CdO							
IIIB	Tl_2O							
IVB	PbO							

The method of presenting molecular formulae of glazes in general use is that originated by Seger (**S52e**) somewhat modified. The oxides are listed in three or more columns, the first containing those of monovalent and divalent elements, the second those of trivalent and the third those of tetravalent elements. The sum of the first column is brought to one.

The division of the elements into columns by their valencies is not as arbitrary as it might seem. Seger's (**S52e**) original idea was to have the three columns 'basic', 'alumina' and 'acidic', and he placed boric oxide with silica. Later boric oxide was moved to the second column because of its similarity to alumina in its valency and certain effects (**P75, B54, S80**).

Present knowledge of the atomic structure of glasses (*see* p. 201) verifies some of Seger and his successors' work, and it is found that all the network modifiers fall in the first column, the network formers in the second, third and fourth columns, and the intermediate oxides are spread across the first four. However, if the elements in each column are listed according to their group, and place in that group, of the Periodic System, it is found that the intermediates tend to be at the bottom of the columns. This method will therefore be adopted in this book as far as possible.

An example is given of the calculations required to present the composition of a typical leadless earthenware glaze in different ways (Table 121).

In the works the most convenient method of noting a glaze is according to the raw materials required and is known as the 'batch composition'. Those compounds that are water soluble are fritted with a certain amount of alumina and silica to give insoluble aluminosilicates before grinding. As the same frit may be used in different mill mixtures for various glazes it is convenient to calculate its composition independently of the total glaze. This is shown in section III of Table 121. In column (2) the percentage of the raw material is multiplied by the fraction of it that will enter the frit, e.g. whiting

TABLE 121

Example of Calculation of Percentage Composition and Molecular Formula from Batch Composition of a Typical Leadless Earthenware Glaze

I. *Batch Composition*

(1) of frit		whiting	12%
		borax	19
		boric acid	4
		feldspar	57
		flint	8
			100
(2) of glaze		frit	88%
		china clay	12
			100

II. *Data required*

Raw material	Assumed formula	M.W.	Oxides entering the glaze	M.W.
Whiting	$CaCO_3$	100·09	CaO	56·08
Borax	$Na_2B_4O_7 \cdot 10H_2O$	381·43	Na_2O	61·99
			B_2O_3	69·64
Boric acid	H_3BO_3	61·84	B_2O_3	69·64
Feldspar	$K_2O \cdot Al_2O_3 \cdot 6SiO_2$	556·49	K_2O	94·19
			Al_2O_3	101·94
			SiO_2	60·06
Flint	SiO_2	60·06	SiO_2	60·06
China clay	$Al_2O_3 \cdot 2SiO_2 \cdot 2H_2O$	258·09	Al_2O_3	101·94
			SiO_2	60·06

TABLE 121—(contd.)

III. *Frit*

(1) Raw material	(2) Nature and quantity of oxide supplied to frit	(3) Percentage composition		(4) Ratio of molecules derived from column (2)	(5) Ratio of molecules rearranged according to Table 120	(6) Divided by 0·27	(7) Molecular formula
Whiting	$CaO \quad 12 \times \frac{[56\cdot08]}{100\cdot09} = 6\cdot724$	CaO	8·0	0·12	$Na_2O \;\; 0\cdot05$	$Na_2O \;\; 0\cdot19$	$\left.\begin{array}{l} Na_2O \;\; 0\cdot19 \\ K_2O \;\;\;\; 0\cdot37 \\ CaO \;\;\;\; 0\cdot44 \end{array}\right\} \left.\begin{array}{l} B_2O_3 \;\; 0\cdot48 \\ Al_2O_3 \;\; 0\cdot37 \end{array}\right\} SiO_2 \;\; 2\cdot74$
Borax	$Na_2O \;\; 19 \times \frac{[61\cdot99]}{381\cdot43} = 3\cdot088$	Na_2O	3·7	0·05	$\left.\begin{array}{l} K_2O \;\; 0\cdot10 \\ CaO \;\; 0\cdot12 \end{array}\right\} 0\cdot27$	$K_2O \;\; 0\cdot37$	
	$B_2O_3 \;\; 19 \times \frac{[69\cdot64] \times 2}{381\cdot43} = 6\cdot937 \left.\vphantom{\begin{array}{l}1\\1\end{array}}\right\} 9\cdot189$	B_2O_3	10·9	0·10	$B_2O_3 \;\; 0\cdot13$	$CaO \;\; 0\cdot44$	
Boric acid	$B_2O_3 \;\; 4 \times \frac{[69\cdot64]}{61\cdot84 \times 2} = 2\cdot252$			0·03	$Al_2O_3 \;\; 0\cdot10$	$B_2O_3 \;\; 0\cdot48$	
Feldspar	$K_2O \;\; 57 \times \frac{[94\cdot19]}{556\cdot49} = 9\cdot645$	K_2O	11·5	0·10	$SiO_2 \;\; 0\cdot74$	$Al_2O_3 \;\; 0\cdot37$	
	$Al_2O_3 \;\; 57 \times \frac{[101\cdot94]}{556\cdot49} = 10\cdot439$	Al_2O_3	12·4	0·10		$SiO_2 \;\; 2\cdot74$	
	$SiO_2 \;\; 57 \times \frac{[60\cdot06] \times 6}{556\cdot49} = 36\cdot901 \left.\vphantom{\begin{array}{l}1\\1\end{array}}\right\} 44\cdot901$	SiO_2	53·5	0·61			
Flint	$SiO_2 \;\; 8$			0·13			
	Total $\;\; 83\cdot986 \sim 84$		100·0				

TABLE 121—(contd.)

IV. Glaze. (Quick method)

It is convenient to make use of the previous calculation for those constituents supplied to the glaze by the frit. The sum of their parts by weight, found in column 2 is 84. The proportion of china clay to give 12% is then

$$\frac{84}{88} \times 12 = 11{\cdot}45$$

(1) Raw material	(2) Nature and quantity of oxides supplied to glaze			(3) Percentage composition	(4) Ratio of molecules		(5) ÷ by 0·27
China clay	Al_2O_3	$11{\cdot}45 \times \dfrac{101{\cdot}94}{258{\cdot}09} =$	4·516				
	SiO_2	$11{\cdot}45 \times \dfrac{60{\cdot}06 \times 2}{258{\cdot}09} =$	5·321				
Frit	Na_2O		3·088		Na_2O	3·3 0·05	0·19
	K_2O		9·645		K_2O	10·3 0·10	0·37
	CaO		6·724		CaO	7·2 0·12	0·44
	B_2O_3		9·189		B_2O_3	9·8 0·13	0·48
	Al_2O_3		10·439	14·955	Al_2O_3	15·9 0·15	0·56
	SiO_2		44·901	50·222	SiO_2	53·4 0·84	3·11
			93·823 ~ 94			99·9	

Molecular formula of the glaze

$$\left.\begin{array}{l} Na_2O\ \ 0{\cdot}19 \\ K_2O\ \ 0{\cdot}37 \\ CaO\ \ 0{\cdot}44 \end{array}\right\} \left.\begin{array}{l} B_2O_3\ \ 0{\cdot}48 \\ Al_2O_3\ \ 0{\cdot}56 \end{array}\right\} SiO_2\ \ 3{\cdot}11$$

---

is calcium carbonate which loses carbon dioxide on heating leaving calcium oxide, the conversion factor is therefore

$$\frac{\text{M.W. CaO}}{\text{M.W. CaCO}_3}$$

The figures obtained for the oxides contributed to the frit may then be brought to a percentage (column 3).

The 'percentage composition' is the usual way of expressing a glass composition in the glass industry. However, as has already been mentioned, in the ceramic industry the more usual way of expressing a glaze composition is by the molecular formula. The molar composition, or relative number of molecules, can be obtained from a weight composition or percentage composition by dividing by the molecular weight, *e.g.* the relative number of molecules of $CaCO_3$ in the frit derived from 12% whiting is

$$\frac{12}{100{\cdot}09} = 0{\cdot}12$$

One $CaCO_3$ contributes one CaO, so the relative number of CaO molecules is 0·12.

In an instance where the number of molecules contributed to the glaze by one molecule of the raw material is not one, the dividing molecular weight must be appropriately multiplied or divided, *e.g.* borax, $Na_2B_4O_7.10H_2O$, contributes two B_2O_3 so the molar ratio for B_2O_3 is

$$\frac{19 \times 2}{381 \cdot 43}$$

In this instance the calculations in column (2) can be used to give column (4) by omitting the molecular weight that is on top (shown in square brackets). The result is then rearranged to the order given in Table 120, p. 527, and those oxides falling in the first column added (column 5). All the figures are then divided by this number so that the first column adds up to 1 (columns (6) and (7)).

To obtain the molecular formula or percentage composition of the glaze itself the molecular weight of the frit is worked out from its molecular formula and an exactly similar calculation made.

It is often possible, however, to by-pass some of the lengthy calculation by noting how many of the oxide constituents are present only in the frit. In this example it can be assumed that only alumina and silica are added to the frit so that four of the six items in the molecular formula will eventually work out to the same as the frit. In Section IV it is shown how all the calculation involving the frit contents can be taken over from Section III.

For accurate work the formulae derived from chemical analyses of the raw materials should be used instead of assumed ideal formulae.

By suitably reversing the calculation* a batch composition can be obtained from a molecular formula. In the case of lead glazes where two frits are necessary careful dividing of the constituents is required which must be considered specially (*see* p. 569).

Use of the molecular formula as a means of comparing glaze compositions has a drawback. It assumes basically that a mixture of ingredients that adds up to a given formula will give the same glaze, however the ingredients are chosen. Even if it is true to assume that the actual fused glaze is in equilibrium this by-passes the fact that the glaze constituents react with the surface of the body. Different raw materials react at different temperatures and in different ways so that a given formula made up in various ways will lose different constituents by reaction with the body before complete fusion and mixing occurs. The batch composition is therefore not to be ignored totally.

The difficulty of deriving an expression for the glaze composition from the

---

* For longer explanations of such calculations *see*
A. HEATH, *A Handbook of Ceramic Calculations with many Problems*. Webberley Ltd., Stoke-on-Trent (1937).
A. I. ANDREWS, *Ceramic Tests and Calculations*. John Wiley and Sons, New York (1928).
W. PUKALL, *Keramisches Rechnen auf Chemischer Grundlage an Beispielen Erläutert*. Ferdinand Hirt, Breslau (1927).

batch composition becomes much greater when fluorine compounds are included. Fluorine can take the place of oxygen in the structure of a glass or glaze; since it is only monovalent it weakens the structure and therefore lowers the melting point. It can also be used to produce opacity. Some workers find it a useful glaze constituent. Its drawback, however, lies in the volatility of some of its compounds, e.g. SiF_4. The amount of fluorine and other constituents that will be lost during firing cannot be exactly predicted and does not stay constant even for a given batch.

As any fluorine that does remain in the glaze is not necessarily still combined directly with the cation it was introduced with, e.g. CaF_2, it is best to convert any fluorides to oxides for the calculation and list the fluorine separately. The molecular formula so derived does not truly represent the finished glaze.

[*Note*: matt, opaque, crystalline, coloured and salt glazes will be considered separately.]

GLAZE COMPOSITIONS WITH REGARD TO GLAZE PROPERTIES

Glazes are glasses and hence their most indispensable constituents are glass-forming oxides seconded by network modifying and intermediate oxides, the functions of which are described in connection with the structure of glasses, pp. 201–215. The adjustment of the numerous possible glaze constituents to suit particular purposes must be the next consideration. The compositions of most glazes, however, fall within the limits shown in Table 122 (**S108**).

TABLE 122

Low softening-point glazes:
   $RO.1·5SiO_2$ to $RO.3SiO_2$
   With RO chiefly PbO or combinations of alkalies

Harder glazes:

$RO.0·1Al_2O_3.2SiO_2$ to $RO\begin{Bmatrix}0·5B_2O_3\\0·4Al_2O_3\end{Bmatrix}.4·5SiO_2$

   RO combinations of PbO, alkalies, alkaline earths

High softening-point glazes:
   $RO.0·5Al_2O_3.5SiO_2$ to $RO.1·6Al_2O_3.14SiO_2$
   RO combinations of alkalies and alkaline earths.

Other oxides, in particular *zinc oxide*, may be used in most RO combinations.

DESIRED PROPERTIES OF GLAZES

(1) *Fusibility* must be such that maximum liquid glass is formed at the desired maturing temperature.

(2) *Viscosity* should be moderate at peak firing temperature so that surfaces even out but no over-all flow occurs down inclined or vertical surfaces.

(3) *Surface tension* should be low to avoid crawling.

(4) *Volatilisation* of glaze components during firing should be minimised.

GLAZES

(5) *Reaction with the body* should be moderate to give good fit without too much change in composition of either glaze or body.

(6) *Absorption into the body* of glaze constituents or eutectics formed during firing should not occur.

(7) *Devitrification* should not occur in transparent glossy glazes.

(8) *Expansion coefficient* and *Young's modulus of elasticity* should relate to those of the body in such a way that maximum strength is achieved.

(9) *Homogeneity, smoothness* and *hardness* to resist abrasion, scratching, etc.

(10) *Chemical durability*.

(11) *Colour* for aesthetic or thermal reasons.

(12) *Electrical properties, e.g.* low power factor.

1. Fusibility

Consideration of the structure of glasses shows that in ceramic glazes the basic glass former is silica, and that its properties are varied by the addition of other glass-forming oxides, *i.e.* B_2O_3 and P_2O_5, network-modifying oxides, *i.e.* the alkalies and alkaline earths, and by intermediate oxides, *i.e.* ZnO, PbO, Al_2O_3, Cr_2O_3, Fe_2O_3, Bi_2O_3, CeO_2, TiO_2, ZrO_2, SnO_2. Also, study of phase rule data shows that addition of one substance to another always lowers the melting point, a minimum being reached in most systems; the more different substances present the greater the lowering. By combination of structural and phase rule data sufficient directive can be obtained for practical tests to produce a glaze of desired fusibility.

Kreidl and Weyl (**K81**) have derived five principal changes based on structural considerations for making a glass more fusible. These are:

(*a*) The replacement of the tetrahedral network of silica by the triangular network of boric oxide. In the absence of alkali the addition of boric oxide to a silica glass loosens it, thereby increasing the expansion coefficient and lowering the melting temperature. In the presence of alkali the boron goes over to tetrahedral coordination and lowers the expansion coefficient, but although no longer complying with the theoretically desired structural change it is nevertheless an admirable flux.

(*b*) Variation of the oxygen-silicon ratio. Excluding other considerations, increase of the oxygen ratio loosens the silica network and lowers the melting point. This is achieved most usually by introduction of alkalies, alkaline earths and other metallic oxides and can also be done by adding P_2O_5.

Replacement of SiO_2 by glass formers of lower valency, *i.e.* Al_2O_3 when in tetrahedral coordination lowers the oxygen ratio thereby increasing the melting point. In the case of alumina this is counteracted by change (*c*) below and only becomes apparent at high alumina content.

(*c*) Replacement of SiO_2 by network co-formers, *e.g.* Al_2O_3, TiO_2, PbO. These oxides, unable to form glasses by themselves, obviously hold the network together less strongly and hence lower the melting point.

In the case of network co-formers of lower valency than four the lowering of the oxygen ratio (change (b) above) might increase the melting point, e.g. with high concentration of alumina. But usually the cation–oxygen bond is weaker than the Si—O bond and hence the melting point is lowered. The fluxing effect of a large tetravalent network co-former, e.g. TiO_2, ZrO_2 and SnO_2 is theoretically the most effective in this class.

The outstanding effect of PbO on glass made it one of the traditional fluxes. Lead can be added to silica in quantities much larger than is expected of a divalent metal, without causing devitrification, in fact $2 \cdot 5PbO.SiO_2$ can form a glass. This has led to its classification with network co-formers. Unfortunately its poisonous properties make it desirable to minimise its use, a matter that will be discussed again.

(d) Variation of network modifiers.

Size Factor. The smaller the modifying cation the less viscous is the glass, so that the maturing temperature of a glaze is lowered by replacing K^+ by Na^+ or Li^+, and Na^+ by Li^+. (This is effective only where the alkali concentration is relatively low but is generally applicable in the range of glaze compositions.)

Complexity. It has already been mentioned that the replacement of one component by two or more will almost certainly lower the melting point. A complex mixture also has a wider maturing range.

In the case of divalent network modifiers only Ca^{2+}, Sr^{2+} and Ba^{2+} can be compared even qualitatively with the alkali series, and Mg^{2+} and Zn^{2+} may show intermediate or co-forming properties in the network.

(e) Monovalent anions. The replacement of the divalent oxygen ion by the monovalent halogen ions F^-, Cl^- and Br^- weakens the structure. This method is not generally applicable to ceramic glazes.

Application of these methods of altering the structure of a glass or glaze to increase or decrease its fusibility is generally done by several of the first four methods at the same time. Thus, comparing two glaze compositions taken at random:

Cone	Temp. (°C)	Temp. (°F)		Na_2O	K_2O	CaO	PbO	B_2O_3	Al_2O_3	SiO_2	Ref.
01	1080	1976	M.F.	0·16	0·06	0·42	0·36	0·6	0·3	2·8	S181
			%	2·8	1·5	6·6	22·6	11·6	8·4	46·5	
14	1410	2570	M.F.	—	0·3	0·7	—	—	0·8	8·0	G67
			%		4·5	6·2			12·9	76·4	F36

The lower melting one has:

(1) More B_2O_3 relative to SiO_2 (change (a)).

(2) A higher oxygen–silicon ratio achieved chiefly by higher proportion of network modifiers (change (b)).

(3) A relatively higher proportion of Al_2O_3 to SiO_2 (change (c)).

(4) A more complex mixture of network modifiers and inclusion of the small ion Na^+ (change (d)).

On the other hand if a small adjustment of the fusibility is required a single one of the listed changes may effect it satisfactorily.

The following two glazes have a maturing point differing by only 30° C (54° F) and their molecular compositions are changed by only 0·1 mol. B_2O_3.

Cone No.	Temp.		Composition						Ref.
			K_2O	MgO	CaO	B_2O_3	Al_2O_3	SiO_2	
6	1200° C	M.F.	0·2	0·1	0·7	0·1	0·4	3·5	S176
	2192° F	%	5·9	1·3	12·2	2·2	12·7	65·7	
7	1230° C	M.F.	0·2	0·1	0·7	—	0·4	3·5	P64
	2246° F	%	6·0	1·3	12·5	—	13·0	67·2	

Knowledge of the eutectic composition of mixtures containing alumina can be particularly useful. Alumina is a desirable glaze component for a number of reasons, but is in itself highly refractory (M.P. 2050° C, 3722° F). On the addition of alumina to another reasonably fusible substance or mixture the melting point is at first lowered until the eutectic point is reached and thereafter rises again. At higher contents of alumina the melting point may be well above that of the original substance or mixture. It is therefore very important that the total mixture on addition of alumina should be near the composition of the eutectic. Unfortunately the amount of alumina required to give the eutectic mixture, *i.e.* the lowest melting point, varies for every system and cannot be predicted accurately.

However, it seems that:

(a) Two minerals of high melting point and relatively large latent heat of fusion have a eutectic point lying closer to the refractory component.

(b) Two minerals of considerably different but high melting points and large heats of fusion have a eutectic point nearer the lower melting substance.

(c) Two minerals of approximately equal latent heat of fusion have a eutectic point nearer the component of higher molecular weight.

Thus, for instance, the eutectic in a system 'low-melting lead-glaze'—alumina is very near the lead glaze end, whereas more alumina can be added to an alkali glaze before reaching and passing the minimum melting point (S108).

Knowledge of the minimum melting point of a system is not only helpful in developing glazes maturing at low temperatures, but also a great aid in finding glaze mixtures that will not soften at temperatures at which the glazed product will be used. Hence Kraner (**K73**) makes the following statement:

> In the case of chemical porcelain or sparking plugs, it would seem that a glaze having a high temperature of initial melting is more desirable than the usual type of glaze which is apt to be soft at the temperature at which it is to be used. The possibility of utilising such a glaze of eutectic composition in this connection is very favourable due to its possible high temperature of melting with rapid rate of maturing.
>
> Where a glaze free of or low in bubble content is desired, one having a late melting point may provide sufficient time for the decomposition of the glaze materials such as carbonates, nitrates, etc., which might ordinarily decompose at temperatures where the long-range glaze has already become somewhat fluid. The result is that in the eutectic composition the probable large volume of gas has been easily liberated from the unfused glaze coating, and thereby does not interfere when melting begins.

The composition and melting points of eutectic mixtures and of cones is also of interest in developing glazes although it is unusual for a cone mixture to be directly applicable as a glaze. They do, however, show what mixtures soften under given firing conditions and show very clearly the fluxing action of boric oxide and how the addition of silica increases the softening point (**S108**).*

2. Viscosity

The viscosity of glazes is an important property governing the success of several processes carried out at different temperatures. Its value at the maturing temperature (largely governed by the fusibility) determines the extent to which the glaze can flow over the body to form a uniform layer without, however, running off inclined or vertical surfaces. The viscosity during the formation of the glaze from its components also determines the ease of escape of gas bubbles evolved during these reactions.

The viscosity of a glaze which is to receive on-glaze decoration is also important at lower temperatures, in the range 500–800° C (932–1472° F). During the third firing which is at about 750° C (1382° F) gas bubbles may be evolved in the body which tend to force their way out through the glaze. If the viscosity is high and the firing short these gas bubbles cause no difficulties but if the glaze has a lower viscosity or the firing is longer bubbles appear in the glaze and they may burst at the surface, leaving jagged-edged craters. Decoration firing is frequently insufficiently prolonged to allow these faults to heal over and the blemishes are known as 'spit-out' (*see* p. 881).

James and Norris (**J17**) have developed two glaze viscosity measuring

---

* see also *Jahrbuch für Keramik, Glas, Email*. Schmelzpunkte. Verlag Sprechsaal, Coburg.
Ceramic Glazes, by Felix Singer. Borax consolidated Ltd. (1940). Appendix I: Glaze Components, Eutectic Mixtures and Melting Points. Appendix II: Seger cones.

techniques for low and high temperatures, respectively (*see* p. 370). The results they obtained on the following glaze types are summarised in the tables:

Eleven lead earthenware glazes.
Four lead china glazes.
One lead sanitary earthenware glaze.
Two lead tile glazes.
Six leadless earthenware glazes.
Four leadless china glazes.
One leadless sanitary earthenware glaze.

TABLE 123

Summary of Low-temperature Viscosity Data on Pottery Glazes (J17)

Temperature		$\text{Log}_{10} \eta$					
		Lead (low-solubility)			Leadless		
(°C)	(°F)	Mean	No. averaged	Range	Mean	No. averaged	Range
600	1112	10.8₅	7	9.5₅–11.4	—		
650	1202	9.6₅	19	8.2₅–10.7	—		
700	1292	8.0	19	7.2–9.1₅	10.1	10	9.2–10.8₅
750	1382	7.4	10	7.1–8.0₅	8.6	11	7.9–9.1
800	1472	—		—	7.5	4	7.2₅–7.9₅

TABLE 124

High-temperature Viscosities of Some Pottery Glazes

No.	Type	$\text{Log}_{10} \eta$ at, °C (°F)					
		800° C (1472° F)	900° C (1652° F)	1000° C (1832° F)	1100° C (2012° F)	1200° C (2192° F)	1300° C (2372° F)
1	Leadless	—	—	5.2₅	4.3	3.6₅	3.3₅
2		—	5.6₅	4.4₅	3.7₅	3.3	3.0₅
3		—	—	4.4	3.8	3.3	2.9
4		6.8	5.4	4.5	3.8₅	3.3	—
5	Low solubility lead	5.5₅	4.4	2.6	3.0	—	—

From their results James and Norris (**J17**) conclude that clear glazes, like glasses, behave as Newtonian liquids, as does also an opaque glaze with 5% suspended matter. There is no definite temperature below which viscous flow ceases and elastic deformation occurs, but instead there is an intermediate range, around 550° C (1022° F), where viscoelastic deformations take place. This stage helps in relieving strains in the glaze that would occur if solidification was instantaneous.

It is also seen that lead glazes are very much less viscous than leadless ones, at 700° C (1292° F) the mean log viscosity for lead glazes is 8·0 whereas that for leadless is 10·1, *i.e.* a hundredfold increase of viscosity, at 750° C (1382° F) the respective figures are 7·4 and 8·6, showing the leadless glazes to be twenty times as viscous as the low solubility lead ones.

Brémond (**B112**) suggests that a glaze would be immature at log viscosity 5, form a matt enamel type of finish at log viscosity 4, and on further reduction pass through maturity until at $\log \eta = 2\cdot 6$ it would be over-fired and full of bubbles. Thus the firing range of a glaze is related to the rate of change of viscosity with temperature.

3. Surface Tension

The flow of glazes during firing is also considerably influenced by their surface tension and angle of contact. Very little work has been done on this, and that is mostly comparative and empirical. Bloor (**B81**) summarises it.

Glass-workers have pointed out the influence of surface tension on reactions between ingredients and the disappearance of inhomogeneities in glass melts. Low surface tension favours elimination of gas bubbles during glass firing, while high surface tension favours reabsorption of the bubbles during the cooling of the glass (**J42, S60, S78**). Morey (**M94**) gives values for the surface tension of silicate glasses around 300 dyn/cm but ranging from 130–500 dyn/cm, the temperature coefficient is usually low at 0·01% to 0·04% per °C. Surface tension is a more or less additive function with regard to the oxide composition of the glass and proposed factors are shown in Tables 125 and 126.

The wetting of ceramic surfaces by glazes has received little attention, but there is little doubt that the composition of the body is important, as well as that of the glaze. Lampman (**L14, 15**) used four methods for investigating this matter:

(1) Inclined flow of glaze on body.
(2) Horizontal flow of glaze on body.
(3) Angle of contact of fused buttons of glaze.
(4) Radius of curvature of fused buttons of glaze.

The angle of contact method (3) proved the most satisfactory and he quotes a case of the same glaze requiring firing four cones apart to achieve the same angle of contact on two different bodies.

High surface tension can lead to the fault known as 'crawling'. Too low surface tension is a subsidiary factor combined with too low viscosity in causing excessive running of glazes (**J17**).

TABLE 125
Effect (in Dyn/cm) of 0·1 Gram-molecule of Each Oxide on Surface Tension of Glass (L51)

Group	Dietzel. At 900° C	Lyon. For glasses in which $SiO_2:Na_2O$ exceeds 3·25	
		At 1200° C	At 1400° C
Li_2O	13·7		
Na_2O	9·3	7·9	6·9
K_2O	9·4	(0·0)	(−7·6)
MgO	26·6	23·3	22·1
CaO	26·9	27·6	27·6
BaO	56·7	(56·7)	(58·3)
ZnO	38·2		
B_2O_3	5·5	1·6	−1·6
Al_2O_3	63·2	61·0	59·6
SiO_2	20·4	19·5	19·5
TiO_2	23·0		
ZrO_2	50·5		
PbO	26·8		
FeO	32·2		
Fe_2O_3		(71·9)	70·3
CoO	33·7		
NiO	33·6		
MnO	31·9		

Results in brackets are given by Lyon as a guide only.

TABLE 126
Factors for Calculating Surface Tension of Glazes at 900° C (in Dyn/cm) from % Composition
(after Dietzel (K92))

MgO	6·6	Fe_2O_3	4·5	BaO	3·7
Al_2O_3	6·2	CoO	4·5	SiO_2	3·4
V_2O_5	6·1	NiO	4·5	TiO_2	3·0
CaO	4·8	MnO	4·5	Na_2O	1·5
ZnO	4·7	ZrO_2	4·1	PbO	1·2
Li_2O	4·6	CaF_2	3·7	B_2O_3	0·8
				K_2O	0·1

For each further 100° C, 4 units should be subtracted

Example:
		%	Factor
Glaze	SiO_2	68%	68 × 3·4 = 231
	CaO	9	9 × 4·8 = 43
	Na_2O	14	14 × 1·2 = 21
	PbO	9	9 × 1·2 = 11
		100	Surface tension at 900° C = 306 dyn/cm
			at 1200° C = 306−12 = 294 dyn/cm

4. Loss of Glaze Constituents by Volatilisation

The composition of glazes tends to alter during firing due to the fact that some of their constituents volatilise more readily than others. The worst offender is lead oxide, followed by boric oxide and the alkalies. The alkaline earths, alumina, silica, etc., are not volatile under normal kiln conditions. However, although the tendency to volatilise is related to high vapour pressure of the oxides concerned when pure, it is frequently not proportional to the quantity of them in the glaze. The nature and quantity of the other constituents in the glaze has a strong influence on the volatilisation. The amount of the loss by volatilisation from a given glaze is also influenced by the time and temperature of firing, and the receptiveness of the surroundings to condensation of the vapour. Porous saggars, etc., readily take up the vaporised oxides and so encourage further loss from the glaze, whereas a glaze-covered environment and a heavily vapour-laden atmosphere prevent volatilisation.

This unpredictable loss of constituents from a glaze not only harms the glaze but also spoils the interior of kilns and in particular the condensate attacks electrical heating elements. Volatile coloured oxides also spoil neighbouring ware by condensing on it, the main offenders being chromium and cobalt oxides, but trouble has also been reported to occur with manganese and rutile.

Kerstan (**K23**) has investigated the composition of deposits on kiln heating elements derived from three known glazes and found the following results.

TABLE 127

Condensates Derived from Known Glazes (K23)

	Glaze A	Cond. from A	Glaze B	Cond. from B	Glaze C	Cond. from C
SiO_2	27.5%	20.60%	33.3%	20.14%	56.7%	*%
TiO_2	4.7	*	—	—	—	—
Al_2O_3	7.7	*	1.3	0.12	8.1	0.30
PbO	43.6	34.24	52.3	48.27	10.7	62.10
ZnO	10.7	3.50	—	—	2.9	*
$Na_2O + K_2O$	1.5	0.30	1.0	*	5.6	3.40
BaO	2.3	0.35	—	—	—	—
SnO_2	2.0	*	—	—	—	—
B_2O_3	—	—	6.6	0.17	7.9	0.11
CaO	—	—	5.5	*	8.1	*
	100.0%		100.0%		100.0%	

* = not determined.

The lead was introduced into all three glazes in a combined form. The results show clearly that the relative amount of lead in the condensate is not related to its percentage in the glaze.

Kerstan (**K23**) therefore went on to find the weight losses of various lead combinations when 0·2 cm³ were heated for one hour at 1000° C (1832° F)

TABLE 128

Weight Loss by Volatilisation of Lead Combinations
(Kerstan (**K23**))

	% Composition	Molecular formula	Weight loss after 1 hr at 1000° C (excluding CO_2 lost)
I	PbO 78·8 SiO_2 21·2	$PbO.SiO_2$	0·39%
	PbO 71·9 SiO_2 28·1	$PbO.1.5\ SiO_2$	0·63%
	PbO 65·0 SiO_2 35·0	$PbO.2SiO_2$	0·71%
	PbO 76·1 Al_2O_3 3·15 SiO_2 20·50	$PbO.0.1Al_2O_3.SiO_2$	0·28%
	PbO 69·00 Al_2O_3 3·15 SiO_2 27·85	$PbO.0.1Al_2O_3.1.5SiO_2$	0·40%
II		0·9 PbO 0·1 CaO } SiO_2	0·60%
		0·8 PbO 0·2 CaO } SiO_2	0·79%
		0·7 PbO 0·3 CaO } SiO_2	0·83%
		0·6 PbO 0·4 CaO } SiO_2	1·73%
III	$PbO.SiO_2 + 3\%\ CaCO_3$ $PbO.SiO_2 + 5\%\ CaCO_3$ $PbO.SiO_2 + 8\%\ CaCO_3$		1·20% 1·78% 2·10%

in a platinum crucible. The surface area of all the samples was the same. The results of three sets of experiments are given in Table 128. It should be noted that in the pure lead silicates the loss of weight increases as the ratio of SiO_2:PbO increases although the opposite might be expected. The addition of 0·1 Al_2O_3 per mol PbO reduces the weight loss. CaO noticeably increases volatilisation and the addition of whiting to a glossy lead glaze to make it matt achieves this end by driving out the lead, a very unsatisfactory arrangement. Other work showed that in spite of its own volatility boric oxide reduces the weight loss when it replaces part of the lead, thus increasing the acidity of the glaze.

Norris (**N24**) described experiments on the volatilisation of a leadless earthenware glaze on a small piece of platinum foil (10 cm^2). The specimen was heated in a small furnace at a constant temperature for 36 hr, being removed, cooled and weighed every 2 hr. The experiment was repeated at temperatures between 1000–1200° C (1832–2192° F) although at the upper limit crawling occurred. Curves of weight loss/unit weight and weight loss/unit area each against time were plotted. A further plot of log weight loss after 8 hr/unit weight against $1/T$ the absolute temperature gives a straight line with an equation of the form:

$$\log(\text{loss}) = A - B/T.$$

This is of the same form as the equation relating vapour pressure to temperature. It may therefore be deduced that the loss by volatilisation is proportional to the vapour pressure of the volatilising material. No linear relationship is found between log(loss/unit area) and $1/T$.

The total weight losses of this leadless glaze could not, however, be predicted from the quantity of volatile matter (Na_2O, K_2O and B_2O_3) in the glaze.

5. *The Interaction between Glaze and Body*

Adequate interaction between the glaze and the surface of the body is essential for keeping the two together. Where the two are well suited to each other in respect of thermal expansion and elasticity the interaction zone forms the bond between them. Where they are not well suited a good zone of interaction will often prevent glaze defects of crazing and peeling that would otherwise occur, because it is of intermediate composition and properties.

Interaction between body and glaze is prevented by deposits of soluble salts or dust on the surface of the body before glazing. The salts may have migrated to the body surface during drying or the earlier stages of biscuit firing. In both cases the glaze fails to get an adequate grip on the body and appears to show the defect of crawling or in certain cases edge-chipping. In contrast to this Campbell (**C4**) has shown that application of a thin fusible mineral layer (borax, feldspars, nepheline syenite, certain frits and frit-talc mixtures) to the biscuit-fired body before glazing promotes good glaze-body interaction giving a more craze-resisting product.

Mellor (**M55**) recommends as long a period of glost firing at as high a temperature as possible to promote good interaction, and for this reason believes once-fired ware to be more satisfactory:

Once-fired ware ... is well known to be less liable to craze than twice-fired. Apart from glaze composition there are at least three favourable conditions in the once-fired ware:

(1) The body in the earlier stages (of firing) is more susceptible to attack by the glaze than is the case with a body previously matured in the biscuit-oven;

(2) The longer period of time required to mature the body gives the glaze more time to develop the interfacial layer;

(3) The higher maturing temperature of the body also favours a more rapid attack by the glaze.

The chemical composition and the method of compounding both the glaze and the body, of course play an important part. Fritted glazes are usually better than raw ones, and fine grinding ensures thorough mixing and a more regular product (over-grinding may, however, make the glaze curl or crawl in firing) (**M55**). The type of body naturally makes a considerable difference, in particular whether it is porous or not.

Purdy and Fox (**P75**) attribute the fine mesh crazing found in certain glazes to chemical incompatibility of glaze and body and in particular to a shortage of alumina in the glaze and its subsequent removal from the body by the glaze. The actual crazing is due to lack of homogeneity in the upper and lower glaze layers, so that whereas in lower-temperature glazes where the alumina content is low, it can be remedied by adding alumina, in those maturing above cone 5 addition of alumina makes them more viscous, which is detrimental. Insley's (**I5**) investigation of the microstructure of the body, glaze and their junction showed that mullite crystals form both in the body and at the body-glaze interface, which accounts for the necessity of alumina.

Parmelee and Buckles (**P18**) made a study of the glaze and body interface with a series of twelve glazes comprising three sets of RO compositions used with the four possible combinations of:

$$RO \begin{cases} 0.1 \ Al_2O_3 & 1.2 \ SiO_2 \\ 0.35 \ Al_2O_3 & 2.4 \ SiO_2 \end{cases}$$

They were fired at cones 2–3, 4–5, and 5–6. Their results again show that the alumina content of the glaze is connected inversely with the attack of the body by the glaze. In the low-Al_2O_3 and low-SiO_2 glazes an interfacial layer containing crystals of anorthite, quartz (from body) and wollastonite was formed. Low-Al_2O_3 and high-SiO_2, and high-Al_2O_3 and low-SiO_2 glazes showed the glassiest portion near the body while the outer glaze had undissolved quartz crystals, bubbles and craters. Attack on the body was less than in the first case and less in the case of the high-Al_2O_3 than the low-Al_2O_3. Anorthite and wollastonite crystals were found at the interface. The high-Al_2O_3 and high-SiO_2 glazes did not appear to form an interfacial layer although some of the glaze was absorbed into the body.

Thomas, Tuttle and Miller (**T23**) studied the effect of a number of

544   INDUSTRIAL CERAMICS

different constituents replacing part of the CaO in a basic glaze, on the penetration of the glaze into several different bodies. The basic glaze was:

	M.F.	%
Na_2O	0·140	3·05
K_2O	0·090	3·00
CaO	0·625	12·40
PbO	0·145	11·45
Al_2O_3	0·300	10·80
SiO_2	2·800	59·30
		100·00

0.325 equivalents of the CaO were replaced in turn by:

Li_2O, MgO, CaF_2, $Ca_3(PO_4)_2$, SrO, BaO, TiO_2, ZrO_2, Cr_2O_3, MoO_2, $NaUO_4$, UO_3, MnO_2, Fe_2O_3, CoO, NiO, CuO, ZnO, CdO, B_2O_3, SnO_2, PbO, Sb_2O_3, Bi_2O_3, SeO_2.

The tabulated results (Table 129) show that no particular substance can

TABLE

Substitutes Arranged in Order of Decreasing

(Vertical brackets include all those with equal

Semi-vitreous body		Talc wall-tile body		Terra cotta body		Hotel china body
One fire	Two fire	One fire	Two fire	One fire	Two fire	One fire
CdO	B_2O_3	B_2O_3	$Ca_3(PO_4)_2$	ZrO_2	PbO	PbO
SrO	ZrO_2	PbO	Li_2O	B_2O_3	ZrO_2	B_2O_3
TiO_2	PbO	Sb_2O_3	TiO_2	$NaUO_4$	BaO	MoO_2
MoO_2	TiO_2	$Ca_3(PO_4)_2$	ZnO	MgO	SnO_2	TiO_2
Sb_2O_3	MgO	ZrO_2	Bi_2O_3	PbO	CuO*	$Ca_3(PO_4)_2$
ZrO_2	$NaUO_4$	CuO	CdO	Cr_2O_3	B_2O_3	MgO
Cr_2O_3	MoO_2	CdO	UO_3	Fe_2O_3	Cr_2O_3	BaO
NiO	SnO_2	TiO_2	NiO	$Ca_3(PO_4)_2$	$NaUO_4$	SrO
CoO	SrO	MoO_2†	MoO_2	MoO_3	CoO	Sb_2O_3
SeO_2	CdO	Bi_2O_3	SrO	SrO	TiO_2	ZrO_2
ZnO	$Ca_3(PO_4)_2$	Cr_2O_3	Fe_2O_3	CaO	UO_3	CuO
CuO	BaO	MoO_2†	B_2O_3	UO_3	$Ca_3(PO_4)_2$	UO_3
B_2O_3	CuO	MgO	PbO	Li_2O	CaF_2	SnO_2
Bi_2O_3	Cr_2O_3	SnO_2	CaF_2	CuO	SrO	MnO_2
$Ca_3(PO_4)_2$	Fe_2O_3	ZnO	CuO	Sb_2O_3	MoO_2	Fe_2O_3
PbO	ZnO	Li_2O	CaO	TiO_2	CdO	ZnO
UO_3	SeO_2	$NaUO_4$	MgO	CaF_2	Fe_2O_3	Li_2O
CaO	Bi_2O_3	SrO	MnO_2	SnO_2	MnO_2	Bi_2O_3
$NaUO_4$	UO_3	NiO	BaO	BaO	SeO_2	CdO
MgO	CaF_2	UO_3	Sb_2O_3	CdO	Sb_2O_3	Cr_2O_3
CaF_2	Li_2O	CaO	ZrO_2	MnO_2	CuO*	NiO
MnO_2	Sb_2O_3	CoO	SnO_2	NiO	NiO	CaF_2
SnO_2	CaO	BaO	CoO	CoO	MgO	CaO
Fe_2O_3	NiO	CaF_2	Cr_2O_3	SeO_2	Li_2O	$NaUO_4$
BaO	MnO_2	Fe_2O_3	$NaUO_4$	ZnO	Bi_2O_3	CoO
Li_2O	CoO	SeO_2	SeO_2	Bi_2O_3	ZnO	SeO_2

* One of these items CuO should presumably read CaO.
† One of these items MoO_2 should presumably read MnO_2.

be recommended for increasing glaze penetration in every instance. In fact they demonstrate very well how the nature of the body and the firing system influence the interaction. However, it is worth noting that PbO comes in the first place six times and in the second twice; B_2O_3 comes in first place twice and second place six times; ZrO_2 comes first five times, second twice, third three times; TiO_2 second three times, third twice, fourth three times; Li_2O first twice, second three times and third twice. The other fluxes that do not impart colour to the glaze, *i.e.* MgO, CaO, CaF_2, $Ca_3(PO_4)_2$, SrO, BaO, ZnO, all have rather varied effects.

Smith's (**S133, S134**) work on the effect of the body composition on the glaze-body interaction has been summarised by Norris (**N24**). A lead glaze of the following approximate percentage composition was applied to bodies in the MgO–Al_2O_3–SiO_2 system:

$$\text{Glaze: } PbO \quad 60.7\%$$
$$Al_2O_3 \quad 6.0\%$$
$$SiO_2 \quad 33.2\%$$

$$99.9\%$$

129

Penetration of Glazes Made Containing Them

effect; underscoring indicates an average) (**T23**)

Hotel china body	Cordierite body		Pyrophyllite body		Steatite body	
Two fire	One fire	Two fire	One fire	Two fire	One fire	Two fire
ZrO_2	PbO	ZrO_2	ZrO_2	PbO	ZrO_2	BaO
BaO	B_2O_3	PbO	CdO	Sb_2O_3	BaO	PbO
Cr_2O_3	Sb_2O_3	TiO_2	Cr_2O_3	MoO_2	Na_2UO_4	Li_2O
Na_2UO_4	MoO_2	CaO	UO_3	CdO	Li_2O	B_2O_3
Fe_2O_3	ZrO_2	CdO	MnO_2	UO_3	CaF_2	MgO
MnO_2	SeO_2	Li_2O	Fe_2O_3	SnO_2	$Ca_3(PO_4)_2$	Sb_2O_3
Li_2O	SrO	Bi_2O_3	ZnO	Li_2O	Fe_2O_3	UO_3
TiO_2	BaO	UO_3	B_2O_3	Bi_2O_3	MnO_2	ZnO
SrO	CdO	SnO_2	Li_2O	Cr_2O_3	SnO_2	Bi_2O_3
MoO_2	NiO	MnO_2	Bi_2O_3	TiO_2	CdO	Na_2UO_4
CdO	CoO	CoO	MoO	Na_2UO_4	Cr_2O_3	ZrO_2
UO_3	MnO_2	Sb_2O_3	SrO	$Ca_3(PO_4)_2$	CuO	CdO
NiO	CuO	CuO	SeO_2	SrO	MgO	$Ca_3(PO_4)_2$
SnO_2	CaO	B_2O_3	CoO	NiO	PbO	Fe_2O_3
ZnO	CaF_2	Cr_2O_3	PbO	ZnO	B_2O_3	CoO
CuO	MgO	$Ca_3(PO_4)_2$	CaO	Bi_2O_3	UO_3	SeO_2
CoO	SnO_2	MgO	CuO	CaO	NiO	NiO
SeO_2	$Ca_3(PO_4)_2$	ZnO	Na_2UO_4	MgO	CoO	MoO_2
$Ca_3(PO_4)_2$	ZnO	Na_2UO_4	BaO	CaF_2	Sb_2O_3	SnO_2
B_2O_3	Cr_2O_3	MoO_2	TiO_2	ZrO_2	TiO_2	SrO
Bi_2O_3	Bi_2O_3	Fe_2O_3	$Ca_3(PO_4)_2$	SeO_2	CaO	CuO
MgO	Na_2UO_4	SrO	NiO	BaO	MoO_2	TiO_2
CaF_2	UO_3	NiO	SnO_2	CuO	SrO	CaF_2
Sb_2O_3	Li_2O	SeO_2	Sb_2O_3	Fe_2O_3	ZnO	CaO
CaO	TiO_2	BaO	CaF_2	MnO_2	Bi_2O_3	Cr_2O_3
PbO	Fe_2O_3	CaF_2	MgO	CoO	SeO_2	MnO_2

The results can be tabulated as in Table 130.

TABLE 130

Body composition			Body–glaze interaction
MgO	Al$_2$O$_3$	SiO$_2$	
<9%	11–23%	69–80%	Intermediate glassy layer containing mullite crystals <0·02 mm long perpendicular to the body surface.
<9%	>23%	<75%	Similar mullite crystals but 0·04–0·08 mm long forming a continuous band. A crack through the centre was always found.
>9%	<26%	<80%	Preferential attack of body leaving sponge-like crystalline network full of glass.
>9%	>26%	<80%	As above with mullite crystals intergrowing in the spongy layer.
<20%		45–65%	Spongy layer formed but tendency to devitrify because glaze dissolves MgO.

The nature of the intermediate layer can have a profound effect on the nature of the ware. For example excessive mullite development may cause peeling because its expansion coefficient is lower than that of either body or glaze. On cooling the mullite is compressed from all sides and eventually cracks through the centre to relieve the stress, thereby separating the glaze from the body (**S134**).

Body–glaze interaction is also influenced by the batch composition of the glaze and whether the body is raw or biscuited. Although the finished glaze is assumed to be a homogeneous glass of the over-all composition deduced from the raw materials, it is applied to the body and initially heated whilst a mixture of any number of different materials. These fuse and interact at different temperatures and may combine permanently with the body surface before complete fusion and mixing of the glaze occurs. As a given glaze molecular formula can be derived from several different batch compositions so the actual composition and properties of the finished glaze may vary.

7. Guards against Devitrification

A number of the suggested methods for lowering the melting point of a glaze can only be applied to a limited extent because compositions may result that devitrify more readily. Also compositions that make glasses on rapid cooling may crystallise when cooled slowly enough to prevent the dunting of large pieces.

On the whole the simpler a mixture the more it tends to devitrify when the oxygen ratio becomes high. Thus a soda–lime–silica glass with more than 20% CaO may devitrify but this can be prevented by replacing 2–3% of the CaO by MgO, B$_2$O$_3$ or Al$_2$O$_3$. Also lithium silicates in the absence of

GLAZES 547

other alkalies crystallise unusually easily, so that lithium is best always used in conjunction with other alkalies, and generally not exceeding one-fifth of the total alkali content (**K81**).

Vogt (**V21**), and Watts (**W26**) found that eutectic mixtures generally give very good glasses. In general, complexity of the mixture ensures a vitreous product and this is assisted in particular by the presence of even small amounts of alumina.

8. *Coefficient of Expansion and Young's Modulus of Elasticity*

These are two basic physical properties of glazes and bodies and are of great importance with regard to 'glaze fit'. For most purposes the mechanical strength of a finished piece is increased if the glaze is in a state of compression (*see* p. 549) and this is brought about if the glaze has a lower coefficient of expansion than the body it is applied to, *i.e.* shrinks less during cooling after the glost firing. Too great a difference between the expansion of body and glaze leads to faults (*see* p. 551).

The linear coefficient of expansion is defined as the increase in length per unit length, caused by a rise in temperature of $1°$ C. It varies with the temperature at which it is measured and this or the range covered should be quoted with values. In the pottery industry expansion data are often expressed differently, namely as the total expansion, in percentage, from $20°$ C up to the lower critical temperature. This assumes that on cooling a glaze suddenly becomes a rigid elastic solid when it passes through the lower critical temperature, and that therefore the stress of the cold piece is determined by the differences of the two total contractions from that temperature down to room temperature. The shapes of the two expansion curves do not affect the ultimate cold stress although they determine the value at any intermediate temperature. As the critical temperature of many glazes is in the region 500–$550°$ C (932–$1022°$ F) the total expansion from 20–$500°$ C (68–$932°$ F) is used as a convenient measure.

For earthenware, bone china and glazed tiles the total glaze expansion from 20–$500°$ C (68–$932°$ F) falls in the range $0·3\%$ to $0·4\%$ and needs to be known to an accuracy of $0·01\%$. For sanitary earthenware and vitreous china the total expansion should be nearer $0·30\%$. For sanitary fireclay the glazes have total expansions of $0·28\%$ up to $500°$ C ($932°$ F) or $0·35\%$ up to $600°$ C ($1112°$ F) (**J17**).

Factors for calculating the coefficient of expansion of glazes from their compositions have been proposed by several workers on the basis of the mean of changes in it found on making substitutions in various glass types.

Table 131 gives a summary of such factors for the common constituents. Experience has shown that the factors do not apply accurately to all types or at all temperature ranges. It is seen that, in general, the weaker the bond the higher the expansion, thus the alkalies allow the highest expansion, the network-formers the lowest and Zn^{2+}, Mg^{2+}, Be^{2+} are intermediate.

TABLE 131

Factors for the Linear Expansion Coefficient (B81)

$$\left(\text{Given as weight \% factor} \times \frac{\text{molecular weight}}{100}\right)$$

Group	Winkelmann and Schott	English and Turner (25–90° C)	Hall (H8) (20°—lower critical point)	Waterton and Turner (0–150° C)	Others (G60)	Takahashi (T1) (cationic not molecular percentages)	
						0–100° C	0–400° C
P_2O_5	9·46						
SiO_2	1·60	0·15	60% 2·40 70% 2·04 80% 1·56 90% 0·90			0·50	0·60
TiO_2 ZrO_2					0·80 6·6 1·4	−3·5 −14·5	
B_2O_3 Al_2O_3 Fe_2O_3	0·23 17·0	−4·6 1·43	1·4 5·1		−1·5 0·47	$AlO_{1·5}$ 2·7 $FeO_{1·5}$ 3·2	3·1 5·0
BeO					0·0 0·5 1·0		5·5
MgO CaO SrO	0·13 9·3	1·8 9·1	0·81 8·4	9·0	13·8 19·2	3·3 13·0	6·0 14·5
BaO	15·3	21·5	18·4		29·1	23·0	25·0
ZnO PbO FeO MnO	4·9 22·3	5·7 23·7	8·1 16·7		5·0	5·0 16·0 10·0 10·0	8·0 18·0 13·0 13·4
Li_2O Na_2O K_2O Rb_2O	20·7 26·7	26·8 36·7	23·6 28·3	14·6 25·9 32·0 48·6	25·8 50·5	$LiO_{0·5}$ $NaO_{0·5}$ 26·9 $KO_{0·5}$	24·5 29·5 33·3

Expansion calculated from $\dfrac{E}{100} = \alpha_1 p_1 + \alpha_2 p_2 \ldots + \alpha_n p_n = \dfrac{10^6 \Delta l}{l \Delta t}$ where α is the factor, p the molecular or cationic percentages.

Another factor determining the ultimate strain in a glaze is the difference between the elastic constants of body and glaze. For any given set of conditions the strains in the glaze and the body are determined by the ratio of their Young's moduli, as is also the distortion produced in pieces that are not glazed on both sides. The higher the Young's modulus of the glaze, other things being equal, the lower the strain in the glaze and the greater that in the body. The majority of pottery glazes have values within the range of $8–12 \times 10^6$ lb/in^2 and most porous pottery bodies have values $1–4 \times 10^6$ lb/in^2 but vitreous bodies have 10×10^6 lb/in^2. Thus glazes usually have higher values, *i.e.* are 'stiffer' than their respective bodies but the reverse may occur (**J17**).

THE INFLUENCE OF GLAZES ON THE MECHANICAL STRENGTH OF CERAMIC BODIES

Rowland (**R60, R61**) has pointed out the importance of correct glazing of porcelain insulators to enhance their mechanical strength. Fracture of a body begins at a surface scratch or crack so that the smoother the surface the more resistant it is to breakage. In a ductile material, *e.g.* metals, small cracks may fill up by flow of adjoining matter, but in a non-ductile one such as ceramic ware the crack will not smooth out, and so once present it is a constant source of danger (Fig. 6.1).

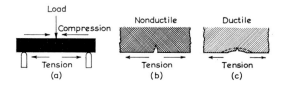

FIG. 6.1. *Behaviour of cracks in non-ductile and ductile materials in tension* (Rowland, **R61**)

Unglazed porcelain always has tiny irregularities in its surface which can offer focal points for incipient cracks. Varnishing of the surface can raise the modulus of rupture 10 to 15%.

Good glazing therefore gives a smooth surface with fewer nuclei for cracks. The effect of the smoothness may, however, be counteracted by lack of fit of the glaze to the body. This is shown by the series of experiments on the transverse strength of porcelain rods either unglazed or with one of seven different glazes (Table 132). Glazing may increase (Group 5) or decrease

TABLE 132
The Transverse Strength of Porcelain Rods $1\frac{1}{8}$ by 6 in. with less than $\frac{1}{64}$ in. Glaze

Group	(1) Un-glazed	(2) Glazed	(3) Glazed	(4) Glazed	(5) Glazed	(6) Glazed	(7) Glazed	(8) Badly fitting glaze
Average strength (lb)	1061	1016	1299	1299	1716	989	957	384
Percent mean variation	4·6	3·6	9·35	8·9	3·8	6·25	4·36	5·5

(Group 8) the strength of the body by about 60%. This depends largely on whether the glaze is in a state of compression or tension before a load is applied. The load tends to bend the rod bringing the lower side into tension. This is more likely to start cracking if it is already in tension than if it is in compression, as is shown in Fig. 6.2.

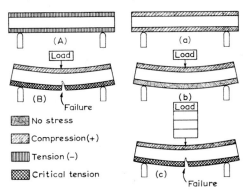

FIG. 6.2. *Comparison of strengths of glazes initially in tension and initially in compression* (Rowland, **R61**)

The tensile strength of a non-vitreous body can also be improved by suitable glazing but usually not to such a large extent as a vitreous body.

The actual tensile strength of the glaze is a determining factor in the extent to which the glaze can improve a body. The tensile strength depends largely on the composition of the glaze. The effect of minor constituents in the glaze composition on its tensile strength is shown in Table 133. Considering the chemical composition of the glaze as a whole it is found that high alkali and lime contents have detrimental effect on mechanical strength (**T19**).

TABLE 133

Minor Glass Constituents in Order of Their Favourable Influence on the Tensile Strength (G16)

(1) Boric oxide.
(2) Barium oxide.
(3) Alumina.
(4) Magnesia.
(5) Zinc oxide.

Another important factor is the firing treatment. For instance a given porcelain glaze fired at cone 14 may improve the mechanical strength about 100%, whereas another at cone 7 only improved it about 50%.

It has been seen how mechanical breakage of a piece usually causes tensile stress and so most work on 'mechanical strength' has been devoted to tensile strength. Some ware is, however, submitted to compressive forces. Compression of a body calls into action forces exactly opposite to those brought into play in tension. In order to increase the resistance of a body to compression it is therefore necessary to apply a glaze that will be initially in tension (**B45**).

INFLUENCE OF GLAZE COMPOSITION ON THE SHRINKAGE OF PORCELAIN

Hard porcelain is biscuit fired at a lower temperature than its glost firing and considerable shrinkage occurs during the latter. It has been found that the amount of shrinkage is affected by the glaze composition. For instance porcelain crucibles were fired at cone 12 with the two following glazes (Table 134). Glaze I produced 5 to 6% greater shrinkage than glaze II (**A120**).

TABLE 134

	I	II
	$\left.\begin{array}{l}0\cdot473K_2O\\0\cdot019MgO\\0\cdot476CaO\\0\cdot032FeO\end{array}\right\}1\cdot739Al_2O_3\left\{\begin{array}{l}14\cdot091SiO_2\\0\cdot039TiO_2\end{array}\right.$	$\left.\begin{array}{l}0\cdot004Na_2O\\0\cdot042K_2O\\0\cdot016MgO\\0\cdot928CaO\\0\cdot010FeO\end{array}\right\}\begin{array}{l}0\cdot827Al_2O_3-\\6\cdot387SiO_2\end{array}$
	%	%
Na_2O	—	0·05
K_2O	4·03	0·74
MgO	0·07	0·12
CaO	2·42	9·89
Fe_2O_3	0·23	0·16
Al_2O_3	16·10	16·04
SiO_2	77·08	73·23
TiO_2	0·28	—
	100·21	100·23

A number of other experiments on porcelain fired between cones 10 and 14 has shown that glazes of higher lime content produce larger porcelain articles. Furthermore the high-lime glazes lead to more cracks at the joins of cup handles than low-lime ones.

With regard to earthenware there is no marked difference of shrinkage as the biscuit firing is higher than the glost but a greater tendency for handles to crack off has been observed with higher-lime glazes.

CRAZING, PEELING AND DUNTING

Incompatibility of the coefficients of thermal expansion of body and glaze together with unsuitable moduli of elasticity are the causes of a number of defects. The glaze fuses on to the body at elevated temperatures and the whole is then cooled to room temperature. If the coefficients of thermal expansion of the body and the glaze, respectively, now two component parts of a single whole, differ widely great stresses are set up between the two, which may eventually cause failure in one or even both of them. If the expansion coefficient of the glaze is higher than that of the body, the glaze contracts more than the body and hence is in tension. This may cause it to break in fine irregular cracks known as 'crazing'. In the reverse case of the expansion coefficient of the glaze being too small the cooled glaze is under compression and chips of it may break off. This is known as 'peeling', or because it happens most readily at sharply curved surfaces 'edge-chipping', and also 'shelling' or 'jinking'.

Schwabe and Syska (**S40**) have investigated the influence of various fluxes on the stresses set up in earthenware glazes. The body used was of rational composition:

60% clay substance;
35% quartz;
5% feldspar.

Twelve glazes were used in which only 0·5 mol. of flux was changed (the representation is not quite traditional because of using fluxes that are not in

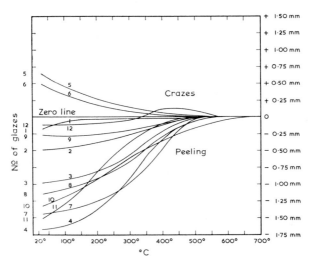

Fig. 6.3. *Stress curves for various earthenware glazes* (**S40**)

the RO group). The relative conditions of these glazes at 20° C (68° F) are listed in Table 135 but Fig. 6.3 shows that the change of stress with increasing temperature is different for the different glazes. The stress in most of them is completely relieved by 575° C (1067° F) but No. 11 remains in compression until 650° C (1202° F).

Crazing and peeling are best prevented by adjustment of the coefficients of thermal expansion of the body and/or the glaze. They are also counteracted by a good interaction layer between the body and the glaze, as has already been mentioned. Greater elasticity of the glaze also assists. The stresses set up between body and glaze may be overcome by annealing. These factors are often automatically adjusted when altering the expansion coefficients of body and glaze. As 'crazing' and 'peeling' although appearing as glaze faults are due to incompatibility of body and glaze, the nature of the body and/or that of the glaze can therefore be adjusted when trying to eliminate them.

GLAZES

TABLE 135

Glaze No.	Molecular Formula		
5	0·5 Na$_2$O 0·5 PbO	0·1 Al$_2$O$_3$.1·5 SiO$_2$	↑ Increasing tension, therefore crazing.
6	0·5 K$_2$O 0·5 PbO	0·1 Al$_2$O$_3$.1·5 SiO$_2$	
12	1·0 PbO	0·1 Al$_2$O$_3$.1·5 SiO$_2$	
1	0·5 CaO 0·5 PbO	0·1 Al$_2$O$_3$.1·5 SiO$_2$	
9	0·5 FeO 0·5 PbO	0·1 Al$_2$O$_3$.1·5 SiO$_2$	
2	0·5 SrO 0·5 PbO	0·1 Al$_2$O$_3$.1·5 SiO$_2$	Increasing compression, therefore peeling.
3	0·5 BaO 0·5 PbO	0·1 Al$_2$O$_3$.1·5 SiO$_2$	
8	0·5 SnO$_2$ 0·5 PbO	0·1 Al$_2$O$_3$.1·5 SiO$_2$	
10	0·5 ZrO$_2$ 0·5 PbO	0·1 Al$_2$O$_3$.1·5 SiO$_2$	
7	0·5 ZnO 0·5 PbO	0·1 Al$_2$O$_3$.1·5 SiO$_2$	
11	0·5 B$_2$O$_3$ 0·5 PbO	0·1 Al$_2$O$_3$.1·5 SiO$_2$	
4	0·5 MgO 0·5 PbO	0·1 Al$_2$O$_3$.1·5 SiO$_2$	↓

Seger (**S52d**) makes the following recommendations:

Crazing. A. Changes in the body to increase its expansion coefficient:

(*a*) reduction of the plastic clay content and increase of the quartz content;
(*b*) partial substitution of china clay by plastic ball clay;
(*c*) reduction of the feldspar content;
(*d*) finer grinding of the silica materials: quartz, sand, flint, etc.;
(*e*) firing of the ceramic body at a higher temperature.

B. Changes in the glaze to decrease its expansion coefficient:

(*a*) increase in the amount of silica or decrease in the amount of fluxing agents;
(*b*) increase of the boric oxide content at the expense of silica;
(*c*) substitution of a flux of high molecular weight by one of lower molecular weight.

C. Changes in the glaze to increase its modulus of elasticity, *e.g.* replacing sodium by lithium (**J17**).

Peeling. A. Changes in the body:

(*a*) increase of the plastic material with simultaneous decrease of the quartz content;
(*b*) replacement of the clay substance in the form of plastic clay by china clay;

(c) increase of the feldspar content;
(d) lowering of the temperature of biscuit firing.

B. Changes in the glaze:

(a) reduction of the silica content or an increase of the fluxing agents;
(b) substitution of a flux of low molecular weight by one of high molecular weight, whereby the percentage silica content is reduced.

The apparent paradox of adding silica to a body to increase its expansion coefficient and adding it to a glaze to reduce the expansion can be explained by the different forms the silica takes. In the body it is crystalline and hence has a high coefficient, higher than the rest of the body; in the glaze it is amorphous and has a low expansion coefficient (**B147**). Alterations of glaze composition must of course be made within limits:

> ratio silica:flux between 2:1 and 3:1;
> ratio alumina:silica between 1:4·2 and 1:5·5 (**K37**).

Mellor (**M55**) qualifies Seger's suggestions on the alteration of firing temperature as applicable to non-vitreous bodies but not to vitreous bodies. In the case of crazing where it is desirable to increase the thermal expansion of the body the non-vitreous body requires higher firing whereas the vitreous body needs lower firing.

Unfortunately the expansion coefficient of a glass of the same composition as a glaze is not a direct measure of the behaviour of the glaze when on a body. Steger (**S176, 179, 180**) showed that the expansion coefficient of a glaze under stress may be 2% higher than that of one free from stress. Further it is influenced by the thermal history, *e.g.* rate of cooling after firing, of the glazed body and also changes with temperature. It is therefore desirable, although not usually possible to achieve, that the body and glaze have compatible expansion coefficients over the whole range from the lowest annealing temperature of the glaze to room temperature (17° C, 62·6° F) (**S108**).

The annealing temperature of a glaze gives the temperature below which the glaze solidifies so rigidly that it becomes a single unit with the body. The critical annealing temperature for average earthenware glazes lies somewhere in between 340 and 480° C (644–896° F), for soft porcelain between 530 and 570° C (986–1058° F), and for hard paste porcelain glazes between 670 and 700° C (1238–1292° F). Above the annealing temperature irregular expansion should not cause glaze defects (**S108**). In this connection Gehlhoff and Thomas (**G15**) came to the following conclusions:

(a) Alkalies lower the annealing temperature in a very pronounced manner, soda more so than potash.

(b) Lime and zinc oxide raise the annealing temperature, the former very considerably; barium oxide lowers it a little, and lead oxide much more so.

(c) Alumina causes a sharp rise in the annealing temperature, while boric oxide produces a pronounced maximum at 20% B_2O_3.

GLAZES   555

(*d*) There is no linear relationship between chemical composition and annealing temperature.

Keeping a glazed body at the annealing temperature helps a great deal in reducing the strains between body and glaze and hence reduces crazing and peeling in the finished ware. The annealing both gives the glaze a better chance to accommodate itself to the body, and also lowers the coefficient of expansion of the glaze by releasing strains. This is seen in the smaller tendency for thick pieces of once-fired ware to craze when compared with thinner pieces of the same composition, for thick pieces inevitably cool more slowly than thin ones. It is therefore most desirable for a kiln to be cooled as slowly as possible through the annealing range (**M55**).

Steger (**S177**) has measured the annealing temperature and lower and upper critical temperatures of twenty-four pottery glazes (Table 136).

Dunting is not necessarily a glaze defect but is most easily considered together with crazing and peeling. Dunting is the sudden fracture of the glazed (or more rarely an unglazed) body during cooling either in the kiln or after its removal. It seems to occur either at about 600° C (1112° F) or below about 200° C (392° F). The surface of the fracture is mostly very smooth, more so than the surface produced by mechanical breakage. Dunting tends to occur more often in thick pieces than in thin ones.

The causes of dunting are probably various. The temperatures at which it most usually occurs coincide with rapid inversions with volume change of crystalline silica; namely the β- to α-quartz inversion at 573° C (1063° F) with 2% volume change and the β- to α-cristobalite inversion at 220° to 275° C (392° to 527° F) with 5% volume change. Such volume changes of one of the body constituents would set up strains which are further aggravated by small thermal conductivity and strains at the body–glaze interface due to incompatible coefficients of thermal expansion of the body and the glaze (or if not glazed the surface layer).

Dunting can be prevented in almost every case by increasing the flux content of the body. If this is done with feldspar the body will become less porous, if with lime or magnesia the porosity will not necessarily be changed, but the dunting is nevertheless prevented, presumably because the silica is combined instead of free.

Other suggested methods are the replacing of some of the non-plastic materials in the body by plastic ones (**M22**) or, controversially, care in not using too much ball clay or feldspar but adding flint and firing higher (**W36**). Dunting is a complex failure and depends on several factors so that one alteration may prevent it in one case whereas another may be required in a different instance.

DELAYED CRAZING

In certain cases a glaze may begin to craze some weeks or even several years after the ware it is on was fired. This occurs more frequently when

TABLE 136
Annealing Temperatures for Pottery Glazes
(Steger (S177))

Glaze type	Beginning of annealing range		Critical zone				'Equalising' temperature		Noticeable softening	
			Lower limit		Upper limit					
	(°C)	(°F)	(°C)	(°F)	(°C)	(°F)	(°C)	(°F)	(°C)	(°F)
Earthenware glazes	340–480	644–896	390–470	734–878	430–560	806–1040	410–515	770–959	430–600	806–1112
Soft porcelain glazes	530–570	986–1058	540–670	1004–1238	620–700	1148–1292	580–680	1076–1256	760–770	1400–1418
Hard porcelain glazes	670	1238	680	1256	760	1400	720	1328	780	1436

the body is porous, and is due to the body gradually absorbing moisture and expanding. It can be prevented by adjusting the body composition (*e.g.* by adding feldspar) and the firing schedule, to make the body vitrify (**A77**). Alternatively, if the porosity of the body is to remain the same, increase of the silica and sometimes lime and magnesia contents prevents moisture expansion. High-silica bodies (with more than 72% SiO_2) may absorb water but they do not combine with it.

9. *Hardness*

Hardness is a property which is defined and measured in several different ways so that materials may not even be placed in the same relative order when tested by the different methods. The two main types of hardness are: (1) resistance to scratching, which is dependent on the nature, pressure and speed of the scratcher; and (2) the resistance to indentation. With regard to glasses and glazes the different tests for hardness may involve any of the following:

(1) friction with consequent heating and perhaps even softening;
(2) elastic deformations;
(3) tensile strength;
(4) resistance to impact;
(5) resistance to scratching;
(6) bond strength and elastic limit.

One of the earliest criteria of hardness was the ability of a material to scratch or be scratched by another. This gave rise to Mohs' scale of hardness in which the test material is classified according to its position in a scale of ten minerals of increasing scratch hardness. This is, however, not sensitive enough for measuring the relatively small differences between various glazes and glasses which usually occupy the position between apatite No. 5 and quartz No. 7.

A measure of relative resistance to scratching may be obtained by using an abrasive wheel for a fixed time at fixed pressure and measuring the scratch produced. Although this and similar methods give no absolute values, useful data relating to the effect of composition on scratch hardness of dinnerware glazes may be obtained. These show that scratch hardness is increased by high silica content with its consequent higher firing temperature; of the minor constituents an optimum quantity of boric oxide has the most favourable effect, and increases of alumina, barium oxide, magnesia and zinc oxide are also useful (**G89**).

Table 137 lists the effect of the minor constituents on scratch hardness. It differs only slightly from that on p. 550 concerning tensile strength.

Ainsworth (**A177**) describes the use of the diamond pyramid indentation test. In this a four-sided diamond pyramid indenter with 136° between opposite faces is brought in contact with the specimen under controlled

TABLE 137

Minor Glass Constituents in Order of Their Favourable Influence on the Resistance to Scratching (G16, S109)

(1) Boric oxide.
(2) Alumina.
(3) Barium oxide.
(4) Magnesia.
(5) Zinc oxide.

conditions and load. The size of the resulting impression gives a measure of the 'hardness' of the surface under test. Tests are made with several different loads and a graph plotted of the square of the diagonal length of indentation against the load. This is linear if the composition is constant through the total depth tested and its slope gives the D.P.H. number. The D.P.H. test can be applied to minute volumes of glass or glaze, the spot being chosen by the microscope incorporated in the equipment. Any flaw or fault can therefore be avoided and a true basic property of the surface measured. Macro tests on glass rarely show any plastic flow because breakage at a flaw almost invariably occurs before flow sets it. In the D.P.H. test the flaws are avoided and the indentations produced at low loads are due to plastic flow in the glass. The D.P.H. test therefore provides a measure of the yield point or elastic limit of the glass, the relationship being approximately:

$$\text{D.P.H. number} = 2\cdot7 \times \text{yield point}.$$

TABLE 138

Effects of Various Oxides on D.P.H. Number of Silicate Glasses (A177)

Glass composition (mol. %)			Ionic radii		D.P.H. number (kg/mm^2)
Na$_2$O	Third. oxide	SiO$_2$			
—	—	100	Si^{4+}	0·41	710
12	8 K$_2$O	80	Na$^+$	0·95	429
20	—	80	K$^+$	1·33	405
—	20 K$_2$O	80			364
18	10 Al$_2$O$_3$	72	Al^{3+}	0·50	578
18	10 CaO	72	Ca^{2+}	0·99	562
18	10 B$_2$O$_3$	72	B^{3+}	0·20	554
18	10 SrO	72	Sr^{2+}	1·13	550
18	10 BaO	72	Ba^{2+}	1·35	522
18	10 ZnO	72	Zn^{2+}	0·74	510
18	10 MgO	72	Mg^{2+}	0·65	498
18	10 PbO	72	Pb^{2+}	1·21	445
18	(SiO$_2$)	82	Si^{4+}	0·41	426
28	(Na$_2$O)	72	Na$^+$	0·95	362

GLAZES

Incidentally the D.P.H. test places the minerals used for Mohs' scale in the same order as the normal scratching test.

With regard to glasses (and glazes) the D.P.H. number has been found to be a measure of the structural conditions in the glass. In general, if oxides are added to a given glass which decrease bond strength, the D.P.H. number is reduced. In some cases the variation is opposite to that found in the tensile strength. For instance adding Na_2O or K_2O to a pure silica glass breaks Si—O—Si bonds, the hardness number is decreased but the tensile strength is initially increased. This latter effect is due to the alkali reducing the number of flaws in the glass.

Some results are shown in Table 138. They illustrate the complex effects that the different oxides may have on the glass structure and may be interpreted as follows. Considering the series based on 18 mols. % Na_2O, 72 mols. % SiO_2 and 10 mols. % other oxide, it is seen that in general small size and high charge increase bond strength. But this may be overshadowed by the fact that if the added ion is a network former it increases the volume more than a network modifier entering an interstice. Thus Ca^{2+} gives greater strength than $\overset{\diagdown\;\diagup}{B}$ and Ca^{2+}, Sr^{2+} and Ba^{2+} lie above $-\overset{|^{2-}}{\underset{|}{Zn}}-$, $-\overset{|^{2-}}{\underset{|}{Mg}}-$ and $-\overset{|^{2-}}{\underset{|}{Pb}}-$. On the other hand if Zn, Mg, Pb are to become four-bonded network formers they must each have two Na^+ ions associated with them, which do not then break Si—O—Si bonds, and hence the glasses with these three intermediate oxides are stronger than the pure sodium silicate one.

10. Chemical Resistance

The amount of chemical attack a glaze must withstand varies, of course, with the use the finished ware will be put to. The range is from artware that must resist the atmosphere and an occasional wash, through dinnerware, tiling, outdoor tiling, etc., chemical stoneware to chemical porcelain which must stand up to concentrated acids and alkalies at elevated temperatures. So when stating that a glaze is resistant it must be added to what it is resistant.

It might be thought that as chemical resistance is bound up with structural strength it would be parallel with the refractoriness. Although this is the general tendency it is fortunately not entirely so, or all low-melting glazes would have lower chemical resistances than higher melting ones.

Kreidl and Weyl (**K81**) describe the two different mechanisms by which reagents can attack glasses:

(1) The attack by acids other than hydrofluoric. Acids have readily available and very mobile hydrogen ions which may change places with alkali ions in holes in the network, gradually converting it to hydrated silica.

The general structure of the glass remains unaltered but the change of composition lowers the refractive index, producing interference colours.

(2) The attack by alkalies, hydrofluoric acid and other reagents that can form soluble compounds with silica. These chemicals actually attack the silica network, gradually dissolving it away. The primary step in attack by alkali, for instance, is the adsorption of $(OH)^-$ onto an oxygen of Si—O—Si bridge. It is therefore strongly influenced by the composition of the surface as well as the structural strength.

In the first case, attack by acids, the resistance is bettered by substitution of potassium and lithium for sodium; CaO, SrO, BaO, MgO, PbO for alkalies, and best by substituting ZnO for other mono- or di-valent oxides as would be expected from structural considerations. Where alkalies are to be eliminated without raising the melting point B_2O_3 can be substituted, but if a given content of B_2O_3 is exceeded it has low-alkali resistance. Similarly TiO_2 increases acid but decreases alkali resistivity. ZrO_2, however, is said to increase resistance towards acid, alkali and neutral media (**K81**).

In the second instance, of solution by alkalies, etc., the matter is much more complex, depending on the combination of major and minor constituents of the glaze. In general, however, alumina is very beneficial. High-silica glazes are resistant but of course have high melting points.

Mellor (**M56**) considers the action of water and general weathering agents. In general complexity favours durability. The alkalies favour absorption of water, potash more so than soda. The effectiveness of other oxides in order of increasing durability, on a molecular basis is shown in the following list: BaO, CaO, PbO, MgO, ZnO, TiO_2, Al_2O_3, ZrO_2.

Corrosion of a glaze under strain is higher than that of a well-annealed strain-free one.

11. Coloured Glazes

The section on colouring agents (p. 143), and that on colour (p. 227), have shown what numberless possibilities there are for making coloured glazes. The chief points that affect the colour are:

(1) Colouring agent.
(2) Remaining composition of the glaze.
(3) Kiln atmosphere.
(4) Firing temperature.

By keeping any one, two or three of these constant and varying the remainder the colour can be changed.

The usual method of making a coloured glaze is to select a good colourless glaze and substitute a colouring oxide or compound colourant (*see* p. 614) for about 1 to 10% by equivalent, of one or more of the colourless oxides. This is fritted and milled and then used in varying proportions with the base glaze. Very fine grinding of the coloured frit is very important to get an

GLAZES

even distribution of the colouring oxide and hence a uniformly coloured glaze. The mixture of coloured and colourless frits also has to be kept well stirred because the colouring agent is often heavier and tends to sink to the bottom.

In other instances the colouring agent is calcined only with alumina, silica, or mixtures not directly related to the glaze batch, as described on p. 606. The remaining glaze must then be adjusted to allow for the extra Al_2O_3, SiO_2, etc. Coloured glazes are discussed in greater detail under 'Decoration' (p. 803).

12. Glazes for Electrical Ceramics

The specialised ceramic bodies developed for the electrical industry have necessitated the development of glazes with properties not normally encountered for earthenware, bone china and porcelain tableware. Bloor (**B81**) summarises the properties of these bodies (Table 139).

TABLE 139

Properties of Electrical Ceramics Relevant in Glazing Problems

Material	Ingredients	Coefficient of expansion over 20°–800° C ($\times 10^6$)	Tensile strength at 20° C (lb/in^2)	Modulus of elasticity at 20° C (lb/in^2) ($\times 10^{-6}$)
(a) Electrical porcelain	Clay Feldspar Quartz	4·5–6·5	3000–6000	10
(b) Aluminous porcelain	Clay Feldspar Alumina	5·5	10 000	17–21
(c) Steatite (low-loss)	Steatite Clay Feldspar or barium carbonate	8·0	8000	15
(d) Zircon porcelain	Zircon Clay Alkaline-earth fluxes	4·9	12 000	22
(e) Cordierite	Clay Talc or serpentine Alumina (Feldspar)	3–4	5000	12
(f) High-titania	Titania Minor proportions of clay and fluxes	7–9	10 000	39
(g) Magnesium titanate	Magnesia Titania Minor proportions of clay and fluxes	10	10 000	

(a) ELECTRICAL PORCELAIN

Many larger electrical porcelain pieces cannot be refired for maturing the glaze because of the high risk of dunting. The glaze must therefore be applied to the raw piece and must not mature too rapidly in the single firing. Raw glazes which remain open for a long time and only slowly fuse to a glass are therefore the most suitable.

Many porcelain insulator glazes are coloured a rich brown with calcined mixtures of iron, manganese, chromium, nickel, and zinc oxides. A suitable addition of these staining oxides, in particular that of manganese, can increase the mechanical strength of the glazed piece. The colour also plays a role in the thermal properties of the ware, *e.g.* the rate at which it will warm up.

(b) SEMI-CONDUCTING GLAZES

It is desirable that the glaze on high-tension insulators should be a semi-conductor of electricity when it can fulfil two functions:

(1) to distribute the voltage evenly over a string of insulators and thereby prevent flashover in heavily contaminated industrial atmospheres;
(2) to minimise radio interference.

A glaze is made semi-conducting by including in it 20–40% of semi-conducting oxides, frequently similar mixtures to the brown stains mentioned above. The bulk of this stain does not dissolve in the glaze and forms a crystalline network which is semi-conducting and is embedded in a normal stained glaze. Unfortunately these oxide networks are very easily attacked electrolytically at points where they are not fully protected by the glaze and can set off serious glaze deterioration particularly near the 'pin' where the current density is high (*e.g.* one of ten insulators supporting a 76-kV line has a current density near the pin of 0·1 mA per in. of glaze).

The electrical properties of some semi-conducting glazes are given by Forrest (**F15**). The surface resistivity, *i.e.* the resistance between opposite faces of a unit square, varies from 0·2 to 50 MΩ at 20° C (68° F). The variation with temperature over the range 0–150° C (32–302° F) is given by

$$\rho = \rho_0\, e^{b/T}$$

where ρ = surface resistivity at temperature $T°$ K;
ρ_0 has values of 25 to 5000 Ω;
b is approximately 3000 corresponding to an activation energy of 0·25 eV.

The contact resistance between an electrode and a semi-conducting glaze decreases as the applied voltage increases. If a voltage greater than a certain limiting value is applied the glaze is destroyed by the heating effect from the cumulatively increasing current. The cause of the disintegration of the glaze near electrodes is due to local heating effects resulting from contact resistance and non-uniformities in the structure of the glaze.

GLAZES

(c) LOW-LOSS STEATITE

Steatite has an unusually high coefficient of thermal expansion and also a tendency to react unfavourably with some glazes. Some suggested glazes are given in Table 140.

(d) ZIRCON PORCELAIN

Zircon porcelain has a low expansion coefficient and a tendency to react with certain types of glazes. Luttrell (**L49**) has developed suitable white opaque low expansion glazes and found that the general basis is as follows:

$$\left.\begin{array}{l} KNaO < 0{\cdot}20 \\ MgO \quad 0{\cdot}40 \text{ to } 0{\cdot}50 \\ \left.\begin{array}{l} CaO \\ ZnO \end{array}\right\} \text{the rest up to } 1{\cdot}00 \end{array}\right\} \left\{\begin{array}{l} B_2O_3 < 0{\cdot}2 \\ Al_2O_3 \ 0{\cdot}3\text{--}0{\cdot}4 \end{array}\right. \left\{\begin{array}{l} SiO_2 \ 3{\cdot}0\text{--}4{\cdot}0 \\ ZrO_2 \ 0{\cdot}5\text{--}0{\cdot}7 \end{array}\right.$$

Other examples of zircon porcelain glazes are:

$$\left.\begin{array}{ll} Na_2O & 0{\cdot}07 \\ K_2O & 0{\cdot}13 \\ MgO & 0{\cdot}65 \\ CaO & 0{\cdot}15 \end{array}\right\} Al_2O_3 \ 0{\cdot}58, \quad SiO_2 \ 5{\cdot}4\text{--}5{\cdot}9 \quad \text{cone } 12$$

$$\left.\begin{array}{ll} Na_2O & 0{\cdot}06 \\ K_2O & 0{\cdot}12 \\ MgO & 0{\cdot}56 \\ CaO & 0{\cdot}26 \end{array}\right\} Al_2O_3 \ 0{\cdot}57 \left\{\begin{array}{ll} SiO_2 \ 5{\cdot}15 & \text{Glost firing } 1200°\text{ C} \\ ZrO_2 \ 0{\cdot}25 & \text{Fritting} \quad 1400°\text{ C} \end{array}\right.$$

(G59)

(e) CORDIERITE

Cordierite has an exceptionally low expansion coefficient and it is difficult to find glazes that will 'fit' as well as maturing with the body at cone 13 on once-fired ware. Thiess (**T22**) found that some cordierite bodies can be self-glazed by making a slip of the body mixture together with some colouring oxides. These bodies all have high sodium feldspar or nepheline syenite contents and it is possible that the self-glazing arises from the sodium evaporating and subsequently reacting with silica at the ware surface. Suitable bodies for this technique are:

$$\left.\begin{array}{l} 0{\cdot}029 Na_2O \\ 0{\cdot}065 K_2O \\ 0{\cdot}911 MgO \end{array}\right\} 1{\cdot}16 Al_2O_3 . 3{\cdot}67 SiO_2$$

$$\left.\begin{array}{l} 0{\cdot}017 Na_2O \\ 0{\cdot}050 K_2O \\ 0{\cdot}809 MgO \\ 0{\cdot}124 ZnO \end{array}\right\} 1{\cdot}09 Al_2O_3 . 3{\cdot}55 SiO_2$$

$$\left.\begin{array}{l} 0{\cdot}029 Na_2O \\ 0{\cdot}093 K_2O \\ 0{\cdot}877 MgO \end{array}\right\} 0{\cdot}982 Al_2O_3 . 2{\cdot}5 SiO_2$$

TABLE

Glazes for

No.	Ref.	Cone	Type	Na_2O	K_2O	BeO	MgO	CaO	ZnO	PbO
1	—	14			0·12		0·88			
2		14			0·3		0·7			
3		14			0·2		0·8			
4		14			0·2		0·8			
5		14			0·1		0·9			
6		14					1·0			
7		14					1·0			
8		14								
9		14								
10	T20	12–13			0·122–0·233		0·446–0·539	0·266–0·339		
11	T21	10–12	once-fired	0·10	0·19		0·33	0·19		0·03
12				0·30	0·24		0·44	0·02		—
13				0·10	0·18		—	0·72		—
14				—	0·2		0·5	0·3		—
15	S197	10–12	once-fired							
16	G20	2/3	glost							1·00
17	S197	07	Frits GL5		0·317					0·683
18	T21	—		0·19						0·81
19			6	—						1·00
20			7	—						1·00
21			8	—						1·00
22			batches							
23										
24										
25		0·15								1·0
26										1·0
27									0·5	0·5
28	G14	014	Frit at			0·67			0·25	0·08
29	to	1260° C			0·67			0·25	0·08	
30	[B81]	012	(2300° F)			0·70			0·21	0·09
31			glost fire			0·74			0·19	0·07
32						0·70			0·21	0·09
33						0·70			0·21	0·09
34						0·70			0·21	0·09
35						0·68			0·24	0·08
36						0·68			0·24	0·08
37						0·68			0·24	0·08

Steatite

No.	B_2O_3	Al_2O_3	SiO_2	Miscellaneous	P_2O_5	Batch compositions	Properties
1		0·40	3·9				
2		0·40	4·0				
3		0·30	4·0				
4		0·30	3·0				
5		0·20	2·7				
6		0·5	1·5				
7		0·5	2·0				
8						Feldspar 95%, ball clay 5%.	
9						Norwegian feldspar 23·2% Steatite pitchers 9·4% Witherite 8·4% Calcspar 6·4% Zettlitz kaolin 18·3% Quartz 34·3%	
10		0·387–0·517	3·67–4·30				Calculated coefficient of expansion 22–550° C ($\times 10^{-6}$)
11		0·64	3·35	Li_2O 0·16		Potash feldspar 50%, spodumene 20%, Californian talc 12%, whiting 5%, ball clay 10%, lead bisilicate 3%	6·92
12		0·67	3·78			Potash feldspar 44%, nepheline syenite 35%, cal. talc 15%, ball clay 6%	7·41
13		0·49	3·5			Potash feldspar 45%, whiting 19%, flint 22·2%, ball clay 13·8%	7·29
14		0·4	4·0				
15						Potash feldspar 75%, lime talc 20%, clay 5%	
16	3·13	0·48	7·82				
17		0·131	2·39	SnO_2 0·230			
18	1·774	0·041	3·372				
19	1·631	—	3·014				
20	0·238	—	0·780				
21	—	0·254	1·910				Calculated coeff. of exp. 22–550° C ($\times 10^{-6}$)
22						Frit GL6 43·8%, GL8 50·2%, bentonite 3%, Florida kaolin 3%	5·67
23						Frit GL5 94%, bentonite 3%, Florida kaolin 3%	5·73
24						Frit GL5 30%, GL7 64%, bentonite 3%, Florida kaolin 3%	6·27
25			1·00				
26	1·0		1·00				
27	1·0		1·00				

						Electrical properties				
No.	B_2O_3	Al_2O_3	SiO_2	Miscellaneous	P_2O_5	Power factor % dry	Dielectric constant dry	Loss factor: fast firing		Loss factor: slow firing	
								dry	wet	dry	wet
28		0·41			1·56	0·071	6·69	0·69	0·73	0·59	1·28
29		0·31			1·69	0·168	6·53	1·11	1·01	0·60	0·86
30		0·31			1·74	0·105	6·60	0·65	0·74	0·53	0·86
31		0·28			1·54	0·099	6·57	0·66	0·69	0·54	0·85
32		0·22			1·23	0·116	6·47	0·75	0·70	0·62	0·79
33		0·16			0·91	0·179	5·72	1·01	3·82	0·69	1·09
34		0·12			0·69	—	—	—	—	—	—
35		0·43		Bi_2O_2 0·05	2·39	0·122	6·54	0·79	1·28	0·59	1·63
36		0·43		Ce_2O_3 0·03	2·39	0·158	6·58	0·97	0·87	0·58	0·90
37		0·43		TiO_3 0·06	2·39	0·267	6·47	1·74	1·67	1·65	1·60

Thiess (**T22**) also developed a successful slip glaze of the following molecular formula

$$0.773\text{MgO} \atop 0.227\text{FeO} \left\{ {0.093\text{B}_2\text{O}_3 \atop 0.266\text{Al}_2\text{O}_3 \atop 0.086\text{Cr}_2\text{O}_3} \right\} 2.25\text{SiO}_2$$

Thomas, Tuttle and Miller (**T23**) found that there is considerable interaction of most glazes with cordierite and glaze fit can be improved by allowing deep penetration and the formation of an intermediate layer. Their list of the effect of oxides on glaze penetration has been given (Table 129).

Glazes have also been proposed in which a crystalline phase is allowed to separate. This has a higher specific volume than the residual glass so that the expansion occurring on crystallisation opposes the cooling contraction of the glass. The following glaze has a reversible coefficient of expansion of 7.3×10^{-6} per °C

$$\left. {0.1636\text{MnO} \atop 0.1067\text{CoO} \atop 0.2683\text{CuO} \atop 0.4613\text{PbO}} \right\} 0.55\text{Al}_2\text{O}_3.0.4904\text{SiO}_2 \quad (\textbf{H30}).$$

(f) HIGH-TITANIA BODIES

Titania and titanate bodies react readily with glazes, absorbing them, so that soft fritted glazes with a fast firing schedule must be used (**B81**).

FRITTING

It has already been mentioned that if water soluble raw materials are to be used in a glaze they must first be combined with other materials in such a way that an insoluble product is obtained. This is the primary reason for fritting, but the process serves a number of other purposes with regard to the production of a good glaze.

In raw glazes the prepared slop contains various different materials of different specific gravities and particle sizes which may sediment and separate. In a fritted glaze many of the constituents are combined to a single material of uniform density so that layered sedimentation does not occur. A ground frit does, of course, tend to sediment and may form a very hard deposit in so doing. This is counteracted by retaining some of the insoluble parts of the glaze batch from the frit and adding them in the mill; usually 5% of clay or occasionally sand, feldspar or lime; addition of dilute acids or organic sizes also helps. Fritting therefore helps in applying a uniform glaze to the ware.

During firing of a glaze its constituents melt according to their own melting points and gradually mix and combine with each other, until the most refractory of the raw materials melts. During the interval in which only some constituents have fused these may react with the body or volatilise so that the end product will not be the same as if the whole glaze had melted simultaneously. Once melted together the softening point of the resulting glaze is considerably lower than the temperature for melting the most refractory constituent, so that a fritted glaze can be matured at a lower temperature and in less time than the same glaze applied raw. Naturally fuel and time are required in melting the ingredients to make the frit. Frits composed of low-melting materials can be prepared more readily than those including the more refractory ones, but any of the latter not melted in the frit kiln will have to be melted on the ware surface in the glost kiln. So that for fast glost firing as much of the glaze as is compatible with mixing and application considerations should be fritted. For once-fired ware that is to receive a slow firing, a raw glaze that remains porous for a longer time is preferable.

The main considerations affecting the division of the glaze batch between frit and mill additions are:

(1) The frit or frits must make the soluble ingredients insoluble.
(2) The mill addition should bring the ground frit in a suitable condition for applying the glaze.
(3) The combination of frit and mill addition should fuse at a rate suitable to the method of firing.

1. *Frit Composition*

Newcomb (**N11**) gives the following rules for compounding a frit.

(*a*) Hold the ratio of acids to bases within the range of easy fusion—from 1:1 to 1:3.

(*b*) Alkalies cannot be used as the only bases in the RO group of the frit, as the alkaline silicates are readily soluble. One or more insoluble bases must be added; they should total at least 50% of the equivalent weight of the RO group.

(*c*) If B_2O_3 is present, its total of molecular equivalents should not be more than one-half that of the silica in the frit.

(*d*) The ratio of the total number of equivalents of oxygen in the RO and R_2O_3 groups combined, to the equivalents of oxygen in the RO_2 group should fall between 1:2 and 1:6.

Purdy and Fox (**P76**) state that the frit should contain at least 50% by weight of the glaze and should include all the feldspar, borates, boric acid, all the clay except 0·05 mol., all the lime except 0·10 mol., all the alumina introduced as such but only enough silica to make the oxygen ratio (*see* (*d*) above) of the frit the same as that of the glaze.

2. *Mill Addition*

The most usual mill addition for keeping the glaze in suspension is a portion of the clay in the batch, varying from 4 to 12% of the weight of the frit.

Pukall (**P71**) recommends retaining 20% of the glaze batch unfritted and this may include any insoluble material that will readily fuse into the frit during the glost firing, *e.g.* kaolin and feldspar. Red lead, marble, chalk, magnesite, white lead, witherite, calcite could also be used on this basis but nowadays all lead materials are fritted. Marble, chalk and calcite will not fuse into the glaze if the kiln temperature is less than cone 02 (1060° C, 1940° F).

3. *Maturing the Glaze*

Mellor (**M51**) states that the relation of frit and mill batch should be such that the glaze shall have completed its action on the body at the same time that the glaze surface has attained its best gloss.

Sortwell (**S151**) found that if a basic glaze was made up in different ways the resultant glaze varied. Increasing the proportion in the frit improved the fusibility, and the appearance of the finished glaze. With the same proportion of frit in the glaze the most easily fusible and best looking glazes were obtained when the refractory ingredients, sand and part of the china clay, had been fritted.

Steger (**S178**) found that varying the division of the glaze between frit and mill can change the stress between an earthenware body and its glaze by 100% and materially alter the craze resistance. He found that if the frit was low melting it had a longer opportunity to react with the body surface to form an intermediate layer that relieves stresses.

In this connection it is worth noting that the form in which a single ingredient such as silica is added also affects the maturing temperature and range, surface texture, defects and glaze fit. In this respect it has been found that diatomaceous silica is superior to other forms of silica as a raw constituent of glazes. Next to this come chalcedony and fused silica while potter's flint and pyrophyllite are less satisfactory sources of silica (**R72**).

THE PROBLEM OF THE USE OF LEAD

It could generally be considered that a glaze that is sufficiently durable not to contaminate anything in contact with it noticeably, or to wear off before the ware breaks would be adequate. Unfortunately, however, a common constituent of glazes is lead, and this has an insidious poisonous nature. So that any vessel for food, etc., ought to have a glaze that either does not dissolve at all or that does not contain lead. There is further the even greater danger to the health of the workers producing lead glazed ware.

Complete exclusion of lead from the glaze composition is not easy. Lead imparts brilliance, hardness and smoothness to glazes not readily equalled by leadless ones. Furthermore lead glazes are much less susceptible to minor irregularities of make-up and firing conditions than leadless ones. This is chiefly because of the high viscosity of lead aluminosilicates which decreases comparatively little with increasing temperatures. Many of the older kilns cannot be regulated more accurately than to 60° to 100° C of the optimum. Under these conditions an alkali containing, leadless glaze may show incomplete fusion in the coolest part, and too low viscosity in the hottest, whereas the lead glaze will give a good result throughout.

There are therefore two possible ways of dealing with the situation, one is to develop lead glazes whose solubility is so low that they are no danger to the consumer, and whose manufacture is so arranged to eliminate danger to the workers, the other is to produce satisfactory leadless glazes. Both ends can be achieved.

Legislation

The danger of lead poisoning is of course far greater for the workers in the potteries than the consumer, and formerly constituted an occupational disease. In Great Britain legislation has therefore been mainly directed at protection of the workers. It had been found that women and young persons are more susceptible than men to lead poisoning so it is now not permissible to employ them in any process, such as dipping the ware in the glaze, where raw lead is used, or for cleaning rooms where raw lead is used. Those workers who do handle raw lead must be protected as follows: by wearing respirators when handling or mixing dry lead materials, by having efficient exhaust draughts in such places as dust is unavoidably created, by being supplied with clean overalls and head coverings at least once a week and washrooms with lockers for outdoor clothing, forbidding eating and smoking in the workshops and arranging regular medical inspections.

By law, Pottery (Health) Special Regulations 1947, it is now only permissible to use 'low solubility' lead glazes whose solubility when measured in the standard way laid down in the Factory Acts shall not exceed 5%. This ruling automatically safeguards the consumer as well as the worker.

In other countries the legal position may be different, but as the desired result is always to eliminate risk to both worker and consumer the technical aspect remains the same.

Low-solubility Glazes

In the workshops the danger from lead compounds is high because of the very small particle size. Lead-containing dust is easily inhaled and gets into the nose and mouth and into the stomach. Here many lead compounds, although almost water insoluble, are dissolved by the hydrochloric acid of the gastric juices. The lead chloride is absorbed and accumulated until there is sufficient for symptoms to appear.

The technical problem then is to convert the lead raw materials to compounds that are insoluble in the gastric juices and to do this as early as possible in the over-all manufacture. This is achieved by fritting. Certain lead silicates are sufficiently insoluble in 0·25% HCl to be harmless when used as a glaze constituent. If then the raw lead is fritted with enough silica and certain other compounds only the workers involved in charging the frit kiln are in any danger of lead poisoning.

The effect of the different glaze components on the solubility of the lead must be carefully considered before making up a lead-containing frit.

SILICA

The higher the ratio $SiO_2:PbO$ the lower the solubility of the lead, but as high silica compounds have too high a melting point the most suitable ratio is about 2:1. $PbO.3SiO_2$ and $PbO.4SiO_2$ have too high fusion temperatures for consideration.

TABLE 141

Solubility and Melting Temperature of Lead Silicates

Compound	% lead	Lead solubility		Melting temperatures	
		% of frit	% of lead content	Initial glass formation	Complete fusion
$2PbO.SiO_2$	81·82	73·11	89·25	—	765° C
$PbO.SiO_2$	73·15	22·80	31·20	700° C	890° C
$PbO.2SiO_2$	60·02	4·51	7·61	700° C	

ALUMINA

Alumina decreases the solubility of lead silicates very sharply, thus Harkort (**H24**) found that an addition of 0·1 equivalents Al_2O_3 to $PbO.SiO_2$ reduces the solubility from about 40% to 10%, whereas with silica the same result can only be achieved with 0·8 equivalents SiO_2.

Addition of alumina to lead bisilicate not only greatly improves its acid resistance but lowers the melting point until the eutectic mixture is reached, further addition raises it again. The mixture with the eutectic composition is therefore very useful.

TABLE 142

Comparisons of Lead Bisilicate and the Eutectic Mixture
$PbO.0·254Al_2O_3.1·91SiO_2$ (K54)

Molecular formula	Percentage composition	Melting temperature			Visible viscosity	Lead solubility % of total frit	Percentage loss of lead content	
		Initial glass formation	Complete fusion	Fritting			In acid	In water
1·00PbO 2·00SiO_2	65·01 34·99 100·00	700° C 1292° C	890° C 1634° F	1250° C 2282° F	Viscous, rather like heavy oil	4·59	7·61	0·12
1·00PbO 0·254Al_2O_3 1·91SiO_2	61·35 7·12 31·53 100·00	730° C 1346° F	770° C 1418° F	1150° C 2102° F	Fluid, similar to a light oil	0·70	1·23	0·20

BORIC OXIDE

Lead borate $PbO.0·9B_2O_3.0·7H_2O$ is completely soluble in acids and 44·5% of the lead in lead borosilicate $PbO.B_2O_3.0·6SiO_2$ is soluble. Addition of boric oxide to a lead frit increases the solubility markedly and can only be counteracted by increase of alumina or larger increase of silica.

THE ALKALIES

The alkali oxides increase the solubility of a lead frit but not as much as boric oxide, K_2O does so more than Na_2O. This can be counteracted by adding Al_2O_3.

GROUP II OXIDES

MgO seems to have practically no effect on lead solubility. ZnO, BaO and CaO decrease the solubility, CaO most. BeO does so too.

GROUP IV OXIDES

TiO_2 and ZrO_2 decrease lead solubility (**K54**).

There are three main ways of incorporating these facts to build up low solubility lead frits and glazes.

(1) The use of a double frit. The lead is fritted with those other raw materials which make a good acid-resisting frit, *i.e.* alumina and silica, and the remaining soluble raw materials, particularly those containing boron, are fritted separately. This is highly recommendable.

(2) The use of a single frit incorporating all the soluble raw materials and large amounts of alumina, lime and silica. Although insoluble frits can be prepared in this way it is not a generally satisfactory method.

(3) The use of a single highly acid-resistant lead frit together with an insoluble boron mineral in the mill addition. This is only practicable where the minerals are available but should then be taken advantage of more than is now usual.

The first method of the double frit is the only one that is universally applicable to lead glazes.

Standard low-solubility lead frits are frequently made in separate works and delivered in granulated form to the potteries. Others are made on the spot in special frit kilns and the raw lead compounds not allowed into general circulation space.

The actual solubility of the frit depends, of course, on the grain size. The statutory test is carried out on samples that pass B.S. 170-mesh sieve but are retained on B.S. 240-mesh sieve. As the frit is ground very much more finely before incorporation in the glaze its actual solubility becomes higher than the measured one and may eventually become higher than is safe. Excessive grinding should therefore be avoided.

The solubility of a lead frit must also be well within the limit allowed because when the highly acid-resistant lead frit combines with the leadless boron-containing frit in the glost firing, the product may be less acid resistant. But the conditions are not comparable. Before firing the lead compounds are in a finely powdered form which can easily become airborne and be taken in through the nose and mouth. The surface area is large and the time spent

TABLE 143

Examples of Low-solubility Lead Frits

(1) *Lead–alumina–silica Eutectic*

	M.F.	%
PbO	1·0	61·2
Al_2O_3	0·254	7·1
SiO_2	1·91	31·7

Solubility 0·70% of frit
1·23% of lead content

(2) *Commercial 'Lead Bisilicate Frit'* (A194, G56)

	M.F.	%
PbO	1·0	65·0
Al_2O_3	0·074	2·2
SiO_2	1·82	31·8

Solubility 1%

(3) *Commercial frit with high lead content* (A194, G56)

	M.F.	%
PbO	1·0	70·0
Al_2O_3	0·078	2·5
SiO_2	1·30	24·5
TiO_2	0·080	2·0

Solubility 2%

(4) (K54)

	M.F.	%
CaO	0·15	2·45
PbO	0·85	55·19
Al_2O_3	0·25	7·42
SiO_2	2·00	34·94

Solubility 0·71% of frit
1·38% of lead content

(5) (K54)

	M.F.	%
K_2O	0·07	0·21
CaO	0·10	1·77
PbO	0·83	58·44
Al_2O_3	0·20	6·43
SiO_2	1·75	33·15

Solubility 0·95% of frit
1·75% lead content

(6) (K54)

	M.F.	%
PbO	1·0	54·15
Al_2O_3	0·20	4·95
SiO_2	2·50	36·43
ZrO_2	0·15	4·47

Solubility 0·5% of frit
1·07% of lead content

GLAZES 573

in the stomach long. After firing the total surface area is vastly smaller and the attacking acids can only work outside the body on the vessel. Successive washing of a lead glazed vessel with dilute acid removes successively smaller quantities of lead; so that a glaze produced by the double frit method, if it is sufficiently resistant to wear and tear is of no danger to the consumer. By law the maximum permissible solubility of a lead glaze tested under given conditions is 5%. By making the lead frits with a maximum solubility of 2% any lowering of resistance when they combine to form the glaze is allowed for.

Examples of six low-solubility lead frits are given in Table 143. Two glazes using them are given in Table 144, the first shows the normal British practice of fritting almost all the components and retaining only 5 to 10% china clay for the mill addition, the second shows a common American practice of keeping back a number of insoluble raw materials for the mill addition.

The solubility of frits (2) and (3) is determined by samples passing B.S. sieve No. 170 and retained on B.S. sieve No. 240 and 0·25% HCl at 20° C (68° F) (*see* p. 373).

The solubility of frits (1), (4), (5) and (6), is determined using samples passing B.S. 150-mesh sieve and retained on B.S. 200-mesh sieve and boiling 4% acetic acid (**K54**).

TABLE 144

Examples of Glazes Using Low Solubility Lead Frits

I. Giving method of deriving batch composition from glaze formula

Glaze formula

$$\left. \begin{array}{l} Na_2O \ 0{\cdot}16 \\ K_2O \ \ 0{\cdot}06 \\ CaO \ \ 0{\cdot}42 \\ PbO \ \ 0{\cdot}36 \end{array} \right\} \left. \begin{array}{l} B_2O_3 \ 0{\cdot}6 \\ Al_2O_3 \ 0{\cdot}3 \end{array} \right\} SiO_2 \ 2{\cdot}8 \qquad \text{(S181)}$$

Frit No. 1

PbO 1·0, Al_2O_3 0·254, SiO_2 1·91, M.W. 363·81

Raw materials	Formula	M.W.	Parts by weight	Batch %
Minium	Pb_3O_4	685·63	(685·63/3) × 1·0 = 211·88	58·61
China clay	$Al_2O_3.2SiO_2.2H_2O$	258·09	258·09 × 0·254 = 65·55	18·13
Flint	SiO_2	60·06	60·06 × (1·91 − 0·508) = 84·08	23·26

Parts by weight required to provide 0·36 PbO for glaze: 0·36 × 363·81 = 130·97

TABLE 144—(contd.)

Leadless frit

Considering the K_2O: Raw material: feldspar, $K_2O.Al_2O_3.6SiO_2$, M.W. 556·49.
  Pts. by wt. required $0·06 \times 556·49 = 33·39$.
  This also contributes $0·06$ Al_2O_3, $0·36$ SiO_2.

Considering Na_2O supplied by borax, $Na_2O.2B_2O_3.10H_2O$, M.W. 381·43.
  Pts. by wt. required $0·16 \times 381·43 = 61·03$
  This also contributes $0·32$ B_2O_3.

Considering B_2O_3 supplied by borax and boric acid, H_3BO_3, M.W. 61·84
  Pts. by wt. boric acid $(0·6 - 0·32) \times 61·84 \times 2 = 34·63$

Considering the CaO supplied by whiting $CaCO_3$, M.W. 100·09
  $0·42 \times 100·09 = 42·04$

Considering the Al_2O_3 supplied by frit (1)   0·09
                             by feldspar   0·06
  China clay $Al_2O_3.2SiO_2$ M.W. 258·09
  $(0·3 - 0·15) \times 258·09 = 38·71$

This china clay is not fritted but used as mill addition.

Considering the SiO_2 supplied by frit (1)   0·69
                          by feldspar in leadless frit   0·36
                          by china clay in mill   0·30
                                                         ——
                                                         1·35

Flint for leadless frit $(2·8 - 1·35) \times 60·06 = 87·08$

	Parts by weight	%
Leadless frit		
Feldspar	33·39	
Borax	61·03	
Boric acid	34·63	
Whiting	42·04	
Flint	87·08	
	258·17	
Glaze		
Lead frit No. 1	130·97	30·61
Leadless frit	258·17	60·34
China clay	38·71	9·05
	427·85	100·00

II. *Glaze formula*

$$\left.\begin{array}{l} Na_2O \ 0·06 \\ K_2O \ 0·12 \\ CaO \ 0·43 \\ ZnO \ 0·13 \\ PbO \ 0·26 \end{array}\right\} \left.\begin{array}{l} B_2O_3 \ 0·31 \\ Al_2O_3 \ 0·27 \end{array}\right\} SiO_2 \ 2·60$$

(K54)

TABLE 144—(contd.)

II.—(contd.)

Lead frit No. 5

K_2O 0·07
CaO 0·10
PbO 0·83
Al_2O_3 0·20
SiO_2 1·75   M.W. 312·75

Parts by weight required to give 0·26 PbO = 81·31

Mols. supplied to glaze $\left(\times \text{frit M.F. by } \dfrac{0\cdot 26}{0\cdot 83} \right)$

K_2O 0·02
CaO 0·03
PbO 0·26
Al_2O_3 0·06
SiO_2 0·55

Leadless frit (**K54**)

Na_2O 0·149
K_2O 0·207
CaO 0·644
B_2O_3 0·800
Al_2O_3 0·200
SiO_2 1·900   M.W. 255·07

Parts by weight required to give 0·31 B_2O_3 = 79·07

Mols. supplied to glaze ($\times$ frit M.F. by 0·31/0·8)

Na_2O 0·06
K_2O 0·08
CaO 0·25
B_2O_3 0·31
Al_2O_3 0·08
SiO_2 0·73

Mill addition

This must supply:

	Mols.		Parts by weight	M.F.
K_2O	0·02 ⎫		1·88	0·07
CaO	0·15 ⎬ 0·30		8·41	0·50
ZnO	0·13 ⎭		10·58	0·43
Al_2O_3	0·13		13·25	0·43
SiO_2	1·32		79·28	4·40
			113·40	

TABLE 144—(*contd.*)

II.—(*contd*).
Total glaze

	Parts by weight	Batch, %
Lead frit	81·31	29·70
Leadless frit	79·07	28·88
Mill addition	113·40	41·42
	273·78	100·00

LEADLESS GLAZES

The search for leadless earthenware glazes began as far back as 1884 when Seger (**S51**) examined the problem. It was hoped not only to prevent the danger to health connected with lead glazes but also to eliminate their other disadvantages which are their mechanical softness, tendency to dull in sulphurous atmospheric conditions, sensitiveness to reducing gases and yellowish tinge.

There is now a range of highly satisfactory leadless earthenware glazes available. In general low melting is achieved by using a combination of alkalies, alkaline earths, magnesia and zinc oxide, and almost invariably boric oxide. Parmelee and Lyon (**P12**) give the following limits for frits containing Na_2O, CaO, B_2O_3, Al_2O_3 and SiO_2:

Na_2O 0·3 to 0·7 equivalents;
CaO 0·7 to 0·3 equivalents;
B_2O_3 not more than 0·5 equivalents;
Al_2O_3 up to 0·1 equivalents;
SiO_2 up to 3·0 equivalents.

Experience has also shown that not less than 2·0 equivalents SiO_2 should be used. The solubility of the frit in water is roughly proportional to the molecular ratio:

$$\frac{Na_2O + B_2O_3}{CaO + SiO_2}$$

with small additions of Al_2O_3, *i.e.* 0·1 equivalents reducing it by about half. A frit of low solubility can nevertheless be made without Al_2O_3.

When a wider range of components is used it should be remembered that barium compounds, if soluble, are poisonous and although there is no legislation about their use the same care ought to be exercised as in the case of lead.

A few examples of leadless earthenware frits and glazes are given in Tables 145–147 and the batch compositions of some of them in Table 148. Examples

TABLE 145

Leadless Commercial Frit Compositions

(compiled by Parmelee and Lyon (**P12**))

Na_2O	K_2O	MgO	CaO	B_2O_3	Al_2O_3	SiO_2
0·239	0·225	—	0·536	0·478	0·226	2·48
0·279	0·023	—	0·699	0·548	0·242	2·372
0·284	0·017	—	0·699	0·549	0·298	2·54
0·360	0·088	—	0·552	4·691	0·160	6·265
0·365	0·067	0·091	0·477	1·000	0·149	3·000
0·267	0·027	0·010	0·680	0·545	0·266	2·576

TABLE 146

Examples of Satisfactory Leadless Frits

(1) Na_2O 0·035, MgO 0·118, CaO 0·471, SrO 0·059, BaO 0·118, ZnO 0·200 ; B_2O_3 0·365, Al_2O_3 0·192 ; SiO_2 2·706   (**O7**)

(2) Na_2O 0·18, K_2O 0·37, CaO 0·45 ; B_2O_3 0·47, Al_2O_3 0·37 ; SiO_2 2·72

(3) Na_2O 0·23, K_2O 0·32, CaO 0·45 ; B_2O_3 0·47, Al_2O_3 0·32 ; SiO_2 2·52

(4) [NaK]O 0·424, CaO 0·274, BaO 0·301, FeO 0·001 ; B_2O_3 0·473, Al_2O_3 0·093 ; SiO_2 1·645   (**H17**)

(5) Na_2O 0·50, CaO 0·50 ; B_2O_3 1·00, Al_2O_3 0·25 ; SiO_2 2·50

(6) Li_2O 0·143, Na_2O 0·086, K_2O 0·029, MgO 0·214, CaO 0·314, SrO 0·214 ; B_2O_3 0·329, Al_2O_3 0·200 ; SiO_2 2·500   (**O7**)

using lithium have been included although its use has so far only proved economical in the United States. So also have some glazes containing magnesia which is used successfully in Continental Europe and America but not in Great Britain. Glazes (1) and (2) are for unusually low temperatures.

The batch compositions given for glazes (7) and (9) do not define the type of clay used. It is usual to use china clay in the frit, where absence of coloured impurities is most important, but ball clay or bentonite is sometimes used in the mill addition where the colloidal properties are of prime importance.

TABLE

Leadless Earthenware

No.	Ref.	Cone No.		Li_2O	Na_2O	K_2O	MgO	CaO
1	B121	021	M.F. %		0·157 5·6	0·103 5·6		0·062 2·0
2	R13	012 to 08	M.F. %		0·38 8·6	0·13 4·4		
3	R13	05 to 1	M.F. %		0·33 6·1	0·12 3·4		0·05 0·8
4	R68	03	M.F. %		0·23 3·3	0·22 5·1		0·55 7·4
5	M65	02–2	M.F. %		0·20–0·30			0·40–0·50
6	M66	4	M.F. %			0·20 6·8	0·15 2·2	0·25 5·0
7	O7	5–6	M.F. %	0·10 1·0	0·12 2·5	0·04 1·3	0·15 2·1	0·44 8·4
8	W27	6	M.F. %			0·20 5·0	0·05 0·5	0·40 6·0
9	M29	5–7	M.F. %	0·006 0·1	0·110 2·4	0·101 3·3	0·104 1·4	0·490 9·6
10	O7	6–7	M.F. %	0·041 0·4	0·138 2·8	0·093 2·9	0·153 2·0	0·423 7·8
11	O7	6–7	M.F. %		0·06 1·3	0·12 4·0	0·10 1·4	0·40 8·0

TABLE

Batch Compositions of Certain

Glaze No. 5 This is a raw glaze and can be used in studio pottery and as a base for coloured glazes.

Raw materials: Soda-potash feldspar, $0·64Na_2O, 0·36K_2O.Al_2O_3.6SiO_2$
Colemanite, $2CaO.3B_2O_3.5H_2O$
Calcined zinc oxide
Barium carbonate
Flint
Georgia kaolin or English china clay (**M65**)
Batch according to glaze formula chosen

Glaze No. 6 This is a raw glaze and is suitable for studio potteries and as a base for coloured glazes.

Raw materials:	%
Buckingham feldspar	38·2
Steatite	6·4
Colemanite	17·3
Calcium nitrate (to counteract deflocculation by colemanite)	1·0
Barium carbonate	6·6
Calcined zinc oxide	8·2
Flint.	22·3
	100·0 (**M66**)

147

Glazes

No.	SrO	BaO	ZnO	B$_2$O$_3$	Al$_2$O$_3$	SiO$_2$	TiO$_2$	F$_2$	
1		0·070 6·2	0·608 28·6	0·556 22·4	0·043 2·6	0·776 27·0			
2		0·20 11·2	0·25 7·4	—	0·1 3·7	2·83 61·8	0·10 2·9		
3		0·20 9·1	0·30 7·2	—	0·22 6·7	3·6 64·3	0·10 2·4		
4				1·20 20·1	0·55 13·5	3·50 50·6			
5		0·40–0·00	0·0–0·40	0·60–0·75	0·25–0·35	2·50–3·75			Batch given
6		0·10 5·5	0·30 8·8	0·375 9·5	0·22 8·1	2·50 54·1			Batch given
7	0·15 5·3			0·23 5·5	0·35 12·2	3·00 61·7			Batch given
8		0·05 2·0	0·30 6·5	0·10 1·9	0·50 13·6	4·00 64·5			
9		0·077 4·1	0·112 3·2	0·287 6·9	0·315 11·2	2·764 57·8			Batch given
10	0·152 5·1			0·235 5·3	0·347 11·5	3·146 61·7		0·041 0·5	
11	0·05 1·8	0·10 5·5	0·17 4·9	0·31 7·7	0·27 9·8	2·60 55·6			

148

Leadless Glazes in Table 147

Glaze No. 7

Frit batch	%		Glaze batch	%
Hydrated boric acid	13·4		Frit	58·1
Lithium carbonate	3·5		Whiting	7·2
Strontium carbonate	10·4		Nepheline syenite	11·8
Talc	8·8		Clay	10·3
Nepheline syenite	17·1		Flint	12·6
Pyrophyllite	8·7			
Whiting	10·3			100·0
Flint	27·8			
	100·0			(**O7**)

Glaze No. 9

Frit batch	%		Glaze batch	%
Borax	4·5		Frit	65·1
Boric acid	11·4		Nepheline syenite	15·5
KNO$_3$	6·1		*Lithospar	3·1
Dolomite	7·7		Clay	8·0
Whiting	15·6		Flint	8·3
BaCO$_3$	6·1			
ZnO	3·7			100·0
Clay	9·9			
Flint	35·0			
	100·0			(**M29**)

Lithospar: Trade name for a mixture of potash and soda feldspars with spodumene having P.C.E. value 2–3 (**M28**).

The lowest melting mixtures of these three minerals lie within the following limits: spodumene 35–15%, potash feldspar 50–25%, soda feldspar 50–20% (**B102**).

Fig. 6.4. *Snakeskin glaze*

GLAZES

Crackle Glazes

A glaze that will craze noticeably is sometimes applied for decorative effect. It must have a large expansion coefficient and if applied thickly some unusual effects may be seen. An example is:

Frit		Glaze	
Borax	52·4%	Frit	40%
Stone	16·1%	Whiting	45%
Feldspar	5·6%	Stone	15%
Whiting	3·5%		
Soda ash	22·4%		

(A47)

Assuming Cornish stone of composition

$$Al_2O_3.2SiO_2.2H_2O.5(K_2O.Al_2O_3.6SiO_2).10SiO_2$$

and potash feldspar, the M.F. of the frit and glaze are:

Frit

$$\left.\begin{array}{l}Na_2O \quad 0·842\\K_2O \quad 0·074\\CaO \quad 0·084\end{array}\right\} \left.\begin{array}{l}B_2O_3 \quad 0·067\\Al_2O_3 \quad 0·084\end{array}\right\} SiO_2 \quad 0·567$$

Glaze

$$\left.\begin{array}{l}Na_2O \quad 0·530\\K_2O \quad 0·154\\CaO \quad 0·316\end{array}\right\} \left.\begin{array}{l}B_2O_3 \quad 0·042\\Al_2O_3 \quad 0·181\end{array}\right\} SiO_2 \quad 1·257$$

Snakeskin Glazes

Glazes with high surface tension leading to crawling can also be developed for decorative effects. Kwederawitsch (**K92**) describes experiments with them. Suitable glazes of this type will split into small islands and give a snakeskin effect. If an opaque glaze is used the islands are opaque and a darker transparent phase usually fills the gaps between them.

Surface tension is governed by the chemical composition (*see* table of factors, p. 538), magnesia and alumina being the best components for increasing it, and also by the temperature. Higher temperatures decrease surface tension. Under average conditions a snakeskin glaze can be expected at 1000° C (1832° F) if the surface tension exceeds 298 dyn/cm. At 1140° C (2084° F) a minimum surface tension of 360 dyn/cm is necessary. The snakeskin effect is also dependent on the thickness of the glaze application, the islands increasing in size with greater glaze thickness, and on the stage at which separation into islands occurs. The use of a large content of raw plastic clay in the glaze will lead to cracking of the dry raw glaze and gives rise to a satisfactory snakeskin. To counteract the poor adherence of such glazes to the body, adhesives should be added to them. Less trouble is experienced on flat articles (Fig. 6.4).

TABLE 149
Miscellaneous Glazes for the Lower Temperature Range

No.	Application	Ref.	Cone		Na_2O	K_2O	MgO	CaO	—	PbO	B_2O_3	Al_2O_3	SiO_2	—	Batch
1	Cordierite		015/014 800°C	M.F. %					ZnO 0·216 4·76	0·784 47·43	0·433 8·22	0·199 5·52	1·693 27·58	ZrO_2 0·194 6·48	
2	High-titania bodies	A16	08/07 950°C	M.F. %	0·08 0·16	0·02 0·60	0·02 0·23	0·30 4·65		0·58 36·50	0·48 9·49	0·32 9·50	2·18 37·04	Fe_2O_3 0·15 TiO_2 trace Ign. loss 0·52	
3	Once-fired German stove tiles	A8	07	M.F. %		0·09 2·58				0·91 61·82		0·35 10·90	1·35 24·70		Lead oxide 60 Stoneware clay 20 Feldspar 15 Quartz 5
4	Building bricks	H123	05–03 1000°C– 1048°C	M.F. %	Li_2O Na_2O 0·065 0·072 0·65 1·49	0·095 2·99	0·096 1·29	0·228 4·28	BaO 0·387 19·84		0·432 10·11	0·279 9·54	2·479 49·80		Flint 25·0 Kaolin 3·6 Talc 3·6 $BaCO_3$ 22·6 Nepheline syenite 13·5
5	Albany slip glaze for stoneware	S90	(4–12)	M.F. %		0·23 4·24	0·21 1·66	0·42 4·61	FeO 0·14 1·97			0·80 15·99	6·09 71·54		Colemanite 17·6 Lepidolite 14·1
6	Albany slip glaze for stoneware	S90	(4–12)	M.F. %		0·19 5·24	0·18 2·13	0·36 5·91	FeO 0·27 5·68			0·51 15·26	3·74 65·78		
7	Feldspathic glaze for stoneware	S90	(4–12)	M.F.		0·2	0·1	0·7				0·4–1·0	3·5–9·0		

TABLE 150
Glazes General (mostly from S108)

Type	Ref.	Cone		Na₂O	K₂O	MgO	CaO	ZnO	SrO	B₂O₃	Al₂O₃	SiO₂	Batch
		7	M.F.		0.2	0.1	0.7				0.4	3.5	
			%		6.0	1.3	12.5				13.0	67.2	
		7	M.F.		0.2	0.1	0.7			0.1	0.45	3.90	
			%		5.5	1.2	11.2			2.0	13.1	67.0	
		8	M.F.		0.2	0.1	0.7				0.45	3.90	
			%		5.5	1.2	11.4				13.4	68.5	
		8	M.F.		0.2	0.1	0.7			0.2	0.6	5.0	
			%		4.3	0.9	9.0			3.2	13.9	68.7	
		9	M.F.		0.2	0.1	0.7				0.5	4.3	
			%		5.1	1.1	10.5				13.7	69.6	
		9	M.F.	0.2	0.3		0.5			1.0	0.6	5.2	
			%	2.4	5.5		5.5			13.6	11.9	61.1	
		10	M.F.		0.2	0.1	0.7			0.1	0.65	5.5	
			%		4.0	0.9	8.4			1.5	14.2	71.0	
Alkali-free		11	M.F.		0.2	0.22	0.78				0.3	3.0	
			%		2.9	3.4	16.6				11.6	68.4	
		11	M.F.			0.1	0.7			0.2	0.9	8.0	
			%			0.6	6.0			2.1	14.2	74.2	
English porcelain	M88	9–14	M.F.	0.13	0.19		0.66				0.48	3.85	
			%	2.3	5.2		10.8				14.3	67.4	
		12	M.F.		0.2	0.1	0.7			0.1	0.85	7.5	
			%		3.1	0.7	6.6			1.1	14.2	74.3	
		12	M.F.		0.2	0.1	0.7				0.75	6.5	
			%		3.5	0.8	7.4				14.4	73.9	
		13	M.F.		0.2	0.1	0.7				0.85	7.5	
			%		3.1	0.7	6.5				14.5	75.2	
Alkali-free		14	M.F.				1.0				0.6	5.0	
			%				13.4				14.6	72.0	
Alkali-free		14	M.F.				0.5				0.6	5.0	
			%				6.8				14.9	73.5	
		14	M.F.	0.4	0.4	0.5	0.2				1.1	8.0	
			%	3.7	5.7	4.8	1.7				16.8	72.1	

TABLE 150—(contd.)

Glazes General (mostly from S108)—(contd.)

Application	Ref.	Cone	Na₂O	K₂O	MgO	CaO	ZnO	SrO	B₂O₃	Al₂O₃	SiO	Batch
Rosenthal laboratory porcelain		14	M.F.		0·15	0·15	0·60		0·10	1·1	10·0	
Porcelain		14	%		1·8	0·8	4·3		1·3	14·5	77·3	
Porcelain dinnerware	A3	15	M.F.		0·17	0·22	0·61			0·77	6·8	
			%		2·9	1·6	6·3			14·4	74·8	
Berlin laboratory porcelain		15	M.F.		0·111	0·222	0·667			1·111	11·111	
	A8		%		1·2	1·1	4·5			13·5	79·7	
Berlin porcelain		15/16 (1450° C)	M.F.		0·3	0·1	0·6			1·0	10·0	
			%		3·7	0·5	4·4			13·3	78·1	
High-alumina sparking plugs	A5		M.F.		0·658		0·342			1·82	14·22	
			%		5·5		1·7			16·6	76·2	
		16	M.F.		0·2	0·1	0·7			1·2	10·0	Chalk 3·0
			%		2·4	0·5	5·0			15·5	76·6	Kaolin 4·0
		16	M.F.		0·2	0·1	0·7			1·4	12·0	Quartz 41·8
			%		2·0	0·4	4·2		0·2	15·2	76·7	Feldspar 32·0
									1·5			Fired kaolin 19·2

INHOMOGENEOUS GLAZES

Homogeneous glazes with only one vitreous phase are transparent. If, however, the body is an undesirable grey or yellowish colour a transparent glaze is often not suitable. By incorporating matter that produces a second phase in the glass, opaque, matt or crystalline glazes can be made. Matt and crystal glazes are only used where their special attractive appearance is on show, e.g. wall tiles, artware. Opaque glazes are used frequently where a uniform glossy surface is required, e.g. sanitary ware.

Opaque Glazes

A vitreous phase is rendered opaque if the path of light is suitably broken up and made diffuse. This is brought about by fine particles insoluble in the glass and of different refractive index. The opacity of the glaze is thus determined by the relative refractive indices of the vitreous phase and the disperse phase, and by the particle size and shape of the latter.

Stevels (**S187**) describes the three ways by which dispersed phases are produced:

I. Crystals formed by crystallisation of the melt.
II. Introduction of crystals insoluble in the melt into the batch.
III. Two (vitreous) phases formed by immiscibility.

The first method is not practicable commercially and has only scientific significance. Method II is by far the most usual and is the cause of opacity obtained by adding fluorides, tin, zirconium, titanium, ceric and magnesium oxides. In the case of fluorides the mill addition may be cryolite (Na_3AlF_6), sodium silicofluoride (Na_2SiF_6), fluorspar (CaF_2), uverite ($7CaO.CaF_2.6TiO_2.2Sb_2O_5$) but the final opacifying agents if reaction is complete are always tiny crystals of alkali or alkaline earth fluorides. The use of fluorides is, however, not very satisfactory for glazes as the reaction may be interrupted at any stage of completion, whereas in glasses it can be brought to completion and in enamels it can be arranged to have hardly started. The oxides, however, remain stable in the glaze and hence their final particle size is limited by the minimum size obtainable by milling, i.e. about $2\,\mu$.

The third method is shown by phosphate and silicate glasses which, although miscible to a certain extent at high temperatures, tend to separate on cooling. The phosphate forms tiny particles of disperse phase and may crystallise. The particles are much smaller than those produced by milling for method II.

Opacity depends on the particle size of the disperse phase, but not in a regular way. For particles of the order 1 to $2\,\mu$ opacity, i.e. diffuse light in all directions, depends on the path of light being broken by refraction or reflection as often as possible, and hence increases as the particle size decreases and the surface increases. This trend, however, reaches a maximum when the particle diameter is about $0.4\,\mu$. Very small particles whose size is of the same order as the wavelength of light are unable to arrest the light rays and hence opacity falls off with particle size decrease in the finest range.

TABLE 151
Refractive Indices of Important Opacifiers

Opacifier	R.I.
Sodium fluoride	1·33
Cryolite	1·34
Calcium fluoride	1·44
Average value for vitreous glaze	1·50 to 1·55
Arsenic trioxide	1·73
Zinc spinel	1·90
Tin oxide	2·04
Antimony trioxide	2·09
Zirconium oxide	2·13–2·20
Cerium oxide	2·33
Titanium oxide	2·50

Opacity increases as the difference in the refractive indices of the vitreous and the disperse phases becomes larger.

The R.I. of fluorides is lower than an average vitreous base. For a wavelength of 5893 Å the R.I. of NaF is 1·326, of CaF_2 is 1·434 as against an average value for the vitreous phase of 1·50 to 1·55. The presence of lead increases the R.I. of a glass and hence the opacity of this type of glaze.

The oxides have higher refractive indices than silicate glasses, so that the presence of lead in the vitreous phase will reduce the difference and with it the opacity. The difference of R.I.'s between vitreous and disperse phase is greater for the oxides than the fluorides and they are better opacifiers.

Calcium phosphate, the most probable opacifier in phosphate opaque glazes has a R.I. of from 1·58 to 1·62 which thereby differs little from that of the vitreous phase. In this case, however, the method of formation from a homogeneous melt makes the particles much smaller, evidently approaching the optimum size for opacity.

It is interesting to note that opacifiers of R.I. both lower and higher than that of the vitreous phase can be successfully used together.

Some of the constituents of the glassy phase have a marked effect on the opacifier. The presence of alumina has been shown to be imperative, it stabilises the vitreous phase and by forming the stable complex anion AlF_6^{3-} prevents volatilisation of any fluorine present.

Zinc ions increase the formation of nuclei so that a large number of small particles can form and produce high scattering power.

The effect of lead in increasing the R.I. of the vitreous phase has been discussed; barium, calcium and magnesium have the reverse effect and hence are detrimental to fluoride opacifiers (**S187**).

The effectiveness of any particular opacifier (p. 143) is determined by:

(1) Its insolubility in the glass, depending both on the composition of the glass and the opacifier.
(2) Its final particle size and shape.

TABLE 152
Opaque Glazes

Ref.	Cone		Na₂O	NaKO	K₂O	BeO	MgO	CaO	BaO	ZnO	PbO	B₂O₃	Al₂O₃	Fe₂O₃	SiO₂	TiO₂	ZrO₂	SnO₂
Opacifier SnO₂																		
L14	08	MF	0·035		0·157		0·101	0·105		0·096	0·505		0·241		1·90			0·10
P44	06	%	0·72		4·90		1·35	1·93		2·64	37·28		8·16		38·02			5·00
		MF			0·166		0·166	—		0·109	0·559		0·268		2·122			0·192
L15	02	%			4·60		1·97			2·61	36·60		8·04		37·64			8·54
		MF	0·070		0·054		0·090	0·176		0·087	0·522		0·223		2·250			0·088
L15	2	%	1·34		1·57		1·12	4·78		2·19	35·91		7·04		41·95			4·10
		MF	0·035		0·147			0·248	0·248	0·285		0·252	0·210		2·87			0·14
H60	6	%	0·66		4·20			4·22	13·25	6·90		5·34	6·50		52·51			0·15
		MF			0·30			0·50	0·10	0·10			0·35		3·20			6·42
W54	7	%			8·50			8·45	4·62	2·45			10·76		58·39			6·82
		MF			0·286		0·074	0·403	0·146	0·146			0·446		2·331			0·104
M36	8	%			9·60		1·06	8·05	4·97	4·23			16·22		50·28			5·59
		MF			0·41			0·38		0·20			0·44		2·98			0·11
M83	9	%			12·20			6·73		5·14			14·19		56·50			5·24
		MF	0·05		0·36		0·04	0·35	0·06	0·23			0·54		4·00			0·10
M30	11	%	1·18		8·63		0·61	5·00	2·34	4·76			14·03		61·41		0·05	3·80
		MF			0·12			0·42	0·04	0·33			0·32		2·40		2·32	0·05
		%			4·29			8·93	2·32	10·18			12·39		54·92			2·85
Opacifier ZrO₂																		
P25	08	MF	0·196		0·113		0·38	0·057		0·221	0·375	0·337	0·095	0·004	1·430	0·146	0·650	
		%	3·55		3·12		0·45	0·94		5·48	24·47	6·91	2·84	0·19	25·29	3·43	23·33	
L14	07	MF	0·150		0·150				0·243	0·457		0·531	0·039		0·925		0·082	
		%	4·54		6·90				18·19	18·15		18·13	1·94		27·25		4·90	
L14	06	MF	0·221		0·063			0·216	0·200	0·100	0·200	0·390	0·212		2·890		0·250	
D8	05	%	3·72		1·61			3·29	8·32	2·21	12·09	7·22	5·88		47·35		8·31	
		MF	0·221		0·063			0·216	0·10	0·30	0·10	0·39	0·212		2·89		0·25	
L14	04	%	3·92		1·70			3·47	4·40	7·00	6·39	7·82	6·49		50·02		8·78	
		MF	0·091		0·060			0·224	0·224	0·428	0·197		0·041		1·507		0·260	
D9	02	%	2·25		2·25		0·035	0·422	13·66	13·85	17·46		1·67		36·19		12·67	
		MF	0·125		0·075		0·035	0·243	0·100	0·243		0·193	0·210		2·215		0·138	
M68	01	%	2·99		2·72		0·54	9·11	5·90	7·61		4·93	8·25		51·45		6·50	
		MF	0·179		0·068			0·292		0·319	0·142	0·262	0·287		2·298		0·329	
M68	8	%	3·49		2·01			5·15		8·16	9·94	5·76	9·22		43·60		12·67	
		MF	0·048		0·152		0·150	0·500	0·075	0·075			0·330		3·00		0·280	
L43	9	%	0·94		4·50		1·90	8·82	3·62	1·92			10·60		56·92		10·78	
		MF	0·0652		0·1265		0·1533	0·4860		0·1690			0·287		3·090		0·357	
		%	1·25		3·69		1·92	8·45		4·26			9·08		57·80		13·55	

TABLE 152—(contd.)

Ref.	Cone		Na_2O	NaKO	K_2O	BeO	MgO	CaO	BaO	ZnO	PbO	B_2O_3	Al_2O_3	Fe_2O_3	SiO_2	TiO_2	ZrO_2	Sb_2O
Opacifier ZrO_2—(contd.)																		
M30	11	MF	0·048		0·152		0·150	0·40	0·10	0·150			0·33		3·00		0·28	
		%	0·92		4·44		1·88	6·96	4·76	3·79			10·45		56·16		10·64	
M68	12	MF	0·57		0·143		0·25	0·15	0·10	0·10			0·40		4·00		0·30	
		%	0·93		3·56		2·67	2·23	4·06	2·15			10·81		63·87		9·72	
Opacifier TiO_2																		
B148	6	MF	0·15		0·19			0·19		0·37			0·19		2·73	0·12		
		%	5·79		6·68			3·98		11·23			7·24		61·49	3·59		
Opacifier P_2O_5																	P_2O_5	
K78	014	MF	0·190		0·577					0·235		0·304	0·706				0·425	
		%	4·95		22·77					7·92		8·91	30·20				25·25	
G51	04	MF			0·20					0·15	0·40		0·33		1·75		11·47	
		%			6·10					3·95	28·84		10·91		34·19			
S212	6	MF			0·20					0·20		1·00	0·55		2·30		0·20	
		%			5·20					4·49		19·32	15·51		38·34		7·84	
G14	7	MF								0·251	0·076		0·413				1·568	
		%								6·4	5·3		13·2				69·8	
S212	7	MF						0·8571		0·1429			0·357		2·30		0·3857	
		%						17·45		4·22			13·23		50·39		14·71	
G14	7	MF				0·676				0·248	0·076		0·431				2·385	
		%				3·7				4·4	3·7		9·6				73·9	
A80	8 or higher		$Mg_3(PO_4)_2$ is recommended for replacing SnO_2 in raw leadless sanitary ware glazes.															
Opacifier CeO_2																	CeO_2	
M67	9	MF	0·10		0·10			0·40			0·40	0·20	0·30		2·80		0·128	
		%	1·65		2·50			5·96			26·31	3·71	8·14		44·87		4·8	
Opacifier ZnO																		
C62	04	MF				0·673											0·15	
		%				5·3											6·86	
P80	7	MF		0·20				0·20		0·55	0·45		0·35		1·60			
		%		3·90				2·80		16·13	36·15		12·89		34·83			
P80	7	MF		0·333						0·60			0·70		4·20			
		%		5·76						12·18			17·84		63·28			
P80	7	MF		0·50						0·666			0·80		4·80			
		%		8·66						12·00			18·09		64·15			
B22	10	MF			0·30					0·50			0·80		4·80			
		%			7·48					9·01			18·11		64·22			
		MF								0·70			0·50		4·00			
		%								15·08			13·52		63·92			

											SO₃		As₂O₃
Z₃ 02–1	MF	0·204			0·003	0·005	0·788		0·073	0·0008	2·141	0·0006	
	%	5·92			0·06	0·13	30·03		3·49	0·06	60·20	0·03	
	MF	0·263			0·006	0·160	0·571		0·152	0·0009	2·206	0·002	
	%	7·39			0·11	4·08	21·11		7·03	0·06	60·15	0·07	
	MF	0·275			0·002	0·147	0·576		0·063	0·0007	1·815	0·0006	
	%	9·06			0·05	4·40	24·95		3·42	0·07	58·02	0·03	
Opacifier As₂O₃ D₇ 08	MF	0·083		0·124		0·002		0·791			2·158		0·918
	%	1·5		3·4		0·4		51·4			28·0		5·3
Opacifier Sb₂O₃ and fluorides Si63 07	MF	0·387	0·203				0·410	0·706	0·203		1·218	Na₃AlF₆ 0·127	0·070
	%	8·99	7·16				12·49	18·49	7·76		27·34	10·01	7·56

(3) The difference between its refractive index and that of the vitreous phase.

In practice, tin oxide has proved to be the most useful opacifier. Although a certain amount dissolves in the vitreous phase, incidentally often improving its fluidity, gloss and mechanical strength, the undissolved portion appears to remain as very small crystals which are near the optimum size (**A205, 206**). The normal amount required is 4 to 5% although the quantity may vary in accordance with the glaze composition from 2 to 10%. Unfortunately the high price of tin oxide has made it necessary to look for cheaper alternatives.

Zirconium compounds have proved to be good opacifiers. American practice seems to prefer double silicates, etc., *e.g.* barium zirconium silicate; calcium zirconium silicate; magnesium zirconium silicate; zinc zirconium silicate; zirconium spinel $1\cdot26ZnO.Al_2O_3.1\cdot7ZrO_2.2\cdot2SiO_2$ (**A89**). But recent British work by Booth and Peel (**B92**) has shown that the modern commercial zircon with its consistent physical and chemical properties has many merits as an opacifier. Some of their glazes are given in Table 153.

Titanium oxide is the opacifier whose R.I. differs most from that of glass, but as it dissolves in alkali and hence to a certain extent in the glaze it is often not as effective as might be hoped. It is, however, useful when used in conjunction with tin oxide, zircon and zinc oxide.

Cerium oxide makes an excellent opacifier if very pure and not exposed to temperatures above 1050° C (1922° F).

The use of antimony and arsenic trioxides as opacifiers has to be undertaken with care because of their poisonous natures. Pentavalent antimony is non-poisonous, so danger is eliminated by using pentavalent raw materials and oxidising firing. It can be used in the frit or mill addition and is a very good opacifier. If lead is present Naples yellow is usually produced. A white lead–antimony glaze can, however, be made by special methods. Arsenic oxides must be fritted, but if due care is taken make good opacifiers. They are mainly used in Italy and Spain.

Sodium, calcium and magnesium phosphates can be used to give the double vitreous phase glazes, but magnesium phosphate can only be used for glazes maturing at cone 8 or above.

Vanadium pentoxide develops whiteness and opacity in lead borosilicate enamels maturing at 610–650° C (1125–1200° F) (**S129**).

Zinc oxide plays a remarkable role in all glazes and can also produce opacity. It is particularly useful in increasing the effect obtained with any single opacifier.

Vielhaber (**V17**) stated that opacity increases with the amount of opacifier only to a certain limit; after this has been reached greater opacity can only be obtained by using a combination of agents.

A few examples are given of opaque glazes using different opacifiers and maturing at various temperatures (Table 152).

TABLE 153

Zircon Opacified Glazes
(by Booth and Peel (**B92**))

(As the correct formulation of the base glaze is very important in zircon opacification the molecular formulae given omit the zircon introduced, including the silica contained in it.)

Glaze type	Z/215 Leadless tile	Z/221 Leadless tile	Z/500 Leadless tile	Z/226 Leadless tile	Z/238/31 Leadless tile	Z/215 Lead tile	2/8 Lead tile	H.33 Milled, earthenware sanitary ware	VC51 Milled, vitreous china sanitary ware	136 Raw, sanitary ware	731 Blunged, sanitary fireclay
Maturing temperature °C	900–950	1000–1050	1050–1100	1100	1120–1180	900–950	1000–1050	1080	1170	1200–1250	1150–1220
Batch composition of frits:											
Potash feldspar	15·8	21·1	21·1			15·8	46·9	40·6	19·8		29·8
Flint	28·1	32·2	35·8			28·1	14·6	28·2	47·7		20·3
Whiting	7·0	12·7	9·7			7·0	15·1	13·0	12·6		15·8
Barium carbonate	9·4	5·2	—			9·4	3·1				
Zinc oxide	3·8	2·1	—			3·8	—	7·8	8·7		16·1
Boric acid	16·5	9·5	14·2			16·5	13·0	7·9			
Sodium carbonate	—	—	—			—	—	2·5			
Anhydrous borax	9·3	7·7	11·5			9·3	4·7		7·6		6·9
Kaolin	10·3	9·5	7·7			10·3	2·6		3·6		11·1
Mill charge:											
Frit (from above)	84·1	76·5	80·7			71·4	71·4	84·8	83·4		
Standard borax frit					18·1						11·8
Lead bisilicate frit				—							
Kaolin	4·5	11·5	7·3	6·0	9·1	14·3	14·3	6·1	6·8	8·8	9·6
Milled zircon	11·4	12·0	12·0	12·3	10·6	3·6	3·6	9·1	9·8	10·6	10·8
Potash feldspar				37·4	14·7	10·7	10·7			26·7	23·3
Flint				24·0	28·5					26·7	26·6
Whiting				10·5	7·6					11·3	11·3
Talc										5·3	
Zinc oxide				6·5	5·7	9·3	4·7			3·4	6·6
Barium carbonate				3·3	5·7	10·3	2·6			7·2	
Molecular formula:											
$Na_2O + K_2O$	0·3	0·3	0·5	0·25	0·2	0·262	0·31	0·3	0·35	0·175	0·2
MgO										0·15	
CaO	0·3	0·5	0·5	0·45	0·45	0·264	0·48	0·4	0·35	0·4	0·5
BaO	0·3	0·1	—	0·05	0·1	0·176	0·05			0·125	
ZnO	0·2	0·1	—	0·25	0·25	0·175		0·3	0·3	0·15	0·3
PbO	0·2					0·123	0·16				
B_2O_3	1·0	0·6	1·2	—	0·15	0·818	0·49	0·2	0·1	0·3	0·1
Al_2O_3	0·35	0·35	0·5	0·3	0·28	0·313	0·36	0·3	0·175		0·325
SiO_2	3·25	3·5	5·0	3·0	3·0	3·12	2·92	3·0	3·0	3·0	3·25

Matt Glazes

Matt glazes differ from transparent glossy glazes in having a mass of tiny crystals embedded in a glassy matrix. They are formed when a completely fused glaze cools and part of it crystallises out. The crystals must be so minute and regularly dispersed that the surface of the glaze is smooth and velvety to touch. It should be possible to write on a matt glaze with an ordinary pencil and then to rub the mark off with the finger.

The crystals in a matt glaze break up light rays making the glaze more or less opaque. Of course, opaque glazes can be made matt as well.

The same components with occasional minor additions are used to make matt glazes as are used for glossy ones. The proportions are altered. The silica is reduced and replaced by alumina, or if this makes the mixture too refractory by boric oxide. Suitable ratios of $Al_2O_3:SiO_2$ lie in the range 1:3 to 1:6 (**P13**). Some of the sodium is replaced by calcium or magnesium, or some of the potassium and calcium is replaced by magnesium. Introduction of small quantities of barium sulphate, pure titania, or zinc oxide and tin oxide often produces very fine velvet glazes. In the case of the barium salt it should be noted that although the molecular formula shows BaO which could theoretically be introduced as $BaCO_3$ or $BaSO_4$, etc., in the glaze batch, the sulphate is much more effective than the carbonate. In all probability it does not decompose entirely to BaO and SO_3, and the $BaSO_4$ forms small insoluble nuclei for the crystals to build on during cooling (*see* example No. 23, Table 154). Titania must be used with care as even if introduced pure it very easily forms a yellow stain with any iron impurity present. Zinc oxide has unwanted effects on chromium colours especially green ones but is very useful in white glazes, and non-chromium coloured glazes. It must not be used in quantities larger than 0·3 equivalents (**P13**) as zinc silicate might then crystallise out in large crystals.

The actual crystals formed in the matt glazes if lime is present seem to be mainly anorthite ($CaO.Al_2O_3.2SiO_2$) and occasionally wollastonite ($CaO.SiO_2$). If barium is present the crystals formed are thought to be the barium feldspar, celsian ($BaO.Al_2O_3.2SiO_2$) (**P13**).

The formation of a good matt glaze depends on the rate of cooling after firing. Rapid cooling almost invariably produces a glossy glaze. The higher the silica content compared with the RO and R_2O_3 contents the slower must be the cooling for crystallisation to take place (**H83**).

In coloured matt glazes the colouring agent is often taken up preferentially by either the crystals or the vitreous phase (**S112**) giving numberless very effective possibilities.

Certain matt glazes are more resistant to crazing than similar glossy ones and may give the ware very good mechanical properties.

Batch composition of velvet matt glaze for cone 8, No. 23, Table 154:

Chalk	8·8
BaSO$_4$	27·5
Feldspar	49·1
China clay	8·3
Quartz	6·3
	100·0

Crystalline Glazes

The use of crystalline glazes is confined to artistic purposes. They have sizeable shapely crystals set in a vitreous matrix which may be a different colour, and by careful work can be made to look very beautiful.

These large crystals are produced when a glaze that is supersaturated with a compound that crystallises easily is cooled very slowly. As the thickness of the glaze restricts growth in one dimension to about 0·5 mm only those substances whose crystals grow in two directions only will give good crystalline glazes. The hexagonal crystal system seems best suited to the purpose (**N30**). Glaze constituents that can produce crystalline glazes include zinc silicate in both leadless (Table 155 examples 2 and 5) and lead glazes, zinc titanate in leadless (example 1) and lead glazes (example 3), manganese silicate (example 4), calcium and magnesium silicates and titanates, and occasionally mullite, $3Al_2O_3.2SiO_2$.

Purdy and Krehbiel (**P78**) found that manganese oxide had the greatest crystallising tendency, producing large and varied crystals. Zinc oxide gives large crystals in certain areas as though the crystallising substance had segregated. Titania produces small but evenly distributed crystals. They found that sodium is more conducive to crystal formation than potassium, and except in high titania glazes give the following limits for alkali and zinc

$$\left. \begin{array}{l} 0{\cdot}7KNaO \\ 0{\cdot}3ZnO \end{array} \right\} \text{ to } \left\{ \begin{array}{l} 0{\cdot}4KNaO \\ 0{\cdot}6ZnO. \end{array} \right.$$

Colouring agents are usually taken up preferentially by the crystals or the vitreous phase and so can produce very fine effects. A typical example is seen when trade nickel oxide, which invariably contains a trace of cobalt oxide, is used to colour a zinc silicate crystalline glaze buff. The vitreous part of the glaze is in fact buff, but the crystals are blue, having attracted the entire cobalt content of the glaze (**S112**) (Fig. 6.5).

Crystallisation of a vitreous phase occurs when nuclei are formed and are thereafter maintained at a temperature suitable for crystal growth. The growth temperature may be the same as, or higher than, the nuclei formation temperature. The growth depends on there being sufficient thermal agitation for bond breaking and making, so that the correct atoms can become attached

TABLE 154

Matt Glazes

No.	Ref.	Cone		Na_2O	K_2O	MgO	CaO	SrO	BaO	ZnO	PbO	B_2O_3	Al_2O_3	SiO_2	TiO_2	ZrO_2
1	A116	08 to 07	M.F. %		0·20 8·0	0·39 6·6	0·09 1·6	0·41 18·0					0·20 8·6	1·8 45·6		
2		03	M.F. %							0·14 3·7	0·77 55·4	0·45 13·2		1·5 29·2	0·4 10·3	
3	B56	03 to 5	M.F. %		0·225 7·2		0·200 3·8				0·575 43·9		0·35 12·2	1·6 32·9		
4	C50	1	M.F. %	0·294 5·2	0·048 1·2		0·266 4·0				0·392 23·8	0·605 11·5	0·190 5·2	2·997 49·1		
5	P81	2–3	M.F. %		0·15 4·6					0·10 2·7	0·75 54·5		0·15 5·0	1·7 33·2		
6	H83	2–3	M.F. %	0·175 4·2			0·075 1·3		0·250 11·9		0·500 34·7		0·33 10·5	2·0 37·4		
7	S162	2–3	M.F. %	0·126 2·9	0·124 4·4		0·500 10·5		0·250 14·3			0·394 10·3	0·35 13·3	1·97 44·3		
8	O3	3–4	M.F. %		0·05 1·5	0·10 1·3	0·2 4·0			0·35 9·3	0·50 36·2		0·50 16·6	1·8 35·1		
9	B52	3–4	M.F. %		0·1 3·3		0·2 3·9			0·2 5·7	0·5 39·5		0·375 13·5	1·60 34·0		
10	B52	3–4	M.F. %				0·2 3·9		0·2 10·5	0·1 2·8	0·5 38·4		0·325 11·4	1·60 33·0		
11	B52	3–4	M.F. %		0·1 3·3				0·8 10·8		0·5 39·2		0·25 9·0	1·60 33·8		
12	R11	4	M.F. %		0·184 6·0				0·195 10·4		0·621 48·2		0·246 8·7	1·28 26·7		
13	R11	4	M.F. %		0·120 4·2				0·144 8·3		0·747 62·4		0·157 6·0	0·85 19·1		
14	A129	4	M.F. %		0·15 5·4	0·20 3·1	0·35 7·5				0·30 25·7		0·28 11·0	1·56 36·0	0·37 11·3	
15	S162	4	M.F. %	0·15 5·6			0·30 8·0			0·55 21·3			0·4 19·4	1·6 45·7		

No.	Code	Range		C1	C2	C3	C4	C5	C6	C7	C8
16	O12	4	M.F.	0.15		0.35	0.30	0.10	0.10	0.233	1.90
			%	5.7		7.9	18.5	3.3	9.0	9.6	46.0
17	H83	4	M.F.	0.225		0.200			0.575	0.35	1.6
			%	7.2		3.8			43.9	12.2	32.9
18	S162	4–5	M.F.	0.15		0.35	0.30	0.10	0.10	0.233	1.90
			%	5.7		7.9	18.5	3.3	9.0	9.6	46.0
19	B10	4–5	M.F.	0.20		0.45			0.35	0.29	1.87
			%	7.1		9.6			29.6	11.2	42.5
20	H83	5–6	M.F.	0.15		0.3			0.55	0.35	1.67
			%	4.9		5.8			42.4	12.3	34.6
21	H83	7	M.F.	0.16		0.44				0.37	1.9
			%	6.1		9.9				15.1	45.8
22	A129	7	M.F.	0.25	0.15	0.60	0.35	0.05		0.40	2.20
			%	10.0	2.6	14.2	21.5	1.6		17.3	55.9
23	—	8	M.F.	0.3		0.3	0.4		See batch composition (p. 593)	0.4	2.4
			%	9.7		5.8	21.0			14.0	49.5
24	A129	9	M.F.	0.30	0.05	0.65				0.45	2.50
			%	10.7	0.8	13.9				17.5	57.1
25	H83	9–10	M.F.	0.30		0.70				0.7	2.7
			%	9.4		13.0				23.7	53.9
26	B53	9–10	M.F.	0.3		0.7				0.65	2.48
			%	10.0		13.9				23.4	52.7
27	—	13–14	M.F.	0.1	0.2	0.7		0.2	0.4	1.4	5.2
			%	1.8	1.6	7.7		6.6	19.9	27.9	61.0
28	—	13–14	M.F.							0.23	2.1
			%							9.5	51.0
29	—	14	M.F.	0.2	0.8	0.7				1.4	5.2
			%	3.6	13.0	7.6				27.6	60.4
30	—	14	M.F.		0.1					0.236	2.962
			%		0.8					9.9	73.4
31	—	14	M.F.	0.1	1.0	0.7		0.5		1.6	5.8
			%	1.7	16.7	6.9		13.0		28.7	61.3
32	—	14	M.F.	0.2	0.3					0.24	3.6
			%	6.0	3.9					7.9	69.2
33	—	14	M.F.	0.1	0.9					0.24	3.6
			%	3.3	12.7					8.5	75.5

TABLE 155

Crystalline Glazes

No.	Ref.	Cone		Na$_2$O	K$_2$O	CaO	BaO	MnO	ZnO	PbO	B$_2$O$_3$	Al$_2$O$_3$	SiO$_2$	TiO$_2$
(1)	N30	Various	M.F.	0·0513	0·235	0·088	0·0513		0·575			0·162	1·70	0·202
			%	1·4	10·1	2·3	3·6		21·3			7·5	46·5	7·3
(2)	W87	5	M.F.	0·33					0·66		0·20	0·05	1·6	
			%	10·79					28·28		7·37	2·69	50·87	
(3)	P63	7	M.F.		0·10	0·20			0·20	0·50		0·10	2·00	0·50
			%		2·95	3·51			5·10	34·89		3·20	37·81	12·54
(4)	P78	10–11	M.F.	0·1				0·9					1·0	
			%	4·7				49·1					46·2	
(5)	K32	11	M.F.	0·20					0·80				2·30	
			%	5·74					30·08				64·18	

to the nucleus, and unwanted atoms removed; but the thermal agitation must not be so large as to break up nuclei before they grow. In his experiments using the glaze shown as example (1), Table 155, Norton (**N30**)

FIG. 6.5. *Crystalline glaze showing preferential absorption of colouring matter*

showed that a growing period of an hour at fixed temperature, somewhat lower than the maturing temperature of the glazed body, could produce crystal growths. He also showed that growth periods at different temperatures produced different crystal types. By applying 'seeds' crystal growth can be induced in a specific spot. The kiln atmosphere throughout the firing should be oxidising.

Crystalline glazes are not mechanically as strong as homogeneous vitreous, or matt glazes. The glaze surface is not always quite smooth and there are weak spots at the junctions between patches of crystals and glassy areas.

Aventurine Glazes

These are also crystalline glazes but instead of having long needle-like crystals, often growing in patches, they have a mass of thin plate-like crystals or spangles precipitated in the glaze.

Aventurine glazes may be produced when the glaze contains iron, chromium or copper oxides, iron being the most usual. The order of the beneficial effect of the RO constituents is soda, potash, lead and lime. Careful firing and cooling is necessary (**P9**) (Table 156).

TABLE 156

Aventurine Glaze

No.	Ref.	Cone		Na_2O	FeO	B_2O_3	Al_2O_3	Fe_2O_3	SiO_2
As given with Fe_2O_3	P9	4	M.F.	1·0		1·25	0·15	0·75	7·0
			%	19·0		26·7	4·7	36·7	12·9
Recalculated, with FeO			M.F.	0·4	0·6	0·5	0·06	—	2·8

SALT GLAZES

Salt glazing differs fundamentally from slip glazing. Instead of coating the raw or biscuited ware with a water-suspended glaze slip and then fusing this to form a glaze in the glost fire, the salt glaze is formed in the kiln by the action of vapours, produced from salt, water and other substances on the hot, fired body.

Salt glazing is used in the manufacture of general stoneware, especially sewer pipes, chemical stoneware, glazed bricks and tiles, and certain artware.

Successful salt glazing is influenced by three interdependent factors:

(1) Composition of the body.
(2) Firing schedule.
(3) Composition of the salting mixture.

Composition of the Body

The sodium ions in the glazing vapours react primarily with silica in the body surface, and it is found that the ratio $SiO_2:Al_2O_3$ may not be too small, although opinions about the best value differ. Barringer (**B11**) gives it as 4·6 to 12·5 whereas Mäckler (**M21**) says that a good salt glaze can be obtained with as low a value as 3·3. In practice a ratio of about 5 is normal. Clays that will not take a good salt glaze because of silica deficiency may have finely ground sand or sandstone added.

Although white-burning, iron-free, clays may be salt glazed, as seen in the old Staffordshire ware, the presence of iron in the clay helps in the formation of the salt glaze. Clays with 0 to 2% iron oxide take white to tan glazes, with 3·5 to 4·75% browns and with 4·75 to 8·2% mahogany are produced. The actual colour being influenced by the presence of other constituents and very largely by the firing schedule. The iron compounds in the body assist in the salt glaze formation in two ways: (a) by reacting with the salt vapours during the glaze formation; (b) by acting as a flux to vitrify the body which can then better take the glaze. It is the ferrous compounds that assist in body vitrification, although the iron in the raw body is usually in the ferric state. Ferric compounds are decomposed to ferrous ones at about 1200° C (2190° F) so that in bodies fired above this temperature the iron is in the beneficial condition. In bodies fired to lower temperatures the fluxing action is obtained by using a reducing kiln atmosphere. It is, in fact, traditional in many works to maintain a reducing atmosphere towards the end of the firing, before salt glazing, whether it is necessary to assist in the reduction of the iron or not.

Lime makes a salt glaze dull at low temperatures and at high ones it turns an iron rich glaze yellow if cooled slowly or greenish black if cooled fast. 1·5% magnesia assists in salt glazing but above 3% makes the glaze dull. 1 to 5% titania in otherwise suitable clays tends to make the glaze more brilliant.

If the clay contains more than 0·1% soluble calcium and magnesium salts these come to the surface during drying, producing a scum that prevents proper reaction with the body by the vapours during the glazing. This is prevented by faster drying where possible and incorporating barium compounds, e.g. $BaCO_3$, $Ba(OH)_2$, black ash (impure barium sulphide) with soda ash.

Treatment of the clay to remove large hard particles which give the fired body a rough surface improves the finished ware considerably. This can be done by grinding, screening and pugging. The most objectionable lumps are those of iron minerals, as the glaze builds up on any such specks in the body surface, making larger irregularities.

Firing Schedule

The vaporisation of salt and its interaction with steam, etc., is considered not to take place satisfactorily below cone 4 (1160° C, 2120° F). Normal salt glazing cannot, therefore, be undertaken below that temperature. Certain bodies that mature earlier and have a high alkali and iron oxide content can, however, be salt glazed at cone 03 to 02 (i.e. 1050° C, 1922° F). Other low-temperature bodies can be salt glazed by mixing borax with the salt.

The actual application of the salt glaze occurs towards or at the end of the firing. With red burning shales and clays this is at about cones 3 to 6 (1140 to 1200° C, 2084 to 2192° F); with fireclays and stoneware clays the usual temperatures are cones 8 to 11 (1250° C to 1320° C, 2282 to 2408° F). The ware is usually fire flashed before salting, *i.e.* kept in a reducing atmosphere so that the iron is reduced to the ferrous state (**S38**).

Salt glazing is done in periodic kilns fired with coal, gas or oil. When setting the ware in the kiln attention must be paid to the necessity of good circulation of the vapours. Pipes must be raised so that gases can enter at the bottom and solid bottom articles such as traps and gullies must be set on blocks or setting pieces.

When the maturing temperature has been held for sufficiently long to make the ware dense the salting mixture and fuel are thrown on to the fires and the damper closed for a short time, and then reopened. When the vapours have cleared and the temperature has been regained the next salting is undertaken. From four to eight saltings may be made.

The fires must be well stoked to prevent drop of temperature due to the extra heat required to volatilise the salt and for the reactions at the body surface. The salts vaporise and interact with each other, water and the body surface producing a grey (ferrous) glaze. The kiln is then cooled slowly with oxidising atmosphere when the ferrous iron is reoxidised to the ferric state and the glaze becomes brown. The practice of slow oxidising cooling in order to make the glaze brown is brought about by the general belief amongst consumers that a good salt glaze should be brown, and so only the brown ware is saleable. Rapid cooling produces a greenish black glaze which is not technically inferior to a brown one. Furthermore a lime-containing body gives a yellow glaze if the iron reoxidises during slow cooling but a greenish one if the cooling is rapid and it remains in the ferrous state. Also the higher the firing temperature, the more of the iron in the body reacts and the darker the colour.

The question of salt glazing in tunnel kilns is a controversial one. Even in periodic kilns the kiln lining is badly attacked by the fumes given off during the glazing but this becomes far worse in a tunnel kiln where the attack is continuous. Attempts to salt glaze in a tunnel kiln at the higher temperatures made the structure collapse in about a fortnight. By using special refractories and lower-burning clays tunnel kilns appear to be successfully operated in the United States. In one instance a body made up of low-grade fireclay, soft shale, fine-grained and plastic clay, hard shale and grog is fired and salt glazed at 1108° C (2030° F) (**A24**). In Europe when a change is made to tunnel kilns it is probable that there will also be a change over to slip glazes.

The Salting Mixture

The actual mechanism of the chemical reaction during salt glazing is not completely clear, but essentially the sodium chloride is vaporised, and then meets hot steam which reacts with it, either directly or indirectly, in presence of silica, to give hydrogen chloride gas and sodium oxide. All coal-fired kilns contain a certain amount of steam but salt glazing is assisted by adding some water with the salt. The sodium oxide immediately reacts with the silica in the body to give sodium silicate and then further with other body constituents to give a glaze of about the following composition (**B11**):

$$\left.\begin{array}{ll} Na_2O & 0{\cdot}777 \\ K_2O & 0{\cdot}002 \\ MgO & 0{\cdot}002 \\ CaO & 0{\cdot}174 \\ FeO & 0{\cdot}045 \end{array}\right\} Al_2O_3 \ \ 0{\cdot}586, \quad SiO_2 \ \ 2{\cdot}588.$$

This has $SiO_2:Al_2O_3$ ratio of 4·418.

The iron oxides also play a part in the reaction probably via their chloride, e.g.

$$(Al,Fe)_2O_3.xSiO_2 + 6NaCl \rightarrow (Al,Na_3)_2O_3.xSiO_2 + 2FeCl_3$$
$$2FeCl_3 + 3H_2O \rightarrow Fe_2O_3 + 6HCl \qquad (\mathbf{K44})$$

In general the thinnest salt glazes, up to 0·001 in. (0·0254 mm) thick, are the most satisfactory and do not craze (**F18**).

The salting mixture used has become a subject of investigation in recent years. The old methods employed only salt (about 300 lb/ton of pipes, ~136 kg/tonne (**A60**), or 1·5–2 kg/m³ kiln capacity, 2·5–3·4 lb/yd³ (**T36**). This had the disadvantage of setting the lower temperature limit for salt glazing at cone 4 and also that there were fairly high losses to the manufacturer due to imperfect ware. By introducing borax and/or boric oxide sometimes with sodium nitrate many faults can be overcome.

Schurecht and Wood (**S38**) made both laboratory and plant investigations leading to the following recommendations (Table 157). The best method of salting with boron compounds was found to be by salt glazing with salt alone for the first three saltings while a mixture of salt and boron compounds was used for the fourth salting and the boron compounds were used alone for the final salting. The poorest method was that in which the boron compounds alone were used for the first salting and salt alone for the last three saltings. When mixtures of the boron compounds and salt were used, the results ranked between the other two methods described above.

Artware can be made in various brilliant colours. All-over colouring can be achieved by adding certain colouring compounds to the salt, e.g. using cobalt chloride with the sodium chloride on an iron-free body a blue glaze is produced, and with zinc greens can be obtained. Where only a portion of the piece is to be coloured this may be done by painting on a suitable pigment containing colouring metallic oxides when maroon, turquoise blue and darkblue, green, iron-red and black can be produced at will (**B73**). If the surface of the body is not suitable for direct decoration and glazing a slip made up of mixtures of china clay, flint, alkali silicates, frits, ball clay, and colouring

TABLE 157

Possible Improvements in Salt Glazing

Improvement to be made	Mixtures which produce better results than salt alone	Mixtures which produce the best results	Improvement made
(1) To decrease salt glazing temperature.	(a) Use mixtures with 1 to 15% boric acid and 99 to 85% salt.	Use 4 to 8% boric acid and 96 to 92% salt.	The salt glazing temperature was decreased 180° F (100° C).
	(b) Use mixtures of 1 to 15% borax and 99 to 85% salt.	Use 6 to 10% borax and 94 to 90% salt.	The salt glazing temperature was decreased 144° F (80° C).
(2) To increase thickness of glaze.	(a) Use mixtures with 1 to 15% boric acid and 99 to 85% salt.	Use 4 to 8% boric acid and 96 to 92% salt.	At cone 03 the thickness was increased 29·2 to 45·2%; at cone 2, 23·4%; and at cone 6, 10·9%.
	(b) Use mixtures of 1 to 15% borax and 99 to 85% salt.	Use 7 to 12% borax and 93 to 88% salt.	At cone 03 the thickness was increased 42·9 to 114·2%; at cone 2, 48·7 to 88·4% and at cone 6, 35·8 to 40·8%.
(3) To overcome roughness due to 'pig popping'.	(a) Use mixtures of 1 to 15% boric acid and 99 to 85% salt.	Use 8 to 15% boric acid and 92 to 85% salt.	A very smooth glaze is obtained without roughness.
	(b) Use mixtures of 1 to 15% borax and 99 to 85% salt.	Use 15% borax and 85% salt.	Although this glaze was smoother than that obtained with salt alone it was not as smooth as glazes obtained with boric acid mixtures.
(4) To overcome 'pig skinning' or waviness of glaze.	(a) Use mixtures of 1 to 15% boric acid and 99 to 85% salt.	Use 8 to 15% boric acid and 92 to 85% salt.	A very smooth glaze practically free from 'pig skinning' was produced.
	(b) Use mixtures of 1 to 15% borax and 99 to 85% salt.	Use 15% borax and 85% salt.	Although borax reduces 'pig skinning' somewhat it was not nearly as effective as boric acid in overcoming this defect.
(5) To darken the colour of salt glazes	(a) Use 4 to 15% boric acid with 96 to 85% salt.	Use 15% boric acid with 85% salt.	The colour was changed from a brown to a dark mahogany colour.
	(b) Use 4 to 15% borax with 96 to 85% salt.	Use 15% borax with 85% salt.	The colour was changed from a brown to a darker shade but was not as effective as boric acid.

(6) To overcome dull spots in salt glazes.	(a) Use 4 to 15% boric acid with 96 to 85% salt.	Use 10 to 15% boric acid with 90 to 85% salt.	The dull spots disappeared and the glaze became brighter.
	(b) Use 4 to 15% borax with 96 to 85% salt.	Use 10 to 15% borax with 90 to 85% salt.	The dull spots disappeared but glaze was not as bright as when boric acid was employed.
(7) To increase the resistance of the glazes to chemicals.	(a) Use 4 to 8% boric acid with 96 to 92% salt.	Use 4% boric acid with 96% salt.	The glaze was very resistant to hydrochloric acid, acetic acid, sulphuric acid, nitric acid, and sodium hydroxide.
	(b) Use 4 to 8% borax with 96 to 92% salt.	Use 4 to 8% borax with 96 to 92% salt.	Glazes produced with borax were not as resistant to chemicals as those produced with boric acid.
(8) To reduce crazing of salt glazes.	(a) Use 1 to 15% boric acid with 99 to 85% salt.	Use 4% boric acid with 96% salt.	Glazes under high tension were changed to glazes with practically no tension.
	(b) Use 1 to 15% borax with 99 to 85% salt.	Use 4% borax with 96% salt.	Glazes under high tension were changed to glazes with practically no tension.
(9) To make glazed ware more resistant to leakage of water under low pressures.	(a) Use 4 to 15% boric acid with 96 to 85% salt.	Use 4% boric acid with 96% salt.	The resistance to water pressure before leakage occurred was increased 38 to 40%.
	(b) Use 4 to 15% borax with 96 to 85% salt.	Use 4 to 8% borax with 96 to 92% salt.	The resistance to water pressure before leakage occurred was increased 26 to 50%.
(10) To make glazed ware more resistant to leakage of gas.	(a) Use 1 to 15% boric acid with 99 to 85% salt.	Use 4% boric acid with 96% salt.	The resistance to gas leakage was improved 54 to 70%.
	(b) Use 4 to 12% borax with 96 to 88% salt.	Use 8% borax with 92% salt.	The resistance to gas leakage was improved 32 to 60%.
(11) To improve the strength of glazed ware.	(a) Use 1 to 10% boric acid with 99 to 90% salt.	Use 5% boric acid with 95% salt.	The strength was increased about 10%.
	(b) Use 1 to 15% borax with 99 to 85% salt.	Use 4% borax with 96% salt.	The strength was increased about 12%.
(12) To improve the average properties of the salt glaze.	(a) Use 2 to 14·0% boric acid with 98·0 to 86·0% salt.	Use 7·0 to 9·0% boric acid with 93·0 to 91·0% salt.	All of the properties were improved including increased thickness of glaze, smoother glazes, better colour, brighter glazes, increased resistance to chemicals, less crazing, greater resistance to leakage of water and gas, and greater strength.
	(b) Use 2·5 to 14·0% borax and 97·5 to 86% salt.	Use 8·5 to 10·5% borax with 91·5 to 89·5% salt.	All of the properties were improved but not as much as they were with boric acid.

metallic oxides may be applied by painting, spraying or dipping (Fig. 6.6).

The finished salt-glazed ware usually has a pitted surface. A smooth glossy surface can, however, be obtained on artware. In some cases very fine crystals occur giving a kind of aventurine glaze.

FIG. 6.6. *A coloured salt-glazed plaque in blue and white*

General Considerations

The advantages of a good salt glaze when compared with slip glazes are the elimination of the dip process, reducing handling, and the good resistance of the glaze to chemical attack and mechanical shock because of its high alumina content (about 20% Al_2O_3). There is also a considerable range of colour and texture available.

Nevertheless a good salt glaze cannot improve a body as much as a good slip glaze. Even a good salt glaze increases the mechanical strength of the body only very little but a good slip glaze can do so from 50 to 100%. But salt glazes on silica-poor bodies are very prone to crazing. When this occurs the strength of a body may be reduced by 20 to 40% and the surface so roughened that it readily absorbs water, etc., whereas a vitrified unglazed surface does not (**J44**). Remedying of this and other defects must always be done by altering the body, which may put up the cost of production too much. A slip glaze, of course, can be adapted to the body as well as the body to the glaze.

The difficulties also lie in the necessity for skilled control of the kiln atmosphere and temperature, high fuel consumption and rapid attack of the kiln linings. There are also often higher losses of salt glazed ware when compared with slip glazed due to irregular glazing in the various parts of the kiln, non-uniformity of different batches, warpage, etc., due to higher kiln temperature. Comparisons of the cost of salt and slip glazes show that salt glazes are somewhat cheaper to produce, the raw materials being very much cheaper but the labour costs relatively high. But when the quality of the glazed ware is considered salt glazing is not necessarily the most economical (**A29**).

Change from salt to slip glazing for sewer pipes is, however, hampered by byelaws and standard specifications. The old British byelaws insisted on brown salt-glazed pipes and the same applies in all the United States except California. In this State the industry has convinced the appropriate governing bodies that salt glaze of unspecified quality is not necessarily the best. The Californian specifications (**P1**) have therefore for some time laid down the crushing and hydrostatic tests for pipes regardless of whether salt glazed, slip glazed or unglazed. The new British Model Byelaws 1952 (**H112**) make similar stipulations of quality rather than precise type of materials, so that the way now lies open for the development of better drain pipes.

Other Vapour Glazing

Vapour glazing can also be done with materials other than sodium chloride, *e.g.* with zinc salts. A variety of colours can be achieved according to the composition and absorption of the body. For instance with clays of 50–70% silica and 18·5–27·5% alumina and given quantities of iron and alkalies the following colours are obtained (Table 158). Higher firing encourages formation of greens, especially when followed by rapid cooling. Slow cooling favours brick reds, brown and tans (**L13**).

TABLE 158

Colour	Absorption	Iron oxide	Alkalies
Greens	<2%	8–18%	>6%
Greys	<13%	1–3%	0–9%
Browns	1–16·6%	3–24%	<9%
Brick-red	7·2–19·8%	1–9%	<4%

CERAMIC COLOURANTS

The common raw materials for making ceramic coloured stains have been listed (p. 143) and the causes of and influences on the appearance of colour discussed (p. 227). The composition and preparation of actual ceramic stains or colourants is influenced by a number of considerations derived from the decorative method for which they are to be used and the possible action that may occur on them in the finished article. The decorative methods are described in detail in Chapter 9 but their technical distinctions will be given briefly here.

Body Stains. Here the prepared colourant is mixed intimately with the ceramic body and fired with it. It must therefore withstand any treatment given to the body.

Glaze Stains. These have already been mentioned in 'coloured glazes' (p. 560) and are colourants suspended in a glaze batch. They must withstand any treatment given to the glaze.

Underglaze Stains or Colourants. These are used for decorative work on the biscuit body, are then covered with the raw glaze and undergo the glost firing. They must withstand the temperature of the glost firing without reacting with the glaze and thereby appearing to run.

On-glaze Colourants. These colourants are used for decorative work on a glazed article. They are fixed to the glaze by a further low-temperature (700–900° C, 1292–1652° F) " enamel " firing, and are assisted in this by being mixed with fluxes so that they sink into the glaze, but no running may occur. On-glaze decoration presents the problem of durability as the larger the temperature difference between the enamel firing and the glost firing the more exposed is the decoration to abrasion and chemical action.

In preparing coloured stains for ceramic decoration the aim is to produce a uniform, reproducible, material in a chemically inert form which will neither decompose by itself, nor react with, nor dissolve in, the bodies and glazes they are to colour. Most colouring agents by themselves do not comply with this requirement so that they must usually be combined with other materials which may themselves be colourless and may or may not influence the colour. An exception to this is found in the under-glaze colour solutions (p. 644).

The main processes involved in preparing colourants are: (1) mixing; (2) calcining; (3) washing; (4) grinding.

1. MIXING

The final colour obtained from a colouring agent is considerably influenced by the other constituents of the compounded stain. If successive batches are to be the same it is therefore very important that reliable uniform raw materials of known composition be carefully weighed out and mixed. It is for this reason that many mineral sources of colouring agents which may give very beautiful colours cannot easily be used in mass production.

The best method of mixing is by wet grinding, drying and lawning. Other methods are heaping the ingredients together on a 60-mesh lawn and passing them through, or in a mechanical dry mixer.

2. CALCINATION

Calcination is the main part of the preparation of ceramic colourants, its purpose being to make them stable. Several different reactions may occur, depending on the nature of the raw materials and of the desired end product. The temperature of the calcination should be at least equal to that at which the stain will subsequently be fired on the ware and in some cases should be as high as possible. The two reaction types occurring during calcination are: (*a*) decomposition, and (*b*) combination.

(*a*) *Decomposition.* Many raw materials decompose on heating to give off gases, *e.g.* carbonates, sulphates, manganese and cobaltic oxides. If this should happen when the colourant has been applied to ware the gas evolution might cause bubbling or spitting, leaving white spots on the decorated surface and transferring colourant to undecorated places. Such decompositions must therefore be *completed* in the calcining.

(*b*) *Combination.* The coloured compound actually used to stain the ware is frequently made during this calcination when coloured and colourless materials are combined to give a product that is inert towards the solvent action of both body and glaze. Such compounds are the spinels which are formed at high temperatures by solid-state reactions and can then be used for body and glaze stains even at high temperatures. Very thorough calcination is particularly important when several colouring agents are mixed together as is often done to obtain black. Unless they undergo solid-state reactions and sinter well they can easily be separated by differential solution in the glaze and so give the decoration coloured haloes.

Calcination was formerly always done in well sealed saggars placed in certain positions in the normal production biscuit or glost kilns. Although this is satisfactory for the more robust colours it does not give sufficient control for the more delicate ones and independent small muffles are recommended.

3. WASHING

After calcination the colourant is crushed and then washed to remove all soluble matter, particularly if it is coloured (this may be due to excess soluble raw material or a soluble reaction by-product). Hot water must be used to remove dichromate and chromates which otherwise make haloes round a print of a chrome colour. Borates, if not removed, form a skin on the colourant.

4. GRINDING

The insoluble residue is then wet ground in pan or ball mill to the desired fineness. There is no rule about the grain size of colourants although the usual requirement is that it must all pass through a 300-mesh lawn. Underground colourants are gritty, difficult to apply, have low covering power and appear spotty after firing. Over-ground colourants dust and may not take a glaze well, causing 'crawling' (**M41, A44**).

Applicability of Ceramic Colourants

The application of each ceramic colourant is restricted in some way by temperature and environmental conditions. Thus many colourants decompose or volatilise at temperatures below those at which some ceramic processes are carried out and can only be used for lower temperature applications. The range of colourants available for on-glaze decoration is therefore much wider than for under-glaze or glaze and body stains; the smallest colour range is that for the highest temperature work, namely hard porcelain body and glaze stains and under-glaze decoration. Other colourants react with certain glaze or other colourant constituents, *e.g.* sulphide colourants are turned black by lead glazes. Another point to consider when choosing on-glaze colourants is their resistance to acid and alkali attack; some colourants are completely destroyed by organic food acids whereas the same colour produced by a different colourant may be resistant.

Auxiliary Materials for applying Ceramic Colourants

UNDER-GLAZE DILUENTS AND FLUXES

Under-glaze colourants may need to be mixed with colourless materials to dilute them and to fix them to the surface so that they are not damaged by the application of the glaze. The latter function is performed by so-called 'fluxes' although these must not be confused with the low-melting on-glaze fluxes. They should fuse around the colourant particles during the hardening on fire without melting sufficiently to form a vitreous patch on which the glaze will not take. On the other hand a flux of too high melting point does not fix the colourant sufficiently and may lead to subsequent shelling off of the decoration together with the glaze above it.

Some typical under-glaze fluxes are:

(1) Glost pitchers or mixtures of pitchers and stone.
(2) Mixtures of stone, feldspar and flint.
(3) Mixtures of borax and flint, *e.g.* borax 54%, flint 46%.
(4) Mixtures containing glass, *e.g.* flint glass 53%, sand 47%.
(5) Mixtures containing lead, borax and flint similar to the No. 8 flux for on-glaze colourants (No. 13, Table 159), *e.g.* red lead 61%, borax 8%, flint 31% (**B57, A44**).

However, it is possible that the stone used in examples (1) and (2) would give rise to 'spit out', especially if on-glaze decoration is applied to the same piece.

ON-GLAZE FLUXES

Colourants destined to be used in on-glaze decoration are mixed with low-melting glasses or frits, termed a 'flux' in this connection. When heated to 700–850° C (1292–1562° F) in the enamel kiln this fuses into the glaze, thereby fixing the decoration. A number of properties are required of these fluxes which may in practice act against each other. Firstly they must mature at as low a temperature as possible, as this reheating of the glazed article may cause gas formation in the body and blemishes on the surface known as 'spit-out' (*see* p. 881). There is usually a maximum safe reheating temperature for any particular glaze–body combination. Secondly good mechanical integration of the flux–colourant mixture with the glaze is required. Without this the decoration may peel off because of incompatible expansion coefficients or be easily removed by abrasion in use. The best combination of flux and glaze is achieved when the decoration is matured near the softening point of the glaze, *i.e.* at a relatively high temperature, when spit-out will readily occur, and so contradict condition (1). The third requirement is adequate resistance to the alkalies in washing-up water, particularly washing-up machines. These have been found to attack the flux, reducing the gloss and gradually exposing the colourant to removal by mechanical means. Fourthly, the flux must be resistant to the attack by food acids. If it contains lead the same considerations apply as for lead glazes (*see* p. 568). As regards the acid resistance of the decoration it has been found that attack by food acids seems to be specific to certain colourants rather than to the fluxes. More careful selection of the colourant is therefore necessary.

The actual conditions to which the ware will be subjected in use vary with its nature. The harshest conditions are applied to crockery used in institutions where there are washing-up machines. Such ware receives the abrasive action of the cutlery and of other plates, etc., when stacked; it receives normal attack by food acids but the attack by alkaline detergents may be high due to high concentration and/or high temperature, when compared with domestic use. Washing-up machines use detergent concentrations between the limits 0·2% and 1·5% with a general average of 0·333%, the temperature in the cleaning chamber varies between 40 and 80° C, although temperatures above 60° C render all egg white-containing foods insoluble and subsequently very hard to remove, rinsing is done at 70–100° C, averaging at 90° C. Various laboratory tests have been evolved to simulate the action of both food acids and washing-up conditions on on-glaze decorations (*see* p. 378) (**K47**), *see also* (**A126**).

Dale and Francis (**D5**) have reviewed the literature relating to the factors influencing the durability of on-glaze decoration, and this has been followed by practical work on the influence of the firing treatment, carried out by Sharratt and Francis (**S58**). They describe first the types of deterioration of the decoration:

(1) The gloss may diminish.

(2) The colour of the decoration may fade or even change completely.

(3) The flux may be dissolved by chemical action; or rendered more susceptible to erosion by friction, vibration or impact.

Those parts of the decoration which are appreciably raised above the general level of the glaze will probably suffer most from friction and impact during handling and stacking.

TABLE
On-glaze

No.	Ref.	Na_2O	K_2O	MgO	CaO	BaO	ZnO	PbO	B_2O_3	Al_2O_3	SiO_2	F_2
1	Nos.1–24							1·00			1·25	
2	S88	0·53						0·47	1·06		1·78	
3	after	0·32	0·11					0·57	0·65		1·08	
4	A153	0·45	0·14					0·40	0·90		1·43	
5		0·35	0·50					0·15	0·69		2·88	
6		0·22						0·78	0·44		0·98	
7		0·87						0·13	1·75		0·99	
8		0·13						0·87	0·26		0·88	
9								1·00			0·69	
10		0·36	0·01					0·63	0·71		1·19	
11		0·29	0·03				0·16	0·52	0·64		0·98	
12		0·17	0·02				0·05	0·76	0·34		0·86	
13		0·16						0·84	0·32		1·06	
14								1·00	0·15		0·94	
15		0·25						0·15	1·70		1·71	
16								1·00	2·37		1·26	
17								1·00	0·73		0·28	
18								1·00	0·25		1·08	
19								1·00	0·40		0·42	
20								1·00	3·31			
21								1·00	0·60		1·26	
22								1·00	0·50		0·60	
23		0·60	0·10					0·30	2·50		0·70	
24		0·30						0·60	1·10		0·80	
25	E10		0·1					0·9		0·2	2·3	
26	E10		0·1					0·9		0·2	1·9	
27	E10		0·1					0·9		0·2	1·5	
28	E10	0·1	0·1	0·1	0·5	0·2			0·75	0·2	2·5	
29	E10	0·1	0·1	0·1	0·5	0·2			0·75	0·2	3·0	
30	E10	0·1	0·1	0·1	0·5	0·2			0·75	0·2	3·5	
31	E9	0·1						0·9	0·1		1·5	
32	E9	0·3						0·7			1·3 to 2·0	
33	P10	0·32		0·49				0·19			1·90	
		0·52		0·27				0·21			2·08	
		0·37			0·41			0·22			2·21	
		0·57			0·21			0·22			2·25	
34	—						0·2	0·8	0·2	0·2	1·5	
35	A154							1·0	0·5		0·5	
36	A163	0·3						0·7	0·5		0·5	
37	A44	0·29						0·71	0·57		0·91	
38	A44	not calculated because of the number of kinds of flint glass									
39	A44											
40	A44	0·60			0·40				0·34	0·03	1·26	0·58
41	P3							1·0	0·51		0·53	
42	H16	0·56	0·25				0·18		1·40	0·25	3·36	
43	H125	0·89	0·11						0·45	0·072	3·39	0·43

Fluxes

No.	Batch	Softening temp. (°C)	Softening temp. (°F)	Remarks
1		~540	~1004	Lead flux or so-called Rocaille flux
2				
3				
4				For blue.
5				For purple
6				
7				
8				
9		~530	~986	
10				
11				
12				
13				For grey, also for blue and iron red
14				For grey, also for iron red and yellow
15				For crimson, also for green
16				For purple
17				For violet
18				For green (for Meissen green)
19				For green (after Salvéta)
20				Special flux for light green
21				Flux for turquoise blue
22				Neutral flux for yellow, iron red, brown
23				For purple
24				For violet
25		730	1736	mg dissolved in HCl in 5 hr from 1000 cm² surface: 32
26		720	1328	59
27		710	1310	96
28		770	1418	Good acid resistance can be achieved with low softening temperature: 261
29		820	1508	267
30		860	1580	267
31				Very good acid resistance
32		800	1472	Very good acid resistance
33	SiO₂ 58%, PbO 22%, CaO or MgO 5–10%, Na₂O 10–15%	<570	<1058	Good acid and alkali resistance
34	Eutectic PbO.0·254Al₂O₃.1·91SiO₂: 80%, ZnO.2B₂O₃: 20%	650	1202	Recommended for porcelain without crazing
35				General flux
36				Flux for purple
37	Flint: 1, red lead: 3, borax: 2			Flux No. 8. General purpose up to 80% of colour mixture
38	Red lead: 10, flint glass: 9, borax: 6			
39	Red lead: 11, flint glass: 2, boric acid: 4			
40	Sand: 34, borax: 29, fluorspar: 14, cryolite: 6, soda ash: 15, nitre: 2%			Lead-free flux for selenium red
41	Quartz sand: 10, litharge: 70, boric acid: 20%			Frit at 1250–1450° C
42	Feldspar: 9·00, silica: 7·08, borax: 13·74, boric acid: 2·22, zinc oxide: 0·96			Flux for selenium red, use 120 parts to 20 parts red
43	Borax (anhyd.): 22·1, feldspar: 40·0, silica: 23·5, soda ash: 4·1, cryolite: 7·6, kaolin: 2·7 Na₂O: 11·1, Na₂F₂: 4·7, K₂O: 2·8, B₂O₃: 8·2, AlF₃: 3·2, SiO₂: 54·1%			Flux for selenium reds

TABLE

No.	Ref.	Na$_2$O	K$_2$O	MgO	CaO	BaO	ZnO	PbO	B$_2$O$_3$	Al$_2$O$_3$	SiO$_2$	F$_2$
44	I1	0·20	0·05		0·20		0·10	0·45	0·70	0·025	1·50	
45	S207	0·54	0·46						1·09	0·46	4·92	
46	S207	0·67	0·33						0·73	0·43	3·79	
47	S207	0·58	0·07		0·35							0·61
48	S207	0·74	0·03		0·23							0·35
49	H107							1·0			1·24	
50	H107							1·0		0·91	1·86	
51	H107							1·0		0·29	0·59	
52	H107							1·0		0·53	0·66	
53	H107	0·69						0·31		0·81	1·52	
54	H107	0·63						0·37		0·74	1·64	
55	H107	0·61	0·05					0·34		0·59	1·30	
56	H107	0·48					0·52			0·53	1·24	
57	L40	0·63				0·12	0·25		1·44		0·25	
58	A119							1·0			0·5	
59	A119							1·0	0·25		0·6	
60	A119							1·0	0·15	0·10	1·12	
61	A119							1·0	0·125	0·125	1·25	
62	A119							1·0		0·25	1·9	
63	A119							1·0		0·15	1·00	
64	A119							1·0		0·15	1·50	
65	A119							1·0		0·15	1·75	
66	A119							1·0		0·15	2·00	
67	A119							1·0		0·10	2·00	
68	A119							1·0		0·254	1·91	
69	A119							1·0		0·30	1·91	
70	A119							1·0		0·35	1·91	
71	N36							1·0		0·69	1·0	

(4) The layer of decoration may become completely detached from the glaze as a consequence of imperfect attachment, or by preferential attack by a corrosive agent between the surfaces of contact, or by reason of differences in thermal expansion between the decoration and the glaze.'

159—(contd.)

No.	Batch	Softening temp. (°C)	Softening temp. (°F)	Remarks	
44				Flux for chrome pink. Use 100 parts to 25 parts colourant	
45	Borax: 26·6, feldspar, 32·6, quartz: 16·8				
46	Borax: 23·0, feldspar: 30·0, quartz: 18·0, cryolite: 7·0			Fluxes for cadmium yellow	
47	Fluorspar: 8·4, soda: 11·2, saltpetre: 4·2,				
48	Calcite: 4·5, soda: 9·0, saltpetre: 1·0				
49	Lead oxide: 75, silica: 25				
50	Lead oxide: 50, silica: 25, boric acid: 25			For porcelain, high enamel firing. The rest can be matured at 580–620° C (1076–1148° F)	
51	Lead oxide: 76, silica: 12, boric acid: 12				
52	Lead oxide: 68, silica: 12, boric acid: 20				
53	Lead oxide: 23, silica: 30, boric acid: 33, soda: 14				
54	Lead oxide: 25, silica: 35, boric acid: 28, soda: 12				
55	Lead oxide: 28, silica: 29, boric acid: 27, soda: 14, potash: 2				
56	Zinc oxide: 20, silica: 35, boric acid: 31, soda: 14				
57	SiO_2 7·8%, B_2O_3 51·8%, BaO 9·7%, ZnO 10·4%, Na_2O 20·3%	470	878	For selenium red, maturing at 570° C (1058° F) for 1 to 2 hr	

No.				Matured at	Resistance to trisodium phosphate
58				704 °C	Strong corrosion
59				732	
60				788	Weak etching
61				788	Iridescent spots
62				843	No attack
63				816	Etching
				982	Weak etching
64				816	Weak etching
				982	Very weak etching
65				816	Trace of etching
				982	No etching, but a little loss of gloss
66				816	Little loss of gloss
				982	No action whatsoever
67				927	No etching, but gloss reduction
				1038	No action
68				927	No action, but bad flow of glaze
				1038	Good
69				927	No action, but immature
				1038	Good
70				927	No action, but more immature
71	Flint : 16, Red lead : 61, boric acid : 23			1038	No action, but signs of immaturity

The factors influencing the durability of the decoration can be grouped as follows:

(1) Nature of the glazed surface to be decorated.

(a) Constitution of the glaze.

    (b) Degree of maturity of the glaze.
    (c) Presence of glaze vapours in the kiln atmosphere.
    (d) Presence of oils, sulphurous and other gases in the kiln atmosphere; even those present in the glost firing may ultimately affect the success of the enamel firing.
    (e) Thickness of the glaze.
    (f) Faults in the glaze, particularly bubbles.

(2) Application of the decoration.
    (a) Composition of the colourant.
    (b) Composition of the flux.
    (c) Admixture of oils, sizes, etc., in applying the colour.

(3) Nature of the enamel firing.
    (a) Heat-work done in enamel kiln.
    (b) Maximum temperature reached in enamel kiln.
    (c) Presence of various gases due to item (2c).

1. *Nature of the glazed surface to be decorated.*

The composition of the glaze is usually determined by factors other than the on-glaze decoration and the decorating technique must be adapted to it rather than vice versa. However, the success of the decoration will obviously be greater if the glaze is of even thickness, properly matured and without bubbles, pinholes and other faults.

2. *Application of the decoration.*

(a) Although recorded investigations on the durability of on-glaze colourants have repeatedly shown that a given colour has different resistances to attack according to its origin, no data on the nature of the various colourants is given. As the direct attack of the colourant is chemical it is reasonable to expect best results from colourants carefully prepared to be as inert as possible (*see* p. 606).

(b) Many fluxes are lead–boron–soda–silica glasses, whose solubility has already been discussed. The proportions of the constituents of such glasses are very important. Other empirical work shows that there is a critical maximum of B_2O_3 if fluxes are to be acid resistant (**B26**), and that small additions of alumina are very beneficial (**H23**). Peddle (**P23**) recommends using the soda and potash in the ratio of 7 to 3 for the alkali portion of the flux for percentages up to 20. Koenig (**K53**) describes work aimed at producing a suitable lead frit insoluble in both acids and alkalies. Starting with lead oxide the decreasing order in which other oxides reduce solubility is: Al, Si, Zn, Ba, Ca, Be, Ti and Zr. MgO is neutral. If only one frit is used much Al_2O_3 has to be added to compensate for the effect of boric acid and alkalies on the lead oxide. Other more specific results are given in Table 159, Nos. 25–33 and 58–70 inclusive.

GLAZES

(*c*) A number of organic materials are frequently added to the on-glaze colourant to render it suitable for application by transfer, brush, screen printing, etc. These either volatilise or burn out during the early stages of the enamel firing. Slow heating with good ventilation is necessary to remove all the gases before any fusion of the flux starts.

3. *The enamel firing.*

(*a*) In the experiments described by Sharratt and Francis (**S58**) identical pieces were fired in three different parts of six different enamel kilns corresponding to 'hard', 'medium' and 'easy' firing. Heat-work was measured by Holdcroft's bars No. 7 (down at about 790° C, 1454° F) which showed a maximum difference of sag between easy and hard of 1·37 cm. (The maximum sag of a bar is 1·50 cm when it touches the base.) It was found that the harder firing increased the resistance of the decoration to acid attack. Over-firing is of course detrimental and it is found that under-firing reduces resistance more than slight over-firing increases it. The hardness of the firing within the small range used does not greatly or consistently affect the resistance to alkali.

(*b*) Sharratt and Francis (**S58**) also found that the development of certain colours, particularly pink, in the enamel firing is more dependent on the maximum temperature reached than on the heat-work performed, and further that the acid resistance of these pinks increases linearly with the degree of maturity of the colour. Alkali resistance also increases with better colour development but in a less regular manner.

OIL, ETC., FOR APPLICATION

Various oils and other media are worked up with a ceramic stain to bring it into a condition suitable for painting, transferring, printing, etc. (**M41, A44**). These are listed in Chapter 1.

RECIPES FOR CERAMIC COLOURANTS

Although the preparation of ceramic colourants is becoming more scientific and explicable the necessity for adaptation to particular production conditions makes general instructions impossible. It is still the normal practice to use a 'recipe' tied to particular raw materials, milling, precipitating and calcining conditions. In the tables that follow we have abstracted a few of such recipes and comments from available experience and literature. (Compilations in other books have largely been avoided.) The data given are often incomplete and where possible we have added suggestions in brackets in the columns 'application' and 'temperature'. These recipes can probably only be applied directly in rare instances but they should act as starting points for experiments to develop colourants to suit individual requirements. They are listed alphabetically according to the metallic ion present, as seen in the summary, except for the under-glaze colour solutions which are given together at the end.

TABLE 160

Ceramic Colourants

(1) Cadmium: yellows, oranges, reds.
(2) Cerium: ivory.
(3) Chromium: (a) greens;
    (b) browns and yellows;
    (c) with tin, pinks to crimsons;
    (d) blacks.
(4) Cobalt: (a) blue;
    (b) black.
(5) Copper: (a) blues and turquoise;
    (b) red.
    Gold: (a) pinks, purples, browns;
    (b) metallic.
(6) Iridium.
(7) Iron: yellows to reds.
(8a) Iron–chromium: browns and yellows;
    (b) green.
(9a) Lead antimonate: Naples yellow;
    (b) basic lead chromate: coral red.
(10) Manganese, pinks, violets and browns.
(11) Metallic surfaces.
(12) Neodymium: pale violet.
(13) Uranium: (a) red;
    (b) brown;
    (c) black.
(14) Vanadium: yellows, greens, blues.
(15) Composite colourants: (a) whites;
    (b) ivories;
    (c) blacks.

Ref.	Colour	Application	Temperature	Composition and preparation
1. Cadmium yellow and oranges				
K65b W41 S206	Yellow to deep orange	[On-glaze and earthenware glaze stain]	Max. 800° C to 835° C	Cadmium sulphide. Colour depends on grain size and crystalline form, and therefore on the method of preparation α-CdS yellow, β-CdS red. Sublimes 980° C. Dry preparation $CdCO_3 + S(\text{flowers}) \xrightarrow{340-600°} $ mixture α- and β-CdS. Wet preparation Soluble Cd salt $+ H_2S \rightarrow$ cadmium yellow. $CdSO_4 +$ freshly prepared $Na_2S \rightarrow$ cadmium yellow or orange. $CdCO_3 + Na_2S$ or $NaSH \rightarrow \beta$-CdS, suitable for preparing cadmium red (see below). Cadmium yellow may be mixed in any proportions with cadmium–selenium red. It is destroyed by most other colourants especially Fe and Cr and also by high lead glazes.
A14	Cadmium yellows	[On-glaze]	[*ca.* 700–800° C]	$CdCO_3$ 68·7 72·3 80·0 S 16·9 17·3 20·0 ZnS 14·5 10·4 — Calcined for 1 hr at 625° C 615° C 600° C temp. given

Cadmium—Selenium oranges and reds

Code	Colour	Application	Temperature	Composition and notes
P3	Yellow	Glass enamel	550° C	Flux 86.4% CdS 13.6% Flux: Quartz sand 10, Litharge 70, Boric acid 20 — frit at 1250–1450° C. Oxidising atmosphere is essential to prevent the formation of black PbS.
K65a	Light bright reds and oranges	On-glaze and glaze stain	Max. cone 013 [ca. 835° C]	Mixed crystals of CdSe and α-CdS prepared by heating cadmium sulphide with selenium or cadmium carbonate with selenium and sulphur to about 650° C. The redness depends on the selenium content: dark red 20% Se fire red 16–18% Se orange-red 12–15% Se Colour destroyed by most other colourants particularly Fe and Cr.
H34	Cadmium-selenium reds	On-glaze and enamel	[ca. 800° C]		Dark	Medium	Light	 CdO 23.1 / 8.0 / 49.1 CdS 57.7 / 72.0 / — Se 19.2 / 20.0 / 19.7 S — / — / 31.2 Calcine the mixtures at 500° C for 5 min and use 1–4% additions to the enamel.
H16	Cadmium–selenium red	[On-glaze and glaze stain]	[ca. 800° C]	Colourant: CdS 65.4 Se 18.2 NH₄NO₃ 16.4 Flux: feldspar 27.3, silica 21.4, borax 41.6, boric acid 6.7, zinc oxide 3.0 Red colour 14.3 Flux 85.7 remelted at 800° C to give more stable colourant.
A14	Cadmium–selenium reds and oranges	[On-glaze]	[ca. 700–800° C]		Red	Red	Red	Red	Orange	Orange	Orange	Orange	Orange	Orange | CdCO₃: 76.3 / 72.2 / 72.5 / 78.6 / 77.9 / 76.6 / 76.5 / 72.2 S: 14.0 / 13.7 / 11.6 / 16.7 / 15.9 / 14.1 / 13.5 / 13.7 Se: 9.7 / 14.1 / 15.9 / 4.7 / 6.2 / 9.3 / 10.0 / 14.1 1 hr calcination at given temp. (°C): 600–700 / 530 / 530 / 580 / 570 / 560 / 555 / 530

TABLE 160—(contd.)

Ref.	Colour	Application	Temperature	Composition and preparation
2. *Cerium*				
S158	Yellow	Under-glaze	[Max. cone 15]	Cerium tantalate together with colourless oxides. See also 15. *Ivory*, below.
3a. *Chromium greens*				
A136	Greens	Under-glaze, glaze stains, on glaze		The colouring is obtained by additions of Cr_2O_3. It may be used alone to obtain dark green. Cr_2O_3 is a good under-glaze colourant as it is heat stable (slight volatility at 1100–1200° C) and relatively insoluble in glazes.
M61				Chrome-greens tend to non-uniformity, being affected by the kiln atmosphere and glaze composition. Reducing atmosphere gives the best greens. SnO_2 must be absent or some 'pink' forms, making the green look dirty. B_2O_3 is detrimental and should be less than the Al_2O_3 content, it is counteracted by SiO_2 and CaO. In oxidising fire chromates (yellow) are formed with bases, frits must therefore be thoroughly washed.
—	Green	Under-glaze	Cone 13 reducing	Chromium oxide (pure) 84% Glaze 16% } wet grind 300 hr, no calcining.
—	Sèvres green (bluish green)	Under-glaze	Cone 14	Chromium oxide 17·8% Cobalt oxide 22·2% Aluminium hydroxide 60·0% } wet grind together 300 hr, dry and then calcine at cone 14.
—	New Sèvres green (slightly darker than above)	Under-glaze	Cone 14	Chromium oxide 25·8% Cobalt oxide 35·5% Aluminium hydroxide 38·7% } wet grind together 300 hr, dry and then calcine at cone 14.
—	Strong light green	Under-glaze	Cone 14	Cr_2O_3 30% SiO_2 70% } wet grind 300 hr, dry, calcine at cone 14.
H90	Victoria green	Under-glaze		Segers composition: $K_2Cr_2O_7$ 36% $CaCl_2$ (fused) 12 Quartz 20 Marble 20 Fluorspar 12 Quartz, marble and fluorspar are finely ground. Treat with hot saturated solution of $K_2Cr_2O_7$ and $CaCl_2$ and evaporate to dryness. Mix in mortar. Calcine as high as possible. *This is difficult to prepare in large batches because of the deliquescent nature of $CaCl_2$ and the irritant properties of $K_2Cr_2O_7$.*

Substitute compositions:

		(1)	(2)	(3)	(4)	
—	$K_2Cr_2O_7$	36	—	—	—	Mix dry. Calcine at cones 6 to 7.
	Flint	30	20	20	20	
	Fluorspar	13	15.5	15.5	15.5	
	Whiting	21	25.9	25.9	25.9	
	Cr_2O_3	—	18.6	18.6	18.6	
	NaCl	—	20	—	—	
	KNO_3	—	—	20	—	
	White lead	—	—	—	20	
Victoria green	Cr_2O_3	25	—	Wet grind, calcine, wash out.		
Glaze stain, under-glaze, on-glaze	Whiting	50	34			
	Quartz	25	—			
	$BaCrO_4$	—	46			
	Boric acid	—	20			

Victoria green is destroyed by zinc glazes unless 15% Bi_2O_3 is added to the frit.

Green (G7)	Potassium dichromate	35.0%	Mix dry and pass through 40-mesh sieve. Calcine in closed saggars to cone 10. Grind wet to pass 200-mesh and wash free of soluble salts.
	Plaster of Paris	12.5	
	Flint	20.0	
	Whiting	20.0	
	Fluorspar	12.5	
Blue (BE2)	Cobalt oxide	3.0%	Grind wet, dry, calcine at cone 10 in closed saggars, grind wet to pass 200-mesh, wash.
	Zinc oxide	12.5	
	English china clay	50.0	
	Flint	34.5	

K8 — Green with blue cast opaque U.S. standard green No. 11 — Vitreous china sanitary ware — Cone 7

Basic glaze 0.30 Na_2O } 0.40 Al_2O_3 3.20 SiO_2.
0.35 CaO
0.25 PbO
0.10 BaO

Batch			
	Soda feldspar	42.0%	Plus Borax 2.0
	English ball clay	6.9	Zirconium oxide 7.5
	Flint	19.2	G7 stain 3.0
	Whiting	9.4	BE2 stain 2.5
	White lead	17.2	
	Barium carbonate	5.3	
		100.0	

TABLE 160—(contd.)

Ref.	Colour	Application	Temperature	Composition and preparation		
3a. A136	*Chromium greens*—(contd.)					
	Green	Under-glaze for earthenware General on-glaze		Cr_2O_3 Feldspar or stone Flint	30% 20% 50%	25% 15% 60%
	Brighter green			Cr_2O_3 Co_2O_3 Feldspar Flint	21% 4 15 60	Calcine to earthenware biscuit temp.
	Blue-green (Russian green)			Cr_2O_3 Co_2O_3 Feldspar Flint	15% 10 15 60	Calcine to earthenware biscuit temp.
	Yellow-green	On-glaze		Cr_2O_3 NiO Flint Borax	12% 6 42 40	Calcine to earthenware biscuit temp.
	Olive greens	On-glaze		Cr_2O_3 NiO Calc. borax Flint	16% 8 34 42	Calcine to earthenware biscuit temp.
	Brown			Cr_2O_3 NiO Stone Flint	27% 40 15 28	Calcine to earthenware biscuit temp.

			Dark green	Chrome green	Light green	Light yellow-green
S88	Greens	Earthenware under-glaze				
		Cr_2O_3	34·6	100·0	25·0	—
		$BaCrO_4$	—	—	—	46·0
		NiO	65·4	—	—	—
		$CaCO_3$	—	—	50·0	34·0
		SiO_2	—	—	25·0	—
		H_3BO_3	—	—	—	20·0
		Calcine at cone	14 reducing	—	14	1050° C

Code	Name	Type	Temp	Recipe / Notes
A136	Victoria green	General [earthenware under-glaze, on-glaze]	[1000–1200° C]	$K_2Cr_2O_7$ 38% Whiting 20 Fluorspar 22 Flint 22 Dissolve $K_2Cr_2O_7$ in water and mix with other ingredients. Dry paste, calcine in ghost kiln. Grind, wash with hot water and HCl or vinegar until all traces of chromate are removed. Dry and mix with flux.
M61				There is a tendency to get unevenness and browning due to: (1) traces of Al_2O_3, SnO_2, ZnO in the colour; (2) presence in the glaze of ZnO, BaO, MgO; (3) glazes of the type 0·2–0·3 NaKO $\left.\begin{array}{l}0{\cdot}3{-}0{\cdot}42 B_2O_3 \\ 0{\cdot}22{-}0{\cdot}35 Al_2O_3\end{array}\right\}$ 2·42–3·15 SiO_2 0·25–0·4 CaO 0·4 PbO Good colour is obtained with the glaze type: 0·15 NaKO $\left.\begin{array}{l}0{\cdot}03{-}0{\cdot}2 B_2O_3 \\ 0{\cdot}2 Al_2O_3\end{array}\right\}$ 2·3–2·8 SiO_2 0·45 CaO 0·4 PbO
P3	—	Glass enamels	550° C	Colourless, clear flux: Quartz sand 10% Litharge 70 } frit at 1250–1450° C, wet or dry grind. Boric acid 20
	Green with yellow tinge	Glass enamels	550° C	Flux 90% Cr_2O_3 10%
	Pure green	Glass enamels	550° C	Flux. Cr_2O_3. CoO or CuO.
	Green	Glass enamels	550° C	Sand 8·7 Boric acid 23·1 Litharge 60·9 Cr_2O_3 4·4 Cobalt oxide 2·9
C59	Muddy green	Body stain		Chromic oxide stain. This is made brighter by CuO, but the latter may be dissolved by the glaze.
S88	Green	Body stain		Cr_2O_3.
	Green-blues	Body stain		Varied mixtures of Cr_2O_3 and CoO.
—	Grass-green	Body stain		Wollastonite ($CaO.SiO_2$). Zirconia bodies with Cr_2O_3.

TABLE 160—(contd.)

Ref.	Colour	Application	Temperature	Composition and preparation					
					Greenish brown	Light brown	Chinese brown	Bright brown	Dark brown

3b. *Chrome browns, yellows, etc.*

Ref.	Colour	Application	Temperature	Composition and preparation
A136	Browns	[Earthenware under-glaze, on-glaze general]	[1000–1200° C]	Cr_2O_3 — Greenish brown 45·4, Light brown 33·4, Chinese brown 14·2, Bright brown 14·0, Dark brown 48·9 ZnO — 18·2, 38, 73·0 Stone — 36·4, 26·6 Fe_2O_3 — —, —, —, 63·3, 51·1 calcine cone 7 Boric acid — —, —, 12·8, 22·7 Mn_2O_3 —
				Cr_2O_3 2–5%; } Use 2–3% in body mixture. Sb_2O_5 10–20%. } TiO_2 the rest. Cr_2O_3 with FeO or MnO_2.
H15	Yellow	Body stain		
S88	Browns and greys	Body stains		

See also Iron–Chrome browns and blacks.

3c. *Chrome–Tin pinks to crimsons*

Ref.	Colour	Application	Temperature	Composition and preparation
M60	Pinks to crimson	Under-glaze, glaze stain [On-glaze]	[Max. 1200–1250° C cones 6–8 oxidising atmosphere]	The chrome–tin pink stain is probably a deposit of highly dispersed chromic oxide on stannic oxide acting as a mordant. Variations in tint are determined: (1) by the temperature of deposition, which, in turn, probably determines the average sizes of the colloidal particles. (2) by the amount of chromic oxide precipitated upon the stannic oxide. It is reported that proportions of Cr_2O_3 to SnO_2 give colours as follows: $Cr_2O_3:SnO_2::1:17$ deep red. 1:25 pink. 1:>25 palening. 1:>17 purpling. 1:5 green.
I1	Pink	[On-glaze]		$CaO.SnO_2$ 65·0 SnO_2 6·7 SiO_2 28·3 } calcine at max. $ZnCrO_4$ 3·0 } 1250° C Borax 3·0 —— 106·0 Flux: Na_2O 0·20 K_2O 0·05 CaO 0·20 B_2O_3 0·70 } SiO_2 1·50 ZnO 0·10 Al_2O_3 0·25 } PbO 0·45 Colourant 25 parts to flux 100 parts.

				Pink	Pink	Pink	Pink	Pink	Lilac
A136	Pinks and lilac	[Underglaze Glaze stain On-glaze]	[Max. 1200–1250°C cones 6–8 oxidising atmosphere]	SnO_2	52	65	57	84	50	70
				$CaCO_3$	26	28	40	9	25	20
				SiO_2	19	4	—	6	18	—
				$Na_2B_4O_7$	3	3	—	—	4	4
				$K_2Cr_2O_7$	—	—	3	1	3	3
				$PbCrO_4$	—	—	—	—	—	—
				$PbCO_3$	—	—	—	—	4	—
					100	100	100	100	104	97

It is recommended to have 25% CaO, in order to avoid purple discoloration. Wet grind insoluble ingredients, dry out to a paste. Dissolve soluble ingredients and add slowly to the paste, dry, grind, calcine to cone 8–10, grind and wash until all soluble chromate removed.

The colour is affected by glaze constituents:

    alkalies   →purple tint.
    ZnO   →browns.
    low-lime   →purple.
    high-B_2O_3→purple.
    high-silica→purple.

J8	Pink	Under-glaze	[Max. cone 7–8 oxidising atmosphere]	SnO_2 60 $\\$ Whiting 31 $\\$ China clay 3 $\\$ $K_2Cr_2O_7$ 5	Suitable glaze type: $\\$ 0·15 KNaO $\\$ 0·45 CaO } 0·03–0·2 B_2O_3 } 2·3–2·8 SiO_2 $\\$ 0·40 PbO 0·25 Al_2O_3 $\\$ Glaze that causes purpling: $\\$ 0·2 –0·3 KNaO $\\$ 0·25–0·4 CaO } 0·03–0·2 B_2O_3 } 2·42–3·15 SiO_2 $\\$ 0·4 PbO 0·22–0·35 Al_2O_3

					Chrome pink	Chrome pink	Dark red	Light lilac	Lilac
S88	Pinks and reds	Earthenware under-glaze	Cone 9 oxidising	Cr_2O_3	0·6	—	0·5	4·7	—
				$BaCrO_4$	—	1·1	—	—	4·1
				$CaCO_3$	23·2	24·2	34·8	9·4	—
				SiO_2	23·2	24·7	19·7	9·4	68·5
				SnO_2	53·0	50·0	45·0	76·5	27·4
				$Na_2B_4O_7$	—	—	—	—	—
				Calcine at cone	9 oxid.	9 oxid.	9 oxid.	9 oxid.	1050°C oxid.

May not be fired above cone 9 and atmosphere must be oxidising.

TABLE 160—(contd.)

Ref.	Colour	Application	Temperature	Composition and preparation
3c.	Chrome—Tin Pinks to crimson—(contd.)			
K8	U.S. Standard Orchid No. 20 Light bluish-pink, opaque	Vitreous china sanitary glaze stain	Cone 7	Orchid stain L1 Tin oxide 86.0% Borax 8.6 $K_2Cr_2O_7$ pulverised 5.4 } Mix dry, pass through 40-mesh. Calcine cone 10, grind, wash. Glaze: 0.20 K_2O, 0.20 CaO, 0.40 PbO, 0.20 ZnO } 0.30 Al_2O_3 3.00 SiO_2. Batch: Potash feldspar 30.0% English ball clay 6.9 Flint 25.8 Whiting 5.3 White lead 27.7 Zinc oxide (calcined) 4.3 ——— 100.0 Plus: Borax 2.0% Tin oxide 3.0 ZrO_2 3.0 L1 stain 5.0
K65a	Red	Glaze stain	Max cone 6 [oxid. atm.]	SnO_2 50 parts by wt. $CaCO_3$ 25 $K_2Cr_2O_7$ 3 Quartz 18 Borax — Calcine at least cone 6 and wash out any soluble remainder. Glaze must have high lead and calcium and low boron content.
K65a	Pink	Glaze stain	Max. cone 6 [oxid. atm.]	SnO_2 50 parts by wt. $CaCO_3$ 36 $K_2Cr_2O_7$ 4 Calcine at minimum cone 6 and wash out any remaining soluble material. Use with high lead and calcium, low boron glaze.
S112	Pink	Earthenware body stain	Max. cone 7 [oxid. atm.]	Chrome–tin pinks.
A160	Pink	Porcelain body stain	Max. cone 12 oxidising throughout	

	SnO_2	Marble or chalk	Borax	Quartz	Cr_2O_3 or $K_2Cr_2O_7$
(a)	73.5%	19.4%	—	5.9%	1.2–1.4% or 2.4–2.8
(b)	50.5%	24.5%	4.0%	18.0%	3.0–5.0%
(c)	45–50%	29–27%	4.5–0%	20.4–20.0%	1.1–3.0%

Wet grind and add potassium dichromate in solution. Evaporate to dryness, crush, calcine cone 5–10, oxidising, grind, wash. Use 5 to 12% in body.

Other chrome pinks				
K65a	Pink	Under-glaze	All temps. if oxid.	$Cr_2O_3 + Al_2O_3$.
Y1	Violet-pink	[Glaze stain] Porcelain body stain	[Oxid. atm.] Cone 14 with reduction	$CrPO_4 + Al(OH)_3$ or $AlPO_4$, when $Al_2O_3/Cr_2O_3 < 6$. Calcine at 1300–1400° C.
S112	Pink	Porcelain body stain	Cone 12/13	$Cr_2O_3 + Al_2O_3$.
A160	Pink			Alumina 70–80% or 105–120 parts hydrated alumina. Boric acid (cryst.) 15–10% $K_2Cr_2O_7$ 15–10% Mix wet, dry, crush, calcine cone 5–10, oxidising, grind, wash. Use 5–12% in body.

3*d.* *Chromium oxide in blacks, see No. 15*

4*a.* *Cobalt blues*

S88	Blue	General		Blues are achieved by additions of cobalt oxides or carbonate. There are two main colourants formed—cobalt silicate and cobalt aluminate. $CoO.Al_2O_3$. Cobalt oxide or silicate oxidises on heating and at higher temperatures gives up this extra oxygen which then gives rise to bubbling or spitting. In the presence of alumina this reversible reaction does not occur.
M62	Cobalt blue	[Under-glaze]	[Cone 12–14 1350–1410° C]	Analysis of Chinese mineral used for blue stain: SiO_2 14.90 Al_2O_3 29.64 Fe_2O_3 4.24 NiO 2.40 Co_3O_4 9.75 CuO 1.23 MnO 24.91 CaO 0.66 K_2O 0.11 Na_2O 1.02 HgO 0.55 It is questioned whether a high degree of chemical purity of cobalt oxide is necessary, or whether the large content of manganese does not in fact enhance the colour. The mineral is calcined in the ware kiln before grinding and use.
—	Thenards blue	Under-glaze porcelain	Cone 14	Cobalt oxide 37.27 Aluminium hydroxide 62.73 } Wet grind 300 hr, dry, calcine cone 14, regrind 300 hr.
—	Blue	Under-glaze porcelain	Cone 13 reducing	Cobalt phosphate 24.5 Zinc carbonate 39.0 Alumina 36.5 } Wet grind 300 hr, dry, fire in glost kiln as high as possible (cone 14) regrind 300 hr.

TABLE 160—(contd.)

4a. Cobalt blues—(contd.)

Ref.	Colour	Application	Temperature	Composition and preparation							
					Dark blue	Dark blue	Blue	Light blue	Dark blue-green	Blue-green	Light blue-green
S88	Blues	Earthenware under-glaze		CoO	—	44·6	—	—	41·8	—	—
				$Co_3(PO_4)_2$	64·0	—	32·5	9·0	—	—	—
				$CoCrO_4$	—	—	—	—	—	28·2	7·4
				Cr_2O_3	—	—	—	—	19·4	—	—
				Al_2O_3	36·0	55·4	48·4	53·9	39·0	51·6	54·8
				ZnO	—	—	19·1	37·1	—	20·2	37·8
				Calcine at cone	7	14	7	7	14	14	14
						reducing	reducing	reducing	reducing	reducing	reducing

S213	Purple	[Under-glaze]	Cone 2–3	Na_2O 0·2 MgO 0·6 $\}$ B_2O_3 1·4 SiO_2 1·0 CoO 0·2

M59	Matt-blue	Under-glaze	Cone 01	Example: Cobalt oxide 20 $\}$ Mix, calcine cone 10, wet grind, sieve through 200's lawn, ZnO 20 $\}$ dry. Al_2O_3 60 This colour is turned to ultramarine by lead glazes but remains more stable under a leadless one. There is always a tendency to convert to the ultramarine cobalt silicate, therefore minimum temperature and time should be used.

—	Blue	On-glaze	[Cone 014]	Cobalt oxide 9·4% Zinc oxide 15·0 Red lead 49·1 Melt at cone 1, grind. Boric acid (cryst.) 6·5 SiO_2 20·0

S88				CoO 0·24 ZnO 0·35 $\}$ B_2O_3 0·1 SiO_2 0·65 PbO 0·41 Cobalt on-glaze colourants should be used with alkali-free fluxes.

Code	Color	Type	Temperature	Composition / Notes
P3	Blue	Glass enamel	550° C	Sand 9.3% Boric acid 18.5 Litharge 65.0 Cobalt oxide 5.4 ZnO 1.8
P3	Light blue	Glass enamel	550° C	Thenard's blue. Mix soluble cobalt and aluminium salts in proportions to give $CoO:Al_2O_3::1:3\cdot55$. Ppt. hydroxides with NH_4OH, not used in excess. Filter, dry, calcine 1250–1450° C. Flux 79·92% Thenard's blue 23·08
K8	U.S. Standard Blue No. 40 Light greenish-blue, opaque	Vitreous china sanitary glaze stain	Cone 7	Blue stain BE7 Cobalt oxide 3·0 } Grind wet, dry, calcine cone 10, in Zinc oxide 12·5 } closed saggars, grind wet to pass English china clay 50·0 } 200-mesh, wash. Flint 34·5 Glaze: 0·10 K_2O 0·10 Na_2O } 0·20 B_2O_3 } 2·50 SiO_2 0·30 CaO } 0·25 Al_2O_3 0·40 PbO 0·10 ZnO Frit No. 3: $2Na_2O.2CaO.Al_2O_3.4B_2O_3.10SiO_2$ 44·9% Borax 11·7 Whiting 15·2 Georgia china clay (pulverised) Flint 28·2 100·0 Glaze batch: Frit No. 3 17·8% Potash feldspar 16·2 English ball clay 7·4 Flint 20·5 Whiting 5·8 White lead 30·0 Zinc oxide (calc.) 2·3 100·0 Plus: Borax 2·0 Tin oxide 8·0 BE2 stain 8·0 Copper carbonate 0·75
S112	Blue	Body stain	min. 600° C	Cobalt silicate. This will develop in the body by using cobalt oxide or carbonate.
S112	Blue	Body stain	min. around 1200° C	Cobalt aluminate is more stable than the silicate but will only develop in the body at higher temperatures. It may also be added in the precalcined form, when less trouble from uneven colouring will result.

TABLE 160—*(contd.)*

Ref.	Colour	Application	Temperature	Composition and preparation
4a.	*Cobalt blues*—(contd.)			
C59	White	Body stain	All temps.	Slightly yellow bodies may be made to appear white by adding about 0·01% cobalt in the form of its carbonate. Specking easily occurs and the raw material should first be ground at 10–20% strength with clay and silica, brought to a fixed specific gravity and then added to the body in the blunger or ball mill.
4b.	*Cobalt oxide is used in blacks, see No. 15c*			
5a.	*Copper greens, turquoise and blues*			
S126			Oxid. atm.	The cupric ion is very sensitive to its environment and a variety of colours can be obtained according to the glaze composition. The intensity of colour is greatly affected by aluminium, zinc, lead, magnesium and zirconium. PbO 1·0. Al_2O_3 0·2. SiO_2 2·0, $+ 1–2\%$ CuO.
	Emerald green	Glaze stain	900–1050° C	In general green is obtained in high PbO and B_2O_3 glazes.
	Turquoise	Glaze stain	Oxid. atm. 900–1050° C	$\begin{array}{l} Li_2O \ 0{\cdot}1 \\ Na_2O \ 0{\cdot}3 \\ K_2O \ 0{\cdot}3 \\ CaO \ 0{\cdot}3 \end{array} \left\{ \begin{array}{l} B_2O_3 \ 0{\cdot}3 \\ Al_2O_3 \ 0{\cdot}1 \end{array} \right\} SiO_2 \ 1{\cdot}8, +1–2\% \ CuO$
	Blue	Glaze stain	Oxid. atm. 800–1000°	In general turquoise is obtained in the presence of K_2O and B_2O_3. Generally with CuO in presence of alkalies particularly Na_2O and absence or minimum of B_2O_3 and Al_2O_3.
S88	Smaragd green	On-glaze	Oxid. atm.	$\left. \begin{array}{ll} CuO & 2{\cdot}4 \\ \text{Antimonic acid} & 24{\cdot}4 \\ \text{Flux No. 1 (Table 159)} & 73{\cdot}2 \end{array} \right\}$ Wet grind but do not frit together. Very susceptible to overheating.
C34	Greens and blues	Body stains	Oxid. atm.	Copper oxide is rarely used as a body stain as it is easily leached out by the glaze. It can only be used where this does not matter.
5b.	*Copper reds*			
S126	Red, flame, purple, grey	Glaze stain	Max. cone 8 reducing conditions	Chinese red, sang de boeuf, rouge flambé, are art colours not easily mass produced. The effect is of irregularly varying red streaked with purple or grey and with blue or green spots where the glaze is thin, it is due to colloidal metallic copper. It is produced in glazes maturing below cone 8 by the addition of 0·1 to 0·5% CuO in the presence of iron and/or tin oxides, the latter must not exceed 1%. The kiln atmosphere must be reducing during the period after all the deposited carbon has burnt out until the glaze is sealed.

K92	Red	Glaze stain on unfired stoneware or glost porcelain	Cone 6–7 1200–1230° C	Example: $\left.\begin{array}{l}0{\cdot}3\ K_2O\\0{\cdot}7\ CaO\end{array}\right\}0{\cdot}4\ Al_2O_3{\cdot}3{\cdot}5\ SiO_2 + 0{\cdot}1\text{–}0{\cdot}2\%\ CuO$. Fire normally until 800–900° C, reduce until 1000° C, mature in neutral atmosphere and cool normally.
		In-glaze		Glazes: $\begin{array}{l}0{\cdot}2\text{–}0{\cdot}7\ KNaO\\0\ \ \ \text{–}0{\cdot}7\ PbO\\0\ \ \ \text{–}0{\cdot}3\ CaO\\0\ \ \ \text{–}0{\cdot}3\ MgO\\0\ \ \ \text{–}0{\cdot}2\ BaO\\0\ \ \ \text{–}0{\cdot}3\ ZnO\end{array}\left.\begin{array}{l}0{\cdot}1\text{–}0{\cdot}35\ Al_2O_3\\0{\cdot}2\text{–}1{\cdot}0\ B_2O_3\end{array}\right\}1{\cdot}0\text{–}4{\cdot}0\ SiO_2.$ Colourant $CuO + SnO_2$. Reduction obtained by: (1) reducing firing, (2) addition of a reducing agent, (3) reduction during cooling, (4) subsequent reducing firing, and (5) pot reduction, *i.e.* the fired ware is packed in clay pots filled with reducing agents.

Gold
(a) *Under-glaze, in-glaze and on-glaze: pinks, purples, browns, etc., see* 'Colloidal Gold'
(b) *Metallic gold, see* 'Liquid Gold' *and* 'Burnish Gold'

6. Iridium			
S88	Black	On-glaze	Max. temp. as little as possible above temp. of preparation	1 part iridium sesquioxide. 3 parts flux No. 13 (Table 159).
7. Iron-yellows to reds			
S88	Red	Earthenware under-glaze		$\left.\begin{array}{l}Fe_2O_3\ \ 50{\cdot}0\\Al_2O_3\ \ 50{\cdot}0\end{array}\right\}$ Calcine at 950° C.
M47	Orange-red to deep violet-red	On-glaze		Ferruginous earths have been used as raw materials for rich browns and reds, *e.g.* ochres, siennas, umbers, Rhodian red, Indian red, Armenian bole, Lemnian earth; but the uniformity of the source is always doubtful. *Undissolved* Fe_2O_3 dissolved in a glaze colours it yellow. *Undissolved* Fe_2O_3 retains the colour imparted in its preparation, *e.g.* $FeSO_4$ heat very gradually to 580° C and maintain at 600° C for 3 hr →orange red. Heat to 700° C→predominantly red. Heat to 800° C→some blue appears. Heat to 900° C→purple. Using $FeCO_3$ less bright colours are produced. Using ferric alums higher temperatures are required to give the same colour. Calcination of ferrous oxalate and tartrate gives darker colours. The difference in colour is due to changes of particle size, finest yellow, larger red.

TABLE 160—(contd.)

Ref.	Colour	Application	Temperature	Composition and preparation
7. *Iron-yellows to reds*—(contd.)				
M47	Iron reds	On-glaze		Low temperature iron colours can withstand higher temperatures when suitably fluxed. Fluxes for iron reds:
				Flint: 21·1, 12·5, 14·3, 16·7
				Red lead: 63·2, 37·5, 42·9, 50·1
				Borax: 15·7, —, —, 33·4
				Boric acid: —, 50·0, 42·9, —
				$\begin{Bmatrix}0·13\\0·87\end{Bmatrix}\begin{matrix}Na_2O\\PbO\end{matrix}\begin{Bmatrix}0·26\ B_2O_3\\1·10\ SiO_2\end{Bmatrix}$ $\begin{Bmatrix}0·44\\0·56\end{Bmatrix}\begin{matrix}Na_2O\\PbO\end{matrix}\begin{Bmatrix}0·87\ B_2O_3\\0·71\ SiO_2\end{Bmatrix}$ $PbO\,\begin{Bmatrix}1·84\ B_2O_3\\1·27\ SiO_2\end{Bmatrix}$ $\begin{Bmatrix}0·28\\0·72\end{Bmatrix}\begin{matrix}Na_2O\\PbO\end{matrix}\begin{Bmatrix}0·57\ B_2O_3\\0·91\ SiO_2\end{Bmatrix}$
K65a	Yellowy-red Blood red Violet to black	On-glaze On-glaze On-glaze		Fe_2O_3 chemically precipitated and calcined to $\begin{cases}400-500°\ C.\\600-800°\ C.\\above\ 900°\ C.\end{cases}$
P3	Red	Glass enamel	550° C	Flux: Quartz sand 10, Litharge 70, Boric acid 20 } frit at 1250–1450° C. Enamel: Flux 90%, Fe_2O_3 10%
M47	Red-brown Aventurine	Glaze stain	China glost slow cooling	Flint 44·0, Borax 33·0, Ferric oxide 15·0, Feldspar 1·4, Nitre 3·8, Barium carbonate 2·7. Firing must be high enough for the Fe_2O_3 to dissolve completely in the glaze, followed by slow cooling to allow the Fe_2O_3 to crystallise out in sizeable spangles. [The commercial method is to use two glazes, the lower being the iron one, and the upper being lower melting, generally a lead glaze, with a high solvent action on the iron.]
K8	U.S. Standard brown No. 51	Vitreous china sanitary glaze stain	Cone 7	Brown stain T7 Ferric oxide 6·0, Chromium oxide 4·0, Alumina 40·0, Zinc oxide 50·0 } Mix dry, pass through 40-mesh, calcine to cone 10 in closed saggars, wet grind to 200-mesh, wash.

Code	Colour	Type	Recipe / Composition	Firing
	Light brown, buff, tan, autumn brown or sand. Opaque		Glaze: 0·20 K₂O, 0·10 Na₂O, 0·40 CaO, 0·30 ZnO {0·20 B₂O₃, 0·35 Al₂O₃} 2·80 SiO₂. Frit No. 3: 2Na₂O.2CaO.Al₂O₃.4B₂O₃.10SiO₂. Borax 44·9% Whiting 11·7 Georgia china clay (pulverised) 15·2 Flint 28·2 —— 100·0 Batch: Frit No. 3 20·0% Potash feldspar 36·4 English ball clay 8·4 Flint 17·6 Whiting 9·8 Zinc oxide (calcined) 7·8 —— 100·0 Plus: Borax 2·0% ZrO₂ 8·0 T7 stain 6·0	
K65b	Brownish yellow	Glaze stain	Fe₂O₃ dissolved in glaze.	Max. cone 1 [cooling must be slow enough to allow for reoxidation]
S88	Browns or grey	Body and glaze stains		
A160	Rouge de Thiviers	Porcelain body stain	Brown ferric oxide decomposes above 1200° C to give grey ferrous compounds. The brown ones are reformed on cooling only if this is slow and in oxidising atmosphere. SiO₂ 83·00 Fe₂O₃ 9·87 Al₂O₃ 2·35 MnO 0·05 CaO 0·81 MgO 0·10 KNaO 0·72 Ignition loss 2·80 2–5% of colourant.	Cone 14 oxidising throughout
	Pink Red		5–12% of colourant. Fluxes or glazes containing boric oxide destroy this colour.	

TABLE 160—(contd.)

Ref.	Colour	Application	Temperature	Composition and preparation
7. *Iron-yellow to reds*—(contd.)				
K65b	Yellow	Body stain	Max. cone 10 (1300° C) Max. 500–600° C	Fe_2O_3 as part of clay molecule in presence of calcium. Free Fe_2O_3 in presence of calcium.
Iron greens				
S169	Seladon greens	Coloured glaze on white stoneware	Cone 6 with reduction	Frit: $\left.\begin{array}{l}0.7\ Na_2O\\0.3\ CaO\end{array}\right\} 2.5\ SiO_2$

	Glossy	Half matt	M.F. of half-matt glaze
Frit	46.5	47.4	
China clay	9.3	9.5	$Na_2O\ 0.28$
Calcspar	9.3	16.6	$K_2O\ 0.03$
Zinc oxide	18.6	9.5	$MgO\ 0.07$ $\left.\right\} 0.10\ Al_2O_3\ 1.8\ SiO_2$
Talc	4.6	4.7	$CaO\ 0.41$
Feldspar	9.3	9.5	$ZnO\ 0.21$
Iron oxide	2.3	2.3	$+Fe_2O_3\ 2.36\%$
Tin oxide	—	0.5	$SnO_2\ 0.47\%$

8. *Browns Fe_2O_3–Cr_2O_3 and $Fe_2O_3 + Mn_2O_3$ mixtures*

Ref.	Colour	Application	Temperature
A138	Browns	[General on-glaze and earthenware under-glaze]	[Max. 1100° C] [oxid. atm.]

	Red-brown	Pale red-brown	Orange brown	Yellow brown	Chocolate	Salmon	Chestnut	Yellow brown
Fe_2O_3	34	18	23	28	24	12	24	27
Cr_2O_3	34	24	11	12	20	10	35	26
ZnO	32	30	29	14	53	50	17	12
China clay	—	28	37	46	—	—	—	—
Al_2O_3	—	—	—	—	3	28	24	35

Code	Colour	Type			Dark brown	Reddish brown	Light red-brown	Yellowish brown
S88	Browns	Earthenware under-glaze		Fe_2O_3	22.1	22.8	17.8	13.7
				Cr_2O_3	21.0	21.7	16.9	13.1
				Al_2O_3	56.9	—	11.4	17.6
				ZnO	—	55.5	53.9	55.6

Calcine at cone 14.

Code	Colour	Type			Enamel. French brown	Under-glaze. Dark brown	Under-glaze. Vandyke brown	Enamel. Sepia
B57	Browns	[General, on-glaze and earthenware under-glaze] [Oxid. atm.]		Zinc oxide	47.8	—	9.8	38.4
				Iron chromate	34.8	45.4	58.9	—
				Red lead	17.4	—	—	—
				Manganese dioxide	—	54.6	31.4	46.2
				Cr_2O_3	—	—	—	15.4

Calcine to earthenware glost temperature. Calcine to hard earthenware glost.

		Enamel [on-glaze]			
B57		Yellow lead chromate	10.3		
		Zinc oxide	41.4		
		Iron chromate	20.7		
		Sienna (Fe_2O_3 mineral)	27.6		

Code	Colour	Type		1	2	3	4	5	6	7	8	9	10	11	12
S88	Brown	On-glaze	Iron oxide	39	45	49	51	51	49	45	39	52	53	77	87
			Zinc oxide	33	39	42	44	44	42	39	33	45	46	—	—
			Cobalt oxide	28	16	9	5	—	—	—	—	—	—	—	—
			Manganese oxide	—	—	—	—	5	9	16	28	—	—	—	—
			Nickel oxide	—	—	—	—	—	—	—	—	3	1	23	13

Use with Flux No. 13. 1 part colour to 3 parts flux.

Flux: Quartz sand 10, Litharge 70, Boric acid 20 } frit at 1250–1450° C.

Code	Colour	Type				
P3	Red	Glass enamel [on-glaze]	Fe_2O_3	8.39%		550° C
			$PbCrO_4$	7.96%		
			Flux	83.92%		
			$PbCrO_3$	16.7%	} Melt at less than 1000° C.	
			Flux	83.3%		
	Orange					

TABLE 160—(contd.)

Ref.	Colour	Application	Temperature	Composition and preparation

9a. Naples yellow. Lead antimonate

S88 — Yellows — Earthenware under-glaze

	Yellow	Light yellow	Orange
Sb_2O_3	35·0	26·0	26·2
PbO	51·0	38·0	39·7
Fe_2O_3	—	—	14·3
Al_2O_3	14·0	11·0	—
SnO_2	—	12·0	—
$CaCO_3$	—	13·0	—
KNO_3	—	—	19·8
Calcine at	959° C	950° C	1000° C

S208 — Naples yellow — Under-glaze — Cone 2, Cone 5

Lead pyroantimonate with suitable quantities ZnO and Al_2O_3.
Lead pyroantimonate, zinc oxide, alumina plus ferric oxide.
Lead antimonate 75%
Stannic oxide 25%

K65b — Naples yellow — [On-glaze and earthenware under-glaze] — Max. cone 6

" — Max. cone 5–6

Lead antimonate with Fe_2O_3, SnO_2 and Al_2O_3.

A137 — Dark yellow to orange — " — Max. cone 03/02 1050° C

Yellows — On-glaze earthenware under-glaze best with lead glazes

	Naples yellow			Bright yellow	Yellow	Orange yellow
Potassium antimonate	60	40	40	40	40	30
Red lead	20	—	40	56	40	45
Tin oxide	20	60	20	—	10	—
Antimony oxide	—	40	—	—	8	—
Soda ash	—	—	—	4	—	25
Ferric oxide	(5)	(5)	(5)	—	—	—
Calcine at	cone 09	cone 09	cone 09	cone 09	cone 09	cone 09

Code	Colour	Application	Temperature	Composition / Notes
A14	Naples yellow		Max. 1000° C cone 05	Antimony oxide 27·3%, Lead nitrate 60·6, ZnO 10·6, KNO$_3$ 1·6. Above plus 15% SnO$_2$. Heat in covered crucible in tunnel kiln up to 1080° C total time 30 hr. Grind, wash, dry at 100° C.
S88	Yellow	On-glaze	Max. 1050° C cone 03/02	0·339 ZnO, 0·116 NiO, 0·545 PbO } 0·044 Sb$_2$O$_3$, 0·173 Fe$_2$O$_3$ } 2·057 SiO$_2$. Frit in with flux No. 14 or 22 (Table 159).
K65b	Naples yellow	On-glaze	Max. cone 1	Lead antimonate prepared by grinding and calcining: Tartar emetic 14·3% (Antimony oxide.), Lead nitrate 28·5 (Litharge.), Common salt 57·2. Can be used highly concentrated.
—	Yellows	In-glaze		Naples yellow in-glaze is very susceptible to glaze composition. Better results are obtained with RO as $\{0{\cdot}2\text{–}0{\cdot}1\ \text{NaKO} / 0{\cdot}8\text{–}0{\cdot}9\ \text{PbO}\}$ than as 1·0 PbO. Colour is destroyed by CaO, BaO, SrO. Colour is unaffected by MgO, ZnO, B$_2$O$_3$, Al$_2$O$_3$, SiO$_2$.

9b. Basic lead chromate. Coral red

Code	Colour	Application	Temperature	Composition / Notes
K65a	Coral red with very small black content	On-glaze for art ware only as readily soluble in acids	Up to max. cone 013	Pb(PbO)CrO$_4$. Must be applied on glazes of high lead and other base contents. Short, cool firing. Readily soluble in acids therefore may not be used on crockery. May now be completely replaced by the cadmium–selenium/cadmium yellow range.

10. Manganese pinks, violets and browns

Code	Colour	Application	Temperature	Composition / Notes
—	Pink	General	Cone 14 with some reduction	0·15 MnO.Al$_2$O$_3$–0·2 MnO.Al$_2$O$_3$.
Y1	Pink			
Y1	Brown			Calcine at 1100° C.
W80	Pinks to reds	Porcelain and earthenware body stains, glaze stains, under- and on-glaze	Cone 14 with some reduction	4MnHPO$_4$.3H$_2$O + 6Al(OH)$_3$. Manganese spinel MnO.Al$_2$O$_3$. Manganese pink is prepared from manganous phosphate and alumina in such a way that together with the glaze a high alumina content is reached. It will withstand some reduction during firing but with too much becomes paler until it is colourless. It can only be fired in certain parts of periodic hard porcelain kilns. Manganese pink causes casting slip to separate and must be counteracted. Manganese compounds in alkaline glazes.
A130	Purple, Brownish purple	Glaze stains		Manganese compounds in lead glazes. The mineral pyrolusite containing MnO$_2$ is used as a raw material for manganese colours but it must be well calcined to decompose it to Mn$_3$O$_4$ by losing oxygen, before use on ware.

TABLE 160—(contd.)

Ref.	Colour	Application	Temperature	Composition and Preparation
10.	*Manganese pinks, violets and browns*—(contd.)			
W80	Pink	Porcelain under-glaze	Cone 14 reducing	Al_2O_3 81·16; P_2O_5 9·44; MnO 9·57
W80	Brown	Earthenware under-glaze		Potash alum 52·0, Manganous sulphate 15·2, Sodium carbonate 32·8 } dissolve } mix solutions. Wash precipitate, dry, calcine.
W80	Brownish violet	Earthenware under-glaze		$Mn_3(PO_4)_2$ 70·3%; Stannic oxide 29·7
A207	Pink	Porcelain		Al_2O_3 is mixed with $MnCl_2 + Na_2HPO_4$ or $Mn_3(PO_4)_2$, heated to 1200–1400° C treated with aqua regia. The pigment is then separated, washed and dried.
S112	Pink or brown	Body stains	1100–1700° C cone 1–31	Pink stained bodies will only remain so if they are glazed. An unglazed manganese stained body oxidises to dark brown manganese compounds during the cooling. To obtain pink, alumina must be present, and in earthenware bodies alumina plus phosphoric acid are necessary to prevent brown.
S88	Violet	Body stain		Manganese oxide 2; Quartz 32; Potash 66 } frit together and grind.
A160	Pink	Porcelain body stain	Cone 12/13 oxidising or slightly reducing	17–20% manganous phosphate; 67–61% alumina; 12–14% ammonium phosphate; 4–5% ammonium nitrate } mix and calcine in oxidising atmosphere at cone 2/3. Grind and use 5–12% in the body.
				8–10% manganese carbonate; 68–86% alumina; 12–2% phosphoric acid (50%); 12–2% ammonium nitrate }
	Brownish pink			25% manganous phosphate; 65% alumina; 10% potassium nitrate }
A160	Pinks	Porcelain body or glaze stain	Cone 12/13	77% aluminium hydroxide, 23% hydrated manganese sulphate or 26% manganese nitrate dissolved in water and mixed. Calcined at cone 1. To 100 parts calcined product add aqueous solution of 7–7·5 parts ammonium phosphate. Calcine cone 1. Repeat treatments with 23% aqueous manganese sulphate and then 7–10% ammonium phosphate. Grind.

Code	Color	Application	Cone	Recipe / Notes
A160	Pinks	Porcelain body stain	Cone 12/13	66% Al_2O_3 or 100% hydrated alumina. 14% aluminium phosphate. 20% $MnSO_4$ or $Mn(NO_3)_2$ in aqueous solution. Mix, dry, calcine at cone 10.
A160	Pinks	Porcelain body stain	Cone 12/13	86% Al_2O_3. 14% $MnSO_4$ in aqueous solution. Mix, add a little ammonia to precipitate Mn^{2+}, mix, dry, grind. Use 5–12%.
A160	Pinks	Porcelain body stain	Cone 12/13	1–2 parts $KMnO_4$ dissolved in 10% HCl. 10–15 parts hydrated alumina. Evaporate to dryness excluding air as much as possible. Mix well. Calcine cone 12/13. Grind. Use 3–7%.
A160	Blacks			See Blacks No. 15c.

11. *Metallic applications. See special section for:*
Aluminium, Antimony, Bismuth, Bronze, Cobalt, Copper, Gold, Iron, Indium, Lead, Nickel, Palladium, Platinum, Rhodium, Ruthenium, Silver, Tin

Code	Color	Application	Cone	Recipe / Notes
12. K65a	*Neodymium* Pale violet		Including cone 14	Neodymium phosphate.
13a. K65a	*Uranium red* Tomato red	Glaze stain Under-glaze	Max. cone 05	Sodium uranate ($Na_2U_2O_7$) 10–20%. Lead glaze with little or no boron. Overheating gives progressive (sometimes desirable) colour changes to yellow, green and then black.
D31	Tomato red	Glaze stain	Cone 07	Litharge 52·6% This glaze must be fritted. Quartz 18·4 Feldspar 9·7 Uranium oxide UO_3 19·3
D31	Tomato red	Glaze stain	Cone 07	Frit: Litharge 6·9% Tin oxide 1 Kaolin 2 Quartz 22 Feldspar 6 Mill batch: Frit 72 Kaolin 6·4 Ground quartz 1·4 Zinc oxide 0·8 UO_3 19·4

TABLE 160—(contd.)

Ref.	Colour	Application	Temperature	Composition and Preparation
13a.	*Uranium red*—(contd.)			
B58	Vermilion	Glaze stain	Cone 02	Red lead 57% Feldspar 20 Zinc oxide 2 Flint 12 Sodium uranate 9 The glaze is fritted.
13b.	*Uranium browns and yellows*			
K8	U.S. Standard ivory No. 30 Light yellow-brown, opaque	Vitreous china sanitary glaze stain	Cone 7	Glaze: 0·20 K_2O 0·20 CaO 0·40 PbO 0·20 ZnO $\Big\}$ 0·30 Al_2O_3 3·00 SiO_2. Batch: Potash feldspar 30·0% English ball clay 6·9 Flint 25·8 Whiting 5·3 White lead 27·7 Zinc oxide (calc.) 4·3 —— 100·0 Plus: Borax 2·0% Tin oxide 6·0 Sodium uranate 1·4 ($Na_2O.2UO_3.6H_2O$)
D31	Yellowish brown	Porcelain body stain	Oxid. atm.	Uranium oxide: UO_3.
	Blackish brown	Porcelain body stain	Reducing atm.	Uranium oxide: UO_3.
—	Light to dark brown	Porcelain under-glaze	Cone 14 with reduction	UO_3 30–40% Porcelain body (finely ground) 70–60 $\Big\}$ Wet grind, dry, calcine at cone 13, regrind with 10% kaolin. Uranium browns palen in oxidising fire.
S88	Yellow	On-glaze		1 part uranium oxide. 2–3 parts flux: 0·43 Na_2O 0·57 PbO $\Big\}$ 0·86 B_2O_3.0·72 SiO_2.
H5	Greenish-yellow lustre	On-glaze		Resin soap: Powdered pine resin 62·5 Caustic soda in water 12·5 25·0 $\Big\}$ Stir, cool, break up, wash, dry. Uranium resinate: Resin soap in water 28·6% 57·1 Uranium nitrate 14·3

13c. *Uranium blacks (see also composite blacks 15c)*

—	Black	Porcelain under-glaze	Cone 13 reducing	Black uranium oxide 50% Porcelain body 50 Frit at cone 13.
—	Black	Porcelain under-glaze	Cone 14 reducing	Black uranium oxide 91% Porcelain glaze 9 Grind and calcine at cone 14.

14. *Vanadium yellows, greens and blues*

D22	Yellow	Glaze stain, body stain	Max. cone 12 1350° C (2462° F) oxid. atm.	Vanadium–tin combinations.
	Ice blue	Glaze stain, body stain	Max. cone 12 1350° C (2462° F) oxid. atm.	Vanadium–zirconium combinations. turned greenish black by reducing atmosphere.
	Lime, green and sea green, turquoise	Glaze stain, body stain	Max. cone 12 oxid. atm.	Mixtures of vanadium–tin yellow and vanadium–zirconium blue, turned dirty yellow by reducing atmosphere as the blue is more affected than the yellow. These colourants are made paler by the presence of unfritted titanium, calcium and magnesium compounds (including using high calcium water). Glazes containing these colourants tend to settle and must be suitably treated, special additions may be supplied by the colour manufacturer.
H15	Dark yellow	Body stain Glaze stain Under-glaze		V_2O_3 2–3%. SnO_2 the rest.
H36	Green	Under-glaze Body stains Engobes Glaze stains On-glaze As above		V_2O_5, 3 to 10, most desirably 3 to 5 parts by wt. The ZrO_2 and SiO_2 ZrO_2 40 to 85, most desirably 60 to 70 parts by wt. must not be introduced in combination. SiO_2 50 to 10, most desirable 37 to 25 parts by wt. Calcine at 550° C to 1200° C, preferably 700° C to 900° C for 1 to 8 or more hours, preferably 2 to 6 hr.
	Greenish-blue to blue	As above		As above plus ½ to 5% alkali. As above except when potassium present when minimum temp. is 750° C, but maximum 1300° C.
V14	Yellow with greenish tinge, opaque	Glaze stain		Titanic acid, anhydrous 1·5% Quartz flour 13·5 Stannic oxide 83·5 } Wet grind, calcine at cone 9–10, grind. Vanadic acid, hydrated, techn. 1·5

TABLE 160—(contd.)

Ref.	Colour	Application	Temperature	Composition and Preparation
14.	*Vanadium yellows, greens and blues*—(contd.)			
V14	Yellowish orange	Glaze stain	Cones 6–9	Anhydrous titanic acid 1·5% Techn. vanadic acid hydrate 11·5 Zirconium oxide 87·0 } Wet grind, dry, calcine at cones 8–10.
	Leaf green	Glaze stain	Cones 6–9	Anhydrous titanic acid 1·5% Techn. vanadic acid hydrate 15·0 Zirconium oxide 65·0 Quartz flour 18·5 } Wet grind, dry, calcine at cones 5–7.
	Turkish blue	Glaze stain	Cones 6–9	Barium carbonate 8·5% Techn. vanadic acid hydrate 12·5 Zirconium oxide 48·0 Quartz flour 31·0 } Wet grind, dry, calcine at cones 5–7. The above three colourants were selected from a large test series as the most useful in industrial practice for ware to be glost fired at cones 6 to 9. They are not recommended for low temperatures.
15a.	*Whites (see also opacifiers)*			
P3	White	Glass enamel	550° C	Sand 8·8% Boric acid 17·6 Litharge 62·9 Clay 3·4 Cryolite 4·0 Arsenic 2·7 Blue colourant 0·6 Blue colourant (see cobalt blue, same ref. No.) } Melt together. Flux: Quartz sand 10 Litharge 70 } frit 1250–1450° C Boric acid 20 } and crush
A143	White satin vellum			Glaze plus: SnO$_2$ 4% ZnO 4% TiO$_2$ (pure) 4%
	Cream satin vellum			Glaze plus: SnO$_2$ 4% ZnO 4% Rutile mineral 4%
	'Break-up'			Glaze plus: Rutile about 8%. The titania crystallises irregularly in the glaze and the patches have coloured edges of titanates.
				Batch: Flux (2–3mm) 88·5% Cryolite 4·4 Clay 3·7 Arsenic 2·9 Blue colourant 0·6

15b. *Ivories*

Code	Colour	Type	Firing	Notes
K65b	Ivory	Glaze stain for earthenware		2% of vanadium–tin yellow. Exceptionally unaffected by variations of kiln temperature. Iron oxide is less satisfactory. Cerium oxide gives a weaker but very clear ivory, it must be cooled slowly to allow for reoxidation, otherwise it is colourless.
K65b	Ivory	Glaze stain for hard porcelain	Above cone 8 up to cone 14 with reduction, and *slow* cooling with oxidation	Pyrolusite (MnO_2). The formation of the ivory colour is dependent on the action of oxygen at high temperatures, *i.e.* very slow initial cooling. The output of intermittent kilns is therefore variable, depending on the position of the ware. Manganese coloured ivory has a bad effect on most on-glaze colours particularly iron.
K65b	Ivory	Glaze stain for porcelain	Cone 14 with reduction	Cerium oxide gives a clear ivory colour with the minimum of black. It is somewhat affected by the cooling rate which must be slow, but uniform colouring can be achieved. Good results can be got with all normal on-glaze colours. $CeO_2:TiO_2$ $\quad CeO_2\ 68.2\%$ / $TiO_2\ 31.8\%$ Wet grind, calcine, regrind.
K65b	Ivory	In-glaze		$CeO_2:TiO_2::2:3.5$ fritted and used as 3% of glaze. Colour intensity increases with content of MgO, CaO and SrO, the effect being least with MgO and greatest with SrO, spottiness increases in the same way.
K65b	Deep cream ivory	Glaze stain on-glaze	Above cone 03	Uranium salts in the presence of boron and alumina.
C59	Orange	Body stain		Chrome–alumina pink or manganese–alumina pink: blended with chrome–iron–zinc browns. A zinc glaze brightens these colours.
C59	Blue-green	Body stain		Cobalt–chrome colourants.
C59	Dove-grey and blue-grey	Body stain		Tin–antimony combinations.
C59	Yellow and buff	Body stain		Tin–antimony combinations.
C59	Buffs	Body stain		Rutile, natural titania, with various impurities, *e.g.* 99% TiO_2, 1% Fe_2O_3
S88	Yellow	Earthenware under-glaze		ZnO 50.0 / 99% pure rutile 50.0 Calcine at 1000° C.
K65b	Leather coloured	Glaze stains		Natural rutile which frequently has a Fe_2O_3 content.

TABLE 160—(contd.)

Ref.	Colour	Application	Temperature	Composition and Preparation
15b.	*Ivories*—(contd.)			
K65b	Light yellowy green	On-glaze		Barium chromate.
K65b	Yellowy green	Body stain	Cone 14	Porcelain body plus 5% praesodymium oxide and 5% praesodymium phosphate.
K65b	Brownish yellow	Body stain	Cone 6	Heat together oxidised compounds of antimony, titanium and chromium in the presence of sodium. Use 4% of stain in body.
15c.	*Blacks (mixtures of oxides of complementary colours)*			
—	Blacks or in thin layers grey	Porcelain under-glaze	Cone 14	Alumina 25 25 Wet grind 300 hr, calcine cone 14, Zinc oxide 25 25 regrind 300 hr. Cobalt oxide 7 6 3 Chromium oxide 3 3 3 Uranium oxide 40 41 44
S88	Blacks	Earthenware under-glaze		Calcine at cone 14 reducing.
A138	Black	[Earthenware] under-glaze		Iron oxide 41·1 36·3 — Chrome oxide 32·4 6·8 90 Black cobalt oxide Co$_3$O$_4$ 20·6 31·8 10 Manganese oxide 5·9 12·2 — Nickel oxide — 12·9 — If coloured edges due to preferential solution of one component are to be avoided hard calcination and a suitable glaze are important. Calcine to earthenware biscuit temperature.

For the "—" row (Blacks or in thin layers grey):

	Greenish black	Brownish black	Bluish black
Fe$_2$O$_3$	34·4	51·2	45·4
Cr$_2$O$_3$	65·5	48·8	43·2
CoO	—	—	11·4

15c. *Blacks (Mixtures of oxides of complementary colours)—(contd.)*

	Black		
P3	Glass enamel 550° C	Flux: Quartz sand 10⎫ Litharge 70⎬ frit at 1250–1450° C Boric acid 20⎭	Colourant: Flux 77·5 85·00 Cr_2O_3 3·9 2·59 Co_3O_4 18·6 12·41
K8	Vitreous china sanitary glaze stain	Glaze: 0·20 K_2O ⎫ 0·20 CaO ⎬ 0·30 Al_2O_3 3·00 SiO_2. 0·40 PbO ⎪ 0·20 ZnO ⎭ Batch: Potash feldspar 30·0% Georgia china clay 6·9 Flint 25·8 Whiting 5·3 White lead 27·7 Zinc oxide (calcined) 4·3 —— 100·0	plus: Borax 6·6% Cobalt oxide 4·0 Ferric oxide 2·0 Cupric oxide 1·0

U.S. Standard black No. 60 bluish black high gloss   Cone 7

Underglaze Colour Solutions

A more unusual method of applying under-glaze colours to biscuited bodies, is to use aqueous solutions of soluble salts of the colouring oxide, either the chloride or the nitrate. The solution is brought to a useful consistency for spraying or brushing without running by adding solutions of sugar, syrup or glycerin. Drying is assisted by adding alcohol. Viehweger (**V16**) gives the following examples of: (1) metal salt solutions, (2) auxiliary solutions and (3) colourant mixtures, although the proportion and type of auxiliary in the final mixture has to be adjusted to the nature of the biscuit body.

TABLE 161

Metal Salt Solutions

(1) *Chrome*	
chromium nitrate	75%
water	25
(2) *Cobalt*	
cobalt nitrate	75
water	25
(3) *Copper*	
copper nitrate	75
water	25
(4) *Gold*	
(*a*) gold chloride	50
water	50
(*b*) gold chloride	30
water	70
(5) *Iron*	
Ferric nitrate	75
water	25
(6) *Manganese*	
manganese nitrate	86
water	14
(7) *Nickel*	
nickel nitrate	75
water	25
(8) *Platinum*	
platinum chloride	95
water	5
(9) *Uranium*	
uranium nitrate	75
water	25

GLAZES

TABLE 162
Auxiliary Solutions

No. 1	syrup	65%
	water	35
No. 2	spirit	25
	glycerin	75
No. 3	*Standard type*	
	syrup	50
	spirit	25
	water	25

The solutions may be coloured with an aniline dye to make them more readily visible during application.

TABLE 163
Colourant Mixtures (V16)

No.	Colour	Metal salt solution		%	Auxiliary solution	%
1	Dark blue	No. 2	Cobalt	71	No. 3	29
2	Dark yellow-brown	No. 7	Nickel	71	No. 3	29
3	Buff	No. 5	Iron	71	No. 3	29
4	Light stone grey	No. 6	Manganese	71	No. 3	29
5	Pink	No. 4b	Gold	71	No. 3	29
6	Chocolate	No. 4a	Gold	57	No. 3	43
7	Light blue-grey	No. 6	Manganese	57	No. 3	29
		No. 2	Cobalt	14		
8	Dark blue-grey	No. 6	Manganese	47	No. 3	29
		No. 2	Cobalt	24		
9	Light yellowy brown	No. 7	Nickel	24	No. 3	29
		No. 5	Iron	47		
10	Light brown	No. 1	Chromium	47	No. 3	29
		No. 7	Nickel	24		
11	Dark brown	No. 1	Chromium	71	No. 3	29
12	Light grey-green	No. 2	Cobalt	18	No. 3	29
		No. 1	Chromium	53		
13	Dark grey-green	No. 1	Chromium	57	No. 3	29
		No. 2	Cobalt	14		
14	Light blue-green	No. 1	Chromium	20	No. 3	60
		No. 2	Cobalt	20		
15	Dark blue-green	No. 1	Chromium	35·5	No. 3	29
		No. 2	Cobalt	35·5		
16	Violet-red	No. 4a	Gold	57	No. 3	29
		No. 2	Cobalt	14		
17	Sap green	No. 3	Copper	71	No. 3	29
18	Platinum grey	No. 8	Platinum	71	No. 3	29
19	Yellow-orange	No. 9	Uranium	71	No. 3	29

The Effect of Glaze Composition on Ceramic Colourants

It is sometimes not enough to make ceramic colourants as inert as possible and some adjustments of the glaze and/or flux they are used with may be

TABLE 164

Causes of Fading of Under-glaze Colours and Recommendations for Improvements (D51)

Direct cause of fading	Indirect cause of fading	Recommendations for improvement
Melt formation at low temperature or excessive amount of initial melt causing solution of colourants.	Excess frit content. Use of low-melting frit. Raw lead compounds (white lead) used. Chemical composition too high in PbO, B_2O_3 and/or KNaO.	Decrease amount of frit. Substitute a higher melting frit. Frit all lead. Replace all or part of PbO with SrO; reduce B_2O_3 and KNaO.
Chemical alteration by glaze constituents.	Alteration by ZnO, MgO, SnO_2. Alteration by Al_2O_3.	Exclude ZnO, MgO, SnO_2. Reduce Al_2O_3 to minimum consistent with acceptable glaze properties.
Low viscosity of maturing glaze.	Improper chemical composition of glaze. Fast glost fire.	Increase SiO_2; reduce B_2O_3; increase K_2O; reduce Na_2O; increase CaO; increase SrO; reduce PbO. Adopt slower rate of firing glost ware.
Glost firing schedule incorrect.	Extended soaking period at elevated temperature; excessive solution of colourant. Long firing schedule necessitated by high surface tension of glaze. Volatilisation of colourant.	Reduce soaking time. Decrease Al_2O_3; substitute K_2O for Na_2O. Adjust rate of firing.
Deficiency of colourant-supporting oxide. Reaction with kiln atmosphere.	Low free whiting in glaze batch. Reducing atmosphere	Include 10% or more whiting in glaze batch. Provide oxidising kiln atmosphere.

necessary as well. For instance, cadmium yellow and cadmium–selenium red cannot be used in conjunction with lead glazes or fluxes. Many colourants are unfavourably influenced by tin, zinc and magnesium oxides in the glaze.

Under-glaze colours may suffer from a number of circumstances peculiar to the method of applying a thin layer of colourant under the glaze. Donahey and Russell (**D51**) have investigated this and summarise their results in tabular form (Table 164).

DECORATION WITH GOLD AND OTHER NOBLE METALS

Noble metals can be used for decoration either in a colloidal, highly coloured form, or to give a metallic surface. Metallic gold is applied in two ways, liquid or brilliant gold which makes a very thin layer but is immediately shiny, and burnish gold which must be burnished with agate after firing but has greater depth and lasting qualities. Other metals are usually applied in the 'brilliant' form. Metallic surfacing of ceramics has many applications apart from decoration, *e.g.* in printed electrical circuits for V.H.F. amplifiers, detectors, oscillators, trigger circuits, hearing aids, electrical instruments, radio sets, radar, sub-miniature radio transmitters; and other electrical contacts, etc., in condensers, resistors, cathode-ray tubes, high-frequency condensers, inductance coils, insulators, soldering adjoinments, etc. (**A191**).

Colloidal Gold

Several very beautiful colours can be achieved by precipitating gold in various ways. Different under-glaze, in-glaze and on-glaze colourants are known which are all coloured by the colloidal gold they contain. The actual colour seen is probably dependent on the particle size (*see* p. 234).

UNDER-GLAZE COLOURS WITH COLLOIDAL GOLD

Brown (suitable for porcelain fired in reducing atmosphere cone 13):

Zettlitz kaolin	250·0 g;
Silver carbonate	12·5 g;
Basic bismuth nitrate	2·5 g;
Glycerin (chemically pure)	150·0 ml;
Gold chloride	100·0 g;
Sodium hydroxide	200·0 g.

All the constituents except the gold chloride are mixed in water, and the gold chloride solution placed in a dropping funnel with the stem leading below the surface of the liquid. It is then slowly run in. The mixture is warmed, filtered, washed and calcined at cone 13. After calcination the stain is wet ground with 10% plastic clay.

Pink (suitable for porcelain fired in reducing atmosphere cone 13):

Porcelain slip (S.G. 1·70)	660 g;
Gold chloride	20 g;
Glycerin	35 g;
Sodium hydroxide	42 g.

The glycerin and alkali are added to the porcelain slip and gold chloride solution run in slowly from a dropping funnel whose stem reaches below the surface of the liquid in the main mixing vessel. The mixture is then warmed, filtered, washed, dried and calcined at cone 13. The product is wet ground with 20% of hard porcelain, cone 13, glaze mixture. (This can also be used as a porcelain body stain (**A160**).)

Purple-red:

Kaolin	90 parts;
Auric chloride solution, containing	10 parts gold;
Sodium carbonate to make alkaline;	
Glucose	20 parts.

Boil kaolin in water and pass through a fine sieve. Add auric chloride solution slowly, with constant stirring, and then make alkaline with sodium carbonate. A glucose solution is then added and the whole boiled for 30 min, adding fresh water to replace that evaporated. Filter, wash and dry (**H5**).

Rose-pink. As for purple-red using 2 parts gold to 98 parts kaolin (**H5**).

Purple of Cassius. Although usually described for on-glaze decoration (*see* below), may also be used for under-glaze work, and as a body stain. For pink porcelain the precipitated purple of Cassius is used wet and added to the body in such a way that it contains 0·02–0·10% gold (**A160**).

COLLOIDAL GOLD AS A GLAZE STAIN FOR ROSE AND RED COLOURS

Mix gold chloride solution with the slop glaze, preferably leadless, make it alkaline with soda ash and then add a reducing agent such as glucose solution.

Rose-coloured glaze requires	0·01% gold;
Dark-red coloured glaze requires	0·1% gold (**A142**).

PURPLE OF CASSIUS. COLLOIDAL GOLD

When gold chloride solution is reduced by stannous chloride the gold produced remains in the colloidal form, being stabilised by the stannic oxide which is the other product of the reaction. The actual colour exhibited by this sol depends on the particle size of the gold and is altered by other constituents in the reaction mixture and the conditions used. The largest particles are yellow and with decreasing size turn to purple and crimson. The purple and crimson will not form in the absence of a little stannic chloride in the stannous chloride. The proportion of stannic to stannous chloride has a considerable effect on the colour.

GLAZES

As with other colloidal sols exact reproduction is difficult and depends on very careful repetition of the conditions as well as quantities used in the preparation. The 'recipes' therefore tend to appear complicated. A few are given below.

RECIPES FOR PURPLE OF CASSIUS (A142 AFTER B57)

Purple Base

      1 dwt.* of fine gold dissolved in 10 dwt aqua regia;
      6 dwt of tin dissolved in 30 dwt aqua regia.

Filter each through glass wool. Dilute gold solution with three pints of cold water, add the tin solution and stir for 10 min. Add 5 pints hot water, stir, and then allow to settle out. Filter off the precipitate and wash with warm water.

Maroon Base. As above using ratio of gold to tin 1:4.
Rose Base. As above using ratio of gold to tin 1:5.
Light Purple Base. As above using ratio of gold to tin 1:10.

These bases are then mixed with a small amount of silver carbonate to enhance the purple cast, although excess must be avoided because the silver tarnishes, together with a suitable flux as tabulated below (Table 165).

TABLE 165
Purple of Cassius Stains (B57) for On-glaze Decoration

Colour	Purple of Cassius base	Ag_2CO_3	Flux quantity and type	Flux composition	
Maroon	Maroon base containing 1 dwt gold	16 grains	1 lb maroon	Maroon flux Red lead Flint glass Boric acid	6 parts 2½ parts 8 parts
Purple	Purple base (1 dwt gold)	1 dwt	4 oz purple	Purple flux Red lead Flint Borax	2 parts 2¾ parts 8 parts
Ruby	Maroon base (1 dwt gold)	6 grains	4 oz ruby	Ruby flux Red lead Flint Borax	2 parts 5¼ parts 9 parts
Rose	Rose base (1 dwt gold)	12 grains	1 lb rose	Rose flux Red lead Flint Borax Stone Nitre	6 parts 5½ parts 6¼ parts 2 parts 1 part
Pink	Rose (1 dwt gold)		½ lb purple ½ lb rose	see above	

* 1 dwt = one pennyweight = 24 grains troy. 437.5 grains = 1 oz. (Av.).

The mixtures are fritted to easy glost temperature (1000–1050°C, 1832–1922° F).

Some manufacturers vary the colour by the addition of different quantities of metallic silver to a standard purple of Cassius base (**A142**).

These stains can be used up to 850°–1000° C, depending on the flux used **K65a**).

Metallic Gold

PREPARATION OF PURE GOLD FROM THE COMMERCIAL PRODUCT, AS THE STARTING POINT FOR MAKING LIQUID GOLD AND BURNISH GOLD

Commercial gold contains traces of copper and silver which if present in a metallic gold decoration eventually oxidise and impair the colour. It is therefore dissolved in aqua regia, diluted, filtered and evaporated to form crystals of gold chloride. These are purified by recrystallisation and then dissolved in water. Gold is precipitated from this solution by the addition of ferrous sulphate solution. It is then filtered off, washed and dried (**A65**). The precipitate is usually dark brown and is hence known as 'brown gold', it may also appear violet or green according to its particle size.

LIQUID GOLD OR BRIGHT GOLD

This is the cheaper form of applying gold on-glaze decoration and is a dark coloured liquid containing about 11·5% of gold in the form of a soluble compound in an oily medium. It can then be brushed, printed, etc., on to the ware. On heating the oil, etc., evaporates and/or burns away by about 300° C, the gold compound decomposes and a thin shiny layer of gold is left on the ware; this is heated to about 750° C to fix it on to the glaze (**C23**). If the glaze surface to which liquid gold is applied happens to be rough the gold will appear matt after the firing. This can be used deliberately to produce both matt and shiny gold decoration simultaneously. Special lithographs containing roughening materials can be applied to the glaze before the liquid gold in order to give a matt pattern in a bright background (**A65**).

The preparation of this colourant is usually done by specialist firms but some of the ingredients are given below.

Liquid gold is useful for lining and stamping but as it is very thin any defect will show as white spots if it is used for bands or larger areas. Antimony colours tend to darken it (**A142**).

The quantities of liquid gold required by a skilled hand painter for decoration in the form of a 3 mm (0·118 in.) wide border and trimmed handles have been given variously as follows:

6 × 26 piece dinner services for 6, 10 g (0·35274 oz.)
18 × 9 piece coffee sets for 6, 10 g (0·35274 oz.)
one 26 piece dinner service for 6, 5 g (0·176 oz.)
one 9 piece coffee service for 6, 3 g (0·106 oz.) (**A155**)

GLAZES

A great deal therefore depends on the actual conditions of working.

The basic ingredients of liquid gold are:

(1) Gold usually in the form of the chloride.
(2) Rhodium, about 1% of the weight of gold (or some other metals, *e.g.* iridium, aluminium, thorium, tin) in order to prevent the gold crystallising, when it would appear matt.
(3) A metallic flux to fix the gold to the glaze surface, this is usually bismuth nitrate but may be a salt of one or more of the following: lead, antimony, tin, uranium, chromium, cadmium and bismuth.
(4) Sulphur balsam, also known as sulphuretted balsam, sulphonated wood resins. *See below* for preparation.

Any or none of the following ingredients may also be included in recipes:

Mercury to amalgamate the gold before treatment.
Resin.
Lavender oil.
Rosemary oil.
Nitrobenzene.
Fenchel oil.
Spike oil.
Camphor oil.
Sassafras oil.
Clove oil.
Syrian asphalt.
Toluene.
Colophony (**A142, 111, C23, 24**).

Sulphur balsam is made by boiling flowers of sulphur with turpentine until no sulphur is precipitated on cooling. The type of turpentine used varies in different recipes. In some a mixture of French and Venetian turpentine is specified (**A142**). Venetian turpentine is described as being prepared by slow evaporation of turpentine oil in open dishes without the application of heat. It may be used alone for making the sulphur balsam *e.g.* 100 parts Venetian turpentine are heated with 27·1 parts flowers of sulphur to about 180° C (maximum 190° C) for 2–3 hr. On cooling it is dissolved in ether and must give a clear solution (**A118**).

A recent (1956) (**A159**) detailed description of the commonest methods of making brilliant gold gives the following recipes.

METHOD BASED ON THAT FIRST PATENTED BY THE DUTERTRÉ BROTHERS

Solution 1. Aqua regia based on 128 g nitric acid. Dissolve in this, in turn, 32 g gold, 0·12 g metallic tin, 0·12 g antimony chloride, $SbCl_3$. When these are completely dissolved dilute with 500 ml distilled water.

Solution 2. 16 g sulphur are dissolved by warming in 16 g Venetian turpentine and 80 g turpentine oil. After solution is complete, 50 g lavender oil are added.

Solutions 1 and 2 are now mixed and stirred over a water bath until all the gold has passed into the oily layer. The acid aqueous layer is then decanted off and the oily one washed several times with warm distilled water. It is next dried. Further additions are made: 5 g basic bismuth nitrate, 100 g ordinary thick turpentine oil, and the whole is then diluted with 85 g lavender oil. This mixture is now ready for application which is assisted by warming it.

CARRÉ'S METHOD FOR BRILLIANT GOLD

Solution 1. 100 g aqua regia (HNO_3:HCl::1:3) are used to dissolve 10 g fine gold; this is diluted with 150 ml distilled water. This solution is shaken with 100 g rectified ether until all the gold has been removed from the aqueous layer which is then poured off.

Solution 2. 20 g potassium sulphide dissolved in 1000 ml distilled water is decomposed by 200 g nitric acid. Sulphur is precipitated and this is washed and dried. The dry sulphur is dissolved in a mixture of 25 g turpentine oil, 5 g nut oil, and diluted with 25 g lavender oil.

Solutions 1 and 2 are now mixed and concentrated on a water bath until a syrupy consistency is reached when 1·5 g bismuth oxide and 1·5 g lead borate are added. This mixture is diluted for use as required with a mixture of equal parts of lavender and turpentine oils (**A159**).

Meyer (**M71**) also describes the preparation of the different materials required for the final mixture he gives.

Gold Resinate:

(1) Dissolve gold in aqua regia in presence of KCl to obtain $KAuCl_4$; dissolve this in methyl alcohol to give 0·25 g gold/1 ml.
(2) Dissolve flowers of sulphur in turpentine oil to give sulphur balsam.
(3) Add the $KAuCl_4$ solution to the sulphur balsam, heat on water bath to form precipitate of gold resinate.
(4) Dissolve gold resinate in chloroform to give 45% solution.

Rhodium Resinate:

(1) Aqueous solution of Na_3RhCl_6 + $Ba(OH)_2$ gives $Rh_2(OH)_6$ precipitate.
(2) Dissolve precipitate in HCl and methyl alcohol.
(3) Add solution to sulphur balsam.

Bismuth Solution. Boil bismuth oxide with sulphur balsam.
Chromium Solution. Add sulphur balsam to an aqueous solution of CrO_3.
Asphalt Solution. Dissolve Syrian asphalt in nitrobenzene and toluene.
Colophony Solution. Dissolve colophony in fenchel oil.

The 12% liquid gold prepared from the above is given in Table 166.

GLAZES

TABLE 166

	%
Gold resinate, 45% Au	26·7
Spike oil	26·0
Asphalt solution	14·0
Fenchel oil	11·1
Bismuth solution diluted with spike oil to 6% Bi	7·0
Clove oil	1·5
Rhodium solution diluted with spike oil to 3·5% Rh	1·4
Chromium solution diluted with spike oil to 8% Cr_2O_3	0·6
Mixture of 1 part asphalt solution to 2 parts colophony solution	11·7

A number of similar recipes appear in the literature. They all differ in detail (**A131, A118, H5**).

BURNISHED GOLD

For quality products a thicker, more uniform, and more lasting application of gold is made with mixtures incorporating powdered or amalgamated gold. Suitable fluxes are added and the decoration fired on. It then appears dull or matt and has to be polished or burnished with agates. Silver may be added to the gold to brighten it.

'Powdered gold' or 'brown gold' is made by dissolving the metal in aqua regia and evaporating to dryness to drive off excess acid. The gold chloride is then dissolved in water and ferrous sulphate or mercurous nitrate solution added, when powdered gold precipitates. This is thoroughly washed and then dried at a temperature not exceeding 100° C.

Mercury either in the metallic form or as the easily decomposed oxides or halides is often added in quantities of up to half the weight of the gold to increase the covering power of the gold. It vaporises away during the firing. 5–10% flux are used, often bismuth nitrate with a little borax or high borate flux.

Up to 25% silver chloride or nitrate may be added to brighten the gold colour which is otherwise reddish. These ingredients are ground together for up to two weeks. The whole is then mixed with a suitable oil for application by decalomania, brushing, stamping, etc.

Gold dust may sometimes be applied directly to ware by dusting on to an oil printed on the ware in the required pattern (**A142, A131, A65**).

The actual mixture used for applying burnished gold varies slightly according to the type of pattern, method of application, etc., even within one works. The four examples shown in Table 167 are different mixtures used for slightly different purposes in one porcelain factory. Binns (**B57**) also gives examples (Table 168).

TABLE 167
Burnished Gold

	I	II	III	IV
Precipitated gold powder	100	100	120	120
Black mercuric oxide	22·5	30	15	15
Red mercuric oxide	37·5	50	15	25
Silver nitrate	—	12·5	15	15
Bismuth nitrate	—	2·5	15	5

TABLE 168
Burnished Gold (Binns, B57)

	Medium burnish	Printing	Bronze	Chasing
Brown gold (precipitated gold)	20	50	2·5	20
Mercury	10	50	1·0	14
Silver chloride	1·5	—	—	1·5
Bismuth	1·0	—	—	0·5
No. 8 flux	—	7	—	1·0
Copper oxide	—	—	2	—
Gold flux	—	—	0·25	—

THE RECOVERY OF GOLD FROM SCRAP AND UTENSILS

The gold decoration from pieces that have proved unsaleable after decoration and from brushes, sponges, etc., can be recovered wholly or partially in the works. Discarded decorated ware is broken up to pieces about the size of a sixpence and those with gold on placed in a stoneware vessel with a well-fitting lid and a stopcock at the bottom. Brushes, etc., are burnt and the mineral ash treated with or similarly to the crockery. Aqua regia is then poured over the scrap, the lid put on and left for about a day. After this the acid is drawn off and may be re-used until saturated with gold chloride. The scrap is rinsed with successive small quantities of distilled water until a particular rinsing water contains no further gold chloride. These solutions may then be sent away for the further extraction.

Alternatively powdered gold may be precipitated with ferrous salts from the solutions obtained, or they are evaporated to dryness and the residue which may contain lead chloride, silver chloride, silica, iron oxides, etc., fused with soda ash, potash and saltpetre in a refractory crucible when metallic gold is obtained (**P66**).

Platinum Grey

A very fine grey for under-glaze use on very expensive ware can be derived from platinum. The metal is dissolved in aqua regia and the solution evaporated to dryness. An equal weight of ammonium chloride is dissolved in water and added to the platinum chloride to form the double salt ammonium chloroplatinate which separates on evaporating to dryness. It is then dissolved in water and mixed with ground pitchers. On heating the chloroplatinate decomposes, leaving finely divided platinum intimately mixed with the pitchers (**A65**). This can be painted, sprayed, etc., as under-glaze decoration.

Bright Platinum

Since silver tarnishes in air it is not normally suitable for metallic on-glaze decoration. So-called 'silver' bands, etc., are therefore achieved with platinum to which gold is added to reduce the cost. The liquid or bright platinum is prepared similarly to liquid gold (**A157, C22**). Platinic chloride and a small percentage of rhodium chloride are each dissolved in sulphur resin, bismuth and chromium resinates are added and the whole dispersed in a mixture of benzene, turpentine, nitro benzene, lavender oil, rosemary oil, and alcohol. This final solution contains 7% platinum. Four parts of it are mixed with one of liquid gold and after application it is fired to 800° C (**A16**).

Metallic Silver

Although unsuitable for decoration because it tarnishes, silver is coming into extensive use for printed electrical circuits. Silver is a good conductor not only in the metallic form but also when it has tarnished to oxide or sulphide forms. Other, cheaper, good metallic conductors such as copper and aluminium oxidise and corrode to non-conducting products and so cannot be used in the thin layers required in printed circuit techniques.

Correctly matured on-glaze silver circuits can be used as a base for soldering on other parts, the tensile bond strength between the silver and its ceramic base being 3000 lb/in^2 (211 kg/cm^2).

The silver may be applied in several ways as shown in Table 169. The methods of preparing silver resinate are similar to those for gold resinate or liquid gold and can be varied considerably, one example only is given.

TABLE 169

On-glaze Silver Inks and Paints

(after **A66**)

Form of silver	Preparation	Solvent	Remarks
Colloidal metallic silver	Suspend in binder	Butyl acetate, amyl acetate, toluol, turpentine, pine oil, etc.	Air drying Fire at about 700° C
Silver chloride	Mix with turpentine	—	For glass and porcelain
Silver chloride	Mix with fluxes and essential oils to paste.	Oils, etc.	—
Silver oxide	Mix with gum and glacial acetic acid.	Water	—

INDUSTRIAL CERAMICS

TABLE 169—(contd.)

Former of silver	Preparation	Solvent	Remarks
Silver carbonate	Mix with flux and essential oil to paste	Oils	Long firing time
Metallic silver	Mix with molten wax	—	Heat 200–300° C, then 400–600° C
Silver oxide	Mix with gum dammar and linseed oil, to paste	Linseed oil, glycerol, castor oil	For rubber stamp printing
Silver resinate (A16)	8 parts silver carbonate dissolved in 5 parts resin at 126–135° C. Powder and dissolve. Prepare sulphur resin from 7¾ parts sulphur and 35 parts Venetian turpentine at 170–175° C, and make 60% solution in turpentine. Add sulphur resin to silver resinate together with manganese naphthenate and bismuth resinate	Mixture of nitrocellulose, methylated spirit, benzyl alcohol, nitrobenzene, xylol, methylcyclohexanol monoethyl glycol ether, butyl acetate and citronella oil	Fire on mica 560–580° C Fire on glass 550–600° C Fire on porcelain 700–730° C

The use of low-alkali flux in minimum quantities is important if connections are to be soldered to the silver layer. Tin–lead solder is often reluctant to adhere to a fired silver coating unless this is first copper-plated but a tin–silver (95:5) solder is recommended for direct use (**A66**). Fluxes used may be of a glassy nature (0·5–1·0% alkali, **P38**) alone or together with compounds of barium, bismuth, strontium and titanium.

TABLE 170

Summary of Non-decorative Metallic Applications to Ceramic Surfaces
(after **A66**)

Metal	Form in paint or ink	Application
Aluminium	Powder or flake plus binder Deposited film—powder	Protective coating On glass for infra-red heating On glass for electrical resistors
Antimony	Sprayed metal (in reducing atmosphere)	On porcelain insulators
Bismuth	Metallic paint—fired	Light filters (glass)
Bronze	Sintering powder—bronze 10%, glass 50%, water 35%, filler 2%	Base metal cannot be reduced to metallic state by firing
Cobalt	Cobalt powder paint—fired	After tinning with solder, wires, etc., can be attached by soldering

TABLE 170—(contd.)

Metal	Form in paint or ink	Application
Copper	Oxide paste—fired Oxide paste—fired and then zinc plated	Coating porcelain insulators Soldering sleeves on porcelain
Gold	See 'Liquid Gold', p. 650	Printed circuits exposed to chemical vapours that attack silver
Indium	Indium–gold alloy	Glass-to-metal seals (inserts)
Lead	Lead powder with varnish and thinner—fired in hydrogen atm.	
Nickel	Nickel flakes in binder—fired	
Palladium	As resinate in oils, etc.—fired	Alternative to platinum for high-duty printed circuits
Platinum	Chloride, or resinate, see 'Bright Platinum', p. 655—fired	High-duty printed circuits
Rhodium Ruthenium	Similar to platinum	Printed circuits (experimental)
Silver	See special Table 169	Standard for printed circuits
Tin	Boiling tin chloride	Forms conducting film
Iron and/or nickel with copper (**J45**)	Metallic powders and ceramic flux—fired 1000–1400° C, reducing atmosphere	Readily wetted by silver solders, e.g., silver–copper eutectic May be furnace brazed to metal components at 800° C in atmosphere of hydrogen or cracked ammonia, without use of solders
Molybdenum (**A16**)		Better than iron for hard soldering

LUSTRES

Glazes of various colours can be made lustrous and iridescent by special treatment. This effect is due to the presence of very thin layers of metal or metal oxide particles and may be achieved by five different methods:

(1) On-glaze application of resinates, etc., of noble metals.
(2) Reduction in a third firing of lead glazes containing colouring oxides.
(3) Special firing technique with certain glaze composition.
(4) Spraying of hot glost ware with solutions of metal salts.
(5) Application of ochres and metals to tin glazes.

(1) On-glaze application of dilute oily or ethereal solutions of organic compounds of the noble metals, copper, nickel, cobalt, cadmium, uranium, iron, silicon, titanium, phosphorus, arsenic, antimony chloride, bismuth

nitrate. These are prepared similarly to the bright noble metal solutions, *e.g.* colourless bismuth lustre.

Method 1. 26·1% colophony are melted on a sand bath and 8·7% crystalline basic bismuth nitrate stirred in, 34·8% lavender oil are gradually added when the mixture begins to brown. The mixture is allowed to cool. 30·4% lavender oil are then added and any undissolved portions allowed to settle out. The residue is left to thicken in the air.

Method 2. 4·6% basic bismuth nitrate and 30·6% colophony are ground together and heated on a sand bath until almost completely dissolved. 64% lavender oil are then added (**A115**).

Coloured lustres of this type may be based on 'brilliant noble metal' preparations: *e.g.* gold lustre may be 1 part brilliant gold, 3 parts bismuth lustre.

Very thin applications of brilliant gold appear pink, and silver is yellow. If a fired piece of brilliant silver is covered with brilliant gold and refired, a moiréd or marbled effect with reddish edges is produced. If bright silver is applied over blue ware very glossy green lustres result. Alternatively various oxides are made into resinates and dissolved in ether or oils and mixed with a portion of noble metal resinate. Basic bismuth nitrate is added in most cases as a flux. Colours obtained are:

Iron oxide in absence of bismuth, red.
Iron oxide in presence of bismuth, golden.
Uranium, greenish-yellow with high gloss.
Mixture of uranium and bismuth, yellow mother-of-pearl effect.
Copper, reddish-brown.
Nickel, light brown.
Cadmium, reddish yellow.
Cobalt, chocolate to black (**S89**).

Marble and crackle effects can be produced in connection with these on-glaze lustres by applying a further solution over the dry but not fired lustre preparation (**D21**).

Viehweger (**V15**) gives some recipes of lustres of this type for cones 021–019.

(2) Lead glazes containing colouring oxides are given a third firing up to 800° C in a strongly reducing atmosphere.

Particularly suitable are copper, cobalt and manganese glazes as well as those containing silver, iron, uranium, bismuth and molybdenum. Alternatively the lustre is achieved in one firing by cooling the glost kiln until the glaze has just solidified and then introducing a strongly reducing atmosphere.

(3) Iridescent particles may be developed within the glaze by firing to above cone 7 in neutral atmosphere followed by rapid cooling. As well as the usual coloured oxides titanium, molybdenum, tungsten or vanadium oxides may be added.

Franchet (**F25**) describes a number of such iridescent lustres based on the following frit:

Quartz	12
Pegmatite	10·5
Kaolin	2
Sand	20
Litharge	30
Borax	19·2
Boric acid	2
Pearl ash	2
Salt	1·8
	99·5

The frit is melted at 970° C (1778° F) and used in the glazes set out in Table 171. The fired glaze is cooled to dull red heat before reducing.

TABLE 171

Iridescent Lustre Glazes (Franchet F25)

	1a	2a	3a	4a	5a	6a
Frit	100	100	100	100	100	100
Kaolin	10	10	10	10	10	10
Zinc oxide	—	1	—	—	—	1
Tin oxide	—	1	—	—	—	4
Silver carbonate	2	—	½	2	2	—
Copper oxide	—	—	3	—	—	—
Bismuth sub nitrate	—	—	—	4	—	—
Copper carbonate	—	—	—	1	—	—
Copper sulphide	—	—	—	2	0·3	½
Silver sulphide	—	—	—	—	2	—
	increasing reduction time ↓	whitish gold metallic brown black metallic black non-metallic	coppery with bright iridescences	blue with 1a green	bright multiple iridescences	

(4) The fourth method involves spraying the ware emerging hot from the glost kiln with solutions of metal chlorides or nitrates. The solvent evaporates and the salt is then decomposed leaving a very thin deposit of the metal which is thereafter fixed by refiring in a decorating muffle. The glazes most suitable to this treatment are lead ones with a small zinc content. Some of the materials used and colours obtainable are:

Copper oxide: red, violet and blue.
Iron oxide: blue, black and violet.
Titania (rutile): blue, cloudy white and violet.
Molybdenum and vanadium compounds: blueish iridescence.

Best results are obtained if silver oxide is added to the above.

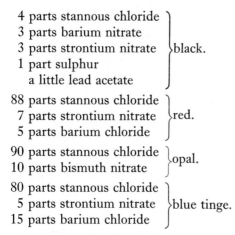

4 parts stannous chloride
3 parts barium nitrate
3 parts strontium nitrate } black.
1 part sulphur
a little lead acetate

88 parts stannous chloride
7 parts strontium nitrate } red.
5 parts barium chloride

90 parts stannous chloride
10 parts bismuth nitrate } opal.

80 parts stannous chloride
5 parts strontium nitrate } blue tinge.
15 parts barium chloride

(5) Tin glazes with alkali and copper or combinations of metal oxides may be glost fired in oxidising conditions and then decorated with a mixture of ochre or other earths with copper, silver and bismuth suspended in vinegar and refired in reducing conditions (**S89**). Franchet (**F25**) gives some examples (Table 172). The vinegar formerly used to mix the paste is now often replaced by ammonium oxalate solution, this breaks up during the heating, giving off carbon monoxide which helps in the reduction.

TABLE 172
Mixtures applicable to tin glazes to obtain lustres (F25)

	1	2	3	4	5	6	7	8
Copper carbonate	30	28	—	—	—	95	—	—
Silver carbonate	—	2	3	1	—	5	—	—
Bismuth sub nitrate	—	—	12	10	—	—	—	17
Copper oxalate	—	—	—	5	—	—	—	—
Copper sulphide	—	—	—	—	20	—	—	2
Tin oxide	—	—	—	—	25	—	—	—
Silver sulphide	—	—	—	—	—	—	5	1
Red ochre	70	70	85	84	55	—	95	80
	gold iridescence		greenish blue	bluish green	gold iridescence	whitish yellow	yellow, gold and brown	bluish green

Chapter 7
The Mechanical Preparation of Ceramic Bodies and Glazes

The processes that are undertaken in ceramic works are very varied. The preliminary ones depend on the condition of the raw materials. It has been mentioned that some raw materials are crushed, ground, washed, purified, and dried before sale from the mining firm. Others are, however, delivered in a coarse condition and are dealt with either by a manufacturer specialising in purification and grinding or by the ceramic works themselves.

The non-plastic raw materials require crushing or disintegrating followed by dry or wet grinding to various degrees of fineness. The individual materials need different machines for this according to their size, hardness and type of cleavage.

Plastic raw materials are treated in either the dry, the plastic or the wet state. Dry preparation may involve drying, crushing, grinding and air separation similar to that used for some raw materials (p. 251). Plastic preparation may also include crushing or disintegration followed by further kneading and mixing. Wet preparation is only used for high quality bodies. The clay is brought into aqueous suspension of a liquid consistency, when impurities can be removed most efficiently by settling, sieving and magneting. It can be ground wet. The other body constituents previously brought to the right sizes are added and the well-mixed body filter-pressed. It is then dried, or worked up into a plastic mass, or made into a casting slip according to the shaping method to be used.

It is proposed in this section to consider the various processes which might be applied in the preparation of raw materials and of ceramic bodies. The actual combination of methods necessary for the production of any one type of ware will be considered subsequently.

CRUSHING AND GRINDING

Many raw materials need to have the size of their lumps, aggregates, grains, particles, etc., reduced before they can be used in ceramic manufacture. The various crushing and grinding processes aim to do this by mechanical as opposed to chemical means.

Several terms are used in this connection, the difference between them being one of application and scope rather than one of principle. In general 'crushing' refers to reduction of large lumps to a convenient size for secondary reduction. 'Pulverising' is usually employed if the product is a fine powder. 'Grinding' is often used in a general sense but in other instances implies producing a fine powder.

The basic principles of mechanical reduction processes are the three following:

(1) A hammer blow.
(2) The crunch of compression.
(3) Tearing or shearing action.

These may or may not be combined.

The 'blow' is obtained by striking the material with a moving hammer or falling ball, or throwing it against a breaking plate. The 'crunch' is achieved in jaw crushers or with rotating drums which force the material through a confined space. Shearing action occurs when a tooth or claw is dragged through the stationary material (**A96**).

When choosing the type and size of crushing and grinding equipment the following points must be considered (**A27**):

(1) Hardness and toughness of raw material.
(2) Size of lumps to be fed in.
(3) Moisture content of material.
(4) Desired size of product.
(5) Quantity of product required.
(6) Possible impurities and whether they should be rejected or crushed.

Another point to consider in connection with crushing and grinding equipment is whether it is intended for batches, or for continuous operation. In the latter case grinding may be undertaken in open circuit or in closed circuit. The older open-circuit method involves using a feed rate that is slow enough for all particles to be reduced below the permitted maximum size. In many machines the fines produced early on have a cushioning effect, and so prolong the time and power consumed to reduce the last particles. If such grinding machines are connected to a classifier which removes those particles that are sufficiently fine, and returns to the grinder those that are not, much power can be saved and greater feed rates used. Grinding in closed circuit can be done with both wet and dry grinding, hot air being supplied to damp material. Ball, pebble, tube, rod and hammer mills can be connected in closed circuits.

Primary Crushers

Miller and Sarjant (**M80**) have contributed a comprehensive description of the evolution of various types of crushers for hard materials. To these must be added a number of machines more especially suited to clays (**A27**).

(1) Double and single toggle jaw crushers will reduce any hard but free-breaking rocks, etc., so long as they do not pack or stick. The single toggle machines can also be used to give a finer granular product (Fig. 7.1).

(2) The gyratory crushers of long and short shaft types and the cone crusher are used for hard materials such as silica, flints and hard fireclay.

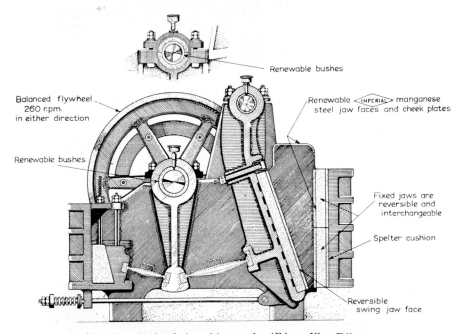

FIG. 7.1. *Sectional view of jaw crusher (Edgar Allen, **E4**)*

The gyratory crushers consist primarily of four parts: lower shell, upper shell, spider crushing head and hopper. The lower shell, which has a removable bottom supports the machine, the upper shell is securely bolted on to it and often has the spider cast on to it in one piece. The hopper encircles the upper end of the top shell. The conical crushing head is pivoted from the spider and is made to gyrate in a circular or rolling movement by an eccentric, for instance, of the ball and socket type. A bevel gear wheel is keyed to this eccentric, and motion is imparted to it via a countershaft and pinion. The crushing action is therefore intermittent like the jaw crusher but a greater throughput can be obtained although more fines are often produced.

(3) Crushing rolls either gear or belt driven, smooth, fluted, grooved or toothed, in pairs or single, perform a large number of duties in the reduction of materials.

(a)

(b)

Fig. 7.2. Crushing rolls: (a) heavy duty toothed rolls; (b) stone separator roll (Bradley and Craven, **B105**)

Geared rolls take the heaviest work, they may be run at the same or at different speeds. Toothed rolls at different speeds tear apart tenacious clays. For softer clays containing stones a cylindrical or conical roller with a helical groove or thread allows the clay to be broken down and the stones to be moved along the roller and discarded at one end. Smooth or fluted gear-driven rolls, sometimes in sets of two pairs above each other, can produce fine material. Single roll breakers will handle sticky materials of varying hardness effectively (Fig. 7.2).

MECHANICAL PREPARATION

Shuttle crushing rolls, in which the rollers move to and fro along each other while turning are popular in Europe for crushing harder lumps and stones in softer clays.

(4) Hammer mills or hammer pulverisers have increased in popularity a great deal. They can deal with hard or soft, dry or moist material and effectively reduce fairly large lumps to dust in one operation (Fig. 7.3). The illustration shows the method of working of the most usual hammer mills. The striking media may be fixed beaters or loose or pivoted hammers of various shapes. The feed may enter under the descending hammers, fall on the ascending hammers, or fall in a stream where the descending hammers hit it during the fall. In all the cases the incoming material is struck by the hammers and thrown against the side. In one of these impacts the lumps generally explode into a multitude of small fragments. These receive further strokes from the hammers before falling out at the bottom. A grid to retain oversize pieces may or may not be incorporated in the base of the mill. As the lumps are broken into fragments at a single stroke the size and grading of the product cannot be so accurately controlled as in roller crushers.

A mill, 'Original Pallman', for very fine grinding of minerals of medium hardness such as clay, gypsum, bentonite, feldspar, kaolin, chalk, talc and colours also operates on the hammer mill principle (**A162**).

CLAY SHAVING MACHINE

Certain tough, plastic clays that are mined in a more or less wet condition and are to be worked in a plastic or slip state respond better to a clay shaver than to crushing machines. The mined clay is dumped into a hopper and is then acted on by carbide tipped shaving tools that will produce shavings as fine as 0·04 in. (1 mm) according to the setting. The shavings drop on to a collecting plate and are discharged by a scraper. The machine can therefore be used as a continuous feeder of subsequent machines. Stones are rejected (**S149**) (Fig. 7.4).

Pan Grinding

The most widespread secondary crushing and fine grinding methods in the ceramic industry are those using a 'runner(s)' in a 'pan'. In their different types and sizes they are used for wet, moist, or dry clays, calcined flint, feldspars, frits, colours, etc. Some of these machines are also used for tempering and mixing.

The essential features of pan mills are the circular pan holding the material and the moving runner(s) which do the grinding. The pan may be solid and stationary, or the base may revolve and/or be perforated. In the simplest instances the runner is cylindrical and only slightly smaller than the pan, it is rotated about a central axis. In other similar cases the runner may be elongated, its longest diameter being just less than that of the pan.

Fig. 7.4. *Clay shaver or cutter (Edgar Allen, E4)*

Fig. 7.3. *Cross-section of a hammer mill pulveriser showing the principle of operation (Edgar Allen, E4)*

A type in general use, especially for flint in slop form has the runners fitted to the arms of a suitable revolving framework.

There is always an optimum speed for running such grinding pans, up to which efficiency increases with speed, but after which it drops. Its actual value depends on the hardness and consistency of the material to be ground and the type of pan and relation of grinding surface to its capacity and cannot be derived theoretically. Mellor (**M54**) has, however, worked out the relation between the speeds of machines of the same type but different sizes, grinding the same material, whose pans have runners extending from the edge to the pivot. He shows that the best speed of revolution varies inversely as the diameter of the pan:

$$Nd = N'd'$$

where N and N' are the speeds in revolutions per minute, and d and d' are the diameters of the pans. So if the optimum speed N of grinding of a particular material in a given pan is known, the optimum speed N' for another pan of the same type can be deduced.

The materials used for the grinding surfaces vary considerably according to the nature of the matter to be ground or mixed; steel, granite and chert are examples and stoneware has also been applied successfully.

The wet-pan grinding mill (for the grinding of flint) was first introduced some 150 years ago. Its design has changed very little since. The pan and runners are chert or hornstone. The pan is some 14 ft (4·3 m) diameter and 3 ft (0·9 m) high and the bottom is lined with chert blocks about 1 ft (0·3 m) thick. The chert runners, weighing about 16 cwt (813 kg), are fixed to four arms at right angles revolving at about 15 r.p.m. The calcined flint is ground with water to produce a slop. This is often done by specialist milling firms who deliver the ready-made slop to the ceramic manufacturers. The chert pan could be superseded by other grinding methods which have smaller initial outlay and use less power, but the industry often specifies that the flint slop should be ground in the chert pan, instead of laying down the properties (purity and grain size, etc.) of the slop (**A100**).

For the working of dry, moist or plastic clays the most widespread machines are the edge-runner mills. There are numerous variations of this type of machine and they are often ambiguously termed 'dry pans', 'wet pans' and 'grinding mills'. They consist essentially of a 'pan' and two (also very occasionally one) 'runners' which are rotating wheels. The axis of the wheels is horizontal and joins the central vertical shaft running through the pan like the arms of the simpler 'pan' just described. The runners either revolve about a fixed centre while the pan turns under them, or they 'roll' round the pan. The pan may or may not be perforated. The positive driving power is usually applied to the pan if it turns, or otherwise it is connected directly to the runners. Scrapers push the material under the runners.

The illustration (Fig. 7.5) shows the general appearance. A few of the different combinations that can make up edge-runner mills will show their applications.

Marls, shales and similar fairly dry and friable materials can be ground in mills with perforated bottoms. The rollers turn about fixed positions while the pan revolves. The portion of the pan coming under the rollers is of solid hard-wearing material and is called the 'dead plate'; the outer part contains grids or sieves. Centrifugal force and scrapers take the crushed material from the rollers to the grids and other scrapers return the oversize

FIG. 7.5. *Edge runner mill, 'dry pan', grinding fletton brick clay. This passes through the perforated base on to a bucket elevator which takes it to screens (Fig. 7.14) the oversize being returned (London Brick, L41)*

lumps to the rollers. The finer material passing through the sieves either falls down a shute on to an elevator in the open base machines, or on to a dish from which scrapers on the underside of the pan push it round to an opening. The latter type requires less depth, and can be more easily completely cleaned for use with different material. The normal open-based mill has to be driven from above, the one with a dish and underscraper can be driven from underneath also. The under-driven mill is particularly advantageous when grinding abrasive materials for refractories as the wearing parts can be very quickly renewed.

For working plastic clays it is usually better for the pan to be stationary and the rollers to move round it. Some have partially perforated bases, so that the clay gets a certain amount of working, tempering, mixing, etc., before reaching a grid, others have solid bottoms and are used for batches, they have scoops, shovels, or sliding doors for emptying.

A specially designed edge-runner grinding mill for plastic clays has the rollers set at different distances from the shaft so that almost all the pan bottom is covered, thereby reducing the work of the scrapers. The rollers themselves are each carried on crank-pins so that they have independent action. If one is lifted somewhat by an uneven thickness of clay, as may easily happen to the leading roller on the feeder side, it does not affect the other (**G55**). The pans and runners or chasers are made of cast iron, steel, granite or stoneware.

Until the last decade grinding pans were usually driven by steam engines because the starting torque is very heavy. Electrical motors are now being introduced and often save power, space and maintenance costs. Another advantage of electrical power is that each pan has its own motor and can be run quite independently. For instance seven pans formerly driven by a 250 h.p. tandem compound steam engine with massive shafts, are now driven by two 60 h.p. and five 40 h.p. individual motors (**A97**).

Ring Roll Mills

These mills consist of a heavy steel anvil ring with a concave face secured in a head and supported and rotated. Inside it three rolls with convex faces are uniformly pressed against it. The material to be crushed is introduced at the inner face of the ring. Centrifugal force holds it on the ring surface and it is then brought under the rolls, turning them by friction. The crushing action forces the finer material over the edge of the ring, when it is conveyed to a screen or air classifier and the oversize returned. These mills exert pressures of up to 40 000 lb. and can effectively crush very hard material (**S215**).

Materials used for the wearing parts of crushing and grinding machinery

The various forms of primary crushers, etc., are usually of steel (p. 163). Manganese steel (14%) is mainly used for the faces actually in contact with the stone, etc. This alloy has the property of work-hardening more than any other. It is, however, difficult to machine, and also somewhat ductile.

Chromium steel is often used for large forged roll-shells as it is comparatively easily machined. In fact all roll shells that have to be machined periodically to maintain their true alignment should be made of chromium or high-carbon steels. They should be heated up when being fixed into position to forestall the effects of the temperature increases that occur under severe working conditions. Chromium steels are greatly improved by heat treatment.

Crusher-shafts that are liable to fatigue under frequent reversals of stress are best made from heat-treated steels, nickel steel being used in some of the recent high-speed crushers. The frames and other parts not submitted to direct abrasion are made from a good grade of carbon steel, suitably annealed to resist shock.

It is most important to know not only the hardness and toughness but also the abrasiveness of the material to be crushed when choosing the type of crushing surface (**M80**).

Rollers for preparing brick clays are often made of chilled iron. They last some 15 to 18 months when grinding 120 tons of crushed shale and being reground weekly.

Secondary Grinding

GRINDING IN BALL, PEBBLE, ROD AND TUBE MILLS

An increasingly large proportion of the fine grinding of ceramic materials is carried out in rotating mills containing hard spheres or rods. The mills are rotated at a speed at which the balls, etc., are carried up the side and then roll over each other to the bottom. The grinding is therefore effected by impact and rubbing.

There are a number of designs of these mills with different relative dimensions, types of lining and of the balls and pebbles. They are used either as batch grinders, or continuously in closed or open circuits.

TUBE MILLS

In its simplest form the tube mill consists of a long steel cylinder lined, if necessary, with quartz, silex, or porcelain, supported on hollow trunnions and charged with flint pebbles. The wet or dry feed enters through one trunnion and the ground product is discharged through the other. The diameter varies from 3 to 8 ft ($\sim$ 90 to 210 cm) and the length from 12 to 30 ft ($\sim$ 3.5 to 9 m) but is usually 22 to 26 ft ($\sim$ 6.5–8.0 m). The output can be varied from 1 cwt ($\sim$ 50 kg) to 20 tons per hr. The tube mill finds its greatest application where a very finely ground product is needed with little regard for the power consumed. This is because 85% of the grinding is achieved in the first 5 ft (1.5 m) of the mill, the remaining 13 to 15 ft (3.9–4.5 m) being used for the other 15%, the previously produced fines having a strong cushioning effect. However, by regulating the feed rate the size of the output can be considerably adjusted and up to 98% through 200-mesh can be achieved in open circuit. Where less fines are required shorter tube mills in closed circuit with a classifier would seem to be more economical.

The tube mill is nevertheless more economical than the 'chert pan' for grinding flint and other free-grinding materials, being initially cheaper and less bulky to install and once started requires less power. It is being used increasingly for this purpose (**A100**) (Fig. 7.6).

Ball tube mills differ by having metal liners and steel balls. Compartment ball tube mills are divided by one or more steel grates with metal lifters, and each compartment has different-sized balls so that the particle size reduction occurs in stages (**M69**).

CYLINDRICAL BALL AND PEBBLE MILLS

The essential difference between 'tube' and 'cylindrical' mills is that in the latter the length is the same as the diameter. The balls and linings may be steel, flint, porcelain, alumina or quartzite. Rubber linings are also highly recommended. They are used as either batch or trunnion overflow mills, in open or closed circuits. They lend themselves very well to this last (**M69**). The variety of their applications is illustrated by the large size range. For laboratory use and for grinding colours and enamels porcelain pot or jar mills are used. Their diameter (and length) is from about 6 to 12 in. (15–30 cm), and they are rotated in motor- or belt-driven frames holding one to twelve jars (Fig. 7.7).

There are a number of variants on the simple jar mill. In one springs and weights make the axis take an elliptical path so that each point on the circumference makes a small circle, while the whole thing vibrates at high frequency. The balls then not only cascade but also jump across the jar. Very hard material is therefore quickly reduced by the violent impacts (**A152**).

Another, the planetary ball mill, is set with its axis upright. This executes a circular path, while a guide rod prevents the jar itself rotating. The balls inside are compelled to follow a definite course and grind by impact more than by friction. This mill is designed for rapid very fine grinding of small quantities, particularly in laboratories (**H77**).

At the other end of the range are the vast steel grinding cylinders of several feet diameter and length, taking a charge measured in hundredweights, *e.g.* a 7 ft by 7 ft (2 m by 2 m) ball mill will grind 30 cwt (1524 kg) of pre-crushed calcined flint in about eleven hours to pass 160-mesh. These are lined with porcelain, silex, rubber or other suitable materials and charged with flint pebbles. They are used for batch grinding of calcined flint, Cornish stone, feldspar, fritted glazes, bone and colours (Fig. 7.8) (**B94**). Before use the particle size distribution of each finished batch should be determined, *e.g.* by hydrometer, elutriation, etc.

CONICAL MILLS

Conical mills have a steep 60° cone at the feed end, then a cylindrical part and then a 30° cone at the discharge end. The diameter is $1\frac{1}{4}$ to 3 times the length of the straight section. This shape may approach that of a sphere with short chords giving the cones and cylinder. It therefore has almost the minimum internal surface area for a given volume, which reduces the amount of wear on the liners.

Fig. 7.6. *Compound ball-tube mill for fine reduction (Johnson, J54)*

Fig. 7.7. *Motor-driven unit of twelve porcelain pot mills (Steel-Shaw, S173)*

Various sizes of ball or pebble are used. During grinding the centrifugal force classifies the balls and the material being ground, so that energy is not unnecessarily expended. The finer material is gradually brought up the shallow cone to the discharge end. Conical mills can therefore be used for open-circuit grinding.

It is more efficient, however, in instances where no oversize particles are permissible to grind dry material in a closed circuit in conjunction with a classifier. (Damp material can be dried during the process if the air used is suitably heated.) Material is circulated on the air-swept principle. Air is blown into the discharge end of the mills where it picks up material small

Fig. 7.8. *Large grinding cylinder* (*Boulton*, **B94**)

enough to be airborne, reverses its direction and conveys it out to a classifier or collector. After sizing on (1) sieves for down to 35-mesh, or (2) in a loop classifier for down to 90% through 200-mesh, or (3) a conical superfine classifier for down to 99·5% through 325-mesh, the oversize material is returned to the mill with the feed. The conical mill is able to take large circulating loads (**H20**) (Fig. 7.9).

ROD MILLS

The simplest type of rod mill is cylindrical, with the length about twice the diameter. It is supported on trunnions. The grinding medium consists of high-carbon steel rods the same length as the cylindrical section of the mill and $1\frac{1}{2}$ to 4 in. (3·8–10·2 cm) in diameter. Line rather than point contact occurs and this produces a granular product without excessive fines, *e.g.* $\frac{1}{2}$ to 1 in. feed is reduced to 60-mesh grade. Rod mills may have trunnion overflow for wet grinding and either trunnion or peripheral outlets for dry material. They are better used with a smaller charge than a ball mill usually takes (**M69**). Sizes made have outputs of from 7 to 10 cwt/hr (355 to 508 kg/hr) to 10 to 12 tons/hr (**J54**).

Rod mills also make good mixers.

SELECTION OF GRINDING MILLS

The particle shape, grading, and some chemical properties of ground materials depend on the type of grinding, batch grinding producing the largest quantity of very fine material for a given maximum particle size.

FIG. 7.9. Conical ball mills (Hardinge, H20)

Open-circuit grinding in tube mills also gives a large quantity of fines. Closed-circuit grinding can give a product of fairly uniform size. Wet and dry grinding, respectively, may give particles of different shapes, and wet grinding may in addition alter the chemical nature of the product, e.g. feldspar, where some of the alkali dissolves, or flint which eventually combines with water to form a silica gel.

In changing from one grinding system to another it is important to bear in mind the particle shapes and the size grading, as mixtures of the same maximum size but different grading give totally different results when used in the ceramic body. The traditional grinding methods do not necessarily give the best size mixtures, but any change must be experimented with before use in the main production. Ceramic manufacturers frequently specify that their flint must be ground in the chert wet pan. Such grinding can be done much more economically in tube, or pebble mills. By research into the optimum particle size grading it may be found that a better product can so be achieved.

Metz (**M69**) makes the following suggestions for the choice of grinding mills for ceramic materials.

(1) *Tube Mills*

(*a*) For wet or dry grinding from feed sizes below $\frac{1}{8}$ in. to produce a finished fine-ground product in one passage through mill.

(*b*) Where an excess of superfine powder is needed or where several materials are ground together requiring intimate mixing and a finished product in one passage through the mill.

(*c*) For wet or dry grinding in closed circuit with wet or air classifier with low circulating load to produce very fine-ground products with maximum superfines; moisture in feed for dry grinding should not be more than 1%.

(2) *Cylindrical Pebble Mills*

(*a*) For wet or dry grinding in closed circuit with auxiliary wet or air classifiers to produce fine-ground products with nominal amount of superfines; feed size should not be more than $\frac{1}{2}$ in.

(*b*) For wet or dry coarse grinding to -4- to 10-mesh in open circuit or as fine as -35-mesh in closed circuit with vibrating screens; feed size should not be more than 1 in. except on very friable material, and the moisture content should not be more than 2% for dry grinding.

(3) *Conical Pebble Mills*

(*a*) For wet or dry grinding in closed circuit with auxiliary wet or air classifiers to produce fine-ground products with nominal amount of superfines; feed size should not be more than $\frac{3}{4}$ in.

(*b*) For dry grinding in closed circuit with vibrating screens to produce medium fine products with uniform grading down to finer meshes or with minimum fines; high circulating loads.

(c) When equipped with superfine air classifier for dry grinding to produce uniform fine-ground products where a self-contained unit of mill and classifier is needed, eliminating elevators or conveyors.

(d) With superfine air classifier where material being fed contains a small amount of moisture up to 4%.

(e) With superfine air classifier and mill drying with air heater, where material contains more than 4% and not over 10% moisture; feed size for air-classifier pebble mills should not be more than 1 in.

(f) For wet or dry grinding to −4- to 10-mesh in open circuit or to as fine as −35-mesh in closed circuit with vibrating screens; feed size should not be more than 1 in. except on very friable material and the moisture content should not be more than 2% for dry grinding.

(4) *Cylindrical Ball, Conical Ball, aud Ball-tube Mills*

(a) The same general rules apply as under pebble mills, with or without classifiers, or conical mills with or without superfine air classifiers, except that these mills cannot be used on material where iron contamination is detrimental to the finished product; iron contamination hardly ever exceeds 0·2% when steel balls and lining are used; ball mills usually have 2 to $2\frac{1}{2}$ times the capacity of the equivalent sized pebble mills; the feed size to ball mills should not be more than $1\frac{1}{2}$ in. and preferably $\frac{3}{4}$ in.; ball mills will handle higher circulating loads due to the greater weight of grinding media.

(b) Alumina balls. The advantage of heavier balls without the disadvantage of iron contamination can be attained by using alumina balls, density 3·4.

(5) *Rod Mills*

(a) For wet or dry grinding materials where slight iron contamination is not detrimental and where a uniformly graded product with a minimum fines below 100-mesh is needed.

(b) Should not be used for grinding finer than 48-mesh.

(c) Feed size, unless material is very friable, should not be more than $\frac{3}{4}$ in.; larger diameter mills can handle feed up to $1\frac{1}{2}$ in.

(d) For products between 16- and 48-mesh, the mills should be operated in closed circuit with vibrating screens.

(e) If feed is slightly damp, the mill should be of the peripheral-discharge type where the mill is dry grinding (**M69**).

RUNNING OF PEBBLE AND BALL GRINDING MILLS

There are six factors affecting the efficient running of ball and pebble mills (**G6**):

MECHANICAL PREPARATION

(1) Speed of the mill.
(2) Quantity of balls.
(3) Size of balls.
(4) Quantity of material.
(5) Consistency of material for wet grinding.
(6) Initial particle size.

These are dependent on each other.

The Speed of Grinding Mills. The essential feature of any of these rotating mills with the grinding medium loose inside is that the grinding should be achieved by the rolling of the pebbles, balls or rods and not by their impact

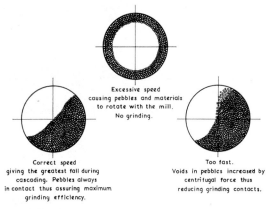

FIG. 7.10. *Speed of grinding mills and ball action (Steele, S175)*

on falling freely. The mill must therefore rotate at a speed at which the grinding medium is carried up the side enough to roll down again over itself but not so great that it tends to be carried clear of the general mass and then falls. If the grinding medium is allowed to fall and therefore disintegrate the material to be ground by impact, it wears out very much faster and contaminates the product more. Of course, as the rotation speed is increased even further a speed is reached when the balls, etc., are carried right to the top of the circle and do not fall but continue round on the wall of the mill. No grinding occurs at all, then. This is known as the critical speed (Fig. 7.10).

Good results are achieved when the angle between the radius to the point where the outer balls break from the periphery, and the horizontal is about 45°. Under these conditions there is a large amount of relative motion of the balls in contact with each other as the upper ones roll down and the lower ones are carried up. The speed required for achieving this 'angle of break' depends somewhat on the mill contents. For instance, water makes the balls slip more than when they are dry (**G6**).

678　INDUSTRIAL CERAMICS

The actual linear speed of the periphery of a mill rotating at a given number of revolutions per minute is greater the larger the diameter. The larger mills are therefore run more slowly than the small ones. This is seen in the curve marked 'theoretical carry-over of outer layer of balls' in Fig. 7.11. The graph shows grinding curves derived by different authors for somewhat different conditions. It can be seen that in every case the practical curve is considerably below the theoretical critical speed.

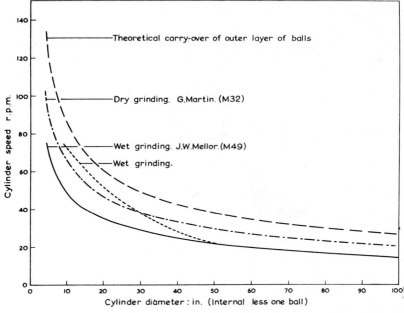

FIG. 7.11. *Grinding curves*

Quantity of Balls. The ball charge should be at least 45% of the volume of the mill but should not exceed 55%. There is an optimum quantity for particular conditions. Some operators introduce a few new balls with every charge. This must not be done without periodic checks as the balls do not always wear down at the same rate.

Size of Balls. It is the surface of the balls or pebbles that do the grinding by their contacts with others. Smaller balls have larger surfaces for given weights and volumes than large ones and so are more efficient. On the whole if 2 in. (5 cm) balls are put in the mill and the worn balls removed when they get less than 1½ in. (3·8 cm) better results are achieved than with 3 in. (7·6 cm) balls. However, use of just a few larger balls is sometimes beneficial.

For grinding enamel frits, Biddulph (**B48**) suggests using 55% by volume of pebbles of the sizes shown in Table 173. Use of too small balls is not good as their size then becomes comparable to the raw material and they wear down much more quickly.

TABLE 173

Mill diameter		Speed	Pebble sizes
(ft)	(cm)	(r.p.m.)	
1	30·5	68	80% at 1 in. (2·54 cm); 20% at 2 in. (5·08 cm)
2	61	48	80% at 1½ in. (3·81 cm); 20% at 2 in. (5·08 cm)
3	91	37	75% at 1½ in. (3·81 cm); 25% at 2½ in. (6·35 cm)
4	122	30	40% at 1½ in. (3·81 cm); 40% at 2 in. (5·08 cm); 20% at 2½ in. (6·35 cm)
5	152	24	
6	183	21	

Quantity of Material. Theoretically the most efficient use of the grinding balls is made when all the voids are filled with the material to be ground, and the balls are just covered with it. The voids usually make up 40% of the volume of the balls. Where the balls occupy 45% of the mill volume this is 18%, and if the balls are to be covered 21·5% of the mill volume (**S175**). For a mill with 55% by volume of balls the theoretical charge would be about 25%. In the writers' experience when grinding plastic material for a ceramic body somewhat less than these quantities are the best, *i.e.* about 20% by volume with 55% of pebbles.

Consistency of the Material. The consistency of the mixture for wet grinding also affects the results. A thick slurry mixture makes the balls clog together, carry-over, or 'float', that is, not make contact with each other. A very thin slip may cause slipping so that higher mill speeds have to be used to carry the balls up far enough.

In theory the best contact should be achieved if the voids in the raw material are just filled with water, this would mean 8·5% by volume of a mill with 45% by volume of balls (**S175**).

Incorporation of air in materials being ground for bodies is, however, most undesirable and such mills are best filled with water.

The senior writer's curve in Fig. 7.11 is for optimum conditions when grinding traditional materials without incorporating air. It is for a 1000-litre (35·32 ft³, or 220 gall.) mill charged with pebbles occupying 55% of its volume (about 550 kg, 1212 lb) and 400 kg (882 lb) plastic (about 22% by volume) or 500 kg (1102 lb) non-plastic (some 25% by volume) material, of S.G. 2·5 and with the mill filled with water.

In Denmark and Sweden the set-up of grinding mills that is much favoured is rather in contradiction with what has been said here. However, as it appears to be successful it is worth mentioning. The recommended conditions have higher grinding speeds, a larger quantity of pebbles, a smaller quantity of material to be ground and a shorter grinding period. An example is shown compared approximately with the writer's experience of optimum conditions (Table 174). This method of grinding produces small batches of material in short periods of time but the more frequent filling and emptying operations involve more labour and the over-all output is not necessarily better.

TABLE 174

	Scandinavian	Writer's
Diameter of mill	150 cm	
Revolutions per minute	30	20
Contents	3000 l.	
Pebbles	1800 kg	1650 kg
Material to be ground:		
Quartz and/or feldspar	500 kg	1500 kg
Water	500 l.	~1500 l.
Grinding time	$2\frac{1}{4}$ hr	12 hr

Initial Size of Material to be Ground. As has been mentioned, if the feed is fairly large it wears down the balls and mill unnecessarily. A fine feed leads to efficient grinding and uniform and uncontaminated products.

Power Input, Efficiency and Time of Grinding Mills. When a particle is broken the energy used is proportional to the area sheared, that is to the new surface produced. This relationship, Rittinger's law (**R37**), generally holds for the fine grinding of ceramic materials so that the energy input in grinding mills is usually proportional to the increase in specific surface of the material being ground.

Of the power used by, say, an electric motor driving a ball mill through a reduction gear, only 15% does actual work in grinding the product and grinding medium, and only 0·5% of the input energy actually breaks ionic links. About 25% of the energy used is lost as heat by the motor, the reduction gear and the main bearings of the cylinder, and some 60% is dissipated as heat from the ball mill. The grinding efficiency is given by:

$$\frac{(S_2 - S_1)C}{kWh}$$

Where S_1 is the original and S_2 the final specific surface (cm²/g);
C is the weight of the charge (tons);
and kWh is the energy input in (kWh).

If kWh is the energy supplied to the prime mover, then the figure obtained is the 'over-all efficiency', if it is the actual energy imparted to the mill, *i.e.* as measured by a dynamometer the result is the 'mill efficiency'.

The efficiency of grinding rather coarse or very fine materials is less than for intermediate sizes because there are wrong size ratios between the grinding medium and the product (**P45**).

MATERIALS FOR MILL LININGS AND BALLS AND PEBBLES

Apart from the small jar mills which are porcelain or stoneware throughout, all the grinding cylinders, etc., have cast-iron or steel shells.

The liners fall into two main categories: iron or steel, and materials that do not cause iron contamination. Various nickel, manganese and chromium steels fall in the first class. They can be made in numerous patterns to fit each other and the mill shell very precisely. They are used with steel balls.

Garms and Stevens (**G7**) investigated the wear of forged balls when compared with cast ones. They found that a definite thickness wears off in a given time regardless of the size although the classifying centrifugal action makes the large balls at the outside do most of the impact crushing whereas the smaller ones work by attrition. When more than a certain amount has been worn off a ball it loses its spherical shape, *e.g.* a 2 in. ball at 1·25 in.

In a ceramic works aiming at a fine and white product contamination by iron during the grinding cannot be tolerated and the mills must be lined with non-ferrous material. Both hard white natural rocks such as Belgian silex, quartzite, 'Jasper adamant silica', and some granite, and also the ceramic products porcelain and stoneware are used. Rubber linings are also coming increasingly into the picture, they are much more resistant to abrasion and much quieter (**W70**). The manufactured liners have the advantage of having a known consistent composition and hardness and of being made in exact sizes so that narrow joints can be used. Both stone and porcelain liners are set in white Portland cement.

The pebbles used in these mills are usually French or Danish flint pebbles. Porcelain and stoneware balls are also used.

When these materials wear during use they only contaminate the product with matter that is normally included in ceramic bodies. (Stoneware liners are only to be recommended for off-white bodies, *i.e.* stoneware itself.) They must, however, be replaced when the surfaces become rough or pitted when material to be ground or even balls may get lodged. Pebbles must be removed when they cease to be spherical.

Trade restrictions during World War II necessitated the search for substitutes for the traditional lining materials and flint pebbles, particularly in the United States. Some of the results obtained are of lasting interest.

For instance, before World War II the United States imported flint pebbles and silex mill linings from Europe. Cut off from this supply during the war they developed their own products which they now claim to be better than the imported materials. Pebbles are produced from a quartzite deposit of hardness 7 occurring in Wisconsin. The rock is disintegrated by an oxyacetylene flame process and then screened. The $1\frac{1}{2}$- to $3\frac{1}{2}$-in. product is fed to a ball mill and ground for some 5 hr. When these more abrasive pebbles are used in combination with granite liners the mill efficiency is increased from 5 to 20% over that achieved with smooth pebbles. The quartzite pebbles grind white, giving no colour contamination (**T39**).

Comparison between various American substitutes for imported Belgian silex were made. It was concluded that quartzite (Minnesota) and granites (Carolinas) were satisfactory for mill linings, although the latter may colour contaminate wet-ground white pigments. From this point of view porcelain is the best substitute (**B13**).

In another instance the following materials were tried: porcelain balls, Newfoundland, North Carolina and Texas pebbles, and jasper and adamant blocks and cubes from Minnesota. The adamant proved better than imported pebbles and silex. Zirconia balls are also found to be a very good heavy ball suitable for replacing steel if they were less expensive (**M70**). Alumina balls of specific gravity 3·4 (also known as 'high density grinding media') are very useful for working with a shorter grinding period at a higher load without any iron contamination.

With the development of pure oxide ceramics contamination of any sort even if it is not iron is found to be deleterious. Schofield (**S28**) therefore suggests that the mill linings and balls should be one of the following:

(1) The same material as that to be ground.
(2) That metal which will be converted to the oxide being ground, on oxidation during the firing.
(3) Organic material that will burn out completely. Suitable examples of his and others are given in Table 175.

Replacement of spherical balls and pebbles by cylindrical rods of equal length and diameter has also been recommended because of the greater grinding surface. Such rods are made of alumina or other high specific gravity material (**D38**).

TABLE 175

Ball Mill Grinding Without Contamination (S28)

Material to be ground	Mill lining	Balls
Fused beryllium oxide	Beryllium	Beryllium
Fused stabilised zirconia	Rubber	Zirconia or zirconium
Zirconium carbide	Rubber	Zirconium
Fused magnesia	Rubber	Magnesia
Graphite	Carbon	Carbon
Alumina	Rubber	Sinter alumina

SPECIAL MACHINES FOR VERY FINE GRINDING

There are a number of machines for fast fine grinding embodying various principles, which find uses in certain sections of the industry, *e.g.* colours, laboratories, etc.

The 'Attritor' (**A43**), the 'Mikro pulveriser' and 'Mikro-atomiser' (**P73**), are dry grinders which can be supplied with hot air to dry damp material. The grinding action takes place in an airstream which together with centrifugal force automatically classifies. A product of definite size can therefore be obtained in one operation.

Some small grinders and blenders work at very high speeds of rotation, making use of centrifugal forces and the turbulence caused by obstructions.

A number of small mills are operated by high-frequency vibrations. Equipped with electromagnetic vibrators minerals of hardness Mohs' 9 can be easily and speedily ground to a specific surface of 35 000 cm²/g for a power input one-twentieth that of a ball mill of equal capacity. A larger mill of this type, the 'Podmore-Boulton Vibro Energy Mill', has now been developed. The grinding chamber is supported on high-tensile springs, eliminating heavy bearings, so that power can be applied directly. 95% of the volume of the grinding chamber can be used for wet or dry grinding. Alumina cylindrical grinding media are recommended, and suffer practically no wear in this machine (**B94, P46**).

Fast fine grinding can be achieved in mills containing a rotating and a stationary 'carborundum' stone. The material is fed through the axis of the stationary stone and spreads out between it and the 'rotor'. The two stones can be set touching or a small amount apart. The ground material collects in a trough around and below the rotor and is discharged down a chute. These machines are very compact (**M91**).

Summary

TABLE 176

Normal Values for Surface Factors of Ground Raw Materials

Application	Flint	Stone	Feldspar	Whiting	Bone
Earthenware	230 ± 10	230 ± 10			
Tiles	210 ± 10	230 ± 10			
Bone china		245 ± 5			270 with 75 to 80% < 0·01 mm
General			240 ± 10	200–250	

Blunging

The clays for fine ceramic ware are naturally fine grained. They nevertheless need treatment to separate the particles and bring them into optimum condition for mixing with the other materials to be incorporated in the same body. Except where air-floated pulverised clays are available the most usual process involves blunging. In this the clay is stirred with water in large vessels until a homogeneous slip is produced. The mechanically driven stirrer has paddles, blades, knives, screws or propellers on a shaft. The old type of blunger had large paddles or gates that swept through the whole of the vessel, and were run at about 15 r.p.m. The newer types have relatively

TABLE 177

The Application of Different Types of Crushing and Grinding Equipment

(after Hendryx (**H81**))

Particularly in the Refractories Industry

Type of material	Hard	Medium		Soft
Specifications	Breaks with few fines	Breaks into cubes	Pulverises with many fines	Breaks up easily into fines
Examples	Quartzite, ganister rock or hard flint clays	Flint clays, some shales	Some flint clays, many shales	Soft shales, plastic fireclays
Jaw crusher	Primary crushing	Seldom	Seldom	Never
Gyratory crusher	Primary and secondary crushing	Seldom	Seldom	Never
Single roll crusher	Never	Good	Good	Good
Double roll toothed crusher	Never	Good if not too hard	Good if not too hard	Excellent, especially for frozen or wet clay
Flat head cone or gyrasphere crusher	Good, produces minimum fines	Never	Never	Never
Double smooth rolls	Not good for abrasive materials	Produce minimum fines	Produce minimum fines	Sometimes used
Dry pan with screen plates	Satisfactory, produces some fines	Satisfactory, produces some fines	Produces many fines	Many fines, not good if clay is too wet
Grinder or dry pan with raised rim	Not good for abrasive materials	Produces less fines than dry pan	Usually gives enough fines	Good, especially if material is wet or frozen
Hammer mill with grates	Never	Satisfactory, but may wear rapidly	Usually satisfactory	Not good for damp materials
Impactor mill without grates	Sometimes, but not good for abrasive materials	Satisfactory, usually makes less fines than dry pan	Satisfactory	Can be used on damp materials if not too sticky
Ball mill (material must be dry, or made fluid with water)	Excellent for grinding finer than 40-mesh. Good for abrasive material.	Satisfactory for fines if material is dry	Not needed	Not needed
Rod mill (material must be dry, or made fluid with water)	Does not produce many fines	Produces less fines than dry pan	Not needed	Not needed
Ring roll	Never	Seldom. Never use on abrasives	Satisfactory if dry	Satisfactory if dry

small propeller blades run fast enough (150 to 300 r.p.m.) for the whole mixture to be kept in constant movement. These may be arranged so that the initial thrust on the liquid is downwards, thereby supporting the weight of the propellor and shaft on the liquid instead of dragging on the rotating mechanism. The propeller blunger not only agitates and mixes better than the old gate type but can also be started up in a slip that has been allowed to settle, getting it well mixed quickly, whereas the gate type had to be kept going continuously (Fig. 7.12).

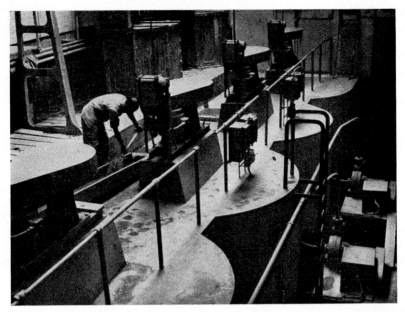

Fig. 7.12. *Blunging. Ball clay and china clay are each separately mixed with water in an enclosed blunger. Each agitator has independent drive (Wedgwood, **W35**)*

Fig. 7.13. *Triple-deck vibrating screen, equipped with stainless steel woven wire screening surfaces of 36-mesh, 40-mesh and 50-mesh respectively (Niagara Screens, **N14**)*

SIZE GRADING

The particle size of the products of crushing and grinding operations must either be tested by taking a sample to the laboratory or graded in some way to remove oversize material. The three main methods are screening, wet classifying by gravity, and air separating. For wet material vibrating screens are usually quicker than classifiers and give a product of more uniform size.

Sieves and Screens

Sieves and screens perform several different functions in the ceramic industry. They are used for testing samples for particle size (p. 279), for grading material and for cleaning it. The essential parts of a sieve or screen are the mesh and the mounting connected to charging and discharging devices.

The mesh is woven in stainless steel, phosphor bronze, brass, mild steel, monel metal, aluminium, copper, nickel, various alloys, nylon or silk. The diameter of the wire, method of weave, size and tolerance of apertures, for standard sieves is laid down in British Standard Specification 410:1943 (for U.S.A. American Society for Testing Materials: Standard E 11–39). Coarse test sieves (aperture from $\frac{3}{16}$ in., 4·76 mm, to 4 in., 101·60 mm) are made of perforated plate. The aperture sizes of coarse, medium and fine standard sieves are given in the tables in the Appendix (p. 1301). The last are given together with the older series from the Institute of Mining and Metallurgy and the standards of other countries.

In the I.M.M. series the wire diameter is the same as the size of the aperture, so that there is always an effective screening area of 25%.

The finest U.S.A. Tyler and the standard American fine series of sieves are numerically related to each other in the ratio $\sqrt[4]{2}:1$ or 1·189:1. In the coarse series the $\sqrt[4]{2}:1$ ratio can be continued by substituting the 4·24 in., 2·12 in., 1·06 in., 0·530 in. and 0·265 in. sieves (which are all shown in brackets) for the 4, 2, 1, $\frac{1}{2}$ and $\frac{1}{4}$ in. ones. The Afnor series is in the proportions $\sqrt[20]{10}:1$.

It is normal practice to designate a sieve by the mesh number, in the English series that means the number of apertures per linear inch. But unless either the standard being adhered to, or the diameter of the wire, is stated the mesh number alone cannot indicate the aperture size.

Test sieves should be made of plain weave mesh except for the finest ones (240 and 300 B.S.S.) which may be twilled. They are usually used in nests of some half-dozen or more of graded mesh size, fitted with a lid and a receiver. There are also small machines for vibrating these stacks of test sieves in a reproducible manner.

The type of weave best suited for screening in the works varies. Slip or dry material will pass through plain weave successfully but clay that is slightly damp clogs it very quickly. It will, however, often pass through a vibrated piano wire screen. As the name implies these screens are made up of wires lying parallel along the length.

The mesh is mounted on circular, rectangular or cylindrical frames and must be correctly tensioned for use. Cylindrical screens are revolved in use, and can be fitted with bushes or rappers to keep the mesh free.

Flat screens and sieves are used largely for slips and have to be vibrated for material to pass through, and in some instances electrical heating of the mesh is advantageous. The screens are set in tiers of two to four and sloped so that the oversize particles will move off. The vibrating mechanism must be arranged to vibrate only the screens so that no vibration is transmitted to foundations and hence to other machinery when unpredictable collapses may occur. (Fig. 7.13)

Granular material is frequently screened through steeply inclined screens with wires running lengthways only. Where such clay is of moisture content of more than 5 or 6% considerable difficulty is often experienced as the screen clogs up very quickly and a lot of the fines then get discarded with the coarse material. The screens therefore have to be brushed by hand at frequent intervals, which not only loses time but means they wear out fast. Blinding can be prevented by heating the screens, and the best method of doing this is by a low-voltage electric current. (Fig. 7.14).

A heavy current is supplied to the screens through a step-down transformer with tappings allowing a choice of current ratings. The current used depends on the size and mesh of the screen, and the moisture content of the clay. It must be such that all the wires are uniformly heated just enough to keep entirely free of material throughout the working. For example, one tile company that has recently installed such screen heating is now able to work its screens continuously unattended with an increased output of 25%, whereas previously the screens had to be cleaned several times an hour (**A40**).

Combinations of rotary screens can be arranged to sort raw materials by virtue of their different crystallographic shapes, *e.g.* feldspar can be sorted from mica, clay, etc. (**B25**).

Wet Classifiers

The methods of classifying waterborne particles have already been met in the purification of china clay (p. 239). Different sized particles can be separated by arranging the flow of the mixture to be such that only the large ones settle. However, they usually entrap a certain amount of the fine ones so that where both fractions are required some washing action of the settled coarse material must take place (p. 245). This occurs in rake classifiers. The slip is fed into a pool formed at the lower end of a sloping tank, from which it can overflow to a collector for fines. One or two sets of rakes move up the slope of the tank pulling the material that settles gradually up through the water and then out into the air for delivery at the upper end. The gradual movement allows any fines caught up with the coarse material to be washed out again (**S196**).

Fig. 7.14. *Heated piano-wire screen. The material successfully passing is seen on the near side, the oversize slides down the inclined screen on to a conveyor which returns it to the dry pan (London Brick, **L41**)*

Fig. 7.15. *Cutaway view of Hardinge counter-current classifier (Hardinge, **H19**)*

In other classifiers clear water is added to remove the fines from the coarse material. The illustration (Fig. 7.15) shows such a counter-current classifier. The mixture to be separated is fed in by the chute on the right and any deposited material is gradually rolled uphill against the stream of wash water. The fines are discharged on the far right. This is effective in a range from 10-mesh to 100-mesh and can be used in closed circuit with a wet grinding mill.

Air Separators

One type of conical air separator was described in connection with kaolin preparation (p. 251). There are other similar separators with two cones separated by an annular gap which is connected to the space inside the inner cone by adjustable ports or vanes. In all of them centrifugal motion makes the oversize particles drop out of the airstream. Products of over 99% through 325-mesh can be obtained.

Where a less fine product of about 90% through 200-mesh is required the simpler loop classifier is adequate. The airborne material is guided in a loop path, which throws the larger (heavier) particles out. They collect at the bottom on an adjustable damper and are cleansed of fines by an airstream. These combine with the fines completing the loop to the upper outlet (**H20**).

Both types of classifier are well suited for inclusion in closed-circuit dry grinding installations.

STORAGE

After grinding or blunging and grading many materials are stored for short or long periods.

Dry materials are delivered into hoppers or bunkers from which they can be withdrawn at will. But much of the work of sieves and classifiers can be undone in a storage hopper if the necessary precautions are not taken. If a material with particles of different sizes is allowed to fall down a chute into a hopper, a cone will form and the larger particles will roll down the outside while the small ones remain in the centre. Whether the withdrawal point is in the centre or at the side of the bottom an unrepresentative sample will be withdrawn. Segregation by cone formation must therefore be avoided either by moving the filling chute about, or raking the material flat as it enters the hopper, etc. Many modern hoppers have baffles, air inlets, etc., to prevent segregation.

Level indicators which actuate switches to stop material flow or operate warning devices can be fitted to storage vessels for liquids, powders and granular and flaked materials (**B79**).

Clay slips, slop ground flint, etc., mixed body slips or glazes may be stored in arks which are kept agitated just enough to prevent separation. They operate much like blungers, but less vigorously.

IMPURITIES

Removal of Iron and its Compounds

The presence of iron compounds in most ceramic bodies both lowers the maturing point, decreases the melting point and colours the burnt product. This is useful in production of bricks, etc., where the red or yellow product is often pleasing to the eye and the lower maturing temperature saves fuel, but it is most undesirable where a white and/or refractory product is required.

Iron contamination of the raw material may occur from various mechanical inclusions of iron oxides, sulphides, etc., or because iron may be a chemical constituent of the clay mineral itself by replacement of aluminium in the crystal lattice.

During the course of mining, the clay may also pick up tramp iron such as nuts, bolts, chains, etc., which could seriously damage crushing and grinding equipment. Also, during processing, clays may become contaminated by fine iron particles from the wear of the machinery through which it passes.

Such free iron, together with certain of the iron minerals can be removed with magnets. Schurecht (**S35**) gives the decreasing order of magnetic susceptibility of iron minerals as follows: magnetite, siderite, haematite, limonite, marcasite and pyrites. Those at the latter end cannot be separated from clay by magnetic methods. The susceptibility is increased after heating to dull red heat. This is not normally practicable or economical, but it is exploited on quite a large scale at the Neurode clay pits in Germany where the main output is in any case burnt before sale (**N10**).

If the clay is transported from the pit to the plant by conveyor belt a magnetic separator in the form of the terminal pulley or an overhead magnetised belt can be incorporated in the system. A typical magnetic pulley of 24 in. (61 cm) diameter and 39 in. (99 cm) face, capable of treating 7000 ft^3 (198 m^3) per hr, consumes 2·1 kW. If the material passes through a chute, a portion of the chute itself may be replaced by a hinged magnetised unit consisting of mild steel pole pieces energised by horseshoe-shaped permanent magnets. The smallest of these 'Magnaplates' effectively cleans a feed of $2\frac{1}{2}$ in. (6·35 cm) depth at 200 ft (61 m) per min (**H50**).

For cheap products where the crude clay from the pit is not normally treated in any way, further removal of the iron contamination is too expensive to be applicable. For higher grade products where the clay is treated, certain extra precautions to remove the iron and thus improve the product are economical.

Where the inclusion of larger or harder grains of non-magnetic iron compounds occurs, hand sorting of the mined product, blunging and screening, and sedimentation will remove most of the impurity. Where the total iron content is not high but because of local concentrations produces spots in the ware, fine grinding and good mixing will give a homogeneous light-coloured product.

Grinding and processing the clay releases smaller magnetic particles not mechanically free to be removed in the initial magnetic separation; it also introduces fine iron particles. It is therefore necessary to magnetise either the dry ground product or the slip. The 'V'-type separator is very efficient for removing up to 3% of fine iron from dry feed. It consists of a series of stationary magnets above which a series of wedge-shaped pole pieces is agitated. The tapered edges of the poles attract the iron, which then

FIG. 7.16 *Super intensity agitated type electromagnetic chute separator* (*Electromagnets*, **E13**)

becomes brushed into the recesses by the oncoming flow of material. A collapsible shute at the outlet is held in place by the magnets so that in the event of a power failure no contaminated material passes out along the normal channels (**H50**). Another type is shown in the illustration (Fig. 7.16), and is suitable both for dry feeds and also the sticky powders so often encountered in the ceramic industry.

For wet feeds some type of percolator separator is advisable. The magnet either has a corrugated surface, or, is a grid through which the slip flows, bringing any iron present into close contact with the metal. A very

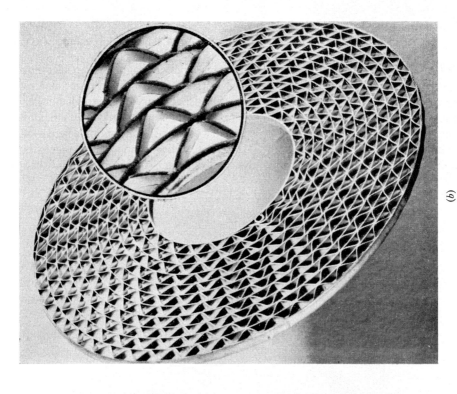

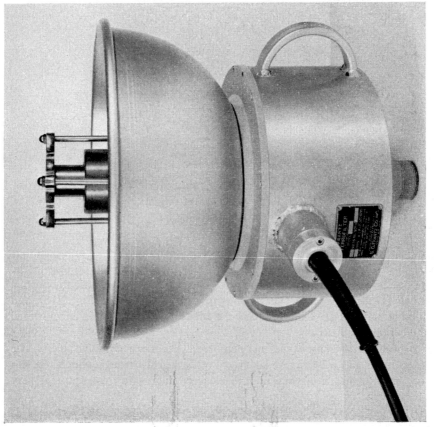

Fig. 7.17. *Frantz ferrofilters:* (a) *gravity percolator for slips;* (b) *a ferrofilter grid (electromagnets use from sixteen to thirty of these grids in the stack).* (c) *section of the dry ferrofilter (Frantz* **F28**). (For (c) see facing page)

popular percolating magnetic separator in the United States is the Frantz Ferrofilter (**F28**). This is circular and is produced in three models for use with slips under gravity, slips being forced upwards or for complete enclosure in a pipeline (there is also one for dry powders). The slip passes through an electromagnetised screen fine enough for any magnetic particle to be attracted out of the liquid and held (Fig. 7.17).

The use of three magnetic filters in each production line has been advocated; first an electromagnet followed by a 'policeman' permanent magnet. This latter is to catch any particles washed off the electromagnet when it has collected a heavy layer. Such particles are usually aggregates of those originally present in the mixture to be purified and will show worse

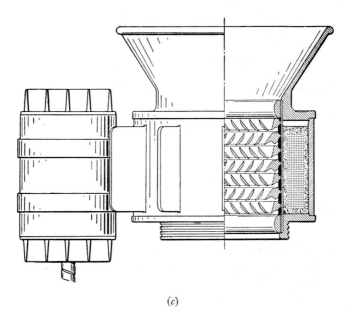

(c)

although fewer spots than the unpurified body. Then thirdly a line permanent magnet at the entrance to the filter press to remove any contamination that has occurred during processing (**E23**).

Recently there has been a tendency to return to permanent magnets instead of electromagnets as certain alloys of nickel, aluminium, cobalt, copper and iron, 'Alnico', 'Ticonal' and 'Alcomax' can be made into very strong and durable magnets.

Water and air act on iron sulphides, giving the soluble sulphates. Weathering or the action of hot water produces this effect but must be continued sufficiently long to leach out the iron sulphates as by introducing a soluble salt which is also a flocculating agent more harm may be caused than by the original iron sulphide.

All the iron impurities cannot, however, be removed by these methods, and so numerous chemical methods have been suggested, but they have not proved to be sufficiently inexpensive to be applied industrially.

Calcium Carbonate

The presence of coarse calcium carbonate in a ceramic body can be very harmful. During the firing carbon dioxide is driven off and calcium oxide, or lime, remains in the body. If this is porous, any water that penetrates to the lime will slake it, evolving heat and causing expansion. If the lime grain is large this may shatter the piece, if smaller splintering occurs.

When a clay that is to undergo blunging and grinding treatment is known to be calcareous it is well worth trying to remove the larger 'pebbles' and grains by passing a blunged slip through sieves, etc., before grinding. Those grains that are too small for mechanical removal can be made less harmful by very fine grinding or crushing between rollers and uniform distribution through the body. They become harmless once they are fine enough to pass a sieve number 52 (**K14**).

Complete removal of carbonates by treatment with acids is usually too costly.

In products (bricks, etc.) where pretreatment of the raw material cannot be considered the after-effect of leaving the lime in the clay may be prevented by the common practice of immersing the burnt product in water. The lime grains will then become attacked by considerably larger quantities of water than under normal atmospheric conditions and will not only slake and expand, but will also be broken up by the uncombined water and the expansion taken up by the pores surrounding the original grain of lime.

Magnetic Purification of Certain Minerals

Certain minerals can be purified by magnetic methods although they are not ferromagnetic. The susceptibility to intense magnetic fields of such feebly magnetic materials is insufficient to attract them against the force of gravity, as is done in the removal of strongly magnetic material. But by applying a very high magnetic field intensity horizontally to a falling stream the more susceptible particles can be deflected sufficiently to effect a separation.

The material must be crushed and ground to a size where the minerals to be separated are no longer mechanically bound together. However, this should not be finer than 100-mesh because air friction disturbs the flow of very fine particles. The ground material should be dried to ensure free flow and if possible screened so that only grains of about the same size pass through the separator together.

Materials that can be purified by this magnetic method are:

(1) Feldspar freed of muscovite mica.
(2) Kyanite freed of iron and then silica.

(3) Separation of ilmenite ore into ilmenite, zircon and other minerals.
(4) Bauxite (**J40**).

The Removal of Soluble Salts

It has already been mentioned that certain salts often present in raw clays interfere with the preparation of casting slips. Soluble salts are also the cause of scumming and efflorescence. When they are present in the shaped ware they tend to migrate to the surface with the water as it dries out and are left behind as it evaporates. During the firing this accumulation of salts acts as a local flux giving irregular ware. Those salts that do not come to the surface during the drying of the green ware do so during subsequent wettings and dryings of the fired pieces giving the familiar sight of efflorescence on bricks. The commonest salts present are calcium and magnesium sulphates followed by iron, potassium and sodium sulphates.

Preparation of bodies by the wet method of blunging, mixing and filter-pressing removes a lot of the undesirable soluble salts. But if the raw materials are not purified in this way the best treatment is with barium carbonate. This can fortunately be done sufficiently cheaply to be applicable even to the production of common bricks.

Barium carbonate reacts with soluble sulphates to give the very insoluble barium sulphate and only slightly soluble carbonates. Barium carbonate is itself only slightly soluble so that when used in excess it does not reintroduce the soluble salt problem. The reaction with calcium sulphate and the solubilities in 100 ml water at 18° C of the substances involved is as follows:

$$BaCO_3 + CaSO_4 = BaSO_4 + CaCO_3$$
2·4 mg   201·6 mg   0·23 mg   1·3 mg

The reaction is of course relatively slow because of the low solubility of the barium carbonate, and it is usual to allow a mixture containing it to mature for several days. The reaction is quicker with the very finely divided chemically precipitated powder than with the ground mineral.

As the barium carbonate is required only in small quantities, of the order of 0·05 to 0·15% of the dry clay, its homogeneous distribution may present some difficulties, particularly if the dry powder is used. The actual quantity required may also vary from batch to batch or day to day. It is therefore normal to use rather more than is really necessary, although rapid testing of each batch is quite simple (p. 322).

Very much easier and better distribution can be achieved with a barium carbonate emulsion in water, stabilised with clays, or organic binders. The finely divided state of the carbonate makes it more reactive and less need be used. Its solubility can be further increased by adding carbonic acid, especially if the solubility of the carbonic acid is made greater by the presence of humic acid. This highly reactive emulsion can be added with the mixing water in pans, mixing troughs, etc. (**S102**).

SUCCESSION OF PROCESSES FOR THE COMMONEST CERAMIC HARD RAW MATERIALS

In the fine ceramic industry some hard raw materials are bought in ready ground condition from specialist milling firms. This applies particularly to flint, which because of its hazard to health is always handled in a watery suspension, 'slop', once it has been ground. Some of the usual processes in the preparation, whether at the mill, or in the works are listed.

FLINT

Calcining to give a S.G. of 2·50–25·2, in intermittent bottle kilns or continuous gas fired vertical shaft kiln → one- or two-stage crushing → sieves ────────── max ¾ ins. ──→ pan or cylinder grinding, batch or continuous (viscosity may be reduced by adding soda ash, so long as calcium chloride is subsequently added to prevent setting) → classifiers → thickeners → adjustment of pint weight.

CHINA STONE

Two-stage crushing → screen → pan or cylinder grinding, batch or continuous → classifier → thickener → adjustment of pint weight (**R35**).

FELDSPAR

Mining or quarrying → primary jaw crusher → 1 in. (2·54 cm) diameter → rod mill → selective rotary screening

↓ ↓ ↓
feldspar mica waste
20 mesh to ¾ in.

↓
grinding to specification (**B25**).

MIXING

Theoretically mixing problems can be divided up under the headings: gas–gas, gas–liquid, gas–solid, liquid–liquid, etc., but in ceramics it is probably more expedient to use the divisions: wet, plastic and dry. The term 'wet mixing' then covers mixing a substance with water to form a suspension that will flow, and mixing together of such watery suspensions. 'Plastic mixing' is concerned with mixtures that are both too dry to flow like wet mixtures but too wet to flow freely like granular powders; they are sticky and easily form lumps. 'Dry mixing' covers materials that will flow freely, even though they may contain a certain amount of moisture.

In this country and on the Continent wet mixing is one of the most important parts of the preparation of fine bodies. The individual raw materials are 'blunged' and the slips obtained mixed in the correct proportions. This is done in large vessels fitted with stirrers or propellers.

Wet or dry mixing of batches can also be carried out in revolving cylinders, tubes, etc. If grinding and mixing are to be carried out together a pebble mill is used; if only mixing is required the pebbles, balls, rods, etc., are left out.

Plastic mixing can be carried out in batch or continuous edge-runner mills as used for crushing and grinding and in similar machines designed more expressly for mixing. These are equipped with ploughs as well as runners, and suitable emptying mechanism. Such machines can also be used for adding water to dry material to produce a uniform plastic mass. They are used for heavy clay products and refractories.

Plastic mixing is very frequently done in a trough or an enclosed cylindrical vessel by some form of rotating worm, screw, knives, blades, etc., either in batches or continuously. The design of the blades is very varied. In some types they are arranged so that the material is swept alternately from the middle to the ends and back as well as revolving and being cut. One type has two paddles with two blades each, one turning at twice the speed of the other, and so that each one is scraped clean by the other during each revolution. This is of very great importance when mixing sticky clay material (**B28**) (Fig. 7.18). This type of mixer is also very much more satisfactory than a pan-type for mixing materials where the size grading should not be altered during mixing, *e.g.* refractory grog.

'Dry mixing' is usually simpler than wet or plastic mixing and can be done in revolving cylinders with or without pebbles, in continuous worm troughs, and batch troughs with turning bladed agitators, etc.

BLENDING AND FEEDING

Before considering the blending of different raw materials to make up the body mixture the question of the uniformity of the individual constituents arises.

Probably the most variable raw materials as regards composition, moisture content, ceramic properties, etc., are the clays. Even when care is taken at the clay pits to deliver uniform products fluctuations are in the very nature of clay formations.

Ideally every load arriving at a works should be tested and the batch composition of the body calculated accordingly. But this cumbersome procedure can be avoided in most cases by using several clays and a number of storage compartments. If possible similar clays are obtained from

(a)

(b)

Fig. 7.18. *Mixers for plastic bodies:* (a) double-shafted trough mixer for brick clays, marls, etc. (*Johnson*, **J54**); (b) plan view of batch mixer for plastic bodies (*Baker Perkins*, **B5**)

different sources and stored separately. Failing this each new load from the same source is stored in a separate compartment, six or eight giving very reproducible mixtures. When the body is to be made up fixed quantities are taken from every compartment. The idea that the various clays will not all change composition at the same time and in the same way holds in most instances.

Dry Materials

Different materials can be fed continuously in known volumes on to a conveyor belt by means of box, rotary, drum, disk, etc., feeders.

The box feeders consist of compartments holding the various materials

FIG. 7.19. *Proportionating machine in use for the clay, sand and grog used for vitrified stoneware pipes (Doulton, **D56**)*

with openings which are controlled by gates or slides which can be raised or lowered by means of racks and pinions. These can be adjusted to let through the required amount of each material. Below the openings there is a trough which either has a mixing worm, or a conveyor belt at the end of which a cutter takes off regular vertical slices. If the material is lumpy or sticky a cutter can be placed inside the box compartment. Batching can be accurate to $\pm 5\%$ (**A211**) (Fig. 7.19).

If a dial is geared up to the gate-moving mechanism adjustments to produce mixtures of different compositions can be very easily made.

Other mechanisms involve rotating cutters or disks that give a definite volume of material per turn. They have adjustable speed controls to vary the output rate. Granular materials can be passed through funnel-shaped feeders of adjustable outlet on to a revolving disk with a discharge plough, accuracy being $\pm 3\%$ (**A211**).

It would often, however, be much more accurate to measure materials of known composition and humidity by weight. This applies particularly to lumpy ones that do not flow freely and regularly. Newer machines are now available for feeding continuously by weight at a predetermined rate (**H20**). Free-flowing materials can also be batch weighed on a registering conveyor before delivery to the main conveyor (**A211**).

Continuous instead of batch feeding is very advantageous for edge-runner mills. Dumping large amounts of material in the pan at once forces the first runner up, with consequent unnecessary wear on the bearings, etc. A recent patent (**L42**) deals with this trouble even more fully by introducing a switch mechanism that stops the feed when the depth of material in the pan exceeds a given amount.

Wet Materials

An essential part of mixing processes is the correct measurement of the different components that are to make up a mixture.

The standard method for measuring materials that are in slip form is by the specific gravity and then by volume. That is, each raw material is blunged or agitated with water in proportions that will give a slip of predetermined standard weight in ounces per pint (pint weight). Each batch is tested with standard quart or pint measures, and must be adjusted until correct. A small difference on the weight of a pint of slip is a very large difference when hundreds of gallons are concerned. It is therefore advisable to check the volume of the measuring can periodically; it may be oversize when new as the Government stamp guarantees a minimum volume, and after being knocked over several times it may be sufficiently dented to be undersize.

Meredith (**M64**) points out the consequence of this as follows:

'It may be argued that as all slips are checked in the same can, no difference is caused, but a few minutes' thought will show that this is only true if all the slips are at the same weights. When there are relatively large differences in weight, *e.g.* ball clay at 24 oz and flint at 32 oz, the increase in the dry content of the material at the lighter weight is out of all proportion to that at the heavier, and may be enough to alter the behaviour of the body. For example if the quart can hold only 39 oz of water instead of 40, the dry content of ball clay slip which apparently weighed 48 oz would have increased by about 15%, while the dry content of the flint which apparently weighed

MECHANICAL PREPARATION

64 oz would have increased by only 7%. From this it is also apparent that a small discrepancy in the weights of the lighter body constituents are much more serious than similar errors in the weights of the heavier ones.'

When each constituent is correctly made up they are run in turn into a mixing ark. This has parallel vertical sides and a vertical measuring stick marked at different heights for the level to be reached as each constituent is added. Hence although the slip is mixed by volume it is measured in 'inches'. Here again a small mistake on the 'height' of any material added may be a large volume if the mixing ark is wide. However, this method, already in use for many years, gives good results if care is taken.

FIG. 7.20. *Slip blending by volume–weight machine* (*Wedgwood*, **W35**)

The difficulties of sampling and of small errors leading to large fluctuations is overcome if a modern machine for measuring by the volume–weight method is used. In these several hundred gallons of the blunged raw material are pumped into a graduated tank on a weighing machine. This records both the volume and the weight so that the average pint–weight for the whole is found. If it is not exactly as required suitable tables inform the operator what adjustment he must make in the volume he delivers from the weighing tank to the mixing ark (Fig. 7.20).

FILTERING

The clay slurries and slips used for purifying and mixing can be dewatered by a number of ways. The commonest method is the filter press, but various vacuum or centrifugal continuous methods are gradually being introduced (**S55**).

Filter Presses

A filter press consists essentially of a series of grooved wooden or metal plates or trays which can be fitted together to make thin circular or rectangular chambers. They have an inlet which connects all the chambers and individual outlets. Each chamber is lined with a filter cloth in such a

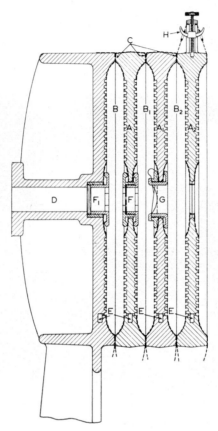

FIG. 7.21. *Cross-section of a few chambers of a filter press (Johnson, J53)*
A, $A1$, $A2$, *filter plates forming chambers* B, $B1$, $B2$; D, *inlet*; E, *outlet ports*; C, *machined rims of plates. Cloth fixings*: $F1$, F, *screw unions*; G, *instantaneous clip*; H, *adjustable hook for use when cloth is sewn round centre hole*

way that the inlet opens inside the cloth but the outlet outside it (Fig. 7.21). In Great Britain rectangular trays are favoured whereas round ones are often used on the Continent and in the United States.

There are two main types of filter press: the recessed plate type, and the flush plate and distance frame type.

In the recessed plate type there is a raised rim round both sides of the flat grooved plate so that when placed together chambers are enclosed. The feed inlet is usually central although it may be placed at a top or a bottom corner. The outlets pass from the bottom of the grooved surface through the plates to a bib and cock, a connecting passage or a simple opening. Each plate is covered on both sides by a filter cloth. This is either hung over the top with a hole in each half for the inlet, the joint round which being made with a cloth clip, or the two halves are sewn together round the hole, eliminating the need for the clip, and then held together at the top of the plate by suitable hooks. The plates are then screwed together with considerable pressure to form the filtering chambers.

In the flush plate and distance frame type of press the grooved plates have the faced joint surfaces practically flush with the filtering surfaces (**B39**). They are separated by hollow frames with faced surfaces corresponding with those of the plates. The thickness of the distance frames determines the depth of the chambers. The feed inlet passages may be through the centre or in the corners like in the recessed plate type, or through lugs external to the joint surfaces and connecting to the chambers through the distance frames. The latter have the distinct advantage of not requiring holes in the filter cloth. The jointing of these passages is done with independent cloth cuffs or hydraulic lipped rubber collars (**J53**).

The earliest filter presses, introduced about 1860–1870, were constructed entirely of wood. This practice is continued for production of clay and bodies for bone china due to the fear of rust contamination from cast iron ones. The most usual ones now are those with cast iron trays. A recent development is the use of aluminium trays which are becoming popular in the bone china industry (**H53**).

A very important part of a filter press is the tightening mechanism. For small presses this can be worked by hand and consists of combinations of tightening screws with or without spur wheels, ratchet tightening gears with or without lever bars, hand hydraulic tightening gear, etc. Large presses have hydraulic tightening and opening gear. Hydraulic gear has the advantage of being able to exert a known pressure so that over-tightening causing unnecessary strain can be eliminated. Electrical tightening gear can also be set to exert a predetermined pressure.

It is the normal practice to use double filter cloths inside the presses: a fine cloth next to the material being filtered and a coarser backing cloth to protect it from wear and rust on the plate surface. Until recently the filtering cloth has always been cotton twill cloth weighing about $12\frac{1}{2}$ oz per yd^2 (423 g/m^2), backed by jute tarpaulin of about 16 to 18 oz per yd^2 (542 to 610 g/m^2) in plain weave. These give a good filtering medium, a good seal between plates, and stand up to the mechanical wear and tear sufficiently well. But they have the great disadvantage of being attacked by mildew which rots the fibre. This was at first combated by treating filter cloths with creosote, but this made them stick to the filter cake and leave nap in it. The cuprammonium process introduced in the 1920's is far more satisfactory. The filter cloth is run through a bath of cuprammonium and then dried on a steam-heated drum, before being made up. The chemical dissolves off surface fibres of cotton, imparting a cellulosing effect to the surface, which makes it come away from the filter cake cleanly. In use, however, the proofing gradually washes off (impregnating the jute backing cloth to good purpose on the way) and after a certain time must be repeated if mildew is to be avoided.

A new process still in the trial stage is the acetylation of cotton fibre converting it entirely to a cellulose mono-acetate which is not attacked by mildew.

Meanwhile, however, cotton is gradually being replaced by synthetic fibres, in particular nylon. Nylon cloth woven from filament yarn is much stronger than cotton and so fabric of half to one-third the weight of the usual cotton cloth can be used. The delivery of the filter cake is much cleaner as there are no stray fibres as from staple spun yarn, and also less blinding of the filter cloth. Nylon cloths can be used for much longer without washing, and they can be washed easily in lukewarm water and soap, rinsed and then re-used immediately.

The disadvantage of nylon filter cloths is their lack of resilience, which makes them easily bruised, and otherwise unsuitable for making good seals in between the plates. This can be overcome by fitting a specially developed rubber rim to the plates to make the seal in place of the cloth. This is U-shaped and is made to fit various types of trays and plates rather like a tyre. Once on, they become an integral part of the plate and need not be removed.

The best backing cloth for use with nylon filter cloths also presents some difficulties. Jute cloths like those for cotton filter cloths become prone to mildew as the proofing solution no longer gets washed through them, and so have to be rot-proofed themselves. They will then last about six months. Synthetic backing cloths and perforated sheets of polyvinyl chloride are being developed (**H53**).

Terylene and P.V.C. fabrics are now in use for filter pressing (1956).

The method of forcing the slip into the filter press has considerable bearing on the uniformity of the filter cake. If a pump is used there is a tendency for some constituents to settle out of a mixture between the strokes. The resulting filter cake may then have zones of different materials. This can be seen in petrographic investigations of the cake and of fired sections of it. The effect is reduced by using four-stroke instead of two-stroke pumps, or by doubling the length of the press and halving the pressure.

For controlled uniform filling of filter presses the pneumatic system is best. The slip is pumped into closed vessels termed 'forcing receivers', 'forcing rams' or 'Montejns'. From these it is displaced continuously by compressed air. Automatic twin rams can be installed in such a way that continuous output is achieved, one ram being filled while the other empties and the compressed air being re-used (**J53**). The time required to fill a press is less by the pneumatic method than by direct pumping (**A87**).

The pressure applied to the slip being pumped into a filter press is regulated to suit the purpose for which the press cake is required, different making methods needing different water contents and therefore different pressures. Correctly filter-pressed bodies are wedged or pugged and are then ready for plastic making.

When sufficient water has been removed the compression on the press is released and the trays separated. The press centres are individually stripped off and dropped on to trolleys under the filter press (Fig. 7.22, *see* also Fig. 7.24). Press cakes are further dried on steam-heated pipes, etc., if a drier body is required, *e.g.* for dust pressing. The dry cakes must then be disintegrated in a dry pan.

Fig. 7.22. *Filter press being emptied. Each filter cake weighs about 1 cwt (Wedgwood, **W35**)*

Vacuum Filtration

The method of the filter press is here reversed, the water being drawn away from the solids instead of being forced out. Both rotary drums and disks are used successfully in the United States. Manufacture of the drum type began in England towards the end of 1952. The clay slip, containing about 70% moisture, is dewatered while passing round the evacuated drum and then discharged by a scraper or by endless strings passing over a roller placed away from the drum, as is shown in Fig. 7.23. Its water content is

FIG. 7.23. *Vacuum filter with string discharge:* (a) *photograph;* (b) *diagram of unit incorporating both filter and dryer* (Stockdale Engineering, **S191**)

then 25 to 30%. The filter can also be directly combined with a dryer when the moisture content is reduced to 10%, as shown in the diagram. The filtering rate is about 8 lb of dry solids per ft^2 of filtering area (39 kg/m^2) per hour, and warmed slip, up to 70° C (160° F), gives the best results.

Centrifugal Dewatering

Although centrifuges are proving efficient for the purifying and dewatering of clays at the pits it is doubtful if they can be adapted for use with clay bodies as segregation of the mixture occurs according to the size and weight of the particles.

Heated Drums

A body can be dried directly to the right water content for dust pressing, about 7%, on a rotating steam-heated drum. The drum dips in the slip and dries the thin layer taken up on its surface during some three-quarters of a revolution, when it is scraped off in flakes on to a conveyor belt.

To be economical a deflocculated slip of maximum pint weight must be used, 32 oz/pint being considered a minimum, while more than 34 oz/pint being impracticable. The process is particularly useful for bodies that have a high content of plastic clay and are therefore slow to filter in a filter press.

Spray Drying

There have been trials of the spray drying of slips to replace the labour-consuming method of producing 'dust' for dry pressing by filter pressing, drying of the press cakes, followed by grinding. The method consists of spraying the slip either up or down a tower with a down blast of hot air. Humid granules ready for pressing are collected at the bottom and conveyed to storage silos. The process is continuous, requires only one attendant and saves space; its installation and running costs are said to be lower than for the traditional method (**B21**), but the latter is probably only so where the cost of heating the air is unusually low.

WET PREPARATION METHODS WITHOUT DEWATERING

A short cut to wet preparation of plastic bodies is described by Appl and Lachmann (**A211**). The moist raw materials are cut up in a clay shaver and then deflocculated with soda ash, sodium silicate, humic acid or other suitable chemical in solution and blunged to a thick slip. The water content should not exceed 30%, giving a pint weight of 37 to 39 oz (1850–1950 g/litre). This is then run through a sieve to remove the main impurities, the system having been particularly devised for removing the paramagnetic pyrites and marcasite (FeS_2). Calcium silicates in quantities of 0·6 to 0·8% are then added to the slip and mixing continued for 3 to 5 min when a plastic body will have been formed.

More detailed description of the successive superimposed blungers used to give an intermediate, deflocculated slip of $38\frac{1}{2}$ oz/pint is given by Kleinpaul (**K33**). This is followed by injection of 'Adsorba' which converts the slip to a plastic body ready to go into a pugmill, etc. A similar process for refractory bodies with small amounts of bond clay is also described.

WEDGING

However it is produced, whether by dewatering a fluid slip or by moistening drier clay, the plastic mass lacks complete uniformity and contains entrapped air. Some form of working, kneading, wedging, etc., is necessary. Traditionally the heavy stoneware bodies were worked by being trodden under foot. Pottery bodies are wedged. A lump is knocked or thrown on a smooth surface on each of its sides, cut through, reunited in a different way, knocked together, and so on until when cut no flaws show. Much of the labour of hand kneading or wedging has been taken over by machines but the final wedging is often still done personally by hand throwers.

The Kneading Table

A machine that had great popularity in Continental Europe and is still used there is the kneading table. By means of horizontal and vertical, cylindrical and conical rollers it simulates the action of hand kneading. The body is alternately spread out and piled up. It is claimed that this action de-airs the body as well as mixing it thoroughly but this is not entirely true. The de-airing action is not complete nor very regular, and furthermore if a body is worked on the kneading table for an excessively long time air is actually re-introduced.

Pugmills and De-airing Pugmills

The function of pugmills is to improve the uniformity of a plastic clay body giving it greater workability because of the proper coating of clay particles with water. The de-airing machines do this more thoroughly as they also remove any air bubbles.

Steele (**S174**) points out that the term 'pugging' means 'mixing with water' and not 'mixing', the latter implying the mixing of two or more solid materials. The pugmill is sometimes expected to mix solids more intimately and it can be shown that it does not do this.

The pugmill consists essentially of knives set as a screw on one or two rotating shafts moving in a trough. The machine is a continuous one and is fitted with a suitable feeding hopper, and usually the discharge end has an auger and a mouthpiece for extrusion of a solid column of pugged material. The knife blades and the auger screw blades may be set on the same shaft.

The amount of material that should be present in the pugmill while it is working seems to cause some controversy. Steele (**S174**) says that the time a given lump of clay stays in the pugmill is as important as the actual agitation provided and hence the volume of clay present is the governing factor rather than the length or inside diameter of the pugmill. He therefore suggests that it is beneficial to a clay if the pugmill is sufficiently full to cover the shaft so that each individual part spends longer in it, even though the power consumption is then greater. On the other hand Seanor (**S47**) maintains that a pugmill cannot work efficiently if the shaft is covered and that adjustments of the knife angles or the shaft speed must be made to keep the material passing through sufficiently quickly. It would seem that for the desirable long pugging a long or a large pugmill is necessary.

The objection to the normal open pugmill is that it tends to incorporate air bubbles in the clay which have an adverse effect on its plasticity and other properties. Straight (**S202**) has shown that this is due to the shape of the pugmill blades. These are usually blunt at the back so that the clay on either side of the blade cannot reunite directly the knife has past, and so air is drawn in. By streamlining the pug blades and auger wings so that the clay joins smoothly after their passage a very much better result is achieved.

DE-AIRING

Pugging of a clay or clay body under reduced pressure increases its plasticity considerably. The process is called 'de-airing' but actually achieves more than merely removal of air bubbles.

To become plastic a clay needs to be homogeneous and to contain as few large non-plastic particles as possible. Air bubbles are non-plastic and so their removal by evacuation improves the plasticity. This de-airing is, however, no longer considered to be the major cause of the benefits derived from passing the clay through a vacuum.

Reduction of the air pressure reduces the boiling point of the water thereby increasing the vapour pressure. Its reaction is therefore accelerated, so that the effect of lengthy ageing can be produced on passing the clay through a de-airing pugmill.

The first de-airing patent was taken out in 1902 but it was not until 1920 that its application was under serious consideration. De-airing pugmills for the heavy clay industry came into production in 1927, and are now in fairly universal use. Machines in a large range of sizes from a laboratory model run with less than one horse power, some four feet long (**J48**) to 150 horse power machines some 25 ft long (**F3**). They are made up of three essential parts. First a pugmill through which the material is introduced. At the end of this it is forced through a small opening into the vacuum chamber, the clay therefore making the seal. Inside, the column is cut up by revolving knives and then falls through the vacuum chamber on to an auger which consolidates it and finally extrudes it. This extrusion may or may not be a major part of the shaping process and the extruding section may be either horizontal or vertical, in the latter case it is continuous with the vacuum chamber (**R6a S168**) (Fig. 7.24).

There are various aspects to be considered when installing a de-airing pugmill if optimum results are to be obtained. Everhart (**E27**) considers the factors that affect the sealing of the inlet to the vacuum chamber, which must be complete but without unnecessary compression of the body. These are:

(1) the material used, as each one has different inherent sealing qualities;
(2) variations in the feed rate;
(3) change of moisture content;
(4) change of vacuum;
(5) wear of machine parts;

it is therefore best if the sealing part of the system is adjustable.

Fig. 7.24. *De-airing pugmill. The trolley of press cakes is seen on the right* (*Wedgwood*, **W35**)

Once in the vacuum chamber the clay or body must be disintegrated and opened up thoroughly to the greatest degree of fineness with the minimum amount of applied pressure, and then permitted to remain in an open, unpacked condition for the greatest length of time.

The shape of the de-airing chamber, and the relative positions of the shredder at the inlet and the compression screw at the outlet is very important. For satisfactory continuous action the clay must fall from the former to the latter, which therefore cannot be on the same shaft. Economies are sometimes made by using a single continuous shaft through both parts of the pugmill, but they are false as such machines clog up very often holding up the production process badly.

The advantages found when a body is passed through a de-airing pugmill are numerous. Firstly it greatly increases the workability, in some cases the blending of clays can be dispensed with, in others 20 or 24 ft of open pugmills can be replaced by 7 or 8 ft of de-airing pugmills. Secondly the green, dry and fired ware is denser and stronger.

Surveys of the experiences of plants which have introduced de-airing equipment were made in the U.S.A. by Garve (**G8**) and in Great Britain by Clews (**C39**), and show that every clay or body must be given individual treatment. Many plants found that a higher water content must be used; the optimum vacuum varies but is usually high, around 28 in. of mercury (0·967 kg/cm^2); the angle of the knives and blades, and the shape of the dies must sometimes be altered from that used without de-airing; drying is often somewhat slower; firing of carbonaceous clays must be done more carefully. The power consumption is, however, the same or less than with non-de-airing and the product always more plastic, which is usually beneficial.

De-airing of Casting Slips

It is obvious that a number of benefits could be derived from a de-aired clay slip. There would be the direct advantage of eliminating blockage of pipe lines due to air pockets in the pumps. But a more far-reaching result would be the possibility of producing absolutely dense ware by casting. At present if a given electrical porcelain body is made up in the plastic state and passed through a de-airing pugmill the product can have an absolute porosity of down to 1·5%. The same body shaped by casting has an absolute porosity of 4 to 8%. There have been attempts to de-air casting slips by allowing them to flow in a thin stream through evacuated vessels, but in general they have been ineffective. The trouble with a slip as against a plastic clay is that there is not only the clay which holds air but also the water which dissolves 20 cm^3 air per 1000 cm^3 water at 15° C (59° F) and very easily takes it up again after de-airing (**S118**).

MATURING OR AGEING

This process is akin to weathering (p. 263) and consists of storing a ready mixed ceramic body in a moist cellar. The Chinese formerly matured some of their porcelain bodies for up to 100 years. Here again the clay–water reaction is allowed to proceed (and other slow bacteriological reactions), and chemical reactions can take place. The clay's plasticity is improved. It is often thought that the water content becomes more uniform because of capillary action but Ford and Noble (**F14.1**) have shown that this does not necessarily occur.

Maturing can be considerably accelerated by increasing the temperature, thereby raising the water vapour pressure, as the rate of maturing of a clay is a function of the water vapour pressure (**S94, 116**). For instance a stoneware body that was normally matured for four weeks at 15° C (59° F) was equally successfully matured in half the time by increasing the temperature by 10° C to 25° C (77° F).

With increased mechanisation in the heavy clay industry the labour required to spread and collect up a plastic body to be matured becomes disproportionate. It has been shown that for a number of clays and bodies most of the benefit derived from maturing is achieved in the first 12 to 24 hr. Such bodies, mostly brick bodies but also roofing tile and stoneware ones can be matured in a 'maturing tower'. This free-standing tower stands about $16\frac{1}{2}$ ft (5 m) high, and has a capacity of up to 3531 ft^3 (100 m^3), a

Fig. 7.25. *Maturing tower* (*Rieterwerke*, **R30**)

normal day's output. It is fed into the top by conveyor, where a cone causes the material to spread out. The base of the tower is made up of a slowly turning plate, powered by 12 horse power. The gravity build-up keeps the material under pressure, while slow mixing occurs. After maturing overnight augers remove the material from the bottom, the revolving base plate preventing bridging (**R30**) (Fig. 7.25).

Such towers can relatively easily be equipped with a heating device to accelerate the maturing (**A211**).

RE-USING OF PRESS-WATER, BODY SCRAP, ETC.

When the body is prepared by the wet method large quantities of water are required, a certain amount of which may be recovered in liquid form at the filter presses, etc. The demand on the water supply can be considerably reduced if the recovered water can safely be re-used for the blunging. This is, however, not always the case. If, as is normal, the raw materials contained soluble salts the press water has a higher salt content than the mains water originally introduced. It also probably has a different pH value. It will therefore have a different effect on the clays it is mixed with, so that the properties of each successive batch may vary.

In many works, however, it is considered that the over-all composition of the water is sufficiently constant if a given proportion of the mixing water always comes in fresh from the mains.

Rather similar considerations apply to the re-using of body scrap. If the body contained no soluble salts and distilled water was used for mixing every time, there would be no difference between the reworked body and the fresh one. But these ideal circumstances are very rare. Not only does the body have salts which increase at every remixing with salt-containing water, but they also tend to pick up bits of plaster, dirt, etc. The best way to reincorporate scrap is as early as possible in the preparation process, *e.g.* in a wet method it is best to blunge the scrap and add it in fixed quantity to the mixing ark. The shorter the circuit of the scrap the more dangers there are. Many works only use scrap in second grade ware.

It is essential that scrap from casting which has had deflocculants added should only be re-used for casting. Scrap from plastic shaping can be re-used for the same purpose or for casting.

SLIPHOUSE PUMPS

During the course of preparation of the body the slip has to be pumped on several occasions. For transporting purposes, only the head of liquid has to be counteracted, but in the filling of filter presses high pressures have to be used and they have to be applied uniformly throughout the operation.

There are numerous types of pump being used: dead weight and other ram pumps, diaphragm pumps, gland and glandless centrifugal pumps, hydraulic pumps, plunger pumps, constant pressure plunger pumps, rotary pumps, compressed air elevators, controlled flow pumps, and Moyno pumps.

Dead weight pumps are very widely used for filter presses and for casting slip. They have the advantage that a tap can be fitted on the delivery pipe which when turned off automatically stops the pump. For filter presses they have the disadvantage of pulsating flow. In their modernised form dead weight pumps can be made entirely self-contained with their own motor (**E6**). The capacity and required power for one make of dead weight pumps is given in Table 178 (**G61**).

TABLE 178
Capacities of Dead Weight Pumps and Horse Powers Required to Drive
(G61)

Size of pump		Capacity per pump		Single unit	Double unit
(in.)	(cm)	(gal/min)	(l./min)		
2	5·08	1½	6·8	1½ h.p.	3 h.p.
3	7·62	4¼	19·3	3 h.p.	5 h.p.
4	10·16	8	36·4	3 h.p.	5 h.p.
5	12·70	15	68·2	3 h.p.	5 h.p.

The constant-pressure slip pump is an improvement on the normal plunger pump and does not require a safety valve. The two plungers are so weighted that they will stop pumping when the required pressure has been achieved and restart when it drops again. The ratings given by the makers of an American machine are (**C67**):

Horse power, 3; pressure range, 50 to 100 lb/in² (3·5–7·0 kg/cm²); capacity, 19 gal/min (86 l./min) when operating a normal pinion shaft speed of 140 r.p.m., which gives 28 strokes/min of the pistons.

A further step is the four-cylinder pump which gives a constant flow with no pulsation (**C67**).

Wathey (**W19**) describes the Willett 'flow-control' pump. The pump is driven by fluid transmission. It is operated by a 1¾ h.p. electric motor coupled to a centrifugal oil pump which operates a 2 in. (5 cm) bore oil–hydraulic ram. The main feature of the pump is that the stroke is always the full length even when it is slowed down by increased resistance, and the return stroke is always very quick. This is claimed to reduce the time normally required to fill a filter press with a dead weight pump by 15%.

Centrifugal pumps with glands have the disadvantage of wearing very quickly and thereby reducing the effective lift and capacity of the pump. In the newer glandless pump the omission of the gland makes it unnecessary to have very fine clearances on parts that wear. All wearing parts can therefore be covered with abrasion-resisting rubber. However, it is necessary to have the inlet above the impeller, or a special priming device must be added (**S194**).

Rubber diaphragm pumps can replace metal (or porcelain) pistons for raising thickened pulps or slurries and hence avoid the metallic contamination and quick breakdown that occurs when abrasive slurries are handled by metal moving parts. These pumps, however, are not intended for pumping coarse material or against a head. They are best used as suction pumps for the control of discharge from thickeners or settling tanks (**S195**).

HANDLING EQUIPMENT

Both in the preparation of ceramic bodies and glazes and in many subsequent stages of manufacture there are uses for numerous types of mechanised handling equipment. Its application depends very largely on the particular works. The main types may be grouped as follows:

DISCONTINUOUS

Trackless carriers: (1) wheelbarrows, (2) rigid trucks, (3) hand lift trucks, (4) power-operated lift trucks, (5) dumpers.

Tracked carriers: (1) narrow gauge railways, (2) overhead tramrail (monorail) systems, (3) skip hoists, (4) lifts.

Carrierless: (1) friction spiral chutes, (2) roller conveyors, (3) spiral and straight chutes, (4) wheel conveyors.

CONTINUOUS

Chains: carrierless, *e.g.* chain conveyors, wire belt conveyors; with carriers, *e.g.* pallet conveyors.

Belts: belt conveyors.

Screws or worms.

Pneumatic and hydraulic.

Dust Control

Dust in the potteries is a constantly occurring hazard to health. Extractor hoods have to be fitted over any operation causing dust and the final deposition and disposal of this dust also causes difficulties. A new wet method of doing this has been described (**A125**). (Legislation, *see* p. 1322, appendix.)

Chapter 8
Shaping

THE shaping of traditional clay ceramic bodies depends on the plastic and flow properties of clay. The ease with which a clay–water mixture changes its shape depends on the water content (*see* graph, p. 66). Thus when the water content is about 50% a slurry forms that will flow like a liquid to fill all spaces in vessels and moulds at only 0·1 atm pressure. When the water content is dropped to 40% 0·4 atm is required to produce the same flow, at 35% it is 1 atm and at 30% it is $2\frac{1}{2}$ atm. At the pressures mentioned two separate portions of the same body will unite to form a homogeneous whole. Drier bodies will also do so but at much higher pressures, namely:

25% moisture	10 atm
20%	40
17%	100
15%	200
10%	infinite (**S90**)

The methods of shaping are therefore divided up according to the condition of the body, as follows: (*a*) liquid; (*b*) thick slurry; (*c*) plastic; (*d*) semi-dry; (*e*) dry.

Both (*a*) and (*b*) flow under gravity or low pressure. In the plastic and semi-dry condition flow is achieved only with considerable pressure but the new shape is retained when the pressure is released. In the dry condition flow cannot usually be induced.

The oldest methods of shaping (hand-modelling and throwing on a wheel) require the clay bodies to be in a plastic condition. Plastic bodies are also used for jolleying and jiggering, extrusion, and pressing in plaster and other moulds either by hand or with mechanical or hydraulic presses.

The casting of clay in a condition where it will flow under gravity and hence not retain an imposed shape began in the eighteenth century. At first oiled metal moulds were used, but the plaster of Paris ones were introduced in England in 1745 (**A54**).

Shaping of bodies too dry to flow under pressure by 'dry pressing' is

SHAPING

comparatively new. As, however, there is almost continuous transition from plastic hand shaping through to dry pressing under high pressures, this sequence will be taken right through in the following discussion and casting dealt with separately thereafter.

SHAPING OF PLASTIC BODIES

The most characteristic condition of clay is the plastic state and it is on this that all primitive methods of shaping were based. These were first hand modelling and building up of vessels from coils, and subsequently the potter's wheel. They are still used for individual art pieces, for very large pieces and for some commercial production, although decreasingly so, but on them are based many of the modern methods of mass production.

Although a variety of shaping methods depend on the ability to deform clay by relatively little pressure after which it will retain its shape the actual consistency or stiffness of the body varies for different methods. A high water content means easy working but high shrinkage and therefore greater drying losses. Where a machine replaces manual work the body can usually be made stiffer and therefore easier to dry.

Soft Plastic Shaping of Bricks and Tiles

The oldest methods of hand moulding bricks used a soft plastic body. Only certain types of clay of relatively recent geological origin can be brought into a suitable condition; however, they do produce a very pleasing facing brick.

Hand making is a skilled operation. The maker takes a lump of the prepared body of the correct size, forms it into a clot in his hands and then throws it into a mould. The impact of the clay must be sufficient to spread it evenly throughout the mould. The shaped brick is at first very soft and cannot be handled before it has partially dried.

Because of the aesthetic appeal of these bricks for facing work machines have been developed for their speedier production.

Soft plastic moulding is also used for fireclay and other refractory bricks and shapes. These are often much larger than building bricks, a large 'walk' has to be prepared and is then thrown into the moulds. The 'drop machine' has been developed in U.S.A. for moulding silica bricks in a manner similar to the hand-moulded process. The wad or walk is prepared by machine and then dropped automatically through a gate into the mould some 20 ft (6 m) below (**C64**).

Plain roofing tiles and pantiles are also often hand moulded. As the tile is thinner a rather stiffer clay must be used. The mould is sanded. After release from the mould the nib of a plain tile is formed by hand. A certain amount of drying on flat trays follows and then the tiles are 'horsed' to give them the correct curvature. For this six tiles stacked with the nibs at alternate ends are placed on a three-legged stool with a top cambered to about a 10 ft radius, here they are given several blows with a wooden block of the same curvature.

Any distortion arising in pantiles during the first stage of drying is corrected by 'thwacking' them when half dry. Here the tiles are placed on a block of the correct curvature and beaten with a bevelled wooden stave (**C41**).

SEMI-STIFF MUD

The soft mud process has the disadvantage that the bricks are difficult to remove from the moulds and cannot be handled until considerable drying has taken place. With the greater power available in machine shaping a somewhat stiffer consistency can be used. For example, the 'Lancaster' Autobrik Machine (**P49**), with automatic pallet car loader, mass produces bricks from clays that could be hand moulded. The body used is, however, slightly stiffer than for a hand-made soft mud brick. It is fed through an auger machine into a row of sanded moulds, struck off and then pressed. The moulds move on and are bumped sufficiently to loosen the bricks which can then be tipped on to a pallet. The pallets of bricks are mechanically loaded on to rack cars and transferred to the drying chambers or tunnels. The brick moulds continue in a circuit under the machine for resanding before refilling with clay.

A new type of stiff-plastic brick machine has recently (1958) been brought into operation in Great Britain, the Hewitt machine. In this the pugmill filling the moulds with clots is offset so that the centre boss is not on the centre line of the brick mould. The pug is also run more slowly so that each clot is the result of a single sweep of the pug blade, instead of a number of independent layers which have to be pressed together. Bricks are produced at about 2550 per hour. As the machine avoids cores and laminations it is also recommended for firebricks (**A41**).

Hand Modelling and Hand Moulding

Hand modelling is of course used by the artist craftsman for making individual pieces, for making prototypes from which moulds are made, etc., in the fine ceramic industry. It is also used considerably in the highly skilled production of large pieces in stoneware or refractory bodies.

For instance, the large refractory glasshouse pots have their base moulded on a baseboard covered with grog or a damp sack and the sides may then be built up by hand modelling. Small pieces are thrown into position and then modelled into place. On the first day about 9 in. (23 cm) may be built up before leaving the piece some 24 hr to stiffen. On successive days 2–3 in. (5–7$\frac{1}{2}$ cm) only are added, so that a large covered pot may take about four months to build up.

Hand moulding, by throwing, pressing and beating successive pieces of

SHAPING

the body into position against mould walls is also used for making glasshouse pots. Large flat pieces of stoneware are made by hand spreading or smearing. Here successive layers of body are spread quite thinly on top of each other, each layer being roughened by drawing bent fingers across it before applying the next.

Finer-grained stoneware or faience bodies are hand moulded in plaster of Paris moulds to make commemorative plaques, brewers' signs and other unsymmetrical shapes.

Plastic Pressing

It is possible to mechanise the moulding of certain bodies by using metal dies and mechanical or hydraulic presses. This is applicable to large pieces, *e.g.* saggars, crucibles, insulators. The dies must be designed to allow no

FIG. 8.1 *Saggar press* (*Dorst*, **D54.1**)

oozing out of the body and the exact amount of body required must be measured for each pressing. The dies should not be oiled, but heating assists the release of the pressed piece (Fig. 8.1).

Mechanisation of plastic pressing enough for mass production is a recent introduction of the 1950's. The *ram process* is described by Blackburn

(B71). It consists basically of pressing the plastic body in between gypsum cement moulds, the backs of which are evacuated so that water is readily removed. A hand-operated hydraulic press is used. Compressed air is then applied to either the upper or the lower die as the press is opened, leaving the formed piece on the other one, from which it is subsequently released by the same means. The set-up, then, is a press, with one pair of gypsum cement dies set in metal rings for each product. The prepared body is cut off the extruded column in the correct size and may be spread into a bat (*see* jiggering) before placing on the lower die. During the pressing the water content can be reduced any desired amount but 1 to 3% is normal. The process sets up remarkably little strain in the pressed piece which results in very little warping. Very many shapes can be made, small pieces at a rate of some 2200 per day, or if the operator pugs his own clay, etc., a self-contained unit with one man makes about 1200 to 1800 pieces per day, or with two men 2400 to 3600. The method should therefore be compared with the longer established processes of jiggering and casting. In both cases the ram process has the advantage of only requiring one pair of moulds whereas the others need a series which must be dried between using. Furthermore, jiggering is limited to circular or at most oval pieces whereas the ram process can make many shapes. However, the extra precision required on the presses means that for circular pieces little is saved when compared with jiggering. For complex pieces the saving when compared with casting is considerable due to the economy of moulds, space, drying heat, time, and in the strain-free ware.

The Potter's Wheel

The revolving wheel is the basis for the shaping of most articles of circular horizontal sections, either, by free-hand throwing, or by jiggering, in every degree of automatic mechanisation.

The flat wheel head, or a specially shaped one for taking moulds is mounted on a spindle. In a foot-operated machine this is mounted on a large and heavy balance or flywheel. The outer edge of this is either directly motivated by the thrower's foot or by a foot-operated treadle. The weight of the balance wheel converts the jolting foot movement to regular revolutions. The spindle may also be run by a belt drive or directly from an electric motor. Suitable arrangements for varying the speed and for braking must be incorporated.

Hand Throwing

Hand throwing on the potter's wheel is an ancient, highly skilled and fascinating operation. It is used nowadays mainly for individual art pieces. It is, however, also used for making some cylindrical insulator blanks which are subsequently shaped by turning.

Hand throwing by a skilled potter appears to be very simple and no verbal description can do justice to this shaping method requiring so much 'feel'. A well-prepared lump of plastic clay of the required size is thrown on to the centre of the wheelhead. It is 'centred' by pressure with the hand held completely steady. Its shape is then altered a number of times by applying pressure at its sides or on top to draw it up and down and thus knead it through thoroughly. This initial working on the wheel is considered by many to be essential for the production of strain-free ware. The desired shape is then gradually made by hollowing out the piece and then slowly drawing up the sides. Overhaste leads to strained ware that cracks during drying or firing. (Fig. 8.2.)

To assist in achieving desired dimensions a graduated stand with pointers and profiles may be fixed behind the wheel (Fig. 8.3) (**A172**).

Various methods of catching the water and slip thrown off the potter's wheel during work include the semi-circular trough found round many of them. A device claimed to prevent splashing more completely is described by Böttner (**B87**).

Jiggering and Jolleying

Where numbers of identical articles are required the wheel is used in conjunction with plaster moulds and metal profiles.

[There appears to be some confusion about the use of the words 'jigger', 'jolley', 'jiggering' and 'jolleying'. It is proposed here to use the two terms to differentiate between shaping the concave or the convex surface on the mould. Thus 'jiggering' (Fig. 8.4) is the process of shaping an article *on* a convex plaster mould, *i.e.* the *inside* of a plate; and 'jolleying' (Fig. 8.5) is the process of shaping an article *in* a concave mould *i.e.* the *outside* of a cup.]

The flat wheelhead used in hand throwing is replaced by a holder for a mould, usually with a central conical hole and is known as a 'jigger-head' or 'jolley-head'. The moulds are made of hard plaster and give the shape of either the inside, as for plates, or the outside of the piece to be made.

A series of identical moulds have to be available to each spindle. The required amount of body is placed in or on the mould either directly from the pugmill column or after being spread into a bat (p. 724). A metal profile of the second surface of the piece is then gradually brought down on to it. The profile is fitted either on a lever or it moves up and down vertically on a shaft, being counter-balanced over a pulley. The shape of the piece may determine which type must be used but frequently the choice is a matter of taste and expediency. Undercut vessels have to be made with special tools of the vertical type on which the profile can be moved horizontally when it is down. The profile is applied to the body on the mould slowly, a certain amount of kneading work being performed on it as in throwing. It presses the body down on to the mould and also scrapes and pushes any surplus away. Many profiles are fitted with stops to prevent them being pushed too far on to

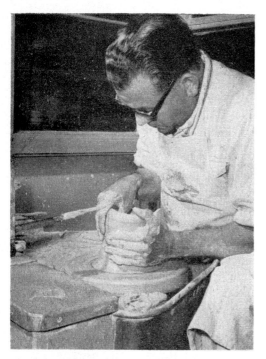

FIG. 8.2. *Hand throwing; opening out and beginning to draw up the sides* (*Wedgwood*, **W35**)

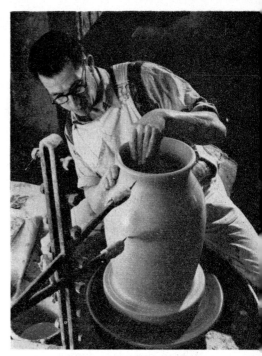

FIG. 8.3. *Hand throwing of a large stoneware jar almost complete; note the pointers to assist with the dimensions* (*Doulton*, **D56**)

FIG. 8.4. *Jiggering a plate* (*Wedgwood*, **W35**)

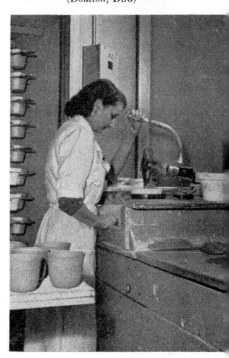

FIG. 8.5 *Jolleying bone china cups, moulds with shaped cups are seen in the dryer, left.* (*Wedgwood*, **W35**)

the mould. When the piece is finally shaped the profile is raised, the spindle stopped, and the mould removed and placed to dry with the piece on it. A fresh dry mould is then inserted and the process repeated. Some spindles are fitted with a treadle-operated device for lifting off the moulds, which become firmly wedged during the working.

FIG. 8.6. *Forming a large cylindrical stoneware vessel on a vertical jolley machine* (*Doulton*, **D56**)

The speed of production by jiggering or jolleying is dependent on the plasticity of the body and on the size of the pieces. Earthenware bodies are very plastic and plates can be jiggered almost four times as fast as those made of the much less plastic hard porcelain body. Bone china also has to be shaped slowly. For large insulators, etc., made of less plastic porcelain body considerable skill and time are necessary to produce strain-free pieces that will not subsequently crack or warp. (Fig. 8.6)

The plaster of Paris moulds used for jiggering have a porosity of about 30%. Careful control of the porosity of the plaster for each shape is necessary, adjustments being made for different shapes. Uneven drying of plates, the rim drying first, may be avoided by periodic rubbing of oil into the rim of the plaster mould. Moulds last about 190 to 200 cycles.

The most usual material for the profiles is mild steel. This is relatively easy to work, file, regrind, etc., but wears down sufficiently to require readjustment in about a fortnight or after jiggering 200 to 300 dozen pieces. Tool tips of sintered carbide, in particular tungsten carbide, sinter alumina, certain hard alloys, etc., have been introduced instead of the mild steel and have the tremendous advantage of much longer wear. However, once they do require adjustment the filing and grinding is very much harder.

The jigger profiles are sometimes heated.

Faults found in a good deal of jiggered flatware are uneven heights of the rim and concave or convex bottoms. Although these faults may be due to the preparation of the body, the drying or the firing, they are very often initiated during the jiggering process. They are much more frequent in vitreous ware than in earthenware and most particularly in hard porcelain.

Investigation of such ware during various stages of the firing process shows that the densities of different parts are not the same. This arises during the shaping when some parts are more compressed than others, so that some have a higher water content and are subsequently more porous on drying and have a higher shrinkage on firing, causing strains. This leads to a rising or sinking of the centre. Alteration of the jigger design and angle can remedy this trouble. Most jigger tools work by compression as well as cutting, A in Fig. 8.7, by reversing the tool the action is almost entirely by cutting, B. By judicial combinations a more uniform piece of jiggered ware can be obtained; *e.g.* Gould (**G62**) found that for hard porcelain plates very much better results were achieved by having a compression tool for the flat centre and a cutting tool for the rim.

Similar results may be obtained by altering the angle the jigger tool makes with the ware, making it larger on some parts of the piece than on others, thus exerting different pressures.

Round ware firing to a non-round shape with a wavy edge may be traced to the transfer of the bat to the jigger-head. If the centre of the bat does not fall on the centre of the mould the strains set up during the two operations are not concentric. With normal hand transfer the centres rarely coincide but if the drum-head method of batting is used this can be achieved (Fig. 8.8).

Batting

The jiggering of plates, dishes and other flatware involves placing the body on the mould in the form of a flat, circular 'bat'. Correct forming of these bats to a uniform thickness and consistency is essential to good jiggering.

A bat may be formed on a plaster base covered with filter cloth by smashing it flat with a plaster maul. This is the old manual method, but can also be done by machine. Alternatively the plaster disk is revolved and a spreader tool brought down on to it. These methods work very well for plastic bodies, in particular earthenware even at high speeds but when the much less plastic porcelain body is used, difficulty is experienced at all but low speeds. Furthermore a porcelain bat will not hold together during manual transfer to the jigger on the cloth. It is therefore made on a rigid drum, transferred to the

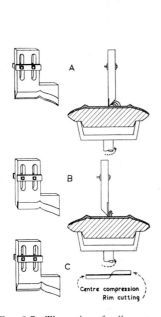

FIG. 8.7. *The action of a jigger profile:*

A, compression tool, clay forced under tool
B, cutting tool, clay shaved off
C, combination tool (Gould **G62**)

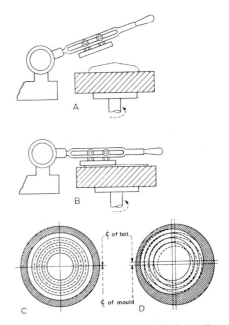

FIG. 8.8. *Batting out and transferring to jigger*

A, clay on whirler
B, tool spreading bat
C, clay centred on mould
D, clay off centre on mould
——— *line of strains induced in clay by bat forming*
- - - *line of strains induced in clay by the jigger tools* (Gould, **G62**)

jigger and then cut off it. The drum method also has the considerable advantage of ensuring that the centre of the bat comes central on the jigger mould.

The batting method evolved for the Miller automatic jiggering machine has considerable merits of its own, and has been subsequently used for a separate batting-out machine (**M78**). Instead of starting with a ball of the body a thin circular slice is taken off the end of a pugmill column, so that part of the deformation required during batting is eliminated from the start.

This slice is then hydraulically raised against a heated die, high pressure being exerted on the body.

For larger pieces of hollow-ware to be made in a jolley the clay is first spread slightly to make a thicker smaller bat and then pulled over an upright post of plaster covered with a cloth. This can be removed by hand and transferred into the jolley mould and the cloth removed.

Semi- and Fully-automatic Jiggering

The jiggering of small ware and in particular tableware can be made either semi- or fully-automatic if the body is sufficiently plastic. In both cases skilled jiggerers are not required and the speed of production per man is increased. Such machines were first developed in the United States, there being three types of semi-automatic machines and the Miller fully-automatic one.

The twin unit jiggers have two spindles side by side serviced by one unskilled operator. These are either both used for shaping ware or one is used for batting out. The operator throws a ball of clay in the mould, places it on the spindle and then sets the machine in motion. This then performs the entire jiggering operation, including the application of water and trimming, and stops. The mould is removed by hand and set in the dryer and a freshly charged one put in its place. Figures given for the Allen semi-automatic jigger (A182) state that whereas a batter out and hand-jiggerman can make some 20 dozen 7 in. diameter plates in an hour a single operator can make 23 dozen similar plates per hour on the semi-automatic jigger. The uniform action of this machine also produces more strain-free ware. Cups can be made at 34 doz./hr.

The second type of semi-automatic jigger has a slowly revolving table with three or more spindles distributed round the circumference. Each spindle has a corresponding profile-holder above it which moves round with it. At one point during its journey round with the table the spindle stops for manual unloading and loading. For the rest of the time it revolves and the profile comes down on it automatically, spraying water at certain points and trimming the ware. A machine with eight heads can make about twenty-four plates per minute (120 doz./hr). On some of these machines a different shape can be made on each head.

In the third type the roles of jigger-head and stationary profile are reversed. A revolving table with six mould holders is loaded with fresh moulds with balls of body and unloaded manually. It takes each mould to the single shaping head where it is raised gradually and the work done on the body by a revolving head holding *three* profiles in a hood. The speed of this head is 700 r.p.m. as opposed to 250 r.p.m. of a normal spindle for similar ware, so that there is an equivalent of 2100 passages of the ware under the profile per minute. Water is sprayed on as required and surplus clay extracted through the hoods. The shaping of a plate takes 4 to 6 sec, production rate being 540–600 pieces per hour. This machine is particularly suitable for jiggered pieces. As these are shaped on the outside of the plaster mould they are subjected to centrifugal forces tending to lift them off the mould. This is eliminated when the profile revolves instead of the mould. The three profiles instead of the usual one make for more even pressure as well as greater speed, and hence less strain in the ware.

Of the fully automatic jiggering machines, the Miller (**M78**) machine is the best known. This has from one to twelve production lines. It is fed directly by a Fate-Root-Heath de-airing pugmill with 3, 5 or 7 in. (7·6, 12·7 or 17·8 cm) nozzles fitted with an elbow so that the column is delivered vertically. The emerging slug or wad is cut automatically at the right thickness by a piano wire cutter and falls on to a mould below, which is on an intermittently moving slide. The mould moves on to a plunger head which raises it against a heated die where the approximate shape is arrived at, it is then sprayed and jiggered or jolleyed to its final shape. It passes on automatically into and out of a dryer after which the ware must be removed from the moulds by hand. The moulds move on to the starting point again automatically.

The Miller machine itself is made to make tableware, except ovals, up to 12 in. (30·5 cm) diameter and 5 in. (12·7 cm) height. Each line can make 70 to 80 dozen flat ware per hour and up to 120 dozen hollow-ware per hour. Its labour requirements are one girl for every two lines for removing ware, and one, two or three men to service the single, eight- or twelve-line machines, respectively (Fig. 8.9).

Automatic jiggers require plastic bodies, they work very well with earthenware, slightly less well with vitreous china and probably cannot be directly applied to hard porcelain or bone china. This is most noticeable in the production of flat ware which requires batting out. Most batting-out machines are unsuitable for vitreous bodies, but the Miller method of slicing a pugged column could probably be used successfully. The less plastic bodies could be adapted for automatic jiggering by adding organic binders or small quantities of certain chemicals, *e.g.* tannic acid, probably more easily than the machines could be altered to suit the bodies.

Semi-automatic jiggers and jolleys have been adapted for making bone china ware by using special cams to allow for adequate lubrication while the piece is being made.

The Roller Machine

A plate-making machine differing considerably from the jigger has been recently introduced in Great Britain. This rolls the clay on to the mould, the roller taking the place of the jigger profile. It is claimed that this does not strain the clay as a conventional profile may do. A big advantage is that a plate can be made directly from the wad of clay fed to the machine and does not require batting out first. Operatives therefore need not learn to transfer the bat correctly. Small flatware can be produced at 44–50 dozen/hr. Larger ware is less successful.

FIG. 8.9. *A multiple-line Miller jigger machine with eight lines producing flatware at a rate of 80 doz. pieces per line per hour, i.e. 640 doz. pieces total per hour* (**M78**)

Fig. 8.10. *Extrusion*

Extrusion

The extrusion process is used for the shaping of pieces with regular cross-sections. It is also the normal method of withdrawing plastic clay from the pugmill in the form of a uniform column. The column can then be cut into regular lengths containing the correct quantity of body for each piece to be made from it *e.g.* for hand throwing, jiggering, etc.

Where extrusion is the main shaping process the body is usually stiffer, having less water than a plastic body for jiggering, etc. Extrusion is used in the manufacture of wire-cut bricks, hollow bricks, roofing and floor tiles, drain-pipes, tubes, perforated plates, insulators, sparking plugs, etc. For many objects the extruded pieces are merely blanks for further shaping by turning and cutting when they are in leather-hard condition.

Basically the extrusion process consists of pushing a column of clay through a die. Various methods have been evolved for forcing the clay through although the auger is the most widespread. Of the others, there is the piston method, where a piston forces clay from an intermittently charged box through the die. This finds some application in sewer pipe manufacture. Alternatively clay may be forced continuously through a die by expression rolls, a pair of rollers set one above the other which force an oversize column into a tapering die box. This is used in the *roller bat-machine* for roofing tiles where stiff-plastic clay is passed between rolls or 'bat-wheels' about 36 in. (91 cm) in diameter from which the clay is delivered as a continuous ribbon of uniform thickness. The column is automatically dried and cut into lengths. The output is 5000 to 7000 bats per hour (**C41**).

With an auger clay columns can be extruded intermittently or continuously, the latter being the commonest method and is frequently attached to the de-airing pugmill (Fig. 8.10). The whole combination has to be very carefully designed if good results are to be achieved. Extruded columns from pugmills often show serious defects during drying and firing. They may show waviness, S-shaped cracks in the centre, longitudinal cracks, circular cracks breaking off all the sides, concave ends, diagonal twist, sunken faces, bow, depressed and bulged ends, etc. The direct causes are the auger action and the die design, but Seanor (**S47**) suggests going back right to the beginning when looking for the trouble. The following factors should be considered:

(1) Machine design.
(2) Auger design and angle.
(3) Selection of raw materials, and their adequate mixing.
(4) Water content of body (*see* Robinson and Keilen, **R44**).
(5) pH value of body.
(6) Grinding. Even a textured brick must have enough fines to bind the coarse material.
(7) Feeding the clay. The feed to the auger must be regular.
(8) Pugging must be uniform and is best if the pugmill is not overloaded.
(9) Auger speed.
(10) Die design.

The degree to which it is desirable to eliminate faults depends, of course, on the type of ware to be produced.

The auger and die design and the auger speed are best considered together. The faults are caused in the ware first by the cutting action of the auger. These cuts together with the screw action give the body a screw structure, and the ware distorts accordingly as it dries. Further distortions are caused by the auger pushing the material in the centre of the column at a different rate to the sides, and by the dragging action of the die.

Seanor (**S47**) states that the auger should run fast enough for every lump of clay dropping on to its ends to be taken into the barrel immediately. If the clay tends to back out of the auger it is overfull and will exert too high pressure on the clay in it and produce pressure laminations. An indication is given by the temperature of the barrel, this heats up if the auger is too slow but remains cool if it is fast enough.

The cuts caused by the auger are sealed better the longer the stretch between the auger tip and the mouthpiece. This also reduces the screw distortions. The speed is, however, lowered. The S-cracks are also due to the auger action. The theory by Koller which is generally but not universally accepted in Europe and America is that a circular hole forms opposite the boss of the auger which during subsequent movement becomes first oval and then S-shaped. Tarasenko (**T12**) maintains that they are due to the irregular application of pressure by auger blades, with the greatest variations at the outside and the least at the centre. The ability of the body to seal together after the auger action and lose the screw structure depends on its tendency to stick to itself even when surfaces have been polished by the auger blades. This is largely determined by the water content and it is assisted by grog.

The length and design of the spacer between the auger and the die is most important. If there is a sharp decrease of cross-section there will be considerable differential flow in the column. The centre will flow readily as it can take up the space in front of the auger shaft, while the circumference has to change direction and overcome the friction. The spacer should therefore be as long as possible.

The flow of material can also be straightened by using a tapered auger in the spacer. The distance between the blades is increased as the section decreases so that the volume of material held between adjacent ones is the same. The pitch is increased towards the die, too. This type of auger is good on plastic clays that knit together easily (**A37**).

The S-cracks, wavy columns, etc., which are derived from the screw-cutting action of the auger can be eliminated by radically altering the direction of flow of the clay after it leaves the auger, *i.e.* by setting the spacer and mouthpiece perpendicular to the auger barrel. As the auger will tend to push the clay forward to the bend and cause a deficiency at the back of the perpendicular column it must be tapered to counteract this, as shown in the diagram (Fig. 8.11). The auger is further cut down in any position directly in line with a bulge in the extruded column, and must be adjusted to suit each particular, body–spacer–die set-up.

For particularly strain-free pieces of, say, circular cross-section the auger set-up with the perpendicular mouthpiece is used to extrude a wide ribbon-like column of thickness equal to the length of the required cylinders. These are then cut out of the flat piece, perpendicular to the direction of extrusion.

A method of overcoming the inhomogeneity of extruded columns has recently been worked out in Germany, and is first described by Hagen (**H3**) in 1955. It consists of interposing an oscillating screen after the end of the auger. The speed of oscillation acts on the thixotropic nature of clay and makes it flow through more readily. Stiffening re-occurs in the spacer and die. Brückner (**B125**) further describes a better method of motivating the screen by eccentric drive on a shaft running through the centre of the auger, which is hollow. This uses relatively little power and can be adjusted as regards frequency and amplitude. The method is particularly recommended

732   INDUSTRIAL CERAMICS

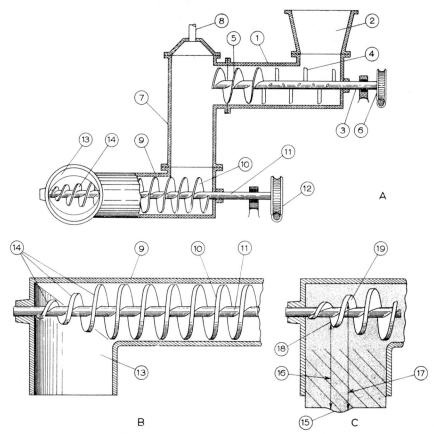

FIG. 8.11. *Elimination of extrusion faults*

A, elevation of de-airing pugmill set up, with tapering extrusion auger and the mouth-piece perpendicular to the auger

B, plan of extrusion portion of machine

1. mixer barrel;   2. hopper;   3. shaft;   4. knives;   5. auger blades;   6. power supply;   7. vacuum chamber;   8. to vacuum pump;   9. auger barrel;   10. auger blades;   11. shaft;   12. power;   13. mouthpiece;   14. tapering of auger

C, detail of *B* shown in operation. The bulge (15) shows that the portion of the auger between points (18) and (19) found by the connecting lines (16) and (17), should be filed down a little

for the extrusion of porcelain insulators where laminations and other structures are so particularly detrimental.

From the end of the auger the clay passes directly or indirectly to the die. For high-grade ware which should have a smooth surface, a long slowly tapering die is used, sometimes preceded by a wider expansion chamber in which the uneven strains set up in the auger can be released. However, this may in itself set up other strains due to the greater surface for friction. Cheaper products may have a fairly sharply tapering die set close to the auger. The low friction surfaces allow greater speed of production but the ware is more likely to have a wavy surface and structure.

A certain amount depends on the composition of the body, large pieces being easier to extrude when they have a higher grog content. Schurecht (**S39**) has, however, considered the problems of extrusion of brick using a fine-grained body. Here there is easily a tendency for bowing, twisting, sunken faces, bulging ends, and under-fired centres. The bricks investigated showed considerable strain between the central and the outer portions, resulting in a centre that had a lower specific gravity after drying and firing and tending to be under-fired. This together with the twist were largely relieved by inserting two pairs of cross-wires at 45° to each other across the sleeve immediately after the end of the auger. Bowing can be prevented by adjustment of the level of the cutting table with respect to the mouth-piece.

The design of dies for hollow-ware is much more complicated than for solid columns because of the difficulty of supporting the central parts of the die without leaving a permanent blemish in the ware. The 'bridges' which hold the central part should be as near the auger as possible so that there is the maximum possible pressure to knit the clay together again after passing round them. As the flow near these bridges tends to be slower than in the unobstructed part it is sometimes necessary to decrease the flow in the latter by lengthening the die to constrict it over a longer period or similar means to produce a strain-free column.

Heating of the dies by steam, gas or electricity makes extrusion much easier.

To facilitate the movement of the column through the die, this is usually designed with means of introducing lubricants. Water, steam or various oils are used. In general water or oil-in-water emulsions are suitable for bricks and roofing tiles. More complex tiles, and thin-walled pipes require more positive lubrication. The choice of oils must be undertaken with care, as well as adequate lubrication, they must not enter the body itself, and must not coke on the surface or induce local sintering. Mineral oils used pure or with up to 5% vegetable oil of medium light, light or translucent qualities, suitably bleached, were considered to be the best for this purpose. Oils obtained from the distillation of coke or lignite are usually too highly coloured, and animal and vegetable oils, apart from their high price, coke too easily.

The viscosity required depends largely on the pressure employed, varying from 2·5–3·5 degrees Engler at 20° C (**A168**).

More recently water-soluble oils and waxes in emulsion form are gaining ground in all fields of lubrication.

EXTRUDED AND WIRE-CUT BRICKS AND TILES IN MEDIUM PLASTIC, 'STIFF-MUD', CONDITION

Large quantities of bricks and tiles are produced by extrusion methods. The cheapest method for producing common building bricks from plastic clay is by the 'wire-cut' process. For this the preparation and shaping machinery is combined into a continuous set-up requiring only feeding of the raw material and removal of the bricks. The largest such machines will produce up to 14 000 bricks per hour, although an average figure is more like 30 000 to 40 000 per day.

FIG. 8.12. *Wire-cut brickmaking plant, with clay preparation at the top feeding the pugmill, extruding die, and finally the cutting-off table,* bottom right (*Shenton,* S65)

The machinery is usually set up vertically, taking up about three stories. The clay or clays, grog, etc., are fed in at the top and pass through crushing rollers and/or edge runner mills, mixing troughs and pugmills which may or may not be de-aired. The prepared body is then extruded with a rectangular cross-section on to a cutting table. A grid of wires (18-gauge piano wire of 0·8/0·85% carbon steel) then cuts it into bricks which are pushed off on to

FIG. 8.13. *Extrusion of hollow clay block* (*London Brick*, **L41**)

the side (Fig. 8.12). For building bricks the extruded column has the cross-section of the largest face, *i.e.* in Great Britain 9 × 4½ in. (23 × 11½ cm) when fired. Other sorts of brick may be extruded along their length.

When a sand-faced brick is being made a stream of sand is run on to the top and sides of the extruded column and distributed evenly with brushes.

The extrusion process is also used for the increasingly popular hollow bricks and blocks, and for agricultural drain pipes. The difficulties of manufacture have already been mentioned and careful design of, and experiments with, the dies cannot be over-emphasised (Fig. 8.13).

Tiles of various shapes can also be extruded. They are made from finer grade material to give a more uniform and impervious product than common bricks. An L-shaped die may be used to give a continuous nib or several 'ribbons' may be extruded at once, which are then cut, and partially dried.

Wire-cut common bricks are dried and fired in the condition in which they leave the cutting table. They show a certain number of surface blemishes, cracks, etc., and may be laminated. Facing bricks and tiles are further processed after preliminary drying by repressing. Hand or power presses taking one or several bricks, etc., at a time are used. Where a sand-faced brick is being made more sand is added at this stage. The lips and holes in roofing tiles are made at this stage.

EXTRUSION OF STONEWARE PIPES

Large quantities of stoneware pipes of different diameters and for different purposes are made by extrusion. Both horizontal and vertical machines are used. Horizontal machines are used for rapid production of the smaller diameter pipes. The vertical machines were formerly steam-driven intermittent piston machines, but except where hydraulic pressure is used to drive the pistons they are largely superseded by auger machines.

Where auger machines are used it is possible to incorporate a de-airing system. This increases the plasticity of the clay and makes the extruded pipe less brittle, a very great advantage where bends are to be produced. De-airing can, however, slow down by about 10% the production rate of small pipes, up to 8 in. (20 cm) diameter, which can be broken off from the mouthpiece if the body is not de-aired, but must be cut off, if it is. De-airing benefits some clays very much more than others.

Most drain-pipes are of the socket and spigot type. Pipe machines are designed to make these in one piece, the socket being made integrally with the pipe by means of equipment fitted below the mouth of the die plate. This consists of a table that moves up and down and on which is bolted a mould for the socket. When the table is locked up against the die plate the extruded clay is forced into the socket mould, after this has been filled the table is released and the extruded pipe gradually presses the table down to the required depth when the auger is stopped and the pipe cut off from the die plate and removed from the table (**C41**).

SHAPING

This method of pipe making may be manually controlled, but new machines can now do it automatically. Hydraulically operated automatic de-airing pipe machines (Fig. 8.14) make an average of 750 pipes an hour (**B105**), increasing output per machine by 100% and output per man-shift by 200%.

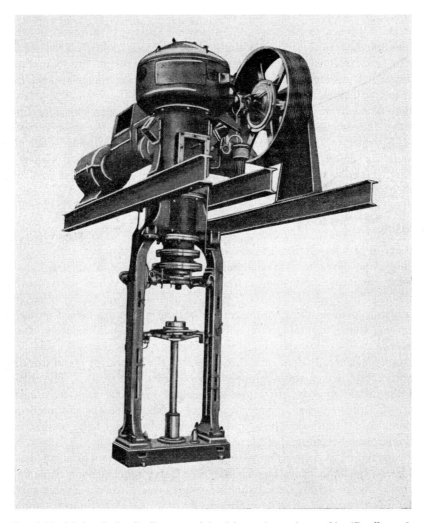

Fig. 8.14. *Modern hydraulically operated de-airing sanitary pipe machine (Bradley and Craven,* **B105**)

In an American plant using automatic grinding, feeding and auger extrusion for 4, 6 and 8 in. pipes, only three men are required between the clay bin and the dryer for an output of 25 000 tons per year (**A32**).

Curved pipes of large radius of curvature are made by manually pulling the pipe to one side with the aid of a template, during the extrusion.

CHEMICAL, ELECTRICAL AND ENGINEERING WARE

Extrusion is used for a variety of products using fine ceramic bodies, chemical stoneware, chemical and electrical porcelain, and other electrical bodies. Here, optimum conditions for the body, and the best designs for auger and dies, are of utmost importance. De-airing is usually necessary (Fig. 8.15).

For short bodies to be extruded into small and very thin-walled tubes hand wedging, if necessary with the addition of tragacanth, potato starch, glycerin, etc., followed by intermittent piston extrusion are sometimes recommended. Tiny tubes 3 mm (0·112 in.) outside diameter and 0·2 mm (0·008 in.) wall thickness can be so made.

For some types of very small thin-walled tubes, especially where the body is costly, *dust extrusion* is used. The body is mixed in a pan mixer with only 15% water to a humid powder. It is then placed in the cylinder of a hydraulically driven piston pump and de-aired. It may then either be extruded vertically directly in the required shape or it is extruded as a solid cylinder and then extruded to shape in a horizontal cylinder.

Turning

A variety of articles of circular cross-section are shaped in two stages. The first stage may give the general shape or only produce a 'blank'; it may be throwing, jiggering, jolleying, extrusion or casting. When partially dry, leather-hard or cheese-hard, 10–16% moisture, or more rarely when dry, 2% moisture or even after the biscuit firing, the pieces are *turned* on lathes to the final shape, *e.g.* tableware, artware, sparking plugs, insulators, drainpipes, etc.

Turning is generally done on lathes similar to those used for metal working. They are motor driven and equipped with variable speed gears. The cutting tools should be tipped with abrasion-resistant materials such as tungsten carbide or sinter alumina. The piece to be turned is usually mounted on a horizontal spindle although large and heavy insulators have to be worked on a horizontal wheel turning about a vertical axis.

The machining is done either with a manually operated tool or mechanically with specially shaped tools or wheels. The heavy foot on bone china cups, which cannot be made directly in the jolley, was formerly always turned by hand but can now be done mechanically at a rate of 160 dozen cups per day. The same machine can also be adapted for use with earthenware.

Turning finds particular application in the cutting of screw threads inside or outside ware, which should have a moisture content of 12–14%. For this a special lathe turns the blank while moving it horizontally past a hand-operated tool. Alternatively specially designed machines now make the operation semi- or fully-automatic.

If the blanks are in the dry state much dust is produced and must be appropriately removed before it can be inhaled. (Fig. 8.16)

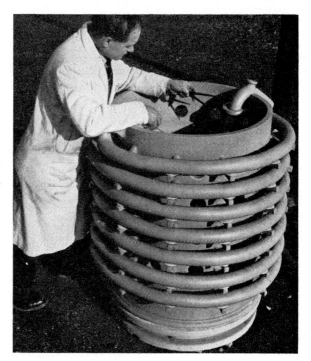

FIG. 8.15. *Chemical stoneware cooling coils are extruded on to the slowly revolving rack* (*Doulton,* **D56**)

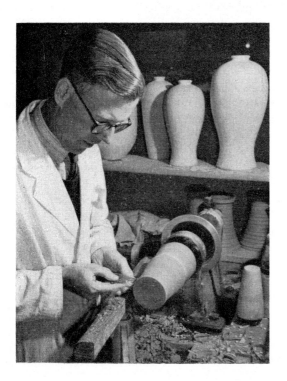

FIG. 8.16. *Turning* (*Wedgwood,* **W35**)

Scalloping

Shaping of the rims of tableware into curves of various kinds, by scalloping, is a form of decoration. Flatware can be produced directly with a scalloped edge but cups, basins, etc., have to be cut to shape when they are leather-hard or dry. This was until recently a skilled manual operation. A centrally holed circular template marked with the positions of the scallops fits inside the vessel to be scalloped. The scallops are cut with a knife. The template is removed with the aid of the hole and the article is sponged.

Machines have now been developed to perform this task reliably and rapidly. In one the cup is placed, on a chuck, upside down on the requisite number of tools of hardened felt. These rotate partly in a tray of moist sponge and act by an abrasing action (**C54**). Another machine uses rotating knives (**A178**).

STIFF-PLASTIC EXTRUSION-PRESSING SHAPING OF BRICKS

In brickworks using soft-plastic and medium-plastic bodies the factors which most often limit the output are the space, labour and time taken up by the drying process. This can be eliminated if the body used has a lower moisture content, when, however, it requires greater pressure to shape it. The stiff-plastic processes are particularly applicable to clays that are hard and fairly dry and would need weathering, tempering and working to bring them in a truly plastic condition.

The stiff-plastic brickmaking machines consist essentially of mixing troughs followed by a vertical pugmill. From this the clay is filled into moulds to form a clot. In some machines (**N13**) there are two clot boxes which come under the pugmill alternately for filling and then discharge the clot which moves on to the press. In others (**B105**) a number of brick moulds are arranged on a rotating table and the clots are made directly in them. The clot is then pressed into the shape of a brick. Facing and engineering bricks may be conveyed from the press straight into a repress where the second pressing gives them a denser texture and sharper outline. Rustic and other decorative surfaces can be produced on these pressed bricks and a large variety of special shapes can be made.

The pressed bricks are sufficiently firm to withstand stacking several feet high and their relatively low water content can be driven off during the preheating period of the firing.

Roofing tiles are also made by stiff-plastic methods. Plain tiles are largely made by feeding extruded tile bats into hand-operated screw presses or power presses. In the latter the output is about thirty tiles per minute per die. Steel moulds are generally used.

For the more complicated shapes of interlocking tiles somewhat softer clay in the form of a clot rather than a bat is pressed in plaster moulds. A common type of press used on the Continent for such tiles is the revolver press. A horizontally revolving drum carries five or six copies of the lower press mould. These are filled with clots in turn and move intermittently into position under the plunger with the upper mould, where they are pressed into shape. The finished tile is then removed by hand or conveyor as it moves clear of the press. The pressing itself is a double action. When the final size has almost been reached the plunger is raised $\frac{1}{8}$ in. (3 mm) allowing trapped air to escape, before being brought down to the final position. Some of these presses are fitted with trimmers to remove the fringe of surplus clay forced out between the two halves of the mould.

DRY AND SEMI-DRY PRESSING

Ceramic bodies in a granular condition can be shaped in dies by applying pressure. The process is often called 'dry-pressing' but as conditions vary at least the two extreme forms of the range of methods should be differentiated.

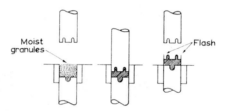

FIG. 8.17. *Basic dust or semi-dry pressing*
(*Thurnauer*, **T33**)

Semi-dry or *dust pressing* is the end point of the series of clay–water combinations requiring more pressure, the lower the water content. The bodies contain clay and have about 10–15% moisture. Under sufficient pressure plastic flow into the die cavities occurs, in fact, some of these bodies although appearing granular will form a compact when pressed in the hand. The pressed pieces are rather fragile and have a flash where excess body has squeezed out of the die. There is still a drying shrinkage. The fired pieces of vitrified bodies shaped this way do not attain the density of a truly dry-pressed piece or one made by a wet process. Electrical porcelain can therefore only be made for low-voltage, low-frequency applications. Dust pressing is universally used for making wall tiles. Semi-dry pressing can rarely be made fully automatic. (Fig. 8.17.)

True *dry pressing*, although up to 4% water may be present, makes no use of the natural plasticity that could be developed in any clay by water that is present. By suitable additions of lubricants and binders dry-pressed ware can be produced from any raw materials that will become hard on firing, whether they possess any plasticity or not. Dry-pressed ware has little or no drying shrinkage and so cuts out drying time. It can be made with high dimensional accuracy, completely vitrified, with high dielectric strength, and with special precautions vacuum-tight ware can be made. Dry pressing can be made fully automatic, as there is no stickiness and the shaped pieces are strong. However, it must be remembered that no flow of the body will occur under pressure, so that the granules must be correctly distributed in the die before pressing starts, and also no excess can squeeze out between the parts of the die (**T33**).

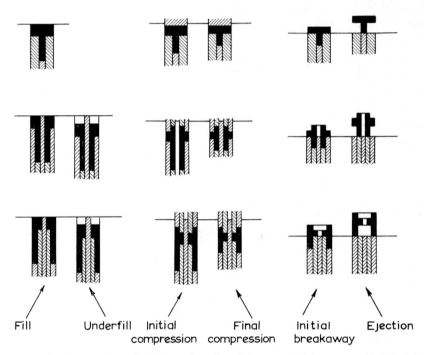

FIG. 8.18. *Dry pressing. Typical punch motions (Thurnauer,* **T33**, *courtesy F. J. Stokes Machine Co.*)

The dies must therefore be designed in such a way that when they are filled with loose powder they have exactly the right quantity of material to make the finished piece, and the different depths of dust in the various parts of the open die must be in the same ratios to their respective thicknesses of pressed product. The pressing of tiles and bricks can therefore be done in straightforward two-piece dies on automatic or semi-automatic presses. Where more complex shapes are required it is often necessary to have several moving parts in the lower part of the die. It is usually necessary to lubricate the metal dies to prevent the pressed pieces sticking to them, fatty oils with a greater affinity for the metal than the body are suitable (**S119**). (Fig. 8.18.)

As the dry-pressing method is in other ways very suitable for making electrical insulators, means have had to be devised for overcoming the flow difficulties so that pieces of complicated shapes can be made. Foremost amongst these is proper design of shapes with regard to the method by which they are made, avoiding unnecessary sharp corners, excrescences, weaknesses, etc.

Thurnauer (**T33**) indicates some of the points to be observed in design of pieces and dies for dry pressing:

(1) Top and bottom punches must not touch each other during the pressing cycle, and care should always be taken that sufficient space exists between the two punches to avoid the danger of damage or breaking of punches.

(2) Holes should not be located too close to the outside, to avoid cracking of thin walls. The same applies to countersinks.

(3) Uniform cross-sections aid in making uniformly dense compacts. Tapered compacts tend to result in varying density.

(4) Side holes can be formed by a pin in the die, if they are open on either the top or the bottom of the compact. Indentations on the inside are not practical.

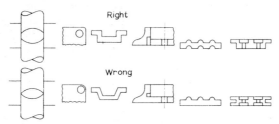

FIG. 8.19. '*Right*' and '*wrong*' ceramic designs for automatic dry pressing (Thurnauer, **T33**)

(5) Grooves on the outside of a compact require an additional machining operation after firing, but may be replaced by a shoulder, at less cost.

(6) Elevations on the top surface of a compact are difficult to form, but if necessary they should be compensated by equivalent indentations at the bottom, to obtain uniform density. (Fig. 8.19.)

Small quantities of binders, lubricants and anti-sticking agents, and plasticisers are frequently added. Stone (**S199**) summarises their functions as follows:

Binders have two functions in dry-press bodies, namely: (1) to produce a high plastic strength at the mixer (supplementing the clay) in order to facilitate granulating; and (2) to produce sufficient mechanical strength in the pressed pieces to prevent damage in subsequent handling.

The anti-sticking agents have three functions, namely: (1) to act as lubricants in aiding flow of the particles of the body in the die while the forming pressure is applied; (2) to prevent the body from sticking to the punches when the forming pressure is released; and (3) to provide the 'slip' necessary to prevent the pressed pieces from sticking to the die walls and core pins when ejected.

In functions (1) and (3), the anti-sticking agents act as lubricants.

Water-soluble solid binders fulfil the first required function but do not act for the second when the water has been eliminated. They must therefore be used with a plasticiser or a second binder which will function in the absence of water.

The function of the plasticiser is to develop the adhesiveness of the binder in the absence of water. Preferably it should be of such nature that it can be added along with the mixing water in a muller or other mixer. It should be non-volatile so that it will not be removed when the water is evaporated from the granules.

Stone (**S199**) investigated the usefulness as binders of a number of dextrines and starches, natural gums, synthetic materials (of types related to plastics) and waxes, and the application as anti-sticking agents of various stearates and oils. He recommends a number of combinations for use with steatite bodies pressed in different shapes. In general he found that better results are obtained by using water-soluble binders (3%) together with high melting microcrystalline waxes (3%) than by having only one type of binder. Waxes can be combined with fats and fatty acid soaps as anti-sticking agents but not with oils.

Lecuir (**L20**) recommends the use of binders that not only have low surface tension but also low melting points, for example:

Urea, M.P. 138° C (280° F). Surface tension 40 dyn/cm at 150° C (302° F).

Stearine, M.P. 69° C (156° F). Surface tension 35 dyn/cm at 150° C (302° F).

These can be mixed with the body above their melting point and are suitable for electrical porcelain and refractory oxides.

Uniform distribution of small quantities of oils and other matter insoluble in water, through an almost dry batch presents certain difficulties. It cannot be achieved by mere mixing or mulling. The oil, etc., forms a film on the outside of the granules and on pressing separates them somewhat. It then evaporates or burns out leaving tiny cracks. This is not noticeable in a porous body, but in porcelain, which is itself completely vitreous the dust-pressed pieces have water absorptions of 0·5 to 2·0%. This can be overcome by saponifying the oil with alkali, for instance by adding oil to a casting slip of the same body and using this to moisten the dry powder to make the pressing dust. It is for this reason that wax emulsions, which are water soluble, are finding increasing favour (**A88**).

Certain wetting agents, detergents and deflocculants are also sometimes added to eliminate the need for ageing the batch. Matthews and Shonk (**M38**) have shown that the addition of either up to 6·0% waste sulphite lye, 0·25% sodium carbonate with 0·5% teepol, or 0·5% teepol alone, increase the dry strength of an unaged body more than 24 hr ageing did; the 6·0% sulphite lye doing so more than three times as much.

The lubricants, binders, etc., must be of such a nature that they are not detrimental to the finished product. Most of them are organic, and must be of the sort that can be easily eliminated by oxidation or sublimation. The inorganic ones must not be detrimental to the electrical properties of insulators made in this way.

The best quantities for any of these additions have to be determined experimentally for every particular case. Ungewiss (**U2**) has developed a small machine for testing the suitability of oils for dry-pressing insulators. The pressure required to eject a test piece pressed under standard conditions is measured. Once a suitable lubricant and optimum composition have been found they should be adhered to. However, as these lubricant oils are often mixtures it is advisable to test each batch delivered by the supplier. Ungewiss (**U2**) suggests finding the specific gravity, the viscosity and the saponification number in these routine tests.

Preparation of Bodies for Dry Pressing and Dust Pressing

Most normal earthenware, porcelain, steatite, brick or refractory bodies can be adapted for shaping by dry pressing. The bodies are prepared by wet or dry grinding and mixing and brought to the correct water content by controlled drying or by complete drying followed by the addition of the exact quantity of water. The necessary quantity of wetting agents, binders, anti-sticking agents, lubricants, etc., may be added at this stage.

If no compensating measure is taken the body is then aged for from 24 to 72 hr, when the moisture content should become evenly distributed. Without this the products may show cracks and indentations. Tests made with a porcelain body showed that 24 hr ageing increased the dry strength from 235 lb/in^2 to 340 lb/in^2 (16·5 to 23·9 kg/cm^2), some 40% (**M38**).

The body, now of correct composition, water, lubricant and binder content, must be brought into a suitable granular form. A fine dust is difficult to move through conveyors and tubes and to fill dies with because it will not flow like a granular material. On pressing a very fine dust, lamination often occurs. It also has a greater tendency to become airborne, with consequent health problems and unbalancing of the body composition. Not only must the body be in a granular form that flows well but the grain sizes must be such that they pack and consolidate well in the die. The size range must also be so reproducible that successive fillings of the die always hold the same weight of material. If the die is filled first with a large-grained material and then with a fine-grained one the second article will be heavier and on firing it will be larger.

If the body has been prepared by the wet method and then filter pressed and dried the press cakes can be granulated directly in small edge runner mills with perforated bottoms, or disintegrators of the hammer mill type. If the body has been prepared dry, or a wet-mixed body has been completely dried and powdered a small amount of water and usually an organic binder are added and the whole pressed into bricks which are then granulated.

The granules so formed are denser than those from filter cakes (**S119**). Dry body may also be mixed with the binder in a muller in such a way that granules form, which are then screened and dried, or a wet body may be spray-dried directly to granular form. The latter produces smaller granules of lower density but of more uniform particle size distribution than the former (**T33**).

Presses

The choice of the type of press, its speed of action and the pressure depends on both the body and the shape. Crank, eccentric, toggle, screw (pressures up to 10 000 lb/in^2, 703 kg/cm^2), pneumatic and hydraulic presses (pressures up to 100 000 lb/in^2, 7030 kg/cm^2) all find their place for different purposes. The last named, which had lost favour earlier on in this century, are now having a dramatic come-back for fast and quiet production. For small pieces numerous types of tabulating machines used in other industries are also applicable in ceramics.

Because of the bad transmission of pressure through dust bodies it is important to apply it from below as well as above to achieve greater uniformity. Where the shape is very complicated and pressure transmission very poor a special hydraulically operated rubber mould has been suggested (**J43**). The rubber die, which is sufficiently stiff to hold the required shape, is set within a metal mould. It is filled, sealed and then hydraulic pressure applied between the rubber and the metal moulds, so that the piece inside comes under identical pressure from all directions.

Dry Pressing of Ceramic Insulators

Dry pressing finds the widest application in the shaping of steatite insulators which can be produced to tolerances of $\pm 1\%$, with a minimum of ± 0.005 in. (0·127 mm). In steatite there is little plastic clay but the talc is soft and has a lubricating effect on the die. Any other body can also be dry pressed including the pure oxide bodies. To produce dense vitrified bodies the particle size of the softer raw materials such as talc or pyrophyllite must be less than 200-mesh, and for non-plastics such as alumina and titanates it must be less than 1–20 μ.

The prepared body should have shrinkage and vitrification temperature checked before granulation. The shaping process on tableting machines or specially designed ceramic presses can be fully automatic and very fast.

Dust or Semi-dry Pressing of Wall Tiles

Plain tiles can be made on automatic presses, those with curved or moulded edges are better done on semi-automatic machines and complicated shapes on hand-operated ones.

Where a normal clay earthenware body is used, it is brought to 8–9%

SHAPING

moisture and then granulated. The lubricating oil is applied direct to the dies.

In an automatic press the prepared body dust is fed by conveyor through a hopper to a feed box. This fills the die with loose dust. The tiles are pressed in two strokes allowing air to escape after the initial compression, and are then delivered on to a conveyor which takes them through the fettling machine where the edges are fettled and the faces softly brushed. The operative merely has to lift the tiles off and check their condition before sending them on to the next process (**H18.1**) (Fig. 8.20).

FIG. 8.20. *Automatic 40-ton press for dust pressing of wall tiles* (*Sheepbridge*, **S62.1**)

Semi-dry Pressing of Bricks and Refractories

The terms semi-plastic, semi-dry and dry are all applied somewhat confusingly to the type of heavy ceramic body composition which is prepared in a dust form that will flow over itself, but on pressing in the hand will produce a ball. The moisture content is in fact somewhere under 15%. The advantages of dry pressing are the uniformly well-shaped and dense products, requiring no drying, etc., while the snags lie in the need for close control of grain size gradations, moisture content, etc., together with the need to add lubricants and binders.

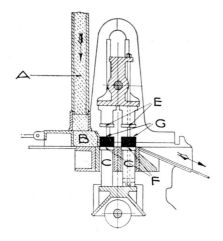

Fig. 8.21. *Semi-plastic repress brickmaking machine. A, raw material being fed into sliding carriage, B, The left-hand piston plate, F, drops to reveal the mould, C, which is filled by B, while this also pushes the back brick, G, on to the right-hand piston plate F, and the front brick out on to the delivery table. The second piston plate is dropped and the top pistons, E, lowered for the pressing* (Whittaker, **W61**)

Fig. 8.22. *Toggle brickmaking and pressing machine for bricks, blocks and shapes* (Johnson, **J54**)

SHAPING

BRICKS

The process is therefore only used for making building bricks in those areas where the clays naturally contain organic matter that acts as lubricant and binder, that is in Great Britain the Fletton districts.

Large automatic machines can make 2000 repressed bricks an hour, average-sized machines produce 12 000 a day. The mined clay is fed to a perforated pan grinder and water added if the moisture content is below the requisite 8%. It is then screened and the oversize particles returned to the

FIG. 8.23. *Semi-dry pressing of fletton bricks on repress machines* (*London Brick*, **L41**)

grinder before passing on to the feeding chute. The empty moulds are fed from the chute by gravity and then pass in between pistons that apply two quick pressure bumps. (Plastic pressing is a slow continuous movement.) The pressed brick may or may not then be repressed (Figs. 8.21–8.23).

PAVING BRICKS

Everhard (**E26**), working with several American commercial bodies, showed that better paving bricks could be produced by the dry-press process than by the stiff-mud one. The four main variables are: (1) forming pressure; (2) rate of pressing; (3) moisture content, and (4) grain size. The first three are interrelated; high pressure in a fast cycle at high moisture content producing pressure cracking, while the best ware is made at high pressures if the moisture content is low and the cycle not too rapid. Where only low pressures (500 lb/in², 35 kg/cm²) can be obtained a high moisture content is necessary. In all cases evacuation of the mould gives better bricks.

As the normal brick body does not give perfect pressure transmission it is important for both the upper and the lower half of the die to move during the pressing. It is found that where only one part moves the resistance to abrasion of the brick is higher on the side of the moving die than on the side of the stationary one.

FLOOR TILES

Dust pressing is largely used for the production of quarry and other floor tiles.

REFRACTORIES

Dry pressing (6 to 8% moisture) is the most satisfactory method of producing high-grade firebricks of regular size and shape. Certain important sections of industry specify that their firebricks should be made this way (**M20**).

There are, however, a number of difficulties encountered in the manufacture of dry-pressed firebricks, namely, pressure cracking, irregular density, spalling, slag penetration, lamination, etc. Birch (**B60**), Rueckel (**R67**), Dodd and Holmes (**D48**) and others (**A147**) describe work on the correct application of the pressure to eliminate these faults and McWilliam and Gilmour (**M20**) have recently done work on the preparation of the body.

It has been found that the proper grading of the raw material is very important as the bonding in the green brick is largely mechanical and is obtained by forcing the particles into and against one another so that they interlock. Proper tempering and mixing is of great importance to develop the workability of any plastic constituents present and distribute them to form a coating of non-plastic ones.

McWilliam and Gilmour (**M20**) found that a number of small additions to the tempering water increase the green and fired modulus of rupture, *e.g.* 3% sulphite lye in the water increases the green strength from 75 to 130 lb/in² (5·3 to 9·1 kg/cm²), while 3% sulphuric acid increases the fired strength from 953 to 1505 lb/in² (67 to 105·8 kg/cm²).

The best method for mixing the batches is frequently discussed. There are the advocates of using only a mixing action with blades and ploughs or in a cylinder, and those that recommend a mulling action. Small edge runner mills with solid bottoms and comparatively light rollers slightly raised are used. Mulling changes the size grading of the batch which must therefore be allowed for; it may also squeeze water out of the material. Under favourable conditions, however, it distributes the water uniformly through the batch, making it easier to 'pour' in the press moulds, and can more than double green modulus of rupture, increase fired modulus of rupture by 25%, and increase density and resistance to abrasion (**M20**). It is important to note that every body will react differently when treated by admixtures to the tempering water and mulling. There are also distinct possibilities in the application of vacuum to the batch during mulling.

With regard to the pressures applied in shaping the bricks, the actual values achieved in commercial mechanical presses can only be obtained by comparison of the properties of the pressed brick with those made in hydraulic presses where the pressures can be measured. Birch (**B60**) showed that commercial pressures were only about 1500 lb/in^2 (105·5 kg/cm^2), the limit often being due to pressure cracking in the brick rather than mechanical limitations. He found that this could be overcome by reducing the moisture content. When this is done the bulk density and transverse strength of the green and dry pressed bricks and their firing shrinkage, fired density, crushing strength and resistance to abrasion increase with increasing forming pressure, the greater increments being at the lower pressures. There is a theoretical limiting pressure at which complete compactness is reached.

Use of high pressures has a greater tendency to produce pressure-cracking the faster the pressing; evacuation of the mould does, however, increase the pressure-cracking limit of clays, the effect being more pronounced on plastic than on non-plastic clays.

Adequate pressure transmission through large pieces to give a product of uniform density does not always occur. Fine-grained clays transmit worse than coarse ones, and the best results are obtained when there is a large (up to 50%) proportion of grog.

Lamination and cracking are due to air entrapped in a raw feed of irregular grain size distribution and blending. De-airing of the feed and/or the mould reduce cracking due to air alone but uniform filling is important.

The manufacture of refractories, especially large blocks, necessitates the use of large machines which are less fully automatic than those for building bricks.

As the mould linings and dies are subject to severe abrasion they are made renewable. The steel used is hardened and stress relieved high-carbon–high-chromium steel or a lower-carbon–lower-chromium steel (p. 165).

TAMPING

Bodies containing a large proportion, up to 90%, of non-plastic material can be shaped by pneumatic tamping into moulds. The 10% of plastic clay used as a bond is made up into a deflocculated slip. This is mixed with the non-plastic in such a way that every grain becomes coated with the clay. The mixed body then has a water content of only 5·0%. For minimum porosity of the product the non-plastic must be graded to give dense packing.

The shaping is done in strong metal moulds. Material is gradually introduced and tamped down with a pneumatic tamp. A strong bond is formed immediately so that the piece can be unmoulded at once. There is little drying or firing shrinkage so that accurate shapes can be made.

The method is largely used for refractory blocks of silica, highly grogged fireclay, corundum, silicon carbide, sillimanite, etc. (**C40**).

HOT PRESSING

The use of heated dies in plastic pressing and extrusion, and of heated profile tools in jiggering have been mentioned. Heat plays a more dominant role in shaping by pressure, using softening or sintering as part of the process:

(*a*) using organic binders that are solid at ordinary pressures but can easily be softened, *e.g.* bakelite;
(*b*) heating the body to its softening point and then pressing it into shape;
(*c*) heating single component bodies to sintering point under pressure. This combines shaping and firing.

Heating to Softening Point

Harman (**H32**) describes experiments with (1) clay of small plasticity and low softening point, (2) plastic clay with a long softening range, and (3) shale with a high softening temperature but long range. These were preheated, respectively, to 1260–1290° C (2300–2354° F), 1233° C (2251° F), 1344° C (2451° F) and then pressed into shape under some 2987 lb/in^2 (210 kg/cm^2). In this way thin dense plates can be made successfully, the clay flows readily to fill the die. The pressed piece is held at temperatures from 700–750° C (1292–1382° F) before being cooled off. Other examples show that in general the correct preheating temperature for clays lies between 1230–1340° C (2246–2444° F) (**A167**).

Garbisch (**G4**) has patented a process for making bricks, tiles, etc., from the waste fine sand and glass mixture resulting from the polishing of glass. The purified and dried by-product is heated to $\sim$ 870–925° C (1600–1700° F) and can then be pressed. However, it readily balls up and will not subsequently flow to fill the mould properly like the heated clays, and so it is best heated in a tray which is dropped into the press and forms part of the mould.

Hot Pressing and Sintering

The sintering of solid materials at temperatures below their melting point can be greatly accelerated if pressure is applied. This principle is used in the method of 'hot pressing' of pieces by which shaping and firing are combined. It is used for making high density accurately shaped pieces of refractory oxides, carbides, nitrides or other single material bodies. The chief difficulty in the method is that the powder tends to react with the mould at the temperatures at which it sinters readily.

Norton (**N35**) describes the most usual method of hot pressing non-metals in heated moulds, which are usually of graphite. The heating is done electrically either by resistance or by induction. An example of the resistance method is given by the Ridgeway and Bailey (**R20**) patent shown in Fig. 8.24.

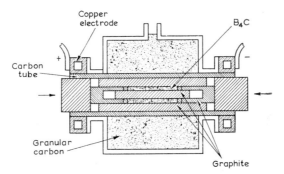

Fig. 8.24. *Ridgeway's method of hot pressing (electrical resistance heater) (Norton,* **N35**)

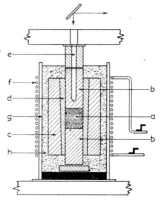

Fig. 8.25. *Hot pressing by induction heating (Norton,* **N35**) :

(a) *powder to be compacted between plungers ;*
(b) *plungers, forced together with a hydraulic press ;*
(c) *graphite susceptor in which heat is induced ;*
(d) *removable graphite liner ;*
(f) *water-cooled copper inductor ;*
(g) *quartz tube ;*
(h) *powdered graphite insulation.*

The heating cylinder is made of graphite and fitted with two water-cooled electrodes through which the heavy current needed for heating is supplied. The powder to be pressed, in this case boron carbide of 200-mesh, is loaded into the mould. The floating core, mould walls, and plungers are also made of graphite. The load is applied by a lever and weight, although an air cylinder could be used, with pressures up to 5000 lb/in^2 (351·5 kg/cm^2) and temperatures up to 2350° C ($\sim$4260° F). This furnace can be used continuously or periodically. The principle of hot pressing by induction is illustrated in Fig. 8.25.

For commercial production graphite has proved to be the only suitable liner, and so temperatures are limited by its reaction with refractory materials as follows:

TABLE 179

Reaction Temperatures Between Pure Graphite and Pure Refractory Materials

Refractory	Temperature	
	(°C)	(°F)
MgO	1800	3270
BeO	2300	4170
ZrO_2	1800	3270
ThO_2	1950	3540
W	1500	2730
Mo	1600	2910

Hot pressing can only be applied to materials that sinter at a temperature below that at which they react vigorously with graphite but good results have been obtained with the following oxides, nitrides and carbides whose properties are tabulated on p. 1150.

Oxides: alumina; beryllia; magnesia; thoria; zirconia; chrysoberyl; spinel.

Carbides: boron carbide; beryllium carbide; silicon carbide; tantalum carbide; titanium carbide; vanadium carbide; tungsten carbide; zirconium carbide.

Nitrides: beryllium nitride; titanium nitride; zirconium nitride.

CASTING

The method of shaping ceramic articles by pouring a liquid slip into a porous mould was invented some one hundred and fifty years ago. At that time the deflocculating action of sodium salts was not known, so that slips containing 40 to 60% water had to be used, and drying must have been a lengthy process involving large shrinkages and risk of cracking. The process was nevertheless better and quicker in man-hours than shaping from plastic clay.

By the middle of the nineteenth century the use of sodium carbonate for making liquid slips of low water content was known, and we now understand the mechanism of deflocculation, flocculation and protection of colloids. The casting process is used for making numberless articles from both traditional clay bodies and new bodies without clay.

Basically the casting process consists of the judicious addition of chemicals to the body to produce a slip of good flow properties with a minimum water content, this is conveyed to and poured into plaster moulds where the double action of water removal and flocculation, by the calcium sulphate of the mould, makes the body set. In due course the cast piece dries and shrinks away from the mould. It is then removed, trimmed, joined if necessary, and further processed like plastic shaped ware.

The casting process is used for shaping pieces in almost any type of body from fine earthenware, bone china and porcelain to refractory articles with a high content of grog, etc. It can also be used for non-traditional, non-clay bodies by suitable additions of binders, etc., *e.g.* pure oxide bodies: zirconia, alumina, magnesia, titania; mixtures of spinels, silicates, titanates or aluminates (**H4**); silicon carbide (**B36**).

Deflocculation

The preparation of a good casting slip from bodies containing clay is not always an easy matter, nor is it one on which any definite rules or methods can be laid down. Not only is every clay different but it, together with the other mineral raw materials and in particular the water are constantly varying. This means that the reaction to added chemicals will also be variable, and yet, the prepared slip must have constant viscosity, setting time, produce a cast of uniform properties, etc.

The optimum conditions for making a given body into a good casting slip have to be found by experiment. In laboratory trials to make new casting slips the consistency to be aimed at should be such, that, when it is thoroughly stirred with a glass rod the groove left behind the rod must close up immediately. If a little is taken up on the rod it should run off it in long strings at first and then in regular rod-shaped drops, none adhering to the glass rod. The slip running off the rod on to the general surface of slip must be taken up into it immediately and not remain as a pile on the surface.

It should be possible to produce a slip with good flow and casting properties with a water content not more than 2% higher than the plastic body of similar composition. In general these amounts fall within the following ranges:

Fireclay	16–18%
Stoneware	20–22%
Earthenware and vitreous china	23–28%
Hard porcelain	28–30%
Bone china (without deflocculants)	50·5%
(with sodium silicate)	35·5%

The traditional deflocculants* are sodium carbonate (soda ash) and sodium silicate (of composition $Na_2O.3\cdot3SiO_2$), and most tests are started using varying quantities of these.

---

* It is unfortunately widespread practice to term deflocculating agents 'electrolytes'. This term correctly means: 'Compound, which, in solution or in the molten state, conducts an electric current and is simultaneously decomposed by it. Electrolytes may be acids, bases or salts' (**U6**). It therefore covers all the inorganic acids, bases and salts we encounter in connection with clays whether they are flocculants or deflocculants. The writers will therefore try to avoid confusion by not using the term in either sense.

The basic factor in the deflocculation of a clay or clay body is, of course, the presence of colloidal clay particles which are susceptible to the action of the added chemicals. The fluidity of a slip depends on the repulsion between these particles, and the minimum viscosity is determined by the total amount of clay colloids present, and on their nature. Phelps (**P33**) gives curves for the viscosity/deflocculant relationship of bodies with different clay colloid contents. It can be seen that the slip of higher clay colloid content can be deflocculated to a lower viscosity although it requires more deflocculant (Fig. 8.26).

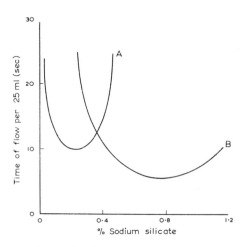

FIG. 8.26. *Viscosity–electrolyte curves showing the effect of clay colloid content upon dope requirement and deflocculation range. Slip A contained 7·5% and slip B had 15% clay colloids* (Phelps, **P33**)

The need for variations in the composition of the optimum deflocculant arises from the differences in the natures of the clay colloids and other constituents presents.

Webb and his co-workers (**W30**) did work on the relative merits of the two basic deflocculants sodium carbonate and silicate. They deduced that the underlying cause for their different action was that the soda ash hydrolyses to give the deflocculant sodium hydroxide and carbonic acid, whereas the sodium silicate hydrolyses to give both the free alkali and also silicic acid which is a protective colloid. They and others found that the deflocculation of certain clay bodies is made considerably easier by the addition of certain 'protective colloids' (*see* p. 154, Chapter 1) which decrease the tendency to flocculate, but that there is an optimum quantity of such substances after which further additions are detrimental. Other clays, however, already contain organic matter of the protective colloid type. This leads to a simple basic rule on the choice of deflocculant:

'If a body is of low plasticity and has little protective colloid, the main deflocculant must be sodium silicate or some similar salt yielding protective colloids on hydrolysis. If the body is of relatively high plasticity and has a high protective colloid content, the main deflocculant must be soda ash or some similar salt yielding no protective colloid on hydrolysis' (**W30**).

It is frequently expedient to use a mixture of soda ash and sodium silicate, their relative proportions and total quantity being determined by experiment. Where such a mixture is used the order in which the different deflocculants are introduced is of importance. Where they are used in the proportion of 3 parts sodium carbonate to 4 parts silicate the most fluid slip is obtained when the carbonate is added first, the highest viscosity occurs when both are added together, and an intermediate value is found when the silicate is added first.

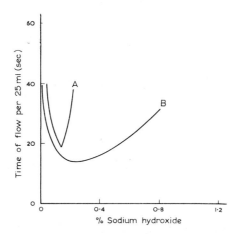

FIG. 8.27. *Viscosity–electrolyte curves showing the effect of a protective colloid upon the deflocculating power of sodium hydroxide. Slip A contained no organic material while Slip B had 0·5% added ground lignite (Phelps, **P33**)*

A number of other inorganic salts, *e.g.* sodium aluminate, have also come into use. These all, however, have the disadvantage, often negligibly small, of introducing a flux into the body, which moreover tends to concentrate on that surface which was against the mould, causing a fluxed skin on the ware. Some salts, *e.g.* sodium ones, also decompose the plaster mould by forming the soluble sodium sulphate. These two disadvantages are not present when organic deflocculants are used. These do not attack plaster and are completely removed during drying or firing leaving no ash. However, they frequently smell unpleasant and may be relatively costly.

Deflocculants can be assisted by the direct addition of protective colloids. These may not act as deflocculants themselves, in fact in certain circumstances many of them are flocculants. But they stabilise a suspension once it is made (Fig. 8.27).

It is not, however, always possible to produce a good casting slip with deflocculants and protective colloids alone. A common reason for this is the presence of flocculating ions: sulphate, calcium, magnesium, ferrous iron and aluminium. These are frequently present in natural clays and in almost any water. Sulphates are the worst disturbers of deflocculation and their presence in natural raw materials is given by Shell and Cartelyou (**S64**) as follows (calculated on dry weight basis):

Ball clays 0·01–0·28% (usually nearer the lower value).
China clays 0·002–0·015%.
Bentonites up to 0·194% (usually relatively high).
Miscellaneous raw materials 0·000–0·007%. (Fig. 8.28.)

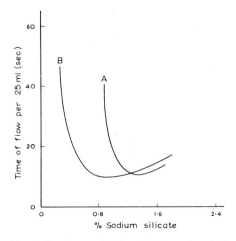

Fig. 8.28. *Curves showing the effect of variable sulphate upon the viscosity–electrolyte relationship for an English ball clay. A represents high, and B low sulphate clay* (Phelps, **P33**)

Sulphates can be precipitated and rendered harmless by the addition of barium ions. Barium sulphate has a solubility of 0·23 mg in 100 ml water at 18° C. If in addition the barium is added in the form of its carbonate certain cations are removed at the same time, *e.g.* Ca, Fe, Mg, by double decomposition.

$$BaCO_3 + CaSO_4 = BaSO_4 + CaCO_3$$
$$2\cdot4 \quad\quad 201\cdot6 \quad\quad 0\cdot23 \quad\quad 1\cdot3$$
solubility in mg/100 ml H_2O at 18° C.

Barium carbonate is itself only slightly soluble. This means that it reacts only slowly with the soluble unwanted salts but that any small excess added will not in itself disturb deflocculation (barium ions are flocculants) (**S102**,

J49). It is usual therefore to add 0·02 to 0·1% (by weight of the dry clay) barium carbonate to a clay before attempting to deflocculate it. Freshly precipitated barium carbonate is more active than the ground mineral, but in either case the powder should be wet ground with some clay before addition to the body and allowed to remain in contact for as long as possible.

Having removed the sulphates it is frequently advantageous to use a dual-purpose deflocculant whose anion forms an insoluble salt with flocculating cations that are present. These are sodium and lithium carbonates, oxalates, phosphates, silicates, etc., which are hydrolysed to the hydroxide required for deflocculation (**J49**). A list of deflocculants and their most direct functions is given on p. 152, Chap. 1. However, as each clay reacts differently to the various deflocculants it is always worth trying out a large number of them in different quantities and combinations when preparing a new process. The quantities required are usually 0·05 to 0·20% of the weight of dry clay, and are largely determined by the type of clay mineral. Kaolins should not have more than 0·1% deflocculant, whereas other clays need up to 0·2% and should in no case have more than 0·3% added.

When no combination of deflocculants and their auxiliaries will give a usable casting slip the clay can generally be shown to contain some montmorillonite. The peculiar swelling properties of this in the presence of water and electrolytes have been described (p. 56). Where such a clay has to be used the montmorillonite must be deactivated, which in the case of calcium montmorillonite can be done with carbonic acid (**K7**). This evidently attaches itself to the positions that would otherwise take up alkali cations and water. If a sodium montmorillonite has to be deactivated the sodium must first be replaced by calcium. Preheating also can make a highly colloidal clay usable in casting slips (**C63**).

There are a number of hindrances to effective regular deflocculation in the factory even when tests have shown that a good casting slip can be made from a given body. These are mainly the irregularity of the raw materials.

Firstly, the clays have variable amounts of soluble salts. Furthermore some clays contain iron pyrites, which is itself insoluble, but is oxidised by moist air to iron sulphates which are soluble and active flocculants. Their content will vary with the mining and weathering conditions. Washed china clays sometimes contain variable amounts of flocculants (alum) added in the recovery process.

Secondly, the water, unless specially demineralised, will always be a variable factor. This applies in particular to any water that is re-used within the factory as it will have more dissolved matter in it after each occasion of use.

Ground non-plastic body constituents may also be the source of soluble matter affecting the deflocculation of the slip. The feldspars and nepheline syenite always show an appreciable alkalinity, the more so the finer they are ground, and therefore act as deflocculants. Talcs may have enough soluble calcium sulphate to retard deflocculation, as does whiting, with its soluble calcium ions. Prepared fluxes and frits are particularly trying with their sometimes large solubility (**P33**) (Fig. 8.29).

The literature of the deflocculation of clay slips is vast and as no result is directly applicable to another problem we are not going to describe it at length. The salient features of a number of papers are given in Table 180, which can be used in conjunction with the lists of auxiliary raw materials (Chap. 1, p. 152).

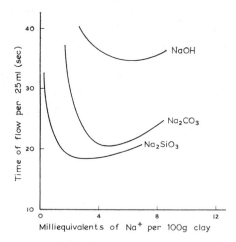

FIG. 8.29. *Viscosity–electrolyte curves indicating the difference in the deflocculating effect of sodium hydroxide, sodium carbonate and sodium metasilicate in the presence of calcium ions* (Phelps, **P33**)

Non-clay Casting Slips

Bodies containing no clay particles may also be deflocculated and made into casting slips. The colloidal content is produced artificially by fine grinding. Pure oxides are found to be deflocculated by acid media, *e.g. zirconia*, 70–80% by weight in 1% aqueous solution of polyvinyl alcohol at pH 1·5–2·0 (St. Pierre, **S2**); *alumina*, slip at pH 6–3 (Shiraki, **S69**); *magnesia*, 1:1HCl, 0·0423–0·0576% (Stoddard and Allison, **S192**).

Casting Behaviour

A good casting slip has a number of well-defined characteristics in addition to that of high fluidity at maximum solids content. In fact a completely deflocculated slip with all flocculating ions removed does not always make a good casting slip.

After the fluidity and viscosity of a slip have been ascertained its action with time must be investigated. For instance, the reactions occurring when an apparently good slip is left standing for 24 hr, whether it forms a skin or stiffens throughout and if the latter, then whether it liquefies again on stirring.

TABLE 180
Preparation of Casting Slips

Chemical	Clay or body type	Action	Comment	Ref.
Sodium aluminate		Deflocculant	Found to eliminate some of the troubles encountered with sodium carbonate	C15
Sodium or ammonium phosphates or pyro-phosphates	Any clay	Removal of flocculants and deflocculant	Precipitates magnesium and calcium	—
Calgon		Deflocculant		A179
Sodium oxalate	Any clay	Removal of flocculants, and deflocculant	Precipitates calcium	—
Ammonium oxalate		Deflocculant and removal of flocculants	This burns out during the firing and does not leave a skin of the soluble salt on the surface	—
Ammonium and sodium sulphur cyanides	Fireclay	Deflocculant	A small optimum quantity greatly reduces the viscosity of fireclay casting slips	V8
Lithium citrate		Deflocculant	Tests comparing this with sodium citrate showed it to be more effective and it does not produce flocculation on overtreating	—
'Foamacrex liquid'	China clay, fireclay or china and ball clays	Deflocculant and wetting agent	Useful agent where organic matter is to be mixed in the body	P31
'Pantarin'		Deflocculant	When used instead of soda in a body that showed streaks, eliminated them	V3
'Clay deflocculant'	Various	Organic deflocculant	Volatilises, excess not detrimental, produces fast cast and does not attack mould, produce skin or introduce flux	K1
Polyvinylamine Ethylamine, piperidine, tetramethyl-ammonium hydroxide		Deflocculant Deflocculants	Concentration required 0·002 to 0·005% of dry body weight Attack plaster moulds less than sodium carbonate and silicate	P11
Lignin, humic acid, tannic acid, colloidal silicic acid	Any clays	Protective colloids	These materials may occur in the natural clay. There is usually an optimum quantity of protective colloid and so the nature of the deflocculant used must be governed somewhat by the organic matter already present	W30
Totanin, peat, tannic acid	Ball and/or china clay	Peptising protective colloids		M39
Quebracho extract	All clays	Peptising protective colloid and prevents thixotropy		
Tannin 0·5%	Clay 10%, bentonite 10%, grog 80%		Bentonite is used to produce a porous body of high mechanical strength (? see remarks about bentonite)	W59
Soluble sulphates	Any clay	Flocculant	Some grinding firms add plaster to the material being ground to keep it in suspension. This is very detrimental	
Plaster	Any clay	Flocculant		
Barium carbonate	Any clay	Removal of flocculants	Precipitates calcium, magnesium, ferrous iron and sulphate	

A thixotropic slip that stiffens quickly when left unstirred can be very disturbing. It requires a lot of agitation during storage, it tends to clog up pipes conveying it to the casters, and stiffens in parts of the mould from which it should be poured out, and appears to set when it has not really done so and jarring will liquefy it again. On the other hand a slight thixotropy in the casting slip can be advantageous in the actual setting of the cast, as will be seen.

Next come the setting properties of the slip. There are two main methods of casting. The first, *hollow or drain casting*, uses moulds giving the shape of the outside of hollow-ware. The casting slip is poured in, and the mould takes up water from it so that a layer of drier body builds up on the mould surface. When this layer is of the required thickness the remaining slip is poured out and the cast allowed to dry further. The second method, *solid casting*, uses moulds dictating the shape of all the surfaces, so that all the space in them is filled with the cast. In either case it is desirable for the slip to become sufficiently solid for removal from the mould as soon as possible, but it is particularly in the first method that thixotropic slips are objectionable.

There are several factors involved in the solidifying of casting slips in plaster moulds:

(1) The flocculating action of calcium sulphate. This is considered to be the main factor by many authorities (**J50**), and is shown by the fact that a casting slip will set even in a wet plaster mould, and does not set so well in a porous mould of other material.

(2) The removal of water by the plaster mould. This is undoubtedly an important factor, and hence it is of great importance that the moulds be well made, with regular porosity (*see* p. 161).

(3) Thixotropy, which enables a clay body to be rigid even when a water layer separates all the solid particles. This is applicable only in certain cases. The thixotropy of the slip determines the amount of water that must be removed before a cast that can be handled is obtained. If it is high the piece is relatively plastic and easy to trim and join. If it is low the cast is very hard and trimming necessitates cutting through solid particles and may result in 'chattering' and 'checking' (**P33**).

(4) The rate at which water can pass through the layer of clay already formed between the slip and the mould which determines the rate at which a cast will build up. Fine-grained clays resist the passage of water more than coarse-grained ones, so that generally primary china clays and kaolins are better casting clays than fine sedimentary clays.

(5) The deflocculant used. Webb (**W30**) describes the different effects of sodium carbonate and sodium silicate on the casting properties of the slips they have deflocculated. For an average commercial body deflocculated with a normal quantity of chemical he concludes that: (1) Sodium silicate always gives a more fluid slip than sodium carbonate for a given amount of soda. (2) Over the whole range of slips of different fluidities even when the soda

ash slip was more fluid than the silicate slip, nevertheless the soda ash slip casts more quickly. This seems to be due to it forming a more permeable cast. (3) That soda ash used alone gives a flabby cast and a tendency to the fault known as 'balling', whereas sodium silicate produces a hard cast with a tendency to stick to the mould and the slip becomes stringy and drains badly. In the case of bone china bodies the difference between the two slips is more marked. Soda ash alone will not deflocculate these effectively and only very dilute slips giving soft casts can be obtained. Sodium silicate on the other hand will give concentrated slips and hard casts, but three to four times as much of it is required than is used for earthenware. It can therefore be concluded that a better castability can be obtained by using sodium carbonate and silicate together. For instance a slip deflocculated with the carbonate alone used for drain-casting may leave streaks on the inside of the piece and dry badly. When a small quantity of the carbonate is replaced by silicate a good cast is obtained. But when too much silicate is used the initial layer of the cast becomes impermeable and the casting rate reduced.

The rate of cast can be divided into two parts:

(a) The initial casting rate during the build up of a thin layer on the mould surface.
(b) The rate at which the casting velocity falls off with increasing thickness of cast. This determines the rate of casting of thick articles.

Three main factors affect the rate of casting:

(i) The clay–water ratio (pint-weight) of the slip.
(ii) The grain size of slip constituents, which for the clay is directly related with the extent of deflocculation.
(iii) The specific nature of the deflocculant.

Another factor to consider is the ease of removal of the cast from the mould. An overdeflocculated slip tends to stick to the mould. It is therefore better to produce a slip just short of the minimum viscosity (**N36**).

Phelps (**P32**) suggests a method of testing a slip for 'castability'.

(1) Particle size (*see* p. 278).
(2) pH (*see* p. 319).
(3) Casting rate. The deflocculated slip is blunged for 2 hr, screened through an 85-mesh sieve and stored for 24 hr before testing. It is then investigated for settling, livering, etc., then stirred and poured into test moulds. These are simple cone shaped with a plugged drain in the bottom. The plug is removed after periods of 5, 10 and 20 min. The casts are removed from the moulds in leather-hard condition, and cut vertically into halves and air dried. Casting rates can then be computed by plotting the thickness against time.
(4) Green strength. The slip is cast in simple bar moulds and allowed

to stand for 12 hr before removal The bars are then air dried for 2 days and oven dried for 2 hr at 105° C, cooled in a desiccator and broken in a transverse breaking machine (p. 335).

(5) Shrinkage. The slip is cast in a solid disk mould marked with 5 cm shrinkage marks.

The correct time for drain-casting any particular shape has to be determined in the works by some trial casts allowed to stand for different lengths of time and then pierced to find their thickness. These casts are of course unsuitable for further processing.

Preparation and use of Casting Slip

The casting slip may be prepared from a wet-mixed filter-pressed body, or a ball-milled slip, or a dry-mixed body, by the addition of the deflocculant and its auxiliaries and water. In every case it is essential that it should be efficiently blunged. Johnson and Steele (**J51**) suggest that where press cakes or a pug column are being incorporated a propeller-type blunger be used. It should be remembered that a body made up with water to a slip of low solids content, and subsequently dewatered by a filtration method, will have less soluble salts than if it is made up directly to the casting slip. Re-use of press water in the blungers, however, may increase the salt content as the water dissolves more in each circuit. Where accurate control is required so that the casting can be mechanised distilled water is sometimes used.

The prepared casting slip is usually kept some 24 hr before use (in U.S.A. it is gently stirred for 5 days). It may be agitated slowly during this time, which not only prevents settling of heavier components but allows the escape of air bubbles. Complete de-airing of a prepared casting slip has proved to be impracticable as mere evacuation is not effective. Some slips may be de-aired by boiling for several hours but this is both costly and not always successful (**S118**).

A method of producing a de-aired casting slip has, however, been suggested (**S184**). Dry raw materials are mixed in an evacuated rotary chamber with a mixing device in it, and water containing the necessary chemicals is added without admitting air. The slip is then discharged through pipes, the outlet end of which is always kept immersed in slip.

A stored slip must usually be well stirred before actual casting begins.

The moulds may be filled by a primitive method of ladling the slip into them by buckets or jugs dipped into the storage vessel or by modern pipeline systems worked by gravity or pressure. Pumping tends to introduce air bubbles into the slip leading to pinholing, gravity systems are therefore to be preferred. If the slip is de-aired by slow agitation a gravity pipeline system is essential. Air bubbles are also avoided by the use of a continuous slip pumping system, all surplus being returned through a pre-loaded valve to the agitated stock ark, so that all pipes, valves, pumps, etc., are always full.

A compressed air pressure system suitable for a slip that does not require agitation after the initial blunging has been applied (**L52**). The prepared slip is passed by gravity to one of two tanks, filling it. The feed is then shut off and transferred to the second tank, allowing the first to be used. Compressed air is passed into this through a suitable valve at a pressure of about 28 lb/in^2 (2 kg/cm^2) and the slip forced into pipes of 2 to $2\frac{1}{2}$ in. (5 to 6·4 cm) diameter leading to the casting benches. $1\frac{1}{4}$ in. (3·1 cm) pipe

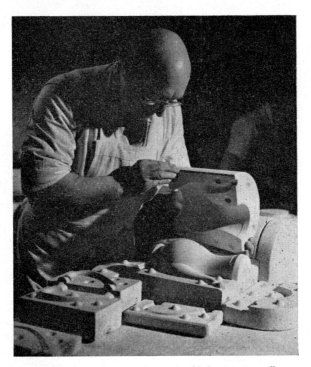

Fig. 8.30. *Preparing the plaster mould for a new coffee pot design* (Doulton, **D56**)

drops lead from these to each caster. It is found that no more air becomes entrapped in the slip than by other methods so long as the pressure tanks are not agitated.

The plaster moulds in which the casting is done can only very rarely be of one piece only. Generally three or more pieces are required to give a well-finished shape that can easily be removed from the mould. The pieces of the mould are made to interlock accurately and are tied together during casting. In spite of the advantages of casting for difficult shapes many pieces of ware are made up of several casts subsequently joined together, *e.g.* teapots are sometimes made with the pot, spout, and handle separate, whilst elsewhere a more complicated mould is used to make them in one casting (Figs. 8.33 and 8.30).

A funnel, either ceramic, zinc or rubber, is placed in the top of the mould if the hole is small, and the slip is poured in, either direct from the pipe line or by ladling. The moulds of circular objects may be rotated slowly during

FIG. 8.31. *Casting : filling moulds with the slip* (*Doulton*, **D56**)

filling. The mould must always be designed with a 'spare' at the top which acts as a reservoir as the level of the slip drops during casting (Fig. 8.31).

In *drain casting* the caster must estimate when sufficient thickness of stiffened body has built up against the mould walls. The mould is then inverted over a trough, which collects the slip for re-use, and the cast drains. Incorrect handling during the pouring off of this surplus slip results in uneven thickness on the inside of the ware (Fig. 8.32).

The piece is left to dry in the mould until it is stiff and has shrunk away from the mould a little. It is then removed and the seams smoothed off and any joins that are necessary made (Fig. 8.33).

Solid casting is used for thick pieces such as sanitary ware, chemical stoneware and refractory blocks and hollow-ware. In the case of refractory bodies with a high grog content the casting may be assisted by vibration. When the cast piece is sufficiently rigid to retain its shape the core of the mould is drawn. Finishing operations can then be carried out on the exposed surface while the piece dries further, until it can safely be removed from the mould (Fig. 8.34).

Special methods for speeding up the casting process and obtaining other advantages are described in the literature and patents, *e.g.*:

(1) Rapid rotation of moulds so that the centrifugal force ejects the water thereby lessening casting time and by making a denser cast lowering the firing shrinkage (**H6, A198**).
(2) The filled mould is sealed with a rubber diaphragm and pressure applied so that water is expelled (**M95**).

In another process the plaster mould is replaced by a perforated galvanised mould covered with canvas. Slip with about 35% water is placed between the two portions of such moulds and pressure applied to these until the water content of the body has been reduced to 12%. This is particularly applicable to making thin flat disks, etc.

Casting Defects

A number of defects in ware occur solely in pieces produced by the casting process, such as casting spots, streaks, fissures and air bubbles. They are due to incorrect compounding of the body, unsuitable deflocculants, wrong viscosity, haphazard methods of filling the moulds, and badly made or dried moulds (p. 161).

'*Casting spots or stains*' or '*flashing*' are vitreous annular rings which will not take a glaze properly. Pfefferkorn (**P29**) investigated this fault in detail. It occurs at the point of impact of the stream of casting slip on the empty plaster mould, and again on the sides if there is any splashing. The greater the absorbing power of the mould the worse the fault. It can be reduced by bringing the outlet of the slip container as near the bottom of the mould as possible and by moistening the mould at the point where the first slip will touch it. Rotating the mould while the slip is poured in also helps.

The cause of casting spots can, however, be traced back to the body composition, and therefore can in many instances be eliminated at the source. This fault occurs almost exclusively when the body contains mica, this is

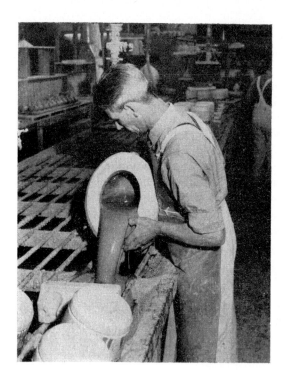

Fig. 8.32. *Drain casting: pouring off the surplus slip* (Wedgwood, **W35**)

Fig. 8.33. *Casting: removing cast teapot from mould and preparing to fix separately cast spout and handle* (Wedgwood, **W35**)

difficult to keep in homogeneous admixture with the other constituents and tends to separate out as an upper layer. If the body is freed from mica the casting spots cease. Where this cannot be done, reduction of the proportion of clay colloid and increase of quartz helps.

'*Streaks*' usually show most on the inside of drain cast ware. Pfefferkorn (**P30**) attributes this to separation of the different body constituents brought on by the electrochemical properties of the finely divided deflocculated clay

FIG. 8.34. *Casting lavatory basins:* Centre left, *newly filled closed mould;* centre, *core removed;* right, *removing casting seams* (Doulton, **D56**)

colloids. It therefore depends largely on the type and quantity of deflocculant. Pressure used in transport of the slip and filling of the moulds may also cause this separation. Streaking can be considerably reduced by rotating the moulds during filling and stirring the slip that has not yet set.

Air bubbles forming small cavities in the ware are more often due to air bubbles entrapped in the slip than air dissolved in the water or absorbed on the clay, the latter may, however, be the cause. Complete de-airing of the casting slip is impossible but there are various useful measures. Subsequent handling must then be done with care to prevent entrapping more air, *e.g.*

by turbulent discharge through air, etc., or using a pumped pipeline system instead of a gravity one.

'*Wreathing*' is the appearance of fine horizontal lines on the piece showing how the slip level rose up the mould at an irregular rate. Fast pouring and vibration of the mould eliminate this (**N36**).

Ware cast in moulds made up of several pieces shows '*seams*' when it is removed. These may be readily sponged off in earthenware, but in vitreous ware the seams reappear during firing even if they have been completely levelled off beforehand. This is due to the fact that clay mineral platelets orientate themselves parallel to each other and the mould surface during the casting, at a seam, however, the additional removal of water makes them tend to lie perpendicular to the others. During firing there is more shrinkage perpendicular to the plane of the platelets than within it, and hence the seam is left standing out. This can be overcome in bone china by hammering the seam and its immediate neighbourhood to induce random orientation. Addition of some 1% water-soluble oil to the slip reduces the appearance of seams. With hard porcelain a good method is to fire to just below maturing point, grind down the seam with an abrasive wheel and then refire (**N36**).

Casting Control

There are so many factors that can change the properties of a casting slip, making the output rate and the product irregular that effective control is of extreme importance. Here again, the sooner deviations from the normal are spotted, the easier and often more effective the remedy. Many different methods of control have been suggested; the best ones for any particular case have to be found experimentally.

RAW MATERIALS

(1) Uniformity can be approached if a large stock is kept and batches are made up with a little from each delivery (*see* p. 697).

(2) The raw materials particularly the clays should be tested:

    (i) chemical composition;
    (ii) grain size (*see* p. 278);
    (iii) amount and nature of soluble salts (p. 322);
    (iv) organic material present.

It should then be possible to adjust the amounts of deflocculants and their auxiliaries.

CASTING SLIP

It is usually found best to keep a constant check on the viscosity of the slip using the quick method of timing the flow of a definite volume through an orifice (p. 323). If an abnormality is found further tests must be made:

SHAPING

(1) Specific gravity.
(2) pH.
(3) Effect of more deflocculant.
(4) Temperature.

CASTING PROPERTIES

Rate of cast.
Setting up or gelling properties.
Drain characteristics.
$\begin{cases} \text{'Trim'.} \\ \text{'Feel'.} \\ \text{Plasticity.} \end{cases}$
Temperature and humidity of casting shop.
Water retention of the cast.
Shrinkage.
Dry strength.

ETHYL SILICATE CASTING PROCESS

Pure refractory non-plastic materials may be given a temporary bond for shaping purposes which converts on firing to a permanent bond with minimum undesirable additions by the ethyl silicate process.

Ethyl silicate is used to moisten the dry graded refractory grain to make a slurry which is vibrated or tamped into moulds. From 1 to 5% ethyl silicate is used. It then gelates and forms a green bond and the mould can be stripped. Complicated shapes can readily be made.

During firing the organic portion of the binder volatilises leaving finely divided highly active silica which causes the body to sinter rapidly. The actual quantity of silica thus added to the refractory is very small, certainly very much smaller than that of bond clays frequently used for this purpose, so that a much higher quality refractory results.

The process can be used for the following bodies:

(1) Alumino–silicate: molochite; sillimanite; mullite.
(2) Alumina–silica–zirconia.
(3) Silicon carbide.
(4) Alumina.
(5) Magnesia.
(6) Zirconia.
(7) Zircon.
(8) Forsterite (**Q2**).

HANDLES

Numerous items of crockery etc., require handles. These are made by casting in plaster moulds, pressing in plaster moulds, or extrusion from a 'dod' machine. They are allowed to become leather-hard and are then trimmed to fit the profile of the vessel they are to be stuck on to. This is now frequently done on small machines (**A178, K11**).

FINISHING

Sponging and Smoothing

The casting seams and other surface blemishes that are present after turning, joining, etc., are removed by sponging of fine ware. Heavy bodies containing grog are smoothed with iron, wood or rubber scrapers and spatulas (Figs. 8.35 and 8.36).

FIG. 8.35. *Sponging (Wedgwood,* **W35***)*

FIG. 8.36. *Finishing a large stoneware vessel (Doulton,* **D56***)*

Fettling

Jiggered ware after drying and removal from the mould requires rounding off at the point where the profile met the mould. This is done by hand on a small turntable under a dust extractor hood (Fig. 8.37).

Dust-pressed ware also requires fettling to remove the flash that appears between the two halves of the die. This can be made automatic in tile pressing.

FIG. 8.37. *Fettling* (*Doulton*, **D56**)

Joining or " Sticking up "

Several types of ware are best made in two or more separate pieces and joined up when partially dry. The most well-known example is, of course, the handles of cups, jugs, etc. The pieces are joined with a slip of the same body as they are made of. This can further be suitably adjusted to be a good adhesive. The body is deflocculated with 0·1 to 0·2% deflocculant, as if to make a casting slip; 0·1 to 0·4% $MgSO_4$ in aqueous solution is then added. This is a flocculant and makes the mixture very sticky and tends to solidify it, up to the total weight of the slip in water must then be added to bring it into a workable consistency. Occasionally it is better to use ammonium acetate instead of magnesium sulphate, and in rare cases it is necessary to add dextrine, but this should be avoided as far as possible.

The test of a good joining slip is to break a joined piece after it has dried. The join should be mechanically stronger than the rest of the piece, and hence the fracture should occur at any point other than the join.

FIG. 8.38. *Handling* (*Wedgwood*, **W35**)

There are many successful instances where the addition of magnesium sulphate to thicken the slip is not done, however, this method will almost invariably overcome any difficulties.

The joining is done with either dry or leather-hard pieces. In France the parts of large leather-hard pieces that are to be joined are first softened with a slurry for a few hours.

Very little of the joining slip is actually required for making a good join. The practice of dipping handles, etc., into a vat of slip means that the excess has to be cleaned off the joined cups afterwards, and the handles are easily distorted.

Application of the required amount only can be done by a simple device.

SHAPING

A glass roller half submerged in a trough of joining slip is rotated so that it carries over the slip. The piece to be joined is touched on to it to pick up slip and then pressed on to the other half. The roller is made with the same curvature as the cup, etc., on to which the handle is to go, so that the latter does not get distorted. The consistency of the slip is adjusted until the thickness carried over by the roller is exactly that required for the join, so that no excess has to be subsequently removed.

The pieces to be joined must be pressed together gently but with a definite pressure, the operation therefore requiring a certain amount of skill and training (Fig. 8.38).

FIG. 8.39. *The Strasser multi-head cup handling machine* (**S205**)

In recent years machines have been developed to undertake the task of joining up mass-produced ware. The Strasser (**S205**) multi-head cup handling machine has four pairs of arms set round a central motor-driven shaft. The upper one of each pair is charged with a sponged cup by one operator, who also places the handle in a cradle in the lower arm. An automatic applicator applies joining slip to it. The arms revolve together and the lower one being hinged gradually brings the handle into contact with the cup under predetermined pressure and releases it. A second operator removes the cup, sponges it if necessary and places it in the dryer. A team of two can handle from fourteen to eighteen cups per minute (Fig. 8.39).

A single-headed version of this machine, working at half the speed with one operative, has also been developed for the smaller works that do not manufacture enough of one shape to use the multi-headed one economically.

A second cup-handling machine made by Airey is now available (**A178**).

Fig. 8.40. *Preparation for sticking on branch to make a stoneware pipe junction* (Doulton, **D56**)

Joins in large and thick-walled pieces that are expected to undergo strain in use, *e.g.* sanitary ware, stoneware pipe junctions, chemical ware, are not made just by sticking smooth pieces together with slip. In some instances the surfaces are considerably roughened before joining so that with the pressure applied the two pieces become knit together. In others, a temporary slip join is made but is progressively cut out and replaced by a hand-moulded join made by hand spreading (Fig. 8.40).

Chapter 9
Glazing and Decorating

PREPARATION OF GLAZES

THE batch composition of glazes is normally worked out to contain only water-insoluble ingredients, whether these are the raw materials themselves or prepared frits. These are ground down to the necessary fineness for application of the glaze, normally in a ball mill in presence of water. In order to keep the ground material in suspension there is usually about 5% of raw china clay in the batch. Under particular circumstances this may be replaced by ball clay or by a small quantity of bentonite, or clay may be eliminated in favour of organic adhesives.

The glaze slip must then be brought to the correct consistency for application. This requires adjustment of the water/solids ratio and of the viscosity and may involve addition of soluble matter to arrive at a useful product. [The actual consistency required depends on the method of application and will be discussed below.] The consistency of glaze slip is considerably influenced by water-soluble matter. This is automatically included in the water used and it is often necessary to keep a check on the pH and the nature and quantity of soluble matter. Soluble matter may also be derived from the glaze batch which, although nominally insoluble in water, generally gives up small quantities of alkalies, boron, etc., during the wet-grinding process. The consistency of the slip is largely dependent on the clay present which is very greatly affected by the pH value and the nature and quantity of soluble matter.

Grinding of Glazes

Glazes are normally ground in pebble or ball mills to particle sizes that will entirely pass through the finest standard sieve. The actual fineness of the resultant glaze slip is usually expressed in hours of grinding or thousands of revolutions of the mill, etc., and rarely as an absolute grain size value. That the average sizes and size distribution of these sub-sieve particles is very important is generally acknowledged, but most experimental work is rather empirical.

In most instances greater fineness, *i.e.* longer grinding, of glazes improves them and this is usually only restricted by economic factors. There are, however, examples of faults that are due to over-grinding.

Koenig and Henderson (**K55**) have investigated the particle size of commercial earthenware and vitreous china glazes with an Andreasen pipette.

TABLE 181

Particle-size Distribution of Some Commercial Semi-vitreous China Glazes (K55)

Glaze No.	Diameter (μ)								
	> 30	30–20	20–15	15–10	10–5	5–2	2–1	1–0.5	< 0.5
	Weight (%)								
A	2.4	7.2	8.5	14.8	25.5	21.1	7.7	6.8	6.0
B	1.9	2.0	1.9	12.3	31.4	28.9	10.1	5.6	5.9
C	2.2	2.2	4.1	12.3	30.7	22.0	9.8	6.6	10.1
D	0.7	2.8	2.3	10.5	33.8	30.5	9.0	4.1	6.3
E	0.3	1.9	2.0	11.5	24.7	22.3	18.4	7.0	11.9

TABLE 182

Particle-size Distribution of Some Commercial Vitreous China Glazes (K55)

Glaze No.	Diameter (μ)							
	> 30	30–20	20–15	15–10	10–5	5–2	2–1	1–0.5
	Weight (%)							
A	4.9	7.8	9.5	15.0	24.6	17.2	6.7	4.1
B	3.5	8.4	6.5	11.7	23.9	25.7	7.9	5.8
C	1.4	0.5	0.9	3.2	28.4	35.1	14.7	7.1
D	1.0	0.9	1.3	16.5	30.3	19.8	11.2	8.5
E	0.6	0.9	2.0	8.9	27.3	25.0	10.0	12.4

In connection with the earthenware glazes they also investigated the relationship between the solubility of the lead frits and the particle size.

TABLE 183

Solubility Characteristics of Commercial Lead Frits Versus Particle Size (K55)

Frit No.	Lead in frit (%)	Percentage of lead dissolved*	
		14–20 Mesh (%)	150–200 Mesh (%)
1	73·15	28·52	91·36
2	45·00	12·51	59·49
3	31·56	12·96	54·53
4	40·20	1·82	8·56
$2PbO.SiO_2$	81·82	65·58	96·01
$PbO.SiO_2$	73·15	28·52	50·35
$PbO.2SiO_2$	60·02	1·47	9·67

* The frit or glaze was subjected to the action of 4% acetic acid and the dissolved lead determined by the molybdate method.

The grinding treatment of a particular glaze was varied from 12 to 28 hr and the solubility of the lead together with the general properties of the glaze tested. The lead solubility against milling time is a curve (**K55**) (Fig. 9.1).

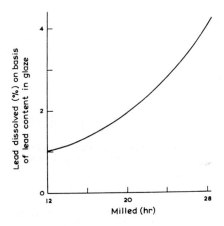

FIG. 9.1. *Lead solubility of semi-vitreous glaze versus reduction of particle size* (Koenig and Henderson, **K55**)

Although the glazes milled for only 12 and 16 hr were slightly inferior in texture to the other three, milled for 20, 24 and 28 hr, respectively, these latter were of about the same quality. All of the glazes passed autoclave and thermal shock tests which would show them as suitable for commercial glazes. It therefore seems that where a lead frit is used care should be taken not to overstep the necessary diminution of particle size.

Koenig and Henderson (**K55**) also did detailed work on the relationships between milling time, particle size distribution and various properties of both a raw and a fritted leadless porcelain glaze.

The raw glaze had the following formula and batch:

$$\left.\begin{array}{l} 0\cdot079\,Na_2O \\ 0\cdot184\,K_2O \\ 0\cdot737\,CaO \end{array}\right\} 0\cdot488\,Al_2O_3 . 4\cdot401\,SiO_2$$

Feldspar	48·08
Flint	27·24
Fla. clay	13·01
Whiting	17·66
Stain	5·00
	104·99

The fritted glaze:

Frit $\left.\begin{array}{l} 0\cdot2782\,Na_2O \\ 0\cdot0236\,K_2O \\ 0\cdot6981\,CaO \end{array}\right\} \left.\begin{array}{l} 0\cdot548\,B_2O_3 \\ 0\cdot268\,Al_2O_3 \end{array}\right\} 2\cdot555\,SiO_2$

Glaze $\left.\begin{array}{l} 0\cdot133\,Na_2O \\ 0\cdot140\,K_2O \\ 0\cdot726\,CaO \end{array}\right\} \left.\begin{array}{l} 0\cdot149\,B_2O_3 \\ 0\cdot427\,Al_2O_3 \end{array}\right\} 3\cdot875\,SiO_2 + 5\%$ stain

Batch
Feldspar	33·66
Flint	21·79
Fla. clay	10·41
Whiting	14·13
Frit	20·00
Victoria Green	5·00
	104·99

TABLE 184
Particle-size Distribution of Raw Glaze for Different Lengths of Milling Time (K55)

Milling (hr)	Diameter (μ)								
	>30	30–20	20–15	15–10	10–5	5–2	2–1	1–0·5	<0·5
	Weight (%)								
4	2·2	4·0	7·7	19·1	29·5	18·5	6·6	4·2	8·2
8	0·9	1·9	2·3	11·4	34·3	25·7	9·0	4·5	10·0
12	0·4	0·7	1·0	2·5	30·4	33·8	13·0	6·5	11·7
18	0·7	0·6	0·8	1·6	21·3	38·1	14·4	7·0	15·5
24	0·1	0·3	0·3	0·8	12·5	41·7	17·7	9·1	17·5
36	0·2	0·4	0·4	0·2	5·4	37·8	23·1	13·7	18·8
48	0·0	0·1	0·1	0·5	3·3	32·0	25·0	18·2	20·8

TABLE 185

Particle-size Distribution of Fritted Glaze for Different Lengths of Milling Time (K55)

Milling (hr)	Diameter (μ)								
	>30	30–20	20–15	15–10	10–5	5–2	2–1	1–0.5	<0.5
	Weight (%)								
4	0.6	6.6	9.8	21.4	26.0	17.7	6.8	3.5	7.6
8	0.5	5.8	5.9	14.6	32.9	20.6	7.9	4.0	7.8
12	0.3	1.0	2.7	9.3	38.1	23.0	10.6	5.7	9.3
18	0.7	0.9	1.1	1.3	28.5	36.5	13.1	7.7	10.2
24	0.2	0.7	0.5	0.2	21.7	39.9	16.4	10.5	9.9
36	0.1	0.3	0.1	0.4	12.3	41.9	21.3	12.6	11.0
48	0.0	0.1	0.0	0.3	4.5	41.6	25.1	14.9	13.5

Each batch had the following tests applied:

I. On the glaze, as milled.
   (1) pH.
   (2) Fluidity (Marriotte tube).

II. On the glaze, stood for up to one week, then stirred:
   (1) pH.
   (2) Fluidity.
   (3) Glaze taken up by standard specimen in 'in–out' dip test, immersion 2 sec.

III. Glaze adjusted to even practical feel for dipping by adding N. Brand Na Silicate.
   (1) Deflocculant quantity required.
   (2) pH.
   (3) Fluidity.
   (4) S.G.
   (5) Glaze taken up by standard specimen.

They gave the following general results. Both the raw and the fritted glaze showed greatly increased viscosity as grinding time increased. When dipped, the glazes showed progressive difficulty in draining, and cracking on drying. The pH of the raw glaze increased at first but fell off for the longest times whereas that of the fritted glaze increased throughout.

A progressive increase in the amount of deflocculant was necessary to achieve the same practical feel for dipping, when a nearly uniform amount of glaze was taken up by the standard specimen. The pH and viscosity of these adjusted glazes increased with milling time.

Comparable fritted glazes were more fluid and had higher pH values than the raw glazes, before any deflocculant was added.

Firing tests were also made and showed that increased grinding always improves texture but that the longer ground glazes tend to crack on drying and crawl on firing. Finer grinding increases fluidity under equal firing conditions.

Scratch hardness tests on the fired glazes showed finer grinds to increase resistance to scratching and generally to have less defects. It is probable that more glaze–body surface interaction occurs with a finer-ground glaze, which alters the composition of the fired glazes.

Schramm and Sherwood (**S30**) found a linear relationship between viscosity measured by flow timing and grinding time. They also found considerable improvements in appearance of a glaze of fixed fired composition when it was compounded with a softer frit but ground for the same length of time. The old glaze contained feldspar with 7·50% potash and 3·21% soda. This was replaced by a feldspar with 9·86% potash and 3·21% soda which necessitated reducing the feldspar content and increasing the borax. Particle size analysis of the old glaze ground for the standard 36 000 revolutions were compared with that for the new glaze ground under the same conditions for different lengths of time.

TABLE 186

Particle Size Analysis of Two Glazes of the Same M.F. but Different Batch Compositions (S30)

Radius (μ)	Glaze I	Glaze II		
	36 000 revs.	36 000 revs.	32 000 revs.	30 000 revs.
Over 8·3	13·3	12·3	12·4	16·0
8·3–2·5	43·1	35·3	41·8	40·6
Below 2·5	43·6	52·4	45·8	43·4

Parmelee (**P16**) gives a survey of a good deal of work on the grinding of glazes. He too emphasises the need to either know the condition of the material placed in the mill, or to be sure that it is always in the same condition. The grain size of zinc oxide has a marked influence on the viscosity of glaze slip and is affected by the thoroughness with which it has been calcined.

It is suggested that very finely ground glazes can be applied more thinly and thereby have less likelihood of crazing.

Water-jacketed pebble mills which are cooled prevent excessive solution of the frit during milling.

Brown (**B124**) recommends stepwise addition of a glaze batch to the ball mills in order to use them efficiently for the hard materials and to avoid over-grinding soft ones. Thus in a fritted glaze the frit, feldspar, flint and other hard material, together with a small amount of the plastic material should be ground first. From 70 to 90% of the water required to bring the slip to the required application consistency should be added at this stage. It is found that these materials are ground to almost the required fineness in from one-quarter to three-quarters of the time normally required for the complete batch. The soft materials and remaining clay and water should then be added and grinding continued for a short time, *e.g.* one hour. By this method grinding time is saved and some faults due to overgrinding are eliminated.

Koenig (**K52**) cites instances where glazes were shivering, peeling or cracking on curves. This was cured by using coarser flint.

In general therefore it is probably true to say that most glazes are improved by finer grinding but that this must not be overdone.

Additives for Glaze Slips and their Ageing

The normal glaze slip contains a relatively small proportion, about 5%, of raw clay which, however, plays a very important role in the consistency of the slip. The colloidal nature of this clay thickens the slip and helps prevent the settling out of the ground non-colloidal constituents. But it is very susceptible to changes in its ionic environment, *i.e.* in the water-soluble matter. The fine grinding of frits and other glaze raw materials exposes such large surface areas that noticeable solution occurs. Blanchard and Andrews (**B77**) have listed thirty-six soluble salts that may occur in enamel liquors together with their properties and probable effects on the slip. They also point out that even the so-called 'deflocculants' act as flocculants when present in large amount. The consistency of a slip of given specific gravity (or pint-weight) may therefore vary considerably because of small differences in its soluble matter and can be altered by such additions. A number of methods of regulating slip consistency depend on altering the quantity of flocculant or deflocculant present, by suitable additions of:

Soda.   Calcined magnesia (**A144**).
Borax.   Calgon.
Boric acid.   Nitric acid.
Calcium chloride.   Sulphuric acid.
Calcium borate.   Hydrochloric acid.
Sodium nitrite.   Magnesium sulphate (**M40**).
Potassium carbonate.   Sodium metasilicate.
Vinegar.   Soda ash.
Ammonium carbonate.   Ammonium oxalate (**B124**).

Other methods of slip control depend on the addition of more colloidal matter. Additions of up to 5% of bentonite, ball clay or fuller's earth (a corresponding reduction of china clay content can be made if the firing temperature increases noticeably) assist greatly in achieving a smooth glaze that will go on to the ware evenly. Organic colloidal materials are frequently used:

    Starch.   Carragheen.
    Dextrin.   Quince seed (**A156**).
    Gum arabic (**F17**).   Syrup.
    Gum tragacanth (**A144**). Milk.
    Gelatin (**G35**).   Animal blood.
    Methyl cellulose (**A144**). Agar (**D4**).
    Cellulose ethers free of methyl groups (**C65**).

The final condition of the clay should in any case be one of part flocculation in order to obtain high viscosity which prevents separation of constituents by settling and running when applied to ware (**S30**).

Slips containing clay are always affected by ageing and almost invariably they are improved by ageing of at least two to three days, preferably a week. In order to reduce variations in the ageing process it is recommended to use three clays of differing properties. Proper choice and blending of raw clays can obviate to a large extent the additions of soluble matter to bring the slip to the right consistency (**P16**).

Glaze Consistency

Consistency is an ill-defined word and may cover specific gravity, mobility, adhesion, etc. With regard to glazes the specific gravity is checked by finding the pint-weight; the normal practice is to make up glazes slightly too heavy and then dilute them to obtain the precise pint-weight required. Parmelee (**P16**) reports that hydrometers do not give consistent values for the same glaze before and after agitation whereas pint-weights are the same, and the difference between the specific gravity given by hydrometer and pint-weight measurements may amount to 11%.

The yield value of a glaze determines the thickness of glaze adhering to dipped ware, while the mobility governs the rate of drainage of the excess glaze from the ware. The yield value of glazes to be sprayed should be sufficiently high for the glaze to flow on the ware but not so high as to cause non-uniform glaze thicknesses with poor levelling qualities. The yield value is largely dependent on the colloid content of the slip and on its degree of flocculation (**P16**).

Harman and his colleagues (**H26–29**) have developed a method for control of glaze slip to be used for dipping, by defining easily measured slip properties. These are:

(1) The 'coherence value', which is the weight of slip adhering to a unit area of glass plate immersed in a slip and withdrawn therefrom at a constant rate of 1 cm per sec. It is expressed as weight in grammes of slip per square centimetre of plate area.

(2) The 'receptivity', which is the weight of water picked up and absorbed per square centimetre surface of a test sample of ware after immersion under standard conditions. This weight per square centimetre is arbitrarily multiplied by 10^4 to eliminate cumbersome decimals.

(3) The 'pickup', which is the weight of solids, in grammes, deposited per square centimetre area of the test sample of ware after immersion in slip under standard conditions.

For constant receptivity (R) the plot of coherence value (C) against pickup (P) is linear and leads to the following equation:

$$P = K\left(\frac{R}{100}\right)(C - 0{\cdot}001)$$

where P = glaze pickup (g/cm²);
C = coherence value (g/cm²);
K = deposition constant;
R = body receptivity (gH$_2$O × 10⁴/cm²);
$0{\cdot}001$ = coherence value of water (g/cm²).

Plots of coherence value against the clay content, and therefore viscosity, of a given type of slip reach a maximum and then fall away. The test cannot be applied to conditions beyond this maximum but such glazes have a tendency to be faulty in practice appearing 'livery' or 'clabbery'. Coherence values change noticeably when a slip is aged, the maximum moving towards lower clay contents. The maximum is pushed to the right, *i.e.* higher clay content, by the addition of deflocculants, and such treated glazes give good results on the ware. The quality of glazed and fired ware is definitely related to the coherence value, so that by checking this considerable control can be exercised.

Graphs can be plotted of R-value against P for constant C. These are linear, but may show a break and change of slope. In the latter cases specimens glazed under conditions represented by points on the section further from the origin showed a much larger percentage of defects. Such defects appear to be due to: (1) glaze slip too thick for body, (2) too long dipping time, (3) over-milled glaze containing too much superfine material. Such graphs can therefore prove useful in defining the range of conditions under which glaze dipping is most successful, and for determining the required coherence value to give the desired coating for any particular receptivity.

By plotting P against C for constant R 'iso–R charts' are obtained. These are linear and meet the C axis at a point designated by the coherence value of water, $0{\cdot}001$ g/cm². These charts may serve as control charts for glazes, as the slope of the iso–R line shows the deposition properties of the glaze. The recommended procedure is:

(1) Determine the R-values of the body by using the same dipping time as is used in applying the glaze.

(2) Choose the desired weight or thickness of glaze to be applied to the body surface.
(3) Follow the weight line on the chart until it cuts the iso–R line corresponding to the R-value of the body.
(4) Read off the corresponding coherence value.
(5) Set the glaze to this coherence value and dip the ware.
(6) Use the coherence value as a control test to check the glaze from time to time.
(7) Check the receptivity of the ware from time to time. The R-value is a more useful check for the uniformity of ware than porosity or density and can be found more rapidly.

Once of a correct consistency a glaze must be kept so by gentle agitation. Glazes otherwise tend to settle in one of the following ways:

(*a*) Slow sedimentation to a thicker slip at the bottom of the vessel and a thinner one at the top.
(*b*) Setting to a hard deposit which is difficult to redisperse.
(*c*) Differential sedimentation of the heavier constituents.

Additives may be necessary to prevent (*b*) and (*c*).

BRUSHING OFF AND BACKSTAMPING

Before glazing every precaution must be taken to see that the surface of the ware is free from loose particles and dust. This applies particularly to biscuited ware to which bedding sand and dust adhere. This has to be removed by brushing, which was formerly done by hand but can now in most cases be done by machine. After brushing, the ware is backstamped with the necessary trade and distinguishing marks (*see* Decoration).

Hand brushing, although it has the advantage that each piece of ware is given individual attention, has a number of disadvantages. Firstly, it is difficult to extract efficiently the dust produced, so that there is both a danger to the workers' health, and also the possibility that dust will resettle on the cleaned ware stacked near the worker. Secondly, in a mechanised works this process becomes a bottleneck holding up glazing operations.

Brushing machines need to be of two different types for flat ware and hollow-ware, respectively. The Roller Flat Ware Brushing Machine requires a team of three operatives. One feeds single pieces to the machine. Here they pass between a pair of shaped, revolving brushes placed under a hood and dust-extractor fan. From there the ware is delivered bottom up on to a chute from which it slides on to a conveyor belt passing at a suitable speed a worker who back-stamps the ware. It then continues to the taker-off who stacks it as necessary for the dipping process. 2000 doz or more of flatware can be treated on one such machine per day.

The Vertical Double Brushing Machine is designed for 'fruits' and

GLAZING AND DECORATING

'soups', which cannot be passed through the roller machine. Here the piece of ware is held on a lower flat revolving brush while an upper brush on a moving spindle is lowered on to it. This holds the piece during the necessary brushing time (**W21**).

GLAZE APPLICATION

Dipping

The hand-dipping of ware is a skilled operation. The dipper picks up the ware in a way that will leave the minimum mark on it, and he may attach wire 'fingers' to the thumb and two fingers of his working hand in

FIG. 9.2. *Glazing by dipping (Doulton*, **D56**)

order to increase his span. The piece is immersed and then 'handled-out' with a suitable twist or jerk so that the glaze distributes itself as evenly as possible and the surplus is thrown off back into the dipping tank. A good dipper knows how best to handle each type and shape of ware in order to produce uniform results. He also knows by the 'feel' whether his glaze slip is of the correct consistency. Dippers of even twenty-five to thirty years' experience may nevertheless turn out ware with weights of glaze differing by 100% and the glaze thickness variations on different parts of a plate, etc., may be 100%. (Fig. 9.2)

Dale and Francis (**D4**) list the specific faults found in dipped ware:

(*a*) Uneven glaze-layer on a single article, short-glazed edges, over-glazed centres of plates, etc., with a tendency towards in-glaze bubbles. With ivory glazed ware, uneven thickness of glaze will cause inequality of tint on a single article, while different thicknesses of glaze on different articles will cause variation from the desired standard colour. Excessive thicknesses of glaze may lead to crazing.
(*b*) Fringing.
(*c*) Run-back.
(*d*) Flouriness, dustiness, or undue sensitivity of dried dipped ware to handling.
(*e*) Difficulty of supplying a sufficiently heavy glaze coating to vitreous or leather-hard articles in a single dip, and at the same time avoiding running or dripping after dipping.
(*f*) Glaze wasted and time occupied in 'tapping' after dipping.

The main factors influencing the condition of the applied glaze are:

(*a*) Body surface condition (receptivity).
(*b*) Glaze consistency.
(*c*) Mechanical distribution of the glaze on the body.

It has already been seen that the nature, size, frequency, etc., of the pores in the surface of the ware as measured by the receptivity are related linearly to the glaze pickup. The appearance of blisters on some types of porous ware is also related to the nature of the pores. Where these are an even diameter the glaze is drawn in regularly and air may become permanently trapped, but where they are of different sizes air may escape up small tributary capillaries that the glaze does not penetrate so readily. This occurs more easily with totally immersed ware than with tiles, etc., glazed on one side only.

Irregular receptivity giving irregular pickup is counteracted by the inclusion in the glaze slip of matter smaller than the pore sizes. This rapidly clogs up the pores so that the glaze is thereafter built upon the equivalent of a vitreous body. Bentonite has been especially recommended as a glaze addition to overcome unevennesses due to this cause. The thickness of glaze adhering to a porous body due to dewatering by the body is greater if the particle size is large, or the colloids are flocculated, and small if the colloids are deflocculated and the particle size small.

Glaze will, however, adhere even to vitreous bodies, the thickness being governed by the fluidity of the glaze slip. Low fluidity gives greater thickness. The thickness is also dependent on the rate of withdrawal of the article from the dipping tub, fast withdrawal giving much heavier glaze application than slow, *e.g.* hotel ware 9 in. plates dipped in glaze slip at 31·25 oz/pint:

	Quick withdrawal 2 sec	Slow withdrawal 4 to 6 sec	Slower withdrawal 6 to 8 sec
Weight of dry glaze (g)	61·39	40·03	38·41

In considering glaze slip consistency all the factors discussed in connection with the flow properties of clays, thixotropy, flocculation and deflocculation, ageing, etc., come into play.

A glaze slip containing highly thixotropic materials, such as bentonite, gives an adhered glaze layer that can be roughly handled when dry. Such a slip also has much less tendency for differential sedimentation of the heavier frit particles. The thixotropy may be considerably influenced by soluble matter present in the slip and can therefore be regulated, within limits.

Again the benefits to be obtained by adequate ageing cannot be over-emphasised. Not only is the slip improved thereby but it becomes more stable with respect to time. The effect of soluble additions, etc., does not wear off as it does with fresh slip.

Dale and Francis (**D4**) conclude with practical recommendations for avoiding the eight following difficulties.

(1) *Effect of uneven porosity on dipped weight.* (a) Use coagulated slip of low pint weight, thus reducing the amount of dry glaze picked up for the same water absorbed. (b) Use dispersed high colloid slip that chokes the pores.

(2) *Over-glazing.* Use coagulated slip at lowest possible pint weight to give required thickness.

(3) *Running after dipping.* Use slip of moderate fluidity and thixotropy.

(4) *Settling-out.* Increase the thixotropy.

(5) *Too slow drying.* Use deflocculated slip of maximum pint weight.

(6) *Difficulties with green ware.* As water is not absorbed use high pint weight and adequate deflocculated colloid.

(7) *Dusting off of dry dipped ware.* Add colloids, *e.g.* bentonite, gelatin, agar, dextrin, gum, etc. The organic colloids can be preserved by small quantities of formaldehyde or other bacterial poison.

(8) *Livering.* Increase the maximum stirred fluidity by adding water or a deflocculant. Decrease the thixotropy if this is shown by the torsion viscometer to be excessive.

Coloured glazes are more difficult to apply successfully by dipping as the last drops of glaze to be removed cause dark spots. The addition of glue or dextrin to the glaze retards drying a little so that the dipper can distribute the glaze better by shaking.

MACHINE DIPPING

Machine dipping finds ready application in the tile industry where the ware is glazed on one flat surface only. Two types of machine are used. Firstly, the roller type, in which the tiles are placed face downwards and pass between two rollers. The lower one picks up glaze from a trough and transfers it to the tile surface. Secondly, the cascade type, in which the tiles are placed face up and pass through a cascade of the glaze. This may be coupled to an automatic fettler which scrapes off surplus glaze (**R5**) (Fig. 9.3).

FIG. 9.3. *Tile dipping machine (Gosling and Gatensbury*, **G61**)

Automatic dipping of tableware was first developed for saucers (reported, **R5**, 1953) but new machines for all flatware came into use for the new English translucent china in 1959 (**D56**).

DIPPING OF LARGE WARE

The dipping of large and heavy pieces may require the use of various mechanical aids in the form of hoists, pulleys, etc.

PARTIAL GLAZING OF WARE

Many pieces of ware are required to be kept free of glaze on certain specified portions. This is done by partial immersion, poured slip, stoppered mouths or ends, etc. (Fig. 9.4).

Brushing

Glaze may be applied to ware by hand with a brush. In the past this has been normal practice for large and once-fired pieces, *e.g.* sanitary ware and insulators, although spraying can now be used. It is also the best method for parti-coloured ware that cannot be successfully glazed by partial dipping.

For brushing, the glaze slip should be sufficiently viscous for a reasonable quantity to be picked up on the brush. If the clay content of the glaze does not give the required consistency gelatin, glue, gum, starch, cellulose products or wax emulsions may be added.

Brushing usually gives a thicker application of glaze than is possible with spraying. (Fig. 9.5.)

Spraying

The spraying of glazes eliminates many of the irregularities of application encountered in dipping, particularly when automatic spraying machines are used. In the latter case the process is not only quick and labour saving but also saves glaze by the elimination of over-glazing.

The consistency of glaze slip for application by spraying must be controlled within very close limits and every batch should be tested, and a trial run made before it is placed in the tank. Glaze slip may be gravity fed or pumped from the tank to the gun. Here it meets the air jet, supplied at from 30 to 40 lb/in^2 (2·1–2·8 kg/cm^2) pressure and is atomised. By suitable manipulation of the ware with respect to the spray gun, an absolutely even layer of glaze can be applied. This is particularly noteworthy in case of uneven porosity of the ware which leads to uneven pickup of glaze when dipped, and also with embossed, grooved and sharply angled ware which readily takes up surplus glaze both spoiling the decoration or shape and leading to crazing. Spraying is also the best way of regulating the area to be covered by glaze if this is not the whole piece, stencils being employed for delicate work (**W68**).

Hand spraying is done in a spraying booth with means for ventilation and for recovering over-sprayed material. The worker must wear a mask, and the atmosphere becomes very humid and unpleasant. The gun must always be held perpendicular to the surface of the ware and a uniform distance from it. Glazing by hand spraying is much slower than dipping but much more satisfactory for coloured glazes (Fig. 9.6).

Automatic spraying can be carried out with high speed and efficiency. The machines are usually specially designed for each job and have either straight or circular conveyors with or without rotating spindles. The spraying booths are equipped with recovery facilities for over-sprayed glaze. The spray guns may be stationary, oscillating, reciprocating or nodding. The ware may pass through the spraying booth continuously or stop under a spray gun. Round ware is usually placed on a slowly turning spindle.

For example a machine required to glaze 350 dozen pieces of miscellaneous

Fig. 9.5. *Glazing by brushing. (Salamander super basin used for melting and maintaining aluminium alloys.) (Morgan Crucible, M96)*

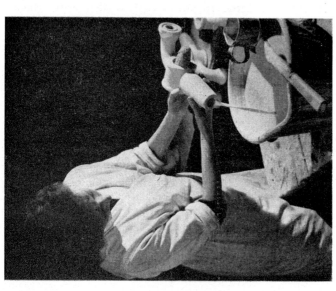

Fig. 9.4. *Partial glazing, inside of tap only (Doulton, D56)*

shapes of biscuit earthenware tableware with four unskilled workers is described as follows. The machine is circular, the moving rim carrying spindles at 8 in. (20 cm) centres. On to these fit three-pronged ware holders for flatware, or cup, or other holders. The holders can be easily exchanged by the loader while the machine is running. Two loaders place the ware in appropriate holders. The spindles then start spinning. The ware passes successively through: a gas-heated preheating tunnel, the automatic spraying booth and an after-heat tunnel; it then reaches the unloading station and is removed by a single worker. The empty spindles pass on through a water wash tunnel to the loading station.

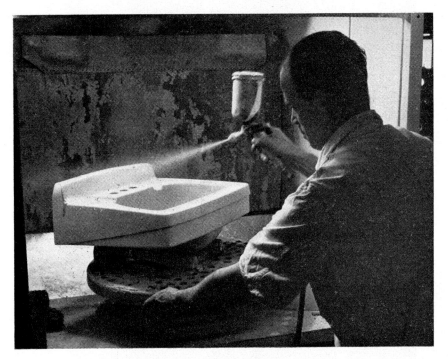

Fig. 9.6. *Glazing vitreous china lavatory basin by spraying (Doulton,* **D56**)

The glaze supplied to the machine is pumped from the glaze cellar to a self-contained pumping unit which keeps all the glaze within the machine in circulation. The spray guns are gravity fed and are so situated to spray all surfaces of the ware simultaneously. They can be adjusted to give different glaze thicknesses on the upper and lower surfaces if this is required. Over-sprayed material mostly collects on the back of the booth which is automatically moistened with water so that the excess glaze flows down to a pump tank beneath the spraying booth. From there it is pumped back to the glaze cellar for reprocessing. The exhaust from the booth is passed through water-washed baffles and then a vertical water scrubber to remove all

FIG. 9.7. *Automatic spraying of table ware. Girl in foreground is loading, girls near window are unloading (Schweitzer Equipment, S44)*

further glaze particles. These washings drain into a settling tank and the comparatively clean water that can be drawn off this is used for the washing of the ware holders, etc. (**A74**) (Fig. 9.7).

Meeka (**M42**) describes the machine glazing of heavy clay tiles. These are glazed raw either immediately after extrusion and cutting, or after drying.

In the former case where the wet column is to be glazed the spraying booth, etc., have to be aligned to the extruding mouthpiece. The column is lightly oiled, then cut, any crumbs made blown-off and rough edges rolled flat. It then slides on to stainless steel plates just narrower than the column and proceeds into the spraying booth. Here the top and one side are sprayed. Five guns spraying from the top and five from the side will glaze the output of a large extrusion machine making up to about 200 tons per day. These can be stationary except for those applying a stipple coat on mottled tiles.

Air is supplied from a water-cooled single stage compressor with a 75–100 lb/in^2 ($5 \cdot 3$–$7 \cdot 0$ kg/cm^2) shut off pressure. It must be free of water, oil or any other foreign matter. This is most successfully achieved by using firstly a compressor with either a graphite cylinder lining or graphite piston rings which entirely eliminate the need for oil in the compression chamber, and secondly, water-cooled after-coolers, water separation tanks, blow-off valves in low places in the air line, and highly efficient felt pad air filters.

The glaze slip used for glazing wet columns should contain twice as much organic binder or gum as that used on dried tiles, in order to anchor the glaze on the tile as it dries. After being tested for density and viscosity it is screened into a pump tank. The pump mounted in this tank keeps a pressure of 25 lb/in^2 ($1 \cdot 75$ kg/cm^2) in the glaze lines and through a by-pass valve, thus keeping circulation going.

The spraying booth must be arranged so that accumulations of over-sprayed material cannot drip on to the ware, but instead flow into a collecting tank for re-use. Suitable baffles prevent too much glaze reaching the exhaust system. An exhaust fan and appropriate ducts and stacks are fitted to the booth.

In glazing a dry column the ware must be placed on a conveyor belt. The ware, preferably still hot from the drying ovens is placed on one end, passed through one or more spraying booths and then through a small drying tunnel so that the glaze is dry enough for easy handling. The ware can then be placed directly on kiln cars.

Christensen and Schramm (**C32**) describe the teething troubles of an automatic spraying machine for vitreous ware. The machine consisted of preheating burners, spraying booth and after-heating burners. The first mistake that had to be eliminated was over-quick drying on of the glaze in the patterns left by the spray guns. Less preheating and a more fluid slip had to be used. It was also found necessary to regulate the slip pressure for each gun individually. As over-sprayed material was double the actual quantity used a very efficient recovery system had to be installed with a wash tower and a maze with an 80 ft long settling path.

Sanitary ware can also be sprayed automatically. It is usually done in a non-conveyorised single or double spindle machine employing moving guns. As much as 120 pieces per hour can be sprayed with a saving of one-third of the glaze used in hand glazing (**V7**).

The reclamation of over-sprayed material both because it is too expensive to waste and because of air pollution, can prove to take up much labour, space and time. A large factory with five automatic glazing machines found that 4 tons of material could be salvaged per week. For this ten electrostatic precipitators were installed to replace settling chambers, which had previously taken up ten times as much floor space and cleaning time as the new equipment (**A76**).

Loss of glaze by over-spraying could be reduced if the piece to be glazed attracted the glaze spray. In electrostatic glaze spraying the ware is earthed and the glaze spray is polarised in a 100 000 V high-tension field, the glaze particles are then attracted by the ware (**D40**). With electronic equipment adapted from standard radio transformers and vacuum tubes the ware can safely and inexpensively be made to attract glaze spray, even round bends (**A39**).

SPONGING

After glazing of vitreous ware the glaze is sponged off the foot, so that the ware may be placed directly on a saggar base or batt. This is done on a small machine with a moistened belt moving round two rollers, one at each end.

The Crawling of Glazes

Irregular contraction of a glaze into thick lumps and islands leaving bare unglazed patches is known as 'crawling'. The fault may originate in several different stages of manufacture but the adverse end-result is similar.

I. CRAWLING ARISING FROM THE COMPOSITION, PREPARATION AND APPLICATION OF THE UNFIRED GLAZE

(*a*) Bad adherence of the glaze to the ware surface: A glaze adheres better to a porous surface than to a dense one, and is particularly difficult to apply to any oily or greasy surface. The shape of the surface, *e.g.* sharp curvature, may also increase a tendency for the glaze to contract. Better adherence may be achieved by reducing the surface tension of the glaze slip by soluble additives, *e.g.* inspissated ox gall (plus sodium benzoate to prevent fermentation), turkey red oil (sulphonated castor oil) (**S156**).

(*b*) Excessive shrinkage leading to cracking of the raw glaze: This may be due to one of the following:
(1) Over-fine grinding.
(2) Too-thick glaze layer.
(3) Presence of constituents with large drying shrinkage, such as plastic clays, uncalcined zinc oxide, organic binders.

GLAZING AND DECORATING

(4) Reabsorption of moisture either before setting or in the preheating stage of firing.
(5) Moist ware preheated too rapidly.
(6) Irregularities in the body surface such as large grains of grog, giving irregular shrinkage.

(c) Lack of internal cohesion in the raw glaze which leads either to its dusting off when dry, or to subsequent crawling. It is largely due to the presence of large quantities of very finely ground material and may be counteracted as follows:

(1) Less fine grinding.
(2) Introduction of binding materials, either by replacing china clay by ball clay or bentonite, or by adding one of these, or by adding organic binders.
(3) Maturing of the made-up glaze for some weeks or months to improve the binding power of any clay present (**A104**).

If the additions necessary to prevent crawling change the consistency of the glaze while maintaining the same solids/water ratio then some deflocculant or flocculant must be added to bring it back to the desired conditions, *e.g.* gum arabic is a very useful addition against dusting but it makes the slip runnier, this can then be counteracted by adding ammonium acetate, when a good glaze showing no crawling may be achieved.

II. CRAWLING ARISING FROM THE COMPOSITION OF THE FIRED GLAZE

High surface tension of the fused glaze draws it together and prevents it wetting the body surface. The surface tension is dependent of the glaze composition and can be calculated approximately by use of factors (p. 538). It is not too strongly dependent on temperature but does decrease with increasing temperature. As some constituents have a much larger influence on the surface tension than others small replacements may appreciably alter the surface tension and so prevent crawling.

DECORATION

There are seven chief methods of colouring and decorating ceramic ware; three concerned with the body and four with the glaze. They are applied at different stages of the manufacture.

Body Decoration

(1) COLOURED BODIES AND BODY STAINS

Very many ceramic raw materials contain colouring agents which give products made with them characteristic colours. This is seen particularly in red building bricks and tiles where the colouring agent is iron, and in yellow bricks where it is iron in the presence of calcium. Coloured clays are also used for finer ware, for figures, crockery, etc. Fine grinding of the raw material is important if the colour is to be uniform.

Deliberate colouring of white or light coloured bodies is also undertaken in some instances either for decorative reasons or to make a permanent distinguishing mark for technical purposes.

Coloured bodies may be left unglazed, or covered with a colourless glaze. Interesting effects are also achieved by using a slightly opacified white or light-coloured glaze.

(2) ENGOBES

Coloured or stained body material may be used to decorate the surface of a piece. For instance thick ware may be made from a body, whose colour is undesirable, and then covered with an engobe, or body slip, of higher quality and good colour. Engobes containing barium carbonate can be applied over a sulphate-bearing body to arrest efflorescence or trouble with the glaze (**S211**). Engobes of contrasting colour are also used to paint or spray designs on to ware.

(3) RELIEF WORK AND INLAYING

A number of decorative methods depend on surface work on the leather-hard pieces including scratching, modelling, and piercing. Applied relief motifs are also used in the same or contrasting body colour, and contrasting body may be inlaid.

Glaze Decoration

(4) COLOURED, OPAQUE AND CRYSTALLINE GLAZES, ALSO CALLED 'IN-GLAZE COLOURING'

The surface of a piece may be uniformly coloured by using a glaze to which a colouring agent has been added. If the body itself is coloured an opaque glaze may have to be used. An allover uneven but not predetermined effect is achieved by mottling and also by the use of crystalline glazes.

(5) UNDER-GLAZE DECORATION

The ceramic colouring agents are processed into the form of paint or ink with fillers, carriers, oils, etc., and can then be applied to the ware by spraying, brushing, dusting, decalomania, stamping, transfer or screen printing, in any desired design and colour combinations. The glaze is subsequently applied over this decoration and the piece then undergoes the glost firing. This is the most durable way of applying design but the interaction of the colouring agents with both body and glaze and the high temperature of glost firing restrict the available colour range.

(6) ON-GLAZE DECORATION

Here the colouring agents mixed with a low-melting glassy flux are applied to ware that has already received its glost firing and are fixed to it by a further firing at a lower temperature. Again any design and colour combination can be achieved by brushing, spraying, dusting, stamping, transfer and screen printing. As the firing temperature and duration are lower than in a glost firing the range of colours for on-glaze decoration is much wider than for under-glaze. The mechanical durability of on-glaze decoration is naturally less than that of under-glaze but varies widely according to the temperature interval between the decorating and the glost firings. Thus for bone china this interval is only about 350° C (630° F) and considerable interaction occurs between the colours and the glaze and the decoration is therefore sufficiently durable. But for hard porcelain the glost–decoration temperature interval is about 700° C (1260° F) and so little interaction occurs, the on-glaze decorations are therefore relatively easily scratched or worn off. Bone china on-glaze decorations also have softer appearance and greater depth of colour than hard porcelain on-glaze decorations.

(7) IN-GLAZE DECORATION

This is an unusual form of decoration. The ware is glazed and when this is dry but not fired the decoration is hand painted on to it. A characteristic soft effect is produced (**C16**).

BODY DECORATION

(1) and (2) *Coloured Bodies and Engobes*

The uniform colouring of a ceramic body presents several problems which can only be overcome by exercising considerable care. The general principle used is to mix a finely ground colouring agent with the body, but it frequently happens that even with the most careful grinding and mixing the fired body appears spotty or blotchy. This can be considered to be due to a number of causes. One is the tendency for the body constituents to react partially with the colouring agent without the reaction going to completion so that the colouring ions occur in different states of combination in the fired body, and hence appear to have different colours. Another cause is the unhomogeneity of a ceramic body, some constituents of which may react with colourants more readily than others so that the colour concentrates in certain favourable phases (as seen in Crystal Glazes, p. 593). A third cause is the unavoidable relatively small variations of body composition, in particular the fluctuations of minor constituents which are normally of no consequence to ceramic production, *e.g.* the gallium in aluminium minerals, rubidium in potassium feldspars, fluorspar in Cornish stone, cobalt in nickel preparations, etc.

All these troubles may be overcome if the reaction between colourant and body can be prevented by using chemically inert colouring agents. A type of compound that has proved to be exceptionally resistant to attack by silicates and borates is the 'spinel' type (*see* p. 229). Many colouring agents can be brought into this state of combination by reaction with suitable other coloured or colourless materials. The product is then finely ground and used as the stain for the body. Another highly resistant colourant is chrome–tin pink (**S112**).

The common synthetic body stains are mentioned in the general table of ceramic colourants (p. 616). It is also of interest to investigate the colourants in naturally coloured clays.

The colouring agent in many of these is iron oxide. It is very often in an extremely finely divided state and on firing produces an even colouring. McVay (**M18**) has investigated the artificial addition of iron as a colourant, the best method of introducing it and the colours obtainable. The experiments were made with engobes of different combinations of clay, feldspar and flint and fired to cone 10–11. The iron was added in various ways: in the form of ferric hydroxide it was introduced first by absorption from sols, then by precipitation in the body slip, in a third set of trials red iron oxide was added directly; ferric hydroxide precipitated in silica gel, and calcines of ferric oxide with feldspar or aluminium hydroxide were also tried. The most satisfactory method for obtaining uniform colours was found to be the precipitation of the hydroxide in the engobe. For this a weighed quantity of the engobe is blunged with water to which a standard solution of ferric chloride is added. Ammonia is then added to precipitate the ferric hydroxide. The solids are allowed to settle and must be filtered off and dried at 125° C (257° F) to destroy the gelatinous water retention properties of the hydroxide. The dried engobe is then blunged with more water and used. Without the intermediate drying the freshly precipitated ferric hydroxide causes serious drying and firing cracks. The composition and grain size of the engobe itself were found to affect the colour though practically no fusion or chemical reaction with the ferric oxide took place at the temperature used. For instance, coarse grained clays gave a more intense coloration than fine ones, as did commercial potter's flint compared with finely ground flint. Reds are produced when the feldspar content is high (*e.g.* 60% feldspar, 30% clays, 10% flint) and the ferric oxide content is from 3 to 5%. Buffs are produced with a high clay content (*e.g.* 60% clays, 30% feldspar, 10% flint) and 1 to 3% ferric oxide.

PREPARING COLOURED BODIES

Owing to the cost of colourants it is usual to use only small quantities when colouring a body throughout, although there are exceptions. The normal maximum is 5% colourant, but there are many instances where less than 1% is used. In these cases the colourant does not noticeably upset the composition and reactions of the body, and more particularly is this the case if an inert colourant has been selected. The procedure for its addition is to wet grind it at 10 to 20% strength with clay and silica or some of the body mixture itself and bring it to a known specific gravity. It can then be added by volume to the rest of the body in the blunger or ball mill.

Cobalt blue colourants are used in bodies that have a natural yellowish tint and are required to appear white. About 0·01% Co_2O_3 is required. Very careful milling, measuring and blending are necessary if the result is to appear uniform and free from specks (**C59**).

PREPARING COLOURED ENGOBES

The composition of engobes is frequently deliberately made different to that of the bodies they cover because higher quality materials are used to cover lower grade ones. Even if this is not the case and the coloured engobe used is based on the same body, the relatively large proportion of colourant may alter the properties of the engobe.

The chief points to bear in mind when developing an engobe for a given body are the relative drying and firing shrinkages and maturing temperatures. All of these must be similar for the two parts. As the engobe is applied as a slip to the raw but dry and therefore shrunk body, its drying shrinkage should be slightly less than that of the body so that it does not crack on drying. The engobe should therefore be less plastic than the body itself. This must be remembered when wet milling a portion of the base body for use as the engobe as this will alter the plasticity which must therefore be adjusted accordingly. The maturing temperatures of the two parts must also be similar if the engobe is neither to fuse, nor to dust off through immaturity. The maturing point of the engobe must be adjusted to suit the ground body by altering the flint/flux ratio.

The engobe is prepared by fine grinding the colourant with the non-plastic materials for about 24 hr. A wetting agent may need to be added to the colourant. The clay constituents are then added and blunged or further milled for another 12 hr. The slip is aged for several days to allow entrapped air to escape. It is then adjusted to the correct specific gravity, generally around 1·60 for spraying and 1·30 for dipping, although the viscosity plays a more important role than the specific gravity. Some colourants have a deflocculating or flocculating effect on the engobe slip and may have to be counteracted by other additions. In that case only inorganic substances should be used and organic additives that may ferment or decompose in the ageing period should be avoided (**C59**).

APPLYING SLIPS AND ENGOBES

Coloured or white engobes are widely used to cover completely a large article made of low-grade materials. The engobe is applied to the ground body in the leather-hard condition by dipping, spraying or brushing. (A brush of contrasting colour to the engobe is recommended so that any hairs that become detached are seen and removed.)

Where the ground body is of sufficient quality for it to appear on the surface a number of decorative effects can be achieved with engobes in slip form.

An engobe of contrasting colour may be brushed on to the piece, either free-hand or with the assistance of a dotted outline powdered on through a perforated stencil (*see* p. 811). Slip may also be applied decoratively with a bulb and quill, the method is known as *slip-trailing* and is used for earthenware and stoneware. A more refined form of this method is seen in the *pâte sur pâte* decoration of porcelain. Here a low relief of white porcelain body is built up in successive layers on a coloured base. The natural translucency of the porcelain makes special effects available to the skilled artist.

Where the engobe is to cover most of the piece the portion of the body to remain clean may be covered with a resist of wax, fat and colophony and then dipped, or a stencil may be held against it and the engobe sprayed on. An alternative to this is the *Sgraffito* method in which the complete piece is covered with the engobe and when it is nearly or wholly dry the unwanted part is scraped away. A clean, sharp edge to the decoration is thus obtained. For repetitive work a zinc stencil is made and placed on the ware, and the unwanted engobe removed with a sandblaster (*see* p. 812) (**A117, N37**).

Parti-coloured pieces can be slip-cast by brushing the engobe slip on to the desired part of the mould before filling it with the casting slip (**N37**).

(3) Body Relief Work and Inlaying

The surface decoration of cast ware is usually incorporated in the mould but ware shaped on the wheel, either freehand or by jiggering and jolleying, etc., may have further work done on its surface.

SCRATCHED OUTLINES

These may be made as a decoration in themselves or combined with regional application of coloured glazes.

SURFACE MODELLING

Surface modelling such as fluting can be cut into fairly thick-walled pieces.

PIERCING

This was already done by the ancient Chinese and although painstaking can be very effective. The thrown piece is turned to an even wall thickness and the design drawn or traced on. The hole is started with a twist drill and then worked with a thin-bladed knife.

GLAZING AND DECORATING

SPRIG WORK

Reliefs may be applied from a mould by this means. Plastic clay of the same or contrasting body colour to the main piece is pressed into a mould and the whole applied to the surface of the plastic piece until the relief sticks to the body surface and comes away from the mould.

APPLIED RELIEFS OR 'ORNAMENTING'

Unmoulded decorative pieces pressed on to leather-hard ware are used. This is the more popular method. The plastic clay is pressed into plaster or biscuit moulds, the surplus scraped off and the surface then rubbed with a rounded tool to make the relief curl enough to fall out of the mould. The relief is then placed on a damp plaster bat.

The leather-hard pieces to be decorated are brushed with water and the reliefs picked up with a spatula and pressed on to the surface. As the water dries out the relief is sucked on to the surface. Considerable skill is required to press the delicate shapes firmly on to the ware without damaging them.

Applied reliefs may be in a matching or a contrasting colour to the ground body (Fig. 9.8).

INLAY

Inlaying has been widely practised on floor tiles. The design is cut out of the plain tile surface and the contrasting plastic body pressed in. The surface is then scraped smooth so that sharp outlines appear (**N37**).

SANDING

Many building bricks are given textured and coloured surface effects by sanding. Both natural sands of suitable colour, and artificially coloured sands and grogs are used. Three faces of the green bricks, one side and both ends, are moistened by spraying and then have the sand sprayed and brushed on. This can be done mechanically (Fig. 9.9).

GLAZE DECORATION

(4) *In-glaze Decoration with Coloured, Opaque, and Crystalline Glazes*

Where a decorative glaze is used to cover all the portions of the piece that are to be glazed the procedure is identical with ordinary glazing except that a thicker application is usually necessary. The use of more than one glaze on the same piece requires special methods. For instance if the inside and outside of a piece of hollow-ware are to be differently glazed, the inside is glazed first by pouring, and the outside then covered by careful dipping, or spraying the piece with the mouth stopped.

For more complicated arrangements spraying with stencils or brush

Fig. 9.9. Sanding machine (London Brick, **L41**)

Fig. 9.8. Ornamenting, applying bas reliefs to leather-hard ware (Wedgwood, **W35**)

GLAZING AND DECORATING

application must be used. Some arrangement must always be made to stop the glazes running into each other, this may be slip-trailed relief or a scratched outline.

(5, 6, 7) *The Application of Under-glaze, In-glaze or On-glaze Coloured Designs*

The diversity of physical and chemical properties of ceramic colourants together with the variety of surfaces and shapes they are to be used on has led to a very wide selection of methods of applying the coloured designs. Some of these, although used in industry, differ little from artists' methods, others can be truly regarded as mass-production methods. A preliminary classification can be made by considering the origin of the design that is applied to the ware.

(a) DIRECT APPLICATION OF DESIGN

Various means can be used for the ultimate design to be produced directly on the ware: hand painting; crayoning; spraying; banding and lining; ground laying.

(b) SEMI-DIRECT APPLICATION OF DESIGN

The design is prepared in a permanent form which is used to apply the colourant to the ware: stamping; stencilling; screen printing.

(c) INDIRECT APPLICATION OF DESIGN

The design and colourant are produced on a thin sheet of paper from which it is transferred to the ware: offset; paper transfers: intaglio printing or engraving, lithography, screen printing; slide-off collodion-backed transfers.

THE INDIVIDUAL PHYSICAL PROPERTIES OF CERAMIC COLOURANTS

In every method of applying coloured design the physical properties of individual colourant preparations play an important part and will, in general, be different for every one. Thus one will have more tendency to run, settle out, clog up, dust off, etc., than another. It is therefore necessary to get the right knack for every colourant. It is also very important to derive the best sequence for applying a number of colourants to a single piece as the final effect will differ with the order used.

DIRECT DECORATION METHODS

Free-hand Painting or Brushing

Both under- and on-glaze colourants can be mixed with suitable oily media for application with a brush. Under-glaze colourants may also be mixed with glycerin and thus avoid the hardening on firing. Most in-glaze decoration on the raw glaze is applied by hand painting. Good brushes suited to the work are essential, an arm or hand rest is also often necessary.

A free-hand design may be sketched on in pencil before starting to paint. For repetitive work the outline may be transferred to the work with a perforated stencil and a talc or charcoal dusting bag (*see* Stencils, p. 811) (Fig. 9.10).

Ceramic Crayons

Ceramic under- and on-glaze (**A128**) colourants may be made into crayons. These allow delightful casual appearances to be achieved by a skilled artist. The crayons were first made by mixing the colourant with beeswax when a very soft and crumbly crayon resulted. Harder pencils with which fine lines can be drawn are achieved by mixing the colourant with a plastic white-burning clay, shaping and firing to 740° C (1364° F). The proportion of colourant should be varied according to its colouring power, *e.g.* chrome–tin pink, 70% with 30% clay, but 10% cobalt blue with 90% clay and 50% chrome green with 50% clay (**S135**).

Spraying

Uniform applications of coloured engobes on green ware or of colours on unfired, biscuit or glost ware can be obtained by spraying with compressed-air pistols.

There is now a semi-automatic machine for applying coloured bands to tableware in this way, at a rate of about eight pieces per minute. The operator places a piece in position and sets the machine going, when the piece becomes automatically centred on an inclined faceplate and is held there by vacuum. It then rotates whilst being sprayed, the centre of the piece is protected by a stencil plate from which any surplus colour drains away, as does also the over-spray caught in a collector behind the piece (**S174**).

UNDERGLAZE SPRAYING ON HARD PORCELAIN

It is widely considered that the spraying technique for under-glaze decoration of hard porcelain produces the most beautiful results with soft edges and every gradation of colour intensity. It is, however, a skilled and slow operation and it is rather difficult to produce identical pieces.

The process is restricted to hard porcelain as this is the only type of body that is sufficiently soft to be scratched by hand tools in the biscuit state. The outline of the design is drawn on to the piece with the aid of stencils, transfers, etc., and then scratched into the surface just enough to show through the sprayed colourant but insufficient to be visible on the finished piece.

The colourants to be used are ground and brought into a watery suspension suitable for spraying. They must then be arranged in order of the intensity and obscuring powers of the final colour. The darkest is applied first and

Fig. 9.12. *Ground laying. The plates in the foreground have the resist showing as the dark pattern (Wedgwood, W35)*

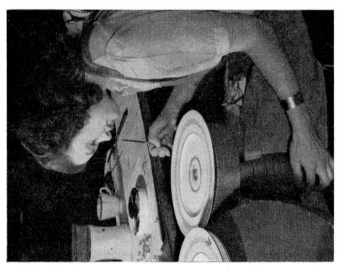

Fig. 9.11. *Hand lining (Wedgwood, W35)*

Fig. 9.10. *Hand painting on-glaze decoration. The hand paintress must space and adapt the design to fit the various pieces of a set (Wedgwood, W35)*

the most delicate tones last. The first colourant is then sprayed on to the piece, to cover as nearly as possible only the required part of it. The edge of a sprayed section will always appear lighter than the centre. This is frequently desired and is the correct application of the spraying technique as it produces the typical 'soft edge' to the design. Where the coloured section is to appear uniformly intense an area considerably larger than that finally required has to be sprayed. When the application has dried all the colourant that has adhered to portions of the ware where it is not required must be carefully scraped off, in particular all parts that are eventually to appear white must be made scrupulously clean. The next and subsequent colourants are then applied in the same way, overlapping as little as possible with the previous ones unless the mixed tone is desired.

Both the colourant spray and the dust made when scraping surplus off are easily inhaled. At best they are lung irritants and at worst they are poisonous, *e.g.* uranium. It is therefore imperative that the whole process be carried out under extractor hoods.

Banding and Lining

Bands or lines of under- or on-glaze decoration may readily be applied to any article that is in some way circular about an axis. For hand application the article is mounted in such a way that it can be revolved about its axis. An arm-rest is arranged above or beside it so that the hand can be kept quite steady. The colourant is then applied by brush while the free hand turns the article. This basic principle has been mechanised in varying degrees. (Fig. 9.11)

One example for hollow-ware is hand operated and designed for small producers and is readily adapted to various sizes. The machine comprises a base plate on which are mounted a headstock, tailstock and saddle. The headstock has a spindle, and a faceplate on which are mounted four spring-loaded radial fibre pads. The pads are actuated through the tapered spindle which is fitted with a quick release. The spindle is from a hand wheel. The vessel to be decorated is largely held by the sprung pads pressing inside it but it is also steadied by a rubber pad mounted on the tailstock spindle and bearing on the outside of the vessel bottom.

The lining instrument itself is held on the saddle and applied to the vessel where required. It consists of a cylindrical box with colour container through which runs a wheel that then deposits colour from its periphery. It can be adjusted to give single or double lines of varying thickness (**A64**).

Larger motorised machines which deal with flat or hollow-ware have the rotation axis vertical. Adjustable chucks to take the different articles can be interchanged on the spindle. The applicator consists of a colour reservoir metering the ink on to a steel drum from which it is taken off by a rubber roller which applies it to the ware. It is this applicator roller that actually turns the ware on its spindle so that the speed of application is in linear quantities of line, not in revolutions. Large diameter ware therefore takes longer than small. For edge-lining plain circular ware the Ryckman machine works at 600 linear inches per min (15 m/min) but for difficult shapes (scallops, etc.) speed must be reduced to 400 in./min (10 m/min) (**W18**).

Machines for producing lines on the surface of ware have now also been developed, one using brushes and the other rubber rollers (**A174**).

Ground Laying

This is the method of producing evenly coloured areas of on-glaze decoration. The parts not to be coloured are first covered with a resist made up of a strong sugar solution containing a water-soluble dye. The remainder of the piece is then painted with a drying oil, *e.g.* linseed oil and turpentine, and the surface patted with cotton wool to remove all brush marks and make the surface absolutely regular, on this depends the evenness of the final colour. It is a skilled operation. When the oil has dried to the right tackiness the powdered colourant is dusted on, and the excess shaken or dusted off. This must be done under an extractor hood.

The piece is dried and then washed in warm water to remove the resist. (9.12.)

SEMI-DIRECT DECORATION METHODS

Rubber Stamping

Monochrome decoration can be applied by a rubber stamp, usually mounted on soft rubber to increase the 'give'. The usual method is to spread the colourant evenly on a glass plate and then press the stamp in it before application to the ware. Only a very thin layer of ink is transferred in this way so that the ratio of colourant to medium must be kept as high as possible and only colourants of high intensity used. A method of obtaining a thicker deposit is to print a fatty oil with the stamp and then dust on the colourant in powder form to adhere to the oil (**D61, C55**).

Stamping can be used for both under- and on-glaze decoration. The under-glaze colourants may be mixed with various media for direct application, as seen in the following examples (Table 187).

TABLE 187
Under-glaze Colourants Prepared for Stamping

'Prepared stain' including 1 to 10% white ball clay	50%	39·4%	30·3%	28·4%
Water	—	40·6%	41·2%	51·6%
Wood vinegar	—	16·0%	21·4%	—
White syrup	—	4·0%	—	—
Glycerin	50%	—	7·1%	20·0%

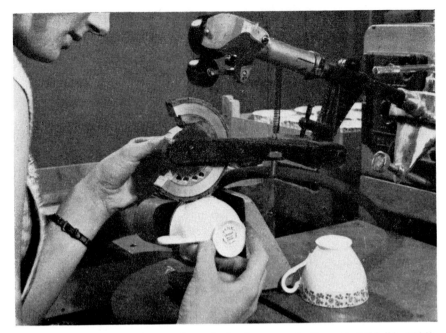

Fig. 9.13. *Applying a border-type decoration by the segmental roller method* (*Malkin*, **M23**)

A girl can stamp 6000–8000 porcelain bottle stoppers in 8 hr with under-glaze chrome green prepared in this way. Similar stoppers stamped with oil and then powdered for on-glaze decoration are worked at a rate of 4000 stamped only, in 8 hr, or 2000 stamped and dusted, in 8 hr.

Stamping is popular for applying gold decoration and fully automatic machines dealing with several shapes of ware simultaneously have been evolved which permit speeds of more than 250 dozen/hr (**R74**).

A number of semi-automatic machines are now available for decorating hollow-ware and flatware chiefly with gold on-glaze patterns but also with under-glaze colours. For bowls a border pattern can be applied from a small roller with rubber stamps fixed round it. The operative has to stop the machine before overlapping of the ends of the pattern occurs. With cups, there is the difficulty of the handle. This is overcome by using a roller of larger diameter than the cup with a segment cut out so that it can pass over the handle. The remaining section carries exactly the required amount of pattern to circumscribe the cup from one side of the handle to the other. The cup is placed in a vacuum chuck. The pattern-roller is allowed to rotate over an inking roller, and then pressed down on to the cup in position at one side of the handle. The cup is then rotated by hand until the pattern is completely applied. (Fig 9.13)

In another machine stamped patterns can be applied over most of the surface of a cup. Here the stamp pad, set with a complete set of rubber stamps for one cup, is revolved in a horizontal plane. Again a start is made against one side of the handle.

Flat ware can also be completely stamped in one operation by building up the stamps on a rubber pad. If this requires to be noticeably curved it cannot be inked directly from a roller. Instead a flexible belt is inked and passed over a plate with a circular hole into which the stamp pad presses it to obtain even inking (**A174**).

Stencilling

The different types of stencils and their application in ceramic decoration have been described in detail by Viehweger (**V13**). The main types are:

(1) Tissue or parchment paper prepared by pricking and used to transfer a guide on to ware to be hand painted with engobe, or under-, in- or on-glaze decoration.

(2) Tinfoil of various thickness hand cut with special knives or etched, and used with a stencil brush or a spray for under-glaze colours.

(3) Zinc sheeting of various thickness hammered to correspond to the shape of the ware and cut out with a fretsaw. This is used with the sand-blasting method of removing small areas of over-all engobe or on-glaze colour.

(1) PAPER STENCILS

Stiff parchment paper may be used when the article to be decorated is flat but thinner paper is required when it must be shaped to conform to the article. The required design outline is traced on to the chosen paper. This is then placed on a felt pad and holes pricked along the lines at $\frac{1}{5}$ in. (5 mm) intervals. The back of the stencil is then rubbed smooth with very fine sandpaper.

These perforated stencils are used to transfer the outline of repetitive work to ware that is to be hand decorated. The stencil is held against the ware and a linen bag, containing talcum powder if the piece is coloured or charcoal powder if the piece is white, is dabbed on, so that the outline appears on the ware in the chosen powder. The design can then be painted in whether it is engobe or, under-, in- or on-glaze colourant. Neither talc nor charcoal produce any faults on the ware during the firing.

(2) TINFOIL STENCILS

Tinfoil stencils may be prepared in two ways, firstly by hand-cutting or secondly by etching.

Before hand-cutting the tinfoil is coated with a thin layer of kaolin mixed with a gum-arabic solution to a creamy consistency and brushed on. When this coating has dried it should not readily wipe off. On it the required design can be clearly drawn. It also serves a useful function during the stencilling operation.

The coated tinfoil is placed on a matt glass plate and cut with special steel knives. A steel ruler may be used to produce any straight lines. A sharpening stone should be at hand to keep the knives in good condition. The portion cut away is that required coloured. Care must of course be taken to leave suitable bars to make the stencil hold together. Long cuts even where no part falls out are undesirable as the stencil easily lifts and gives blurred edges to the design.

For the etching process the tinfoil is first coated with a heated mixture of 41% beeswax, 40% stearine candle wax and 19% thin machine oil, and allowed to cool and become solid. It is then placed on the matt glass cutting plate and a ready-cut stencil placed on it. The wax is then brushed off all the exposed parts with a stiff short-haired brush. It is next immersed in a nitric acid bath (50% technical nitric acid and 50% water) until the unwanted parts have dissolved. The acid is then washed off thoroughly, followed by removal of the wax with turpentine.

Tinfoil stencils prepared in either way are used for brushed and sprayed under-glaze decoration. The stencil is lightly vaselined underneath and held on to the article with the left hand while the colourant is brushed or sprayed on with the right. The vaseline assists adherence to the piece and so reduces the amount of colourant accidentally getting under it. It should be renewed whenever there is evidence of this happening. After each application the stencil should be dried before re-use. A number of stencils and a heat-controlled hot plate for drying them are therefore necessary for continuous production. When too much colourant builds up on the front of the stencil it should be tapped off. The colourant may require mixing with suitable vehicles for stencilling, *e.g.*

Stain	80%	71·4%
Wood vinegar	16%	21·4%
White syrup	4%	—
Glycerin	—	7·2%

(3) ZINC SHEET STENCILS

The sheet is coated with a similar kaolin–gum arabic mixture as used on the tin. It is shaped to fit the ware with the aid of a wooden mallet and incisions. The design is then traced on and subsequently cut out with a fret-saw.

These strong stencils are used for the sandblasting process. This is not a widely applied but nevertheless noteworthy method of decoration.

One application is to ware which is to be largely covered with an engobe with small portions of the base body showing through. The ware is entirely covered with the engobe and allowed to become completely dry. The stencil is then placed in position and very fine sand blown on to it with a minimum of air until the base body is cleaned of engobe in the required places. Care must be exercised not to use coarse sand or too great an air blast as the body itself would be worn away.

Similar techniques can be applied to on-glaze decoration. The colourant is applied to the ware, mixed with turpentine oil and just sufficient balsam to make it impossible to wipe the dried application off by hand. Too much balsam would make the next operation unnecessarily difficult. The stencil is placed on the fully dry ware and very fine sand blown on with a suitably adapted spray. This must be done under an extraction hood to remove the sand as soon as it leaves the ware face (**V13**).

Screen Printing

Screen printing can be used to produce mono- or poly-chrome decoration either directly on the ware or to make transfers. The basic method of applying the colourant is to mix it with oily substances and then force it through stretched fabric lying above the surface to be printed. The pattern is achieved by making the fabric impermeable to the ink in all places where no printing is required. The set up then, is a suitable fabric stretched over a wooden or metal frame on which the negative of the desired pattern has been applied in one of a number of dense materials. This is placed over the piece of ware or transfer paper and the colourant is pressed through it with a rubber blade, known as a 'squeegee'. The process has the advantage over intaglio and lithographic printing in the speed and relative cheapness of preparation of screens. Over lithography it has the added advantage of being able to deposit a much thicker layer of colourant.

PREPARING THE SCREENS

Dubuit (**D61**) has described the process in detail and himself developed many machines for it. The original fabric used for screen printing was silk and the process is often known as 'silk screen printing'. It is now occasionally replaced by nylon or metal. The material used for silk screens is bolting cloth which is very strong and can be stretched considerably without losing shape. It is woven in three different ways: full gauze, in which every thread has a double twist; half gauze, in which every other thread has a double twist; plain weave, in which the threads are as in ordinary material. Nylon thread, although much stronger than silk, cannot be woven into full gauze. Instead, the plain weave is heated slightly to melt the strands into each other and fix them. Metal screens are of course much stronger than either textile and more resistant to alkali; if, however, they should accidentally get out of shape they are difficult to repair. Stainless steel mesh has proved the most durable.

The choice of material is not only governed by the wear and tear the screen will get. On the type of weave, the mesh number and the nature of the thread depends the density of the colourant deposit. Metal threads are thinner than silk ones, so that for the same mesh number the holes through which ink passes are larger in a metal screen. The tiny squares of colourant deposited from the metal screen are not only larger but also closer together so that when they flow together they give a thicker coating of colourant.

The chosen fabric has to be evenly stretched over wooden or metal frames and special types of frame have been evolved to achieve this.

The application of the design to the prepared screen can be done in a number of ways the choice of which depends on the nature of the design (whether it is bold or fine), the nature of the colourant (whether oil or water-based) and the equipment available. The three main groups are: direct, cutting out or stencil, photochemical.

In the direct method the negative of the design is painted on to the screen in shellac. As this material is rather viscous it is not possible to make very fine designs in this way.

In the cutting or stencilling method special coated paper is used. It may be transparent or opaque and is coated with shellac, cellulose acetate, gum-lac, etc. The mirror image of the required design is cut in the coating film and removed, leaving the backing paper intact. This stencil is then placed paper uppermost on the silk screen and the film fixed to it by a hot iron or a solvent according to its nature. The backing paper may then be removed. Again only bold design is readily produced.

In the photographic methods use is made of the light sensitivity of potassium dichromate in combination with gelatine, gum-lac, fish-glue, albumen, polyvinyl alcohol, and a number of other colloids. (This is also used in the photographic preparation of engravings and lithographic plates.) The methods have the advantage of speed and perfect reproduction. For the direct method the chosen colloid is made up in solution together with the potassium dichromate, coated on to the screen and dried in the dark. The design positive is placed on this and the whole exposed to strong light. The screen is then developed in warm water if gelatine or glue has been used or cold water for polyvinyl alcohol, when the unexposed portions will wash away. The edge of the design produced in this way may not be quite straight and particularly where it is on the bias of the cloth it may be jagged. To obtain true edges the indirect method must be used. The light-sensitive film is applied to a sheet of paper, exposed and then placed face down on to a sheet of celluloid. It is then developed in water when the paper and the unexposed portions come away. The celluloid with the gelatine uppermost is placed on a table and the silk screen pressed on to it when the gelatine transfers to the screen. In this way the design edges are independent of the weave of the cloth.

Numerous further methods have been worked out for more accurate reproduction or special effects.

For under-glaze printing water-based colourants may be used and the screen must then be made waterproof. In the direct and the stencilling methods of screen covering waterproof lacquers can be used. For the photographic method the gelatine deposit may be subsequently covered on

GLAZING AND DECORATING

both sides with lacquer but water eventually always gets into the gelatine and makes it swell and deform. Alternatively, the screen may be prepared with the negative of the design instead of the positive. After developing it is entirely covered with a fatty or cellulose lacquer, dried, and then boiled in water. The gelatine swells and peels off taking its covering of lacquer with it and leaving the lacquer on the clean parts of the screen in readiness for printing.

DESIGN TYPES SUITABLE FOR SCREEN PRINTING

Generally all line designs are suitable for screen printing but half-tone effects require considerable care and skill. In the first place it is difficult to arrange that dots on the negative (or positive) register on the screen mesh in such a way that they appear as dots in the prepared screen. If the negative is too fine there is a danger that the dots will fall across the stitches and disappear, if it is too coarse the result will only be recognisable when seen from a distance. The best copy is achieved when the mesh of the silk screen is twice as fine as that of the half-tone screen. Plain weave of the silk screen gives better results than the full or half gauze as here the twists and knots in the thread nullify the reproduction of small dots. Another trouble arising from the superimposition of two half-tone screens (the original and the silk screen itself) is the *moiré* effect, which is the appearance of regular geometrical designs when the two screens are in certain relative positions. This is most effectively eliminated when the lines of the two screens form an angle of 30° between them. This means either stretching the silk on the frame in bias or making the photo negatives with the screen at 30° to the frame.

Even with a successfully prepared screen, half-tone designs are difficult to print on-glaze as the oily colourant dots easily run together. Very good results can, however, be achieved under-glaze on porous biscuit as the print dries so quickly.

MEDIA FOR APPLYING THE COLOURANT

The normal medium for applying colourant by screen printing is an oily one, known as squeegee oil. The ground colourant is mixed intimately with the oil to a consistency at which it will not drip through the screen unassisted but can be readily pressed through with the squeegee. Prints made with this oily medium must be carefully dried at 175–200° C (347–392° F) before a second colour can be applied, so that polychrome prints require a piece to pass through a drying oven between each printing. With under-glaze printing there is the further disadvantage that the final print must be hardened on at about 400° C (752° F) to burn out all the oil before a glaze can be applied.

More satisfactory working on biscuit ware is obtained by using glycerine alone or with water as the base for the colourant. In this case the screen masking must be made waterproof with lacquer, etc. A glycerine and water-based ink is immediately absorbed by porous biscuit ware enabling successive printings to be made without drying and the glaze can also be applied without a hardening-on firing.

For polychrome on-glaze printing new thermoplastic media make rapid printing possible. These media are waxes or resins that are solid at room temperature but can easily be melted at moderately elevated temperatures 50–100° C (122–212° F). The colourant is mixed in when the vehicle is liquid and kept in a thermostat-controlled bath. The screen is heated either by using a stainless steel screen as a resistance heater as well as a printing screen, or by infra-red lamps, or a combination of these. The hot liquid ink can then be squeezed through the screen on to the cold ware surface when it immediately solidifies. There are a number of advantages in addition to that of being able to make successive superimposed prints without intermediate drying and these have been enumerated by Remington (**R9**). The chief advantages are probably found in automatic printing of mass-produced articles. The machines for multicolour printing can be made much simpler. Furthermore a design can be printed completely round a cylindrical object where formerly the screen would have smudged the beginning when arriving at the end. The prints are also much sharper and clearer, less smudges and mistakes occur so that a higher percentage of good prints are obtained.

The thermoplastic media were first developed for the polychrome printing on bottles and other cylindrical objects where most trouble had been experienced with traditional squeegee oils. All these objects having a curved surface always have very brief contact with the printing screen. In applying the technique to printing flat surfaces there may be difficulties due to the screen and surface being in contact for a longer period so that the screen may freeze on. Remington (**R9**) recommends that the screen be placed not less than $\frac{1}{8}$ in. (3 mm) above the ware, and a fast squeegee stroke be used on a taut mesh, so that there is instant release of the screen after passage of the squeegee.

SCREEN PRINTING DIRECTLY ON TO THE WARE

The screen-printing process can be applied to biscuit or glost ware with varying degrees of mechanisation. The basic principle is to lay the screen over the ware, put some colourant on it and press it firmly over the whole screen with a rubber-bladed squeegee. The squeegee action can be mechanised.

For flat surfaces it is usual to have the screen hinged along one side and have stops underneath it to register the ware against. Machines can screen print glazed tiles at a rate of 2500–3000 per hour. Cylindrical objects are arranged to revolve about their axis and the squeegee presses on the point of contact of screen and ware. The screen moves in a straight line at the same speed as the surface under it. Conical objects can also be revolved about their axis but the screen must be made to move around the vanishing point of the cone it is printing. Hand-operated machines will do these operations at the rate of about 500 per hr; semi-automatic ones where the operator only needs to feed the pieces to the roller, and the movement of screen, article and squeegee are automatic, print about 1000 per hr; whilst machines with continuous turntable operation produce 2500–3000 per hr.

Concave articles such as plates can be screen printed on special machines. The screen is placed over the plate and a special squeegee comes down to print the centre pattern, a different and suitably shaped squeegee then prints the border by pivoting round the axis through the centre of the plate.

Various odd-shaped articles may be printed by machine by using slack screens and squeegees shaped to the article. For very awkward shapes where only a few articles are required in comparison to the number of straightforward ones, e.g. tureens, sauce boats, etc., the design is printed on to transfer paper and applied from that to the ware (**D61**).

Another method of printing odd-shaped articles was invented by Valiela of Argentina. The required design is printed in the normal way on to a deflated flattened rubber bladder. This is then inflated against the ware to be printed (**C55**).

Printing several colours in registration is quite straightforward on flat pieces. Shaped pieces are always complicated owing to small variations of size. Fortunately a screen will allow slight slipping of the ware without smudging so that if the pattern is not continuous the two can be brought into alignment. Machines have been developed with suitable gears, etc., to make a given point on the chuck or other ware support correspond to a certain point on the screen. By correctly placing the ware on the machine perfect registrations can be achieved.

Printing with thermoplastic media has many advantages and eliminates drying between printings, etc., but the actual printing machine is rather more complicated. The best main source of heat is the screen itself. This should be stainless steel mesh (copper and bronze have too low resistance to heat up properly). The most usual mesh size is 165 although 180-mesh and 200-mesh are also used, the wire for the first of these is 0·002 in. (0·05 mm) diameter. Low-voltage current is fed to this from a step-down transformer to heat it up to about 175–200° C (347–392° F) (**R9**). The screen mesh must be insulated from the screen frame if this is metal. Very precise temperature control for each particular medium is required, involving thermocouples operating the switches by relay. The heated screen is frequently supplemented by infra-red radiation and a heating element in the squeegee. The colour pot must also be heated to a particular temperature and kept at this accurately. The colourant should be dispensed automatically in quantities just sufficient for each print.

The possibility of melting and therefore smudging a first print when making the second is overcome either by keeping the screen at some distance from the ware or by using vehicles of successively lower melting points for the various superimposed printings (**D61**).

SCREEN PRINTING OF GOLD DECORATION

The development of screen printing has opened up a new field of application of gold decoration hitherto confined to brushing and allied methods. Some adaptations must be made in the technique to allow for the chemical nature of gold preparations and the fact that the application of gold must be very thin. Metal screens are ruled out both because they are too thick and because they may react with the solvents in the gold preparation. For the latter reason also synthetic fabrics are excluded. The best screens are made from natural silk in hand-woven full-gauze weave of about 200×168-mesh. In applying brilliant gold it is not only economy that dictates thin applications, as the brightest mirror finishes are only then obtained. It is therefore tremendously important that material applied to the screen to prevent printing unwanted portions be of negligible thickness. A film thickness of 0·0010 in. (0·0254 mm) is the maximum for producing sharp bright prints (**S21, M79**).

Schmidt (**S21**) points out that interesting results can be obtained with screen printed gold in combination with polychrome over-glaze decoration. The question of correct sequence must be borne in mind, however. Firstly, the other colours must always be applied before the gold. Gold contains less flux than other on-glaze colourants and so makes a weaker bond with the glaze, to over-print on it would make the decoration wear badly. The other colourants should also receive their enamel firing before applying the gold. The gold preparation should then be arranged to require a decoration firing at 20–30° C (36–54° F) less than the other colourants so that it can safely be fired on without marring the previous work.

INDIRECT DECORATION METHODS

Offset Printing from Raised Type and the Rejafix Machine

To use raised type for curved surfaces an offset process may be employed. An impression is taken off the inked or oiled block on to a rubber pad and the article to be printed pressed or rolled on to the pad. If an oil base has been used the colourant is then dusted on. If ink is used directly it must have very high colour intensity as so little is applied (*see* also 'Rubber Stamping') (**D61**).

This process is used in the Rejafix machine applicable particularly for back stamping or printing labels on bottles, shell cases, etc. The machine incorporates a small muffle furnace, to which ware passes automatically after printing, for the decorating firing, at a rate upwards of 5000 articles per hour. The inks used have been specially developed for the machine and the working of the furnace made to the accurate temperature limits required for them (**C55, A63**).

THE MURRAY CURVEX PRINTING MACHINE

In this process the colour is offset on to flatware, which is of course not flat but includes anything from plates to soups, by means of a convex pad of gelatine. The pattern to be applied is engraved, or etched, on to a flat copper plate, with the working surface protected by a deposit of hard chromium (see p. 166). Gold-printing colour, developed especially for this process, is applied to the plate and any surplus scraped off with a 'doctor knife'. The gelatine pad is made by melting and casting, and is fixed facing down below the operating mechanism. This causes it to make one downstroke on to the inked engraved plate, then to move along the table until it is over the automatically centred plate to be printed, and then to make a second down-stroke on to this (Fig. 9.14).

Paper Transfers

The commonest indirect method of applying designs to ware is by paper transfers. These may give an outline in one colour only or may have a complete multi-coloured pattern on them. Simple transfers pulled off engraved copper plates may be made as required in the decorating department but complicated polychrome transfers are usually made by specialist firms. In this case the relative numbers of complete designs required for the different pieces in a set are worked out, and the correct proportions of the various centres, sprays, borders, etc., crammed on to a single sheet.

The colourants adhere to the transfer paper by means of various sticky oils. After being rubbed on the ware they are held in place by means of size. This may be applied over the printed design on the transfer or direct to the ware surface. Both oils and size evaporate or burn out when heated to about 700° C (1292° F).

Transfers can be used for both under- and on-glaze decoration although until recently the former found little favour. When used for under-glaze decoration the ware must undergo a special heating or 'hardening-on' process to remove the oils and size before a glaze will adhere properly. In on-glaze decoration the auxiliary matter is removed automatically in the enamel kiln.

TRANSFER PAPERS

The paper used for printing transfers should be a thin, strong, smooth, hairless and non-absorbent tissue paper. The most suitable type is a hard-sized tissue with a smooth but not highly glazed finish (**S48**). In this country duplex paper is used for polychrome prints prepared by specialist firms, allowing the use of very thin tissue backed by a stronger paper which is removed before applying the transfer. In Continental Europe thicker simplex tissues are used, but they are less easy to apply to the ware.

FIG. 9.14. *The Murray Curvex printing machine (Wedgwood,* **W35**)

MONOCHROME TRANSFERS BY INTAGLIO PRINTING

Transfers for decorating in one colour only are made by printing from an engraved copper plate in a press or from engraved rollers in a printing machine. This is done on the premises where the prints are to be used, and the design is transferred to the ware before the printing ink dries. The older method, which is still applicable where the demand is small, is to engrave or etch the design on a copper plate. This is filled with oil-based colourant and then printed on to tissue paper by passing it through a press. Many of the larger works now have printing machines with the design engraved on to rollers which are automatically coloured and cleaned off before transferring the pattern to a continuous strip of paper, which is cut up into convenient lengths and thereafter into the various parts required for decorating different pieces (**A134**).

The engraving of plates or rollers is a highly skilled craft that is being revived by the introduction of the Murray Curvex Machine (**A127**). The work is done with a number of chisel-like 'gravers', and 'punches' which can produce a variety of lines and dots, respectively. Intensity of solid colour varies with the thickness of colourant and is regulated by the depth of the engraved line. Wide bands of continuous colour cannot be achieved directly as the plate cannot hold colourant in wide cuts when the unwanted material is wiped off, so that cross-hatched areas must be used. The light tones of the design are obtained by using the punches to give dots of varying size and density (Fig. 9.15).

Because of the shortage of engravers and the time taken by engraving, many printing plates and rollers are prepared by etching usually combined with a photographic method of applying the design to the plate. The plate or roller is covered with gum or albumen containing ammonium dichromate and dried in the dark. A clear photopositive of the design is then laid over it and exposed to a bright light. This hardens the exposed part of the dichromate–gum film and renders it insoluble. The unexposed part can then be washed out with warm water. A resist is applied to the remaining protective film. The plate is then etched either in iron perchloride solution or by electrolysis. In both cases if the depth of etching is to vary in different parts, in order to obtain different thicknesses of colourant in the print, the etching process must be interrupted and resist applied to the parts which are to have the shallowest cuts. Even with such additional work an etched plate will always give prints differing in texture from a hand engraved one (**W39**).

Ceramic colourants are very abrasive and wear away printing plates very quickly and copper plates are quickly ruined. Their life can be very considerably lengthened by chromium or steel plating the engraved or etched plate. About 2000 prints can then be taken off before the plating starts wearing thin, when it is stripped off and the undamaged copper engraving replated for further use.

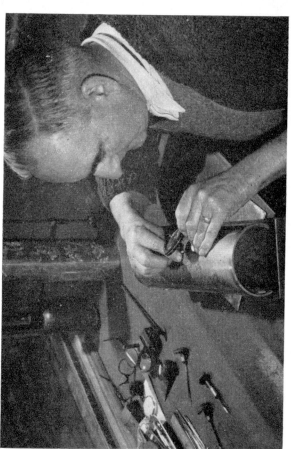

Fig. 9.15. *Engraving a roller for transfer printing* (Wedgwood, **W35**)

Fig. 9.16. *Intaglio printing. Stripping the print from the engraved plate* (Wedgwood, **W35**)

The traditional medium for printing ceramic colourants is based on linseed oil boiled with resin and Stockholm tar and other additives. The Stockholm tar retards the surface drying of prints to form skins which would prevent the colourant sticking to the ware, it is a natural anti-oxidant containing complex phenolic substances that are more effective than simple synthetic anti-oxidants such as β-naphthol and pyrogallol. Red lead and litharge may be added and are assumed to react with fatty acids and to become dissolved as the acid salts, in which form lead acts as a polymerisation catalyst and assists in the bodying of the oil. This printing oil has the disadvantage of unpleasant smell and such high viscosity that it must be heated for use, but the high wetting power that linseed oil has for ceramic pigments, and other good properties of the mixture, have not so far been attained by synthetic products.

Efficient dispersion of the ceramic colourant in the printing oil is very important. It is recommended that the colourant powder aggregates should first be broken up in a disintegrator mill and then mixed with the oil in a triple roll mill or a colloid mill (**C55, A134**).

The method of printing from engraved plates in a hand press is straightforward. The plate is inked while on a small heating stove kept at about 100° C (212° F) so that the ink attains a workable consistency. It is rubbed into all the depressions of the engraving and the excess is wiped off with a cloth. This 'bossing' should be done across and not along lines or ink will be removed from the engraving. The transfer paper is sized with soft soap solution and applied to the plate while still moist and rubbed out smooth.

For intaglio printing of transfers the traditional sizing material is a solution of soft soap, but many alternatives, including a number of water-soluble wetting agents, may be used. Of these the most successful is a solution of polyvinyl methyl ether whose solubility is lowered by increasing temperature, so that when the wet sized paper makes contact with the heated copper printing plate, or roller, the size precipitates to form a film between the paper and the printing ink and prevents its penetration into the fibre of the paper. This paper can therefore be sponged off more readily (**C55**). The plate and paper are then slid between the rollers of the printing press, the upper one being covered with five layers of rough flannel and one of smooth flannel. When it comes out the paper is stripped off and taken to the rubbing-down tables (Fig. 9.16).

In the printing machine the process is automatic and internally heated by steam, gas or electricity. Here the upper roller is engraved and it presses on the lower roller which is rubber. The paper passes between them, being drawn through by the rotation from a roll of paper weighted to prevent any slack. The paper is unsized as this would make it stretch and warp. The inking and cleaning of the roller is done continuously by a steel knife pressing on the top. The ink is fed behind this so that as the roller turns the knife forces the ink into the engraving and simultaneously scrapes it off the non-printing surface. This action must be borne in mind when laying out the design on the roller for engraving or etching. No straight line may run parallel with the axis of the roller as the pressure on the knife may force it into the cut and damage the engraving. The abrasive nature of some ceramic colourants necessitates frequent attention to the knife to keep its edge sharp and true, considerable experience is necessary to keep the pressure on it correct throughout a printing run. The temperature of the roller should be fixed for each colourant and kept constant. The range used is 80 to 95° C (176–203° F).

The continuous sheet of wet printed paper coming off the end of the printing machine is cut into lengths which must then be circulated to the tables where they are further cut up and applied to the ware. As the sheets are of thin tissue paper which must neither be torn nor allowed to fold back on themselves to smudge the wet prints this is not as easy as it might be. Wells (**W39**) describes a convenient print-carrying trolley with a number of hinged leaves fitted with separating strips of wood at the ends. This is filled from the bottom upwards at the printing press and then wheeled to the rubbing-down team having an empty trolley, who use the transfers from the top first. They are then not rushed by the supply of further wet prints for which they have no room on their table.

The cutting up of wet prints without smudging them is again not easy and is not satisfactorily carried out with scissors as this necessitates picking up the work. The use of a razor blade obviates this difficulty, but the complicated shapes to be cut usually mean awkward twists of the hand. This is overcome by using a cutting wheel mounted on a shaft with a ball-bearing joint in it.

After cutting up, the print is placed ink side down on to the ware, rubbed into position and allowed to dry for a short while. It is important when applying these prints to biscuit ware that it should have a uniform porosity approaching that for which the printing medium has been adjusted. Difficulty may be found in decorating the polished surfaces of thrown and turned ware, or fettled instead of sponged ware.

When the print has adhered to the ware properly it is washed, either by hand dipping in water until the paper comes away by shaking, or by placing the pieces on a conveyor belt passing through a trough of water or under a series of waterfalls. The last method, in which no piece uses water coming off another is the best way of avoiding chance specks of colour. After the removal of the paper the ink is allowed to dry completely (**W39, D64**).

PREPARING DESIGNS FOR MAKING POLYCHROME TRANSFERS

Multicoloured transfers were originally always made by lithographic printing but are now prepared by other methods as well, notably screen printing. Many people still term the result a litho irrespective of the printing method used. The American term is 'decalomania', known as 'decal'. They are usually made by specialist firms.

The first step in preparing polychrome transfers is to analyse the artist's

GLAZING AND DECORATING

design for the decoration and to decide how many colourants will be required. As each colourant has to be printed on to the paper separately the cost of the transfer is proportional to the number of colourants required. Any adjustments required to bring the cost of the transfers within the limits set must be made at this stage.

The number of different sections and sizes of decoration needed for balanced production are then worked out and arranged in the correct proportions on a sheet, getting as many as possible in.

The design must next be broken down into the component colours. This was formerly always, and is still frequently, done by hand by an artist and may be drawn directly on to the lithographic plate or stone, or the silk screen. Different intensities of colour are obtained by stippled dots set more or less closely together. Alternatively the breakdown into separate colours can be achieved photographically by using coloured filters in front of the camera and colour-sensitive film in it. Thus with a red filter and red-sensitive film a photograph showing only the red parts of the design is obtained and similarly for the other colours. The colour separation negatives are developed and dried and then used to make a positive directly on to light-sensitised litho plates or silk screens. In order to produce a stippled effect to show changes in intensity a fine screen is interposed between negative and positive.

LITHOGRAPHED TRANSFERS

Polychrome transfer sheets may be prepared by lithographic printing both from true lithographic stones, and from zinc plates, which are gradually superseding the stones. The method originally used special absorbent stones which can be cleaned off and re-used, but which are heavy and difficult to store. They are being replaced by zinc plates which have their surface grained by scouring with sand and porcelain balls in a vibrating trough so that they too will retain moisture.

The design is drawn on to these moist stones or plates in a special lithographic ink which contains an oily substance which will not mix with water. Plates are prepared for each of the colours to be printed, great care being taken that they register properly. They are then clamped into printing presses.

The hand drawing of designs on to the lithographic stones or plates can now be replaced by photolithography. For this process grained zinc plates are almost always used, due to their greater ease of handling. They are coated with a solution of gum or albumen containing potassium dichromate, and dried by rotating in a special machine. This coating is light sensitive and the plates must be kept in a dark room thereafter and handled only with a 'safe light'. The colour-separation negatives for the design to be reproduced are then placed on the plates and exposed to the powerful light of a carbon arc. This exposure makes the dichromate react with the gum to make it water insoluble. The whole is inked with greasy lithographic ink and then placed in warm water when the unexposed portions wash out. The plate is dried and the inked gum-bichromate design dusted with powdered resin and chalk to act as a resist. The remainder of the plate is then etched to remove any tendency to pick up grease. Finally the plate is gummed over with gum arabic solution, and dried ready for use in the printing press.

The printing is done by keeping the lithographic plates damp and then applying a greasy substance or sticky varnish by roller. The paper is put on and the press applied. The sheet of paper with the design printed on it in grease is then removed and passed to a machine where it is dusted with the colourant in dry powder form, which will adhere to the greasy print. After twenty-four hours the next machine removes any powder remaining on the white paper, before the sheet passes on to the next printing press to receive the grease outline for the next colour, and so on. The sheets are scrutinised at each stage for register and other possible faults. The finished litho is then coated with a layer of special size to protect the colourants and hold the pattern together on the ware when the paper has been sponged off (**A135**).

SCREEN-PRINTED (PAPER-BACKED) TRANSFERS

Polychrome transfers can now also be prepared by screen printing (described in greater detail, p. 813). The cost of preparing the printing screen is less than that for lithographic plates. The screens are prepared by hand drawing and filling of the unwanted parts, by ironing-on film-coated stencils or by the same gum–dichromate photographic process as used for photo-lithography but colour separation positives must be used. A screen is prepared for each colour wanted on the finished transfer. The ceramic colourant is mixed with squeegee oil for printing directly on to the transfer paper. By careful choice of the relative size of hole to fibre in the material of the screens the thickness of colourant deposited on the transfer can be altered. Much thicker deposits can be achieved than with lithographic printing, giving bolder colours. Screen-printed transfers can therefore replace hand-painted on-glaze 'enamels', and are also very suitable for gold decorations.

Paper-backed screen-printed transfers are applied to the ware in the same way as the lithographed ones, but if the colours are thick there is difficulty in sticking them down so that they have largely been replaced by waterslide transfers (p. 830).

APPLYING PAPER-BACKED TRANSFERS

The printed sheets of transfers have to be cut up and if they have a mixture of patterns and shapes on them they must be sorted out. This is usually done with scissors but a modern improvement is to use a glass-cutting wheel. The transfer paper is laid on a turntable and covered with Perspex and the designs cut out by running round them with a glass cutter's wheel (**C55**).

The ware to which transfers are to be applied whether in the biscuit or glost state must be suitable sized to make it tacky. In the case of under-glaze transfers the size may be applied by the printers to the transfer itself, but these transfers must be used within a month of delivery (**A134**). The sizing of biscuit ware for under-glaze decoration is particularly important if it shows signs of surface vitrification so that the colourant cannot penetrate into its pores. In this case the size should form a slightly tacky continuous surface film and numerous solvent–resin combinations have been tried. Copeland (**C55**) recommends a 1 in 10 solution of Shell Epikote 7007 resin in acetone. On the other hand very porous biscuit ware must first be sealed with a hydrophilic organic compound so that the varnish or size does not soak in. This must be dried before applying the size (**B75**).

The traditional sizes appear to be based on either turpentine, *e.g.* 4 kg colophony, 10 l turpentine and 15–20 g spike oil (8·8 lb colophony, 2 gal turpentine, 0·53–0·71 oz spike oil), or benzene. It is important to thin the sizes with the same solvent as has been used in their preparation, and makers' instructions should be followed carefully in this respect.

Glaze will not adhere to sized biscuit ware so that this has to be heated to about 540–650° C (1000–1200° F) to burn out the seal and size before it can be glazed. This hardening-on firing is cumbersome and substances are being sought that will stick transfers to the ware while remaining easily wettable to the glaze slip (**A134**). In this connection Blair (**B75**) has developed a special size for applying under-glaze transfers to porous biscuit ware. It is a wax–water emulsion which will act as both seal and size and give a surface that remains tacky for 30 to 60 days. Unlike ordinary size which must be applied evenly, the amount applied does not affect the result so that it may be sprayed on. When dry the transfer is applied in the ordinary way and the backing paper sponged off. The wax is then removed from the ware surface by heating it to 82–105° C (180–220° F), by a gas flame, radiant lamp, etc., when it sinks into the ware leaving the surface ready for glazing in the ordinary way. During the glost firing the wax volatilises at about 205° C (400° F) while the glaze is still porous, without leaving a carbon deposit. In industrial trials this size not only eliminated the sealing and hardening-on processes but produced a higher proportion of saleable ware.

The normal size for applying on-glaze transfers has to be dried on to the ware at 70–100° C (158–212° F) before use. Quick-drying sizes are now available consisting essentially of a film-forming resin and a tackifier resin dissolved in a volatile solvent, *e.g.* Pentalyn-X (film former), Aroclor 1254 (tackifier) dissolved in xylene (**C55**).

The transfers cut to the correct size are then placed print side down on to the sized ware and rubbed down with considerable vigour, first with a flannel and then with a hard brush (Fig. 9.17). This hand process is very laborious but was not easily replaced by machine. One type of machine will rub down five prints a minute on to flat or nearly flat (saucers, deep soups, etc.) ware by means of a pad of sorbor rubber on a steel plate in a simple automatic press. Each transfer is pressed several times (**A134**). Other machines work by air pressure above the transfer which is protected from air penetration by an impervious film of synthetic material placed over it (**R74**), but these machines only work well on centres and certain bolder types of decoration and fail where there is any embossing or the printing is more delicate. For this work a newer machine using both vacuum and pressure has been introduced (**C55**). In this each article is placed between moistened sponge rubber faced rubber diaphragms. These are flat for plates, etc., but shaped into male and female upper and lower parts for cups and other hollow-ware.

FIG. 9.17. *Hand application of paper-backed transfers. The printed sheet is seen in the background (Wedgwood,* **W35**)

Spouted articles cannot be placed in the machine. After placing the ware, two levers are pulled to bring the upper diaphragm beam into position. If either lever is accidentally released the beam movement is instantly stopped, so that the operator can never get a hand trapped. The upper and lower diaphragms make a sealed chamber around the ware except for one outlet. To this a vacuum is briefly applied to remove all air trapped between the transfer and the ware. Behind the diaphragms both below and above the ware lie pressure chambers, these interconnect when the machine is closed. After evacuation, pressure is applied at about 100 lb/in^2 (7 kg/cm^2) which presses the transfer on to the ware while equalising the pressure from below so that breakage cannot occur. The time taken is 2 to 4 sec each for vacuum and pressure so that output is 1400 to 2000 dozen per week on the standard two-headed machine (**A48**, **B150**) (Fig. 9.18).

After rubbing down the transfer may be allowed to dry on for a short while. The paper backing is then removed with water. Glost ware may be dipped in a water trough or placed under running water but porous biscuit ware would take up too much water in this way and a damp sponge is used instead.

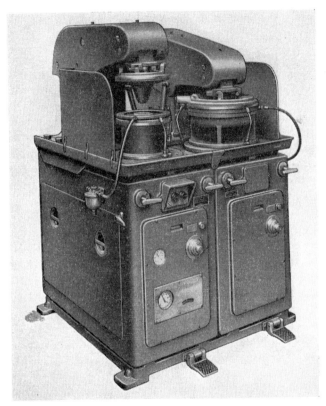

FIG. 9.18. *Transfer pressing machine* (*B.U. Supplies*, **B150**)

Trouble is occasionally experienced in the form of disfiguring spots on the surface of ware that has been washed to remove transfer backing papers. They are usually in the positions where the last drops of water evaporated and consist largely of the salts dissolved in mains water. The trouble is therefore termed 'water-spotting' or 'water marking'. Another source of it may be dust, etc., settling on the ware as it dries. Suggested remedies are: rinsing the ware in distilled water after it leaves the washing off tank (a 1000-gal dip tank will rinse 14 000 dozen pieces of ware before becoming contaminated), eliminating all dust from the drying atmosphere, using suitable wetting agents in the immersion rinse and placing the ware upright in such a way that the last drop is left on the back (**S154**), by rinsing the ware in a 1 to 4000 solution of saponin after removing the paper or by adding a de-foaming agent like turkey red oil to the saponin solution in the paper-rinsing tank (**B88**).

Occasionally the backing paper is left on glost ware and burnt off in the enamel kiln but it tends to fill this with ash and lead to faulty ware. It is to be discouraged.

Screen-printed Waterslide Collodion-backed Transfers

A new type of transfer has recently become available which is applied to the ware with the printed side uppermost, unlike the normal type where the printing is placed face down on the ware. In these the colourants are printed on to a film of collodion which is mounted with gum on a backing sheet of paper which serves only to protect it.

A number of these transfers are placed in a row on a cloth or pad kept saturated with water. After a short period the first will have uncurled and become limp so that the backing paper can be slid off. The collodion is then moistened all over and the transfer applied to the unsized ware printed side uppermost. It can be moved about easily until the desired position is reached so long as this is done before the water begins to dry. It is then pressed down with a rubber squeegee for most flat ware, a roller for large pieces, or a pad of blotting paper on curved surfaces. In every case it is important to work from the centre towards the edge to press out all air and water bubbles. In the case of pronounced curvature of the ware, especially concave ware, it is difficult to secure good contact. Radial cuts should be made in any suitable spaces in the transfer. This must be remembered at the design stage.

The ware is ready for the decorating kiln almost immediately after the transfer has been applied. As in all forms of decoration in organic media good ventilation is essential during the early stages of the firing (**B100, A183, J46**).

Waterslide transfers can also be produced by lithographic printing.

Cover-coat Transfers

Cover-coat transfers have been developed to overcome the difficulty of applying waterslide transfers to sharply curved surfaces and also to eliminate the time spent in cutting up sheets of transfers. For these transfers a water absorbent gummed backing sheet is used as for the waterslide ones, but the colours are printed directly on to it. A flexible supporting film is then applied on top of the colours in patches just larger than each individual motif on the sheet.

For applying these transfers a whole sheet can be soaked in water and is then placed on a suitable platform. Each motif can then be slid off at will

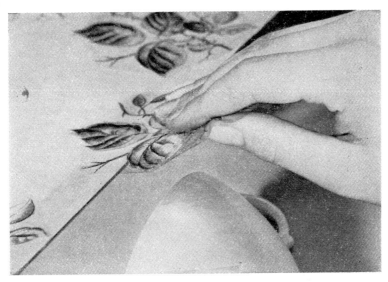

Fig. 9.19. *Applying cover-coat transfers (Johnson Matthey, **J46**)*

and applied to the ware. The cover coat is thick enough to be easily handled and manipulated into position, but flexible enough to allow it to be stretched sufficiently to fit any shape of ware without creasing.

The cover coat burns away readily in the enamel firing, without any of the difficulties that can arise from the layer of size or of collodion that is between colour and body in litho or waterslide transfers, respectively (Fig. 9.19) (**B101**).

Chapter 10

Drying of Shaped Ware

WATER forms an essential part, in greater or lesser quantity, in most ceramic shaping processes. But, thereafter, its job in the body is done and it must be removed as completely as possible before the firing of the ware.

It has been shown (p. 67) that the drying process is complex, taking place in two main stages. Firstly the constant rate period when the water loss is from the surface and shrinkage occurs equal in volume to that of the water lost. Secondly the falling rate period(s) when water evaporates inside the body and little or no shrinkage occurs. The first is of far greater concern to the ceramist, as it is during this that losses through cracking or warping occur.

The primitive potter probably learnt very quickly that it was no good trying to dry his ware fast and the traditional methods of drying goods on open shelves in the workshops or in ventilated sheds, etc., grew up. Large ware may take weeks or even months to dry in the natural way, nevertheless it is still being done.

If therefore the drying is not going to determine the speed of other processes by being a serious bottleneck it must be speeded up and made more economical of labour. It is not, however, an easy process to accelerate. Conditions of hot, fast airstreams which would remove water quickly immediately set up large moisture gradients in the body, making it liable to crack. In order to avoid this the best aim is to accelerate the movement of moisture in the body by warming it, but only to remove water vapour from the surface at the rate at which it is replaced from the interior of the piece. This is the basis of the 'humidity drying' system in which the ware is first heated in a humid atmosphere and then allowed to dry. This is applied more or less strictly in all artificial drying methods.

CONSIDERATION OF BODY COMPOSITION IN CONNECTION WITH DRYING

The nature of the body to be dried is often determined by considerations unconnected with drying it. A little thought and co-operation can, however, overcome many drying troubles.

As shrinkage, which causes cracking, is equal in volume to that of the water removed during the constant rate period, *i.e.* most of it, it is clear that the more water in the body the more occasion there is for drying cracks. The water is chiefly bound up with plastic clays and hence they must be used discriminately, replacing them with non-plastic raw materials or grog.

In highly colloidal clays considerable drying difficulties arise not only due to shrinkage but also due to impermeability (**P39**). Whereas clays with a high drying shrinkage develop long deep cracks if dried too quickly, certain other clays produce innumerable small cracks running in all directions. Here the colloidal clay particles make up an impervious structure throughout so that water cannot migrate to the surface, the surface layer therefore dries and shrinks.

One of the most effective methods of treatment of these tender-drying clays is preheating. It is, however, expensive and requires special plant.

Where cheaper treatment is necessary, chemical means have been recommended (**P39**). For lower-grade ware if the clay contains free silicic acid treatment with lime is helpful as it combines with the gelatinous acid. More generally a coagulant is used. The choice of this depends somewhat on the grade of ware to be produced, and it is of little use adding a substance of which any excess will remain as a soluble salt causing further clogging. Suitable reagents are acids, particularly organic acids, aluminium salts where the excess will hydrolyse to aluminum hydroxide; and for heavy clay products ferric chloride (used with a trace of sodium chloride) (**P39**). They are added in the tempering water. In the case of acids (usually HCl) clays vary considerably in their response to such treatment but a sedimentation test gives definite indications of both whether a clay will respond favourably, and also approximately the amount required. The actual amount required is small in all cases, never reaching the total quantity the clay can absorb and never exceeding 0.25% of dry weight (**M5**).

A tabulated guide of recommendations for adding electrolytes to difficult drying clays is given by Schurecht (**S37**) but the need for careful tests in each case cannot be overemphasised.

Another drying trouble is brought about by soluble matter in the body, which was often deliberately added to facilitate shaping. This together with very fine colloidal particles migrate to the surface with the water but remain behind when the latter evaporates. They then clog up the pores so that the interior will not dry properly. Soluble organic binders such as dextrin and gelatin are particularly obstructive.

The clogging of the pores can be prevented by incorporating a soluble substance that will decompose to volatile products on warming. Thus if ammonium carbonate or bicarbonate is added to the extent of 0.1 to 0.5% of the wet body, drying is accelerated very much. On drying the ammonium carbonate decomposes to the three volatile substances, ammonia, carbon dioxide, and water. Results obtainable are shown by the following examples:

Example 1

Refractory blocks for glass furnaces weighing about 1 ton, composed of $33\frac{1}{3}\%$ refractory plastic ball clay, and $66\frac{2}{3}\%$ refractory grog, containing 16–18% of water, normally need up to nine months for proper drying. If $\frac{1}{2}\%$ of ammonium carbonate is added, the drying time, under the same drying conditions, can be reduced to a fortnight.

Example 2

Sillimanite with no inorganic binder, but shaped with the addition of 3% of dextrin and 10–12% of water, cannot be dried in any length of time, due to the gastight dextrin skin developed on the surface during the first part of drying. If $\frac{1}{2}\%$ of ammonium carbonate is added, pieces of 1 cwt (51 kg) can be dried without risk of breakage in four to five days.

Example 3

Thin dinnerware, jiggered in the form of plates on to plaster moulds, need a few hours to dry in the usual humidity dryers. If 0·4% ammonium bicarbonate is introduced into the dinnerware body, the quality of the ware is improved, and the drying time reduced by 80–90%. A considerable advantage obtained thereby is that the plaster moulds remain in the dryer for a relatively short time and therefore disintegrate less (**S121**).

DRYERS

The factors to consider when installing drying equipment are:

(1) Space. Artificial dryers take up far less floor space than natural or hot floor drying.

(2) Handling. The old drying methods inevitably involved a lot of handling. Planning of the dryers as an integral part of the works should eliminate much of it.

(3) Manipulation and control. There is naturally more of this on artificial dryers.

(4) Maintenance, as for (3).

(5) Fuel efficiency. This is the ratio of the amount of fuel theoretically necessary to drive off the water, to the amount actually used. Here, the enclosed artificial dryers score highly against a hot floor. Insulating the dryer increases the fuel efficiency.

(6) Time Efficiency. This is the ratio of the minimum time in which particular goods can be dried safely under ideal conditions to the actual time taken in a given dryer. Time efficiency is not easy to measure as the minimum time has to be found experimentally for each shape and type of ware. Only in special cases is it possible to calculate the maximum safe drying rate of one article from another of different size made from the same body (**M4**).

(7) Maximum rate of safe drying and its optimum temperature. This depends not only on the body but also on the size and shape of the piece.

Data gained from one piece cannot generally be transferred to another, and so the drying rate used is usually arrived at by a hit and miss process. This is most especially the case for large pieces, sanitary ware, refractory blocks, etc., where tedious experiments would have to be made with full-size samples to find the highest safe rate. It is of course just these pieces, which, if dried at a high time efficiency, could make a large difference. Macey (**M6**) did full-size experiments with fireclay blocks $10 \times 10 \times 6$ in. ($25 \cdot 4 \times 25 \cdot 4 \times 15 \cdot 24$ cm) (fired size). He found that for blocks of equal initial moisture content there is an increase of the maximum safe drying rate with temperature increase until a maximum is reached, after which the rate falls off. Maturing of the moulded piece, even if the body has been matured, allows faster drying This is thought to be due to strains set up during shaping being relieved. Slight initial air drying before artificial drying allows faster drying. This may be allowed to occur during the maturing period. Faster drying conditions are permissible towards the end of the process.

Because of the difference of optimum drying schedule for each type of ware dryers can be run most economically for single types at each setting.

The maximum temperature that may be used in dryers for cast or jiggered ware on/in plaster moulds is determined by the plaster. This deteriorates if heated above 45° C (113° F). In the initial stages where the humidity is high the air entering the dryer may be somewhat hotter without danger but as drying proceeds the safe temperature must not be exceeded.

Drying ware on/in moulds is also complicated by the fact that it should dry evenly from both sides to avoid warping. This means that the air stream passing over the exposed surface should remove water at the same rate as does the plaster mould from the other side. Drying can therefore often not be as fast as for the same piece without a mould. Artificial drying is nevertheless extremely beneficial for ware on moulds allowing the moulds to return to the maker much more quickly and better dried than by natural methods. Thus fewer moulds are required and they have longer useful lives.

Unheated Dryers

In warm dry climates much heavy clay ware will dry easily when exposed to the atmosphere; in fact it may be necessary to prevent too rapid drying by placing damp cloths over the surface.

In northern climates there must be at least a roof with overhanging eaves over the ware. Special drying sheds, as used for instance for bricks, have louvred walls, the openings of which are adjusted according to the weather. The bricks are stacked in honeycombed walls, and after some time further channels at right angles are opened through them by partial restacking. Drying takes a number of weeks, depending on the nature of the clay, so that a large amount of space has to be given over to it. These older installations also require a lot of labour. Where fuel is very expensive it is nevertheless economical to use natural drying methods coupled with efficient handling equipment similar to that used in filling chamber dryers.

The main difficulty with drying thin-walled ware in the workshops is again space and handling, but here also the health of the workers must be considered as the workshop may be made unnecessarily warm and dusty.

Heated Dryers

Heat can be supplied to the ware by convection, conduction and radiation. *Convection* is the usual method, especially as air movement is necessary to carry the water vapour away. Convection is induced by placing steam pipes in the floors and walls of dryers, or by blowing in hot air or kiln waste gases. Ware is heated by *conduction* by placing it directly on a heated floor in 'hot floor' dryers but as it is a thermal insulator uniform heating cannot occur. Heating by *radiant heat* is being investigated and introduced at present (1950's). As radiant heat is that section of electromagnetic wave band that is termed 'infra-red' various terms such as 'infra-red drying' are commonly used.

Radiant heat is the name given to the energy which a body emits by virtue of its temperature, and the nature of the radiation depends on the temperature of the source. The radiation emitted by a source at 500° C (932° F) is all in the infra-red region of the spectrum, whilst a source at 2500° C (4532° F) emits an appreciable amount of visible light as well as infra-red rays. The radiation travels through air without appreciable absorption and is capable of being reflected, refracted or absorbed at a surface; a body heats when it absorbs radiation but is not affected by any radiation which it reflects. Absorption takes place at the surface (except in a few substances which are semi-translucent to infra-red rays) and generates heat there, any heating effect elsewhere is due to conduction from the surface. The relative amounts absorbed, reflected and transmitted depend on the physical properties of the receiver and emitter and on the wavelength of the source. All practical sources of radiation are an integration of many wavelengths, but the peaks differ with the source. Gas-heated medium temperature surfaces (450° F to 650° F, 232° C to 343° C) emit longer wavelengths than electric ones at 2227° C (4040° F) where the peak is at about 17 000 Å. Absorption of the former is less affected by the colour of the receiver than in the latter case (**A204**).

Radiant heat has been applied so successfully to drying and stoving of paints, curing of plastics, drying of textiles, etc., where tremendous shortening of the drying time occurs, that its possible use for ceramics must be thoroughly investigated. As the heat transmission is rapid but to the surface only, its most obvious application is to thin ware, *i.e.* crockery, small insulators, etc.; Hayman (**H51**) (in 1946) quotes the maximum total thickness at 1 in. to $1\frac{1}{4}$ in. ($2\frac{1}{2}$ to 3 cm). In 1958 Roberts (**R40**) describes successful drying of thick pieces. Radiant heat may be produced by gas-heated panels or by electric light bulbs set in front of reflectors. The Osram infra-red industrial

DRYING OF SHAPED WARE

lamps (**G33**) run at 115 V, 250 W and are set in rows in rhodium-plated reflector troughs. The working temperature is attained as soon as the current is switched on. The flux densities obtainable by gas and electricity, respectively, are:

gas-fired, medium-temperature 'black emitter' at 600° F (315° C):
  3000–4000 B.t.u./ft² per hr;
  8137–10 850 kcal/m² per hr;

250 W electric lamps:
  1420–2900 B.t.u./ft² per hr;
  3852–7866 kcal/m² per hr (**A196**).

Atkin (**A196**) also gives the following comparison of gas and electric tunnels for identical work, which shows that gas is more satisfactory for speed and cost. For ceramics the highest speeds may not, however, be the most satisfactory.

TABLE 188

Comparison of Gas-fired Radiant Tunnel and Electric Infra-red Tunnel (A196)

	Gas	Electric
Output of articles per hour	600	600
Length of heating zone	9 ft	32 ft
Time to pass through heating zone	1·7 min	6·6 min
Cost of installation	£135	£1920
Energy consumption per hour	480 ft³ (2·16 therms)	96 kWh
Running cost per hour for 600 articles	1s.	4s. 9d.

TESTS SHOWING TIME EFFICIENCY OF RADIANT HEAT DRYING

TABLE 189 (A196)

Small articles 3 in. long and ¾ in. in diameter (7·62 cm long, 1·7 cm diameter) usually dried in forced convection dryer in 5 hr.

Panel temperature		Time to reach constant weight
(°C)	(°F)	(hr)
157	315	After 2
229	445	1
260	500	½
293	560	¼
309	588	½
327	622	½

Drying of cups with gas-fired radiant panels at 500–600° F (260–315° C).

(1) Before mould release, the mould being cooled with an air stream of 1·6 ft²/min (0·045 m³/min): 3 min for cups of 0·1 in. thickness; 5 min for hotel ware.

(2) After removal from mould and after sponging: 5 min normal cups; 13 min hotel ware (**A109**).

As, however, the fuel and equipment costs of radiant heat drying are much higher than convection drying it must justify itself on a broader economic basis.

Gould, Evans and Flannigan (**G63**) have considered a large number of factors in the case of a particular pottery works and have found it an economic proposition to introduce a radiant heat dryer for the following reasons.

If an automatic jiggering machine making 80 to 100 dozen pieces per hour is used with a convection dryer taking 90 to 150 min to dry ware enough to take it off the mould some 150 to 200 dozen moulds are required to keep the jigger going. The usual life cycle of a mould is 200 fillings, so that to use the set of moulds economically 35 000 dozen pieces of one sort would have to be made in one run unless large storage space is available. With a radiant heat dryer the drying time is reduced to 12 to 15 min so that only 25 dozen moulds are required.

In addition to the saving of moulds and storage space it was found that defective ware cracked during the drying rather than the biscuit firing stage, the body then being re-usable and some handling saved.

They found that radiant heat was most effective in the initial stage of drying prior to mould release whereas convection heating was more satisfactory thereafter. They used an electric source of the radiant heat and found that the colour of the ware affected the rate of drying, dark bodies drying faster. Radiant heat is very suitable for drying the glaze on dipped or sprayed biscuit ware.

The application of *high frequency alternating current* across a non-conductor causes heating inside it. As the heating arises in the centre of the piece the method would seem to be very helpful in the drying of large and thick pieces. It has, however, proved too expensive for other than experimental work so far.

Some ware is dried in the kiln in which it is to be fired especially where this is a continuous chamber kiln but generally a separate dryer is much to be preferred.

Dryers can be classified according to the way the heat is introduced:
    (1) Hot floor.
    (2) Steam pipes.
    (3) Hot air.
    (4) Radiant heat (also called infra-red).
    (5) High frequency.

They may also be divided into:
- (1) Batch dryers.
- (2) Continuous dryers.

Batch Dryers

HOT FLOOR

Drying of heavy clay products, bricks, refractories, etc., on a 'hot floor' has become an established practice. The bricks or blocks are spread out in a single layer on a heated floor using steam pipes or kiln waste heat. As they approach leather-hardness they are stacked in an open arrangement at one end of the floor to make room for more wet ware, and left until they are as dry as is possible in the atmosphere prevailing. The advantages of the hot floor are its easy maintenance and the ease of observing the condition of the goods. The disadvantages, which in themselves are more numerous and are only outweighed if the hot floor is already built, are the space, labour, low fuel efficiency, uneven drying due to the face actually on the floor being 10 to 20° hotter than the upper one and hence frequently a high wastage, and usually haphazard ventilation (**T40**). Air generally only heats up a little by contact with the floor and the bricks and then escapes upwards at a fairly low temperature. Restricting this escape of heat by placing curtains round the filled part of the floor or by placing slow, large, propeller, downward forcing fans over the ware can assist considerably.

CHAMBER AND CORRIDOR DRYERS

Enclosed batch dryers with temperature and humidity control can be constructed to take any size, quantity and nature of ware. A big consideration with them is the handling and placing of the ware. Large ware that takes a relatively long time to make can reasonably be placed in a walk-in dryer by the maker. Such dryers are then placed alongside the making stations. Small, quickly produced pieces are best placed on trays, racks, pallets, etc., and mechanically conveyed to the dryer. This is well worked out in the Keller and Ivo systems for drying bricks (Fig. 10.1).

These chamber or corridor dryers are constructed with doors at each end, a central track and ledges at, say, 9 in. (15 cm) intervals up both sides along their whole length. The bricks are placed on pallets of the same width as the corridor, as soon as they are made. These are then loaded mechanically on a finger car which carries ten pallets above each other at the height intervals of the ledges in the dryer. The car is driven into the corridor with the pallets just clear of the ledges and then lowers them into place. After drying, unloading is undertaken in the same way from the other end of the dryer.

Other corridor dryers load the ware on trucks which are left in during the process. For small ware there are various cabinet tray dryers.

It is probably more possible to approach ideal drying conditions with a properly controlled chamber dryer than with a continuous method. Neilson

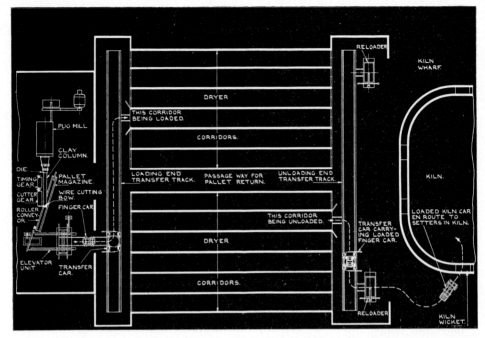

Fig. 10.1. *Ivo chamber humidity brick dryer: (a) typical layout*

(N7) describes some of the drying methods used for bricks. The simplest modern chamber or corridor dryer has a flat roof gently rising towards one end where there is an exhaust stack. Parallel steam pipes, often gilled, are installed on the floor below the level of the car rails, and hot air may be admitted through the floor at the end of the corridor furthest from the stack. The main drying action comes from the convection currents induced by the steam pipes, so that the atmosphere is perforce moist at the beginning of the process. The exhaust air may be of high temperature and saturation indicating high thermal efficiency, but constant supervision and control either by men or instruments is necessary.

Alternatively the source of heat is concentrated in the hot air supplied. If, however, this is done the simplest way, *i.e.* by admitting it through the floor at the lower end of the chamber, there are two main snags:

(1) The hot air tends to rise immediately and then move along under the roof so that the bricks at the bottom of the portion of the chamber near the exhaust stack do not get any.
(2) When hot air comes directly into contact with cold damp bricks there is a risk of cracking them.

The first fault can be avoided by placing a number of inlets and outlets along the corridor, the second by using first moist and then dry air streams; the whole being much more economically achieved in units of at least two chambers.

FIG. 10.1. *Ivo chamber humidity brick dryer:* (b) *Photograph of loaded finger car; a transfer car is also seen* (*Ivo*, **I11**)

Neilson (**N7**) goes on to describe a successful multi-chamber arrangement. The chambers or corridors are about 4 ft 6 in. (1·37 m) wide, 6 ft (1·83 m) high and 40 ft (12·2 m) long. The arched roof contains a number of inlet valves. These are baffled to give even air distribution by a perforated ceiling fitted at the base of the arch. The perforations are small and close near the centre under the valves and larger at either end. Because hot air tends to rise it will spread out along the roof before going through the ceiling. A similar duct is formed under the plates that make up the floor from which the outlet valves open. It is therefore possible to arrive at a regular air flow throughout the dryer.

The type of ware to be dried decides whether two or more stages are required, and hence the number of inlet and outlet flues and the number of chambers that make up a unit. For instance for a three-stage drying process each chamber has three independent valved inlet and outlet flues and three chambers form a unit.

Hot waste air from the kilns, or if necessary specially heated at about 105° C (220° F), is forced by fan (1) into the flue above the chamber ready for the third stage of drying and the valve gradually opened. The outlet flue leading to fan (2) is then opened, and withdraws air at about 38° C (100° F) and 50% humidity. This then passes down the neighbouring chamber in the middle stage of drying and is withdrawn by fan (3) at about 24° C (75° F) and 75% saturated. The volume of air and therefore rate of flow decreases at each stage so that the moist warm air passed into the first-stage chamber by fan (3) flows very gently. Fan (4) finally withdraws and ejects the air from the stage (1) chamber.

It may be argued that an updraught drying chamber would need less powerful fans. But as the even distribution of hot air along the length of the chamber in a down-draught one depends on its disinclination to flow downwards, this would be far more difficult to achieve in an up-draught dryer.

The temperature and humidity of the air can be checked between each stage and any necessary additions of hot air or steam made.

Thor (**T28**) describes a Dutch two-chamber brick dryer where both up- and down-draught are used. Hot air enters the preheated, stage two, chamber at the bottom and is allowed to rise through the goods. When cooled down and with a higher moisture content its natural tendency is to fall, which fact is exploited in its down passage through the stage one chamber to preheat it.

The fuel requirements of chamber dryers have been considered by a number of authors. Sachse (**S1**) makes detailed calculations of the heat requirements of the dryer and the heat output of cooling chambers of a periodic kiln. Estimating the velocity of the air through the latter at 7·2 ft/sec (2·2 m/sec) the available waste heat is 85% of the requirements of the dryer.

Intermittent drying stoves are also used for the thinner ware of the fine ceramic industry. These may have shelves, or the ware may be placed on trucks with shelves that fit the stove. Formerly the stoves were heated by steam pipes, etc. They can now be fitted with a jet drying system. In this, warm air at high speed is caused to impinge on the ware surface, at right angles to it. The air jets are carefully distributed over the surface of each piece to give uniform drying. High speeds can be attained (*see* Table 190).

Such stoves are very useful for oval dishes, etc., which form only a small proportion of dinner services and so are treated separately from the plates. The trucks that fit the stoves are placed near the maker who places the jiggered ware on its mould, in two rows on each shelf. The filled truck is then pushed into the stove. Each side of this carries projecting arms with downward facing air jets, under which the pieces on the truck locate. The doors are shut, and drying can commence. The system is also suitable for the smaller type of suspension insulators 6 in. to 10 in. (15–25 cm) (**H13**).

Continuous Dryers

There are a large variety of continuous dryers for a number of applications. The main feature they have in common is that the ware is always placed and taken out from the same respective external positions. Many types can be arranged to take a major part in the general transport of a well-laid-out factory.

DOBBINS OR POTTERS STOVES

The ware moves through these stoves in a horizontal circle about a central vertical axis. Most dobbins are octagonal with six closed and two open sides. They are divided into eight compartments, one for loading, six in the drying zone, and one for unloading. These are fitted with shelves suitable for the ware in question. The rotation is done manually by the operator when the loading section is full, by moving the whole on one section. Traditional dobbins are heated by hot air.

Underneath the drying racks there is a distribution chamber for the hot air which is forced down a duct to it by the fan which also extracts air from the top of the heating chambers. Air is passed into the six drying chambers. Its turbulence causes it to take a different path in each so that as a batch moves on from one chamber to the next every piece in it gets treatment. Some of the air removed is exhausted, the remainder is reheated and recirculated. The damper regulating the proportion of air removed controls the humidity of the air in the dryer and must be correctly adjusted. (Fig. 10.2.)

Dobbins are very popular for drying plates and other flatware on plaster moulds, as they can be placed near each maker with a fettler beside him receiving the dried ware. Similar units are built for drying the glaze on biscuited ware (**C17**).

The difficulty about dobbins is that they cannot easily be adjusted to a different making rate if the drying rate and time are to remain the same. So that installation of a faster making machine involves a new dryer as well.

Nevertheless they have regained considerable popularity in a new form incorporating jet drying. Jet drying dobbins are circular instead of octagonal and have twelve sections. The shelves are hollow, and except when they are at the loading or unloading stations they are supplied with hot air. The

TABLE 190

Intermittent Drying Stoves for Jiggered Ware

(after Hancock (**H13**))

Type of heating	Size	Floor area		Drying time (hr)	Moulds in dryer	Steam				Power		Total	
						Pressure	Consumption			3-phase (A/phase) (416V)	(kWh)	(B.t.u./lb water)	(kcal/kg water)
		(ft²)	(m²)			(lb/in²)	(kg/cm²)	(lb/hr)	(kg/hr)				
									Per unit water evap.				
Steam: Hot air jet	4½ × 7 × 10 ft 1·4 × 2·1 × 3·0m	31·5	2·93	1¾	72	5–10	0·35–0·7	26·6	12 1·67	1·75	0·76	1833	1017
Steam pipes 'traditional'	3½ × 7 × 10 ft 1·0 × 2·1 × 3·0m	24·5	2·28	6	60	5–10	0·35–0·7	20	9 6·70	—	—	6700	3718

TABLE 191

Comparative Data for Jet, Dobbin and Mangle Dryers

(Hancock (**H13**))

Type	Floor area		Drying time (hr)	Moulds per dryer	Steam			Electricity			Total (B.t.u./lb water)	Total (kcal/kg water)
	(ft²)	(m²)			Consumption (lb/hr)	(Consumption/lb water evap.)	(B.t.u./lb of water evap.)	(A per phase, 3-phase)	(kWh)	(power as B.t.u./lb water evap.)		
Jet	33	3·0	¾	240	20	0·86	860	1·5	0·66	95	955	530
Dobbin	97·2	9·0	1¼	600	45	1·80	1800	4·5	1·94	264	2064	1145
Mangle	32	2·9	1¼	480	60	2·13	2130	4·5	1·94	236	2366	1313

first jet drying dobbin was the Bloor cup dryer (1943) (**B84**) and jets of hot air were directed into each cup, so that the rim did not dry prematurely. For flatware the air jets are arranged in concentric circles above each plate. In order to avoid 'whirlers', *i.e.* too fast drying of the centre, and 'humpers', *i.e.* too fast drying of the flange, these air holes must be carefully placed and their sizes graded, *e.g.* for large flatware the central 2 in. has no jets, there are then three circles of $\frac{1}{8}$ in. diameter holes and then three of $\frac{1}{4}$ in. diameter

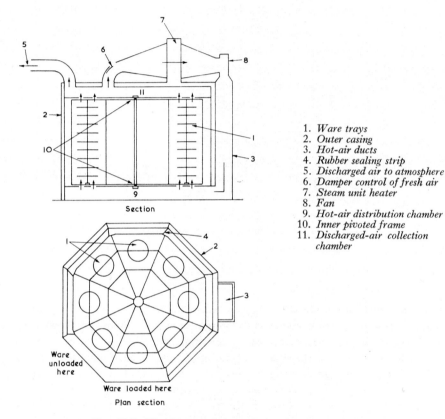

1. *Ware trays*
2. *Outer casing*
3. *Hot-air ducts*
4. *Rubber sealing strip*
5. *Discharged air to atmosphere*
6. *Damper control of fresh air*
7. *Steam unit heater*
8. *Fan*
9. *Hot-air distribution chamber*
10. *Inner pivoted frame*
11. *Discharged-air collection chamber*

FIG. 10.2. *Diagrammatic layout of potters stove* (Hind, **H97**)

holes. Ware can be dried in $\frac{3}{4}$ hr in the jet dryer that takes $1\frac{3}{4}$ to 2 hr in the traditional dobbin (**H13**) (Table 191).

MANGLES

Here a conveyor system carries trays of ware from a loading station, through a drying chamber to an unloading station, and then continues with empty moulds, etc., back to the loading point. They can be made a number of shapes to fit in with the available space and will deliver ware to towers, etc., some distance away from the makers. They are also more easily adapted to an increased making rate than dobbins.

The arrangement of the conveyors may be considered in two main types, the single-loop ones running to considerable height but economical of floor space and the more compact multi-loop ones (Fig. 10.3).

The drying is usually achieved by circulating heated air through the main chamber of the mangle. Baffles are arranged to prevent its excessive escape into the workshops. Steam pipes are sometimes installed in the chamber as well. A good modern mangle dryer should have: (*a*) good controllability

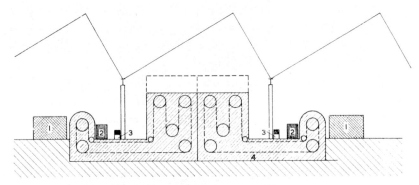

Fig. 10.3. *Mangle dryer (Hind,* **H97**)

1. *Maker*
2. *Tower*
3. *Belt conveyor*
4. *Holds 80 doz. 8-in. moulds*

of hot air temperature and of air humidity, (*b*) the use of larger air velocities, (*c*) application of closely controlled air recirculation, (*d*) possibility of altering the drying cycle, *e.g.* for drying ware after dipping (**W5**).

In order to economise the space in a mangle dryer it is customary to suspend sets of shelves instead of single ones. In the case of plates, for which mangles are widely used, three shelves constitute a satisfactory unit. This means, however, that the ware on the two lower shelves is somewhat less open to the circulating air and hence dries a little more slowly. The individual pieces do not therefore emerge with exactly the same moisture content as each other. Fortunately in the case of plates this is not critical (**F27**).

As in all dryers adequate thermal insulation is of importance and can be combined with visual control of the process by encasing the conveyor system in double glazing (**K21**). (Table 191)

TUNNEL DRYERS

Tunnel dryers find most application for heavy pieces that take a considerable time to dry, *e.g.* bricks, refractories, sanitary ware, etc., they are also used for mass-produced small articles such as tiles. The ware is moved through the tunnel on track trucks or by monorail. Earlier tunnel dryers used other means (**H110**), but the usefulness of having the ware on a means of general transport usually outweighs the extra cost, in fact it is often convenient to set the ware on the kiln cars if a tunnel kiln is used. The heating is now generally by hot air circulated by fans, which may be augmented by steam pipes in the dryer. Holland and Gardner (**H110**) in 1924 describe a tunnel dryer heated through its walls by hot gases drawn along underneath it and then returned along the sides.

It is very important that the construction of a tunnel dryer be sound; the dimensional tolerances for good working are considerably smaller than in a chamber dryer. As in the corridor dryers the construction must be air-tight, or the air cannot be uniformly distributed, and should be well insulated. The ware must fill the internal cross-section as nearly as possible to prevent the hot air by-passing it; the more precise the kiln construction is the larger the setting of ware on the trucks can be. Well-fitting doors can be difficult to achieve. Garve (**G9**) proposes a galvanised iron-clad wooden-framed door which combines lightness and rigidity with fire resistance. The tunnel, flues, ducts, etc., should be drained to keep them free of condensation water.

The direction of the air flow may be either counter- or concurrent with the car motion. The former is the more usual and cheaper to install; the latter is thermally more efficient and better for tender drying clays. Neilson (**N7**) describes the two types as applied to drying bricks. The simplest counter-flow dryers have an air inlet where the bricks are removed and an exhaust stack where they are put in. Steam pipes are laid on or near the tunnel floor. Pipes on the floor do not have a large effective area, so that a large number are required. They are also easily fouled by fallen bricks, etc. In order to make the freshly taken in air effective as soon as possible the largest heating areas must be near the brick exit. The hottest air passes over the bricks that enter, and as it is not necessarily of high humidity may cause cracking.

In the simple concurrent air flow tunnel dryer the bricks enter into an atmosphere almost identical with that outside and are gradually warmed up, reaching the highest temperature when they are driest. This is a far better treatment for difficult clays.

In modern installations the air is distributed along ducts outside the dryer and injected in such a way as to ensure even evaporation from all surfaces.

Forced convection has been introduced in a number of tunnel dryers by using fans to move the air to and fro across the tunnel or up and down alternate cars. The most notable example is the Proctor (**P57**) dryer.

Temperature, time and humidity control can be effectively maintained by various dampers and can even be made automatic to comply with a predetermined drying curve (**W5**).

TABLE
Types of Dryer used in
(Hind

Type of dryer	Material or ware normally treated	Process of production	Source of heat
Waste-heat tunnel	Press-clay	Drying for dust	Tunnel-kiln waste heat
	Tiles in saggars	Drying before biscuit fire	
Tunnel	Tiles in bats	Drying before biscuit fire	Exhaust or live steam
Walk-in stove (1)	Tiles in bungs	Drying out	Exhaust- or live-steam pipes
(2)	Insulator blanks	Hardening for turning	Hot clean air from kilns or special stoves
(3)	Plaster moulds	Initial and service drying (week-ends)	
Potters' dobbin (old type)	Clayware in moulds	Between maker and tower	Live-steam pipes below ware
Potters' stove (new type)	Clayware in moulds (chiefly flat-ware)	Between maker and tower	Exhaust steam in heat exchanger
Chamber-dryers (Carrier and similar types)	Turned insulators	Between turning and clay-glazing	Exhaust steam in heat exchanger
	Plaster moulds (pottery)	Initial drying	
Rotating table (1) Multiple	Cups in moulds	Between maker and sponger	Hot air from exhaust steam pipes
(2) Single	Cups in moulds	Between maker and sponger	Hot air from exhaust steam pipes
Conveyor belt (1)	Cups after handling	Final drying	Hot air from exhaust steam pipes
(2)	Dipped china, special assortment	Before china-glost fire	Electricity
(3)	Small flat and cups	Glaze drying	Gas (or electricity)
Chain conveyor (1) ('Mangles')	Glaze-dipped ware	Drying after dipping	Live steam pipes
(2)	Glaze-dipped ware (up to 8 in. flat)	Drying after dipping	Exhaust steam pipes
(3)	Flat and hollow-ware in moulds	Initial drying	Exhaust steam pipes or (direct) Town's gas

Method of application of heat	Method of moving ware	Time of drying (approx.)	Max. temp. (approx.) (°C) (°F)		Origin
Direct or via heat exchanger	On light metal trucks	16 hr	75	167	
	On tunnel-kiln cars	51 hr			Bricesco
Fan-driven air with recirculation control	Three lines of 14 trucks	24–36 hr	—		Mitchell
Pipes below racks					
Distributing ducts at high level (sometimes fitted with recirculating fans)	Manual, on boards	2–7 days	49	120	
Natural convection	Manual rotation, work on boards	24 hr	49	120	Boulton
Fan-driven circulation with recirculation control	Manual rotation, placed one at a time	2–3 hr	71	160	Boulton
Fan-driven circulation with recirculation control	On movable stillages, wheeled or lifting units	40 hr	88	190	Mitchell
		2–6 hr	71	160	
A jet into each mould	Manual rotation	20 min	71	160	Bloore patent
Downward hot-air blast	Manual rotation	15 min	68	155	Unsworth
Downward hot-air blast, recirculated	Motor-driven	30 min	77	170	Victoria
Radiants above and below belt	Motor-driven	15 min	—		Chandos
Radiants and fan	Motor-driven	3–4 min	—		Mitchell
Radiation and natural convection	Motor-driven	16 hr	105	220	
Fan circulation and recirculation, with blast into ware if required	Motor-driven	15 min	60	140	Mitchell
Fan circulation and recirculation, with blast into ware if required	Motor-driven	Cups 20 min	77	170	Service (Engrs.), Mitchell, Williams-Gardner, Boulton Hopol

Fig. 10.4. *Twin-tunnel dryer built in association with a car tunnel kiln for wall tiles* (Gibbons, **G39**)

Tunnel dryers for more expensive ware may be divided into sections of controlled humidity:

(1) Humid preheating.
(2) Gradual drying.
(3) Final drying.

A tunnel dryer using radiant heat and a system of ventilation similar to jet drying, especially designed for hollow-ware is described by Poutroie and Sobole (**P52**).

Where a tunnel dryer is used in conjunction with a tunnel kiln a number of labour and heat economies can be made. The dryer should be built to take the loaded kiln cars. It should be of such a length that the cars move through it at the same rate as through the kiln which is usually between half and fully the length of the kiln. Not only should the dryer use the waste heat from the kiln but the ware should pass from the dryer to the kiln directly without allowing time for it to lose the heat acquired in the dryer. (Fig. 10.4)

Dryer Efficiencies

The efficiency of a dryer can be regarded from several viewpoints, namely: (1) fuel; (2) time; (3) cost.

(1) FUEL

Most dryers can be at least partially heated by waste heat from the kilns. Furnace gases, etc., should be passed through heat exchangers but hot clean air can be used direct. This may or may not need to be boosted by specially produced steam for heated pipes or electrical heating in fan systems. Adequate thermostatic control is necessary.

Dryers with a system of re-using the exhaust air, either directly or through a heat exchanger can have a higher fuel efficiency. As the quantity of moisture needed to saturate air is considerably more at high temperatures than at low ones it is expedient to remove the air from the dryer when it is hot. This would, however, appear to waste heat. In certain humidity dryers a proportion of this hot, humid exhaust is mixed with the fresh dry intake. In others the exhaust can be passed through a heat exchanger to warm the fresh intake.

Tunnel dryers should work out more economical of heat as the structure does not have to be reheated for each batch.

In addition to the heat consumed the power required to drive fans must be taken into account. There is a tendency to use unnecessarily powerful fans and/or to make their work harder by using too narrow ducts with sharp bends. It should also be remembered that a portion of the power supplied to a fan is converted into heat. This can sometimes be usefully employed to supplement the direct heating methods.

Jones (**J56**) and Westman and Mills (**W47**) consider the questions of fuel efficiency and use of waste heat in some detail.

(2) TIME

An example of some time efficiencies is given by Macey (**M4**) using figures obtained by Bodin and Gaillard (**B86**) on six distinct tile clays (Table 193).

TABLE 193

Drying Time Efficiencies of Tile Clays
(**M4** after **B86**)

Clay	Drying time (hr)		Time efficiency (%)
	At works	In tests	
A	62	29·9	48
B	144	22·4	16
C	120	13·9	12
D	168	179·3	106
E	168	21·3	13
F	30	15·5	52

TABLE 194

Drying of Die-pressed Tile Bats

(Bullin (**B141**))

No.	Type of dryer	Drying medium	Extent of drying (% H$_2$O)	Time (hr)	Wt. of drying medium		Electricity, kWh/ton dry tiles	Labour, man hr per ton dry tiles	Thermal effic. (%)	Cost/ton dry tiles			
					Per lb dry tiles	Per lb H$_2$O evapd.				Fuel, pence	Electricity at 1·0d./kWh	Labour at 3s. per hr, pence	Total cost, pence
13	Open stillage stove	Exhaust steam	9·5 to 5·5	96–144	0·25	4·85	nil	2·5	22	Steam at 0·07d./lb 34·0	nil	90·0	124·0
14	Partially enclosed stillage stove	Exhaust steam	9·5 to 5·5	96–144	0·11	2·08	nil	2·5	51	ditto 13·5	nil	90·0	103·5
14a	Small totally enclosed stillage stove	Exhaust steam	8·5 to 0·5	96	0·21	2·4	nil	2·5	44	ditto 32·7	nil	90·0	122·7
15	Low-temperature tunnel dryer	Steam and air movement	8·0 to 3·0	24	nil to 0·034 lb	nil to 0·63 lb	23·5	2·13	—	ditto nil to 5·3	23·5	76·0	99·5 to 104·8
16	Intermittent oven (wall tiles), 20 ft diam.	Coal	5·5 to nil	48	0·081 lb coal (996 B.t.u.)	1·35 lb coal (16 505 B.t.u.)	nil	0·5	6·7	Coal at 75s. per ton 74·0	nil	18·0	92·0
17	Intermittent oven (floor tiles) 16 ft 6 in. diam.	Coal	5·5 to nil	60	0·30 lb coal (3690 B.t.u.)	4·9 lb coal (60 270 B.t.u.)	nil	1·0	2·4	ditto 270·0	nil	36·0	306·0
18	Tunnel dryer (wall tiles)	Waste hot air	8·0 to nil	36	465 B.t.u.	5400 B.t.u.	23·0	0·1	19·6	Waste heat?	23·0	3·6	26·6

(3) COST

This cannot really be expressed as an efficiency as there is no ideal to use in the ratio. The factors concerned are:

(1) Cost of heat used, *i.e.* type of fuel and manner of convertion into heat.
(2) Cost of power.
(3) Labour.
(4) Breakages and rejects due to drying defects.
(5) Maintenance and repairs.
(6) Interest and depreciation on the total cost of installed plant.
(7) Interest on the value of the factory space used.
(8) Interest on the articles in process of drying, and consequential costs.
(9) Other overhead charges, such as supervision, administration, rates, insurance, depreciation and obsolescence (**J56**).

Bullen (**B141**) gives comparative figures for various efficiency factors for drying dry-pressed tiles in different types of dryer (Table 194).

Chapter 11

Firing and Kilns

FIRING OF CERAMIC WARE

CERAMIC ware must by definition undergo at least one firing, which converts the shaped ware irreversibly into a hard product, resistant to water and chemicals. Unglazed ware is fired only the once.

Glazed ware is traditionally fired twice. Firstly the *biscuit firing* when all bodies except hard porcelain are fully matured. The *biscuit ware* is then glazed and *glost fired* at a lower temperature, for the glaze to mature. In the case of hard porcelain the biscuit firing does not mature the body, a porous article being produced. The glost firing is then at a higher temperature, body and glaze maturing together.

Modern tendency is to eliminate the second firing and glaze raw ware so that it can be finished in one firing, such ware is termed *once-fired*. Both body and glaze compositions must be suitably adjusted for this method to be successful.

Decorated ware may have to undergo even more heating processes. Under-glaze decoration is frequently applied with oils or varnishes which must be burnt off in the *hardening-on fire* at about 700–800° C (1292–1472° F) before a glaze is applied. On-glaze decoration is fixed to the ware by firing to 600–900° C (1112–1652° F), usually 750–850° C (1382–1562° F), in a *decorating kiln* (enamel kiln) which must be a muffle kiln or an electric kiln. Different colours requiring different decorating temperatures may necessitate a number of such firings.

FIRING THE BODY

The physical and chemical changes brought about by heat on the various raw materials and some of their mixtures have already been discussed (Chap. 2). The firing of ceramic bodies is generally more complex as they are mixtures of these materials, and involve both complete and incomplete reactions, fast and slow ones, etc. The geometry of the ware is also an important factor.

FIRING AND KILNS

In particular the firing of ceramic ware does not merely involve bringing it to a desired high temperature, the rate of heating and cooling are always important.

The best firing schedule for a body is governed by a number of different types of reaction which occur in successive changes. These in turn are affected by other factors:

Factors Due to Body Composition

(1) Removal of free, hygroscopic, and combined water.
(2) Combustion and removal of organic impurities and admixtures.
(3) Combustion and removal of sulphurous impurities.
(4) Reduction or oxidation of body constituents.
(5) Gradual volume changes.
(6) Sudden volume changes due to inversions during both heating and cooling.
(7) Maturing temperature.

Factors Due to Body Preparation

(1) Grain sizes of constituents.
(2) Geometry of ware.
(3) Permeability to escaping gases, heat conductivity and elasticity at different temperatures.

Factors Due to Firing Methods

(1) Time and heat needed to heat up the kiln structure and furniture.
(2) Time lag between first and last pieces of the setting to attain a given temperature.
(3) Controllability of heating method.

FIRING THE GLAZE

This is more straightforward than firing bodies, the glaze layer being thin and the reaction being required to go to completion and involving a liquid phase. Chief factors to consider are:

(1) Uniform heating and cooling of the piece.
(2) Oxidation of carbon deposits before the glaze seals the surface.
(3) Proper maturing of the glaze without reducing its viscosity until it flows down the ware.

WATER SMOKING PERIOD

The removal of water from the ware during the initial firing period is usually divided into three sections:

(1) Mechanical water.
(2) Hygroscopic water.
(3) Chemically held water.

However, there is no sharp definition between hygroscopic and constitutional water as regards the clay minerals.

MECHANICAL WATER

The amount of water present depends firstly on the minerals and secondly on the efficiency of the drying method used for the ware. In fact there are still instances where no independent drying is undertaken, the initial stages of the firing being used to dry the ware.

Most kilns are not suitable dryers. It is desirable to increase the temperature faster than is right for drying, thus setting up strains, uneven shrinkage and cracks. Also the water evaporated from the hotter part of the setting easily condenses out on the cooler part making it moist, causing first expansion and then shrinkage, scumming, etc.

Even where drying has been undertaken considerable variations occur. Ware that has been dried under 'natural' conditions with the prevailing atmospheric conditions of temperature and humidity contains more mechanical and therefore relatively free water than ware dried in a heated and ventilated dryer, furthermore the water content of the former will vary with the weather. The work of a good dryer can also be partially undone by allowing ware to stand around after leaving it before setting in the kiln, as it will absorb atmospheric water again.

It is therefore desirable to set thoroughly dried ware. The less dry it is, the longer must be an initial period of slow temperature increase and rapid air circulation. Even well-dried ware has hygroscopic water and this is preferably driven off at as low a temperature as is practicable with as much turbulent air movement as possible to remove the water vapour. A large amount of low-temperature heat is required to produce even temperatures throughout the kiln and within the ware itself (**A199**).

HYGROSCOPIC WATER

The hygroscopic water held by the clay minerals is generally not driven off in the drying process. The temperature and heating rate for its removal depend on the mineralogical and mechanical make-up of the body and also on the kiln. Russell (**R70**) has reported that it is eliminated at 149° C (300° F). Geller (**G17**) describes circumstances for which the best heating rate is 20° C (36° F) per hr up to 301° C (575° F) taking some 16 hr.

CHEMICALLY HELD WATER

There then remains the chemically combined water. By the time decomposition of clay minerals to give off water vapour begins (around 400° C, 752° F) the whole setting is too hot for there to be a danger of condensation. The removal of the hygroscopic water leaves a relatively porous body, and it has been found that the chemically combined water can be driven off relatively more quickly. Grim and Johns (**G83**) working on bricks found that it can be done in 3 to $3\frac{1}{2}$ hr of rapid firing with little control. However, Robinson (**R43**) showed that very rapid firing of bricks during this period, although not disintegrating them, markedly reduced their fired strength and Austin (**A199**) points out that any unevenness of heating produces high-pressure steam in some parts of a piece which may disrupt it, so that care must still be taken.

The complete removal of water at as low a temperature as possible is of major importance as the water vapour formed blocks the oxidation reactions that must take place to remove certain impurities. These reactions could begin around 400° C (752° F).

OXIDATION AND DECOMPOSITION REACTIONS

I. *Carbon*

Many raw clays contain a certain amount of carbonaceous matter which usually must be removed by decomposition and oxidation during the firing. In certain instances organic matter is actually added to the body to facilitate shaping, but it too must normally be burnt out. Furthermore during the initial firing period the ware is cooler than the flame and so carbon is actually deposited even in oxidising conditions.

The organic compounds begin to decompose at about 400° C (752° F) giving off CO_2, CO, H_2O, etc., and leaving free carbon in the body (**P67**). When the water vapour has been swept away and oxygen can penetrate the the pores, combustion of the carbon sets in.

As the necessary oxygen must penetrate into the body for the carbon, etc., to burn and the carbon dioxide and water evolved must escape it is essential that all the combustible impurities are removed before sintering starts. If this is not achieved the body will either be left with a dark centre and/or bubbles and craters will be left on the surface where the gases have escaped through the partially molten or sintered body. The porosity curve for the body has to be studied to find the highest temperature at which gases can move through it. The accompanying curves give examples of porcelain bodies and various glazes (Figs. 11.1, 11.2). Those for the bodies show that oxidation must probably be completed by 950° C (1742° F).

Another deciding factor is the rate of reaction. In general a 10° C (18° F) rise of temperature doubles the rate of a reaction. So that a body that can be oxidised successfully in 1 hr at 900° C (1652° F) would theoretically take about 6 weeks at 800° C (1472° F) (excluding the local heating effect due to the actual combustion), or $3\frac{1}{2}$ sec at 1000° C (1832° F), as set out in Table 195.

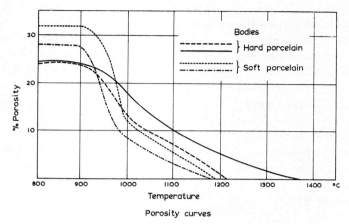

Fig. 11.1. *Porosity curves for bodies*

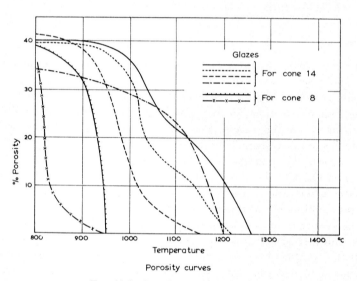

Fig. 11.2. *Porosity curves for glazes*

Lower temperature combustion takes advantage of better porosity conditions in the body and, if it does not take too long, is more economical of fuel, as a good draught and excess air are required which make flue losses greater the higher the temperature. On the other hand a temperature must be chosen at which the reaction proceeds at a reasonable rate.

At high temperatures, apart from the risk of disrupting the body because the porosity is too low, the reaction can occur so rapidly that the gas evolved explodes the piece.

In practice the oxidation period is undertaken somewhere between 800 and 1100° C (1472–2012° F), when it can be completed in about one hour. In

TABLE 195

Theoretical Time Taken to Remove Carbon by Combustion Deduced from an Optimum of 1 hr at 900° C (1652° F)

°C	°F	Time
800	1472	1024 hr (about 6 weeks)
810	1490	512 hr
820	1508	256 hr
830	1526	128 hr
840	1544	64 hr
850	1562	32 hr
860	1580	16 hr
870	1598	8 hr
880	1616	4 hr
890	1634	2 hr
900	1652	1 hr
910	1670	30 min
920	1688	15 min
930	1706	7·5 min
940	1724	3·75 min
950	1742	1·9 min
960	1760	56 sec
970	1778	28 sec
980	1796	14 sec
990	1814	7 sec
1000	1832	3·5 sec

periodic kilns this is about the time taken to clean all the fireboxes, when excess air automatically enters and temperatures remain fairly stationary or drop.

In fact, in the traditional centres of ceramic production it is usually found that the factor determining the temperature for the oxidation is the burning properties of the coal rather than the body itself. In periodic kilns the fireboxes gradually fill with ash, clinker, etc., and have to be cleaned. The temperature at which the ash content necessitates this cleaning varies with the type of coal and so the oxidation period varies with it.

Where this empirically-found oxidation temperature has proved satisfactory it is generally used again in a tunnel kiln schedule without further investigation.

There is nevertheless a move to ascertain whether the oxidation could not be carried out more economically at lower temperatures. Kraner and Fritz (**K74**) therefore both reviewed the foregoing literature on the subject and then did careful experiments to determine at what temperatures and at which heating rates the oxidation can be completed. The literature they cite suggests that oxidation may begin at 350° C (662° F), becomes rapid above 650° C (1202° F), and may not be completed until 800° C–900° C (1472–1652° F). From their own work, however, they conclude that rapid oxidation occurs at lower temperatures than generally believed. A porcelain or ball clay–flint body may be completely oxidised at 600 to 650° C (1112–1202° F) and a close grained flint body is 90% oxidised by 650° C (1202° F), and completely oxidised by 700–800° C (1292–1472° F). They therefore suggest that it would be more economical to maintain an oxidising atmosphere up to 650° C (1202° F) than to have an oxidising period at some higher temperature when heat losses up the flue will be greater. The rate of oxidation seems to depend more on the grain size of the body than on the actual content of combustible material.

An alternative method of removing carbon from bodies to be fired at high temperatures has been suggested by Bernardaud (**B42**). Use is made of the following reactions which prevent the presence of free carbon in mixtures containing steam and carbon dioxide by the time 1000° C (1832° F) has been reached:

$$C + H_2O = CO + H_2$$
$$C + 2H_2O = CO_2 + 2H_2$$
$$C + CO_2 = 2CO$$

The fuel consumption and firing time can therefore be considerably reduced.

Highly bituminous clays must be fired with care and the minimum of fuel because once ignited the bituminous matter will act as fuel and may even burn the ware too fast. Some clays contain so much combustible matter that only 0·5 cwt of coal (25 kg) is needed to burn 1000 bricks. If bloating, through sintering before combustion is complete, is to be avoided the kilns must be fitted with good dampers and vents to control the temperature (**A38**).

II. *Carbonates*

Magnesium carbonate begins to give off carbon dioxide at 408° C (766° F) and calcium carbonate does so at 894° C (1641° F). The gas given off is generally not of sufficient quantity to cause disturbance if the heating is slow enough to take account of the other reactions that occur at the same temperature.

III. *Sulphates*

Table 196 shows the temperatures at which various sulphates decompose. In the presence of the other body constituents the reaction may, however, be accelerated or retarded, and may depend on firing conditions. For instance magnesium sulphate, which is very harmful to glazes, is decomposed if the firing around 500° C (932° F) is fast, but not if it is slow. Similarly sulphur dioxide in the kiln gases is more harmful if the firing is slow around this temperature than when it is fast.

If decomposition occurs after vitrification has started, bloating and blistering take place. (Fig. 11.3)

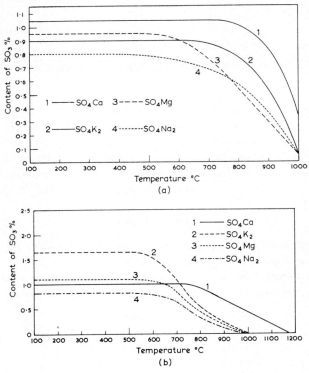

FIG. 11.3. *Decomposition of sulphates in clays:*
(a) *a kaolin without lime;*
(b) *a kaolin (Munier,* **M100**)

TABLE 196

Decomposition Temperatures of Sulphates
(after Searle (**S48**))

Sulphate	Decomposition begins at		Rapid decomposition		Product of decomposition	Colour of product
	°C	°F	°C	°F		
$FeSO_4$	167	333	480	896	$Fe_2O_3.2SO_3$	Yellow brown
$Fe_2O_3.2SO_3$	492	918	560	1040	Fe_2O_3	Red
$Al_2(SO_4)_3$	590	1094	639	1182	Al_2O_3	White
$CuSO_4$	653	1207	670	1238	$2CuO.SO_3$	Orange
$MnSO_4$	699	1290	790	1454	Mn_3O_4	Red
$ZnSO_4$	702	1296	720	1328	$3ZnO.2SO_3$	White
$2CuO.SO_3$	702	1296	736	1357	CuO	Black
$NiSO_4$	702	1296	764	1407	NiO	Green
$CoSO_4$	720	1328	770	1418	CoO	Brown
$3ZnO.2SO_3$	755	1391	767	1412	ZnO	White
$MgSO_4$	890	1634	972	1782	MgO	White
$CaSO_4$	1200	2192	—	—	CaO	White
$BaSO_4$	1510	2750	—	—	BaO	White

IV. *Iron Pyrites*

Iron pyrites decompose in heat and in the presence of sufficient oxygen will give ferric oxide and sulphur dioxide.

$$4FeS_2 + 11O_2 = 2Fe_2O_3 + 8SO_2$$
$$4FeS_2 + 15O_2 = 2Fe_2O_3 + 8SO_3$$

Jackson (**J2, 4**) has shown that with slow heating and an abundant, turbulent supply of oxygen the reaction takes place at about 425° C (797° F) and by 510° C (950° F) 95% of the sulphur content is removed.

If, however, heating is too rapid and the draught insufficient this oxidation and removal of the sulphur is not completed and different reactions occur. At about 425° C (797° F) pyrites decompose even in a non-oxidising atmosphere:

$$FeS_2 = FeS + S$$

giving off sulphur vapour and leaving ferrous sulphide dispersed in the body. The sulphur vapour will combine with any available oxygen in the kiln atmosphere before this can penetrate the body. The ferrous sulphide will react further in different ways according to the oxygen available:

$$FeS + 2O_2 = FeSO_4$$
$$6FeS + 13O_2 = 2Fe_2(SO_4)_3 + 2FeO$$
$$4FeS + 7O_2 = 2Fe_2O_3 + 4SO_2$$
$$4FeS + 9O_2 = 2Fe_2O_3 + 4SO_3$$

Ferric sulphate decomposes at a higher temperature to give ferric oxide and sulphur trioxide. The ferric oxide combines with the silicates, but the resulting compound is less highly coloured than that formed by ferric oxide free to react at lower temperatures. This gives rise to 'pink spot' and 'pink core'.

Ferrous sulphate is stable enough to be formed above 500° C (932° F) and easily combines with silica to form a black viscous glass which is contributory to 'black core'. If the glass forms an envelope around more ferrous sulphate and the body is heated until this decomposes spontaneously (unfortunately below the finishing temperature), bloating and distortion occur (**J4**).

The decomposition of iron pyrites can be assisted in certain cases by a small addition of ammonium chloride. For instance, a mixture of shales and fireclay with a high carbon and pyrites content, used for facing bricks, normally had a tendency to have black cores. After the addition of 0·25% ammonium chloride to the dry clay (or in solution in the mixing water), firing could be accelerated by 7 to 8% without any sign of black cores (**A31**).

CARBON AND SULPHUR TOGETHER

When both carbon and sulphur are present in a clay the carbon burns first although the pyrites ignite first. The sulphur will not get a share of available oxygen until the carbon is oxidised. Meanwhile temperatures usually rise

FIRING AND KILNS 863

which increases the chance of forming sulphates with all their attendant troubles (**J4**).

V. *Iron Oxides*

Practically all ceramic bodies contain some ferric oxide. Its possible colour changes and evolution of oxygen during the firing must therefore be taken into consideration.

Ferric oxide decomposes, giving off oxygen, at various temperatures above 1100° C (2012° F), depending on the other body constituents. If a piece containing it is already sintered by this time it will become bloated by the gas given off; 0·1% Fe_2O_3 gives off a volume of oxygen equal to 30% of the body it is in. It is therefore necessary to reduce the ferric oxide before sintering occurs, by regulating the kiln atmosphere, if the ware is to be fired above the decomposition temperature. This is so whether the desired final condition of the iron oxide is in the reduced state or not. (There are certain exceptions to this.)

Ware that is to appear 'brick red' or brown, etc., must then be reoxidised during cooling once the temperature is below 1300° C (2372° F). For relatively dense ware the oxidising cooling between 1300° C (2372° F) and 1000° C (1832° F) must be slow to allow oxygen to penetrate as far as possible. For example, a particular instance has been given of engineering bricks fired at cone 8–9 to a final water absorption of 1–3% for which the oxidising cooling stage takes 16 hr (**T43**).

Ware of low ferric oxide content, up to about 0·5%, can be made to appear white by preventing reoxidation of the ferrous oxide. The maximum reduction must be brought about by the kiln gases before sintering sets in. A neutral atmosphere is then used for the soaking period and cooling must be relatively rapid to prevent reoxidation (**D29**).

In principle, therefore, it is always desirable to reduce ferric oxide at a reasonably low temperature. The lowest temperature at which reduction can begin is, however, determined by the necessary oxidation of combustible matter which in turn is influenced by the removal of water from the body.

KILN ATMOSPHERE

It has already been seen that the nature of the kiln atmosphere is of very great importance. It is required to be dry and fast moving during periods when water and decomposition gases are given off from the ware, and it has to be oxidising or reducing at definite stages of the burning process. In the traditional coal-fired kilns these requirements are fulfilled only by skilful adjustment of air inlet at the stoke holes, dampers, etc. In more modern grates and with gas or oil burners more automatic regulation can be achieved. In electrically heated kilns the atmosphere is introduced independently of the heating method and so can be regulated at will.

As well as the oxidising, reducing and water vapour saturation powers of kiln gases the sulphur dioxide they may contain is of considerable importance. Most industrial coals and oils contain some sulphur and many brick clays, etc., have some sulphur compounds, both of which give rise to sulphur dioxide. In the presence of water vapour and oxygen this can combine with bases in bodies and glazes to form sulphates.

It is interesting to note that the troubles that can arise from these sulphates became worse as more modern firing methods came into use some 50 years ago. Seger (**S52b**) found that the change from wood, which is sulphur-free, to sulphur-bearing peat, lignite and coal, together with the abandonment of reducing, smoky and wasteful burning of the fuel, gave rise to bricks of lowered resistance to weathering.

The sulphuric gases combine with the bases on the surface of the ware, doing so more readily with any carbonates, e.g. those of iron, calcium and magnesium, than with silicates. Jackson (**J3**) has shown that contrary to the general assumption the amount of sulphur dioxide absorbed is not proportional to the percentage of it in the kiln atmosphere, a high-sulphur coal producing little more damage than a low one.

The formation of the sulphates removes the bases from their rightful duties in the body. For instance in bricks normally coloured red or rust by iron–calcium silicates the calcium is removed as sulphate and the colour is lighter. In those that are usually yellow removal of the lime leaves the iron free to colour the brick red, often only on the surface. Thus actual examples with yellow cores and red surfaces had sulphate contents inside and outside as follows:

Core	Surface
0·61%	8·49%
0·74%	19·58%
0·88%	11·07% (**K22**)

Reduction of sulphates after they have formed never completely eliminates the trouble as the sequence of the reactions has been upset and a mark always remains. They may furthermore decompose after vitrification has set in causing bloating, blistering, etc. Thorough oxidation during slow heating from 648 to 815° C (1200–1500° F) should prevent this (**R49**).

Sulphates that are not decomposed during the firing, whether they were present in the raw material or formed from sulphurous kiln gases, cause trouble in finished bricks. They are all more or less soluble in water and will therefore be carried by water moving through the pores of the brick and then deposited when it evaporates. There is then a tendency for crystal growth, the force of which eventually disrupts the brick. This is found more with sodium and magnesium sulphates than with potassium and calcium sulphates, where the sulphates come out on to the surface in sufficient quantity an unsightly white scum appears, known as *efflorescence*.

Sulphurous kiln gases also have a detrimental effect on glazes combining with bases to form sulphates which decompose only at higher temperatures than those at which the base would have combined with silicates, etc. The glaze then appears matt or egg-shelled and shows pin-pricks, blistering and other faults. To avoid this Robson (**R49**) states that the sulphur content of oil must be below $\frac{1}{2}\%$ and that of gas below $\frac{1}{4}$ grain per 100 ft^3 (0·016 g/ 2·8 m^3).

MATURING OF THE BODY

During the initial firing stages while the main attention is being devoted to removal or alteration of impurities some of the reactions leading to the final product are beginning to take place. The removal of the chemically bound water is the first stage of the transformation of kaolinite to mullite (*see* p. 77), etc.

Above about 900° C (1652° F) the firing may be accelerated and most of the important reactions already described in detail (Chap. 2) take place; *e.g.* fusion, sintering, inversion, solution, chemical reaction, etc. It must again be emphasised that the course which these reactions take and the product achieved depend not only on the raw materials and their proportions but also very largely on their grain size and related factors.

The body is said to be matured when the desired final condition is achieved. This is irrespective of whether this is a porous or a vitrified state, more or less approaching equilibrium.

The proper maturing of ceramic bodies is dependent on the highest *temperature* achieved and the *time* spent reaching it, held at it and cooling from it. Although certain reactions require a given heat input and would theoretically be brought about equally by short heating at a higher temperature or a longer heating at a lower one, the properties of ceramic bodies produced in this way are not the same. There are generally optimum values for the temperature and the time, respectively.

Hursh and Morgan (**H126**) investigated the effects of variations of soaking time and temperature on structural clay products. Vitrification will only set in above a certain minimum temperature regardless of the length of soaking time. Above this, increases of temperature are more effective in reducing porosity than are increases of soaking period; similarly an increase in temperature will reduce the necessary soaking time to obtain a given porosity. They found that seven out of eight clays brought to a suitable temperature will mature in 8 to 12 hr.

In the production of porous ware generally, increasing temperature produces denser ware, higher shrinkage, and greater strength. Prolonged soaking at the maximum temperature also leads to greater strength.

In the case of fully vitrified ware, in particular porcelain, variations of temperature and time have a marked effect not only on the strength, but also on the colour and translucency of the product. The whitest porcelain of best mechanical strength is achieved at high temperatures for short periods.

TABLE 197
Maturing Temperatures of Some Ceramic Bodies
(given in order in which discussed in Chap. 5)

Body	Cone	°C	°F
(1) Brickware, general	010–05	900–1000	1652–1832
fine	5–6	1180–1200	2156–2192
(2) Refractories			
Silica	14/15–18	~1430–1500	2606–2732
Fireclay	1–13/14	1100–1400	2012–2552
Aluminous fireclay, etc.	10–35	1300–1770	2372–3218
Zircon	10–31	1300–1690	2372–3074
Zirconia	35–42	1770–2000	3218–3632
Silicon carbide	10–11	1300–1320	2372–2408
Carbon	05	1000	1832
Magnesite	18–27	1500–1610	2732–2930
Dolomite	12–16	1350–1460	2462–2660
Chrome	6 upwards	>1200	>2192
Chrome magnesite	13/14–29	1400–1650	2552–3002
Forsterite	20–32	1530–1710	2786–3110
Barium anorthite	13/14	1400	2552
(3) Thermal insulating	015/014 upwards	800	1472
(4) Stoneware			
engineering bricks	03–8	1040–1250	1904–2282
(5)–(8) general	4–12	1160–1350	2120–2462
(9) Fireclay, tiles	07	960	1760
sanitary	7–10	1230–1300	2246–2372
(10) Coloured earthenware	010–03/02	900–1050	1652–1922
(11) and (12) White earthenware	08–10	940–1300	1724–2372
(13) Vitreous china	5–12	1180–1350	2156–2462
(14) Soft porcelain	5–12	1180–1350	2156–2462
(15) Bone china	6–11	1200–1320	2192–2408
(16) Hard porcelain	14–15	1410–1435	2570–2615
(17) Electrical porcelain	4–14	1160–1410	2120–2570
(18) Chemical porcelain	14–16	1410–1460	2570–2660
(19) and (20) Mullite and high alumina porcelain	10–~36	1300–1800	2372–3272
(21) Zircon porcelain	6–26	1200–1580	2192–2876
(22) Grinding wheels, traditional	8–12	1250–1350	2280–2462
fritted	02–1	1060–1300	1940–2012
(23) Cordierite	6/7–14	1210–1410	2210–2570
(24) Steatite, block	05	1000	1832
body	6–14	1200–1410	2192–2570
(25) Forsterite	14	1410	2570
(26) Spinel	14/15–40/41	1425–1947	2600–3540
(28) Wollastonite	4	1160	2120
(29) Lithia porcelains	04–2	1020–1120	1868–2048
(30) High magnesia	—	1540	2804
(31) High beryllia	10–~27	1300–1600	2372–2912
(32) Single component, e.g. Al_2O_3		1300–1950	2372–3542
BeO		>1370	>2498
ZrO_2		1500–1700	2732–3092
ThO_2		>1500	>2732
ZnO		1400	2552
$ZrSiO_4$		>1400	>2552
(33) Cermets	17–31	1480–1690	2696–3074
(34) Rutile	12	1350	2462
(35) Titanates	7/8–26/27	1400–1600	2552–2912
(36) Ferrites	8–13/14	1250–1400	2282–2570

A similar degree of vitrification can be produced at a somewhat lower temperature for a longer time but the appearance is not so white and the structure contains more glassy matter and is not so strong. This is due to the longer time at elevated temperature allowing a greater approach to equilibrium, and may even lead to decompositions giving off gases and making the porcelain somewhat porous again.

The question of the colour in porcelain dinnerware is of prime importance. A porcelain body fired to maturity at cone 13/14 is white whereas the identical body fired to the same density, etc., at cone 12 has a greyish tinge.

Maturing temperature is affected not only by chemical composition of the constituents of the body but also by their grain size and grading and their packing density. This may be determined by the shaping method. A dry shaping method under high pressure achieves much better contact than a wet low-pressure method, and so reactions occur more easily. This is very important in the production of dense articles of single compounds where sintering is the only means of densification. Even in low-firing brick clays Straight (**S203**) found considerable variation of maturing temperature with making method. Soft mud bricks need to be fired some 110° C (200° F) higher than stiff mud ones made from the same clay, because of their more open structure in the dry condition, after losing the greater water amount. Deflocculation, although most deflocculants are fluxes, increases the required firing temperature.

No definite data can therefore be given about maturing temperatures. In Table 197 an indication is given of the range of temperatures used for the various body types. Frequently the lower limit is considerably below normal usage and may relate to rather special conditions. Generally the better ware is produced near the upper limit and this would cover most commercial production. Maturing cone numbers have been mentioned individually with the batch compositions of bodies in Chapter 5, and with respect to ceramic products in Chapters 12 to 18.

RATE OF SHRINKAGE

The rate of volume change of ceramic bodies varies with the stage of the firing process. Unlike the drying shrinkage which occurs essentially during the initial stages of drying when the main weight loss is occurring, the firing shrinkage does not parallel the weight loss and occurs chiefly towards the end of the firing (Fig. 11.4).

The length or volume changes during firing are observed on test pieces either by periodic removal of samples from a furnace or by continuous measurement throughout firing. Van der Beck and Everhart (**V2**) describe such a method and applied it to clays used for structural clay products. Their general results are:

small expansion, up to about 750° C (1382° F);
rapid shrinkage, above 875° C (1607° F).

Stevens and Birch (**S189**) found three critical stages of rapid shrinkage in refractory bricks:

(1) 480–600° C (896–1112° F).
(2) 850–938° C (1562–1720° F).
(3) 927–1296° C (1700–2300° F).

It is during rapid expansion or contraction of pieces that strains are set up that may rupture them. It is therefore desirable that a good firing schedule must aim at producing a constant rate of volume change by adjusting the rate of temperature change. It is also very important for the production of uniform-sized ware that the entire setting attains a temperature above the main shrinkage period and is given time for this to be completed. (Fig. 11.5)

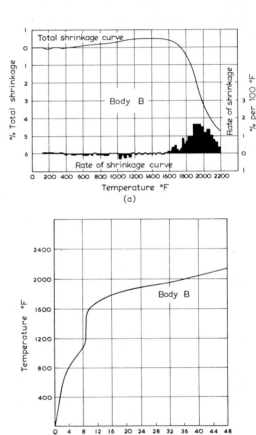

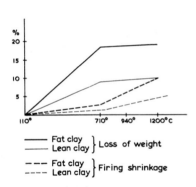

FIG. 11.4. *The weight loss and shrinkage of two typical clays during firing* (**S83**)

FIG. 11.5. (*a*) *Firing shrinkage behaviour for a given structural clay body.* (*b*) *Forty-eight-hour firing schedule for the same body to give approximately constant length change* (Van der Beck and Everhart, **V2**)

COOLING

Successful cooling of matured ware entails consideration of any physico-chemical changes that may occur. For instance the quartz inversion involves a considerable volume change and cooling should be very slow at this point.

Too rapid cooling contraction may set up strains that cause cracking either immediately or in the first few days after removal from the kiln. It is also at this stage that the strains between a body and an ill-fitting glaze are set up, and crazing will follow. Austin (**A199**) suggests that annealing of the cooling ware at a suitable temperature might relieve many of these strains.

The cooling rate may also be influenced by oxidation reactions which may be desired, in which case cooling at the correct temperatures should be slow, or they may be undesirable in which case cooling through the dangerous period must be fast.

HEAT REQUIREMENTS IN FIRING CERAMIC CLAYS AND BODIES

It is desirable in the designing of kilns, in the calculation of the efficiency of existing kilns, etc., to know the heat required to transform a ceramic clay or body from the raw to the fired state.

Heat is required in a number of different ways:

(1) Physically, to raise the temperature of the body and is determined by its specific heat. This depends on the individual specific heats of its constituents, each of which is temperature dependent, allowance being made for changes due to reactions (Table 198). Heat is again given up on cooling and is governed by the specific heat of the fired body.

(2) Chemical and physical reactions which either require heat or give out heat:

(*a*) Dehydration.
(*b*) Decomposition.
(*c*) Combustion.
(*d*) Dissociation.
(*e*) Fusion.
(*f*) Solution.
(*g*) Reaction.
(*h*) Changes of structure.

Shorter (**S70**) has reviewed methods for measuring such heat requirements for clays and carried out similar work. His conclusion (in 1948) was that it is not yet possible to make accurate quantitative measurements. However, his values for the 'specific heats' in the various temperature ranges, which therefore also indicate 'heats of reaction' can act as useful guides in calculations (Table 198 (*a*)–(*j*)).

TABLE

Specific Heats of Ceramic Materials

(compiled by

(a) Specific Heat of Calcined Alumina

Temperature interval (°C)	Specific heat (Shorter, **S70**)	Temperature interval (°C)	Specific heat (Calorimetric, Wilkes, **W69**)
—	—	30– 100	0·2060
100–219	0·19	—	—
—	—	30– 300	0·2260
100–403	0·23	—	—
100–540	0·24	30– 500	0·2395
100–641	0·25	—	—
100–725	0·26	30– 700	0·2500
100–797	0·265	—	—
100–859	0·27	—	—
100–913	0·27	30– 900	0·2580
100–958	0·28	—	—
100–998	0·28	—	—
—	—	30–1100	0·2645

(b) Specific Heat of Quartz

Temperature interval (°C)	Specific heat (Shorter, **S70**)	Temperature interval (°C)	Specific heat (White, **W60**)
102– 235	0·18	0– 100	0·1868
102– 446	0·205	0– 300	0·2168
102– 591	0·23	0– 500	0·2379
102– 705	0·23	0– 700	0·2543
102– 794	0·23	—	—
102– 870	0·24	—	—
102– 929	0·24	0– 900	0·2596
102– 981	0·25	—	—
102–1013½	0·27	—	—
—	—	0–1100	0·2641

Thermal value of inversion at 573° C equals 3·6 cal/g approx.

Over Various Temperature Ranges
Shorter (**S70**))

(c) *Specific Heat of Silicon Carbide*

Temperature interval (°C)	Specific heat
140– 254	0·18
140– 460	0·17
140– 602	0·17
140– 707	0·165
140– 790	0·16
140– 858	0·16
140– 915	0·16
140– 962	0·155
140–1004	0·1575

(d) *Specific Heat of Magnesia*

Temperature interval (°C)	Specific heat (Shorter, **S70**)	Specific heat (Wilkes, **W69**)
93–186	0·27	0·240
93–355	0·29	0·250
93–483	0·30	0·259
93–579	0·30	—
93–655	0·31	—
93–719	0·31	—
93–774	0·32	—
93–823	0·32	—
93–868	0·32	—
93–909	0·325	0·2765

(e) *Specific Heat of Calcined Clays*

Calcined china clay		Calcined ball clay	
Temperature interval (°C)	Specific heat	Temperature interval (°C)	Specific heat
100–225	0·18	100–218	0·18
100–412	0·21	100–400	0·20
100–547	0·23	100–536	0·21
100–649	0·24	100–638	0·21
100–732	0·25	100–721	0·21
100–799	0·26	100–789	0·22
100–855	0·275	—	—
100–903	0·29	—	—
100–944	0·30	—	—
100–979	0·315	—	—

TABLE 198—(contd.)

(f) Specific Heat of Alumina–China Clay Mixture
1·30 g Al$_2$O$_3$ + 0·65 g china clay

Temperature interval (°C)	Specific heat mixture	Specific heat of clay (calc.)
98– 214	0·23	0·25
98– 414	0·24	0·26
98– 540	0·28	0·365
98– 659	0·352	0·56
98– 747	0·349	0·54
98– 816	0·347	0·53
98– 873	0·348	0·53
98– 921	0·350	0·532
98– 961	0·353	0·537
98– 995	0·350	0·527
98–1025	0·353	0·535

(g) Specific Heat of a Ball Clay

Temperature interval (°C)	Specific heat
96–218	0·30
96–418	0·26
96–528	0·36
96–597	0·46
96–657	0·45
96–740	0·43
96–809	0·41
96–868	0·395
96–916	0·39
96–959	0·38
96–990	0·37

(h) Specific Heat of a Yorkshire Fireclay

Temperature interval (°C)	Specific heat
103–225	0·22
103–424	0·195
103–539	0·23
103–652	0·34
103–743	0·39
103–811	0·335
103–868	0·336
103–913	0·338
103–956	0·334
103–990	0·335

(i) Specific Heat of a Scottish Aluminous Fireclay

Temperature interval (°C)	Specific heat
93–215	0·285
93–438	0·23
93–532	0·29
93–655	0·42
93–744	0·415
93–813	0·405
93–871	0·400
93–919	0·397
93–965	0·377
93–996	0·38

(j) Specific Heat of a French Aluminous Fireclay

Temperature interval (°C)	Specific heat
98–210	0·29
98–398	0·24
98–507	0·29
98–608	0·47
98–696	0·455
98–762	0·436
98–816	0·420
98–864	0·407
98–903	0·40
98–937	0·39
98–966	0·38

FIRING AND KILNS 873

(k) *Interval Specific Heats* (Dinsdale (**D44**))

Range (°C)	Clay substance	Fired silica brick
0– 100	0·380	0·198
0– 200	0·390	0·206
0– 300	0·430	0·214
0– 400	0·448	0·222
0– 500	0·484	0·229
0– 600	0·557	0·236
0– 700	0·520	0·242
0– 800	0·501	0·248
0– 900	0·502	0·253
0–1000	0·497	0·259
0–1100	0·502	0·264

Calculating the Heat Requirements for Firing Ware

The theoretical heat requirements for firing ware of known composition to a definite temperature can be calculated. Dinsdale (**D44**) gives an example of such a calculation:

It is desired to biscuit fire 100 lb of earthenware to 1100° C (2012° F). The mechanical water content is 1%. The combined water content is 5%.

I. HEAT REQUIRED FOR DRYING STARTING WITH BODY AT 0°C

Heat required to raise temp. of 1 lb water to 100° C	180 B.t.u.
Latent heat of water at 100° C	970 B.t.u.
Heat required to raise temp. of steam from 100° C to 150° C	40 B.t.u.
	1190 B.t.u.

II. TO FIND THE OVER-ALL SPECIFIC HEAT OF THE DRY BODY

It is assumed that the body is made up of pure clay substance and of other materials whose specific heat resembles that of fired silica brick. The clay substance content is deduced from the combined water content.

$Al_2O_3.2SiO_2.2H_2O$ has a water content of 13·95%; a body containing 5% therefore has a clay content of:

$$\frac{5 \times 100}{13.95} = 35.8\%$$

From Table 198 (k) it can be seen that:

Sp. ht. clay substance from 0° to 1100° C = 0·502
Sp. ht. fired silica from 0 to 1100° C = 0·264

Therefore specific heat of body:

$$(0.358 \times 0.502) + (0.642 \times 0.264) = 0.349$$

Hence the amount of heat required to fire the dry body:

$$99 \times 1100 \times 1.8 \times 0.349 = 68\ 400 \text{ B.t.u.}$$

The actual heat content of the body at 1100° C which will be given up on cooling, assuming sp. ht. of 0.264, is

$$94.05 \times 1100 \times 1.8 \times 0.264 = 49\ 200 \text{ B.t.u.}$$

Thus the amount of heat actually required for the chemical process of firing and therefore irrecoverable is

$$68\ 400 - 49\ 200 = 19\ 200 \text{ B.t.u.}$$

this is about 28% of the total supplied.

To summarise:

Total heat supplied: 69 590 $\begin{cases} \text{drying: 1190.} \\ \text{irrecoverable: 19 200.} \\ \text{Given up on cooling: 49 200.} \end{cases}$

(*See also* B.S.S. 1388:1947.)

DERIVATION OF THE FIRING SCHEDULE FROM THE RELEVANT FACTORS

Munier (**M100**) describes a method of deriving a theoretical firing schedule from the curve of volume changes experienced by a particular body. He then adapts this to allow for the other factors that affect the firing (Fig. 11.6).

The experimental expansion, shrinkage and contraction curve is plotted on the axes OX/OY and is represented by $OABCDEFGHIJKLMN$. The portion $OA\ldots HI$ corresponds to the heating and the portion $IJ\ldots MN$ to the cooling of the piece. In order to obtain the cumulative curve of dimensional change with temperature the portion $O'RSTD$ is drawn symmetric to $OABCD$ about the axis ZZ'. The new curve $O'RSTDEFGHIJKLMN$ is now referred to the axes O_1X_1 graduated in length changes and $O'Y'$ in temperature.

The aim is to produce a heating schedule that brings about length changes proportional to the temperature changes:

$$\Delta L = k \times t$$

If O_1Y'' is graduated in time, then the curve for the length variation as a function of time is represented by the straight line $O_1\alpha$. At any point P on O_1Y'' representing a length of time of n hr of the firing the length change is given by O_1V and the temperature by $O'W$. If a new X-axis O_2X_2 is drawn to represent time, its divisions corresponding to those of O_1X_1 times the factor

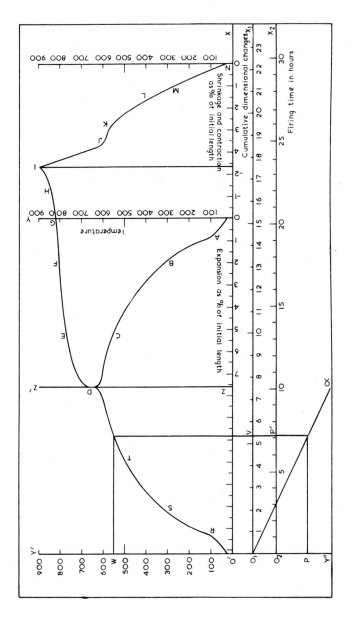

Fig. 11.6. *Derivation of theoretical firing curve (Munier, M100)*

k, then at any point P' on it representing n hours of firing the temperature should be $O'W$. In fact the curve $O'RSTDEFGHIJKLM$ referred to the axis O_2X_2 (time) and $O'Y'$ (temperature) is the desired theoretical time–temperature schedule.

This schedule can be applied directly to ware being fired for a second or third time (glost, decoration, etc.) where no side reactions evolving gases, etc., occur.

For a raw body the following adaptations are necessary:

HEATING

(a) *ORS*. The samples used for length change tests are thoroughly dried whereas commercial ware is not necessarily so. This stage must therefore be fired slower than shown by the curve.

(b) *ST*. Less water is now leaving the body so that the curve can be adhered to.

(c) *TDE*. The curve indicates slow firing because of the α-β quartz change and the loss of chemically bound water. It is, however, advisable to go even slower than the slope shows to ensure the completion of these solid state reactions.

(d) *EF*. A little shrinkage takes place, but more rapid firing can nevertheless be applied.

(e) *FG*. Large shrinkage during the period when the clay minerals have lost their original structure and not yet attained their new one. The body is at its weakest at this stage and should be fired very slowly.

(f) *GHI*. New crystalline structures are now forming giving the body sufficient strength to fire it faster than indicated by the curve.

(g) *I*. The highest temperature is maintained for a certain length of time to ensure an even temperature throughout and allow reactions to reach the desired state.

More flat portions may have to be inserted into the curve for individual bodies to allow for oxidation of carbon, decompositions of sulphates, carbonates, etc., and the reduction of iron compounds.

COOLING

The whole of the cooling should take longer than the theoretical curve shows (*see* table below).

(h) *IJ*. Little shrinkage takes place during the initial cooling which can be undertaken rapidly.

(i) *JK*. This kink represents the quartz inversion and is larger the more quartz is present and the cooling must be very slow.

(j) *KL*. Rapid cooling.

(k) *LM*. If there is a noticeable cristobalite inversion which would show as a bend in the curve, slow cooling is necessary.

FIRING AND KILNS

TABLE 199

Relative Duration of Heating and Cooling in Efficient Tunnel Kilns, the Total being taken as 100

	Heating	Cooling
Refractories	58	42
Paving bricks	50	50
Earthenware biscuit	57	43
Earthenware glost	57	43
Vitreous china	53	47
Electrical porcelain and large pieces	55	45
Red building brick	45	55

Munier points out that the temperatures shown on the graph are those of the ware itself, *i.e.* inside any muffles or saggars and *not* flame temperatures, or temperatures in flues, ducts, etc.

FACTORS DETERMINING THE TOTAL FIRING TIME

The total firing time is inserted on the O_2X_2 axis of the theoretical curve and is dependent on:

(*a*) Type of body.
(*b*) Density of setting. The denser the setting the more time must be allowed for equalisation of temperature.
(*c*) Thickness of ware. Same considerations as (*b*). (Fig. 11.7.)

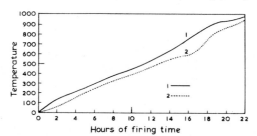

FIG. 11.7. *Temperature at the surface (curve 1) and the centre (curve 2) of a thick porcelain cylinder (Munier, M100, after Hegemann-Dettmer)*

(*d*) The cross-section of the kiln and hence the time taken for temperature changes to reach the centre. The graph (Fig. 11.8) shows the effect of the cross-section of tunnel kilns on the speed of firing of different bodies.

Munier gives a further graph showing three curves (Fig. 11.9).
(1) The expansion–contraction curve for a biscuited body to be fired glost. This is taken directly as the theoretical firing schedule.
(2) The recommended practical firing schedule derived from (1).
(3) The firing schedule of a typical tunnel kiln.

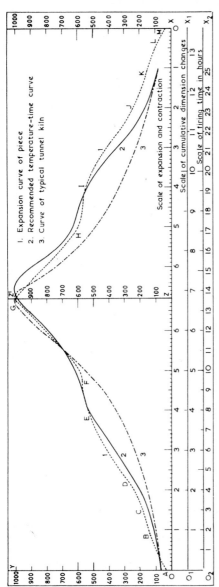

FIG. 11.9. *Temperature–time curve for glost firing of feldspathic earthenware tableware (Munier, M100)*

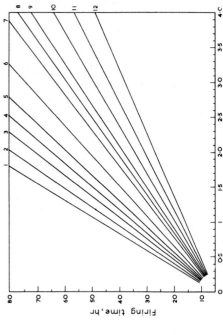

FIG. 11.8. *Combined influence of the nature of the piece being fired and the cross-section of the kiln on the firing time (Munier, M100)*

(1) Floor tiles
(2) Engineering bricks
(3) Biscuit earthenware tiles
(4) Bricks and tiles
(5) Glost earthenware tiles
(6) Grogged refractory
(7) Biscuit sanitary earthenware
(8) Main fire porcelain
(9) Biscuit earthenware tableware
(10) Glost sanitary earthenware
(11) Biscuit porcelain
(12) Glost earthenware tableware

TABLE 200

(Note: Section A gives total firing times, Section B gives time to reach maximum temperature.)

A. Time of Firing and Fuel used for Different Products in Periodic and Tunnel Kilns Respectively
(after Gatzke (**G11**))

Product	Total firing time (hr)		Fuel consumption (kg coal/kg of fired product)	
	Periodic	Tunnel	Periodic	Tunnel
Porcelain dinnerware				
glost	76	28–40	3·20–4·00	1·43–1·71
Electrical porcelain	80	90–100	2·07–2·37	0·95–1·25
Earthenware dinnerware				
biscuit	72	27–37	0·85–1·00	0·20–0·23
glost	72	17–30	1·00–1·80	0·30–0·34
Sanitary earthenware				
biscuit	86	34–50	1·20–1·45	0·49–0·62
glost	76	23–36	1·40–2·10	0·50–0·63
Earthenware wall tiles				
biscuit	180	80–100	0·50–0·62	0·13–0·27
glost	160	30–50	0·40–0·48	0·24–0·35
Stoneware floor tiles	230	100–120	0·30–0·38	0·32–0·37
Refractories	110	70–90	0·20–0·23	0·08–0·10

TABLE 200

B. Some Typical Firing Times
(Dinsdale (**D46**))

Product	Time (hr) to reach maximum temperature		
	Intermittent	Tunnel	
Earthenware biscuit	75	Open-flame gas	50
		Electric	28
Earthenware and china glost	28	Gas-fired radiant tube muffle	11
		Gas-fired Dressler muffle	13
		Gas-fired Revergen	16
		Electric	15
Decorating	14	Electric	6
China biscuit	55	Gas-fired open flame	35
Tile biscuit	100	Gas-fired	65
Sanitary earthenware biscuit	55	Oil-fired muffle	27
Sanitary fireclay	200	Gas-fired muffle	75
Jet and Rockingham	24	Gas-fired	13

The deviation of (3) from (2) is considerable. The heating is unnecessarily slow at the beginning of the firing, and too fast at the crucial quartz inversion stage, the same applying to the cooling.

FAULTS IN FIRED WARE

Faults in fired ware may or may not be due to faulty firing. The firing shows up any stresses, strains and blemishes introduced during preparation, shaping and drying, as well as being a source of further mistakes.

Hielscher (**H86**) has listed and discussed the sources of such faults in connection with bricks.

A. FAULTS DUE TO RAW MATERIALS

(1) Stony, lumpy impurities in the clay lead to an unhomogeneous product tending to crack (*see* p. 664 for removal of stones).

(2) Crystals and grains of gypsum and iron pyrites decompose causing bloating and the latter leave a highly fluxing deposit (*see* pp. 239, 686 for washing and classifying).

(3) Limestone, calcium carbonate in any form becomes calcined to quicklime during the firing, and thereafter recombines with water disrupting the brick (*see* p. 694 for treatment of limestone-containing clays).

(4) Surface discolorations, these may be due to soluble salts that migrate to the surface during drying (p. 695) (they may also be due to kiln gases).

B. FAULTS DUE TO SHAPING METHODS

The main source of cracks, bowed bricks, etc., is the extrusion process (p. 730). Other faulty shaping processes that show up after firing are:

(1) Incorrect transfer of bats on to jiggering mould (p. 724).

(2) Joining of parts of different moisture content or with wrong jointing slip (p. 773).

C. STRAINS SET UP BY INCORRECT DRYING

Too rapid drying of the surface while the inside of a piece is still moist will set up strains that may produce drying cracks but the piece may hold together until firing and then crack.

D. TRUE FIRING FAULTS

(1) Watersmoking period, disruption of ware by too rapid removal of water, and spoiling of the surface of other pieces by impurity-laden condensation.

(2) Bloating due to sintering before various gas-evolving decompositions have occurred.

(3) Cracking occurs during the water-smoking period, during periods of rapid volume changes and during cooling. In a coal-fired kiln of open set ware the later the cracking occurs the cleaner the cracks appear, so that those started during the initial stages have particles of ash, etc., in them whereas cooling cracks are clean.

(4) Black core is due to incomplete burning out of carbon and/or incomplete

oxidation of iron below the decomposition temperature of ferric oxide when more fusible ferrous silicates vitrify the centre of the ware (**A199**).

'SPIT-OUT' OR 'SPITTING' OF ON-GLAZE DECORATED WARE

This fault actually appears after the enamel or decorating firing of on-glaze decorated ware but its sources are to be found both in the composition of body and glaze and in the method of conducting the biscuit and the glost firing as well as the decorating firing. The defect is characterised by small craters with sharp or somewhat rounded-off edges showing that gas bubbles have forced themselves through the surface of the glaze.

A great deal of empirical data exist on conditions favouring or curing spitting and this has been collected by Mellor (**M57**) together with his own work on the subject. Koenig (**K48**) has continued this work. It now seems evident that gas bubbles of various origins are able to force their way through the glaze even though it does not undergo over-all softening during the enamel firing. Having released the pressure the glaze is not sufficiently fluid to heal over the crater which is therefore still present when the piece emerges from the kiln. The reason that spit-out occurs only at the lower temperature of the enamel kiln is that although gas bubbles may push their way through the glaze in both the glost and the enamel firing the glost kiln is hot enough for the glaze to flow over the escape points whereas the enamel kiln is not. If on the other hand the glaze resists gas pressure throughout the decorating firing no spit-out occurs even though gases may have been present in the body. The occurrence or presence of spit-out is therefore dependent on two opposing forces. The aggressor is the gas pressure building up in the body during the heating, which is opposed by the mechanical resistance of the glaze to the escape of gas bubbles.

The above papers discuss two sources of the gas bubbles, moisture and carbon and hydrocarbons, to this must be added fluorine from Cornish stone.

Moisture is an ever-present factor. Unglazed biscuit absorbs moisture from the atmosphere and so also does glazed ware with any imperfections in the glaze such as stilt marks or an unglazed foot. Once in the body the moisture gradually migrates and may thereafter occur in regions distant from the point of entry and therefore also far from a means of exit. If such pieces are heated rapidly steam is developed and sets up gas pressure. This may then force an exit through the glaze. If the glaze is still hard when the gas pressure overcomes its mechanical resistance, whole pieces of it may be forced off, known as 'spurting'; if the glaze has softened bubbles force their way through it as 'spits'. It is due to this cause that it is sometimes found that ware decorated and refired immediately after leaving the glost kiln is perfect whereas ware stored for a long time in the glost state produces spitting when it undergoes the decorating firing.

Long storage of glazed ware in a moist atmosphere can also result in moisture being absorbed by the glaze itself. On reheating, superficial spitting occurs leaving the surface very rough.

The second common cause of spit-out is unburnt carbon in the body, supplemented by absorbed hydrocarbon gases. Deposits of carbon in a ceramic body arise from both the raw materials, particularly ball clays and calcined bone, and the fuel gases and are normally largely or completely removed during the biscuit and glost firing. If, however, some should remain in the glost ware they will ignite if sufficient oxygen and heat are supplied. Hydrocarbon gases arising from the fuel or from organic decorating media and vehicles can in favourable circumstances be absorbed and given passage by glazes. Similarly oxygen can gain access to the body through glaze imperfections, etc. In favourable circumstances the oxygen will reach the combustibles at relatively low temperatures so that slow burning results. The volume of carbon dioxide produced by the combustion of carbon being equal to that of the oxygen used up no gas pressure should be set up. It is possible, however, that the oxygen only reaches the carbon when it is already very hot and then instantaneous combustion occurs with evolution of heat, causing the carbon dioxide to expand and set up gas pressure, while also softening the glaze locally so that its resistance to the pressure is lessened.

The third source of gas pressure, not mentioned in the available literature, is the decomposition of fluorides present in Cornish stone. This is a slow reaction and may not be completed in the biscuit and glost firings where it is harmless. It is for this reason that spit-out is much more common in the British Isles than in other countries where raw materials containing fluorine, particularly Cornish stone, are not used in the body.

All these sources of gas bubbles may be effectively opposed by a glaze that is mechanically strong and well integrated with the body surface and having a softening temperature high in relation to that of the decorating kiln.

With these explanations of spit-out it is possible to interpret all the empirical data about conditions that affect it and therefore to use them to help cure the defect. (Table 201)

SILVER MARKING

Certain glazed or more particularly glazed and decorated ware is found to be easily marked by soft metals such as table silver. Some of these marks can be washed off easily whereas others are more or less permanent. The marking may occur on new ware or may only begin after the ware has been in use for some time.

Microscopic investigation has shown that the actual cause of the marking is a roughness of the surface due to tiny crystals imbedded in the glaze. The origins of these crystals appear to vary and it is not always clear in the literature whether different causes lead to the same fault or whether a combination of circumstances always occurs and only one of these is traced and recorded. Some points are probably generally applicable.

Silver marking usually occurs only on ware that has been reheated after

FIRING AND KILNS

TABLE 201

Spit-out

Factors Concerning the Removal of Combustible Material Before the Enamel Firing

Easy biscuit-fired ware is more liable	Because it may contain more carbon through having been in a less ventilated part of the kiln.
Vitreous ware is more liable than non-vitreous	Because the glost firing has not removed combustible material so efficiently.
A glaze compounded from a hard frit leads to less spitting than one with a soft frit even if the final glaze is the same	Fusion occurs later in the glost kiln so that carbon can be burnt out for a longer period.

Factors affecting Combustion in the Enamel Kiln

Ground-laid ware in enamel kiln favours Lithographed ware is prone	} Because of the hydrocarbon gases evolved at the initial heating which are absorbed by the glaze and subsequently burn.
Ware fired quickly in the enamel kiln is more liable	Because gases formed get less opportunity to diffuse harmlessly.

Factors Relating to the Resistance of the Glaze to Gas Pressure

Whiting added to glazes cures.	It sometimes reduces fusibility of glaze.
High boric acid in the glaze favours.	It increases fusibility of glaze.
Matt and semi-matt are less liable than bright.	Because they are less fusible and less fluid when fused.
Leadless glazes are less liable than lead ones.	Because they are less fusible and less fluid when fused.
Once-fired ware is less liable than twice-fired.	Because it generally has a harder glaze.
Easy-fired glost ware is more liable than normal or hard-fired ware.	Because there is less interaction between body and glaze and so the glaze is more fusible.
Thinly dipped ware is less liable than thickly dipped.	Because there has been better interaction between glaze and body.
Ware fired harder in the enamel kiln is more liable.	Because the glaze is softened.

---

the glost firing, for the purpose of on-glaze decoration. The marking may then occur even on undecorated portions of the ware, in fact reheating of a glost piece in a decorating kiln, even without any decoration, may make it prone to silver marking.

Silver marking appears to occur only if lead is present, *i.e.* on ware with a lead glaze, ware with lead on-glaze flux, or leadless ware fired in kilns or saggars where lead fumes have once been present. This last instance is frequently the cause of silver marking on bone china with a leadless glaze and it really means that once a kiln has fired ware with a lead glaze or lead on-glaze flux it will thereafter make leadless ware prone to silver marking.

Geller and Creamer (**G19**) have shown that the presence of sulphur dioxide in decorating kiln gases can make ware prone to silver marking. Concentrations of 3% SO_2 and over did this in their experiments. With higher concentrations a slight scum became visible and the gloss was reduced, this scum was composed at least in part of lead sulphate.

Other investigations have shown that higher glost firing or the reheating in the decorating kiln may lead to greater interaction between glaze and body allowing either some devitrification of the glaze or migration of crystals from the body. The result may be a slightly roughened surface which will silver-mark immediately or the crystals may be below the surface. In the latter case silver marking may develop after use in which some of the glaze has become dissolved.

KILNS FOR FIRING CERAMIC WARE

It has been seen that drying of ceramic ware can be done in the open by natural air circulation, at prevailing temperatures. Use of enclosing structures and applied heat accelerates the process. Firing the ware, however, must always be done in enclosed structures with applied heat, and kilns of some sort have been built since pottery began, and are sometimes recognised in excavations of prehistoric sites.

The simplest method of burning is the clamp, which, ancient as it is, is still used occasionally for hand-made bricks. The bricks are stacked alternately with fuel to form a mound, or often up a hillside, and then covered with earth, etc. The bottom is then set alight and the fire allowed to travel through the clamp, drawing cooling air after it. When cold the whole is pulled down.

The next step is the periodic or intermittent kiln with a permanent structure (sometimes the roof is temporary). These are usually round, but may be rectangular. They have an inner lining of suitable refractory and an outer one of protective building brick. Periodic kilns may be worked on the up-draught, horizontal draught or the down-draught principle, the last being much more satisfactory. Ware is placed in the kiln and it is then gradually heated up, kept at maximum temperature for a given period and allowed to cool. The ware can then be removed and a fresh setting put in.

It can readily be seen that apart from inefficiency by heat losses through walls and up the flue, etc., a periodic kiln must use a large amount of fuel to heat up the structure for each batch of ware, all of which is lost during the cooling. Furthermore the continual heating and cooling of the structure wears it out much faster than a constant high temperature would.

Continuous kilns, *e.g.* Hoffmann kilns, make use of the waste heat given off during cooling. They consist essentially of a number of intermittent kilns connected in a circuit. The air flow is regulated to pass first through ware that has been fired and is cooling, and then, in its heated condition it moves on to the kiln that is being fired. The hot waste gases pass on to ware about to be fired, preheating it, so that it needs less fuel in its actual firing. The main principle is that the fire is always alight and keeps moving round the kiln circuit. Waste heat is used but the kiln structure still has to be heated and allowed to cool for each batch.

In the tunnel kiln the reverse process occurs. A tunnel structure has zones at constant, different temperatures corresponding with a firing schedule and the ware is pushed through on trucks or refractory slabs. In theory this is the ideal method of firing where maximum fuel efficiency can be achieved. In practice the tunnel kiln is therefore rapidly becoming recognised as the best firing method for mass production, although improved intermittent kilns will probably continue to be used for small or individual batches.

The placing or setting of the ware in the kilns is also subject to a number of considerations and varies considerably. Factors that must be considered are firstly, whether or not the ware can be subjected to direct contact with the flames and combustion gases produced by burning the fuel. Where no contact is permissible there are two alternative methods of protecting the ware. The first is to use a muffle kiln which consists of a more-or-less sealed chamber inside the kiln, round which the flames and hot gases are led; the second consists of placing the ware in refractory vessels, saggars, before setting it in the kiln.

The second important factor is whether or not the ware can be stacked on top of itself to the full height of the kiln. Bricks, etc., can usually be so stacked. Much other ware can be stacked a little and so requires some intermediate support. Glazed ware receiving its glost firing may not touch at all, or sticking occurs.

Before describing the various kiln types in detail it is proposed to discuss some of the constructional details that are common to more than one type of kiln, *e.g.* grates, burners, refractories, insulation, construction, setting methods, etc.

KILN CONSTRUCTION MATERIALS AND METHODS

The choice of materials for kiln construction is difficult as there is no universally 'best' product. The conditions of service must be assessed carefully, such as: maximum service temperature; rate of temperature change; maximum load; abrasion by ash, etc.; attack by fumes, vapours, slag, oxidising or reducing atmosphere. These show that more than one property of constructional materials must be known and the most suitable combination chosen to give long service.

The basic materials used are themselves ceramic, being refractory (P.C.E. above 26), insulating, and common bricks, blocks and shapes. To these must be added mortars, ramming mixtures and concrete. The properties of refractories used for kiln construction together with other related ceramic products are tabulated in Chap. 17, p. 1212.

The requirements of ceramic kiln building generally involve knowledge about the following properties:

Refractoriness in service, that is the maximum temperature which the material will withstand for prolonged periods of time, on repeated occasions and under the load of the structure above it, plus a safety factor. Figures are usually available for the pyrometric cone equivalent, which is not directly useful, and the more helpful ones on the deformation temperature under a given load, and the shrinkage at a given temperature. The coefficient of thermal expansion must also be known in order to make sufficient expansion joints to give the structure stability.

Thermal spalling resistance or *resistance to thermal shock* is an important factor as a material that withstands high temperatures well but cracks if cooled rapidly is useless in, for instance, a short-cycle periodic kiln. The safe cooling of the kiln structure may, in fact, sometimes dictate a firing schedule of length unnecessary to the ware being fired. Spalling resistance of materials is often only given in qualitative terms such as 'excellent', 'good', 'fair', etc., but details of repeated rapid coolings may be available.

Abrasion and *impingement resistance:* different parts of a kiln undergo very varied treatment during use, including solid fuel stoking, air-borne ash and dust, etc. Only comparative tests can be made, giving qualitative results.

Resistance to slag, fumes, etc.: refractories are largely classed as acid, neutral or basic according to their resistance to acid and basic slags, neutral ones being at least partly resistant to both. For ceramic kilns it must further be known what resistance the materials have to combustion gases, particularly sulphurous gases in the presence of water vapour, fumes given off from glazes, particularly lead ones, sodium chloride vapour, etc.

Constructional properties: the load-bearing properties both cold and hot, together with the apparent density, possible shrinkage, reversible expansion, thermal conductivity and specific heat should all be known in order to decide on wall thicknesses and methods of construction.

Materials most commonly used in Brick Form for Kiln Construction, after A33 *and* S218

A6. High-fired super duty fireclay. Used as lining of periodic and tunnel kilns, cones 8–12.

A1. Super duty fireclay. Useful up to cone 29. Used as generally high temperature kiln lining and kiln car tops, *e.g.* crowns of muffle and direct-fired kilns, burner surrounds and bag walls.

A2. High duty fireclay. Up to cone 20. General lining for lower temperature kilns and lining of hot end of preheating and cooling zones in tunnels; side walls of periodic kilns.

TABLE 202

Refractories Classification List

(after Swain (S218))

(Temperatures have been read off from the chart given by Swain, in °F to the nearest 10° F. The nearest British cone is given.)

Temperature of Application								Refractory
Normal application range						Light service condition extension		
From			To					
°C	°F	Approx. cone	°C	°F	Approx. cone	°C	°F	Approx. cone
981	1800	06	1647	3000	29	1703	3100	32	A. FIRECLAY
815	1500	014	1525	2780	20	1647	3000	29	(1) Super duty.
648	1200	021	1259	2300	8	1592	2900	26	(2) High duty.
537	1000	<022	1148	2100	4	1531	2790	20	(3) Intermediate duty.
									(4) Low duty.
									(5) Abrasion resistant.
									(6) High fired, super duty.
									(7) Glass tank block.
									B. HIGH ALUMINA
981	1800	06	1675	3050	30	1730	3150	33	(1) 50% alumina.
1092	2000	1	1703	3100	32	1758	3200	35	(2) 60% alumina.
1203	2200	6	1730	3150	33	1786	3250	36	(3) 70% alumina.
1253	2290	8	1758	3200	35	1814	3300	37	(4) 80% alumina.
1314	2400	11	1786	3250	36	1814	3300	37	(5) 90% alumina.
1314	2400	11	1814	3300	37	1925	3500	40	(6) 99% alumina.
1314	2400	11	1758	3200	35	1869	3400	39	fused alumina.
									C. INSULATING FIRE BRICK
537	1000	<022	870	1600	011	—			(1) Group 16.
648	1200	021	1092	2000	1	—			(2) Group 20.
759	1400	016	1259	2300	8	—			(3) Group 23.
870	1600	011	1414	2580	14	—			(4) Group 26.
981	1800	06	1525	2780	20	—			(5) Group 28.
537	1000	<022	1592	2900	26	—			(6) Lightweight silica.
									(7) Loose insulation.
1092	2000	1	1647	3000	29	—			(8) Group 30.
1092	2000	1	1703	3100	32	—			(9) Alumina insulating brick.
815	1500	014	1436	2620	15	1592	2900	26	D. SEMI SILICA 80–82% SiO_2 bound with clay.
									E. KAOLIN
									F. MULLITE
1248	2280	8	1703	3100	32	1758	3200	35	(1) Mullite or sillimanite.
									(2) Molten cast mullite.
									G. SILICA BRICK
648	1200	021	1658	3020	29/30	—			(1) Regular.
648	1200	021	1703	3100	32	—			(2) Super.
648	1200	021	970	1780	07/06	—			Spalling resistant.
815	1500	014	1758	3200	35	1814	3300	37	H. SILICON CARBIDE
									I. ZIRCON
									J. ALUMINA ZIRCONIA
1314	2400	11	1758	3200	35	1897?	3450?	39	K. CHROME
1314	2400	11	1814	3300	37	1925?	3500?	40	L. MAGNESITE
									(1) Dead burned 85–90% MgO.
									(2) 90–95% MgO (periclase).

TABLE 202—(contd.)

Temperature of Application								Refractory
Normal application range					Light service condition extension			
From			To						
°C	°F	Approx. cone	°C	°F	Approx. cone	°C	°F	Approx. Cone	
1092	2000	1	1758	3200	35	1897?	3450?	39	M. MAGNESITE–CHROME
1092	2000	1	1758	3200	35	1897?	3450?	39	N. CHROME–MAGNESITE
1092	2000	1	1592	2900	26	1647	3000	29	O. FORSTERITE
			1650	3002	29				— BARIUM ANORTHITE
									P. PLASTIC FIREBRICK
									(1) Standard.
1092	2000	1	1647	3000	29	1703	3100	32	(2) Super.
									Q. RAMMING MIXTURES
									(1) Magnesite.
									(2) Magnesite–chrome.
									(3) Chrome–magnesite.
									(4) Forsterite.
									(5) High alumina.
									(6) Fireclay.
									(7) Insulating.
									R. CASTABLES
									(1) Fireclay.
									(2) Basic.
(−18)	0	—	1092	2000	1	—			(3) Insulating 501b.
									(4) High alumina.
(−18)	0	—	1314	2400	11	—			Group 24.
815	1500	014	1481	2700	17	—			Group 27.
815	1500	014	1647	3000	29	—			Group 30.
(−18)	0	—	1536	2800	20	1758	3200	35	Chrome.
(−18)	0	—	1508	2750	19	1647	3000	29	Mullite.
(−18)	0	—	1342	2450	12				Insulating 701b.
1092	2000	1	1536	2800	20	1647	3000	29	P. (3′) High duty.
1092	2000	1	1647	3000	29	1703	3100	32	(4′) Super slag resistant.
(−18)	0	—	1647	3000	29	1758	3200	35	(5′) Chrome.
1092	2000	1	1730	3150	33	—			(6′) Mullite.
									MORTARS
815	1500	014	1481	2700	17	1703	3100	32	Fireclay.
815	1500	014	1536	2800	20	1703	3100	32	Fireclay–calcine mix.
(−18)	0	—	1370	2500	13	1481	2700	17	Hydraulic setting.
(−18)	0	—	1647	3000	29	—			Air-setting high temp.
1092	2000	1	1703	3100	32	—			Heat-setting high temp.
815	1500	014	1536	2800	20	1592	2900	26	Silica cement.
981	1800	06	1703	3100	32	—			Super silica.
1092	2000	1	1758	3200	35	—			High alumina.
(−18)	0	—	1758	3200	35	—			Air-setting mullite.
981	1800	06	1758	3200	35	—			Heat-setting mullite.
(−18)	0	—	1647	3000	29	1758	3200	35	Air-setting chrome.
1092	2000	1	1758	3200	35	—			Heat-setting chrome.

A3. Intermediate duty fireclay. Cone 8. Backing for hot zones behind super or high duty, and lining for intermediate sections of preheating and cooling zones of tunnel kilns; side walls of periodic kilns.

A4. Low duty fireclay. Cone 4. General backing material and a cool portion of kilns, entrance, exit and below kiln cars.

B2. High alumina containing about 60% Al_2O_3. Up to cone 32. Surrounds of grates, burners, bag walls and other places where direct flames impinge and there is abrasion, *e.g.* crowns of muffle kilns.

B4 and 5. High alumina containing 80–90% Al_2O_3. Up to cones 35/36. Hollow front walls of muffle combustion chambers. Abrasion resistant.

B5 and 6. High alumina containing 90–99% Al_2O_3. Up to cones 36/37. Floor and back of muffle combustion chambers. Abrasion resistant.

D. Semi-silica. Up to cone 15. Crowns and linings of kilns.

E. Kaolin. Up to cone 16 to 20. Crowns and linings of kilns also periodic ones used for salt glazing.

F1. Mullite. Up to cone 32. Floor, back and hollow front of combustion chamber of muffle tunnel kiln. Ports for oil and gas burners. Lining of electric kiln with resistor grooves.

G1. Regular silica brick. Up to cone 29/30. Hot zone kiln lining.

G2. Super silica brick. Up to cone 32. Floor and back of combustion chamber of muffle kiln. Abrasion resistant.

H. Silicon carbide. Up to cone 35. Front hollow wall of combustion chamber of muffle kiln. Bag wall where flame impinges. Abrasion resistant.

TABLE 203

Melting Points (°C) (Birch (B63))

Element	Oxide	Aluminate or mixture of oxide with Al_2O_3	Compound or mixture of oxide with MgO	Zirconate or mixture of oxide with ZrO_2	Titanate or mixture of oxide with TiO_2	Silicate or mixture of oxide with SiO_2
Magnesium	2800	2135	—	2150	1840	1890
Aluminum	2050	—	2135	1910*	1860	1810
Zinc	1975†	1950	No data	1810*	1830	(1510)
Titanium	1825	1860	1830	(1750)*	—	(1540)*
Zirconium	2690 P	1910*	2150	—	(1750)*	2430
Barium	1925 H	2000 H	(1500)*	2650	No data	(1700)
Strontium	2430 H	2015 H	1935*	2800	No data	(1580)
Calcium	2570 H	(1765) H	2360*	2550	1840	2130
Chromium	3000 R	2800*	3000*	2320*	No data	No data
Nickel	1990 R	2020	1990*	No data	No data	(1725)*
Cobalt	1935 PR	1960	1935*	(1730)*P	No data	(1725)*
Beryllium	2530 P	1820(?)	1955*	2200*P	1810	(2000)
Cerium	2600 P	1885*	2250*	2400*P	(1500)*	No data
Thorium	3050 P	1910*P	No data	2690*P	(1640)*	No data
Hafnium	2810 P	No data	No data	No data P	No data	No data
Lanthanum	2320 P	1870*	2030	2300	—	—
Uranium	2180 P	—	—	—	—	—
Yttrium	2410 P	—	—	—	—	—

Notes:
    The letters P, H and R are to suggest that the principal limitations are price, hydration and reducibility, respectively.
    Melting points falling below 1800° C are given in parentheses.
    If two oxides combine to form more than one compound, the data represent the more refractory one.
    * No chemical compound. In such instances the minimum melting point is given if known.
    † Volatilises readily.
    The chief reference paper for melting points was 'Melting and Transformation Temperatures of Mineral and Allied Substances', by F. C. Kracek, Section II of *Special Papers No. 36*, Geological Society of America, *Handbook of Physical Constants*, edited by Francis Birch, pp. 139–74 (1942).

VERY HIGH TEMPERATURE REFRACTORIES (ABOVE 1800°C (3272°F))

Only very pure materials consisting of a single compound resist very high temperatures, *e.g.* pure oxides, aluminates, zirconates, titanates and silicates, and also carbon, carbides, nitrides and borides. Birch (**B63**) has listed the eighteen elements forming oxides and some other compounds with melting points above 1800° C (3272° F) as shown in Tables 203 and 204.

TABLE 204

Data on Refractory Oxides

(prices current 1945)

Element	Density of oxide	Cost per lb $	Estimated weight of 9 in. brick (lb)	Cost of material in 9 in. brick $
Beryllium	3·0	4·00	8·2	25·00
Magnesium	3·6	0.05	10·0	0.50
Zinc	5·5	0.75	15·0	1.10
Strontium	4·7	0.50	12·9	6.50
Barium	5·7	0.10	15·6	1.56
Calcium	3·3	0.005	9·0	0.05
Aluminium	4·0	0.055	10·0	0.55
Cerium	7·6	0.82	20·8	17.00
Thorium	9·7	7.00	26·5	185.00
Titanium	4·9	0.09	12·4	1.10
Zirconium	5·7	0.90*	15·6	14.00
Yttrium	4·8	12.50	13·1	164.00
Lanthanum	6·5	14.00	17·8	250.00
Hafnium	9·7	—	26·5	—
Chromium	5·0	0.25	13·7	3.40
Cobalt	5·7	1.84	15·6	28.60
Nickel	7·4	0.35	20·0	7.00
Uranium	5·6	3.00	15·3	46.00

* Zircon, the silicate, is much cheaper than the oxide.

He then discusses the potentialities of these materials. Some of them, namely alumina, magnesia, dolomite ($MgCO_3.CaCO_3$), chrome spinel, zircon, zirconia are already used for commercial refractories. Of the remainder price or stability eliminates them all for large-scale use. However, where the quantity is small enough for price to be less important some other materials have been or could be used, *e.g.* for burner tips, crucibles, sample tubes, etc. The oxides suitable are those of beryllium, thorium, zirconium, cerium.

Refractory Mortars, Ramming Mixtures and Renderings

The various mortars, ramming mixtures, spraying and patching mixtures, etc., necessary for the building and maintenance of kilns and furnaces must be chosen with the same care as the refractories themselves. Miehr (**M74**) describes the requirements for and nature of such materials and attempts to clarify some misconceptions.

Firstly, the term 'refractory' as applied to mortars, etc., has the same meaning as when it is applied to bricks, blocks, etc., namely that the material shows no melting below cone 26 (1580° C, 2876° F). Certain mortars which depend on their partial fusion and the formation of a glassy phase to make the bond are therefore not refractory and might be termed 'thermoadhesives'.

Secondly, refractory mortars, etc., set by ceramic mechanisms quite unrelated to the hydraulic setting of cement and lime mortars and unnecessary confusion arises by the inclusion of the word 'cement' in their nomenclature. Refractory mortars are in fact raw ceramic bodies, consisting of plastic binding materials made workable with water, and non-plastic grog. They become hard and form bonds by the normal process of firing in their final position.

Refractory mortars are used in kiln construction to form a bond and a seal between refractory bricks and blocks. They have to take up any irregularities in the surfaces they are joining and completely prevent the passage of gas along the joints while having a minimum total thickness. The grain size may therefore not exceed 0·04 in. (1 mm). The proportion of binder to non-plastic material must be such that a plastic workable material can be made by mixing with water which is easy to apply with a trowel, and which on drying does not crack, and on firing does not shrink too much or fuse and flow away. In general the quality, composition and refractoriness of the mortar should resemble that of the bricks it is to bind. Its expansion properties must also be comparable.

Thermoadhesives are materials which can be applied like mortars but contain fluxes so that they fuse during service. They are used for repairing cracks and holes, sealing lids, doors, etc.

Ramming mixtures are used for monolithic kiln and furnace linings, to replace complicated special shapes in arches and around ports, etc., for repairing damage of any size or shape without the necessity of removing all damaged bricks. Their use does not require skilled labour and the finished work is jointless and therefore less open to attack and disintegration (**R26**). To fulfil these duties the requirements for the mixtures are very stringent regarding their refractoriness, refractoriness under load, shrinkage, vitrification, porosity, mechanical strength, resistance to thermal spalling and attack by fumes and slags.

Ramming mixtures may be wholly ceramic consisting of grog and clay or they may be bound by aluminous cement. Additional initial binding may be afforded by organic binders and silicones. The grog used is considerably coarser than that for mortars but should be graded to give close packing using sizes from 0–1·2 in. (0 to 30 mm). To achieve the maximum refractoriness the minimum of binding clay should be used and any fluxes, however much they can help in binding, must be rigorously excluded.

As such rammed construction may have considerable thickness expansion joints must be made occasionally and the first heating must be slow and carefully regulated. Small cracks may appear nevertheless and these should be patched.

Refractory coatings and renderings may be brushed or sprayed over hot or cold masonry surfaces to protect them from slag or abrasive attack and in particular to keep the joints intact. The general considerations given for mortars are applicable although it may be necessary to use a smaller grain size for spraying and less binding clay to avoid shrinkage. In order to obtain immediate adhesion when spraying on to hot surfaces a small amount of fusible material may be included. Application of a protective coating on a kiln or furnace lining before it is used and patching or replacing it whenever necessary may be an economical method of prolonging the life of the lining itself.

RAW MATERIALS FOR REFRACTORY MORTARS, RAMMING MIXTURES, ETC.

A. Binding

(1) Ceramic binding set up during service. (*a*) clays: fireclays, stoneware clays, kaolins, bentonites; (*b*) colloids: waterglass, alumina sol; (*c*) frits, glazes, etc.: low-melting silicates.

(2) Hydraulic cold setting followed by vitrification in service. Aluminous cement, magnesium hydroxide.

(3) Organic binders setting cold and burning out.

B. Grog

Burnt fireclay, calcined bauxite, mullite, corundum, sinter alumina, talc, olivine, forsterite, cordierite, sinter magnesite, sinter dolomite, chromite, zirconia, spinels, quartz, silica, shale, sandstone, etc., coke, graphite, silicon carbide, tungsten carbide, boron nitride, etc. (**M74**).

Selection of Refractories

A finished kiln is a composite structure made up of several materials chosen primarily for their positive qualities and secondarily as the most economical for the various parts of the kiln. Thus highly refractory bricks are only used where very high temperatures are anticipated and abrasion-resistant dense bricks only where really necessary, etc. The outermost skin of the kiln is often common building brick.

Apart from the constructional considerations of combining sections built of different materials, *i.e.* where bonded construction may or may not be used, the physicochemical interactions must be allowed for. Certain pairs of refractories react with each other at kiln temperatures, damaging one or both of them. The extent of the damage sometimes depends on which of the two is on top. Birch (**B61**) and McGill and McDowell (**M10**) have investigated the reactions between pairs of common refractory materials at various temperatures by placing half bricks on each other in reversed pairs in suitably heated test furnaces. Their results are summarised in Table 205

FIRING AND KILNS

and although the actual temperatures of reaction found in the experiments may not be identical with those at which reactions occur under the load and atmospheric conditions of a kiln they are indications of where caution should be observed.

Swain (**S218**) points out that kiln constructional materials rarely have to be selected without the opportunity of studying past experience of the same or similar applications. Systematic evaluation of such findings can lead to suitable adjustments based on complete service conditions. The points he suggests looking out for are as follows:

(1) Correct zoning, *i.e.* use of high quality materials where conditions are severe but of cheaper ones where they are less maltreated.

(2) Abrasion shown as a rough dry surface can be counteracted by the use of harder fired bricks of a slightly less refractory nature or if necessary silicon carbide or fused alumina may be used.

(3) Slagging giving the refractory a smooth glassy surface due to attack by slags or fluxes. Mild attacks may be counteracted by using denser bricks but severe trouble requires the use of bricks of different chemical composition.

(4) Melting can be differentiated from slagging by analysis of the fused portion which then is identical with the original brick. It occurs rarely and means that a brick that is more refractory under load must be used.

(5) Mechanical spalling is the loss of portions of brick by mechanical stresses. The fracture is flat and often the spalls are deep at the edge of the brick, decreasing to a thin layer at the centre. The cause is in the constructional method rather than the refractory and is generally due to inadequate expansion joints.

(6) Structural spalling is due to vitrification of the surface by flux or slag action or melting, which then shrinks and sets up strains until it splits off.

(7) Thermal spalling due to too rapid temperature changes shows as random curved cracks in the bricks. It occurs more easily in kiln car tops where considerable compressive forces exist.

(8) Rigidity. Highly refractory materials begin to soften at much lower temperatures when under heavy loads and when this is found to occur better methods of transferring the load to cooler portions of the structure are the best solution.

(9) Disintegration due to chemical action reducing the brick to a powdery mass of no strength.

(10) Heat properties, conductivity, specific heat as discussed (p. 886).

In evaluating service he suggests making use of the chart on 'Utility of Refractories' developed by Phelps (**P34**) and improved by McDowell (**M2**) (Fig. 11.10).

TABLE

Reaction Temperatures between different Refractories

(The temperature at which reaction was first observed is given under T_1, and that at particular conditions of the test and may be lower in furnace and kiln conditions. Series B

Refractory	Series	Silica		High-duty fireclay		70% alumina		90% alumina	
		T_1	T_2	T_1	T_2	T_1	T_2	T_1	T_2
		°C °F	°C °F	°C °F	°C °F	°C °F	°C °F	°C °F	°C °F
Silica	A	—	—	1500 2732	*	1600 2912	*	—	—
	B	—	—	1497 2730	1647 3000	1597 2910	*	1647 3000	*
High-duty fireclay	A	1500 2732	*	—	—	No reaction		No reaction	
	B	1497 2730	1647 3000	—	—	No reaction		No reaction	
70% high alumina	A	1600 2912	*	No reaction		—	—	No reaction	
	B	1597 2910	*	No reaction		—	—	No reaction	
90% high alumina	B	1647 3000	*	No reaction		—	—	—	—
Chrome	A	No reaction		1600 2912	*	1600 2912	1600 2912	—	—
	B	1647 3000	*	1597 2910	*	1597 2910	*	1647 3000	*
Magnesite, low iron	A	1500 2732	1600 2912	1500 2732	1500 2732	1700 3092	*	—	—
Magnesite, regular	A	1500 2732	1600 2912	1400 2552	1500 2732	1500 2732	*	—	—
	B	1497 2730	1597 2910	1397 2550	1497 2730	1497 2730	1703 3100	1597 2910	1647 3000
Forsterite	A	1700 3092	1700 3092	1500 2732	1600 2912	1700 3092	1700 3092	—	—
	B	1597 2910	1703 3100	1497 2730	1597 2910	1647 3000	1703 3100	1647 3000	1703 3100
Chrome–magnesite	B	1597 2910	*	1597 2910	*	1597 2910	1703 3100	1703 3100	*
Magnesite–chrome	B	1597 2910	1647 3000	—	—	1597 2910	1703 3100	1597 2910	*

* Indicates that the reaction was not damaging up to 1700° C.

Insulation of Kilns

The need to minimise the waste of fuel in any process that is to be economic is always present. A large amount of heat is lost through the walls of kilns, built in the traditional way of firebrick and building brick, by conduction through the masonry followed by radiation. Proper insulation of the structure can reduce this loss to a negligible amount.

In the case of the tunnel kiln where the inside face is always kept at the same temperature heat losses through the structure occur at a constant rate and it would appear well worth preventing them. Furthermore, the temperature of the structure itself remains constant so that increasing its bulk by adding insulation does not in itself increase the fuel consumption. It is therefore an essential part of the construction of a tunnel kiln to incorporate efficient insulation.

In the case of the intermittent kiln there are two opposing factors to consider. Firstly the heat losses by conduction and radiation and secondly the heat required to heat up the structure for every firing. Decreasing the conduction losses by increasing the total bulk of the structure may eventually use up as much fuel as it saves. Insulation of periodic kilns has therefore lagged until the recent introduction of refractory insulating materials.

Proper insulation not only reduces the fuel consumption but assists in keeping the kiln at a more uniform temperature and makes working conditions in the vicinity of the kilns very much more pleasant.

Proper insulation of the floor of a kiln must not be overlooked. The concrete normally placed across the site has a higher conductivity than either bricks or soil. Soil, too, has a marked conductivity. So heat losses through

after (B61) Series A and (M10) Series B

which it became damaging under T_2. The actual temperatures probably apply only to the appears to have used an approximation in the temperature conversion.)

Chrome				Low iron magnesite				Regular magnesite				Forsterite				Chrome–magnesite			
T_1		T_2		T_1		T_2		T_1		T_2		T_1		T_2		T_1		T_2	
°C	°F	°C	°F	°C	°F	°C	°F	°C	°F	°C	°F	°C	°F	°C	°F	°C	°F	°C	°F
No reaction				1500	2732	1600	2912	1500	2732	1600	2912	1700	3092	1700	3092	—		—	
1647	3000		*	—		—		1497	2730	1597	2910	1597	2910	1703	3100	1597	2910		
1600	2912		*	1500	2732	1500	2732	1400	2552	1500	2732	1500	2732	1600	2912	—		—	*
1597	2910		*	—		—		1397	2550	1497	2730	1497	2730	1597	2910	1647	3000		*
1600	2912	1600	2912	1700	3092		*	1500	2732		*	1700	3092	1700	3092	—		—	
1597	2910		*	—		—		1497	2730	1703	3100	1647	3000	1703	3100	1597	2910	1703	3100
1647	3000		*	—		—		1597	2910	1647	3000					1647	3000		*
—		—		No reaction				No reaction				No reaction				—		—	
—		—		—		—		1647	3000		*	1597	2910		*	—		—	
No reaction				No reaction				No reaction				No reaction				—		—	
No reaction				—		—		—		—		No reaction							
1647	3000		*	No reaction				—		—				1703	3100	No reaction			
No reaction				No reaction				No reaction				—		—					
1597	2910		*	—		—				1703	3100	—		—		1703	3100		*
—		—		—		—		No reaction				No reaction				—		—	
—		—		—		—		—		—		1703	3100		*	—		—	

Compositions of the Refractories

Refractory	Series	SiO$_2$	Al$_2$O$_3$	TiO$_2$	Fe$_2$O$_3$	CaO	MgO	Cr$_2$O$_3$	Others	Total
Silica	A	95·9	1·0		1·0	2·0	0·1	—		100·0
	B	96·3	0·8	0·6		2·2			0·2	100·1
High-duty fireclay	A	52·9	42·7		2·3	0·3	0·4	—		98·6
	B	54·0	39·0	2·3					4·7	100·0
High alumina 70%	A	22·7	74·5		1·8	0·1	0·3	—		99·4
	B	23·2	69·8	3·5					3·5	100·0
High alumina 90%	B	8·5	90·0	trace					1·5	100·0
Chrome	A	5·0	29·5		13·3†	—	17·8	32·7		98·3
	B	4·5	28·0		14·5		18·0	34·0	1·0	100·0
Magnesite, low iron	A	2·0	0·7		2·1	1·7	93·6	—		100·1
Magnesite, regular	A	3·5	1·0		6·0	3·7	85·7			99·9
	B	4·0	1·5		3·0		87·0	trace	4·5	100·0
Forsterite	A	31·7	1·1		6·4†	2·8	57·2	0·1		99·3
	B	33·0	1·5		9·0		53·0	1·5	2·0	100·0

† Iron oxide reported as FeO.

the base can be considerable. Furthermore the heat thus led into the ground burns the soil, causing shrinkage, so that the entire kiln may subside by as much as a foot. As it is also important to keep moisture away from the kiln floor it is suggested that the whole kiln site should be excavated down to 1 ft (0·3 m) below the footings of the walls. At this level it should be concreted over, the walls then built and insulating concrete placed under the floor (**A35**).

INSULATING MATERIALS

The principle of insulation is based on the poor conductivity of air, and all insulating materials aim to incorporate as much air as other considerations

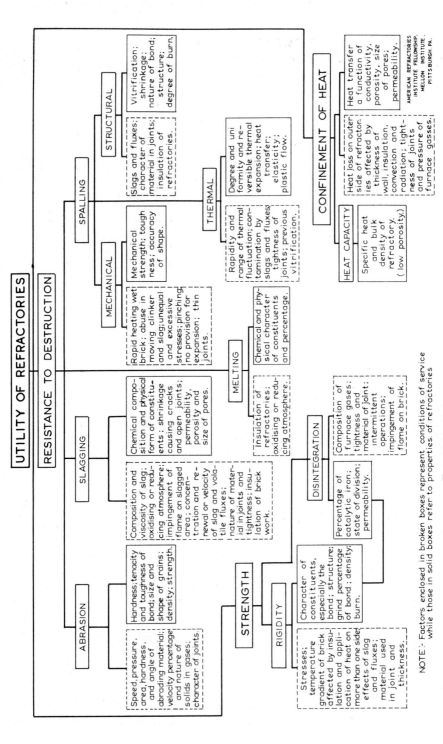

Fig. 11.10. *Refractory utility chart* (P34, M2)

will permit. Heat transfer through solids is by conduction, but as soon as large air spaces are introduced convection and radiation play a part. Convection is prevented by making the air spaces too small to allow appreciable air movement. Radiation depends on the temperature differences between two opposite solid faces. The smaller the air gap, the less the temperature difference across it and the less the radiation loss (**S124**).

The relative importance of the three factors changes with the temperature. At low temperatures conduction is the most important and the lighter the material the better it insulates. At high temperatures convection and radiation convey so much heat that the air spaces must be very small. For instance, at 1027° C (1881° F) a pore of 0·1 mm (0·039 in.) transmits half as much by radiation as by conduction whereas one of 5 mm (0·197 in.) diameter transmits thirty times as much (**H11**).

As well as the pore size a matter of very great importance to kiln insulation is the temperature which a porous material will withstand without the pores collapsing and heavy shrinkage occurring. The most readily available porous material of very high insulating value is diatomaceous earth. This is used either as a loose infilling or bonded and made into bricks, or as the aggregate in insulating concrete. But it will not normally withstand temperatures above 800 to 900° C (1472–1652° F). It therefore has to be shielded by a layer of refractory brick thick enough to reduce the temperature of the face against the diatomaceous earth to less than 900° C (1652° F). This makes the total structure bulky, and although very suitable for tunnel kilns it cannot be used for really efficient insulation of a periodic kiln because of the total weight.

Ceramic refractory insulating bricks are now being made which withstand temperatures up to 1540° C (2804° F). These can be used directly as the kiln lining, not only insulating but actually reducing the weight of the structure. With them real fuel savings can be made in periodic kilns.

Bullin (**B140**) shows the relative thicknesses of various structures necessary to give the same insulation (Table 206). Table 207 lists some insulating materials. The properties of some commercially available insulating materials are given in Chapter 18 (p. 1286).

The more porous surface of insulating bricks makes them more liable to crumble and produce dust. In order to overcome this they are often washed with a zircon preparation to seal the surface layer and improve its strength.

Construction Methods for Kilns

The walls of kilns are of masonry construction with expansion joints. They are usually built of different materials in skins or layers. Sometimes the bonding goes throughout whether the materials are the same or not. In others the refractory lining and the outer structure are independent and may have a gap between them with loose insulating infilling. Except in the Scotch kiln with its temporary roof, these walls must bear the load of the roof of traditionally built kilns. These roofs are of arched masonry structure,

TABLE 206

Table of Equivalent Insulators (B140)

Materials	Firebrick 18 in. (45·7 cm)	Red brick 13 in. (33·0 cm)	Firebrick 9 in. (22·8 cm)	Diatomaceous insulation 4½ in. (11·4 cm)	Refractory insulation 9 in. (22·8 cm)	Refractory insulation 4½ in. (11·4 cm)	Diatomaceous insulation 2½ in. (6·3 cm)
Total thickness, in. (cm)		31 (78·7)		13½ (34·2)	9 (22·8)		7 (17·7)
Approx. wt./ft², lb (kg)		340 (154)		112 (51)	40 (18)		27 (12)
Approx. Storage at equilibrium, B.t.u., kcal		78 000 19 656		40 000 10 080	7800 1966		7400 1864

the refractories being made in wide ranges of shapes and sizes to suit the various shapes of arch. This construction must be arranged to divert the maximum possible load from the hottest refractories.

In the newer type of suspended construction the refractories themselves bear little or no load, as they are suspended individually or in small groups from a steel structure. The method is particularly useful for ceilings, which can be flat, giving a more convenient and larger setting space with better heat distribution. (Flat kiln or furnace ceilings are nevertheless called 'arches'.)

There are a number of designs for suspended flat arches in which specially shaped bricks are hung individually from steel bars, either by tie rods hooked into the bricks or by clamps fitting over projections in the bricks, or by water-cooled tubes passing through the bricks. In some designs several bricks are supported from a single hanger, or make a double layer by hooking one over another, and a further modification is to have interlocking surfaces on the bricks so that broken ones do not fall out (**L37**). A monolithic arch can also be formed by use of a suitable ramming mixture on hangers (**J55**).

Suspended bricks can span a kiln of any desired width without any load bearing on the walls, which can therefore be more economically constructed. Suspended arches can be made any shape and in particular need not be arched. The main disadvantage of suspended roofs is that if insulation is placed above the refractories and therefore round the metal hangers they will overheat, while if these hangers are water-cooled the insulation is less effective. In the type of roof where the bricks are suspended by projecting pieces which are clamped it is possible to lay insulation between the projections, thus insulating about half the area. Suspended roofs are suitable for use with hot face insulating bricks (**T29**).

CHOICE OF FUEL

The following sources of heat can be used for ceramic firing:

(1) Wood and charcoal.
(2) Peat.
(3) Lignite.
(4) Coal.
(5) Coke.
(6) Anthracite.
(7) Natural gas.
(8) Town gas.
(9) Propane.
(10) Producer gas.
(11) Oil.
(12) Electricity.

The choice of it is governed by:

(1) Availability and price per unit of heat.
(2) Capital and maintenance cost of equipment necessary to burn or use it.
(3) Labour requirements.
(4) Nature of gases discharged into kiln.

TABLE

Insulating

No.	Ref.	Material	Range		Maximum temperature	
			(°C)	(°F)	(°C)	(°F)
1 2 3 4	G5	*Loose fill* of diatomite, expanded vermiculite, etc. *Classes of ceramic bricks and blocks* (1) (2) (3) (4)	up to 900 900–1200 1200–1350 1100–1450	1652 1652–2192 2192–2462 2012–2642		
		Examples:				
5	A71	Low-temperature diatomite slabs, as mined			900	1652
6	A71	Solid grade diatomite bricks			851	1564
7	A71	Light grade diatomite bricks			850	1562
8	M85, S128, C29	Moler bricks			850–900	1562–1652
9	A71	Intermediate temp. bricks with diatomite base			1100	2012
10	A71	Vermiculite bricks			1100	2012
11	A71	Vermiculite concrete (cement fondu)			1200	2192
12	M85	Fosalsil super			1350	2462
13	S75	Silex products, limits			1400 to 1250	2552 to 2282
14	A71	Hot face insulating firebrick I			1350	2462
15	A71	Hot face insulating firebrick II			1450	2642
16	A71	Hot face insulating firebrick III			1550	2822
17	M96	Refractory insulator			1538	2800

Materials

No.	Thermal conductivity at mean temperature		Refractoriness		Bulk density		After contraction (% after 2 hr at temperature)
	(kcal) (°C)	(B.t.u.) (°F)	(°C)	(°F)	(g/cm³)	(lb/ft³)	
1	0·699	1·26@ 760° C			0·56/0·64	35/40	
2	1·365	2·46@ 870° C			0·67/0·75	42/47	
3	1·942	3·50@1000° C			0·80/0·88	50/55	
4	2·497/3·052	4·5/5·5@1000° C			0·80/1·12	50/70	
5	0·305 0·321 0·338	0·55@300° C 0·58@400° C 0·61@450° C					
6	0·660 0·677 0·693	1·19@300° C 1·22@400° C 1·25@450° C					
7	0·466 0·488 0·510	0·84@400° C 0·88@450° C 0·92@500° C					
8	0·333–0·666	0·6–1·2			0·40–0·80	25–50	
9	0·627 0·649 0·682	1·13@350° C 1·17@400° C 1·23@500° C					
10	0·577 0·821	1·04@200° C 1·48@600° C					
11	1·093	1·97@78–667° C					
12	0·832	1·5 @625° C			0·85	53	
13	1·665 to 0·555	3·0 to 1·0	1650 to 1600	3002 to 2912	1·15 to 0·45	72 to 28	
14	0·821 0·888 0·954 1·032 1·148	1·48@200° C 1·60@300° C 1·72@400° C 1·86@500° C 2·07@600° C	1690	3074	0·76	47·4	−0·03@1200° C −0·37@1300° C −2·69@1400° C —
15	1·276 1·376 1·476 1·570 1·670	2·30@200° C 2·48@300° C 2·66@400° C 2·83@500° C 3·01@600° C	1730	3146	0·94		+0·03@1200° C +0·02@1300° C −0·61@1400° C −3·66@1500° C
16	1·148 1·581 1·637 1·720 1·831	2·07@200° C 2·85@300° C 2·95@400° C 3·10@500° C 3·30@600° C	over 1770	over 3218	1·02		+0·01@1200° C −0·10@1300° C −0·16@1400° C −0·99@1500° C
17	1·332 1·609	2·4@538° C 2·9@816° C	1710	3110	0·76	47·5	

In general it is considered that in inefficient kilns the only possible fuel is the cheapest, namely coal in Gt. Britain, whatever its disadvantages in labour costs and the dirt and impurities it carries into the kiln and up the flue into the atmosphere. But in efficient kilns with good recuperation and insulation the extra cost of a clean fuel is offset by the higher percentage of good ware and by being able to eliminate saggers.

Town gas is finding increasing favour for firing tunnel kilns. It is clean, generally constant in supply and calorific value, requires no storage and is trouble-free in operation.

Producer gas, either hot and raw, or cooled and cleaned works out cheaper than town gas but requires the capital, space and labour for the producer equipment. Small manufacturers prefer town gas, whereas it pays a large manufacturer to have a gas producer.

Oil has many of the advantages of reliability shown by town gas except that it nearly always contains sulphur. However, it requires storage with the accompanying fire risk.

Electricity is the cleanest and most easily controlled source of heat. The thermal efficiency of an electrically heated kiln is about double that of any other type but this in itself does not compensate for its high cost per unit. Its convenience is, however, found to do so, for high quality ware, and particularly for decorating kilns. In our further discussion of the use of electricity no allowance will be made for power supplies subject to frequent failure as are still prevalent in underdeveloped countries.

It is difficult to give prices or even relative prices for fuels as these change rapidly and also vary with the district, the season and the amount used. Table 208A gives costs that are now out of date, but it shows for which fuels additional cost due to labour and handling, etc., occur. Table 208B gives 1956 values.

Wood and Charcoal

Wood and charcoal are the ancient traditional materials for firing pottery and are still used extensively in countries where wood is more available than coal. Wood has a much higher volatile content than coal and requires grates of somewhat different design. The ash is fine and does not contain stones, clinker, etc., found in coal ash. Wood contains no sulphur compounds, which is a great advantage.

In Australia a method has been developed to use up waste eucalyptus wood by mixing it with brown coal and carbonising it in the presence of steam, when a high-grade charcoal is produced (**A151**).

Wood can also be used successfully and economically in gas producers.

Lignite

Lignite or brown coal which occurs extensively in central and south-eastern Europe, where there is little true coal, is usually pressed into

TABLE 208

A. 1939 Fuel Costs (F22)

Fuel	Net cost per therm	Gross cost, i.e. including capital charges, labour, handling, power, etc.
Coal	0·75d. to 1·00d.	1·10d. to 1·60d.
Oil	2·50d.	2·75d. to 3·00d.
Town gas	2·50d. to 4·00d.	2·50d. to 4·00d.
Hot producer gas (coke)	1·00d. to 1·30d.	1·45d. to 1·80d.
Clean producer gas (coke)	1·00d. to 1·30d.	1·80d. to 2·20d.
Electricity	7d. to 10·5d.	7d. to 10·5d.

B. 1956 Fuel Costs (A146)

Fuel	Calorific value	Cost	Cost (pence)/therm*
Electricity	3412 B.t.u./kWh	1¼d. unit†	36·6
Gas	470 B.t.u./ft³	—	12·75 to 17‡
Gas oil	166 000 B.t.u./gal	1–1⅞d./gal	8·4
Fuel oil, S.G. 0·968	18 900 B.t.u./lb	1–1½d./gal	7·4
Coke	11 964 B.t.u./lb	£7 5s. per ton	6·5
Coal	12 857 B.t.u./lb	£7 per ton	5·8

* Therm = 100 000 B.t.u.
† Actual charge depends on tariff adopted and may involve maximum demand charges.
‡ Depends on amount consumed, Stoke-on-Trent (may be higher elsewhere).

briquettes before use. It is fired in grates similar to those for coal, although an inclined one is preferred to the flat type. A grate that has 50% free space is the most successful (**S95**).

Coal

Coal has replaced wood as the main fuel in most parts of the industrial world today. It has been defined as 'a combustible rock which had its origin in the accumulation and partial decomposition of vegetation'.

Adams (**A175**) describes the origin and evolution of the different coals now present at various depths in the earth's crust. The first factor determining the type of coal is the nature of the vegetable debris accumulating, whether it has the usual predominance of carbohydrates or an unusual enrichment of hydrocarbons. Secondly comes the type of biochemical decay, whether 'peat-forming' in the presence of some air, or 'putrefaction' in deeper water, or one of the many graduations between these two extremes. During these processes the oxygen content is reduced and the deposit gradually enriched

TABLE

Average Properties of

(An attempt has been made in the accompanying table to indicate what may be regarded
tion. It is to be understood that wide variations will be encountered among fuels belonging

		Wood	Peat	Lignite
Moisture content as found (or made)		20	93	50
Moisture content of good commercial fuel		15	20	15
Proximate analysis (dry fuel):				
Volatile matter		80	65	50
'Fixed carbon'		20	30	45
Ash		Trace	5	5
Ultimate analysis (dry ash-free fuel)	C	50·0	57·5	70·0
	H	6·0	5·5	5·0
	N + S	1·0	2·0	2·0
	O	43·0	35·0	23·0
Ultimate analysis (good commercial fuel with ash and water content as above)	C	42·50	43·70	56·52
	H	6·78	6·42	5·72
	O	49·87	44·36	31·89
	N+S	0·85	1·52	1·62
	Ash	—	4·00	4·25
Theoretical air requirements per kilogram of good commercial fuel:				
kg air		5·08	5·33	7·11
m³ air at 0° C and 760 mm		3·93	4·12	5·50
Waste gases per kg of good commercial fuel:				
m³ of wet waste gas at 0° C and 760 mm		4·65	4·79	6·04
CO_2 content of dry waste gases (%)		20·4	20·1	19·5
Composition of wet waste gases:				
CO_2 %		17·1	17·1	17·5
H_2O %		16·1	14·9	10·5
N_2 %		66·8	68·0	72·0
Dew point of waste gases (°C)		56	54	47
Calorific value (dry ash-free fuel): (cal/g)	gross	4450	5000	6400
	net	4130	4710	6140
Good commercial fuel: (cal/g)	gross	3780	3800	5170
	net	3420	3460	4870

Combustion data refer to the fuels having a moisture content indicated as characteristic of good
air is assumed to be dry.
The air requirements (in ft³/lb of fuel) are obtainable by multiplying by 16·018 the air requirements
Net calorific values have been calculated by deducting from the gross values 586 cal/g of water
ultimate analysis.
C.V. in B.t.u./lb = 1·8 × C.V. (in cal).
The calorific value of pure amorphous carbon is 8080 cal/g.

Solid Fuels (Speirs (S153))

as the typical composition of fuels falling into each of the classes included in the classifica-
to each of the groups.)

	Bituminous coals			Semi-bituminous coals	Semi-anthracite	Anthracite	Manufactured fuels		
I	II	III	IV				Charcoal	Coke	Semi-coke
10	5	2	2	1	1	1			
10	5	2	2	1	1	1	2	2	2
40	35	30	25	18	10	4	10	1	8
55	60	65	70	78	86	93	89	92	85
5	5	5	5	4	4	3	1	7	7
77·0	82·0	86·0	88·0	90·5	93·0	94·0	93·0	95·0	93·0
5·5	6·0	5·5	5·0	4·5	3·5	3·0	2·5	1·0	3·0
2·5	2·5	2·5	2·5	2·0	1·5	1·5	1·5	2·0	2·0
15·0	9·5	6·0	4·5	3·0	2·0	1·5	3·0	2·0	2·0
65·83	74·00	80·07	81·93	86·01	88·39	90·27	90·23	86·58	84·76
5·82	5·98	5·34	4·87	4·39	3·45	3·00	2·64	1·13	2·96
21·71	13·01	7·37	5·98	3·74	2·78	2·32	4·69	3·61	3·60
2·14	2·26	2·32	2·32	1·90	1·42	1·44	1·46	1·82	1·82
4·50	4·75	4·90	4·90	3·96	3·96	2·97	0·98	6·86	6·86
8·68	10·04	10·75	10·87	11·28	11·25	11·34	11·11	10·22	10·59
6·71	7·76	8·31	8·40	8·72	8·70	8·77	8·59	7·90	8·19
7·18	8·18	8·67	8·71	8·98	8·91	8·95	8·77	7·99	8·39
18·9	18·4	18·6	18·8	19·0	19·4	19·6	19·9	20·5	19·6
17·2	16·9	17·3	17·6	18·0	18·6	18·9	19·3	20·2	19·0
9·0	8·1	6·9	6·2	5·4	4·3	3·7	3·3	1·6	3·9
73·8	75·0	75·8	76·2	76·6	77·1	77·4	77·4	78·2	77·1
44	42	39	37	35	31	28	26	15	29
7200	8300	8600	8600	8750	8600	8600	8300	8050	8200
6910	7890	8310	8340	8510	8420	8440	8170	8000	8040
6160	7490	8000	8000	8310	8170	8260	8050	7340	7470
5850	7170	7720	7740	8080	7990	8110	7910	7280	7310

commercial fuel and an ash content equivalent to the figure shown for the dry fuel. The combustion
(in m³/kg of fuel). The same factor applies to the volume of waste gases.
contained in the waste gases. This is equivalent to 52·7 cal for each 1% of hydrogen shown in the

in carbon. As the deposit becomes covered by successive layers bacterial action ceases and further changes are of a physicochemical nature.

The increasing pressure and temperature compress the peat, expelling water and volatile compounds containing oxygen. The carbon-to-oxygen ratio increases and in due course lignitic coals and brown coals are formed. These have a high water content, above 10%, and yield 40% to 55% of hydrocarbons on heating to 900° C (1652° F).

Later on sub-bituminous and then bituminous coals formed. The latter are the commonest types of coal in general use. Low-ranking ones have a volatile content of 30% to 45% and are the most suitable for making coal gas and producer gas. High-ranking varieties contain only 20% to 30% volatile matter and provide high-grade coke suitable for metallurgical processes.

The highest ranking coals derived from peat are the anthracites with up to 95% carbon and only 5% volatiles. They are very hard and burn slowly and almost without a flame.

The different origins and conditions of decomposition and pressure, etc., have perforce produced various types of coals, even amongst those of the same age. From the user's point of view these are divided into 'brights' and 'hards'.

The *bright*, shiny-black, well-laminated, soft type is termed *clarain* and is made up of the shiny black *vitrinite*, minor quantities of the duller *micrinite* and a little of the powdery *fusinite*. It is frequently banded.

Durain is the term applied to the *hard*, lustreless coal that shows little or no banding. It is largely made up of *micrinite* and *fusinite* (**A175**).

Although clarain and durain coals may have very similar compositions their burning properties differ considerably. Clarain tends to have a high tarry resin content that holds the solid particles together in cakes. The volatile matter given off on heating is comparatively heavy and requires a large amount of air well mixed with it and high temperatures for complete combustion. Smoke forms very easily unless care is taken. Fortunately most 'bright' coals have bands of durain which counteract the caking and make them easier to burn.

Durain coals give off lighter volatile matter which burns with a longer, less luminous flame and makes less smoke than clarains. The hard coal breaks open gradually during burning without disintegrating into dust and so presents large glowing surfaces to the air stream (**G2**).

All coals and lignites contain a certain amount of sulphur, varying from 0·5 to 3%. It is present in at least three forms. The largest quantity occurs as iron pyrites, FeS_2. During the burning of the coal this is oxidised to ferric oxide and sulphur dioxide. During destructive distillation it is decomposed so that some of the sulphur goes off in the gas and some stays in the coke. A second portion of the sulphur is present as calcium sulphate which becomes reduced to calcium sulphide, and a third occurs in various organic compounds.

FIRING AND KILNS

Most of this sulphur appears at some time as sulphur dioxide in the kiln gases, and as such is highly deleterious to ware of all types (*see* Kiln Atmosphere, p. 863) (**S57**).

As well as the combustible matter of plant origin coals contain more or less inorganic incombustible matter which remains as ash after burning the coal. The ash content of British coals varies between 1% and 20% (**G2**).

The quantity and fusibility of the ash must be taken into account when choosing a coal for a particular purpose. Some firing methods depend on a bed of ash to support the fuel and will not work if there is too little ash or if it fuses into a solid cake or clinker.

In general if the ash is light coloured it is mainly silica and/or alumina and is refractory, if it is dark it may contain iron oxides which lower the fusion point particularly under reducing conditions. Analysis of ash does not, however, give a true picture of the behaviour of a coal when fired in industrial grates. Factors to consider when selecting coal are:

(1) Quantity and nature of volatile combustible matter.
(2) Caking properties of non-volatile matter.
(3) Sulphur content.
(4) Ash quantity and refractoriness.

GRATES AND FIRING METHODS FOR COAL

Firing Holes in the Kiln Roof

In some kilns and in particular continuous brick kilns where the air is preheated sufficiently to ignite the coal the fuel is dropped into the kiln through holes in the roof. The setting of bricks is placed so that a space is left under the fuel holes, so that most of it drops down to the bottom. Some bricks are, however, allowed to project into this chute and catch small amounts of coal and then act as intermediate hearths (*see* Fig. 11.23, p. 957).

Where the ware will stand such direct contact with the fuel this method gives more even heat distribution in a large kiln than some constructions using hearths at the outside walls can achieve.

Fire Holes in the Kiln Walls

The most usual method of firing the coal is through holes in the kiln walls. These are from 14 to 18 in. ($35\frac{1}{2}$–$45\frac{3}{4}$ cm) wide and the arch over them slopes or steps downwards and inwards, the lowest point being the 'glut arch'. Beyond this a vertical baffle confines the gases into a space called the 'bag', opening on to the vault of the kiln. Other baffles may also drive some or all the heat along underfloor flues (**G2**).

The fuel is burnt either on horizontal or inwardly inclined bars or directly on the brick floor of the fire hole. In the latter case a coal forming suitable supporting clinker is required.

Most or all of the air for combustion enters under the fuel bed. It has been shown that when air passes through a solid porous layer of burning carbon all the available oxygen is used up in a bed thickness of 4 in. (10 cm). So that if a thicker fuel bed is used, as is frequently the case, secondary air must be admitted to complete combustion. If the fire hole is closed between refuellings, 'baitings', with a firedoor, this must have an air inlet that can be regulated. Where there is no firedoor a skilled fireman can regulate the secondary air by the height between the fuel bed and the glut arch. A continuous recording CO_2-meter to show the fireman what conditions prevail inside the kiln would allow him to find the best setting for secondary air.

Variable draught due to short chimneys and changing wind conditions alter the rate of drawing in the primary air which must be counteracted with dampers, etc., if efficient combustion without excessive smoke is to be achieved. Forced draught achieved with fans eliminates the variable chimney factor.

In most cases there is an optimum width and length for the furnace and an optimum rate of combustion. Too high a burning rate causes overheating of the furnace, damage to the brickwork, firebars, etc., and slagging. Too low a rate gives a sluggish fire and incomplete combustion. It has been proved in boiler practice that the most efficient combustion takes place when the burning rate is between 20 and 30 lb of coal per square foot of grate surface per hour (98–146 kg/m^2).

The normal method of firing a kiln with such fireboxes is for the fireman to shovel fuel on each grate in turn and clean them as necessary at intervals of several hours. The fire therefore goes through cycles of temperature fluctuations, being cooled by the new fuel and gradually rising and then falling before the next baiting. Similarly the tendency to emit smoke is greatest after each baiting and the amount of excess air passing through the fuel bed is greatest shortly before baiting and least at about half time between them. Regular heating and known atmospheric conditions can therefore never be achieved with this firing system.

Greater uniformity can be attained by baiting alternate fireholes at one time and the remainder at half time between these fuellings.

Step or Cascade Grates

A considerable improvement as regards fuel consumption, smoke emission, temperature fluctuation, and effective control of the air input can be achieved by replacing the open type of firebox by a step or a cascade grate (**B140, S95**).

The general construction of the two types is similar but the step grate has a low arch 9 in. or 18 in. (23 cm or 46 cm) thick extending below the glut arch almost to the level of the base of the bottom flue. The front of the fire mouth is cut off by a step arrangement of iron bars sloping towards the inside of the kiln. Water drips on to the top bar and thence on to the others to keep them cool and prevent clinkering. The lower door, behind the steps, has the primary air regulator door (Fig. 11.11).

The fuelling procedure consists of raking the burned fuel down the steps to below the arch and adding fresh fuel through the upper door until the lower end of the slope just fills the gap under the arch.

With this method a steady temperature rise can be achieved instead of the lag before a baiting and sudden rise after one, found with flat open grates.

There is a definite fuel saving of from 10 to 15% and cheaper grades of fuel can be used (**I8**).

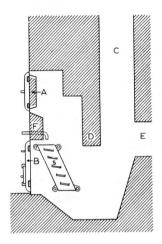

Fig. 11.11. *Side elevation of step grate* (*Bullin*, **B140**)

A, fuelling door, lined with fireclay to resist heat
B, door with sliding primary air regulator
C, bag
D, low arch
E, bottom flue
F, glut arch
S, steps

Automatic Stoking

Hand stoking of kiln fireholes and fireboxes has been shown to be extravagant of both labour and coal, as well as producing unsteady temperature increases and gas conditions within the kiln. Some form of continuous feeding of the grates is therefore highly desirable. A variety of automatic stokers were first applied extensively to brick and tile kilns and more recently to kilns for higher grade ware.

Over-feed Stoking

The simplest form of continuous stoking is by gravity from a hopper fitted to the firemouth with a sloping grate. Air, or air and steam are blown in through the ash pit door and up through the fuel bed, making the firebox into a kind of gas producer. The carbon monoxide and hydrogen formed are burnt in the secondary air, the flame length being regulated by the air input (**G2**).

Other stokers deliver a semi-continuous supply of fuel which can be easily regulated. They consist essentially of hoppers fitted with baffles to regulate flow and mechanically operated shovels, shoes, compartments, wheels, etc., which deliver the fuel. Various sizes and grades of fuel, slack and dust, up to 3 in. (7·6 cm) nuts, can be catered for. The power required to drive the motors is small, and one motor will frequently drive several stokers. Many of these devices are easily moved, either in their entirety or their more expensive motor parts alone, from one firehole to another so that they need not be fitted to every firehole.

Regular stoking with small quantities of fuel has shown economies of fuel and better ware (**B140, B34, C68, R6b**).

Under-feed Stoking

Under-feed stoking involves rather different firing methods. Small-size fuel is carried along a tube by a worm or screw to a retort under the firebed. Here it is mixed with a forced airstream and moves up to the fire zone. The forced draught ensures complete combustion of the coal with practically no smoke. As the fuel becomes spent new fuel pushes it aside and slag and ash collect around the fire opening. Such firing methods have been found

FIG. 11.12. *Under-feed stoker* (*Ashwell and Nesbit*, **A214**)

to work very well for boilers where the heat is constantly being removed. In furnaces and kilns where far higher static temperatures are reached trouble from slagging may occur if the fuel is not carefully selected. However, if this is done 25 to 30% fuel saving can be achieved.

Under-feed stokers are either situated around the kiln in fireboxes, in which case the fire can be cleaned during firing, or placed centrally in the kiln. In the latter case the heat distribution is more effective but as the fire cannot be reached for cleaning a non-caking, non-slagging, low-ash content coal must be used. Enough space must be left around the burner for all the ash formed in an entire firing to accumulate there. Under-feed stokers have been successfully installed in both intermittent and continuous kilns (**A192**) (Fig. 11.12).

NUMBER OF FIREBOXES FOR PERIODIC KILNS

The number of fireboxes and therefore the total area of grate surfaces is determined by the size of the kiln and the rate at which it is to be fired and the final kiln temperature.

For average-sized kilns where it is usually desirable to complete the heating and soaking stages in about 24 hr there is an empirical formula to give the total grate area:

$$\text{grate area} : \text{volume of contents} :: 1 : 10$$

Kilns of extreme sizes do not follow this rule. Very small kilns have a relatively far greater heat loss from the surface and may need the ratio to be 1:2 or 1:3. In very large kilns slower firing and more flues combine to make larger grate areas necessary, the ratio being about 1:8.

PULVERISED COAL

Coal can be burnt completely, without emitting black smoke or leaving ash that has a combustible content, if it is first pulverised. Firing with pulverised coal is being increasingly introduced.

The coal is either pulverised on the spot, one pulveriser often feeding several burners, or bought in the prepared condition. (Coal pulverisers may be of the hammermill, ball mill, ring roll mill or attritor types coupled with air classifiers in closed circuit.) The coal dust is forced to the burner by fans introducing the primary air. Secondary air is led to the burner. The general firing principles are similar to those for oil and the rate of burning is as easily controlled.

Although firing with pulverised coal is reputed to combine the lower cost of coal with the higher efficiency and ease of control of gas or electricity, it has a big disadvantage in the ceramic industry because of its ash content. The ash is naturally very fine and is carried into the kiln by the combustion gases and penetrates everywhere, causing blemishes on the ware. The system is therefore difficult to use for high-grade ware.

Gas

The use of gaseous fuel has great advantages over solid fuel. These are largely derived from its ease of control and its cleanliness.

(1) The burning of gaseous fuel is much more easily controlled than that of solid fuel. The flow of fuel to the burner is continuous instead of by intermittent stoking. It is controlled by a valve instead of requiring heavy shovelling. It can in fact be automatically controlled to give constant temperatures (for tunnel kilns) or to follow a predetermined heating and cooling curve (for periodic kilns).

(2) Gaseous fuel leads to much greater cleanliness both inside and outside the kiln. The flame and hot gases produced by its combustion are true gases and do not, like coal, carry ash, tar, etc., with them into the kiln where they damage ware, saggars and kiln linings. Town gas is also freed of sulphur compounds by passing it through hydrated ferric oxide, and natural gas often has a very low sulphur content, which therefore eliminates most of the troubles arising from damp sulphurous kiln gases.

Outside the kiln the totally enclosed pipe system for gas distribution is often the deciding factor when compared with the dirt and labour of solid fuel.

These two considerations lead to the following possible economies and advantages when gas firing is suitably installed:

(1) Fuel economy.
(2) Better ware.
(3) Reduced labour costs.
(4) Better working conditions.
(5) Reduced expenditure on saggars, etc.
(6) Smoke abatement (**S164**).

The application of gas firing is limited because of its price. The actual cost per unit of heat is more than for that obtained from coal, so that direct conversion of kilns from coal to gas firing is generally uneconomical. Gas is best reserved for the firing of modern kilns where insulation, shortest schedules, optimum cross-section, minimum kiln furniture and all the other factors concerned with economical firing are applied. In particular gas is suitable for laboratory furnaces and tunnel kilns.

There are three main types of gaseous fuel:

(1) Coal gas produced by the destructive distillation of coal at town gas works and piped to consumers.
(2a) Producer gas, made by the action of air on red-hot coal, anthracite or coke and used hot within the same works.
(2b) Semi-water gas made by passing air and steam through coke.
(2c) Water-gas made by passing steam through white-hot coke.

FIRING AND KILNS

(3) Natural gas, which is usually associated with oil deposits and is piped over considerable distances overland.

Natural and coal gases are, of course, much more convenient to use than producer gas. The gas producer introduces a certain amount of the mess and labour of solid fuel that use of gas for kiln firing is sought to eliminate. But as the over-all cost may be considerably less where a large amount of heat is required, it is well worth considering.

TABLE 210

Fuel Gases (S66)

	H_2	CH_4	Other hydrocarbons	CO	CO_2	N_2
Coal gas	45–50	30–35	4	5–10	—	8
Blue water gas	45	0·5	—	44	4	7
Producer gas	5	2	—	29	2	62
Blast furnace gas	4	1	—	30	10	55
Natural gas	3–30	65–95	traces	—	—	traces

Most descriptions of gaseous fuels omit to mention the sulphur content. This is, however, very important in ceramic firing and must be ascertained before using the gases. Perrins (**P26**) makes a comparison between producer and town gases including average data on sulphur compounds (Table 211). The usually accepted values are 0·04% for the sulphur content of town gas and that of producer gas as 1·5%.

TABLE 211

	Producer gas	Town gas
Calorific value (B.t.u./ft³)	130–150	450–500
(kcal/m³)	1157–1335	4004–4449
Flame temp. (°C)	1600	2300
(°F)	2912	4172
Dust (oz/1000 ft³)	0·02–0·03	Nil
(g/1000 m³)	20–30	
Sulphur content		
(a) with H_2S removed (grains/100 ft³)	15–20	10–30
(g/100 m³)	34–46	23–67
(b) without H_2S removed (grains/100 ft³)	55–70	H_2S must be
(g/100 m³)	126–160	removed
Specific gravity (air = 1)	0·85–0·95	0·42–0·48
Air required for combustion	0·9–1·0	4·25–4·5

Coal Gas or Town Gas

When coal is destructively distilled at a gas works each ton will, on average, give the following products:

Coal gas	83 therms
8·5 cwt saleable coke	119 therms
1·0 cwt saleable breeze	11 therms ⎫
15 gal tar	26·5 therms ⎬ not necessarily used as fuels.
3 gal benzole	4·4 therms ⎭
	243·9 therms

On average, if gas is used to fire a modern tunnel kiln 1 ton of coal must be carbonised to mature the same amount of ware as 1·2 tons of coal fired in a periodic kiln would produce. This furthermore leaves the remaining products of the distillation for other uses. There is therefore a considerable fuel saving, although this may not show directly in the cost of the fuel (*see* Table 212).

TABLE 212

Heat Requirements and Weight of Coal used for Firing Various Ceramic Products in a Coal-fired Periodic Kiln and a Gas-fired Tunnel Kiln Respectively (S164)

Class of ware	Therms per ton of ware		Tons of coal required	
	Solid fuel in periodic kiln	Coal gas in tunnel kiln	Per ton of ware in periodic kiln	For production of gas/ton of ware in continuous kiln*
Biscuit tiles	215	83	0·74	1·00
Glost tiles	436	50	1·50	0·60
Sanitary biscuit	519	163	1·76	1·95
General earthenware biscuit, saggared	262	120	0·90	1·44
Ditto, open placed	—	94	—	1·14
General earthenware, glost	355	78	1·22	0·94
Sanitary fireclay	847	120	2·91	1·44
Electric porcelain	372	115	1·27	1·39

* Each ton of coal gives 83 therms of gas and by-products which if used as fuel amount to a further 160·9 therms.

The first gas-fired kiln to be installed in Stoke-on-Trent was started in 1932. Between 1936 and 1939 the number was rapidly increased to over fifty. During World War II there was a drop, but by 1946 sixty kilns were being fired by gas using some 2200 million cubic feet (62·3 million m^3) of gas. At that time it was estimated that some 20% of the North Staffordshire

FIRING AND KILNS

ceramic and allied industry (excluding brick and roofing tile trades) had changed from coal to gas firing (**S164**).

Natural Gas

Natural gas is a mixture of several paraffin hydrocarbons diluted with nitrogen and oxygen occurring in conjunction with mineral oil deposits. It generally has a high calorific value, United States natural gas ranges from 950 to 1150 B.t.u./ft^3 (8453–10 233 kcal/m^3) and has a specific gravity, compared with air, of 0·57 to 0·65. It generally has low sulphur and moisture contents.

Natural gas differs from manufactured gas by requiring a larger volume of air for its complete combustion, namely about ten times its own volume (**G64**).

HEAT AVAILABLE IN PRODUCTS OF COMBUSTION OF 1 FT^3 OF NATURAL GAS (**A84**)

B.t.u. per Cubic Foot of Gas Burned

Gross B.t.u. value = 1010 B.t.u./ft^3 (8987 kcal/m^3).
Net B.t.u. value = 810 B.t.u./ft^3 (7208 kcal/m^3).
Products of combustion contain 8% CO_2.
Sensible heat in air, preheated to 1500° F, required for combustion of 1 ft^3 of natural gas = 371·9 B.t.u. (3309 kcal/m^3).

Producer Gas

Basically there are three ways of converting coal, coke, anthracite, peat, lignite, wood, or vegetable waste completely to gaseous fuel and incombustible ash.

(1) True producer gas is made by blowing air through hot coal, coke or anthracite, when the carbon is converted to carbon monoxide and the volatile constituents are distilled off:

$$C + O_2 \rightarrow CO_2$$
$$CO_2 + C \rightarrow 2CO$$

The product is always diluted with the nitrogen present in the air used.

(2) Water gas is made by forcing steam through white-hot coke when carbon monoxide and hydrogen result:

$$C + H_2O \rightarrow CO + H_2$$

This reaction absorbs heat so that alternate blasts of air and steam are used, the former causing combustion and heating of the coke and its products being allowed to escape. Water gas is not diluted with nitrogen. It has higher calorific value than producer gas.

(3) Semi-water gas is made by blowing air and steam through coal or coke when the cooling effect of the water–gas reaction counteracts the tendency of overheating in the producer gas reaction. This is the type of gas generally made in the modern gas producer and termed 'producer gas'.

In all three types a certain amount of the calorific value of the solid fuel appears in the actual temperature of the gas (for commercial producer gas this is 5·4%) (**H95**). In order not to lose this, the gas must be used as near the producer as possible, if possible without scrubbing, cleaning, etc. In many instances nevertheless cleaned, cold producer gas proves economical.

The commercial gas producer (general references **A103, W38b, K21**) consists essentially of a vertical shaft with a means of supplying fuel to the top without allowing gas to escape, a gas draw-off near the top, steam-saturated air inlet and ash-removal device at the bottom.

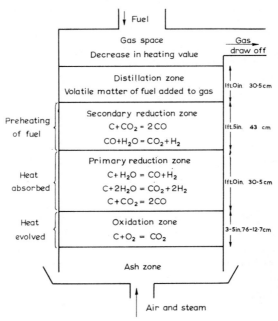

FIG. 11.13. *Reactions in gas producers (after* **A103**)

The main reactions taking place are summarised as follows. The air entering at the bottom becomes preheated as it passes through the ash layer. It then reaches incandescent carbon above this and burns it to form carbon dioxide. A large amount of heat is given out here and is carried up with the gases to initiate further reactions. In the next zone the hot carbon dioxide is reduced to carbon monoxide by the carbon. Towards the top of the fuel layer the hot gases distil the volatile matter out of fresh fuel and are eventually drawn off for use.

The initial combustion zone becomes very hot and tends to melt the ash which then forms clinkers and obstructs the proper working of the producer. Steam is therefore introduced with the air, in proportions decided by the nature of the particular ash, to cool the ash enough to prevent clinkering. Having performed this function the steam goes on to react with carbon and carbon monoxide to give carbon monoxide, hydrogen and carbon dioxide, the first two being fuels (Fig. 11.13).

The nature of the tars produced in the distillation zone is dependent on its depth. If the gas is to be cleaned and used cold a deep distillation zone is needed in order to obtain tars in a fluid state that are easy to remove. With a shallower one the temperature of the distillation is such that the tars are solid at atmospheric temperatures. The nature of the tars is immaterial where the gas is to be used raw and hot.

Three main types of fuel must be catered for separately in gas producer design.

(1) Hard fuels (coke, anthracite, etc.) and feebly caking coals.

(2) Strongly caking coals. The caking of many coals entirely defeats the object of passing the air, etc., over large areas of solid fuel. They can only be gasified by special techniques.

(3) Soft fuels such as vegetable waste, squeezed sugar cane, cotton-seed husks, olive refuse, wood, sawdust, etc.

Individual producers are more highly specialised as regards the size, ash type, volatile content, etc., of particular fuels. They are also divided into hand-operated or 'static' and mechanical types.

Variation of the fuel and the process gives gases of different compositions, as shown in Table 213.

TABLE 213

Various Producer Gases (A103)

Type	Fuel	CO_2	CO	H_2	CH_4	N_2	Calorific value (gross at 30 in. and 60° F sat.)
Hand-operated	Coke (gas works)	6·0	25·0	12·5	0·6	55·9	125
Mechanical	Coke (coke oven)	5·5	28·5	10·5	0·6	55·0	129
Mechanical	Coal	4·0	29·0	15·0	2·7	49·3	167
Mechanical	Anthracite	5·5	25·5	17·0	0·9	51·4	145
Hand-operated	Wood	7·0	28·0	13·0	3·5	48·5	165
Hand-operated	Vegetable refuse			Variable			130–160
Mond system	Coal	16·0	11·0	27·0	3·0	43·0	151

CONVERSION OF HEAT VALUE OF COAL WHEN GASIFIED **(H95)**

Producer gas, at least	88%
Remaining in ash	1%
Sensible heat of gas	5·4%
Excess hot water	1·5%
Hot ashes, radiation and unaccountable losses	4·1%
	100·0%

HEAT AVAILABLE IN PRODUCTS OF COMBUSTION OF 1 FT³ OF PRODUCER GAS (**A84**)

B.t.u. per Cubic Foot of Gas Burned

Gross B.t.u. value = 135 B.t.u./ft³ (1201 kcal/m³).
Net B.t.u. value = 120 B.t.u./ft³ (1068 kcal/m³).
Products of combustion contain 18% CO_2.
Sensible heat in air, preheated to 1500° F (815° C) required for combustion of 1 ft³ of producer gas = 34 B.t.u. (302·5 kcal/m³).
Sensible heat in 1 ft³ of producer gas, measured at 62° F when preheated to 600° F (315° C) = 10 B.t.u. (89 kcal/m³).

THE MECHANICAL GAS PRODUCER

The desirability of eliminating both heavy manual labour and constant skilled attendance is bringing the mechanical gas producer into increasing favour. Its parts will be described in general here. Various combinations of hand and mechanical operation also exist, a good deal depending on the size of producer and the type of fuel.

Having followed the path of the gases from bottom to top in the previous discussion we will now follow the solid fuel from top to bottom.

The fuel should be graded and only that between set limits fed to the producer adjusted to that size. Such grading is given in Table 214 (**W38b**).

TABLE 214

Anthracite	Coke
Grains $\tfrac{3}{16}$–$\tfrac{3}{8}$ in. (4·75–9·5 mm)	Breeze $\tfrac{1}{4}$–$\tfrac{1}{2}$ in. (6·35–12·7 mm)
Peas $\tfrac{3}{8}$–$\tfrac{5}{8}$ in. (9·5–15·9 mm)	Peas $\tfrac{1}{2}$–$\tfrac{3}{4}$ in. (12·7–19·05 mm)
Beans $\tfrac{5}{8}$–$\tfrac{7}{8}$ in. (15·9–22·2 mm)	Beans $\tfrac{3}{4}$–$1\tfrac{1}{4}$ in. (19·05–31·75 mm)

Many producers will not take coke breeze, but suitable adjustment to the fuel charging and other parts make it possible (**P55**).

The coal, etc., is brought by truck, skip or conveyor belt to fill storage hoppers above each producer. From there it is periodically run into an intermediate hopper. During this operation valves between the intermediate hopper and the producer are closed. In between these fillings the valve at the top of the intermediate hopper is closed and the lower one is open so that the fuel is fed by gravity down a chute extending on to the fuel bed in the producer. The fuel therefore does not fall through the gas in the gap at the top of the producer which would pick up any dust, etc., in the fuel. The fuel bed is kept at a constant level and shape.

The producer shaft itself is constructed of firebrick-lined masonry for small installations but large ones have at least the hot zone in a steel water

FIRING AND KILNS

jacket. This latter cools the ash and prevents clinkers sticking to the walls and supplies the steam required for saturating the air blast.

The base of the shaft is set in a water seal through which the ash is withdrawn. Where the diameter is small the ash can be shovelled out by hand without a rotating grate. In general, however, the grate and water seal bowl are rotated and fitted with hooks, bars, ploughs, rings, etc., to grind up any clinker and push the ash out.

CLEANING PRODUCER GAS

Where the tars, etc., are not objectionable the gas leaving the producer is passed through a vortex dust catcher and past baffles to remove the larger dust particles and then through insulated pipes to the burners.

If the dust and tars are to be removed thoroughly the gas will inevitably cool down. It is therefore desirable to use the heat in the gas deliberately, by waste heat boilers or some other heat exchanger.

The usual cleaning method is by water spray which removes the high-boiling tars and larger dust particles, followed by a fine mist of water removing the lighter tars, the latter may be repeated. This method has the disadvantage of producing water–tar mixtures which are not useful by-products and and also a certain amount of the gas heat is wasted.

More recently a process has been introduced by which it is claimed the gas is made 99·9% clean, the by-products are a good quality pitch and water-free light tar oils and 80% of the heat contained in the gas is recovered as clean hot water at about 90° C (194° F). The raw gas passes through a dust catcher to remove the larger particles and then up a tower against a spray of tar oil which dissolves the heaviest tar fractions as pitch. In the following tower the now rather cooler gas meets a counter-flow of light oil which absorbs the light tar fractions. The pitch and tar oil from the bottoms of the towers flow to tanks where they give up their heat to water coils and are then pumped back to the tower or to storage tanks for sale.

The gas passes on for final cleaning to a spray separator and either a direct or an indirect cooler (**A55**).

Another very effective cleaning method is the electrostatic dust precipitator although precautions must be taken and alterations must be made to counteract building up of tarry deposits (**D55**).

APPLICATION OF PRODUCER GAS

Producer gas is generally the cheapest gaseous fuel per B.t.u. that can be obtained from solid fuel, so that it finds application firstly in those industries where a large bulk of relatively cheap ware is produced and an improvement on coal firing is desired without the need for a highly refined fuel and firing technique.

Hendryx (**H80**) points out its suitability for firing refractories, because of its improvement on coal firing as regards temperature and draught variations, ash deposits on ware and localised over-heating. With producer gas, firing can be regular and delayed combustion can be readily maintained even with highly preheated air so that the whole kiln is filled with a soft luminous flame with high radiation characteristics. The theoretical flame temperature of producer gas is about 1703° C (3100° F) as compared with 2036° C (3700° F) for natural gas and 2091° C (3800° F) for oil.

Bituminous coal is usually a cheaper fuel source than coke or anthracite but the gas obtained is more difficult to handle. It comes off with a temperature of 676–787° C (1250–1450° F), whereas that from coke or anthracite is 315° C (600° F), so that it requires large mains to accommodate the necessary volume. These must be lined with insulating material to prevent heat loss both because of the waste of heat and because the tars will condense and clog the pipes. Raw bituminous producer gas is therefore chiefly used for large furnaces, kilns, etc., located immediately adjacent to the producer, although very recent distribution plant designs do make it possible to pipe it some distance.

Raw bituminous producer gas requires special burners (p. 921), whereas coke or anthracite gas and cleaned cold bituminous gas can be burnt in standard ones.

Propane and Butane

These two hydrocarbons are obtained from natural gas, oil and certain coal processes. They are readily liquefied under proper conditions of temperature and pressure and can be transported in the liquid state. Release of pressure, however, evaporates them so that they are burnt as gases.

TABLE 215

	Propane	Butane
B.P., °C (°F)	C_3H_8 −42 (−43·6)	C_4H_{10} 10 (50)
Calorific value (kcal/m³)	22 270–24 240	28 100–30 500
(B.t.u./ft³)	2503–2724	3158–3428
Abundance in natural gas (%)	25	6
Sulphur content	Nil	Nil

These fuels combine the advantage of the easy transport of oil, without its attendant pumps, heating coils, sludge, grease, etc., and the easy firing of gas without any troubles of sulphur contamination. They are generally only used in isolated works or where local town gas supplies are insufficient (**J18, B126**).

GAS BURNERS

In general coal gas, natural gas and clean producer gas can be fired through the same types of burners. The variety of burners is large and although the

FIRING AND KILNS

main principles of design can be divided up into the groups described below the commercial burners do not necessarily fall into any particular group.

(1) *Piped gas, and air by kiln draught.* Gas pipes with drilled caps are placed in each firebox and the air for combustion is pulled in by the natural draught of the kiln. The system is cheap to install but unsatisfactory in its lack of control of the air intake which is therefore likely to be too large, causing waste of heat (**G64**).

(2) *Piped gas under varying degrees of pressure entraining the necessary amount of air from a piped supply at lower or atmospheric pressure.* In such systems the products of combustion can be properly controlled. But as the volume of air required is always several times that of the gas (two to six times for manufactured gas and ten times for natural gas) the resulting mixture has very little pressure and does not move away from the area of the burner fast enough to keep a large kiln at a uniform temperature.

(3) *Air under pressure is used to entrain the correct proportion of gas which is supplied under low or no pressure.* In these burners as the large volume of air is under pressure the reduction of velocity caused by the entrained gas is small and the products of combustion move away from the burner. The burners embody numerous devices such as spirals, many instead of one outlet, etc., to ensure rapid and complete mixing of the gas and air. The flame is then very short, and with the best designs only hot gases enter the kiln so that only the minimum muffles, baffles, etc., are needed to distribute these well. One valve only is required to regulate the gas and air flow so that these are always in the correct proportion to each other (Fig. 11.14).

(4) *Air and gas are mixed in a chamber and then delivered to the burner by a blower.* This ensures the correct proportioning and the immediate combustion. The burners must be designed never to allow the flame to pass back through them and there is always a risk when piping an inflammable mixture.

(5) *Radiant burners.* The flame of the burning gas (however the mixture is made) can be directed by baffles or its own rotation on to a suitable refractory surface which heats up sufficiently to make it radiate heat. Combustion itself is complete within the radiant cavity. This has many advantages over flame burners because no muffles or saggars are required to protect ware from direct flames and the heat can be directed exactly. The burners can be set in patterns up the side of the kiln wall to produce a completely uniform temperature (Fig. 11.15).

BURNERS FOR RAW BITUMINOUS PRODUCER GAS

The tar in raw producer gas clogs up many types of burners and special designs have to be used, although some of the general types are suitable, *e.g.* Basequip K. The burner must be so designed that it can be cleaned during use.

Hendryx (**H80**) in discussing the application of producer gas to firing of

FIG. 11.14. *Gas burner, Eddy ray, for clean gases* (Gibbons, **G39**)

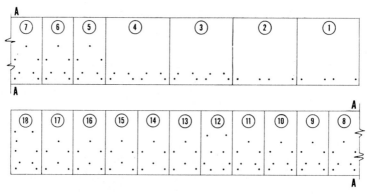

FIG. 11.15. *Scheme of location of radiant gas burners in the preheating and firing zones of a tunnel kiln* (Gradiation, **S53**)

refractories, says that where the fuel requirements are large the gas and highly preheated air are simply admitted into the furnace through ports in its walls.

Where a number of smaller burners are required a cast-iron burner with a fireclay burner block is used. The air, usually preheated, may be used to inject or draw the gas into the burner block and a very short oxidising flame can be obtained by using enough air and moving the air pipe away from the block. A reducing or delayed combustion flame is achieved by using less air and keeping the air pipe close to the throat of the burner. The slide or valve controlling the gas flow must be opened and shut about once every hour to keep it clean and the burner itself cleaned every twenty-four hours.

RADIANT HEAT FROM RECIRCULATING GAS TUBES

A method of using gas to give radiant heat whilst keeping the products of combustion completely separated from the kiln atmosphere is to fire the gas in metal tubes. The maximum heat is extracted from this by having each tube in the form of a loop through which the combustion gases are recirculated.

The system consists firstly of an efficient proportional mixer to give an intimate mixture of gas and air in exactly the right ratio. This mixture is delivered through suitable valves, pressure controls, and backfire arresters to the individual burners. The tube radiating unit fitted to each burner is made of a high-grade refractory alloy. Alloys for use up to kiln temperatures of 1850° F (1008° C) are available. The tube is made without joints within the kiln in a closed loop with parallel entrance and exit. The burner is fixed through the entrance tube so that the gases emerge from it at the edge of the circuit. This draws hot gases round from the discharge end to recirculate. Amounts of combustion gases equal to the input are allowed to discharge. The intimate mixing of gas and air before delivery to the burner is essential if combustion is to be complete in an otherwise inert atmosphere of combustion products. The high velocity of the gas jet entraining hot combustion gases to give rapid circulation ensures even distribution of the heat throughout the tube without any local heating which is detrimental to the tubes and causes irregular heat output.

The heat given out by these radiation tubes is similar to that produced by electric heating elements, consisting solely of radiation from their hot surfaces augmented by the natural convection of the kiln atmosphere. This heating method has been found very suitable for decorating kilns (**D58**).

GAS FOR HIGH-TEMPERATURE KILNS

Heating kilns to above 1600° C (2912° F) presents a number of problems not met with at lower temperatures, amongst these is that of achieving a sufficiently high flame temperature. Gas, especially if the combustion air is preheated, fulfils this need very well. Schüffler (**S36**) has tabulated data for using gas to heat a kiln to 1800° C (3272° F).

TABLE 216

Theoretical and Practical Flame Temperature, Flue Gas Losses and Gas Consumption for a Kiln at 1800° C (3272° F) using Coke Oven Gas with a Calorific Value of 4006 kcal/m^3 with and without Preheating of the Gas and Air (S36)

Air temp.	Gas temp.	Flame temperature			Losses in combustion gas removed at 1800° C			Gas consumption at 1800° C for heating kiln and ware			
		Theor.	Prac.	$\eta t\%$	kcal/m^3	B.t.u./ft^3	η/m^3	m^3 gas/ 1000 kcal	ft^3/1000 B.t.u.	m^3 gas/860 kcal = 1kWh	ft^3/kWh
20° C	20° C	2110° C	1800° C	85	3340	375	83	1·45	12·9	1·25	44
68° F	68° F	3830° F	3272° F								
400° C	20° C	2300° C	1930° C	84	3340	375	74	0·85	7·5	0·73	26
752° F	68° F	4172° F	3506° F								
400° C	400° C	2340° C	1960° C	84	3340	375	71	0·74	6·6	0·63	22
752° F	752° F	4244° F	3560° F								
800° C	20° C	2520° C	2120° C	84	3340	375	66	0·58	5·1	0·50	18
1472° F	68° F	4568° F	3848° F								

It can be seen that preheating of the gas brings less reward than preheating air as the volume of air used is much larger, also gas may not be heated above 400° C (752° F) because of possible decomposition whereas air can be heated to any temperature, 950° C (1742° F) being the practical limit. Preheating of the gas has a marked effect on the gas consumption for maintaining a steady temperature of 1800° C (3272° F).

Schüffler (**S36**) goes on to describe individual periodic kilns for use at 1800° C (3272° F) using surface combustion or multiple Eddy ray burners, and a tunnel kiln for sparking plugs in single file.

Oil

Industrial kilns, furnaces, boilers, etc., may be economically fired with 'heavy fuel oils'. These are the oils that together with bitumen make up the residue after crude oil has been distilled to give off petrol, paraffin oil, gas and diesel oils. Their characteristics may vary considerably according to the crude oil from which they are produced and the extent to which distillation has been taken (**B2**). Unfortunately, fuel oils nearly all contain sulphur, up to 3%, which can be detrimental to their use in ceramic kilns.

Benda (**B33**) gives the following characteristics for the three main grades of fuel oil (Table 217).

TABLE 217

Fuel oils	Heavy	Medium	Light
Specific gravity at 15° C (59° F)	1·10–0·94	0·93–0·91	0·90–0·81
Flash point (°C)	~200	~150	<120
(°F)	~390	~300	<248
Viscosity, Engler degrees			
at 20° C (68° F)	—	—	8–10
50° C (122° F)	50–100	8–10	4–5
100° C (212° F)	3–7	1·8–2·0	—
Solidifying point (°C)	−10 to +30	~−10	below −10
(°F)	+14 to 86	~+14	below +14
Calorific value kcal/kg	10 000–9000	10 000–9000	10 000
B.t.u./lb	18 000–16 200	18 000–16 200	18 000

Bailey, Battershill and Bressey (**B2**) give descriptions of the characteristics and the applications of heavy fuel oils for high-temperature processes. Some of their figures have been assembled in Table 218, together with their comments.

(1) *Specific Gravity.* No direct indication of the quality of a fuel oil can be ascertained from the specific gravity. The relationship between it and viscosity is only linear for oils derived from the same crude oil.

(2) *Flash Point.* This applies to oil in the bulk and is always well above the regulation minimum of 65·5° C (150° F). When properly stored and handled, therefore, there is no greater fire hazard than with other fuel.

(3) *Viscosity.* Of great importance for the actual use of the oil is its viscosity. By suitable preheating a more viscous oil can be made to flow as well as a less viscous one.

(4) *Calorific Value.* As oil is bought by volume the calorific value per gallon (or per litre) is of more interest than the weight one, except for comparative purposes.

(5) *Storage.* Fuel oils must be stored in properly designed tanks combined with proper handling equipment, pumps, valves, etc. The tanks must be equipped with steam coils or electric immersion heaters to keep the oil at a temperature at which it will be easy to handle. The oil should be delivered to the burners by a ring main designed to have branch lines of minimum length.

TABLE 218

Characteristics of Three Typical Grades of Heavy Fuel Oil

(after (**B2**))

	Pool fuel oil	Pool heavy fuel oil	A 1500 sec fuel oil
(1) Specific gravity at 60° F / 16° C	about 0·935	about 0·95	about 0·96
(2) Flash point (closed) (°F) / (°C)	150 minimum / 65·5	150 minimum / 65·5	150 minimum / 65·5
(3) *Viscosity:* Redwood I at 100° F (38° C)	220 max.	950 max.	1500
Kinematic viscosity (stokes)	0·54	2·35	3·71
Saybolt Universal (sec)	249	1084	1712
Engler degrees	7·1	31·0	49·0
(4) Gross calorific value:			
(B.t.u./lb)	about 18 900	about 18 750	about 18 700
(kcal/kg)	about 10 489	about 10 406	about 10 378
(Therms per gal at 60° F)	about 1·77	about 1·78	about 1·80
(kcal/l. at 16° C)	about 9811	about 9866	about 9977
(5) Oil storage temperatures (°F)	45	70	80
(°C)	7	21	27
(6) Oil temperature at burners:			
(°F)	110–140	170–200	190–230
(°C)	43–60	77–93	88–110

(6) *Oil Burners.* For high temperature processes all oil burners work on the principle of forming a fine mist or spray of oil particles, each of which can come into contact with the combustion air. There are numerous makes of burner achieving this which fall into one of the following categories:

(i) Pressure jet.
(ii) Blast atomising type: (*a*) air jet, including low, medium and high pressure air; (*b*) steam jet.
(iii) Rotary cup burners (**A30**).

(i) *Pressure jet* burners are used largely for boiler firing rather than industrial furnaces. The oil is pumped at high pressure through a specially designed orifice, coming out as a fine spray in the form of a cone (**B2**). As the atomising depends on oil pressure the rate of burning can only be varied over a small range (**A30**).

(iia) In *air jet* burners the atomisation is brought about by a stream of relatively high velocity air which strikes the oil. The pressure of air may vary between wide limits, so that three main group types are recognised:

(1) Low-pressure air 0.5 lb/in^2 to 1.5 lb/in^2 (0.035–0.105 kg/cm^2) (at 1 lb/in^2 (0.07 kg/cm^2) the atomisation uses about 20% of the combustion air).

(2) Medium-pressure air 1.5 lb/in^2 to 10 lb/in^2 (0.105–0.703 kg/cm^2).

(3) High-pressure air 10 lb/in^2 to 150 lb/in^2 (0.703–10.55 kg/cm^2) (at 50 lb/in^2 (3.52 kg/cm^2) the atomisation requires about 2.5% of the combustion air).

Air passing through the burner for the jet should not be preheated above 500° F (260° C). As it is desirable to preheat as much of the combustion air as possible by recuperation and regeneration, the high-pressure burners, where 97–98% of it can be heated, find most application in high-temperature work.

(iib) The *steam jet* burner works on the same principle as the air jet but produces slightly lower flame temperature and so is only used for furnace work when cheap steam is available.

(iii) *Rotary cup burners* atomise by the action of a high-speed rotating metal cup, the oil being thrown from its edge by centrifugal force into a stream of low pressure air entering around the cup (**A30**).

The quantity of air supplied to the burners should barely exceed that required for complete combustion. That is for 1 lb (0.45 kg) of oil burnt about 180 ft^3 (51 m^3) of air at 60° F (16° C) at 30 in. of mercury (1.03 kg/cm^2) are required. The volume of wet combustion products under the same conditions is then 192 ft^3 (5.44 m^3) and the volume percentage of CO_2 in the dry products of combustion is ideally 16.0. Such combustion can only be achieved if the atomising is really good, the use of properly designed burners cannot be overemphasised.

Modern burners not only bring about complete atomisation but they are also fitted with a number of safety devices. For example the air and oil flows can be controlled by a single valve so that they are kept at their optimum proportions. Turning down the air without doing the same to the oil causes the emission of thick black smoke. Photoelectric cells connected to switching and valve controls will stop the oil flow should it fail to ignite properly at the burner, etc.

The advantages of using liquid fuel are summarised by the authors quoted previously (**B2**):

(i) Uniform quality of the fuel. Modern fuel oils vary very little from day to day or even year to year and adjustments of firing procedure need only rarely be made.
(ii) Cleanliness. The fuel is handled throughout in a closed-pipe system producing no dust or ash.
(iii) Ease of storage. A million B.t.u. are produced from 0·9 ft^3 (0·025 m^3) of fuel oil, 1·5 ft^3 (0·042 m^3) of coal and 3·0 ft^3 (0·085 m^3) of coke.
(iv) Labour saving. Conveyance, stoking, etc., are all fully automatic.
(v) Controllability and flexibility, alterations of burning rate being achieved with the turning of valves.
(vi) Facility of handling.
(vii) Absence of ash.
(viii) High flame temperatures (*see* Table 219).
(ix) Luminous flames and therefore high radiating power.
(x) High furnace efficiencies.
(xi) Greater outputs from furnaces because of fast heat transfer.

TABLE 219

Flame Temperatures (B2)

Fuel	Flame temperatures	
	(°C)	(°F)
Fuel oil	2080	3780
Coal gas	2045	3710
Producer gas	1600	2910
Blast furnace gas	1460	2660

To take full advantage of these special features of fuel oil Benda (**B33**) points out some of the precautions that must be taken when installing an oil-burning installation. The high flame temperature of fuel oil can lead to uneven firing unless a number of small burners are carefully distributed. The high flame temperature combined with the unfortunately high sulphur content can lead to more damage of ware than the same sulphur content in other fuels because of the much greater speed of reaction. Very careful control of oil and air pressure leading to a small excess of air in the finishing stages of the full fire can eliminate any damage due to sulphur.

It had been thought by some users of oil burners that the use of oil was the direct cause of rapid disintegration of the kiln lining. This has now been shown to be due to faulty atomisation of the oil which is not completely burnt and therefore condenses on to the lining. It is subsequently charred and leads to disintegration. With properly adjusted burners where complete combustion of the oil occurs, the lifetime of the kiln structure is comparable with that of kilns using other fuels.

GAS–OIL BURNERS

Convertible burners taking oil or gas are finding considerable application where exact control is desired and continuity is essential. The fuel used can be changed at will to produce flames of different type and temperature

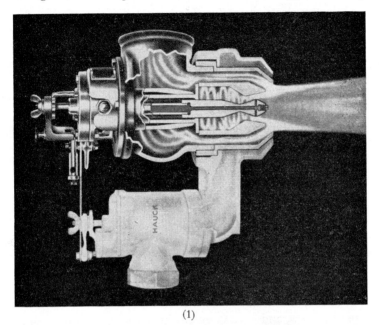

(1)

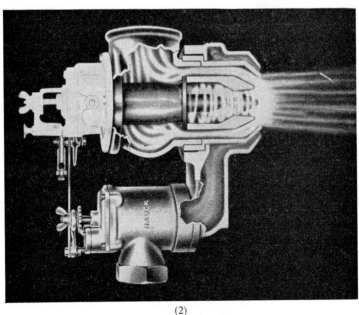

(2)

Fig. 11.16. *Gas–oil burner shown* (1) *burning oil;* (2) *burning gas.* (*Hauck*, **H41**)

and the dual system prevents shortage of either fuel from interfering with production (**L48**). (Fig. 11.16.)

Electricity

Heating kilns by electricity differs fundamentally from all the other fuels. Whereas they depend on chemical energy generated by combustion, electrical heating arises from the physical property of resistance. Its limitations are chiefly the temperature which different types of resistors will withstand and their period of usefulness.

NICHROME AND KANTHAL

The easiest type of resistor to handle and operate is the metallic one. Two alloys are in common use for ceramic kilns. For temperatures up to about 1050° C (1922° F) a nickel–chromium alloy is used, *e.g.* nichrome which is 80% nickel and 20% chromium. A cheaper alloy is 65% nickel, 15% chromium, 20% iron and is useful to up 900° C (1652° F). For higher temperatures alloys of iron, chromium, aluminium and cobalt can be used, *e.g.* kanthal, 37·5% chromium, 7·5% aluminium and 55% iron, or 2–3% cobalt, 4–6% aluminium, 22–25% chromium and the rest iron. The second type has about three times the life of nichrome at the same temperature or can be used at temperatures up to 250° C (450° F) greater, but it is brittle and must be very carefully supported. The average life for nichrome elements is 5–10 years at 800° C (1472° F); 2–4 years at 1050° C (1922° F) if the glaze is leadless, and 9–12 months with a lead glaze at this temperature; kanthal has an average life of 2 years at 1250° C (2282° F) (**H95**).

The metal alloy elements are usually wound round refractory fireclay formers which act as supports. In other instances the element is made self supporting by using heavy gauge nichrome tape. Lengths of about 40 ft (12 m) of this are bent into a regular zigzag and connected to nichrome leads of larger cross-section to prevent heat losses. Such an element is capable of carrying 3kW, but its life is prolonged if the maximum loading is not applied (**A98**). However they are supported it is important that metal elements are free to expand and contract and also to radiate their heat well to prevent overheating.

Nickel–chromium alloy may also be cast into a grid which is then suspended in the kiln. By using this spread-out heat source it need not be heated very much higher than the desired temperature in the kiln (**A78**).

Sulphur compounds attack nickel–chromium but not kanthal. Reducing conditions should be avoided with kanthal (**A146**).

SILICON CARBIDE

For higher temperatures electrical resistance heaters are made of silicon carbide (carborundum) rods, tubes or plates. In general the highest temperature for continuous use is 1400° C (2552° F) but some specially prepared resistors can be used up to 1575° C (2867° F). Silicon carbide is more difficult to use than metal resistors. One snag is the connection to the incoming leads which cannot be made direct to the silicon carbide as this is too hot. The ends of the rods therefore have free silicon incorporated in them, this conducts better than its carbide and hence becomes less hot. Where the carbide has been prepared in unusually strong form, under the trade name 'crusilite', metal terminal caps can be shrunk on to the 'metallised' ends and leads can be connected without difficulty (**A91**).

Silicon carbide is gradually attacked by oxygen during use which converts it to silica and carbon dioxide, and leads to two undesirable consequences. Firstly, the resistance increases and a higher voltage must be applied to give the same heat output. Secondly, in changing silicon carbide to cristobalite there is a volume increase calculated as 105%. This causes considerable mechanical weakness and eventual failure. The volume effect is aggravated if an element containing cristobalite is allowed to cool below 600° C (1112° F) when the high–low inversion will take place.

The life of an ordinary silicon carbide element used in a tunnel kiln at 1300–1350° C (2372–2462° F) is from 1500–2800 hr, *i.e.* from 2 to 4 months (**G38**). The new crusilite made in a single piece with no joints to fail, and by a special process which retards oxidation is said to last considerably longer. This material can be made over a range of resistances, and sets of elements do not vary by more than 5% from the mean (**M96, A91**).

The 'Silit three-phase elements' have three silicon carbide rods connected by a bridge. These require only one entry to the kiln. They take up the power supplied very evenly and so have a good lifetime even at 1500° C (2732° F) (**K34**).

MOLYBDENUM SILICIDE $MoSi_2$

A new element (1958) based on molybdenum silicide will reach higher temperatures. Small elements especially for laboratory kilns can be used up to 1650° C (3002° F). For larger kilns the useful range is 1350 to 1500° C (2372–2732° F) (**K34**).

Molybdenum silicide is also an important constituent of the new *kanthal super* cermet elements, which also contain silica and other metallic and ceramic materials. They can be operated up to 1600° C (2912° F). The elements are shaped as hairpin rods with straight heating sections, allowing high surface loads of 10–20 W/cm^2. Electrical resistance rises rapidly as temperature increases. Power consumption falls with rising temperature. It is necessary to use a voltage regulator in order to start at a low voltage when resistance is low, and then increase it.

Oxidising conditions are recommended, although reducing ones can be tolerated if they are followed by oxidation. Sulphur and its compounds, and any substance forming fusible silicates must not be allowed contact (**A146**).

CARBORUNDUM WITH GRANULAR CARBON

Another non-metallic element consists of a carborundum trough and airtight lid filled with granular carbon. Electrodes lead into the ends of the trough. The whole is elevated on refractory piers away from the kiln wall to allow good air circulation (**E11**).

STABILISED ZIRCONIA

This material has the composition:

ZrO_2	94·10%
CaO	4·96
SiO_2	0·20
Fe_2O_3	0·52
TiO_2	0·22
	100·00

and has the property of a negative resistance characteristic (Table 220).

TABLE 220

Temperature		Specific resistivity (Ω/cm)
(°C)	(°F)	
700	1292	2300·0
1200	2192	77·0
1300	2372	9·4
1700	3092	1·6
2000	3632	0·59
2200	3992	0·37

It is used in two types of furnace. In the ultra-high-frequency furnace an insulated stabilised zirconia muffle is surrounded by an induction coil supplied with ultra-high frequencies up to 10 Mc/s. The muffle is heated initially by gas until it becomes a conductor and is further heated by the induction coil. In resistance furnaces the stabilised zirconia tube is wound with a coil of pure molybdenum. The two are insulated with zirconia grain and connected in parallel. When current is switched on it flows through the molybdenum which heats the zirconia until it takes an increasing amount of the current itself. Eventually the molybdenum coil carries only 9% of the current. Such furnaces can operate continuously at 2000° C (3632° F) (**G38**).

MOLTEN TIN ELECTRODE

Tin melts at 231° C (449·4° F) but does not boil until 2300° C (4175° F). It has a remarkably long liquid range. Its resistivity at 21° C (70° F) is 11·5 $\mu\Omega$/cm³. Molten tin is used as an electric heating element by enclosing it in a narrow quartz tube. The ends of the tube are wider so that less heating occurs and a cool join can be made to the tungsten electrodes dipping in the tin and eventually joined to copper leads.

Such an element has been in use for nearly a year at 1500° C (2732° F) without the slightest change in pressure drop across it (**G38**).

Protective Atmospheres

In traditional ceramic firing the atmosphere surrounding the ware is a mixture of air with combustion gases. This can be regulated to have: (i) more air than is required for combustion, when it is oxidising; (ii) the right amount of air—neutral; or (iii) too little air–reducing. Generally conditions will vary during the course of the firing.

Certain new bodies require firing in a controlled non-air atmosphere. These must be fired in muffle kilns or electric kilns supplied with the right gas from a separate 'gas generator'.

Meid (**M46**) describes the production of some of the gases that find application in special heat treatments:

(1) *Hydrogen*, the greatest reducing power of any of the protective molecular gases, is bought in cylinders containing about 200 ft^3 (5·6 m^3) of gas.

(2) *Dissociated ammonia* is used as a low-cost substitute for hydrogen, the mixture obtained being 75% hydrogen and 25% nitrogen. Raw, dry ammonia is passed over a heated catalyst, which in a well-designed dissociator, will dissociate it with 99·95% effectiveness.

(3) *Exothermic atmosphere*, of neutral or reducing nature is produced by the exothermic cracking of readily available hydrocarbon gases such as coke-oven gas, natural gas, propane, butane. The gas and air are accurately metered in the proper proportions, then mixed and pumped to a combustion chamber where the gas is partially burned to raise the temperature to ($\sim 1095°$ C) 2000° F. The catalyst present then enables the products of combustion to react with the unburned hydrocarbon.

The most reducing atmosphere that can be obtained is:

Hydrogen, maximum	18%.
Carbon monoxide, maximum	11%.
Carbon dioxide, minimum	40%.
Nitrogen	balance.

A neutral atmosphere can be obtained consisting of nitrogen, carbon dioxide and water vapour. An oxidising atmosphere of oxygen content controllable to 0·1% can also be obtained.

(4) *Endothermic atmosphere*, of reducing power intermediate between dissociated ammonia and the exothermic atmosphere. Here gas and air are mixed and compressed and then passed over an externally heated catalyst. Little or no combustion occurs so that carbon dioxide and water vapour are

not present in the product. The result is about 60% maximum reducing constituents as against the 29% in the exothermic atmosphere.

Hydrogen	~40%.
Carbon monoxide	~20%.
Carbon dioxide	0–3%.
Methane	0–1%.
Nitrogen	balance.

(5) *Nitrogen* is produced cheaply from gas and air by passing the mixture first through an exothermic unit, where the gas is burned, and then through an absorption unit to remove carbon dioxide, and a drying unit.

INSTRUMENTS FOR OBSERVING, RECORDING AND CONTROLLING KILN AND OTHER WORKS CONDITIONS

The most important property of the kiln interior that must be constantly under observation is the temperature. This may be variously observed, or continuously recorded, furthermore it can now be automatically controlled. Other measurements are made on the draught, moisture and CO_2 content of the flue gases, air and gas flow to burners, etc.

Temperature and Heat-work Measurement

Most of the changes occurring in ceramic ware during firing are slow ones and they are frequently not allowed to go to completion, *e.g.* shrinkage, vitrification, fusion, sintering, chemical interaction, some inversions, etc. These changes are as much dependent on time as on temperature, they take a different course according to the rate of temperature increase and the time held at any one temperature. Few changes are dependent only on temperature and quickly reach equilibrium, *e.g.* melting of eutectics, α–β quartz inversion at 573° C (1063° F), melting of silica at 1710° C (3110° F), melting of some decoration colours (**D30**). Mere measurement of temperature will therefore not give a true indication of the amount of 'heat-work' done on the ware and of its degree of maturity. It is often more helpful to follow the progress within the kiln with some method that responds to time and temperature in a similar way to the ware. Such heat-work recorders or thermoscopes were, in fact, the only devices available to potters until the development of electrical methods of temperature measurements which can now be used in conjunction with them. The advantages of electrical instruments are that they give a precise reading, can be made continuously recording and can be made to operate automatic control gear.

Thermoscopes or Heat-work Recorders

Heat-work recorders have been developed from two different thermal changes of ceramic bodies: (*a*) shrinkage, (*b*) fusion with deformation; the latter is the more widely adopted.

It is often stated that a ring contraction number or cone number is equivalent to a certain temperature. The very nature of the rings and cones makes such statements unwise unless the heating rate is stipulated. Properly made standard heat-work recorders will repeatedly give the same reading at the same temperature after the same rate of heating but if the heating is more rapid this same reading will be given at a higher temperature and if it is slower at a lower one. In irregular firing these recorders will give the best possible idea of the condition of the ware.

The Table III (p. 1304) in the Appendix gives the standardised temperature equivalents of thermoscopes.

PYROMETRIC CONES

Mixtures of kaolin, feldspar, quartz, marble, etc., of various softening points are made into thin three-sided pyramids. These are set at a slight (standard) angle to the vertical in a refractory base which is placed amongst

Fig. 11.17. *A typical set of fired pyrometric cones showing the 'signal' cones H1 and H2, the 'critical' cone H3 with tip just touching the base, and the 'witness' cone H4 remaining upright* (Harrison, **H33**)

the ware where it can be observed through a spy hole. With increasing temperature the mixture softens and the tip of the cone bends down. A cone is said to be 'down' when the tip is level with the base. The normal practice in using cones is to set three or four different ones spanning the desired finishing temperature on each refractory holder, warning is then given of the approaching temperature as the lowest goes down and a check on overfiring is obtained if the highest remains standing (Fig. 11.17).

The first standard and regularly spaced series of pyrometric cones was introduced by Seger in 1886 (**S52c**). It was a great success and his numbering became so well established that extensions and additions to the range have been given new numbers in order not to disturb the well-known numbers. The result is an apparently very curious series of numbers. The original series of Seger cones were numbered 1 to 36. It was subsequently extended to lower temperatures twice 01 to 09 and 010 to 022, and to higher temperatures 37 to 42. The compositions of cones 022 to 6 inclusive were altered to eliminate lead and iron oxides and these new cones designated with an 'a'. In Great Britain production was started during the 1914–1918 war but the same numbers and temperature equivalents adhered to. A new series are now also available designated with an 'H' which includes intermediate cones making an almost constant interval of 10° C between cones. The old numbers are adhered to and the insertions have an 'A' added (**H33**). The French cones are like the Seger cones (**R12**).

The United States Orton cones (**O13**) although working on the same principle, have a different temperature scale as seen in the Table III on p. 1304.

Of the Staffordshire Seger cones the following cones are recommended for the different uses:

Glass colours	022 to 016
Enamel kiln (soft)	019 to 015
Enamel kiln (ordinary)	014 to 011
Enamel kiln (hard)	012 to 07
Red brick kiln	010 to 01
Tile glost and majolica kilns	07 to 01
Glost oven (soft glazes, Rockingham)	05 to 1
Glost oven (earthenware and bone china)	03 to 5
Earthenware biscuit	3 to 8
Vitreous tile and granite biscuit	4 to 9
China biscuit and sanitary	7 to 12
Hard porcelain	13 to 17
Fireclay and silica bricks	7 to 15
Special refractories	17 and over
Refractoriness tests	26 and over (**H33**)

Seger cones do not react normally in very sharp firing or reducing and sulphurous atmospheres. In the latter case a refractory crust of carbon and sulphates is built up which prevents the cone bending even when the unaffected material inside has flowed out of the base. Irregular behaviour of cones, e.g. bloating, cracking, crusting, bending in wrong order or in abnormal way can, however, give an idea of what has gone wrong with the firing and help with the post mortem more than a pure temperature chart can.

Shipley (**S68**) describes an experimental series of cones with a regular temperature interval between members of 10° C. They have been compounded in such a way that they will always fall in the correct order even if abnormal heating rates are applied.

Voldsevich, Gerasimova and Lyutsareva (**V22**) describe Russian work on producing satisfactory cones for reducing conditions. They found that if the standard cones were made up from pure synthetic oxide components instead of natural minerals the blackening, bloating and losing of shape in reducing conditions is eliminated. Such cones have squatting temperatures 40–50° lower in nitrogen–hydrogen and hydrogen conditions than in oxidising ones. Alternatively a new series was tried using the synthetic glass, CaO 23·3%, Al_2O_3 14·7%, SiO_2 62·0% which softens at 1170° C (2138° F) and is fritted at 1300° C (2372° F) in varying proportions with calcined alumina.

HOLDCROFT'S BARS

Like the Seger cones Holdcroft's bars are mixtures with differing softening points. They are dust-pressed into bars of rectangular cross-section and supported at both ends in a special refractory holder. The progress of the firing is followed by the amount successive bars have sagged. Like cones they are used in sets of four, two for preliminary warning, one for confirmation of the maturing point, and one as a check for possible over-firing. They must not be directly in the way of the flame or of a cold draught (Fig. 11.18).

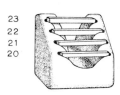

Before using

After using

FIG. 11.18. *Holdcroft's bars before and after use* (*Harrison*, **H33**)

The approximate bending temperatures are given in Table III, p. 1304. The manufacturers' recommendations for application are as follows:

Enamel kilns	4– 8
Hard kilns	8–12
Majolica glaze kilns	15–20
Earthenware and china glost	20–24
Earthenware biscuit	24–28
China biscuit	28–32
Red brick and roofing tiles	20–24
Vitreous floor tile biscuit	25–29
Blue bricks	24–28
Firebricks	25–30
Hard paste porcelain	30–40 (**H108**)

WATKINS RECORDERS

These are a series of numbered pellets which fuse at different temperatures. Sets of five of them are placed by the makers in ¼ in. (6·3 mm) cavities in small refractory bars $2 \times \frac{1}{2} \times \frac{1}{2}$ in. (5·08 × 1·3 × 1·3 cm) and held in place with paper strips. They can be placed straight into the kiln without further preparation as the paper burns off during the firing. After the firing the bar is inverted and the plugs which have not fused fall out. The heat-work done is assessed from the number of the last plug to fuse (see Table III, p. 1304).

The progress of the firing cannot be watched through a spyhole if Watkins recorders are used but they are very compact for placing inside Saggars where the inspection can only be made after the firing in any case (**W22**).

BULLERS RINGS

Measurement of heat-work on this system depends on the progressive shrinkage occurring during firing. A set of identical rings of ceramic material is placed in the kiln near each trial hole and one is withdrawn with a hooked rod at regular intervals. It is rapidly cooled and then measured on a gauge.

In order to give reliable results a most careful check must be kept on the composition and grain size of the raw materials and on every stage of manufacture of the rings. Their body composition is of the porcelain type and they are dust pressed. It has been found useful to make four grades of ring to suit different temperature ranges and these are coloured with vegetable dyes for ready identification.

The No. 27 or original standard rings are coloured green. They are recommended for use within the range 960° C (1760° F) to 1275° C (2327° F) and are mainly used for the firing of earthenware and tiles (biscuit and glost) building, engineering and fire bricks.

The No. 72 rings are coloured natural and are intended for higher temperatures up to 1310° C (2390° F). Their additional uses over the No. 27 include electrical porcelain, china biscuit, grinding wheels, firebricks and refractories.

No. 55 rings are coloured brown and are intended for the range 960° C (1760° F) to 1200° C (2192° F). They have a rapid shrinkage between 1000° C (1832° F) and 1100° C (2012° F). They are very useful for firing glost ware and building bricks of low finishing temperature.

No. 26 pink rings are an extension of the high temperature range to 1400° C (2552° F) especially for slow firing as used for high-grade firebricks and heavy refractories.

Below 960° C (1760° F) when the combined water is being driven off from both ware and rings the latter expand a little and the scale of the gauge makes allowance for this. The rings are therefore useful in following the critical water-smoking period. If the firing is oxidising up to 850° C (1562° F) a reducing atmosphere thereafter does not affect the readings given.

The brass gauge for measuring the shrinkage of the rings is a simple device consisting of a plate with two studs to rest the ring against and a pivoted pointer which rests on the free side of the ring and moves over a scale marked on the plate. The markings are 0–5 to the left for expansion and 0–60 on the right for contraction. It is a good idea to keep a new gauge in a secure place for periodic checks on those in use at the kiln.

The rings should be used in two ways. Firstly as a check on progress by placing about six at each trial hole and withdrawing them at regular intervals. Secondly by placing them singly in various parts of the kiln inaccessible from the trial holes they can assist in a post mortem on the finished kiln (Fig. 11.19) (**B139c**).

The Measurement of Temperature

The nature of the property 'temperature' makes it impossible to measure it directly. Instead some of the energy causing the temperature of a body must be transferred to an instrument which responds to it in a known and repetitive way. Different types of response are suitable for the different temperature ranges as shown in Table 221.

(1) LIQUID-IN-GLASS THERMOMETERS

The well-known mercury-in-glass thermometer depending on the large thermal expansion of mercury relative to that of glass finds widespread use. Its limitations are those of the temperature maximum of 500° C (932° F), fragility and difficulty of reading it in dark or inaccessible positions. However, with the help of protective guards and a right-angled bend between the bulb and the scale part of the stem, widespread use for the following purposes does occur:

(*a*) Room temperature.
(*b*) Drying chamber temperatures.
(*c*) Maximum and minimum registration.
(*d*) Flue gas temperatures.
(*e*) Steam pipe temperatures.
(*f*) Humidity measurements by the wet and dry bulb method.

These thermometers cannot be made to record the temperature they indicate.

(2) MERCURY-IN-STEEL THERMOMETERS

These have a slightly longer range than the mercury-in-glass type, namely up to 600° C (1112° F) and have the advantages of robustness and of transmitting readings up to 120 ft (36·6 m) where they can be automatically and continuously recorded. The bulb is connected by a long capillary tube to a Bourdon pressure gauge which operates a pointer. The pointer can be equipped with a pen under which a chart is rotated. By suitable connections two such thermometers can be made to record on the same scale.

(a)

Fig. 11.19. *Bullers rings*

above: (a) *rings in position both for drawing through the wicket during the firing (upright) and for checking after firing (flat);*

top right: (b) *drawing a ring from an intermittent brick kiln ((a) and (b) courtesy Berryhill Brickworks);*

bottom right: (c) *measuring a ring* (**B139c**)

(b)

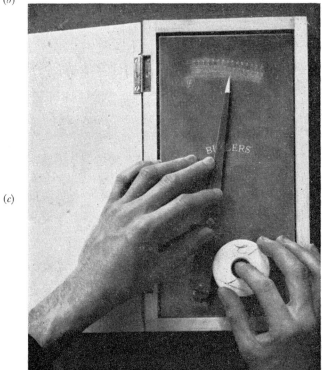

(c)

TABLE 221

Main Types of Temperature Measuring Instruments (C5)

	Type	Description	Range and remarks	
(1)	Liquid expansion	Toluol or pentane in glass	$-200°$ C to $+30°$ C	Indicating
		Mercury-in-glass	$-30°$ C to $+500°$ C	
(2)	Liquid expansion	Mercury-in-steel	$-50°$ C to $600°$ C. Readings can be transmitted up to distances of 120 ft	Indicating and recording
(3)	Vapour pressure	Ether or other organic liquid	$-50°$ C to $400°$ C. Readings can be transmitted up to distances of 200 ft	Indicating and recording
(4)	Vapour pressure	Mercury-in-steel	$350°$ C to $800°$ C. Readings can be transmitted up to distances of 120 ft	Indicating and recording
(5)	Electric resistance (change in resistance of a platinum wire with temperature)	Mica frame (in which the wire is wound on mica and has a resistance of $2·5\Omega$ or 25Ω at $0°$ C)	$-200°$ C to $+1000°$ C. Suitable for accurate laboratory work	Subdivided into two groups of direct deflection or null measurement, can be indicating or recording
		Steatite spool (in which the wire is wound on steatite and then glazed and has a resistance of 100Ω at $0°$ C)	$-100°$ C to $500°$ C. Particularly suitable for the direct reading of air temperatures in buildings, power plants, ships, etc.	Subdivided into two groups of direct deflection or null measurement, can be indicating or recording
		Copper tube (in which the wire covered with silk is drawn through a thin walled copper tube)	$-100°$ C to $140°$ C for the quick measurement of temperature of liquids or for obtaining the mean temperatures of large enclosures	Subdivided into two groups of direct deflection or null measurement, can be indicating or recording
(6)	Thermo-electric	Iridium, iridium–rhodium	$-200°$ C to $1927°$ C (**W77**)	Subdivided into groups of direct deflection or null measurement, either indicating or recording
		Platinum, platinum–rhodium	$-200°$ C to $1800°$ C (**W77**)	
		Base-metal titan wires	$-200°$ C to $1200°$ C	
		Base-metal iron constantan	$-200°$ C to $800°$ C	
		Base-metal copper constantan	$-200°$ C to $500°$ C	
(7)	Radiation Total	Féry pyrometer	$600°$ C upwards. Particularly useful for recording temperatures above $1000°$ C	Indicating or recording
(8)	Optical	Wanner type (polarized light)	$700°$ C upwards without limit	Indicating
		Disappearing filament	$700°$ C–$4000°$ C	Indicating
		Photo cell		Indicating or recording

FIRING AND KILNS

The fact that the expanding liquid in these thermometers is mercury makes it possible to make them operate switches in electrical circuits on reaching given temperatures, *e.g.* alarm bells, valves of gas, air, steam, oil or water. Control mechanism can therefore be automatically operated.

Applications are in the preparation, mixing, shaping and especially drying of ceramic bodies.

(3) and (4) VAPOUR PRESSURE THERMOMETERS

These are similar in design and operation to the liquid mercury-in-steel ones.

(5) ELECTRICAL RESISTANCE THERMOMETERS

These thermometers depend on the regular change of resistance with temperature of pure metals. The metal chosen is pure platinum. The change of resistance is usually measured on a Wheatstone Bridge circuit either by the galvanometer deflection or by the change of resistance in the opposite arm of the bridge necessary to give a null reading on the galvanometer. In certain circumstances the resistance change may be measured potentiometrically.

The number of precautions required regarding temperature changes of the leads to the resistance coil, type of circuit, etc., depend very largely on the accuracy of the results required. For very accurate laboratory work a null method must be used together with suitable compensating leads or a Smith's Bridge which eliminates any lead effect by connection reversals. For general use where readings are required only to the nearest 1° C a direct deflection instrument with the galvanometer graduated in degrees finds very widespread use. Recording arrangements can be fitted.

Although electrical resistance and thermoelectric thermometers both cover the temperature ranges up to 1000° C (1832° F), the resistance type is to be preferred for the lower ranges up to 500° C (932° F). They are very suitable for centralised readings from scattered points. They can be arranged to indicate temperature in one place and simultaneously record it in another, etc.

(6) THERMOELECTRIC THERMOMETERS

Here the temperature is measured by the e.m.f. that is set up when two dissimilar conductors are joined into a circuit in such a way that one junction is hotter than the other. The metals normally used for the thermocouples are:

copper/constantan (60% copper + 40% nickel) up to 500° C (932° F);
iron/constantan up to 850° C (1562° F);
chromel (90% nickel + 10% chromium)/alumel (98% nickel + 2% aluminium) up to 1100° C (2012° F);
platinum/87% platinum + 13% rhodium up to 1500° C (2732° F).

Wisely (W77) reports new developments in thermocouples for higher temperatures, these are:

94% platinum + 6% rhodium/70% platinum + 30% rhodium up to 1800° C (3270° F);
iridium/60% iridium + 40% rhodium up to 1927° C (3500° F);
or iridium/75% iridium + 25% rhodium up to 1927° C (3500° F);
tungsten/iridium up to 2100° C (3812° F) in neutral atmosphere or vacuum.

Of the well-known types the chromel/alumel and platinum/platinum–rhodium thermocouples can be used to measure the temperatures attained in many ceramic kilns and are therefore of great importance in the industry. The measurement of the e.m.f. they develop has been the subject of considerable research and progress.

The platinum/platinum–rhodium thermocouples are the most reliable for high-temperature work, suffering less from corrosion and other alterations. The first question is therefore how to have the maximum temperature difference between the hot and cold junctions without using excessive quantities of these very expensive metals. This is achieved by using compensating leads of alloys having similar thermoelectric constants to the thermocouple. They are joined to the head of the thermocouple just outside the kiln and lead to a convenient cold junction which can either be kept at constant temperature, *e.g.* by burial several feet in the ground or immersion in a thermostat bath, or where proper compensation for variation of the cold junction temperature is made from tables or automatically in the instrument measuring the e.m.f. it may be permitted to fluctuate.

The precious metal wires are protected by refractory sheaths so that only the junction between them receives the full effect of the kiln heat.

The measurement of the e.m.f. set up is either by direct deflection of a galvanometer arranged as a millivoltmeter or by potentiometric measurement. The latter is very much more accurate.

Prior **P56** describes the advantages and disadvantages of the millivoltmeter, mechanical potentiometer and the electronic potentiometer, respectively.

Direct Deflection Millivoltmeter

Advantages. (a) Simplicity. (b) Relatively long-life. (c) Low production cost.

Disadvantages. (a) Rather poor accuracy rarely more than 1% of the full-scale reading and dependent on stable field distribution and strength. (b) Accuracy affected by external resistances in the circuit. (c) Suppressed ranges are difficult to arrange due to the requirement of accurately cutting the hair springs. (d) Automatic cold junction compensation requires the insertion of a bimetal spiral between the hair spring and the frame. (e) The instrument is comparatively delicate, requiring regular maintenance to remove dust and ensure that pivots are sharp, etc. (f) The instrument is

FIRING AND KILNS 945

not stable over long periods of time mainly due to magnet strength falling off so that frequent recalibration is necessary.

The simplicity of the circuit and low initial cost nevertheless make this the commonest instrument in industry. The modern instruments are designed with galvanometers of high resistance so that the resistance of the connecting leads and the thermocouple, which should be allowed for in calibration, form only a small percentage of the total resistance of the circuit. The resistance of the galvanometer is made temperature independent by making the galvanometer coil of copper wire with a manganin coil in series with it.

Portable table models and wall models of various types are available as are also recording galvanometers.

The calibration and checking of thermocouple millivoltmeter circuits is done with a portable potentiometer (**C5**).

The mechanical potentiometer, introduced to the pottery industry for measuring e.m.f. about 1930, overcame many of the difficulties inherent in the millivoltmeter.

Advantages. (*a*) Reasonably independent of external resistance; (*b*) easily provides for suppressed ranges; (*c*) easily provides for automatic cold junction compensation; (*d*) null-balance eliminates all troubles due to irreproducible galvanometer movement allowing accuracy of $\pm \frac{1}{4}\%$; (*e*) wider scales and charts possible; (*f*) generally more robust.

Disadvantages. There are nevertheless certain disadvantages which are inherent in the mechanical movement of the contact along the slide wire and the fact that a sensitive galvanometer still has to be incorporated in the circuit; (*a*) slow response; (*b*) stepwise response to continuous changes transmitting stepwise alterations to control mechanisms; (*c*) high maintenance (**P56**).

Within these shortcomings very good instruments of this type are available to the industry (**K16, F23**).

In the modern electronic potentiometer many of the disadvantages of the mechanical potentiometer are overcome while its advantages are retained. These instruments give rapid response in a continuous manner and can easily be harnessed to control mechanisms.

The *advantages* listed by Prior (**P56**) are: (*a*) Continuous measurement allowing for linear corrective control action to input changes. (*b*) Sensitivity limited only by amplifier and slidewire design. (*c*) By simply changing the motor speed it provides the ability to follow rapidly changing variables and by balanced printing, the ability to record large numbers (up to 16) of curves with very small intervals between each. (*d*) The maintenance requirements are reduced by the elimination of the galvanometer and cyclic balancing mechanisms. The instrument is generally more robust and independent of attitude or movement. (*e*) It is easily connected to control mechanisms.

Multiple-point instruments of this type that can run through 16 points in a minute are available (**L22**).

The installing of both the hot junction or 'sensing element' and of the measuring instruments must be done with care. It is desirable that the sensing element be in a position where it will measure the temperature to which the ware is being subjected. But it is undesirable to expose long lengths of the precious metals to harsh conditions both of temperature and atmosphere and also the risk of mechanical knocks, etc., during placing and drawing. In a tunnel kiln permanently installed thermocouples must obviously be on the surface of the lining and of such a size as not to foul the moving ware. Special kiln cars carrying thermocouples amongst the ware, usually at three levels, are periodically passed through the kiln to obtain a complete firing curve. Where possible the thermocouple wires are passed

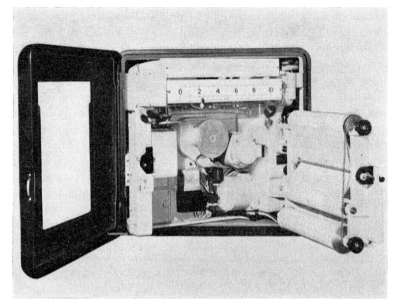

FIG. 11.20. *Electronic strip chart recorder (Kent,* **K16**)

through the car top to the cooler undercar space where compensating leads are paid out.

It is frequently found advisable to place all the measuring and recording instruments for a kiln or a battery of kilns on a single well-illuminated panel away from a main thoroughfare. If the cold junctions are at the instruments a steady temperature is very helpful and total enclosure of the panel in a small room helps to keep out dust, etc. Automatic cold junction compensators consisting of temperature variable resistances can be installed at the measuring instrument. (Fig. 11.20)

(7) TOTAL RADIATION PYROMETERS

For measuring temperatures above those that thermocouples will directly withstand radiation pyrometers must be used.

Under 'black-body' conditions when a body receives as much energy as it loses by radiation the emitted radiation is a measure of the temperature of the body. This is stated in the Stefan–Boltzmann law as:

$$E = \sigma(T^4 - T_0^4)$$

where E is the total energy radiated at absolute temperature T to the surroundings at absolute temperature T_0, and σ is a constant. T_0 is generally very small compared with T, when the equation becomes

$$E = \sigma T^4$$

In the majority of modern kilns the ware approaches black-body conditions, so that if the radiation it emits is measured through a small opening its temperature can be found. The same body at the same temperature but in the open does not emit black-body radiation and a radiation pyrometer will not give a true indication of its temperature although the reading given is reproducible under the same conditions.

The Féry radiation pyrometer measures total radiation emitted by a hot body. In it, a concave mirror focuses the radiation on to a thermocouple set up at its focus, an e.m.f. is set up and read on a millivoltmeter or potentiometer as required. The instrument can be used at considerable distances from the hot object so long as the image of the hot body formed by the mirror is large enough to cover the thermocouple. In the Cambridge instrument (No. 13195) this means the distance of pyrometer from hot object may not exceed 2 ft for every 1 in. (1 m for every 40 mm) diameter of the hot body (C5).

The temperature values obtained can be indicated, or recorded, and transmitted to control mechanism.

(8) OPTICAL RADIATION PYROMETERS

The intensity of the visible radiation emitted by hot bodies may also be used as a measure of their temperature. Two types of instrument are available for such measurement. In the one the light from the hot body is optically matched with that from a constant comparison lamp, this type was invented by Wanner about 1894 and has been improved upon. In the second type the brightness of the comparison lamp is varied to match that of the hot body and was originally due to Holborn and Kurlbaum about 1901.

In the Wanner type of optical pyrometer light from the hot source and from a comparison lamp are passed through monochromatic red glass to give beams of a single wavelength. Both are then polarised, the one in a plane at right angles to the other, and then passed through a Nicol prism. Rotation of this diminishes the brightness of one beam while increasing that of the other until the brilliancy of the two fields matches. The temperature can be determined from the angular position of the Nicol prism.

For practical purposes the Wanner type is being replaced by the disappearing filament type of optical pyrometer because of its greater ease in use and standardisation. In this the filament of a small incandescent lamp is placed at the focus of a telescope objective receiving light from the hot source. Red glass is fitted in the eyepiece so that only monochromatic light passes to the observer. The current passing through the filament is varied by means of a rheostat until it is of the same brightness as the background of light from the hot body when it disappears against it. The current then passing is read on an ammeter which is calibrated to read directly in temperature degrees.

Both instruments can be made portable or even pocket size but are always only indicating (**C5, B111, F16, E15**).

Further optical pyrometers using photocells connected in Wheatstone Bridge circuits with galvanometer or electronic balancing arrangements can be indication or recording (**W4**).

MAINTENANCE, TESTING AND CHECKING OF TEMPERATURE MEASURING INSTRUMENTS

Instrument makers usually supply instructions with their instruments for maintenance and checking. Certain supplementary instruments may be necessary to carry this out. It is very important to follow these instructions and carry out regular checks of all instruments, which may have to be as frequent as once a week for thermocouples in constant use at high temperatures.

Automatic Temperature Control

The logical sequel to accurate electrical temperature measurement is the use of the circuits to operate automatic controls to maintain the desired temperature. Instruments are now available to maintain a given temperature or to follow a definite time–temperature programme. The latter is achieved by inserting a cam which moves the setting of the controller which is otherwise the same as for a steady temperature (**W6**) (Fig. 11.21). Automatic controls can operate electric switches and valves in air, gas, steam or oil pipes. They are therefore applicable to many tunnel kilns and some intermittent ones.

The whole temperature control device consists of three parts:

(*a*) temperature-measuring instrument;
(*b*) controller;
(*c*) mechanism to alter the heat input.

In electrically heated kilns (*b*) and (*c*) are one.

(*a*) The temperature-measuring instruments that can operate controllers have already been described, those usually used are the mercury-in-steel thermometer for low temperatures, the resistance thermometer and Wheatstone Bridge for intermediate temperatures and the thermocouple and Féry radiation pyrometer with potentiometers for high temperatures.

Correct placing of the sensing element with respect to the burners or heating elements it controls is very important.

FIRING AND KILNS 949

The controller must only be coupled to burners which act on its sensing element. It is best if the reaches of adjacent elements overlap a little. In muffle kilns thermocouples should be not more than two car-lengths apart. In direct-fired ones they should be more frequent and in those with radiant panels or burners it is best to have one for each heat source. It must also be remembered that whereas a radiation pyrometer focused on the ware gives the temperature of the ware, a thermocouple on the wall or roof of the kiln gives the temperature of the atmosphere to which the ware temperature is proportional but not necessarily equal, even in a tunnel kiln.

FIG. 11.21. *Programme controller* (*Kent*, **K16**)

(*b*) Bennett (**B35**) enumerates and describes the five main categories of automatic controllers:

(1) Two-position or 'on–off', or 'high–low'.
(2) Floating.
(3) Proportional.
(4) Proportional plus reset.
(5) Proportional, or proportional plus reset, plus rate action.

The choice between them depends on both the amount of variation the ware will stand, and the method of transmission of the heat together with the time lag between adjustments and their recording by the sensing element. For instance in a muffle kiln the high heat storage of the walls counteracts temperature fluctuations before they reach the ware while radiant burners transmit all variations directly to the ware.

(1) TWO POSITION CONTROL

This is almost self-explanatory. The valve, etc., has two settings giving, respectively, too much and too little heat which are turned to according to whether the measured temperature is below or above that desired. The action is therefore cyclic. The instrument is the cheapest to install but should only be used where muffle or bag walls with high heat capacities take up the variations or where the ware is unaffected by them, *e.g.* some structural products and biscuit whiteware.

In some instruments there are three available positions for the valve (**F24**).

(2) FLOATING CONTROL

In the single-speed floating control the valve adjustment is at a fixed but relatively slow speed and any particular setting can be held. The control mechanism has a dead or neutral zone covering the desired temperature. When the temperature registered enters within this zone the element or valve 'floats' in whatever position it was at that moment and remains so until the temperature moves out of the neutral zone again.

This type of control is good for cases where the heat transfer lag between burner and ware is short, and the load and schedule in the kiln are constant. With an added refinement to allow adjustment of the speed of action on the valve, the rate of heat transfer can be allowed for.

(3) PROPORTIONAL CONTROL

In this system there is a continuous linear relationship between the temperature registered and the position of the valve, etc. A certain deviation of temperature will repeatedly give the same setting of the control valve for any particular setting of the proportional band. Response is therefore quicker and more effective in the case of large deviations than with the floating control.

When the load is at the desired temperature it is periodically necessary to adjust the valve to maintain it there. This is done manually on the controller that shifts the position of the proportional band. An experienced operator can also make adjustments to allow for variations in quantity of ware in the kiln and other disturbing factors. This type of control allows for more errors in valve size (**F24, B120**).

(4) PROPORTIONAL PLUS RESET CONTROL

In this type the resulting action also is automatic and gradual. The proportional and the resetting devices may operate at the same time. The rate of resetting can be adjusted.

This type of control is almost universally applicable in tunnel kilns the only exception being when there is considerable 'dead time', that is a long

FIRING AND KILNS

time lapse between an adjustment of a valve and registration of it on the thermocouple or pyrometer.

(5) RATE-ACTION SUPPLEMENTARY CONTROL

This is used in conjunction with (4) and (5) when these have been set to act slowly for a system where considerable dead time occurs, and cannot therefore respond quickly to a change of load. The rate action part of the instrument takes into account the speed at which the temperature deviates.

(c) The controllers act electrically or pneumatically or by a combination of both, on valves or switches and so can be used for gas or oil-fired or electrically heated kilns. They are not normally applicable to coal firing.

In electrically heated kilns the heat input can be altered directly by the control instrument by switching the heating circuit on or off or by switching a resistance into or out of the heating circuit.

For gas or oil firing the controller may operate a small motor which turns valves in the fuel or air pipelines. The circuit is arranged so that the motor operates in alternate directions on alternate occasions. Bullin (**B140**) suggests that the most satisfactory gas installation for such control is one using high-pressure air which inspirates low-pressure gas into the burner. The automatically controlled valve is in the air supply. In other systems the motor controls valves in both the gas and the airlines arranged to keep the proportion of gas to air constant (Fig. 11.22).

Alternative valve operating mechanisms are hydrostatic or pneumatic. The valve itself has a diaphragm on which water or air pressure acts to alter the flow through the pipe. The controller alters this water and air pressure. The method is particularly applicable when liquid or vapour pressure thermometers are used in which the pressure is altered by temperature (**C5b, B120**).

Measurement and Control of Gas and Air Pressure and Flow, and of Draught

A number of processes in a ceramic kiln are dependent on correct movement of the gases present in it. Firstly, the proper combustion of fuel depends on an adequate air supply coupled with removal of the combustion products from the burning zone. Secondly the distribution of the heat of combustion is carried out by the circulation of the hot gases. Thirdly the spent gases must be removed in a flue, taking with them water vapour, sulphurous gases, carbon dioxide, etc., removed from the ware.

Measurement of the gas flow, *i.e.* draught, both in the flues of any kiln, and within a tunnel kiln and whenever practicable inside other kilns, allows for proper regulation and control, so that enough draught is present to achieve the desired results without an excess which wastes fuel.

Draught measurement is carried out with gas flowmeters. These consist of two parts, one being fitted into the flue in such a way that a pressure difference is set up between the two sampling tubes, the second part measures and indicates and/or records this pressure difference. There are several types of sampling devices using nozzles, orifices, sampling tubes, and Venturi tubes. The pressure difference set up by the gas flow is measured by a diaphragm or a hydrostatic meter. These instruments can be made to operate switches in circuits regulating the positions of dampers on the air inlet or the flue (**B1d, B119**).

Fig. 11.22. *Motorised butterfly valve* (*Kent*, **K16**)

Flue Gas Analysis

Information about the efficiency of fuel combustion can be derived from the composition of the flue gases. Usually only one or two constituents are measured according to the general firing conditions.

In oxidising firing there is a general tendency to use too much excess air which then wastes fuel by absorbing heat that could be used for the ware. The quantity of excess air is measured by finding either the oxygen content or more usually the carbon dioxide content of the flue gases.

In a well-regulated firing process, attempting utmost fuel efficiency by cutting down the excess air, tests must be made for incompletely oxidised fuel, carbon monoxide and hydrogen. This is done in addition to the carbon dioxide measurement.

In dusty operations, such as calcining of raw materials, it may be necessary to measure the quantity of dust and fumes being carried up a chimney stack. A self-contained apparatus for filtering metered samples of flue gas has been developed by B.I.S.R.A. primarily for use in steelworks but is applicable also in the ceramic industry (**A53**).

CHEMICAL GAS ANALYSIS

Most chemical methods of gas analysis are based on the Orsat apparatus. In this a sample of gas is measured in a gas burette under known conditions of temperature, pressure and water vapour saturation. It is passed through suitable three-way taps and by manipulation of the mercury reservoir into a caustic potash solution where carbon dioxide is absorbed. The gas is returned to the burette and remeasured. By similar manipulations the oxygen is then absorbed in alkaline pyrogallol solution. The remaining gas may then either be assumed to be nitrogen and inert gases or it may be further treated to find the content of combustible gases (CO and H_2). In this case the carbon dioxide is removed from a second sample of the gas and the rest is passed into an electrically heated combustion tube and the resulting mixture measured before and after absorption of the carbon dioxide from it. If insufficient oxygen is present in the mixture for this combustion a measured quantity must be added.

A continuously recording instrument depending on the chemical absorption of carbon dioxide is available. A measured sample of gas is passed through caustic absorption solution and then remeasured. The percentage of volume loss under the same pressure conditions is automatically recorded as the percentage of CO_2 (**G58**).

ELECTRICAL GAS ANALYSIS

Carbon Dioxide. The electrical estimation of carbon dioxide content is based on the different thermal conductivity of that gas when compared with air or other possible flue gas constituents. The test is made in a narrow horizontal tube with a central electrically heated resistance wire kept slightly hotter than the walls. Under these conditions heat transmission by convection and radiation is negligible. The temperature, and therefore resistance, of the central wire is dependent on the conductivity of the gas surrounding it. Changes in the temperature of the gas and the instrument generally are compensated for by a sealed duplicate thermal conductivity measuring cell containing air of known composition. The measurement actually made is the difference of resistance of the wires passing through air and flue gas, respectively, and is made with a Wheatstone Bridge circuit.

The result in percentage carbon dioxide may be indicated or recorded, or presumably could be made to work an automatic air inlet control (**C5c, B1b**).

Carbon Monoxide, Hydrogen and Other Combustibles Measurement. Instruments to measure any combustible gases present can be made independent of, or be fitted to work in conjunction with the electrical carbon dioxide meters. The principle used is the heat evolved to complete combustion with an excess of air. This is done in a tube fitted with an electrically heated wire whose resistance changes with the increased combustion temperature. The wire is connected into one arm of a Wheatstone Bridge and external temperature changes are compensated by placing a similar but sealed tube of air in the other arm. The result in percentage $CO + H_2$ is indicated or recorded (**C5c, B1a and c**).

Oxygen Measurement. For processes where carbon dioxide is evolved independently of the fuel combustion a better test for completeness of the combustion is the presence of oxygen in the flue gases. This is measured by the change of thermal conductivity when the gases are passed over hot carbon, thus converting any oxygen present into carbon dioxide. The instrument is similar to that for measuring CO_2 with a pair of thermal conductivity cells on each side of a Wheatstone Bridge. The flue gases are passed through one pair then through a small furnace containing hot carbon rods and then on to the other pair. The change in thermal conductivity and therefore increase of CO_2 content is measured. The result can be indicated or recorded (**C5d**).

SAGGARS AND KILN FURNITURE

Although some types of ware can support themselves when stacked in the kiln, *e.g.* bricks, many kinds require various forms of support and protection. These are given by saggars and kiln furniture. Saggars are round, oval or rectangular vessels with flat bottoms and straight vertical sides. Ware is placed in them outside the kiln and they are then stacked on top of each other in the kiln or on the kiln car. The ware is thereby supported in stacks far higher than it could be placed upon itself and also protected from flames, ashes, sulphurous gases, etc. Within the saggars some ware is further supported by small props or rings to prevent distortion and glost ware is prevented from touching by point supports. Alternatively, where the presence of a muffle or clean firing makes it possible to dispense with the protection of saggars, a system of shelves is erected for placing the ware.

Saggars and most kiln furniture are themselves ceramic products and are described in Chapter 17.

In Europe it is usual for saggars to be made in the works where they are to be used whereas the more intricate shapes of kiln furniture are made by specialists. In the United States it is considered better to have the saggars made by specialist firms as well, so that the manufacturer of fine tableware, etc., need not encumber himself with the separate grinding, mixing, shaping, drying and firing equipment required for good saggars.

It is generally considered now that a good saggar should have a useful life of at least a hundred firings (formerly it was necessary to make do with only twenty to thirty cycles). Good saggars for earthenware at cone 4 to 5 have a life of 200–300 cycles and good ones for porcelain fired at cone 14 to 15 last 200 times.

Now that technical understanding makes bad saggars inexcusable the further question of reducing the weight of kiln auxiliaries arises. The firing of ware in saggars or on bats, etc., means that kiln capacity and fuel are used up for articles that are not saleable. In the intermittent kiln the heat required to bring the saggars, etc., to the firing temperature of the ware is lost completely. In a good tunnel kiln a large amount of it is recuperated but kiln capacity is nevertheless taken up. There is therefore every incentive to reduce the weight and volume of firing auxiliaries. This has, in fact, been found possible in tunnel kilns where the stack of saggars is only 6 ft (1·8 m) high instead of the 15 ft (4·6 m) normal in intermittent kilns and so the load on the saggars is reduced and thinner sections can be used (**D57**).

The most important factor, however, is cleaner and flameless firing which makes the protective function of saggars unnecessary so that open methods of supporting ware can be adopted. This can immediately reduce the ratio of kiln furniture to ware from 5:1 to 2:1.

Further reduction of the ratio can be achieved by making the bats and props stronger so that thinner sections can be used, or greater loads carried. Walker (**W1**) has shown that if a bat is made twice as strong its maximum safe load is increased $2\frac{1}{2}$ times. To reduce the thickness of a fully loaded bat it has to be made very much stronger as the ratio by which thickness can be reduced is only about the square root of the ratio of critical stresses, *e.g.* a fourfold increase of critical stress from 100 to 400 lb/in² (7 to 28 kg/cm²) permits a thickness reduction from 0·68 to 0·3 in. (1·7 to 7·6 mm) while the ratio of payload to furniture increases from 1·95 to 4·3.

Walker also showed that higher firing of kiln furniture will give better hot strengths, as seen in Table 222, where a silicon carbide body is shown for

TABLE 222
Effect of Firing Temperature on Clay-bonded Sillimanite
(after (**W1**))

Material	Firing temp.		Breaking strength		Average permanent deformation
	(°C)	(°F)	(lb/in²)	(kg/cm²)	(10^{-4} in./in. strain)
Clay-bonded sillimanite	1500	2732	1070	75·2	17
Clay-bonded sillimanite	1550	2822	1110	78·0	11
Clay-bonded sillimanite	1650	3002	2310	162·4	3
85% silicon carbide mix			3730	262·2	3

comparison. It is probable that great benefits would be found in controlled firing of kiln furniture in tunnel kilns.

He also enumerates three ways in which the pottery industry itself can contribute to the further development of kiln furniture: (1) by maintaining precise records of furniture performance and its conditions of use, (2) by reducing to a minimum the risk of mechanical damage, (3) by facilitating a reasonable degree of standardisation so that the pieces can be more economically produced.

THE SETTING OF WARE IN KILNS

The method of setting ware in the kiln is dependent on a number of factors related both to the ware and to the kiln.

WARE

(1) Biscuit *or* glost.
(2) Load bearing when hot *or* more or less liable to warp.
(3) Unharmed by flames and kiln atmosphere *or* requiring more or less protection from it.

KILN

(1) Open flame *or* muffle *or* electric.
(2) Relatively dirty (coal) *or* clean (gas) fuels.
(3) Intermittent, continuous *or* tunnel types.

In general, unglazed ware can be stacked in contact whereas each piece of glazed ware must be carefully kept apart from all others or they will stick together. The lower grades of ware, bricks, roofing tiles, etc., may be fired in open contact with kiln gases, whereas quality ware, particularly if it should be white, has to be protected from direct contact with flames, smoke, ash, etc., by placing it in saggars or using muffle kilns.

The placing of either the individual pieces, or the saggars in the kilns must be arranged to achieve the optimum circulation of kiln gases for maximum regularity of temperature and heat exchange. Too open a setting allows the gases to pass through easily and wastefully. On the other hand too dense a method of stacking will prevent the gases from reaching all pieces evenly. One air gap larger than others will draw off the hot gases from them, so the gaps between large uneven shaped pieces (sanitary ware) should be filled in with smaller ones to prevent the large ones becoming over-fired. (Fig. 11.23)

Each particular type of ware will deform during the firing if more than a certain weight is bearing down on it. This maximum must be determined and not exceeded. If higher stacks are required a structure of refractory posts and shelves must be introduced (Fig. 11.24).

Setting Tableware and Wall Tiles

This ware can never be stacked directly into kilns or on to kiln cars. It requires both support and protection. There are two basic methods of placing: (a) in saggars, formerly the only method; (b) in open settings on bats and props. In each, the individual pieces are placed differently according to their stage in manufacture and their body type.

Saggars both support and protect the ware. For tableware they are generally round or oval and for tiles square or rectangular (Fig. 11.25).

FIG. 11.23. *Setting fletton bricks in top-fired continuous chamber kiln by fork-lift truck. The bricks coming from the presses are set on pallets with the lowest layer arranged to fit the inflatable prongs of the fork lift truck. Successive layers have optimum layout for gas circulation, note that the lowest one has the bricks on end. Note also the occasional line of bricks jutting out into the path of the fuel fed through the roof* (London Brick, **L41**)

Subsidiary support may be given to each pile of ware placed in the saggar depending on the ware. There is also a tendency to make individual setter-saggars for ware that must have complete, individual support, *e.g.* biscuit firing of American fine china and glost firing of European hard porcelain.

Shaw (**S61**) describes the design of economical individual setters for American fine china:

Firstly the correct body composition must be selected and its drying and firing shrinkage ascertained. The method of shaping it is then chosen, and although pressing results in less warpage and shrinkage than other methods the cost of dies may be prohibitive; jiggering also lessens warpage but many shapes cannot be made this way; casting is therefore the most generally applicable for difficult shapes. The design of the saggar must therefore be such that it is easily released from a mould.

Fig. 11.24. *Setting salt-glazed stoneware pipes including bends, traps, junctions and other shapes. Pipes can be superimposed but the shapes are individually supported* (*Doulton*, **D56**)

The next requirement is samples of the green dry ware and of the fired ware, each cut in half. The design of the biscuit saggar is then made to conform to them both as shown in the diagram (Fig. 11.26).

Using the Fired Half Section. A is a straight portion from the centre of the plate to within about ¼ in. of the foot. *B* is horizontal and about ¼ in. below the foot. *C* is at 30° or 45° to *A* and *B* depending on their distance apart, being smaller if the distance is great to facilitate mould release. *D* follows the contour of the side of the plate.

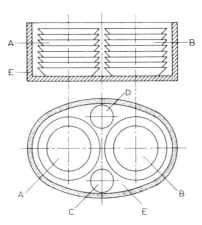

FIG. 11.25. *Oval saggar for biscuit tableware:*

A and *B*, plates
C and *D*, saucers, etc.
E, base and wall of saggar (**A105**)

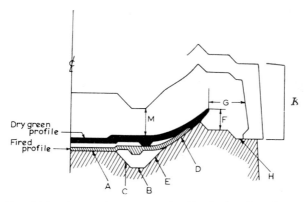

FIG. 11.26. *Layout for a saggar for biscuit American fine china (Shaw, S61)*

Using the Green Half Section. *E* is sloped to allow clearance for the foot which lies further out in the green than in fired ware. *F* is the clearance that must be allowed for placing the ware in the saggar by hand. *G* must be at least ⅜ in. (9·5 mm) to allow the next saggar to be set on top without danger to the ware. *H* is the step for the next saggar, usually about ½ in. (1·27 cm). *K* the height of the leg is determined by the saggar shape as *M* the minimum clearance should be at least $\frac{3}{8}$ in. (9·5 mm). The angle of the leg from the vertical must be enough for good mould release, about 5°.

In general, however, although saggars have the advantage of easy production, of being filled away from the kilns and of being suitable for any kiln, their disadvantages are causing their elimination where possible. The main points against saggars are that they are heavy and require male labour to handle them. It is also in their nature that they must be handled for every filling and emptying and this has been shown to cause more wear and tear on them than the repeated firing. They tend also to dirty the ware. The greatest objection to the use of saggars is the heat uselessly consumed in heating them up as well as the ware.

Open setting. There is therefore an increasing tendency to use a clean firing method and an open setting for the ware. For this the ware is supported on refractory bats which form shelves supported by props. The arrangement may be made very flexible for easy mounting and dismantling to accommodate different ware sizes, or a fairly permanent structure may be made for a single type of ware such as glost wall tiles. It is desirable that the bats be free to expand and contract and also that any that fail can be replaced without dismantling a great deal of sound work. The design of the props varies widely. Some have iron posts set in the base of the setting over which refractory tubes are slipped which act as spacers for bats with holes. Others are all ceramic with various interlocking slots, etc. Others are flat-topped pillars on which the bat rests. This allows free expansion.

The aim is to make the weight ratio of kiln furniture to ware as low as possible. Even with very economical structures for biscuit earthenware it is difficult to reduce it below 1:1. When thinning down the bats and props to reduce their weight it becomes increasingly important to know their strength. Gautby (**G12**) discusses and gives graphs relating to the load-bearing capacities of bats and props.

Considerable weight can be saved if the ceramic refractory can be replaced by metal, as this can be made in thinner sections. Rohn (**R52**) describes a heat-resisting metal alloy that can be used up to 1200° C (2192° F). This not only allows thin sections for the kiln furniture and therefore the placing of more ware, but also its better thermal conductivity allows a shorter firing time and a lower soaking temperature to finish the ware successfully, *e.g* in an electric kiln the following results are obtained:

	Fireclay furniture	Metal alloy furniture
Weight kiln furniture per car (kg)	786	150
Weight ware per car (kg)	550	740
Firing time (hr)	$16\frac{1}{2}$	$8\frac{1}{2}$
Power used (kWh)	760	560
Power used/kg ware	1·38	0·83
Power saved (%)		40

FIRING AND KILNS

BISCUIT SETTING

In the biscuit firing earthenware and hard porcelain should not deform and so flatware can be stacked in bungs twelve to fifteen pieces high with a refractory setting ring only at the top, and hollow-ware is boxed to keep the circular shape. Large and special shapes must, however, be supported. Hotel ware does deform a little so that a filler must be put between plates in a stack. Sand is strewed on the bat or saggar bottom to allow all such ware to contract freely. (Fig. 11.27.)

FIG. 11.27. *Biscuit placing of earthenware on an open prop and bat structure, to be fired in an electric tunnel kiln (Wedgwood,* **W35**)

Bone china and soft or frit porcelain definitely sag and deform in the biscuit firing and specially shaped supports must be used for each piece. In Europe this is done by placing each piece in a shaped bed of dry inert powder. The powder formerly used was flint but this is being replaced by alumina because of the danger of the dry flint causing silicosis. A plaster bat is pressed into the bed to make it the shape of the finished piece. In U.S.A. soft porcelain is set in specially shaped saggars.

Setting of hard porcelain figures, bowls, etc., requires skilled propping with green setters made of the same body.

GLOST SETTING

During the glost firing the glaze will stick to any piece with which it is in contact, whether it is the body it is required to cover or a neighbouring one. Each piece must therefore be supported independently of any others on the minimum surface.

Earthenware should not deform in the firing and so can be supported on the smallest possible points. Flatware is either arranged in vertical stacks with some means of separating the pieces, 'dottling', or alternatively, if saggars are in any case to be used, the pieces may be 'reared' on edge and supported at the top and bottom. The simplest methods of dottling use

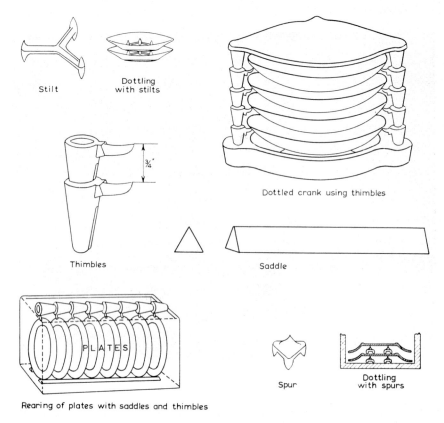

FIG. 11.28. *Firing supports for glost earthenware* (*Bullers*, **B139a**, *and Gimson*, **G50**)

stilts and spurs placed between the plates. Both these supports are made in various sizes and different relative sizes of point and arm. In both cases marks are left on the upper as well as the lower side of the ware. It is therefore sometimes advantageous to place the ware upside down so that any saggar strew damages the underside only.

By using a socketed crank base or individual sockets, 'claws', and thimbles each plate, etc., is independently suspended and therefore retains marks only on its underside. The cover at the top of each crank (normally about fifteen pieces) holds the pillars of thimbles together and protects the top plate from strew, etc. (Fig. 11.28).

The kiln supports used for all three of the above methods of dottling have complicated shapes and so cost precludes that they should be used once only. The support point gradually wears down with repeated use and so increases the size of the blemish left on the plate it has supported. The newer dottling methods eliminate this trouble by supporting the ware on very cheap triangular pins produced by extrusion. These can be discarded after a single using. The cranks here are built up on a slotted base with perforated pillars. The pins which actually support the ware are placed into the holes in the pillars (Fig. 11.29).

A special machine has been developed for loading the pillars with pins. With this one girl can do the work formerly done by four women loading

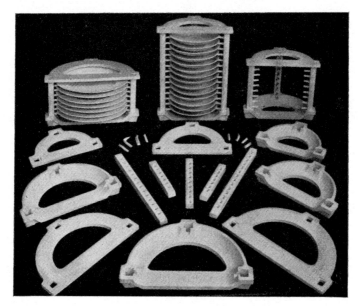

Fig. 11.29. *Pinplacing equipment for glost earthenware* (Bullers, **B139a**)

pins by hand. Loose pins are placed in a hopper with a slowly revolving drum at its base. This has circumferential grooves into which the pins are swept by a second revolving drum with rubber wipers. 'Scavenging' grooves reject dust and small particles. From the drum the pins drop by gravity in a series of tubes from which a foot-operated mechanism releases a full set into a waiting pillar. The machine will load the crank pillars as fast as they can be presented to it by the operator (**W20**).

The individual cranks of flatware may be placed in saggars or on open shelving, although the method of placing was developed largely for the latter case.

Rearing of plates must be done in saggars and so is losing favour as open-set tunnel kilns replace periodic kilns that use saggars. Two long 'saddles' of triangular or curved three-cornered cross-section are placed along the bottom of the saggar. Plates, etc., are propped vertically and at right angles to these and held apart at the top by thimbles (Fig. 11.28).

Hollow-ware does not lend itself readily to stacking. Bowls may be dottled in saggars with the aid of 'dumps' (Fig. 11.30). Small objects may be placed inside larger ones with the aid of stilts or butterflies (tall three-legged supports), but the arrangement is usually too insecure for more than

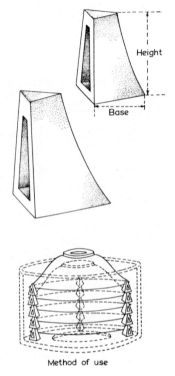

FIG. 11.30. *Hand basin props (dumps)*
(*Gimson*, **G50**)

two pieces. Cups, etc., are therefore usually placed singly for the glost firing either on stilts or with most of the glaze removed from the foot directly on the saggar bottom or shelf (Fig. 11.31).

Bone china needs a little support even in the glost firing. Cranks can be built up from interlocking triangles which hold longer pins supporting more of the foot of the ware rather than the rim as in earthenware cranks (Fig. 11.32). Pin cranks can also be used with longer pins supporting the foot. These are therefore further apart than earthenware ones.

In U.S.A. *fine china* (soft porcelain) is set in individual saggars for the glost as well as the biscuit firing, although as Shaw (**S61**) points out the design is simpler as the ware does not change shape in this firing. It is supported

Fig. 11.31. *Kiln cars with glost ware* (*Doulton*, **D56**)

on from three to six small buttons or landings in the bottom of the saggar (Fig. 11.33a).

Hard porcelain receives its high firing only during the second and glost firing. It is during this that it softens and easily deforms. It is therefore placed in individual specially designed saggars. As the foot is not glazed it does not require point supports but can rest on the bottom of the saggar. The type shown is designed to economise in kiln space (Fig. 11.33b).

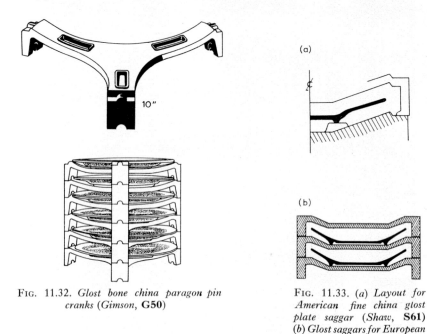

Fig. 11.32. *Glost bone china paragon pin cranks* (Gimson, **G50**)

Fig. 11.33. (*a*) *Layout for American fine china glost plate saggar* (Shaw, **S61**) (*b*) *Glost saggars for European hard porcelain* (Larchevêque, **L17**)

DECORATED SETTING

The firing of decoration is less severe than the preceding biscuit and glost firing. The ware must be kept apart but there is less tendency for it to fuse on to its supports. Cranks may therefore be constructed from cast iron or nickel alloy bases and pillars with thimbles threaded on to the latter.

TEST CODES FOR PERFORMANCE AND EFFICIENCY OF KILNS

The British Standards Institute has compiled two test codes for ceramic kilns, one for the heavy clay industry and the other for the fine section as follows:

B.S.S. 1081:1942 (Amended 1953). *Test code for kilns for heavy clay ware, including refractory materials.* Deals with testing for evaluating the performance and efficiency of intermittent, continuous car tunnel, ring tunnel and chamber continuous kilns used in the heavy clay and refractories industries and fired by solid, liquid or gaseous fuel. The codes cover both comprehensive and less detailed commercial tests for general industrial use. Sections are included in each code with full explanatory notes dealing with general information and design data in regard to the plant being tested, the test conditions, the information to be reported derived from log sheets and preliminary deductions, the thermal statement and the summary of the kiln performance.

B.S.S. 1388:1947. *Test code for ovens and kilns firing pottery*. Provides for evaluating the performance and efficiency of pottery firing ovens and kilns of the intermittent and continuous types, fired by solid fuel, liquid fuel, gas or electric heating. The code covers a comprehensive test and a less detailed test for commercial purposes. The sections of the code cover the following broad headings: general information, and design data; test conditions; mean observations and preliminary deductions; thermal statement; kiln performance. Under each of these sections items are listed together with explanatory notes.

KILNS INDIVIDUALLY

Periodic or Intermittent Kilns

There are now two main classes of intermittent kilns. Firstly the traditional ones, for which there was formerly no alternative, with types developed for every class of ware. Secondly the modern intermittent kiln.

The *traditional* intermittent kilns, especially the up-draught bottle kilns, compare so unfavourably with modern kilns as regards performance and efficiency that they are no longer being built. However, they have a long life, and will continue to be used by small works already possessing them until the Clean Air Act forces their conversion or removal, and so they are described below.

Modern intermittent kilns, generally gas-fired or electrically heated, have a number of points in their favour and have recently regained popularity in preference to tunnel kilns. Some reasons being:

(1) There may not be space available on a congested site for a tunnel oven, but several intermittent electric or gas kilns can be put into the space formerly occupied by a couple of bottle ovens.

(2) The capital investment is only a fraction of that required for a tunnel kiln.

(3) Fluctuations in trade can easily be dealt with by reducing the number of ovens in use. A large tunnel kiln presents serious difficulties in similar circumstances.

(4) Small ovens give a rapid rate of fire and the firing can be mechanised by programme controllers so that it can be done overnight with occasional supervision from a night watchman (**A146**).

(5) A wider variety of products to be fired at different temperatures can be handled.

(6) Maintenance can usually be accomplished without an interruption of production (**N3**).

TRADITIONAL INTERMITTENT KILNS

The traditional industrial intermittent kilns were originally all fired with solid fuel, in Great Britain almost always coal. The floor plan may be round, oval, square or rectangular.

Richardson (**R17**) lists the advantages and disadvantages of the two main shapes. The round kiln has the advantage of low construction and maintenance cost compared with its capacity, and more even temperature distribution; its snags are the uneconomical use of the site, more awkward attendance and setting, and that it is less adaptable to mechanical stoking. It is used extensively where the ware is placed in round saggars. The rectangular kiln has the advantage of full use of a site, easy transport arrangements for fuel and ware and the possibility of using suspended arches. The main disadvantages of higher building and maintenance costs apply only when vaulted roofs are used and can be eliminated with suspended arches. The rectangular kiln is used for stoneware, sanitary ware, etc., that cannot be economically packed into a round one. With the more economical construction now possible it will probably take precedence over round kilns.

The walls and crown may be built either in a double layer of refractories inside and building brick outside or a triple-layered arrangement with insulators backing the refractories. With the recent introduction of refractory insulators very considerable economies in fuel consumption, weight of construction and firing time can be achieved by their use in place of dense refractories.

Each kiln has one or two doorways or 'wickets' for access to place and draw the ware. They are bricked up during firing. The fireholes, fireboxes, burners, etc., are distributed evenly around the circumference of round kilns and along one or two sides of rectangular ones.

ESTABLISHMENT OF DRAUGHT IN KILNS

The heating of kilns by burning fuels depends on establishing a draught. The combustion process heats the air immediately around it causing its expansion and a reduction of its density. This makes it rise above the cold denser air around it. The upward movement in an enclosed space leads to reduced pressure at the bottom so that cold air is sucked in, and increased pressure at the top forcing the air out through any apertures. A neutral zone where the internal pressure equals atmospheric occurs between them. The pressure difference increases with the height of the enclosed column of air, so that for a large pressure difference a tall chimney is required.

The air movement due to the local heating therefore sets up a vertical

FIRING AND KILNS

pressure gradient which induces the draught required for the combustion and for the distribution of the hot combustion products.

Up-draught Kilns

In the simplest type of kiln the hot products of combustion are allowed to enter the main part of the kiln directly from the firebox, surrounded by only a low bag wall, and to make their way upwards towards the flue(s) which open(s) out of the top of the kiln. Skilful placing of the ware in the path of these gases assists in distributing them and taking up the heat from them, but the bottom centre of a sizeable kiln never gets hot enough unless some of the combustion gases are led off through an under-floor flue and released into the centre well-hole. The introduction of dampers as well as the main central flue in the crown, if used skilfully, makes the firing of an up-draught kiln much less crude than it seems. Nevertheless very much heat is allowed to go up the chimney in the gases and the fuel consumption and smoke emission is very high. It is therefore not often worth converting these kilns to modern firing methods.

Down-draught Kilns

One way to reduce the heat losses in the waste gases which account for so much of the fuel consumption in the up-draught kiln is to force the hot gases to take longer paths through the ware and to give up more of their heat to it. In the down-draught kiln the flues drawing off the spent gases open from the base of the kiln. They may then pass individually up the sides of the kiln thereby assisting in heating up the masonry or they lead to a collecting chamber and a single flue. The latter may give a better draught.

In the Minton oven, which is the only one working completely on the down-draught system, all the hot gases from the firebox are forced up to the kiln roof by a high bag wall, usually reaching to the bottom of the arches. They are then free to percolate the setting on the way down to the floor flues.

A more even distribution of the hot gases is achieved by replacing the individual 'bags' round each firehole by a continuous internal screen wall, tied to the outer wall only where necessary for structural reasons.

In the true down-draught kiln the top of the setting heats up before the bottom. If there is an appreciable moisture content in the ware there is a danger of this drying out of the top and condensing on the bottom of the setting. A number of kilns therefore use an up- and down-draught principle, e.g. in the Wilkinson and Robey types some of the hot gases from each fire can be led under the floor and released through the well-hole into the bottom of the kiln (**D46**). In the Thomas kiln supplementary fires placed in between the fireboxes have bags that are nearly closed at the top but open at the front. They heat the bottom 3 ft (~ 1 m) of the kiln only, thereby preventing condensation and making it possible to finish the bottom of the kiln at the same time as the top (**S49**). It is usual to use coal with a high content of volatile combustibles and to arrange the secondary air so that these gases burn in the midst of the setting rather than the firebox.

The down-draught system produces much more uniform heating of the setting than the up-draught, but complete regularity cannot be achieved so that careful placing according to the needs of the ware is still necessary.

Horizontal Draught Kilns

Certain kilns are so arranged that the hot gases move through the ware more or less horizontally. They are a narrow rectangular shape with fireboxes and a screen wall at one end and the outlet to the chimney at the bottom of the other end wall.

Muffle Kilns

In some instances even saggars are insufficient protection for the ware and so the kiln is made up of an inner gas-tight chamber containing the ware and completely surrounded by an outer chamber in which the hot gases circulate.

A gas inlet and outlet are connected to the inner chamber and a suitable atmosphere is passed through. This is kept at a slightly higher pressure than that in the combustion chamber in order to counteract the inevitable leaks in the muffle walling.

THE FIRING SCHEDULE

It is impossible even to approach the ideal firing schedule (p. 874) in an industrial periodic kiln. The ideal curve indicates the temperatures at which firing must be, respectively, slow, fast, oxidising, reducing, etc., but its time factor cannot be followed. In firing a periodic kiln allowance must be made for those parts of it which receive heat last to spend sufficient time in each important temperature range. This means that the parts that receive heat sooner either spend much longer than necessary at a certain temperature, or attain higher temperatures. As it is unlikely that the entire contents will reach the same maturing temperature it is important that this is chosen to lie on a part of the expansion–shrinkage curve where there is little volume change for increasing temperature.

In a large kiln it is impossible to follow the firing process in the centre even though temperature and maturity measurements (*see* p. 934) can be made round the circumference. In spite of theoretical calculations and allowances, only full-scale trials will show what readings are shown in the accessible parts of the kiln when maturity is achieved in the inaccessible and ill-favoured parts. The final adjustments of the firing schedule can only then be made (**O4**).

Experience has shown that the optimum *time* for firing a kiln for average ware from the moment of lighting the fires to that when the fireholes are closed divides into three equal parts, corresponding to three stages in the firing.

(1) Normal temperature to *ca.* 110° C (230° F). Removal of mechanical water.

(2) 110° C (230° F) to 800° C (1472° F). (*a*) Removal of hygroscopic water. (*b*) Removal of constitutional water. (*c*) Decomposition of carbonates. (*d*) Decomposition of sulphurous compounds. (*e*) Destructive distillation of organic compounds.

(3) 800° C (1472° F) to final temperature. (*a*) Combustion of carbon. (*b*) Reduction of iron compounds. (*c*) Fusion of some components. (*d*) Sintering.

Kilns that produce faulty ware, or seem to take an uneconomically long time to fire are frequently found to have a schedule that does not divide into thirds of the total time. For instance in a kiln where water-smoking has been found to be successfully completed in 24 hr but the remaining firing takes 3 days, this can often be speeded up to 2 days. In another the total firing may take 3 days but only 12 hr be devoted to water-smoking, and faulty ware be too numerous. This can be corrected by lengthening the initial stage to 1 day.

FUEL CONSUMPTION OF THE PERIODIC KILN

The fuel consumption of the periodic kiln is very large. Unless special heat recovery methods are used it is therefore uneconomical to use any but the cheapest fuel, namely coal, with all its attendant disadvantages. A traditional 'bottle' kiln for biscuit earthenware in saggars consumes at least twice as much fuel as the weight of goods it contains. The average heat balance is given by different authorities in Table 223.

TABLE 223

	H111	D46
Heat used for firing the ware	2%	4%
Heat lost in brickwork and through radiation	36%	44%
Heat lost in saggars (weighing about five times their total contents)	10%	8%
Waste gases	52%	44%

A test to compare the thermal efficiency of the up-draught and down-draught systems was carried out on two kilns in the same factory. They were the same size, the ware was identical and set to the same density, the same fuel, fireman, firing time and finishing temperature were employed (**D46**). (*See* Table 224.)

TABLE 224

	Up-draught	Down-draught	Improvement as % of up-draught figure
Fuel consumption:			
Tons	19·0	15·1	21
Therms	5570	4210	24
Cwt/ton ware	24·7	19·0	23
Therms/ton ware	358	265	26
Cwt/ton setting	7·1	5·5	23
Therms/ton setting	103	77	25
Sensible heat lost in flue gases (therms)	2090	1320	37
Heat loss by radiation, convection, conduction from structure (therms)	2360	1910	19

Some of the heat in the waste gases can be used for drying or boilers. There is also a prospect of lighter, stronger saggars although their capital cost might be more. However, much of the heat wasted on the brickwork of the kiln is inherent in the working of a periodic kiln. The complete structure of the kiln has to be heated up for every firing and this heat is wasted every time the kiln is cooled. It is therefore desirable to have proportionately as little structural matter to waste heat on as possible, and so large kilns with less surface area per unit of volume are more economical than small ones in this respect. The size of kiln is, however, limited by the increasing difficulty of achieving uniform temperatures and of placing the ware.

Radiation losses can be considerably reduced by insulation of the kiln structure. As has already been pointed out (p. 894), insulation of periodic kilns was not always a useful means of fuel saving so long as it involved increasing the total bulk of the structure. With the introduction of refractory insulating bricks in all the shapes required for building the crown as well as the walls of a kiln, considerable fuel savings can be achieved for two reasons. Firstly the insulation reduces radiation losses, and secondly the total weight of the structure that has to be heated during each firing is reduced. In new kilns built in these refractories the firing installation can actually be made smaller than for an old kiln of the same size (**S124**). Because of the smaller heat content of a refractory insulator structure the cooling of the kiln is also accelerated, e.g. one normally taking 72 hr to cool can now be drawn after 48 hr (**B140**).

Bullin (**B140**) gives some figures obtained from actual works statistics over a considerable period, which show the order of efficiency of refractory insulation on strictly comparative terms (Table 225).

TABLE 225

1. *Brick kiln*	
R.I. lining and crown	2·7 cwt coal/ton bricks
Firebrick lining and crown	5·3 cwt coal/ton bricks
Maximum temperature	1300° C (2372° F)
% fuel saving	49%
2. *Red brick kiln*	
R.I. lining and crown	13·3 cwt/1000 bricks
Firebrick lining and crown	19·5 cwt/1000 bricks
% fuel saving	32%
3. *Blue brick kiln, rectangular*	
R.I. lining and crown	10·5 cwt/1000 bricks
Firebrick lining and crown	14·0 cwt/1000 bricks
% fuel saving	25%
4. *Pottery biscuit oven, Wilkinson down-draught*	
R.I. crown and lining to within 4 ft of bags	4·0 cwt/ton ware and saggars
Firebrick crown and lining	4·5 cwt/ton ware and saggars
% fuel saving	11%

The actual fuel consumption varies with the type of ware because of the various final temperatures, soaking periods, presence or absence of saggars, etc. Miehr (**M75**) gives the figures for traditional periodic kilns (Table 226).

TABLE 226

Type of ware	10^6 kcal/ton
Bricks	0·7 to 1·0
Flower pots	0·7 to 0·85
Refractory bricks	1·0 to 1·4 at cone 8–10
Silica bricks	1·25 to 1·55 at cone 12–14
Stoneware	2·1 to 3·5

Individual Types of Traditional Periodic Kiln

Even if traditional periodic kilns are no longer considered suitable when new construction is undertaken, so many will continue to be used that it is worth giving a brief description of some of the commoner types.

THE UP-DRAUGHT BOTTLE KILN

The round up-draught kiln with its vaulted ceiling is surmounted by an exterior cone to collect the gases from the various smoke holes and lead them to the tall central stack. The external shape is therefore that of a bottle and hence the name given to the kiln type (Fig. 11.34). The diameter is from 14 to 24 ft (4·2 to 7·3 m) (**N11**), generally 18 ft (5·5 m), the height then being about 20 ft (6·1 m) (**H111**). The fireholes are set at regular intervals around the circumference. The gases pass up the kiln and out through damper-controlled holes in the vaulting (Fig. 11.35).

Fig. 11.34. *Bottle kilns* (*Wedgwood*, **W35**)

The operation of a typical round up-draught bottle kiln is executed by a number of controls. The air intake at the base may be through: (*a*) firebars, this is primary air going through the fuel bed; (*b*) fire door, this is secondary air and its amount depends on the proximity of the fuel bed to the glut arch; (*c*) regulator hole, this is secondary air entering beyond the glut arch. Intake (*a*) can be regulated only in certain types of firebox, (*b*) and (*c*) can be regulated at will by the fireman.

The flue outlets are placed one in the crown and four (or more) about halfway along the radii of the vaulting. Each has a damper. The ring dampers are opened during water-smoking and baiting to take away the steam and heavy smoke, respectively. They are also used individually to check or accelerate the progress of different parts of the kiln. In the up-draught system if a section of the kiln gets hotter than the rest it creates more draught and hence overheats further. The ring dampers serve to regulate this.

During the early stages of the firing when the top dampers are open the hot gases rise from the firemouth straight up the bag to the arches. The outermost (first) ring and the top of the setting heat up much more rapidly than the remainder. As the temperature rises the region of increased pressure gets larger and gradually pushes the neutral line downwards. The fireman usually tries to keep it level with the spy hole above the firebox (**D46**). During the water-smoking period it is, in fact, advantageous to have the top

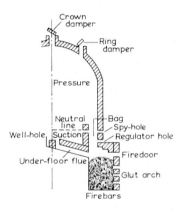

FIG. 11.35. *Vertical section through intermittent up-draught kiln showing controls* (*Dinsdale*, **D46**)

of the setting hotter than the bottom. The water is driven off the top first and the ware then becomes too hot for condensation of the steam coming from the bottom to occur on it. A bright fire must be kept up with sufficient but not too much excess air passing through the kiln (**B140**).

After the water-smoking the true firing begins. The kiln tends to heat up at the top first, the regulator is then opened so that cold air is allowed to enter and be pulled up to the top, holding it back, while the reduced draught through the firebox allows more of the gases to pass along the underfloor flue to the well hole to heat up the bottom centre of the kiln.

In spite of these controls, that in themselves need a skilled fireman, it is impossible to obtain a uniform temperature in an up-draught kiln. Ware that requires different temperatures and times of firing must be carefully selected and placed in the various heat zones of the kiln. This needs considerable skill and even then leads to high losses.

The fuel efficiency of up-draught kilns is the lowest of any kiln, a very large proportion of the heat being lost up the chimney. (*See* Table 228, examples (1) and (2).)

THE TWO-TIER PORCELAIN KILN

It has already been pointed out that a large amount of the useful heat produced by burning fuel in an up-draught kiln is lost up the chimney. The Continental porcelain industry therefore developed a double-decker round kiln for firing at two different temperatures. The biscuit firing of hard porcelain occurs at cone 010 ($\sim 900°$ C, 1652° F), whereas the glost firing is at cones 12 to 16 (~ 1350–1460° C, 2462–2660° F). If the glost setting is in the lower part of a round kiln the waste gases are often sufficiently hot to biscuit fire the raw ware which is set in an identical round kiln above the glost one. The lower kiln may be worked on the up-draught or the down-draught principle. The upper kiln is an up-draught one (**S83**) (Fig. 11.36).

THE SCOTCH KILN

The Scotch kiln is a rectangular up-draught kiln for burning bricks. It has a simple design and structure and can be built as a temporary kiln in unburnt bricks at a new site to obtain a supply of bricks for building the works (**A35**).

The internal dimensions are 30 ft × 11 to 12 ft (9·1 × 3·3 to 3·6 m) and will hold 35 000 standard bricks or 32 000 3 in. bricks. The end walls are 2 ft 3 in. (68 cm) thick and each has a 3 ft (91 cm) wide wicket. The side walls are thicker, 4 ft 6 in. (1 m 28 cm) and have narrow 9 in. (22 cm) fireholes set 2 ft 3 in. (68 cm) apart along both sides. The first 3 ft (91 cm) of the setting are placed in blocks of three bricks leaving channels corresponding to the fireholes. A 9 in. (22 cm) wall of green bricks 3 ft (91 cm) high divides the kiln longitudinally. Above this an even setting is made, the whole totalling about 11·0 ft (3·3 m). The top of the setting is covered with a 'platting course' of underburned bricks with $\frac{1}{8}$ in. (3·17 mm) wide open joint. These are covered with earth as and when required to regulate the draught.

The narrow fireholes have no grate. As the ash increases and the fuel bed is required deeper, firebricks are inserted in the mouth in an open structure to build it up. Firing is started slowly with wood chips and coke and the main fire is achieved with coal. Firing takes $2\frac{1}{2}$ to 3 days.

The Scotch kiln can fairly easily be converted to a down-draught kiln by adding a roof, underfloor flues and stack, and bag walls (**A35**).

THE GERMAN KILN

This kiln is rectangular with a vaulted roof. There is a wicket at each end. Firing is achieved by dropping fuel through holes in the roof on to grates set in the floor. The combustion gases rise up through the ware and out through holes in the roof. There is no chimney. Because of the un-

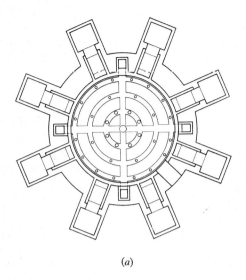

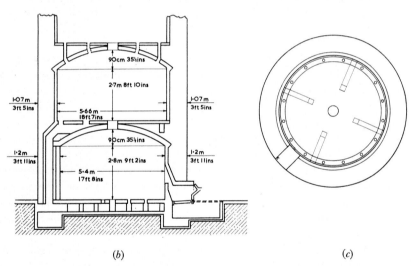

Fig. 11.36. *Two-tier hard porcelain kiln* 18 ft (5.5 m) *diameter*
(a) *plan of lower kiln*
(b) *cross-section*
(c) *plan of upper kiln*

even firing and direct damage to the ware by ashes and clinkers such kilns can only be used for bricks. They are furthermore very extravagant of fuel, requiring 2·8 to $4·2 \times 10^6$ kcal per 1000 bricks (11·1 to $16·66 \times 10^6$ B.t.u./ 1000 bricks (**S20**)). In other instances this kiln has fireholes and resembles the Scotch kiln.

THE NEWCASTLE KILN

The 'Newcastle' is a horizontal draught kiln suitable for bricks and firebricks requiring higher temperatures than are reached in other brick kilns. It has a rectangular plan, up to 15 ft (4·57 m) wide and up to 20 to 30 ft (6·09 to 9·14 m) long. One end has one or more flues leading to the chimney stack. The other has two fireboxes and a wicket into which a third firebox is built.

A space 3 to 4 ft (91 cm to 1·21 m) wide is left inside the firing wall before the first bricks are set, to allow completion of the combustion process. This space may be defined by a screen wall or perforated flash wall.

The capacity for bricks is usually made 25 000 to 30 000 (**S49**) (*See* Table 228, example 14.)

THE KASSELER KILN

This kiln is similar to the Newcastle. Dimensions are given as 5 to 8 m ($\sim 16\frac{1}{2}$ to $26\frac{1}{4}$ ft) long, 2 to 3·5 m ($6\frac{1}{2}$ to $11\frac{1}{2}$ ft) wide and 2·6 to 3·2 m ($8\frac{1}{2}$ to $10\frac{1}{2}$ ft) high, narrowing a little towards the chimney end. These kilns, although wasteful of fuel and producing non-uniform ware, are used for bricks, roofing tiles, irrigation pipes and flowerpots. The heat required for 1000 bricks is given as $2·1$ to $3·0 \times 10^6$ kcal ($8·3$ to $11·9 \times 10^6$ B.t.u.) (**M75**).

DOWN-DRAUGHT KILNS

These have proved to be the best of the large coal-fired intermittent kilns and are now the most popular and the only ones being newly built. They have also been improved in a number of ways during this century to give better draught control, to take automatic stoking, or to be fired with oil or gas.

For firing fine ware in saggars the kilns are normally round and of about the same size as the up-draught bottle kilns, namely about 18 ft (5·5 m) diameter. Some well-known types have already been mentioned on p. 969, *e.g.* Minton, working only on down-draught, and Wilkinson, Robey and Thomas using up- and down-draught. For examples of Wilkinson kilns, *see* Table 228, Nos. 3, 4. Example 6 is an oil-fired Beehive down-draught kiln.

The field in which these kilns are most suitable is for heavy clay ware, salt-glazed pipes and refractories, most especially for special shapes and large pieces; certain classes of ware needing special conditions of firing, kiln atmosphere or cooling; and for works with a small or a variable output. They are used predominantly for silica bricks, basic bricks, fireclay shapes, glass tank blocks; salt-glazed pipes; blue, brindled and multi-coloured bricks, roofing-tiles and quarries; red roofing tiles, floor tiles and quarries. Such ware may be fired in round or in rectangular kilns, much progress having been made in their design using definite ratios between the areas of the stack, main flue, subsidiary flues, floor openings, grates, etc., and the internal floor area. The round kilns are larger than those for fine ware with 30 to 40 ft (9·1 to 12·2 m) diameter, but with five damper-controlled well holes connecting to the flue and stack instead of the former single central well hole (Fig. 11.37). Rectangular kilns have two, three or four off-take flues each controlled by a damper. There is a tendency to replace built-in flues by a single external stack.

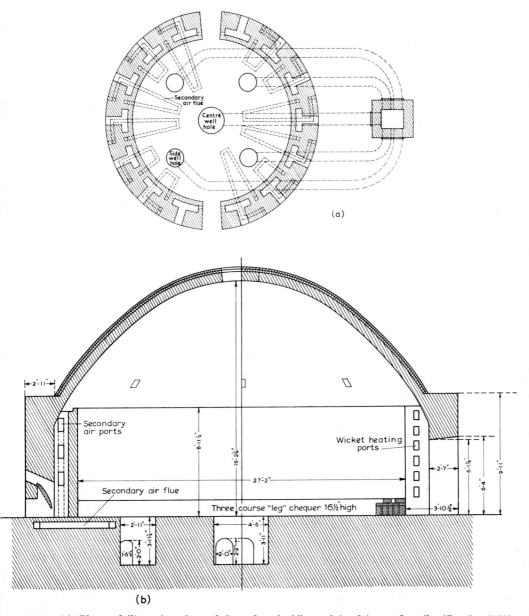

FIG. 11.37. (a) Plan and (b) section of round down-draught kiln used for firing roofing tiles (Rowden, **R59**)

For uniform gas flow different methods of floor construction have been developed for different products: chequers of bricks for roofing tiles; holey-boy perforated floor blocks for salt-glazed pipes; perforated or slotted fire-clay blocks for fire-clay goods; solid floors with centre well-hole only, or with radial or ring flues for silica goods (**R59**). It is advisable to protect such floors during setting and drawing from fallen dust and scrap and from serious damage from wheels, etc., by using tarpaulins and steel plates as necessary (**N3**). (*See* Table 228, examples 7 to 13, 15.)

Modern Intermittent Kilns

Apart from modernisation of large down-draught kilns, converting them to automatic stoking, or gas or oil firing, the main tendency in new intermittent kilns is for small ones for fine ceramic ware. These are usually electrically heated or gas-fired. Although using much more expensive fuel than the coal-fired bottle kilns they have been shown to cost just less to run because of a number of factors: (*a*) elimination of saggars; (*b*) saving of male labour in handling saggars; (*c*) more even firing, more 'firsts' in the output; (*d*) automatic programme-controlled firing; (*e*) frequently no need for shift work; (*f*) generally pleasanter conditions of work.

The self-contained periodic kilns have a hinged door which may also be fitted with heaters, and removable shelves or normal open shelf type of kiln furniture. The smaller ones are usually electrically heated and may be free standing and therefore movable. The larger ones, especially the gas-fired ones are of masonry, *in situ*, structure. Such kilns are used not only in industry but also in laboratories, studios, colleges and schools.

Some of the heat and time economies of the tunnel kiln are obtained in two new types of intermittent kilns, the truck kiln and the top hat kiln. In these the base of the setting is independent of the heated kiln walls.

The *truck kiln* or *trolley hearth kiln*, has kiln cars which may have one end wall of the rectangular kiln built up on each. The ware is set on this, away from the kiln, and then kept in readiness at a kiln being cooled. The cooling truck can be withdrawn from a kiln before it is quite cold and the fresh one inserted to use the heat remaining in the kiln structure. When in position connections are made to heating elements in the end wall and the deck of the truck (Fig. 11.38).

In the *top hat kiln* the setting base is stationary thus eliminating any losses due to vibration on a movable truck. The heating mantle which makes up the walls and roof is lowered on to the base for each firing. Again ware can be set in comfort during day shifts and the heating mantles re-used as fast as the cooling ware can stand atmospheric conditions (Fig. 11.39).

These kilns, first developed for enamel firing, glost firing and then biscuit firing of earthenware and bone china, now also find application for firing electrical porcelain, abrasive wheels and sanitary ware. (Table 228, examples 16–23 incl.)

FIG. 11.38. *Trolley hearth or truck kiln (Kilns and Furnaces,* **K24.1**)

FIG. 11.39. *Top hat kilns showing alternate bases being fired or open for setting and drawing (Shelley,* **S64.1**)

SMALL SPECIALISED PERIODIC KILNS AND FURNACES

Specialised kilns and furnaces are required in laboratories for testing and development, and also for firing new bodies, single-component bodies, oxides, carbides, nitrides, etc., titanates, ferrites, etc. Very high temperatures, very exact temperatures, controlled atmospheres, application of loads or pressure at high temperatures may be required.

FIG. 11.40. *Gas-fired under load testing furnace (Brayshaw,* **B110**)

The designs of small kilns and furnaces are very numerous and cannot be entered into in detail. It is of interest, however, to note the conditions that can be achieved and the specialised types of furnace made for certain tests.

I. *Atmosphere*

Kilns are available for working with: (*a*) oxidising atmosphere; (*b*) reducing atmosphere; (*c*) neutral atmosphere; (*d*) vacuum (**M37**); (*e*) nitrogen.

II. *Temperature*

A kiln is described capable of maintaining temperatures in the range of 700–1000° C (1292–1832° F) for long periods, *e.g.* 4 months with fluctuations of only $\pm 1°$ C (**A208**).

A gas-fired test furnace capable of maintaining temperatures above 1725° C (3137° F) is described (**T25**), as are also electrical resistor furnaces which can be used even with an oxidising atmosphere up to 2000° C (3632° F) (**G21**), 2400° C (4352° F) (**D11**), a small research furnace in which a tungsten crucible large enough to hold a walnut is heated by radiant heat and high frequency induction is hoped to reach 2600° C (4712° F) (**A85**).

In another test furnace the heating zone can be moved hydraulically with respect to the stationary sample. Different atmospheres can be maintained (**S54**).

Most of these furnaces for very high temperatures are very small, making the testing of larger samples impossible. The research department of the Société Française de Céramique were therefore set the task of building a larger high-temperature kiln. The one they developed is capable of temperatures up to 1700° C (3092° F) and is 475 mm (18·7 in.) high, 450 mm (17·7 in.) wide and 450 mm (17·7 in.) deep. In order to maintain such high temperatures without enormous fuel consumptions the construction of the kiln and of the gas burners appears slightly complicated but is found to achieve the desired results using largely commercially available materials (**S148**).

III. *Special Testing Furnaces* (Fig. 11.40)

These are available for testing the following: (*a*) refractoriness; (*b*) refractoriness under load; (*c*) shrinkage; (*d*) transverse strength up to 1500° C

TABLE 227

Power or Gas Requirements of Certain Small Kilns

Ref.	Internal size						kW	V	A	ft³ gas/hr	Max. temp. achieved
	Width in	Height in	Length in	Width cm	Height cm	Length cm					
C52	10	8	18	25·4	20·3	45·7	4·8	230	21		
	8	6	14	20·3	15·2	35·5	3·4	230	14·8		
	10	6	18	25·4	15·2	45·7	4·5	230	19·6		
F35	15·7	15·7	27·5	40	40	70	12				
	27·5	21·6	29·5	70	55	75	24				
	47·2	25·5	40·1	120	65	102	48				
							kVA				
W31	12	10	18	30·4	25·4	45·7	5·5				
	15	13	18	38·1	33·0	45·7	9·5				
	18	16	27	45·7	40·6	68·6	18				
	24	22	27	60·9	55·9	68·6	32				
										360	1100° C in 3 hr
W71	14	14	18	35·5	35·5	45·7				400	1300° C in 5 hr

(2732° F) (**M13**); (*e*) temperature gradient method for determining firing range by observing shrinkage, hardness, bloating, crazing and colour of 8 in. (20·3 cm) specimens with an applied temperature gradient of several hundred degrees (**S200**); (*f*) spalling test by heating twelve standard refractory bricks simultaneously up to $1350° \pm 20°$ C ($2462 \pm 36°$ F) and then removing one brick from each side of the furnace at a time and subjecting it to an air blast for 10 min (**B23**).

Data on Ceramic Kilns

Much data on kilns in table form has been published and republished. Rather than re-use this, some of which may be out of date, the writers have tried to obtain fresh data from present-day users of kilns. Most ceramic manufacturers in Great Britain were asked (in 1956) to cooperate with this work and some thirty have supplied information on the questionnaires sent to them. This has been collated in Table 228, intermittent kilns, Table 229, continuous kilns, and Table 236, tunnel kilns, with the minimum of editing. Obviously many aspects have not been covered, but the tables probably do show the diversity of ceramic firing methods. (As the majority of contributors preferred to remain anonymous code numbers have been used.)

Interconnecting Intermittent Kilns

Heat is wasted in intermittent kilns in two main ways: (*a*) by allowing hot gases to escape up the flue: (*b*) by dissipating the heat given up by ware and structure during cooling.

The superimposition of two kilns, one for ware requiring less heat than the other uses some of the heat otherwise lost up the chimney (p. 976). By linking intermittent kilns by underground flues both the hot combustion gases and the clean hot air given off during cooling can be led into other kilns to help preheat the ware. A number of systems for firing interconnected round or rectangular down-draught kilns in a cycle have been tried but they have not come into widespread use (**R59**).

There are, however, interesting examples of linked kilns being built as seen in Table 229, examples 1 and 2. The kiln given in example 2 not only uses waste heat from the high-temperature kiln to fire the low-temperature one, but also incorporates an arrangement for burning, the black smoke from the chamber being reduced. Reduction of stoneware and porcelain above 800–1000° C (1472–1832° F) inevitably produces black smoke. This consists of finely divided carbon which does not ignite unless the air with which it is subsequently mixed is preheated to about 850° C (1562° F). Alterations to the connecting underfloor flue made in 1957–1958 incorporate an extra firebox under one of the main kiln fireboxes which can be heated by directing one of the oil burners downwards for about half an hour. The tertiary air is then drawn through this and mixed with the smoke coming from the kiln during the reducing period, full combustion then occurs and black smoke is emitted only during a short time while the change from oxidation to reduction is being made, instead of throughout the reducing period (**D14**).

Another way of using up waste heat from intermittent kilns is to withdraw it for use in dryers. Combustion gases must be kept in flues but clean air from cooling kilns can be led directly into the dryers.

Continuous Kilns (Moving Fire)

Consideration of intermittent kilns always leads to the conclusion that a very large proportion of the fuel is inevitably used up for purposes other than firing the ware. It is therefore natural that means should be sought to lessen this, and it can be readily deduced that by continuous instead of intermittent firing the heat given off by ware that is cooling can be used to heat raw ware and thus save fuel. There are two main methods of achieving continuous firing, the first has the ware stationary in a series of interconnected chambers or an annular ring and the fire progresses round the circuit, in the second the ware passes through a tunnel with appropriate preheating, main fire, and cooling zones.

The moving-fire type came into general use first and will be considered here.

THE HOFFMANN CONTINUOUS ANNULAR BRICK KILN

The first successful continuous kiln was invented by Friedrich Hoffmann in 1856. Many subsequent kilns have been modelled on it with the addition of some refinements. The original Hoffmann kiln consisted of a circular gallery* of average length 144 ft (~ 44 m) with twelve wickets set equidistantly round the outer wall, and twelve damper-controlled flue inlets in the inner wall. The flues lead to a smoke-collecting chamber and then a stack. Temporary paper partitions can be placed between each wicket and flue pair, dividing the kiln into twelve 'chambers' as desired. This kiln was designed to fire bricks which can be allowed to come into intimate contact with burning fuel and was fuelled through holes in the roof (Fig. 11.41).

The firing is as follows. When chamber No. 1 is empty, No. 2 is being emptied, Nos. 3, 4 and 5 are cooling with the assistance of air coming in from the open chambers (Nos. 12, 1 and 2), this heated air passes on to chamber No. 6 where the firing is almost complete, and Nos. 7 and 8 under full fire with fuel being dropped in from the roof. Hot gases from these fired chambers pass on to chambers Nos. 9, 10 and 11 bringing them successively to red heat when fuel will be added and ignite immediately. When the gases are cooled to about 150° C (302° F) they pass into flue No. 11. Between chambers Nos. 11 and 12 there is a paper partition to keep the draught in the right direction. Chamber No. 12 is being filled. When this is completed a partition will be erected between No. 12 and 1, the one between Nos. 11

---

* *Gallery*: It is proposed to use this term for the arched tunnel-like structure of a continuous kiln in order to avoid confusion with 'tunnel kilns'.

TABLE
Intermittent

No.	1	2
Code No.	13	19
I. (a) Type of kiln	Bottle	
(b) Up-draught, down-draught or horizontal draught	Up-draught	Up-draught
(c) Rectangular or round	Round	Round
(d) Area internally (ft²)	165 d. ~14½ ft	78¼ d. 10 ft
(e) Capacity (ft³)	2310	h. 10 ft
II. (a) Fuel used	Coal (seams: 10 ft, mixture Bowling-alley, Hardmine, Bulhurst)	Coal
(b) Heating capacity of fuel (B.t.u./lb or per ft³)	—	—
(c) Type of grate or burner	Hob grate	5 open fireboxes
III. (a) Nature of ware	Green china	Earthenware
(b) Moisture in green ware as set (%)	<1	5
(c) Nature of setting, saggars or open, with or without auxiliary supports	Saggars and setters	Saggars and open with supports
(d) Maturing temperature or cone No.	1265° C	1000° C
IV. (a) Weight of setting: ware	3572 lb	8–9 tons
(b) kiln furniture	35 317 lb (including green)	?
total	44 248 lb (plus 5359 lb alumina)	
(c) Quantity of setting	447 dozen (5364 pieces)	5000 to 8000 pieces according to size
V. (a) Time of firing: preheat and full fire	34 hr preheat, 57 hr full fire	30 hr preheat, 24–28 hr full fire
(b) cooling	72 hr	48 hr
(c) total	129 hr	4 days
VI. (a) Fuel consumption: for complete firing	12–14 ton per oven	3 tons
(b) per ton/lb ware	8 lb per lb ware	⅓ ton per ton ware
(c) per quantity of ware	67 lb per doz	~½ ton per 1000 pieces
VII. (a) Labour requirements: skilled	24 shifts (8 hr)	2 men
(b) unskilled	9 shifts (8 hr)	1 man

Kilns

3	4	5	6
9	1	25	18
Wilkinson	Wilkinson's (bottle)	—	Beehive
Down-draught	Down-draught	Down-draught	Down-draught
Round	Round	Round	Round
270 d. ~18½ ft	300 d. ~19½ ft	201 d. 16 ft	78·5 d. 10 ft
3462	6300	1474	~490 h. 5 ft 6 in. to shoulder 8 ft to dome
Coal	2 in. nuts (graded)	Coal	Shell gas oil 4103
12 700	13 920 with 3·8% moisture and 4·3% ash	As for steam hards	?
10 firemouths, horizontal grate 30 in. long × 18 in. wide × 24 in. deep	Inclined (hob type) bar	Fire bars	4 Massey burners with separate air and fuel controls
White wall tile biscuit 7–8	Clayware biscuit 1–2 max.	Plant pots 10–15	Earthenware 1–1½
Saggars	Saggars	Open with supports	Saggars
1120° C	B.R. 38–40 ~1150° C	960° C	1120° C
42 tons (green) 40 tons (burnt)	22 tons	15 tons approx.	3400–3700 lb
54 tons (burnt)	34 tons	5 tons approx.	12 000–13 000 lb
96 tons	56 tons	20 tons approx.	15 400–16 700 lb
3200 yards super	7000 dozens (84 000 pieces)	30 000–40 000 plant pots	~5000 pieces
144 hr	63 hr	48 hr	22–24 hr
120 hr	19–72 hr	63 hr	4 days
264 hr	132 hr	111 hr	5 days
20 tons	23 tons	4½ tons approx.	330 gal
¼ ton per ton ware	1·05 lb/lb ware	6 cwt	~$\frac{1}{10}$ gal per lb
6¼ ton per 1000 yd super	736 lb/doz ware	~7½ tons/1000	~1 gal per 15 pieces
27 } to operate 10 kilns 2 } in rotation	skilled	51·5 man-hr 102·0 man-hr	skilled and/or semi-skilled; 100–130 man-hr for setting and drawing plus 10 man-hr for firing

TABLE

Intermittent

No.	7	8
Code No.	11	26
I. (a) Type of kiln	Beehive	—
(b) Up-draught, down-draught or horizontal draught	Down-draught	Down-draught
(c) Rectangular or round	Round	Round
(d) Area internally (ft^2)	314 d. ~20 ft	270 d. ~18½ ft
(e) Capacity (ft^3)	2200	2200
II. (a) Fuel used	Coal (unscreened)	Coal
(b) Heating capacity of fuel (B.t.u./lb or per ft^3)	10 000 as received basis	13 000
(c) Type of grate or burner	Open grate	4 under-feed stokers
III. (a) Nature of ware	Fire bricks	Machine-made quarries and red hand-made tile
(b) Moisture in green ware as set (%)	5	10
(c) Nature of setting, saggars or open, with or without auxiliary supports	Open setting, no supports	Open, without auxiliary supports
(d) Maturing temperature or cone No.	1300° C	1080° C; Buller's ring values: Top 25 Bottom 14,
IV. (a) Weight of setting: ware	75 tons	54 tons
(b) kiln furniture	Nil	None
total	75 tons	54 tons
(c) Quantity of setting	15 000, 3 in. firebricks	—
V. (a) Time of firing: Preheat and full fire	Drying 2 days, preheat 10 days, Topheat 2 days	Preheat 200 hr approx Full fire 100 hr
(b) cooling	Cooling 3 days	240 hr
(c) total	17 days	540 hr
VI. (a) Fuel consumption: for complete firing	—	18 tons
(b) per ton/lb ware	7·5 cwt per ton	6⅔ cwt fuel (ton fired product)
(c) per quantity of ware	37·5 cwt/1000	—
VII. (a) Labour requirements: skilled	3 setting, 3 drawing	35 man-hr per oven
(b) unskilled	Nil	28 man-hr per oven 10 semi-skilled per oven
VIII. (a) Salt glazing: consumption in one firing		
(b) per ton/lb ware		
(c) per quantity ware		

228—continued

Kilns—continued

9	10	11	12
24	11	15	15
Intermittent round beehive	Rectangular	—	—
Down-draught	Down-draught	Down-draught	Down-draught
Round	Rectangular	Rectangular	Rectangular
210 d. ~16ft.	—	674	520
3238	—	7583	6407
Coal (washed nuts)	Piece or washed coal	Coal	Coal
	—	13 400	13 400
Inclined	Automatic stokers	Inclined bar grate	Prior stokers
Red clay hollow-ware	Bricks, stiff plastic (pressed)	Stoneware pipes and fittings	Stoneware pipes and fittings
—	12	3	3
Open with bricks and tiles	Open setting (five on two)	Open on rings with rings to tie top, topped off with odd stuff	Open on rings with rings to tie top, topped off with odd stuff
900° C	1020° C	1200° C	1200° C
24 tons	320 000 lb	39 ton 2 cwt	34 ton 16 cwt
27·4 tons	—	2 ton 8 cwt	2 ton 16 cwt
51·4 tons	320 000 lb	41 ton 10 cwt	37 ton 12 cwt
—	32 000 bricks	—	—
48 hr	⎧ 7 days preheat	96 hr	96 hr
58–66 hr	⎨ 7 days full fire	96 hr	96 hr
140 hr	4 days	192 hr	192 hr
	18 days		
	—		
	35 tons	29 ton 0 cwt	24 ton 0 cwt
930 lb fresh fuel per ton ware	640 lb per ton	14·8 cwt/ton	13·8 cwt/ton
—	1 ton for 1000 bricks	—	—
3 men	3 setters and 3 burners	⎧ 1 man/2 kilns slow fire	⎧ 1 man/2 kilns slow fire
3 men	2 runners and 3 drawers	⎨ 1 man/1 kiln mid fire	⎨ 1 man/1 kiln mid fire
		⎩ 2 men/1 kiln full fire finishing	⎩ 2 men/1 kiln full fire finishing
		500 lb	450 lb
		13 lb/ton	13 lb/ton

TABLE

Intermittent

No.	13	14
Code No.	22	11
I. (a) Type of kiln	Sherwin and Cotton	Newcastle
(b) Up-draught, down-draught or horizontal draught	—	Combination
(c) Rectangular or round	Rectangular	Rectangular
(d) Area internally (ft^3)	44	160
(e) Capacity (ft^3)	330	1360
II. (a) Fuel used	Coal	Unscreened coal
(b) Heating capacity of fuel (B.t.u./lb or per ft^3)	—	12 500
(c) Type of grate or burner	Plain mouths	C.I. bars
III. (a) Nature of ware	Glazed tiles glost	Best quality facing bricks
(b) Moisture in green ware as set (%)	—	12
(c) Nature of setting, saggars or open, with or without auxiliary supports	In cranks	Open
(d) Maturing temperature or cone No.	1040° C	1060° C
IV. (a) Weight of setting: ware	3 ton	35 ton 9 cwt
(b) kiln furniture	50 cwt	Nil
total	5 ton 10 cwt	—
(c) Quantity of setting	100 yd^2 tile + 120 doz. 8 × 4 block	10 750
V. (a) Time of firing: preheat and full fire	24 hr	8 days
(b) cooling	24 hr	3·4 days
(c) total	48 hr	11 to 12 days
VI. (a) Fuel consumption: for complete firing	2 ton 5 cwt	13 ton 8 cwt
(b) per ton/lb ware	15 cwt/ton ware	7·5 cwt/ton ware
(c) per quantity of ware	—	1 ton 5 cwt per 1000
VII. (a) Labour requirements: skilled	2 ⎫ for drawing, setting	3 men
(b) unskilled	1 ⎭ and firing	1 part-time

228—continued

Kilns—continued

15		16	17
8		27	16
Recuperative oil fired		Webcot electric	Electric
Down-draught with horizontal draught		—	—
Rectangular		Rectangular	Rectangular
4½		4	15 and 30
9		10	—
Paraffin oil, tractor vaporising oil (TVO) or light diesel oil		Electricity	Electricity
19 600		—	—
Single jet high-pressure atomising oil burner		—	—
Stoneware biscuit	Glost	Stoneware biscuit	Biscuit earthenware and bone china
Well dried	—	6–7	
Open setting in semi-muffle, with supports for glost		Open on batt	
855	1350–1400	1200	Earthenware 1170° C, bone china 1250° C
80 lb	30 lb		
Nil	67¼ lb		
~80	~97¼ lb		
—	—		
9–10 hr	12–13 hr	24 hr (4 hr to 400° C)	18–24 hr
36 hr	36–48 hr	24 hr	16 hr
46 hr	48–61 hr	48 hr	34–40 hr
	22 gal		About £1 worth of electricity
8¼ gal/80 lb	13½ gal/30 lb		
1 or 2 operators for 13 hr	—	1 man	No special labour requirements

TABLE

Intermittent

	No.	18	19
	Code No.	2	13
I.	(a) Type of kiln	Muffle	Two trucks on track forming complete box when pushed together from one end of small blank-ended tunnel
	(b) Up-draught, down-draught or horizontal draught	None	—
	(c) Rectangular or round	Rectangular	Rectangular
	(d) Area internally (ft²)	7·9	20
	(e) Capacity (ft³)	18	70·5
II.	(a) Fuel used	Electricity	Electricity
	(b) Heating capacity of fuel (B.t.u./lb or per ft³)	—	—
	(c) Type of grate or burner	30 kW elements	Kanthal 'A' wire
III.	(a) Nature of ware	Biscuit, glost or enamel	Green china
	(b) Moisture in green ware as set (%)	?	<1
	(c) Nature of setting, saggars or open, with or without auxiliary supports	Open in shelves	Setters for flat. Open with normal kiln furniture for hollow-ware
	(d) Maturing temperature or cone No.	Biscuit C, 985° Glost 1070° C	1265° C
IV.	(a) Weight of setting: ware	5 cwt 3½ cwt	414 lb av.
	(b) kiln furniture	1 cwt 1½ cwt	1800 lb av. (alumina included)
	total	6 cwt 4 cwt	Flat ware 2900 lb, Hollow-ware 1800 lb
	(c) Quantity of setting	2500 pieces 700 pieces	68 doz. av. (flat and hollow-ware)
V.	(a) Time of firing: preheat and full fire	9 hr	preheat 11 hr, full fire 32 hr
	(b) cooling	15 hr	14 hr
	(c) total	24 hr	46 hr
VI.	(a) Fuel consumption: for complete firing	~270 units	2050 kW
	(b) per ton/lb ware	54 units/cwt 79 units/cwt	5 kW/lb
	(c) per quantity of ware		30 kW/doz.
VII.	(a) Labour requirements: skilled	1	1 shift (8 hr)
	(b) unskilled	—	

228—*continued*

Kilns —*continued*

20 3 Gas kilns converted to electricity by boxing		21 23,1 Electric. 2 superimposed chambers, upper taking waste heat	
Nil			
Square 9 30 Electricity		Rectangular 4 6·3 Electricity	1·8 1·7
Spiral elements Brown clay (Staffordshire) Biscuit 2 approx. Glost Open with auxiliary supports		Raw one fire glazed 5 shelves	Decorating

Thermoscope bar
  No. 18, 985° C
8 cwt
2 cwt
	Thermoscope bar
	  No. 22, 1080° C
	4 cwt
	2 cwt
		1040° C
		
		1 cwt
		7 lb
			800° C

10 cwt	6 cwt		
100 doz.	60 doz.	All-size pottery	
Preheat 4 hr, firing 9 hr		14 hr	
24 hr 33 hr 432 units of electricity 54 units/cwt 4·32 units/doz.	 108 units/cwt 7·2 units/doz.	24 hr 38 hr 170 units 170 units/cwt	
2		Automatic switch gear	

TABLE 228—*continued*

Intermittent Kilns—*continued*

No. Code No.			22 23 II	23 23 III
I.	(a)	Type of kiln	Electric	Electric
	(b)	Up-draught, down-draught or horizontal draught		
	(c)	Rectangular or round	Rectangular	Arched rectangular
	(d)	Area internally (ft^2)	3	4
	(e)	Capacity (ft^3)	5·6	7·5
II.	(a)	Fuel used	Electricity	Electricity
	(b)	Heating capacity of fuel (B.t.u./lb or per ft^3)		
	(c)	Type of grate or burner		
III.	(a)	Nature of ware	Raw clay ware, glazed hot	Raw clay ware, glazed hot
	(b)	Moisture in green ware as set (%)		
	(c)	Nature of setting, saggars or open, with or without auxiliary supports		Open with 5 shelves
	(d)	Maturing temperature or cone No.	1040° C	1040° C
IV.	(a)	Weight of setting: ware	1 cwt	1 cwt
	(b)	kiln furniture		7 lb
		total		119 lb
	(c)	Quantity of setting		
V.	(a)	Time of firing: preheat and full fire	12 hr	14–15 hr
	(b)	cooling	24 hr	24 hr
	(c)	total	36 hr	38 hr
VI.	(a)	Fuel consumption: for complete firing	150 units	170 units
	(b)	per ton/lb ware	150 units per cwt	170 units per cwt
	(c)	per quantity of ware		
VII.	(a)	Labour requirements: skilled	Automatic switch gear	½ hr all told
	(b)	unskilled		

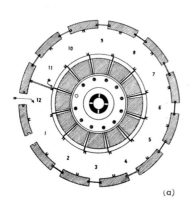 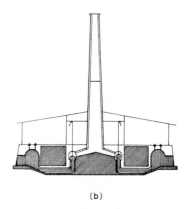

FIG. 22.41. (a) *Plan and* (b) *cross-section of the original Hoffmann kiln*

and 12 pierced and No. 12 wicket bricked up. By this time No. 2 should be empty and No. 3 cool enough to draw, while the main fire will be in Nos. 7, 8 and 9.

Variants on the Hoffmann kiln involve some or all of the following factors:

(1) Oblong instead of circular plan.
(2) Zigzag plan.
(3) Greater length of gallery to allow slower burning of tender clays.
(4) Omission of arches.
(5) Flue and damper arrangements to use hot air, from outside or cooling chambers, instead of kiln gases for the water-smoking of green bricks.
(6) Grates instead of roof fuelling to pillars to prevent direct contact of fuel with the ware.
(7) Structural instead of temporary division into chambers.
(8) Arrangements of bags, screen walls and flues to give either up-draught, down-draught or horizontal draught firing.
(9) Fans instead of chimney stacks.
(10) Automatic stokers.
(11) Gas or oil instead of solid fuel.

(1) THE OBLONG ANNULAR KILN (Fig. 11.42)

Instead of the original circular gallery most continuous kilns now have two straight series of chambers joined at the ends either by a semicircle of chambers (see Fig. 11.42), rectangularly placed chambers, or by connecting flues only. This is more economical of site space, easier to attend to by truck and lorry and gives most chambers a rectangular shape which is easier to set.

(2) THE ZIGZAG KILN

To be effective a continuous kiln must be in constant use at full capacity. The annular Hoffmann kiln has a minimum economic size. Where smaller but continuous production is desired the zigzag kiln should be considered. Here the firing gallery is narrow and is doubled back on itself in a double zigzag. The wickets in the side walls of the kiln as a whole now come in the ends of the individual 'chambers'. Fig. 11.43 shows a twelve 'chamber' example.

Compared with the annular kiln the zigzag kiln is more economical of fuel, partly because there is relatively less masonry and partly because the fire front travels so fast that the masonry is not necessarily heated up to equilibrium conditions.

If the output from a zigzag kiln with its narrow firing gallery is to equal that of an annular kiln the length of gallery must be longer and the fire travel faster. Thus where a normal annular kiln is 70 m (230 ft) long the zigzag one must have a total length of 90 to 120 m (295–394 ft) (**A166**).

Although the kiln can be satisfactorily served by a natural draught chimney

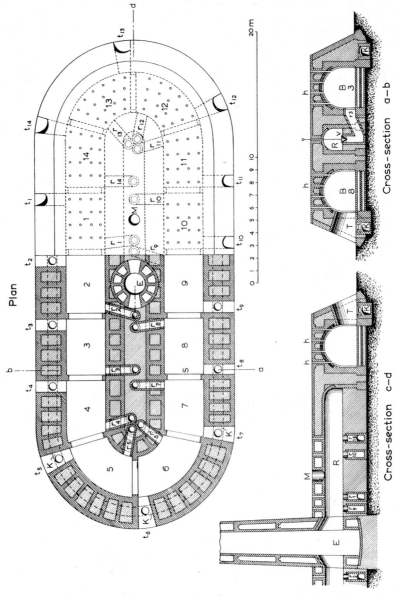

FIG. 11.42. *The annular kiln:*

B, firing gallery divided into sections (1) to (14)
T, wicket, t_1–t_{14} individual wickets to sections (1)–(41)
r_1–r_{14}, flues
R, smoke-collecting chamber
v, damper connecting individual flues r to smoke-collecting chamber R
E, chimney stack
S, movable partition between chambers
h, firing holes
K, hot air flues

stack it is very well suited to fan draught, when the tall stack can be omitted.

The kiln is suitable for firing common, facing and engineering bricks. In an average twenty-eight chamber zigzag kiln the weekly output is about 250 000 bricks (**H89**).

The narrow gallery width is very well suited for the setting of roofing tiles which are less easily placed in the long rows required by a wider firing gallery (**A166**).

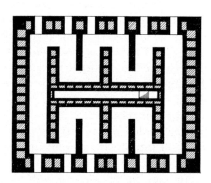

FIG. 11.43. *The zigzag kiln*

(3) THE LENGTH OF THE GALLERY

The early continuous kilns had twelve chambers of at least 12 ft (3·6 ft) length, giving a total of 144 ft (44 m). Greater length is really desirable to give slow enough preheating and cooling. Searle (**S49**) suggests at least 225 ft (68·6 m) total length and at least sixteen chambers.

Hielscher (**H86**) points out that the fuel consumption of a continuous kiln falls as the speed of fire travel increases. So that a long gallery allowing rapid fire travel is an economy, so long as it is kept in full production and is not too narrow.

Thor (**T27**) confirms this in his study of kiln lengths and firing times. He found a fire travel speed of 15 m (49 ft) per day with a whole circuit taking 6 days, and therefore a total length of 90 m (295 ft) gave the best results.

(4) THE ARCHLESS KILN

Continuous gallery kilns are sometimes constructed without the arch or crown, *i.e.* with walls and flues only. The arches are replaced by a temporary covering of two courses of bricks and layer of ashes placed on top of the setting. The kiln is fired in the same way as other continuous kilns, the flues being arranged to give up- or down-draught as desired. An indication of the maturity of a chamber is given by the amount that the level of the covering bricks has sunk below those on unburnt chambers.

The ashes are removed from cooling chambers allowing rapid cooling. The wickets are made large enough to take a lorry, and loading can be done by crane through the open roof (**S49**).

Although cheaper to build than complete kilns they involve the hot and dusty job of placing and removing the temporary roof. There are also considerable heat losses through the roof making it hot for the burners servicing the fireholes in it (**H89**).

(5) THE WATER-SMOKING OF A FRESH SETTING

In the simplest system of air and gas movement in a continuous kiln fresh air enters from the empty chamber, receives heat from the cooling ware, becomes mixed with combustion products in the chambers under fire, and then picks up moisture in the chambers being dried and preheated. When it has cooled to 120° C (248° F) there is a serious risk that it will deposit water, salts, etc., on any further cooler ware it meets and cause scumming, etc. It should therefore be diverted to the flue at this point.

The preheating of ware to 120° C (248° F) without scumming, etc., must be brought about by other means.

(*a*) Direct heating by small independent fires in the wicket or roof fireholes. This has the disadvantage of introducing combustion gases.

(*b*) Diverting heated air from the cooling chambers through special flues. This is clean hot air but may be insufficient. Primary air must then be admitted to the firing chambers to make up for the diverted air.

(*c*) Hot air from heat exchangers, either independent or built into the roof of the kiln and heated by the chambers in full fire.

Chambers being water-smoked individually must be isolated from the main gallery by partitions, etc., and the damper to their own flue opened.

(6) GRATES, FIREBOXES, ETC.

In the original Hoffmann kiln the fuelling is through the arched roof into pillars built up in the green ware. One or, better, two bricks with $\frac{1}{8}$ in. (3 mm) gap between them, stick out into the pillar at intervals to catch the fuel so that it burns at various levels. In this way, however, a considerable number of bricks are spoilt either by discoloration or by over-firing.

In many other continuous brick kilns a flat grate, or a trough or gutter is built right across each chamber with a primary air supply under it. One such grate will serve 15 ft (4·6 m) of the gallery for normal bricks. It is fuelled through fireholes either in the roof or in one side of the kiln. Far fewer bricks are discoloured or damaged.

In chamber continuous kilns, particularly those with down-draught the firing is with fireboxes and grates as found in periodic kilns. Automatic stoking can be introduced and usually results in a saving of fuel and more 'firsts'.

(7) CHAMBER KILNS

For better control of the movement of gases and the fire front itself it is often considered advisable to make masonry divisions between the chambers.

FIRING AND KILNS

These may be complete so that communication between adjacent chambers is limited to flues, or partial, with or without controllable dampers over the openings.

From the structural point of view this allows deeper chambers as the arches can spring from the partition walls instead of the front and back.

(8) DOWN-DRAUGHT

Chamber kilns can be constructed with fireboxes, bag or screen walls and underfloor flues to operate like an intermittent down-draught kiln to which are added the fuel savings of a continuous kiln. In this way high-class ware can be fired.

(9) CHIMNEY DRAUGHT VERSUS FAN DRAUGHT

The idea that a fan uses power whereas a chimney only necessitates the initial cost of construction is a false one. A chimney stack will only cause a draught when hotter gases enter at the bottom than surround it at the top. So that a chimney will not draw properly if only completely spent gases with no remaining heat, are fed to it. In effect fuel is consumed to keep the chimney draught going, and this is generally an irregular and unknown quantity.

In order to obtain a uniform draught, whatever the condition of the combustion gases, at a known expenditure of power, many engineers now recommend using a fan (**H89**).

(10) AUTOMATIC STOKERS

Various kinds of automatic stokers can be fitted to the roof or side fireholes as required.

OTHER NARROW KILNS (*see* Table 229, example 3)

Apart from the original and improved *Hoffmann* kiln and the *zigzag* kiln, both of German nineteenth-century origin, there are a number of other 'narrow' kilns with a long narrow barrel arched gallery. These differ from the Hoffmann in the details of their ground plan, the disposition of the flues and the introduction of grates instead of pillar firing.

The 'Ideal' kiln, of English design, has an oblong gallery. The main flue to the fan or chimney runs underground along the length of each side of the kiln. This is a source of heat loss. The sub-flues leading to it have a dual role, first as an exit for vapour and combustion products and later as a channel furnace supplied with primary air. Each 'chamber' has a transverse grate fuelled through holes in the roof. Each wicket has a furnace with a drop arch 3 ft (0·91 m) back, reducing the capacity of the kiln by one-third.

The kiln is suitable for facing bricks.

The 'Belgian' kiln may be a continuous gallery or it may be partitioned into chambers. It is fired by a single transverse grate in each chamber fuelled from one end at ground level. Used as a continuous gallery it makes high-quality firebricks and engineering bricks (*see* Table 229, example 4). This kiln can be successfully converted to automatic stoking with considerable fuel saving (**A105**).

'*Brown's Patent*' *kiln* may be a continuous gallery or divided into chambers. It is fuelled through roof fireholes into a trough grate supplied with heated primary air. It will fire roofing tiles, land drains, engineering, facing and rustic bricks in various colours (**H89**).

CONTINUOUS CHAMBER KILNS WITH INCOMPLETE PARTITIONS

The Staffordshire and Super Staffordshire Kilns

These are two of the most elaborate continuous kilns with arrangements for complete control of the temperature and atmosphere prevailing at any point. The Super Staffordshire is very suitable for high-quality facing bricks. Its features are as follows:

(1) Division into chambers by perforated walls over which dampers can be lowered. Any chamber can therefore be completely isolated if desired but the fire can progress through the partition when the dampers are off. This allows controlled water-smoking, preheating, soaking and cooling.

(2) In the Staffordshire kiln firing is through holes in the roof into charge holes in the ware as in the Hoffmann. In the Super Staffordshire kiln the fuelling is through the roof but on to grates. Under these grates there is a deep passage to admit air and collect ash. The air may be fresh or from the hot air flue. There is no contamination of the bricks with ash, etc.

(3) Each chamber works on up-draught. Damper control allows the gases either into the flue system leading to the stack or to that leading to a main hot air flue. This hot air is then available at any of the chambers for combustion of the fuel, drying or preheating of green ware.

(4) Fresh cold air can be admitted to any chamber as desired, and mixed with the hot air where necessary.

(5) The volume of air admitted to the chambers under fire can be adjusted to produce reducing or oxidising conditions (*see* Table 229, examples 5, 6, 7).

The Manchester Kiln has many features similar to the Staffordshire kiln.

The '*English*' *Kiln*

This chamber kiln is used largely for fletton bricks. The chambers are divided by perforated walls through which the fire can travel. The arches are transverse, allowing large chambers to be built. The chief feature is the ingenious system of hot air flues placed in two parallel rows on top of the arches. The circular plate dampers regulate the direction of flow and the amount of hot air entering or escaping from each chamber. Subsidiary flues supply the hot air to the bottom of the chambers and collect it from the top.

The Lancashire continuous chamber kiln is designed for mass production of common bricks. The large transverse-arched chambers have wickets big enough to allow lorries to enter for loading. These wickets also allow rapid cooling. Production may be 250 000 to 300 000 per week from one fire face (**H89**).

CONTINUOUS CHAMBER KILNS (COMPLETE PARTITIONS)

A number of kiln types consist of a series of individual chambers connected only by flues. They may or may not have a common wall.

The Barnett and Hadlington Kiln is the only well-known English true continuous chamber kiln. It consists of a series of individual chambers with transverse arches separated by a small gap containing flues, dampers, etc. The hot gases within the chamber are drawn off through the floor at one side, and enter the next chamber at the side opposite to the offtake, also from the floor. Long grates fed from the front of the kiln are used, and when fuel is being burnt the main flue opening in the middle of the floor is brought into use. The combination of horizontal draught for preheating and up- and down-draught for firing is considered to give the most satisfactory heat distribution.

Water-smoking is achieved with hot air from the cooling chambers and the latter also is used as primary air for the fires. These kilns are more popular on the Continent where well-known designs are by Bock, Diesener, Mendheim and Dannenberg.

Dannenberg's Kiln has a built-in system of flues to collect hot air from the top of a cooling chamber and deliver it to the top of a preheating one. It is drawn off the latter at the bottom.

GAS-FIRED CONTINUOUS CHAMBER KILNS

The more refined continuous kilns can very successfully be fired by gas. Very much greater cleanliness is thereby achieved making the kilns suitable for ware other than bricks. It may also be possible to use a coal (coke, etc.) in the gas producer that would be unsuitable for direct firing of a kiln.

For very high temperatures gas is more economical than coal and it is also more suitable for producing reducing atmospheres at elevated temperatures.

Richardson (**R15**) lists a number of advantages of the gas-fired continuous chamber kiln as compared with coal firing, intermittent and tunnel kilns:

(1) The fuel saving as compared with intermittent kilns is 60 to 76%.

(2) The kiln plant is compact and uses much less space than intermittent kilns.

(3) No coal or ashes are distributed round the plant. Mechanical unloading of coal and removal of ashes is all at one place.

(4) The firing is rapid and requires fuel only after the hot gases from previous chambers have brought the ware to 'red' heat, or 'white' heat.

(5) Burning can be better controlled than in periodic kilns.

(6) Heat distribution is uniform, giving uniform products.

(7) Greater flexibility and adaptability to varied ware than in the tunnel kiln as the fire can be hastened or retarded by damper control.

(8) No sulphuring of goods as pure hot air is used for the water-smoking.

(9) Less cost for repairs than the periodic kiln and greater accessibility for repairs during production than in the tunnel kiln.

The Mendheim Gas Chamber Kiln

This consists of a connected series of gas-fired down-draught chamber kilns. In the earlier designs the burner(s) was situated behind a bag wall at one side of the chamber, more recently one burner has been placed in each corner. The hot gases are drawn off through a perforated floor (old system) or a single central hole in the floor (newer system) and passed on to the bags of the next chamber or to the central flue.

The Shaw Producer Gas Fired Kiln

This too is a connected series of gas-fired down-draught chambers and is used for firebricks, silica bricks, glazed bricks, fireclay sanitary ware, glost earthenware, etc. (**R59**) (*see* Table 229, example 8).

The Dunnachie Kiln is especially designed for high-temperature firing of firebricks, but is of course also suitable for lower temperatures.

The two rows of five chambers are set some 20 ft (6 m) apart with the main flue running outside them to a stack at one end. The central space has the gas producer at one end of it with the delivery pipes running the shortest route to each chamber to utilise the heat in the gas as much as possible. (The space may be used as a drying chamber.) Sub-flues connect the chambers so that air heated in the cooling chambers is used for the combustion of the gas. The hot combustion gases then pass through succeeding chambers to preheat the ware. The water-smoking is done independently with any surplus hot air from cooling chambers or by burning some gas with cold air, openings in the roof and round the walls allow the steam out.

Oil firing can be directly introduced into any kiln designed for burning the fuel on grates, preferably with the stoking from ground level (Fig. 11.44).

Tunnel Kilns

In the discussion of periodic kilns it was concluded that for mass production a more continuous method of firing that re-used the heat given off by cooling ware would be more economical. The first successful improvement was the Hoffmann continuous kiln, where the fire front moves through the stationary ware. It can be noted, however, that although this kiln re-uses the heat in the ware, fuel must still be used to heat up successive portions of the masonry of the kiln as the fire moves round the circuit.

In the tunnel kiln the alternative method of continuous firing is applied, namely the ware is moved through a heated tunnel. The temperature encountered by the ware as it moves through the tunnel increases and then decreases gradually as in an intermittent kiln but the kiln structure at any

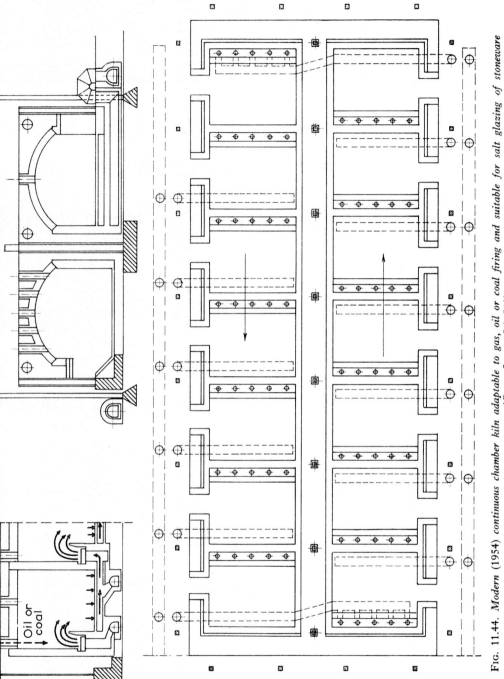

Fig. 11.44. Modern (1954) continuous chamber kiln adaptable to gas, oil or coal firing and suitable for salt glazing of stoneware (Ooms, Ittner, **O6, I10**)

TABLE

Continuous

(see introductory note

No.	1	2	
Code No.	20	Ref. **D14**	
I. (a) Type of kiln	3-chamber semi-continuous down-draught 'climbing' kiln.	2-chamber connected kiln	
(b) Open gallery or chamber	Chamber	Chambers	
(c) No. of chambers	3, 2 glost, 1 biscuit	2, 1 glost, 1 biscuit	
(d) No. of chambers per circuit	3	2	
(e) Area of chamber (ft^2)	$19\frac{1}{4}$	—	
(f) Volume of chamber (ft^3)	110	260 round	240 rectangular
II. (a) Fuel used	Fuel oil atomised with wood finishing	Gas oil (diesel oil)	
(b) Heating capacity of fuel (B.t.u./lb or per ft^3)	?	—	
(c) Type of grate or burner	Massey burner with large fan to provide an ample volume of air at about lung pressure	3 Massey low pressure burners	—
III. (a) Nature of ware	Reduced or oxidised stoneware and porcelain	Domestic stoneware and porcelain	
		Glost	Biscuit
(b) Moisture in green ware as set (%)	Bone dry	—	
(c) Nature of setting, saggared or open, with or without auxiliary supports	Saggars in front facing the flames and open shelves behind	Saggars and kiln furniture	Saggars and kiln furniture
(d) Maturing temperature or cone No.	Glost cone 10 $\sim 1300°$ C Biscuit cone 010 $\sim 900°$ C	Cone 10–12 reducing	09A
IV. (a) Setting: wt. ware/chamber	—	1800 lb	1800 lb
(b) wt. kiln furniture/chamber	—	3950 lb	3580 lb
(c) No. of pieces/chamber	750 of average 1 pint beer size	—	—
V. (a) Time of firing: time in kiln	Average 25 hr start to finish	44 hr	41 hr
rate of finishing chambers	—	—	—
VI. (a) Fuel consumption: per chamber per wt. ware per quantity ware	Per kiln firing 130 gal	300 gal	
VII. (a) Labour requirements: skilled	2 at a time in shifts		
unskilled	—		

229

Kilns

to Table 228, p. 984)

3	4	5	6
12	5	11	11
Hoffmann	Belgian	Staffordshire	Staffordshire
Open	Open	Chamber	Chamber
14	(Setting spaces) 28	20	20
14	(Setting spaces) 28	10	20
225	(Setting spaces) 98	—	336
1300	(Setting spaces) 750	—	2519
Coke breeze, preferably under $\tfrac{3}{8}$ in.	Coal	Fine or ground coal, also D.S. nuts for wicket ends	Doubles
ca. 11 000 as received	13 000 approx.	—	13 000
Kiln bottom via automatic feeders	Usual Belgian kiln horizontal firebar	Automatic Riley stokers	Fire-clay grids on one side
Facing bricks and agricultural drainpipes	Firebrick	Stiff plastic pressed (shales)	Best quality facing bricks
~4	~8 average	12	12
Open	Open	Open setting (five on two)	Open (5 on 2)
1020° C	Cone 12–14	1020° C	1060° C
31 tons, fired wt.	(Setting space) ~22 tons	180 000 lb	61 tons (fired wt.)
0	—	—	Nil
—	(Setting space) ~5800 ($9 \times 4\tfrac{1}{2} \times 3$ in.)	18 000	18 500
14 days	280 hr, paper to drawing, 42 hr. firing	20 days (five days firing)	26 days
1 per day		1 per day	4·84 chambers per week
50 cwt	—	114 cwt	6 ton 10 cwt
1·6 cwt/ton (fired wt.)	3·1 cwt/ton (fired wt.)	6·2 cwt per $3\tfrac{1}{2}$ ton bricks	2·13 cwt per ton ware
$3\tfrac{1}{2}$ cwt/1000 ($2\tfrac{5}{8} \times 4\tfrac{3}{8} \times 8\tfrac{3}{4}$ in.)	12 cwt/1000 ($9 \times 4\tfrac{1}{2} \times 3$ in.)	6·2 cwt per 1000 brick	7·03 cwt per 1000
4 (including burners) including lorry 6 loading	1 kiln burner 1 operative responsible for coal and ashes	5 { 3 setters; 3 burners; 1 coal wheeler; 4 drawers }	3; 1 part-time

TABLE 229—*continued*

Continuous Kilns—*continued*

No.	7	8
Code No.	11	29
I. (a) Type of kiln	Super Staffordshire transverse arch	Shaw
(b) Open gallery or chamber	Chamber	Chamber
(c) No. of chambers	16	14
(d) No. of chambers per circuit	3 burning, 3 pre-heating, 4 drying (6 cooling, drawing and setting)	14
(e) Area of chamber (ft^2)	294	316
(f) Volume of chamber (ft^3)	2190	2928
II. (a) Fuel used	Coal (unscreened)	Crude producer gas
(b) Heating capacity of fuel (B.t.u./lb or per ft^3)	12 000 as received basis	—
(c) Type of grate or burner	Open grate	—
III. (a) Nature of ware	Building bricks	Diatomite insulating bricks
(b) Moisture in green ware as set (%)	15	21
(c) Nature of setting, saggared or open, with or without auxiliary supports	Open, no supports	Open
(d) Maturing temperature or cone No.	1000° C	Holdcroft bars 8–10 ($\sim$840–875° C)
IV. (a) Setting: wt. ware/chamber	72 tons	20 tons approx.
(b) wt. kiln furniture/chamber	Nil	—
(c) No. of pieces/chamber	18 000	11 000
V. (a) Time of firing: time in kiln	18 days (2 days firing)	7 days
rate of finishing chambers	1 chamber per day	9 chambers per week
VI. (a) Fuel consumption: per chamber	3 ton 12 cwt–4½ tons	6 tons coal
per wt. ware		3 tons coal/ton ware
per quantity ware	4–5 cwt/1000	5 tons coal/1000 bricks
VII. (a) Labour requirements: skilled	3 setting, 3 drawing	6
unskilled	Nil	9

point is always at the same temperature. Thus the nearest approach is made to the ideal firing method where heat energy is only consumed for the irreversible chemical changes. The heat used merely to change the temperature of the ware is recovered as completely as possible during the cooling.

A number of features of the tunnel kiln distinguish it from other firing methods. It consists essentially of a long straight or circular tunnel of relatively small internal area. Extreme dimensions of kilns so far erected have been given as lengths 2·2 m (7·2 ft) to 210 m (689 ft), widths 2 cm (0·8 in.) to 300 cm (9·8 ft) and heights 8 cm (3 in.) to 200 cm (6·6 ft). The smaller limit is exceptional. Through this tunnel passes a means of conveying the ware. This is usually a rail system with trucks or cars. The top of the truck is protected from the heat by a refractory slab and a sand seal on either side prevents the heat passing to the wheels and rails. The ware travels against a counter-current of air which takes up heat from the cooling ware, and transmits it to the heating ware. In the centre of the kiln direct heating is applied by burning fuel or using electricity. The ware is placed on the trucks outside the kiln to give a structure corresponding to the inside cross-section of the kiln. This is relatively much smaller than that of a traditional intermittent or continuous kiln so that a desired temperature can be attained in the centre of the setting very much more quickly and without the outside having meanwhile become much hotter. This is the basis of the very rapid firing obtainable with a tunnel kiln. The actual firing schedule can more nearly approach the 'ideal' one (p. 874) than in any other production kiln. The firing schedule can be very finely adjusted and controlled, so that a tunnel kiln is purpose made for a given production rate and firing schedule That is the reason for the very wide variation of dimensions.

The fact that the tunnel kiln fires the ware on a 'conveyor' belt system fits in with the general present-day trend in factory layout. In every case it reduces the difficult and often arduous task of setting and drawing ware from larger kilns, and in factories laid out round the kiln the movement and handling of ware is minimised. In some works a tunnel dryer is adjacent to the kiln and if it uses the same trucks cuts out the handling between dryer and kiln. If ware is transferred directly from the dryer to the kiln heat is also saved.

The history of the tunnel kiln is very short compared with that of ceramic firing itself, but it is probably true to say that, after the teething troubles in the first quarter of this century, it has now been developed to fire any regular production better than any other kiln. The chief limitation on its use is that once it has been started and adjusted it will only fire ware of the same body and similar dimensions. However, as small tunnel kilns are now being successfully constructed, wherever the production is regular, however small, a tunnel kiln could be used.

In general it is true to say that a tunnel kiln can be designed to fulfil the following needs in the ceramic industry:

(1) It can be made to fire any type of ware for any particular process, *i.e.* biscuit, glost, once-fired, salt glazing, decorating and enamelling. The results will be as good as, or better than, those in periodic kilns with a higher percentage of uniform, first-quality ware.

(2) The firing schedule can be finely adjusted to allow for any peculiarities of the ware while being as fast as possible. The 'ideal firing schedule' is more nearly approached than in any other production kiln (*see* Table 200, p. 879) comparing times in intermittent and tunnel kilns).

(3) The kiln can be made to cater for any size of production so long as this is fairly regular.

(4) The fuel consumption is less than in other kilns and firing requires less skill, in fact it may be fully automatic, *e.g.* 1000 bricks in a periodic kiln require 8 to 10 cwt coal, in a Hoffmann it is 4 to 5 cwt and in a tunnel kiln it is 3 cwt.

(5) Clean fuels are usually used, which may eliminate the need for saggars.

(6) Waste heat can be supplied to dryers, etc.

(7) The labour required for handling of ware and firing the kiln is less than any other kiln.

(8) Repairs are less frequently necessary. Only a small portion of the

kiln is at high temperatures and can therefore be built of the most suitable refractories, and as each part is at constant temperature there is not the repeated expansion and contraction found in other kilns.

The chief disadvantage of the tunnel kiln is the considerable capital outlay which cannot even be done piecemeal as it could for a number of intermittent kilns.

As each tunnel kiln is designed individually for every particular set of circumstances there are probably no two identical ones. There are, however, a number of initial decisions to make, and a number of recurring features which will be discussed first. This will be followed by brief descriptions of the main types of tunnel kiln.

DESIGN FEATURES

One Large or Two Small Kilns? For a steady production of identical or similar ware the more economical firing method is by a single tunnel kiln. In some cases, however, the more flexible arrangement of two small kilns is preferable although the total capital outlay and the running costs are larger. This is particularly so where an existing works is switching over from periodic to tunnel kilns. Some periodic kilns can be kept going while the first tunnel kiln is being built and tried out. The experience gained on this can be used on the second one. The capital outlay is thereby staggered.

Two kilns allow much more economical working if there is either a fluctuation of production or a change of ware type.

Single or Multi-kiln? Although the firing schedule in a small tunnel kiln can be made to approach the ideal, as soon as a large output is required the cross-section is increased and time for equalisation of temperature must be allowed. The result is a long kiln of large cross-section taking perhaps three times as long as the minimum time to fire the ware.

During and since the World War II there has been considerable development and application of the multi-kiln. In its most refined form it consists of sixteen, twenty-four or thirty-two passages placed above and next to each other to make a compact whole. Each passage is about 15 in. (38 cm) wide by 6 in. (15 cm) high or a size suitable just to take the largest object to be fired. The ware is placed on refractory bats which are pushed through the passages. The ware is passed through alternate passages from alternate ends in equal amounts to allow maximum heat interchange between cooling and preheating ware. Each passage is independently heated in the main firing zone.

In these tiny tunnels the heat distribution through the ware is so regular that the firing schedule can be brought as near to the 'ideal' as is liked. The kilns are therefore only 30 to 40 ft (9·1 to 12·2 m) long, although the over-all cross-section may be 10 ft by 8 ft (3·0 by 2·4 m) (**G37**).

Nelson and Wilson (**N8**) in considering the replacement of one large by two or four smaller adjacent car tunnel kilns operated in opposite directions in 1941, list the following advantages. These give even greater rewards in the multi-kilns of from sixteen to thirty-two passages subsequently brought into use.

(1) More rapid firing cycle because the heat penetrates more quickly to all parts of the load.

(2) Greater uniformity of heat distribution, for the same reason.

(3) Higher efficiency within the kiln unit owing to simpler and more direct heat recuperation between incoming and outgoing ware.

(4) Greater compactness, as less length is required for installation and operation (this is especially desirable for small factories).

(5) Lower initial cost of construction, indicated by smaller volume and savings in refractories.

(6) Lower cost in fuel or energy because heat requirements are lower, indicated by less radiating surface.

To these must be added the advantage that:

(7) Less expensive foundations are necessary which is a considerable advantage in areas subject to mining subsidence (**G37**).

A matter that should not be overlooked, however, when installing any counterflow type of tunnel kiln is the fact that green ware must be transported to *both* ends of the kiln and fired ware removed from both. The arrangement of conveyors, tracks, etc., must be made accordingly.

Straight or Circular Tunnel Kiln? The main factor deciding whether the tunnel should be straight or circular is the shape of the available space. The straight tunnel requires a long straight stretch, whereas if only a short but wide site is available a circular one can be fitted in. Both, of course, take up the same total area if everything else is to be equal, but in fact the space in the middle of a circular one may be wasted.

In general, the straight kiln is preferred for the following reasons:

(1) It is possible to see straight along the kiln and observe obstructions; and to withdraw the cars in case of a breakdown.
(2) Doors can more easily be fitted to the straight kiln.
(3) There is less distortion and warpage of the track.
(4) There is more allowance for leisurely placing and drawing, and a supply of filled trucks can be built up during the day for the night and week-end firings.

Its disadvantage as compared with the circular kiln is:

Arrangements must be made for transferring the trucks from the discharging end to the charging end. The cost of transfer trucks, rails, and spare trucks is thus added to the otherwise similar costs of the two kiln types of the same capacity.

The chief points to bear in mind about a circular kiln are that the conveying mechanism is more complicated and therefore more likely to go wrong; and that the drawing and setting has to be a streamlined continuous operation as cars are only out of the kiln for a short time (**D45**). The disadvantages of circular kilns which are due to their being run at high temperatures, *i.e.* warping and uneven heating, are less marked at the temperatures required for decorating firing and they have a certain popularity for this purpose.

Open or Closed Ends? It is difficult to arrange for circular kilns to have doors at the ends of the enclosed part of the circuit, but with the straight kiln there is a choice of having the ends closed with intermittent propulsion or open and continuous movement of the cars.

In the open-ended system the entering ware suffers no abrupt temperature change on entering the kiln. If it is not properly dry it will not be so easily damaged. The charging and discharging of cars is continuous.

With the closed-end system there is much greater control of air intake and output and therefore of the air-borne heat losses. There is, however, a need for more care in the simultaneous charging and discharging of the cars.

Some kilns have a small chamber, the length of one car, at each end of the main tunnel. These can act as air locks.

Siting the Tunnel Kiln. The continuous firing method of the tunnel kiln is an essential feature of a modern factory layout with continuous production methods. The making shops, drying chambers or tunnels, and kiln must be designed to complement each other to eliminate handling and waste of space and equipment. The quality of the ware can be considerably improved if the number of times it is handled or bumped about is reduced to a minimum. If the choice arises of transporting either the green or the biscuited ware a longer distance, it is always advisable for it to be the latter.

There should be plenty of space and good light at each end of the kiln (or the open bay of a circular one) to give the best conditions for careful setting and drawing.

The Dimensions of the Kiln. A number of interdependent factors determine the dimensions of the kiln.

(1) Total output required.
(2) Ideal firing schedule for the body concerned.
(3) Size and thickness of the pieces.
(4) Delicacy of shape of the ware.
(5) Protection required for the ware.
(6) Maturing temperature.

The ideal firing schedule for the body, modified by consideration for the thickness and shape of the ware gives the minimum time each piece must spend in the kiln. This can be achieved by slow passage through a short kiln or fast passage through a long one, assuming that the cross-section is so small that no time need be allowed for the heat to reach the centre of the setting.

The same total output can be achieved by slow movement of a setting of large cross-section or fast movement of a small cross-section. The larger the cross-section the more time must be added to the ideal firing schedule to allow for equalisation of temperature through the setting.

Thus for a large production with long minimum firing time a long kiln of large cross-section is required. For productions of smaller quantities or shorter schedules the kiln can be long and narrow, shorter and wide, or short and narrow.

Korach (**K67**) has shown that in kiln working under ideal conditions the heat losses are proportional to the cross-section of the kiln and hence increased output can be achieved more economically by lengthening a kiln than by increasing the cross-section. The cross-section should be the minimum permissible by the dimensions of the ware to be fired.

The arrangements for the counterflow of air through the kiln depend on the speed of throughput of the ware. Fast car movement means rapid removal of hot ware from the firing zone, and rapid entry of cool ware into it. The first requires more cold air to cool the ware before it reaches the end of the kiln and this helps to counteract the second by supplying more hot air for combustion and therefore more hot combustion products for preheating. In general the increase of fuel consumption is 30% of a given increase of throughput rate.

It has been found that for very high maturing temperatures, *e.g.* 1800° C (3272° F) required for sparking plugs, a very shallow setting space is the most economical. The ware is set in a single layer on the top of the kiln car, eliminating all kiln furniture, which has to be of a very high quality and is therefore expensive. The heat wasted on heating the kiln furniture is saved, and the setting can be brought to the correct maturing temperature with greater precision.

Direct or Indirect Heating (Open Flame or Muffle)? In heating the tunnel kiln the choice lies between a direct flame method, involving the protection of the ware by saggars and/or the cleaning of the fuel, and the muffle kiln with the fuel completely separated from the ware.

In the muffle kiln the burners discharge into chambers separated from the ware tunnel by refractory walls. The heat is transmitted primarily by conduction through the refractory, then by radiation and only subsequently by convection. It therefore necessitates a higher combustion temperature for the same heat transfer, involving higher fuel consumption, greater heat losses or insulation costs and more wear on the refractories. But there is *no* contamination of the ware so that saggars can be dispensed with, and up to twice as much ware can be placed on each car.

The incorporation of the muffle chambers necessitates a greater over-all width of the kiln for a given width of setting space, and in addition the slower equalisation of heat through the setting means the muffle kiln must be longer than the direct-fired one.

One suggestion for improving the heat economy of the muffle kiln is to use a heat recuperator harnessed to the flue gases to heat clean air which is then passed into the ware tunnel (**W84**).

In the direct-fired kiln the amount of saggars, etc., necessary to protect the ware is inversely proportional to the cleanliness of the fuel. In the early coal-fired tunnel kilns ware was spoilt not only by the combustion gases but also by fly ash, and everything had to be most carefully protected. The cheap fuel was therefore offset by the need to use a large proportion of setting space and fuel unproductively. Even with normal gas and oil fuels, saggaring is usually necessary. Under these circumstances the economics of open flame and muffle kilns are similar.

The most recent developments have, however, given the open flame kiln the advantage. If the gas used is purified to remove all sulphur compounds open placed ware can be fired directly. Both biscuit and glost can be fired in this way (*see* Table 230).

It should be remembered in all discussions about elimination of saggars by clean firing that porcelain and bone china flatware always requires the physical support of saggars and must therefore always be fired in them, however clean the kiln atmosphere.

TABLE 230

Typical Examples of Weights of Ware and Auxiliaries in Saggared and Open Settings (B32)

	Ware		Saggars		Pillar and shelf	
	(lb)	(kg)	(lb)	(kg)	(lb)	(kg)
Tableware	419	190	838	380	573	260
Tableware	617	280				
Sanitary ware	141	64	705	320	419	190
Sanitary ware	212	96				

Choice of Fuel

The early tunnel kilns were fired with coal, either dropped through the roof or through side fireboxes. But it is not possible to take full advantage of all the superior qualities of tunnel kiln firing with a fuel that cannot be fired continuously and under precise control. Raw coal is therefore only used in a few brick kilns. Coke and anthracite although smokeless still have the disadvantage of intermittent stoking lacking fine control.

Pulverised coal (coke, etc.) has been tried in tunnel kilns as it allows continuous and controlled firing but the dust and ash carried into the kiln has a very detrimental effect. It acts as an abrasive on the refractories, causing them to wear too fast; it clogs muffles; and it deposits on the ware, possibly combining as slag if the temperature is high enough. Pulverised coal firing is therefore usually only applied to bricks fired at too low a temperature for the ash to melt.

For most ware the choice, therefore, lies between gas, natural, town or producer, oil and electricity. When choosing between them both the technical and the economic aspects must be considered.

Natural gas when available is very suitable for tunnel kilns, particularly if the sulphur content is low or is eliminated.

Town gas is probably the most popular fuel for new tunnel kilns in Great Britain. The supply is regular, clean, and of known calorific value, and lower tariffs are frequently quoted for the industry.

Producer gas may be cheaper for a large concern but it has low calorific value and so high flame temperatures cannot be attained.

Oil may be installed as an alternative to producer gas in case of breakdown, or in its own right.

Special purification of fuels to remove sulphur compounds is often worthwhile for firing tunnel kilns if all muffles and saggars can then be omitted.

Combustion of any of these fuels automatically induces circulation of the kiln atmosphere which carries the heat with it. It also involves certain heat losses in the flue gases.

Electricity is the most expensive source of heat per unit but its conversion to useful heat in the ware has a much higher efficiency than combustion of fuels. The heating elements can be more suitably disposed about the ware and there are no flue losses. Saggars, muffles, etc., can all be dispensed with. Fox (**F22**) estimates that electricity prices per therm should be reduced by about 30% when comparing them with fuels to compensate for the higher conversion efficiency.

Sharp (**S59**) has made a comparison of gas and electricity consumption in tunnel kilns for glost firing of tableware. Six kilns for each heat source were considered; of the gas ones two were open-flame, two were muffle, one a radiant tube and the sixth a Revergen kiln; the electric kilns were two of the conventional type, one twin-track glost, one twin glost and biscuit, one walking-beam and one continuous-belt type. Table 231 shows that there is a considerable variation in consumption but that the higher setting density possible with electricity does much to compensate for its higher price.

Heating the Kiln to a Fixed Temperature Curve

The predetermined temperature curve is achieved in the kiln by one or more of the following methods:

(1) Spacing and number of burners or heating elements.
(2) Regulation of heating power of individual units.
(3) Regulation of the rate of air flow bringing in preheated air from cooling zone and removing hot air from main firing zone.

The actual heat input required is determined by the length of the kiln, the

TABLE 231

Fuel Consumption for Firing Glost Ware

(after (S59))

		Fuel consumption (therms)	
		Per 1000 ft³ of setting space	Per ton ware
Gas	A	254	78
	B	343	100
	C	452	120
	D	287	125
	E	310	143
	F	357	232
Mean		334	133
Electricity	A	118	27
	B	274	45
	C	203	46
	D	274	54
	E	168	69
	F	249	88
Mean		214	55

temperature needed, the rate of progress of the cars and the nature and density of the setting, together with the rate of air flow.

In spite of all previous experiments, calculations and experience the heating unit arrangement must be flexible enough to allow for some adjustments in the heat input.

Distribution of Heat through the Setting. As in any other kiln or oven there is a tendency for the heating to be irregular. The sides of the setting naturally receive direct heat from the burners, etc., in preference to the centre, and the hot gases tend to rise to the top of the kiln. Once there, the draught will pull them out without the heat being further utilised unless the setting is very near the arch; over-firing of the top may then occur.

Uneven heat distribution is, of course, more pronounced the larger the cross-sectional area of the setting and is also proportional to the rate at which the cars move through the kiln. The trouble is most obvious in the part of the preheating zone immediately before the main firing zone when the top of the setting may be 300° C (540° F) hotter than the bottom. The bottom will then catch up rapidly as the car comes to the firing zone, and this high heating rate is detrimental and leads to failure.

Uniform heat distribution is achieved by the correct spacing of the burners and turbulent air and gas movements.

In the main firing zone the burners must be arranged to counteract the

FIRING AND KILNS

tendency for heat to rise as much as possible. Most of them are near the bottom of the kiln. In an electrically heated kiln there may be a heating element actually in the top of the car to assist in heating the centre of the bottom of the setting, which is usually the coldest spot. This hot zone is also frequently kept under pressure to help distribute the heat better from top to bottom, side to side and also longitudinally along several cars.

In the preheating zone the air movement occasioned by draught or fans is the main means of equalising the temperature. This may be supplemented by a few small burners or heating elements near the bottom.

Down-draught Tunnel Kilns

The troubles arising from uneven distribution of heat through the setting are felt acutely in kilns designed to fire large quantities of cheap goods, particularly bricks. For these a narrow tunnel has too small an output and so one with a wide cross-section may be desired. Its inherent disadvantages have already been discussed. The principle of down-draught firing has therefore been applied to such kilns.

In order to fire a tunnel kiln on the down-draught principle it must be possible to divide it temporarily into chambers by incorporating baffles at intervals. The firebox is then shielded by a high bag wall so that the gases pass straight up to the arches. They are then pulled through the ware and into flues in the trucks until they reach a baffle, where vertical flues lead them to the top of the next compartment. The movement of the cars is intermittent so that the baffles always register with the drop arches, dampers, etc., in the kiln construction that make the sealing of the chambers complete.

In this way a combination is achieved of the conveyor belt firing of the tunnel kiln with the uniform firing of large quantities possible in the down-draught continuous chamber kiln. A further advantage that is claimed is that the baffles supersede the shape of the setting in controlling the draught so that a smaller output may be passed through the kiln without adjusting the controls (**R16**).

Kiln Atmosphere

Proper adjustment of the fuel/air ratio at the burners makes it possible to have oxidising or reducing conditions in the different zones of the kiln. The atmosphere can be kept under far better control than in other kilns. In electrically heated kilns the atmosphere can be completely controlled and non-air protective atmospheres can be introduced at will.

Salt Glazing in Tunnel Kilns

Although there are well-known successes in this field it has been generally considered inadvisable to attempt salt glazing in tunnel kilns. There are two main problems to be solved. Firstly, the section of the kiln where the salting takes place must be isolated by baffles, dampers, suction, air pressure, etc., so that salt vapour does not diffuse into other parts of the kiln. Secondly, special refractories must be used that will resist the vapour. It is this latter factor which usually makes the process uneconomical. A further difficulty arises in the case of drain pipes, which are one of the commonest types of mass-produced salt-glazed ware. These are sometimes difficult to fire in tunnel kilns because of their shape as they are set vertically whereas the draught is horizontal so that it is always difficult to avoid temperature differences between the two ends of a pipe and between its inside and outside.

A further snag which became apparent in the early experiments described below is that in the faster cooling of a tunnel kiln the glaze does not develop the traditional deep brown colour generally accepted by the consumer as the mark of quality; instead a dark grey or greenish grey to black is produced. The different colour does not, in fact, indicate any technical disadvantage but good publicity is required to convince users of this.

One of the writers took part in one of the first experiments on salt glazing in tunnel kilns in Münsterberg in 1929. An experimental kiln for drain pipes was constructed 116 ft (35 m) long, 37 in. (94 cm) wide and 45 in. (114 cm) high, fired by coal through six fireholes on a combination of step and horizontal grates. Refractory dampers were inserted to isolate at will a single car near the centre of the firing zone.

In the first instance the firing zone was lined with first-quality refractories for which a life of from 6000 to 8000 hours was calculated from their performance in periodic salt-glazing kilns. But the dampers were found to be wearing badly and in need of replacement after only a few days, and in a fortnight the kiln lining had become useless. Obviously a tunnel kiln cannot be run on these lines. A new special refractory was then developed which would resist the salt vapour at 1250–1300° C (2282–2372° F). It was based on the composition of barium anorthite, $BaO.Al_2O_3.2SiO_2$, and fulfilled the required purpose perfectly (**S101**).

With regard to the actual introduction of the salt three methods were tried, the traditional shovelling of salt on to the fires, blowing powdered salt into the kiln, or spraying a concentrated solution of it. The last proved the most satisfactory.

With the improved kiln, good salt-glazed pipes could be produced continuously but a number of economic reasons, some of which were connected with local conditions, made the method of firing uneconomical.

In the United States a successful and commercially applicable method of salt-glazing pipes in a tunnel kiln has been developed, although it has not become very widespread because of the trend to replace salt glazes in any case. In one case a circular kiln is used on a 48 hr cycle. The circumference is 220 ft (67 m), diameter 70 ft (21 m) and cross-section 6 ft 2 in. (1·9 m) wide and 4 ft (1·2 m) high. Pipes 6 in. (15 cm) and 8 in. (20 cm) diameter are placed together on a refractory platform moving at 4 ft (1·2 m) per hour. The kiln is divided into the usual preheating, full fire and cooling

FIRING AND KILNS

zones. The maximum temperature is 1108° C (2030° F) for the materials used, which include low-grade fireclay, soft shale, fine-grained and plastic clay, hard shale and grog. The hot zone has twelve burners for natural gas, six on each side. Of these two are set tangentially to heat the structure and thence through checker work to the sides and top of the kiln, four are used for introducing the salt and others heat under the refractory platform. Salt of specified fineness is treated with sodium carbonate and is fed through fluted hoppers directly into the burners. These hoppers are arranged to feed salt for 4 min in every hour. The dampers on a special exhaust stack work automatically in time with the salting mechanism, and salt fumes are prevented from entering the cooling zone by a region of high air pressure built up at the end of the hot zone. A second tunnel kiln of different dimensions, fires 3 ft (0·91 m) pipes (**A24**). Maintenance on the car tops and kiln linings is found to be fairly high.

A straight kiln fired with oil has been described (**S38**). Here the salt and borax are vaporised at the burning zone by means of a small vaporising chamber. An exhaust stack with entrances at the top and bottom of the kiln is situated at the end of the hot zone to prevent fumes entering the cooling zone (**A28**).

To prevent excessive corrosion on the kiln lining by incompletely vaporised salt settling on it, it is recommended to have an auxiliary vaporiser outside the kiln instead of passing solid salt into the burners. This can be replaced without interruption of the firing (**F9**).

Temperature and Heat-work Measurement

The course of the firing in tunnel kilns is followed with both thermoscopes (heat-work measurers) and pyrometers. Seger cones, Bullers rings, Holdcroft's bars, etc., according to choice are placed on the cars at the top, centre and bottom for observation through spy holes or as a record to be inspected when the car is finished. Thermocouples are fitted in the kiln walls and on occasional cars with extending leads being paid out under the car. Nickel–nichrome couples are used in the preheating and cooling zones and rare metal ones in the hot zone. Radiation pyrometers may be focused on the ware in the hot zone.

The electrical temperature measuring instruments are connected to gear for indicating, or indicating and recording, or recording only on an instrument panel. Multi-point instruments connected to several sensing elements are frequently used.

Automatic Control

Tunnel kilns where the temperature at any point should remain constant indefinitely are particularly well suited to the application of automatic control mechanisms (p. 948). The type of controller required depends on the amount of temperature variation permissible, the time lag between adjustment of the heat input and its registration by the sensing instrument, and the amount that loads may vary and thus upset the heat input adjustment (**B35**).

CONSTRUCTION OF THE TUNNEL KILN

The detailed design and construction of a tunnel kiln is usually divided into three zones, although the transition from one to another must be at least in part progressive.

The Pre-heating Zone

In this zone the ware is brought from room temperature (or ex-dryer temperature) to the temperature at which directly generated heat must be applied.

There are two technical snags in the way of strict adherence to the firing schedule with the maximum of heat recuperation using the kiln atmosphere moving against the ware throughout. The one, as has been mentioned, is the difficulty of heating the bottom of the setting at the same time as the top. This can be assisted by fans or by low-power burners near the bottom. The second is the water content of the kiln gases. If the combustion gases are used for preheating they are bound to carry with them the water formed by the combustion of hydrocarbon fuels, even before they take up the water given off by the ware. On no account may these gases be allowed to take up so much water that, at the lower temperature they later fall to, they become saturated and therefore deposit water on the cooler ware. This water causes scumming, etc., and may even soften the ware sufficiently to cause deformation. Either the ware must enter the tunnel kiln direct from the dryer at a temperature above 100° C (212° F) or the damp gases must be removed from the kiln before they reach cold ware, and dry warm gases used in their stead.

Constructional methods to overcome these difficulties may be chosen from the following, a combination sometimes being used:

(1) The combustion gases from the main fire zone are extracted at a point in the preheating zone before they become harmful. They are then passed into iron pipes on either side of the kiln, whence they radiate their heat to the incoming ware. When all useful heat has been extracted the gases are led (blown) to the flue.

(2) The efficiency of heat exchange from piped hot gases is improved by a fan-assisted circulation of the gases in the kiln to pass alternatively over the pipes and right through the setting (**G40**).

(3) After extraction of the combustion gases near the hot end of the preheating zone, hot clean air removed from the cooling zone is blown in. In some kiln systems all the heated air from the hot end of the cooling zone is removed by fans, part of it is used for the combustion of the fuel and the remainder is used for the preheating, possibly in conjunction with radiant heating (*see* (1) above).

The preheating zone is constructed of common bricks near the entrance, low-temperature insulating blocks along most of its length and more refractory insulating blocks or firebrick backed with low-temperature insulation as the firing zone is approached.

The Burning, or Main Fire, or Hot Zone

In the burning zone the already hot ware is gradually raised to and held at its final firing temperature. In this zone the heat is supplied by the burning of fuel or the use of electric power, although in the former case recuperated hot air may be used for the combustion. The firing curve is achieved either by the spacing of the burners, etc., or by the rate at which they burn fuel and hence emit heat, or by a combination of these methods.

The materials and method of construction of the hot zone of the tunnel kiln are of extreme importance. It is vital that only the best regarding both suitability and durability are used. The importance of durability is due to the fact that there can be no routine overhauling and maintenance of the inside of a tunnel kiln or its burners or fireboxes, as there can with an intermittently fired kiln. The need for quality arises from that for durability and also because of the exact control of temperature, atmosphere, fuel consumption, heat loss, etc., that form an integral part of tunnel kiln firing.

One advantage is to be gained from the fact that the structure is continuously at a high temperature, it is not damaged by the repeated expansion and contraction caused by intermittent firing.

The inner face of the kiln is built of the best refractories for the particular temperature to be used. These are backed by insulating blocks able to withstand the temperature transmitted by the firebricks and then further by low-temperature insulating blocks or loose filling. Alternatively the new refractory insulating blocks can be used directly as the kiln lining, thus reducing the weight, thickness and heat storage capacity of the structure.

Coal firing of Tunnel Kilns

In early tunnel kilns there was only one firebox on each side with a wide baffle wall to spread out the combustion gas. Subsequently a number of fireboxes on each side were introduced, the siting being staggered so that no two are directly opposite each other. If side firing is to be by intermittent hand stoking, cascade grates or deep grates being fired as gas producers are to be preferred because they give out more regular heat. Automatically stoked grates of under- or over-feed types give more regular firing. The mechanism for all the grates along each side may be run off a single shaft.

In some brick kilns top firing similar to that in continuous kilns has been adopted. The crown of the kiln is pierced with from twelve to fifteen rows of five fireholes set one-third of a car length apart. Coal is fed through these mechanically. The rate of feed to the different rows can be varied independently. The bricks are set on the cars in three blocks with 9 in. spaces left to act as combustion chambers for the fuel. The cars are pushed intermittently, one-third of their length at a time (**C41**).

Many of the advantages of the tunnel kiln over intermittent kilns as regards the possibility of a high proportion of first-class ware are lost if coal firing is used. If the stoking is manual the temperature tends to fluctuate, and if it is automatic, the pulverised coal used carries too much fly ash into the kiln so that complete saggaring is essential and the ash tends to build up in the kiln. By 1959 it appeared that there were no longer any coal-fired tunnel kilns in Great Britain (**A146**).

Gas-fired Tunnel Kilns

Gas firing is probably the most popular form of heating tunnel kilns and gives scope for considerable variation. In choosing burners it is wise to select those that have been deliberately designed for continuous use. Use of burners intended for intermittent use may cause frequent breakdowns. It should be possible to clean burners for producer gas with the minimum of interruption to the firing.

Gas firing is used for direct-fired, muffle, or radiant tube firing. In the direct fired kiln the burner has to be fired against the pressure of the firing zone of the kiln which is kept high to give better equalisation of heat. If sulphur-free gas is used an open setting can often be used where saggars or muffles were previously necessary.

Gas which is fired in tubes which radiate heat can be used as a means of regulating the vertical temperature distribution in the firing zone, by placing a series of horizontal tubes on each side of the kiln. The gas and air supply to the burners in each one is independently controlled so that the lower tubes can be made hotter to counteract the tendency for the top to accumulate heat.

There are a number of possible methods of heat recuperation and fuel economy applicable to fuel-fired tunnel kilns and gas-fired ones in particular.

Firstly the hot air obtained from the cooling zone may be used directly for the fuel combustion or it may be removed and used for preheating or other purposes. In the latter case the combustion gases may be passed through a heat exchanger to heat the gas and/or air. Direct use of the hot air is to be preferred for the over-all economy of the kiln. Ideally, relatively little air flowing against the movement of the ware should be able to take up the heat and itself attain a high temperature when it reaches the combustion zone. If high-speed gas with little or no air is then injected into it combustion occurs without a localised flame, and with the minimum of gas being used to heat air, etc. Efficient heat exchange from cooling ware to give a little very hot air is difficult to achieve and can only be done with an open setting of thin ware. With denser settings of thick ware successful cooling usually produces a large amount of only moderately hot air. The volume of this is too great for direct use in combustion and to allow it all to enter the firing zone would dilute the effect of the burning gas. At least a part must therefore be removed and used elsewhere in the works, or in the preheating zone. In many tunnel kilns all the cooling air is removed before reaching the combustion zone. Some of it may be passed into the kiln again through the burners, or it may all be used elsewhere. In this latter case the air used for combustion is cold, so that more fuel has to be burnt and hence more CO_2 and H_2O are released into the kiln atmosphere. In general then, the more the recuperated heat is used directly in the main fire zone the less combustion products are liberated in the kiln (**L25**).

Producer Gas. Orth (**O10**) describes the advantages of producer gas for burning firebricks in an open flame kiln. One of the difficulties of firing firebrick is the need for a large amount of heat in the early stages in order to drive off volatile matter and overcome the initial shrinkage before temperatures are too high, and achieve uniform heating of large masses of ware before it reaches the main fire zone. As producer gas has low calorific value it means that a large volume of it is used in the firing zone and thereby a large volume of hot combustion gases are available for preheating. A second advantage is that the flame temperature of producer gas is only a little above that required to finish the ware so that there is little danger of local overheating. Kilns are described producing 36 000 firebricks per day, with cars being charged and discharged once an hour, maturing occurs at cone 14. The fuel consumption is 1100 to 1200 lb per thousand bricks compared with 1800 to 2200 lb per thousand in intermittent kilns.

Recirculating Radiant Gas Tubes. P. Dressler (**D58**) describes the use of recirculating radiant gas tube firing for decorating kilns. This method has the same advantages as electricity over a gas-fired muffle furnace in that there is no leakage of combustion gases into the kiln atmosphere, whereas refractory muffle walls are always somewhat porous. The kiln described has the ware set on alloy trays moving over spaced rollers. The setting is 1 ft 3 in. (38 cm) high. Twenty recirculating tubes are used to heat the kiln, fourteen being set below the conveyor and six above the ware. The exhaust from the burners is led under the incoming ware, dampers controlling the amount of heat applied to it.

It is also possible to use gas recirculating tubes to heat car-type decorating kilns if they are placed as much as possible round the ware, both alongside it and partially below it.

Oil-firing of Tunnel Kilns

The choice between gas or oil for firing tunnel kilns is usually one of cost and handling, which is common to all kiln firing (pp. 829, 925). Oil generally has a content of sulphur and so saggars or muffles are necessary for delicate ware.

Oil firing is sometimes installed as an alternative to producer gas, for use in case of a breakdown at the producer and during the regular cleaning and burning out of the gas system (**O10**).

In Great Britain oil has usually been found to be more expensive than a fuel derived from native coal. It is now becoming increasingly clear, however, that fuel will have to be imported, and as the transport of oil is easier than that of coal its use may be much more widespread in the future.

Electrically Heated Tunnel Kilns

Instances have arisen in different parts of the world for electricity to be used for firing kilns of almost any size for the full variety of articles made in the ceramic industry. Temperatures used vary from 680° C (1112° F) for on-glaze decoration to 1410° C (2570° F) for glost porcelain.

There are three main types of electrically heated tunnel kilns. The conventional car, or metal sole, single-track, straight or circular kiln. These are similar to fuel-fired kilns in their general layout, dimensions and construction except for arrangements for movement of air or gases. Secondly there are smaller kilns where the ware is carried on wire belts and is usually placed in a single layer and fired fast. Thirdly there are U-shaped, twin, or multi-passage counter-flow kilns where heat recuperation from the cooling to the preheating ware is direct and efficient. A manufacturer of both single- and twin-track kilns estimates that the latter uses 35% less energy (**E12**).

Although one of the big advantages of the electrically heated kiln with respect to utilisation of fuel is the fact that no heat is lost in flue gases, there are several attendant problems. In the basic tunnel kiln air is drawn in at the exit, cools the fired ware, is used for combustion in the main zone, preheats the incoming ware and is then exhausted taking with it not only the products of combustion of the fuel but also the water vapour and decomposition products of the ware. In all subsequent refinements there is always air movement to recuperate heat and remove unwanted gases. In the electric kiln air movement is undesirable because it wastes heat and the counter-flow of air and ware is not introduced. It is therefore necessary to ventilate the kiln artificially to remove decomposition products. Heat recuperation from the cooling ware is also very often inefficient. In some of the mesh conveyor belt kilns water-cooled tubes effect the cooling and heat energy has to be expended for the preheating. Their power consumption is therefore high.

Nevertheless, in spite of the fact that an electric kiln has no draught the heat from the hot zone does become dissipated to the preheating and cooling zones, retarding the work of the latter, and wasting heat energy. The Gibbons 'Controvec' system is designed to counteract this by blowing low-velocity air into the ends of the kiln in a carefully regulated manner. Tests have shown that its application in a glost kiln operating at 1050° C (1922° F) makes a current saving of 20 to 25% (**A99**).

From the point of view of heat economy there is much to be said in favour of twin- or multi-passage counter-flow kilns. These may have a thin dividing wall or be open throughout. Herbst (**H82**) also describes a small U-shaped decorating kiln for obtaining the heat recuperative action of a twin kiln when quantities only warrant a single one. The preheating and the cooling zones of this kiln lie alongside each other. Half-way through the hot zone the ware comes under the control of a second set of pushing gear at right angles to its previous movement which pushes it on to the return track. Here a third pusher takes over. The kiln is economical of space and fuel but requires three pushers and often involves excessive jarring of the ware. In general, straight twin kilns are to be preferred. Electricity is probably the best means of heating multi-passage kilns and the efficient heat recovery achieved by the counter-flow makes it an economical application of this more expensive heat source. German (**G38**) gives comparative performance figures to demonstrate this, they are for earthenware and tile glost (Table 232).

TABLE 232

Kiln type	Finishing temperature		kWh/ton ware
	(°C)	(°F)	
Metal belt	1080	1976	1200
Twin tunnel	1070	1958	800
Multi-passage	1100	2012	268

Another advantage of the electric kiln with its own inherent snags is the siting of the heating elements. These are placed on the inside of the kiln wall, in suitable grooves, etc., and can be arranged on its top, bottom and sides to give absolutely uniform heat distribution, while in the fuel-fired kiln some consideration must be given to the supplying of fuel to the burners which may restrict their siting. In the latter case heat is lost through each firebox or firehole; but the burners, etc., are reasonably accessible from outside. In the electric kiln there is neither this heat loss nor the means of access to the elements. Special methods of dealing with defective resistors have, therefore, to be devised. This is done by one of three ways:

(1) Extra heating elements are installed but not connected. If a resistor burns out it is disconnected and a new one connected in its place. During the annual overhaul defunct elements are replaced.

(2) A transformer is connected into the power supply and the voltage on the good resistors is increased when one (or more) drops out to keep the over-all heat output the same.

(3) Arrangements are made to change individual resistors from outside the kiln without cooling it. Some types have the element in holes in the kiln refractory and can be slipped out. Others have them mounted in small sections on the refractory blocks making up the kiln wall. These are only lightly cemented in and can be removed complete and replaced.

The most usual types of electric resistance heaters used in tunnel kilns are: nichrome for decorating kilns and for the cooler parts of others, up to 1050° C (1922° F); kanthal for the hotter parts up to 1250° C (2282° F); and silicon carbide for porcelain and bone china, firing up to 1400° C (2552° F).

They may be mounted in horizontal grooves on the inside refractories of the kiln in such a way that heat can easily be dissipated and avoiding local heating and damaging of the element (and the ware). Alternatively they are inserted in holes through the refractory. As heat is less easily dissipated thermocouples are enclosed alongside them to keep a check on possible overheating. The latter type can be exchanged without stopping the firing.

German (**G38**) gives some data about the dimensions, output and consumption of the different types of electric tunnel kilns (Table 233).

TABLE 233

I. *Straight Decorating Kilns*
    Length: about 100 ft (30 m).
    Cars: about 1 ft 4 in. wide (40 cm).
              4 ft 6 in. long (137 cm).
    Setting height: 2 ft to 3 ft (60 to 91 cm).
    Consumption: 10 000 to 13 000 kWh per 168 hr week.
    Output: 3500 to 7000 doz. of ware.
                 1·5–2·5 units/per doz.

II. *Circular Decorating Kiln*
    Space required: 35 ft × 30 ft (10·6 × 9·1 m).
    Outside diameter: 27 ft 6 in. (8·4 m).
    Mean diameter: 22 ft (6·7 m).
    Sole width: 1 ft 4 in. (40 cm).
    Setting height: 2 ft 11 in. (89 cm).
    Consumption: 9650 kWh per week.
    Output: 3000–4500 doz. per week.
               2–2·7 units/doz.

III. *Twin Biscuit and Glost Earthenware*
    Length: 272 ft (83 m).
    Carload: 6 cwt (305 kg) biscuit + *ca*. 12 cwt (610 kg) kiln furniture.
        or 2 cwt (102 kg) glost + *ca*. 4 cwt (203 kg) kiln furniture.
    Temperature: Biscuit 1140° C (2084° F).
                 Glost 1070° C (1958° F).
    Time of cycle: biscuit 60 hr.
               glost 25 hr.
    Output: 15 000 doz. biscuit/week.
               12 500 doz. glost/week.
    Consumption: 1135 kWh/ton biscuit.
              800 kWh/ton glost.

IV. *One-way Metal Belt Kiln for Glost Earthenware*
    Length: 70 ft (21 m).
    Furnace zone: 22 ft (6·7 m).
    Width of belt: 36 in. (91 cm); 18 in. (46 cm).
    Overall width: 9 ft (2·7 m); 7 ft (2·1 m).
    Working height: 1 ft (30 cm).
    Time of cycle: 3–4 hr.

FIRING AND KILNS

TABLE 233—(contd.)

IV. One-way Metal Belt Kiln for Glost Earthware—(contd.)

Loading and belt life: 5 lb/ft² (0·35 kg/cm²) 9 months.
7 lb/ft² (0·49 kg/cm²) 7 months.

Output: at 4 hr schedule   3 hr schedule.
Saucers, 7500 doz.   cups 11 000 doz.
Mixed cups and
  saucers 9400 doz.
10 in. (25 cm) plates 2500 doz.
8 in. (20 cm) plates 6300 doz.

This kiln has no air circulation, the heating elements are in three independently controlled groups to give, say, 860° C (1580° F), 1050° C (1922° F), 1080° C (1976° F), respectively and water cooling is used at the end of the cooling zone. The short belt life makes this kiln more expensive to run than might be expected from the very rapid output.

V. Single-way Walking Beam Kiln for Glost Earthenware

In this kiln some bottom heat can be applied to the ware through the supporting refractory bats.

Length 98 ft (30 m).
Fuel consumption: 3·5 units/doz. dinnerware.
2·5 units/doz. teaware.

TABLE 234

Consumption in Electric Tunnel Kilns

Ref.	Ware type	kWh/lb	kWh/doz.	Kiln type
G43	Decorating pottery	0·25		
R1	Porcelain 1380°	2·5		
G38	Decorating kilns		1·5–2·7	One-way
G38	Earthenware biscuit 1140° C	0·51		Twin
G38	Earthenware glost 1070° C	0·36		
G38	Earthenware glost		2·5–3·5	One-way
G38	Earthenware glost	0·12		Multi-passage
G38	Earthenware glost	0·54		One-way
G38	Earthenware glost		0·02	4-passage
G38	Earthenware biscuit	0·001		24-passage

Electrical Firing of Porcelain. It is essential for the production of white porcelain to reduce the ware at a certain point in the firing schedule. In order to make the atmosphere in an electrically heated kiln reducing small blocks of wood (or coal, etc.) are introduced through openings in the kiln walls in a manner calculated to give sufficient reducing gases without harmful smoke or ashes.

The Cooling Zone

The cooling zone of the tunnel kiln generally plays the dual role of cooling the fired ware according to a predetermined schedule and converting the heat given off into a useful form of hot air or water. The heat transfer occurs either directly to air flowing against the travel of the ware, or indirectly by radiation to air or water in pipes in the kiln walls. In both cases the flow of coolant must be regulated to give the required cooling rate. In closed-end kilns the air intake can be regulated, in open-end ones the draught has to be regulated at offtakes and flues. Thin ware in open setting can be cooled by a swift air current but thick ware or dense settings require slower treatment if irregular cooling, setting up strains in the ware, is to be avoided.

The heated air may pass directly into the firing zone as combustion air or it may be partially or completely extracted into ducts at one or more points along the cooling zone.

In the electrically heated kiln air currents in the cooling zone are undesirable and the cooling of the ware is generally achieved indirectly. But as heat from the hot zone tends to diffuse along the cooling zone a small amount of air is sometimes blown in near the kiln exit to counteract it. No actual draught is created thereby (**G42**).

In twin counter-flow kilns the heat from the cooling ware is required to pass to preheating ware adjacent to it. Some of this heat exchange is by direct radiation but the remainder depends on air movement and can therefore be greatly assisted by fans and ducts which induce the air to move through adjacent cars in turn, *e.g.* air is extracted from the side of the kiln in the preheating zone, is passed through ducts and a blower to the cooling side and then forced through first the hot and then the cool ware, alternatively a fan situated in the top of the tunnel can make the air move down the centre and up the sides and thus equalise the temperature right across (**G41**).

From the heat economy aspect the ware should reach room temperature before it leaves the tunnel kiln. But in most kilns it is found best to draw the cars when they are at about 125° C (257° F) and not to recover the heat fully from the ware. If maximum heat recovery is not deemed important, the cooling zone can be a little shorter, so long as the ware is not passed out so hot that cracking due to the sudden temperature drop ensues.

The construction of the cooling zone is similar to that of the preheating one, in reverse.

HOW THE WARE IS TRANSPORTED THROUGH THE KILN

Cars or Trucks

The normal method of carrying the ware through the larger types of tunnel kiln, in particular straight ones, is on cars or trucks travelling on rails. These consist of metal frames mounted on two pairs of wheels. On this is constructed a superstructure of insulating and refractory slabs sufficiently thick to protect the metalwork from the heat of the kiln. These slabs may be ceramic refractories or refractory concrete (p. 885). At the sides of the cars metal flanges dip into troughs of sand running down each side of the kiln, thus aiming to divide the kiln into two separate compartments. The effectiveness of the sand seal is augmented by tongued and grooved construction of the refractory slab and the corresponding kiln wall.

The kiln cars undergo more wear and tear both thermally and mechanically than any other part of the tunnel kiln; in fact more even than the structure of a periodic kiln because of the faster cycle. The greatest care in their design and maintenance is therefore essential to the smooth running of the kiln. The main problems are keeping the metal parts cool and lubricated and making a superstructure that withstands the steep thermal gradient necessary when the top is at the main firing temperature and the bottom is to be cool enough for the metal parts.

In connection with keeping the metal parts cool straight tunnel kilns have been designed both open-bottomed and closed-bottomed. The former was evolved to make the car bearings and wheels cooler. It depends on accurate installation and good seals at the side of the cars. The closed-bottom kiln is easier to construct and can be equipped with forced cold air circulation under the cars. Church (**C33**) ran a kiln of each type over a period of four years firing heavy clay products and found that the controlled air-cooling in the closed-bottom kiln was more effective than the natural air-cooling in the open-bottom one.

Deliberate air-cooling of the track and wheels can, however, be the source of undesirable cold draughts in the upper part of the kiln, unless this is under pressure, because cold air can seep into it through inefficient sand seals and between the cars.

Spurrier (**S157**) describes an early method of checking the temperatures reached by the car wheels in order to find a suitable lubricant for the bearings. A number of substances of different melting points were sealed into short sections of Pyrex test tubes and placed in an iron tube slung under the kiln car. These located the temperature reached between $271°$ C ($520°$ F) and $320°$ C ($608°$ F). The lubricant then chosen was pure fine graphite with less than $\frac{1}{2}\%$ ash content suspended in an oil with a flash point of $300°$ C ($572°$ F) and fire point of $340°$ C ($644°$ F), the proportions by weight being 74% oil and 26% graphite. A saving of about 40% in the power required to move the cars was found.

More recently Orth and Oberst (**O11**) reported on lubricating of bearings of cars with an under-car temperature of $\sim 150°$ C ($300°$ F). While using petroleum oil and graphite lubricants it had been necessary to inspect and clean the bearings at 6-month intervals to remove the hard, abrasive, carbonaceous residue, and even so the cars were difficult to push manually and the train pusher required pressures of 100 to 150 lb/in^2 (7·0 to 10·5 kg/cm^2). On introducing a new synthetic lubricant (**C7**) whose oxidation and thermal decomposition products are largely volatile, low-molecular weight materials which vaporise leaving no significant carbon residue, cleaning became unnecessary and the pressure required on the pusher fell to 60 lb/in^2 (4·2 kg/cm^2).

The refractory kiln car tops have to withstand both mechanical wear from the goods stacked on them and severe thermal strains. The top of the structure undergoes the full firing cycle with the ware and must have high thermal shock resistance while the bottom remains considerably cooler. A laminated structure is therefore often adopted which combines a more selective use of materials, *i.e.* high temperature refractories at the top and lower temperature insulators at the bottom, with prevention of too great temperature gradients within single pieces. Expansion joints have to be left between adjacent blocks, but loose material must be prevented from entering the joints as this will eventually force the blocks apart and make the over-all dimensions of the car top too great.

The general trends for car tops is described as follows (**A33**):

Cone 4 to 7, Pyrophyllite or pyrophyllite over firebrick (U.S.A. only).

Up to cone 11, refractory concrete made with aluminous cement and possibly incorporating expanded clay aggregate.

Up to cone 12, combinations of high-duty alumina and high- and low-duty fireclay refractories, and refractory insulators.

Cone 13 to 26–27, combinations of super- and high-duty firebrick, insulating firebrick and mullite. The life of a car top at these high temperatures is inevitably shorter than that used at lower ones.

Many types of ware can be stacked directly on to the car top, *e.g.* bricks, others are placed in saggars which are stacked, while a large amount of small pieces are placed single in a pillar and bat superstructure on the cars (see p. 961, Fig. 11.27 and p. 365, Fig. 11.31).

The movement of the cars through the kiln may be continuous or intermittent, the latter being the only possible method if the kiln has doors, or is divided internally by partitions made by certain structures on the car, or is fuelled through the roof. The means of propulsion may be applied either to the end car only or to all of them.

Ram Propulsion worked from an Oil Pump

A $\frac{1}{2}$ h.p. motor drives a small oil pump delivering oil into a cylindrical ram of from 3 to 6 in. (7·6–15·2 cm) diameter. The amount of oil passing from the pump to the ram can be finely adjusted and controls the speed at which the ram moves forward. The length of each stroke is equal to the length of a car. The pump is connected to a limit switch which stops the ram at the end of a stroke and allows it to be returned to the starting position by a counterweight. The total time taken over each stroke can be adjusted to anything between 5 min and 5 hr.

The pressure required to move a normal train of cars is from 300 to 400 lb/in^2 (21 to 28 kg/cm^2). If the pressure rises above this it is a sign of an obstruction. The pump is set to cut out at from 800 to 900 lb/in^2 (56 to 63 kg/cm^2) when there must be some serious obstruction (**C33**).

Chain Propulsion

A 1 h.p. motor running at 1420 r.p.m. working through a reduction gear drives a continuous propelling chain for the tunnel kiln cars (**A98**).

Cars may be propelled through a kiln by a continuously running gear situated beneath the track. This engages in each car independently, moving the whole train simultaneously without any jarring. Further mishaps from fallen ware can be avoided by inclining the car decks slightly inwards towards the axial line (**A34**).

Walking Beam

The ware is placed on refractory bats which are lifted and moved forward at regular intervals by a walking beam underneath the kiln floor. This method is cheaper in capital outlay than a car system and requires less foundations. It also takes up less fuel for unprofitable purposes.

Pusher

In many smaller kilns the ware is placed on refractory bats or slabs which move on rails or on the kiln floor. The end slab is mechanically pushed and the subsequent slabs bear on and transmit motion to the one in front. The system is used in some smaller single-tunnel kilns and very extensively in multi-kilns.

French patents, intended to prevent opening of joints and jamming in the kiln, describe the use of two interlocking courses of refractory slabs and the placing of a compressible material, *e.g.* a deal block, between the ram and the slab being pushed into the kiln (**A185a**).

Berns and Milligan (**B43**) describe the properties required of the kiln floors and the refractory bats for efficient application of this system. There are two schools of thought regarding the relationship between the slabs and the kiln floor. The one prefers a very hard dense floor and a softer slab so that the wear takes place predominantly on the slab, the other aims at a reasonably hard floor and a very hard slab, usually silicon carbide. The most general practice is to use a hard but not vitrified fireclay brick for the floor and a silicon carbide pusher slab, this normally results in a floor life of two years and a longer life for the slabs which eventually fail for reasons other than wear.

Essential properties for the flooring materials including the mortar are that they should not soften or produce any sticky matter during the operation of the kiln.

The properties required of the pusher slabs are very exacting and they must be specially compounded and made for this purpose.

(1) Thermal shock resistance. The firing cycles in pusher-type kilns are usually short and so a considerable thermal shock is applied to the slabs. It is vital that they should not warp or crack in the kiln if wrecks and complete stoppages are to be prevented.

(2) Resistance to abrasion. This is essential because of the method of propulsion. The product of abrasion must also be dry and mobile so that it moves out of the kiln.

(3) Good mechanical strength both at room temperature and firing temperatures. The slabs have to bear considerable loads in the kiln and to withstand handling outside it.

(4) Resistance to chemical and physical change which would cause alteration of size, strength, etc.

(5) They should be and remain true to dimension, level and square. This is most important if movement is to be transmitted successfully.

Three main classes of material are used:

(1) Fireclay; the cheapest material and all right if thermal shock and temperature conditions are not too severe.

(2) Mullite; used where the slab rather than the floor is to undergo wear and of higher resistance to warpage and thermal shock than fireclay.

(3) Silicon carbide; the most commonly used, its long life compensating for the higher cost.

The maximum size of pusher slabs is dependent on the kiln cycle; larger slabs do not withstand such fast cycles as smaller ones of the same material.

Conveyor Belts

In some kilns an endless conveyor belt is passed through the tunnel and returned over rollers or drums under the kiln, *e.g.* for a kiln of maximum temperature 1150° C (2102° F) the belt may be of nickel–chromium alloy wire mesh. In its passage through the heating zone it is supported on nickel–chromium alloy skid rails which rest on a refractory hearth. The motion of the belt is activated by a motor driving the drum at the loading end. This has a variable-speed gearbox to allow for adjustments in the speed of operation. A self-tensioning device may be incorporated in the driving mechanism to compensate for expansion and contraction of the belt (**B64**).

In decorating kilns, where high temperatures are not required, light-weight alloy conveyor belts can be used for the ware. This makes it possible to heat the ware from below as well as the sides and above and so uniform heating can more easily be achieved.

One type consists of a series of alloy rollers carried on alloy supports on the kiln floor. Over these pass alloy trays carrying the ware (**D58**).

Basket Conveyors

In Germany the basket method of conveying the ware has found considerable favour. The ware is placed in heat-resisting chrome–nickel steel baskets and then pushed or pulled over rails or rollers. In this way a large ratio of ware to support can be established, which is particularly important in electrically heated kilns. Although these baskets can be used for porcelain biscuit and glost firing their use is more usually for decorating kilns. Above 1000° C (1832° F) there is a tendency for the steel to spit glowing sparks which cause black spots on white glazes (**B131**).

Individual Types of Tunnel Kiln

Brief mention will now be made of individual types of tunnel kiln which have proved successful. In general the types that have been superseded will not be described. On the whole the kilns are discussed in the order, open flame, muffle, electric. The same applies to Table 236 showing the recently collected data.

THE FAUGERON TUNNEL KILN

The Faugeron tunnel kiln is one of the oldest kiln systems, and the first to be successfully direct-fired. It is distinguished from all other directly fired kilns by the fact that in the preheating zone and also in part of the firing zone, a reduced pressure is produced in the arch of the kiln. For this purpose, openings are arranged in the sides of the brickwork, in order to allow the sideways escape of the combustion gas in the places where the reduced pressure is required.

By this construction, it can be arranged that the combustion gases take a serpentine path through the preheating zone, and because of this guided motion a better and uniform preheating of the ware is obtained.

Before the change-over to gas heating, the Faugeron kiln was mostly built with the customary grate and semi-producer firing systems. The original kilns were built in France for biscuit and glost firing of china, etc. A later development with extra grates for reduction was built in Germany for hard porcelain. In U.S.A. Faugeron kilns were built to fire firebricks, refractory products and porcelain insulators.

HARROP KILNS

C. B. Harrop took out a number of patents between 1919 and 1930. The improvements he brought forward include:

(1) Staggered clearance spaces in the side walls of the kilns, increasing towards the bottom of the setting, enabling the gases to travel more readily through the lower parts of the clearance spaces.

(2) Introducing air to the oxidation zone independent of the products of combustion of the fires.

(3) Applying mechanical stokers at the sides of car-tunnel kilns.

(4) Staggering the furnaces to give greater uniformity of heating in the firing zone, introducing cold air into the far end of the cooling zone and removing the greater part as hot air for drying purposes before it reaches the firing zone, and providing staggered inset panels on both sides of the preheating zone so as to distribute the gases laterally into the setting (**R59**).

Harrop kilns are built in the U.S.A.

THE BRICESCO–HARROP KILN

These kilns are designed for the direct firing with coal, oil or gas for large outputs of bricks or refractories. The output of bricks is 15 000–20 000 per day with kilns 300–400 ft (91–122 m) long and about 6 ft × 6 ft (1·83 × 1·83 m) in cross-section. They may be coupled with dryers using the same cars and setting and heated by waste heat from the cooling zone.

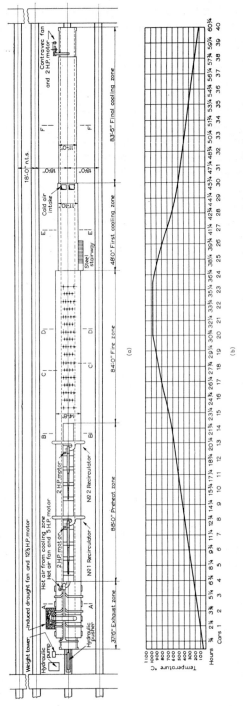

Fig. 11.45. *Modern* (1953) *Gibbons top-fired tunnel kiln using fuel-oil* (*Gibbons*, **G39**):

(a) plan
(b) approximate heat curve

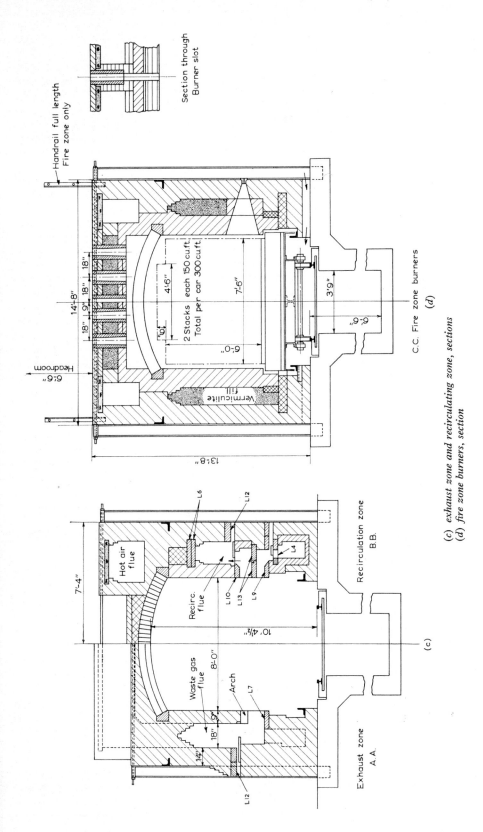

(c) *exhaust zone and recirculating zone, sections*
(d) *fire zone burners, section*

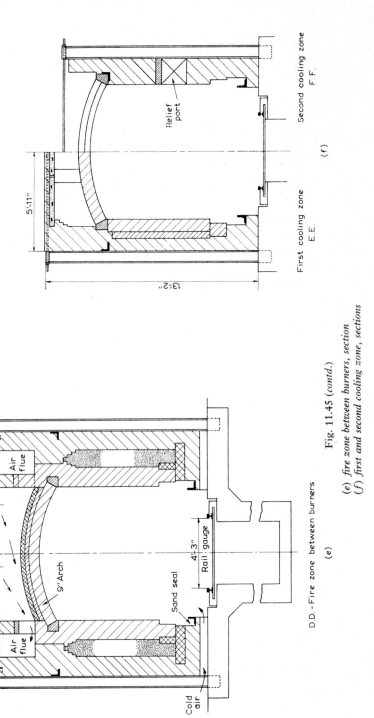

Fig. 11.45 (contd.)
(e) fire zone between burners, section
(f) first and second cooling zone, sections

FIRING AND KILNS

In the British Isles the brick kilns are usually fired with coal by mechanical stokers of the ram type at 6 or 7 fireholes on either side of the kiln. Such stoking results in relatively poor heat distribution within the kiln limiting the width to about 5 ft 6 in. (1·67 m). Much better results are achieved by firing with producer gas.

The refractory kilns are fired either by producer gas or by side coal firing supplemented by oil burners in the sides and top (**R59**).

Bricesco kilns are given in Table 236, examples 1 and 2.

THE MONNIER OR TOP-FIRED TUNNEL KILN

The Monnier kiln has been developed especially for firing bricks, roofing tiles, etc., which have formerly been fired in Hoffmann kilns. It is a direct fired kiln with a top feed of coal falling into gaps left in the setting (*see* p. 1019). Modern top-fired kilns are fired with gas or oil (Fig. 11.45).

TABLE 235

Outputs and Dimensions of Monnier Top-Fired Car-Tunnel Kilns (R59)

	1	2	3	4
Length of kiln (ft)	327	327	297	297
(m)	100	100	90	90
Width (ft)	6 ft 6¾ in.	5 ft 11½ in.	5 ft 11¼ in.	5 ft 11¼ in.
(m)	2	1·8	1·8	1·8
No. of cars in kiln	32	32	29	29
Car travel (ft/24 hr)	141	105¼	112	105
(m/24 hr)	43	32	34	32
Output per week bricks	210 000	160 000	167 000	156 000
Output (tons/24 hr)	87	62	67	71
Fuel consumption cwt coal/1000	3·06	4·0	4·2	3·17
kg coal/1000	155	203	213	161
Fuel consumption				
lb coal/ton output	116	154	164	123
kg coal/ton output	52·6	69·8	74·4	55·8

THE 'WILLIAMSON' TUNNEL KILN

The firing zone can be fired by down-draught, up-draught or horizontal draught. In the down-draught type the gas (or oil) is admitted to the bottom of combustion chambers on either side of the kiln with only a small amount of air. The chamber is fan-shaped, spreading the gas out to a long strip by the time it reaches the top of the setting. Here a stream of hot secondary air is admitted and the mixture ignites as it reaches the crown of the kiln and the opposite gas stream. It then passes down the setting and is extracted through the car platform into side flues. These then lead the hot gases down either side of the preheating zone where they radiate heat.

The kiln is so arranged that only air without combustion gases contacts the ware in both the cooling and the preheating zones whereas the firing zone is directly fired. The kiln was designed and has been used for salt-glazing pipes (**R59**).

This type of kiln has the following advantages:

(1) Air and gas are thoroughly mixed before meeting the ware.
(2) The flame intensity from each pair of opposite burners is equalised by the latter meeting at the crown of the arch, so that an even temperature is obtained on each side of the cars.
(3) Since the flame passes under the bottom of the car, in addition to traversing the sides and centre, every part of the load is at the same temperature as the flame.
(4) The fact that each pair of burners can be regulated to give equal temperatures makes a long soaking heat possible (*see* Table 236, example 4).

THE 'DRAYTON' DOWN-DRAUGHT KILN

This kiln was designed for direct firing, on the down-draught principle, of high-grade facing bricks. It is divided into 34 ft (10·36 m) compartments, each consisting of three 10 ft (3·04 m) kiln cars and one 4 ft (1·21 m) baffle car. These are moved intermittently the whole 34 ft (10·36 m). The baffles register with dampers in the side and top of the kiln at these points. The dampers are pressed against the baffles when in position and withdrawn when movement is to take place. The kiln gases are taken to the arch of the kiln either by the bag walls shielding the fireboxes, or by the flues in the baffles. They then pass through the ware into flues in the kiln cars which lead them to a vertical flue in the baffle, which opens into the top of the preceding chamber (**R16**).

THE 'RICHARDSON' DOWN-DRAUGHT TUNNEL KILN (Fig. 11.46)

The kiln is divided into compartments by a baffle constructed on the rear end of every third kiln car. This has a tilting damper on it which registers automatically on drop arches in the kiln structure. The main structure has a flat suspended arch and the bags, although circular where they surround the

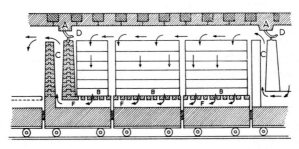

FIG. 11.46. *Richardson down-draught tunnel kiln, internal side elevation* (Bullin, **B140**)

fireboxes, present a straight wall to the inside of the tunnel. The down-draught principle is applied to the firing and preheating zones except the water-smoking zone. The combustion gases are not allowed to enter this but are passed along its sides in iron pipes. This radiant heat is boosted by hot air derived from the cooling zone (**R16**).

THE KOPPERS TWIN-TRACK TUNNEL KILN

This kiln was especially designed for firing silica bricks which required slow and careful treatment; it incorporates cross draught and heat regenerators in the firing zone together with the counter-flow twin-track method of trans-fering heat from the cooling to the preheating ware. Twin tracks taking ware moving in opposite directions are placed in a single relatively wide tunnel kiln.

Refractory doors are lowered from closed damper chambers to cut the central firing zone off from the end zones. Inside there are two complete sets of eighteen burners with hot-air supplied from adjacent regenerators and the hot gas passing into regenerators on the opposite side. Firing is by cold, clean producer gas and the direction of use of the equipment is reversed every twenty minutes.

Baffles and dampers are let down into the two preheating zones allowing air to pass under them near the car top and in the cooling zones forcing air to pass over them. The adjacent preheating and cooling zones are separated by a thin partition (through which heat can radiate) except near the firing zone. Air is therefore drawn in at the exit, passes in an up and down motion through the cooling zone and then similarly back through the preheating one. The kiln was especially designed in 1925 for firing silica bricks which must have a long firing cycle. The aim was to have a reasonable output without an excessively long kiln. It held fifty-four cars each 8 ft $2\frac{1}{2}$ in. square (2·5 m) with a setting height of 6 ft 3 in. (1·9 m) and holding 11 tons. Daily output per track was from four to five cars (**R59**).

THE 'REVERGEN' KILN

This is a direct-fired kiln in which the flame itself does not contact the ware (Fig. 11.47). On each side of the kiln horizontal and vertical burners occur alternately, there are also heat regenerators of chequered brickwork on each side. Gas is led to only one side at a time, and burnt with air heated by drawing it through the regenerator on that side. The combustion gases are pulled quickly under the setting or up the side, over the crown and down the other side, by a strong draught taking them into the regenerator on the other side which they heat up. The combustion gases themselves therefore do not percolate the setting but they set up convection currents and heat the kiln structure which radiates heat. Every twenty minutes valves reverse the direction of firing (**D46**).

This kiln is suitable for biscuit and glost tiles and tableware, and grinding wheels (*see* Table 236, example 10).

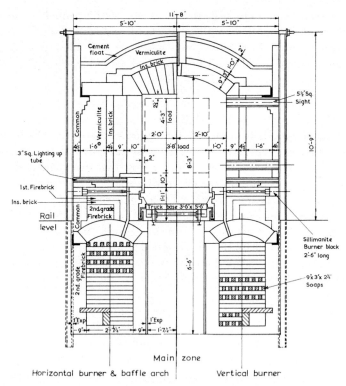

FIG. 11.47. *Gibbons Revergen tunnel kiln* (1955). *Cross-section of main firing zone* (**G39**)

SEMI-MUFFLE TUNNEL KILN, OOMS-ITTNER

Like the Revergen the flame is not allowed to play on the ware but the combustion gases do make contact with it (Fig.11.48). Gas and air are burnt in a firing chamber behind a muffle wall and then forced under pressure to enter the tunnel near the bottom of the setting. Good temperature equalisation results with no over-firing of the sides of the setting. The hot gases are drawn through the preheating zones and withdrawn at a number of points as necessary to keep to the correct firing schedule. The entering ware passes through a roller shutter air lock in order not to upset this carefully regulated draught. Cooling is indirect by air passed through cavities in the kiln walls (**O6**).

DRESSLER KILNS

Conrad Dressler was the first to build a successful muffle tunnel kiln, in 1910, and a number of his patents have led to a wide range of Dressler kilns both muffle and open flame. They are built individually and with particular refinements by firms in England, U.S.A., France and Germany.

Multi-burner Open Flame Dressler. With this type, a number of burners are arranged on both sides of the kiln firing directly into the tunnel, and

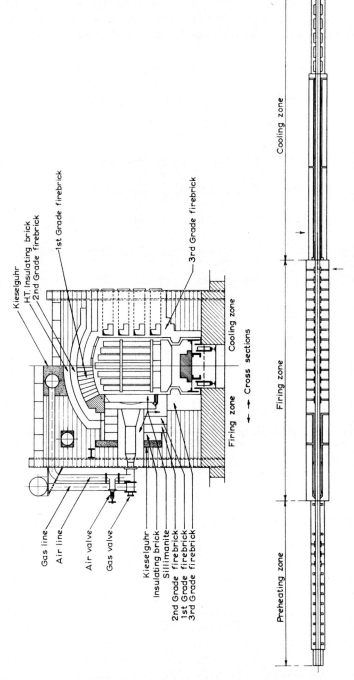

Fig. 11.48. *Ooms Ittner semi-muffle kiln for coloured floor tiles* (1956) (**O6**)

preheated air for combustion is drawn from the cooling zone. In order to regulate the rate of increase in temperature, the waste gases are extracted at a number of different points in the preheating zone, and steps are taken to equalise the temperature over the load both here and in the cooling zone. A Phillips recirculator is sometimes used to make the gases travel through the ware instead of going along the sides of the tunnel. This comprises two units, one for low temperatures near the entrance, and another for high temperatures near the firing zone (*see* Table 236, example 11).

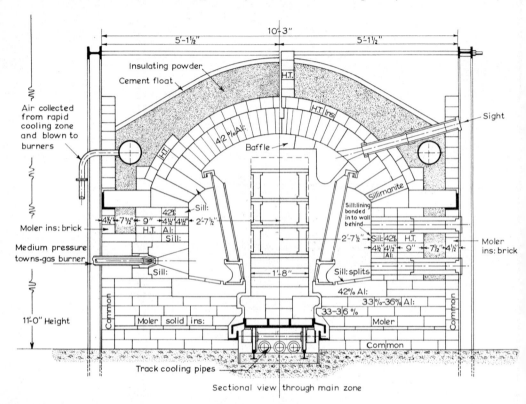

FIG. 11.49. *Dressler thrift muffle tunnel kiln. Cross-section of main firing zone* (Gibbons, **G39**)

Regenerative Dressler Kilns. These work similarly to the Revergen kilns with firing from alternate sides and the hot gases passing across the kiln to regenerators. A twin track counter movement of the ware may also be incorporated.

Dressler Thrift Muffle Kiln. The best known of the Dressler kilns and developed from the original Dressler muffle kiln. Triangular combustion chambers are formed on either side of the firing zone within the kiln structure with the side towards the ware formed by straight hollow slabs of alumina, sillimanite, carborundum or fireclay. The combustion gases are led alongside the preheating zone in similar chambers (Fig. 11.49).

These kilns are used chiefly for biscuit and glost earthenware tiles and tableware, glost bone china and other fine ware (*see* Table 236, examples 14–18 inclusive).

ALLIED KILNS

A large number of allied kilns are circular. Designs of both straight and circular ones cater for all types of ware and fuel. Where sulphur-free gas is available ware that would normally have to be saggared or muffled is direct-fired with radiant burners as these have no effective flame to impinge on the ware.

An indirect cooling system is incorporated to cool the ware emerging from the hot zone quickly through the range where this is safe. No cold air is thereby introduced into the kiln which would otherwise use more fuel because of its diffusion into the hot zone (**A184**).

Allied have developed a successful method of salt-glazing in tunnel kilns by vaporising the salt in an independent vaporiser(s) outside the kiln. Corrosion of the kiln refractories is largely due to unvaporised salt and this is prevented thereby (**B15**).

THE KERABEDARF KILNS

There are four main types of Kerabedarf kiln, the open-flame, muffle, convertible open-muffle, and electric kilns. Each type is adaptable to a considerable range of goods, output, etc. The non-electric kilns are usually fired by gas although some may be fitted with gas–oil burners. All 'Kera-kilns' have the following features:

(1) An inspection channel under the track which allows supervision and checking of the cars, their chassis, the underside of the platforms and the effectiveness of the packing strips at the ends. The channel also allows for rapid clearance of a wreck should this occur.

(2) The use of the highest quality refractories on the inside of the kiln backed by highly efficient insulation with a corrugated metal or concrete outer casing on all but the firing zone which is brick.

(3) Electrically controlled air locks at the entrance and exit of the kiln keep the kiln atmosphere under complete control without contamination from outside. The safety devices prevent accidents due to careless manipulation. Propulsion of the train of cars is hydraulic and continuous. The ram is set to come back to its starting position about five times as quickly as it moves forward.

(4) Waste heat is fully recovered from the cooling ware and from the combustion gases and after all the kiln requirements have been fulfilled there is still some available in the form of clean air at 80–120° C (176–248° F) for other purposes. The main heat recuperation takes place as the ware leaves the hot zone. The initial cooling of most ware can safely be very rapid, so that here cold air is blown through the ware and extracted hot for use in the

gas burners. The remainder of the cooling zone is very carefully regulated for temperature by adjusting the air flow in sections, air being drawn into the cavity walling as required.

(5) The combustion gases are extracted before there is danger of either water being deposited on cold ware or of the sulphur dioxide combining with water to give acid that attacks fans and flues, *i.e.* at 150° C (302° F). The use of a tunnel dryer in line with the kiln using clean hot air and delivering the ware at 120–150° C (248–302° F) is recommended.

(6) Very careful graphic record is kept of the temperatures of all parts of the kiln including the flue gases. It is obtained both by static thermocouples and pyrometers in the kiln walls and by passing a special recording truck through the kiln. This truck has three pyrometers at three heights, which make sliding contact with underfloor cables leading to the control panel. The ware is set round the instruments in as normal a manner as possible so that a true picture is gained of the conditions in different parts of the kiln car setting. Graphic record of the draught is also made.

(7) Burners set both at the bottom of the kiln and in the walls and roof can be adjusted to give temperature variations in a cross-section of not more than $\frac{1}{2}$ to 1 cone. The burners are designed to have no metal parts in regions where they can become very hot. Preheated air and gas delivered to the burner in the correct proportions are mixed by turbulence enacted by the shape of the burner throat. Low pressures are used throughout so that there is no flame jet. Anti-explosion measures are taken by making the vertical section of the gas pipes leading to each burner open at the bottom. This is immersed in water and collects tar as well as being an explosion outlet which is easily put in order again.

(8) The kiln cars have roller bearings lubricated by dry colloidal graphite. Special transfer cars eliminate the need for a set of rails at a lower level than the rest of the factory floor.

(9) In the muffle kilns the muffle itself is divided into two or three sections to give better temperature control.

(10) The convertible muffle–open kiln can be changed from one type to the other in a few minutes and can thus be used alternately for biscuit or glost firing.

(11) Electric tunnel kilns resemble other types, twin- and multi-track kilns being recommended, individual designs have been developed to fulfil the following tasks:

(*a*) Decorating and hardening on.
(*b*) Biscuit porcelain and coloured earthenware.
(*c*) Low-temperature earthenware biscuit.
(*d*) High-temperature earthenware and fireclay biscuit.
(*e*) Low-temperature earthenware glost.
(*f*) High-temperature earthenware glost.
(*g*) Glost electrical porcelain (**K21**).

FIRING AND KILNS

THE MOORE CAMPBELL ELECTRIC TUNNEL KILN

This is a truck kiln for decorating of pottery. The heating elements are metallic, mounted in grooves in the refractory wall. Excellent insulation is used throughout, so that the outer face of the hot zone is only a few degrees above that of the cooler ends. The trucks are propelled by rack-operated gear with mechanical cut-out arrangements when the desired travel has been completed or when a finished truck is a few inches from the closed exit door.

The pyrometric control equipment consists of four thermocouples and a dial-reading thermometer, a three-colour thread recorder, and automatic temperature regulator with the necessary transformer and relay keeping the kiln temperature within the range of 5° C.

Power consumption for earthenware decoration is 0·25 kWh/lb (**G43**). (*See* Table 236, examples 20, 21).

ROTOLEC ELECTRIC ENAMEL KILN

This is a circular kiln for firing decoration on china and earthenware (Fig. 11.50). With mean diameters of 16 ft (4·87 m) or 22 ft (6·70 m) and loading heights of 2 ft 4 in. (71 cm) or 2 ft 11 in. (89 cm) the cycle is 22–14 hr and the output from 3000 to 5000 dozen per week. The electrical consumption per unit dozen of mixed teaware and dinnerware varies between 1·43 and 2·75 kWh, depending on the density of the setting and the cycle (**G39**) (*see* Table 236, example 22).

THE GOTTIGNIES MULTI-PASSAGE KILN

This multi-kiln has sixteen, twenty-four or thirty-two passages worked in counter-flow with the ware on bats pushed from the ends (Fig. 11.51). The passages are about 15 in. (38 cm) wide and $4\frac{1}{2}$ to 6 in. (11·4–15·2 cm) high, and the kiln length is 30 to 40 ft (9·1 to 12·2 m). The over-all cross-section is up to 10 ft × 8 ft (3·0 × 2·4 m).

The kiln is heated by electric resistance coil elements of nichrome or kanthal inserted together with thermocouples into holes moulded into the refractory slabs which form the roof and floor of superimposed passages. The elements can be withdrawn and replaced without stopping and cooling the kiln.

At each end of the kiln a steel structure has platforms aligned to the passages for loading and unloading and the hydraulic propelling gear is incorporated (**R59**).

This kiln is used successfully for wall tiles, floor tiles, grinding wheels, tableware (*see* Table 236, example 23).

SPARKING PLUG KILN

Sparking plugs require very high firing temperatures but their small size makes special kiln design possible. Schüffler (**S36**) describes such a gas-fired kiln for firing up to 1800° C (3272° F) with the sparking plugs moving in single file (Fig. 11.52).

Fig. 11.50. *Rotolec decorating kiln* (*Gibbons*, **G39**)

Fig. 11.51. *Gibbons-Gottignies multi-passage kiln*:
(a) *photograph*

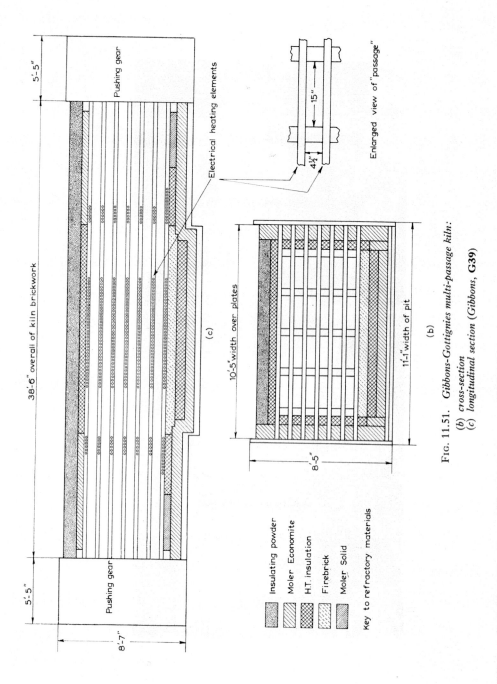

Fig. 11.51. *Gibbons-Gottignies multi-passage kiln:*
(b) *cross-section*
(c) *longitudinal section (Gibbons,* **G39**)

FIRING AND KILNS

FIG. 11.52. *High-temperature sparking plug kiln* (*Schüffler*, **S36**)

MISCELLANEOUS

Further kiln types that have proved reliable in their own fields but have not found widespread application are the Lengersdorff, Mendheim which is a muffle kiln including muffle chambers in the car superstructure, Padelt, Möller and Pfeifer.

Table 236 gives data on tunnel kilns collected in 1956, *see* notes on Table 228, p. 984.

The *Deutsche Keramische Gesellschaft* have published at least two special numbers with detailed data on kilns:

Fachausschussbericht Nr. 5. Anhaltzahlen für Brennstoffverbrauch und Leistung keramischer Öfen (Feb. 1954).
Fachausschussbericht Nr. 6. Anhaltzahlen für keramische Öfen (Aug. 1955).

Frit-kilns

The fritting of glaze constituents is a quite different process from any other ceramic firing and therefore requires different kilns. Fritting involves heating a mixture of powdered materials until it melts, gives off various gases and eventually makes a homogeneous liquid. This is then run off into a tank of water where the sudden temperature drop makes it shatter into

TABLE

Tunnel

	No.		1	2
	Code No.		30	30
I.	(a)	Type of kiln	Bricesco	Bricesco
	(b)	Straight or circular	Straight	Straight
	(c)	Open-flame or muffle	Open flame	Open flame
	(d)	Length: pre-heating zone	92 ft	92 ft
	(e)	firing zone	32 ft 6 in.	32 ft 6 in.
	(f)	cooling zone	83 ft 6 in.	83 ft 6 in.
	(g)	total	208 ft	208 ft
	(h)	width	3 ft internal	3 ft internal
	(i)	height above car or slab top	2 ft 8 in.	2 ft 8 in.
	(j)	Number of cars in kiln	33	33
II.	(a)	Fuel used	Town gas	Town gas
	(b)	Heating capacity of fuel (B.t.u./ft^3)	470	470
	(c)	Type of grate or burner	Bricesco	Schieldrop No. 5
III.	(a)	Nature of ware	Diatomite insulating bricks	Insulating refractory
	(b)	Moisture in green ware as set (%)	21	18
	(c)	Nature of setting, saggared or open, with or without auxiliary supports	Open	Open
	(d)	Maturing temperature or cone numbers	Holdcroft bar Nos. 8–10 840–875° C	Holdcroft bar Nos. 31–33 1350–1430° C
IV.	(a)	Setting per car: weight ware	1300 lb	1800 lb
	(b)	kiln furniture	Approx. 2240 lb	Approx 2240 lb
	(c)	total	Approx. 5000 lb	Approx. 5400 lb
	(d)	quantity of ware	291 pieces	200 pieces
V.	(a)	Total time in kiln	40 hr	58 hr
	(b)	Travel of cars (ft/24 hr)	115	82
	(c)	Output per day: quantity	5590 pieces	2750 pieces
	(d)	weight	15 500 lb	14 600 lb
VI.	(a)	Fuel consumption: per wt. ware	2 ft^3/lb	4·5 ft^3/lb
	(b)	per quantity ware	8 ft^3/piece	40 ft^3/piece
	(c)	per day	47 000 ft^3	111 000 ft^3
VII.	Labour requirements:			
	(a)	Skilled per car	3/4 man-hr	1 man-hr
	(b)	per weight	1 man-hr/1700 lb	1 man-hr/1800 lb
	(c)	Unskilled per car	3/4 man-hr	1 man-hr
	(d)	per weight	1 man-hr/1700 lb	1 man-hr/1800 lb

Kilns

3	4	5	6
12	7	14	10
—	Williamson	British Ceramic Service Co. Ltd.	Built by user
Straight	Straight	Straight	Straight
Open flame	Open flame heat zone, muffle preheat zone	Open flame	Open flame
82 ft	146 ft	41 ft 3½ in.	90 ft
28 ft	30 ft	29 ft 7 in.	28 ft
54 ft	130 ft	38 ft 1½ in.	28 ft
166 ft	306 ft	109 ft 0 in.	146 ft
40 in.	4 ft 6 in.	20 in.	30 in.
48 in.	3 ft	27½ in.	36 in.
30	53 plus 22 in dryer	18	28
Producer gas from coke breeze ⅛ in. +	Town gas	Britoleum 200° gas oil	Town gas
1150 approx. as received	475	?	440
Wall pockets	Own design	Nu-way 'Y' type burner	Open flame, naturally aspirated
Facing bricks and agricultural drain pipes	Floor tiles	Red earthenware and white earthenware, biscuit and glost	Fireclay tiles, biscuit
4 approx.	1	2 approx.	8
Open	Saggared	All saggered, glost flat ware with cranks	Open with auxiliary supports
1020° C	Bullers rings 18–46	1030° C	Bullers ring 32
1½ ton fired wt.	1400 lb	Approx. 96–120 lb	16 cwt
Nil	2520 lb	Approx. 440 lb	Nil
1¼ ton	3920 lb	Approx. 560 lb	16 cwt
670 (2⅝ × 4 3/16 × 8¾ in.)	26·2 yd super	Approx. 16 doz. biscuit and glost	50 yd, ⅜ in. tiles
48 hr	65 hr	12 hr	96 hr
83	108	220	30
10 000	500 sq yd	570 doz., biscuit and glost	350 yd
22·5 tons fired weight	12 tons	3600 lb approx.	5½ tons
1·35 cwt/ton ⎫ average	94 therms/ton	6¼ gal/150 lb ware	34 therms/ton
3 cwt/1000 ⎬ as	2·06 therms/sq yd	6¼ gal/23 doz. ware	
30 cwt ⎭ received	216 000 ft³/dy 1030 therms/dy	156 gal	50 000 ft³
All heat for dryer also supplied			
0·66 hr ⎫	Nil	1 man/6 cars	3 men/24 hr
0·44 hr/ton ⎬ including		1 man/600 lb	(1 man per shift)
1·6 hr ⎪ lorry	18·8 man-hr	None	Nil
1·07 hr/ton ⎭ loading	30 man-hr/ton to load and unload		

TABLE

Tunnel

	No.	7	8
	Code No.	10	4
I. (a)	Type of kiln	Davis	—
(b)	Straight or circular	Straight	Straight
(c)	Open-flame or muffle	Open flame	Open flame
(d)	Length: pre-heating zone	36 ft	100 ft
(e)	firing zone	15 ft	45 ft
(f)	cooling zone	39 ft	75 ft
(g)	total	90 ft	220 ft
(h)	width	40 in.	4 ft internal, 11 ft 6 in. external
(i)	height above car or slab top	26 in.	7 ft
(j)	Number of cars in kiln	16	40
II. (a)	Fuel used	Town gas	Town gas
(b)	Heating capacity of fuel (B.t.u./ft^3)	440	470
(c)	Type of grate or burner	Open flame, naturally aspirated	'Amal' gas jets
III. (a)	Nature of ware	Glost tiles	Biscuit tiles
(b)	Moisture in green ware as set (%)	—	1
(c)	Nature of setting, saggared or open, with or without auxiliary supports	Open in cranks	Saggared
(d)	Maturing temperature or cone numbers	1050° C	Buller's rings 23–33
IV. (a)	Setting per car: weight ware	521 lb	1600 lb approx.
(b)	kiln furniture	666 lb	2000 lb approx.
(c)	total	1187 lb	3600 lb approx.
(d)	quantity of ware	13 yd	45 yd of 6 in. tiles approx.
V. (a)	Total time in kiln	18¼ hr	7 days
(b)	Travel of cars (ft/24 hr)	100	31
(c)	Output per day: quantity	250 yd	270 yd of 6 in. tiles
(d)	weight	4¼ tons	7900 lb
VI. (a)	Fuel consumption: per wt. ware	14 000 ft^3/ton	—
(b)	per quantity ware		—
(c)	per day	60 000 ft^3	5000 ft^3
VII.	Labour requirements:		
(a)	Skilled per car	⎫	Total, 6 placers
(b)	per weight	⎬ 7 men in shifts	
(c)	Unskilled per car	⎨ (2 per shift + 1 floater)	Total, 2 emptiers
(d)	per weight	⎭	

236—continued

Kilns—continued

9	10	11	12
4	13	21	21
Davis	Davis 'Revergen'	Gibbons Dressler	Gibbons
Straight	Straight	Straight	Straight
Open flame	Open flame	Open	Muffle
45 ft	45 ft	110 ft	74 ft
44 ft	17 ft	109 ft	66 ft
36 ft	45 ft	111 ft	70 ft
125 ft	107 ft	330 ft	210 ft
3 ft 3 in. int., 8 ft ext.	4 ft 6 in. int., 11 ft 6 in. ext.	42 in.	26 in.
4 ft 6 in.	2 ft 8 in.	51 in.	42 in.
24	20	49	43
Town gas	Town gas	Town gas	Town gas
470	475	470	470
Regenerating type gas jets	Firing zone: Independent supplies of gas and air mixed inside kiln Preheating zone: Amal type burners	Amal burner (induced primary air)	Low-pressure air and gas
Glazed tiles	Glost earthenware	Biscuit earthenware	Glost, white and coloured glaze
(Open with auxiliary supports)	—	2	$< \frac{1}{2}$
	Saggars	Open on decks	Open, Ceramic bats and cranks
Bullers rings 11–16	1090° C	1130° C	1080° C
693 lb approx.	—	1309 lb	280 lb
495 lb approx.	—	2400 lb	560 lb
1188 lb approx.	—	4500 lb	840 lb
22 yd of 6 in. tiles approx.	—	220 doz.	50 doz.
24 hr	32 hr	Related to data above taking into account variety of speeds down to approx. 60–70 min service	12–60
125	74		420–85
528 yd of 6 in. tiles approx.	750 doz.		4200–850 doz.
16 632 lb approx.	—		
—	—	940 ft³/cwt	210–42·5 cwt
—	—	4700 ft³/100 doz.	1000 ft³/cwt
2500 ft³	77 400 ft³	168 000 ft³	5000 ft³/100 doz.
			192 000
		3 attendants, 8 hr each per day (any speed)	
Total, 9 skilled placers	3 men fill 7½ cars per shift, 2 shifts		
Total, 2 unskilled emptiers	2 men empty 7½ cars per shift, 2 shifts	6–9 depending on speed for loading	

TABLE

Tunnel

	No.	13	14
	Code No.	4	9
I. (a)	Type of kiln	Bricesco	Dressler
(b)	Straight or circular	Straight	Straight
(c)	Open-flame or muffle	Muffle	Muffle
(d)	Length: pre-heating zone	45 ft	100 ft
(e)	firing zone	40 ft	20 ft
(f)	cooling zone	35 ft	80 ft
(g)	total	120 ft	203 ft
(h)	width	2 ft int., 8 ft ext.	3 ft 2 in.
(i)	height above car or slab top	4 ft 6 in.	30 in.
(j)	Number of cars in kiln	30	36
II. (a)	Fuel used	Town gas	Town gas
(b)	Heating capacity of fuel (B.t.u./ft^3)	470	475
(c)	Type of grate or burner	Premixed gas and air jets	Multi-orifice Williams burner
III. (a)	Nature of ware	Glost tiles	Glost wall tiles
(b)	Moisture in green ware as set (%)	—	<2
(c)	Nature of setting, saggared or open, with or without auxiliary supports	Open, in cranks	Cranks
(d)	Maturing temperature or cone numbers	Bullers rings 6–10	Bullers rings No. 55 contraction, 32–34 ~1050–1060° C
IV. (a)	Setting per car: weight ware	380 lb approx.	5 cwt
(b)	kiln furniture	270 lb approx.	6 cwt
(c)	total	650 lb approx.	11 cwt
(d)	quantity of ware	12 yd of 6 in. tiles approx.	22 yd^2
V. (a)	Total time in kiln	30 hr	32 hr
(b)	Travel of cars (ft/24 hr)	96	144
(c)	Output per day: quantity	288 yd of 6 in. tiles approx.	594 yd^2
(d)	weight	9070 lb approx.	—
VI. (a)	Fuel consumption: per wt. ware	—	—
(b)	per quantity ware	—	—
(c)	per day	2200 ft^3	96 000 ft^3
VII.	Labour requirements:		
(a) Skilled	per car	3 skilled placers	3
(b)	per weight		—
(c) Unskilled	per car	1 unskilled emptier	Nil
(d)	per weight		—

236—*continued*

Kilns—*continued*

15	16	17	18
13	17	17	6
Gibbons Muffle 'Thrift'	Gibbons Dressler 'Thrift'	Gibbons Dressler 'Thrift'	Dressler
Straight	Straight	Straight	Straight
Muffle	Muffle	Muffle	Muffle
41 ft	165 ft	161 ft	140 ft
43 ft 2 in.	68 ft	65 ft	6 ft
41 ft 10 in.	145 ft	92 ft	96 ft
126 ft	378 ft	318 ft	242 ft
21 in.	4 ft 8 in.	4 ft 8 in.	3 ft 6 in.
2 ft 8½ in.	4 ft 6 in.	30 in.	6 ft 6 in.
26	60	50	42
Town gas	Coke producer gas	Coke producer gas	Producer gas
475	142	142	?
Thrift burner	—	—	Step grate open port burner
Earthenware	Sanitary fireclay	Sanitary fireclay	Stoneware, biscuit and glost
2	0·6	0·6	2
Sillimanite bats 19 × 13 × ¾ in. with fireclay props	Open, some trucks have furniture	Open, no support	Open
1140° C	1180° C	1180° C	Bar No. 26
214 lb	2020 lb	1397 lb	7 cwt
409 lb	1295 lb	Nil	12 cwt
627 lb	3315 lb	1397 lb	Incl. chassis 2 tons 8 cwt
40 doz.	20 pieces	12/14 pieces	50 pieces biscuit, 350 pieces glost
45 hr 30 min	131 hr	125 hr	42/28 hr
62	68	62	132/200
520 doz.	220 pieces	120/140 pieces	24/36 trucks
2561 lb	22 220 lb	13 970 lb	Variable
—	0·382 lb coke	0·587 lb coke	Not determined
—	41 lb coke	63 lb coke	Not determined
62 400 ft³	8200 lb coke	8200 lb coke	5 tons
⅕ shift per 8 hr	0·027	0·027 men	⅙ men
—	—	—	—
—	0·068	0·068 men	1/12 men

TABLE

Tunnel

	No.	19	20
	Code No.	13	4
I. (a)	Type of kiln	Gibbons	Moore
(b)	Straight or circular	Straight	Straight
(c)	Open-flame or muffle	Open	Open
(d)	Length: pre-heating zone	16 ft 1 in.	6 ft
(e)	firing zone	29 ft	38 ft
(f)	cooling zone	34 ft 6 in.	36 ft
(g)	total	79 ft 7 in.	80 ft
(h)	width	18 in.	2 ft
(i)	height above car or slab top	2 ft 4 in.	4 ft 6 in.
(j)	Number of cars in kiln	16	18
II. (a)	Fuel used	Electricity	Electricity
(b)	Heating capacity of fuel (B.t.u./ft^3)	—	—
(c)	Type of grate or burner	—	Elements
III. (a)	Nature of ware	China glost	Glost earthenware
(b)	Moisture in green ware as set (%)	—	—
(c)	Nature of setting, saggared or open, with or without auxiliary supports	Sillimanite bats $18 \times 17 \times \frac{3}{4}$ in. with fireclay props	Open with supports
(d)	Maturing temperature or cone numbers	1080° C	Bullers rings 4–12
IV. (a)	Setting per car: weight ware	—	—
(b)	kiln furniture	—	—
(c)	total	—	—
(d)	quantity of ware	—	50 doz. pieces
V. (a)	Total time in kiln	34 hr 40 min	18 hr
(b)	Travel of cars (ft/24 hr)	50	Right through kiln (80 ft)
(c)	Output per day quantity	125 doz.	1100 doz.
(d)	weight	—	—
VI. (a)	Fuel consumption: per wt. ware	—	—
(b)	per quantity ware	—	—
(c)	per day	—	3000 units
VII.	Labour requirements:		
(a) Skilled	per car	$\frac{1}{4}$ shift emptying and filling	5 total
(b)	per weight		
(c) Unskilled	per car		1 total
(d)	per weight		

236—*continued*

Kilns—*continued*

21	22	
21	13	
Gibbons Moore Campbell	Rotolec	
Straight	Circular	
Open	Open	
45 ft	—	
36 ft	—	
35 ft	—	
116 ft	—	
16 in.	—	
32 in.	—	
24	18	
Electricity	Electricity	
—	—	
Elements in the walls	Nichrome	
On-glaze decorated ware	Earthenware and china enamel	
Trace		
Open. Iron bats and cranks for cups, heat resistant steel	Alloy steel plates with sillimanite bats with steel rods and supports also refractory supports	Sillimanite bats $17\frac{1}{2} \times 16 \times \frac{5}{8}$ in. with refractory supports
735° C	740° C	800° C
179 lb		
246 lb for flat ware, 582 lb for cups		
517 lb		
32 doz.	3 doz.	4 doz.
10–38 hr	15 hr	18 hr
280–73		
1800–480 doz.		
—		
50 kWh/cwt		
250 kWh/100 doz.		
1780 kWh/day		
	1 man per 8 hr shift	1 man per 8 hr shift

TABLE 236—*continued*

Tunnel Kilns—*continued*

		23	24
	No.		
	Code No.	4	4
I.	(a) Type of kiln	Gottignie passage	Heurty passage
	(b) Straight or circular	Straight 24 passages	Straight 16 passages
	(c) Open-flame or muffle	Open	
	(d) Length: pre-heating zone		
	(e) firing zone		
	(f) cooling zone		
	(g) total	Varying	38 ft 6 in.
	(h) width	1 ft	15 in.
	(i) height above car or slab top		8 in.
	(j) Number of cars in kiln		
II.	(a) Fuel used	Electricity	Electricity
	(b) Heating capacity of fuel (B.t.u./ft^3)	—	
	(c) Type of grate or burner	Kanthal A1 elements	Kanthal A1 elements
III.	(a) Nature of ware	Biscuit earthenware	Glost earthenware and clay-dipped earthenware
	(b) Moisture in green ware as set (%)	1	
	(c) Nature of setting, saggared or open, with or without auxiliary supports	Open, on bats	Open, on bats
	(d) Maturing temperature or cone numbers	Bullers ring, 25–30	Glost, Bullers ring 10 Clay-dipped, Bullers ring 28
IV.	(a) Setting per car: weight ware	Approx. 5 lb/bat	5 lb/bat
	(b) kiln furniture		
	(c) total		
	(d) quantity of ware		
V.	(a) Total time in kiln	24 hr	12 hr
	(b) Travel of cars (ft/24 hr)		
	(c) Output per day: quantity		
	(d) weight		
VI.	(a) Fuel consumption: per wt. ware		
	(b) per quantity ware		
	(c) per day	1500 kW	2000 kW
VII.	Labour requirements:		
	(a) Skilled per car	5 shift work placers	7 shift work placers
	(b) per weight		
	(c) Unskilled per car		
	(d) per weight		

glassy granules. The purpose of the fritting process is twofold (*see* p. 566): to make soluble ingredients combine with other materials to produce insoluble compounds, and to decompose any raw materials that give off gases when heated. This second action although harmful in a glaze applied to a body is, in fact, very helpful in the fritting process, as gas evolution helps to keep the molten mass stirred. Raw materials with water of crystallisation are sometimes deliberately chosen for this purpose.

Completion of the fritting process is tested by withdrawing samples on a poker. These must be homogeneous and clear, quite free from gas bubbles or solid material.

Fritting is usually done intermittently in reverberatory kilns or rotary kilns. Where large quantities of the same frit are required continuously a continuous process may be used.

REVERBERATORY OR STATIONARY HEARTH FRIT KILN

This is the old type of frit kiln. It is a rectangular masonry structure with a shallow hearth built of aluminous firebrick. It may be fired by coal, gas or oil. In the first case a firebox is separated from the melting hearth by a bridge wall. In all cases the charge is heated from above by the passage of the combustion gases and by radiation from the arched kiln roof. The charge is introduced through the top of the kiln and the molten product tapped off into a water tank from the bottom. A new charge is introduced immediately to make use of the stored heat in the structure.

The disadvantages of this type of frit kiln are the difficulty of heating the charge uniformly and the very severe conditions applied to the kiln lining. The curved roof radiating heat on to the rectangular hearth tends to concentrate the heat on the centre. The charge in the corners therefore takes longer to melt than that in the centre and may not become uniform with it. The central portion of the charge may become over-heated while waiting for the corners to melt. The molten frit readily penetrates the joints between the bricks of the floor of the hearth and eventually causes this to collapse even when highly resistant bricks have been used, and the great heat of the flame on the arch makes this comparatively short lived (**H94**).

SEMI-ROTARY AND ROTARY FRIT KILNS

A number of the disadvantages of the stationary hearth kiln can be overcome by using a rotating drum. This can, however, not be fired by coal, it requires gas or oil.

The rotary kilns consist of steel drums lined with suitable refractory materials. They can be rotated about a horizontal axis and some are mounted on a trunnion so that they can be tipped for pouring and filling through the ends. Heating is by gas or oil burner in one end with a flue outlet under a cowl leading to the chimney stack at the other end.

The charge is filled into the drum and it is brought into a horizontal position if it is not permanently so. The burner is lit and the drum rotated very slowly so that the charge lies at its angle of repose up one side of it. As melting occurs the liquid flows off the charge to the bottom of the incline and a new surface is exposed. In this way the melting of the lower parts of the charge does not depend on the conduction of heat through the upper parts as in the stationary-hearth type of kiln, and so is very much quicker. When the charge is completely molten the furnace is rotated a little faster so that mixing occurs. Throughout the operation each part of the furnace lining is in turn heated directly by the flame and then cooled by the charge. A higher proportion of the combustion heat is thereby transmitted to the charge and less strain imposed on the kiln lining (**H94, G44**).

In semi-rotary furnaces the drum is only rocked to and fro during the initial melting of the charge and then rotated to give a mixing action during the maturing period (**M86**).

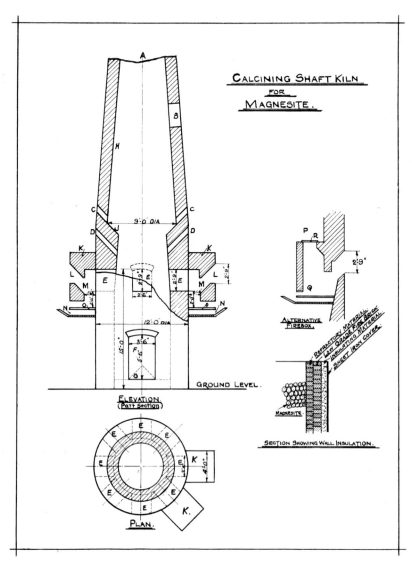

FIG. 11.53. *Shaft calcining kiln for magnesite*

Both types of rotary furnaces may be hand or motor operated.

The initial cost of rotary frit melters is considerably more than stationary reverberatory kilns but their operation is more economical and a more uniform frit can be produced.

CONTINUOUS FRITTING

Fritting can be made a continuous process where the constant demand makes this desirable. For instance the charge may be fed on to the top of a flight of shallow steps with a burner directed up them. It then flows downwards in a shallow stream eventually dropping into a water tank (**A59**). This process is found to corrode the refractory lining very much less, thereby giving a purer frit and also allowing further improvement by making it economical to use more expensive iron-free refractories. There is also waste heat available for drying the quenched frit, etc. (**P47**).

FIG. 11.54. *Polishing glost-fired tableware after removing the points of the kiln furniture with a chisel* (*Wedgwood*, **W35**)

TABLE 237

Glass/Refractory Reactions (H133)

	Soda–lime (commercial, as used for electric light bulbs)	Lead (57% SiO_2.29% PbO. 5·5% Na_2O.8·5% K_2O)	Borosilicate (commercial heat-resisting glass)
Good ↑ ↓ Bad	Chrome–magnesite Magnesite Alumina Firebrick (42% Al_2O_3) Kyanite Chrome Forsterite Firebrick (76% Al_2O_3) Mullite (fusion cast) Silica (95% SiO_2) Zircon Clay tank block	Alumina Firebrick (42% Al_2O_3) Chrome–magnesite Magnesite Kyanite Forsterite Zircon Mullite (fusion cast) Silica (95% SiO_2) Firebrick (76% Al_2O_3) Clay tank block Chrome	Alumina Zircon Firebrick (76% Al_2O_3) Firebrick (42% Al_2O_3) Mullite (fusion cast) Kyanite Clay tank block Silica (95% SiO_2) Chrome–magnesite Magnesite Chrome Forsterite

FRIT KILN LINING

Frit kiln linings undergo very severe physical and chemical conditions of abrasion, heat, sudden cooling and chemical attack. The possible preventive measures are to a small extent contradictory. For instance chemical attack is greater on a porous body than on a vitreous body whereas thermal shock resistance is higher with the porous one.

FIG. 11.55. *Grinding a fired porcelain insulator to accurate dimensions*
(*Doulton*, **D56**)

Hyslop, Stewart and Burns (**H133**) describe a test to find the extent to which a glass or slag will react with a refractory brick to form low-melting, fluid and therefore corrosive products. Three glasses were used and the relative resistances of various refractories are tabulated. This test does not take the physical nature of the refractory very much into account, but some interesting information can be obtained.

Kilns for Calcining Raw Materials

A number of raw materials require heat treatment before incorporation in the ceramic body either by calcining to drive off combined water, carbon dioxide, etc., and perhaps change the crystal form, or by firing to make grog.

If clay in largish lumps is to be fired to give grog this may be done in batches in periodic down-draught or continuous chamber kilns but there is a tendency for gas circulation and therefore heat distribution to be poor and for flues to become blocked. Labour requirements are heavy. Tunnel kilns are generally too costly for firing grog, they also need to be wide and shallow with retaining walls built on the trucks to prevent grog falling off and fouling the track.

It is therefore usual to use *shaft kilns* (Fig. 11.53), or *rotary kilns* not otherwise used in the ceramic industry, for firing powdery or granular materials (**M72**).

Inspection and Finishing

Ware coming out of ceramic kilns is stored for inspection before it undergoes any further processing. This may involve visual inspection, sorting into first, seconds, etc., sorting out blemished ware that must have the points of kiln furniture ground off, or that can be otherwise put right by polishing (**S109**) (Fig. 11.54), sorting out blemished ware with pinholes, etc., that can be treated and refired, etc. Ware that is fired more than once is inspected after each firing and useless pieces culled immediately. In classes of ware where grog or pitchers are required in the mixture, faulty pieces are crushed and ground for re-use.

Much technical ware must undergo rigorous sizing. Some ware is ground to size after firing (Fig. 11.55).

PART II
Ceramic Products

CERAMIC products have here been grouped according to their uses and described very briefly under standard headings. Relevant physical properties are then given. Chapter 12 deals with 'heavy' or 'coarse' ceramics and the general sequences for preparation, drying, etc., are given in greater detail at the beginning under 'bricks' and 'tiles' and subsequently referred back to from Chapter 17 as well. Fine ceramic preparation sequences are given in most detail in Chapters 12 and 13 and subsequently referred back to. Stoneware is described in Chapter 14.

Chapter 12
Ceramic Building Materials

COMMON BRICKS

Characteristics. Mass-produced bricks in which no special care is taken in manufacture to avoid surface blemishes.
Uses. Construction where appearance is of no account, *i.e.* footings, backing up, surface for rendering, etc.
Body type. Brickware.
Raw Materials. Local easily obtained clays together with sand, grog, ash.
Preparation and Shaping. According to the nature of the clay and the consistency required for shaping, these are largely general methods for 'heavy ceramics'.

(1) *Easily worked clay* to make 'wirecuts': box or worm feeder→wet pan→finishing rolls→pugmill and auger→cutting-off table.

(2) *Harder clay* to make 'wire-cuts': jaw crusher→box feeder→dry pan→bucket elevator→vibrating screen→storage bins→disk feeder→open-trough mixer→pugmill and auger→cutting-off table.

(3) *Shale* for stiff plastic making: pan feeder→perforated grinding pan→ [oversize loop] screening→storage hopper→revolving disk feeder→double-shafted mixer, water added→brick machine, where clots are formed and pressed into bricks.

(4) *Hard shaly clay* for semi-dry pressing (largely confined to Flettons in Great Britain): trough feeder→perforated dry pan→screens→moisture [oversize loop] content adjustment→presses.

(4a) As applied in largest most mechanised works:
dragline→conveyor→grizzly screen, coarse grading→ large size, manual removal of stones
small
storage trough←conveyor←kibbler preliminary crushing
dry pans with automatic feed control→bucket elevator
[oversize loop]
→heated piano-wire screens→chutes→presses→setting on gridded pallet ready for fork lift truck to remove complete to kiln.

(5) Special shapes are made by cutting down standard bricks when in the leather-hard condition or by pressing in suitable moulds.

Additional preparation sometimes necessary:

(a) magnets for removing tramp iron;
(b) crushing rollers with helical thread to remove stones;
(c) prolonged weathering.

Drying.

(1) Hack drying in open or in unheated sheds.
(2) Hot floor.
(3) Chamber dryers.
(4) Tunnel dryers.

Firing. 960–1180° C (1760–2156° F) cone 07–5.

(1) Hoffmann, round, oblong and zigzag.
(2) Transverse arch kilns, Staffordshire, etc.
(3) Tunnel kilns, direct fired with coal, producer gas or fuel oil, e.g. Faugeron, Bricesco-Harrop, Top-fired (Monnier) (**C41**).

(*See also The Firing of Common Bricks*, W. Noble and A. T. Green, H.M.S.O., 1950.)

FACING BRICKS AND STOCK BRICKS

Characteristics. Bricks produced with a view to appearance as well as structural use; the natural brick of a particular district, when not sand faced, engobed, etc., is known as the stock brick.

Uses. Constructional work where appearance counts.

Body type. Brickware.

Raw materials. Selected local clays together with sand, grog, etc.

Preparation and Shaping. Determined by the nature of the clay and the rate of production required. These include general methods for 'heavy ceramics'.

(1) *Easily worked clay* for 'soft-mud' hand moulding: box or worm feeder →wet pan→finishing rolls→assistant making clots→slop or sand moulder→ drying floor→mould removed.

(2) *Readily worked clay* for 'semi-stiff mud' machine moulding: box or worm feeder→wet pan or crushing rolls→automatic brick machine making clots and pressing them.

(3) *Soft* or *hard clays* for 'wire-cuts': see preparation of common wire cuts as far as cutting off table→sanding→partial drying→re-pressing.

(4) *Shale* for stiff plastic making: as for common bricks plus re-pressing.

(5) *Shale* for semi-dry pressing: as for commons then sanding or engobing or surface scratching to give rustic bricks.

(6) Special shapes are made by cutting down standard bricks when in the leather-hard condition or by pressing in suitable moulds.

Additional processes included:
(a) magnetic removal of tramp iron;
(b) crushing rollers with helical thread to remove stones;
(c) prolonged weathering.

Drying.
(1) Hack drying in the open or in unheated sheds.
(2) Hot floor.
(3) Chamber.
(4) Tunnel.

Firing. 960°–1180° C (1760–2156° F) cone 07–5.
(1) Clamp.
(2) Scotch, round or rectangular down-draught, Newcastle.
(3) Most types of continuous kiln.
(4) Tunnel kilns, e.g. Faugeron, Bricesco-Harrop, Monnier (**C41**).

Standards for Building Bricks. B.S. 1257:1945; B.S. 657:1950.
Tests. Size. Compressive strength of single bricks, piers and walls. Water absorption. Chemical analysis. Soluble salts. Efflorescence. Moisture movement, rarely above 0·01% should not exceed 0·025%. Frost resistance.
Labour Requirements. These vary with the output and with the nature of the clay and the machinery available. Most machines require a minimum of attendance regardless of their output so that works with larger outputs tend to need less man-hours per thousand saleable bricks. Tables 238 (a)–(c) give summaries of minimum labour requirements in a number of brickworks.

TABLE 238(a)
Minimum Labour Requirement for Wirecut Brick Manufacture (C41)

Works Kind of dryer Output (bricks per week)	1 Hot-floor 96 000		2 Hot-floor 192 000		3 Tunnel 96 000		4 Tunnel 192 000		5 Tunnel 480 000		6 Chamber 120 000	
	Men	*m.h.t.	Men	m.h.t.	Men	m.h.t.	Men	m.h.t.	Men	m.h.t.	Men	m.h.t.
Excavator and cleaning	2	1·0	2	0·5	2	1·0	2	0·5	3	0·3	2	0·8
Haulage	—	—	1	0·25	—	—	1	0·25	2	0·2	1	0·4
Making	8	4·0	14	3·5	7	3·5	9	2·25	19	1·9	10	4·0
Setting	2	1·0	4	1·0	2	1·0	4	1·0	9	0·9	3	1·2
Burning	3	1·75	3	0·87	3	1·75	3	0·87	6	0·7	3	1·4
Other kiln labour (wheeling coal)	—	—	1	0·25	—	—	1	0·25	2	0·2	1	0·4
Drawing	3	1·5	6	1·5	3	1·5	6	1·5	13	1·3	4	1·6
Foreman	1	0·5	1	0·25	1	0·5	1	0·25	1	0·1	1	0·4
Boiler man	1	0·75	1	0·37	—	—	1	0·37	2	0·2	1	0·5
Engine man	—	—	1	0·25	—	—	1	0·25	1	0·1	1	0·4
General labourers	1	0·5	2	0·5	2	1·0	2	0·5	4	0·4	2	0·8
Maintenance	—	0·25	2	0·5	—	0·25	2	0·5	4	0·4	1	0·4
Totals	21	11·25	38	9·75	20	10·5	33	8·5	66	6·7	30	12·3

* Man-hours per thousand saleable bricks.

TABLE 238(b)

Minimum Labour Requirements for Stiff-plastic Brick Manufacture (C41)

Works Output (bricks per week) Making-machines	1 106 000 2 single-mould		2 156 000 3 single-mould		3 86 500 1 single-mould		4 173 000 2 double-mould	
	Men	*m.h.t.	Men	m.h.t.	Men	m.h.t.	Men	m.h.t.
Excavator and cleaning	2	0·92	2	0·61	2	1·12	2	0·56
Haulage	1	0·46	1	0·31	—	—	1	0·28
Making	8	3·68	10	3·07	7	3·92	13	3·64
Setting	2	0·92	3	0·92	2	1·12	4	1·12
Burning	3	1·61	3	1·07	3	1·96	3	0·98
Other kiln labour (wheeling coal)	—	—	1	0·31	—	—	1	0·28
Drawing	4	1·84	5	1·53	3	1·68	6	1·68
Foreman	1	0·46	1	0·31	1	0·56	1	0·28
General labourers	1	0·46	2	0·61	1	0·56	2	0·56
Maintenance	1	0·46	2	0·61	1	0·56	2	0·56
Totals	23	10·81	30	9·35	20	11·48	35	9·94

* Man-hours per thousand saleable bricks

TABLE 238(c)

Minimum Labour Requirements for Semi-dry Press Brick Manufacture (C41)

Works Output (bricks per week)	1 96 000		2 384 000		3 Very large
	Men	*m.h.t.	Men	m.h.t.	m.h.t.
Excavator and cleaning	2	1·0	3	0·38	0·2
Haulage	1	0·5	2	0·25	0·2
Making	5	2·5	17	2·13	1·65
Setting	2	1·0	8	1·0	1·0
Burning	3	1·75	6	0·89	0·3
Other kiln labour (wheeling coal)	—	—	2	0·25	0·3
Drawing	3	1·5	9	1·12	1·0
Foreman	1	0·5	1	0·12	0·2
General labourer	1	0·5	4	0·5	0·5
Maintenance	2	1·0	6	0·75	1·0
Totals	20	10·25	58	7·39	6·35

* Man-hours per thousand saleable bricks.

CERAMIC BUILDING MATERIALS 1069

ENGINEERING BRICKS, BLUE BRICKS

Characteristics. Harder and denser bricks of bluish-grey colour. (In Great Britain water absorption may be up to 10%, in Continental Europe it is 3–4%.)

Uses. Civil Engineering, *e.g.* bridge construction. In North Europe where high frost resistance is essential they are also used for buildings.

Body Type. Brickware approaching stoneware.

Raw Materials. Natural clays containing iron, and free of lime and magnesia.

Preparation and Shaping. Generally as for common bricks. About 20% are produced by the semi-dry or dry (3–12% moisture) pressing methods. This gives the best results. About 80% are made by plastic extrusion, cutting-off and re-pressing or by stiff plastic pressing methods.

Drying. As for common bricks.

Firing. 1040–1250° C (1904–2282° F) cone 03–8.

   (1) Intermittent. Up-draught and down-draught, Newcastle.
   (2) Continuous gallery and transverse arch chamber types.
   (3) Tunnel kilns. Direct fired. Faugeron. Bricesco-Harrop. Monnier.

Standards for Engineering Bricks. B.S.S. 1301:1946.

Tests. Compressive strength. Water absorption. Chemical analysis. Soluble salts. Efflorescence. Moisture movement.

HOLLOW BRICKS, PERFORATED BRICKS, HOLLOW TILES

Characteristics. Bricks or blocks with large regular air spaces surrounded and separated by relatively thin walls. Vertical slots, holes or perforations are preferred to horizontal ones as no specials are then needed at openings and corners. It has now been shown that these are quite as resistant to water penetration as the horizontally perforated ones. In recent years there has been much development of these bricks in order to produce more economical building methods, better insulation, and for the ceramic brick to meet the competition from concrete blocks, etc. In Europe a large number of types and shapes have evolved. In Great Britain a recent development (1958/59) is a perforated 9 × 9 in. (23 × 23 cm) brick that effectively replaces the two separate bricks of an 11 in. (28 cm) cavity wall.

Uses. Wall construction, especially where light weight and/or good thermal insulation are required. Floor construction of reinforced concrete 'rib and tile' floor (**B145, H12**).

Body Type. Brickware.

Production. As for extruded wire cut bricks, although better preparation of the body may be necessary.

Standards for Hollow Bricks. B.S. 1190:1951.

TABLE

Analyses of Bricks made from Clays from Different Geological

	Alluvial	London stock	Glacial	Oligo-cene	London clay	Gault	Weald	Oxford clay (Fletton)
SiO_2	64.7	68.7	62.5	77.8	64.4	47.2	68.4	56.2
TiO_2	1.6	0.7	0.9	2.2	1.6	0.8	1.3	0.5
Al_2O_3	12.7	11.0	18.6	15.8	15.8	19.4	17.2	20.9
Total iron as Fe_2O_3	8.3	7.0	6.6	0.8	7.9	6.1	6.3	6.0
CaO	7.9	8.1	4.1	0.3	1.1	19.2	1.9	8.1
MgO	1.9	0.8	3.4	0.4	2.4	1.9	1.2	1.7
K_2O	1.5	2.0	2.9	1.8	3.2	3.1	2.2	3.6
Na_2O	0.4	0.8	0.5	0.5	0.5	0.6	0.5	0.5
SO_3	1.4	0.6	0.4	0.5	2.3	1.4	0.7	1.9
Loss on ignition	0.3	Nil	0.3	0.3	1.1	0.4	0.3	0.6

TABLE 240

Typical Test Results on Bricks made from Clays of Various Geological Formations (Bonnell and Butterworth *Clay Building Bricks of the United Kingdom*) (B91)

Brick	Density (g/cm³)	Water absorption				Saturation coefficient C/B	Compressive strength (lb/in²)	Total soluble salts (%)	Liability to efflorescence
		C (24 hr soak)		B (5 hr boil)					
		By wt. %	By vol. %	By wt. %	By vol. %				
London Stock* 'first grade'	1.52	14.7	22.1	26.2	40.0	0.55	2070	0.94	Nil
Glacial W.C. common	1.82	13.4	24.3	17.0	30.9	0.78	4570	0.86	Slight
Oligocene W.C., engineering	2.06	2.8	5.2	4.5	9.3	0.56	9450	0.15	—
London Clay W.C. common	1.80	12.8	23.0	16.9	30.4	0.76		2.11	
Gault S.D. common	1.56	23.9	37.2	28.5	44.3	0.84	1520	1.19	Nil
Weald Red W.C. engineering	2.07	7.2	14.7	9.6	19.8	0.74	8160	0.49	Nil
Weald S.P. engineering	2.09	4.8	10.4	6.4	19.0	0.72	11 540	—	
Fletton S.D.	1.65	18.4	30.4	21.5	35.6	0.85	4040	2.38	Slight
Middle Lias S.D. common	1.96	10.9	21.3	13.7	26.7	0.79	5580	0.68	Slight
Keuper Marl, Top W.C. common	1.75	16.1	28.2	20.4	35.7	0.79	4320	1.29	Fairly slight
Middle W.C. common	1.67	21.0	35.0	24.9	41.5	0.84	2780	5.70	Slight/heavy
Bottom W.C. facing	1.85	12.4	22.8	16.4	30.3	0.75	4900	0.94	Slight
Permian W.C. facing	1.78	14.9	26.6	18.4	32.8	0.81	4030	0.94	Slight
Etruria Marl, brown W.C. engineering	2.31	0.4	1.0	1.8	4.2	0.23	11 750	0.20	—
Coal Measure, Buff S.P. common	1.97	5.5	10.8	9.8	19.2	0.56	5610	0.14	Slight
Red, S.P. engineering	2.42	1.7	4.1	3.0	7.3	0.53	18 800	0.15	
Red S.P. common	2.24	5.9	13.1	7.4	16.5	0.79	8360	0.25	Slight
Devonian, S.P. common	2.04	10.2	20.7	1.27	25.8	0.80	5880	0.15	Nil

* Methods of manufacture: W.C. = wire-cut; S.D. = semi-dry; S.P. = stiff plastic.

Formations (Bonnell and Butterworth) (B91)

Middle Lias	Keuper Marl within 300 ft of top	Keuper Marl (middle)	Keuper Marl less than 300 ft above base	Permian	Etruria Marl	Coal measure	Coal measure	Coal measure	Devonian
57·8	58·3	46·2	66·0	60·1	62·7	54·9	61·7	61·9	59·6
1·2	0·7	0·8	0·6	0·5	1·2	0·6	1·2	1·1	1·2
23·2	15·3	13·7	13·9	16·7	23·1	34·9	21·6	24·0	19·9
9·3	6·0	6·0	6·8	5·8	8·4	3·4	8·0	8·7	11·4
1·0	6·2	11·4	3·5	6·5	0·9	2·1	0·6	0·6	0·2
2·5	7·3	12·8	2·9	4·2	1·2	0·7	1·0	1·1	1·2
2·9	4·7	3·3	4·1	3·3	2·6	2·6	3·1	1·6	4·2
0·9	0·7	0·3	0·6	1·2	0·4	0·1	1·2	0·2	1·0
0·3	0·5	5·6	1·0	0·9	0·7	0·1	1·5	—	0·1
1·0	0·4	0·5	0·2	0·3	Nil	0·6	0·1	0·5	0·5

TABLE 241

Range of Results for Measurements of Compressive Strength and Water Absorption for Different Kinds of Clay Building Brick (C41)

	Mean compressive strength (lb/in^2)	Water absorption (5 hr boiling)	
		(% wt.)	(% vol.)
Engineering Bricks			
British Standard, Grade A	10 000–c. 20 000		
British Standard, Grade B	7000–10 000		
Facing Bricks			
Hand-made—Extreme Range	1000– 8500	9–28	19–42
Machine-made			
Extreme Range	500–15 000	1–37	2–50
London Stocks (1st Grade)	500– 2500	22–37	36–50
Boulder Clay W.C.	4000– 7500	9–19	19–34
White Gault W.C.	2000– 3000	22–28	38–44
Keuper Marl W.C. (Leicester)	4000– 6500	12–21	24–37
Coal Measure Shale (S.P. or W.C.)	5000–15 000	1–16	2–30
Flettons S.D.	2000– 4500	17–25	30–40
Common Bricks			
London Stocks	500– 1500	27–38	42–51
Boulder Clay W.C.	3000– 6000	14–25	26–40
Keuper Marl W.C. (Birmingham, Leicester, Nottingham)	3000– 4000	17–26	31–42
Coal Measure Shale S.P.	1500–10 000	4–18	10–34
Flettons	2000– 4500	17–25	30–40

GLAZED BRICKS

Characteristics. Off-white bricks bearing white opaque engobe and glaze on one face.
Uses. Constructional, where light requirements or the need to wash the walls down make a glazed surface necessary.
Body Type. Some high-grade brickware or low-grade refractory.
Raw Materials. Body: fireclay of low refractoriness, with grog or sand (loess). Engobe: Ball clay, china clay, flint, stone.
Preparation and Shaping. Brick: as for facing bricks. Engobe: prepared as a pottery body. Raw brick in leather-hard or dry condition is engobed by brushing or spraying.
Glazing. Opaque white glaze is applied by brushing or spraying on top of the engobe.
Drying. As for common bricks, taking care to place the glazed side vertical out of contact with other bricks.
Firing. 1100–1200° C (2012–2192° F) cone 1–6.
  (1) Intermittent. Up-draught, down-draught, direct-fired or muffle kilns.
  (2) Continuous gallery and chamber types where grate fired.
  (3) Tunnel direct-fired or muffle kilns.

ROOFING TILES, PLAIN OR INTERLOCKING

Characteristics. Thin pieces shaped to overlap and/or interlock on sloping surfaces. Appearance also counts.
Uses. Roofing or exterior wall cladding.
Body Type. Brickware of close-grained and denser texture than bricks. Colouring of the body or engobing of the tile are not uncommon.
Raw Materials. Selected local clays together with sand if necessary.
Preparation and Shaping. This is very largely dependent on local conditions and varies widely. For example:
  (1) *Soft plastic* hand making of plain tiles: haulage from pit→watering and weathering in heaps→crushing through three pairs of rollers, the last being $\frac{1}{16}$ in. (1·5 mm) apart→extrusion from pug to form clots→souring→hand moulding, sanding, colouring→cambering.
  (2) *Plastic extrusion* of plain tiles: haulage from pit→watering and weathering→wet pan→tempering through two pairs of rollers $\frac{1}{4}$ in. (6·3 cm) and $\frac{3}{32}$ in. (2·4 mm) apart→pugmill→clots→storage and souring→extrusion auger giving continuous rib tiles→cutting-off table→sanding→cambering.
  (3) *Stiff plastic* method for plain tiles: haulage from pit→dry-pan→elevating→screening→mixing, water being added→wet grinding→pugmill→expression rollers→bat making→tile pressing→sanding→cambering.
  (4) *Semi-stiff plastic* method for interlocking tiles: preparation as (1), (2) or (3), →pugmill→clots→revolver press.

CERAMIC BUILDING MATERIALS

Drying. Soft tiles are dried on trays or racks. Stiffer tiles are dried in bungs on drying floors or on stillages in chamber dryers. Some leather-hard tiles are set in the kiln and finished there.

Firing. Temperature *ca.* 900–1160° C (1652–2120° F) cone 010–4.

(1) Round or rectangular down-draught kiln.
(2) Gallery type continuous kilns, Hoffmann, zigzag, Belgian, etc.
(3) Tunnel kilns, direct-fired with coal, producer gas or oil, *e.g.* Faugeron, Bricesco-Harrop, Monnier (**C41**).

Standards for Roofing Tiles. B.S. 402:1945; B.S. 1424:1948; B.S. 2717:1956.

Tests. Transverse strength. Water absorption. Frost resistance.

TABLE 242

Range of Values of Properties of Roofing-tiles (C41)

	Plain tiles		Pantiles
	Etruria Marls (9 works)	Tertiary and recent clays (6 works)	(7 works)
Water absorption			
Range (%)	3–8	4–17	12–17
Mean (%)	5·2	9·9	14·2
Breaking load			
Range (lb)	150–290	140–280	—
Mean (lb)	220	200	—
Mean thickness (in.)	0·44	0·52	—
Transverse strength			
Range (lb/in^2)	1500–2500	700–2000	—
Mean (lb/in^2)	1900	1000	—

CHIMNEY POTS, FLOWER POTS

Characteristics. Round, hollow, red, porous pieces.
Uses. *See* heading.
Body Type. Brickware similar to that used for roofing tiles.
Preparation. As for roofing tiles.
Shaping. Hand throwing. Jolleying, by hand or automatic. Cylindrical chimney pots, extruded and cut off.
Drying. As for bricks.
Firing. 950–1100° C (1742–2012° F) cone (08/07–1). Kilns as for common bricks.
Standards for Chimney Pots. B.S. 1181:1944.

DRAINAGE PIPES, FIELD DRAINS, AGRICULTURAL DRAINS

Characteristics. Porous pipes which can be butt jointed and will withstand compression.
Uses. Land drainage.
Body Type. Brickware.
Raw Materials. Local easily obtained clays together with sand, grog, etc.
Preparation and Shaping. As for common wire-cut bricks.
Drying.
  (1) In open or in unheated sheds.
  (2) Hot floor.
  (3) Chamber dryer.
  (4) Tunnel dryer.
Firing. 960–1180° C (1760–2156° F) cone 07–5.
  (1) Round or rectangular down-draught intermittent.
  (2) Continuous, Hoffmann type.
  (3) Continuous, transverse arch, Staffordshire, etc.
  (4) Tunnel kilns, direct-fired with coal, producer gas or fuel oil, *e.g.* Faugeron, Bricesco-Harrop.
Standards for Drainage Pipes. B.S. 1196:1944.
Tests. Compressive strength.

SALT-GLAZED (OR VITREOUS UNGLAZED) STONEWARE PIPES, FITTINGS AND OTHER SHAPES

Characteristics. Complete resistance to chemical attack, neither decomposing, corroding nor contaminating (except HF), resistant to bacterial attack, smooth and non-absorbent, resistant to erosion, alternating wet and dry conditions, frost, rodents, roots, etc.
Shapes and Uses.
  (1) Spigot and socket pipes up to large diameters, bends, junctions, traps, gulleys, etc. For sewage, chemical and surface water disposal. For supplying acidic mains water.
  (2) Perforated pipes. Drainage of land used for roads, airports, buildings, etc. Agricultural irrigation.
  (3) Septic tanks.
  (4) Well linings.
  (5) Split pipes for gutters and open surface water drains.
  (6) Flue linings, round or square.
  (7) Wall copings.
  (8) Lining tiles for larger containers for sewage works, etc.
Body Type. Common stoneware.
Raw Materials. Selected local natural 'stoneware' clays, grog, sand.

Preparation. Mined clay→(crushed)→sorted→stored or weathered→ dried→worm feeder→dry pan or hammer mill→screens→disc feeders→
↓
oversize→pulveriser
conveyors→dry mixers→mixing auger or wet pan, water added →pugmill or de-airing pugmill (may be part of shaping machine).

Shaping.
 (1) Spigot and socket pipes by vertical pipe presses→when leather hard to finishing machine where ends cut and grooved.
 (2) Straight small diameter pipes by horizontal extrusion.
 (3) Bends made by hand during extrusion.
 (4) Junctions, etc., made by cutting and joining leather-hard pieces.
 (5) Hoppers, gullies, traps by hand moulding in plaster moulds, casting or casting round extruded parts.
 (6) Tiles by dry pressing.

Drying.
 (1) Open sheds.
 (2) Hot floors.
 (3) Chamber or corridor dryers ⎫
 (4) Tunnel dryers ⎬ Humidity for preference.

Firing. 1120–1280° C (2048–2336° F) cone 2–9.
 (1) Intermittent round or rectangular down-draught 4–10 day schedule.
 (2) Continuous gallery and chamber kilns.
 (3) Tunnel kilns with isolation chamber for the salting which must have special lining (primarily U.S.A.).

Glazing. Salt glazing during firing. (Fig. 12.1.) In enlightened communities where the bye-laws lay down the degree of imperviousness and strength of sewage pipes but do not insist on salt glazing or actual colour, glazing is omitted or a slip glaze is used (**S62**).

Standards for Common Stoneware. B.S. 65:1952; B.S. 539:1951, Parts 1 and 2; B.S. 540:1952; B.S. 1143:1955.

Tests. Hydraulic internal pressure. Water absorption. Acid resistance.

FLOOR TILES, QUARRY TILES

Characteristics. Dense, fully vitrified coloured tiles, frequently the colour of natural clays, but also in various colours and textures achieved by additions. Many are smooth surfaced and some slightly self-glazed, others are made 'non-slip' by addition of abrasive grain or by a corrugated surface. High resistance to abrasion, weather, stains, etc. 'Quarry' tiles are thicker and frequently larger than 'floor tiles' and of slightly coarser texture, water absorption 2–5%. Shapes are precise.

Uses. Floors of kitchens; dairies; slaughterhouses; public buildings; balconies; terraces; porches; yards; sills; wall linings.

Fig. 12.1. *Salt-glazed stoneware sewage pipes* (Doulton, **D56**)

Body Type. Stoneware.
Raw Materials. Ball clays, stoneware clays firing to various colours, sintering well but without high shrinkage. (Stains.) China clay. Cornish stone, feldspars, pegmatites, mica. Grog.
Preparation.
Single Clay Bodies. (1) Wet preparation: (as for roofing tiles) watering and weathering→wet pan→expression rollers→pugmill→extrusion.

(2) Dry preparation: dry clay→dry pan→crushing rolls→sieving→ adjustment of water content to 6–9%.
Composite bodies. (1) Wet preparation using grinding mills, for lean bodies: wash feldspar→wet grinding 20 hr. Weigh clays→wet grinding 8 hr (batch grinding therefore requires two feldspar mills for each clay mill) →mixing ark→pumps→filterpress→drying tunnel→cooling→dry pan with water spray, perforations 1–1·5 mm (0·04–0·06 in.)→press dust of 6–8% water content→silo for few days maturing.

(2) Wet preparation using blunger and grinding mill, for lean bodies: weigh hard materials→wet grind in mill. Weigh clay→blunger plus 0·1% soda to give slip of 1·46–1·55 S.G.→mixing ark→sieve→storage ark, HCl added to eliminate soda→pumps→filterpress →then as in (1) above.

(3) Half-wet preparation, for fat bodies using clean clays: moist clay→ shredder→drying drum→8% moisture→various clays weighed, spread out in superimposed layers and then vertical sections shovelled into the dry pan

(or box feeders)→conveyor→storage→mixing ark. Hard materials→wet grinding 20 hr→mixing ark, water added→plastic body→auger→extrusion→bats→trucks through drying oven→and then as in (1) above.

(4) Dry preparation: moist clays→shredder→drying drum→crushing rolls →storage. Hard materials→either dry or allow for moisture and use hot air in the tube mill. Tile grog→crushing rolls. Saggar grog: the fines not usable for making saggars can be used here. Each material may either be ground individually or the whole batch is weighed into tube or conical dry pebble mills in closed circuit with air classifiers→conveyor with water spray→dry pan, perforations 1–1·5 mm.

(*Note:* It is recommended to use dry pans for the final granulation rather than hammer mills because the latter aerate the press dust and cause laminations in the pressed tile.)

Parti-coloured tile bodies are prepared by mixing two or more coloured bodies:

(*a*) if colour gradation is required, 'Porphyry', by placing both bodies in the dry pan together;
(*b*) if sharp colour contrast, 'Granite', is required by mixing in body of the secondary colour that has been separately drypanned and had the fines removed by sifting.

Shaping. (1) Dry pressing, 6–8% moisture.
(2) Extrusion and cutting off.

Drying. Complete drying is essential because of the high content of fine-grained stoneware clays. Insufficiently dried ware marks at the points of contact during firing. On removal from the press mould tiles are placed on small racks with upstands at the ends so that they can be stacked.

Drying chambers. If tunnel dryer is used in conjunction with tunnel kiln the moist tiles are set in the saggars (open-flame kiln), or in open setting with layer of sand between tiles (muffle kiln) before drying.

Firing. Cone 6–10.

Open flame kilns for quarry tiles, muffle kilns or saggars for floor tiles.
(1) Intermittent round kilns (coal-fired).
(i) Water-smoking up to 120° C; temperature 5° C/hr for 24 hr, good ventilation.
(ii) Decomposition 120–710° C; temperature rise 10° C/hr for 24 hr; temperature rise 15° C/hr for further 24 hr. Cleaning of fireholes takes place before→.
(iii) Full fire 710° C to finishing temperature; 2 days. Clean atmosphere throughout.
(iv) Soaking. At least 1 hr.
(v) Cooling; 5 hr after finishing the fire holes are closed up, and roof dampers opened. After 48 hr one wicket is opened, after 72 hr four courses are removed from the second wicket and this repeated every 8 hr.

Too rapid initial cooling (down to cherry-red heat) prevents crystallisation and leads to lack of elasticity. Too rapid subsequent cooling leads to cooling cracks. (Cooling cracks differentiate from drying and firing cracks by being curved and having a smooth surface.)

(2) Continuous gas-fired chamber kiln. Similar schedule as for intermittent kiln. Care must be taken not to connect a freshly set cold chamber until the previous one has reached 120° C and finished water-smoking, or condensation will occur.

(3) Tunnel kilns.

Relative costs of firing per square unit of tiling in the three kiln types producing approximately same output (calculated from costs and wages in 1937): Coal round intermittent 2·01: gas continuous chamber 1·35: producer gas open flame tunnel 1·0.

Standards for Floor Tiles. B.S. 1286:1945.

Tests. Tendency to sag during firing. Moisture content of press dust. Water absorption. Abrasion resistance (**V5**). Compressive strength. Bending strength. Acid resistance. Frost resistance. Light reflection (**L26**).

Output. Examples:

		yd²/month	No. of employees
German firms	I	24 000	250–280 (40% female, 15% juvenile)
	II	100 000	800 (10% female, 5% juvenile)
	III	48 000	500 (**A17**)
	IV	136 000	600
	V	40 000	350 (53% female)

DECORATED FLOOR TILES

Characteristics. Dense unglazed tiles with designs in various colours of clays.

Uses. As floor tiles.

Body Type. As floor tiles. The shrinkage of the various coloured bodies used in the same tile must be compatible.

Body Preparation. As for floor tiles.

Shaping. Dry pressing. A team to make decorated tiles needs a presser and an assistant to remove the finished tile from the mould *plus* one assistant for each colour used.

(1) Remove finished tile, clean it and set aside, pass empty mould to→

(2) Clean mould, place master outline stencil for complete pattern in mould→

(3) Place stencil blanking off all sections except those required for one colour, fill in body of that colour to depth of approximately 0·12 in., 3 mm. remove stencil→
(4) As in (3) for second colour.
(5) etc., for subsequent colours.
(6) Fill up mould with background body, gently remove master stencil→
(7) Pressman.

Drying and Firing. As floor tiles (V5).

CERAMIC MOSAIC

Characteristics. Small dense multi-coloured tiles in squares, oblongs, hexagons, rhomboids, diamonds and triangles. These are either unglazed and coloured throughout or glazed. There is a wider range of bright colours in the glazed type.
Uses. Decorative wall finishes.
Body Type. Stoneware.
Manufacture. Similar to floor tiles.

WALL TILES

Characteristics. Mass-produced thin tiles with a porous body and a white or coloured glossy or matt glaze. Shapes and sizes are relatively few, the great majority of the tiles being square, with rectangles, coves, beads, etc., for trimmings.
Uses. Tiling of domestic and institutional kitchens, bathrooms, washrooms, etc., also decorative work such as tabletops and fireplace surrounds but not in places liable to frost attack or severe abrasion.
Body Type. Earthenware, wall tiles.
Raw Materials.
  British: Ball clays. China clays. Cornish stone (ground). Ground flint slop.
  German: Kaolins (usually as press cakes). Various plastic clays, including highly feldspathic clays. Feldspar (ready ground or crushed to $\frac{1}{4}$ mm). Quartz sand (either ground or unground). Ground pitchers. Chalk. Dolomite.
  U.S.A.: China clay or kaolin. Ball clay. Flint. Feldspar. Whiting. Talc.
Preparation. There are a number of general methods of preparation all eventually producing granular 'press dust' of controlled moisture content of 5–10%. The methods are also applicable wherever dry pressing is used.
British Wet Mixing Method (using ready ground hard materials). Blunging of each component and adjustment to correct pint weight→ mixing, by volume→blunging→screening→magneting→de-watering, usually by filter press→drying of press cake on steam heated stillages or in waste heat dryers (formerly in slip kilns) *either* to correct moisture content 9–10% *or* to complete dryness followed by accurate water addition→if left moist

mature for 24–48 hr→reduction to dust, in dry pan with perforated base or hammer disintegrator→(screening or grain size check, should pass 14-mesh and leave little on 18-mesh, or better should pass 18-mesh with small residue on 25-mesh)→stock ark, where no evaporation or condensation on cold walls may occur (**M64**).

British Dry Mixing Method. Raw materials→dry→weigh and proportionate→blunge together→screen→magnet→de-water, etc., as above (**M64**).

American Dry Mixing Method. Raw materials obtained purified, ground and air-floated to pass 200-mesh. Weigh→batch mixing pan with rotating stars and rollers, mixing dry then spraying on of water and mixing to uniformity→dust mill.

German Combined Wet and Dry Method. Feldspar and quartz if unground→drying→dry grinding to pass 200-mesh (or wet grinding to pass 150-mesh).

Ground quartz, ground feldspar, hard siliceous clays, ground pitchers, etc., weighed (with allowance for moisture content) into cylinder mill 6–7 ft diameter→wet grinding 24 hr.

Kaolins and clays→blunger (0·05% $BaCO_3$ sometimes added).

Ground slip + blunged slip→mixing ark, blunged→rotary sieves 130–200-mesh→magneting→filter presses→drying to 5–10% moisture→dust making in perforated edge runner mill (**A12**).

German Dry Mixing Methods. Individual dry grinding of raw materials to required fineness (depending on surface of tile and impurities in materials).

(*a*) Dry method: batch weighing→addition of 5–8% water→mixing trough →sieve→dry pan consisting of oppositely driven pan and rolls with latter set eccentrically, this tears apart any lumps formed with higher moisture content and gives uniform distribution of moisture together with maximum removal of entrapped air→silos.

(*b*) Briquetting method for hard granules: proportionating→addition of 20% water→mixing→extrusion→cutting off→drying→grinding→addition of 5–7% water in perforated dry pan.

(*c*) granule method for hard granules, more economical than briquetting method: proportionating→addition of 14–15% moisture→mixing and granulation in special type of Lancaster mixer with rakes as well as ploughs, the star set eccentrically to the pan→conveyor belt through dryer→perforated dry pan (**R29**).

Applicability of the Different Methods. Fix (**F12**) made a survey of the application of the dry-mixing method of body preparation in electrical porcelain, electrical refractories, floor and wall tiles, chemical stoneware, vitreous china and earthenware, heavy and light refractories and saggars; German, Ratcliffe and others (**G26**) consider the situation for white tiles. In general, if bagged purified raw materials of known grain size are available and any necessary alterations in body composition are made the method is both successful and saves time and labour. The chief advantages would seem to be:

(1) Better control of composition and particularly of moisture content (to 0·2%).
(2) More uniform mixture where fine clays and coarse material are used together, giving regular shrinkage, etc.
(3) Less equipment, power, labour and time for preparing body.
(4) Greater ease in changing from one body to another.

The main disadvantages are:

(1) The iron and lignite impurities that may cause specking.
(2) The lower plasticity of some clays due to the shorter time in contact with water.
(3) The body composition may need adjusting.

Where a new slip house is being installed this method saves equipment, labour, space and time. But it has nevertheless not found favour in this country where it is still considered that blunging purifies the clays and develops the plasticity best and this opinion is now being accepted in the U.S.A. British tile manufacturers that have tried the method for dust-pressed ware have found it successful for talc bodies such as used in U.S.A. but not for non-talc ones (**A133**).

Shaping. Automatic presses for standard straight-edged tiles with automatic fettling.

Output: 4×4 in., two per pressing, 175 yd^2 per 8 hr
6×6 in., one per pressing, 200 yd^2 per 8 hr (**H18.1**).

Semi-automatic presses for round-edged tiles followed by hand fettling and brushing. Hand presses for specials. Tiles may be placed on slightly concave bats to counteract subsequent sagging.

Drying. Chamber dryers, steam heated. Tunnel dryers, tunnel kiln waste heat. Dry to 'white hard'.

Biscuit Firing. ($2\frac{1}{2}$–6 days.) Formerly in saggars in down-draught intermittent kilns. Increasingly in tunnel kilns either open-flame using saggars, or muffle kilns. Town or producer gas firing is favoured.

Biscuit Properties. Water absorption 12–15%. Wet to fired contraction 0–3%.

Selection. Underglaze decoration if any.

Glazing. Using white and coloured low-solubility glazes. White glazes applied thin, e.g. 7·8 oz/yd^2. Coloured glazes thicker, e.g. 32·7 oz/yd^2. Waterfall glazing. Mottling by hand on longer conveyors from waterfall machine.

Glost setting. White and cream, reared in saggars or setters. Coloured, flat in cranks or setters.

Glost firing. 2–2½ days. Increasingly in muffle tunnel kilns.
Finishing. Sizing and packing.
Output Examples. (1947): 200 tons white wall tiles per week with 300 men and girls. (1958, different works): 3 000 000 tiles per week with 1400 people.
Standards for Wall Tiles. B.S. 1281:1945.

DECORATED WALL TILES

Characteristics. Wall tiles with colourless or pastel shade glazes with on-glaze decoration.
Uses. Decorative treatment of walls, both those normally tiled for utilitarian reasons and others, window sills, fireplace surrounds. Teapot stands, etc. Souvenir and commemorative plaques.
General Manufacture. As for plain white tiles.
Decoration. This may be carried out by firms buying finished tiles from the large-scale manufacturer. On-glaze: Hand painting; screen printing (this is especially well suited to tile decoration).
Enamel Firing.

FAIENCE TILES

Characteristics. Decorative tiles frequently in a number of shapes and with relief designs and in a range of sizes. The body may be coloured and porous or almost dense, the latter being suitable for outside use. The glaze is opaque, sometimes mottled, often matt.
Uses. Fireplace surrounds. External facing of buildings; halls and lobbies of public buildings; underground stations.
Body Type. Earthenware; fireclay; fireclay and flint; ball clay and flint (**P32**).

EXTERIOR WALL TILES, OR FROSTPROOF TILES
(Fig. 12.2)

Characteristics. Glazed tiles with a vitreous or semi-vitreous tile body which resist frost action. They are frequently larger and thicker than interior wall tiles.

STOVE TILES (OFENKACHELN)

Characteristics. Large and thick tiles of considerable porosity rendering them into thermal shock resistant, heat insulators and heat storers. Although made of cheap raw materials, etc., the surface is made attractive with engobes, clear or opaque glazes, decoration, etc.

Fig. 12.2. *External tiles. A tile mural in Stevenage New Town approx.* 25 × 20 *ft using vitrified stoneware tiles painted with coloured, opaque low-temperature glazes* (Bajo and Hévézi, **B4.1**)

Uses. These tiles are an integral part of the Continental stove, *kachelofen*, they are heated by the fire and form a low-temperature radiant heat source. They retain heat after the fire is out. They also have high decorative value.

Body Types. Coloured earthenware (with opaque glazes). Fireclay (with engobe and clear glaze). Earthenware (clear glazes).

Raw Materials. As large quantities are required local clays, etc., are often used. For the coloured earthenware or majolica type a highly calcareous clay leads to less crazing. Velten clay is particularly suitable and needs no additions. Fireclays. Kaolins. Stoneware clays. Quartz. Feldspar. Sand. Grog.

Preparation.

Majolica. Velten clay→blunger→settling tanks→water run off→air drying to plastic state→mixer or pugmill.

Fireclay. Preparation of grog→crushing→pebble mill grinding; other hard materials→pebble mill grinding; clays→ground, dry pan; mixing→water addition→pugmill; or clays blunged with electrolyte, other materials added →casting slip.

Earthenware. General earthenware pottery methods.
Shaping. Hand pressing into plaster moulds, extrusion of back pieces and sticking up. Machine plastic pressing. Casting.
Drying. Chamber dryers using steam and kiln waste heat.
Engobing and Glazing of fireclay tiles.
Biscuit Firing. Coloured earthenware, 900–1000° C (1652–1832° F) cone 010–05; fireclay, 960° C (1760° F) cone 07.

Intermittent kilns: Wood fired, open flame; coal-fired, muffle; coal-fired round down-draught kilns, ware in saggars.
Glazing. Face ground flat. Opaque or clear or coloured glazes applied.
Glost Firing. In same kiln as biscuit by careful arrangement of the setting.
Decoration. On-glaze decoration.
Enamel Firing. (S83, A8).

SANITARY EARTHENWARE

Characteristics. Sanitary ware, wash basins and closets, *e.g.* with a porous body and less strength than vitreous china ware but more easily made.
Uses. General domestic use.
Body Type. Sanitary earthenware.
Raw Materials. China clays or kaolins. Ball clays. Ground quartz sand. Ground feldspars or pegmatites. Pitchers. Chalk.
Preparation (General German Methods).
(*a*) **Direct Method.** Measurement of moisture content of samples of raw materials→direct weighing out of raw materials as received allowing for moisture content→feldspar, sand, pitchers, hard clays, large slow pebble mill for 24 hr→screens; kaolins and soft clay→blunge, $BaCO_3$ added to remove soluble sulphates, some deflocculant added→screens→Two slips mixed in an ark and remaining deflocculant added→100–150-mesh screen. Casting slip of 35–36 oz per pint.
(*b*) **Indirect Method.** Slip of lower pint weight made as in (*a*)→filter presses→press cakes blunged with water and deflocculants. Casting slip 35–36 oz per pint. Indirect method is claimed to produce slip of lower thixotropy than direct. Casting scrap is blunged and added to casting slip using 10–50% of scrap.
Shaping. Slip casting in plaster moulds. Closet $\frac{3}{8}$ in. thick takes 2 hr. Moulds may be emptied in evening following morning cast. Hand fettling if possible before ware 'white hard'.

CERAMIC BUILDING MATERIALS

Drying.
(1) Open racks or benches in making shop up to 5 days.
(2) Batch dryers, 24 hr.
(3) Tunnel dryers, waste heat from kiln, 12 hr.
Biscuit Firing. Cone 7, 1230° C (2246° F) 27 hr.
  Tunnel kilns. Muffle kilns with open bat and prop setting or open flame kilns with saggars.
  Intermittent round down-draught kilns.
Biscuit Properties. Water absorption, 3, 10, or 12–14%.
Glazing. Spraying in booths fitted with exhausts. Several layers are applied, especially of coloured glazes. Gelatine or glue in some glazes.
Glost Firing. Cone 5–6, 1180–1200° C (2156–2192° F).
  Intermittent down-draught kilns.
  Muffle Dressler tunnel kilns (**A12**).

VITREOUS CHINA SANITARY WARE
(Fig. 12.3)

Characteristics. Highly vitrified body, water absorption less than 1%, of high strength. It requires higher firing than earthenware and there is often some distortion.
Uses. Better grade domestic use.
Body Type. Vitreous china sanitary ware.
Raw Materials. China clays. Ball clays. Quartz sands. Feldspar. Pitchers.
Preparation. As for sanitary earthenware.
  (*a*) Direct method.
  (*b*) Indirect method: casting slips have 35–36 oz/pint.
Shaping. Slip casting in plaster moulds, *e.g.* closet ½ in. thick 1½–2 hr. Hand fettling.
Drying. On benches or racks in making shop, 5 days.
  Chamber dryer, 1 day.
  Tunnel dryer, 12 hr.
Setting. Careful setting on properly shaped setters or with props, etc., is necessary to minimise distortion.
Biscuit Firing. Cone 9–10, 1280–1300° C (2336–2372° F).
  Intermittent down-draught kilns.
  Tunnel kilns. 48–120 hr. Kerabedarf open flame: saggars. Dressler muffle: bat and prop.
Biscuit Properties. Water absorption <1%. Drying contraction 3%. Firing contraction 7–8%.
Glazing. Spraying in booths with exhaust.
Glost Firing. Cone 5, 1180° C (2156° F).

Intermittent down-draught kilns.
Muffle tunnel kilns. Dressler. 45 hr.
Finishing. Grinding flat of foot of closets, etc., with revolving steel plate fed with carborundum powder (**A12, W82**).

(a)

(b)

FIG. 12.3. *Sanitary ware (a) lavatory basin; (b) low-level W.C. suite (Leeds fireclay,* **L21**)

FIRECLAY SANITARY WARE

Characteristics. Strong ware that can be made into large pieces, *e.g.* sinks and urinal stalls, without undue warping. There is a greater tendency for coloured blemishes, etc.

Uses. Public lavatories; institutions; kitchen sinks.

Body Type. Fireclay. Sanitary ware.

Raw Materials. Plastic fireclays. Siliceous fireclays. Kaolin. Grog: saggar sherds, etc., or specially made. Sawdust. Quartz sand.

Preparation.

Casting Slip.

Grog. Briquettes of raw clays or clays with ground pitchers are fired with the ware in intermittent and tunnel kilns, the former developed a higher cristobalite content. Amount of each grog in body adjusted to eliminate crazing (more intermittent kiln grog) or peeling (more tunnel kiln grog). Grog briquettes or saggar sherds, etc.→jaw crusher→dry pan or grinding cylinders→centrifugal separator→magnets→sieves→storage silos (fines frequently discarded).

Clay. Water, deflocculants, barium carbonate and then fist sized lumps of clay added in order to blunger→blunging 12 hr→slip 34 oz/pint→screening gyratory screens to remove pyrites→storage→smaller mixing blungers→ grog added→Casting slip pint weight 40 oz; or, dry clays→dry pan→ screen→weigh→complete batch into blunger; or clays wet ground in cylinder→mixing blunger.

For Plastic Pressing. Wet pan: grog then clay and water added→20 min mixing→pug→hand wedging.

Shaping. Basins, sinks, double sinks, closets, urinal stalls: Solid slip casting. Casting time 6–7 hr, after about 12 hr cover can be lifted. Ware removed from mould about 20 hr later. Some shapes, case removed first, and core after, others core first, case after.

Urinal stalls, large sinks, baths, slop hoppers and other large pieces: Plastic pressing in plaster moulds.

Finishing. Patching of casting holes, immediately after stripping mould. Fettling with knife and palette and beating bases flat with wooden board.

Drying. Large pieces in making shop for first 10 days, small pieces in making shop for first day. Drying chamber at 60–80° C, no humidity control: large pieces further 4–5 days, small pieces further 2 days. Drying contraction 3–4%.

Inspection. Dry ware carefully inspected with aid of shaded lamps and sponges, surface cracks repaired with dried slip.

Engobing and Glazing. Both engobe and glaze wet ground 24 hr→100-mesh sieve→magnets. 32 oz/pint. 3% gelatine added to engobe. Vegetable dyes added to distinguish layers. Five to six coats engobe brushed on. Three to five coats glaze brushed on.

Firing. Cone 8–10, 1250–1300° C (2282–2372° F).

Large pieces: Intermittent gas or coal muffle kilns. Firing time to top temperature 150–180 hr.

Small pieces: Tunnel kiln. Dressler muffle. Gas fired. 130 hr.

Continuous chamber kilns, muffle, producer gas-fired. 8–10 days. Intermittent kiln fired ware has less tendency to craze, or show 'blow-out'. Firing contraction, 6–7%.

Finishing. Cutting: carborundum tipped disks revolving at 1450 r.p.m. Grinding: circular steel table with sand abrasive, carborundum table would be better.

Tests. Routine chemical analysis of raw materials. Daily thermal expansion tests on body and glaze up to 600° C. Water absorption (normally 20%). Crazing test by quenching from 100° C, 110° C, 120° C, etc. (**A12**, **A17**).

STONEWARE SANITARY WARE

Characteristics. Very strong ware, *e.g.*, sinks, basins, wash tubs, urinal stalls, closets, taking a lot of punishment, suitable for institutions, factories, etc. Dense throughout so that no danger results from chipping the pieces. Factories, etc., use the common brown stoneware, hospitals use white chemical stoneware.

Uses. Factories; hospitals; public lavatories; farms.

Body Type. Stoneware. Common brown stoneware. White chemical stoneware.

Manufacture. General stoneware methods (*see* p. 1103, Chemical Ceramics, Chapter 14).

Standards for Sanitary Ware. B.S. 1188:1944; B.S. 1206:1945.

TABLE 243
Properties of Sanitary Ware Bodies

Name or Type		Earthenware	Fireclay
Ref.		**S185**	**S185**
Water Absorption		5–10	12–15
Bulk density	(lb/ft^3)	119–131	112–119
	(g/cm^3)	1·9–2·1	1·8–1·9
Tensile strength	(lb/in^2)	711–1422	995–1138
	(kg/cm^2)	50–100	70–80
Compressive strength	(lb/in^2)	8534–14 223	4978–5689
	(kg/cm^2)	600–1000	350–400
Impact bending strength (cm kg/cm^2)		1·2–1·5	1·0–1·2
Modulus of elasticity	(lb/in^2)	2 844 660–4 266 990	
	(kg/cm^2)	200 000–300 000	
Specific heat 20–100° C (kcal/kg °C)		0·19	
Transverse strength	(lb/in^2)	2845–4267	2845–3556
	(kg/cm^2)	200–300	200–250

Chapter 13
Ceramics in the Home

The main application of non-constructional ceramics in the home is for tableware, others are kitchenware, and ornaments and vases. A large number of body types can be used for tableware, different ones finding more or less favour in different countries.

TABLE 244

Tableware Bodies

No.	Name	Countries generally making	Absorption % (W28)	Mechanical shock resistance (W28)	Translucency (W28)
1	Majolica. Coloured pottery	Most	over 15	very low	none
2	Fine earthenware	Gt. Britain and Europe	10–15	low	none
3	Semi-vitreous china	U.S.A.	4–10	medium to high	little or none
4	Semi-vitreous porcelain	U.S.A.	0·3–4	high	low
5	Hotel china	U.S.A.	under 0·3	very high	medium
6	Household china	U.S.A.	under 0·1	very high	high
7	Bone china	Gt. Britain	0·3–2	medium	very high
8	Beleek china	U.S.A. and Ireland	none	medium to high	high
9	Hard porcelain	Europe and Scandinavia	none	medium to high	high
10	Stoneware	Gt. Britain and Europe	none	medium to high	none

STONEWARE TABLEWARE

Characteristics. Relatively thick and heavy ware deriving beauty from simple shapes and coloured glossy or matt glazes. The entire base or the foot is left unglazed. If suitable bodies are used ovenproof ware can be made.
Uses. If heat-resisting body used, casseroles, cocottes, etc. Dinner services. Informal jug and mug sets. Children's sets. Covered jars.
Body Type. Stoneware.

Raw Materials. Stoneware clays. China clay. Ball clay. Feldspar or Cornish stone. Flint. Minor constituents (*see* General stoneware in Chemical Ceramics).

Preparation. Fine ceramic wet methods, *e.g.* wet or dry grinding of hard materials, blunging of clays→proportionating→sieves→magnets→filter press →pugmill→maturing (some weeks).

Shaping. Hand throwing. Jiggering. Jolleying. Casting.

Drying. Must be very careful.

Biscuit Firing. Cone 015/014–05, 800–1000° C (1472–1832° F).

Glazing. Albany slip. Feldspathic or lowlead, glossy or matt, opaque coloured or clear colourless.

Decoration. Under-glaze. On-glaze. In-glaze. By use of more than one coloured glaze.

Glost Firing. About cone 8, 1250° C (2282° F).

Enamel Firing if necessary.

EARTHENWARE TABLEWARE, VASES, ETC.

Figs. 13.1, 13.2.

Characteristics. General purpose ware produced on a large scale by bulk production methods but nevertheless available in a wide variety of shapes, colours and decorated designs.

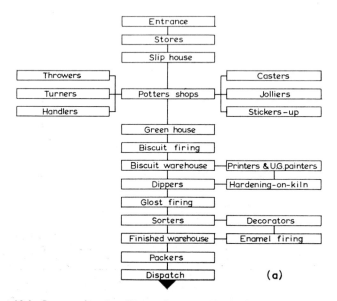

FIG. 13.1. Layout of modern 'Pottery' or general tableware works (*Upright*, **U3**)
(a) Skeleton flow of production

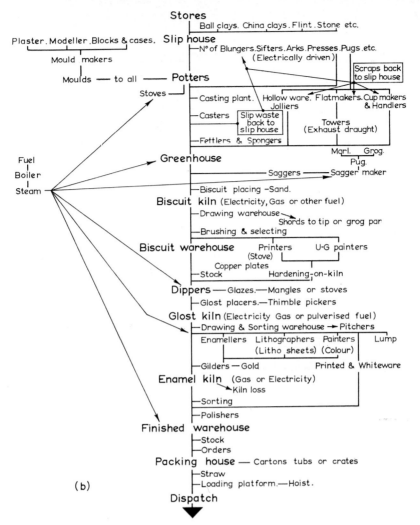

(b) *Detailed diagram of the flow of production*

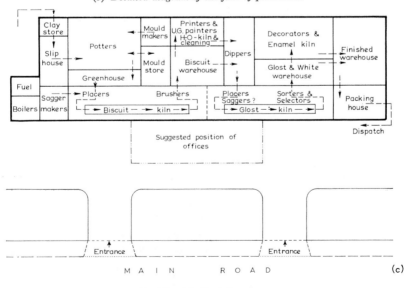

(c) *'Straight line' layout type*

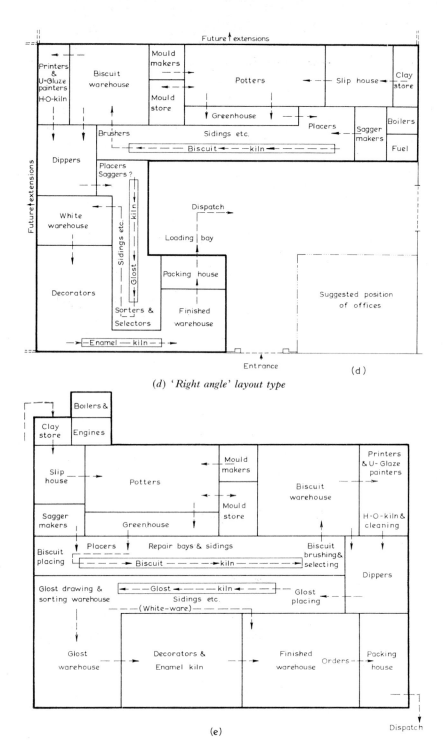

Fig 13.1 (contd.)
(e) 'Complete cycle of production' layout

CERAMICS IN THE HOME 1093

Uses. General tableware. Artware.
Body Type. Earthenware, dinnerware.
Raw Materials. China clays. Ball clays. China stone. Flint. Feldspar. Whiting.
Preparation. General Methods. There are four different general methods originating in: England, Europe, Scandinavia, U.S.A., respectively, and various modern methods using the best of each.

(1) *English Method.* This consists basically of individual blunging of raw materials and adjustment to a required 'pint-weight' before mixing by

Fig. 13.2. *Earthenware tableware in blue and white, coloured bodies with transparent glaze being used (Wedgwood,* **W35**)

volume, followed by cleansing, filter pressing, etc. It obviates any need for accurate weighing and for knowledge of, and allowance for, the water content of the raw materials.

China Clay blunge, *e.g.* propeller circular blunger, 4 ft 8 in. (1·41 m), 300 r.p.m. 30–60 min→agitated stock ark→test pint weight.

Ball Clay blunge, *e.g.* knife type blunger with baffles, 6 ft (1·82 m), 45–50 r.p.m. 4 times as long as china clay→agitated stock ark→test pint weight. Best to have blungers with independent drives.

Flint and Stone. Generally bought ready ground. Flint, once ground, must be kept in slop state. Usual fineness is 200-mesh although stone or feldspar may be 300-mesh. Blunged→stock ark and test for pint weight.

Where own grinding done: primary jaw crushers or roller crushers→ gyratory crushers→wet pan or silex or rubber-lined cylindrical or conical ball or tube mills.

Scrap. Blunged in propeller blunger.

→*Mixing.* Either by adding slip of correct pint weight up to respective notches on the measuring stick in the stock ark *or* by weighing and measuring machine. Addition of 0·01 to 0·05% cobalt stain to give dead white body. Addition of organic stains to differentiate various bodies during processing, *e.g.* fuchsine, methyl violet and brilliant green.→Thorough agitation→ screening 100-, 130-, 150-mesh to remove organic matter introduced with ball clay→magneting→stock ark. (Rubber-lined centrifugal or diaphragm pumps may have to be interposed.) Propeller agitation at 150 r.p.m. in stock ark.

→Filter pressing. Sixty to seventy chamber presses. Deadweight ram pumps or constant-pressure ram pumps. Pressure depends on purpose for which body is required, throwing, jiggering, pressing, etc., and is in the range 80 to 110 lb/in^2 (5·6 to 7·7 kg/cm^2), exact adherence to right pressure is essential.→Pugging→maturing→pugging or de-airing pug (**J51**).

(One disadvantage with the slop method is found when measuring flint, because its specific gravity varies with the degree of calcination.)

(2) **European Method.** (Czechoslovakia, Austria, Italy, Switzerland, France, Belgium, Holland and Germany.)

(Kaolin may arrive in unpurified pit condition, toothed differential rollers→blungers→settling tanks→sand troughs→rakes, etc. Slurry used directly in blunger (**S83**)).

All materials dried and then weighed out. Hard materials such as flint, feldspar and whiting→wet grinding with small addition of ball clay in slow pebble mills. China clay and ball clay→blungers→all materials mixed→ blunger→sieves→magnets→filter press 100–200 lb/in^2 (7·0–14·0 kg/cm^2).

(3) **Scandinavian Method.** (Sweden, Norway, Denmark, Finland.) As in Europe except that fast rotating grinding cylinders are used with relatively smaller charges. This grinding seems to be more efficient.

Both methods (2) and (3) require efficient drying of the raw materials or accurate measurement of moisture content.

(4) **U.S.A. Methods.** If materials obtained dry and ground: Weigh→mix dry→wet blunging→sieving→magneting→filter pressing.

└→or add enough water to make plastic and mix well.

If ball clay damp→blunge to correct pint weight→add other ingredients, blunge, etc.

U.S.A. flint is normally 300-mesh but feldspar is only 200-mesh. Such fine flint requires more feldspar in the body than that at 160–200-mesh, whereas less feldspar is required if it has been ground to greater fineness, 230–300-mesh.

Wet grinding of fine materials is advantageous except that feldspar should

not be wet ground too long. This should be dry ground first and then wet ground and used immediately.
Shaping. Flatware (plates, saucers, etc.): Batting→jiggering→drying stove→towing or fettling→stacking.

Ovals: Batting→special jigger→as above.

Hollow-ware: Batting→jolleying→as above. Some hollow-ware requires turning.

Teapots, jugs, vegetable dishes, etc: casting→drying→sponging.

Vases, mugs, etc.: Hand throwing→dry to leather-hard→turning.

Sticking up, handling, etc., in leather-hard condition. Scalloping.
Drying. Dobbins. Mangles. Open shelves. Green house inspection of ware and removal of imperfect pieces.
Decoration of green ware if any, *e.g.* body relief work, inlaying, sprig, engobing, slip trailing.
Biscuit Firing. Cone 08–10, 940–1000° C (1724–2372° F).
Intermittent bottle kilns, up-draught and down-draught, saggars essential. Tunnel kilns, open-flame, muffle, electric.
Biscuit warehouse inspection.
Under-glaze Decoration, *e.g.* intaglio printing of outlines, followed by hand colouring. Transfers. Hand painting.
Glazing. Hand or machine dipping. Spraying.
In-glaze Decoration, *e.g.* majolica type of hand-painted design.
Glost Firing. About cone 03/02 1050° C (1922° F).
Similar kilns to biscuit. Tendency to use twin tunnel kilns for biscuit and glost, respectively, especially where electric.
Glost warehouse inspection.
On-glaze Decoration. Hand painting, banding, etc. Transfers. Screen printing.
Enamel Firing. This may need to be repeated several times for various colours.
Final inspection, packing.

SEMI-VITREOUS CHINA

Characteristics. Commonest type of lower priced tableware in U.S.A. Of good strength.
Uses. General household use.
Body Type. Earthenware of low porosity.
Raw Materials. China and ball clays, flint, feldspar, whiting, talc.
Preparation. American dry mix methods.
Shaping. This body type is well suited to highly mechanical shaping.
Firing. Biscuit, cone 5–7, 1180–1230° C (2156–2246° F). Glost, cone 2–4, 1120–1160° C (2048–2120° F). There is increasing tendency to decorate and glaze the green ware and fire once only.
Decoration. Under-glaze. On-glaze. Coloured glazes.
Output Example (1942). 14 000 dozens of ware produced on automatic

jiggers per day, using $3\frac{1}{2}$ miles of conveyor. One operator makes 100 doz. cereal bowls per hr. Sandblast machine handles 700 doz/hr. Seven circular tunnel kilns produce 16 000 doz. daily.

AMERICAN HOTEL CHINA

Characteristics. Opaque white vitrified ware of very high strength. It is made in three grades based on wall thickness.
Uses. Grade (1), 'Thick china' $\frac{5}{16}$ to $\frac{3}{8}$ in. (7·9–9·5 mm) walls: lunch counters, army messes. Grade (2), 'Hotel China' $\frac{5}{32}$ to $\frac{1}{4}$ in. (3·9–6·3 mm) walls: hotels, restaurants. Grade (3), 'Medium-weight China', less than $\frac{1}{4}$ in. (6·3 mm) walls: high-class eating places, home use, also for numerous jars, trays, etc., in hospitals.
Body Type. Vitreous china.
Raw Materials. Kaolin. Ball clay. Flint. Feldspar. Whiting. Dolomite.
Preparation. Hard materials→ball milling to 325-mesh. Clays→blunged →magnets→admixture of non-plastics→blunged→magnets→filter press.
$$\downarrow$$
casting slip.

Shaping. Casting. Jiggering. Jolleying, etc.
Firing. Biscuit, cone 11, 1320° C (2408° F).
Under-glaze Decoration.
Glazing.
Glost Firing.
On-glaze decoration.
Enamel Firing.

AMERICAN HOUSEHOLD CHINA

Characteristics. Fine ware of high strength and translucency but tending to cream instead of white.
Uses. High grade domestic tableware.
Body Type. Soft porcelain (**N11**).
Raw Materials. Kaolin. Ball clay. Flint. Feldspar.
General methods of manufacture similar to bone china.

ENGLISH TRANSLUCENT CHINA

Characteristics. Fine ware of exceptional translucency and whiteness, and very good strength. First produced in 1959 to cost about half the price of bone china, and only a little more than fine earthenware. Will take a wide range of colours and decorative treatments including the use of gold and platinum.
Uses. Fine tableware.

CERAMICS IN THE HOME 1097

Body Type. Soft porcelain.
Raw Materials. Include: china clay, feldspars.
Shaping. The body is more plastic than bone china and so quicker shaping methods can be used.
Biscuit Firing. 1250–1300° C (2282–2372° F). Special refractory setters are required to maintain the shape.
Glazing. Automatic dipping machine has been developed especially for the flatware.
Glost Firing. Lower than biscuit (**D56**).

BONE CHINA DINNERWARE (AND ARTWARE)
(Fig. 13.3)

Characteristics. Fine ware of exceptional whiteness and translucency, capable of bearing brilliant under-glaze and soft on-glaze decoration. Very high mechanical strength and resistance to chipping, although thermal shock resistance is not always as good as hard porcelain.
Uses. Fine tableware, dinner services, tea services, figurines, ornaments.
Body Type. Bone china.
Raw Materials. 25% Cornish china clay. 25% Cornish stone. 50% Beef bone (after extraction of glue, etc., but with minimum of alkalies and other contaminants). (Ball clay, bentonite, flint, frits, etc.)

Fig. 13.3. *Bone China* (*Wedgwood*, **W35**)

Preparation. Bone→slow calcination to 900° C (1652° F)→wet grinding in edge runner mill to 80% $<10\,\mu$. China clay→washed. Cornish stone →wet grind→screen. Mixing in blunger→screen→magnets→filter press→ de-airing pugmill→plastic making
                        └→casting slip 33 oz/pint.
Shaping. (1) Flatware, batmaking→jigger. (2) Hollow-ware, handthrown blanks→jolley. (3) Tall pieces or re-entrant shapes, slip casting.

The much smaller plasticity of bone china body compared with earthenware reduces the speed of making and increases losses at almost every stage. A platemaker who can jigger 1000 earthenware plates per day can make only 400 bone china ones.

Drying. Mangles or dobbins. The lean body makes this easy.
Biscuit Firing. 1250–1300° C (2282–2372° F), cone 8–10 (to almost complete vitrification, 0·3–2% porosity).

(1) Intermittent bottle kilns.

(2) Tunnel kilns. Average 250 ft long, setting space 3 ft × 3 ft, 60 hr. Even where saggars are not required each piece must be individually supported, every plate must be placed in a tray with alumina powder, impressed to the correct shape. (One bone china plate therefore takes up the space of twelve earthenware plates.)

Decoration. Under-glaze decoration. Intaglio printed outlines with hand painted colours. Transfers. Hand painting.
Glazing. Usually transparent colourless glazes. Dipping. Spraying.
Glost Firing. 1050–1100° C (1922–2012° F) cone H03A–1.
    Intermittent bottle kilns.

Tunnel kilns, electric with open setting, 30 hr. Setting, as for earthenware glost, with thimbles and pins in cranks, etc.

Decoration and Enamel Firing. 700–800° C (1292–1472° F).

Much decoration is on-glaze. Ground laying. Transfers. Hand painting. Gold banding. Different colours need to be enamel-fired separately so that a highly ornate piece may pass through the enamel kiln twenty times (**S4**).

BONE CHINA HOTEL WARE (SWEDISH)

Characteristics. A bone china made partly of non-British materials with high mechanical strength.
Uses. Tea ware. Fine hotel ware.
Body Type. Bone china.
Raw Materials. Zettlitz kaolin. English china clay. English ball clay. High purity Swedish orthoclase feldspar. Bone ash.
Preparation. Clays→blungers. Bone ash and hard minerals→wet ground

in large cylinders (charge about 1 ton), 16 hr (grain size check). Mixing ark→electromagnets→screen (180-mesh)→filter press→pug.
Shaping. Jiggering. Jolleying. Slip casting.
Drying.
Biscuit Firing. Conc 10, 1300° C (2372° F).
Flatware bedded upside down in purest quality fired sand in round saggars, ten per saggar.
Glazing. Scouring off of setting material. Glazing by dipping.
Glost Firing. Electric, 1120° C (2048° F). Flatware reared.
Finishing. Pin marks, etc., ground off (**O1**).

HARD PORCELAIN TABLEWARE

Characteristics. High-quality ware of purest whiteness and considerable translucency, good mechanical strength hard scratch resisting glaze and thermal shock resistance. Wide range of decoration possible including the characteristic soft-edged under-glaze decoration.
Uses. Quality tableware. Artware.
Body Type. Hard porcelain.
Raw Materials. Best quality: High purity washed kaolin. Pure rock quartz. Pure feldspar. (Berlin Staatliche Porzellan use controlled kaolin washing process that leaves sufficient of the finest fraction of silica in the kaolin (**A8**).) Second quality: Kaolin. Quartz sand. Pegmatite. Minor constituents: Calcspar. Dolomite. Magnesite. Talc. Frits, etc.
Preparation. Rock quartz→calcined. All hard materials crushed in jaw crushers, and ground in edge runner mills and finally wet milled in pebble mills. All machinery designed so that no contact is made with iron. Each batch tested for grain size. Kaolin→blunger.

                 deflocculent   casting slip

Feldspar, quartz and kaolin slurries→mixing ark→filter presses→30% moisture→maturing cellars (assisted by warmth→de-airing pugmill.
                                   ↘kneading machine.
Dry press dust requires 11–14% water and 1–4% vegetable oil.
Shaping. Throwing. Jiggering. Jolleying. Casting. Dry pressing.
Drying. Slow. Around and above two-tier coal-fired porcelain kiln. Steam heated chamber dryers.
Biscuit Firing. 800–900° C (1472–1652° F). Small shrinkage.
Top chamber of two-tier kiln.
Under-glaze Decoration. Hand painting. Stencil. Spraying.
Glazing.
Glost Firing. 1400–1435° C (2552–2615° F), cone 14–15. 15–20% shrinkage. Plates placed in individual shallow saggars made to conform with the under surface of plate in moistened mixture of coarse silica with clay. Cups

have glaze wiped off rim and are boxed or placed rim down on low-fired biscuited ring-shaped setters. All in saggars.
  Lower chamber of two-tiered coal-fired porcelain kiln.
  Tunnel kilns.

On-glaze Decoration. Hand painting. Transfers. Engraving. Spraying. Banding.

Enamel Firing. 700–800° C (1292–1472° F), except for cobalt blue when 1400° C.

Finishing. Grinding and polishing the unglazed foot and rim. Polishing of gold decoration.

Output. Examples: Hand jiggering of plates: two men + one girl on two wheels 800 of 24 cm ($9\frac{1}{2}$ in.) in 8 hr (the more plastic earthenware body would allow the same team to make 3000 plates in 8 hr). Semi-automatic cup jolley: 4000/8 hr.

HEAT RESISTANT WARE, OR OVEN-PROOF WARE. 'FLAME-RESISTANT' WARE

Characteristics. Ware in which cooking can be done in the oven or on the stove with an asbestos mat and which can then be brought to the table. Some ware is also claimed to be flame-resistant so that it can be used on a cooking stove without an asbestos mat.

Uses. Slow cooking, casseroles, cocottes, etc.

Body Types. Porcelain. Cordierite. Stoneware.
  General methods of manufacture as for porcelain or stoneware.

STONEWARE KITCHEN WARE

Characteristics. The general properties of chemical stoneware, particularly acid resistance, resistance to abrasion, mechanical strength find considerable application in the kitchen. Both salt-glazed and brown slip-glazed ware is used.

Uses. Preserving and pickling jars for salting meat and vegetables, preserving eggs. Large jars for jams. Mixing bowls. Wine jars. Cider jars.

Preparation. General stoneware methods for moderately fine ware.

Shaping. Throwing. Jiggering. Jolleying. Casting.

Drying and Firing and Salt Glazing. General stoneware methods (p. 1103).

ARTWARE

Vases, Bowls, Ashtrays, Lampbases, Figurines, Covered Bowls and Boxes, Decorative Plates, Trays, Individual Coffee or Tea Sets, Mugs, Jugs, Teapots, Egg Cups, Cruets, Souvenirs

(Fig. 13.4)

Characteristics. Articles of every variety of artistic and ceramic quality made either very largely by hand in small studio potteries, or with considerable mechanisation in a large pottery, but usually bearing a certain individuality.

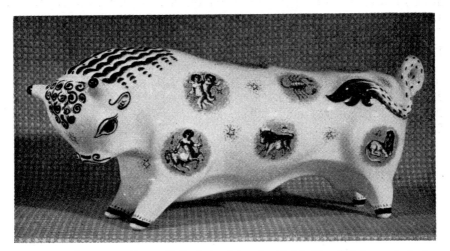

Fig. 13.4. *Art ware. Taurus Bull, designed by Arnold Machin, A.R.A., made in fine earthenware with lithographed decoration* (*Wedgwood*, **W35**)

Body Types. All fine ceramic bodies: Hard porcelain. Soft porcelain. Bone china. American household and hotel china. Stoneware. Beleek. Jasper. Fine earthenware. Also red earthenware bodies, majolica, etc.
Raw Materials and Preparation. According to body type.
Shaping. Hand throwing. Hand modelling. Pressing in plaster moulds. Shaping on plaster bats. Casting. Joining, sticking up, etc.
Drying. Open shelves, or factory methods.
Decoration. On green pieces: Slip trailing. Incision. Scraffito. Hand painting. Sprig.
 Under-glaze: Painting. Screen printing. Spraying.
 In-glaze: 'Delft' hand painting. Use of several different coloured glazes, etc.
 On-glaze: Hand painting. Screen printing. Banding, etc.
Firing. Temperature according to body. Studio potters have largely intermittent kilns: electric for lower temperatures, gas semi-muffle or muffle

for high temperatures. Large firms pack artware in between other pieces, either to use tunnel kiln space more economically or to utilise cooler parts of intermittent kilns, etc.

DENTAL PORCELAIN

Characteristics. Maximum simulation of natural teeth as regards colour, surface texture and translucency combined with resistance to abrasion, saliva, food acids and thermal shock.
Uses. Artificial teeth. Additions for broken teeth.
Body Type. Soft porcelain.
Raw Materials. Feldspar or nepheline syenite. Kaolin. Silica. Chalk. Colouring.
Preparation. Very fine grinding→fritting→grinding perhaps two to three times, plus very careful colour control (**N11**).
Shaping. In bronze dies different coloured layers being placed in turn. The die is then clamped together.
Biscuit Firing. In the die. Fettling.
Main Firing. In gas muffle kilns.
Test. Breaking strength (**S83**).

Chapter 14
Chemical and Technical Ceramics

STONEWARE. GENERAL CONSIDERATIONS

Characteristics. See individual sections in Chapters 12, 13, 14 and 16.
Body Type. Stoneware.
Raw Materials. *Major*: Selected secondary clays. China clay or kaolin. Bentonite. Quartz. Feldspar. Cornish stone. Grog: stoneware, porcelain, fire clay. *Minor*: Granite. Basalt. Porphyry. Whiting. Chalk. Dolomite. Magnesite. Corundum. Graphite. Talc. Rutile. Barium sulphate. Plaster. Zirconia. Zircon. Sillimanite. Cobalt ore. Chrome ore. Fused quartz. Silicon carbide. Ferrosilicon. Ceria. Chrome oxide. Cobalt oxide. Manganese oxide. Enstatite. Clinoenstatite. Forsterite. Plagioclase. Nepheline. Cordierite.
Preparation. A wide variety and combination of methods are used depending partly on the quality of product and partly on the nature of the raw materials.
Clays. When dug individual layers are sorted or they may be systematically mixed: if for wet preparation→outside storage for weathering; if for dry preparation→drying sheds, hot floors, drying drums.
Wet Preparation. Used for impure clays and/or high quality ware.

(1) Clay→blunger→vibrating sieves ⟨quartz→cleaned→ground. / clay slip.

Hard materials individually dry ground in pebble mills→magnets→with addition of clay slip batch mixes in wet grinding pebble mills→Mixing ark with remaining clay slip→filter press→possible addition of coarser grog→pug mill→maturing cellar.

(2) Clay blunged→sieve
    Hard materials dry ground ⟩ batch mixing→continuous wet mill
→filter press→possible addition of coarser grog→pugmill→maturing cellar.

(3) For lower-grade ware. (Impurities ground finely and so evenly distributed.) Single clays or mixture of clays of known moisture content weighed into pit→pre-crushed or ground non-plastics may be added (or because of difficulty in obtaining proper sample when emptying the pit they are added later)→water is added and prolonged soaking allowed. The extreme toughness of stoneware clays makes this a slow process.→Soaked clay or batch shovelled out of pit→wet pan (granite or quartzite runners)→ crushing rollers→pug→maturing cellar.

(4) For bodies where clay impurities, even if ground up, are harmful. Soaking of clay alone→plastic pressing through sieves or slits with help of rams, augers, rollers, etc., removal of granular impurities of more than 2 mm →batch mixing→wet pan→pug→maturing cellar.

Maturing at 22–30° C (71·6–86·0° F) takes about three weeks.

Wet prepared bodies may also be dried and disintegrated to give *press dust* or may be blunged with water and deflocculant to give *casting slip*.

Dry Preparation

(5) Pit clay→dryers→grinding, centrifugal mills for clays without stony impurities, clay need not be quite dry but product finer than in ball or pebble mills which must be used if stony impurities present, air classifier sometimes placed in circuit. Grog→dry pan or ball mills (dry pan gives more desirable jagged particles) size required depends on piece to be made. Quartz, feldspar, etc.→tube or pebble mills to maximum economical fineness. Proportionating by box feeder→mixer→5–8% moisture added for dry pressing.

                                     ↳water added to give plastic body.
                                     ↳water and deflocculants→casting slip.

(6) Raw materials each dried and crushed→box feeder→proportionating→ dry pan→dry press body.

      ↳plastic body.
      ↳casting slip.

Casting slips have 20–35% water and 0·2–0·4% soda deflocculant.

Shaping. All general ceramic shaping methods are used, the choice depending on the size of the piece, its shape and the coarseness of the body used. Further methods especially suited to the highly plastic nature of stoneware clays are also used The methods are listed in order of decreasing water content of the body.

(1) Slip casting.

    (*a*) Drain casting for complicated but thin-walled pieces.
    (*b*) Solid casting for thick-walled vessels as used in the chemical and electrical industries. Heavily grogged bodies are used.
    (*c*) Pressure casting.

(2) Hand throwing of fine-ground body for artware and tableware, ovenware, etc.

CHEMICAL AND TECHNICAL CERAMICS

(3) Jiggering and jolleying of fine-ground body for tableware, ovenware, etc.

(4) Jiggering and jolleying of body containing coarse grog for large pieces (water content is 28%). In some machines clay is gradually hand fed to the scraping blade.

(5) Hand spreading or smearing of coarse-grogged body (water content 22%). This is a special method for large flat stoneware pieces. Successive thin layers of body are spread by hand on top of each other, and then roughened by drawing bent fingers across them, before the next layer is applied.

(6) Extrusion. (Some older piston and roller machines.) Auger machines: Special vertical machines fitted with accessories make socketed sewer pipes, tower sections, round or square sections for fitting to flat bases to make tanks. Horizontal machines for various pipes, normally fitted with cutting-off tables with rollers. Fine-grained plastic body for thin-walled heating and cooling coils, extruded horizontally on to slowly rotating framework of raw stoneware body. The whole is dried and fired together. Solid extrusion for blue bricks.

(7) Plastic pressing for fine-grained body, and relatively small and simple pieces. Re-pressing of extruded bricks.

(8) Semi-dry pressing for engineering bricks and floor tiles.

(9) Dry pressing (3–12% moisture) for engineering bricks, floor and wall tiles.

(10) Joining. Composite pieces are made by joining up more easily shaped pieces. Smaller pieces with joins that will not undergo much strain are joined with clay slip in the ordinary way. Joins that will undergo considerable strain are made temporarily with slip and then by stages cut out and replaced by a hand-moulded join made by the hand-spreading method.

Large rectangular vessels may be made up by placing preformed slabs in metal or wood moulds and hand moulding the joins. Extruded pipes are cut and joined to give junctions, syphons, traps, feeding troughs, etc.

(11) Turning. Extruded, thrown or jolleyed blanks are turned to more intricate shapes, threaded, etc., when leather hard.

(12) Smoothing of surfaces of pieces shaped with a grogged body is done with suitable iron, wood or rubber scrapers and spatulas (**S90, A11**).

Drying. The nature of stoneware clays necessitates the utmost care in the drying of pieces. Too fast or uneven drying can lead to warping and cracking. Large pieces must be set on sand or grog to allow the base to move when shrinking. Edges and corners tending to dry faster than thicker parts of a piece must be retarded with damp cloths, clay or paper. Pieces should be turned during drying to equalise any uneven air currents.

Drying sheds. Open shelves. Drying chambers heated by kiln waste heat. Humidity dryers (suitable for small mass-produced pieces). Tunnel dryers (suitable for small mass-produced pieces).

Firing. Cone 8–9/10, 1250–1290° C (2282–2354° F). Shortest schedule

24 hr, generally several days in intermittent kilns. The high colloidal clay content necessitates very slow and careful firing with considerable excess air and good draught up to 700–1000° C (1292–1832° F) (depending on body). After this, fast firing in a reducing atmosphere is used until ready for salting, after salting an oxidising fire is used to regain full heat. Open-flame kilns for salt-glazed and unglazed ware. Saggars or muffle kilns for slip-glazed or decorated ware.

Periodic kilns, suitable for mixed and fluctuating production. Kasseler. Round up-draught. Round down-draught. Rectangular down-draught. Muffles.

Continuous gas chamber kilns, suitable for steady production of mixed types of ware.

Tunnel kilns, suitable only for steady production of ware of similar sizes, e.g. all tableware; all bricks; all tiles; all pipes; all medium size chemical ware; (not recommended for the largest pieces).

Open flame or muffle according to requirements.

If salt glazing is required a section of the tunnel must be partitioned off and built in specially resistant refractories.

Glazing. Salt glazing: most stoneware is salt glazed during firing.

Slip glazing: tableware, artware and some chemical ware has an applied glaze. Albany slip and raw or fritted, lead or feldspathic glazes are used.

Thick-walled pieces risk dunting if twice fired and can stand the water from the glaze slip. They are therefore glazed by dipping, spraying or brushing when leather hard or bone dry. Thin-walled pieces would soften to much if glazed in green condition. They are biscuit fired to 700–800° C (1292–1472° F) before glazing. As the final pieces are vitreous the foot need not be glazed.

Decoration. Most forms of decoration are used including: Engobes. Carving relief and incision of body. Scraffito. Slip trailing. Under-glaze, in-glaze and on-glaze painting; coloured and/or matt glazes.

Finishing

Grinding. The nature of stoneware makes high dimensional accuracy impossible but complete accuracy can be achieved by grinding the fired pieces. No chemical resistance is lost thereby as it is the body rather than the glaze that is resistant.

Pieces requiring grinding: cocks; taps; valves; lidded vessels; pipes to be butt joined; insulators.

Three types of grinding: planar; internal, including boring holes; external.

Abrasive wheels; natural sandstones; corundum; silicon carbide; diamond; graded for rough through fine to polishing work. Abrasive powder is also used especially for grinding two pieces together.

Special work can also be done with a hammer and chisel, a light hammer is used and only small chips taken. Equipment can be armoured with furan impregnated woven glass, nylon or orlon fabrics (**S90**).

CHEMICAL STONEWARE FOR CHEMICAL ENGINEERING

Characteristics. Corrosion resistance to practically all chemical liquids and gases except hydrofluoric acid, and its derivatives, and hot caustic alkalies. Strong, relatively tough and possible to make into almost any shape and in large sizes (up to about 500 gal, 2273 l). Very low or zero porosity, can be suitably salt or slip glazed to give an easily cleaned surface. There is a tendency towards poor thermal shock resistance and failure in tension which must be taken into consideration when designing and installing plant, but the physical properties can be considerably varied by altering the composition. There is also a tendency to warping during firing and 2% tolerances should be allowed or the extra cost of grinding to accurate dimensions accepted.

Uses. Manufacture of: Acetic acid. Acetic anhydride. Ammonia. Bleaches. Bromine. Chlorine, wet and dry. Dyestuffs. Explosives. Hydrogen peroxide. Hydrochloric acid. Hypochlorite. Magnesium chloride. Naphthalene. Nitric acid, strong and weak. Paint and varnish. Paper. Phenol. Phosphoric acid. Photographic materials and processes. Potassium chloride. Potassium permanganate. Sodium chromate and bichromate. Sodium hydroxide. Sodium persulphate. Sulphide pulp. Sulphuric acid (**A93**).

Apparatus and Equipment. Absorbers. Acid-resisting brick and tile for construction and lining: interlocking shapes, hollow shapes to allow for concrete reinforcing, roughened tiles for two course overlapping work. Agitators. Autoclaves. Ball mills. Condensers. Cookers. Crystallisers. Digesters. Dryers. Drying towers. Evaporators. Fans and blowers. Filter plate rims. Heat exchangers. Cooling coils (Fig. 14.1). Kettles. Pipelines: bell and spigot pipe (cement); conical flange pipe (metal clips, no cement); elbows; tees, Y's, crosses; reducers; blank flange caps; clean-out plugs; traps; spacers. Pumps: centrifugal; vacuum; compressors. Reaction vessels. Separators (Fig. 14.2). Scrubbers. Sinks. Stills. Tanks, basins and jars for: neutralising; settling, storage (Fig. 14.3), wash. Thickeners. Towers, whole, sectional. Tower packings (Fig. 14.4): balls; berl saddles; cones; Lessing rings; partition rings; Raschig rings; splined rings; single, double and treble spiral rings. Tower distribution plates and tower grates. Vacuum filters. Valves, fittings, cocks and dampers: straight-way cocks; three-way cocks; bib cocks; neverstick cocks; right-angle cocks; block cocks; gas faucets; revolving dampers; slide dampers; check valves; safety valves (**A93, S117**).

Design of Vessels and Equipment. As the physical properties can be varied by altering body composition it is essential that users should give manufacturers complete details regarding the purpose for which the stoneware is required, *e.g.* materials to be handled, temperature range and variations, physical loadings. One property may have to be sacrificed in order to obtain the maximum in another.

Thermal shock or spalling resistance is greater for small pieces than for

Fig. 14.1. *Double-ring cooling coil made in heat-resisting acid-proof stoneware (Hathernware,* **H40**)

Fig. 14.2. *Stoneware separator for working under full vacuum. Test pressure (hydraulic) 30 lb/in². 1 in. cone-flanged pipes, 25 mm. bore, with expansion joints. Large vessel* 300 *l. Small vessel* 10 *l.* (*Hathernware,* **H40**)

FIG. 14.3. *Battery of chemical stoneware storage jars of 25–50 gal (114–227 l.) capacity (Doulton, **D56**)*

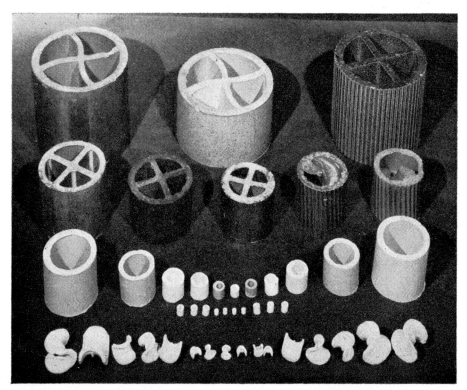

FIG. 14.4. *Various types of tower packing, partition rings, spiral rings, Raschig rings, and Berl saddles (made in both chemical stoneware and chemical porcelain) (Knight, **K45**)*

large ones. In large vessels shape and wall thickness are factors in determining spalling resistance. A vessel of even section thickness and regular circular section will withstand greater thermal shock than a vessel of the same capacity with varying wall thickness and sharp corners or edges. Ideally all vessels which will have to operate under conditions of thermal shock should be circular with hemispherical bases, failing this all corners must have the greatest possible radius. Inlet and exit holes, built-in ducts, etc., should be excluded but if essential they must have an even wall thickness and large radius of curvature at junctions. It is not desirable to attempt to fire a thin baffle into a thick vessel.

It should also be remembered that large articles require grog in their bodies and cannot be fired to so dense a product as small pieces.

Manufacture (*see 'General Stoneware'*).

Standards. B.S. 784:1953; B.S. 1634:1950.

Installation. It should always be remembered that: (1) Stoneware has limited bending and tensile strength although its compressive strength is very high. (2) Thermal shock resistance of stoneware is small.

Pipes. Should be on rigid supports that do not have appreciable heat or moisture movement. They should not be rigidly united to the support but allowed their own expansion movements.

In spigot and socket pipe-lines sockets should face upwards if there is any fall. Pipes should not be run home in the socket, the joint left allows for expansion and also easier replacement of single pipes.

In butt-jointed conical flanged pipes expansion joints must be allowed in long runs. In systems suffering vibration very rigid support is required if joints are not to fail.

Towers. Each section should have individual support. As each section should be vertical it is frequently more economical to machine the ends so that butt joints using little cement and no packing pieces can be quickly assembled (**W8**).

Tanks. It is in the nature of stoneware that the 'flat' bottoms are not quite true and if not evenly supported tensile stresses are set up, leading to failure. The most effective method of bedding, especially if it needs to be acid resistant is to use a 1–2 in. (3–5 cm) deep bed of loose sand constrained in a rim, place the stoneware vessel and move it about a little until unevennesses have been taken up. Adequate support is then achieved without cementing which sets up thermal stresses.

Cements. The correct choice of acid-proof, or other corrosion-resisting cements is very important if chemical stoneware is to serve in optimum conditions. Conditions of use should be compared carefully with manufacturers details of available materials and their instructions followed. (*See also* **A107**.)

Heinel (**H76**) gives figures and graphs concerning the design and installation of stoneware apparatus and machines.

CHEMICAL STONEWARE OF ZERO ABSORPTION

Characteristics. Chemical stoneware of greater mechanical strength and zero absorption. There are no pores in which chemicals can be absorbed either to attack the stoneware or to be lost from the solution or to contaminate the subsequent contents of the vessels, no roughness to catalyse reactions.

Uses. Preparations of valuable chemicals, precious metals, etc. Manufacture of chromic acid. Manufacture of and reactions with hydrogen peroxide (catalytically decomposed in some ordinary stoneware vessels). Foodstuff industries (where traces absorbed in pores can decompose and taint later batches). Grinding mill linings and balls (**S96**). Rollers for bleaching plants and for coating photographic papers. Triturating rolls for preparing oil colours. Crushing and homogenising rollers in provisions industry.

WHITE CHEMICAL STONEWARE FOR THE PHARMACEUTICAL AND FOODSTUFF INDUSTRIES
(Fig. 14.5)

Characteristics. Chemical stoneware with all the advantages of the ordinary type as regards acid resistance, versatility of shape, etc., plus the

FIG. 14.5. *White stoneware laboratory benches*
(*Hathernware*, **H40**)

fact that it is white and therefore shows dirt or impurities and has a smooth glossy glaze that does not set up undesired crystallisation centres, etc. No foreign taste is ever imparted to the contents.

Uses. Pharmaceutical and foodstuff industries (especially soups, meat extracts, chocolate). Hospitals.

Apparatus and Shapes. All larger vessels that are uneconomical in porcelain, *e.g.* troughs, tanks, channels, pipes, machinery, etc.

Manufacture. In general as for ordinary chemical stoneware (**S98**).

AGRICULTURAL SALT-GLAZED STONEWARE
ALSO SOME SLIP-GLAZED AND SOME UNGLAZED WARE

Characteristics. Impervious bodies of smooth surface unattacked by animal feeding stuffs, dairy products, silage, or excrements. Volume stable, no swelling and shrinking as in wood, no absorption of material. Smooth surfaces that do not harm the teeth of farm stock. Easily cleaned.

Uses. Feeding and drinking troughs, mangers, etc., for horses, cattle, sheep, pigs, poultry, rabbits, etc. Dairy vessels. Butter machines. Preserving and salting jars (cabbage, beans, gherkins, etc.). Silage pit linings. Floor tiles for stables, cattle stalls, dairies, etc. Wall tiles for same. Drains, ducts, gulleys, cesspool linings. Irrigation pipes. (B.S.2505:1954.) (**S23**).

ACID-RESISTING BRICKS AND TILES

Characteristics. Bricks and tiles of low water absorption (1%) and high resistance to acids.

Uses. Lining of vats, chambers, towers, etc., in chemical plant. Paving floors subject to acid attack, *e.g.* dairies.

Body Type. Stoneware, with about 80% SiO_2, low lime and iron ($Fe_2O_3 < 1.5\%$ recommended).

Raw Materials. Plastic stoneware clays complying with above specification.

Preparation. Prolonged weathering→adjustment of pH→pugmill→ageing.

Shaping. Hand moulding. Extrusion→cutting off→re-pressing. Casting.

Drying. As for bricks, careful drying is essential.

Firing. 1280° C (2336° F), cone 9.

(1) Intermittent, up-draught, down-draught, open fire and muffle. (2) Continuous chamber kilns, direct fire and muffle. (3) Tunnel kilns direct fire and muffle.

Tests. Acid resistance.

Standards. (B.S. 784:1953).

CHEMICAL STONEWARE FOR THE TEXTILE INDUSTRY

Characteristics. Corrosion and wear resistant pieces up to very large sizes. No contamination of the liquids contained, or absorption of their

CHEMICAL AND TECHNICAL CERAMICS

constituents (very important in dye baths). Easy complete cleaning. Fair thermal shock resistance. Possibility to grind to accurate dimensions, the glaze not being necessary as a protection against corrosion, and hence tight joints in pipe-lines, tiled vessels, etc.

Uses and Shapes

(1) Acid storage: batteries of stoneware jars interconnected with stoneware pipe-lines, cocks, etc.

(2) Dyeing and bleaching: vessels, vats, etc. Pipes, angles, and stopcocks (these can be supplied with ground conical flanges and coupled by metal clips without cement). Pumps, fans. Pot eyes and pot pegs. Rollers.

(3) Rendering wool unshrinkable: high-grade vessels with perforated false bottoms and run-off cocks. Solution storage vessels. Pipe-lines.

(4) Rayon industry: Tanks. Storage jars. Condensing towers. Stills. Vacuum filters. Agitators.

(5) Acetate, cuprammonium, wool-dyeing rayons, new synthetics: various special parts.

(6) Spinning: Thread guides. Baths. Troughs. Rollers. Bobbins. Spools. Precipitation baths. Nozzles, etc.

Body Type. Chemical stoneware, manufactured by general methods (P59, S92).

RUTILE BODY THREAD GUIDES

Characteristics. Excellent mechanical strength and chemical resistance to acids and alkalies, polished surface that remains polished in wear. Some types are semi-conductors and can dissipate static electricity.

Uses. Thread guides for: rayon; nylon; vinyon; silk; glass fibre; jute; linen; press cloth; hard twist cotton.

Body Type. Rutile porcelain.

CHEMICAL PORCELAIN

Characteristics. Dense homogeneous white body. Corrosion proof throughout its thickness to all chemicals except hydrofluoric acid and strong caustic soda. High mechanical and thermal shock resistance. Complete freedom from contamination by iron makes it replace stoneware in a number of applications. Can be produced as thinner walled vessels than stoneware but cannot be made into such large pieces. Has better tensile strength than stoneware and is gas tight up to $1400°$ C ($2552°$ F). It has the advantage over many other engineering materials of being economically made into unusual shapes.

Apparatus and Equipment. Pipes and fittings: elbows; tees; Y's; crosses; spacers; reducers; end caps. Valves and plug cocks: Y-valves; angle valves; diaphragm valves; straight-way cocks; three-way cocks; bib

Fig. 14.6. *Three-piece porcelain vacuum filter* (*U.S. Stoneware*, U5)

cocks; safety valves; check valves; neverstick cocks. Pumps. Exhausters. Towers and tower sections. Tower packings: Raschig rings; partition rings. Jars, tanks, kettles, distillation vessels. Filters for vacuum, pressure or gravity filtration (Fig. 14.6). Tongued and grooved lining tiles. Sinks. Laboratory ware: crucibles; beakers; cells; dishes; tubes; bottles; capsules; pestles and mortars; boats, etc.; funnels; covers (Fig 14.7). Ball mill linings; jars and balls.
Uses. Laboratory ware. HCl plants. Oil refineries. Textile thread guides. Rubber dipping forms for rubber gloves, bathing caps, etc. Vacuum doughnut tubes for Betatron.
Body Type. Hard porcelain. Heat and thermal shock resistance being of more importance than translucency.
Raw Materials. Kaolin or China clay. Feldspar. Quartz. Plastic kaolin. Ball clay. Pitchers.
Preparation. Hard materials→crushing→cylinder grinding until 60% finer than 0·01 mm (finer grinding than for domestic porcelain aims at eliminating free quartz from the fired body). Clays→blunger→160-lawn. Mixing ark→magnets→sieves→filter press→kneading table or deairing pugmill→maturing, up to three weeks.

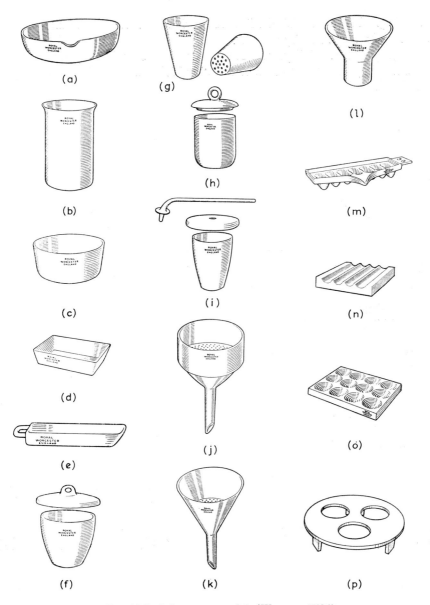

Fig. 14.7. *Laboratory porcelain* (*Worcester* **W86**)

(a) *Flat tipped basin*
(b) *Beaker*
(c) *Round capsule*
(d) *Rectangular capsule*
(e) *Combustion boat*
(f) *Crucible*
(g) *Gooch crucible*
(h) *Liebig crucible*
(i) *Rose's crucible*
(j) *Buchner funnel*
(k) *Hirsh funnel*
(l) *Dental dissolving cup*
(m) *Parting dish*
(n) *Pipette rest*
(o) *Spotting plate*
(p) *Desiccator plate*

Shaping. Hard porcelain bodies are inevitably 'short' and strains leading to crooked ware are easily set up.

(1) Jiggering and jolleying. This is preceded by a batting out machine, the bat being spread on to stretched cloth which can be used to transfer it to the jigger.

(2) Casting.

(3) Extrusion.

(4) Dry pressing.

Drying. Small pieces in dobbins, mangles, open shelves, etc. Large pieces in chamber dryers, corridor dryers. Top deck of three-compartment porcelain kilns.

Fettling

Biscuit Firing. 800–900° C (1472–1652° F).

Intermittent: second deck of two- or three-decker porcelain kilns.

Tunnel kilns. Conveyor belt (12 hr cycle).

Glazing. Raw feldspathic glaze. Dipping.

Glost Firing. 1430–1460° C (2606–2660° F). Reducing atmosphere during final stages.

Intermittent: bottom chamber of two- or three-deck porcelain kilns. Tunnel kilns, gas fired.

Tests.

(1) For porosity or glaze imperfections with eosine dye.

(2) Thermal shock resistance from 250° C to water at 15° C.

(3) Constancy of weight and heat resistance of glaze. Unglazed pieces are heated to 950–1000° C in glazed dish of same make and any loss of weight or sticking looked for.

(4) Resistance to acids and alkalies at 100° C for 4 hr (**A57**).

B.S. 914:1952.

Finishing. Grinding, polishing. Armouring with several layers of glass cloth impregnated with corrosion-resistant resin. This shrinks on to the porcelain when setting, giving a hard durable surface with mechanical properties approaching those of steel. This armouring not only prevents breakage from external mishandling, but if the porcelain should crack it is held together and serious damage due to fire, explosion or toxic substances is reduced (**G32**).

CARBON AND GRAPHITE SHAPES

Characteristics. General chemical inertness, except under oxidising conditions at elevated temperatures (above 350° C, 662° F). Unaffected by fluorine, acids and alkalies. No temperature limit of use except in oxidising conditions and even then as the oxidation products are gaseous contamination does not occur.

A number of shapes can be produced by moulding or extrusion but the material can also be cut, sawn, turned, threaded, planed and polished to any shape.

High resistance to thermal shock. Conductor of electricity. Thermal conductivity depends on porosity and whether amorphous carbon or graphite, can therefore be adjusted to requirements (**D52**).

Resistance to Acids. Complete resistance to: all very dilute acids; HCl all concentrations; HF+dil. H_2SO_4; dil. H_2SO_4; dil. and conc.H_3PO_4; organic acids. Limited resistance to: HF up to 400° C; HF+conc. H_2SO_4 up to 150° C; dil. HNO_3; conc. H_2SO_4 up to 200° C. Not resistant to: conc. HNO_3 even at room temperature.

Uses. Raschig rings and tower packings, *e.g.* alkali scrubbing of hydrocarbon gases to remove sulphur. Brick or tile linings for pickle tanks, vats, digesters, ducts, furnace and ladle linings. Pipes and pipe fittings including valve handles, suitable for hydrofluoric acid. Mould plugs, stopper heads and sintering trays for metallurgical work. Towers and entire plants for gas purification. Oil refinery tower equipment, including grids, distributor plates and trays, bubble caps, heaters, dephlegmating coolers, spurge pipes, etc. Cottrell electrostatic precipitators in sulphuric and phosphoric acid manufacture. Anodes for cathodic protection systems on pipe-lines, tank bottoms, etc. (**D52**). (Fig 14.8)

Uses, Graphite. Impervious graphite for various forms of heat exchanger, *e.g.* atmospheric or cascade cookers, bayonet or candle heaters, countercurrent concentrator tube exchangers, immersion type bundles, tube and shell floating heat-type exchangers. Considerably applied in oil refineries. Pressure limit 60 lb/in² (4·2 kg/cm²). Temperature limit 175° C (347° F). Self-lubricating piston rings and piston rod packings, *e.g.* where pumping liquefied hydrocarbons that wash out lubricating oils, etc. Anodes for cathodic protection systems where chlorine is present (**D52**). Moulds for casting machine tools, special oil well drilling bits, ferrous and non-ferrous ingots (**V23**). HCl absorbers. Towers. Pipes and fittings (**G32**).

Body Type. Refractory, carbon.

General methods of manufacture described in Chapter 17.

'DELANIUM' CARBON

Characteristics. Homogeneous fine-grained carbon product of low (controlled to 2% up to 14% as required) porosity, consisting largely of α-carbon. Various hardness grades, requiring *either* diamond or carborundum wheels, *or* readily sawn, ground or machined by normal methods, are available; the 2% porosity grade being very hard and the 10–14% combining excellent resistance to spalling and cracking while relatively impermeable.

Uses. Tiles for lining tanks, reaction vessels, scrubbing towers, etc. Towers and packings for absorption and distillation plant. Tubes, pipes and fittings for corrosive fluids.

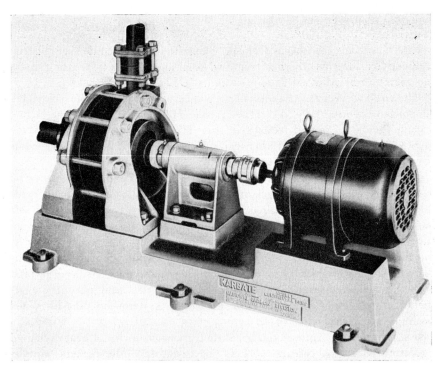

Fig. 14.8. 'Karbate' centrifugal pump (National Carbon, **N5**)

Body Type. Carbon.
Raw Materials. Selected bituminous coal of suitable agglutinating properties. Swelling inhibitors. Lubricants.
Preparation. Cleaning→grinding→grading to size range from 200-mesh to 10μ→blending with additions.
Shaping.
Firing. Preheating circulating ovens and gas tunnel kiln, both taking bogies carrying boxes into which carbon plates are interleaved with metal plates to ensure suitable heat penetration and then topped off with coke fines; 40 hr preheat to 450° C (842° F); 80 hr in tunnel kiln passing through separately controlled zones at 430° C (806° F), 530° C (986° F), 680° C (1256° F), 910° C (1670° F), 975° C (1787° F); controlled cooling down to 200° C (392° F) (**J60**).

'DELANIUM' GRAPHITE
(Fig. 14.9)

Characteristics. Homogeneous properties of Delanium carbon but from ten to fifteen times the thermal conductivity and from four to five times the electrical conductivity.

Fig. 14.9. *Cut-away view of block heat exchanger showing the graphite block (Powell Duffryn, P53)*

Uses. Heat transfer equipment, especially the cubic heat exchanger.
Raw Material. Finished 'Delanium' carbon pieces.
Firing. Up to 3000° C (8600° F). Because of its more crystalline nature 'Delanium' carbon does not graphitize easily and requires a special process (N22).

'KEMITE' LABORATORY TABLE TOPS, AND OTHER PIECES

Characteristics. A material of fine uniform texture that can be made in large slabs of accurately flat and smooth finish, of adequate transverse strength to be able to dispense with excessive support; good resistance to chemicals, impact, abrasion and scratching, and to heat and thermal shock. It consists of a skeleton of porous cordierite (low-expansion) body, the pores of which are filled with coke.
Uses. Laboratory table tops. Laboratory and chemical industry floors. Vats, tanks and other apparatus.
Body Type. Composite body with cordierite (plus silicon carbide) porous skeleton and carbon impregnation.
Raw Materials. *Porous skeleton:* Plastic clay. Porous cordierite grog. Silicon carbide. Paper fibre. *Impregnants:* Pitch and asphalt.

Preparation. All ingredients dry ground. Circular mixing tank with mechanical stirrer, water and paper pulp placed in first and then all the dry ingredients; deflocculant. Stirring continued for 1 hr to remove entrapped air. Water content 30%.
Shaping. Special pressure casting.
Drying. After removal from mould slabs are placed on pallets, allowing air to side in contact. Initial slow drying at room temperature 48 hr; then heat up to 180° C (356° F), 48 hr. Open structure of the body facilitates drying.
Firing. 1190° C (2174° F). Tunnel kiln; slabs placed on edge; 122 hr.
Machining. Grinding to specific thickness on horizontally rotating rubbing bed of heavy iron construction fed with sand and water. Cutting to required length and width by silicon carbide wheels. Drilling, grooving, etc., by suitable stone cutting tools.
Impregnating. Prepared slabs are dried out in the coking oven at 150°C (302° F).

Immersed in impregnating liquid at 210° C (410° F) in autoclave. Soaked 1 hr, then 40–60 lb/in² air pressure applied for 1 hr→drain, excess impregnant removed.
Coking. Gas-fired coking oven, suitably sealed to prevent oxidation and with outlets to recover distillate of volatile matter. Temperature 150° C (302° F) when sealed; 150° C (302° F) to 400° C (752° F) 60–72 hr; soaking until no more distillate coming off; cooling 48 hr down to 50° C (122° F).
Finishing. Hand or machine polishing (**P35**).

KARCITE LABORATORY EQUIPMENT

Characteristics. A material of fine uniform texture consisting of a porous body, whose pores are filled with coke. It has high transverse, crushing and impact strength, high resistance to chemicals, can be made into large and dimensionally accurate pieces (but does not have the thermal shock resistance of 'Kemite').
Uses. Laboratory: sinks, shelves, tanks, pipes and pipe fittings. General building: sanitary ware; partitions; roofing; flooring; stair treads; wainscoting; window sills.
Body Type. Porous structural body with carbon impregnation.
Raw Materials. Grog, made from same plastic mix, 70%; plastic mix, 30%: slip clay 9%; ball clay 20·7%, paper pulp (for pressure casting), or wood flour (for hand pressing) 0·3%. Pitches and asphalts.
Preparation. As Kemite.
Shaping. Slabs, pressure casting as Kemite. Sinks, etc., hand pressing.
Drying. Firing to 1025° C, **machining, impregnating** and **coking** as Kemite (**P36**).

FILTERS
(Fig. 14.10)

Characteristics. Self-supporting filter of good mechanical strength and excellent resistance to heat, thermal shock and corrosion so that drastic cleaning methods may be used. Good abrasion resistance, as required in gas filtration. Available in well-defined grades of known maximum pore size, will generally retain solid particles down to a size approximately one-third that of the measured maximum pore size. Available in numerous shapes. Range of pore size approximately 750 μ to 20 μ.

FIG. 14.10. *Filters. Circular plates for vacuum filter units, with and without non-porous surround. Porous tiles. Standard* 10 × 2 *in.* (250 × 50 *mm.) filter candle with metal fitting* (*Doulton*, **D56**)

Shapes. Tiles. Circular plates, with or without vitreous surround. Candles. Tubes.
Body Type. Permeable porous: diatomite, porcelain, alumina, carbon, graphite.
Uses. Filtration. Liquids, drinking water: diatomite, carbon, porcelain. Bacteriological work: porcelain. Pharmacy: Diatomite, porcelain. Sterilisation of liquids that may not be heated. Gases, removing both solids and liquids. Ducting and pipe-line systems, compressed air systems, cyclones. Plugs to prevent spilling of mercury from manometers (**S123**).

TABLE

Properties of materials

A. Stoneware

Name / Body	1905 Stoneware	1921 Stoneware	1926 Stoneware	1921 Stoneware
Ref.	S86	S86	S86	S117
Country	Germany	Germany	Germany	
Specific gravity				2·479–2·649
Bulk density (lb/ft^3)		137	154	95·0–127·9
(g/cm^3)		2·2	2·473	2·028–2·482
Absolute pore volume (%)		12·9	5·2	6·3–18·2
Water absorption (%)		1·3	0	0·26–2·50
Compressive strength (lb/in^2)	21 334	66 622	82 964	82 727
(kg/cm^2)	1500	4684	5833	5816
Tensile strength (lb/in^2)	640–782	1892	2532	1650
(kg/cm^2)	45–55	133	178	116
Transverse strength (lb/in^2)		5447	12 801	5746
(kg/cm^2)		383	900	404
Modulus of elasticity (lb/in^2)		80 200	59 400	$9·7 \times 10^6$
(kg/mm^2)		56·4	41·75	6850
Specific heat		0·188	0·197	0·191
Thermal conductivity (B.t.u.)		10·0	24·2	10·8859
(kcal hr^{-1}m^{-1} °C^{-1})		1·24	3·00	1·35
Linear thermal expansion $\times 10^{-6}$		4·8	3·9	4·1
Impact strength (ft-lb/in^2)		74 663	111 995	83 996
(cm-kg/cm^2)		1·6	2·4	1·8
Drum test (abrasion resistance) (%)		6·4	3·6	3·6
Torsional strength (lb/in^2)		2916	4594	1570
(kg/cm^2)		205	323	251
Resistance to sand blast (cm^3)		5·0	2·4	3
Hardness scleroscope				62
Hardness, Mohs' scale				
Thermal shock resistance				
Ball test, compressive strength (kg/cm^2)		810	1253	
(lb/in^2)		11 521	17 822	
Electrical resistance (ohm-in^3)				
(ohm-m-mm^2)				
P.C.E.		25	29	
Chemical: Inorganic acids				
Hydrofluoric acid				
Organic liquids				
Corrosive liquids				
Corrosive gases				
Alkalies				
Hot conc. caustics				
Organic solvents				
Oxidising agents				
Phosphoric acid				

for Chemical Plant

1951 Stoneware	1921 Chemical stoneware	1933 Chemical stoneware	Chemical stoneware	Digester lining brick stoneware	Normal stoneware
S117	**S97**	**S97**	**P60** Brit.	Swedish	**M33** Swiss
2·50				2·52	
153·8				124	142
2·463				1·99	2·27
1·5				21·03	
0	0·5	0	0–3	7·55	0·1
116 773	82 723	116 773	45 000–75 000		55 329
8210	5816	8210	3164–5273		3890
7610	1650	7510	900–1500		2404
528	116	528	63–105		169
13 939	5917	13 555	4000–5600		4765
980	416	953	281–394		335
$4·4 \times 10^6$	$5·9 \times 10^6$	$5·9 \times 10^6$			$10·6 \times 10^6$
3080	4175	4175			7480
0·191			0·18–0·19		
31·8512	10·9	31·8			8·1
3·95	1·35	3·95			1·0
0·15	4·1	0·15	4–5		4·3
228 655	88 662	233 322			
4·9	1·9	4·97			
4596	3570	4594			
323	251	323			
	3·0	0·6			
100					
					Max. change 40° C

TABLE

A. Stoneware—(contd.)

Name / Body	Thermal shock resistant cordierite stoneware	General stoneware	Thermal shock resistant stoneware	1934 'Acitherm' heat resisting stoneware
Ref.	M33	G32	G32	U4
Country	Swiss	U.S.A.	U.S.A.	U.S.A.
Specific gravity		2·2	2·2	
Bulk density (lb/ft^3)	141			
(g/cm^3)	2·26			
Absolute pore volume (%)		1·5	1·5	
Water absorption (%)	0·3	0·2	0·2	
Compressive strength (lb/in^2)	49 355	80 000	80 000	118 000
(kg/cm^2)	3470	5625	5625	8296
Tensile strength (lb/in^2)	3101	2500	4000	8000
(kg/cm^2)	218	176	281	562
Transverse strength (lb/in^2)	6528	5000	6500	14 000
(kg/cm^2)	459	352	475	984
Modulus of elasticity (lb/in^2)	12×10^6	10×10^6	10×10^6	
(kg/mm^2)	8650	7030	7030	
Specific heat		0·2	0·2	
Thermal conductivity (B.t.u.)	13·5	8	20	7·5–15
(kcal-hr^{-1}m^{-1} °C^{-1})	1·68			
Linear thermal expansion $\times 10^{-6}$	1·5	per °F 2·7	per °F 2·7	
Impact strength (ft-lb/in^2)				
(cm-kg/cm^2)				
Drum test (abrasion resistance) (%)				
Torsional strength (lb/in^2)				
(kg/cm^2)				
Resistance to sand blast (cm^3)				
Hardness sclerescope				
Hardness, Mohs' scale		8·5	8·5	
Thermal shock resistance	guaranteed over 100° C			
Ball test compressive strength				
(kg/cm^2)				
(lb/in^2)				
Electrical resistance (ohm-in^3)				
(ohm-m-mm^2)				
P.C.E.				
Chemical: Inorganic acids		Resistant		
Hydrofluoric acid		Not resistant		
Organic liquids		Resistant		
Corrosive liquids		Resistant		
Corrosive gases		Resistant		
Alkalies		Resistant		
Hot conc. caustics		Not resistant		
Organic solvents		Resistant		
Oxidising agents				
Phosphoric acid				

245—(contd.)

Drainpipe stoneware	Chemical stoneware	'Embrachit' higher temp. stoneware	'Thermosil' high thermal conductivity stoneware	'Embrit' porcelainlike stoneware for electrical insulators
S185 German	S185 German	S185 German	S185 German	S185 German
134	133	134	131	147
2·15	2·13	2·13	2·1	2·35
3·16	0·85	0·3–1·0	0·2–3·0	0
25 033	38 403	33 709	36 838	70 690–85 340
1760	2700	2370	2590	4970–6000
2034	1252	981	1323	1707–7112
143	88	69	93	120–500
3271	4551	2560	4765	5618–12 801
230	320	180	335	395–900
528 000	7×10^6	4×10^6	9×10^6	$8 \cdot 5 – 11 \times 10^6$
371	4900	3080	6330	6000–8000
0·19	0·19	0·19	0·18	0·19
8·1	8·7	9·7	23·4–~80	10·3
1·0	1·08	1·20	2·9–~10·0	1·28
4·9–5·5	4·9–5·0	2·7	4·1	4·0
65 330–74 663	69 997–88 662	79 329–83 996	65 330–69 997	83 996–107 328
1·4–1·6	1·5–1·9	1·7–1·8	1·4–1·5	1·8–2·3
7–8	7–8	7–8	7–8	7–8
Good	Good	Very good	Good to very good	Good to very good
Resistant	Resistant	Resistant	Resistant	Resistant
	Not resistant	Not resistant	Not resistant	Not resistant
	Resistant	Resistant	Resistant	Resistant
If cold, resistant	If cold, resistant	If cold, resistant	If cold, resistant	If cold, resistant
Not resistant	Not resistant	Not resistant	Not resistant	Not resistant

TABLE

B. Porcelain

Name / Body	General 1920's porcelain	1942 porcelain		Chemical porcelain
Ref.	**S81**	**W3**		**L16**
Country	Germany	Germany		U.S.A.
Specific gravity	2·3–2·5	2·4		2·41
Bulk density (lb/ft^3)				150
(g/cm^3)				2·4
Absolute pore volume (%)				
Water absorption (%)				0
Compressive strength (lb/in^2)	56 893–71 116	Unglazed 56 893–64 005	Glazed 64 005–78 228	100 000
(kg/cm^2)	4000–5000	4000–4500	4500–5500	7030
Tensile strength (lb/in^2)	3700–3900	3600–5000	4300–7000	5000–8000
(kg/cm^2)	261–274	250–350	300–500	351–562
Transverse strength (lb/in^2)	5832–13 939	7112–9956	12 801–14 223	Glazed 15 000 / Unglazed 12 000
(kg/cm^2)	410–980	500–700	900–1000	1055 / 844
Modulus of elasticity (lb/in^2)	117 769	$9·9–11·4 \times 10^6$		$10·4 \times 10^6$
(kg/mm^2)	82·8	7000–8000		7311
Specific heat	0·25	20–100° C 0·19–0·21		
Thermal conductivity (B.t.u.)		10·4–11·3		8·4
(kcal hr^{-1} m^{-1} °C^{-1})		20–100°C 1·3–1·4		1·04
Linear thermal expansion $\times 10^{-6}$	3·66(16–250°C) 4·34(16–1000°C)	20–100° C 3·5–4·5		°F 2·3
Impact strength (ft-lb/in^2)				
(cm-kg/cm^2)				
Drum test (abrasion resistance) (%)				
Torsional strength (lb/in^2)	6116–7112			
(kg/cm^2)	430–500			
Resistance to sand blast (cm^3)	1·7–3·3	0·09–0·14		
Hardness sclerosscope				
Hardness, Mohs' scale	7–8			
Thermal shock resistance		160° C		
Ball test compressive strength				
(kg/cm^2)				
(lb/in^2)				
Electrical resistance (ohm-in^3)				
(ohm-m-mm^2)				
P.C.E.		30		
Chemical: Inorganic acids	Resistant			
Hydrofluoric acid	Not resistant			
Organic liquids				
Corrosive liquids				
Corrosive gases	Resistant			
Alkalies	Resistant			
Hot conc. caustics	Slowly attacked			
Organic solvents				
Oxidising agents				
Phosphoric acid	Hot, slow attack			

245—*(contd.)*

Porcelain	Chemical and electrical porcelain	Porcelain grinding balls	High alumina 'Regalox' grinding balls	AlSiMag 192 TiO$_2$
G32 U.S.A.	**S185** Germany	**W86** Brit.	**W86** Brit.	**A186** U.S.A.
2·78	143–156 2·3–2·5	2·75	3·52	4·0 248·8 4·0
0 0 100 000	0 64 005–71 116	Nil 75 000	Nil 158 000	0–0·2 80 000
7030	4500–5000	5273	11 108	5625
6000–8000	2800–7000			7500
422–562	200–500 5689–14 223			527 20 000
	400–1000			1406
10 400 000 7311 0·25	9·9–11·4 × 10^6 7000–8000 0·19–0·21	11 200 7·87	31 200 22	
10 10 1·24 °F 1·9	10·4 1·3	0·049 0·0062	0·185 0·023	
	3·5–4·5 8400–10 270 1·8–2·2	4·86	7·1	7·3 (25–100° C) 8·7 (25–700° C)
9	7–8 Good to very good		9	
Resistant Not resistant Resistant Resistant Resistant Resistant Not resistant Resistant	Resistant Not resistant Cold, resistant Not resistant			

TABLE

C. Carbon or Graphite

Name ⎫ Body ⎬ Ref. ⎭ Country	Graphite 1921 S117	1921 Retort carbon S117	Hardening carbon 1921 S117	'National' carbon N5 U.S.A.
Specific gravity	2·21	—	2·0	
Bulk density (lb/ft^3)	99	—	97	95–97·8
(g/cm^3)	1·59	—	1·56	1·525–1·57
Absolute pore volume (%)	28·0	—	22·0	
Water absorption (%)				
Compressive strength (lb/in^2)				1910–4100
(kg/cm^2)				134–288
Tensile strength (lb/in^2)				400–800
(kg/cm^2)				28–60
Transverse strength (lb/in^2)				790–1670
(kg/cm^2)				56–117
Modulus of elasticity (lb/in^2)				4·3–9·4 × 10^5
(kg/mm^2)				302–661
Specific heat	0·20	0·3		
Thermal conductivity (B.t.u.)	33·8671	29·8353		4–6
(kcal hr^{-1}m^{-1} °C^{-1})	4·2	3·7		0·47606– 0·74408
Linear thermal expansion × 10^{-6}	7·86	5·5	—	
Impact strength (ft-lb/in^2)				
(cm/kg/cm^2)				
Drum test (abrasion resistance) (%)				
Torsional strength (lb/in^2)				
(kg/cm^2)				
Resistance to sand blast (cm^3)				
Hardness scleroscope				
Hardness, Mohs' scale				
Thermal shock resistance				
Ball test compressive strength				
(kg/cm^2)				
(lb/in^2)				
Electrical resistance (ohm-cm^3)				0·0016
(ohn-m-mm^2)				
P.C.E.				
Chemical: Inorganic acids				
Hydrofluoric acid				
Organic liquids				
Corrosive liquids				
Corrosive gases				
Alkalies				
Hot conc. caustics				
Organic solvents				
Organic solvents				
Oxidising agents				
Phosphoric acid				

245—*(contd.)*

'National' graphite **N5** U.S.A.	'Karbate' carbon **N5** U.S.A.	'Karbate' graphite **N5** U.S.A.	'Impervite' graphite **G32** U.S.A.	Graphite **G32** U.S.A.
95·3–97·3	109·9–110·4	116·1–119·2	109	97
1·53–1·56	1·76–1·77	1·86–1·91	1·75	1·55
			0	30
3050–3420	10 500	9000–11 000	8900	3050
214–240	738	633–733	626	214
440–760	1700–2000	2350–2600	2600	760
31–53	121–141	105–183	183	53
1490–1810	4170–4640	4650–4980	4650	1750
105–127	293–327	327–350	327	123
$6·7–8·8 \times 10^5$	$2·6–2·9 \times 10^6$	$2·1–2·3 \times 10^6$	$2·3 \times 10^6$	$8·8 \times 10^5$
471–619	1828–2309	1476–1617	1617	619
70–94	2·8–3·0	75–95	1020	
8·681–	0·3473–	9·301–10·781	126	
10·173	0·37204		°F with t in °F $(24+0·0039t) \times 10^{-7}$	°F with t in °F $(12+0·0039t) \times 10^{-7}$
0·00036	0·0016	0·00033		
			Resistant to most Resistant to most	
			Resistant to most Not resistant	

TABLE

C. Carbon and Graphite—(contd.)

Name / Body Ref. Country	Carbon blocks **M96** Brit.	Graphite blocks **M96** Brit.	Carbon rods and tubes **M96** Brit.	Graphite rods and tubes **M96** Brit.
Specific gravity				
Bulk density (lb/ft^3)	100	106	100	106
(g/cm^3)	1·6	1·7	1·6	1·7
Absolute pore volume (%)				
Water absorption (%)				
Compressive strength (lb/in^2)	14 000	4000	19 000	5500
(kg/cm^2)	984	281	1336	387
Tensile strength (lb/in^2)	3000	1000	3500	1000
(kg/cm^2)	211	70	246	70
Transverse strength (lb/in^2)	5200	3300	7500	3800
(kg/cm^2)	365	232	527	267
Modulus of elasticity (lb/in^2)	11×10^5	5×10^5	14×10^5	5×10^5
(kg/mm^2)	773	351	984	351
Specific heat	0·2	0·2	0·2	0·2
Thermal conductivity (B.t.u.)	10	60	10	60
(kcal hr^{-1} m^{-1} °C^{-1})	1·24	7·44	1·24	7·44
Linear thermal expansion $\times 10^{-6}$				
Impact strength (ft-lb/in^2)				
(cm-kg/cm^2)				
Drum test (abrasion resistance) (%)				
Torsional strength (lb/in^2)				
(kg/cm^2)				
Resistance to sand blast (cm^3)				
Hardness scleroscope	Shore 36	Shore 35	Shore 60	Shore 40
Hardness, Mohs' scale				
Thermal shock resistance				
Ball test compressive strength				
(kg/cm^2)				
(lb/in^2)				
Electrical resistance (ohm-in^3)	0·0015	0·00045	0·0014	0·0005
(ohm-m-mm^2)				
P.C.E.				
Chemical: Inorganic acids				
Hydrofluoric acid				
Organic liquids				
Corrosive liquids				
Corrosive gases				
Alkalies				
Hot conc. caustics				
Organic solvents				
Oxidising agents				
Phosphoric acid				

245—(contd.)

Impervious carbon blocks **M96** Brit.	Impervious graphite blocks **M96** Brit.	Impervious carbon rods and tubes **M96** Brit.	Impervious graphite rods and tubes **M96** Brit.	'Durabon' carbon **S73** Germ.
106	112	106	112	94–100
1·7	1·8	1·7	1·8	1·5–1·6
				8–50 as required
16 000	7000	19 000	5500	725
1125	492	1336	387	510
3000	2000	3500	1000	853
211	141	246	70	60
5500	5000	7500	3800	
387	351	527	267	
11×10^5	8×10^5	14×10^5	5×10^5	21 335
773	562	984	351	15
0·2	0·2	0·2	0·2	
10	60	10	60	28–36
1·24	7·44	1·24	7·44	3·5–4·5
				3·0
Shore 65	Shore 40	Shore 65	Shore 45	
0·0015	0·0005	0·0014	0·0005	
				40–60

TABLE

C. Carbon and Graphite—(*contd.*)

Name / Body	'Diabon' graphite	'Durabon' impregnated	'Diabon' impregnated
Ref.	**S73**	**S73**	**S73**
Country	Germ.	Germ.	Germ.
Specific gravity			
Bulk density (lb/ft^3)	100–106	109–115	112–118
(g/cm^3)	1·6–1·7	1·75–1·85	1·8–1·9
Absolute pore volume (%)	8–50 as required		
Water absorption (%)			
Compressive strength (lb/in^2)	5689	10 667	10 100
(kg/cm^2)	400	750	710
Tensile strength (lb/in^2)	925	1636	2062
(kg/cm^2)	65	115	145
Transverse strength (lb/in^2)			
(kg/cm^2)			
Modulus of elasticity (lb/in^2)	9956	3 130 000	1 850 000
(kg/mm^2)	7	2200	1300
Specific heat			
Thermal conductivity (B.t.u.)	806–1008	28·22–36·29	806–1008
(kcal hr^{-1} m^{-1} °C^{-1})	100–125	3·5–4·5	100–125
Linear thermal expansion $\times 10^{-6}$	1·8	5·0	3·5
Impact strength (ft-lb/in^2)			
(cm-kg/cm^2)			
Drum test (abrasion resistance) (%)			
Torsional strength (lb/m^2)			
(kg/cm^2)			
Resistance to sand blast (cm^3)			
Hardness scleroscope			
Hardness, Mohs' scale			
Thermal shock resistance			
Ball test compressive strength			
(kg/cm^2)			
(lb/in^2)			
Electrical resistance (ohm-in^3)			
(ohm-m-mm^2)	8–14	40–60	8–14
P.C.E.			
Chemical: Inorganic acids			
Hydrofluoric acid			
Organic liquids			
Corrosive liquids			
Corrosive gases			
Alkalies			
Hot conc. caustics			
Organic solvents			
Oxidising agents			
Phosphoric acid			

'Delanium' carbon	'Delanium' graphite	'Kemite' coke filled porous cordierite	'Karcite' coke filled porous body
P53 Brit.	**P53** Brit.	**P35** U.S.A.	**P36** U.S.A.
—	—	1·87	1·87
97	106	117	117
1·55	1·70	1·87	1·87
		0·75–1·0	0·50–0·75
38 000	20 000	11 400	11 500
2670	1406	801	808
6000	3500		
422	246		
12 000	8000	3000	2500–3000
845	562	211	176–211
3 per °C	40 per °C		
4·5	60		
3·8	3·8	20–450° C	20–450° C
		2·5	6·0–7·5
85	65		
0·0055	0·0008		
			20–32

DIFFUSERS AND AERATORS

Characteristics. Uniform pore size and high permeability. Can be coated with catalysts to give rapid gas reactions.
Uses. Distribution of gases in liquids that absorb them, *e.g.* chlorine or carbon dioxide in alkalies. Carbonation of soft drinks. Organic reactions, *e.g.* chlorination, hydrogenation. Catalytic gas reactions. Oxidation by aeration; aquaria; sewage treatment by activated sludge process (**F10**). Froth flotation methods of mineral dressing. Fluidisation of fine powders, 'Airslide', *e.g.* for cement conveyors, and in storage bins and silos.
Body. Permeable porous: diatomite; porcelain; carbon; alumina; quartz.

POROUS ABSORBERS

Characteristics. Highly porous material of great capillary action can hold a large volume of liquid which cannot be spilt. When used with volatile liquids these evaporate from the surface and are constantly replenished.
Uses. Absorption towers and reactors. Non-spill containers for liquids for absorbing unpleasant odours or releasing bacteriocidal vapours, etc.
Body Type. Porous permeable bodies: silica.

POROUS DRYING PLATES AND SEED GERMINATION SAUCERS

Characteristics. Very fine grained material of high capillary pull.
Uses. Drying of crystals. Photographic processing. Germination of seeds.
Body Type. Permeable porous: Porcelain.

ELECTROLYTIC DIAPHRAGMS

Characteristics. Very fine grained ware of thin but strong walls in a number of shapes. The small pore size is combined with high pore density. Low electrical resistance and high resistance to acids.
Uses. Regulation of chromic ion in chromium plating solutions (**D56**). Porous membrane in cells and batteries to separate anolite and catholite liquids, *e.g.* in plating industry and hydrogen peroxide cells. Porous separator for dialysis cells for purification of clays and colloids. Porous barrier in gaseous diffusion processes, *e.g.* separation of helium from methane or natural gas (**F10**).
Body Type. Permeable porous: Porcelain. High mullite porcelain. Alumina.

Chapter 15
Specialised Laboratory and Engineering Ware

CHEMICAL STONEWARE, PARTICULARLY WHITE CHEMICAL STONEWARE

Characteristics and Manufacture. *See* Chapter 14, p. 1103.
Uses. Laboratory grinding jars. Storage vessels, etc.

CHEMICAL PORCELAIN

Characteristics and Manufacture. *See* Chapter 14, p. 1113.
Uses. Wide range of laboratory ware, usually glazed but also unglazed. Crucibles and lids. Combustion boats and tubes. Bottles. Funnels. Cells. Capsules. Dishes. Pestles and mortars. Grinding balls. Ball mill linings, grinding jars.

MULLITE PORCELAIN WARE (ALSO SILLIMANITE)

Characteristics. Better resistance to high temperatures and thermal shock than normal porcelain, can be used up to 1500° C (2732° F) or even 1700° C (3092° F) according to quality. Impermeable, and capable of maintaining vacuum up to 1400° C (2552° F), high resistance to attack by iron oxides, slags and glasses.
Uses. Grinding jars and linings for grinding cylinders. Grinding balls. Centrifugal pump propellers for acid-resisting pumps. Crucibles, combustion boats, etc., for general laboratory use. Combustion tubes and furnace cores for carbon and sulphur in steel determinations. Pyrometer sheaths. Thermocouple insulation. Burner tips (**T18**).
Body Type. Mullite porcelain.
Raw Materials. Ball clay. China clay. Sillimanite minerals, calcined if necessary. Alumina.

Preparation. Sillimanite minerals are very tough: crushing→dry grinding →wet, cylinder grinding→less than 6 μ. Clays: blunging or wet grinding. Mixing→magneting→screening→filter pressing (→prolonged ageing up to 6 months)→for plastic making→de-airing pugmill; for dry pressing→drying →pulverising→moisture added→briquetting→granulating.
Shaping. (1) Hand rolling (for grinding balls). (2) Jolleying. (3) Extrusion and turning. (4) Dry pressing and turning with diamond-shaped carborundum or corundum tools. (5) Slip casting.
Drying. Chamber dryer.
Firing. 1300–1800° C (2372–3272° F) according to quality.
Gas-fired intermittent or tunnel kilns (**A9**).

ZIRCON PORCELAIN PIECES

Characteristics. Very good mechanical strength, hardness and thermal shock resistance. Completely vitreous.
Uses. Tools for working synthetic materials and light metals. Wire drawing dies. Mill linings. Blasting nozzles. All-ceramic radiant gas burners.

STEATITE

Characteristics. Hard, heat resisting and can be shaped to very accurate dimensions. Lower cost than metals allows more frequent replacement of worn parts.
Uses. Pressing and extrusion dies, *e.g.* for aluminium. Burner tips.
Manufacture. *See* electrical insulator, steatite, Chapter 16, p. 1201.

MOULDED (OR SINTERED) BORON CARBIDE PIECES

Characteristics. Refractory, extremely hard, wear resistant material. Less resistant to oxidation than silicon carbide. Has large ability to absorb thermal neutrons, the interaction giving off α-particles. Chemically inert.
Uses. Gauges, anvils, guides, bearing or bearing liners. Sandblast nozzles, orifice plates. Jet nozzles or venturis. Atomic reactor control rods and shields. Lapping of hard metals. Polishing powder. Arc or electrolytic electrodes.
Body Type. Single component refractory.
Raw Materials. Boron carbide.
Preparation. Grinding.
Shaping. For high-density (2·3–2·5 g/cm^3) wear-resistant pieces hot pressing in graphite moulds below 2375° C (4307° F). For lower densities 1·5–2·3 g/cm^3 normal cold moulding methods.

SINTERED SILICON CARBIDE

Characteristics. Extreme hardness, denseness and abrasion resistance. Good resistance to oxidation, and corrosion. Excellent shock resistance and very high thermal conductivity. Maximum working temperatures are 2300° C ($\sim$ 4200° F) in inert atmosphere and 1650° C ($\sim$ 3000° F) in oxidising atmosphere.
Uses. In atomic energy. Chemical processing. Ore handling, *e.g.* lining of cyclones. High temperature gas handling. Electric heating elements. Electronic resistors. (*See also* silicon carbide refractories.)

SILICON-NITRIDE BONDED SILICON CARBIDE BRICKS AND SHAPES

Characteristics. Extremely high thermal shock resistance. High volume stability permitting accurate complicated shapes. Very high abrasion resistance.
Uses. Nozzles for uncooled rocket motors. Fixtures for brazing silver, copper and aluminium. Rocket combustion chamber liners and exit cones. Parts for handling abrasive slags, slurries and other reactive materials.

TUNGSTEN CARBIDE

Characteristics. Extreme hardness.
Uses. Cutting tools. Wear resisting parts. Heaters and shields in high-frequency furnaces.
Body Type. Single-component refractory.
Raw Materials. Tungsten carbide (for cutting tools: binder 0·5 to 14% cobalt). Lubricant: camphor, cetyl alcohol or waxes.
Preparation. Grinding→grading→mixing.
Shaping. Dry pressing in carbide dies at 100 000 lb/in^2 (7030 kg/cm^2).
Firing. Prefiring to 1050–1100° C (1922–2012° F) for 2 hr. Molybdenum resistance furnace. Sizing. Firing in a vacuum at 1500° C for 2 hr (2732° F).

SINTERED ZIRCONIUM AND TITANIUM CARBIDE SHAPES

Characteristics. Good thermal shock resistance (especially TiC).
Uses. Special refractories (neutral or reducing conditions). Wear resistant parts, *e.g.* anvils, gauges, guides, etc. Bearing or bearing liners. Jet nozzles or venturis. Arc or electrolytic electrodes. Nozzles or orifice plates. Crucibles (**N27**).
Shaping and Firing. Hot pressing.

REFRACTORY NITRIDES

Uses. High-temperature vessels for melting metals in vacuum without contamination. Boron nitride, nuclear reactors.
Body Type. Single-component refractory.
Raw Materials. Pure nitrides, organic binders, *e.g.* carbowax.
Preparation. Grinding TiN, ZrN in steel mills and then acid wash: other nitrides in carbide mills.
Shaping. Slip casting from water or absolute alcohol. Dry pressing.
Firing. In ammonia atmosphere. As electrical conductors heating can be by induction.

SINTER ALUMINA PIECES

Characteristics. Dense refractory material of extremely high mechanical strength and hardness, stable in oxidising and reducing conditions to 2000° C (3632° F).
Uses. Sparking plugs. Cutting tools. Pyrometer tubes. Crucibles, boats, tubes, etc. Grinding balls. D.T.A. apparatus (**T18**). Chemical pumps. Gauges (life twenty-four times that of steel). Cylinder liners for combustion engines and deep well oil pumps. Surface plates for precision checking instruments. Guides and pulleys in textile and wire making industries. Nozzles for sandblasting machines.
Body Type. Single-component refractory.
Raw Materials. High purity calcined α-alumina. Hydrated alumina. 1–3% Cr_2O_3, or MgF_2, or $AlPO_3$ to prevent crystal growth. Bentonite. Organic binders: wheat flour, gum tragacanth, polyvinyl alcohol, waxes, rubber latex.
Preparation. Wet ball milling in steel or alumina mills for 50–75 hr until maximum particle size is 12–15 mμ and 65% is below 5 μ. If in steel mills digest with 10–15% hot HCl for one or more days and wash three to four times. Also sometimes dry grinding (**M70**).

Hydrated alumina treated with acid to make gel→mixed with milled alumina to give plastic body for extrusion or jolleying (**A9**).
Shaping.

(1) Casting in plaster moulds with acid slip (pH3) of S.G. 2·0 (the pH of alumina slips alters with age but equilibrium is reached in times depending on the pH, **H113**). If alumina gel is precipitated on to the alumina particles slips of pH 4·5 and S.G. 3·0 can be made.

(2) Bind dry powdered alumina with 1–5% organic binders and dry press. 10 000 lb/in² (703 kg/cm²).

(3) Hydrostatic pressing at 100 000 lb/in² (7030 kg/cm²).

(4) With high organic binder content or bentonite, extrude.

(5) With alumina gel, extrude or jolley (**A9**).

(6) With thermoplastic resins, injection moulding.
(7) Low-frequency vibratory compacting (**B30**).
Firing. 1800–1950° C (3272–3542° F).
Gas-heated magnesia muffles. Special tunnel kilns.

SAPPHIRE WARE

Characteristics. Sintered high-purity (99·8%) alumina of polycrystalline structure but gem-like quality giving a transparency like that of bone china. High strength, non-porous and chemically inert. Compressive strength 500 000 lb/in^2 (35 154 kg/cm^2) and strength retained up to 1100° C (2012° F) (**A51**).

SINTERED BERYLLIA PIECES

Characteristics. Highly refractory material, useful up to 2500° C (4532° F), a working temperature some 500° C ($\sim$ 930° F) higher than that of alumina. High thermal conductivity and therefore excellent heat–shock resistance, high stability to reducing agents, remains gas-tight at high temperatures, hard, good electrical insulator. Nuclear moderator with low thermal-neutron-absorption cross-section, one of the few efficient moderators available for use at high temperatures.

Unaffected by alkalies, reducing agents, oxidising agents, and metals but attacked by high temperature water vapour, acids and some oxides, Al_2O_3 and MgO.

Note: Dust or vapour of beryllium or its compounds is highly toxic if inhaled, causing serious lung conditions. Adequate control measures have, however, been developed (**B113**).

Uses. Nuclear power, as moderator. Nearest possible approach to theoretical density without the aid of fluxes is necessary (**H130**). Jet propulsion, because of excellent hot strength. Crucibles for melting pure metals, *e.g.* iron, nickel, platinum metals, beryllium, uranium (**A150**). Crucibles for glasses and slags. Cores for tungsten-wound furnaces. Insulation in high-frequency induction furnaces, inert to C and CO up to 2000° C (3632° F) therefore can be used as supports and shields with graphite heaters. Insulators in radio tubes.

Body Type. Single-component refractory.

Raw Materials. Beryl converted to high-purity beryllium sulphate or hydroxide and thence by controlled calcining to beryllia. High purity of materials and cleanliness throughout the operation is essential as the M.P. is rapidly lowered by other refractory oxides.

Preparation. *Controlled calcining* of the hydroxide to give beryllia of different grain sizes. Crystal growth becomes rapid at elevated temperatures owing to high vapour pressure of BeO, *e.g.* calcining to 1200° C (2192° F) gives 0·05–0·2 μ; 1300° C (2372° F) gives 0·05–0·2 μ plus 1·0–2·0 μ; 1400° C

(2552° F) gives few small plus many 2·0–4·0 μ; 1500° C (2732° F) gives all 2·0–4·0 μ (**D65**). The Battelle Institute have found that the most active powder is produced at 998° C (1830° F). The finer powders sinter more readily to high density than the coarser crystals but there is more shrinkage (**Q3, H130, H131**).

Milling and Mixing.
For tamping: 3 parts high-temperature calcine or fused grog plus 1 part low temperature calcined oxide, plus water; mix well. For complicated shapes add 4% solution $BeCl_2$ as binder.

For casting: 200-mesh fused oxide→wet ball mill at 60 r.p.m. for 25 hr until all finer than 50 μ, 80% finer than 6 μ, 25% finer than 3 μ→three acid treatments to remove iron picked up in mill→slip of pH 4·5–5, S.G. 1·9–2·0, water content 37–33%.

For dry pressing: dry ball mill powder, plus lubricant, *e.g.* 14% carbowax 4000 (polyethylene glycol of M.W. 4000) as 20% water solution, plus 0·1% antisticking agent. Mix→partially dry at 80° C→nodulate by rubbing through 14-mesh sieve→dry completely (**N35**); *or* mix BeO powder with camphor dissolved in ethanol, the latter is evaporated before pressing (**H130**).

Shaping. (1) Tamping in paper lined graphite moulds.

(2) Slip casting.

(3) Dry pressing in carbon–chrome steel mould, heat treated at 58 to 62 Rockwell C hardness. Hydraulic press at 20 000 to 30 000 lb/in^2 (1406–2109 kg/cm^2). Weighed amount of mix used to produce uniform size of piece.

Closer tolerances can be achieved by machining the pressed piece with tungsten carbide tools (**N35**).

(4) Hydrostatic pressing in de-aired mould at 100 000 lb/in^2 (7030 kg/cm^2) (**H130**).

(5) Hot pressing.

Firing. Beryllia must be fired separately from other ware to avoid its contamination and also the deposition of beryllia on other articles as it has a high vapour pressure.

Two-Stage Firing
 I. (*a*) Tamped ware heated in mould to 1100° C (2012° F) for 2 hr, then removed from mould.
   (*b*) Cast ware pre-fired to 1000° C (1832° F).
   (*c*) Pressed ware placed in closed insulating firebrick saggar and heated uniformly for 8 hr to maximum of 500° C (932° F) to remove wax. Process requires close control or cracking occurs.

 II. Preheated ware fired to 1800° C (3272° F) and held for 30 min. Heating period depends on size of piece. Small pieces 4 to 5 hr. (Strong but porous pieces are fired at 1500° C (2722° F).)

Catenary kilns of fused-alumina brick fired with gas and cold air. Alumina furnace fired with propane and oxygen (**N35**).

Single-Stage Sintering of Battelle active BeO Powder. Molybdenum-

LABORATORY AND ENGINEERING WARE

wound alumina-tube furnace in hydrogen atmosphere; or globar-heated porcelain-tube furnace in purified argon atmosphere or dried air. Gas-fired furnace. Dry pressed ware (20 000 lb/in², 1406 kg/cm²) of best powder sinters to density of 3·01 at 1536° C (2800° F). Hydrostatic pressed ware sinters to virtually theoretical density at 1370–1758° C (2500–3200° F) (**H130**).

SINTERED MAGNESIA

Characteristics. A plentiful refractory oxide, stable in oxidising conditions to 2796° C (5070° F). Its high thermal expansion makes it sensitive to thermal shocks and it reacts with carbon at elevated temperatures. In vacuum it volatilises rather rapidly above 2000° C (3632° F).
Uses. Crucibles for melting Ni, Fe, Cu, Pt, U and U alloys without contamination. Thermocouple swaging tubes.
Body Type. Single-phase refractory.
Raw Materials. High purity magnesia obtained by controlled calcining of pure compounds. 1–2% silica.
Preparation. Dry grinding. *For casting*, slip made in absolute alcohol (**T26**). *For extrusion* mix with cellulose acetate.
Shaping. Casting. Extrusion. Dry pressing. Hot pressing.
Firing.

SINTERED SPINEL (MAGNESIA–ALUMINA)

Characteristics. Resembles alumina in many respects but is more neutral at high temperatures when alumina reacts as an acid. Spinel has lower thermal shock resistance than alumina as it has low heat conductivity, its hardness and mechanical strength are also less (**K28**). Good resistance to basic slags and coal ash (**T18**).

TITANIA PIECES

Characteristics. Extreme hardness and good tensile, compressive and bending strengths which give good performance where resistance to shock, vibration and accidental damage are required. Low coefficient of friction.
Uses. Thread guides.
Body Type. Single component refractory.
Raw Materials. Titania. Organic binder, polyvinyl alcohol, methyl cellulose.
Shaping. Plastic pressing with organic binder. Hot pressing.
Firing.

SINTERED ZIRCONIA

Characteristics. Good chemical resistance at high temperatures (but *see* special list **S152**), low thermal conductivity, good thermal shock resistance, good resistance to coal ash slags. Becomes an electric conductor at high temperatures. Can be used up to 2400° C (4352° F) in oxidising and moderately reducing atmospheres.

Uses. Gas turbines. Liners for jet and rocket motor tubes. Facing of high temperature furnace walls. Resistance heater for furnaces at 2000° C (3622° F). Laboratory ware for high temperatures. Crucibles for melting metals. High S.G. Grinding Balls.

Body Type. Single-component refractory.

Raw Materials. Pure zirconia, plus CaO alone or with MgO to stabilise, or fused stabilised zirconia. Binders: aluminium chloride, magnesium chloride, zirconium hydroxide, zirconium chloride, dextrin, starch.

Preparation. Grinding→digest with acid; for extrusion: mix with binder and water; for pressing: mix with binder; for casting: bring to pH2 and 80% wt solids.

Shaping. (1) Extrusion. (2) Dry pressing 6000 lb/in^2 (422 kg/cm^2). (3) Slip casting. (4) Ramming and tamping (**T26**). (5) Hot pressing.

Firing. Cast: 1700–1750° C (3092–3182° F). Rammed: 1450° C (2642° F). Lowest porosities achieved by presintering and then sintering at 1800° C (3272° F) for 10 min in an inert atmosphere, *e.g.* argon in the presence of free carbon (**W2**).

THORIA

Characteristics. Extremely high refractoriness, stable in oxidising and slightly reducing conditions to (2996° C) 5430° F. Claimed to be unwetted by molten zirconium, thorium, iron, uranium, osmium. Most resistant of all pure oxide refractories to titanium and its alloys. Radioactive material producing radioactive gases that accumulate in a closed container and are a health hazard.

Uses. Extremely high-temperature uses in oxidising atmosphere or vacuum for melting refractory metals, etc.

Body Type. Single-component refractory.

Raw Materials. Electrically fused thoria.

Preparation. Calcining→wet ball milling→grading (→drying). *For tamping*, moisten with 20% solution of ThCl$_4$. *For casting*, deflocculate slip with P$_2$O$_5$.

Shaping. Tamping. Slip casting (**T26, M101**). Dry pressing. Hot pressing.

Firing. 1800–1900° C (3272–3452° F) (**N35**).

URANIUM DIOXIDE PIECES

Characteristics. Nuclear-fissionable material, dark brown colour, M.P. 2800° C (5072° F) but with considerable vapour pressure below this temperature. Relatively low thermal conductivity, but low specific heat, high thermal

LABORATORY AND ENGINEERING WARE

expansion. Becomes oxidised in air just above room temperature. No reaction with water, HCl or alkalies. Soluble in oxidising acids and bases. Some reaction with carbon. No reaction with hydrogen. No reaction with Al_2O_3, MgO, BeO, SiO_2. Solid solutions and/or compounds ThO_2, ZrO_2, CaO. Reactions with Si, Al, Nb.
Uses. Nuclear reactor fuel for use in very high-temperature reactors or high pressure water reactors.
Body Type. Single component refractory.
Raw Materials. Fine grain uranium dioxide powder.
Shaping. Cold pressing, plus dextrose or wax at 10 000 lb/in² (703 kg/cm²).

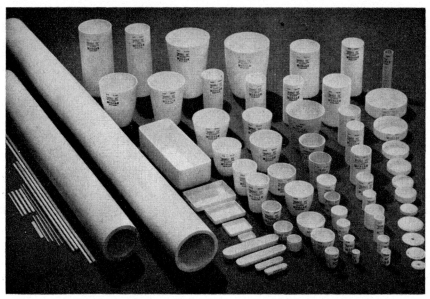

FIG. 15.1. *A selection of pure oxide ware (Morgan Crucible, **M96**)*

Hydrostatic pressing, no binder, 40 000 lb/in² (2812 kg/cm²). Slip casting, with HCl to pH2. Extrusion with organic binders. Hot pressing.
Firing. Up to 1700–1750° C (3092–3182° F). Hydrogen atmosphere required during the temperature range when U_3O_8 is stable. Air or water vapour may be introduced above 1200° C (2192° F) to regulate crystal growth.
Finishing. Grinding with diamond wheel rod grinder (**J52**).

THORIA–URANIA PIECES

Characteristics. A urania containing body that does not require firing *in vacuo* or in hydrogen.
Uses. Nuclear.

Body Type. Oxide refractory.
Raw Materials. Thoria, calcined to 1000° C (1832° F) and screened to −100-mesh. U_3O_8 obtained by oxidising UO_2 or U metal.
Preparation. Proportionating $ThO_2/U_3O_3 \rightarrow$ addition of carbowax dissolved in hot water→wet milling 2 hr→drying→addition of 2% water.
Shaping. Dry pressing.
Firing. Gas-fired kiln, 1700–1750° C (3092–3182° F). During the firing U_3O_8 decomposes to UO_2 and shrinks (**H14**).

ZIRCON WARE

Characteristics. Low coefficient of thermal expansion and high thermal conductivity giving excellent resistance to thermal shock. Resistant to slags, phosphates, metals and acid chemicals.
Uses. Melting of precious metals.

'LEAD' PENCIL

Characteristics. Replacing the old lead/tin alloy pencil in the sixteenth century, the graphite-filled pencil is still known as a 'lead pencil'. Originally natural graphite was used but the modern pencil is a ceramic material made of blends of graphite and clay which can be varied to give a large number of degrees of 'hardness' of the 'lead'.
Uses. Writing. Drawing.
Body Type. Carbon.
Raw Materials. High-quality natural graphite, *e.g.* from Ceylon, Madagascar or Siberia. Highly plastic fine-ground clay maturing at a relatively low temperature, *e.g.* that from Klingenberg, Main, Germany.
Preparation. Prolonged grinding of the graphite for whole weeks or months→mixing.←Clay undergoes prolonged fine preparation.
Shaping. Hydraulic pressing into long strands. Cutting. Straightening.
Drying.
Firing. ∼1200° C (2192° F). In grog saggars covered with graphite powder. Intermittent kilns. Continuous kilns.
Finishing. Impregnation with wax finally determines the hardness on the scale:

Soft ←——————————————————————————→ Hard
EEB, EB, 6B, 5B, 4B, 3B, 2B, B, HB, H, 2H, 3H, 4H, 5H, 6H, 7H, 8H, 9H.

Setting in wood. Best is American cedar, for cheaper pencils dyed and softened alder, lime or ash are used. Varnishing. Stamping (**S210**).

CERMET PIECES

Characteristics. A wide range of properties combining metallic ductility, thermal shock resistance and electric conductivity with ceramic refractoriness, hot strength, oxidation resistance and electric insulation in varying degrees.

Uses. 'Present and potential', after Westbrook (**W43**): Lamp filaments. Gas turbine parts: aircraft jet engines; turbo supercharger; locomotive, marine and stationary turbines. Rocket engine components. Contact materials. Metal oxide magnetics. Cutting, drilling and grinding tools. High-power cathodes. Metal–ceramic vacuum seals. Hard and abrasion resistance applications. Friction parts. Spark plug bodies and electrodes. Laboratory ware. Welding electrodes. Printed circuits. Welding and brazing fixtures. Nuclear power applications. Recoilless rifle parts. Thermoelectric generators. Parts for high-temperature instrumentation. Pumping of liquid metals. Heating elements. Resistors. Current collector brushes. Bearings. Magnetic cores.

Body Type. Cermet.

Raw Materials. Pure metals. Pure oxides, carbides, etc., *see* 'Bodies'.

Preparation. Ball milling.

Shaping and Firing. As listed by Westbrook (**W43**): "Fragmentation pressing. Hydrostatic pressing. Hot pressing. Pressing and sintering. Extrusion. Suspension spraying. Flame spraying. Slip casting. Melting and casting. Dipping. Hot forming. Sintering *in situ*. Cataphoresis. Electroplating. Salt deposition. Chemical reaction. Injection moulding. Vapour deposition." Vapour deposition and sintering (**A69**). Low-frequency vibratory compacting (**B30**). Many of these are specialised methods covered by patents or by security regulations. Available literature is listed by the U.S. Department of Commerce in their *Catalog of Technical Reports, Cermets*.

CHROMIUM–ALUMINA CERMET

Characteristics. High refractoriness, good impact and thermal shock resistance, high modulus of rupture, slow oxidation during service.

Body Type. Cermet.

Raw Materials. Electrolytic chromium. Corundum. Aluminium hydrate.

Preparation. Grinding with tungsten carbide balls in alcohol→mixing.

Shaping. Dry pressing→hydrostatic pressing 55 000 lb/in^2 (3867 kg/cm^2).

Firing. 1703° C (3100° F) for 30 hr in hydrogen atmosphere with controlled amounts of water vapour introduced as an oxidising agent (**B72**).

TABLE
Properties of Pure Refractory

Class	Refractory oxides				
Name	Alumina	Beryllia	Thoria	Magnesia	Zirconia
Formula	Al_2O_3	BeO	ThO_2	MgO	ZrO_2
Molecular wt.	101·92	25·02	264·12	40·32	123·22
S.G.	3·97	3·00	9·69	3·58	6·27
Hardness M, Mohs	9M	9M		6M	6·5M
K, Knoop	2000 K	1250 K	6·5 M		1160 K
Tensile strength; (Mlb/in^2)	36 (20° C) 33 (500° C) 18 (1300° C)	14 (20° C) 8 (500° C)			18 (20° C)
($kg/cm^2 \times 10^3$)	2·53 (20° C) 2·32 (500° C) 1·26 (1300°C)	0·63 (1200° C) 0·98 (20° C) 0·56 (500° C) 0·04 (1200° C)			1·26 (20° C)
Compressive strength; (Mlb/in^2)	413 (20° C) 70 (500° C) 20 (1200° C)	103 (20° C) 60 (500° C) 25 (1200° C)	220 (20° C) 50 (500° C) 25 (1200° C)	112 (20° C)	303 (20° C) 230 (500° C) 100 (1200° C)
($kg/cm^2 \times 10^3$)	29·0 (20° C) 4·92 (500° C) 1·41 (1200° C)	7·24 (20° C) 4·22 (500° C) 1·76 (1200° C)	15·47 (20° C) 3·52 (500° C) 1·76 (1200° C)	7·87 (20° C)	21·30 (20° C) 16·17 (500° C) 7·03 (1200° C)
Young's modulus of elasticity; (Mlb/in^2)	52 (20° C) 50 (800° C) 32 (1400° C)	45 (20° C) 28 (800° C) 12 (1400° C)	21 (20° C) 18 (800° C) 14 (1400° C)	12 (20° C)	36 (20° C) 18 (800° C) 14 (1400° C)
($kg/cm^2 \times 10^3$)	3·65 (20° C) 3·52 (800° C) 2·25 (1400° C)	3·16 (20° C) 1·97 (800° C) 0·84 (1400° C)	1·48 (20° C) 1·26 (800° C) 0·98 (1400° C)	0·84 (20° C)	9·56 (20° C) 1·26 (800° C) 0·98 (1400° C)
M.P. (°C)	2050	2530±30	3050	2800	2690±20
(°F)	3722	4586	5522	5072	4874
B.P. (°C)	2980	3900	4400	2830	4300
(°F)	5396	7052	7952	5126	7772
Sp.ht. (cal/g)	0·21 (20° C) 0·25 (500° C) 28 (1000° C)	0·24 (20° C) 0·50 (500° C)			0·12 (20° C) 0·16 (500° C)
(B.t.u)	0·378 (20° C) 0·450 (500° C) 50·4 (1000° C)	0·432 (20° C) 0·90 (500° C)			0·216 (20° C) 0·288 (500° C)
Linear coeff. thermal exp. °C($\times 10^{-6}$)	6 (20° C) 7 (500° C) 9 (1000° C)	8 (100° C) 8 (500° C) 9 (1000° C)	8 (100° C) 9 (500° C) 9 (1000° C)	11 (100° C) 13 (500° C) 15 (1000° C)	11 (100° C) 13 (500° C) 15 (1000° C)
Coeff. thermal conductivity (B.t.u) kcal-m/m²-hr-°C	17 (20° C) 6 (500° C) 4 (1000° C) 3·978 (20° C) 1·404 (500° C) 0·936 (1000° C)	10·4 (1200° C)		36 (20° C) 12 (500° C) 8·424 (20° C) 2·808 (500° C)	
Resistivity (Ω/cm³)	$2 \times 10^{(15)}$ (100° C) $2 \times 10^{(13)}$ (300° C)	$8 \times 10^{(13)}$ (1000° C) $3·5 \times 10^{(10)}$ (1600° C) 8×10^8 (2100° C)	4×10^{13} (20° C) $1·2 \times 10^{12}$ (500° C)	9×10^{12} (1000° C) 7×10^9 (1600° C) 2×10^8 (2100° C)	
Thermodynamic data:					
F 298° K	−365·9	−132·0		−138·43	−246·0
H 298° K	−388·1	−139·0		−146·1	−258·8
S 298° K	−74·58	−23·42		−25·73	−41·8
Solubility H_2O	sl.s.	sl.s.	i	sl.d.	i
Acids	v.sl.s.	s. conc. H_2SO_4 i.dil	s.h. H_2SO_4 i.dil	s	s. H_2SO_4
Alkalies	v.sl.s.	s. fus. KOH i.dil	i		

Compounds (Norton (N35))

	Titania TiO_2 79.90 4.17 5.5–6.5	Chrome oxide Cr_2O_3 152.02 5.21	Lime CaO 56.08 3.32 4.5M	Baria BaO 153.37 5.72 3.3M	Strontia SrO 103.63 4.7 3.5M	Refractory carbides Boron Carbide B_4C 55.28 2.51 9M 280K
					414 (20° C) 29.11 (20° C)	
	1830 3326	2435±50 4415 3000 5432	2572 4661 2850 5162	1923 3493 2200 3992	2430 4406 3000 5432	2350 4262 3500 6332
	7 (500° C)	7–12 (500° C)				4.5 (20–800° C)
	11.7×10^3 (900° C) 7.5×10^3 (1000° C) 4.4×10^2 (1300° C)	12.7×10^2 (350° C) 7.5×10^1 (750° C)		1×10^6 (300° C) 2.2×10^1 (500° C)		0.45 (20° C) 0.02 (500° C)
	−205.1 −218.0 −43.23 i s. H_2SO_4 i s	−251.1 −270.6 −65.51 i i i	−144.3 −151.7 −24.96 d.s s	d.s s s	−12.5 d	i s. fus.

TABL[E]

Class	Refractory carbides—(contd.)			
Name	Silicon carbide	Tungsten carbide	Titanium carbide	Vanadium carbide
Formula	SiC	WC	TiC	VC
Molecular wt	40·06	195·93	59·91	62·96
S.G.	3·17	15·7	4·25	5·4
Hardness M, Mohs	9–9·5M		8·9M	
K, Knoop	255K	188K	247K	208K
Tensile strength (Mlb/in^2) ($kg/cm^2 \times 10^3$)				
Compressive strength (Mlb/in^2) ($kg/cm^2 \times 10^3$)	82 (20°C) 5·76 (20° C)		109 (20° C) 7·66 (20° C)	89 (20° C) 6·26 (20° C)
Young's modulus of elasticity (Mlb/in^2) ($kg/cm^2 \times 10^3$)				
M.P. (°C)	2700	2777	3140 ± 90	2830
(°F)	4892	5031	5684	5126
B.P. (°C)	2340	6000	4300	3900
(°F)	4244	10 832	7772	7052
Sp.ht (cal/g) (B.t.u.)				
Linear coeff. thermal exp./°C ($\times 10^{-6}$)				
Coeff. thermal conductivity (B.t.u.) (kcal-m/m²-hr-°C)				
Resistivity (Ω/cm^3)	0·5 (20° C)			$1·6 \times 10^{-1}$ (20° C)
Thermodynamic data:				
F 298° K	−26·1		−83·1	
H 298° K	−26·7		−102·2	
S 298° K	−2·0	−1·7	−3·3	
Solubility H_2O	i	i	i	i
Acids	i		s HNO_3	s HNO_3
			s HNO_3 + HCl	i $HCl.H_2SO_4$
Alkalies	d fus KOH			s. fus. KNO_3

Class	Refractory nitrides—(contd.)			Refractory borides	
Name	Tantalum nitride	Vanadium nitride	Silicon nitride	Titanium boride	Zirconium boride
Formula	TaN	VN	Si_3N_4	TiB_2	ZrB_2
Molecular wt.	194·89	64·96	140·21		
S.G.		5·63	3·44	4·50	6·08
Hardness;					
M, Mohs					
K, Knoop				271K	175K
Tensile strength; (Mlb/in^2) ($kg/cm^2 \times 10^3$)					
Compressive strength; (Mlb/in^2) ($kg/cm^2 \times 10^3$)					
Young's modulus of elasticity (Mlb/in^2) ($kg/cm^2 \times 10^3$)					
M.P. (°C)	3360	2050	2170 sub	2900	3060
(°F)	6080	3722	3938	5252	5540
B.P. (°C)			1900		
(°F)			3452		
Sp.ht. (cal/g) (B.t.u)					
Linear coeff. thermal exp. °C ($\times 10^{-6}$)					
Coeff. thermal conductivity (B.t.u.) (kcal-m/m²-hr °C)				0·06	0·057
Resistivity (Ω/cm^3)				$17^{10·3}$	$10^{10·3}$
Thermodynamic data:					
F 298° K	−52·2	−35·1	−154·7		
H 298° K	−58·1	−41·4	−179·3		
S 298° K	−19·9	−21·4	−81·6		
Solubility H_2O	i	i	i		
Acids			s H.F.		
Alkalies			i.a.		

246—(contd.)

Molybdenum carbide Mo_2C 203·91 8·9	Refractory nitrides				
	Titanium nitride TiN 61·91 5·29	Zirconium nitride ZrN 105·22	Boron nitride BN 24·83 2·25	Thorium nitride Th_3N_4 152·39	Beryllium nitride Be_3N_2 55·08
2380 4316 4500 8132	2900 5252	2950 5342	2730 4946 sub	2100 3812	2200 3992 2240 d 4064
2·8 4·2 4·8	−283·7 −310·4 −84·0 i i s aq. reg. s	−75·6 −82·2 −22·0	−27·7 −31·5 −12·9 i d HCl, HF, H_2SO_4	−283·7 −310·4 −79·6 d s	−121·4 −133·5 −40·6 d i d conc

(N27)

Vanadium boride VB_2 5·10	Tantalum boride TaB_2 12·38	Chromium boride CrB_2 5·15	Molybdenum boride MoB_2 7·12	Thorium boride ThB_4 8·45	Uranium diboride UB_2 12·70	Uranium tetraboride UB_4 9·32
—	—	180K	128K	—	139K	—
2100 3812	3000 5432	1850 3362	2000 3632	>2500 >4532	2360 4280	<2400 <4352

TABLE

Single-component

Ref. Body	Carbides				
	N27 Boron carbide	G3 Boron carbide	G3 SiC 85% B_4C 15%	N27 Titanium carbide	G3 Titanium carbide
Making method	Hot pressed	Hot pressed	Hot pressed	Hot pressed	Hot pressed
Structure	Rhombohedral			Cubic	
M.P. (°C)	2450			3140	
(°F)	4442			5680	
Density (g/ml)	2·51	2·50	3·00	4·93	4·74
% density of theoretical		99·2			96·5
Knoop hardness (K100)	2800			2470	
Moh's hardness					
Bending strength (lb/in²)	44 000			68 000	
(kg/cm²)	3094			4781	
Compression strength (lb/in²)	414 000			400 000	
(kg/cm²)	29 107			28 123	
Thermal conductivity c.g.s. (assumed cal-cm/cm²-sec-°C)	0·065			0·041	
(B.t.u.)	[188]			[119]	
Electrical resistivity (micro-Ω-cm)	$3-8 \times 10^5$			70 105 193	
Young's modulus (10^{12} dyn/cm²)	4·5				
Thermal expansion $\times 10^{-6}$/°C	4·5 (20–800° C)	2·61 (24–590° C)	2·34 (24–590° C)	—	4·12 (24–590° C)
Short-time tensile strength at 980° C (lb/in²)		22 550	9950		15 850–17 200
(kg/cm²)		1585	699		1114–1209
1200° C (lb/in²)			7100		8000–9400
(kg/cm²)			499		562–661
Thermal shock resistance, No. of cycles before failure from 980° C		¼	25		25
1090° C			25		25
1200° C			2		25
1310° C					14–21
Porosity	0			3	
Top temperature of use					
(°C)					
(°F)					
Tensile strength (lb/in²)					
(kg/cm²)					
Impact bending strength (ft-lb/in²)					
(Charpy) (cm-kg/cm²)					

Ware

		Nitride	Oxides			
N27 Zirconium carbide Hot pressed Cubic	G3 Zirconium carbide Hot pressed	C8 Boron nitride	G32 Alumina	R75 Alumina Extruded	N27 Dense Alumina	N27 Hot pressed Alumina
3540 6340 6·73	6·30 97·8	2·10	3·4		2000 3632 3·1	2035 3695 3·86–3·99
2090		2	9·0 38 000 2672			2050 22 000 1547
22 500 1582 238 000			100 000	@20° C @1500°C 420 000 14 000		373 000
16 733 0·049			7031	29 529 984	[0·006] 1100° C 17	26 224 [0·010] 1100° C estimated 30
[142] 50 63 75 4·8 3·4 —	3·74 (24–590° C)	high	5·5 (25–100° C) 7·3 (25–700° C)			5·2 7–9

	11 700–14 450
	823–1016
	12 950–15 850
	910–1114
	22 (oxid.)
		excellent		20° C 37 100
				         2608
				1300° C 6500
				         457		

5		Inert Oxid 1650 980 3000 1800	1160 2120 14 000 984 7·0 0·0014		22 1950 3542	0–3 1900 3452

TABLE

Ref. Body	Oxides—(contd.) N27 Insulating alumina	R75 Beryllia	G3 Beryllia	N27 Dense magnesia	G3 Magnesia
Making method			Hot pressed		Hot pressed
Structure					
M.P. (°C)	2000			2700	
(°F)	3632			4892	
Density (g/ml)	1·9		2·96	2·8	3·39
% density of theoretical			97·7		93·6
Knoop hardness (K100)					
Moh's hardness					
Bending strength (lb/in²)					
(kg/cm²)					
Compression strength (lb/in²)		20° C 1500° C			
		112 000 16 800			
(kg/cm²)		7874 1181			
Thermal conductivity c.g.s.	[0·003]			[0·006]	
(assumed cal-cm/cm²-sec-°C)	1100° C			1100° C	
(B.t.u.)	9			18	
Electrical resistivity (micro-Ω-cm)					
Young's modulus (10^{12} dyn/cm²)					
Thermal expansion × 10^{-6}/°C					6·94 (24–590° C)
Short-time tensile strength at					
980° C (lb/in²)			5900–6100		3100
(kg/cm²)			415–429		218
1200° C (lb/in²)		1300° C 630			
(kg/cm²)		44			
Thermal shock resistance, No. of					
cycles before failure from 980° C			25		½
1090° C			3		
1200° C					
1310° C					
Porosity	53			21	
Top temperature of use (°C)	1850			2300	
(°F)	3362			4172	
Tensile strength (lb/in²)					
(kg/cm²)					
Impact bending strength (ft-lb/in²)					
(Charpy) (cm-kg/cm²)					

TABLE

Properties of some

Ref. Oxide system	L6 BeO–Al₂O₃–ZrO₂	G23 BeO–MgO–Al₂O₃	L5 BeO–MgO–ZrO₂
Maturing range (°C)	1500–1600	1500–1725	normal porcelain kilns
(°F)	2732–2912	2732–3137	
Apparent density	2·9–3·4		
Shrinkage (%)	17·8–20·5	9·7–20·5	
Hardness	Knoop (500 g load) 550–830		
Compressive strength (lb/in² × 10³)	238–305	98·56–168	up to 266
(kg/cm² × 10³)	16·7–21·4	6·9–11·8	up to 18·7
Transverse strength	after thermal		
(lb/in² × 10²)	shock 982° C		20° C 982° C
	178–319 151–251		390 320
(kg/cm² × 10²)	12·5–22·4 10·6–17·6		27·4 22·5
Young's modulus (lb/in² × 10⁶)	@ 982° C 28–38		
(kg/cm²)	2·0–2·7		
Thermal shock resistance	good		fair to good
Thermal expansion range		700–800° C	
value		9·4–10·8	
Thermal conductivity (temperature)			
(B.t.u.)			
(kcal m/m² hr °C)			

247—(contd.)

R75 Thoria	N20 Titania	R75 Zirconia	G3 Stabilised zirconia Hot pressed	N27 Dense stabilised zirconia	N27 Insulating stabilised zirconia	R75 Spinel	G3 Zircon Hot pressed
	4·0 8		5·80 95·1	2650 4802 4·2	2650 4802 2·8		4·54 96·2
20° C 1500° C 210 000 1400 14 764 98	130 000 ~9000 0·009 [26]	20° C 1500° C 294 000 2800 20 670 197		[0·002] 1100° C 6	[0·001] 1100° C 4	20° C 1500° C 266 000 14 000 18 702 984	
	6–8		4·95 (24–590)				2·24 (24–590° C)
			5250–6750 369–475			20° C 18 900 1329 1300° C 1120 79	6200–8700 436–612 3600 253
			0				1
	nil			25 2500 4532	50 2000 3632		
	8000 560 2·4 5·1						

48

High Beryllia Bodies

	L5 BeO–MgO–ThO₂	L7 BeO–Al₂O₃–TiO₂	S29 BeO–Al₂O₃– ZrO₂	S29 BeO–Al₂O₃– ZrO₂+MgO	S29 BeO–Al₂O₃– ThO₂	S29 BeO–Al₂O₃– ThO₂+MgO
	some very refractory	1500–1700 2732–3092 3·3–3·7 11–19	1500–1650 2732–3002	1400–1550 2552–2822	1500–1650 2732–3002	1400–1550 2552–2822
	up to 266 up to 18·7	187–280 13·1–19·7				
	20° C 982° C 140 160 9·8 11·2	20° C 982° C 137–250 105–170 9·6–17·6 7·4–11·9 42–47 22–41 2·9–3·3 1·5–2·9				
	poor	poor 25–950° C 0·81–0·89				
			mean 54° C ——— 150° C 93·34—33·26 11·57— 4·12		mean 56° C ——— 143° C 85·36—38·60 10·58— 4·79	

37

TABLE
Physical Properties of some

Composition	Specific gravity (g/cm³)	Rockwell hardness A scale	Transverse rupture (lb/in²) (kg/cm²)		Young's modulus[2] (lb/in² × 10⁶) (kg/cm² × 10⁶)	
97% WC 3% Co	15·25	92·7	170 000	11 952	97·5	6·85
95·5% WC 4·5% Co	15·05	92·3	200 000	14 061	90·5	6·36
94% WC 6% Co	14·85	90 to 92	225 000	15 819	88	6·19
91% WC 9% Co	14·60	89·5 to 91·5	275 000	19 334	83	5·84
87% WC 13% Co	14·15	87·5 to 90	300 000	21 092	80	5·62
80% WC 20% Co	13·55	85 to 87	350 000	24 607	73	5·13
Predominantly WC with TaC and 13% Co	13·90	87 to 88	275 000	19 334	88	6·19
Predominantly WC with TaC and 6% Co	14·70	91 to 92	220 000	15 468	91	6·40
Predominantly WC larger amount of TiC, 7% Co	9·00	92 to 93	150 000	10 546	88	6·19
Predominantly WC with TaC and TiC, 8% Co	11·7	91·5 to 92·5	165 000	11 601	72	5·06
Predominantly WC with TaC and TiC, 11% Co	11·6	90·5 to 91·5	175 000	12 304	81	5·69

1. Values given in this table are representative of properties obtained in good production practice. They are not necessarily the highest obtainable, nor do they represent the lowest which are of practical use.
2. Most values for Young's modulus were obtained by W. H. Davenport of the Norton Co., by the musical pitch method.
3. Most values for compressive strengths are given through the courtesy of P. W. Bridgman of Harvard University.

Properties. (of chromium–alumina cermet, p. 1145)

Firing shrinkage	13·0–14·5%.
Apparent specific gravity	4·60–4·65.
Apparent porosity	<0·5%.
True specific gravity	4·68–4·72.
Thermal expansion	77–1472° F; 4·8 × 10⁻⁶ (25–800° C; 8·65 × 10⁻⁶).
	72–2400° F; 5·25 × 10⁻⁶ (25–1315° C; 9·45 × 10⁻⁶).
Heat transfer	Slightly < sintered alumina.
Thermal conductivity	66·5 ± 20% B.t.u.
Resistance to thermal shock	Good
10 cycles at 2400° F (1314° C)	15–50% gain in strength.
Resistance to oxidation	Excellent up to 2750° F (1453° C).
Hardness	1100–1200 (Vickers pyramid number).
Modulus of elasticity, 75° F (24° C)	5·23 × 10⁷
Impact resistance at 75° F (24° C)	1·05 in.-lb (121 cm-kg) 18·95 in.-lb/in² of area (0·049 cm-kg/cm²).
Compressive strength 75° F (24° C)	320 000 lb/in² (22 498 kg/cm²).
Bending strength (bars) 75° F (24° C)	55 000 lb/in² (3867 kg/cm²)
1600° F (870° C)	43 125 lb/in² (3032 kg/cm²)
2000° F (1092° C)	32 800 lb/in² (2306 kg/cm²)
2400° F (1314° C)	24 400 lb/in² (1715 kg/cm²).
Tensile strength (rods) 75° F (24° C)	35 000 lb/in² (2461 kg/cm²)
1600° F (870° C)	21 560 lb/in² (1516 kg/cm²)
2000° F (1092° C)	18 490 lb/in² (1300 kg/cm²)
2400° F (1314° C)	14 120 lb/in² (993 kg/cm²).

249
Commercial Metal Bonded Carbides

Compressive strength[3] (lb/in²) (kg/cm²)		Proportional limit in compression[4] (lb/in²) (kg/cm²)		Impact strength[5] (ft-lb) (m-kg)		Endurance limit[6] (lb/in²) (kg/cm²)		Coefficient of thermal expansion[7]
815 000	57 300	780 000	54 839					
890 000	62 573	740 000	52 027					
750 000	52 730	600 000	42 184	0·73	0·101	95 000	6679	5·0 × 10⁻⁶
685 000	48 160	540 000	37 966					
625 000	43 942	525 000	36 911	1·10	0·152	105 000	7382	5·9 × 10⁻⁶
550 000	38 669	425 000	29 880	1·75	0·242			
610 000	42 887	475 000	33 396					7·25 × 10⁻⁶
752 000	52 871	670 000	47 106	0·65	0·090	85 000	5976	
725 000	50 973							7·0 × 10⁻⁶
720 000	50 621							6·75 × 10⁻⁶
680 000	47 809			0·60	0·083	85 000	5976	6·0 × 10⁻⁶

4. Proportional limit in compression is the load per unit area at which the increase in strain ceases to be directly proportional to the increase in stress.
5. Impact values are from unnotched specimens of approximately ¼ in. square section. Charpy machine was used.
6. Values for endurance limits are based on 20 000 000 cycles, for specimens of R. R. Moore rotating beam type.
7. Average coefficient of expansion per °C for the range 20 to 700° C (68 to 1290° F).

CEMENTED CARBIDE OR HARD METAL OR METAL-BONDED CARBIDE TOOLS AND SHAPES

Characteristics. Extremely hard pieces suitable for cutting tools giving long life and good wear resistance also resisting abrasive powders, etc., compared with hard steels the tensile strength is low, brittleness high and therefore the method of applying the tools must be different (**S42**).

Uses. Cutting tools for turning, drilling, blanking most materials, including masonry, glass, porcelain, concrete, carbon, marble, slate, limestone, asbestos, tile, plaster, etc., many metals where the chip is short. Less satisfactory with long curling chip (**S42**). Jiggering, jolleying and turning tools for ceramic manufacture. Wear parts for cams and dies in manufacture of wire, nails, tubing, rivets, etc. Metal-forming dies, especially for aluminium. Thread guides, especially for nylon, etc. Matrix for diamonds for wheel dressers, cone drills, centre laps, etc.

Body Type. Cermet.

Raw Materials. Tungsten carbide. Tantalum and titanium carbides. Cobalt. Organic binder (waxes, etc.).

Preparation. Grinding→grading→mixing.

Shaping. Dry pressing in carbide faced dies at 15 to 30 tons/in² (2362–4724 kg/cm²).

Firing. Presintering 650° C (~1200° F). Electric furnace, hydrogen atmosphere. Sintering 1480° C (~2700° F). Electric furnace, hydrogen atmosphere. Shrinkage 40% by volume, 15–20% linear.

Finishing.

(1) Brazing of tips or facings on to steel holders, shafts, etc., using oxy-acetylene torch and silver solder, or shrinkage on steel on to tool, forging, etc., or electronic induction.

(2) Grinding to final shape:

    (*a*) rough grinding silicon carbide wheels 60 to 80;
    (*b*) finish grinding silicon carbide wheels 100 to 120;
    (*c*) finish lapping diamond impregnated wheels 200 to 240.

ABRASIVE GRINDING WHEELS

Characteristics. Ceramic-bonded abrasive grain gives the most satisfactory all-purpose grinding wheel of high strength, high modulus of elasticity, good refractoriness (operating temperatures may be 1000° C, 1832° F).

Uses. Grinding of metals, ceramics.

TABLE 250

Properties of Abrasive Grinding Wheels

(Ricke and Haeberle (R25))
c.g.s. units

Percentage Binder	Properties	Range of values for low fired (cone 1) binders	High fired cone 10 binder
15	Hardness	1·5–2·2	2·2
	Modulus of elasticity	2920–3820	3490
	Bending strength	200–260	182
	Tensile strength	110–130	79
	Compressive strength	420–1000	770
	Impact bending strength	0·54–0·72	0·51
20	Hardness	1·0–1·5	1·5
	Modulus of elasticity	3910–4570	4020
	Bending strength	260–340	200
	Tensile strength	120–145	90
	Compressive strength	540–1210	—
	Impact bending strength	0·59–0·78	0·57
25	Hardness	0·8–1·25	0·88
	Modulus of elasticity	3930–5360	4670
	Bending Strength	245–350	265
	Tensile strength	110–180	130
	Compressive strength	1000–1570	1160
	Impact bending strength	0·57–0·89	0·60

Body Type. Bonded abrasive.
Raw Materials. Abrasive: corundum; silicon carbide. Bond: clay; feldspar; quartz; sodium silicate; frits; Fe_2O_3, $CaCO_3$, ZnO, borax, $MgCO_3$; various phosphates.
Preparation. Bond prepared by normal fine ceramic methods and dried. Weighing of bond and abrasive grain→mixer.
Shaping. Hydraulic dry pressing. Vibration. Casting.
Drying. Batch dryers at uniform temperature for dry-pressed wheels. Cast wheels in humidity dryers or slow heating up.
Shaving.
Firing. Temperature according to bond type. Very slow firing in order to release all internal stresses. Circular down-draught coal-fired kilns, saggars required. Electric intermittent kilns. Tunnel kilns, direct-fired or muffle, 5 to 6 days.
Finishing. Trueing. Balancing. Testing. Grading (**A7, 15**).

SILICON CARBIDE, CORUNDUM OR EMERY GRINDING WHEELS

Body Type. Bonded abrasive.
Raw Materials. Graded abrasive (80%): silicon carbide, corundum, emery. Bond clay, 6·3%. Feldspar 13·7%. Sulphite lye.
Preparation. Dry mixing→addition of water and sulphite lye→further mixing.
Shaping. Hydraulic press, 40 atm.
Drying. Steam-heated racks.
Firing. 1300° C (2372° F). Intermittent round or rectangular, 130–150 hr cycle (**A9**).
Finishing. Dressing and trimming.

FUSED ALUMINA GRINDING WHEELS

(Fig. 15.2)

Characteristics. Suitable for grinding materials of high tensile strength. Available with a variety of grades and grain sizes of the abrasive imbedded in more or less vitrified bond. Porous wheels are faster and cooler cutting and allow more space for chip clearance. Dense wheels are stronger, hold their shape more accurately and wear longer.
Uses. Grinding.
Body Type. Bonded abrasive.
Raw Materials. One of the types of fused alumina grain. Regular: most work. High titania: snagging wheels for heavy stock removal and in

coated products. Finely crystalline: billet grinding and snagging. Pure white: hard, heat sensitive, high alloy tool steels. Clay. Glass frits.
Preparation. Grinding→grading of each raw material. Mixing→water addition.
Shaping. Dry pressing.
Drying.
Firing. Intermittent kilns. Tunnel kilns. Temperature depending on bond mixture.

FIG. 15.2. *Ceramically bonded abrasive wheels in a variety of shapes and specifications (Carborundum, C10)*

Finishing. Dressing by disking or dressing with bell cutters, star dressers, diamond dressers, etc. Trueing. Balancing. After testing, marking with grit, grade, structure, bond type.
Testing. Speed testing. Degree of hardness (**C8a**).

CERAMIC CUTTING TOOLS

Characteristics. High hardness and compressive strength which are retained at elevated temperatures. Compared with metals tensile strength is low and the tools are brittle. Oxide tools do not wet metals and have low friction between oxide and metal. High resistance to abrasion and chemical attack. Dimensional stability (**S219c**). Cutting can frequently be performed at much higher speeds than with 'hard metal' (**S219b**).

Uses. Cutting tough materials, *e.g.* plastics; rubber; wood. Cutting metals, *e.g.* aluminium. B_4C: graphite. Al_2O_3: green and biscuit ceramics (**S219b, c**).

Bodies. Sintered high-purity alumina. Boron carbide. B_4C/SiC. B_4C/TiB_2. SiC re-crystallised (**S219c**). Glass-bonded alumina. Alumina with various additives (**S219f**).

Preparation and Shaping.

Alumina: aluminium hydroxide→controlled calcination →alumina or fused alumina →grinding.

(1) Add water-soluble wax or a wax emulsion→dry pressing 10 000 lb/in² (703 kg/cm²); or pressing plus vibration.

(2) Milling to 10 to 15 μ→acid treatment to remove iron and leave at pH 3 →slip casting.

(3) Hydrostatic pressing (100 000 lb/in², 7030 kg/cm²).

(4) With wax, etc., extrusion.

(5) Hot pressing (**S219a**).

Firing. *Alumina*: 1800–1900° C (3272–3452° F).

Finishing. The brittleness of ceramic tools requires different designs for their shape, cutting angle and tool holders. The cutting angle should be close to 90°, and it is preferable to provide a narrow face on the upper surface to produce a negative cutting angle.

The tool holder must be designed to give maximum support and the clamping forces adjusted to distribute the stress evenly (Fig. 15.3). The lathe also should be free of vibration and cutting be uninterrupted and either completely dry or with sufficient coolant (**S219b, f**).

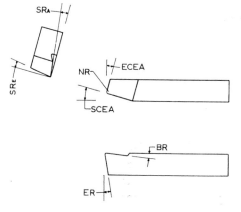

Fig. 15.3. Single-point ceramic cutting tool angles (see Table 251). (*Watertown Arsenal—Rodman Lab.*, **S219f**)

BR, *back rake.*
ER, *end relief.*
SCEA, *side cutting edge angle.*
ECEA, *end cutting edge angle.*
SR_A, *side rake.*
SR_E, *side relief.*
NR, *nose radius.*

Various brazing techniques are also described (**S219b**) and if temperatures are not too high the bit can be bonded to the shank with 'Araldite' (**S219c**) (*see also, Ceramics for Machine Tools* (**G48**)) (Fig. 15.4).

TABLE 251

Utilisation of Ceramics for Metal Cutting Tools
Tool Geometry* (S219f)

Nomenclature	No. 1 Grind	No. 2 Grind	No. 3 Grind
Side cutting edge angle	30°	30°	10°
End cutting edge angle	10°	10°	5°
Back rake angle	0°	−10°	0°
Side rake angle	10°	−10°	10°–7° land
Side clearance angle	10°	10°	10°
End clearance angle	10°	10°	10°
Nose radius (in.)	$\frac{1}{32}$	$\frac{1}{32}$	$\frac{1}{16}$

* See Fig. 15.3 for Nomenclature.

TABLE 252

1954 Turning Tests on Graphite (S219c)

Tool Bit Material	Lineal Inches Turned
SiC/B$_4$C 10/90 wt.%	1950
50/50	1365
60/40	1560
85/15	1443
B$_4$C Regular	1911
B$_4$C/TiB$_2$ (20 vol. %)	1755
Cubic SiC (Alfred)	780
Hexagonal SiC (Norton)	702
Cemented Tungsten Carbide	625
Circle C H.S. tool steel	390

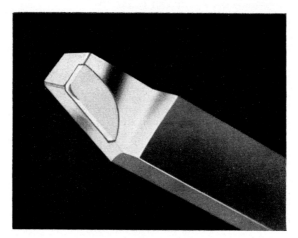

Fig. 15.4. (a) *Sinter alumina cutting tool (B.S.A. tools,* **B129** *and Lodge Plugs,* **L39**).

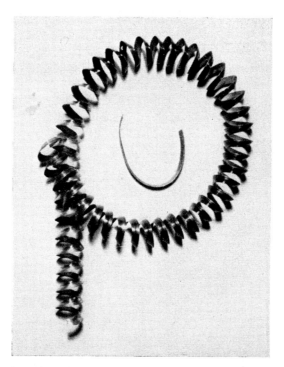

Fig. 15.4. (b) *Steel turnings cut with sinter alumina tool. Maximum cutting $\frac{1}{32}$ in. thickness, 1000 ft per min.*

TABLE 253

Utilisation of Ceramics for Metal Cutting Tools (S219f)
A. Ceramics Investigated

Specimen No.	Material
S-1008	Aluminium oxide with 12% glass.
S-1047	High purity aluminium oxide.
S-1048	Aluminium oxide with chromium oxide.
S-1049	Aluminium oxide with 13% additions.*
S-1050	Aluminium oxide with 5% additions.*
S-1051	Aluminium oxide with 1·5% additions.*
S-1052	High purity aluminium oxide.
S-1053	Metal bonded silicon carbide.
S-1054	Silicon carbide impregnated with alloy steel.
S-1055	Mixed boron and silicon carbides.
S-1056	High purity boron carbide.
S-1057	Metal bonded boron carbide.
S-1058	Aluminium oxide with glass.
S-1059	Aluminium oxide with glass.

*Additions unidentified at the writing of this report.

B. Analyses and Hardnesses of Workpieces

Workpiece material	Composition* (%)					Brinell hardness†
Nonferrous	Cu	Zn	Pb	Sn	Fe	
Brass–commercial grade	62·0 Max.	30·0 Min.	3·0 Max.	0·3 Max.	0·75 Max.	101

	Alloying Elements (%)					
Ferrous	C	Mn	Si	Ni	Cr	Mo	Brinell hardness‡
Cast iron	2·7	1·68	2·13	2·06			174
FS-1020 steel	0·20	0·45	0·18				120
FS-4140 steel	0·40	0·87	0·30		0·95	0·20	183

* Analyses obtained from commercial data.
† 2000 kg load.
‡ 3000 kg load.

C. Machining Tests*

Cutting tool material	Work material	Speed (s.f.m.)†	Feed (Linear) (in.)	Depth of cut (in.)	Travel (in.)
S–1008	Brass	425	0·015	0·062	22
S–1008	Brass	650	0·010	0·125	22
S–1008	Cast iron	300	0·015	0·062	$\frac{3}{16}$
S–1008	Cast iron	450	0·010	0·187	$\frac{1}{4}$
S–1008	FS–1020	385	0·005	0·062	$2\frac{3}{4}$
S–1008	FS–4140	500	0·005	0·062	1
S–1047	Brass	425	0·010	0·062	68
S–1047	Brass	650	0·010	0·125	88
S–1047	Cast iron	460	0·008	0·125	37
S–1047	FS–4140	665	0·010	0·187	12
S–1047	FS–4140	900	0·010	0·062	60
S–1048	Brass	425	0·008	0·093	12
S–1048	Brass	650	0·008	0·093	20
S–1048	Cast iron	300	0·010	0·093	6
S–1048	FS–4140	300	0·010	0·125	—
S–1048	FS–4140	600	0·010	0·062	—
S–1049	Brass	650	0·010	0·062	30
S–1049	Cast iron	300	0·010	0·093	2
S–1049	Steel	400	0·008	0·125	16
S–1050	Brass	650	0·010	0·125	44
S–1050	Cast iron	300	0·010	0·125	12
S–1050	FS–4140	350	0·008	0·093	13
S–1051	Brass	650	0·010	0·093	54
S–1051	Cast iron	300	0·010	0·093	34
S–1051	FS–4140	380	0·008	0·125	38
S–1052	Brass	650	0·010	0·125	88
S–1052	Cast iron	300	0·010	0·125	70
S–1052	FS–4140	350	0·010	0·125	76
S–1052	FS–4140	250	0·008	0·125	184
S–1052	FS–4140	600	0·008	0·160	102
S–1052	FS–4140	950	0·008	0·125	106
S–1052	FS–4140	910	0·008	0·250	86
S–1052	FS–4140	1200	0·008	0·125	80
S–1053	Shape not suitable for cutting tool				
S–1054	Brass	650	0·010	0·125	12
S–1054	Cast iron	300	0·010	0·125	—
S–1054	FS–4140	350	0·010	0·125	—
S–1055	—	—	—	—	—
S–1056	—	—	—	—	—
S–1057	Brass	650	0·010	0·125	24
S–1057	Cast iron	300	0·010	0·125	20
S–1057	FS–4140	250	0·008	0·093	12
S–1057	FS–4140	350	0·008	0·125	$2\frac{1}{2}$
S–1058	Brass	650	0·010	0·125	20
S–1058	Cast iron	310	0·010	0·125	18

continued on next page.

* Machining tests conducted on a standard 18-in. engine lathe.
† Surface feet per minute.

TABLE 253—(contd.)
Utilisation of Ceramics for Metal Cutting Tools (S219f)—(contd.)
C. Machining Tests*—(contd.)

Cutting tool material	Work material	Speed (s.f.m.)†	Feed (Linear) (in.)	Depth of cut (in.)	Travel (in.)
S–1058	FS–4140	250	0·010	0·093	20
S–1058	FS–4140	300	0·008	0·093	3
S–1058	FS–4140	500	0·008	0·062	3½
S–1059	Brass	620	0·010	0·125	80
S–1059	Cast iron	310	0·010	0·125	38
S–1059	FS–4140	300	0·008	0·125	65
S–1059	FS–4140	600	0·008	0·062	30
S–1059	FS–4140	590	0·015	0·125	6

* Machining tests conducted on a standard 18-in. engine lathe.
† Surface feet per minute.

D. Comparative Machining Tests* Data

Item	Cutting Tool Material		
	S–1052‡	C–6§	C–7§
Speed (s.f.m.)†	750	500	500
Feed (linear) (in.)	0·0075	0·0075	0·0075
Depth of Cut (in.)	0·125	0·125	0·125
Travel (in.)	82	80	90
Chip Platform Cratering (in.)	None	$\frac{3}{32}$	Slight
Edge (land) Wear (in.)	$\frac{1}{64}$	$\frac{3}{64}$	$\frac{1}{32}$

* Machining tests conducted on a standard 18-in. engine lathe.
† s.f.m.—Surface feet per minute.
‡ Ceramic material—High purity, high density aluminium oxide.
§ Carbide cutting tools.

CERAMICS FOR GAS TURBINE AND TURBO-JET BLADES

Requirements.

(1) To raise operating temperatures above the 850° C (1562° F) limit imposed by metals to at least 1000° C (1832° F).

(2) High mechanical strength which may be only slightly reduced at operating temperatures.

(3) Freedom from creep (*see* **B146**).

(4) Thermal shock resistance.

(5) High tensile strength.

Certain ceramics possess good high temperature strength but poor thermal shock resistance, low ductility and therefore no ability to distribute stress (**R14**).

Bodies that come under Consideration.

(*a*) Single component: alumina; beryllia; magnesia; zirconia; thoria; spinel (MgO, Al_2O_3).

(b) Multicomponent with high-melting eutectics: silicon carbide–clay; beryllia–clay; BeO–Al$_2$O$_3$–MgO (**C14**); high-alumina porcelain; high-magnesia porcelain; high-mullite porcelain (**G31**).

CERAMIC PARTS FOR ROCKET MOTORS

Requirements. In uncooled liquid propellant motors temperatures rise to 2200–3870° C ($\sim$4000–7000° F), and pressure 300 to 1000 lb/in^2 (21–70 kg/cm^2). Operation time is from 5 sec to 30 min. Exit gases reach sonic velocities at the throat and supersonic velocities downstream from this point.

Ceramic linings have two functions: (1) to maintain the inner contour of the chamber and more particularly the throat and nozzle; (2) to prevent overheating of metal stress-bearing housing.
Bodies. Zirconia. Beryllia. Silicon carbide. Graphite. Silicon-impregnated graphite. Zirconium boride.
Shapes. Either complete chamber, throat and nozzles in the minimum of separate parts or small segment tiles.
Finishing. Most ceramic rocket linings are set in a cast or rammed mixture (light-weight refractory concrete or plaster) in the metal housing (**L56, D62**).

REFRACTORIES FOR THE ATOMIC ENERGY INDUSTRY

Characteristics. This industry will never require large quantities of materials but the requirements regarding quality and purity are extremely stringent.
Uses and Body Types.
A. Metallurgical preparation of uranium, thorium and plutonium: CaF$_2$ crucibles for convertion of UF to U metal, graphite crucibles coated with magnesium zirconate, pure lime, or magnesia, for remelting uranium. Research in pure oxides: BeO, MgO, CaO, Al$_2$O$_3$, ZrO$_2$, UO$_2$, ThO$_2$. Research in pure sulphides: CeS, US, ThS. Research in pure nitride: BN. Research in pure carbides: B$_4$C, SiC, HfC, TaC. Research in pure fluorides: CaF$_2$, MgF$_2$.

B. In reactors: For higher temperature working: research in using fuel elements of UO$_2$, UC, UC$_2$, UN, PuO$_2$, PuC, ThO$_2$, ThC. As moderator, BeO (**C3**). Research in dispersion-type fuel elements: UO$_2$ in BeO, Al$_2$O$_3$, MgO, stainless steel; UC in graphite.

THERMAL SHOCK-RESISTANT CERAMICS

(The work of a number of people has been brought together and summarised by Hummel (**H122**) and by the Soc. Fr. Cer. (**A106**).)

Requirements and Characteristics. Thermal shock resistance is defined by absence of failure under specified conditions of thermal stress in a temperature cycling process. It is a complicated property altered by the conditions of test, such as heat transfer coefficient, size and shape of specimen and interaction with heating and cooling media, and dependent on a number of properties of the material, such as strength, elasticity, coefficient of expansion, thermal conductivity, specific heat and density.

Body texture has considerable influence of thermal shock resistance, porous bodies frequently being much more resistant than dense ones of the same type. Amongst dense bodies the heterogeneity of the body can lead to fatigue phenomena over repeated cycles because of uneven expansion and contraction of the components eventually causing failure under far less stringent conditions than have previously been successfully survived.

The relationship between physical properties and maximum temperature difference that can be withstood by a material is given in the equations:
for low rates of heat transfer

$$\Delta T_{max} = \frac{\kappa \cdot S}{E\alpha}(1-\mu)$$

for high rates of heat transfer

$$\Delta T_{max} = \frac{S}{E\alpha}(1-\mu)$$

where κ = thermal conductivity;
S = strength;
E = modulus of elasticity;
α = coefficient of expansion;
μ = Poisson's ratio.

$$\text{Index of thermal shock resistance} = \frac{S_T}{\alpha E}\sqrt{\frac{\kappa}{Cd}}$$

where S_T = tensile strength (value for porcelains 240–520 kg/cm^2);
α = linear coefficient of expansion (value for porcelain 2–5 × 10^{-6});
E = modulus of elasticity (value for porcelains 0·6–0·8 × 10^6 kg/cm^2);
C = Specific heat (value for porcelains ~0·19);
d = density (value for porcelains 2·4–2·8);
κ = thermal conductivity (value for porcelains 0·004–0·006 c.g.s.).

Figures for these properties for types of ceramic bodies are given in Table 254. Of them, Smoke (**S138**) has shown that the coefficient of thermal expansion is by far the most important.

Body Types. Technical ceramic bodies can be divided into three classes by their coefficient of expansion:

(1) Low expansion: $\alpha < 2\cdot 0 \times 10^{-6}$.
(2) Intermittent expansion: $\alpha = 2\cdot 0$–$8\cdot 0 \times 10^{-6}$.
(3) High expansion: $\alpha > 8\cdot 0 \times 10^{-6}$.

TABLE 254

Physical Properties and Thermal Shock Resistance (H122)

Body type	Relative thermal shock resis. (S140)	Transverse strength (lb/in²)	Modulus of elasticity E (lb/in² × 10⁶)	Poisson's ratio μ	Coefficient of expansion $\alpha \times 10^{-6}$	Thermal conductivity (kcal-cm-sec⁻¹-cm⁻²-°C⁻¹) 100°C	Thermal conductivity 1000°C
Eucryptite					−6·0 (to 700°C)		
β Spodumene					<0·5		
Lithium–alumino–silicate	7 (best)						
Fused silica		7000			0·5	0·0038	—
Aluminium titanate					<1·0 (to 1000°C)		
Zirconium phosphate					<1·0 (to 700°C)		
Cordierite	6	8000			2·0		
Beryl					2·0		
Silicon carbide					3·5		
Mullite	4	13 000	21	0·30	4·0	0·015	0·010
Zircon	5	15 000 and up	20	0·35	4·0	0·016	0·010
Porcelain	2	10 000	10	0·30		0·0041	0·0045
BeO					8·8	0·525	0·049
Al₂O₃	3	45 000	51	0·20	8·0–9·0	0·072	0·015
Spinel			34	0·31	8·0–9·0	0·036	0·014
Graphite						0·426	0·149
TiC cermet		60 000	60	0·30		0·083	
Steatite	1 (worst)				7·7		
Forsterite		20 000					

Group 1. Low expansion, $\alpha = <2\cdot 0 \times 10^{-6}$.

(1) Cordierite. The first low-expansion ceramic. Difficulties are experienced in producing a vitreous body because of the short firing range. There is a tendency to form glass which impairs thermal shock resistance.

(2) Lithium–aluminosilicates. These have excellent thermal shock resistance derived from exceptionally small or negative expansion coefficients, but are not very refractory, melting at about 1400° C (2552° F).

(3) Aluminium titanate. This compound contracts from $-180°$ C ($-292°$ F) to about 20° C (68° F) and then expands increasingly. It dissociates between 1200 and 1300° C (2192–2372° F).

(4) Zirconyl phosphate $2ZrO_2.P_2O_5$. Has interesting expansion curve similar to aluminium titanate.

(7) Tantalum, niobium and vanadium pentoxides are reported to have very low expansion coefficients.

Group 2. Intermediate expansion, $\alpha = 2\cdot 0 – 8\cdot 0 \times 10^{-6}$.

(1) Zircon bodies are suitable for large refractories or fine laboratory ware.

(2) Mullite finds use in refractories, glass tank block, etc.

(3) Feldspathic porcelain although falling within this class because of its relatively small expansion coefficient exhibits poor thermal shock resistance because of the stresses set up between the quartz, glass and mullite present. Much better thermal shock resistance is attained in bodies with no residual quartz.

(4) Silicon carbide has the added advantage of good thermal conductivity and is used for kiln furniture. Its disadvantage is the tendency to oxidise.

(5) Tin oxide has good conductivity, but is costly and very sensitive to reducing atmospheres.

(6) Calcium aluminate, $CaO.2Al_2O_3$, $\alpha = 5\cdot 0 \times 10^{-6}$.

(7) Anorthite, $CaO.Al_2O_3.2SiO_2$, $\alpha = 5\cdot 4 \times 10^{-6}$.

(8) Barium aluminosilicate $BaO.Al_2O_3.2SiO_2$, $\alpha = 4\cdot 0 \times 10^{-6}$.

Group 3. High expansion, $\alpha > 8\cdot 0 \times 10^{-6}$.

(1) Beryllia. It was thought that the very high thermal conductivity would give this material good thermal shock resistance but it only places it above other materials of similar expansion coefficient, *e.g.* corundum.

(2) Corundum is not really very thermal shock resistant but finds applications where its very high strength is of importance.

HIGH THERMAL CONDUCTIVITY BODIES

Requirements and Characteristics. Bodies of high thermal conductivity are rare in the ceramic field and, if also dense and of good dielectric properties would find uses in the field of dielectrics. Figure 15.5 shows conductivities of various ceramics compared with metals and it is seen that all of the traditional bodies fall in the group 0–2.

Bodies. Thermal conductivity (in B.t.u. at 140° F) 2–4. Insulators:

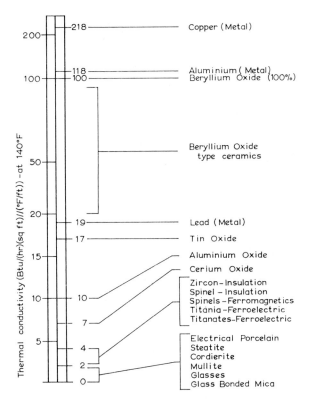

FIG. 15.5. *Thermal conductivity of ceramic dielectrics (Smoke and Koenig,* **S141**).

zircon, spinel. Ferroelectrics: titania; titanates. Ferromagnetic: spinels. Thermal conductivity, 4–20. Cerium oxide: 7, but dielectric constant 22. Alumina: 10, hence its use in sparking plugs. Tin oxide: 17, dielectric constant 9 and low loss.

Thermal conductivity comparable to metals 20–100. BeO: 100, but considerable manufacturing difficulties. Beryllia porcelains: 20–95 (**S141**).

Chapter 16
Ceramics in the Electrical Industry

THE demand for ceramics in the electrical industry has risen phenomenally both in quantity and quality during this century. As the applications have increased so has the need for different combinations of electrical properties, some of which will be briefly discussed.

Dielectric materials are defined by Hippel (**H100**) as the broad expanse of non-metals considered from the standpoint of their interaction with electric, magnetic, or electromagnetic fields.

Dielectric or breakdown strength of a material is its resistance to electrical breakdown and is the property which determines its ability to withstand high voltage. It is measured by applying an increasing voltage of known frequency across a test piece until puncture occurs, the thickness of the piece at the point of puncture is measured and the dielectric strength calculated in volts per mil (thousandth of an inch) or kilovolts per millimetre.

Dielectric strength varies with the thickness of the piece, increasing as it decreases, also with frequency and wave shape of the applied voltage, electrostatic field distribution and temperature of the specimen. Direct current gives 25 to 30% higher values than alternating current at 60 cycles. Bodies other than rutile bodies have decreasing dielectric strength at increasing frequencies.

Dielectric constant or *permittivity*, ϵ, is the ratio of the amount of energy stored in the dielectric to that stored by air occupying the same volume. It is a measure of electric storage ability, or capacitance. The dielectric constant varies with temperature, ceramic dielectrics showing a range of high positive capacity temperature coefficients (porcelain) through low coefficients (steatite) to high negative capacity coefficients (rutile). The dielectric constant may be used to classify dielectric materials.

Loss angle. A theoretically perfect insulator would throw current and voltage waves entirely out of phase, in which case the phase angle (θ) would be 90°. For actual insulators the phase angle is somewhat less than 90°,

The loss angle (δ) is the difference between 90° and the phase angle, *i.e.* 90° − θ = δ. For ceramics δ is usually less than 5°.

Power factor is a measure of the energy lost when an insulator transforms part of the electrical energy into heat. The power factor is given by the size of the loss angle.

$$\text{Power factor} = \sin \delta$$

when $\delta < 5°$ P.F. $\sim \tan \delta$.

Loss factor is the product of the dielectric constant ϵ and the tangent of the loss angle.

$$\text{Loss factor} = \epsilon \tan \delta$$

when $\delta < 5°$ L.F. $\sim \epsilon \sin \delta = \epsilon \times$ P.F.

Loss factor is used in grading high-frequency insulators.

These properties vary considerably with frequency and with temperature, values for more than one frequency usually being quoted.

The loss of electrical energy N, is also given by the following equations:

$$N = E^2 2\pi f C \tan \delta$$

where E is the potential, f the frequency and C the capacitance. The capacitance is also given by the product of the dielectric constant ϵ and a constant K which depends on the shape and field position of the insulator.

$$N = (E^2 2\pi f K) \epsilon \tan \delta$$

Q-value is the ratio of reactance to resistance of an insulator and is given by

$$Q = \tfrac{1}{2} f C R$$

where f is the frequency, C the capacitance and R the resistance. Q is also equal to the reciprocal of the tangent of the loss angle $\tan \delta$. When δ is less than 5° $\tan \delta \sim \sin \delta$ so that $Q \sim 1/\text{power factor}$.

Specific resistance or *volume resistivity* is the electrical resistance per unit volume of a dielectric expressed in ohms per cubic centimetre (Ω/cm^3). Ceramic dielectrics are considered to be insulators if their specific resistance is greater than 1 MΩ ($10^6 \Omega$) per cm³. Many of them have values of from 10^{12} to 10^{19} Ω/cm^3, but these drop considerably with increasing temperature.

Surface resistivity is four times the resistance between electrodes which completely cover opposite faces of a centimetre cube of the insulator when all the current flows through the surface layer. This is very dependent on the nature of the surface, particularly the amount of moisture absorbed there. Insulators to be used in humid conditions are therefore glazed.

Piezoelectric effect is the phenomenon exhibited by certain crystals of expansion along one axis and contraction along another when subjected to an electric field. The converse effect, whereby mechanical strains produce opposite charges on different faces of the crystal, also obtains.

Piezoelectric coupling coefficient is a measure of the strength of the piezoelectric effect. A coupling coefficient of 100% means that all dielectric polarisation would appear as an elastic stress, or, inversely, that an applied external mechanical force would be converted into electric voltage.

Ferroelectric properties are characterised by a high dielectric constant which varies with electric flux density and direction, ferroelectric hysteresis, polarisation saturation, piezo effect and a Curie temperature.

Curie point or *Curie temperature* is the temperature at which a ferroelectric material changes to a non-ferroelectric or at which a ferromagnetic one loses its ferromagnetic properties.

Ferromagnetic materials acquire magnetic properties when placed in a magnetic field. The main magnetic properties are defined by the relationship between induced flux (B) caused by the application of a magnetising force (H). When the magnetising force is raised from zero the increase in induced flux is non-linear, being at first slow, then faster and then slow again as saturation

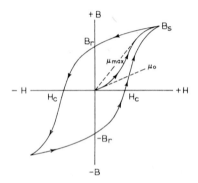

Fig. 16.1. *Typical hysteresis loop of ferromagnetic material showing the relationship between induced flux (B) and magnetising force (H) (after* **D19**)

occurs. After this, cyclic variation of the magnetising force produces a hysteresis curve (Fig. 16.1). When the magnetising force is reduced to zero some flux is retained as the remanent induction (B_r) and a negative magnetising force, the coercive force (H_c), is required to reduce the induction to zero. The area of the hysteresis loop is a measure of the work done in traversing a complete cycle and is a measure of the hysteresis loss when an alternating field is used, the work done appearing as heat.

The *permeability* or ease of magnetisation, μ, of the material is the ratio of B/H and is normally given at two points on the magnetising curve: initial permeability μ_0 at very small values of B, and, maximum permeability μ_{max} at the knee of the curve.

Eddy current loss occurs in materials magnetised by alternating fields because of the induced electrical current. It is inversely proportional to the resistivity and rises proportionately with the frequency.

Residual losses arise from the molecular structure of the magnetic materials and are strongly influenced by impurities and imperfections in the crystal

TABLE 255

Ceramic Materials for the Electrical Industry
(Buessem **B136**)

Application	Ceramic materials	Leading properties	Competitive materials
1. High-voltage application			
(*i*) Production	porcelain	resistivity, dielectric strength	oil organic products
(*ii*) Transmission	porcelain, (glass)	resistivity, dielectric strength	none for outdoor lines
2. Low-voltage application	porcelain, steatite, cordierite, (glass)	resistivity	plastics
3. Electronic components			
(*i*) Tubes	(glass) steatic refractory oxides	dielectric loss factor	(germanium and silicon transistors)
(*ii*) Condensers	barium titanate, titanium dioxide, zirconates, steatite, (glass)	dielectric constant	paper, mica, plastics, Al and Ta electrolyte
(*iii*) Resistors	porcelain, reduced titanates, reduced uranium oxide, silicon carbide and cermets, tin oxide film on porcelain	resistivity	resistance metals, carbon films
(*iv*) Inductances, transformers, high-frequency antennas	magnesium-zinc ferrites, nickel-zinc ferrites	magnetic permeability	iron, nickel, cobalt alloys
(*v*) Memory devices	magnesium-manganese-zinc ferrite	squareness ratio of hysteresis loop	(barium titanate single crystals)
(*vi*) Dielectric amplifier	barium titanate	non-linearity of dielectric constant as function of applied electric field	
(*vii*) Transducers	barium titanate	electro-mechanical coupling coefficient	Rochelle salt, tourmaline lithium sulphate dipotassium tartrate, ethylene diamine tartrate
4. Permanent magnets for instruments and other uses	$BaO \cdot 6Fe_2O_3$ unoriented, oriented	energy product (gauss × oersted)	iron alloys, alnico oriented and unoriented

structure. They cause the induced flux to be out of phase with the magnetising field and are a further source of non-linear behaviour.

Thurnauer (**H100**) gives the following classification for ceramic dielectrics:

(1) Materials with dielectric constant below 12.
   Low frequency: (*a*) low- and high-tension electrical porcelain; (*b*) high-temperature porcelains; (*c*) sparking plug insulators.
   High frequency: (*d*) vitrified materials for high frequency (graded by loss factor); (*e*) refractory ceramics for electronic use.
(2) Materials with dielectric constant above 12; (*a*) class I, ϵ 6 to 500; (*b*) class II, ϵ 500 to 10 000; (*c*) non-linear circuit elements.
(3) Materials with piezoelectric and ferroelectric properties.
(4) Materials with ferromagnetic properties.

LOW-TENSION, LOW-FREQUENCY PORCELAIN INSULATORS

Characteristics. Almost vitreous to vitreous body consisting of mullite and quartz crystals imbedded in a glassy matrix the relative proportions of which are adjustable within limits to give different properties. Able to withstand normal domestic voltages, up to about 440 V. Dielectric strength decreases with increasing temperature (therefore not used above 100° C (212° F)), cast, jiggered and extruded pieces having higher dielectric strength than dry-pressed ones. Both glazed and unglazed ware is made.

Uses and Shapes. *Unglazed* for use in dry conditions. Fittings for open wiring or knob and tube wiring: nail knobs; tubes; cleats; cable racks; bobbins; reels; split insulators; bushes. Outlet boxes. Switch boxes. Connectors. Plugs. Sockets. Ceiling roses. Lamp holders. Fuse holders. Appliance parts in numerous pieces of equipment, using electric resistances to generate heat, *e.g.* electric irons, toasters, fires, sterilisers.

Glazed, for weather-exposed pieces. Farm fencing. Telephone and power supplies to houses. Parts for neon signs. Aerial insulators.

HIGH-TENSION, LOW-FREQUENCY PORCELAIN INSULATORS

Characteristics. Completely vitrified material of zero water absorption covered by a glaze of expansion coefficient slightly lower than that of the body so that the greatest strength improvements are obtained by applying the glaze. This gives complete imperviousness and resistance to freezing, very good mechanical strength in tension as well as compression. Shapeability allows almost any shape to be made.

Corona is reduced and interference with electronic equipment eliminated by using semi-conducting glazes or coatings on areas in contact with metallic surfaces.

Fall of dielectric strength with rising temperature limits use to below 100° C (212° F).

T_e, the temperature at which resistivity drops to 1 MΩ-cm is 300 to 500° C (572–932° F) except where non-alkali fluxes are used, when it can be raised to 800° C (1472° F).

Shapes and Uses. Suspension insulators to be used in 'strings' for voltages over 66 000 V. Pin-type insulators for voltages up to 50 000 to 60 000 V. Tubes. Bushings. Lead-in insulators. Guy-line insulators. Housing for line cut-outs. Fuse cut-outs. Coil forms. Transformer parts. Circuit-breaker parts. Mast bases. Switch parts.

General Production Methods for Electrical Porcelain

Body Type. Porcelain, electrical.
Raw Materials. China clays. Ball clays. Potash feldspars. Pegmatites. Flint. Quartz sand. Quartz crystal. Whiting. Talc. Pressing oils: vegetable oils; mineral oils; oleic acid; emulsified saponified oils; paraffin.

Preparation.

A. Wet Methods.

(1) *Casting slip.* Feldspar→jaw crusher or crushing rolls→wet cylinder mills, perhaps with quartz (particle size test before use).

Flint, quartz, etc.→crushers if necessary→wet cylinder mills, perhaps with feldspar (particle size test before use).

Ball clay → cutters → blunger
China clay → blunger
→ deflocculants added → mixing ark →
ground feldspar and quartz added→casting slip storage.

In the U.S.A. where raw materials obtained ready ground: batch+water+deflocculant→pebble mill.

(2) *Plastic.* Feldspar and quartz → crushing → wet milling.

Clays → blunging.

Scraps→blunged→
all to mixing ark→lawns→magnets→storage (slip may be heated to 40° C or de-aired)→filter press→non de-airing pug→storage cellars→pugging (de-airing for extrusion or throwing, de-airing optional for jolleying or hot pressing) and adjustment of water content.

In the U.S.A. where raw materials obtained ready ground: Batch→blunger→filter press→de-airing pugmill (no maturing).

(3) *Press dust.* Wet grinding, blunging, mixing and filter pressing as in (2)→press cakes to dryer→edge runner or hammer mill→mixer, oil (1–4%) and water (11–18%) added→sieves or special granulators to make dust more solid.

B. Dry Method.

(4) *Press dust.* Feldspar and quartz→dry grinding in tube mill and air floating. Clay→air floated. All to trough or pan mixer, water 11–18% and oil 1–4% added→sieves or special granulators to make dust more solid.

Fig. 16.2. *High-voltage porcelain insulators* (*Doulton*, **D56**)

Shaping.

(1) Casting.

(2) Jolleying→conditioning in humid chambers for large pieces→air drying→trimming→sticking up.

(3) Hot pressed in two stages in plaster mould with heated, revolving tool.

(4) Extrusion (up to 30 in., 76 cm, diameter)→air dried to leather-hard condition→turning.

(5) Throwing (up to 3 ft, 90 cm, high)→air dried to leather-hard condition→turning→sticking up if necessary.

(6) Extrusion of low moisture content (15%) by hydraulic presses. Product leather-hard has no drying warpage.

(7) Dust pressing in hand or semi-automatic hydraulic machines with possible vacuum control in dies to reduce porosity→holes drilled and tapped (if not evacuated only suitable for low-tension use).

(8) Hydrostatic pressing.

Drying. Humidity dryers, chamber or tunnel.
Fettling and sponging.

Glazing. The need for the right 'fitting' glaze that is under compression is graphically illustrated by Rowland (**R61**) (*see* Fig. 6.2, p. 550). Single fire brown, white or semi-conducting glazes applied to all but some small pieces. Sand grit (porcelain grit) applied with aid of glaze adhesives to parts of glazes where joints have later to be made. Spraying or dipping.
Making of glaze joints for very large pieces.

Firing. Low-tension insulators 1150° C (2102° F) upwards; normal 1300–1350° C (2372–2462° F) cone 10–12, high-tension insulators up to 1400° C (2552° F).

Intermittent kilns, coal fired. Very large pieces must always be fired in an intermittent kiln.

Tunnel kilns, gas fired. Unlike table porcelain the atmosphere may be kept oxidising throughout.

Finishing. Grinding of surfaces that must be planar. Assembly: cementing together and curing in steam; soldering and brazing; glass bonding. Inspection and routine testing of every insulator.

Testing. Electrical. Pressure. Load and tension, torsion and bending. Thermal shock. Water absorption. B.S. 137:1941; B.S. 223:1956.

Output. Examples. Weekly: 60–100 tons large insulators with 250 men. Hourly: throwing of $4\frac{1}{2}$ lb insulator blanks, one man with one woman, 300. Turning of leather-hard blanks, two women, 60.

Properties. Table 256, Nos. 1–4.

HIGH-TENSION, LOW-FREQUENCY MULLITE PORCELAINS

Characteristics. Better resistance to temperature and thermal shock than normal porcelains.

Uses. Sparking plugs. Fuse cores. Electric water boiler insulators. Carbon and wire resistor cores.

Body Type. Mullite porcelain.

Raw Materials. Dumortierite. Kyanite. Andalusite. Alumina. Ball clay. China clay.

Preparation. Andalusite and dumortierite are very tough: Crushing→

TABLE

General Properties

No.	1	2	3
Trade name and/or body type	General props of hard porcelain	General hard porcelain	Porcelain
Ref.	N11	N11	L39
Country		U.S.A.	Brit.
Specific gravity	1·90–2·48		2·4
Water absorption (%)	0·2–2·0		
Coefficient of linear thermal expansion:			
Temp. range (°C)			0–300
Value/°C $\times 10^{-6}$	2·5–5·5		4–6
Safe operating temperature (°C)			
Thermal conductivity (c.g.s.)	0·0020–0·0039		0·0024
Tensile strength (lb/in^2)	1500–6000		
(kg/cm^2)	105–422		
Compressive strength (lb/in^2)	30 000–65 000		110 000
(kg/cm^2)	2109–4570		7734
Flexural strength ⎱ (lb/in^2)	3000–8500		14 000
Modulus of rupture ⎰ (kg/cm^2)	211–598		984
Impact strength (ft-lb); ½ in. rod	0·50–0·90		
Impact resistance (lb/in^2)			
Modulus of elasticity (lb/in^2)			
(kg/cm^2)			
Thermal shock resistance	370° C–840° C		
Dielectric strength (V/mil)	90–200	100	70
(V/mm)	3540–7870	3940	2750
Resistivity : Temp. °C			14 | 300
(Ω/cm^3)	1·8–18 $\times 10^{10}$		10^{13}–10^{14} | 10^6–10^7
T_e-value (°C)			
Power factor: c, cycles, or kc, kilocycles (10^3 cycles) or Mc, megacycles (10^6 cycles)	1000 kc		1 Mc
Value	0·6–0·95		0·0055
Loss factor: cycles			
values			
Dielectric constant: cycles		60	1 Mc
value	6·0–7·5	5·6–6·8	5·6
Capacity changes per °C: range (°C)			10–65
Value (p.p.m./°C)			450
Bulk density (lb/in^3)			
(g/cm^3)			
Volume (in^3/lb)			
(cm^3/g)			
Mohs' hardness			
Specific heat Temp. °C			500° C
(B.t.u./lb)			0·25
(c.g.s.)			0·139

of Electrical Ceramics

4 porcelain	5 Mullite porcelain	6 Mullite AlSiMag 203		7 Mullite porcelains	8 Stoneware
S171 Brit. 2·4 0	**N36** U.S.A.	**A186** U.S.A. 2·1 10–15		**R71** 2·4–3·1 0·0	**K27** 2·2 1·7
3·5–4·5	5	25–100 3·5	25–600 4·9	20–700 4·3–5·0	3·6
	0·007	1350 0·003		1250 0·0037–0·0073	
4200–5400		2000		5000–10 000	2000
300–450		140		351–703	140
64 000–68 000	100 000	10 000		50 000–120 000	80 000
45 000–5500	7030	703		3515–8437	5600
10 000–18 000	26 000	6000		13 000–21 000	5000
700–1300	1828	422		914–1476	350
		0·075			
840–980					
	15×10^6				8×10^6
	$1·05 \times 10^6$				$5·6 \times 10^5$
				Good	
640–760				250–400	74
25 000–30 000				9840–15 750	2900
20 300 600		25 300 900			
10^{13-14} 10^{6-7} 10^{4-5}		$>10^{12}$ $7·8 \times 10^7$ $1·0 \times 10^4$		10^{14}–10^{15}	$6·5 \times 10^{12}$
		480		500–650	
300 m 180 m 30 m		1000 kc	10 Mc	1 Mc	
1 Mc 3 Mc 10 Mc					
0·0055 0·0049 0·0063		0·004	0·003	0·004–0·005	
		1000 kc	10 Mc		
	0·03	0·016	0·0117	0·05–0·034	
		1000 kc	10 Mc		
	6	4·0	3·9	6·2–6·8	
450					
		0·076			
		2·10			
		13·20			
		0·48			
				7·5	
					0·2
					0·111

TABLE

No. Trade name and/or body type	9 Porcelain spark plug	10 100% alumina spark plug	11 95% Al_2O_3, 3·5 ZnO_2 1·5 talc spark plug
Ref.	G34	G34	G34
Country			
Specific gravity			
Water absorption (%)			
Coefficient of linear thermal expansion:			
Temp. range (°C)	200–600	200–600	200–600
Value/°C × 10^{-6}	4·5	8·23	8·08
Safe operating temperature (°C)			
Thermal conductivity (c.g.s.)		0·0252	0·0264
Tensile strength (lb/in²)			
(kg/cm²)			
Compressive strength (lb/in²)	64 000	182 600	191 370
(kg/cm²)	4500	12 838	13 455
Flexural strength ⎱ (lb/in²)	17 500	24 600	48 470
Modulus of rupture ⎰ (kg/cm²)	1230	1730	3408
Impact strength (ft-lb); ¼ in. rod			
Impact resistance (lb/in²)			
Modulus of elasticity (lb/in²)			
(kg/cm²)			
Thermal shock resistance	Transverse strength after quenching		
(lb/in²)	3400	11 610	15 340
(kg/cm²)	239	816	1079
Dielectric strength (V/mil)	5868	5860	11 076
(V/mm)	231 020	230 710	436 060
Resistivity: Temp. °C			
(Ω/cm³)			
T_e-value (°C)	557	650	631–647
Power factor : c, cycles, or kc, kilocycles (10^3 cycles), or Mc, megacycles (10^6 cycles)			
Value			
Loss factor			
Dielectric constant: cycles			
value			
Capacity changes per °C: range (°C)			
Value (p.p.m./°C)			
Bulk density (lb/in³)			
(g/cm³)			
Volume (in³/lb)			
(cm³/g)			
Mohs' hardness			
Specific heat (B.t.u./lb)			
(c.g.s.)			

256—(contd.)

12 Natural Indian block talc	13 Natural Yellowstone block talc	14 Phosphate-bonded Yellowstone talc (dry-pressed)	15 Phosphate-bonded Yellowstone talc (hot-pressed)
C49 U.S.A. 2·8 2–3	**C49** U.S.A. 2·6–2·8 2–5	**C49** U.S.A. 2·4–2·6 3·5–9·0	**C49** U.S.A. 2·5–2·6 3·0–5·5
Up to 30 000 Up to 2109 Up to 9000 Up to 633	15 000–30 000 1055–2109 Up to 9000 Up to 633	25 000–40 000 1758–2812 4500–7000 316–492	n.d. — 6000–8000 422–562
1 Mc 20° C | 200° C | 350° C 0·0003 | 0·0013 | 0·007	1 Mc 20° C | 200° C | 350° C 0·0008 | 0·0014 | 0·005	1 Mc 20° C | 200° C | 350° C 0·0006 | 0·0008 | 0·0013	1 Mc 20° C | 200° C | 350° C 0·0003 | 0·0008 | 0·0017
5·8	5·6	5·5	5·5
1	1	1–2	1·5–3

TABLE

	16	17
No. Trade name and/or body type	General clinoenstatite	Clinoenstatite
Ref.	N11	N11
Country	U.S.A.	U.S.A.
Specific gravity		2·6
Water absorption (%)		0–0·05
Coefficient of linear thermal expansion:		
Temp. range (°C)		20–600
Value/°C × 10^{-6}		8·90
Safe operating temperature (°C)		1000
Thermal conductivity (c.g.s.)		
Tensile strength (lb/in^2)		10 500
(kg/cm^2)		738
Compressive strength (lb/in^2)		85 000
(kg/cm^2)		5976
Flexural strength } (lb/in^2)		20 000
Modulus of rupture } (kg/cm^2)		1406
Impact strength (ft-lb); ½ in. rod		0·4
Impact resistance (lb/in^2)		
Modulus of elasticity (lb/in^2)		
(kg/cm^2)		
Thermal shock resistance		
Dielectric strength (V/mil)		60 cycles 240
(V/mm)		9450
Resistivity: Temp. °C		200 V. 60 c: 24 900
(Ω/cm^3)		> 10^{14} 10^4
T_e-value (°C)		
Power factor: c, cycles, or kc, kilocycles,	60 c 1000 kc 10 Mc	60 c 1000 kc 10 Mc
(10^3 cycles), or Mc, megacycles (10^6 cycles)	0·14 0·06 0·04	0·0022 0·0021 0·0015
Loss factor: cycles	60 c 1000 kc 10 Mc	60 c 1000 kc 10 Mc
value	0·91 0·36 0·23	0·013 0·012 0·08
Dielectric constant: cycles	60	60 c 1000 kc 10 Mc
value	5·7–6·5	5·9 5·8 5·7
Capacity changes per °C: range (°C)		
Value (p.p.m./°C)		
Bulk density (lb/in^3)		
(g/cm^3)		
Volume (in^3/lb)		
(cm^3/g)		
Mohs' hardness		
Specific heat (B.t.u./lb)		
(c.g.s.)		

256—(contd.)

18 Steatite porcelain **N36** U.S.A.		19 All-purpose steatite AlSiMag A–35 **A186** U.S.A. 2·5 0–0·05			20 Low-loss steatite AlSiMag 197 **A186** U.S.A. 2·6 0–1			21 Machinable lava AlSiMag 222 **A186** U.S.A. 2·0 14–18		
9		25–100 6·9		25–600 8·7	25–100 7·7		25–600 10·4	25–100 8·0		25–600 9·8
0·008		1000 0·006 8500 598			1000 0·006 8500 598			1300 0·005 2500 176		
100 000 7030 16 000 1125		75 000 5273 18 000 1265 0·37			75 000 5273 20 000 1406 0·15			25 000 1758 8000 562 0·066		
14×10^6 9.8×10^5										
			225 8860			210 8270				
		25 $>10^{14}$	300 6×10^7	900 $<10^4$	25 $>10^{14}$	300 2.5×10^{10}	900 6.8×10^5	25 $>10^{14}$	300 1.8×10^{11}	900 5.0×10^6
			440			840		—		
0·01 6		60 c 0·015 60 c 0·09 60 c 6·1	1000 kc 0·0035 1000 kc 0·021 1000 kc 5·9	10 Mc 0·0030 10 Mc 0·017 10 Mc 5·8	60 c 0·0020 60 c 0·0126 60 c 6·3	1000 kc 0·0012 1000 kc 0·0072 1000 kc 6·0	10 Mc 0·0010 10 Mc 0·0058 10 Mc 5·8		1000 kc 0·0002 1000 kc 0·0009 1000 kc 4·3	10 Mc 0·0001 10 Mc 0·0004 10 Mc 4·2
		+160 0·090 2·49 11·11 0·40			+160 0·094 2·60 10·65 0·38			$+1.0 \times 10^{-4}$ 0·072 1·99 13·85 0·50		

TABLE

	22	23	24
No.	22	23	24
Trade name and/or body type	Steatite Exp. thermal shock resistant No. 2	Steatite No. 11	Steatite
Ref.	S137	S137	L39
Country	U.S.A.	U.S.A.	Brit.
Specific gravity			2·6
Water absorption (%)			
Coefficient of linear thermal expansion:			
Temp. range (°C)	25–700	25–700	0–300
Value/°C × 10^{-6}	6·69	5·67	8
Safe operating temperature (°C)			
Thermal conductivity (c.g.s.)			0·0035
Tensile strength (lb/in^2)			
(kg/cm^2)			
Compressive strength (lb/in^2)			120 000
(kg/cm^2)			8 500
Flexural strength ⎱ (lb/in^2)	—	18 600	18 000
Modulus of rupture ⎰ (kg/cm^2)		1308	1265
Impact strength (ft-lb); ½ in. rod			
Impact resistance (lb/in^2)			
Modulus of elasticity (lb/in^2)			
(kg/cm^2)			
Thermal shock resistance			
Dielectric strength (V/mil)			
(V/mm)			2·00 (kilo?)
Resistivity: Temp. °C			14 | 300
(Ω/cm^3)			10^{14}–10^{15} | 10^8
T_e-value (°C)			
Power factor: c, cycles, or kc, kilocycles, (10^3 cycles), or Mc, megacycles (10^6 cycles)	Dry | Wet	Dry	1 Mc
Value	0·00062 | 0·00089	0·00031	0·0021
Loss factor	Dry | Wet	Dry	
	0·00512 | 0·00731	0·00248	
Dielectric constant: cycles			1 Mc
value	8·2	8·0	6·0
Capacity changes per °C: range (°C)			10–65
Value: (p.p.m./°C)			280
Bulk density (lb/in^3)			
(g/cm^3)			
Volume (in^3/lb)			
(cm^3/g)			
Mohs' hardness			
Specific heat (B.t.u./lb)			
(c.g.s.)			

256—(contd.)

	25 'Frequentite S' steatite			26 'Frequentite' steatite			27 Steatite		
	S171 Brit. 3·0 0			**S171** Brit. 2·7 0			**S171** Brit. 2·6 0		
	7–8			6·2–6·8			7–9		
	5700–7100 400–500 85 000 6000 14 000–17 000 1000–1200			7800–12 000 550–850 12 800–13 500 9000–9500 20 000–22 500 1400–1600			7800–12 000 550–850 120 000–130 000 8500–9200 17 000–20 000 1200–1400		
	1300–1400			1850–2300			1600–2100		
	— —			620–760 24 000–30 000			310–780 12 000–27 000		
			20 10^{15}–10^{17}	300 $2·5 \times 10^{10}$	600 $4·1 \times 10^{7}$	20 10^{14-15}	300 10^{8}	600 3×10^{5}
300 m 1 Mc 0·00012	100 m 3 Mc 0·00010	30 m 10 Mc —	300 m 1 Mc 0·00065	100 m 3 Mc 0·00059	30 m 10 Mc 0·00055	300 m 1 Mc 0·0021	100 m 3 Mc 0·0020	30 m 10 Mc 0·0017
	120			110–130			280		

TABLE

	28	29	30
No.			
Trade name and/or body type	Steatite porcelains	General cordierite	Cordierite
Ref.	**R71**	**N11**	**N11**
Country		U.S.A.	U.S.A.
Specific gravity	2·5–2·8		2·1
Water absorption (%)	0·0–0·1		1–5
Coefficient of linear thermal expansion:			
Temp. range (°C)	20–700		25–600
Value/°C $\times 10^{-6}$	7·8–10·4		2·8
Safe operating temperature (°C)	1000–1100		1250
Thermal conductivity (c.g.s.)	0·0053–0·0063		
Tensile strength (lb/in²)	6200–13 000		3500
(kg/cm²)	436–914		246
Compressive strength (lb/in²)	65 000–130 000		40 000
(kg/cm²)	4570–9140		2812
Flexural strength (lb/in²)	14 000–24 000		8000
Modulus of rupture (kg/cm²)	984–1687		562
Impact strength (ft-lb); ½ in. rod			0·07
Impact resistance (lb/in²)			
Modulus of elasticity (lb/in²)	13–15 $\times 10^6$		
(kg/cm²)	9·1–10·5 $\times 10^5$		
Thermal shock resistance	Poor		
Dielectric strength (V/mil)	200–350		
(V/mm)	7870–13 780		—
Resistivity:			200 V 60 c
Temp. °C			24 900
(Ω/cm³)	10^{13}–10^{15}		$> 10^{14}$ $7·0 \times 10^4$
T_e-value (°C)	500–>700		
Power factor: c, cycles, or kc, kilocycles (10^3 cycles), or Mc, megacycles (10^6 cycles)	1 Mc 0·0002–0·00035	1000 kc 10 Mc 0·40 0·30	1000 kc 10 Mc 0·004 0·003
Loss factor: cycles value	 0·0011–0·026	1000 kc 10 Mc 2·0 1·5	1000 kc 10 Mc 0·020 0·015
Dielectric constant: cycles value	 5·5–7·5	60 4·0–5·0	1000 kc 10 Mc 5·0 5·0
Capacity change per °C: range (°C)			
Value (p.p.m./°C)			
Bulk density (lb/in³)			
(g/cm³)			
Volume (in³/lb)			
(cm³/g)			
Mohs' hardness	7·0–7·5		
Specific heat (B.t.u./lb)			
(c.g.s.)			

31 Cordierite AlSiMag 202	32 Cordierite AlSiMag 72	33 Cordierite Sipa H
A186 U.S.A. 2·1 1–5	**A186** U.S.A. 2·1 5–8	**S170** Germ. 2·1–2·2 0
25–100 | 25–600 1·3 | 2·8 1250 0·003 3500 246 40 000 2812 8000 562 0·075	25–100 | 25–600 1·5 | 2·5 1250 0·003 1500 105 30 000 2109 4000 281 0·075	20–100 1·1 1·7 3400–4800 239–337 41 400–69 000 2910–4851 6900–11 700 485–822 0·15
		508 50 cycles 20 000
25 | 300 | 900 $>10^{14}$ | 8.2×10^8 | 7.0×10^4 600 1000 kc | 10 Mc 0·004 | 0·003 1000 kc | 10 Mc 0·020 | 0·015 1000 kc | 10 Mc 5·0 | 5·0	25 | 300 | 900 $>10^{14}$ | 4.8×10^9 | 2.2×10^5 720 1000 kc | 10 Mc 0·004 | 0·003 1000 kc | 10 Mc 0·0164 | 0·0120 1000 kc | 10 Mc 4·1 | 4·0	20 | 200 | 600 10^{14} | 1.2×10^8 | 2.4×10^4 50 10^6–10^7 | 200 | 40–70 50 5–5·5
0·076 2·10 13·20 0·48	0·076 2·10 13·20 0·48	
		0·36–0·396 0·20–0·22

TABLE

No.	34	35	36
Trade name and/or body type	Cordierite Sipa 14	Cordierite	Forsterite AlSiMag 243
Ref.	S170	R71	A186
Country	Germ.		U.S.A.
Specific gravity	2·1	2·1–2·2	2·8
Water absorption (%)	1·0–5·0	0·0–8·0	0·0–0·05
Coefficient of linear thermal expansion:			
Temp. range (°C)	20–100	20–700	25–100 | 25–700
Value/°C $\times 10^{-6}$	1·2–1·7	2·0–2·8	9·1 | 10·6
Safe operating temperature (°C)		1200	1000
Thermal conductivity (c.g.s.)	2·2	0·0025–0·0061	
Tensile strength (lb/in²)	2000–3400	1500–5000	10 000
(kg/cm²)	140–239	105–351	703
Compressive strength (lb/in²)	41 400–69 000	30 000–70 000	85 000
(kg/cm²)	2910–4851	2109–4921	5976
Flexural strength ⎰ (lb/in²)	6900–9000	4000–12 000	20 000
Modulus of rupture ⎱ (kg/cm²)	485–632	281–843	1406
Impact strength (ft-lb); ¼ in. rod	0·18		0·33
Impact resistance (lb/in²)			
Modulus of elasticity (lb/in²)		7–13 $\times 10^6$	
(kg/cm²)		4·9–9·1 $\times 10^5$	
Thermal shock resistance		Excellent	
Dielectric strength Temp.			
(V/mil)		100–250	
(V/mm)		390–9840	
Resistivity: Temp. °C			
(Ω/cm³)		10^{11}–10^{15}	
T_e-value (°C)		500–700	
Power factor: c, cycles, or kc, kilocycles (10^3 cycles), or Mc, megacycles (10^6 cycles)		1 Mc 0·003–0·007	
Loss factor: cycles			
value		0·012–0·038	
Dielectric constant: cycles			
value		4·1–5·4	
Capacity change per °C: range (°C)			
Value (p.p.m./°C)			
Bulk density (lb/in³)			0·101
(g/cm³)			2·79
Volume (in³/lb)			9·91
(cm³/g)			0·36
Mohs' hardness		6·0–7·5	7·5
Specific heat (B.t.u./lb)	0·36		
(c.g.s.)	0·20		

256 (contd.)

37	38	39	40		
Alumina AlSiMag 393	Sintered alumina	'Heanium' 95% alumina	Coors Al–200 extra high strength alumina		
A186	**N36**		**C53**		
U.S.A.	U.S.A.		U.S.A.		
2·4		3·80	3·57		
12–18		0·0	0·0		
25–100 | 25–600			25–200 | 25–1000		
5·3 | 6·9	6		6·67 | 9·14		
1400		1600	1700		
—	0·03		25° C 0·004 | 800° C 0·01		
			25 000–27 000		
			1758–1898		
	400 000		275 000–290 000		
	28 123		19 334–20 389		
6000	50 000		49 000–50 000		
421	3515		3445–3515		
	52×10^6		$40·2 \times 10^6$		
	$3·6 \times 10^6$	Excellent	$2·8 \times 10^6$		
		215	320° C | 1000° C		
		280–340	20–30		
		8460	11 024–13 386 | 790–1180		
		10^{14}–10^{15}	100° C | 250° C | 500° C		
		> 700	$4·0 \times 10^{14}$ | $5·0 \times 10^{13}$ | $3·3 \times 10^{10}$		
			1070		
1000 kc			100 c | 1000 kc | 10 Mc		
0·0025		0·14	0·0014 | 0·00035 | 0·0003		
1000 kc			100 c | 1000 kc | 10 Mc		
0·014	0·01	0·015	0·0124 | 0·0031 | 0·0026		
1000 kc			100 c | 1000 kc | 10 Mc		
5·5	6	9·4	8·84 | 8·81 | 8·80		
0·087					
2·41					
11·50					
0·42					
		9·5	9		
			0·188		
			0·104		

TABLE

No.	41	42
Trade name and/or body type	Sintox alumina	Sintered 75–100% alumina
Ref.	L39	R71
Country	Brit.	
Specific gravity	3·65	3·1–3·5
Water absorption (%)		0·0
Coefficient of linear thermal expansion:		
Temp. range (°C)	0–300 | 300–800	20–700
Value/°C × 10^{-6}	6·2 | 8·25	5·5–8·1
Safe operating temperature (°C)		1350–1600
Thermal conductivity (c.g.s.)	0·054	0·0073–0·056
Tensile strength (lb/in²)		8000–50 000
(kg/cm²)		562–3515
Compressive strength (lb/in²)	240 000	80 000–410 000
(kg/cm²)	16 874	5624–28 826
Flexural strength ⎫ (lb/in²)	48 400	18 000–65 000
Modulus of rupture ⎭ (kg/cm²)	3403	1265–4570
Impact strength (ft-lb); ½ in. rod		
Impact resistance (lb/in²)		
Modulus of elasticity (lb/in²)	46·7 × 10^6	15–52 × 10^6
(kg/cm²)	3·28 × 10^6	1·0–3·6 × 10^6
Thermal shock resistance		Good
Dielectric strength (V/mil)	122	400–1100
(V/mm)	4800	15 750–43 310
Resistivity: Temp. °C	14 | 100 | 300	
(Ω/cm³)	10^{16} | 2 × 10^{15} | 3 × 10^{13}	10^{14}–10^{15}
T_e-value (°C)		> 700
Power factor: c, cycles, or kc, kilocycles (10^3 cycles), or Mc, megacycles (10^6 cycles)	1 kc | 1 Mc 0·0080 | 0·0006	1 Mc 0·0010–0·0020
Loss factor: cycles		
value		0·007–0·022
Dielectric constant: cycles	1 kc | 1 Mc	
value	10·0 | 9·6	7·3–11·0
Capacity change per °C: range (°C)	10–65	
Value (p.p.m./°C)	125	
Bulk density (lb/in³)		
(g/cm³)		
Volume (in³/lb)		
(cm³/g)		
Mohs' hardness		8·5–9·0
Specific heat (B.t.u./lb)	0·315	
(c.g.s.)	25° C 0·175	

256—(contd.)

	43 Zircon porcelain	44 Zircon porcelain	45 Westinghouse zircon porcelain	46 General zircon porcelain	47 Coors zircon porcelain		
	N11 U.S.A. 3·6–3·8 0	**N36** U.S.A.	**R71** U.S.A. 3·68 0	**R71** U.S.A. 3·0–3·8 0	**C53** U.S.A. 2·85 0		
	20–700		20–700	20–700	25–200	400–600	800–1000
	3·68–4·9 1000–1100	5 0·01	4·9 1050 0·0117	3·5–5·5 1000–1300 0·010–0·015	3·36	5·82 1400	7·70
	12 700 893 90 000 6327 23 000–30 000 1617–2109	 120 000 8437 30 000 2109	12 700 893 90 000 6327 25 000 1757	10 000–15 000 703–1054 80 000–150 000 5624–10 540 20 000–35 000 1406–2460	11 500–13 000 808–914 65 000–75 000 4570–5273 26 000–27 000 1828–1898		
	253	25 × 10⁶ 1·75 × 10⁶	24 × 10⁶ 1·68 × 10⁶ Good	20–30 × 10⁶ 1·4–2·1 × 10⁶ Good	19·4 × 10⁶ 1·36 × 10⁶		
	60 c 280–300 11 024–11 810 100 600 10¹⁴ 1–2 × 10⁶		290 11 417	250–350 9842–13 780	280–320 11 024–12 600 100 250 7·6 × 10¹⁴ 4·0 × 10¹¹		
	1000 kc 10 Mc 0·0005– 0·0005 0·0010		10¹³ > 700 1 Mc 0·0010–0·0014	10¹³–10¹⁵ > 700 1 Mc 0·0002–0·0020	515 1 kc 0·0040		
		0·01	0·009–0·013	0·0016–0·0210	1 kc 0·026 1 kc		
	9·0–10·5	9	9·2	8·0–10·5	6·39		
			8·0	7·5–8·5	8		

TABLE

No. Trade name and/or body type	48 'Stupalith' A2417 lithia body	49 'Stupalith' A2209 lithia porcelain	50 Wollastonite body
Ref.	S214	S214	C2
Country	U.S.A.	U.S.A.	U.S.A.
Specific gravity			
Water absorption (%)	21·5	0·01	
Coefficient of linear thermal expansion:			
Temp. range (°C)	20–500	20–500	
Value/°C × 10^{-6}	0·063	0·85	
Safe operating temperature (°C)		1000	
Thermal conductivity (c.g.s.)			
Tensile strength (lb/in²)			
(kg/cm²)			
Compressive strength (lb/in²)	20 000	60 000	
(kg/cm²)	1406	4218	
Flexural strength ⎧ (lb/in²)	4000	8000	21 000
Modulus of rupture ⎩ (kg/cm²)	281	562	1476
Impact strength (ft-lb); ½ in. rod	0·90	1·40	
Impact resistance (lb/in²)			
Modulus of elasticity (lb/in²)			
(kg/cm²)			
Thermal shock resistance	1260° C to cold water	500° C to cold water	
Dielectric strength (V/mil)		450	
(V/mm)		17 720	
Resistivity: Temp. °C			
(Ω/cm³)			
T_e-value (°C)			
Power factor: c, cycles, or kc, kilocycles (10^3 cycles), or Mc, megacycles (10^6 cycles)		Wet Dry 0·00492 0·00424	
Loss factor: cycles			
value		0·0273 0·0236	0·371
Dielectric constant: cycles			
value		5·56 5·57	5·91
Capacity change per °C: range (°C)			
Value (p.p.m./°C)			
Bulk density (lb/in³)	0·058	0·084	
(g/cm³)	1·60	2·34	
Volume (in³/lb)			
(cm³/g)			
Mohs' hardness			
Specific heat (B.t.u./lb)			
(c.g.s.)			

256—(contd.)

	51 Wollastonite body		52	53	54	55	56	57 General rutile	
			184	183	High magnesia bodies 181	180	179		
	J7 U.S.A.		S147 U.S.A. 3·25 0·10	S147 U.S.A. 3·25 0·14	S147 U.S.A. 3·48 0·0	S147 U.S.A. 3·36 0·0	S147 U.S.A. 3·23 0·062	N11 U.S.A.	
			0·0597	0·0608	0·0917	0·0864	0·0627		
			18 000 1265	15 000 1055	10 800 759	20 000 1406	10 500 738		
Dry 0·0572 Dry 0·360		Wet 0·0622 Wet 0·392	0·044 0·35	0·1013 0·79	0·028 0·22	0·046 0·40	0·044 0·34	1000 kc 0·08 1000 kc 6·8	10 Mc 0·06 10 Mc 5·1
6·3			7·89	7·78	8·21	8·67	7·91	60 85–115	

TABLE

No.	58	59
Trade name and/or body type	Rutile	Titania porcelain
Ref.	**N11**	**N36**
Country	U.S.A.	U.S.A.
Specific gravity	3·9–4·0	
Water absorption (%)	0–0·05	
Coefficient of linear thermal expansion:		
Temp. range (°C)	20–600	
Value/°C × 10^{-6}	7·1–8·6	7
Safe operating temperature (°C)	1000	
Thermal conductivity (c.g.s.)		—
Tensile strength (lb/in²)	6500–8000	
(kg/cm²)	457–562	
Compressive strength (lb/in²)	80 000–120 000	—
(kg/cm²)	5624–8437	
Flexural strength } (lb/in²)	6000–20 000	—
Modulus of rupture } (kg/cm²)	422–1406	
Impact strength (ft-lb); ½ in. rod	0·07–0·1	
Impact resistance (lb/in²)		
Modulus of elasticity (lb/in²)		—
(kg/cm²)		
Thermal shock resistance		
Dielectric strength (V/mil)	60 c 100–200	
(V/mm)	4000–8000	
Resistivity: Temp. °C	24 900	
(Ω/cm³)	$6·1 \times 10^{10}$ $6·1 \times 10^{2}$	
T_e-value (°C)		
Power factor: c, cycles, or kc, kilocycles (10^3 cycles), or Mc, megacycles (10^6 cycles)	1000 kc 10 Mc 0·0008 0·0006	
Loss factor: cycles	1000 kc 10 Mc	
value	0·0068 0·051	0·01
Dielectric constant: cycles		
value	85	100
Capacity change per °C: range (°C)	20–80	
Value (p.p.m./°C)	$-6·8 \times 10^{-4}$	
Bulk density (lb/in³)		
(g/cm³)		
Volume (in³/lb)		
(cm³/g)		
Mohs' hardness		
Specific heat (B.t.u./lb)		
(c.g.s.)		

256—(contd.)

60 'Faradex' rutile **S171** Brit.	61 Rutile **R55**	62 Titania **B80**	
4·0 0	4·0 0·05–0·00	3·6–4·1 0·0–0·1	
		20–100	20–1000
6–8	7·1–7·3	7–8	8–9
		0·007–0·010	
8400–11 600	7500	4000–11 000	
600–800	527	281–773	
80 000–128 000	80 000	40 000–120 000	
6000–9000	5624	2812–8437	
11 600–14 500	20 000	12 000–20 000	
800–1000	1406	843–1406	
	2·4	1·2–2·4	
1400–1500			
		$12·7–15·5 \times 10^6$	
		$0·89–1·09 \times 10^6$	
250–310	100	100–510	
10 000–12 000	4000	$4–20 \times 10^3$	
300	700 · · · 900	20 · · 300 · · 600	
4×10^8	$2·5 \times 10^4$ · · 1×10^4	10^{12} · · $10^8–10^9$ · · 10^5	
		450–500	
300 m · 100 m · 30 m	10 Mc	1 Mc · · 10 Mc	
1 Mc · 3 Mc · 10 Mc	0·06	3–8 · · 2–5	
0·00033 · 0·00030 · 0·00027– 0·00030			
	10 Mc	1 Mc	
	85	80–90 (or down to 32)	
−700 to −800	$−6·8 \times 10^{-4}$	−700 to 800 (for low ϵ −340 to −650)	
		0·126–0·144	
		3·5–4·0	
		0·306–0·378	
		0·17–0·21	

TABLE

		63	64	
No.				
Trade name and/or body type		Titania and titanates	Gen. magnesium orthotitanate	
Ref.		**H100**	**N11**	
Country			U.S.A.	
Specific gravity		3·5–5·5		
Water absorption (%)		0·0		
Coefficient of linear thermal expansion:				
Temp. range (°C)		20–700		
Value/°C $\times 10^{-6}$		7·0–10·0		
Safe operating temperature (°C)				
Thermal conductivity (c.g.s.)		0·008–0·01		
Tensile strength (lb/in^2)		4000–10 000		
(kg/cm^2)		281–703		
Compressive strength (lb/in^2)		40 000–120 000		
(kg/cm^2)		2812–8437		
Flexural strength ⎫ (lb/in^2)		10 000–22 000		
Modulus of rupture ⎭ (kg/cm^2)		703–1547		
Impact strength (ft-lb); ½ in. rod		0·3–0·5		
Impact resistance (lb/in^2)				
Modulus of elasticity (lb/in^2)		10–15 $\times 10^6$		
(kg/cm^2)		0·7–1·05 $\times 10^6$		
Thermal shock resistance		Poor		
Dielectric strength (V/mil)		50–300	500	
(V/mm)		2000–11 800	20 $\times 10^3$	
Resistivity: Temp. °C		Room		
(Ω/cm^3)		10^8–10^{15}		
T_e-value (°C)		200–400		
Power factor: c, cycles, or kc, kilocycles (10^3 cycles), or Mc, megacycles (10^6 cycles)		1 Mc 0·0002–0·050	1000 kc 0·12	10 Mc 0·08
Loss factor: cycles			1000 kc	10 Mc
value			1·8	1·2
Dielectric constant: cycles			60	
value		15–10 000	12·5–19·0	
Capacity change per °C: range (°C)				
Value (p.p.m./°C)				
Bulk density (lb/in^3)				
(g/cm^3)				
Volume (in^3/lb)				
(cm^3/g)				
Mohs' hardness				
Specific heat (B.t.u./lb)				
(c.g.s.)				

256—(contd.)

65 Magnesium orthotitanate	66 Magnesium orthotitanate	67 Magnesium titanate	
N11 U.S.A. 3·1	**R55, N11** 3·1	**B80** 3·2–3·6 0·0–0·1	
20–500 6·2–10	20–500 6·2–10	20–100 6–10	20–1000 8–11
		0·009	
8500–10 000 598–703 70 000–85 000 4921–5976 8500–15 000 598–1055	8500–10 000 600–700 70 000–85 000 5000–6000 8500–15 000 600–1100	8000–10 000 562–703 70 000–80 000 4921–5625 11 000–15 000 773–1055 1·2–1·5	
32–44	32–44		
50 c 375–500 15–20 × 10³	375–500 15–20 × 10³		
10 Mc 0·0008–0·0015	10 Mc 0·0008–0·0015	50 c 1 Mc 10 Mc 10 1 <1	
		1 Mc	
14–16 20–80 (−20 to −30) × 10⁻⁶	14–16 20–80 (−20 to −30) × 10⁻⁶	11–16 20–80 +30 to +50 0·112–0·126 3·1–3·5	
		0·378–0·396 0·21–0·22	

finer crushing→continuous high speed dry grinding cylinder→wet grinding cylinder. Grain size check.

Ball and china clays→wet cylinder grinding. Grain size check.

Mixing→magneting→sieves 140- and 325-mesh→filter pressing.

For dry pressing:→drying of press cake→pulverising→moistening→sieving and grading.

For plastic making: thorough pugging of press cake→de-airing pugmill.

Shaping. Dry pressing followed by fettling and polishing; or followed by turning with carborundum or corundum tools. Extrusion, cutting off, followed by turning when leather-hard.

Drying. Humidity dryers.

Glazing. All over, by spraying. Within defined limits, by rolling over glaze bearing cushion.

Firing. Cone 16–17, 1460–1480° C (2660–2696° F) in saggars. Oil-fired tunnel kilns.

Properties. Table 256, Nos. 5–7.

STONEWARE ELECTRICAL INSULATORS

Characteristics. Special stoneware bodies resemble electrical porcelain in many respects but always have the advantage of better workability and therefore ability to make larger single pieces up to 26 ft (~8 m) high.

Uses. Large one-piece insulators. High-tension conduits. Props, etc., for power stations (**S85**).

Manufacture. *See* Chapter 14, p. 1103.

Properties. Table 256, No. 8.

HIGH-TEMPERATURE INSULATORS

Characteristics. Bodies of incomplete vitrification and the minimum of glassy phase that could soften. The porosity, together with low coefficient of thermal expansion, assists in giving good thermal shock resistances. To retain good resistivity at high temperatures and to prevent attack of the resistor wire no alkalies may be present, fluxes may consist of alkaline earths.

Uses.
(1) Insulating supports for electrical heating elements. Arc chutes.
(2) Vacuum spacers, high-temperature insulation.
(3) High-frequency insulation, vacuum tube spacers.

Body Type. Use (1), porous cordierite. Use (2), alumina, aluminium silicate refractories (mullite). Use (3) massive fired talc, pyrophyllite. Pure magnesia, beryllia, zirconia and thoria.

Shaping. Cordierite heating element supports: extrusion; dry pressing.

IMPERVIOUS RECRYSTALLISED REFRACTORY ALUMINA

Characteristics. Highly refractory, can be used up to 1950° C (3542° F), very high degree of chemical inertness, impervious.
Uses. Tubes for wire-wound furnaces, *e.g.* molybdenum and tungsten. Pyrometer sheaths especially for the rare metal thermocouples.
Body Type. Alumina (**M96**).

Order of increasing Te Value for Ceramic Dielectrics

Feldspathic porcelain	~ 400° C	(752° F).
Feldspathic steatite	~ 450° C	(842° F).
Cordierite bodies	~ 700° C	(1292° F).
Zircon bodies	~ 700° C	(1292° F).
Alumina bodies	~ 750° C	(1382° F).
Quartz glass	~ 900° C	(1652° F).
Spec. steatite	~ 1025° C	(8776° F).
Magnesia bodies	~ 1150° C	(2102° F).

SPARKING PLUG INSULATORS

Requirements (in order of importance as cited by Riddle (**R19**)).

(1) Good hot dielectric properties.

(2) High dielectric strength. The insulator is relatively thin-walled and completely surrounded by metal. To be a reliable insulator it must be in perfect and homogeneous condition. Conductivity may not increase greatly with temperature.

(3) Resistance to thermal shock.

(4) Good thermal conductivity. There is a sharp rise and fall of temperature hundreds of times per minute in an engine and also temperatures up to 1100° C (2012° F) may be generated in hot spots.

(5) Good mechanical strength and resistance to mechanical shock. The insulator is clamped into position in the sparking plug by a gland nut exerting considerable force. The insulator also has to withstand constant vibration.

(6) Resistance to tetra-ethyl-lead corrosion. This additive to petrol forms a lead oxide deposit on the nose of the insulator and will eat into a silicate body to form lead silicate. This glazes the surface, is itself a conductor and forms a gummy surface that collects carbon and dirt and leads to short circuiting (**A61**).

(7) Low dielectric constant.

(8) Resistance to abrasion.

(9) Resistance to attack by carbon.

(10) Low modulus of elasticity.

Uses. Insulators for sparking plugs for all types of internal combustion engine.

Body Type. Sinter alumina, usually with small additions to reduce sintering temperature, *e.g.* 3% SiO_2 or control crystallisation, *e.g.* Cr_2O_3. Also high alumina porcelain, high mullite porcelain, zircon porcelain, steatite.

Raw Material. Pure alumina, fused alumina, controlled crystal size alumina, silica, chrome oxide, manganese oxide, organic binders, clays, zircon, fluxes, etc., *e.g.* magnesia, lime.

Preparation. Wet or dry very fine grinding→mixing. Spray drying to form free-flowing pellets for hydrostatic pressing.

Shaping. Extrusion followed by turning. Dry pressing followed by turning. Hydraulic pressing. Injection moulding with thermoplastic and slowly thermosetting resin mixed with body. Hydrostatic pressing (followed by grinding).

Drying.

Firing. 1600–1800° C (2912–3272° F). Gas muffle tunnel kilns, or open-flame gas tunnel kilns with saggars.

Glazing. Spraying.

Glost Firing. 1450° C (2642° F) (A5).

Properties. Table 256, Nos. 9–11. B.S. 45:1952.

HIGH-FREQUENCY CERAMIC INSULATORS

General Requirements.

(1) Low power loss, therefore low dielectric constant except for capacitors when high dielectric constant required.
(2) High dielectric strength.
(3) High specific resistance.
(4) Small temperature coefficient of items (1) (2), (3).
(5) Smooth non-absorbent surface.
(6) High mechanical strength.
(7) Capable of being easily shaped to close tolerances.
(8) Relatively low thermal expansion.

The relative importance of the requirements varies with the specific application so that a number of different bodies have been developed with particular properties, their grading being according to their loss factor.

Four, more specialised, groupings of high-frequency insulators are:

(1) General insulating purposes. Vitrified, mechanically strong materials, with low dielectric constant and low power factor.
(2) Capacitors. Vitrified materials with high dielectric constant and low power factor.
(3) For temperature independent oscillating circuits. Insulating materials with low coefficient of thermal expansion.
(4) Vacuum tube spacers. Porous or vitrified insulating materials with low dielectric constant and low loss factor.

B.S. 1598:1949.

BLOCK TALC PIECES
(also called 'Lava')

Characteristics. Ware prepared by machining natural steatite talc to the required shape, having very small firing shrinkage and therefore great dimensional accuracy. Although porous (1–3% water absorption) there is no glassy matter giving closed pores so that the material can be readily de-gassed. The electrical properties resemble those of other steatite bodies except that the dielectric loss is somewhat higher.
Uses. Vacuum tube spacers. Filament supports and spacers in electronic power tubes. Gas burner tips. Test and experimental parts and small orders of special shapes, eliminating cost of press dies.
Body Type. Steatite.
Raw Materials. Natural steatite talc free of chlorite, quartz, mica and other contaminants.
Shaping. Machining with lathes, drills, etc.
Firing. 1010° C (1850° F) (**N11**).
Properties. Table 256, Nos. 12, 13.

PHOSPHATE-BONDED TALC BLANKS OR PIECES

Characteristics. A fabricated uniform body that can be machined to close tolerances, and has other of the desirable properties of block talc at a more economical price. The properties are relatively insensitive to normal production variables.
Uses. As for block talc, especially for experimental shapes.
Body Type. Steatite.
Raw Materials. Normal ceramic powdered talcs. Orthophosphoric acid, H_3PO_4 (or aluminium dihydrogen phosphate, $Al(H_2PO_4)_3$, magnesium dihydrogen phosphate $Mg(H_2PO_4)_2$).
Preparation. Grinding→grading→mixing in of phosphate solution.
Shaping. Dry pressing. Hydrostatic pressing. Hot pressing.
Firing. 1000° C (1832° F).
   Gas-fired tunnel kiln.
Finishing. Machining with hardened steel tools (**C49**).
Properties. Table 256, Nos. 14, 15.

LOW-LOSS STEATITE BODIES

Characteristics. Dense, tough and strong low-loss body with high elasticity, which can be economically manufactured to close tolerances and complicated shapes. Power factor remains almost constant from 0–100° C (32–212° F) and rises only slightly up to 300° C (572° F). A highly coherent layer of silver can be deposited on it which can then be electroplated, soldered, etc.

Uses. Most widely used ceramic in electronic applications, *e.g.* radio industry: aerial equipment; wave band switches; tube sockets and supports; trimmer bases; condenser plates, stators and axles; coil formers; variometers; crystal holders; coaxial cable insulators; lead-in and stand-off antennae insulators; resistor shafts; stator supports for air condensers; relay insulators; valve holders, bases and spacers; mast bases; wire resistor cores; ball and socket insulator beads for higher temperature wire insulation.

Preparation.

For Plastic Making. Wet milling→sieves→magnets→filter press→mullers→de-airing pug.

For Dry Pressing.
A. *Wet Methods.*
  (1) Wet grinding in ball mills→filter pressing→drying of filter cake→grinding→moistening with binder solution→mixing→perforated drypan→tabletting machine (5690–7110 lb/in^2) 400–500 kg/cm^2→perforated drypan→size grading (**R29**).
  (2) Wet grinding→sieves→magnets→spray drying→pressing into bricks→disintegrator→size grading.
                                              ⋯⋯⋯⋯⋯⋯⋯⋯⋯⋯⋯⋯fines⋯⋯⋯⋯⋯⋯⋯⋯⋯⋯⋯⋯
  (3) Wet grinding→centrifuge→dry grinding→mixing with melted paraffin 10%→sieve→press (4270 lb/in^2) 300 kg/cm^2→granulate (**A4**).
B. *Dry Methods.*
  (4) Ready ground raw materials→weighed and proportionated→dry ground in pebble mill to mix→Simpson mixer 13% water and 3% dextrine added→granulated by rubbing through vibrating sieve→rotating dryer with direct flame→very hard grains→mixing of hard grains with undried material to give correct water content for different processes.
  (5) Dry body mixture→dry milling→addition of olein, petroleum, water, sulphite lye 0·5–5%→mixer→press (7110 lb/in^2) 500 kg/cm^2→edge runner mill or crushing rolls→size grading sieves.

Shaping. Extrusion, cutting and turning. 20–28% water: horizontal extrusion; 16–20% water: vertical extrusion.

Slip casting (avoided if possible, steatite shows poor castability because of high water content, slow build-up, and lower precision).

Plastic pressing, 12–16% water + 2% vegetable oil (lamination may occur) firing shrinkage 16–20%.

Dry pressing, 2–6 tons/in^2, 0–5% moisture. Bodies of high talc content can be pressed with zero moisture content and therefore require no drying. No lamination occurs. With properly granulated body very high production rates are possible in dry pressing. Firing shrinkage 6–10%.

Drying.
Fettling, etc.

CERAMICS IN THE ELECTRICAL INDUSTRY 1203

Firing. 1250–1400° C (2282–2552° F), cone 8–14. Steatite has very small firing range, 30–40° C, or, for ultra low loss only 10–20° C.

Electric-, gas- or oil-fired tunnel kilns. Small intermittent electric or gas kilns. Coal-fired round intermittent kilns.

Finishing. If fired to only 900° C can be sawn, drilled, ground, shaped, milled, gears and screws cut, etc. Grinding, lapping, honing, tumbling, polishing. Fully vitrified pieces can be highly polished. Metallising and metal assembly.

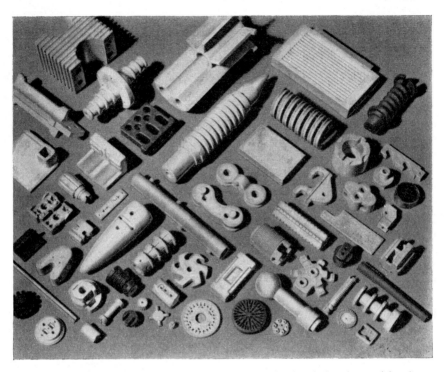

FIG. 16.3. *A selection of precision-made steatite pieces for electrical equipment (American Lava,* **A186**)

Glazing. Spraying.
Glost Firing. Circular tunnel kiln.
Properties. Table 256, Nos. 16–28.

LOW-LOSS CORDIERITE BODIES

Characteristics. Dense bodies of very low thermal expansion and therefore very high thermal shock resistance. They are difficult to glaze successfully. Suitable both for high-frequency and for high-tension applications.

Uses. Resistance wire supports. Coil forms, especially in radio where dimensional stability means constancy of inductance of the coil. Electric water-boiler insulators. Arc chutes. Fuse cores and casings for high currents. Oil burner ignition insulators. Rheostat blocks and dimmer winding cores. Resistor bobbins and spools.
Body Type. Cordierite.
Raw Materials. Talc. Clay. Alumina.
Manufacture. Similar to steatite. The firing range tends to be very small.
Properties. Table 256, Nos. 29–35.

LOW-LOSS FORSTERITE BODIES

Characteristics. Dense ware consisting largely of forsterite crystals with excellent dielectric properties at elevated temperatures. The resistivity falls to 10^6 only above 1000° C (1832° F). The expansion coefficient is high so that thermal shock resistance is low, but expansion is practically linear from room temperature to 1000° C (1832° F). The expansion matches that of some nickel–iron alloys so that strain-free vacuum- or pressure-tight metal–ceramic seals can be made. The power factor remains low at centimetre wavelengths when steatite bodies become quite lossy.
Uses. Low-loss electronic uses.
Body Type. Forsterite.
Raw Materials. Talc. Magnesium hydroxide. Clays. Alkaline earth compounds.
Manufacture. As for steatite.
Firing. 1250–1400° C (2282–2552° F) cone 8–14. Wider firing range than steatite.
Finishing. Sealing into suitable nickel–iron alloy.

(*a*) Small assemblies, not required to insulate high voltages at high frequencies: molybdenum metallising process. (Iron and molybdenum or iron and manganese powders are fused to the ceramic in a hydrogen atmosphere. The conductive coating is electroplated and can then be brazed or soft soldered.)

(*b*) Large assemblies: vacuum–hydride process. (Hydrides of titanium, zirconium, thorium or tantalum are applied as a bond between metal and ceramic and decomposed to the metal by firing *in vacuo* (**H100**)).
Properties. Table 256, No. 36.

LOW-LOSS ALUMINA PIECES

Characteristics. Very low dielectric loss combined with excellent mechanical properties make these bodies especially useful for insulators exposed to mechanical shock or required in small cross-sections.

Uses. Hermetically sealed terminals for transformers, motors, capacitor housings, relays, switches.
Properties. Table 256, Nos. 37–42.

LOW-LOSS ZIRCON PORCELAIN PIECES

Characteristics. Many properties lie between those of porcelain and steatite. Very good heat–shock resistance because of a low expansion coefficient, also good mechanical strength, complete vitrification (impervious to high-pressure dye solutions). Low power factor and low loss factor up to 200° C (392° F).

These bodies contain clay and can be shaped by plastic methods, their firing range is much greater than other low-loss bodies so that they can be mass produced much more easily. However, zircon is very abrasive towards dies.

For many electrical purposes zircon porcelain is superior to all but sinter-alumina which is too expensive to be competitive.

Uses. High-frequency work. Sparking plugs. Circuit breakers (using high strength). Supporting base for tube resistors. High temperature high voltage insulation. Terminal and switch plates.
Body Type. Zircon porcelain.
Raw Materials. Zircon. Double zirconium alkaline earth silicates. Clay. Bentonite.
Preparation. Zircon→grinding. Wet or dry mixing of materials, normal methods.
Shaping. The body is plastic.
  Casting.
  Dry pressing.
  Throwing and turning, extrusion and turning, plastic forming (13–17% moisture). The zircon sand is very abrasive and wears out dies and tools easily. Tungsten carbide tools must be used (**H100**). Wet methods, in which an evacuated die is used, reduce wear (**N11**).
Drying.
Glazing. Normal glazes are absorbed by zircon bodies unless very heavily applied and in any case craze. Special glazes give compression.
Firing. According to body, cones 6 upwards to 26, but normally cone 10–13, 1300–1380° C (2372–2516° F) (**R71**).
Properties. Table 256, Nos. 43–47.

LITHIA PORCELAIN

Characteristics. A body that can be adjusted to have small positive, negative or zero expansion. The last type can be used for completely temperature stable pieces. Very high thermal shock resistance shown by

resistance to 100 repeated cycles between 1090°C and −190° C (1994 to −310° F).
Uses. Temperature stable dielectric pieces.
Body Type. Lithia porcelain.
Raw Materials. Lithium carbonate. Clay. Flint. Alumina.
Preparation, Shaping, Firing. Normal technical porcelain methods.
Properties. Table 256, Nos. 48, 49.

LOW-LOSS WOLLASTONITE PIECES

Characteristics. Very low dielectric losses combined with good mechanical strength. Vitrification occurs at considerably lower temperature than steatite, zircon and other low-loss bodies. High-frequency applications.
Body Type. Wollastonite.
Raw Materials. Wollastonite. Ball clay, (bentonite). Barium carbonate; boron phosphate; zircon; alumina; barium zirconium silicate. Organic binders: methocel; carbowax.
Preparation. Grinding→grading→dry mixing in Lancaster mixer→ solutions of binders added, 17% moisture→mix→comminuting machine to give correct particle size for extrusion. Extrusion scrap dried and graded for dry pressing.
Shaping. Horizontal ram extrusion. Dry pressing.
Drying.
Firing. 1200–1250° C (2192–2282° F), cone 6–8. Dry to fired shrinkage of extruded pieces 14·5–15·8%. Dry to fired shrinkage of power pressed pieces 7–8% in width and 5–6% in depth (**J7**).
Properties. Table 256, Nos. 50, 51.

CERAMIC INSULATORS AS SPACERS IN VACUUM TUBES

Requirements. High electrical resistance and low dielectric loss at high temperature. Easily de-gassed and free from gas inclusions. Must be shaped to very accurate dimensions into intricate shapes.
Bodies. High alumina. High beryllia. High magnesia. High zirconia. Special non-shrinking body: 60% constituents melting above 2000° C, 40% constituents melting below 1000° C, binder; this is fired to slightly above the low-melting constituent (**P72**).
Special Shaping Method for Alumina or Zirconia. Make into thick paste with water, press into mould, dry and fire to 1000° C (1822° F)→ machine to accurate shape with high-speed tools allowing for 10% shrinkage →fire 1700° C (3092° F) (**S19**).

RUTILE BODIES

Characteristics. Bodies of dielectric constant up to about 85, and a negative temperature coefficient of dielectric constant. Small alterations of composition can be made to give predetermined values of ϵ and the temperature coefficient. By suitable blending with materials of positive temperature coefficient a body of zero coefficient can be made. Bodies of very low power factor can be made.

Uses. Capacitors and condensers for trimming circuit components of opposite temperature drift to maintain constant frequency characteristics, *e.g.* tuned wireless circuits (**N11, H100**).

Body Type. Rutile.

Raw Materials. Purified fine titania. Fluxes: titanates; oxides. Clay.

Preparation. Grinding→mixing→sieving→filter pressing→granulating→ size grading.

Shaping.
(1) Dry pressing, with aid of waxes or dextrine.
(2) Hydrostatic pressing and turning.
(3) Extrusion, with plasticisers.
(4) Plastic methods.
(5) Casting, deflocculated with sodium pyrophosphate.

Firing. 1300–1400° C (2372–2552° F), cone 10–13/14. Placed on calcined or fused alumina. Highly oxidising atmosphere with carefully regulated heating and cooling schedule essential (**B80**).

Properties. Table 256, Nos. 57–62.

HIGH PERMITTIVITY RUTILE BODIES

Characteristics. Dielectric constant from 100 to 2000, and positive temperature coefficient. Some samples show different colours in the light and the dark, probably due to some photoelectric effect. Transparent to ultra-violet light.

Uses. Capacitors. Optical and electrical lenses. V.H.F. antennae. Increasing voltage and power output of electrostatic generators, *e.g.* Wimshurst machines. Condenser microphones (**A18**).

General production as for rutile pieces.

Properties. Table 256, No. 63.

MAGNESIUM TITANATE BODIES

Characteristics. Very low power factor below 1×10^{-4} at 1 Mc/s, 20° C, dielectric constant about 10–20 at 25° C and 1 Mc/s, and slightly positive to zero temperature coefficient of capacity.

Uses. Temperature-stable radio equipment.

Making. Low clay and high alkalinity leads to poor plasticity. Small pieces: pressing; extrusion. Larger pieces: pressing followed by turning.

Firing. 1400–1600° C (2552–2912° F) (**B80**).

Properties. Table 256, Nos. 64–67.

HIGH PERMITTIVITY CAPACITORS

Characteristics. High permittivity allowing much smaller capacitors but generally high temperature coefficient and some undesirable ageing effect.
Uses. By-pass and coupling capacitors.
Body Type.

	ϵ at room temperature
Barium titanate bodies	1200–1500
Barium–strontium titanate bodies	10 000
Strontium titanate bodies	225–250
Calcium titanate bodies	150–175 (**H100**)
[Compare heavy grade titania	95–105
Magnesium titanate	13–18 (**K56**)]

Raw Materials. Prepared pure titanates. Bentonite 1–2%. Other binders.
Firing. Pure $BaTiO_3$ 1400–1450° C (2552–2642° F). $BaTiO_3$ with 1–2% clay 1350° C (2462° F). Zirconia as plates, powder and coating slip is the best placing material.

NON-LINEAR CIRCUIT ELEMENTS

Characteristics. Non-linearity of dielectric constant with applied field, unfortunately coupled with large temperature dependence of the dielectric constant.
Uses. Dielectric amplifiers.
Body Type. Barium titanate (**H100**).

PIEZOELECTRIC AND FERROELECTRIC BODIES

Characteristics. Piezoelectric properties comparing very favourably with other materials such as Rochelle salt, but with much better mechanical, chemical and thermal stability.
Uses. Ultrasonic transducers for: emulsification of oil-soluble and water-soluble liquids; stimulation or destruction of bacteria; transformation of crystal structures; mixing of powders and paints; homogenisation of milk; agglomeration of smoke and dust particles; thickness gauges. Microphones. Pickups. Coaxial tweeters. Accelerometers measuring high-frequency shock and vibration. Strain gauges (**H100**).
Body Type. Barium titanate.

PIEZOELECTRIC BODIES

Characteristics. Bodies of better output and temperature operating conditions; with high Curie point permitting the design of transducers for

immersion in boiling liquids and, if necessary, dip and soft soldering; low dielectric loss giving low heating; high electromechanical coupling factor; high dielectric constant; high dielectric breakdown enables the material to be used under conditions approaching its ultimate tensile or compressive strength; high resistivity at high temperature; low coefficient of expansion. Adequate stability of the more important parameters over a wide temperature range. Compared with other ceramics, volume can be reduced for the same power input or power handling capacity increased for the same volume.
Uses. Ultrasonic cleaning of precision parts. Drilling and cutting of hard materials. Descaling of metal. Flaw detection. Prevention of scum in soldering baths. Echo sounders. Sonar systems. Hydrophones. Resonant transformers. High voltage impact generators, 'spark pump' for internal combustion engines. Pressure measurements. Structure vibration analysis. Ultrasonic cold welding. Strain gauges. Accelerometers. Relays. Displacement gauges. Gramophone pickups. Microphones. Filters. I.F. Transformers. Sterilisers. Ultrasonic therapy. Transfilters.
Body Type. Lead zirconate-titanate.
Raw Materials. Solid solutions of lead zirconate-titanate in equimolecular proportions. Plus small amounts of one or more of the oxides of niobium, lanthanum, tantalum, nickel, iron, cobalt, chromium or alkaline earth, to modify properties for the various types of application.
Shaping. Pressing. Extruding. Casting.
Drying.
Firing.
Finishing. Grinding to accurate dimensions in order to achieve the desired resonant frequency.
Testing. All piezoelectric properties (**B128.1**).

FERRITES OR FERROMAGNETIC BODIES

Characteristics. High electrical resistance and so low eddy current losses, combined with permeability high compared with laminated metallic low-loss magnetic cores. Operable at much higher frequencies than metallic components. A number of properties can be varied over a fairly wide range (*see* discussion of 'ferrite bodies', Chapter 5).
Uses. Largely above 1 kc, where metals lose their usefulness. Television receivers, giving larger and brighter pictures for smaller components and therefore lower cost and cabinet size. Radio receivers, especially portables and transistor sets. Remote control of tuning devices, sweep oscillators, travelling wave tubes. Recording equipment and pickup heads. Telephone carrier filter coils. Pulse transformers.
Shapes. Cup or pot-cores; C-cores; E-cores; toroids; circle segments; flat strips; rods; tubes.
Body Type. Ferrite.
Raw Materials. Pure synthetic metal oxides, hydroxides or carbonates of

controlled particle size of from 0·1 to 2 μ. Binders: alginates, starches, gums and polyvinyl alcohol. Press lubricant: stearin.

Preparation. Proportionating→wet ball milling→filter press→dry→prefiring to 1000° C (1832° F)→milling→addition of organic binder→granulation.

Shaping. Dry pressing, mechanical or hydraulic presses 6–10 tons/in², (945–1575 kg/cm²). Tungsten carbide dies. Extrusion.

Firing. Controlled atmosphere required, therefore usually electric continuous kilns with silicon carbide elements. 1250–1400° C (2282–2552° F) depending on compounds. Up to 20% shrinkage.

Finishing. Only limited grinding possible with silicon carbide wheels under water.

ATTENUATOR MATERIALS

Characteristics. Materials that decrease the amplitude of a wave without distortion.

Uses. Microwave devices for radar and communications systems. High-power and wave-guide applications.

Body Type. Steatite or porcelain with varying amounts of silicon carbide or graphite (**T32**).

SILICON CARBIDE ELECTRICAL HEATING ELEMENTS OR ELECTRICAL RESISTORS

Characteristics. Non-metallic electric resistance heater that can be used for temperatures at which metallic resistors fail, *i.e.* up to 1400° C (2552° F) or even, if specially treated, 1575° C (2867° F). Each element must have low-resistance ends to enable metal contact to be made. Expansion coefficient is low, specific resistance high and mechanical strength good. During use slow oxidation takes place leading to increased resistance so that a transformer must be used to step up the voltage accordingly. Eventually mechanical failure occurs due to the expansion taking place during the change.

ARC CARBONS

Uses. Searchlights. Cinema projectors. Photography.

Body Type. Carbon.

Raw Materials. Prefired lampblack. Lampblack. Prefired coke with boric acid. Arc carbon scrap. Pitch. Anthracene oil. Rare earth fluorides for positive cores.

Preparation. Mixing → kneading → rolling → extrusion as 'spaghetti' → broken up in edge runner mill→stamping into slugs.

Shaping. Shells: extrusion; intermittent plunger presses with gas-heated nozzles and steam-heated chambers. Cores: extruded into prefired shells.

Firing. Shells: up to 1000° C (1832° F). Cores: 150 to 900° C (302–1652° F). Ring furnace.
Finishing. Machining: completed shells cut approximately to length on carborundum wheels. After core insertion final grinding to shape. Copper-plating (**A13**).

BATTERY CARBONS

Uses. Batteries.
Body Type. Carbon.
Raw Materials. Anthracite. Battery carbon scrap. Lampblack. Carbon dust.
Preparation. Mixing→rolling→remixing→stamping into slugs.
Shaping. Extrusion. Intermittent heated presses.
Firing. 1000° C (1832° F). Ring furnaces.
Finishing. Breaking to length. Grinding ends. Impregnating with wax (**A13**).

ELECTRIC MOTOR BRUSHES

Body Type. Carbon.
Raw Materials. Hard carbon (containing small amount graphite). Graphite (up to 90% natural graphite). Electrographite. Copper graphite (5 to 60% natural graphite).
Preparation. Mixing.
Shaping. Pressing into blocks. Extrusion.
Firing and electrographitising if necessary.
Finishing. Cutting to size (**A13**).

Chapter 17
Constructional Refractories

THE different types of refractory bodies have been discussed in Chapter 5. The need for this wide range is seen better when some of the demands made on refractory products are considered.

A large number of major industries are dependent on a process using heat and are therefore dependent on heat-resisting structures of some sort. Refractories are of major importance to them and technical advances are frequently influenced by the availability of better refractories. Industries dependent on refractories include:

(1) Iron and steel.
(2) Non-ferrous metals.
(3) Glass.
(4) Gas.
(5) Steam power.
(6) Ceramics.
(7) Gas turbine and jet propulsion.
(8) Nuclear power.
(9) Cement and lime manufacture.
(10) Incinerators.
(11) Paper mills.
(12) Enamelling.
(13) Domestic heating.

The demands made on refractories are not only those of heat resistance but involve a number of other factors derived from their particular uses. These may be divided into three main types:

(1) Strength (resistance to mechanical force).
(2) Resistance to destruction (by physical and chemical forces).
(3) Confinement of heat (**C64**).

Strength

Refractories to be used for construction need to have adequate compressive strength both at normal and elevated temperatures. They also need to have

CONSTRUCTIONAL REFRACTORIES

sufficient dimensional accuracy to be laid with thin joints and also dimensional stability. Properties that must be measured and taken into account when choosing refractories with regard to their strength include the following.

(1) Refractoriness.
(2) Refractoriness under load.
(3) Compressive strength.
(4) Dimensional accuracy.
(5) Dimensional stability, or after-expansion or contraction.
(6) Resistance to abrasion.

Resistance to Destruction

There are two main destructive forces which may act on refractories. The one is the chemical action of slags, dusts and gases. The other is the physical action of expansion or contraction that can lead to spalling.

A. CHEMICAL FORCES

The slags of metallurgical furnaces vary in composition and are classed as acid, basic or neutral. They react least with refractories of the same chemical character as themselves, *i.e.* acid refractories must be used for acid slags. Slag action is also dependent on a number of physical factors. The intensity of corrosive action depends on such factors as the following:

Temperature of the furnace.
Furnace atmosphere.
Chemical composition of the refractory.
Porosity and pore structure of the refractory.
Temperature gradient through the brick, which governs the depth to which a molten flux will penetrate.
Chemical composition and melting temperature of the fluxing agent.
Fluidity of the fluxing agent.
Rate at which fluxing agent is supplied.
Surface tension between flux and the refractory; that is the tendency of the flux to 'wet' the refractory.
Fluidity of the reaction product.
Rate at which the reaction product flows away from the face of the refractory, or is removed in other ways.
Agitation or turbulence of molten fluxing agent, flowing down the face of the furnace wall (the turbulence is usually formed by gases moving at high velocity).
Turbulence of slag floating upon a molten bath, as in a copper converter.
Convection currents, as in a molten bath such as glass (**H18**).

Staerker (**S159**) has investigated the magnitude of the factor derived from the surface tension between the refractory surface and the liquid corrosive agent and found it to be of such importance that special surface treatment with

TABLE 257

Interaction of Refractories

Refractory Types		Temp. at which reaction becomes damaging	
		(°C)	(°F)
Chrome, fired	Forsterite	1650	3000
Chrome, fired	High-Alumina 70%	1600	2910
Chrome–magnesite chem. bond or fired	High-Alumina 70%	1700	3100
Aluminous fireclay	Forsterite	1600	2910
Aluminous fireclay	Magnesite, fired	1500	2730
Aluminous fireclay	Silica	1600	2910
Forsterite	Chrome, fired	1650	3000
Forsterite	Aluminous fireclay	1600	2910
Forsterite	High-alumina 70%	1700	3100
Forsterite	High-alumina 90%	1700	3100
Forsterite	Magnesite, fired	1700	3100
High-alumina 70%	Chrome, fired	1600	2910
High-alumina 70%	Chrome–magnesite	1700	3100
High-alumina 70%	Forsterite	1700	3100
High-alumina 70%	Magnesite, fired	1700	3100
High-alumina 70%	Magnesite–chrome chem. bonded	1700	3100
High-alumina 90%	Forsterite	1700	3100
High-alumina 90%	Magnesite fired	1650	3000
Magnesite, fired	Aluminous fireclay	1500	2730
Magnesite, fired	Forsterite	1700	3100
Magnesite, fired	High-alumina 70%	1700	3100
Magnesite, fired	High-alumina 90%	1650	3000
Magnesite, fired	Silica	1600	2910
Magnesite–chrome chem. bonded	High-alumina 70%	1700	3100
	Silica	1650	3000
Silica	Aluminous fireclay	1600	2910
Silica	Magnesite, fired	1600	2910
Silica	Magnesite–chrome	1650	3000

'Vanal' can make a very appreciable difference. This has been confirmed by Lehmann and Singh (**L29**).

Furnace gases also may have a destructive effect on refractories. Refractories containing iron compounds may be affected by alternate oxidation and reduction which may break bonds and make them friable.

Carbon monoxide can cause severe corrosion by decomposing into carbon dioxide and carbon, the latter being deposited in the brick. The reaction occurs most readily between 370 and 540° C (700–1000° F) and is catalysed by iron oxide. Partially cracked natural gas has a similar effect. Fireclay and high-alumina bricks fired to cone 18 or higher resist this destructive action.

The bond in many kinds of brick is weakened by exposure at furnace temperatures to chlorine, hydrogen chloride, sulphur dioxide and trioxide, volatile metal oxides and carbon monoxide above the temperature at which it deposits carbon.

Refractories also react more or less severely with each other. The pairs shown in Table 257 react damagingly with each other and should therefore be separated in furnace construction by a third refractory that reacts with neither.

B. SPALLING

Norton (**N35**) describes spalling as a fracture of the refractory brick or block resulting from any of the following causes:

'(1) A temperature gradient in the brick, due to uneven heating or cooling, that is sufficient to set up stresses of such magnitude as to cause failure.

'(2) Compression in a structure of refractories, due to expansion of the whole from a rise of temperature, sufficient to cause shear failures.

'(3) Variation in coefficient of thermal expansion between the surface layer and the body of the brick, due to surface slag penetration or to a structural change in service, great enough to shear off the surface layer.'

The first type is considered by many to be the only true form of spalling, the second is due to poor furnace design with insufficient expansion joints, and the third is due to using bricks at temperatures higher than that at which they were burned, or by slag penetration. The first type of spalling is therefore the only one inherent in the nature of the refractory.

Norton (**N35**) goes on to differentiate between spalling occurring, respectively, during uneven heating and cooling of a piece. In the first instance expansion leads to compressive forces and failure due to shear stresses, in the second, contraction leads to tension and failure due to tensile stresses. He derives the following two equations making use of standard physical properties of the material:

α = coefficient of expansion;
κ = conductivity;
ρ = density;
c_p = specific heat;
h^2 = diffusivity, the rate at which a point in a hot body will cool under definite surface conditions

$$h^2 = \frac{\kappa}{\rho c_p}$$

ϕ_b = maximum shearing strain or flexibility;
dx = distance over which the temperature drop dt exists;
Σ_b = maximum tensile strain;
k_1, k_2 = constants.

Tendency to spall due to shear stresses S_s is given by

$$S_s = k_2 \frac{\alpha}{h\phi_b}$$

Tendency to spall due to tensile stresses S_t is given by:

$$S_t = k_1 \cdot \mathrm{d}x \frac{\alpha}{h\Sigma_b}$$

Experimental measurements were made of the coefficient of expansion, flexibility and diffusivity at 500° C (932° F) to give values of S_s. These were in good agreement with the results given by laboratory and service spalling tests.

Generally, experience and laboratory spalling tests have given indications of conditions that can lead to better spalling resistance of refractory bricks. With respect to the composition of the brick, and therefore its coefficient of expansion, fireclay gives good general results, silica is very poor in the low-temperature range but above the inversion temperature it is very good; chrome and magnesite have higher expansions and so are less resistant except when a specially flexible structure is used. Silicon carbide is very good.

The method of manufacture has considerable influence on the spalling resistance of a particular type of brick. The method of shaping, whether hand-made, stiff-mud or dry-press leads to different, though unpredictable, results. The spalling resistance increases with the size of grog, and with the amount of grog up to 60%. It frequently decreases with increased burning temperature. Spalling is greatly aggravated by flaws or laminations in bricks so that apparently identical bricks may give quite different service. Furthermore it is obvious that a highly porous brick with large size grog, that would be very resistant to spalling, will not be mechanically strong enough for constructional use (N35).

Confinement of Heat

Refractories are required under different circumstances to transfer, store or confine heat. The refractory materials are all better conductors of heat than air, so that a brick that is to transfer heat is made as dense as possible and has a crystalline structure. For heat storage, as in chequer work for heat exchangers, again a dense brick of maximum weight per unit volume has the greatest heat capacity (weight × specific heat) as the various refractories have similar specific heats.

For insulation the exactly opposite arrangement, a maximum of air-filled pores, is required. The size and nature of these pores is of considerable importance. Heat transfer through a porous brick occurs by conduction, convection and radiation. Air is a very poor conductor compared to the solid brick so that on this basis the more air incorporated the less heat transfer occurs. However, losses by convection are greater the larger the individual air space and losses by radiation are greater the greater the temperature difference between opposite solid faces, which is greater the larger the air gap. The last factor is of increasing importance the higher the temperature, so that at the service temperatures of insulating refractories a fine pored brick is the

TABLE 258

Relative Factors for Determining the Usefulness of Commercial Refractory Products(N27)

Commercially available product	Refractoriness	Load bearing at high temperatures	Spalling resistance	Heat transfer characteristics	Resistance to oxidation	Resistance to reducing atmospheres	Acid slag resistance	Basic slag resistance	Resistance to slag adherence	Resistance to slag penetration	Electrical resistance at high temperatures	Weight of refractory article	Resistance to mechanical abrasion
Silicon carbide	2	1	1	1	3†	1	1	3	1	1	5	2	1
Aluminous materials:													
Fused alumina	2	2	4	2	1	1	1	2	4	4	2	2	1
Corundum	2	3	3	3	1	1	1	2	4	3	3	2	2
Bauxite	3	3	3	3	1	1	1	2	4	3	3	3	3
Diaspore	3	3	3	3	1	1	1	2	4	3	3	3	3
Magnesium compounds:													
Sintered magnesia	1	4	5	2	1	1	5	1	4	3	1	2	3
Calcined magnesia	1	4	5	3	1	1	5	1	4	3	2	2	3
Sintered MgO with fused alumina	2	4	3	3	1	1	3	2	4	3	2	2	3
Magnesia spinels	2	4	3	3	1	1	3	2	4	3	2	2	3
Forsterite	2	1	3	4	1	1	3	3	4	3	2	2	3
Aluminium silicates:													
Mullite (fused and cast)	2	1	4	3	1	1	1	3	4	1	2	2	1
Sillimanite	2	2	2	3	1	1	1	3	4	2	2	3	2
Andalusite	2	2	2	3	1	1	1	3	4	2	2	3	2
Kyanite	2	2	2	3	1	1	1	3	4	2	2	3	2
Kaolins	2	2	2	3	1	1	1	3	4	2	3	3	2
Fireclay	3	4	5	4	1	4	1	4	4	2	3	3	3
Chromites	2	4	5	3	1	1	1	1	4	2	3	1	2
Silica	2	1	1	4	5	1	1	5	4	1	2	3	2
Graphite	1	1	1	1	1	3	2	1	4	1	5	4	5
Zirconium oxide	1	1	1	4	1	3	2	4	4	3	2	1	3
Zirconium silicate	2	2	1	4	1	3	2	4	4	3	2	1	3

* *Relative classification*: 1. Very high. 2. High. 3. Medium. 4. Low. 5. Very low.
† This figure 3 applies to a temperature range of 800 to 1200° C (1475 to 2200° F); at other temperatures a No. 2 classification applies.

See also Fig. 11.10, refractories utility chart, p. 896.

best. It is also important that the pores should be sealed as far as possible so that hot gases cannot permeate the structure and carry away heat (**S124**).

SILICA BRICK

Characteristics. Acid refractory, capable of carrying a load (50 lb/in^2, 3·5 kg/cm^2) to within a few degrees of its cone fusion point, *i.e.* 1710 to 1730° C (3110–3146° F), and which can safely be used in structures up to 1650° C (3002° F). It is free from shrinkage at temperatures up to its melting point and has a high thermal shock resistance in the range 600–1700° C (1112–3092° F). It has a high resistance to attack by the principal steel furnace fluxes, namely iron oxide, lime, and acid slags generally, and high abrasion resistance. It is readily attacked by basic slags and fluorine. Thermal conductivity at high temperatures is about 25% greater than fireclay. The only serious limitation is the sensitivity to thermal shock below 600° C (1112° F), or below 300° C (572° F) for a well-fired brick.

α-quartz $\longleftrightarrow$ β-quartz 573° C
α-tridymite $\longleftrightarrow$ β-tridymite 117° C
β-tridymite $\longleftrightarrow$ γ-tridymite 163° C
α-cristobalite $\longleftrightarrow$ β-cristobalite 230° C

Uses. Steel plant acid and basic open hearth furnaces where operating temperatures frequently reach the safe maximum of 1650° C (3002° F). Acid Bessemer converters. Electric steel. Coke ovens. Gas works. Glass tanks.
Body Type. Refractory. Silica.
Raw Materials. Quartzite or ganister. Hydrated lime, sulphite lye. Mineraliser, *e.g.* frit containing Na$_2$O, Fe$_2$O$_3$, SiO$_2$ (**S7**).
Preparation. Washing to remove clay impurities→primary jaw crusher reducing to about 2 in. (5·08 cm)→disintegrator or cone crusher

    ↓   oversize ↓
sample for quality control   └──────screens.

or enclosed dry pan plus ball mill for fines;
or perforated ball mill fitted with sieves and supplemented with ball mill ground fines→magnetic separators→bins.

Proportionating of correct size grading, *e.g.* 45% coarse, 10% medium, 45% fine. Average sizes of particles in the silica bricks used in British steel works:

	Limits (%)	
On 6-mesh	13	9
6- to 20-mesh	28	17
20- to 60-mesh	11	25
60- to 120-mesh	14	11
Through 120-mesh	34	38
	100	100 (**S50**)

Salmang (**S8**) quotes 40% at less than 0·06 mm B.S. Sieve 240.

→Addition of hydrated lime, mineralisers, sulphite lye, in revolving-bottom counter-flow mixer with rolls raised to prevent further grinding→sample for grading, moisture, and lime control; *or* stationary pans with revolving mullers and scrapers.

Shaping. Hand soft plastic moulding or Drop machine, 7–8% moisture. Pressing in hydraulic, toggle or table presses, 4–6% moisture. Special shapes by hand tamping or pneumatic ramming in loose lined moulds, 6–7% moisture.

Drying.

(1) Hot floors, this leads to cracking, which is not revealed until after the bricks are fired.

(2) Tunnel dryers for about 24 hr. (Dry-pressed bricks can be set straight on kiln cars and passed first through the tunnel dryer.) Moisture is reduced to 0·2%.

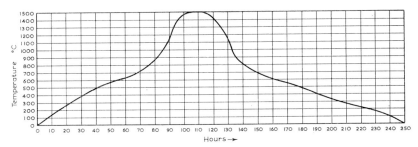

FIG. 17.1. *Ideal firing curve for silica bricks showing the regions where faster firing are permissible* (Miehr, **M73**)

Firing. 1430–1500° C (2606–2732° F) cone 14/15 to 18.

(1) The soaking temperature must be regulated according to the proportion of tridymite to cristobalite desired. For maximum tridymite 1430° C is best, above this, conversion of cristobalite to tridymite slows down and ceases altogether at 1470° C (**B62**).

(2) Very slow firing, including prolonged soaking, is necessary to allow for the solid-state reactions to be completed, and to prevent rapid volume changes during inversions.

Where the firing schedule can be closely regulated, as in a tunnel kiln, the slow (7–10° C per hr) rate of heating or cooling can be confined to below 800° C (1472° F) as shown in the firing curve (Fig. 17.1).

(3) Absolute evenness of temperature is essential.

Kilns: (1) Round down-draught kilns fired by coal, gas or oil, *e.g.* schedule for hand-moulded shapes: 72 hr to 300° C (572° F); 144 hr to 1440/1450° C (2624/2642° F); 24 hr soak at 1450° C (2642° F); about 7 days cooling. Turn-round 20/21 days. Intermittent kilns may have schedules of from 3 to 4 weeks.

(2) Mendheim gas-fired continuous chamber.
(3) Tunnel kilns. These have to be very long, 450–550 ft (137–167 m). Gas or oil fired, *e.g.*

preheating zone	144 ft (44 m)
firing zone	246 ft (75 m) 1450° C (2642° F)
cooling zone	162 ft (49 m)
	552 ft (168 m)

Cars enter at 250° C (482° F)
72 hr to reach 1200° C (2192° F)
72 hr from 1200° C to 1450° C (2642° F)
48 hr soak at 1450° C (264° F)
72 hr cooling to 200° C (392° F) when withdrawn
264 hr (11 days) (**A10**).

Testing.
A. Control. Prepared batch: (1) grading B.S.I. sieves $+30$, $-30+60$, -60; (2) moisture content by rapid calcium carbide method; (3) hydrated lime addition by titration.
Moulded or dried bricks: weighed (for bulk density).
Fired bricks: specific gravity, four bricks from each consignment must have an average S.G. of less than 2·365.
B. General. Porosity. Bulk density. Size. Cold crushing strength. Mechanical strength at high temperatures. Permeability. Microscopic examination. Chemical analysis (alumina content).

SUPER DUTY SILICA BRICKS

Characteristics. Silica bricks of rigidly controlled flux content which can be used under load up to 1705° C (3101° F), 55° C ($\sim$ 100° F) higher than normal silica bricks (**S113**). They also show a lower permeability to gases (**R50**).
Uses. Open-hearth roofs where life increases of 10 to 30% over normal silica bricks occur. High-temperature zones of tunnel kilns especially for firing refractories (**R50**).
Body Type. Refractory. Super duty silica.
Raw Materials. Selected quartzites, ganisters, South African 'silcrete'. Hydrated lime, sulphite lye (**D16, L54**).
Preparation. More careful selection and washing may be required but otherwise as for silica bricks.
Shaping. Drying. Firing. Testing. *See* silica bricks.

Relevant Properties (C26) of silica refractories
Porosity. It is difficult to produce dense silica bricks because the conversion of one crystal form to another leads to complete rearrangement of the grains.

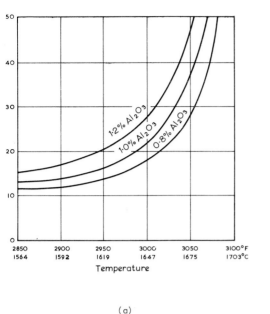
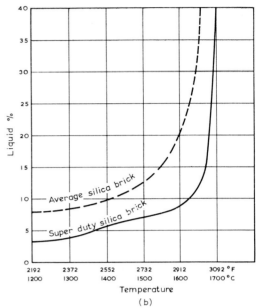

Fig. 17.2. (a) Percentage of liquid in SiO_2–CaO–Al_2O_3 compositions with 2% CaO. (b) Melting behaviour of silica refractories

Normal figures are from 22 to 26%. South African 'Silcrete' bricks may have porosity less than 20%. Miehr (**M73**) states that all the pores of silica brick are open so that absorption measurements give the true porosity.

Bulk Density and True Specific Gravity. This is related to the porosity and the degree of conversion of quartz to cristobalite or tridymite. For raw material S.G. is 2·65 and for completely converted material it is from 2·31 to 2·33. The highest bulk density for fully converted silcrete is about 1·95 g/ml.

Permanent Linear Change on Reheating. Soft fired bricks can show an after-expansion of several per cent when refired in the standard manner. Hard-fired products have little after-expansion.

TABLE 259

Melting Behaviour of Silica Brick based on CaO–Al_2O_3–SiO_2 Equilibrium Diagram (C65)

% Liquid	Conventional Silica Brick		Super-duty Silica Brick	
	(°C)	(°F)	(°C)	(°F)
10	1400	2550	1670	3040
20	1610	2930	1690	3070
30	1650	3000	1695	3085
100	1700	3090	1715	3120

Refractoriness. Silica, 1710–1730° C (3110–3146° F), cone 32 to 33.

Refractoriness under Load. This is very dependent on liquid content as calculated from equilibrium considerations. Flux-free bricks maintain a load of 25 lb/in² (1·75 kg/cm²) at 1600° C (2912° F) but if alumina or alkali content rises from $1\frac{1}{2}$ to 2% then failure occurs well below 1600° C (2912° F). Failure under load of 25 lb/in² (1·75 kg/cm²) appears to occur when one-third of the brick is liquid, and at 18% liquid under 50 lb/in² (3·5 kg/cm²).

Chaklader and Roberts (**C21**) offer a further explanation of the suddenness of collapse of silica refractories. They consider that the non-crystalline transition phase of pure silica which is present to the extent of 20–25% in new silica refractories is highly viscous and is the main bond in the refractory. At temperatures above 1600° C (2912° F) this phase converts increasingly rapidly to cristobalite, in effect destroying the bond, and thereby leading to

TABLE

Properties of Normal Quality Silica Bricks for

Ref.						British
	C26	C26	C26	C26	C26	
Code No.	S.8	S.6	S.5	S.7	S.9	
True porosity (%)						
Apparent porosity (%)	25·4	22·7	22·8	23·6	26·5	
Bulk density (g/cm³)	1·73	1·79	1·83	1·76	1·75	
(lb/ft³)	108	112	114	110	109	
True specific gravity (Rees–Hugill flask)	2·30$_5$	2·33	2·36$_5$	2·32	2·39	
Permeability to air (perp. to working face and through one skin) (c.g.s. units)	0·079	0·049	0·041	0·083	0·105	
After-expansion (2 hr at 1500° C) (%)	0·0	1·1	0·5	0·5	1·7	
Refractoriness under load—maintained at 1600° C for 1 hr, 25 lb/in²	0·6% deformation	0·7% deformation	Sheared in 24 min	No deformation	Sheared after 7 min	
Thermal shock resistance (450° test)	2	2	18	2	3	
Cone melting point	32 (1710° C)	32 (1710° C)	32 (1710° C)	32/33 (1710/ 1730° C)	31 (1690° C)	
Microscopical examination—						
Approximate % quartz	0	30	30	20	30	
Approximate % tridymite	50	20	35	40	10	
Approximate % cristobalite	50	50	35	40	60	
Chemical composition (%)—						
SiO₂						
CaO + MgO						
Al₂O₃						
TiO₂						
Fe₂O₃						
Alkali						
Refractoriness under load:						
Rising temp.						
Coefficient of linear expansion						
Thermal conductivity (B.t.u.)						
(c.g.s.)						
Specific heat (c.g.s.)						
Compressive strength (kg/cm²)						
(lb/in²)						

CONSTRUCTIONAL REFRACTORIES 1223

failure under load without previous viscous flow and giving the 'dry' appearance to the fracture.

Thermal Shock Resistance. This is very poor up to 600° C (1112° F). Soft-fired bricks are more resistant than hard-fired ones. A special test has been developed by which a test brick or piece is heated on one face by an electric hot plate.

Mineral Content. Thin-section microscopy can be used to estimate residual quartz content. X-ray powder diffraction is necessary to differentiate between cristobalite and tridymite. D.T.A. has proved a reliable method (**C21**).

Reversible Thermal Expansion. This varies somewhat with the degree of conversion, being greater for hard-fired bricks. The expansion curve is steeper than other refractories up to 600° C (1112° F) but thereafter almost flat.

260
Furnace Roofs (especially open-hearth roofs)

		American				German	
W55	W55	C26	N27	C26	C26	M73
	'Silcrete'	S.2		S.3	S.4	
24	16·8			23·5	20·7	20–28
1·78	1·90	25·8		1·78	1·85	1·80–1·85
		1·72		111	116	
		107				
2·33	2·335	2·30	2·3–2·4	2·32	2·325	2·32–2·40
		0·189		0·138	0·017	
		0·2	Nil	0·4	0·4	0·0–3·0
at 1650° C and 28 lb/in^2 Fails in 60–70 min		Sheared after 30 min	Low shrinkage at 1500° C and 50 lb/in^2	1·4% deformation	0·4% deformation	
		3		1	8	
32/33		32	30/31 to 33/34	32	32/33	32–34
(1710–1730° C)		(1710° C)	1700–1745° C	(1710° C)	(1710/1730° C)	(1710–1750° C)
		0		5	20	
		50		20	30	
		50		75	50	
95·5	94·8		95–96			92–96
1·9	2·07					1·5–3·0
0·8	0·37					0·5–3·0
0·1	1·7					
0·7	—					0·5–2·5
0·15	0·16					0·3–1·5
50 lb/in^2						28 lb/in^2
1680–1700° C						1600–1700° C
			20–300° C: 430 × 10^{-7}			0–600° C: 0·8–1·1%
			300–1100° C: 30 × 10^{-7}			600–1400° C: 0·3–0·4%
			390–1832° F: 13·0			
			200–1000° C: 0·045			
			21–1000° C: 0·265			150–250
						2133–3556

TABLE

Properties of Open-hearth Furnace

Ref. and Code	British				
	C26(1)	C26(2)	C26(3)	C26(4)	C26(5)
Apparent porosity (%)	20·5	20·7	17·2	29·5	23·7
Bulk density (g/ml.)	1·87	1·83	1·93	1·68	1·77
(lb/ft^3)	117	114	121	105	110
Apparent solid density (g/ml.)	2·32	2·31	2·33	2·37	2·32
Permeability to air, perp. to working face, 1 skin (c.g.s. units)	0·06	0·05	0·03	0·11	0·21
Permanent linear change on reheating (2 hr @ 1500° C)	0·4	0·4	0·2	3·0	0·4
Refractoriness-under-load, 50 lb/in^2, maintained at 1600° C for 1 hr	No deformation	No deformation	No deformation	2·0% exp.	—
Thermal shock resistance—max. safe heating rate (°C/min)	4/5	4/5	3/4	—	—
Refractoriness (°C)	1710	1710	1710	1710	1730
Cone:	>32	>32	>32	>32	33
Chemical analysis (%):					
SiO$_2$	96·6	96·8	94·8	96·0	96·55
Al$_2$O$_3$	0·19	0·49	0·37	0·33	0·39
Fe$_2$O$_3$	0·72	0·70	0·65	0·64	0·88
CaO	1·60	1·59	2·07	2·31	2·05
MgO	0·07	0·11	trace	0·13	0·03
TiO$_2$	0·01	0·02	1·70	0·11	0·07
Na$_2$O	0·12	Not determined	0·10	Nil	Not determined
K$_2$O	0·05		0·16	Nil	
Loss on ignition	0·22	—	0·19	—	—
Refractoriness under load of 50 lb/in^2, rising temp.					
True porosity					
True S.G.					
Cold crushing strength (kg/cm^2)					
(lb/in^2)					

Thermal Conductivity. At a mean temperature of 1000° C (1832° F) this is about 10 B.t.u.-in./ft^2-hr-°F (1·24 kcal-m/m^2-hr-°C). This means that a 12-in. (30-cm) thick roof of inside temperature 1650° C (3002° F) will have an outside temperature of 250° C (482° F).

Slag Resistance. Very high resistance to lime and ferric oxide. Less resistance to ferrous oxide. Attacked by alkalies, ZnO, etc.

Resistance to Gases. Grant and Williamson (**G69**) have shown that silica bricks reheated to 1200° C (2192° F) in hydrogen or carbon monoxide lose strength as shown by impact, crushing, cross-bending and abrasion tests. The effect is, however, not sufficiently marked for a discontinuation of the application of silica bricks in reducing conditions.

SEMI-SILICA BRICKS

Characteristics. Properties intermediate between fireclay and silica, lacking the large after-shrinkage of fireclay and the low thermal shock

261

Roof Silica Bricks "Super-duty" Quality

	British				U.S.A.	Holland	Sweden
	C26(6)	C26(7)	W55	W55	H37	S113	S113
	24.4	25.4				22/24	
	1.75	1.74	1.82	1.84		1.80	
	109	109					
	2.32	2.33					
	0.71	0.04					
	No change	0.3					
	0.4% def.	0.4% def.	28 lb/in² 1650° C 2 hr 0.5% subsidence	0.2% expansion			
	1710	1710					
	> 32	> 32				31	31/32
	94.55	94.50	96.3	97.2	97.02	96/97	96.3
	0.15	0.59	0.45	0.3	0.25	with TiO_2 0.7	with TiO_2 0.8
	0.56	0.80	0.5	0.7	0.50	0.7	
	3.16	2.77	} 2.0	1.6	1.90	2.0	1.7
	0.72	0.25					
	Nil	0.51	0.1	0.1	0.03		
	Not determined	Not determined	} 0.05	0.1	0.08		
	—	—					
			1700–1730° C		25 lb/in² 1705° C	1705° C	1680° C
			21	19		23/25	23
			2.33	2.31		2.35	2.38
						300	500
						4267	7112

resistance of silica. Owing to self-glazing with a highly viscous glass they are highly resistant to slag attack.

Uses. Open-hearth furnace checker settings.

Body Type. Refractory. Semi-silica.

Raw Materials. Sand–clay mixtures. Low-grade quartzites and ganisters together with deflocculated clay slip.

Preparation. Raw materials dried→jaw crusher→cone crusher or disintegrator→grading→addition of clay slip in revolving mixer with knives or mullers.

Shaping, Drying, Firing. As for silica bricks.

Properties. See Table 262. Compared with silica and fireclay.

FIRECLAY BRICK

Characteristics. The cheapest kind of refractory brick, suitable for many varied purposes and only superseded by specialised bricks where extra performance makes the greater outlay worthwhile. Relatively good spalling resistance and thermal insulation. Fair resistance to acid slags and fluxes, lower resistance to basic slags and fluxes.

Uses. 'Dry heat' furnaces. Gas retorts and settings. Gas producers. Annealing furnaces. Reheating furnaces. Boiler settings. Coke ovens. Ladle linings. Cupolas. Blast furnaces tops, inwalls, and stoves. Lime kilns. Aluminium melting furnaces. Billet heating furnaces. Cement kilns. Forge furnaces. Glass tank furnace regenerators. Mill furnace. Non-ferrous furnaces. Open-hearth furnace regenerators and doors. Slab-heating furnace roofs. Soaking pits. Suspended arches. Water gas generator linings. Glass pots. Grates, range blocks, fire backs, stove linings. Casting pits. Ceramic kilns.

Body Type. Refractory. Fireclay.

Raw Materials. Natural flux-free clays. Grog.

Preparation.

Clay. General heavy clay methods (Chapter 12).

For example: (1) Jaw crusher→dry pan→vibrating screen→storage bins→ →disk feeder→batch or continuous mixer together with other clays, grog and water, etc. (2) Toothed or smooth differential crushing rolls→wet pan→ mixer.

Grog. Prepared by crushing in jaw or gyratory crushers→roller, disk, swing hammer, ring roll or fine gyratory crushers→sieves for grading. Fine grog is prepared in hammer mills, ring roll mills or ball mills, preferably in closed circuit with an air separator.

Proportionating and Mixing. For shaping methods (1), (2) and (3): Prepared clay and 15–20% grog, together with the appropriate quantity of water are mixed in trough or Lancaster mixers or in a pug mill.

For shaping method (4): When using clay slip or paste, trough or Lancaster mixers can be used directly. When using damp or dry ground clay it is better to mix the grog and water first and then add the clay (**R29**).

For shaping method (5): 80% graded grog of:

        3 parts 0·0 –0·04 in. (0·0–1·0 mm)
        1 part   0·04 –0·098 in. (1·0–2·5 mm)
        2 parts 0·098–0·236 in. (2·5–6·0 mm)

and 20% by weight of deflocculated ball clay slip of S.G. 1·55 containing:

        Dry clay   96·9%
        Soda ash   1·2%
        Kasseler brown 0·5%
        Dextrine   1·4%

This means that of the dry body about 12% is raw clay.

Grog and slip are mixed in a double arm or Eirich mixer for from 30 to 45 min.

The mixture is used for Scheidhauer and Giessig process and also for hydraulic pressing (**A21**).

For shaping method (6): About 70% grog and 30% clay are deflocculated to make a casting slip with 17% water.

Shaping.

(1) Hand moulding of plastic body; generally only used for special shapes.

(2) Extrusion, from 10 to 15% moisture content→cutting off→re-pressing (the largest proportion are made by this method).

(3) Stiff plastic pressing.

(4) Dry pressing, body containing 5–8% water, de-airing of the mould reduces lamination.

(5) Pneumatic tamping, especially for German high-grog body containing only 5% moisture.

(6) Casting in plaster moulds.

Drying. Hot floors. Chamber dryers. Tunnel dryers.

Firing. Depends on the refractoriness of the body and on the degree of vitrification required.

Range 1100–1400° C (2012–2552° F) cone 1 to 13/14. Careful oxidation of combustible matter before vitrification occurs is very important.

(1) Periodic kilns, coal-fired (now only deal with 50% of British firebricks).

(2) Continuous kilns, Belgian up to 1250 or 1300° C (2282 or 2372° F); Mendheim gas-fired for 1350° C (2462° F) and over; Dunnachie gas-fired.

(3) Tunnel kilns.

Quality Control in a Fireclay Refractories Plant

I. RAW MATERIALS

(1) Drill samples taken during prospecting.

(2) During mining daily samples are taken.

(3) Sampling of each truck-load arriving at works so that it can be dumped on correct stockpile.

(4) Check tests on the stock piles.

Tests for (1) to (4) are for physical properties.

II. PREPARATION

(5) Ground raw material is sampled and checked for grain size distribution, contamination and moisture content.

(6) During mixing and blending of batches samples are tested for moisture content and contamination (visual and physical).

III. SHAPING

(7) Checking of size, shape, weight, moisture, appearance, certain physical properties.

TABLE 262

The Properties of Checker Bricks (C26)

	Semi-silica			Fireclay	Silica
	(1)	(2)	(3)		
Chemical analysis (%)—					
SiO_2	88.7	92.3	88.0	53.8	95.8
Al_2O_3	9.1	5.8	7.9	39.2	0.8
Fe_2O_3	1.0	0.5	1.2	2.7	0.9
TiO_2	0.4	0.7	0.7	1.6	0.0
CaO	0.2	trace	0.4	0.3	1.6
MgO	0.3	0.2	trace	0.7	0.3
Alkalies	0.4	0.5	0.8	1.4	0.5
Loss on ignition (%)	—	—	0.3	0.2	0.2
Cone melting point (°C)	over 1600°	over 1600°	over 1600°	1560°	1690°
Cold crushing strength (lb/in²)	1480	over 8030	1250	2350	4680
Porosity (%)	23.6	23.3	27.7	26.3	26.8
Bulk density (g/cm³)	1.93	1.90	1.85	1.96	1.70
Thermal conductivity (hot face, 700° C)—					
(B.t.u.)	5.6	7.1	5.6	6.2	7.5
(c.g.s. units)	0.0019	0.0024	0.0019	0.0021	0.0026
Specific heat, c.g.s. units (from 750° C)	0.26	0.245	0.27	0.25	0.255
Heat capacity = specific heat × bulk density	0.50	0.47	0.50	0.49	0.43
Average	3.8	0.49	3.8	0.49	0.43
Diffusivity factor = $\dfrac{\text{thermal conductivity}}{\text{bulk density} \times \text{specific heat}}$		5.1		4.3	6.0
Average		4.2		4.3	6.0

CONSTRUCTIONAL REFRACTORIES

IV. DRYING AND BURNING

(8) Check appearance, moisture, size, weight, handling and setting.
(9) Dryers have instruments to control temperature and humidity.
(10) Kilns, instrumented with temperature and atmospheric controllers and operate on prescribed time–temperature curves.

V. FINISHED WARE

(11) Appearance regarding colour and degree of burn.
(12) Size.
(13) P.C.E., modulus of rupture, apparent porosity, bulk density, hot load and reheat behaviour, spalling resistance.

Statistical methods are to be preferred to straight sampling in many cases, in order to be sure that real variations are spotted and chance divergences not corrected for unnecessarily (**H124**).

ALUMINOUS FIREBRICK, SUPER-DUTY FIREBRICK

Characteristics. A more refractory firebrick which can be made with relatively low porosity and therefore high slag resistance. Good volume stability, excellent resistance to thermal spalling. Fair resistance to highly acid slags; lower resistance to basic slags.
Uses. Lining of hot metal mixers. Open-hearth furnace checker chambers and settings.
Body Type. Refractory. Alumino-silicate.
Raw Materials. Aluminous clays, preferably of diaspore type although bauxitic clays are perforce being increasingly used. Grog.
Preparation. Shaping. Drying. Firing. As for fireclay.

MULLITE REFRACTORY BRICKS AND SHAPES

Characteristics. High refractoriness up to 1830° C (3326° F) combined with good slag and glass resistance and thermal and structural spalling resistance.
Uses. Glass tanks, especially to strengthen critical areas or for 'hot repairs', feeder forehearths, port 'jambs', breast walls, furnace crowns. Crucibles. Burner blocks. Frit kilns. Setter tiles for ceramic kilns where high thermal shock and/or high temperatures up to 1700° C (3092° F). Steel industry for tube mills, hottest area of malleable furnace roofs, electric arc furnace roofs, tap hole blocks, pouring area of large ladles and mixers. Non-ferrous metal industry for crucibles and induction furnaces.
Body Type. Refractory. Alumino-silicate.
Raw Materials. Diaspore. Bauxite. Sillimanite, andalusite and kyanite: calcined to 1500–1600° C (2732–2919° F). Dumortierite. Topaz. Corundum. High-alumina fireclays. Aluminous and siliceous materials either

fused or sintered together to give synthetic mullite, 1790–1820° C (3254–3308° F). Bonding clay. Sulphite lye.

Preparation. Crushing→grinding→grading→blending according to shaping method.

Shaping.

(1) Hand moulding of plastic body in wood moulds. Product has poor dimensional accuracy and slag resistance but good spalling resistance because of its open texture.

(2) Slip casting, for blocks and shapes for the glass industry or other applications where extreme uniformity of body and perfect bonding is required. Batch contains 15 to 25% bond clay.

(3) Ethyl silicate casting (**Q2**).

(4) Dry pressing for standard shapes and general use (**K42**).

(5) Pneumatic ramming for shapes too complicated to press.

(6) High-pressure extrusion for tubes, pins and pieces of uniform cross-section, high-clay content bodies being required.

Drying.

Firing. Cone 10, 1300° C (2372° F) to cone 35, 1770° C (3218° F) depending on nature of body mixture and on the product required.

ALUMINA REFRACTORY BRICKS OR CORUNDUM BRICKS

Characteristics. Very high resistance to oxidation and reduction up to 1900° C (3452° F), acid slags and mechanical abrasion; high refractoriness, refractoriness under load, heat transfer, resistance to basic slag, electrical resistance at high temperatures; low resistance to spalling, slag adherence and penetration.

Uses. Muffles for ceramic kilns and vitreous enamelling furnaces. Burner blocks, high temperature gas or oil furnaces.

Body Type. Refractory. Alumino-silicate.

Raw Materials. Electrically fused pure alumina. Bond, refractory clay, up to 20%.

Preparation. Fused alumina→grinding→grading. Clay→grind.

Mixing, *e.g.*

        corundum 0–3 mm (0–0·1181 in.) 40%
        corundum 0–1 mm (0–·03937 in.) 40%
        clay   20%

tempering with water.

Shaping.

(1) Tamping (water content 4–5%).

(2) Ethyl silicate casting (**Q2**).

Firing. Cone 10, 1300° C (2372° F) (**A9**).

CONSTRUCTIONAL REFRACTORIES

GLASS TANK BLOCKS

Requirements.
(1) Large blocks of $12 \times 15 \times 6$ in. ($30 \times 38 \times 15$ cm), $36 \times 18 \times 15$ in. ($91 \times 46 \times 38$ cm) or massive blocks up to 10 cwt (508 kg) in weight.

(2) One face must withstand 1350–1450° C (2462 2642° F) cone 12 16. The other is at air temperature or cooled by air blast. Therefore high resistance to thermal gradient required.

(3) Accurate dimensions and good shape in order to give close joints.

(4) Small permanent volume change when heated.

(5) Any solution in the molten glass shall be slow and uniform, not giving rise to the development of cord, seed or bad colour. It must not be corroded in such a way that pieces break away as 'stones', or if they do they must dissolve completely.

Uses. Construction of sides and bottom of tank, trough or basin used to contain molten glass in a tank furnace.

Body Type. Refractory. Fireclay and alumino-silicate.

Raw Materials. Fireclays with alumina contents between 19 and 45%. P.C.E. about cone 30; softening under load of 2 kg/cm² not less than 1300° C (2372° F); burning to give dense product. Fireclay grog, calcined at 1250 to 1400° C (2282–2552° F) cone 8–13/14, derived from (*a*) lump clay direct from mine, (*b*) prepared and moulded clay, (*c*) wasters and used blocks. Sillimanite, andalusite, kyanite, bauxite, silica.

TABLE 263

Proportions of Materials and Shaping Methods for Glass Tank Blocks (after P7)

Process	% Composition		Water content (%)	Drying time	Shrinkage in manu-facture (%)
	Grog	Raw Clay			
1 Hand moulding of plastic mixtures	50–75	50–25	17–22	3–4 months	4–6
2 Hand moulding ramming	70–85	30–15	7–10	8–14 days	3–4
3 Mechanical ramming or beating	80–95	20–5	5–8	4–5 days	0·2–2·0
4 'Semi-dry' pressing	70–85	30–15	7–10	3 days	2–5
5 Extrusion	50–75	50–25	17–22	3–4 months	4–6
6 Slip-casting	60 min.	40 max.	16–20	2–3 months	3–4
7 Fusion casting	Nil	100	Nil	14 days annealing	10

Preparation.
Clays. Hand sorting→weathering→(removal of iron compounds)→drying →jaw crusher→edge runner mills, chaser mills, roller or ball mills→sieving through 12-mesh (or more recently 40-mesh).
Grog. Jaw crusher to about 2 in. (5 cm)→crushing rolls to pea size→edge runner mill with rollers not touching pan→sieving→grading.

Proportionating (see Table 263)→dry mixer→wet mixing in double-shafted trough mixer followed by a de-airing pugmill, or, in a pan mill→ souring.

Shaping. See Table 263.
Drying. (1) Large blocks: hot floors in moulding rooms or drying sheds. (2) Occasionally chamber dryers.

When leather-hard the face to come in contact with molten glass is smoothed with a palette knife. When dry some blocks are ground with carborundum wheels to true shape.

Firing. 1300–1400° C (2372–2552° F) cone 10–13/14. Care must be taken to oxidise between 500–1000° C (932–1832° F).

Intermittent coal or oil-fired kilns. Heating up: minimum 7 to 8 days. Short soaking. Cooling: 10 to 12 days.

Finishing. Dressing by grinding all faces. (Increasingly unnecessary as shaping methods become more accurate and denser blocks are harder fired.)

Tests.
A. *Physical.* (1) After-contraction. (2) Crushing strength. (3) Density and porosity. (4) Spalling. (5) Refractoriness under load. (6) Tensile strength at elevated temperatures.
B. *Chemical.* (1) Chemical analysis. (2) Resistance to corrosion by molten glass: (*a*) penetration of corroding medium in the refractory; (*b*) solution of refractory in corroding glass; (*c*) examination, after corrosion, of crucibles made of the refractory; (*d*) change of composition of solvent due to contamination by the refractory; (*e*) use of corrosive acid mixtures as solvents; (*f*) corrosion of blocks in small experimental tank firing; (*g*) P.C.E. mullite content, etc. (**P7**).

Properties. *See* Table 264.

Uses of Alumino-Silicate Refractories as referred to in Table 264.
(1) Air furnaces.
(2) Aluminium melting furnaces.
(3) Annealing furnaces.
(4) Billet heating furnaces.
(5) Blast furnaces: (*a*) hearth and bosh; (*b*) inwall; (*c*) top; (*d*) taphole.
(6) Blast furnace stoves.
(7) Boiler settings.
(8) Brass melting furnaces.
(9) Brass reverberatory furnaces.

(10) Burner blocks.
(11) Cement kilns (a) intermediate zone; (b) hot zone.
(11A) Ceramic kilns.
(12) Chemical works furnaces.
(13) Coke ovens.
(14) Continuous slab furnace hearths and roofs.
(15) Combustion chambers of stoker-fired furnaces.
(16) Copper melting and refining furnaces.
(17) Crucible furnaces.
(18) Cupolas.
(19) Desulphuring ladles.
(20) Down-jet furnaces.
(21) 'Dry heat' furnaces.
(22) Destructors.
(23) Electric melting furnaces.
(24) Electric arc furnace crowns.
(25) Enamelling furnaces.
(26) Forge furnaces.
(27) Furnace doors.
(28) Gas producers.
(29) Gas retorts and setting.
(30) Glass tank furnaces: (a) regenerators; (b) uptakes and checkers, ports; (c) superstructures.
(31) Heating furnace walls and roofs.
(31A) Induction furnaces.
(32) Ladle linings.
(33) Lead dross reverberatory furnaces.
(34) Lead melting furnaces.
(35) Lime kilns.
(36) Mill furnaces.
(37) Non-ferrous furnaces.
(38) Oil-fired furnaces.
(39) Open-hearth furnaces: (a) regenerators, walls, crown and checker work; (b) gas ports; (c) furnace doors.
(40) Pottery kiln props, bats and deck slabs.
(41) Powdered fuel furnaces.
(42) Recuperator tubes.
(43) Reheating furnaces.
(44) Reverberatory furnaces.
(45) Soaking pits.
(47) Stills.
(48) Suspended arches, Detrick, Liptak, Karrena, etc., for boilers and industrial furnaces.
(49) Water tube boilers.
(50) Zinc smelting furnaces.

TABLE

Alumino Silicate

1. Firebricks

Name, type	Semi-silica	Semi-silica	Firebrick	Firebrick
Origin	British	British	British	Germ.
Ref.	P22	L21	L21	S93
Analysis: SiO_2	79	80·7	68/72	70·50
Al_2O_3	17	15·0		26·34
TiO_2		1·2		
Fe_2O_3	1·5	2·0	<3·0	2·28
Others		CaO 0·7		
		MgO 0·3		
		NaKO 0·3	<1·0	
Refractoriness: cone	30	30	30	30
(°C)	1670	1670	1670	1670
(°F)	3038	3038	3038	3038
Specific heat (c.g.s.)				
Apparent porosity	23–25	27	22–28	
True porosity				
Water absorption				15·7
Bulk density (g/cm³)	1·92	1·84		1·80
(lb/ft³)	120	114·9		112·4
True S.G.		2·518		
Cold crushing strength (lb/in²)	2303†	3880	>2000	1991
(kg/cm²)	162	273	>140	140
Modulus of rupture (lb/in²)				
(kg/cm²)				
Mean reversible thermal exp. × 10⁻⁶		0·60% @ 1000° C	0·58% @ 1000° C	
Thermal conductivity c.g.s. (for B.t.u. × 2903)				
Permeability				
Linear change on refiring: Temp. (°C)	1410	1410	1350	
Time (hr)	2	2	2	
% change	0·52	−0·5	−1·3	
Refractoriness under load: Load	28 lb/in²†	50 lb/in²	28 lb/in²	2 kg/cm²
Temp. (°C)	1520° C	1500	>1460	1430 | 1590
28 p.s.i. = 2 kg/cm² Time				
% deformation	5	10	5	begins | 40
Thermal shock resistance				
Spalling loss % (High-duty schedule)				
(Low-duty schedule)				
Uses (see List, p. 1232)	21, 29	50, 11a, 29 13	3, 43, 28, 7, 29, 13	

* B.S.S. 1902:1952.
† Tested by B.C.R.A.

Refractories

Low alumina ladle lining Brit. **L1**	Firebrick Germ. **S93**	Firebrick Brit. **S183**	Firebrick Brit. **S183**	Firebrick Brit. **S183**	Firebrick Brit. **S183**	High alumina firebrick Brit. **S183**
62·80	65·42	63–61	60–58	61–57	58–56	52
29·48	31·33	30–32	33–35	33–35	35–37	42
1·20						
2·86	2·34					
CaO 0·40						
MgO 0·82						
NaKO 2·17						
26/27	31	30	31	31	31	34/35
1600	1690	1670	1690	1690	1690	1760
2912	3074	3038	3074	3074	3074	3200
17						
	15·6					
	1·85					
	115·5					
5800	2631					
408	185					
		@ 300° C 0·0031;		@ 700° C 0·0038;		@ 1100° C 0·0046
nil						
1410		1410	1410	1410	1410	1410
2		2	2	2	2	2
nil		0·13	nil	nil	0·2	0·05
28 lb/in²	2 kg/cm²	28 lb/in²	28 lb/in²	28 lb/in²	28 lb/in²	28 lb/in²
1400 | 1500	1465 | 1600	1525	1500	1525	1490	1570
beg. | 10	beg. | 40	5	5	5	5	5
20						
32		13, 18, 28, 32	32	5b, 5c, 6, 35	1, 2, 3, 4, 7, 11, 13, 18, 26, 29, 30a, 32, 36, 37, 39a, c, 45, 48	1, 2, 4, 5a, 5b, 6, 7, 8, 11, 13, 16, 17, 18, 23, 25, 26, 28, 29, 30a, 34, 35, 36, 37, 39a, c, 45, 48

TABLE

1. Firebricks—(*contd.*)   **2. Aluminous**

Name, type	High heat duty firebrick	Medium alumina ladle lining	Fireclay glass tank block	Aluminous firebrick
Origin	Amer.	Brit.	Brit.	Brit.
Ref.	**N27**	**L1**	**P22**	**P22**
Analysis: SiO_2	52–60	55·14	56–57	53·8
Al_2O_3	35–42	37·84	37–38	40·1
TiO_2		1·00		1·4
Fe_2O_3		1·86	2·5–3·0	3·1
Others		CaO 0·30		
		MgO 0·90		
		NaKO 2·59		
Refractoriness: cone	31–33	32/33	32	33
(°C)	1690–1730	1720	1710	1730
(°F)	3074–3146	3128	3110	3146
Specific heat (c.g.s.)	0·26			
Apparent porosity		14	16–18	18–22
True porosity				
Water absorption				
Bulk density (g/cm³)	2·04–2·18			
(lb/ft³)	127·5–136			
True S.G.	2·60–2·70			
Cold crushing strength (lb/in²)		2900		3607†
(kg/cm²)		204		254
Modulus of rupture (lb/in²)				
(kg/cm²)				
Mean reversible thermal exp. $\times 10^{-6}$	20–1200° C			
	5·3			
Thermal conductivity c.g.s. (for B.t.u. $\times 2903$)	204–1314° C			
	0·003			
Permeability		0·004–0·020		
Linear change on refiring: Temp. (°C)	1400	1410	1400	1410
Time (hr)		2		
% change	−2·0	+1·2	0·5	under 0·5
Refractoriness under load: Load	25 lb/in²	28 lb/in²		28 lb/in²†
28 p.s.i. = 2 kg/cm² Temp. (°C)	1350	1150 1520		1540
Time				
% deformation	4·0+	beg 10		5
Thermal shock resistance				
Spalling loss % (High-duty schedule)		30+		
(Low-duty schedule)				
Uses (*see* List, p. 1232)		32	30	5, 6, 16, 18, 22, 36a, 35, 38, 32, 41, 43, 47–8, 50

* B.S.S. 1902:1952.
† Tested by B.C.R.A.

264—(contd.)

Firebrick

Firebrick	Firebrick	Aluminous firebrick	Aluminous firebrick	Firebrick extruded and repressed	Firebrick by S. & G. process	High alumina ladle lining
Germ. **S93**	Germ. **S93**	Brit. **L21**	Brit. **L21**	Germ. **A21**	Germ. **A21**	Brit. **L1**
51·0	54·50		52·4			50·20
41·0	42·80	42	42·6	42	42	43·32
			1·4			1·20
2·17	2·08		2·7			3·28
			CaO 0·4			CaO 0·40
			MgO 0·1			MgO 0·60
			NaKO 0·4			NaKO 0·60
33	34	33	34	33	33	32
1730	1750	1730	1750	1730	1730	1710
3146	3182	3146	3182	3146	3146	3110
			26·29	28	23	10·5
11·6						
1·90	1·85			1·90	2·05	
118·6	115·5			118·6	128·0	
2702	2631		5975	2133	4978	7000
190	185		420	150	350	492
			0·64 @ 1000° C			
						nil–0·015
			1410	1410		1410
			2	2		2
			−0·3	−0·3		nil
2 kg/cm²	2 kg/cm²		50 lb/in²	2 kg/cm²	2 kg/cm²	28 lb/in²
1430 | 1620	1500 | 1680	1580	1645	1380 | 1600	1420 | 1630	1460 | 1620
begins | 40	begins | 40	5	10	begins | 40	begins | 40	begins | 40
good resis.	good resis.	25+		15 cycles	30 cycles	30+
		oil fired 10 up to 1400° C	50, 38, 41 11a			32

TABLE 2. Aluminous Firebricks—(contd.)

Name, type	High alumina 50	Super duty	Super duty firebrick
Origin	Brit.	Brit.	Amer.
Ref.	S183	M96	N27
Analysis: SiO_2		52/53	51–53
Al_2O_3	51–52	43/44	43–44
TiO_2			
Fe_2O_3		<1	
Others		MgO <1, CaO about 1·5, NaKO	
Refractoriness: cone	35/36	35	34
(°C)	1780	1770	1750
(°F)	3236	3218	3182
Specific heat (c.g.s.)			0·26
Apparent porosity	22	13/15	
True porosity		~17	
Water absorption			
Bulk density (g/cm³)		2·16	2·04–2·38
(lb/ft³)		135	127·5–148·75
True S.G.			2·65–2·75
Cold crushing strength (lb/in²)	8000		
(kg/cm²)	570		
Modulus of rupture (lb/in²)			
(kg/cm²)			
Mean reversible thermal exp. × 10⁻⁶		5	20–1200° C 5·3
Thermal conductivity c.g.s. (for B.t.u. × 2903)			0·003
Permeability			
Linear change on reheating: Temp. (°C)	1400 1500 1600	1600	1500
Time (hr)	2 2 2		
% change	Nil +0·1 −0·55	0·5	−1·0
Refractoriness under load: Load	28 lb/in²	25 lb/in²	25 lb/in²
Temp. (°C)	1605	1600–1650 / 1750	1350
28 p.s.i. = 2 kg/cm² Time			
% deformation	5	beg. / 10	2·5–4·0
Thermal shock resistance			
Spalling loss % (High-duty schedule)			
(Low-duty schedule)			
Uses (see List, p. 1232)	6, 45, 31, 44, 39a, 19, 24, 7, 27, 37, 35, 30b, 15, 11a	43, 26, 45, 17, 4, 7, 30b, 39a, 30a, 16, 18, 11a	

* B.S.S. 1902:1952
† Tested by B.C.R.A.

264—(contd.)

High alumina firebrick	Aluminous 50 firebrick	Aluminous firebrick 57	Aluminous firebrick 57–59	Aluminous firebrick	High alumina 63
Amer. **N27**	Brit. **P22**	Brit. **P22**	Brit. **P22**	Brit. **P22**	Brit. **S183**
	40·9	36·5	35·7	30·4	
50–80	52·1	57·7	58·2	63·8	65–67
	1·7	1·8	2·5	2·5	
	3·1	1·9	2·0	2·2	
34–39	34	over 34	36		36/37
1750–1880	1750	1750	1790		1810
3182–3416	3182	3182	3254		3290
~0·26					
	17–20	22–24	24–26	28–31	25
2·18–2·72					
136–170					
2·80–3·40					
	4147†	4098*†	3260*†	3853*	4500
	292	288	229	271	320
20–1425° C					
5·3–6·5					
1600	1400	1400 | 1500†	1400 | 1500†	1400 | 1500	1400 | 1500 | 1600
					2 | 2 | 2
0–3·0	0·1	−0·1 | +0·6	−0·1 | −0·2	−0·1 | −0·3	nil | +0·35 | +0·25
25 lb/in²	28 lb/in²†	28 lb/in²†	28 lb/in²†	28 lb/in²	28 lb/in²
1350	1605	1625	1625	1640 | 1650	1625
2·5–4·0	5	5	5	5 | 10	5
		27	35, 11, 10	35, 11, 21	11 a and b
			43, 16, 50, 23		

TABLE 2. Aluminous Firebricks—(contd.)

Name, type	High alumina 70	High alumina firebrick	High alumina 73
Origin	Brit.	Brit.	Brit.
Ref.	S183	P22	S183
Analysis: SiO_2		25·3	
Al_2O_3	+70	70·1	+73
TiO_2		2·1	
Fe_2O_3		1·7	
Others			
Refractoriness: cone	38		38
(°C)	1850		1850
(°F)	3362		3362
Specific heat (c.g.s.)			
Apparent porosity	26	33–35	24
True porosity			
Water absorption			
Bulk density (g/cm³)			
(lb/ft³)			
True S.G.			
Cold crushing strength (lb/in²)	7500	4299*†	11 000
(kg/cm²)	533	302	773
Modulus of rupture (lb/in²)			
(kg/cm²)			
Mean reversible thermal exp. $\times 10^{-6}$			
Thermal conductivity c.g.s. (for B.t.u. $\times 2903$)			
Permeability			
Linear change on refiring: Temp. (°C)	1400 | 1500 | 1600	1500† | 1600†	1400 | 1500 | 1600
Time (hr)	2 | 2 | 2		2 | 2 | 2
% change	nil | nil | −0·24	+0·32 | −0·25	nil | −0·09 | −0·44
Refractoriness under load: Load	28 lb/in²	28 lb/in²†	28 lb/in²
Temp. (°C)	1655	1660	1640
28 p.s.i. = 2 kg/cm² Time			
% deformation	5	5	5
Thermal shock resistance			
Spalling loss % (High-duty schedule)			
(Low-duty schedule)			
Uses (see List, p. 1232)	11b, 35, 24, 33, 2, 37, 9, 25	35, 11, 43	43, 14, 45, 5d, 39b, 7

* B.S.S. 1902:1952.
† Tested by B.C.R.A.

264—(contd.)

3. Kaolin

4. Sillimanite

Molochite	Kaolin	Kaolin	Glass tank block Kaolin	Sillimanite 70	Sillimanite type		
Brit. **L23**	Amer. **N27**	Amer. **T14**	Brit. **P22**	Brit. **P22**	Brit. **S183**		
54·84	51–53	51–53	52–54	44–46	53		
42·59	44–45	44–45	43–45	51–53			
0·14							
0·61			<1	1			
MgO 0·29							
CaO 0·03							
Na_2O 0·04							
K_2O 1·40							
33–34	31–34	30/31–31/32	34	over 34	36		
1730–1750	1690–1750	1680–1760	1750	1750	1790		
3146–3182	3074–3182	3056–3092	3182	3182	3254		
	0·254	0·254 250–1000° C					
		18–28	15–17	28	21		
	1·72–1·83	2·02					
	107·25–114·4	125·8					
	2·60	2·60–2·70			6000		
					430		
	5·3, 20–1425° C	4·3, 21–1610° C					
	0·005 @ 1314° C	0·0045 200–1000° C					
1410	1650	1650	1400	1410	1400	1500	1600
2				2	2	2	2
0·10	−0·5	0·5	0·2	0·2	nil	−0·1	−0·7
50 lb/in²	25 lb/in²	25 lb/in²		28 lb/in²	28 lb/in²		
1450	1540	1540		1630	1645		
first sign	1	1·0		5	5		
Good resis.		Spalling index 75–85					
49, 43, 26, 18, 41, 38, 39, 31, 13, 11A			30 colourless glass	21, 48, 30c, 11, 20, 42, 23, 12, 10	7, 30b, 4b, 10, 15		

TABLE 4. Sillimanite (contd.)

Name, type	Glass tank block Vacuum slip cast sillimanite	Sillimanite	Sillimanite 85
Origin	Brit.	Brit.	Brit.
Ref.	P22	S183	P22
Analysis: SiO_2	39–41		38
Al_2O_3	55–57	60	60
TiO_2			1
Fe_2O_3	0·5–1·0		0·5
Others			
Refractoriness: cone	36	36/37	over 35†
(°C)	1790	1810	1770
(°F)	3254	3290	3218
Specific heat (c.g.s.)			
Apparent porosity	19–20	21	
True porosity			
Water absorption			
Bulk density (g/cm³)			
(lb/ft³)			
True S.G.			
Cold crushing strength			
(lb/in²)		7000	
(kg/cm²)		500	
Modulus of rupture			
(lb/in²)			
(kg/cm²)			
Mean reversible thermal exp. × 10⁻⁶			
Thermal conductivity, c.g.s. (for B.t.u. × 2903)			
Permeability			
Linear change on refiring: Temp. (°C)	1500	1400 | 1500 | 1600	1410 | 1500†
Time (hr)		2 | 2 | 2	2 | 2
% change	0·5	nil | nil | −0·46	<0·15 | <0·5
Refractoriness under load: Load		28 lb/in²	28 lb/in²†
28 p.s.i. = 2 kg/cm² Temp. (°C)		1680	1750
Time			
% deformation		5	<5
Thermal shock resistance			
Spalling loss % (High-duty schedule)			
(Low-duty schedule)			
Uses (see List, p. 1232)	30 high duty	7, 30b, 40, 10, 38, 25	

* B.S.A. 1902 : 1952
† Tested by B.C.R.A.

264—(contd.)

5. Mullite

Sillimanite	Sillimanite	Sillimanite	Sillimanite		Mullite
Amer. **T14**	Germ. **S93**	Germ. **K66**	Germ. **S93**	**S93**	Amer. **N27**
37·62	34·0	~32	26·30	18·4	38
59·48	63·20	~62	72·14	80·0	62
1·19					
0·81	1·66	max. 1·5	1·56	1·56	
MgO 0·18					
CaO 0·32					
NaKO 0·40					
37	33–39	38/39	39	>40	37/38
1825	1730–1880	1850/1880	1880	>1920	1835
3317	3146–3416	3362/3416	3416	>3488	3335
					0·175
20–23		22			
	16·00		12·20	9·5	
2·34–2·36	2·00	2·4	2·20	2·3	2·31
146·0	124·9	149·8	137·3	143·6	144·5
					3·03
A.S.T.M. C133					
7800	9245–11 378		10 667–14 223	14 223–17 068	
548	650–800		750–1000	1000–2000	
A.S.T.M. C133–39					
2000					
141					
25–1400° C					20–1320° C
5·4					4·5
					0·003
1700					1650
5					
1·0					nil
25 lb/in²	2 kg/cm²	2 kg/cm²	2 kg/cm²	2 kg/cm²	50 lb/in²
1590	1620 | >1800	1630 | >1750	1580 | >1780	1620 | >1800	1455 | 1650
3·0	beg. | 40	beg. | 40	beg. | 40	beg. | 40	0 | 2·0
A.S.T.M. C122–40					
No loss	very good	50 cycles	very good	very good	
24, 10, 42, 31A, 17, 30, 11A, 40					

TABLE

5. Mullite
(contd.)

Name, type	Mullite	Mullite
Origin	Amer.	Brit.
Ref.	T14	S183
Analysis: SiO$_2$	29·39	
Al$_2$O$_3$	68·26	72
TiO$_2$	0·85	
Fe$_2$O$_3$	0·62	
Others	MgO 0·01	
	CaO 0·16	
	NaKO 0·84	
Refractoriness: cone	38	38
(°C)	1850	1850
(°F)	3362	3362
Specific heat (c.g.s.)		
Apparent porosity	20–23	22
True porosity		
Water absorption		
Bulk density (g/cm^3)	2·36–2·38	
(lb/ft^3)	148·0	
True S.G.		
Cold crushing strength	A.S.T.M. C133	
(lb/in^2)	8505+	10 000
(kg/cm^2)	598	703
Modulus of rupture	A.S.T.M. C133–49	
(lb/in^2)	1805	
(kg/cm^2)	127	
Mean reversible thermal exp. ×10^{-6}	5·76	
Thermal conductivity c.g.s. (for B.t.u. ×2903)		
Permeability		
Linear change on refiring: Temp. (°C)	1700	1400 | 1500 | 1600
Time (hr)	5	2 | 2 | 2
% change	−0·0–0·5	nil | +0·12 | +0·41
Refractoriness under load: Load	25 lb/in^2	28 lb/in^2
28 p.s.i. = 2 kg/cm^2 Temp. (°C)	1590	1725
Time		
% deformation	1·8	5
Thermal shock resistance		
Spalling loss % (High-duty schedule)	No loss	
(Low-duty schedule)		
Uses (see List, p. 1232)	24, 10, 42, 31A, 17, 30, 11A, 40	7, 30b, 40 10, 38, 25

* B.S.S. 1902:1952.
† Tested by B.C.R.A.

6. Alumina

	Mullite	Mullite	Bonded fused alumina	Fused alumina	Fused alumina
	Amer. **R10**	Amer. **R10**	Amer. **N27**	Amer. **N27**	Amer. **T14**
	24·27	23·00	0·6		
	74·35	75·65	99·2	90–99·1	88–90
	0·39	0·38			
	0·41	0·39			
	MgO 0·19	MgO 0·18			
	NaKO 0·41	NaKO 0·39			
	39–40	39–40	42	38–41	>42
	1880–1920	1880–1920	2000	1850–1960	2050
	3416–3488	3416–3488	3632	3362–3560	3722
				0·174–0·304	0·272 20–800°C
	22·7	19·8	20–23	19–23	21
	9·4	7·9			
	3·12	3·12	3·06	3·0	3·0
	150·6	156·2	191	187	187
				3·8–4·0	3·90
	A.S.T.M. C.133–49	A.S.T.M. C.133–49			
	6075	9000+			
	427	633			
	A.S.T.M. C.133–49	A.S.T.M. C.133–49			
	1610	2815			
	113	198			
	250–825° C			20–1000° C	21–1000° C
	0·07	6·07		7·0	8·1
	0·005 @ 260–1064° C		>0·006	18–20 ~1100° C	
				2000° F	
	1820	1760 | 1820		1500	1500
	4	4 | 4			
	−0·44	0·00 | −0·03		nil	nil
	25 lb/in²	25 lb/in²	50 lb/in²	50 lb/in²	50 lb/in²
	1700	1700	1500 | 1710	1500	1500
	1½	1½			
	−1·8	−0·2	0·0 | 0·10	1–2	2
	A.S.T.M. C.122–47				spalling index
	0·6%	high	medium		60–70
	30a, b, c, 10,	30a, b, c, 10,			
	40, 25, 24,	40, 25, 24,			
	32, 31A,	32, 31A,			
	37, 1, 48	37, 1, 48			

STABILISED ZIRCONIA BRICKS AND SHAPES

Characteristics. Acid refractory that can be used up to 2400° C (4352° F) under oxidising conditions. Low volatility at high temperatures. Good thermal shock resistance. Safe up to 1600° C (2912° F) in contact with 50–90% alumina, chrome, chrome–magnesite, forsterite, magnesite and silica. Safe up to 1550° C (2822° F) with super-duty firebricks, up to 1900° C (3452° F) with zircon. At about 2200° C (3992° F) decomposed *in vacuo*, carbon, hydrogen and nitrogen. Low electric resistivity at high temperatures.

Uses. Melting of metals with acid slags, provided that oxide formation is prevented. Monolithic furnace linings. Refractories for high temperature gas-fired furnaces (2000° C, 3632° F). Burner blocks; nozzles; glass feeder blocks. Setting or placing media when firing titanates. High-temperature thermal insulation. Catalyst supports (**P24**). High-frequency induction furnaces for melting platinum at 1900° C (3452° F). Nitrogen fixation plant up to 2315° C (4200° F).

Body Type. Refractory. Zirconia.

Raw Materials. Fused stabilised zirconia: ZrO_2 94-95%, CaO 4–5, SiO_2 0·14–0·75, Fe_2O_3 0·2–0·7, TiO_2 0·22–1·0. (This can be prepared direct from siliceous ores by an electric arc fusion process (**W65**).) From 0 to 20% unstabilised zirconia. Organic binder, *e.g.* dextrine; or for hand moulding zirconium sulphate.

Preparation. Crushing→grinding→grading→add binder and water→mix.

Shaping. Dry pressing 2000–6000 lb/in² (140–422 kg/cm²).
Hand moulding for spalling resistant shapes.
Ethyl silicate casting (**Q2**).

Drying.

Firing. Cone 35, 1770° C (3218° F) to cone 42, 2000° C (3632° F) (**C36**).

Properties. *See* Table 265, Nos. 1, 2. Also:

Temperature		Specific Resistivity
(°C)	(°F)	(Ω/cm³)
700	1292	2300
1200	2192	77
1300	2372	9·4
1700	3092	1·6
2000	3632	0·59
2200	3992	0·37

ZIRCON BRICKS AND HOLLOW-WARE

Characteristics. Acid refractory. No crystalline inversions, uniform thermal expansion curve (4.5×10^{-6} per °C from 20–500° C). Good resistance to molten metals except under strongly oxidising conditions. Exceptional resistance to molten aluminium which does not wet it, also little alumina dross forms and this is not strongly adhesive. Resistant to borosilicate glasses, metaphosphates, sodium chloride, sodium chloride–zinc chloride flux, potassium pyrosulphate, phosphorus pentoxide. Attacked by sodium carbonate and sodium fluoride, fluorspar, cryolite, molten barium sulphate, and tetrasodium pyrophosphate (**P24**). Good thermal shock resistance.
Uses. Aluminium remelting furnaces. Calcium phosphate fertiliser furnaces. Glass tanks: for superstructures, feeder parts, feeder forehearths and in the 'balanced' glass furnace. Crucibles for platinum.
Body Type. Single-component refractory.
Raw Materials. Natural zircon sand. Micronised zircon: >60% below 50 μ; >10% below 10 μ; >1% below 2 μ. Calcium or ammonium alginate jelly or starch paste, dextrine and boric acid, goulac and lignin.
Preparation. Natural fine zircon sand→partial sintering→grog→crushing →grading→blending: 2 parts zircon sand: 1 part micronised zircon. *Or* equal parts grog $-14 + 72$ mesh, sand and micronised.
Shaping. Dry pressing. Slip casting. Ethyl silicate casting (**Q2**). Pneumatic tamping (**K43**). Hand moulding. Extrusion.
Firing. 1300° C (2372° F) to over 1500° C (2732° F) according to purity of zircon sand. If impure, resistance to load at high temperatures is increased by firing to 1650–1700° C (3002–3182° F) for at least one hour.
Properties. Table 265, Nos. 3–6.

BARIUM ALUMINIUM SILICATE REFRACTORIES OR BARIUM–ANORTHITE REFRACTORIES

Characteristics. Highly basic refractory, of high density, low expansion coefficient, high impact strength and exceptional resistance to molten aluminium and to salt glaze vapours.
Uses. Aluminium reverberatory and other furnaces. Lining of kilns for salt-glazed stoneware especially tunnel kilns. Protection against X-rays.
Body Type. Barium aluminium silicate.
Raw Materials. Barium carbonate. Suitable pure clay with minimum free quartz.
Preparation. Clay and $BaCO_3$ grind together→fire to make grog 1400–1500° C (2552–2732° F)→crush. Grind finely the same raw mixture of clay and $BaCO_3$ to make bond. Mix 2 parts grog and 1 part bond.

TABLE

Properties of Zircon, Barium Anorthite,

No.	1	2	3
Name, type	Stabilised zirconia	Stabilised zirconia	Zircon
Origin	U.S.A.	Brit.	Brit.
Ref.	N27	P24	P24
Analysis	SiO_2 0·5–1 Fe_2O_3 0·2–5 CaO 4·5–5 TiO_2 0·4–1 HfO_2 <2 ZrO_2 balance	SiO_2 1% Fe_2O_3 max. 0·2 CaO 4–5 TiO_2 max. 0·2 Al_2O_3 0·2 HfO_2 max. 2 ZrO_2 balance	
Refractoriness: cone			
(°C)	2550–2600		
(°F)	4620–4710		
Porosity (app. or tr.)		app. 31	app. 21
Water absorption			
Bulk density (g/cm³)	4–4·4	3·9	
(lb/ft³)	250–275	243	
True S.G.	5·6	5·66	4·59
Specific heat (c.g.s.)	25–1400° C, 0·125		0–400° C 0·13
Cold crushing strength (lb/in²)		9000	12 000
(kg/cm²)		633	844
Modulus of rupture (lb/in²)			
(kg/cm²)			
Mean reversible thermal (exp. × 10⁻⁶ or %)	400–1250° C 6·5	0–900° C | 0–1500° C 10 | 6·0	0–1500° C 4·7
Thermal conductivity: Temp.	1095° C	1000° C | 1600° C	
c.g.s.	0·0021	0·0021 | 0·0022	
B.t.u.	6·2	6·2 | 6·4	
Permeability (c.g.s.)		0·08	0·05
Linear change on refiring: Temp. (°C)			
Time			
% change			
Refractoriness under load: load (lb/in²)	10 | 40	28	28
Temp. (°C)	2110 | 1950	1710	1550 | 1690
Time			
% deformation	collapse	nil	0 | 4
Spalling index			30 cycles
Rising temp. test: load (lb/in²)			
initial softening			
rapid softening			
failure			
Impact strength (cm kg/cm²)			
Modulus of elasticity (kg/mm²)			
(lb/in²)			
Shear strength (lb/in²)			
(kg/cm²)			
Tensile strength (lb/in²)			
(kg/cm²)			
Transverse breaking strength (lb/in²)			
(kg/cm²)			

† Spalling index compared with silicon carbide 100.
§ DIN 1068.

Carbon and Silicon Carbide Refractories

4 Zircon	5 Zircon	6 Zircon	7 Barium anorthite	8 Carbon
Amer. **N27**	Germ. **K60**	Amer. **T14**	Germ. **S101**	Brit. **A56**
ZrO_2 67·1 SiO_2 32·9		ZrO_2 67·3 SiO_2 32·7	BaO 34 Al_2O_3 28 Fe_2O_3 0·8 SiO_2 36·2 29/30	
2424 4400		> 2550 > 4622 20–26	1660 3018	2985 5400 app. 22·3
3 204 4·6 38° C 0·132	0·167	3 204 4·6 —	3·3 2375 167	1·51 94 1·95 6300–15 000 443–1055 $9·3 \times 10^5$ 65 386
4·2 @ 20–1550° C		4·2 @ 21–1550° C	2·4	
0047 @ 200–1000° C	200° C 1400° C 0·0035 0·0049	0·0046 @ 200–1000° C		300°C 600°C 1000°C 0·0055 0·0071 0·0094
13·5 @ 390–1832° F	10 14	13		15·9 20·6 27·2 0·022
1550		1550		1500 2
nil 25 1550–1600		nil 25 1550–1600	28 1410 1620	up to 1 50 4480 1700 1470
fails		failed †70–90	beg. 4 § > 15	nil nil 30+ reversals
			3·0	
				$2·85 \times 10^3$ $4·06 \times 10^6$ 2390 168 4450 323 1920–2420 133–170

TABLE

No. Name, type	9 Carbon	10 Carbon	11 Graphite
Origin		Germ.	Amer.
Ref.	S104	A21	N27
Analysis	C 85–90 Ash <10		C 100
Refractoriness: cone	>42		
(°C)	>2000		2997
(°F)	>3632		>5432
Porosity (app. or tr.)	tr. 30–32	26	
Water absorption	21–23	18	
Bulk density (g/cm^3)	1·35	1·38	2
(lb/ft^3)	84	86	136
True S.G.		1·82	2·25
Specific heat (c.g.s.)			0·29, 21–1000° C
Cold crushing strength (lb/in^2)	1710	3270	
(kg/cm^2)	120	230	
Modulus of rupture			
(lb/in^2)			
(kg/cm^2)			
Mean reversible thermal (exp. × 10^{-6} or %)	0–900° C 5·8		20 @ 20–1000° C
Thermal conductivity c.g.s. (cal cm^{-2} sec^{-1} °C^{-1} cm^{-1})			0·076 @ 1093° C
B.t.u. ft^{-2} hr^{-1} °F in.$^{-1}$			220 @ 2000° F
Permeability (c.g.s.)			
Linear change or refiring: Temp. (°C)			
Time			
% change			nil
Refractoriness under load: load (lb/in^2)			
Temp. (°C)			
Time			
% deformation			none
Spalling index			
Rising temp. test: load (lb/in^2)			
initial softening			
rapid softening			
failure			
Impact strength			
(cm kg/cm^2)			
Modulus of elasticity (kg/mm^2)			
(lb/in^2)			
Shear strength (lb/in^2) (kg/cm^2)			
Tensile strength (lb/in^2)			
(kg/cm^2)			
Transverse breaking strength (lb/in^2)			
(kg/cm^2)			

† Spalling index compared with silicon carbide 100.

265—(contd.)

12 Silicon carbide Brit. **P22** SiC 85	13 Ceramic bonded silicon carbide U.S.A. **C9**	14 Recrystallised silicon carbide Amer. **T14** SiC 98–99	15 Silicon carbide Amer. **N27** SiC 89–91	16 Silicon carbide Germ. **K66**	17 Silicon nitride bonded silicon carbide U.S.A. **C9**
33–34	38			42	
1730–1750	1850	dissoc. 2250	dissoc. 2250	2000	
3146–3182	3362	dissoc. 4082	dissoc. 4082	3632	
app. 18–20	13·2	20	17–20	18–22	7·9
2	2·57	2	3	2·2–2·5	2·87
155	160	153	161·5	137–156	179
		3·17	3·13	3·0–3·15	
	0–1400° C		21–1000° C		0–1400° C
	0·285		0·186		0·288
17 365				11 379–14 223	
1221				800–1000	
	@ 1350° C				@ 1350° C
	2000				5640
	141				397
0–1000° C	25–1400° C	21–1000° C	21– 1100° C		25–1400° C
5·1	4·4	4·5	4·5		4·4
0–1100° C					
0·0161–0·0159	0·036@1206°C	0·225 @	0·023@1093° C		0·039@1206° C
	109@2200° F	21–800° C			
46–47		653	66·0@2000° F		113·5@2200° F
1400		1500	1500		
0·1		+ up to 10% due to oxid.	nil		
28	25	50	50		25
1570	1500	1500	1500		1500
5	0·0	nil † 100	nil		0·0
				1500–1700	

Shaping. Pressing.
Firing. 1400° C (2552° F).
Properties. Table 265, No. 7 (**S101**).

CARBON BLOCKS AND SHAPES

Characteristics. High refractoriness with good volume stability, wide range of resistance to chemical attack, low thermal expansion, high resistance to thermal shock, good strength at high temperature, not wetted by molten slag or iron, relatively high thermal conductivity. Oxidised by air above 350° C (662° F), and reacts with water above 590° C (1094° F) and carbon dioxide above 700° C (1292° F) (**A56**). The alkaline residue from blast furnaces begins to attack carbon brick at 815° C (1500° F) (**N9**).

Uses. In form of interlocking corrugated blocks for base hearth and bosh of blast furnaces, excluding the area around the tap hole. Bricks for stacks of blast furnaces. Trough shapes for run outs of blast furnaces. Boats for sintering tungsten carbide in carbon tube furnaces (**A56**). Refractory lining and cathode of aluminium pots. (Bricks set in ramming mortar of same composition.) Ferro alloy furnaces, *e.g.* ferrosilicon, ferromanganese, also for calcium carbide, phosphorus, ferrophosphorus, phosphoric acid. Moulds for casting ferroalloys. Electrodes (**V23**). Mould plugs in production of steel ingot (**W42**). Melting furnaces for copper and lead (**C64**).

Body Type. Refractory. Carbonaceous.

Raw Materials. High-quality foundry coke, of low ash content, 10–12% maximum; maximum size may also be stipulated, *e.g.* 1½ in. (3·8 cm), 75–82%, (gas and blast furnace cokes are unsuitable). Anthracite replaces 10% of the coke for acid-resisting blocks. Water-free steelworks coke-oven tar, 18–25%, (pitch content, 70%, anthrazenol, 30%). Or up to 10% addition of hard petroleum asphalt. Or mixtures of tar, anthracene oil and bitumen with a little naphthalene.

Preparation.

(1) Coke→dry→jaw breaker→edge runner mill, tube mill or ball mill→ $\frac{1}{25}$ in. (1 mm). Heat coke and tar in steam heated containers→mixing in kneading edge runner mills or heated Baker–Perkins mixer (**S104**).

(2) Coke→dryer→jaw crusher→rolls→screens. Grading 0·0 to 0·2 in. (0 to 0·5 mm)→double-armed steam-heated mixer maintained at 50° C (122° F)→add heated pitch→mix 30–45 min. (coke, 84%, tar, 16%) →Steam-heated metal storage floor at 40° C (104° F) (**A21**).

(3) Coke→screened to remove fines→large to rotary dryer→large dry pan (bottom steam heated) 8 min.→fines returned.

Particle size achieved:

		(%)
0·0 –0·004 in.	(0 –0·1 mm)	24–21
0·004–0·02 in.	(0·1–0·5 mm)	26–24
0·02 –0·04 in.	(0·5–1 mm)	11–10
0·04 –0·08 in.	(1 –2 mm)	7– 7
0·08 –0·12 in.	(2 –3 mm)	5– 6
0·12 –0·16 in.	(3 –4 mm)	1– 3
0·16 –0·20 in.	(4 –5 mm)	2– 2
0·20 –0·24 in.	(5 –6 mm)	7– 8
Plus % tar		15–18

→Tar heated to 80° C (176° F) added and mulling continued for 3 min (**A21**).

Shaping. Ramming with pneumatic hammers in wooden moulds up to $5 \times 5 \times 13$ ft ($1·5 \times 1·5 \times 4·0$ m). Small shapes dry pressing 2850–5700 lb/in^2 (200–400 kg/cm^2). Larger shapes by ramming with heated wedge-shaped rammer.

Hardening. Hardening occurs on cooling. The largest blocks take from 5 to 6 days.

Firing. 1000° C (1832° F). Bricks set edgewise into dense muffles built of firebrick slabs, all interstices being filled with crushed coke and an 8 in. (20 cm) coke layer placed on top of the uppermost layer. On this a flat layer of fireclay bricks is carefully imbedded in refractory mortar or fireclay (**S104**).

Periodic single-chamber coal-fired down-draught kilns.

Continuous gas-heated chamber kilns (**S104**).

Tunnel kiln, oil fired with provision for burning the volatiles given off from the setting. In tunnel kiln each car is fitted with a built-on container to take about 3500 lb ($\sim$1600 kg) products and packing (**A58**).

Firing Schedules
Periodic Kilns.

(1) Heating 700–800° C (1290–1470° F), 3 weeks; cooling, 3 weeks.

(2) Heating to 1000° C (1832° F), 4 days; cooling to removal of flue seals, 10 days; further cooling, 2 days (**A21**).

(3) Preheat to drive off volatiles slowly, 2 days. Heat up to 1000° C (1832° F), 2 days. In order to bring the bricks in the muffle to 1000° C, cone 10 must go down in the kiln. Cool naturally (**S104**).

Tunnel Kiln.

(4) Total time in kiln, 46 hr (**A58**).

Finishing. If necessary, machining to accurate size and shape on a travelling bed grinding machine. This enables the bricks to be set with close joints.

Properties. Table 265, Nos. 8–11.

SILICON CARBIDE BRICKS AND SHAPES

Characteristics. Very high load-bearing cold or hot, resistance to spalling, heat transfer characteristics, resistance to reduction, acid slags, slag adherence and penetration, mechanical abrasion; high refractoriness; resistance to oxidation is only medium in the range 800–1200° C (1475–2200° F) but otherwise is good; the electrical resistance at high temperatures is very low.
Uses. Furnace hearth plates. Lining of forge furnaces. Piers and supports in combustion chambers. Bridge wall and clinker line of boiler furnaces. Lining of water gas generators. Pottery kiln furniture. Pit furnace linings. Slag hole blocks. Skid rails. Roller hearths. Pyrometer tubes. Pumps for abrasive slurries. Spray nozzles for sulphuric acid. Tubes for submerged combustion heating of corrosive liquids. Burner blocks and nozzles. Wire guides. Crucibles. Rocket motor liners. Muffles. Retorts. Radiant heater tubes. Gas generator tubes. Cyclone liners. Hopper liners. Hot blast mains for blast furnaces.
Body Type. Refractory. Silicon carbide.
Raw Materials. Silicon carbide. Plastic fireclay. Soda ash. Kasseler Braun, etc.
Preparation. Jaw crushers→crushing rolls→magnetted→screened→fine, middle and coarse fractions→weighed batches→mixers.

Clay + water + deflocculant→blunger.
Clay slip added to dry graded mixture.

	Fine	Coarse grade
SiC 0–0·1 mm (0–0·004 in.)	42·5%	42·5%
SiC 0–2 mm (0–0·08 in.)	42·5%	—
SiC 0–3 mm (0–0·12 in.)	—	42·5%
Fine clay	15 %	15 %

(A9)

Shaping. Dry pressing. Tamping. Ethyl silicate process (**Q2**).
Firing. Cone 10–11 (1310° C).
Properties. Table 265, Nos. 12–17.

MAGNESITE BRICKS

Characteristics. The main basic refractory brick, of high corrosion resistance and frequently of very dense structure. Extremely high refractoriness and high thermal conductivity. Great resistance to basic slags, poor resistance to slags containing high percentage of silica. Normal brick poor spalling resistance but special types and chemically bonded or steel-encased are better.
Uses. Hearths of basic arc and basic open-hearth furnaces (if silica bricks are used for the roofs of such furnaces the silica and magnesite should be

separated by a course of chrome–magnesite bricks). Lining of barrel-type inactive mixers. Back, front and end walls of steel furnaces. Top courses of checker work. Lining of melting and refining furnaces for copper, tin, lead and antimony. Cement rotary furnaces. Glass melting tanks (low-iron quality bricks). Lower side walls of soaking pits.

Body Type. Refractory. Magnesite.

Raw Materials. Magnesite rock, brucite or sea water magnesia (special types up to 6% alumina) without binder, or with one of the following binders: iron oxide; chromite ore; alkalies; caustic lime; boric acid; coal ashes; clay; cement; magnesium salts; sodium silicate; tar; molasses; caustic magnesia.

Preparation. Mined rock hand sorted, magnesite being recognised by colour and lustre→

or froth flotation separation from talc

crushing ⟨fist size→shaft furnace (12–14 tons per day)
⟨small size→rotary furnace (up to 200 tons per day)

→*burning* at 1600° C (2912° F) (shaft furnace) to 1800° C (3272° F) rotary furnace→slow cooling→hand removal of impurities or magnetic separation (iron content of magnesite makes it magnetic)

→*Test:* At least six tests should be done on burnt magnesite clinker to check its suitability for making magnesite bricks:

(1) Chemical analysis.
(2) Specific gravity.
(3) Crystal size.
(4) Bulk density.
(5) Hydration tendency.
(6) Firing shrinkage.

The specific gravity and bulk density are a measure of the degree of dead-burning when considered in conjunction with the chemical analysis. Roughly speaking a specific gravity correction of 0·01 is necessary for every 1% of iron oxide. Thus the pure Greek material is well burned at a S.G. value of 3·50, whereas the Veitsch material with 8% iron oxide may be as high as 3·65. Well-burned material has a crystal size of 0·03 mm or over. The over-all hydration tendency must stay within a safe limit, nor must there be pockets of highly reactive material. Firing shrinkage test is a comparative one but will show a difference in new supplies, etc. Too large a shrinkage gives cracks and undersize bricks (**C26**).

→*Grinding* in edge runner mills or ball mills with chromium, chromium–nickel, or molybdenum steel balls→grading→medium size only used for brick production→addition of 2 to 3% water and if necessary binder in pan or ball mill→souring (1 to 10 days), necessary only if free lime or other caustic material present.

Shaping. Hydraulic, mechanical or hammer pressing at above 10 000 lb/in^2 (703 kg/cm^2).

Drying. Open sheds followed by humidity drying at 20° C (68° F).
Firing. To about 100° C below sintering temperature 1500–1600° C (2732–2912° F). Magnesite bricks cannot bear heavy load during firing so that in chamber and periodic kilns they must be boxed in load-bearing silica bricks. Fairly fast heating up is permissible, followed by slow cooling.

Down-draught beehive kilns (cycle about 8–14 days).
Chamber kilns, Mendheim, 8-day cycles.
Tunnel kilns, 3 to 5 days (**A21, A10**).

Note: some magnesite bricks are not burned before being built into the user's furnace. If pressed under sufficiently high pressure there is little shrinkage in use.

General Properties. The cold crushing strength is good but the decrease with increasing temperature is greater than for silica bricks. This is due to softening of the matrix which is made up of the various natural or added impurities in the material. Thus many magnesite bricks soften at 1500° C (2732° F), good bricks withstand 1700° C (3092° F) while those made of electrically fused purest natural magnesite are suitable for 2000° C (3632° F) provided that an oxidising atmosphere prevails (**H85**).

Resistance to corrosion is very good. There is absolute resistance to oxidising agents, resistance to reducing agents up to 1500° C (2732° F), complete resistance to molten metals and basic slags, good resistance to acid slags.

Ordinary magnesite bricks tend to have poor thermal shock resistance but a number of brands have overcome this difficulty, some by the addition of 2 to 6% alumina, some by the use of especially pure magnesite (**S105**).

The thermal conductivity of magnesite bricks is very high, 25 B.t.u. (0·0061 cal-cm/cm^2-sec-° C) at 300° C (572° F), falling off with rising temperature.

The electrical conductivity is low at room temperature but becomes substantial at higher temperatures, say, 1500° C (2732° F). In Greaves Etchells type of arc furnace a magnesite bottom can therefore be used as a third electrode.

Thermal expansion is rather high, from 1·2 to 1·4% over the range 20 to 1000° C (68–1832° F).

Properties. Table 266, Nos. 1–20.

PERICLASE BRICK

Characteristics. A basic brick with higher refractoriness up to 2300° C (4172° F) and refractoriness under load than the normal magnesite brick. Above 1700° C (3092° F) cannot be used in reducing atmosphere. High thermal conductivity.

Uses. Embedding resistors in heating units.

Body Type. Refractory. Magnesite.

Raw Materials. Electrically fused magnesia.
Shaping. Ethyl silicate process (**Q2**).

SEMI-STABLE DOLOMITE BRICKS

Characteristics. A cheap basic brick of high refractoriness. It has no natural resistance to water and can only be stored so long as the tar coating is not damaged, and in any case not more than 6 months. The porosity is low, refractoriness under load moderate, thermal shock resistance better than stabilised dolomite as is also resistance to iron oxide–lime slag at 1650° C (3000° F).
Uses. Open-hearth furnace. Basic Bessemer furnace side walls. Side walls of arc furnaces.
Body Type. Refractory. Dolomite.
Raw Materials. Natural dolomite rock. Tar.
Preparation. Burning of rock together with fuel in cupolas or vertical shaft kiln at minimum of 1700° C (3092° F) to reach bulk density of 2·5 g/ml →conveyor belt sorting, removal of soft fired;

or with addition of 5% Fe_2O_3 in rotary kiln→doloma.
  Doloma plus tar→mixer; *or* doloma plus flux and tar→mixer.
Shaping. Ramming.
Firing. Slow as large volume changes at low temperatures.
Finishing. Immediately after firing the bricks must be dipped in boiling tar (**C26**).
Properties. Table 265, No. 24.

STABLE DOLOMITE BRICKS

Characteristics. Basic refractory, cheaper than magnesite, of low porosity, high crushing strength, good refractoriness under load, wide softening range and volume stability but very low thermal shock resistance. The resistance to iron-oxide rich slags is lower than magnesite.
Uses. Sub-hearth of fixed open hearth furnaces. Top layers of sub-hearths in tilting furnaces. Front, back and end banks of fixed furnaces. Reheating furnace hearths. Arc furnace sub-hearths. Backing courses behind chrome–magnesite in front and back linings of open-hearth furnaces. Ladle linings.
Body Type. Refractory. Dolomite.
Raw Materials. Natural dolomite rock. Serpentine. Boric acid or phosphates.
Preparation. Dolomite and serpentine→wet grind in tube mill→stirred slurry tank→rotary kiln 1600° C (2912° F) (loses water, CO_2 and then reacts

TABLE

Magnesite Refractories

	1	2	3
Name, Type	Veitsch magnesite	Burned magnesite	Magnesite
Origin	Austrian	American	German
Ref.	**A21**	**T14**	**K66**
Analysis CaO	2·0–2·7		
MgO	86·0–86·7	83–93	~85–92
Al_2O_3	0·5–0·8		<4
Fe_2O_3	8·4–8·9	2–7	1–9
Cr_2O_3	—		
SiO_2	1·5–2·2		2–7
Refractoriness: cone			42
(°C)		2200	
(°F)		3992	
Porosity (app. or tr.)	17–21	20–28	20–22
Water Absorption			
Bulk Density (g/cm³)	2·90–3·05	2·58	2·8
(lb/ft³)	181–190	161·5	175
True S.G.		3·40–3·60	3·5–3·65
Specific heat (c.g.s.)		0·278 21–1000° C	
Cold Crushing strength (lb/in²)	9250–12 900		5689–8534
(kg/cm²)	650–907		400–600
Modulus of rupture (lb/in²)			
(kg/cm²)			
Mean reversible thermal exp × 10^{-6} or %	1·2% 1000° C	14·7@21–1700° C	
Thermal conductivity Temp.		200–1000°C	
(cal cm^{-2} sec^{-1} °C^{-1} cm^{-1})	0·00114	0·0087	
B.t.u. ft^{-2} hr^{-1} °F^{-1} – in.$^{-1}$	3·3	25·25	
Permeability (c.g.s.)			
Linear change or refiring Temp. (°C)		1600	
Time (hr)			
% change		– begins	
Refractoriness under load: Load (lb/in²)	28		
Temp. (°C)	1700 | 1800		
Time			
% deformation	initial | failure		
Spalling index	0–2‡	20–40†	poor resis.
Rising Temp. test load (lb/in²)			
initial softening			1600–1700
rapid softening			
failure			
Impact Strength			
(cm-kg/cm²)			
Millscale Bursting at 1600° C			
Modulus of Elasticity (kg/mm²)			
(lb/in²)			
Shear Strength (kg/cm²) (lb/in²)			
Tensile Strength (kg/cm²) (lb/in²)			
Transverse breaking Strength (kg/cm²)			
(lb/in²)			
Uses (see key) p. 1270			

‡ Thermal shock resistance at 950° C (1740° F), number of shocks.
† Spalling index compared with SiC 100.

4	5	6	7	8
Magnesite	Magnesite	Magnesite	Magnesite	Fired Styrian magnesite
German **K60**	Austrian **A21**	**P6**	Austrian **A21**	French **L31**
	2·78		3·4	1·5–0
	76·19		85·60	85–89
	4·76		0·00	1–3
	10·42		8·58	4·5–1·2
	4·56		Mn_3O_4 0·50	—
	1·7		7·36	5·5
				42
	20–23	tr. 20–22 app. 19–21	21·0	app. 15–22
	2·81–2·94	2·84	2·82	3·0
	175–184	177	176	187
		3·55	3·57	
3·4–3·7		0·225		0·26
0·263				
	5700	> 6000	5888–7325	5689–11 379
	701	421	414–515	400–800
20–1000° C		12		13·0 20–1000° C
13·74–14·53		500–1000° C		
300° C | 1100° C	—	0·0103		
0·00206 | 0·00096	0·00081	30		
5·98 | 2·78	2·35	0·02		
		1400 | 1500 | 1600		
		2 | 2 | 2		
		0 | −0·6 | −3·0		
	28	25		28
	1530 | 1550	1600	1700	1450
		—		
	begin | fail	4	begin	begin
	> 40‡	25		
		1500		
8000–12 000				
11·4–17·1 × 10^6				

TABLE

Magnesite Refractories—(contd.)

Name, Type	9 Magnesite of high thermal shock resistance	10 Magnesite	11 Magnesite
Origin Ref.	Austrian A21	American N27	Austrian
Analysis CaO	2·4		2·6
MgO	81·0	83–93	83·5
Al_2O_3	7·0		4·1
Fe_2O_3	8·0	2–7	5·2
Cr_2O_3	—		MnO 0·2
SiO_2	1·6		3·6
Refractoriness: cone			
(°C)		2200	
(°F)		3992	
Porosity (app. or tr.)	26		
Water Absorption			
Bulk Density (g/cm³)	2·70	2·57	
(lb/ft³)	169	161	
True S.G.		3·4–3·6	
Specific heat (c.g.s.)		0·278 @ 22–1000° C	
Cold Crushing strength (lb/in²)	2850		
(kg/cm²)	200		
Modulus of rupture (lb/in²)			
(kg/cm²)			
Mean reversible thermal exp × 10^{-6} or %		14·7, 20–1425° C @ 1320° C	
Thermal conductivity.			
(cal cm^{-2} sec^{-1} °C^{-1} cm^{-1})		0·0045	
B.t.u. ft^{-2} hr^{-1} °F^{-1} – in.$^{-1}$		13 @ 2400° F	
Permeability (c.g.s.)			
Linear change or refiring: Temp. (°C)			
Time (hr)			
% change			
Refractoriness under load: Load (lb/in²)	28		
Temp. (°C)	1540 1550		
Time	begin fail		
% deformation	20‡		
Spalling index			
Rising Temp. Test: Load (lb/in²)			
Initial softening			
Rapid softening			
Failure			
Impact Strength			
(cm-kg/cm²)			
Millscale Bursting at 1600° C			
Modulus of Elasticity (kg/mm²) (lb/in²)			
Shear Strength (kg/cm²) (lb/in²)			
Tensile Strength (kg/cm²) (lb/in²)			
Transverse breaking Strength (kg/cm²)			
(lb/in²)			
Uses (see key p. 1270)			1

‡ Thermal shock resistance at 950° C (1740° F), number of shocks.
* Spalling resistance of 3×2×2 in. specimens placed for alternate 10 min. periods in furnace

266—(contd.)

12	13	14	15	16
Magnesite	Electrically fused magnesia	Electrically sintered Grecian magnesite	Magnesite with a little chrome	Austrian Magnesite with alumina addition
British **S183**	**C25**	**C25**	British **S183**	**C25**
91–92				
app. 19·6	21·4	app. 22·2	app. 21·3	app. 18·5
2·82	2·80	2·76	2·94	2·90
176	175	173	184	181
3·55	3·56	app. 3·54	3·67	app. 3·56
11 200	—	1470	6600	1860
787		103	464	131
0–1400° C, 1·74% @1100° C			0–1400° C, 1·38% @1100° C	
0·0072			0·0062	
20·90			17·99	
0·107	0·073	0·088	0·204	0·039
1500	1500	1500	1500	1500
2	2	2	2	2
−0·03	0·0	−0·3	−0·09	−0·3
28	25		28 psi	
1715	1600		1630	
	20 min failed			
10			10	
7*	30 +	30 +	30 +*	30 +
		50		50
		1480		1410
		1600		1570
		1630		1620
2c, d, f, 3, 4a, b, 7, 8a, 9			2c, d, e, n, 4b	

at 1000° C and on cold surface until failure.

TABLE

Magnesite Refractories—*(contd.)*

Name, Type Origin Ref.	17 Unburned magnesite American T14	18 Sea-water magnesite chem. bonded French L31	19 Unfired chem. bonded magnesite C25
Analysis CaO		1·4	
MgO	83–93	86·0	
Al_2O_3		2·7	
Fe_2O_3	2–7	1·7	
Cr_2O_3		—	
SiO_2		5·0	
Refractoriness: cone			
(°C)	2200		
(°F)	3992		
Porosity (app. or tr.)	10–22	app. 12/18	13·2
Water Absorption			
Bulk Density (g/cm³)	2·85	2·8	2·94
(lb/ft³)	178·5	175	184
True S.G.	3·40–3·60		3·39
Specific heat (c.g.s.)	0·278, 21–1000° C		
Cold Crushing strength (lb/in²)		4267	5120
(kg/cm²)		300	360
Modulus of rupture (lb/in²)			
(kg/cm²)			
Mean reversible thermal exp × 10^{-6} or %			
Thermal conductivity			
(cal cm⁻² sec⁻¹ °C⁻¹ cm⁻¹)			
B.t.u. ft⁻² hr⁻¹ °F⁻¹ – in.⁻¹			
Permeability (c.g.s.)			0·0044
Linear change or refiring: Temp. (°C)	1550		1500
Time (hr)			2
% change	−2·6		+0·3
Refractoriness under load: Load (lb/in²)	25	28	25
Temp. (°C)	1480	1450	1600
Time			12 min.
% deformation	3·5	begin	failed
Spalling index	30–50†		2
Rising Temp. Test: Load (lb/in²)			
Initial softening			
Rapid softening			
Failure			
Impact Strength			
(cm-kg/cm²)			
Millscale Bursting at 1600° C			
Modulus of Elasticity (kg/mm²) (lb/in²)			
Shear Strength (kg/cm²) (lb/in²)			
Tensile Strength (kg/cm²) (lb/in²)			
Transverse breaking Strength (kg/cm²)			
(lb/in²)			
Uses (see key, p. 1270)			

† Spalling index compared with SiC 100.
* Spalling resistance of 3 × 2 × 2 in. specimens placed for alternate 10 min. periods in furnace

Dolomite Refractories

20	21	22	23	24
Unburned chem. bonded magnesite	Stable Dolomite	Stable Dolomite	Stable Dolomite	Semi-stable Dolomite
British S183	British C26	British C26	British R7	British C26
	type 40·0		40·0	51·7
	40·4		39·5	38·2
	1·5		2·0	2·2
	3·4		3·5	2·5
	14·4		15·0	4·3
app. 14·9	22·1	24·7	18	22·8
2·89	2·58	2·53	2·78	2·52
180	161	158	174	157
3·45	3·31	3·36		3·27
			0·25	
5400	>8300	5090	8000	3680
380	584	358	565	259
0–1400° C 1·5%			13·0 20–1000° C	
@1100° C			615° C | 870° C	
0·0035			0·0053 | 0·0051	
10·16			15·39 | 14·81	
0·0020	0·088	0·12		0·29
1500	1500	1500		1500
2	2	2		2
−0·19	0·0	0·2		1·2
28	25	25	28	
1510	1600	1600	1500	
	32 min	11 min	2 hr	
	failed	failed	−5	
10			1–3	23
10*				
	1540	1460	1450	1360
	1610	1510		1520
	1680	1600	1650	1610
2b, f, g, k, 4b, 7 8a, b, 10			2p, g, 4, 6, 12	

at 1000° C and on cold surface until failure.

TABLE

Chrome—Magnesite

	25	26	27
Name, Type	Chrome-magnesite	Chrome-magnesite	Chrome-magnesite
Origin	General prop. of British	British	
Ref.	C26	S183	P6
Analysis CaO			
MgO	30		
Al_2O_3			
Fe_2O_3			
Cr_2O_3	70		
SiO_2			
Refractoriness: cone (°C)			
(°F)			
Porosity (app. or tr.)	22–26	app. 18·4	tr. 22–25 app. 22–24
Water Absorption			
Bulk Density (g/cm³)		2·94	2·85
(lb/ft³)		183	178
True S.G.		3·83	3·80
Specific heat (c.g.s.)			0·20
Cold Crushing strength (lb/in²)	~1000	4000	3000
(kg/cm²)	~70	281	211
Modulus of rupture (lb in²)			
(kg/cm²)			
Mean reversible thermal exp × 10⁻⁶ or %		0–1400° C, 1·38	10
Thermal conductivity		@1100° C	500–1000° C
(cal cm⁻² sec⁻¹ °C⁻¹ cm⁻¹)		0·0045	0·0052
B.t.u. ft⁻² hr⁻¹ °F⁻¹ – in.⁻¹		13·06	15
Permeability (c.g.s.)		0·165	0·07
Linear change or refiring: Temp. (°C)	1600 | 1700	1500	1400 | 1500 | 1600
Time (hr)	2 | 2	2	2 | 2 | 2
% change	+small | −2	+0·24	0 | 0 | −0·9
Refractoriness under load: Load (lb/in²)	28	28	25
Temp. (°C)	1600	1730	1600
Time	a few min.		120 min.
% deformation	fail	10	5
Spalling index	30	28+*	>30
Rising Temp. Test: Load (lb/in²)	28		25
Initial softening			1610
Rapid softening			
Failure	1700		
Impact Strength			
(cm-kg/cm²)			
Millscale Bursting at 1600° C			5–10
Modulus of Elasticity (kg/mm²) (lb/in²)			
Shear Strength (kg/cm²) (lb/in²)			
Tensile Strength (kg/cm²) (lb/in²)			
Transverse breaking Strength (kg/cm²)			
(lb/in²)			
Uses (see key p. 1270)		2b, c, d, e, f, g, h, i, k, l, m, n, 4b, 6, 8b, c	

* Spalling resistance of 3 × 2 × 2 in. specimens placed for alternate 10 min. periods in furnace at
‡ Thermal shock resistance at 950° C (1740° F), number of shocks.

266 —(contd.)

Refractories

	28	29	30	31	32
	Chrome magnesite (average of 10 Brands) German **A21**	Chrome magnesite Austrian **A21**	Chrome magnesite German **K66**	Fired chrome magnesite French **L31**	Chem. bond Chrome magnesite French **L31**
	1·26	1·0		3–1	1·5–0
	40·55	48·0		58/65–32/35	70–32
	10·40	13·5	10–13	4/7	5·5–17·5
	13·26	13·0	10–12	9/15–20	4·5–12·0
	28·0	20·5		12/15–32/35	7·5–30
	4·44	4·0	~3	3–6	5·5
			40	42	42
			1920	2000	2000
			3488	3632	3632
	25·1	22–25	24–25	app. 20/25	app. 10/16
	2·92	2·90–3·00	2·8–2·9	3·00	2·95/3·18
	182	181–187	175–181	187	184–99
	3·87		3·8–3·9		
	327–6386	2850	~4266	4266/8534	5689
	23–449	200	~300	300/600	400
		0·88% 1000° C		10, 20–1000° C	
		0·0006			
		1·74			
		28		28	28
1550	1640	1650 1700		1450	1450
big	end	begin fail >30‡	good 1600–1700	begin	begin

1000° C and on cold surface until failure.

TABLE

Chrome—Magnesite Refractories—(*contd.*)

	33	34	35
Name, Type	Unburned magnesite chrome	Magnesite chrome	Magnesite-chrome
Origin Ref.	British **S183**	**P6**	Austrian **A21**
Analysis CaO			1·7
MgO			65·5
Al_2O_3			7·0
Fe_2O_3			11·0
Cr_2O_3			12·0
SiO_2			2·7
Refractoriness: cone (°C)			
(°F)			
Porosity (app. or tr.)	app 11·7	tr. 24–25 app. 22–23	18–21
Water Absorption			
Bulk Density (g/cm³)	3·11	2·75	2·86–3·04
(lb/ft³)	194	172	179–190
True S.G.	3·57	3·65	
Specific heat (c.g.s.)		0·2	
Cold Crushing strength (lb/in²)	9000	3000	4280
(kg/cm²)	633	211	301
Modulus of rupture (lb/in²)			
(kg/cm²)			
Mean reversible thermal exp × 10⁻⁶ or %	0–1400° C, 1·2	10	
Thermal conductivity	@1100° C	500–1000° C	
(cal cm⁻² sec⁻¹ °C⁻¹ cm⁻¹)	0·0023	0052	0·0008 0·0008
B.t.u. ft⁻² hr⁻¹ °F⁻¹ – in.⁻¹	6·68	15	2·35–2·40
Permeability (c.g.s.)	0·0019	0·05	
Linear change or refiring: Temp. (°C)	1500	1400 | 1500 | 1600	
Time (hr)	2	2 | 2 | 2	
% change	+0·57	0 | −0·3 | −3·0	
Refractoriness under load: Load (lb/in²)	28	25	28
Temp. (°C)	1650	1600	1560 | 1600
Time		50	
% deformation	10	>10	begin | fail
Spalling index	23*	>20	>40‡
Rising Temp. Test: Load (lb/in²)		25	
Initial softening		1400	
Rapid softening			
Failure			
Impact Strength (cm-kg/cm²)			
Millscale Bursting at 1600° C		1–3	
Modulus of Elasticity (kg/mm²) (lb/in²)			
Shear Strength (kg/cm²) (lb/in²)			
Tensile Strength (kg/cm²) (lb/in²)			
Transverse breaking Strength (kg/cm²) (lb/in²)			
Uses (see key p. 1270)	2b, g, k, n, 8b		

* Spalling resistance of 3 × 2 × 2 in. specimens placed for alternate 10 min. periods in furnace at
‡ Thermal shock resistance at 950° C (1740° F), number of shocks.

266—(*contd.*)

Chrome, Refractories

36	37	38	39	40	41
Chrome	Chrome	Chrome	Chromite	Chrome	Chrome
British **S183**	American **N27**	British **C26**	**P6**	French **L31**	German **K66**
		nil		0·5	
		18·99		11	
	15–33	11·12		11	10–12
	3–6	21·43		25	12–15
	30–45	39·77		45	
	11–17	8·52		7	~5
	41+			40/42	38–39
	1960–2200			1920–2000	
	3550–4000			3488–3632	
			tr. 22–24		
app. 16·2		app. 22·4	app. 20–22	app. 15/18	20–21
3·00	2·99	3·10	3·00	3·30	2·9–3·1
187	187	194	187	206	181–194
3·95	3·8–4·1	app. 3·99	3·00		3·9–4·0
	0·22@1000° C		0·18	0·2, 0–1000° C	
	@1832° F				
11 000			>6000	14 223	4267–5689
773			422	1000	300–400
0–1400° C, 1·16	8·0@20–1000°C		9·0	9·0@20–1000°C	
@1100° C			500–1000° C	20° C |1000° C	
0·0046	0·0041		0·0041	0·0033 | 0·0042	
13·35	12·1@2400° F		12	9·6 | 12·2	
0·047		0·017	0·03		
1500	1550		1400|1500|1600		
2			2 | 2 | 2		
−0·15	−1·3		0 | −0·6 | —		
28	25		25	28	
1495	1430		1600	1450	
			—		
10	nil		4	begin	
2*		10	25		average
		50	1350		
		1280			1450–1500
		1350			
		1370			
2c, f, j, l, m, o, 5, 6, 11					

1000° C and on cold surface until failure.

TABLE

Chrome—Spinel, Olivine and

	42	43	44
Name, Type	Chrome	Magnesia alumina spinel	Olivine 'A' olivine bonded without magnesite
Origin	American	American	
Ref.	**T14**	**N27**	**S122**
Analysis CaO			
MgO		28·2	
Al_2O_3	15–33	71·8	
Fe_2O_3	3–6		
Cr_2O_3	30–45		
SiO_2	11–17		
Refractoriness: cone			$33\frac{1}{2}$–$35\frac{1}{2}$
(°C)	1950–2200	2100	1730–1800
(°F)	3542–3992	3875	3146–3272
Porosity (app. or tr.)	20–28		20
Water Absorption			
Bulk Density (g/cm³)	2·99	2·72	2·7
(lb/ft³)	187	170	169
True S.G.	3·80–4·10	3·6	3·3
Specific heat (c.g.s.)	0·217, 21–1000° C	0·25@22–1000° C 70–1832° F	
Cold Crushing strength (lb/in²)			
(kg/cm²)			
Modulus of rupture (lb/in²)			
(kg/cm²)			
Mean reversible thermal exp × 10^{-6} or %	10·4@21–1540° C	8·0@20–800° C	
Thermal conductivity	200–1000° C	1315° C	
(cal cm^{-2} sec^{-1} °C^{-1} cm^{-1})	0·0040	0·0499	
B.t.u. ft^{-2}hr^{-1} °F^{-1}–in.$^{-1}$	11·61	145@2400° F	
Permeability (c.g.s.)			
Linear change or refiring: Temp. (°C)	1450 | 1550	1600	
Time (hr)			
% change	nil | –1·3	nil	
Refractoriness under load: Load (lb/in²)	25	any	28
Temp. (°C)	1425	1350	1610 | 1650 | 1687
Time			
% deformation	nil	poor	1 | 3 | fail
Spalling index	25––70†		12–25§
Rising Temp. Test: Load (lb/in²)			
Initial softening			
Rapid softening			
Failure			
Impact Strength (cm-kg/cm²)			
Millscale Bursting at 1600° C			
Modulus of Elasticity (kg/mm²) (lb/in²)			
Shear Strength (kg/cm²) (lb/in²)			
Tensile Strength (kg/cm²) (lb/in²)			
Transverse breaking Strength (kg/cm²) (lb/in²)			
Uses (see key p. 1270)			

* Spalling resistance: Three specimens 3 in. × 2 in. × 2 in. are placed, resting on their 2 in. × 2 in. 2 in. × 2 in. faces, on a cool brick for 10 min. Specimens are then placed in a heated furnace as before.
† Spalling index compared with silicon carbide 100.
‡ Thermal shock resistance at 950° C (1740° F), number of shocks.
§ Number of quenchings by cold air blast from 1600° C.

266—(contd.)

(contd.)

Forsterite Refractories

45	46	47	48	49
Forsterite 'Regular' olivine bonded with magnesite	Forsterite	Forsterite	Forsterite	Forsterite 'B' olivine bonded with chrome magnesite
	American	American	French	
S122	T14	N27	L31	S122
			1·5	
		57·3	55	
			1·5	
			8·0	
			0·5	
		42·7	33·5	
35½–42		40	34/37	35½–42
1800–1900	1910	1920		1800–1900
3272–3452	3470	3488		3272–3452
22–26	24–27		app. 26	12–23
2·7–2·8	2·45	2·45	2·75	2·9–3·4
169–175	153	153	172	181–212
3·4–3·5	3·30 –3·40	3·3–3·4		3·4–3·7
	—	0·22@21–93° C		
		70–200° F		
~2000			4267	
~141			300	
1·5%, 20–1500° C	12·5@21–1500° C	11, 20–1500° C	12·0, 20–1000° C	1·5%, 20–1500° C
0·0035		0·0035		
10·3 up to 1316° C		10·3@2400° F		
1600	1650	1700		
12				
−0·5	nil	nil		
28		50	28	28
1630 | 1675 | 1687		1500	1600	1710 | 1730 | 1737
1 | 3 | fail		nil	begin	1 | 3 | fail
8§	20–40†			20§

faces, in a furnace at 1000° C. After 10 min, specimens are removed and placed, resting on the same Treatment continues until specimens fail.

to give magnesia, $3CaO.SiO_2$, and a small amount of glass)→pea-sized clinkers→cooler→crushers→sieves→hoppers→testing. Mixing of coarse: medium: fine:: 60: 10: 30 plus 4% water in paddle type mixer.

Shaping. Hydraulic pressing 10 000 to 15 000 lb/in^2 (703 to 1055 kg/cm^2).
Drying. In stream of clean dry air below 60° C (140° F).
Firing. 1350–1450° C (2462–2642° F).

Intermittent down-draught kilns, circular or rectangular. Coal, oil or producer gas. Soaking period of at least 24 hr (**S106**).

Tests. Stabilised clinker.
(1) Chemical analysis.
(2) Porosity, bulk density and true specific gravity.
(3) Hydration resistance.
(4) Grading and packing of aggregates.
(5) Sintering tendency.
(6) Angle of rest.
(7) Slag resistance (**C26**).

Properties. Table 266, Nos. 21–23.

Uses of Magnesite, Chrome–Magnesite, Chrome, etc., Refractories as referred to in Table 266.

(1) Rotary cement kiln.
(2) Open-hearth furnace: (*a*) general; (*b*) gas and air ends; (*c*) bath bottom; (*d*) bridge, front and back banks; (*e*) blocks: gas port arch; (*f*) blocks: gas port slope; (*g*) blocks: wing walls; (*h*) blocks: block facings; (*i*) water-cooled ports (basic); (*j*) water-cooled ports (acid); (*k*) back wall: fixed furnaces vertical; (*l*) back wall: fixed furnaces semi sloping; (*m*) back wall: fixed furnaces fully sloping; (*n*) front wall; (*o*) slag pocket, lower courses; (*p*) sub-hearth brickwork; (*q*) backing behind front and back linings.
(3) Hot metal mixers.
(4) Electric steel furnace (basic): (*a*) below slag line; (*b*) above slag line.
(5) Soaking pit, lower courses of side walls.
(6) Reheating furnace: hearth and lower side walls.
(7) Copper smelting reverberatory furnace.
(8) Copper refining furnace: (*a*) bottom up to slag line; (*b*) side walls up to skews; (*c*) roof.
(9) Copper converter.
(10) Dolomite cement kiln.
(11) Soda recovery furnace.
(12) Ladle linings for foundries where desulphurisation by soda ash used.

CHROME-MAGNESITE REFRACTORIES

Characteristics. Mechanical strength and stability at high temperatures. Excellent to good resistance to spalling. High resistance to corrosion by basic slags. Better resistance to torsional and tensional stress at high temperatures than magnesite. There is a tendency to bursting in contact with iron oxide.

Uses. Non-ferrous metallurgical furnaces, *e.g.* copper (sprung arch roofs), converters and reverberatory smelting furnaces; nickel converters; lead reverberatory furnaces lower and upper side walls and roofs; aluminium, bottoms and side walls. Side walls and rib arches in copper anode and wire-bar furnaces. Roofs of all basic open-hearth furnaces. Soaking pits. Reheating furnaces.

Body Type. Refractory. Chrome-magnesite.

Raw Materials. Chrome ore. Dead burnt magnesite.

Preparation. Grinding→grading. Grading has large effect on thermal shock resistance, at least 60% should be coarse and not more than 25% intermediate, the usual British method uses: coarse chrome $+25$ mesh, fine magnesite -72 mesh. But a German mixture for use in vibrating moulds is given as follows:

      5% magnesite dust
    30% magnesite 0·04–0·12 in. (1–3 mm)
    25% chrome ore dust
    40% chrome ore 0·08–0·12 in. (2–3 mm)

→tempering with 6–7% water in mixer for vibrating mould shaping (**A21**).

Shaping. Dry pressing in hydraulic or impact presses. Vibrating moulds (Germ.).

Drying. Moister pieces on pallets in warm room 27–35° C (80–95° F) for 8 days.

Firing. 1400–1650° C (2552–3002° F) (**C26**); 1480–1500° C (2696–2732 °F) (**A10**); 1520–1540° C (2720–2800° F) (**A21**).

  Periodic kilns: If small can have quick cycle; coal-fired.

  Tunnel kilns: gas or oil-fired. Advantage of low setting reducing slumping.

Tests.
  (1) True and apparent porosity.
  (2) Specific gravity and bulk density.
  (3) Shape, size, warpage.
  (4) Spalling index.
  (5) Volume stability at a given temperature.
  (6) Cold crushing strength.
  (7) Refractoriness under load: (*a*) rising temperature; (*b*) maintained temperature.
  (8) Thermal expansion (reversible).
  (9) Thermal conductivity.
  (10) Permeability.
  (11) Specific heat.

(12) Millscale bursting expansion at 1600° C (2912° F) (**P6**).
Properties. Table 266, Nos. 25–31.

MAGNESITE–CHROME

Characteristics. Excellent mechanical strength, stability of volume and resistance to abrasion and erosion at high temperatures. Excellent resistance to spalling. High resistance to corrosion by basic slags.
Manufacture. Generally similar to chrome–magnesite except for the proportions and gradings of the raw materials.
Properties. Table 266, Nos. 34 and 35.

CHROME REFRACTORIES

Characteristics. High resistance to corrosion by basic and moderately acid slags and fluxes. In general, basic slags do not adhere to chrome brick. Under certain unusual conditions, iron oxide is absorbed and causes a damaging expansion. Thermal conductivity lower than that of magnesite but higher than that of fireclay. Relatively low refractoriness under load and thermal shock resistance.
Uses. Basic open hearth furnaces for: facing port blocks, front and back walls at and below the slag line, furnace bottoms. Soaking pits, heating, heat treating, forge and welding furnaces where iron oxide would attack fireclay and silica. Sodium salt recovery furnaces (paper industry). Neutral course between magnesite and silica.
Body Type. Refractory. Chrome.
Raw Materials. Chromite. Organic binders, *e.g.* sulphite lye, molasses, tar, gum arabic, etc., or ceramic bonds, *e.g.* brick clay, loam, fireclay, china clay, bauxite, aluminium hydroxide, etc. (*see* bodies).
Preparation. Jaw crushers or cone stone breakers→roller or ball mills→ wet pan, binders added.
Shaping. Power press. Hydraulic press 4200–5700 lb/in² (295–400 kg/cm²). Hand tamping.
Drying. Green bricks are very fragile Tunnel dryers.
Firing. Straight chrome max. 1420° C (2588° F), with magnesia up to 1760° C (3200° F).

Periodic kilns, slow cooling. Little load may be applied so that chrome bricks must be boxed in silica bricks in periodic kilns.

Tunnel kilns.
Properties. Table 266, Nos. 36–42.
Chrome Bricks in Contact with Iron Oxide. Chrome and to a lesser extent chrome–magnesite bricks tend to expand and burst at surfaces in contact with hot iron oxide. The cause is not fully understood.

OLIVINE REFRACTORY BRICKS

Characteristics. A basic refractory made directly from suitable natural rock without previous calcination, its thermal shock resistance is better than most magnesite, it has practically no firing or after shrinkage, it has exceptional resistance to fused iron oxides and silicates.

Uses. Lining for rotary kilns for calcination of dolomite. Roofs of copper-holding furnaces. Bottoms of metal treatment furnaces. Glass tank checkers. Walls and arches of regenerative chambers (glass tanks). Glass tank ports, uptakes (**H38**).

Body Type. Refractory. Olivine.

Raw Materials. Olivine rock of maximum 6–7% iron content. Bond: *e.g.* sulphite lye, magnesium chloride solution (80 g/l.), 250 g syrup or molasses plus 30 g boric acid per litre.

Preparation. Jaw crushers→mills→sieves. The minimum of fines to be made. Olivine is very brittle and easily breaks down to a fine dust. Three portions are required:

(1) Large grains (0·16–0·02 in.) 4·0 to 0·5 mm.
(2) Dry ground powder to pass sieve 4900 mesh/cm^2.
(3) Wet grinding of some of the fines with the binder solutions 77 parts solids to 23 parts liquid for 6 hr.

Mixing: A. 60 parts coarse-grain olivine are moistened with 1·8 parts liquid. B. 26 parts dry fines are mixed with 18 parts wet-ground slurry (14% solids and 4% liquid). A and B are then thoroughly mixed in mixer.

Shaping. Hydraulic press 3000–6000 lb/in^2 (211–422 kg/cm^2).

Drying. 4 days ordinary temperature. 2 days being heated up to 150° C (302° C).

Firing. 1200° C (2192° F) upwards. No contact must be made with fireclay bricks (**S122**).

Properties. Table 266. No. 44.

FORSTERITE REFRACTORY BRICKS

Characteristics. Basic refractory with the general characteristics of olivine refractory, *i.e.* great volume stability and resistance to iron oxides and silicates but higher refractoriness and excellent strength at high temperatures. Marked resistance to corrosion by alkali compounds, fair resistance to most basic slags; attacked by acid slags.

Uses. Lining for rotary kilns for calcining refractory dolomite and magnesite. Roofs of copper-holding furnaces. Bottoms of metal treatment furnaces. Glass tanks: checkers, walls and arches of regenerative furnaces, ports, uptakes. Lime and cement kilns. Gas-fired open-hearth furnaces, for ports and uptakes (**H18, 38**).

Body Type. Refractory. Forsterite.

Raw Materials. Olivine rock of maximum 6–7% iron content. Dead burnt magnesite. Chrome ore of maximum 12% Al_2O_3 content, or caustic magnesia or magnesium hydroxide. Solution of 250 g syrup and 30 g boric acid per litre.

Preparation. (1) Olivine→jaw crushers→milling and sieving to produce minimum of fines (olivine is very brittle and easily makes fines). Grains from 52- or 180-mesh up to 0·25 or 0·5 to 0·6 in. (6·35 or 12·7 to 15·24 mm) retained. Fines rejected.

(2) Chrome ore and dead burnt magnesite ground finely together, some dry and some wet with binder solution.

Mixing: 60 to 80% olivine grain plus 40 to 20% chrome–magnesite 2:1 bond plus 3 to 6% water.

Shaping. High-pressure automatic hydraulic presses 3000 to 6000 lb/in^2 (211–422 kg/cm^2). Ethyl silicate casting (**Q2**).

Firing. Ordinary forsterite bricks cone 20 (1530° C, 2786° F). Higher qualities forsterite bricks cone 28–32 (1630° C–1710° C 2966–3110° F).

Oil-fired kilns with compressed air.

Testing. Care has to be taken during tests to allow no contact between olivine or forsterite refractories and refractories with which they might react during test.

General Properties of Olivine and Forsterite Refractories

(1) *Chemical.* No reaction with magnesite, chrome–magnesite or chromite refractories. Withstand molten iron up to 1800° C (3272° F). Serious reactions leading to failures occur with: aluminous materials (fireclay, etc.) above 1250° C (2282° F); silica brick or sand above 1450° C (2642° F); lime or very basic compounds of lime above 1400° C (2552° F); reducing agents: carbon, high-grade ferrosilicon, calcium carbide, etc. (**S122**).

(2) *Physical.* See Table 266, Nos. 45–49.

Refractory Hollow-ware

SAGGARS AND OTHER KILN FURNITURE

Characteristics. Refractory hollow-ware and bats, props and other shapes, used for the support and protection of fine ceramic ware during its firing. Requirements include refractoriness adequate to the application, thermal shock resistance, hot strength and dimensional stability, freedom from dusting and long life (*see* p. 408).

Uses. Kiln auxiliaries in the firing of fine ceramic ware (p. 954).

Body Type. Refractory: alumino-silicate, cordieritic, silicon carbide.

Raw Materials. Refractory fireclays, ball clays, kaolins, molochite. Fireclay grog. Broken saggars. Bauxite. Calcined alumina. Sillimanite, kyanite, andalusite. Talc. Silicon carbide.

Preparation.

Grog. Crushing→sieving→grading. Maximum size determined by thick-

CONSTRUCTIONAL REFRACTORIES

ness of piece to be made, up to 5 mm for thick, or sieve mesh-16 for thin.
Clay binder. Fireclays→fine wet grinding.
  Proportionating→mixing→water addition→pugmill→maturing.
Shaping. Hand modelling of saggars. Plastic screw pressing of saggars. Jiggering. Dry pressing: bats, cranks. Extruding of thimbles, saddles, stilts and spurs followed by drying to leather-hard condition, and then pressing to accurate shape. Casting: saggars, cranks.
Drying. According to water content and thickness.
Firing. Ideally to temperatures higher than those of anticipated use. Saggars are usually fired at the top of a bung of fired saggars carrying ware. A few light pieces may be placed in them.

OUTLETS, NOZZLES, STOPPERS, SLEEVES, RUNNERS, PLUGS, ROD COVERS, TUBES, PIPES, TROUGHS, FUNNELS AND TUYERES

Characteristics. Refractory pieces required to have good dimensional accuracy, high resistance to corrosion, abrasion and thermal shock.
Uses. Accessories to heating, melting, etc., furnaces.
Body Type. Refractory. Generally alumino-silicate frequently approximating to 'sillimanite'. Also magnesite, bauxite, chromite.
Raw Materials. Fireclays and grog. Graphite. Or magnesite, bauxite, chromite.
Preparation. General heavy clay methods to give plastic body.
Shaping. Hand moulding. Extrusion. Pressing in hand- or power-driven presses.
Drying. Slow drying.
Firing. Cone 10 to 12.
  (1) Intermittent down-draught kilns.
  (2) Semimuffle tunnel kilns.

CRUCIBLES AND SCORIFIERS OR ROASTING DISHES

(Fig. 17.3)

General Characteristics. *Crucibles* are highly refractory open-mouthed vessels used for heating various materials and particularly for melting metals, alloys, enamels, etc. Sizes vary from 1 in. (2·5 cm) high to about 30 in. (76 cm) holding about 70 to 75 lb (32–34 kg) of iron. There are a number of traditional shapes of crucible, *e.g.* tall narrow 'London' crucible, the shorter 'Cornish', triangular 'Hessian'. Place names also denote a particular body mixture.

Scorifiers or *roasting dishes* are shallow dishes made of the same materials and by similar methods as crucibles. They are used for exposing ores, etc., to air at high temperature in order to oxidise one or more constituent.

(a)

(b)

Fig. 17.3. Crucibles: (a) coal testing crucibles; (b) miscellaneous lift out and tilting crucibles (Morgan crucible, **M96**)

CONSTRUCTIONAL REFRACTORIES

The main requirements for crucibles and scorifiers are: (a) sufficient refractoriness to withstand any temperature to which they are likely to be exposed; (b) thermal shock resistance; (c) sufficient density to cause no loss of contents by absorption; (d) resistance to the corrosive action of the contents, e.g. metallic oxides, slags, ashes; (e) strength to hold the contents when being carried hot.

CLAY CRUCIBLES

Characteristics. Open-topped vessels for containing materials in furnaces, of moderate to high refractoriness, good strength and good resistance to thermal shock. Dimensions vary from 1 in. (2·5 cm) high to 30 in. (76 cm) high by 9 in. (23 cm) diameter.
Uses. The heating or melting of powders, metals, etc., in furnaces.
Body Type. Refractory, fireclay.
Raw Materials. Fireclays, grog, etc.
Preparation. Heavy clay methods to give a plastic body.
Shaping.
   (1) Hand moulding on a revolving wooden block.
   (2) Pressing. The correct-sized ball of plastic clay is placed in a 'flask' the interior of which gives the outside of the crucible, and a plug is rammed in to give the inside.
   (3) Jolleying.
   (4) Throwing (for small orders).
   (5) Slip casting.
Drying. As for firebricks. When firm, fettling or washing with slip gives smooth surface. Drying then completed.
Firing. Some crucibles are used green. Generally as for firebricks.
Chemical Resistance. Moderate resistance to siliceous slags (**S57**). Poor resistance to basic slags and alkalies (**S50**).

SILICEOUS CRUCIBLES
(Hessian Crucibles, Cornish Crucibles)

Body Type. Refractory, semi-silica.
Raw Materials. Silica rock, quartzite, etc. Clay.
   E.g. Hessian crucible: siliceous clay 2 parts, sand 4 to 5 parts; Cornish crucible: ball clays 2 parts, sand 2 parts.

SILLIMANITE CRUCIBLES

Uses. Melting glasses for optical and experimental purposes.
Body Type. Refractory alumino-silicate to give sillimanite.
Raw Materials. As for sillimanite refractory bricks. Kyanite. Andalusite. Sillimanite.
Chemical Resistance. Better resistance to acid than to basic slags.

ALUMINA CRUCIBLES

Characteristics. Very high refractoriness and resistance to boiling or fused alkalies, high thermal conductivity.
Uses. Coal analysis, igniting filters, melting rare metals and high melting point alloys.
Body Type. Refractory. Alumino-silicate, high alumina.
Raw Materials. (*a*) Bauxite, diaspore plus refractory fireclay. (*b*) Sinter corundum and binder. (*c*) Fused alumina and binder. (*d*) Mixtures of (*a*) (*b*) (*c*).
Chemical Resistance. High resistance to acid and basic slags.

GRAPHITE OR PLUMBAGO CRUCIBLES AND SHAPES

Characteristics. Refractory crucibles of high thermal conductivity and thermal shock resistance with little tendency to oxidise their contents. High chemical resistance except to oxidation.
Uses. Crucibles for melting of steels, brasses, copper, aluminium, antimony and various alloys. Ladle nozzles and stoppers. Zinc distillation retorts. Casting moulds for noble metals. Slabs for flattening window glass.
Body Type. Refractory. Carbonaceous.
Raw Materials. Graphite (Ceylon, Madagascar or Alabama U.S.A.) charcoal or coke. Plastic fireclay or ball clay, vitrifying at relatively low temperatures but having no after-contraction, *e.g.* Klingenberg (clay 85%, quartz 9%, feldspar 6%). Grog. (Silicon carbide.) (Sand.) (Feldspar.)
Preparation.
Natural Graphite crushed and separated from impurities by methods giving maximum of large flakes, and depending on the type of admixture. Examples are given below:

Dry methods depending on greater toughness of graphite than admixtures: disintegrators→sieves; or depending on different S.G. in air separator; or depending on different shapes.

Wet methods: sedimentation; flotation using oil to attract the graphite; froth flotation. If the ore is ground with rollers in a suspension of oil in water then as any graphite is released it floats to the surface and so is not unnecessarily crushed up.

Mica usually remains with graphite in these methods of purification.

Crucible strength is influenced by the shape of the graphite particles, flakes being better than grains. Strength is increased if the graphite is ground with a small addition of caustic soda before mixing with the clay.

Clay, Silicon Carbide, Grog, Sand, etc., Dry ground

Bentonite if used must be soaked in from five to ten times its weight in water for from 4 to 8 days and added with the tempering water.

Mix dry and then add tempering water→mix for several hours, edge runner mill or other batch mixer. Maturing, for hand throwing about 8 weeks, otherwise 2–4 weeks.

Pugging in de-airing pug mill.

Shaping.
  (1) Hand throwing.
  (2) Jiggering or jolleying.
  (3) Power pressing of stiff plastic body.
  (4) Dry pressing.
  (5) Hand moulding.
  (6) Ramming.
  (7) Slip casting.

Drying. Particularly difficult as the laminar structure restricts movement of the water.

Humidity dryers recommended.

Engobing. When leather-hard or when dry or when fired, protective engobes are painted on to prevent oxidation, *e.g.* sodium silicate, soda ash, calcium or magnesium chloride, boric acid, borax, etc., various albany slips.

Firing.
  (1) To about 700° C (1292° F) in open-set kilns. The surface becomes oxidised.
  (2) In muffles to 900° C (1652° F) (kiln at 1200° C, 2192° F). The interstices are filled with coke.
  (3) In saggars packed with coke to 1100° C (2012° F) (kiln at 1400° C, 2552° F).

Finishing. Annealing by placing mouth down on cold fresh refuelling of coke in a furnace; surround by coke and burn up slowly until crucible is red hot.

GRAPHITE/SILICON CARBIDE CRUCIBLES

Body. Refractory. Carbonaceous.
Raw Materials. Graphite. Silicon carbide. Tar.
Preparation. Similar to graphite-clay crucibles.
Shaping. Vibration.

SILICON CARBIDE CRUCIBLES

Characteristics. High thermal conductivity and therefore fuel saving, high resistance to thermal shock, and spalling, clean pouring but stain some contents, *e.g.* glass.
Body Type. Refractory. Silicon carbide.
Raw Materials. Silicon carbide. Bond clay or other binders.
Shaping. Dry pressing.
Chemical Resistance. High.

TABLE 267

Properties of Crucibles (S50)

		Clay	Graphite or plumbago	Sillimanite, mullite	Alumina	Magnesia	Spinel	Silicon carbide	Zirconia	Thoria	Beryllia	Silica (fused)
Maximum safe working temp.	(°C)	1700	1700	1750	1950	2000	1900	1300	1750	2200	2200	1300
	(°F)	3092	3092	3182	3542	3632	3452	2372	3182	3992	3992	2372
Maximum safe temp. in contact with carbon	(°C)					2000			1200	1200		
	(°F)					3632			2192	2192		
Volatilisation under reduced pressure	(°C)											
Thermal expansion. Temp. range Coeff. ×10⁻⁶				0–1700 5·3	0–1600 7·7	20–1000 13·7		0–1400 5·0	20–800 7·2	0–600 93·0		
Thermal conductivity (kcal m⁻¹ hr⁻¹ °C⁻¹)					4·9			0·01–0·07				0·72

MAGNESIA CRUCIBLES

Uses. Melting of platinum and other refractory materials and in electric furnaces.
Body Type. Refractory. Magnesite.
Raw Materials. Dead burnt magnesite (fired above 1550° C, 2822° F). Binders: caustic magnesia and magnesium chloride.
Preparation. Crushing and grading as for magnesite bricks.
Shaping.

(1) Pressing or tamping in plumbago mould, prefiring to 1200–1400° C (2192–2552° F) removing magnesia crucible for final firing.

(2) Pressing or tamping in graphite mould, prefiring to 1700° C (3092° F).

(3) Dry pressing 1500 lb/in^2 (105 kg/cm^2).

Firing. To 1900° C (3452° F). Arsem or similar furnace.
Chemical Resistance. Very high resistance to basic slags, low resistance to acid slags.

Tests for Crucibles.

(1) Heat inside another crucible to highest attainable temperature. There should be no sign of fusion.

(2) Heat to redness and plunge in cold water. Even if cracked it should not fall to pieces.

(3) Corrosion: (*a*) half fill with litharge or red lead, melt, cool, break and examine; (*b*) half fill with copper and a little borax and repeat; (*c*) melt together 1 part litharge with 2 parts soda.

GLASS-HOUSE POTS

Characteristics. Large refractory vessels of open or covered shapes, from 3 ft wide by 3 ft high (91 × 91 cm) up to 6 ft diameter (183 cm) and 3 ft high (91 cm). The thickness of the bottom of the pot is one-twelfth the greatest diameter, and the wall thickness one-twentieth the greatest diameter. Although they only need withstand temperatures of 1400–1500° C (2552–2732° F), they must not become so corroded and absorbed by the molten glass as to alter its properties or allow 'stones', etc., to get into the glass.
Uses. Melting of glass. Open pots: leadless glasses, *e.g.* plate glass, blown glass. Covered pots: lead glasses, *e.g.* flint glass, crystal glass, strass, etc. Optical and other high-quality glasses. Plate glass pots, 4–6 ft (123–182 cm) diameter 2 ft 6 in.–3 ft (61–91 cm) high. Bottle glass pots, 3–4 ft (91–123 cm) diameter, 2–3 ft (61–91 cm) high. Pressed glass, hollow-ware, optical or special glasses 14–40 in. (35–102 cm) diameter.
Body Type. Refractory. Alumino-silicate.
Raw Materials. Siliceous fireclays free from coarse particles of quartz,

e.g. Stourbridge and Scottish fireclay. Some aluminous fireclays as long as there is no free alumina. Complete mullite formation is important.

Preparation. Hand picking of obvious impurities→weathering→crushing →drying→grinding, edge runner mill or tube mill→sieving to pass 15-mesh.

$\uparrow$ oversize $\downarrow$

Grog made from specially burned clay at cone 14–16, or from damaged but *unused* glass pots→crush→grind→sieve: 10–20 mesh, 30%; 20–40 mesh, 40%; 40–80-mesh, 17%; through 80-mesh, 13%.

Mixing: up to 1 part grog to 1 part clay. Mix dry in rotary mixing drum or by hand→add water→treading or mixing trough and pugmill→ age→remix→age→remix→de-airing pugmill.

Or blunge to make casting slip, dissolving deflocculants in water first→add clays, when uniform slip→add grog, feldspar, flint, etc.

Shaping.

(1) Hand modelling on a base board with a $\frac{1}{4}$ in. (6·35 mm) bed of coarse grog or a damp sack. The base may be modelled on a separate board mounted on trunnions and then inverted on to the base board. Walls are gradually built up by throwing small pieces on and modelling into place. After 9 in. (23 cm) on the first day and 2–3 in. (5–8 cm) on later days the pot is left to stiffen for 24 hr before the next addition. About 4 months are required for a large covered pot.

(2) Hand moulding. Base hand modelled as in (1). Then wooden boards held by iron bands placed round and covered with strips of wet calico, and walls brought up against mould and beaten daily with wooden beater until dense and hard.

(3) Ramming into cored mould.

(4) Slip casting.

Drying. Chamber dryers, humidity drying. Pots must be placed on $\frac{1}{2}$–1 in. (1·2–2·5 cm) coarse grog, to allow free shrinkage.

Firing. Very careful setting required avoiding proximity to grates or flues. Complete firing rarely carried out by manufacturer. British practice is 'tempering' at cone 05 a, 1000° C (1832° F). U.S.A. practice, firing at cone 12, 1350° C (2462° F).

Firing itself takes place in the glass furnace. Where near enough the pot can be transferred red hot from the tempering kiln to the glass furnace, cooled down to same temperature, and then brought up to full fire.

Properties. Minimum refractoriness, cone 31 1690 °C (3074° F). Minimum refractoriness under load 50 lb/in^2 1250–1300° C (2282–2372° F). Porosity: hand made, 20–28%; cast, 18–23% (**S50**).

RETORTS

Characteristics. Vertical or horizontal refractory vessels made of single pieces or built up from suitably shaped bricks and blocks.

CONSTRUCTIONAL REFRACTORIES

Uses. Distillation of coal, zinc, sodium, etc.
Body Type. Refractory (*a*) Silica. (*b*) Semi-silica. (*c*) Bauxite (55–60% Al_2O_3). (*d*) Graphite. (*e*) Magnesia. (*f*) Silicon carbide. (*g*) Chromite. (*h*) Fireclay.
Raw Materials. (*a*) Ganister, lime bond. (*b*) Natural silica-clay mixtures or fireclay and silica. (*c*) Bauxite and fireclay. (*d*) Graphite 2 parts, fireclay 1 part. (*e*) Sinter magnesite, gelatinous silica bond. (*f*) Silicon carbide, fireclay bond. (*g*) Chromite, fireclay bond. (*h*) Refractory grog, fireclay.
Preparation. Crushing→grinding→screening→proportionating→mixing →tempering→ageing.
Shaping.
(1) Blocks and bricks: As for bricks of corresponding materials.
(2) Whole retorts: (*a*) hand moulding inside a mould; (*b*) hand moulding around a core; (*c*) hand moulding in mould and core; (*d*) pneumatic tamping; (*e*) extrusion in vertical machines with flange attachment, end closure effected by hand; (*f*) slip casting.

All retorts require careful smoothing inside.
Drying. Drying sheds heated to maximum of 25° C (75° F) 2–3 weeks.
Firing. Gas retorts, minimum 1400° C (2552° F) cone 14. Others, 1250° C (2282° F) upwards cone 8. Setting is vertical, as close together as the flanges allow, on bed of sand or grog. Kilns, down-draught round or rectangular 10–14 day.
Glazing. To render retorts impervious from the start, although it will reduce refractoriness, a raw glaze can be applied before firing, *e.g.*

Feldspar   100 parts $\left.\begin{array}{l}0.3\ K_2O \\ 0.7\ CaO\end{array}\right\} 0.5$–$1.0\ Al_2O_3 . 4.0$–$10.0\ SiO_2$
Whiting   42 parts
China clay 30–110 parts
Flint or sand 65–250 parts. Maturing cone 14 (**S50**).

Chapter 18

Thermal Insulators

MOLER BRICKS, 'SILEX' BRICKS, DIATOMITE BRICKS

Characteristics. Light-weight bricks consisting largely of hollow siliceous remains of diatoms (diatomaceous earth or moler) or the unicellular, microporous, siliceous ash remaining after combustion of rice waste (silex).
Uses. Insulation up to the limits of 850–1400° C (1562–2552° F) depending on the particular product, *e.g.* back-up insulation of medium temperature furnaces behind refractories; back-up insulation of high-temperature furnaces behind refractory insulators.
Body Type. Insulating, intermediate temperature, type (1).
Raw Materials. Moler or silex. Plastic clay. Volcanic ash. Combustible material, *e.g.* sawdust or cork.
Preparation. Crushing rolls→mixing trough, water added if necessary→pugmill: ~56% moisture for slop moulding; ~52% moisture for extrusion.
Shaping. Hand slop moulding.
 Extrusion followed by cutting into bricks.
Drying. Considerable care is required because of the high water content, especially if combustibles are included as these must not be accidentally ignited. Tunnel or chamber dryers.
Firing. Cone 015/014–010, 800–900° C (1472–1652° F).
 (1) Intermittent up-draught and down-draught kilns, open-flame.
 (2) Continuous gallery kilns.
 (3) Continuous chamber kilns.
 (4) Tunnel kilns, direct fired.
Finishing. Machining to correct size and shape (**O5**). Single bricks are fragile and have to be transported in cartons. These bricks readily absorb water and are then easily damaged by frost (**S124**).
Tests. Compressive strength. Refractoriness under load. Porosity. Density. Thermal conductivity. P.C.E.
Properties. Table 269, Nos. 2–10.

THERMAL INSULATORS

VERMICULITE INSULATING BRICKS

Characteristics. Intermediate temperature insulator with maximum service temperature between 1100° C (2012° F) and 1150° C (2102° F), and slightly less insulation efficiency than high-grade diatomaceous bricks. Large slabs with good dimensional accuracy are made and can be drilled or cut with wood-working tools for fitting.
Uses. Back-up insulation.
Body Type. Insulating, type (1).
Raw Materials. Vermiculite $\sim 75\%$; eyrite (colloidal magnesium silicate) or bentonite $\sim 25\%$.
Preparation. (1) Exfoliation by heating for from 4 to 8 sec at 900–950° C (1652–1742° F) in special furnace when vermiculite expands to sixteen times its original size. (2) Granulation→mixing with bond and moisture.
Shaping. Special pressing method.
Firing. With eyrite bond 1150° C (2100° F). With bentonite bond 1000° C $\sim$(1800° F) (**O5**).
Properties. Table 269, No.1.

LIGHT-WEIGHT AGGREGATE BUILDING BLOCKS

Characteristics. Light-weight blocks, hollow blocks, etc., with a range of physical properties and good dimensional accuracy.
Uses. Thermal insulation of buildings.
Body Type. Insulating type (7).
Raw Materials. Natural clays and shales with fire-bloating characteristics. Colliery shales. Plus coal dust or other combustible.
Preparation. Dry grinding → mixing → tempering → pelleting → sintering furnace→irregular cake of porous material→crushing→grinding and grading →addition of bond clay→making of bricks by normal methods.
Drying. No shrinkage, can be swift.
Firing. Little shrinkage, can be swift (**S124**).

EXPANDED CLAY BLOCKS

Characteristics. Light-weight blocks with either connecting pores or zero water absorption. Large units can be obtained.
Uses. Thermal insulation at temperatures related to the refractoriness of the clays used.
Body Type. Insulating type (6).
Raw Materials. Natural clays that will fire-bloat; or brick-making or fire clays, plus soda ash and magnesite, or soda ash and dolomite.
Preparation. Dry mill→temper with a little water.

TABLE

Thermal

No. Ref. Name Insulator type Body type Country of origin	1 O5 Vermiculite Intermediate Vermiculite Britain	2 M85 Moler Intermediate Diatomaceous Britain	3 M85 Fosalsil Super High Temp. Diatomaceous Britain
Fusion temperature (°C)			
(°F)			
Maximum safe temperature (°C)	1100–1150	850	1350
(°F)	2012–3001	1565	2462
True porosity (%)			
Bulk density (g/cm^3)	0·46–0·77	0·40–0·80	0·84
(lb/ft^3)	29–48	25–50	53
Permeability (cm^3 sec^{-1} cm^{-3})			
Absorption (%) 24-hr cold soak			
5-hr boil			
Open pores (%)			
Closed pores (%)			
Cold crushing strength (lb/in^2)	120–250	200–1450	750
(kg/cm^2)	8–17	14–102	52
Thermal conductivity at Temp. (°C)	200 | 400 | 600		
(B.t.u. ft^{-2} hr^{-1} – in.$^{-1}$ °F^{-1})	1·32 – | 1·51 – 1·16–1·76 | 1·91 | 2·12	0·6–1·2	1·5
(cal cm^{-2} sec^{-1} cm^{-1} °C^{-1})	0·0003 – | 0·0004 – | 0·0005 – 0·0006 | 0·0007 | 0·0007	0·0002–0·0004	0·0005
After contraction Temp. (°C)	1100–1150		
Time (hr)			
% change	0–<1		
Spalling resistance			
* cycles using small prism test			
Transverse strength (lb/in^2)			
(kg/cm^2)			
Modulus of elasticity (lb/in^2)			
(kg/cm^2)			
Refractoriness under load (lb/in^2)			
Temp. (°C)			
Time (hr)			
Deformation (%)			
Specific heat			

Insulators

4 C37 Diatomite	5 O5 Diatomite Britain	6 M85 Insulex Intermediate Diatomaceous Britain			7 A185 Diatomaceous France			8 S13 P600 Diatomaceous Denmark		
					>1350 >2462					
		930			900			875		
76	66·1	1700			1652			1607		
0·51	0·75	0·28–0·32			0·4			0·600		
32	47·0	18–20			25			37		
0·04	0·25									
130	440	112–182			142			575		
9	31	8–13			10			40		
500	500	57	175	290	100	400	700	100	400	700
1·11	1·70	0·60	0·67	0·78	0·70	1·0	1·3	0·9	1·2	1·4
0·0004	0·0006	0·0002	0·0002	0·0003	0·0002	0·0003	0·0004	0·0003	0·0004	0·0005
900 | 1000 | 1100 2 | 2 | 2 −0·2 | −2·0 | −4·7 >30*	900 | 950 | 1000 2 | 2 | 2 −0·1 | −0·5 | −3·24 >23	870 | | 925 −2·4 | | −3·3								
		36 2·53								
10 900 | 1000 2 | 2 −1·0 | −2·6										

TABLE

	No.	9	10	11
	Ref.	**S76**	**S76**	**C37**
	Name	Sil-3	Silex N550	
	Insulator type	Intermediate	Intermediate	
	Body type	Silex	Silex	Clay
	Country of origin	Italy	Italy	
Fusion temperature	(°C)	1650	1600	
	(°F)	3200	2915	
Maximum safe temperature	(°C)	1400	1250	
	(°F)	2552	2282	
True porosity (%)		52	76	68
Bulk density	(g/cm³)	1·15	0·55	0·83
	(lb/ft³)	72	34·3	52
Permeability (cm³ sec⁻¹ cm⁻³)				0·38
Absorption (%) 24-hr cold soak				
	5-hr boil			
Open pores (%)				
Closed pores (%)				
Cold crushing strength	(lb/in²)	996/1280	142	580
	(kg/cm²)	70/90	10	41
Thermal conductivity at Temp. (°C)				500
(B.t.u. ft⁻² hr⁻¹ – in.⁻¹ °F⁻¹)		3·0	1·0	2·55
(cal cm⁻² sec⁻¹ cm⁻¹ °C⁻¹)		°C × 0·00018 + 0·21	°C × 0·000075 + 0·09	0·0009
After contraction Temp. (°C)				1350 | 1400 | 1450
	Time (hr)			2 | 2 | 2
	% change			−0·4 | −0·9 | −2·0
Spalling resistance				5*
* cycles using small prism test				
Transverse strength	(lb/in²)			
	(kg/cm²)			
Modulus of elasticity	(lb/in²)			
	(kg/cm²)			
Refractoriness under load	(lb/in²)			10 | 10
	Temp. (°C)			1200 | 1300
	Time (hr)			2 | 2
	Deformation (%)			−0·9 | −8·0
Specific heat				

269—*(contd.)*

	12 O5 Fireclay U.S.A.			13 O5 White clay U.S.A.			14 O5 Clay Britain			15 O5 Clay Britain			16 O5 Clay Britain		
	75·5 0·66 41·4 10·6			84·3 0·75 47·0 0·87			80·8 0·51 32·1 1·22			68·3 0·83 52·0 0·38			67·8 0·91 56·6 0·37		
	124 9 500 2·35			90 6 500 1·11			200 14 500 1·67			575 40 500 2·5			633 45 500 2·58		
	0·0008			0·0004			0·0006			0·0008			0·0009		
1250 2 0·43	1350 2 0·47	1450 2 2·48	1150 2 0·0	1250 2 0·58 5	1350 2 1·00	1150 2 −0·2	1200 2 −1·35 8	1250 2 −4·35	1300 2 −0·2	1400 2 −0·9 5	1450 2 −2·0	1300 2 −0·3	1400 2 −1·56 >40	1450 2 −2·83

TABLE

	No.	17	18	19
	Ref.	N16	S217	N16
	Name	Foamclay	Expanded clay	
	Insulator type			
	Body type	Clay	Clay	Expanded clay communic. pores
	Country of origin	U.S.A.	U.S.A.	U.S.A.
Fusion temperature (°C)				
(°F)				
Maximum safe temperature (°C)				
(°F)				
True porosity (%)			70–80	
Bulk density (g/cm^3)		0·80–0·96	0·56	0·75
(lb/ft^3)		50–60	35	47
Permeability (cm^3 sec^{-1} cm^{-3})				
Absorption (%) 24-hr cold soak		35–50		
5-hr boil		70–85		57
Open pores (%)			4–5	
Closed pores (%)			66–76	
Cold crushing strength (lb/in^2)		800–1500	1350–1500	800
(kg/cm^2)		56–105	94–105	56
Thermal conductivity at Temp. (°C)		5		
(B.t.u. ft^{-2} hr^{-1} in.$^{-1}$ °F^{-1})		~1·41	0·8	
(cal cm^{-2} sec^{-1} cm^{-1} °C^{-1})		0·0005	0·0002	
After contraction Temp. (°C)				
Time (hr)				
% change				
Spalling resistance				
* cycles using small prism test				
Transverse strength (lb/in^2)		375–525	350–400	
(kg/cm^2)		26·36–36·91	24·60–28·12	
Modulus of elasticity (lb/in^2)		350 000–870 000		
(kg/cm^2)		24 607–61 167		
Refractoriness under load (lb/in^2)				
Temp. (°C)				
Time (hr)				
Deformation (%)				
Specific heat				

269—(contd.)

20 N16	21 C37	22 O5	23 C37	24 C26	25 C26	26 C37
Expanded clay non-communicating pores U.S.A.	Sillimanite	Sillimanite Britain	Mullite	Silica Britain	Silica Britain	Corundum
				1200 2192	1400 2552	
	67	66·6	73	72	65	69
0·80–1·12	1·02	1·01	0·87	0·69	0·83	0·99
50–70	64	63·6	54	43·3	52·0	62
	0·81	0·81	4·10	2·81	0·09	5·00
0·37–4·10						
1300–4650	450	445	190	330	590	410
91–327	32	31	13	23	41	28
	500	500	500	400	400	500
	2·93	2·93	2·27	1·80	2·75	4·15
	0·0010	0·0010	0·0007	0·0006	0·0009	0·0014
	1400 | 1450 2 | 2 −0·0 | −0·3 9*	1400 | 1450 2 | 2 nil | −0·3 7–10	1500 | 1550 2 | 2 −1·0 | −2·8 10*			1500 | 1550 2 | 2 +0·6 | +0·1 16*
	10 | 10 1200 | 1300 2 | 2 −0·18 | −0·47					10 1300 16 min failed

TABLE

	27	28	29	30
No. Ref.	O5	O5	N27	N27
Name Insulator type Body type	Alumina	Alumina	Alumina	Alumina
Country of origin	Britain	Britain	U.S.A.	U.S.A.
Fusion temperature (°C)				
(°F)				
Maximum safe temperature (°C)				
(°F)				
True porosity (%)	68·6	—	64	53
Bulk density (g/cm^3)	0·99	0·87	1·45	1·89
(lb/ft^3)	62·0	54·0	90	120
Permeability (cm^3 sec^{-1} cm^{-3})	4·96	—		
Absorption (%) 24-hr cold soak				
5-hr boil				
Open pores (%)				
Closed pores (%)				
Cold crushing strength (lb/in^2)	413	190		
(kg/cm^2)	29	13		
Thermal conductivity at Temp. (°C)	500	500		
(B.t.u. ft^{-2} hr^{-1}–in^{-1} °F^{-1})	4·15	2·50		
(cal cm^{-2} sec^{-1} cm^{-1} °C^{-1})	0·0014	0·0008		
After contraction Temp. (°C)	1510 | 1550	1510 | 1550	1900	1900
Time (hr)	2 | 2	2 | 2		
% change	+0·60 | −0·06	0·95 | 2·75	−0·3	−0·1
Spalling resistance	—	—		
* cycles using small prism test				
Transverse strength (lb/in^2)				
(kg/cm^2)				
Modulus of elasticity (lb/in^2)				
(kg/cm^2)				
Refractoriness under load (lb/in^2)				
Temp. (°C)				
Time (hr)				
Deformation (%)				
Specific heat			0·269	

269—(contd.)

31 N27	32 C36	33 C36	34 N27	35 O5	36 O5
Alumina U.S.A.	Grain stabilised Zirconia Britain	Stabilised Zirconia Britain	Stabilised Zirconia U.S.A.	Chrome- magnesite Britain	Chrome- magnesite Britain
	2550–2600 4620–4710	2550–2600 4620–4710			
42 2·33 145	68	51 2·5 156	50 2·80 175	app. 41·8 2·09 130 0·26	app. 51·2 1·74 109 0·10
				570 40 350 5·30 0·0018	165 11 350 3·95 0·0013
	650 | 870 | 1100 2·25 | 2·5 | 2·75 0·0007|0·0008|0·0009	650 | 870 | 1100 3·75 | 4 | 4·4 0·0012|0·0013|0·0015			
1900 −0·2			2300 | 2400 4·0 | 11·5	1500 2 1·90	1500 2 2·2
			0·168		

Shaping. Dry pressing.
Firing. Set in kiln on refractory slabs with separators between each. The refractories are coated with a mixture of crystallised and amorphous silica bonded with bentonite. Large units are made by permitting several small ones to fuse together in the kiln. Fire fast to induce bloating.
Finishing. Trim to shape with silicon carbide saws and planes (**N16**).
Properties. Table 269, Nos. 18–20.

General Grouping of Refractory Insulators

(1) INSULATING FIREBRICK

Characteristics. A porous brick capable of withstanding up to about 1250° C (2282° F) at its hotter face but usually not suitable for exposure to the furnace atmosphere.
Uses. Back-up to dense firebrick in furnace construction.

(2) HIGH-TEMPERATURE INSULATING BRICK

Characteristics. Insulating brick for maximum hot-face temperatures ranging from 1250 to 1600° C (2282–2912° F) but requiring protection from the maximum furnace temperature and atmosphere by a thin layer of dense refractories.

(3) HOT-FACE INSULATION

Characteristics. Insulating brick able to withstand exposure to actual furnace operating conditions.
Uses. Furnace linings. High temperature back up to thin layer of dense refractory (**O5**).

TABLE 268
U.S.A. Grouping of Firebricks
(A.S.T.M. Refractories 1948)

Group Identification	Maximum Safe Temperature		Reheat Change not more than 2% when tested at		Bulk density maximum	
	(°F)	(°C)	(°F)	(°C)	(lb/ft³)	(g/cm³)
Group 16	1600	~870	1550	845	34	0·54
Group 20	2000	~1095	1950	1065	40	0·64
Group 23	2300	~1260	2250	1230	48	0·77
Group 26	2600	~1420	2550	1440	52	0·83
Group 28	2800	~1540	2750	1510	60	0·96

INSULATING BUILDING AND REFRACTORY BRICKS
(COMBUSTIBLE ADDITION METHOD)

Characteristics. Porous, light-weight bricks made of the same materials as solid refractories but acquiring a fluxing ash during manufacture that lowers the refractoriness by about 2 cones for a brick of density 0·9, or about 3 to 4 cones for a density of 0·7.
Uses. Insulation of buildings, backing up of furnaces.
Body Type. Insulation type (3).
Raw Materials. Normal brick or fireclay body mixture plus one or two of the following: coal, anthracite, coke, lignite, pitch, petroleum residues, sawdust (green, weathered, or roasted) charcoal, paper, cork, straw, chaff, husks of cornflower seed, etc. The firing is made easier if two combustibles of different ignition points are used.
Preparation and Shaping. As for dense bricks, *i.e.* slop moulding, extrusion, semi-dry or dry pressing. Allowance may have to be made for the lower plasticity of the body due to the non-plastic content.
Drying. As for dense bricks. The drying shrinkage is smaller.
Firing. Carefully regulated firing is necessary, with a predominantly oxidising atmosphere and good draught. The setting should be such that the bricks get even treatment from all sides, including the inside if they are hollow. As soon as ignition starts the fuel is cut off and a draught of cold air passed through (**S124**).

REFRACTORY INSULATING BRICKS, FINE PORED

Characteristics. Porous bricks with small and uniform spherical pores giving better insulation for the same total pore volume than large-pored bricks. The mechanical strength is better than large-pored bricks of the same density. Good strength for S.G. of 0·5 to 0·6 is possible.
Body Type. Refractory insulating bodies type (5a), made porous by chemical effervescence. They include: general fireclay bodies; high-alumina bodies (up to 95% Al_2O_3) sillimanite bodies (90–95% sillimanite); silica bodies (up to 98% SiO_2); kaolinite bodies.
Raw Materials. One of the following is used: fireclays; aluminous fireclays; calcined sillimanite, kyanite, etc.; ganister, quartzite, etc.; kaolins, etc. Also included are: dolomite or aluminium, sulphuric acid, plaster of Paris or organic foaming agent.
Preparation. (A) Crushing→dry grinding in ball mills for both body and the dolomite. Dry mixing in planetary mixers of body, dolomite and plaster of Paris. (B) Dilute acid in blunger. Dry ingredients added to acid and rapidly mixed.
Shaping. The bubbling slip is poured into plaster of Paris, wood, sheet

metal or cardboard moulds, half filling them. Temperature control is advised.

Drying. Pieces set in conventional dryers in moulds. Drying is fast and shrinkage only 1 to 2%.

Firing. The dry porous pieces are very fragile and cannot be set in tall bungs. Firing itself can be fast and has no difficulties.

Finishing. At least one side has to be machined to shape but this is not difficult (**S107**).

FINE-PORED REFRACTORY INSULATING BRICKS

Characteristics. Porous material with fine spherical pores, low thermal conductivity, relatively high mechanical strength.

Body Type. Insulating type (5b), pores made by the inclusion of foam. General fireclay bodies. Corundum and high alumina bodies. Silica bodies. Chromite bodies. Magnesite bodies. Chrome–magnesite bodies.

Raw Materials. One from each group (A) (B) (C).

A. Plastic, siliceous and aluminous fireclays; corundum; quartzite; chromite; magnesite.

B. Soaps; sodium resinate; saponin; resin.

C. Glue; gum arabic; plaster of Paris.

Preparation and Shaping.

(1) Water foam process (Nicholson and Bole, **N16**) for batch operation with highly plastic bodies. (A) Dry mill raw clay, foamclay grog, calcined gypsum. (B) Whip surface active agent and water to fine foam. Add (A) to (B) in mixer and continue whipping to a uniformly aerated slip→cast in moulds with aluminium base pallet.

(2) Clay foam process (Nicholson and Bole, **N16**) for continuous operation. Raw clay and grog made into deflocculated slip→add wetting agent and whip until foamed→add gypsum→cast.

(3) Water foam process (Pirogov, **P43**). (A) Fluid slip (sodium silicate deflocculant) of 80–90% grog ground to pass 30-mesh sieve, plus 10–20% clay of maximum 3 mm size. Wet grinding must be used to give rounded particles. (B) Beat up foam using sodium resinate as foaming agent and 0·5% glue for stabilising. Bulk density of stable foam 0·07 to 0·09. Mix carefully about 1·0 to 1·50 vols. of foam (B) to 1·0 vol. of slip (A), adjusting quantities to give a mixture of constant bulk density. Cast in collapsible wooden or sheet metal moulds standing on a board or asbestos to form base. Lubricate with oil for sides and wet paper on bottoms.

Drying. This can be rapid. Dried bricks are very fragile.

Firing. Setting height limited to $3\frac{1}{2}$ ft (1 m). Can be set on top of dense refractories of same body and fired at same schedule so long as there is no reduction if plaster has been used as a stabiliser, and no condensation on to

the porous bricks during the early stages, this might cause slumping. Little firing shrinkage.
Finishing. At least one face must be ground to shape.

REFRACTORY INSULATING BRICKS
(Volatile addition method)

Characteristics. Porous, light-weight bricks made of the same material as solid refractories with no addition of fluxing ash.
Uses. Furnace insulation.
Body Type. Refractory insulator, type 4.
Raw Materials. Refractory plastic clay, grog from wasters, raw naphthalene (95% pure).
Preparation. General heavy clay methods.
Shaping.
   (1) Extrusion→cutting off→repressing.
   (2) Hand moulding→repressing.
   (3) Stiff plastic brick machine.
   (4) Semi-dry pressing.
   (5) Casting, for special shapes.
Sublimation and Drying. Sealed chambers with warm air passed through and then led into condensation channels for the recovery of naphthalene.
Firing. Volatilisation of naphthalene requires very slow firing over the range 600–800° C (1112–1472° F). Maturing 1200° C (2192° F).

STABILISED ZIRCONIA INSULATOR

Characteristics. Highly refractory, up to 2400° C (4352° F) insulating material of low thermal conductivity and low specific heat and therefore low heat storage capacity, in spite of high bulk density. Can be used in both oxidising and reducing conditions.

APPENDIX

The Periodic System of the Elements (presentation after Avon Antropoff (A209)). Showing Atomic Number, Symbol, Name, Atomic Weight (1957)

Period	Groups								
	O	I	II	III	IV	V	VI	VII	O
1	0 Nn Neutron 1·0090				I H Hydrogen 1·0080				2 He Helium 4·003
First short period 2	2 He Helium 4·003	3 Li Lithium 6·940	4 Be Beryllium 9·013	5 B Boron 10·82	6 C Carbon 12·011	7 N Nitrogen 14·008	8 O Oxygen 16·000	9 F Fluorine 19·00	10 Ne Neon 20·183
Second short period 3	10 Ne Neon 20·183	11 Na Sodium 22·991	12 Mg Magnesium 24·32	13 Al Aluminium 26·98	14 Si Silicon 28·09	15 P Phosphorus 30·975	16 S Sulphur 32·066±0·003	17 Cl Chlorine 35·457	18 Ar Argon 39·944

Groups	O	Ia	IIa	IIIa	IVa	Va	VIa	VIIa	VIII			Ib	IIb	IIIb	IVb	Vb	VIb	VIIb	O
First Long Period 4	18 Ar or A Argon 39·944	19 K Potassium 39·100	20 Ca Calcium 40·08	21 Sc Scandium 44·96	22 Ti Titanium 47·90	23 V Vanadium 50·95	24 Cr Chromium 52·01	25 Mn Manganese 54·94	26 Fe Iron 55·85	27 Co Cobalt 58·94	28 Ni Nickel 58·71	29 Cu Copper 63·54	30 Zn Zinc 65·38	31 Ga Gallium 69·72	32 Ge Germanium 72·60	33 As Arsenic 74·91	34 Se Selenium 78·96	35 Br Bromine 79·916	36 Kr Krypton 83·80
5	36 Kr Krypton 83·80	37 Rb Rubidium 85·48	38 Sr Strontium 87·63	39 Y Yttrium 88·92	40 Zr Zirconium 91·22	41 Nb Niobium 92·91	42 Mo Molybdenum 95·95	43 Tc Technetium —	44 Ru Ruthenium 101·1	45 Rh Rhodium 102·91	46 Pd Palladium 106·4	47 Ag Silver 107·880	48 Cd Cadmium 112·41	49 In Indium 114·82	50 Sn Tin 118·70	51 Sb Antimony 121·76	52 Te Tellurium 127·61	53 I Iodine 126·91	54 Xe Xenon 131·30
6	54 Xe Xenon 131·30	55 Cs Caesium 132·91	56 Ba Barium 137·36	57 La Lanthanum 138·92	58–71 Rare Earths	72 Hf Hafnium 178·50	73 Ta Tantalum 180·95	76 W Tungsten 183·86	75 Re Rhenium 186·22	76 Os Osmium 190·2	77 Ir Iridium 192·2	78 Pt Platinum 195·09	79 Au Gold 197·0	80 Hg Mercury 200·61	81 Tl Thallium 204·39	82 Pb Lead 207·21	83 Bi Bismuth 209·00	84 Po Polonium —	85 At Astatine —	86 Rn Radon
7	86 Rn Radon	87 Fr Francium	88 Ra Radium	89 Ac Actinium	90 Th Thorium 232·05	91 Pa Protractinium —	92 U Uranium 238·07	93— Trans-uranium elements											

The Rare Earth Elements of Group IIIa

58 Ce Cerium 140·13	59 Pr Praseodymium 140·92	60 Nd Neodymium 144·27	61 Il Illium —	62 Sm Samarium 150·35	63 Eu Europium 152·0	64 Gd Gadolinium 157·26	65 Tb Terbium 158·93	66 Dy Dysprosium 162·51	67 Ho Holmium 164·94	68 Er Erbium 167·27	69 Tm Thulium 168·94	70 Yb Ytterbium 173·04	71 Lu Lutecium 174·99

Trans-Uranium Elements of Group VIa

93 Np Neptunium	94 Pu Plutonium	95 Am Americium	96 Cm Curium	97 Bk Berkelium	98 Cf Californium	99 Es Einsteinium	100 Fm Fermium	101 Md Mendelevium	102 No Nobelium

Elements with more than one name or symbol; beryllium, Be also glucinium, Gl: lutetium, Lu also cassiopeium, Cp: hafnium, Hf also celtium, Ct: niobium, Nb also columbium, Cb: illium, Il also promethium, Pm.

Table Ia
Comparison of Standard Fine Test Sieves
(after B115)

(Openings given on the same horizontal line are within the permissible variations in average opening of the U.S. standard fine series.)

Britain				United States		Germany	France
B.S.I. (1943) (B118) (mm)	(in.)	I.M.M. (1907) (M82) (mm)	(in.)	Tyler (1910) (T45) (mm)	U.S. standard (1940) fine series (A187) (mm)	DIN (1934) (D32) (mm)	AFNOR (1938) (A213) (mm)
						6·000	
				5·613 (3½)	5·66 (3½)	5·000	5·000
				4·699 (4)	4·76 (4)		
				3·962 (5)	4·00 (5)	4·000	4·000
3·353 (5)	0·1320			3·327 (6)	3·36 (6)		
						3·000	3·150
2·812 (6)	0·1107			2·794 (7)	2·83 (7)	2·500	2·500
		2·540 (5)	0·1				
2·411 (7)	0·0949			2·362 (8)	2·38 (8)		
2·057 (8)	0·0810			1·981 (9)	2·00 (10)	2·000	2·000
1·676 (10)	0·0660			1·651 (10)	1·68 (12)		
		1·600 (8)	0·063			1·500 (4)	1·600
1·405 (12)	0·0553			1·397 (12)	1·41 (14)		
		1·270 (10)	0·05				1·250
1·204 (14)	0·0474			1·168 (14)	1·19 (16)	1·200 (5)	
		1·059 (12)	0·0417				
1·003 (16)	0·0395			0·991 (16)	1·00 (18)	1·00 (6)	1·000
0·853 (18)	0·0336			0·833 (20)	0·84 (20)		
		0·795 (16)	0·0313			0·750 (8)	0·800
0·699 (22)	0·0275			0·701 (24)	0·71 (25)		
		0·635 (20)	0·025				0·630
0·599 (25)	0·0236			0·589 (28)	0·59 (30)	0·600 (10)	
0·500 (30)	0·0197			0·495 (32)	0·50 (35)	0·500 (12)	0·500
0·422 (36)	0·0166	0·424 (30)	0·0167	0·471 (35)	0·42 (40)	0·430 (14)	0·400
						0·400 (16)	
0·353 (44)	0·0139			0·351 (42)	0·35 (45)		
		0·317 (40)	0·0125				0·315
0·295 (52)	0·0116			0·295 (48)	0·297 (50)	0·300 (20)	
0·251 (60)	0·0099	0·254 (50)	0·01	0·246 (60)	0·250 (60)	0·250 (24)	0·250
0·211 (72)	0·0083	0·211 (60)	0·0083	0·208 (65)	0·210 (70)	0·200 (30)	0·200
0·178 (85)	0·0070	0·180 (70)	0·0071	0·175 (80)	0·177 (80)		
							0·160
0·152 (100)	0·0060	0·160 (80)	0·0063	0·147 (100)	0·149 (100)	0·150 (40)	
		0·139 (90)	0·0055				
0·124 (120)	0·0049	0·127 (100)	0·005	0·124 (115)	0·125 (120)	0·120 (50)	0·125
0·104 (150)	0·0041	0·104 (120)	0·0041	0·104 (150)	0·105 (140)	0·100 (60)	0·100
0·089 (170)	0·0035	0·084 (150)	0·0033	0·089 (170)	0·088 (170)	0·090 (70)	
							0·080
0·076 (200)	0·0030			0·074 (200)	0·074 (200)	0·075 (80)	
0·066 (240)	0·0026	0·063 (200)	0·0025	0·061 (250)	0·062 (230)	0·060 (100)	0·063
0·053 (300)	0·0021			0·053 (270)	0·053 (270)		0·050
				0·043 (325)	0·044 (325)		
							0·040
				0·038 (400)	0·037 (400)		

Table Ib
British Standard Wire Cloth for Medium Mesh Test Sieves
B.S. 410: 1943

B.S. Mesh size Nominal width of aperture (side of square)		Screening area (approximate) (%)
(in.)	(mm)	
$\frac{1}{32}$	0·79	36
$\frac{1}{16}$	1·60	38
$\frac{1}{8}$	3·18	40
$\frac{3}{16}$	4·76	49
$\frac{1}{4}$	6·35	53
$\frac{3}{8}$	9·53	61
$\frac{1}{2}$	12·70	63

Table II
Table Showing Weight of Solids, % Water, % Solids for a Given Slip Weight per pint

Weight per pint (oz)	Solids per pint (oz)	% Water	% Solids
18	2·17	87·90	12·10
19	3·83	79·84	20·16
20	5·50	72·50	27·50
21	7·17	65·86	34·14
22	8·84	59·82	40·18
23	10·50	54·35	45·65
24	12·17	49·29	50·71
25	13·84	44·64	55·36
26	15·51	40·30	59·70
27	17·18	36·37	63·63
28	18·84	32·72	67·28
29	20·51	29·28	70·72
$29\frac{1}{2}$	21·34	27·66	72·34
30	22·17	26·10	73·90
$30\frac{1}{2}$	23·00	24·59	75·41
31	23·84	23·10	76·90
$31\frac{1}{4}$	24·26	22·37	77·63
$31\frac{1}{2}$	24·67	21·69	78·31
$31\frac{3}{4}$	25·09	20·95	79·05
32	25·50	20·31	79·69
33	27·17	17·67	82·33

Table Ic

British Standard Perforated Plates for Coarse Sieves
B.S. 410: 1943

U.S.A. Standard Coarse Sieves
E11–39

Mesh size Nominal size of aperture (side of square)		Screening area (approximate)	Size of sieve Designation	Sieve opening	
(in.)	(mm)	(%)	(in.)	(mm)	(in.) (approx. equivalents)
$\frac{3}{16}$	4.76	44			
$\frac{1}{4}$	6.35	53	$\frac{1}{4}$ (No. 3)	6.35	0.250
			(0.265)	6.73	0.265
$\frac{5}{16}$	7.94	55	$\frac{5}{16}$	7.93	0.312
$\frac{3}{8}$	9.52	56	$\frac{3}{8}$	9.52	0.375
			$\frac{7}{16}$	11.1	0.438
$\frac{1}{2}$	12.70	64	$\frac{1}{2}$	12.7	0.500
			(0.530)	13.4	0.530
$\frac{5}{8}$	15.88	64	$\frac{5}{8}$	15.9	0.625
$\frac{3}{4}$	19.05	64	$\frac{3}{4}$	19.1	0.750
$\frac{7}{8}$	22.23	64	$\frac{7}{8}$	22.2	0.875
1	25.40	64	1	25.4	1.00
			(1.06)	26.9	1.06
$1\frac{1}{8}$	28.58	64			
$1\frac{1}{4}$	31.75	64	$1\frac{1}{4}$	31.7	1.25
$1\frac{3}{8}$	34.93	64			
$1\frac{1}{2}$	38.10	64	$1\frac{1}{2}$	38.1	1.50
$1\frac{5}{8}$	41.28	64			
$1\frac{3}{4}$	44.45	64	$1\frac{3}{4}$	44.4	1.75
$1\frac{7}{8}$	47.63	64			
2	50.80	64	2	50.8	2.00
			(2.12)	53.8	2.12
$2\frac{1}{4}$	57.15	64			
$2\frac{1}{2}$	63.50	64	$2\frac{1}{2}$	63.5	2.50
$2\frac{3}{4}$	69.85	64			
3	76.20	64	3	76.2	3.00
$3\frac{1}{2}$	88.90	64	$3\frac{1}{2}$	88.9	3.50
4	101.60	64	4	101.6	4.00
			(4.24)	107.6	4.24

Table III
Thermoscopes or Heat-work Recorders

°C	°F	German, Staffordshire and French Seger cones	New 'H' series Staffordshire Seger cones	Buller's rings			Watkin's recorders	United States Orton cones standardisation rate (see footnote)		Holdcroft's bars
				No. 55	No. 27	No. 72		(1)	(2)	
580	1076									
585	1085							022		
590	1094									
595	1103							021		
600	1112	022	H022				1		022	1
605	1121									
610	1130								021	
615	1139									
620	1148									
625	1157							020		
630	1166							019		
640	1184									
650	1202	021	H021				2		020	2
660	1220								019	
670	1238	020	H020				3	018		3
680	1256									
690	1274	019	H019				4			
700	1292									4
710	1310	018	H018				5			
720	1328							017	018	
730	1346	017	H017				6			5
735	1355							016		
740	1364									
750	1382	016	H016				7			
760	1400									6
770	1418							015	017	
780	1436									
790	1454	015a	H015				8			7
795	1463							014	016	
800	1472									
805	1481								015	
810	1490									7a
815	1499	014a	H014				9			
820	1508									
825	1517							013		
830	1526								014	
835	1535	013a	H013				10			
840	1544							012		8
850	1562									
855	1571	012a	H012				11			
860	1580								013	9
870	1598									
875	1607							011	012	10
880	1616	011a	H011				12			
890	1634							010		11
895	1643								010	
900	1652	010a	H010				13			
905	1661								011	12
910	1670									
920	1688	09a	H09				14			13
930	1706							09	09	
935	1715									14
940	1724	08a	H08				15			
945	1733							08		
950	1742								08	15
960	1760	07a	H07	3	0	0	16			16
970	1778		H07A	7	1	1				17
975	1787							07		
980	1796	06a	H06	11	2½	2	17			
985	1805									18
990	1814		H06A	15	4	3			07	

(1) Heat at 20° C/hr.   (2) Heat at 150° C/hr.

Thermoscopes or Heat-work Recorders—(*contd.*)

°C	°F	German, Staffordshire and French Seger cones	New 'H' series Staffordshire Seger cones	Buller's rings			Watkin's recorders	United States Orton cones		Holdcroft's bars
				No. 55	No. 27	No. 72		standardisation rate (1) (*see* footnote)	standardisation rate (2) (*see* footnote)	
1000	1832	05a	H05	18	5½	4	18			19
1005	1841							06		
1010	1850		H05A	21	7	5				
1015	1859								06	
1020	1868	04a	H04	24	8½	6	19			
1030	1886		H04A	27	10	7		05		
1040	1904	03a	H03	30	11½	8½	20		05	20
1050	1922		H03A	32	15	10		04		
1060	1940	02a	H02	34	14	11	21		04	21
1070	1958		H02A	36	15½	12½				
1080	1976	01a	H01	37	17	14	22	03		22
1090	1994		H01A	38	18½	15½				
1095	2003							02		
1100	2012	1a	H1	39	20	17	23			23
1110	2030		H1A	40	21½	18½		01		
1115	2039								03	
1120	2048	2a	H2	41	23	20	24			24
1125	2057							1	02	
1130	2066		H2A	42	24½	21				
1135	2075							2		
1140	2084	3a	H3	43	26	22	25			25
1145	2093							3	01	
1150	2102		H3A	44	27	23				
1160	2120	4a	H4	45	28½	24½	26		1	
1165	2129							4	2	
1170	2138		H4A	46	30	26			3	25a
1180	2156	5a	H5	47	31½	27	27	5		
1190	2174		H5A	48	33	28		6	4	
1200	2192	6a	H6	49	34½	29	28			26
1205	2201								5	
1210	2210			50	36	30		7		
1215	2219		H6A							
1220	2228				37½	31				
1225	2237							8		
1230	2246	7	H7		38½	32	29		6	26a
1240	2264		H7A		40	33				
1250	2282	8	H8		41½	34½	30	9	7	27
1260	2300		H8A		43	37		10	8	
1270	2318		H8B		44½	38½				27a
1280	2336	9	H9		46	40	31			28
1285	2345							11	9	
1290	2354		H9A		47	42				
1300	2372	10	H10		48	44	32			29
1305	2381								10	
1310	2390		H10A			46		12		
1320	2408	11	H11			47	33			
1325	2417								11	30
1330	2426									
1335	2435								12	
1340	2444									
1350	2462	12	H12				34	13	13	31
1360	2480									
1370	2498									
1380	2516	13	H13				35			32
1390	2534							14		
1400	2552								14	
1410	2570	14	H14				36	15		
1420	2588									
1430	2606									33
1435	2615	15	H15				37		15	
1440	2624									
1450	2642							16		
1460	2660	16	H16				38			34

(1) Heat at 20° C/hr. (2) Heat at 150° C/hr.

Thermoscopes or Heat-work Recorders—(contd.)

°C	°F	German, Staffordshire and French Seger cones	New 'H' series Staffordshire Seger cones	Buller's rings No. 55	Buller's rings No. 27	Buller's rings No. 72	Watkin's recorders	United States Orton cones standardisation rate (1)	United States Orton cones standardisation rate (2) (see footnote)	Holdcroft's bars
1465	2669							17	16	
1470	2678									
1475	2687								17	35
1480	2696	17	H17				39			
1485	2705							18		
1490	2714								18	36
1500	2732	18	H18				40			
1505	2741									37
1510	2750									
1515	2759							19		
1520	2768	19	H19				41	20	19	38
1530	2786	20	H20				42		20	
1535	2795									39
1540	2804									
1550	2822									40
								(3)		
1560	2840									
1570	2858									
1580	2876	26	H26				43	23		
1590	2894									
1595	2903							26		
1600	2912									
1605	2921							27		
1610	2930	27	H27				44			
1615	2939							28		
1620	2948									
1630	2966	28	H28				45			
1640	2984									
1650	3002	29	H29				46	29		
1660	3020							30		
1670	3038	30	H30				47			
1680	3056							31		
1690	3074	31	H31				48			
1700	3092							32		
1710	3110	32	H32				49			
1720	3128									
1725	3137							32½		
1730	3146	33	H33				50			
1740	3164									
1745	3173							33		
1750	3182	34	H34				51			
1760	3200							34		
1770	3218	35	H35				52			
1780	3236									
1785	3245							35		
1790	3254	36	H36				53			
1800	3272									
1810	3290							36		
1820	3308							37		
1825	3317	37	H37				54			
1830	3326									
1835	3335							38		
1840	3344									
1850	3362	38	H38				55			
								(4)		
1860	3380									
1865	3389							39		
1870	3398									

(1) Heat at 20° C/hr. (3) Heat at 100° C/hr.
(2) Heat at 150° C/hr. (4) Heat at 600° C/hr.

Thermoscopes or Heat-work Recorders—(contd.)

C°	°F	German, Staffordshire and French Seger cones	New 'H' series Staffordshire Seger cones	Buller's rings			Watkin's recorders	United States Orton cones standardisation rate (4) (see footnote)	Holdcroft's bars
				No. 55	No. 27	No. 72			
1880	3416	39	H39				56		
1885	3425							40	
1890	3434								
1900	3452								
1910	3470								
1920	3488	40	H40				57		
1930	3506								
1940	3524								
1950	3542								
1960	3560	41	H41				58		
1970	3578							41	
1980	3596								
1990	3614								
2000	3632	42	H42				59		
2010	3650								
2015	3659							42	

(4) Heat at 600°C/hr.

Table IV

AMBIGUOUS WORDS

Pottery. This can have various meanings and so has been avoided whenever possible.

Enamel. In the ceramic industry the word enamel is used to signify coloured decoration applied on top of a glaze and then hardened into position in a so-called 'enamel kiln'. As the word enamel is more generally held to mean an opaque siliceous covering on a metal base, we wish to avoid confusion by confining it to this use and replacing it in the ceramic sense by 'on-glaze decoration, colourant, etc.' and enamel kiln by 'decorating kiln'.

Colour. We have tried to restrict the use of this word to the visual sensation received by the human eye and to the physical facts of wavelength. Where the word 'colour' is often applied to materials which appear coloured, or which do so after firing or other treatment, we have replaced it by the word 'colourant'.

Electrolyte. We have used this word only to mean a compound which in solution in the molten state is ionised and will therefore conduct an electric current by which it is simultaneously decomposed. Where the word is used in ceramics to denote a 'deflocculant' we have replaced it by the latter word.

Table V: STANDARD SPECIFICATIONS

British Standard Specifications relating to Ceramics

The following standard specifications are obtainable from the British Standards Institution, 2, Park Street, London, W.1, prices change and so are not quoted.

B.S. 16:1949.	Telegraph material.
B.S. 45:1952.	Sparking plugs.
B.S. 65:1952.	Salt-glazed ware pipes.
B.S. 67:1938.	Two- and three-terminal ceiling roses.
B.S. 137:1941.	Porcelain and toughened glass insulators for overhead power lines.
B.S. 223:1956.	Electrical performance of high-voltage bushing-insulators.
B.S. 402:1945.	Clay plain roofing tiles and fittings.
B.S. 539, Pt 1:1951.	Dimensions of drain fittings in salt-glazed ware and glass (vitreous) enamelled salt-glazed fireclay.
B.S. 539, Pt 2:1951.	Dimensions of drain fittings—Scottish type. Salt-glazed ware and glass (vitreous) enamelled salt-glazed fireclay.
B.S. 540:1952.	Glass (vitreous) enamelled salt-glazed fireclay pipes.
B.S. 620:1954.	Dimensions of grinding wheels and segments of grinding wheels.
B.S. 657:1950.	Dimensions of common building bricks.
B.S. 784:1953.	Methods of test for chemical stoneware.
B.S. 914:1952.	Quality of laboratory porcelain.
B.S. 996:1942.	Test code for the performance of drying ovens.
B.S. 1081:1953.	Test code for kilns for heavy clay ware, including refractory materials.
B.S. 1143:1955.	Special salt-glazed ware pipes with chemically resistant properties.
B.S. 1181:1944.	Clay flue linings and chimney pots suitable for open fires (dimensions and workmanship only).
B.S. 1188:1944.	Ceramic lavatory basins (dimensions and workmanship only).
B.S. 1190:1951.	Hollow clay building blocks.
B.S. 1196:1944.	Clay ware field drain pipes.
B.S. 1206:1945.	Fireclay sinks (dimensions and workmanship).
B.S. 1213:1945.	Ceramic washdown W.C. pans (dimensions and workmanship).
B.S. 1226:1945.	Draining boards.
B.S. 1229:1945.	Fireclay washtubs and tub and sink sets (dimensions and workmanship).
B.S. 1233–35:1945.	Copings of clay ware, cast concrete and natural stone.

B.S. 1236–40:1956. Sills and lintels.
B.S. 1257:1945. Methods of testing clay building bricks.
B.S. 1281:1945. Glazed earthenware wall tiles (dimensions and workmanship only.)
B.S. 1286:1945. Clay tiles for flooring (dimensions and workmanship only).
B.S. 1301:1946. Clay engineering bricks.
B.S. 1388:1947. Test code for ovens and kilns firing pottery.
B.S. 1424:1948. Clay single-lap roofing tiles and fittings (dimensions and workmanship only).
B.S. 1598:1949. Ceramic materials for telecommunication and allied purposes.
B.S. 1614:1949. Brickwork settings for cylindrical boilers.
B.S. 1634:1950. Dimensions for stoneware pipes and pipe fittings for chemical purposes.
B.S. 1752:1952. Sintered disk filters for laboratory use.
B.S. 1758:1951. Fireclay refractories (bricks and shapes) for use in the petroleum industry.
B.S. 1814:1952. Marking system for grinding wheels.
B.S. 1902:1952. Methods of testing refractory materials.
B.S. 1969:1953. Tests for performance characteristics of sintered filters.
B.S. 2064:1953. Dimensions of diamond abrasive wheels and tools.
B.S. 2067:1953. Determination of power factor and permittivity of insulating materials.
B.S. 2484:1954. Cable covers (concrete and earthenware).
B.S. 2496:1954. Preferred sizes of casting pit refractories.
B.S. 2505:1954. Cowhouse equipment.
B.S. 2705:1956. Clay lath.
B.S. 2717:1956. Glossary of terms applicable to roof coverings.
B.S. 2757:1956. Classification of insulating materials for electrical machinery and apparatus on the basis of thermal stability in service.
B.S. 3446:1962. Glossary of terms relating to the manufacture and use of refractory materials.
B.S./M.O.E. 1–7:1947. Sanitary equipment for schools (fireclay).

British Standard Specifications applicable in the Ceramic Industry
B.S. 188:1937. Method for determination of viscosity of liquids in absolute (c.g.s.) units.
B.S. 410:1943. Test sieves.
B.S. 427:1931. Tables of diamond pyramid hardness numbers.
B.S. 526:1933. Definitions of gross and net calorific value.
B.S. 600:1935. Application of statistical methods to industrial standardisation and quality control.
B.S. 600R:1942. Quality control charts.
B.S. 718:1953. Density hydrometers and specific gravity hydrometers.
B.S. 735:1944. Sampling and analysis of coal and coke for performance efficiency tests on industrial plant.

B.S. 742:1947.	Fuel oil for burners.
B.S. 749:1952.	Underfeed stokers (ram or screw type).
B.S. 799:1953.	Oil-burning equipment.
B.S. 823:1938.	Density–composition tables for aqueous solutions of sodium chloride and of calcium chloride for use in conjunction with British Standard density hydrometers.
B.S. 824:1938.	Density composition tables for aqueous solutions of caustic soda for use in conjunction with British Standard density hydrometers.
B.S. 848:1939.	Testing of fans for general purposes (excluding mine fans).
B.S. 860:1939.	Table of approximate comparison of hardness scales.
B.S. 874:1956.	Definitions of heat insulating terms and methods of determining thermal conductivity.
B.S. 878:1939.	Code for comparative commercial tests of coal or coke and appliances in small steam raising plants.
B.S. 891:1940.	Method for direct reading hardness testing (Rockwell principle).
B.S. 893:1940.	Method of testing dust extraction plant and the emission of solids from chimneys of electric power stations.
B.S. 974:1941.	British Standard Symbols for use on diagrams of chemical plant.
B.S. 975:1941.	Density–composition tables for the aqueous solutions of nitric acid for use with British Standard density hydrometers.
B.S. 976:1941.	Density–composition tables for the aqueous solutions of hydrochloric acid for use with British Standard density hydrometers.
B.S. 995:1942.	Test code for gas producers.
B.S. 1016:1942.	Methods for the analysis and testing of coal and coke.
B.S. 1017:1942.	Method for the sampling of coal and coke.
B.S. 1041:1943.	Code for temperature measurement.
B.S. 1042:1943.	Code for flow measurement.
B.S. 1293:1946.	Screen analysis of coal (other than pulverised coal) for performance and efficiency tests on industrial plant.
B.S. 1328:1956.	Methods of sampling water used in industry.
B.S. 1610:1950.	Verification of testing machines. Part 1. Method of load verification; and verification of tensile and compression machines.
B.S. 1611:1953.	Glossary of colour terms used in science and industry.
B.S. 1647:1950.	pH Scale.
B.S. 1733:1955.	Flow cups and methods of use.
B.S. 1794:1952.	Chart ranges for temperature recording instruments.
B.S. 1796:1952.	Methods for the use of British Standard fine-mesh test sieves.
B.S. 1826:1952.	Reference tables for thermocouples. Platinum/rhodium v. platinum.

B.S. 1827:1952. Reference tables for thermocouples. Nickel/chromium v. nickel/aluminium.
B.S. 1904:1952. Commercial platinum resistance thermometer elements.
B.S. 1986:1953. Dimensional features of measuring and control instruments and panels for industrial processes.
B.S. 2074:1954. Size analysis of coke.
B.S. 2082:1954. Code for disappearing-filament optical pyrometers.
B.S. 2643:1955. Terms relating to the performance of measuring instruments.
B.S. 2648:1955. Performance requirements for electrically-heated laboratory drying ovens.
B.S. 2690:1956. Methods of testing water used in industry.
B.S. 2701:1956. Rees–Hugill powder density flask.
B.S. 2765:1956. Dimensioning system and terminology for industrial temperature-detecting elements and pockets.

Liste des normes françaises concernant la céramique (NF) homologuées au 31 Decembre 1958

(List compiled by La Société Française de Céramique)

Indice	Date	Titre
		Produits Refractaires
		Produits Réfractaires :
B. 40–001	Octobre 1939	Définition—Classification.
B. 49–101	Avril 1940	Mésures des variations permanentes de dimensions.
B. 49–102	Avril 1940	Résistance pyroscopique.
B. 49–103	Avril 1940	Essai de compression à température ordinaire.
B. 49–104	Avril 1940	Densités—Porosités—Définitions et déterminations.
B. 49–105	Avril 1940	Essai d'affaissement sous charge à haute température.
		Analyse chimique des produits réfractaires :
B. 49–401	Novembre 1943	Produits à base d'argile (silico–alumineux, alumineux, extra-alumineux) produits de bauxite, de cyanite (sillimanite), de corindon, produits siliceux et leurs matières premières (à l'exclusion des produits de silice).
B. 49–431	Novembre 1943	Produits de silice et leurs matières premières.
B. 49–441	Mars 1948	Produits à base de magnésie, de dolomie et de silicates de magnésie.
B. 49–442	Mai 1950	Réfractaires à base d'oxyde de chrome—Minerais de chrome.
B. 49–443	Juin 1953	Carbure de silicium et produits à base de carbure de silicium.
		Economie Domestique—Equipement Sanitaire—Materiel
D. 10–101	Février 1957	Eviers caractéristiques générales.
D. 11–101	Avril 1944	Lavabo en céramique.
D. 11–102	Juillet 1946	Evier en céramique.
D. 11–105	Juillet 1946	Cuvette de W.C. en céramique à chasse directe et siphon caché à sortie arrière et sortie centrale.
D. 11–106	Juillet 1946	Cuvette de W.C. en céramique à chasse directe et siphon apparent.
D. 11–107	Juillet 1946	Bidet en céramique.
D. 11–109	Juillet 1946	Dessus de cuvette de W.C.
D. 11–110	Juillet 1946	Consoles de lavabo.

Liste des normes françaises concernant la céramique (NF) homologuées au 31 Decembre 1958—(contd.)

Indice	Date	Titre
D. 11–115	Mars 1950	Urinoirs — Encombrement — Raccordement.
D. 11–116	Mars 1950	Bacs profonds en céramique.
D. 25–101	Juin 1942	Assiettes en porcelaine.
D. 25–102	Juin 1942	Tasses et soucoupes en porcelaine.
D. 25–601	Novembre 1944	Assiettes en porcelaine pour collectivités.
D. 25–602	Novembre 1944	Tasses et soucoupes en porcelaine pour collectivités.
D. 25–603	Novembre 1944	Bols à pied en porcelaine.
D. 25–604	Novembre 1944	Pots à lait en porcelaine pour collectivités.
D. 26–101	Mars 1943	Assiettes en faïence.
D. 26–102	Mars 1943	Tasses et soucoupes en faïence.
D. 26–103	Mars 1943	Bols à pied en faïence.
		Bâtiment et Genie Civil
P. 10–301	Mai 1942	Hourdis pour planchers en béton armé.
P. 13–301	Juillet 1937	Briques (qualités).
P. 13–401	Avril 1944	Hourdis en terre cuite pour planchers en béton armé.
P. 13–402	Avril 1944	Hourdis droits et bardeaux en terre cuite.
P. 13–403	Novembre 1944	Briques semi-modulées.
P. 13–404	Juillet 1937	Briques (dimensions).
P. 14–301	Avril 1958	Blocs creux en béton de granulats lourds (sable et gravillon) pour murs et cloisons.
P. 14–302	Décembre 1940	Briques silico–calcaires (qualités).
P. 14–303	Avril 1958	Blocs creux en béton de machefer pour murs et cloisons.
P. 14–304	Avril 1958	Blocs creux en béton de granulats légers (pouzzolane ou laitier expansé).
P. 14–401	Avril 1944	Hourdis en agglomérés pour planchers en béton armé.
P. 14–402	Avril 1944	Agglomérés pour maçonnerie (dimensions).
P. 14–403	Décembre 1940	Briques silico–calcaires (dimensions).
P. 18–301	Mai 1941	Agrégats pour bétons de construction.
P. 18–303	Mai 1941	Eau de gâchage pour bétons de construction.
P. 18–304	Mai 1941	Granulométrie des agrégats.
P. 18–310	Juillet 1957	Pouzzolane.
P. 18–311	Juillet 1957	Laitier expansé.
P. 31–301	Mai 1935	Tuiles mécaniques.
P. 61–401	Mai 1944	Carreaux de grès cérame.
P. 61–402	Mai 1944	Carreaux de faïence.
P. 61–403	Mai 1944	Carreaux de grès émaillé.
		Normes Générales
X. 11–501	Juin 1938	Analyses granulométriques par tamisage.

European Lexicon of technical terms in the Brick and Tile Industries, compiled by the European Brick and Tile Federation (T.B.E.).

Deutsche Normen relating to Ceramics

DIN	Date	DK...	Title
			Prüfung keramischer Roh- und Werkstoffe
51030	December 1954	666.31:620.174	Bestimmung der Trockenbiegefestigkeit.
51031	June 1954	666.31:666.295:620.1	Bestimmung der Bleiabgabe von eingebrannten Glasurfarben, Glasuren und Dekoren.
51063 (Ersatz für DIN 1063)	December 1954	666.31:620.1	Bestimmung des Kegelfallpunktes nach Seger (SK).
51070	April 1957	666.76:543.6	Chemische Analyse von Roh- und Werkstoffen mit 90% oder mehr Kieselsäuregehalt.
51071	May 1957	666.76:543.6	Chemische Analyse von Roh- und Werkstoffen mit Kieselsäuregehalt unter 90%.
51072	June 1956	666.76:543.6:546.722-31	Gerät für die Bestimmung von Eisen-(III)-oxyd nach dem Titan-(III)-chlorid-Verfahren und Titerstellung der Titan-(III)-chlorid-Lösung.
51073	February 1956	666.31:543.6	Chemische Analyse von Roh- und Werkstoffen mit hohem Magnesiumoxydgehalt.
51091	September 1956	691.833:666.75:620.193.41/.42	Bestimmung der Säure- und Laugenbeständigkeit von Fliesen und Platten für Wand- und Bodenbeläge.
51092	September 1956	691.833:666.75.016.5:620.193 41/.42	Bestimmung der Säure- und Laugenbeständigkeit der Glasur von Fliesen und Platten für Wand- und Bodenbeläge.
51093	September 1956	691.833:666.75:620.193:536.48/.49	Bestimmung der Temperaturwechselbeständigkeit von Fliesen und Platten für Wand- und Bodenbeläge.
51094	September 1956	691.833:666.75:620.193.6:535.683	Bestimmung der Lichtechtheit der Färbung von Fliesen und Platten für Wand- und Bodenbeläge.
51100	April 1957	666.31:620.1:543.721	Bestimmung der löslichen Salze (Perkolatorverfahren).
51102 Blatt 1	May 1958	666.31:620.193.41:628.45	Bestimmung der Säurebeständigkeit — Verfahren mit stückigem Prüfgut.
51102 Blatt 2	May 1958	666.31:620.193.41:666.774	Bestimmung der Säurebeständigkeit — Verfahren mit gekörntem Prüfgut.
(Ersatz für DIN 4092 Ausgabe 10.44 Verfahren A.)			
51033	April 1958	666.31:620.1:539.215.4	Bestimmung der Korngrössen durch Siebung und Sedimentation. Verfahren nach Andreasen.
(Einsprüche bis 31 Okt. 1958)			
			Siebe
4186 Blatt 1	February 1957	621.928.028.3	Webedrähte rund, Masse.
4186 Blatt 2	February 1957	621.928.028.3:620.1	Webedrähte rund. Technische Lieferbedingungen und Prüfung.
4188 Blatt 1	February 1957	621.928.028.3:620.168.32	Drahtgewebe für Prüfsiebe. Masse.
4189 Blatt 1	November 1953	621.928.028.3	Drahtgewebe Quadratische Maschenweiten, glatte Bindung.
4189 Blatt 2	October 1953	621.928.028.3	Drahtgewebe Quadratische Maschenweiten, glatte Bindung. Kennlinien gleicher offener Siebflächen.

Deutsche Normen relating to Ceramics—(contd.)

DIN	Date	DK...	Title
			Siebe—(contd.)
4189 Blatt 3	February 1957	621.928.028.3: 620.1	Drahtgewebe Quadratische Maschenweiten, glatte Bindung Technische Lieferbedingungen und Prüfung.
			Prüfsiebung
4190 Blatt 1	October 1955	621.928.028	Doppeltlogarithmisches Körnungsnetz
Blatt 2			Erläuterungen.
24041	November 1954	621.928.028:622.74	Lochbleche. Rundlochung.
			Prüfung fester Brennstoffe
51700	April 1958	662.62.017:620.1	Allgemeines und Übersicht über Untersuchungsverfahren.
51701 (Ersatz für DIN 53711)	August 1950	662.62.017	Probenahme und Probeaufbereitung von körnigen Brennstoffen.
51708	April 1956	662.62.017:536.6	Bestimmung der Verbrennungswärme und des Heizwertes.
51718	August 1950	662.62.017:543.812	Bestimmung des Wassergehaltes
51719	August 1950	662.62.017:543.822	Bestimmung des Aschegehaltes.
51721	August 1950	662.62.017: 543.842/.843	Bestimmung des Gehaltes an Kohlenstoff und Wasserstoff.
51724	February 1955	662.62.017:543.845	Bestimmung des Schwefelgehaltes.
51725	February 1955	662.62.017:543.847	Bestimmung des Phosphorgehaltes.
51722	March 1954	662.62.017:543.846	Bestimmung des Stickstoffgehaltes.
51727	February 1955	662.62.017:543.848	Bestimmung des Chlorgehaltes.
51730	June 1954	662.62.017:662.613.114	Bestimmung des Asche-Schmelzverhaltens.
51739	March 1956	662.62.017:662.74	Bestimmung des Kokungsgrades von Steinkohle nach dem Dilatometer-Verfahren.
51741	March 1956	662.62.017.662.74	Bestimmung des Blähgrades von Steinkohle.
			Prüfung flüssiger Brennstoffe
51603	May 1957	662.753.325.017	Heizöle. Mindestanforderungen.
51795	December 1958	662.75:620.1: 543.82	Bestimmung des Abdampfrückstandes von wirkstoffhaltigen Ottokraftstoffen.
			Keramische Werkstoffe der Elektrotechnik
40685	January 1957	621.315.612: 666.593	Keramische Isolierstoffe für die Elektrotechnik. Gruppeneinteilung und Technische Werte.
40686	January 1952	621.315.612	Oberflächenbehandlung. Richtlinien.
41341	November 1953	621.39:621.319.4	Keramik-Kondensatoren. Elektrische Daten und Aufbau.
			Prüfverfahren für feuerfeste Baustoffe
1064	July 1930	691.4:666.76:620.1	Erweichen bei hohen Temperaturen unter Belastung (Druck-Feuer-Beständigkeit DFB).
1065	March 1942	691.4:666.76:620.1	Spezifisches Gewicht, Raumgewicht, Wasseraufnahmefähigkeit, Porosität.
1066	October 1941	691.4:666.76	Prüfverfahren für das Nachschwinden (NS) und Nachwachsen (NW).
1067	July 1930	691.4:666.76:620.1	Bestimmung der Druckfestigkeit bei Zimmertemperatur.
1068	July 1931	691.4:666.76:620.1	Bestimmung des Widerstandes gegen schroffen Temperaturwechsel. Temperaturwechselbeständigkeit (TWB) *a*. Normalsteinverfahren. *b*. Zylinderverfahren.

Deutsche Normen relating to Ceramics—(contd.)

DIN	Date	DK...	Title
			Prüfverfahren für feuerfeste Baustoffe— (contd.)
1069	July 1931	691.4:666.76:620.1	Beständigkeit gegen den Angriff fester und flüssiger Stoffe bei hoher Temperatur. Verschlackungs-Beständigkeit (VB) *a.* Tiegelverfahren (VBT). *b.* Aufstreuverfahren (VBA).
			Feuerfeste Baustoffe — Feuerfeste Steine
1081	November 1940	691.4.666.76	Normalsteine. Abmessungen.
			Gütenormen für feuerfeste Baustoffe
1086	March 1942	691.4:666.76	Allgemeines und zulässige Abweichungen.
1087	August 1931	691.42:669.162.212	Gütenormen für Hochofensteine.
1087 addition	August 1931	691.42:669.162.2	Hochofensteine. Erläuterungen.
1088	August 1931	666.76:691.42	Siemens-Martin-Ofensteine.
1088 addition	August 1931	691.42	Siemens-Martin-Ofensteine. Erläuterungen.
			Feuerfeste Baustoffe
1089 Blatt 1 (Mit Blatt 2 Ersatz für DIN 1089 und DIN 1089 Bbl.)	December 1956	691.42:666.76	Koksofensteine. Gütewerte von kalkgebundenen Silikasteinen.
1089 Blatt 2 (Mit Blatt 1 Ersatz für DIN 1089 und DIN 1089 Bbl.)	December 1956	691.42:666.76	Koksofensteine. Gütewerte von Quarzschamottesteinen.
1090	March 1939	691.4	Wannesteine aus Schamotte. Gütenormen.
			Baustoffe
105	March 1957	666.71/.72:691.421	Mauerziegel. Vollziegel und Lochziegel.
456 (Zugleich Ersatz für DIN 453, DIN 454 und DIN 52250)	May 1958	666.74:691.424	Dachziegel. Güteeigenschaften und Prüfverfahren.
1180	September 1931	626.862:691.4:666.73	Dränrohre.
1180 addition	September 1931	626.862:691.4:666.73	Dränrohre. Erläuterungen.
52501 (DIN DVM 2501 ist seit Aug. 1955 DIN 52501)	July 1937	691.8	Tonhohlplatten (Hourdis). Prüfverfahren.
278	April 1938	666.725:691.81.431	Tonhohlplatten (Hourdis).
409	May 1938	691.4:697.26:666.76	Kacheln für Tonöfen, quadratisch.
			Freistehende Schornsteine
1056 Blatt 1 (Einsprüche bis 31.Okt.1956)	January 1956	697.84:351.78:662.922.2	Grundlagen für Berechnung und Ausführung.
1057	January 1956	691.82:662.922.2	Mauersteine und Mauerziegel.
			Kabelzubehör
279	September 1937	621.315.687.6	Lieferbedingungen und Prüfvorschriften für Kabelschutzhauben aus Ton für Schwachstromkabel.
457 Blatt 1 (Ersatz für DIN 457)	September 1955	621.315.687.6	Kabelkanal-Formsteine ein- und mehrzügig.

Deutsche Normen relating to Ceramics—(contd.)

DIN	Date	DK...	Title
			Kanalklinker
4051 Blatt 1	November 1955	628.241:666.714.14	Masse. Gütebestimmungen.
4051 Blatt 2	November 1955	628.241:666.714.14	Prüfung.
			Rohre, Formstücke, Sohlschalen und Platten aus Steinzeug
1230 Blatt 1 (Mit Blatt 2 Ersatz für DIN 1230)	August 1957	621.643.2:628.2	Abmessungen und Gütebestimmungen.
1230 Blatt 2	February 1958	621.643.2:628.2	Prüfbestimmungen und Prüfverfahren.
1230 addition	March 1958	621.643.2:628.2	Herstellerzeichen der Steinzeugwerke.
			Prüfverfahren für saurefeste keramische Baustoffe Säurefeste Steine
4091	October 1944	691.4:666.774	Bestimmung des Widerstandes gegen Temperaturwechsel. Temperaturwechselbeständigkeit (TWB)
4092 (Einheitsblatt)	October 1944	691.4:666.76	Bestimmung der Säureloeslichkeit.
12912	December 1955	542.12:691.833:666.75	Fliesen für Laboratoriumstische.
18155 (Ersatz für DIN 1400 Blatt 1 und 2 und DIN 18154)	March 1958	666.75.691.42 −431:003.62	Keramische Wand- und Bodenfliesen.
			Schleifwerkzeuge
69100	February 1949	621.922:001.4	Bezeichnung.
69102	February 1949	621.922	Richtlinien für die Wahl von Schleifkörpern.
69103	February 1949	621.922	Richtwerte und zulässige Höchstwerte für die Umfangsgeschwindigkeit beim Schleifen.
69106	February 1949	621.922	Das Auswuchten von Schleifscheiben.
			Prüfung von Email
51150	March 1957	666.29:620.193. 474.771	Bestimmung der Beständigkeit gegen kalte Zitronensäure.
51151 (Ersatz für DIN 6050)	March 1957	666.29:620.193. 474.771	Bestimmung der Beständigkeit gegen kochende Zitronensäure.
			Baugipse
1168 Blatt 1 (Ersatz für DIN 1168)	March 1955	691.55:001.4	Begriffe und Kennzeichnung.
1168 Blatt 2 (Ersatz für DIN 1168)	March 1955	691.55	Stuckgips und Putzgips Anforderungen, Prüfverfahren und Prüfgeräte.
			Prüfung von Naturstein
52102 (Nachdruck August 1947 von DIN DVM 2102)	February 1940	691.2:620.1	Rohgewichte (Raumgewicht). Reingewichte (spezifisches Gewicht). Dichtigkeitsgrad.
52103 (Ersatz für DIN DVM 2103)	November 1942	691.2:620.193.19	Wasseraufnahme, Wasserabgabe.
52104 (Ersatz für DIN DVM 2104)	November 1942	691.2:620.1	Frostbeständigkeit.
52105 (Ersatz für DIN DVM 2105)	November 1942	691.2:620.1	Druckfestigkeit.
52106	March 1943	691.2:620.1	Wetterbeständigkeit.
52108 (DIN DVM 2108 ist seit Mai 1947 DIN 52108)	October 1939	691.2:620.1	Abnutzbarkeit durch Schleifen.

APPENDIX

American Society for Testing Materials
1916 Race Street, Philadelphia 3, Pa., U.S.A.

Extract from List of Standards, Sept. 1959

Manufactured Masonry Units.

BRICK

Specifications for:

Building Brick (Solid Masonry Units Made from Clay or Shale)	C 62–58
Facing Brick (Solid Masonry Units Made from Clay or Shale)	C 216–57
Industrial Floor Brick (*Tentative*)	C 410–57 T
Masonry Units, Chemical-Resistant	C 279–54
Ceramic Glazed Structural Clay Facing Tile, Facing Brick, and Solid Masonry Units (*Tentative*)	C 126–59 T
Paving Brick	C 7–42
Sewer Brick (Made from Clay or Shale)	C 32–58

Methods of:

Sampling and Testing Brick	C 67–57

STRUCTURAL TILE AND FILTER BLOCK

Specifications for:

Clay Load-Bearing Wall Tile, Structural	C 34–57
Clay Non-Load-Bearing Tile, Structural	C 56–57
Clay Floor Tile, Structural	C 57–57
Clay Facing Tile, Structural	C 212–54
Ceramic Glazed Structural Clay Facing Tile, Facing Brick and Solid Masonry Units (*Tentative*)	C 126–59 T
Vitrified Clay Filter Block for Trickling Filters	C 159–55

Methods of:

Sampling and Testing Structural Clay Tile	C 112–52

Definitions of Terms relating to:

Structural Clay Tile	C 43–55

PIPE AND DRAIN TILE

Specifications for:

Clay Drain Tile (*Tentative*)	C 4–59 T
Ceramic Glazed or Unglazed Clay Sewer Pipe, Standard Strength (*Tentative*)	C 261–57 T
Ceramic Glazed or Unglazed Clay Pipe, Extra Strength (*Tentative*)	C 278–57 T

Clay Flue Linings . . . . C 315–56
Clay Sewer Pipe, Standard Strength (*Tentative*) . . C 13–57 T
Clay Pipe, Extra Strength (*Tentative*) . . . . C 200–57 T
Clay Pipe, Perforated, Standard Strength (*Tentative*) . C 211–57 T
Vitrified Clay Pipe Joints Using Materials Having Resilient
    Properties (*Tentative*) . . . . . . C 425–58 T

Methods of Test for:

Clay Pipe . . . . C 301–54

Recommended Practice for:

Installing Vitrified Clay Sewer Pipe (*Tentative*) . . C 12–58 T

REFRACTORIES

Classification of:

Castable Refractories (*Tentative*) . . . . . C 401–57 T
Fireclay Refractory Brick . . . . . . C 27–58
Fireclay and High-Alumina Refractory Brick (*Tentative*) C 27–58 T
Insulating Fire Brick . . . . C 155–57
Ground Refractory Materials, Single- and Double-
    Screened . . . . C 136–55
Silica Refractory Brick (*Tentative*) . . . . C 146–58 T

Specifications for:

Refractory for Heavy Duty Stationary Boiler Service . C 64–51
Refractories for Moderate Duty Stationary Boiler Service C 153–51
Air-Setting Refractory Mortar (Wet Type) for Boiler and
    Incinerator Services . . . . . . C 178–47
Fireclay Plastic Refractories for Boiler and Incinerator
    Services . . . . C 176–47
Fireclay-Base Castable Refractories for Boiler Furnaces
    and Incinerators . . . . C 213–58
Refractories for Incinerators . . . . . C 106–51
Refractories for Malleable Iron Furnaces with Removable
    Bungs and for Annealing Ovens . . . . C 63–51
Steel Pouring Pit Refractories (*Tentative*) . . . C 435–59 T
Insulating Fire Brick for Linings of Industrial Furnaces
    Operated with a Neutral or Oxidising Atmosphere
    (*Tentative*) . . . . C 434–59 T
Fire Clay, Ground, as a Mortar for Laying-Up Fireclay
    Brick . . . . C 105–47
Quicklime and Hydrated Lime for Silica Brick Manufac-
    ture . . . . C 49–57

APPENDIX

Methods of Test for:

Chemical Analysis of Refractory Materials	C 18–52
Chemical Analysis of Refractory Materials (*Tentative*)	C 18–56 T
Panel Spalling Test for Refractory Brick, Basic Procedure in	C 38–58
Panel Spalling Test for High Duty Fireclay Brick	C 107–52
Panel Spalling Test for Super Duty Fireclay Brick	C 122–52
Refractory Brick, Testing Under Load at High Temperatures	C 16–49
Panel Spalling Test for Fireclay Plastic Refractories	C 180–52
Resistance to Thermal Spalling of Silica Brick (*Tentative*)	C 439–59 T
Drying and Firing Shrinkage, Combined, of Fireclay Plastic Refractories	C 179–46
Workability Index of Fireclay Plastic Refractories	C 181–47
Thermal Conductivity of Refractories	C 201–47
Thermal Conductivity of Castable Refractories (*Tentative*)	C 417–58 T
Thermal Conductivity of Fireclay Refractories	C 202–47
Thermal Conductivity of Plastic Refractories (*Tentative*)	C 438–59 T
Thermal Conductivity of Insulating Fire Brick	C 182–47
Reheat Change of Refractory Brick	C 113–46
Reheat Change of Insulating Fire Brick	C 210–46
Reheat Change of Carbon Brick and Shapes (*Tentative*)	C 436–59 T
Permanent Linear Change on Firing of Castable Refractories (*Tentative*)	C 269–57 T
Pyrometric Cone Equivalent (PCE) of Refractory Materials	C 24–56
Refractoriness of Air-Setting Refractory Mortar (Wet Type)	C 199–47
Cold Bonding Strength of Air-Setting Refractory Mortar (Wet Type)	C 198–47
Cold Crushing Strength and Modulus of Rupture of Refractory Brick and Shapes	C 133–55
Crushing Strength and Modulus of Rupture of Insulating Fire Brick at Room Temperature	C 93–54
Modulus of Rupture of Castable Refractories (*Tentative*)	C 268–55 T
Warpage of Refractory Brick and Tile	C 154–41
Size and Bulk Density of Refractory Brick	C 134–41
Size and Bulk Density of Insulating Fire Brick (*Tentative*)	C 437–59 T
Bulk Density of Granular Refractory Materials	C 357–58

Methods of Test for:

Apparent Porosity, Water Absorption, Apparent Specific Gravity, and Bulk Density of Burned Refractory Brick	C 20–46
True Specific Gravity of Refractory Materials	C 135–47
Sieve Analysis and Water Content of Refractory Materials	C 92–46
Disintegration of Refractories in an Atmosphere of Carbon Monoxide	C 288–56

Definitions of Terms Relating to:

Refractories	C 71–55
Heat Transmission, Symbols for	C 108–46

CERAMIC WHITEWARES

Methods of Test for:

Chemical Analysis of Ceramic Whiteware Clays	C 323–56
Free Moisture in Ceramic Whiteware Clays	C 324–56
Sampling Ceramic Whiteware Clays	C 322–56
Shrinkages, Drying and Firing, of Ceramic Whiteware Clays, Determination of	C 326–56
Wet Sieve Analysis of Ceramic Whiteware Clays	C 325–56
Specific Gravity of Fired Ceramic Whiteware Materials	C 329–56
Sieve Analysis of Nonplastic Pulverised Ceramic Materials	C 371–56
45-deg. Specular Gloss of Ceramic Materials	C 346–59
Impact Resistance of Ceramic Tableware	C 368–56
Compressive (Crushing) Strength of Fired Whiteware Materials	C 407–58
Crazing Resistance of Fired Glazed Whitewares by Autoclave Treatment (*Tentative*)	C 424–58 T
Linear Thermal Expansion of Fired Whiteware Products by the Dilatometer Method	C 372–56
Linear Thermal Expansion of Fired Ceramic Whiteware Materials by the Interferometric Method	C 327–56
Moisture Expansion of Fired Whiteware Products	C 370–56
Modulus of Rupture of Fired Cast or Extruded Whiteware Products	C 369–56
Flexural Properties of Fired Dry-Pressed Whiteware Specimens at Normal Temperature	C 328–56
Water Absorption, Bulk Density, Apparent Porosity, and Apparent Specific Gravity of Fired Porous Whiteware Products	C 373–56
Thermal Conductivity of Whiteware Ceramics	C 408–58
45-deg, 0-deg Directional Reflectance of Opaque Specimens by Filter Photometry	E 97–55

Definitions of Terms Relating to:

Ceramic Whitewares and Related Products	C 242–56
Ceramic Whitewares and Related Products (*Tentative*)	C 242–58 T

THERMAL INSULATING MATERIALS

Diatomaceous Earth Block-Type Thermal Insulation (*Tentative*)	C 333–59 T
Diatomaceous Earth Thermal Insulation for Pipes (*Tentative*)	C 334–59 T

APPENDIX

85 Per Cent Magnesia Block-Type Thermal Insulation	C 319–55
85 Per Cent Magnesia Moulded-Type Thermal Insulation for Pipes	C 320–55
Sampling Preformed Thermal Insulation (*Tentative*)	C 390–57 T
Density of Preformed Block-Type Thermal Insulation	C 303–56
Density of Preformed Pipe-Covering-Type Thermal Insulation	C 302–56
Mechanical Stability of Preformed Thermal Insulation by Tumbling (*Tentative*)	C 421–58 T
Breaking Strength and Calculated Flexural Strength of Preformed Block-Type Thermal Insulation	C 203–58
Breaking Strength and Calculated Modulus of Rupture of Preformed Insulation for Pipes (*Tentative*)	C 446–59 T
Compressive Strength of Preformed Block-Type Thermal Insulation	C 165–54
Thickness and Density of Blanket- or Bat-Type Thermal Insulating Materials	C 167–50
Emittance, Normal Total, of Surfaces of Materials 0·01 in. or Less in Thickness at Approximately Room Temperature (*Tentative*)	C 445–59 T
Hot-Surface Performance of High-Temperature Thermal Insulations (*Tentative*)	C 411–58 T
Linear Shrinkage of Preformed High-Temperature Thermal Insulation Subjected to Soaking Heat (*Tentative*)	C 356–55 T
Mean Specific Heat of Thermal Insulation (*Tentative*)	C 351–59 T
Temperature, Maximum Use, of Preformed High-Temperature Insulation, Determining the (*Tentative*)	C 447–59 T
Thermal Conductivity of Materials by Means of the Guarded Hot Plate.	C 177–45
A detailed description of three typical guarded hot plates, the National Bureau of Standards plate, the National Research Council plate, and the Alundum plate, which comply with the requirements of Method C 177	
Thermal Conductivity of Insulating Materials at Low Temperatures by Means of the Wilkes Calorimeter (*Tentative*)	C 420–58 T
Thermal Conductivity of Pipe Insulation (*Tentative*)	C 335–54 T
Thermal Conductance and Transmittance of Built-Up Sections by Means of the Guarded Hot Box (*Tentative*)	C 236–54 T
Water Vapour Transmission of Materials Used in Building Construction (*Tentative*)	C 365–59 T
Water Vapour Transmission of Materials in Sheet form, Measuring (*Tentative*)	E 96–53 T

Recommended Practice for:

Clearance of Preformed Thermal Pipe Insulation	C 312–55

Definitions of Terms Relating to:

Thermal Insulating Materials	C 168–56

ACOUSTICAL MATERIALS

Methods of Test for:

Impedance and Absorption of Acoustical Materials by the Tube Method	C 384–58
Strength Properties of Prefabricated Architectural Acoustical Materials	C 367–57
Sound Absorption of Acoustical Materials in Reverberation Rooms (*Tentative*)	C 423–58 T

Recommended Practice for:

Laboratory Measurement of Airborne-Sound Transmission Loss of Building Floors and Walls	E 90–55

ELECTRICAL INSULATING MATERIALS

Ceramic Products (Glass, Porcelain, Steatite)

Specifications for:

Aluminium Oxide Powder (*Tentative*)	F 7–58 T

Methods of Test for:

Electrical Porcelain	D 116–44
Steatite Used for Electrical Insulation	D 667–44

Table VI

HEALTH HAZARDS IN THE POTTERY INDUSTRY

(1) Pneumoconiosis resulting from inhalation of silica (silicosis) and silicate dusts.

(2) Poisoning by compounds of: antimony; arsenic; beryllium; cadmium; chromium; cobalt; LEAD; manganese; mercury; selenium; uranium; vanadium. Poisoning also by: nitrobenzene; benzene; acetone; petrol and ether (additional fire hazard).

(3) Burns from hydrofluoric acid. Inhalation of fumes from hydrofluoric acid and other fluorine compounds.

(4) Dermatitis: (*a*) from turpentine, or other organic solvents; (*b*) continuous wetting of hands; (*c*) continuous friction.

(5) Carbon monoxide poisoning from gas appliances.

(6) Over-hot workrooms.

(7) Strains from weight-lifting.

The Dust Hazard in the Ceramic Industry

A number of processes in the ceramic industry produce dust which can be readily inhaled by workers, *e.g.* towing and fettling green ware; brushing biscuit ware; dust-pressing of tiles; setting of bone china and vitreous china; general handling of dry materials; etc.

Any mineral dust entering the lungs causes damage to them by the formation of a fibroid tissue which reduces the area of effective lung tissue, leading to a condition known as *fibrosis*. Silica dust, particularly cristobalite and tridymite, is especially harmful and the resulting condition is known as *silicosis*. Development of the disease is gradual and a long period of years may elapse before marked effects, such as shortness of breath, become apparent. There is also a lowered resistance to respiratory diseases, colds, bronchitis, pneumonia and tuberculosis. Silicosis complicated by tuberculosis is also known as *phthisis*.

Only particles of less than 5 to 10μ diameter enter the lungs. If the damage were due only to mechanical irritation there would probably be a lower size limit below which the particles would not act as irritants. However, the greater toxicity of free silica made further investigations necessary, and these have shown that fine silica particles will dissolve in aqueous fluids, such as are present in the lungs. Colloidal silicic acid could then be formed which apparently tends to destroy the phagocytes or natural scavenging cells, which normally attack and absorb poison, or tubercle bacilli.

It is not known exactly what dust concentration maximum is permissible to avoid all health hazards. It would probably vary between individuals. The need for maximum prevention and removal of dust therefore becomes obvious (**S32**, **B85**).

We have usually indicated in the text the various operations that require effective exhaust hoods and ducts.

The Pottery (Health) Special Regulations, 1947, dated October 7, 1947, made by the Minister of Labour and National Service under section 60 of the Factories Act, 1937 (1 Edw. 8 and I.G.E.O. 6, c. 67) (S.R. and O. 1947, No. 2161). Short title: *The Pottery (Health) Special Regulations, 1947.*

(1) Prohibition of any glaze which is not a leadless glaze or a low solubility glaze.

'Leadless glaze' means a glaze which does not contain more than one per cent of its dry weight of a lead compound calculated as lead monoxide.

'Low solubility glaze' means a glaze which does not yield to dil. hydrochloric acid more than five per cent of its dry weight of a soluble lead compound calculated as lead monoxide when determined in the prescribed manner (*see* p. 373).

(2) Prohibition of ground or powdered flint or quartz for certain purposes:
(*a*) placing of ware for the biscuit fire;
(*b*) polishing of ware;

(c) as an ingredient in a wash for saggars, trucks, bats, cranks or other articles used in supporting ware during firing;

(d) as dusting or supporting powder in potters' shops ('ground or powdered flint or quartz' does not include natural sands).

(3) Prohibition of ground or powdered flint or quartz except in slop or paste with the following exceptions:

(a) rooms or buildings set apart for the manufacture of ground or powdered flint or quartz, the making of frits or glazes, or the making of colours or coloured slips for the decoration of pottery;

(b) processes where the ground or powdered flint, or quartz, is transported in closed bags or containers and handled in enclosures into which no person enters and from which no dust can escape to places where persons are employed.

The Pottery (Health and Welfare) Special Regulations 1950, dated January 16, 1950 (S.I. 1950 No. 65). Short title: *The Pottery (Health and Welfare) Special Regulations, 1950*.

Regulations prohibiting the employment of women and young persons in (a) processes using material that does not pass the low solubility glaze test; (b) dust making processes; (c) heavy lifting or treading.

Regulations regarding medical examination of persons employed on the above tasks, health registers, protective clothing, washing facilities, messrooms and food storage, food, drink and tobacco.

Regulations covering the ventilation of work rooms with particular regard to the drying of ware, the temperature of work rooms.

Twenty-seven processes are listed as requiring an efficient exhaust draught.

Regulations regarding: the nature of and cleaning of floors, work benches. Preparation, manipulation and storage of clay dust, tile presses. Raw lead compounds. Glazing, etc. Colour blowing, etc. Hydrofluoric acid. Lithographic transfer making. Separation of processes. General suppression of dust. Respirators.

The Refractory Materials Regulations 1931, dated April 28, 1931 (S.R. and O. 1931, No. 359). Short title: *The Refractory Materials Regulations, 1931*.

These regulations apply to material containing not less than 80 per cent total silica (SiO_2) and deal with dust hazards.

The Clay Works (Welfare) Special Regulations 1948. (S.I. 1948 No. 1547.) Short title: *The Clay Works (Welfare) Special Regulations, 1948*.

The regulations apply to factories making bricks, tiles, blocks, slabs, pipes, stilts and spurs, nozzles or similar articles. They cover shelters, washing facilities, protective clothing, clothing accommodation, first aid, canteens and messrooms.

Industrial Health. A survey of the Pottery Industry in Stoke-on-Trent. A report by H.M. Factory Inspectorate. Published for the Ministry of Labour and National Service by H.M.S.O. (1959) 5s. 0d.

Bibliography and Authors Index

THERE are two main sources of information and illustrations other than personal experience, that have been used in the preparation of this book. Firstly, published papers and articles that have appeared in the scientific and ceramic presses of a number of countries, and secondly, information supplied by individuals or firms either personally or through leaflets, photographs, diagrams, etc. Acknowledgement to all these sources has been placed together in the following list, which, however, has a few features which it is hoped will make it easy to use and of value in itself.

(1) All authors' and firms' names have been arranged alphabetically. Any authors' names appearing after the first in the title of a paper also appear in the appropriate place in the alphabet on their own. Where full names are printed in a publication this has been followed.

(2) Wherever possible the full titles of papers and articles are given. This gives a reader wishing to look up more about a subject a better idea whether or not a reference is going to be worth looking up.

(3) Where possible the address of firms is given, so that further information can be readily obtained.

(4) Anonymous material, editorial matter, etc., has been arranged together at the beginning according to the alphabetical order of the periodical in which it appeared.

(5) It is hoped that the abbreviations used for the names of periodicals are clear and consistent (*Min.* is used only to mean Mining, and not Mineral, Mineralogy, etc.) The order of the reference given is Volume number, [issue number], page number, month, only if this is not apparent from the issue number (year). Although not universal practice the year is given last because this can be more readily picked out at the end of an entry.

(6) A full list of publications by the writers has been included.

(7) The right-hand column gives the numbers of the pages on which the references appear, pages with figures being in italics.

A1. ANON., Report of Committee C-8 on refractories. *Am. Soc. Testing Materials Proc.* **54**, 344 (1954).   408

A2. ANON., Steatit Werkstoffe. *Ber. D. K. G.* **34** [4], 96 (1957).   493, 496

A3. ANON., *German Hard Paste Porcelain. German Earthenware Bodies.* B.I.O.S. Report on Tableware Industries.   442–447, 470–473, 584

A4. ANON., *Telefunken Metal Ceramic Radio Valves.* B.I.O.S. 496, 1202
   Final Report No. 30, Item No. 1.
A5. ANON., *Bosch Sparking Plug Factories.* B.I.O.S. Final 476, 584, 1200
   Report No. 68, Item No. 19, 22, and 26.
A6. ANON., B.I.O.S. Misc. Report No. 107, pp. 42–60. 413
A7. ANON., *The German Abrasive Industry.* B.I.O.S. Final 1157
   Report No. 113, Item No. 21 and 31.
A8. ANON., *The Ceramic Industry of Germany.* B.I.O.S. Final 438, 472–475,
   Report No. 458, Item No. 22. 582, 584,
   1084, 1099
A9. ANON., *High Temperature Refractories and Ceramics.* 476, 1136,
   B.I.O.S. Final Report No. 465, Item No. 21 and 22. 1138, 1157
   1230, 1254
A10. ANON., *Heat Treatment of Refractory Materials.* B.I.O.S. 1220, 1256
   Final Report No. 831, Item 21 and 22.
A11. ANON., *The Manufacture of Chemical Stoneware in Germany.* 434, 1105
   B.I.O.S. Final Report No. 1087, Item No. 22.
A12. ANON., *German Glazed Wall Tile and Sanitary Ware Fac-* 448–451, 454,
   tories. B.I.O.S. Final Report No. 1187, Item No. 22. 1080, 1085–
   1088
A13. ANON., *The Carbon Industry in Germany.* B.I.O.S. Final 1211
   Report No. 1230, Item No. 22 and 31.
A14. ANON., B.I.O.S. Final Report No. 1402. 616, 635
A15. ANON., *Grinding Wheel Manufacture in Germany.* B.I.O.S. 481, 1157
   Final Report No. 1407, Item No. 21.
A16. ANON., *German Radio Ceramics.* B.I.O.S. Final Report 582, 655–657
   No. 1459, Item No. 22.
A17. ANON., *German Floor Tile Factories with Further Informa-* 436, 438, 440,
   tion on Glazed Wall Tile and Sanitary Industries. 1078, 1088
   B.I.O.S. Final Report No. 1551, Item No. 22.
A18. ANON., *Lutz & Co., Lauf/Pegnitz.* C.I.O.S. File No. 516, 518, 1207
   XXIX–31, Item No. 1.
A19. ANON., *Refractories in Turbine Blades.* Intelligence Objec- 494
   tives Subcommittee, Item 1, 18, 21 and 26, File XXX-
   22.
A20. ANON., *Refractories in Turbine Blades.* Intelligence Objec- 494
   tives, Sub-committee, Item, 1, 18, 21 and 25, File XXXI-
   22, p. 10.
A21. ANON., *The Manufacture of Refractories and Information* 1227, 1237–
   Concerning their Use in the Iron and Steel Industry of 1271
   Western Germany. F.I.A.T. Final Report No. 432.
   B.I.O.S.
A22. ANON., F.I.A.T. Final Report No. 617, pp. 73–74.
   (a) B.I.O.S. Report No. 458, Item 22, p. 27. Steatit 488
      Magnesia A.G.
   (b) B.I.O.S. Report No. 458, Item 22, p. 20. 488
   (c) B.I.O.S. Report No. 1459, Item 22, p. 180. E. Sem- 488
      bach & Co.

(d) B.I.O.S. Report No. 1459, Item 22, p. 42.   488
(e) F.I.A.T. Report No. 617, p. 71. Steatit Magnesia   488
A.G.

A23. ANON., *Manufacture of Technical Ceramics in Germany for the Chemical Industry.* F.I.A.T. Final Report No. 1275.   434, 488

A24. ANON., You can burn sewer pipe in a tunnel kiln—and salt glaze them, too. *Brick & Clay Record* 17 February (1941).   600, 1017

A25. ANON., Scrapers do two operations in one. *Brick & Clay Record* **120** [2], 50, February (1952).   259

A26. ANON., The winning of clay. *Brick & Clay Record* **121** [4], 44, October (1952).   259

A27. ANON., Crushers and crushing. *Brick & Clay Record* **121** [5], 54, November (1952).   662, 663

A28. ANON., Use tunnel kilns. *Brick & Clay Record* **116** [5], 48 (1950).   1017

A29. ANON., Salt glaze or slip glaze? *Brick & Clay Record* **119** [3], September (1951).   605

A30. ANON., Heat. Proper oil atomization boosts kiln efficiency. *Brick & Clay Record* **125** [1], 65, July (1954).   926, 927

A31. ANON., They've reduced black coring with ammonium chloride. *Brick & Clay Record* **126** [2], 40 (1955).   862

A32. ANON., World's most automatic pipe plant in operation. *Brick & Clay Record* **127** [2], 39 (1955).   737

A33. ANON., Refractories. *Brick & Clay Record* **128** [1], 66 (1956).   886, 1028

A34. ANON., A modern brickworks equipment. *Brit. Clywkr.* 17, April (1935).   1029

A35. ANON., The design and burning of kilns—I. *Brit. Clywkr.* 385, February (1938).   895, 976

A36. ANON., Pyrophyllite in refractories. *Brit. Clywkr.* 10, April (1941).   95

A37. ANON., Design of pitch and speed of augers. *Brit. Clywkr.* **55** [655], 133, November (1946).   731

A38. ANON., Burning bituminous clays. *Brit. Clywkr.* **54**, 231 (1946).   860

A39. ANON., Ceramic glazes. *Brit. Clywkr.* **59** [706], 288, February (1951).   796

A40. ANON., Electrical heating of screens—The cure for blinding. *Brit. Clywkr.* **62** [741], 311, January (1954).   687

A41. ANON., The Hewitt stiff-plastic brick machine. *Brit. Clywkr.* **67** [794], 50, June (1958).   718

A42. ANON., Building Research Station Survey: The costs of mechanical plant. *Brit. Clywkr.* **67** [798], 167, October (1958).   263

A43. ANON., The 'Atritor' unit pulveriser: Its application to the ceramic industry. *Ceramics* 397, October (1950).   682

A44. ANON., The manufacture of ceramic colours. *Ceramics* 509, December (1950).   607, 608, 610, 615

A45. ANON., A viscometer for clay slips. *Ceramics* 540, December (1950). 324
A46. ANON., Potter's plaster. *Ceramics* **2** [23], 573, January (1951). 159
A47. ANON., Crazing and craze resistance. *Ceramics* **3** [26], 74, April (1951). 581
A48. ANON., A development in transfer pressing. *Ceramics* **3** [26], 91, April (1951). 829
A49. ANON., China clay—Past and present. *Ceramics* **3** [29], 236, July (1951). 241
A50. ANON., The testing of clays. *Ceramics* **3** [30], 303, August (1951). 299
A51. ANON., Sapphire laboratory crucibles. *Ceramics* **3** [30], 335, August (1951). 1139
A52. ANON., Fluxes for ceramic bodies. *Ceramics* **3** [32], 401 (1951). 440
A53. ANON., Sampling of dust and fume from chimney stacks. *Ceramics* **9** [113], 22, July (1958). 953
A54. ANON., Some developments in the mixing and shaping of pottery bodies. *Ceramics* **3** [34], 504 (1951). 716
A55. ANON., Clean producer gas from bituminous coal. *Ceramics* **3** [34], 533 (1951). 919
A56. ANON., Carbon—A refractory and material of construction. *Ceramics* **4** [37], 7, March (1952). 1249, 1252
A57. ANON., Laboratory porcelain. *Ceramics* **4** [41], 201, July (1952). 472, 1116
A58. ANON., Carblox Limited—New tunnel kiln. *Ceramics* **4** [42], 31, August (1952). 1253
A59. ANON., Frits—Their preparation and applications. *Ceramics* **5** [49], 5, March (1953). 1059
A60. ANON., Salt glazed ware. *Ceramics* **5**, 162 (1953). 601
A61. ANON., A new ceramic insulating material for the manufacture of aero-engine sparking plugs. *Ceramics* **6** [61], 37, March (1954). 1199
A62.—
A63. ANON., Inks for high speed marking. *Ceramics* **6** [64], 184, June (1954). 818
A64. ANON., A lining machine. *Ceramics* **6** [68], 376, October (1954). 808
A65. ANON., The use of noble metals in the ceramics industries. *Ceramics* **6** [72], 536, February (1955). 650, 653
A66. ANON., Inks for ceramic base printed circuits. *Ceramics* **7** [73], 26, March (1955). 655–657
A67. ANON., *Ceramics* 156, June (1955). 436, 446
A68. ANON., Metal-ceramic combinations. *Ceramics* **7** [77], 221, July (1955). 512
A69. ANON., Vapour deposition of metals on ceramic particles. *Ceramics* **7** [78], 271, August (1955). 1145

A70. ANON., Improved method for applying cermets. *Ceramics* 512
7 [80], 357, October (1955).
A71. ANON., Refractories for kilns for the ceramic industries. 900
Ceramics **7** [81], 383, November (1955).
A72. ANON., Electrical porcelain and its manufacture. *Ceramics* 474
8, [85], 525, March (1956).
A73. ANON., Refractory uses of boron and titanium carbides 515
bonded with metals. *Ceramics* **8** [91], 184, September
(1956).
A74. ANON., Spray glazing mechanisation at Johnson Brothers. 795
Ceramics **8** [91], 190, September (1956).
A75. ANON., Improved clay cleaning. *Cer. Age* **38** [6], 167 249
(1941). Reprinted in *Chem. Age* **46** [1181], 89, 14
February (1942).
A76. ANON., Salvaging glaze materials by an electrostatic process. 796
Cer. Age **28**, 143 (1936).
A77. ANON., Delayed crazing. *Cer. Ind.* **19** [5], 527 (1928). 557
A78. ANON., Continuous electric decorating kiln. *Cer. Ind.* 930
March (1932).
A79. ANON., Lithium compounds. *Cer. Ind.* February (1939). 105, 124
A80. ANON., *Cer. Ind.* **38** [1], 85 (1942). 588
A81.—
A82. ANON., Ceramic materials issue. *Cer. Ind.* **40** [1], 114 460
(1943).
A83. ANON., X-ray uncovers hidden defects. *Cer. Ind.* July 341
(1944).
A84. ANON., *Cer. Ind.* **48** [4], 97 (1947). 915, 918
A85. ANON., Develop super-hot furnaces for laboratory use. 983
Cer. Ind. **52** [2], 27 (1948).
A86. ANON., *Cer. Ind.* **77**, March (1950). 452
A87. ANON., Using air to save slip pumps. *Cer. Ind.* **54** [3], 80 704
(1950).
A88. ANON., Makes automatic tile pressing possible. *Cer. Ind.* 744
93, May (1950).
A89. ANON., Ceramic materials. A complete directory. *Cer.* 90–156, 169,
Ind. January (1952). 192, 436, 452,
590
A90. ANON., Colour tests for clays. *Chem. Age* **44** [1645], 136, 307
20 January (1951).
A91. ANON., Crusilite. New silicon carbide furnace element. 931
Chem. Age **71** [32], 377 (1954).
A92. ANON., Metal boride developments. *Chem. Age* **79** [2012], 512
226, February (1958).
A93. ANON., Materials of construction for chemical engineer- 1107
ing equipment. *Chem. & Met. Eng.* 97, September
(1944).
A94. ANON., Pyrophyllite. *Clay Prod. News (Canada)* **19** [1] 95
January (1946).
A95. ANON., *United States Government Master Specification for* 450

Plumbing Fixtures. Dep. Comm. Bureau of Standards. Fed. Spec. Board. No. 448.

A96. ANON., Crushing and grinding simply explained. *Edgar Allen News* **1** [8], 75, Mid-June (1940).   662

A97. ANON., *Electr. Rev.* 11 October (1946).   669

A98. ANON., Firing pottery. *Electr. Rev.* **132** [3408], 375, March (1943).   930, 1028

A99. ANON., Firing china glost. *Electr. Rev.* **134** [3455], 182, February (1944).   1022

A100. ANON., Editorial. Fine grinding in the potteries. *Engineer* **134**, 22 (1922).   667, 671

A101. ANON., Der technische Fortschritt: Auslaufbecher zur Bestimmung der Viskosität. *Euro-Ceramic* **8** [6], 144, June (1958).   324

A102. ANON. Bestimmung der Druckfestigkeit feuerfester Baustoffe. *Euro-Ceramic* **8** [6], 148, June (1958).   357

A103. ANON., *Producer Gas.* Arrow Press Student Publication No. 4. A 'Gas Times' Publication, 30 November (1951).   *916*, 917

A104. ANON., Le 'retirement' des glaçures. *L'Ind. Cér.* [455], 203, July-August (1954)   797

A105. ANON., *L'Ind. Cér.* [470], 313, December (1955).   *959*, 1000

A106. ANON., Service des Recherches Techniques de la S.F.C. Renseignements théoriques et pratiques concernant la résistance au choc thermique des produits vitrifiés. *L'Ind. Cér.* [493], 11, 2 January (1958).   469, 472, 1165

A107. ANON., Jointing of stoneware pipes. *Ind. Chem.* **29** [344], 415, September (1953).   1110

A108. ANON., Reliability of 'workability index' of fire-clay and plastic refractories. *Ind. Heat.* **14**, 1336 (1947).   328

A109. ANON., *Radiant Heat Drying by Gas.* Industrial Uses of Gas Series No. 14.   838

A110. ANON., Ceramic turbine blades. *Iron Age* **77**, 10 February (1949).   140

A111. ANON., Making liquid gold. *Jap. Pat.* 2602 (1924).   651

A112. ANON., Resistance to abrasion of a sillimanite porcelain die when used for extruding clay columns. *J. Franklin Inst.* **213**, 323 (1932).   165

A113. ANON., *J. Franklin Inst.* **244**, 147 (1947).   506

A114. ANON., What effect has warm water upon the washing process? *Ker. Rund.* **31**, 416 (1923).   250

A115. ANON., *Sprechs.* 25 (1926); *Ker. Rund.* 159 (1927).   658

A116. ANON., Matt- und Ausscheidungsglasuren für Ofenkacheln. *Ker. Rund.* **45** [6], 175 (1937).   594

A117. ANON., Über die Verzierung von Steingut. *Ker. Rund.* **45** [37], 419 (1937).   802

A118. ANON., Glanzgoldherstellung. *Ker. Rund.* **47** [19], 215, May (1939).   651, 653

A119. ANON., Versuche mit Schmelzflüssen für Aufglasurfarben. *Ker. Rund.* **49** [17], 171 (1941).   612

A120.—

A121. ANON., *Investigation of Moulding Properties, Mechanical and Electrical Characteristics of Pyrophyllite.* Ministry of Supply Development Contract No. 6/WT/4205/CB. 16 (Carried out by Bullers Ltd. and Admiralty Signals and Radar Establishment, Haslemere).   94

A122. ANON., China clay production in Cornwall. *Pottery Gazette* 1665, November (1950).   241

A123. ANON., Ball clay production in S. Devon. *Pottery Gazette* October (1951).   256

A124. ANON., Ball clay from the Bovey Basin. *Pottery Gazette* 1406, September (1952).   255

A125. ANON., Dust control by new wet system. *Pottery Gazette* 83 [971], 622, May (1958).   715

A126. ANON., The dangers of washing up. *Pottery Gazette* July (1958).   609

A127. ANON., Engraving for pottery. *Pottery Gazette* 84 [980], 253, February (1959).   821

A128. ANON., Crayoning on glazed ware. *Pottery and Glass.* 31 [7], 220 (1953).   806

A129. ANON., Notes on mat glazes. *Ramsden Monthly Bull.* [14], 3, April (1928).   594–595

A130. ANON., Materials and processes of the ceramic industries. Decoration. *Ramsden Monthly Bull.* [98], April (1936).   635

A131. ANON., Materials and processes in the ceramic industries. *Ramsden Monthly Bull.* [100], June (1936).   653

A132.—

A133. ANON., Body mixing. *Ramsden Monthly Bull.* [138], March (1949).   1081

A134. ANON., Decoration. *Ramsden Monthly Bull.* [148], February (1950).   821, 823, 827, 828

A135. ANON., Lithographs. *Ramsden Monthly Bull.* [149], March (1950); *Ibid.* [150], April (1950).   826

A136. ANON., Chromium compounds in the ceramic industries. *Ramsden Monthly Bull.* [154], August–September (1950); *Ibid.* [155], October (1950).   618, 623

A137. ANON., Antimony compounds in the ceramic industries. *Ramsden Monthly Bull.* [156], November (1950).   634

A138. ANON., Iron compounds in the ceramic industries. *Ramsden Monthly Bull.* [160], March (1951).   632, 642

A139. ANON., Electrical porcelain. *Ramsden Monthly Bull.* [165], August–September (1951).   474

A140.—

A141. ANON., Barium compounds in the ceramic industries. *Ramsden Monthly Bull.* [177], October (1952).   456

A142. ANON., Gold compounds in the ceramic industries. *Ramsden Monthly Bull.* [179], December (1952).   648–653

A143. ANON., Some less familiar elements in the ceramic industries. *Ramsden Monthly Bull.* [181], February (1953).   640

A144. Anon., Organic compounds in the ceramic industries. *Ramsden Monthly Bull.* [185], June (1953).   783, 784

A145. Anon., Body preparation. *Ramsden Monthly Bull.* [237], March (1958).   326

A146. Anon., The firing of pottery. *Ramsden Monthly Bull.* [246–251], January–June (1959).   903, 930, 931, 967, 1020

A147. Anon., Dry process refractories—Cause of cracking. *Refr. J.* [3], 84, March (1950).   750

A148. Anon., Clay Convention works visit. English Clays, Lovering, Pochin & Co. Ltd., Lee Moor Works. *Refr. J.* 302, July (1951).   242

A149. Anon., *Refr. J.* **31** [1], 3 (1955)   111

A150. Anon., Refractory materials and atomic energy. *Refr. J.* **33** [12], 550 (1957)   1139

A151. Anon., Charcoal from eucalyptus wood. *S. Afr. Min. & Eng. J.* **54**, Pt. II [2641], 29 (1943).   902

A152. Anon., Ball mill: Vibratory laboratory type (Introduced by Griffin and Tatlock, Kemble St., London). *S. Afr. Min. & Eng. J.* **55**, Pt. II [2700], 231 (1944).   671

A153. Anon., *Sprechs.* 696 (1911).   610

A154. Anon., *Sprechs.* [48], 996 (1911).   610

A155. Anon., Fragekasten. *Sprechs.* [48], 833, November (1933).   650

A156. Anon., *Sprechs.* [38], 560, September (1936).   786

A157. Anon., The decoration of ceramics with noble metals. *Sprechs.* **84**, 127, 150 (1951).   655

A158. Anon., Bedeutung der Gleichgewichtsdiagramme für den Keramiker. *Sprechs.* **86** [10], 251 (1953).   216

A159. Anon., Die Herstellung von Glanzgold. *Sprechs.* **89** [1], 3 (1956).   651, 652

A160. Anon., Rosa gefärbtes Porzellan. *Sprechs.* **89** [24], 563, December (1956).   624, 625, 631, 648

A161. Anon., Ionenaustauscher-Anlagen zur Wasservollentsalzung. *Sprechs.* **91** [4], 73, 20 February (1958).   152

A162. Anon., Die Pralltellermühle 'Original Pallman'. *Sprechs.* **91** [4], 75, 20 February (1958).   665

A163. Anon., *Techn. Rund.* [41], 585 (1911).   610

A164. Anon., Alumina in industry. *The Times Review of Industry* 18 January (1948).   114

A165. Anon., *Rock Products* **29** [21], 60; *Tonind.-Ztg.*, **51**, 520 (1927).   160

A166. Anon., Ringofen-Zickzackofen. *Tonind.-Ztg.* **61** [17], 199 (1937).   995, 997

A167. Anon., Zur Herstellung neuartiger Tonerzeugnisse durch Heisspressung. *Tonind.-Ztg.* **64** [7], 44 (1940).   752

A168. Anon., Mould- and die-oils for the ceramic industry. *Tonind.-Ztg.* **64** [68], 540 (1940).   733

A169.—

A170.—

A171. Anon., *Trans. Brit. Cer. Soc.* **50** [2], 79a, February (1951). 470
A172. Anon., *Formende Hände.* Tonindustrie, Heisterholz (1952). 721
A173. Anon., Resistance des couleur de moufle. *Verre et Silicates Ind.* **6**, 431 (1935). 380
A174. M. W. Abberley, Machines for lining and banding and for stamping. *Trans. Brit. Cer. Soc.* **57** [9], 534 (1958). 806, 809, 811
— L. L. Abernethy, see J. H. Handwerk.
A175. P. J. Adams, The origin and evolution of coal. *Discovery* **16** [5], 190 (1955). 903, 906
A176. B. Aggeryd, *Chalmers Tekn. Högskolas Handl.* (1945). 183
A177. L. Ainsworth, A method for investigating the structure of glazes based on a surface measurement. *Trans. Brit. Cer. Soc.* **55** [10], 661 (1956). 378, 557, 558
A178. K. W. Airey Ltd., Milton, Stoke-on-Trent. 740, 772, 776
A179. Albright and Wilson Ltd., Calgon—Properties and applications. 761
A180. F. H. Aldred and A. E. S. White, Some equipment for ceramic research. *Trans. Brit. Cer. Soc.* **58** [4], 199 (1959). 277, 338
— H. W. Alexander, see A. I. Andrews.
A181.—
— L. T. Alexander, see S. B. Hendricks.
A182. Allen Jigger Corporation, Syracuse 9, New York, U.S.A. 726
— A. W. Allen, see E. D. Lynch.
— E. T. Allen, see A. L. Day.
— Kirby K. Allen, see Ercel B. Hunt.
A183. L. Allen and Johnson, Matthey & Co., Ltd., Pat. 399922, 27 May (1932). 830
A184. Allied Engineering Division of Ferro Enamel Corporation, 4150 East 56a St., Cleveland 5, Ohio, U.S.A. 1041
A185. Soc. des Etablissements René Amand et Cie (France), Improvements in sliding hearths for tunnel furnaces. *Brit. Pat.* 704, 386, 24 February (1954); *Ibid.* 704, 448, 24 February (1954). 1029, 1287
A186. American Lava Corporation, (a) *Data and Property Chart* No. 416. (b) *Alsimag* 72 and 202. (c) *General Information.* Chattanooga, Tennessee, U.S.A. 488, 1127, 1179–1189, *1203*
A187. American Society for Testing Materials (A.S.T.M.), *Sieves for Testing Purposes.* Standard E11-39. 1301
— Adrian G. Allison, see Stephen D. Stoddard.
A188. Noel L. Allport, *Colorimetric Analysis.* Chapman & Hall, London (1947). 267
A189. A. Anable, Dorr classifiers for clay washing. *J. A. Cer. S.* **11**, 791 (1928). 245
A190. A. Anable, Modernized treatment improves results in a non-metallic industry—I. *Eng. and Min. J.* 6 April (1929). 245

— O. ANDERSEN, see N. L. BOWEN.
A191. H. V. ANDERSON, Precious metal paints. *Cer. Age*, August (1949).   647
A192. J. P. ANDERSON, The Belgian kiln and the underfeed stoker. *Brit. Clywkr.* **67** [798], 178, October (1958).   911
— K. ANDERSON, see J. A. HEDVALL.
— N. ANDÖ, see S. TARMARU.
A193. A. H. M. ANDREASEN, Einige Beiträge zur Erörterung der Feinheitsanalyse und ihrer Resultate. *Wiss. Arch. Landwirstsch.* Abt. A. *Arch. Pflanzenbau* **6**, 230 (1931).   284
A194–A202 inclusive follow after A214.
A203. A. H. M. ANDREASEN and S. BERG, Beispiele der Verwendung der Pipettemethode bei der Feinheitsanalyse unter besonderer Berücksichtigung der Feinheitsuntersuchungen von Mineralfarben. Beihefte zu den *Z. Ver. deut. Chemiker* [14]. Verlag Chemie, G.m.b.H, Berlin W35 (1935).   *285*
A204. L. W. ANDREW and E. A. C. CHAMBERLAIN, Radiant heating for industrial processes. *J. Inst. Fuel* **17**, 41 (1944).   836
A205. A. I. ANDREWS, G. L. CLARK and H. W. ALEXANDER, Progress report on determination of crystalline compounds causing opacity in enamels by X-ray methods. *J. A. Cer. S.* **14**, 634 (1931).   590
A206. A. I. ANDREWS, G. L. CLARK and H. W. ALEXANDER, The determination by X-ray methods of crystalline compounds causing opacity in enamels. *J. A. Cer. S.* **16**, 385 (1933).   590
— A. I. ANDREWS, see M. K. BLANCHARD.
A207. N. K. ANTONEVICH, N. E. KOLYUN and K. V. SALAZKINA, Pink pigment for porcelain and similar ware. *Russ. Pat.* 51907, 31 October (1936); *Chem. Abs.* **34**, 3896 (1940); *J. A. Cer. S.* Abs. [9], 218 (1940).   636
A208. A. V. ANTONOVIČ, Ofen zum Erhitzen von Proben bei Kurz- und Langzeitversuchen. *Zavodsk. Labor.* **15**, 618 (1949) (In Russian); *Schriftumsbericht Werkstofforschung* [3], 93 (1951).   983
A209. A. VON ANTROPOFF, Les formes usuelles du systeme périodique des éléments. *Annales Guébard-Séverin* **13**, (1937). Neuchâtel.   *1300*
A210. TAKEO AO, Studies of thermal and chemical changes of Kato kaolin. *J. Soc. Chem. Ind. Japan* **36** [1] 12–133 (1933); *J. A. Cer. S.* Abs. **12** [10–11], 402 (1933).   195
A211. APPL and LACHMANN, Aufbereitung und Aufbereitungsmaschinen. *Euro-Ceramic* **7** [10], 255 (1957); *Ibid.* **8** [1], 3; *Ibid.* **8** [2], 31 (1958).   699, 700, 707, 712
A212. C. A. ARENBERG, H. H. RICE, H. Z. SCHOFIELD and J. H. HANDWERK, Thoria ceramics. *Bull. A. Cer. S.* **36** [8], 302 (1957).   508
— A. ARBO, see K. E. SMITH.

A213. ASSOCIATION FRANÇAISE DE NORMALISATION (A.F.N.O.R.). 1301
Standard Control Sieves (1938).
A214. ASHWELL AND NESBIT LTD., Iron Fireman Automatic Under- 910
feed Stokers. Barkby Rd, Leicester.
A194. ASSOCIATED LEAD MANUFACTURERS LTD., Low Solubility 572
Lead Frits for Ceramic Glazes. London, Newcastle and
Chester (1952).
A195. C. ASTARITA, Contribution a l'étude du choc thermique. 361
Bull. Soc. Fran. Cér. [41], 3 October–December (1958).
A196. F. L. ATKIN, Infra-red drying. Gas J. 20 and 27 Septem- 837
ber (1944).
A197. LEON M. ATLAS, Effect of some lithium compounds on 418
sintering of MgO. J. A. Cer. S. **40** [6], 196 (1957).
A198. C. R. AUSTIN and G. H. DUNCOMBE, JR., Centrifugal slip 767
casting. Bull. A. Cer. S. **20**, 113 (1941).
— CHESTER R. AUSTIN, see JOHN D. SULLIVAN.
A199. J. B. AUSTIN, The physics and chemistry of firing ceramic 856, 857, 869,
ware. Bull. A. Cer. S. **14** [5], 157 (1935). 881
A200. GRUBE 'AUSTRIA', Haidhof Opf, Germany. 49
A201. A. I. AVGUSTINIK and O. P. MCHEDLOV-PETROSYAN, Zhur. 187
Priklad. Khim. (J. Applied Chem. U.S.S.R.) **20**, 1125
(1947).
A202. A. I. AVGUSTINIK and V. S. VIGDERGAUZ, Properties of talc 91, 192
during heating. Ogneupory **13** [5], 218 (1948); J. A.
Cer. S. Abs. 66 (1949).

— R. A. BACH, see J. H. HANDWERK.
B1. BACHARACH INDUSTRIAL INSTRUMENT COMPANY,
   (a) Combustible Gas Analyser, Bull. 255. 954
   (b) Electric CO_2 Meter, Bull. 256. 954
   (c) Electric CO Meter, Bull. 257. 954
   (d) Flow meters, Bull. 90. 952
   7000 Bennett St., Pittsburgh, Pa., U.S.A.
— B. L. BAILEY, se R. R. RIDGEWAY.
B2. T. C. BAILEY, F. J. BATTERSHILL and R. J. BRESSEY, Liquid 925–928
Fuel for High Temperature Processes. Paper read at the
Conference on 'Fuel and the Future', London, 8, 9 and
10 October, 1946, Institute of Petroleum, Manson
House, 26 Portland Place, London, W.1.
B3. G. J. BAIR, Constitution of lead-oxide–silica glasses. I— 214
Atomic arrangement. J. A. Cer. S. **19** [12], 339 (1936).
B4. BAIRD AND TATLOCK (LONDON) LTD., 14 St. Cross St., E.C.1. 289
B4.1. G. J. BAJO and E. HÉVÉZI, Studio for Mural Design, 9, 1083
Edeline Ave., London, S.W.16.
— C. J. W. BAKER, see B. E. VASSILIOU.
B5. BAKER PERKINS LTD., Engineers, Westwood Works, Peter- 698
borough.
— W. J. BALDWIN, see C. A. BEST.

B6. D. Balarew, *Koll.-Z.* **104**, 78 (1943).   183
— G. H. Baldwin, see R. T. Stull.
B7. J. C. Banerjee and D. N. Nandi, Comparative study of spalling test methods for refractories. *Refr. J.* **33** [11], 488 (1957).   358
B8. G. J. Barker and Emil Truog, Improvement of stiff-mud clays through pH control. *J. A. Cer. S.* **21**, 324 (1938); Factors involved in improvement of clays through pH control. *Ibid.* **22**, 308 (1939).   63, 64
B9. George J. Barker and Emil Truog, Method of determining the pH of ball clays. *Bull. A. Cer. S.* **21**, 263 (1942).   63, 320, 321
— L. R. Barrett, see C. W. Parmelee.
B10. W. E. Barret, Glazes colored by molybdenum. *J. A. Cer. S.* **8** [5], 306 (1925).   595
B11. L. E. Barringer, The relation between the constitution of a clay and its ability to take a good salt glaze. *Trans. A. Cer. S.* **4**, 211 (1902).   598, 601
— E. Barnes, see P. Murray.
B12. M. Barnick, *Strukturbericht* **4**, 71, 207 (1936).   5, 119
B13. C. E. Barry, *Wear Resistance Tests on Domestic Materials for Pebble Mill Linings.* Amer. Inst. Min. Met. Eng. Tech. Pub. No. 1948; *Min. Tech.* **10** (2), 8 (1946).   682
B14. W. P. Barsakowski, Neues bei der Untersuchung des Mullits. *Nature (Moscow)* **42** [2] (1933); abs. in *Silikat Technik.*   112
B15. N. Barsby, Salt glazing in tunnel kilns. *Clay Prod. J. (Australia)* **25** [1], 26, November (1957).   1041
— T. F. W. Barth, see L. W. Greig.
B16. J. A. Basmajian and R. C. DeVries, Phase Equilibria in the system $BaTiO_3$–$SrTiO_3$. *J. A. Cer. S.* **40** [11], 373 (1957).   520
B17. B. Bassa, Über den Massenaufbau für Steinzeugplatten. *Ker. Rund.* **46** [22], 243–246 (1928).   436
B18. R. W. Batchelor, Steatite in Electro-Ceramics. *Ceramics* **9** [115], 11, September (1958).   499
— W. H. Bateman, see E. A. Jamison.
B19. Th. F. Bates, *Amer. Mineralogist* **35**, 225 (1950).   *21*
B20. Thomas F. Bates and J. J. Comer, Electron microscopy of clay surfaces. *Proceedings of the Third National Conference on Clays and Clay Minerals* p. 395. National Academy of Sciences—National Research Council, Washington (1955).   *14/15, 17, 18, 23*, 309
— F. J. Battershill, see T. C. Bailey.
B21. A. Baudran, Le séchage continue des barbotines par pulvérisation. *Bull Soc. Fran. Cér.* [21], 29, October–December (1953).   707
B22. Ernst P. Bauer, Ueber den Einfluss von Zinkoxyd auf bleifreie, ungefrittete Glasuren für Sanitätssteingut. *Ber. D. K. G.* **3**, 286 (1922).   588
— Stefan George Bauer, see Raymond Francis Walker.

B23. H. N. BAUMANN, JR., An electric spalling furnace. *J. A. Cer. S.* **13**, 167 (1930).   984

B24. R. J. BEALS and R. L. COOK, Low-expansion cordierite porcelains. *J. A. Cer. S.* **35** [2], 53 (1952).   484, 490

— R. BEARD, see A. DINNIE.

— RICHARD BEAVER, see J. E. WISS.

B25. HUBERT O. DE BECK, Consulting Geological Engineer, Custer, S. Dakota, U.S.A.   687, 696

B26. K. BECK, F. LOWE and P. STEGMULLER, *Arb. Kaiser Ges. Amt.* **33**, 203 (1910).   614

B27. C. A. BEEVERS and M. A. S. ROSS, The crystal structure of 'beta alumina' $Na_2O.11Al_2O_3$. *Z. Krist.* **97**, 59 (1937).   114

B28. LAVINO (LONDON) LTD., *Beken Mixers.* 103 Kingsway, London, W.C.2.   697

B29. W. C. BELL and D. H. MCGINNIS, Large lightweight structural clay building units. *Bull. A. Cer. S.* **30** [10], 333 (1951).   429

B30. WILLIAM C. BELL, RICHARD D. DILLENDER, HAROLD R. LOMINAC and EDWARD G. MANNING, Vibratory compacting of metal and ceramic powders. *J. A. Cer. S.* **38** [11], 396 (1955).   1139, 1145

B31. BELL CLAY & COMPANY, *Bell Sergeant China Clay Data.* Tennessee, U.S.A.   35, 40

B32. LUTZ BENDA, Der Tunnelofen in der Steingut- und Sanitär-Industrie. *Ber. D. K. G.* **23**, 437 (1942).   1012

B33. LUTZ BENDA, Die Ölfeuerung in der Keramik. *Sprechs.* **89** [6], 111, March (1956).   925, 928

— B. BENGSTON, see R. JAGITSCH.

B34. BENNIS COMBUSTION LTD., *Air-draught Stoker and Self Cleaning Furnace.* Little Hulton, Bolton.   910

B35. DONALD C. BENNETT, Automatic controls and tunnel kilns. *Bull. A. Cer. S.* **34** [3], 82 (1955).   949, 1018

B36. DWIGHT G. BENNETT and W. A. GRAFF, Slipcast ceramic combustion tubes. *Cer. Ind.* **48** [4], 118 (1947).   755

B37. H. BENNETT, W. G. HAWLEY and R. P. EARDLEY, Rapid analysis of some silicate materials. *Trans. Brit. Cer. S.* **57** [1], 1, January (1958).   266

B38. BIG HORN BENTONITE COMPANY, Greybull, Wyoming.   260

B39. BERDEL, Praktische Ausführung der technischen rationellen Analyse. *Sprechs.* **36**, 1337, 1407 (1903).   84, 299, 316

— M. BERG, see H. H. GREGER.

B40. SÓREN BERG, The diver method for determination of particle-size distribution. *Transactions of the International Ceramic Congress,* Holland, 1948, p. 152.   288, 289

— S. BERG, see A. H. M. ANDREASEN.

B41. A. S. BERMAN and M. E. SHENINA, New method for determining the moisture content of ceramic masses and controlling their workability. *Steklo i Keram.* **5** [4], 11–15 (1948); Abs. *J. A. Cer. S.* **48** (1949).   328

B42. Société L. Bernardaud and Company, Elimination of the oxidising phase in the firing of glazed whiteware. *Fr. Pat.* 927876, 19 May (1947); 17 April (1946). — 860

B43. M. H. Berns and C. E. Milligan. Refractories and the pusher-type kiln. *Bull. A. Cer. S.* **31** [9], 317 (1952). — 1029

B44. C. A. Best, K. Traub and W. J. Baldwin, Effect of varying firing temperature and furnace atmosphere on the ceramic and electrical properties of dielectric titanate bodies. Nineteenth Symposium on Ceramic Dielectrics, Rutgers University. *Cer. Age* **64** [6], 31 December (1954). — 520

B45. C. Bettany and H. W. Webb, Some physical effects of glazes. I—The effect of the glaze on the mechanical strength of electrical porcelain. *Trans. Brit. Cer. Soc.* **39**, 312 (1939–1940). — 550

— A. G. Betteley, see P. W. Dager.

B46. George C. Betz, Zircon as a Ceramic Material. *Foote-Prints* **9** [2], 15, December (1936). — 478, 479

B47. M. Bichowsky and J. Gingold, Ueber die physikalischen Eigenschaften keramischer Massen aus Speckstein und Ton. *Sprechs.* **64** [37], 679, September (1931). — 488, 496

B48. A. Biddulph, The milling of porcelain enamels. *Enamelist* **22** [15], 7 (1945). — 678

— Hans O. Bielstein, see Berthold C. Weber.

B49. F. M. Biffen, Sodium and potassium determination in refractory materials using flame photometer. *Anal. Chem.* **22** [8], 1014 (1950). — 272

— H. C. Biggs, see J. C. Stein and Company Ltd.

B50. A. Bigot, Volume changes under heat treatment of kaolins, clays, bauxites etc. *Céramique* **26**, 56 (1923). — 80

— W. Bilke, see U. Hofmann.

B51. E. C. Bingham, *Am. Soc., Testing Materials Proc.* **19** [2] (1919); *Ibid.* **20** [2] (1920); *J. A. Cer. S.* **5**, 350 (1922). — 327

B52. Charles F. Binns, The development of the 'Matt' glaze. *Trans. A. Cer. S.* **5**, 50 (1903). — 594

B53. Charles F. Binns, Mat glazes at high temperatures. *Trans. A. Cer. S.* **7**, 115 (1905). — 595

B54. Charles F. Binns, The function of boron in the glaze formula. *Trans. A. Cer. S.* **10**, 158 (1908). — 527

B55. Charles F. Binns, A plea for bone china. *Trans. A. Cer. S.* **12**, 175 (1910). — 460

B56. Charles F. Binns, Contribution to discussion of 'Matt Glazes' by W. N. Claflin. *Trans. A. Cer. S.* **12**, 541 (1910). — 594

B57. Charles F. Binns, *Manual of Practical Potting.* London (1922). — 608, 632, 649, 653, 654

B58. Charles F. Binns and Frobisher Lyttle, A vermilion colour from uranium. *J. A. Cer. S.* **3**, 913 (1920). — 638

B59. D. B. BINNS. Routine ceramic testing. *Claycraft* **23** [10], 495, July (1950). 381

B60. RAYMOND E. BIRCH, Forming pressure of dry pressed refractories. I—The effect of pressure variations on the properties of green and dry bodies. *J. A. Cer. S.* **13**, 242 (1930); II—The effect of pressure variations on fired properties. *Ibid.* **13**, 831 (1930). 750, 751

B61. RAYMOND E. BIRCH, Reaction temperatures between various types of refractory brick. *Amer. Refr. Inst. Tech. Bull.* [52], October (1934). 892, 894–895

B62. RAYMOND E. BIRCH, Phase-equilibrium data in the manufacture of refractories. *J. A. Cer. S.* **24** [9], 271 (1941). 417, 418, 419, 1219

B63. RAYMOND E. BIRCH, Refractories of the future. *Ohio State Univ. Eng. Exp. Stn. News* **17** [2], 3 October (1945). 889, 890

— R. E. BIRCH, see D. K. STEVENS.

— R. E. BIRCH, see F. A. HARVEY.

B64. BIRLEC LTD., Birlec Works, Tyburn Rd, Erdington, Birmingham. 1030

B65. F. VON BISCHOFF, *Z. anorg. Chem.* **250**, 10 (1942). 185

— J. BISCOE, see B. E. WARREN.

B66. J. BISCOE and B. E. WARREN, X-ray diffraction study of soda–boric oxide glass. *J. A. Cer. S.* **21** [8], 287 (1938). 208, *210*

B67. J. BISCOE, C. S. ROBINSON, JR. and B. E. WARREN, X-ray study of boric oxide-silica glass. *J. A. Cer. S.* **22** [6], 180 (1939). 204

B68. J. BISCOE, M. A. A. DRUESNE and B. E. WARREN, X-ray study of potash-silica glass. *J. A. Cer. S.* **24** [3], 100 (1941). 206

B69. J. BISCOE, A. G. PINCUS, C. S. SMITH, Jr. and B. E. WARREN, X-ray study of lime phosphate and lime borate glass. *J. A. Cer. S.* **24** [4], 116 (1941). 208, 210

B70. J. BISCOE, X-ray study of soda-lime-silica-glass. *J. A. Cer. S.* **24** [8], 262 (1941). *207*, 208

— ALF BJORKENGREN, see J. A. HEDVALL.

B71. A. R. BLACKBURN, Plastic pressing, a new production tool for the ceramic industry. *Bull. A. Cer. S.* **29** [6], 230 (1950). 720

B72. A. R. BLACKBURN and T. S. SHEVLING, Fundamental study and equipment for sintering and testing of cermet bodies. V—Fabrication, testing and properties of 30 chromium-70 alumina cermets. *J. A. Cer. S.* **34** [11], 327 (1951). 512, 1145

— A. R. BLACKBURN, see GEORGE A. LOOMIS.

B73. J. F. BLACKER, *The A.B.C. of English Salt-Glaze Stoneware.* Stanley Paul, London (1922). 601

B74. L. C. F. BLACKMAN, Some factors involved in the preparation of a reproducible magnesium–manganese ferrite. *Trans. Brit. Cer. Soc.* **56** [11], 624 (1957). 521

B75. L. BLAIR, Investigation of a new decal-application medium. *Bull. A. Cer. S.* **27**, 374 (1948). 827

B76. L. R. BLAIR and C. H. MOORE, Jr., Method of varying the spectral absorption of atomic transition elements in silicate melts. *Cer. Age* **54** [5], 295, November (1949).   232, *233*

B77. M. K. BLANCHARD and A. I. ANDREWS, Fundamental properties of soluble salts in enamel liquors. *J. A. Cer. S.* **27** [1], 17 (1944).   783

— M. K. BLANCHARD, see C. G. HARMAN.

— W. BLASCHE, see R. FRICKE.

B78. SCHAMOTTEWERKE BLASCHEK Eppertshansen / Hessen, Germany.   51

B79. BLAW KNOX LTD., 94 Brompton Road, London, S.W.3.   689

B80. E. C. BLOOR, The composition and properties of electrical ceramics. *Ceramics, a Symposium.* British Ceramic Society (1953).   479, 488, 493–497, 516–520, 1195–1197, 1207

B81. E. C. BLOOR, Glaze composition, glass structural theory, and its application to glazes. *Trans. Brit. Cer. Soc.* **55** [10], 631 (1956).   538, 548, 561, 564–566

B82. E. C. BLOOR, Plasticity: A critical survey. *Trans. Brit. Cer. Soc.* **56** [9], 423 (1957).   63, 328, 329

B83. E. C. BLOOR, Plasticity in theory and practice. *Trans. Brit. Cer. Soc.* **58** [7–8], 429 (1959).   28, 56

B84. R. B. BLOOR, *Trans. Brit. Cer. Soc.* **43**, 63 (1944); *Brit. Pat.* 565989.   845

B85. W. A. BLOOR, K. L. GOODALL and H. W. WEBB, Dust investigations in the ceramic industry. *Trans. Cer. Soc.* **38** [1], 1 (1939).   1323

— E. B. BLOUNT, see K. FORREST PENCE.

— D. P. BOBROWNIK, see P. P. BUDNIKOW.

B86. BODIN and GAILLARD, *Brit. Clywkr.* **40**, 168 (1931).   851

B87. SIEGFRIED BÖTTNER, Schlickerschutz für Töpferscheiben. *Sprechs.* **91** [1], 2, 5 January (1958).   721

B88. G. A. BOLE, A method for eliminating water marks from tableware. *J. A. Cer. S.* **14**, 454 (1931).   830

— G. A. BOLE, see C. A. COWAN.

— G. A. BOLE, see C. M. NICHOLSON.

B89. HERMANN BOLLENBACH, Studien zur chemischen und rationellen Analyse der Tone. *Sprechs.* **41**, 340, 351 (1908).   299

B90. HELEN BLAIR BONLETT and R. R. THOMAS, JR., A Study of the mineralogical and physical characteristics of two lithia-zirconia bodies. *J. A. Cer. S.* **17** [2], 17–20 (1934).   503

B91. D. G. R. BONNELL and B. BUTTERWORTH, *Clay Building Bricks of the United Kingdom.* Nat. Brick Advis. Council. Paper 5, H.M.S.O., London (1950).   1070–1071

B92. F. T. BOOTH and G. N. PEEL, The principles of glaze opacification with zirconium silicate. *Trans. Brit. Cer. Soc.* **58** [9], 532 (1959).   590, 591

B93. BORAX CONSOLIDATED LTD., '20 mule team'. *Diary*, 1953.   131

B94. WILLIAM BOULTON LTD., Providence Engineering Works, Burslem, Staffs. — 671, *673*, 683

B95. G. BOUVIER and E. KALTNER, Selected examples of the examination of refractory materials with the help of radioactive isotopes. *Sixth International Ceramic Congress, Wiesbaden. Refr. J.* 35 [2], 83 (1959). — 345

— B. BOVARNICK, see *Symposium on Ceramic Cutting Tools*. **S219**

B96. E. J. BOWEN, *The Chemical Aspect of Light*. Clarendon Press, Oxford (1942). — 228

B97. N. L. BOWEN and O. ANDERSEN, *Amer. J. Sci.* 37, 487 (1914). — 198

B98. N. L. BOWEN and J. W. GREIG, The system Al_2O_3–SiO_2. *J. A. Cer. S.* 7 [4], 238 (1924). — 198, *217, 482*

B99. N. L. BOWEN and J. W. GREIG, *J. A. Cer. Soc.* 7, 238 (1924); *Ibid.* 16, 496 (1933). — 406

— N. L. BOWEN, see G. W. MOREY.
— N. L. BOWEN, see J. F. SCHAIRER.

B100. E. R. BOX, Screen printed transfers. *Ceramics* 536, December (1950). — 830

B101. E. R. BOX, Modern types of transfer. *Trans. Brit. Cer. Soc.* 57 [9], 541 (1958). — 831

B102. J. E. BOYD, JR., Pyrometric properties of spodumene–feldspar mixtures. *J. A. Cer. S.* 21 [11], 385–388 (1938). — 104, 579

B103. R. S. BRADLEY, The structural scheme of attapulgite. *Amer. Mineralogist* 25, 405 (1940). — 19, *24*

— R. S. BRADLEY, see GRIM.

B104. R. S. BRADLEY and B. K. MILLER, Prospecting, developing and mining semiplastic fire clay in Missouri. *Min. Tech.* July (1941). — 236, 257

— W. F. BRADLEY, see J. E. COMEFORO.

B105. BRADLEY AND CRAVEN LTD., Stiff plastic brickmaking machinery. *Clayworking Machinery*, Sec. 2. P.O. Box No. 21, Wakefield. — 664, *737, 740*

— CYRIL BRADSHAW, see IVAN B. CUTLER.
— W. L. BRAGG, see B. WARREN.

B106. W. L. BRAGG and J. WEST, *Proc. Roy. Soc.* A 111, 691 (1926). — 5, 139

B107. E. J. BRAJER, Effect of some ceramic techniques on the piezoelectric properties of barium titanate. *Bull. A. Cer. S.* 36 [9], 333 (1957). — 519, 520

B108. A. I. BRANISKI, Refractarité du system serpentine-dolomie. *L'Ind. Cér.* [491], 319, November (1957). — 419, 498

— C. A. BRASHARES, see W. F. ROCHOW.

B109. P. BRASSEUR, *Bull. soc. Chim.* 12, 412 (1945). — 189

— BRAY, see GRIM.

B110. BRAYSHAW FURNACES AND TOOLS LTD., Belle Vue Works, Manchester, 12. — *982*

B111. BRAYSHAW FURNACES AND TOOLS LTD., *Pocket Radiation Pyrometer*. Mulberry St., Hulme, Manchester. — 948

— J. G. BREEDLOVE, see J. E. COMEFORO.

B112. P. Brémond, *Bull. Soc. Fran. Cér.* [11], 4 (1951); *Transactions of the Third International Ceramic Congress*, 1952, p. 41. 538

B113. A. J. Breslin, The control of beryllium oxide in the ceramic industry. *Bull. A. Cer. S.* **30** [11], 395 (1951). 504, 1139

— R. J. Bressey, see T. C. Bailey.

B114. Joseph R. Bressman, Ceramic materials show promise for high temperature mechanical parts. *Materials & Methods* 65, January (1948). 506

B115. R. E. Brewer, *Comparison of Fine Series, Square Mesh Wire Test Sieves of Different Countries.* Bur. of Mines Report of Investigation 3766, July (1944). 1301

B116. G. W. Brindley, Stevensite, a montmorillonite type mineral showing mixed-layer characteristics. *Amer. Mineralogist* **40**, 239 (1955). 17

B117. G. W. Brindley and K. Hunter, The thermal reactions of nacrite and the formation of metakaolin, γ-alumina, and mullite. *Min. Mag.* **30** [228], 574, March (1955). 77

B118. British Standards Institute, *Test Sieves.* British Standard 410 (1943). 1301

B119. British Standards Institute, *Flow Measurement.* British Standard 1042 (1943). 952

B120. The British Thermostat Co. Ltd., *Teddington, Industrial Controls.* Sunbury-on-Thames, Middlesex. 950, 951

B121. The British Thomson-Houston Company Ltd., Brit. Pat. 551606. 578

B122. Bahngrell W. Brown, Preliminary investigation of Wyoming cordierite. *Bull. A. Cer. S.* **27** [11], 443 (1948). 121, 487

— G. G. Brown, see A. A. Klein.

B123. J. F. Brown, Spectrographic analysis of some ceramic materials. *Trans. Brit. Cer. Soc.* **57** [4], 218 (1958). 270

B124. Lawrence H. Brown, Preparing of slips and glazes for best results. Personal Communication. 783

B125. Ernest Brückner, Strukturzerstörer für Schneckenpressen. *Sprechs.* **91** [15], 351 (1958). 731

B126. H. Brückner, Propane and butane as industrial gases. *Gas und Wasserfach* **77** [25], 425 (1934). 920

B127. A. F. Bruins, Removal of detrimental fluorine compounds from the atmosphere in enamelling plants. *Ber. D.K.G. und Ver. deut. Emailfachleute* **29** [2], 52 (1952). 128

B128. B. Brunowski, *Acta physicochim. U.S.S.R.* **5**, 863 (1936). 5

B128.1. Brush Crystal Co. Ltd., Hythe, Southampton. 1209

B129. B.S.A. Tools Ltd., Birmingham. *1161*

B130. I. G. Bubenin, *Stroitel Materialy* [3], 53 (1937). 181

B131. Rudolf Buchkremer, Elektrotunnelöfen mit Korbförderung in der keramischen Industrie. *Sprechs.* **83** [7] (1950). 1030

— P. E. Buckles, see C. W. Parmelee.

B132. O. S. Buckner, Aluminium phosphate bonded diamond abrasive. *U.S. Pat.* 2420859. Bay State Abrasives Products Co.   523

B133. P. P. Budnikow and D. P. Bobrownik, Über die Reaktion im festen Zustand zwischen SiO_2 und CaO, Kaolin und Seinen Brennprodukten und CaO. (Ukrainian). *Ukr. Chem. J.* **12**, 190 (1937).   191

B134. P. P. Budnikov and K. M. Shmukler, Effects of mineralisers on the process of mullitization of clays, kaolins, and synthetic masses. *Zhur. Priklad. Khim.* **19** [10–11], 1029 (1946); *J. A. Cer. S. Abs.* **31** [3], 54 (1948).   195, 196

B135. W. R. Buessem, Thermal shock testing. *J. A. Cer. S.* **38** [1], 15 (1955).   360

B136. W. R. Buessem, Present trends in electroceramics. *Mineral Industries* **26** [3], 1–3 (1956); *Abs. Ceramics* **9** [109], March (1958).   1173

B137. W. R. Buessem and E. A. Bush, Thermal fracture of materials under quasi-static thermal stresses (Ring test). *J. A. Cer. S.* **38** [1], 27 (1955).   361, *362*

B138. W. Bussem and W. Eitel, *Z. Krist.* **95**, 175 (1936).   212

B139. Bullers Ltd.,
(a) *Bullers Placing Equipment.*   962, 963
(b) *Bullers Furniture for Enamel Kilns.*
(c) *Bullers Firing Trial Rings.*   939, *940–941*
Joiners Square Works, Hanley, Stoke-on-Trent.

B140. L. Bullin, Some recent developments in ceramic firing. *Trans. Cer. Soc.* **35**, 53 (1936).   897, 898, 908, 909, 910, 951, 972, 975, *1036*

B141. L. Bullin, Drying in tile manufacture. *J. Inst. Fuel* **27**, 115 (1954).   852, 853

— Jack, J. Bulloff, see Herbert D. Sheets.

— K. Bunde, see W. Jander.

B142. E. N. Bunting, Phase equilibria in the system SiO_2–ZnO. *J. A. Cer. S.* **13**, 5 (1930); *J. Res. Nat. Bur. Stand.* **4**, 134 (1930).   132

B143. E. N. Bunting, *Bur. Stand. J. Res.* **11**, 719 (1933).   170

B144. E. N. Bunting, G. R. Shelton and A. S. Creamer, *J. A. Cer. S.* **30**, 4 (1947).   181

— E. N. Bunting, see R. H. Ewell.

B145. B. Butterworth and D. Foster, The development of the fired-earth brick. *Trans. Brit. Cer. Soc.* **55**, 457, 481 (1956); *Ibid.* **56** [10], 529 (1957); *Ibid.* **57** [8], 469 (1958); *Ibid.* **58** [2], 63 (1959).   1069

— B. Butterworth, see D. G. R. Bonnell.

B146. M. D. Burdick, R. E. Moreland and R. F. Geller, *Strength and Creep Characteristics of Ceramic Bodies at Elevated Temperatures.* N.A.C.A. T.N. No. 1561, National Bureau of Standards, Washington.   1164

— M. D. Burdick, see S. M. Lang.
— N. Burns, see J. F. Hyslop.
B147. S. G. Burt, The coefficient of expansion of silica. *Trans. A. Cer. S.* **5**, 340 (1903).   554
B148. Werner Busch and Hilde Strumm-Bollenbach, Ueber die Weisstrübung bor- und bleifreier Glasuren mit Titanoxyd. *Ber. D. K. G.* **25**, 17 (1944).   588
— E. A. Bush, see W. R. Buessem.
B149. Bhan Bushan and H. N. Roy, Dichtgebrannte Massen aus indischem Talk mit niedriger Wärmeausdehnung. *Sprechs.* **91** [4], 67, 20 February (1958).   487
B150. B.U. Supplies and Machinery Co. Ltd., *Transfer Pressing Machine*. Law St., Leicester.   829

C1. Cabot Minerals (A Division of Cabot Carbon Company), (a) Big Big Future for wollastonite. Reprint from *Business Week*. 12 April (1952). (b) Introducing wollastonite. *Cabot* **5** [2] (W-1), April (1952). (c) William M. Jackson II, Low loss electrical bodies from wollastonite. Paper presented at the *Convention of the American Ceramic Society*, 1952. 77, Franklin St., Boston 10, Mass.   120
C2. Godfrey L. Cabot Inc., 77 Franklin St., Boston 10, Mass.   500, 1192
C3. R. Caillat, L'industrie des réfractaires et le développement industriel de l'énergie atomique. *Bull. Soc. Fran. Cér.* [39], 9, April–June (1958).   1165
C4. R. M. Campbell, Interfacial reactions: Symposium on Glaze Properties. *Cer. Ind.* **32** [5], 39 (1939).   542
C5. Cambridge Instrument Co. Ltd.,   942, 945, 947,
    (a) The accurate measurement of temperature. List No. 198 U.D.C. 536.5.   948
    (b) Cambridge automatic temperature regulators.   951
    (c) Cambridge electrical CO_2 and CO instruments.   954
    (d) Cambridge oxygen recorder.   954
    13, Grosvenor Place, London, S.W.1.
C6. Capitain & Co., Tonbergbau-Mahlwerk, Vallendar am Rhein, Germany.   37, 38
C7. Carbide and Carbon Chemicals Division (Union Carbon and Carbide Corporation), 30 East 42nd St., New York 17, New York. UCON LB-550-X.   1027
C8. Carborundum Co., *Facts about Aluminium Oxide*. Niagara Falls, N.Y. (1953).   1151, 1158
C9. Carborundum Co., Perth Amboy, N.J., U.S.A.   412, 1251
C10. The Carborundum Co. Ltd., Trafford Park, Manchester 11.   *1158*
C11. W. F. Carey and C. J. Stairmand, Size analysis by photographic sedimentation. *Trans. Inst. Chem. Engs.* **16**, 57 (1938).   290

C12. K. Carr, R. W. Grimshaw and A. L. Roberts, The constitution of refractory clays. IV—The constitution of some Yorkshire fireclays. *Trans. Brit. Cer. Soc.* **51**, 334 (1952).   29

C13. K. Carr, J. Hargreaves, R. W. Grimshaw and A. L. Roberts, The Constitution of refractory clays. V—relationship between constitution and properties with special reference to the influence of micaceous minerals. *Trans. Brit. Cer. Soc.* **51**, 345 (1952).   29

— W. P. Cartelyou, see H. R. Shell.

C14. T. G. Carruthers and A. L. Roberts, *Ceramics as Gas Turbine Blade Materials.* Iron and Steel Institute Symposium on High Temperature Steels and Alloys for Gas Turbines. Special Report No. 43 (1951).   1165

C15. W. K. Carter and R. M. King, Sodium aluminate as an electrolyte for casting slip control. *J. A. Cer. S.* **15**, 407 (1932).   761

C16. Carter, Stabler and Adams Ltd., *Potting.* Poole, Dorset.   799

C17. Casburt Ltd., *Casburt Dippers Dryer.* Park Rd., Stoke-on-Trent.   843

C18. H. Cassan and A. Jourdain, Contribution à l'étude expérimentale de la plasticité des argiles. *Céramique* **40**, 117 (1937).   328

C19. H. C. Castell, S. Dilnot and M. Warrington, Reaction between solids. *Nature, Lond.* **153**, 653 (1944).   191

C20. F. Catchpole, Production of lightweight aggregate by the Sinter-Hearth process. *Trans. Brit. Cer. Soc.* **56** [10], 519 (1957).   429

C21–C27 inclusive follow after **C30**.

C28. Cealglske and Houden, *Ind. Eng. Chem.* **29**, 805 (1937).   74

C29. Cellactite and British Uralite Ltd., *The Kimolo (Moler) Technical Handbook* No. 1 (1950).   900

C30. *Ceramic Data Book,* 1950–1951. 5, South Wabash Ave., Chicago 3, Ill., U.S.A.   160

— N. I. Chaĭkovskaya, see M. I. Zaĭtseva.

C21. A. C. D. Chaklader and A. L. Roberts, Relationship between constitution and properties of silica refractories. *Trans. Brit. Cer. Soc.* **57** [3], 115, 126 (1958).   169, 403, 1222, 1223

C22. L. Chalmers, Liquid Silver: its use and manufacture. *Brit. Ind. Finishing* **2** [21], 181 (1931).   655

C23. L. Chalmers, Liquid gold. *Brit. Ind. Finishing* **2** [22], 201 (1931).   650, 651

— E. A. C. Chamberlain, see L. W. Andrew.

C24. Chemnitius, Gold in ceramic work. *Chem. Trade J.* **81**, 260 (1927).   651

C25. J. H. Chesters, Magnesite refractories. *Iron Age* **151**, 46, 75, 3 and 10 June (1943).   1261–1262

C26. J. H. CHESTERS, *Steelplant Refractories* (2nd Ed.). The United Steel Co. Ltd., Sheffield (1957). — 96, 98, 111, 116–118, 357, 363, 402, 407, 420, 426, 1220–1230, 1255–1271, 1291

C27. J. H. CHESTERS and C. W. PARMELEE, The measurement of reaction rates at high temperatures. *J. A. Cer. S.* **17**, 50 (1934). — 191

C28–C30 inclusive follow after **C20**.

C31. R. CHEVATUR, *Bull. soc. Chim. France* 576 (1950). — 189

— CARL J. CHRISTENSEN, see IVAN B. CUTLER.

— CARL J. CHRISTENSEN, see EDMOND P. HYATT.

C32. P. L. CHRISTENSEN and EDWARD SCHRAMM, Application of the Schweitzer spray machine to the glazing of vitreous ware. *Bull. A. Cer. S.* **31** [10], 384 (1952). — 795

C33. MARSHALL T. CHURCH, *Design and Operation of Tunnel Kilns.* Paper read to Ceramic Society of Southwest U.S.A. (1945). — 1027, 1028

C34. V. CIRILLI, *Gazetta Chimica Ital.* **75**, 233 (1945). — 183, 628

— G. L. CLARK, see A. I. ANDREWS.

— J. D. CLARK, see J. W. DONAHEY.

C35. N. O. CLARK, China clay research in Cornwall. *Trans. Brit. Cer. Soc.* **49**, 409 (1950). — 309, *310*

C36. N. CLARKE JONES, Fused stabilised zirconia. *Ceramics* **4** [38], 75, April (1952). — 1246, 1293

C37. J. F. CLEMENTS, High temperature insulating materials, their properties and testing. *Ceramics* **3** [26], 110, April (1951). — 1287–1291

C38. J. F. CLEMENTS and W. R. DAVIS, A sonic spalling test. *Trans. Brit. Cer. Soc.* **57** [10], 624 (1958). — *359*

C39. F. H. CLEWS, Practical experiences in the de-airing of heavy clay goods: A report on a questionnaire. *Trans. Cer. Soc.* **37** [6], 231 (1938). — 711

C40. F. H. CLEWS, Fireclay and alumino-silicate refractories in the Twentieth Century. *Ceramics—A Symposium.* British Ceramic Society (1953). — 752

C41. F. H. CLEWS, *Heavy Clay Technology.* British Ceramic Research Association, Stoke-on-Trent (1955). — 349, 361, 362, 392, 399, 718, 730, 736, 1020, 1066–1068, 1071, 1073

— F. H. CLEWS, see W. NOBLE.

C42. A. G. COCKBAIN and W. JOHNSON, Examination of used chrome-magnesite refractories by petrological and X-ray techniques. *Trans. Brit. Cer. Soc.* **57** [8], 511, August (1958). — 340

C43. T. Tusting Cocking, *pH Values, What They Are and How to Determine Them.* British Drug Houses Ltd., Graham St., London, N.1. — 320

C44. William W. Coffeen, Simple procedure for correction of mineral compositions calculated from felspar analysis. *Cer. Age* **60** [6], 29, December (1952). — 301, 302

— M. Coleman, see D. M. Newitt.

C45. G. C. Collins and H. Polkinhorne, An investigation of anionic interference in the determination of small quantities of potassium and sodium with a new flame photometer. *Analyst* **77** [917], 430 (1952). — 272, 274

C46. P. F. Collins, Study of china bodies of Belleek type. *J. A. Cer. S.* **11**, 706 (1928). — 456

C47. P. F. Collins, Auxiliary fluxes in ceramic bodies. *J. A. Cer. S.* **15**, 17 (1932). — 440, 441

C48. J. E. Comeforo, R. B. Fischer and W. F. Bradley, Mullitization of kaolin. *J. A. Cer. S.* **31** [9], 254 (1948). — 77, 192

C49. J. E. Comeforo, J. G. Breedlove and Hans Thurnauer, Phosphate bonded talc: A superior block-talc substitute. *J. A. Cer. S.* **37** [4], 191 (1954). — 524, 1181, 1201

— J. J. Comer, see Thomas F. Bates.

C50. C. H. Commons, Jr., Effect of zircopax additions on abrasion resistance and various properties of several glazes. *J. A. Cer. S.* **24** [5], 145 (1941). — 594

— R. L. Cook, see R. J. Beals.

C51. Nach B. Cool bearbeitet von H. Harkort. Ungefrittete Borverbindungen in keramischen Massen. *Sprechs.* **91**, [6] 112, 20 March (1958). — 132

C52. Cooley Electric Manufacturing Corporation, 38, Shelby St., Indianapolis, Ind., U.S.A. — 983

C53. Coors Porcelain Co., Golden, Colerado, U.S.A. — 1189–1191

C54. Copeland Bros., Fenton, Stoke-on-Trent. — 740

C55. R. S. C. Copeland, Developments in the printing process as used in the decoration of china and earthenware. *Ceramics* **7** [78], 262, August (1955). — 809, 818, 823, 826–828

C56. C. W. Correns, Die Minerale der Ziegeltone. *Sixth International Ceramic Congress, Wiesbaden.* Abs. *Euro-Ceramic* **8** [10], 256 (1958). — 30

C57. G. Correns, Über die Bestandteile der Tone. *Zeit. Deut. Geol. Ges.* **85**, 706 (1933). — 309

C58. Cornwall Mills Ltd., Yate Mills Ltd., Par Harbour, Cornwall. — 104

C59. William G. Coulter, How to prepare and apply engobes and body stains. *Cer. Ind.* **63** [3], 76, September (1954). — 621–641, 801

C60. Courtaulds Ltd., *Courlose*. Plastics Division, Little Heath, Coventry. — 156

C61. C. A. Cowan, G. A. Bole and R. L. Stone, Spodumene as a flux component in sanitary chinaware bodies. *J. A. Cer. S.* **33** [6], 193 (1950). — 452

C62. PAUL E. COX, Zinc oxide, friend and enemy. *Cer. Age* 195, 588
April (1947).
C63. PAUL E. COX, Preheating makes highly colloidal clays work- 759
able in casting slips. *Cer. Age* **53**, 124 (1949).
C64. JAMES R. COXEY, *Refractories.* State College, Pa., U.S.A. 717, 1212, 1252
(1950).
— A. S. CREAMER, see R. F. GELLER.
— A. S. CREAMER, see E. N. BUNTING.
C65. FRIEDRICH CRÖSSMANN, JULIUS STAWITZ and JULIUS VOSS, 784
Verdickungsmittel in Engoben und Glasuren. *Tonind.-
Ztg.* **85** [59], 582 (1941).
C66. L. J. CRONIN, Refractory cermets. *Bull. A. Cer. S.* **30** [7], 512
234 (1951).
C67. THE CROSSLEY MACHINE COMPANY INC., Trenton, New 714
Jersey, U.S.A.
C68. CROSTHWAITE FURNACES AND SCRIVEN MACHINE TOOLS 910
LTD., *Mechanical Stokers and Self Cleaning Furnaces.*
York Street, Leeds 9.
C69. H. L. CROWLEY and A. M. HOSSENLOPP, Steatite-type body. 496
U.S. Pat. 2382137, 14 August (1945).
C70. A. L. CURTIS, Mining English china clay—I. *Sands, clays and* 241
minerals **1** [3], 13 (1933).
C71. CARL E. CURTIS, The electrical dewatering of clay suspen- 248
sions. *J. A. Cer. S.* **14**, 219 (1931).
C72. CARL E. CURTIS, Applications of electrophoresis and electro- 248, 249
osmosis in the ceramic industries. *Cer. Age* **31** [5], 142,
May (1938).
C73. IVAN B. CUTLER, CYRIL BRADSHAW, CARL J. CHRISTENSEN 505
and EDMOND P. HYATT, Sintering of alumina at tempera-
tures of 1400°C and below. *J. A. Cer. S.* **40** [4], 134
(1957).
— IVAN B. CUTLER, see EDMOND P. HYATT.
C74. CZEDIK-EYSENBERG, Beiträge zur Prüfung Feuerfester 343, 353
Baustoffe. *Actes Cong. Cér. Int., Paris,* 1952, p. 165.

D1. P. W. DAGER and A. G. BETTELEY, Comparative study show- 165
ing the possibilities of sillimanite for use in extrusion
dies. *J. A. Cer. S.* **14**, 706 (1931).
D2. P. H. DAL and H. HARKORT, Die Einwirkung von Soda auf 64
Ton. *Sprechs.* **91** [21], 511, November (1958).
D3. A. J. DALE, The control of silica brick making, based on load- 403
test indications. *Trans. Cer. Soc.* **26**, 203, 210, 217, 225
(1927).
D4. A. J. DALE and M. FRANCIS, The technical control of glazing 784, 788, 789
by dipping and other methods. *Trans. Brit. Cer. Soc.*
41 [10], 167 (1942).
D5. A. J. DALE and MARCUS FRANCIS, The durability of on-glaze 378, 609
decoration—I. *Trans. Brit. Cer. Soc.* **41**, 245 (1942).

D6.—

D7. R. R. Danielson and H. P. Reinecker, The production of some white enamels for copper. *J. A. Cer. S.* **4**, 827 (1921).   589

D8. R. R. Danielson, Low-lead bright opaque glazes at cones 08 to 06. *J. A. Cer. S.* **30**, 247 (1947).   587

D9. R. R. Danielson and D. V. Van Gordon, Leadless opaque glazes at cone 04. *J. A. Cer. S.* **33**, 323 (1950).   587

D10. John Dasher and O. C. Ralston, New methods of cleaning glass sands. *Bull. A. Cer. S.* **20** [6], 187 (1941).   99

— S. T. E. Davenport, see H. H. Macey.

D11. W. H. Davenport, S. S. Kistler, W. M. Wheildon and O. J. Whittemore, Jr., *High Temperature Electric Furnace Using Oxide Resistors*. Norton Co., Worcester, Mass.; Abs. *Bull. A. Cer. S.* **29** [3], 122 (1950).   983

— Ben Davies, see J. C. Hicks.

— D. A. Davies, see Johnson, Matthey & Co. Ltd.

D12.—

D13. C. W. Davis and H. C. Vacher, *Bentonite : Its Properties, Mining, Preparation and Utilization*. Technical Paper 438, Dept. of Commerce, Bureau of Mines, U.S. Govt. Printing Office (1928).   258

D14. Harry Davis, Successful smoke elimination. *Brit. Clywr.* **67** [799], 197, November (1958).   985, 1004

D15. N. B. Davis, Note on apatite substituted for bone ash. *Trans. Cer. Soc.* **18**, 378, 381 (1918–1919).   460

D16. W. Davies, Low-alumina silica bricks. Raw materials. *Trans. Brit. Cer. Soc.* **51** [2], 95 (1952).   1220

— W. R. Davis, see J. F. Clements.

— W. Dawihl, see H. Hirsch.

D17. A. L. Day and E. T. Allen, *Isomorphism and Thermal Properties of Feldspars*. Carnegie Institute of Washington (1905).   198

D18. Daniel L. Deadmore, Sintered alumina for filter purposes. *Cer. Age* **56** [7], 15, January (1951).   521

D19. W. T. Dean, Magnetic ceramics. *Ceramics* **9** [116], 42, October (1958).   520, 521, *1172*

D20. J. Debras and I. A. Voinovitch, Dosage des ions alcalins Li, Na et K dans les silicates, par photométrie de flamme. *Bull. Soc. Fran. Cér.* [38], 77, January–March (1958).   275

D21. Degussa, Abt. Keramische Farben. Degussa-Lüsterfarben. Mitteilung Nr. 18. Frankfurt/M.   658

D22. Degussa, Abt. Keramische Farben. Interessante keramische Farbkörper für hohe Temperaturen. Mitteilung Nr. 26. Frankfurt/M.   639

D23. De Lapparent, Formula and structural scheme of Attapulgite. *Compt. rend. Acad. Sci.* **202**, 1728 (1936).   19

D24. Demarara Bauxite Co. Ltd., British Guiana.   *115*

D25. DEMPSTER BROTHERS, INC., Knoxville 17, Tenn., U.S.A.   260
D26. I. DENTON and P. MURRAY, A note on grain growth in fabri-   507
cated beryllia. *Trans. Brit. Cer. Soc.* **58** [1], 35 (1959).
— I. DENTON, see P. MURRAY.
D27. DENVER EQUIPMENT COMPANY. *Denver 'Sub-A' Flotation.*   250
Bulletin No. F10.B29. 1400, 17th Street, Denver,
Colorado, U.S.A.
D28. DENVER EQUIPMENT COMPANY. *Flotation Feldspar.* Bulletin   102
No. M7-F21; Deco Trefoil **16** [6], 17 (1952).
D29. FRIEDRICH DETTMER, Ueber die Bedeutung des Eisenoxydes   190, 863
für den Brandausfall keramischer Massen. *Sprechs.* [17],
317 (1928).
D30. FRIEDRICH DETTMER, Das Segerkegeldiagramm als Grund-   934
lage für das Brennen und Prüfen fein- und grobkera-
mischer Erzeugnisse. *Sprechs.* **63**, 173 (1930).
D31. FRIEDRICH DETTMER, *Die Herstellung des Porzellans.* Chem.   412, 464–473
Lab. für Tonindustrie, Berlin (1938).   488, 499,
     637, 638
D32. DEUTSCHE INDUSTRIE NORMEN, DIN 1171 (1934).   1301
D33. DEUTSCHE KERAMISCHE GESELLSCHAFT, Vorschläge für   300
chemische Untersuchungen keramischer Rohstoffe und
Erzeugnisse. *Ber. D. K. G.* **15**, 633 (1934).
D34. DEUTSCHE TON- UND STEINZEUG-WERKE A.-G., Mullite.   195
Ger. Pat. 589556, 9 December (1933); *J. A. Cer. S. Abs.*
13 [10], 265 (1934).
D35. DEVON AND COURTENAY CLAY CO. LTD., Kingsteignton,   36, 37, 42, 44
Devon.
D36. R. C. DE VRIES and R. ROY, Phase equilibria in the system   520
$BaTiO_3$–$CaTiO_3$. *J. A. Cer. S.* **38** [4], 142 (1955).
D37. R. C. DE VRIES, R. ROY and E. F. OSBORN, Phase equilibria   513, 520
in the system CaO–TiO_2–SiO_2. *J. A. Cer. S.* **38** [5],
158 (1955).
D38. DIAMONITE PRODUCTS DIVISION, U.S. Ceramic Tile Co.,   682
Canton 2, Ohio, U.S.A.
— E. DI CESARE, see *Symposium on Ceramic Cutting Tools.* **S219**.
D39.—
D40. T. A. DICKINSON, Electrostatic glazing. *Cer. Age* **54** [1],   796
28 (1949).
— H. DIEHL, see H. EISENLOHR.
D41. A. DIETZEL, Die Transparenz von Porzellan in Abhängigkeit   100
von der Quarzsorte. Kurzreferate der Vorträge zur
Hauptversammlung der Deutschen Keramischen Gesell-
schaft in Goslar, 24–26, September, 1952. *Europäische
Tonind*, **2** [11], 308, November (1952).
D42. A. DIETZEL, H. WICKERT and N. KÖPPEN, Zustandsdia-   503
gramm des Systems Li_2O–BaO–SiO_2. *Glastechn. Ber.*
27, 147–151 (1954).
— RICHARD D. DILLENDER, see WILLIAM C. BELL.
— S. DILNOT, see H. C. CASTELL.

BIBLIOGRAPHY AND AUTHORS INDEX 1353

D43. A. Dinnie, R. Beard and R. Richards, Measurements of the thermal transmittance of external walls by a guarded hotplate method. *Trans. Brit. Cer. Soc.* **58** [2], 113 (1959). — 355

D44. A. Dinsdale, *Heat Requirements in Pottery Firing.* Paper read at the Conference on 'Fuel and the Future', London, 8, 9 and 10 October 1946. — 873

D45. A. Dinsdale, *Some Aspects of Pottery Firing.* B.P.R.A., February (1947). — 1010

D46. A. Dinsdale, The development of firing in the pottery industry. *Ceramics, A Symposium.* British Ceramic Society (1953). — 879, 969, 971, 975, 1037

D47. A. E. Dodd, The evaluation of plasticity. *Refr. J.* **13**, 573 (1937). — 326

D48. C. M. Dodd and M. E. Holmes, A study of the effect of grog on pressure transmission in dry pressing. *J. A. Cer. S.* **14**, 899 (1931). — 750

D49. C. Doelter, *Handbuch der Mineralchemie,* Vol. II, Pt. 3, p. 395. Theodor Steinkopf, Dresden and Leipzig (1921). — 102

D50. John W. Donahey and J. D. Clark, A neglected cost cutting flux. *Cer. Ind.* **52–53**, 74, November (1949). — 444

D51. John W. Donahey and Ralston Russell, Jr., Color fading of underglaze decalcomania. *J. A. Cer. S.* **33** [9], 283 (1950). — 646–647

D52. M. J. Dorcas, Carbon as a material of construction. *Refiner and Natural Gasoline Manufacturer* **20** [8], 73 (1941). — 1117

D53. Gebrüder Dorfner, Kaolin- und Krisstallquarzsand-Werke, Hirschau, Germany. — 34

D54. Dorr-Oliver Co., Barry Place, Stamford, Connecticut, U.S.A. — 246

D54.1. Maschinenfabrik Dorst A.G., Oberlind-Sonneberg, Germany. — *719*

D55. G. Dougill, The cleaning of producer gas. *Trans. Inst. Chem. Engs.* **23**, 1 (1945). — 919

D56. Doulton & Co. Ltd., Doulton House, Albert Embankment, Lambeth, London, S.E.1. — *348, 365, 458, 699, 722, 723, 739, 765, 766, 769, 772, 773, 776, 787, 790, 792, 793, 958,* 965, *1060, 1076,* 1097, *1109, 1121,* 1134, *1176*

D57. Philip Dressler, Progress in the elimination of saggers in the firing of glost and bisque general ware and glost wall tile. *J. A. Cer. S.* **13** [2], 143 (1930). — 955

D58. Philip Dressler, Application of the recirculating gas-tube firing system to continuous decorating kilns. *J. A. Cer. S.* **20** [12], 383 (1937).   923, 1021, 1030

D59. M. Drosdoff, The separation at identification of the mineral constituents of colloidal clays. *Soil Science* **39**, 463 (1935).   305, 307

— M. A. A. Druesne, see J. Biscoe.

D60. I. M. Dubrovin, *Zhur. Priklad. Khim.* **9**, 1049 (1937).   187, 189

D61. L. Dubuit, Decorating glass and pottery. *Ceramics* **4** [44, 45, 46], 113, 413, 448, October, November, December (1952).   809, 813–818

D62. W. H. Duckworth and J. J. Krochmal, Ceramics for rockets. *J. A. Ordnance Ass.* May–June (1955).   1165

— W. H. Duckworth, see Herbert D. Sheets.

— W. H. Duckworth, see J. F. Lynch.

— W. H. Duckworth, see A. K. Smalley.

— W. H. Duckworth, see Collin Hyde.

D63. A. F. Dufton and H. Sheard, *J. Inst. Heat. Vent. Eng.* **5**, 82 (1937).   355

— G. H. Duncombe, Jr., see C. R. Austin.

D64. A. Durose, Underglaze printing. *Ceramics* **6** [62], 81, April (1954).   824

D65. Pol Duwez, Francis Odell and Jack L. Taylor. Recrystallisation of beryllium oxide at 2000°C. *J. A. Cer. S.* **32** [1] 1 (1949).   1140

D66. W. Dyckerhoff, Dissertation, Frankfurt (1925).   185

E1. Eagle Iron Works, Des Moines, Iowa, U.S.A.   262

— R. P. Eardley, see H. Bennett.

E2. N. F. Eaton, T. J. G. Glinn and R. Higgins, Effect of thermal treatment on the properties of silica refractories. *Trans. Brit. Cer. Soc.* **58** [2], 92 (1959).   404

E3. C. H. Edelman and J. Ch. L. Faverjee, On the crystal structure of montmorillonite and halloysite. *Z. Krist.* **102**, 417 (1940).   16, 20

E4. Edgar Allen and Co. Ltd., Imperial Steel Works, Sheffield 9.   663, 666

E5. Edgar Allen and Co. Ltd., *Special Alloy Tool-Steels* (3rd Ed.). Imperial Steel Works, Sheffield 9.   166

E6. Edwards and Jones Ltd., Globe Engineering Works, Longton, Stoke-on-Trent.   162, 713

E7. H. Edwards and A. W. Norris, The examination and maturing of glazes. *Trans. Brit. Cer. Soc.* **56** [3], 133 (1957).   374

E8. J. D. Edwards, Alumina nomenclature. *Chem. and Ind.* **66**, 229 (1947).   114

— M. Ehrenberg, see G. F. Hüttig.

E9. H. Eisenlohr, *Sprechs.* **59**, 645 (1926).   610

E10. H. Eisenlohr and H. Diehl, *Sprechs.* **65**, 42 (1932). 610
— W. Eitel, see W. Bussem.
— A. Eldh, see J. A. Hedvall.
E11. Electric Furnace Co., *Ceramic Firing with Electric Heat in the Continuous Recuperative Tunnel Kiln.* Salem, Ohio, U.S.A. 932
E12. Electric Resistance Furnace Co. Ltd. (E.F.C.O.), *E.F.C.O. Electrically Heated Pottery Kilns.* Leaflet P. 17, Victoria St., London, S.W.1. 1022
E13. Electromagnets Ltd., Engineers, Boxmag Works, Bond St., Hockley, Birmingham 19. 691
E14. Hende Eleöd, Versuche zur Herstellung von Knochenporzellan. *Ker. Rund.* **43**, 191 (1935). 462
E15. Elliott Brothers (London) Ltd., '*Siemens*' *Optical Pyrometer (Disappearing Filament Type).* Century Works, Lewisham, London, S.E.13. 948
E16. Friedrich Emich, *Microchemical Laboratory Manual.* John Wiley, New York (1932). 266
E17. K. Endell, Ueber die Wirkung alkalischer Reinigungsmittel auf farbigdekoriertes Porzellan-Hotelgeschirr. *Ber. D. K. G.* **12** [11], 548 (1931). 379
E18. K. Endell, R. Zorn and U. Hofmann, *Angew. Chem.* **54**, 376 (1941). 307
— K. Endell, see U. Hofmann.
— K. Endell, see H. Lehmann.
E19. W. von Engelhardt, Über silikatische Tonminerale. *Forschr. Min. Krist. und Petr.* **21**, 276 (1937). 19, 24
E20. English Clays, Lovering Pochin & Co. Ltd., 14 High Cross St., St. Austell, Cornwall. *239, 240*
— English Clays, Lovering, Pochin & Co. Ltd., see Leefra.
E21. Erbslöh & Co., Geisenheimer Kaolinwerke. Geisenheim am Rhein, Germany. 33, 34
E22. Th. Ernst, W. Forkel, and K. von Gehlen, Zur Benennung natürlicher Tone und keramischer Massen. *Sprechs.* **91** [11], 263, 5 June (1958). 26
E23. R. A. Etherington, Magnetic filters and the pottery industry. *Ceramics* 363, September (1950). 693
E24. Evans Electroselenium Ltd., Colchester Road, Halstead, Essex. *272, 273*
— B. B. Evans, see R. E. Gould.
E25. J. L. Evans and J. White, Further studies of the thermal decomposition of clays. *Trans. Brit. Cer. Soc.* **57** [6], 289 (1958). 77
E26. J. Otis Everhart, The production of paving brick by the dry-press process. *J. A. Cer. S.* **15**, 107 (1932). 749
E27. J. O. Everhart, Some aspects of de-airing generally overlooked. *Cer. Age.* February (1936). 710
E28. J. O. Everhart, Some special wear-resisting materials for clay machinery parts. *J. A. Cer. S.* **21** [2], 69 (1938). 165

— J. O. Everhart, see Roland R. Van der Beck.

E29. R. H. Ewell, E. N. Bunting and R. F. Geller, Thermal decomposition of talc. *J. Res. Nat. Bur. Stand.* **15**, 551, November (1935).   91, *92*, 192

— Raymond H. Ewell, see Herbert Insley.

F1. G. L. Fairs, The use of the microscope in particle size analysis. *Chem. Ind.* **62**, 374 (1943).   290

F2. R. Fahn, Armin Weiss and U. Hofmann, Über die Thixotropie bei Tonen. *Ber. D. K. G.* **30** [2], 21 (1953).   59

F3. Fate-Root-Heath Co., *Bulletin* No. 15. Plymouth, Ohio, U.S.A.   709

— J. Ch. L. Faverjie, see C. H. Edelman.

F4. Albert H. Fay, *A Glossary of the Mining and Mineral Industry*. Bulletin 95, Dept. of Int. Bureau of Mines (1920).   96, 102

— M. S. Federowa, see W. Ja Lokschin.

F5. H. R. Feichter, Vitreous and vitrifiable composition. *Can. Pat.* 465201, 16 May (1950).   476

F6. Clarence N. Fenner, The different forms of silica and their interrelationship. *J. Wash. Acad. Sci.* **2** [20], December (1912); *Amer. J. Sci.* **36** [4], 331 (1913).   194

F7. R. F. Ferguson, Review of literature on laboratory slag tests for refractories. *J. A. Cer. S.* **11** [2], 90 (1928).   363

F8.—

F9. Ferro-Enamel Corporation, East 56th St., Cleveland, Ohio, U.S.A. Salt glazing in continuous kiln. *Pat.* No. 631613; *Brit. Clywkr.* **57** [694], 319, February (1950).   1017

— C. L. Fillmore, see S. M. Lang.

F10. Filtros Inc., 601, Commercial St., East Rochester, N.Y., U.S.A.   1134

— E. Fingas, see C. Kröger.

— R. B. Fischer, see J. E. Comeforo.

F11. Eugene Fisher and Robert Twells, High-temperature glazing of alumina bodies. *J. A. Cer. S.* **40** [11], 385 (1957).   476

F12. F. A. Fix, Application of dry mixing in the whiteware and refractory industries. *Bull. A. Cer. S.* **17** [9], 351 (1938).   1080

— A. M. Flannigan, See R. E. Gould.

F13. O. W. Florke, New aspects on the mineralogical composition of silica bricks, especially concerning the so-called 'Glass-Phase'. *Refr. J.* **34** [11], 510 (1958).   403

F14. Follsain-Wycliffe Foundries Ltd., Lutterworth.   165

F14.1. R. W. Ford and W. Noble, The development of drying-strains in wire cut bricks. *Trans. Brit. Cer. Soc.* **59** [2], 38 (1960). R. W. Ford and W. Noble. Moisture   711 redistribution in plastic clay during storage. *Trans. Brit. Cer. Soc.* **59** [2], 58 (1960).

— W. F. Ford, see H. J. S. Kriek.
— H. Forestier, see J. Lougnet.
— W. Forkel, see Th. Ernst.

F15. J. S. Forrest, The electrical properties of semi-conducting ceramic glazes. *J. Sci. Instr.* **24**, 211(1947). 562

F16. Foster Instrument Co. Ltd., *Foster Optical, Radiation and Photoelectric Pyrometers.* Letchworth, Herts. 948

— D. Foster, see B. Butterworth.

F17. E. S. Foster, Organic agents as aids to adhesion and suspension of glazes. *J. A. Cer. S.* **12**, 264 (1929). 784

F18. H. D. Foster, Salt glazes for structural clay units. *J. A. Cer. S.* **26** [2], 60 (1943). 601

F19. Wilfrid R. Foster, Contribution to the interpretation of phase diagrams by ceramists. *J. A. Cer. S.* **34** [5], 151 (1951). 197, 216–*219*

F20. Wilfrid R. Foster, High temperature X-ray diffraction study of the polymorphism of $MgSiO_3$. *J. A. Cer. S.* **34**, 255 (1951). 91, 492

F21. Wilfrid R. Foster, Solid-state reactions in phase equilibrium research. *Bull. A. Cer. S.* **30**, [8], 267 (1951); *Ibid.* **30** [9], 291 (1951). 215

— H. B. Fox, see R. C. Purdy.

F22. John Fox, Some factors influencing the choice of fuels and kilns in the ceramic industries. *J. Inst. Fuel* **12**, 257, June (1939). 903, 1013

F23. The Foxboro Company, *Potentiometer Recording Pyrometers.* Foxboro, Mass., U.S.A. 945

F24. Foxboro-Yoxall Ltd., *Foxboro Potentiometer Controllers.* Lombard Rd., Morden Rd., London, S.W.19. 950

F25. M. L. Franchet, *Ann. Chim. et Phys.* [8], 9 (1906). 659–660

— E. E. Francis, see Rustum Roy.
— Marcus Francis, see A. L. Dale.
— Marcus Francis, see D. A. Holdridge.
— Marcus Francis, see E. Sharratt.

F26. J. Frank, Ein schreibendes Schwindungs-Messgerät. *Tonind.-Ztg.* **77** [1–2], 8 (1953). 334

F27. C. E. L. Franklin, Changes in the theory and practice of drying, 1900–1950. *Ceramics, A Symposium*, p. 315. British Ceramic Society (1953). 846

F28. S. G. Frantz Co., Inc., Brunswick Pike and Kline Ave., P.O. Box 1138, Trenton 6, N.J., U.S.A. *692–3*

— Felix Fraulini, see C. G. Harman.

F29. V. D. Fréchette, Cermet development, U.S.A. *Cer. Ind.* **63** [3], 79, September (1954). 512

— V. D. Fréchette, see C. W. Parmelee.

F30. I. L. Freeman, Firing-shrinkage of some brick clays. *Trans. Brit. Cer. Soc.* **57** [6], 316 (1958). 83

— H. Frey, see W. Jander.
F31. R. Fricke and W. Blasche, *Naturwissenschaften*, **31**, 326   187
(1943).
— K. S. Fridmann, see A. V. Pamfilov.
F32. K. Friedrich, Beiträge zur thermischen Dissoziation und   192, 193
zu der Konstitution leicht zerlegbare Mineralien.
Zentralblatt f. Mineral. 616–626, 651–660, 684–693,
(1912).
F33. Firma Fries, Heibel and Keil, Die neue Tonstech-   256
maschine. *Europäishe Tonind.* **3** [10], 286 (1953).
— E. H. Fritz, see Hobart M. Kraner.
F34. Fuchs'sche Tongruben K. G., Ransbach im Westerwald,   39, 41, 49
Germany.
— Yoshio Fukami, see Toshio Nakai.
— S. D. Fulkerson, see J. R. Johnson.
F35. Industrieofenbau Fulmina, Friedrich Pfeil, *Fulmina*   983
Brenn-Ofen. Edingen-Mannheim, Germany.
F36. William Funk, *Rohstoffe der Feinkeramik.* Julius   442
Springer, Berlin (1933).

— Gaillard, see Bodin.
— F. Ja. Galachow, see N. A. Toropow.
G1. A. Gallenkamp and Co. Ltd., *Universal Torsion Viscometer.*   324
G2. W. H. Gamble and R. F. Paget, Uses of coal for brickworks.   906, 907, 910
Trans. Cer. Soc. **35**, 209 (1936).
G3. James J. Gangler, Some physical properties of eight re-   1150–1153
fractory oxides and carbides. *J. A. Cer. S.* **33** [12], 367
(1950).
G4. Norbert Samuel Garbisch, Improvements in or relating to   752
process of and apparatus for forming ceramic bodies.
Pat. No. 532781 (1941).
G5. J. S. F. Gard, *Thermal Insulation as Applied to Pottery*   900
Furnaces and Carbonising Plant. Inst. Fuel Wartime
Bulletin, April (1945).
— W. J. Gardner, see J. Holland.
G6. Oakley H. Garlick, Get the best from a ball mill. *Cer.*   676–677
Ind. **55** [5], 59, November (1950).
G7. W. I. Garms and J. L. Stevens, Ball wear and functioning   681
of the ball load in a fine-grinding ball mill. *Amer. Inst.*
Min. Met. Eng. Tech. Pub. 1984; *Min. Tech.* **10** [2], 4 pp.
(1946).
— E. Garping, see J. A. Hedvall.
— B. Garre, see G. Tammann.
G8. T. W. Garve, Factory experience in de-airing. *Bull. A. Cer.*   711
S. **15** [10], 335 (1936).
G9. T. W. Garve, Tunnel drier refinements. *Bull. A. Cer. S.*   847
20 [4], 117 (1941).
G10.—

G11. P. Gatzke, Elektrische Beheizungsmethoden an Keramische Oefen. *Ber. D. K. G.* [6], 297 (1936). — 879

G12. H. R. B. Gautby, Design of modern kiln superstructure. *Ceramics* **9** [102], 10, August (1957). — 960

G13. Kenneth, A. Gebler and Harriet R. Wisely, Dense cordierite bodies. *J. A. Cer. S.* **32** [5], 163 (1949). — 484, 490

G14. Kenneth A. Gebler, P_2O_5 glazes for steatite. *Cer. Age* 15, September (1951). — 564, 588

— K. A. Gebler, see H. R. Wisely.

G15. G. Gehlhoff and M. Thomas, Schnell-Kühlung von Glas. *Z. Techn. Phys.* **6**, 333 (1925). — 554

G16. G. Gehlhoff and M. Thomas, I—Die physikalischen Eigenschaften der Gläser in Abhängigkeit von der Zusammensetzung. II—Die mechanischen Eigenschaften der Gläser. *Z. Techn. Phys.* **7**, 105 (1926). — 550, 558

— K. von Gehlen, see Th. Ernst.

G17. R. F. Geller, Water smoking of clays. *J. A. Cer. S.* **4** [5], 375 (1921). — 856

G18. R. F. Geller, Bureau of Standards investigation of felspar. *J. A. Cer. Sec.* **10**, 423 (1927). — 452

G19. R. F. Geller and A. S. Creamer, Metal marking of whiteware glazes as influenced by sulphur and carbon in kiln gases. *J. A. Cer. S.* **14** [9], 624 (1931). — 884

G20. R. F. Geller, *J. Res. Nat. Bur. Stand.* January (1938). — 564

G21. R. F. Geller, A resistor furnace, with some preliminary results up to 2,000°C. *J. Res. Nat. Bur. Stand.* **27**, Research Paper, R.P. 1443, December (1941). — 983

G22. R. F. Geller and P. J. Yavorsky, *J. Res. Nat. Bur. Stand.* **35**, 87 (1945). — 198

G23. R. F. Geller, P. J. Yavorsky, B. L. Steierman and A. S. Creamer, Studies of binary and ternary combinations of magnesia, calcia, baria, beryllia, alumina, thoria and zirconia in relation to their use in porcelains. *J. Res. Nat. Bur. Stand.* **36** [3], 277 (1946); *Cer. Abs.* 162, September (1946). — 197, 506, 1152

G24–G30 inclusive follow after **G38**.

G31. R. F. Geller, *Ceramic Bodies for Turbo-Jet Blades.* Soc. Automotive Engineers (1948). — 1165

— R. F. Geller, see M. D. Burdick.
— R. F. Geller, see R. H. Ewell.
— R. F. Geller, see S. M. Lang.

G32. General Ceramics Corporation, Keasbey, N.J., U.S.A. — 1116, 1117, 1124, 1127, 1129, 1151

G33. General Electric Company Ltd., *Infra-Red Lamp Heating Equipment.* Magnet House, Kingsway, London, W.C.2. — 837

G34. General Motors Corporation, Improvements in spark plug insulators. *Brit. Pat. Spec.* 546191. — 476, 1180

— V. D. Gerasimova, see G. N. Voldsevich.

G35. Albert C. Gerber, The use of gelatine in glaze application. *J. A. Cer. S.* **7** [6], 494 (1924).   784

G36. W. H. Gerisch, Ceramic insulating material. *U.S. Pat.* 2304562, 8 December (1942).   504

G37. W. L. German, Experimental firing with a short multi-passage kiln. *Ceramics* 441, October (1950).   1008, 1009

G38. W. L. German, Electric tunnel kilns. *Ceramics* **3** [25], 29, March (1951).   931–933, 1023–1025

G24. W. L. German, Bone—Its preparation and use in the bone china body. *Ceramics* **4** [43], 56, September (1952).   108–110

G25. W. L. German, S. W. Ratcliffe et al., Relative translucencies of the system bone–china clay–stone. *Trans. Brit. Cer. Soc.* **53** [3], 165 (1954).   462

G26. W. L. German, S. W. Ratcliffe et al., Experiments in dry-mixing for a white-tile body. *Trans. Brit. Cer. Soc.* **55** [2], 157 (1956).   1080

G27. W. L. German, Small scale equipment and the works laboratory. *Ceramics* **9** [99], 12, May (1957).   265

G28. W. L. German, Chemically prepared phosphates in a bone china body. *Ceramics* **9** [112], 12, June (1958).   463

— W. L. German, see W. Gilbert.

— W. L. German, see S. W. Ratcliffe.

G29. H. Gessner, Der verbesserte Wiegner'sche Schlämmapparat. *Mitteilung aus dem Gebiete der Lebensmitteluntersuchung und Hygiene* **13**, 238 (1922) Bern.   287

G30. H. Gessner, Der Wiegner'sche Schlämmapparat und seine praktische Anwendung. *Koll.-Z.* **38**, 115 (1926).   287

G39. Gibbons Bros. Ltd., Dibdale Works, Dudley, Worcs.   850, 922, 1032–1034, 1038, 1040, 1043, 1044–1046

G40. Gibbons, Bros. Ltd., Improvements in or relating to tunnel ovens and the like. *Pat. Spec.* 312786, 6 June (1929).   1018

G41. Gibbons Bros. Ltd., Improvements in tunnel kilns. *Brit. Pat. Spec.* 461088, 10 February (1937).   1026

G42. Gibbons Bros. Ltd., Improvements in tunnel ovens. *Brit. Pat. Spec.* 485578, 19 May (1938); *Ibid.* 486628, 8 June (1938).   1026

G43. Gibbons Bros. Ltd., *The 'Moore Campbell' Patent Electric Tunnel Kiln.*   1043

G44. Gibbons Bros. Ltd., *Oil Fired Rotary Frit Melters.*   1057

G45. George Gibson and Roland Ward, Reactions in the solid state. III—Reaction between sodium carbonate and quartz. *J. A. Cer. S.* **26** [7], 239 (1943).   191

G46. W. Gilbert, Works control in refractory making. *Ceramics* **9** [99], 22, May (1957).   387–391

G47. W. Gilbert, E. Marriott and W. L. German, The use of coal washery tailings and other waste products in the manufacture of bricks. *Trans. Brit. Cer. Soc.* **57** [5], 258, May (1958).   400

G48. R. M. Gill and G. Spence, Ceramics for machine tools. *Ceramics* **9** [114, 115], 30, 27, August, September (1958).   1159

G49. Gilliland, *Ind. Eng. Chem.* **30**, 506 (1938).   70

— J. McM. Gilmour, see J. M. McWilliam.

G50. J. Gimson and Co. (1919) Ltd., (a) *General Catalogue*. (b) *Pin Placing*. Fenton, Stoke-on-Trent.   962, 964, 966

— J. Gingold, see M. Bichowsky.

G51. L. D. Gittings (Monsanto Chemical Co.), Vitreous glaze and enamels. *U.S. Pat.* 3423971, 15 July (1947).   588

— T. J. G. Glinn, see N. F. Eaton.

G52. V. M. Goldschmidt, Olivine and forsterite refractories in Europe. *Ind. Eng. Chem.* **30**, 32, January (1938).   422

G53. V. M. Goldschmidt, New refractories from Norwegian raw materials. *Teknisk Ukeblad* [1] (1940).   422

G54. G. Gollow, A simple method for obtaining photographic records of time-settling curves with elutriation. *J. A. Cer. S.* **17**, 116 (1934).   287

— K. L. Goodall, see W. A. Bloor.

G55. Clayton, Goodfellow and Co. Ltd., Engineers, Blackburn, Lancs. *Grinding Mills*.   669

G56. Goodlass Wall and Lead Industries Ltd., *Brit. Pat.* 494060.   572

G57. O. E. Gorbunova and L. I. Vaganova, X-ray investigation of the transformation of γ-aluminium oxide into α-aluminium oxide. *Trudy Tsentral. Nauch.-Issledovatel. Lab. Kamnei Samotsvetov Tr. 'Russkie Samotsvety'*, No. 4, 66 (1938); Cer. Abs. *Trans. Brit. Cer. Soc.* 87a (1940).   114

G58. James Gordon and Co. Ltd., *Simplex Mono CO_2 Recorder*. Regent House, Kingsway, London, W.C.2.   953

G59. D. V. van Gordon, High-fire opaque glazes for zircon bodies. *J. A. Cer. S.* **34** [2], 33 (1951).   563

G60. D. V. van Gordon and W. C. Spangenberg, Adjustment of thermal expansion of cone 8 glazes. *J. A. Cer. S.* **38**, 331 (1955).   548

G61. Gosling and Gatensbury Ltd., Hanley, Staffs.   713, 714, *790*

— Hans Gotthardt, see Gustav Keppeler.

— Sidney Gottlieb, see Theron A. Klinefelter.

G62. R. E. Gould, Experimental determination of factors causing strains in true hard-fire porcelain flatware and a practical means for their elimination in production. *J. A. Cer. S.* **20** [12], 389 (1937).   724, *725*

G63. R. E. Gould, B. B. Evans and A. M. Flannigan, The use of infra red in drying pottery. *Bull. A. Cer. S.* **28**, 62 (1945).   838

G64. ROBERT F. GRADY, JR., Conversion of periodic kilns from coal firing to natural gas for firing glazed clay products. *J. A. Cer. S.* **16**, 141 (1933).   915, 921

— WM. A. GRAFF, see DWIGHT G. BENNETT.

G65. ROBERT P. GRAHAM and JOHN D. SULLIVAN, Critical study of methods of determining exchangeable bases in clays. *J. A. Cer. S.* **21** [5], 176 (1938).   318, *319*

G66. ROBERT P. GRAHAM and JOHN D. SULLIVAN, Improved machine shows different forms of failure of clay bodies in torsion. *Bull. A. Cer. S.* **18** [3], 97 (1939).   329

G67. ALBERT GRANGER, *Die Industrielle Keramik* (German Trans. by RAYMOND KELLER), p. 401. Julius Springer, Berlin (1908).   534

G68. A. GRANGER, Les porcelaines européennes et les porcelaines orientales. *Céram., verre, emaill.* **1** [2], 5, 7, 9–10, 45–50.   456

— R. GRAMSS, see F. KRAUSS.

G69. K. GRANT and W. O. WILLIAMSON, Loss in strength of silica bricks reheated in hydrogen or carbon monoxide. *Trans. Brit. Cer. Soc.* **56** [6], 277 (1957).   1224

G70. F. GRASENICK, Beitrag zur topologischen Erfassung und Identifizierung von Gefügebestandteilen in metallischen und feuerfesten Werkstoffen. *Radex Rundschau* [5–6], 843, March (1957).   340

G71. E. I. GREAVES and J. MACKENSIE, High temperature strength of basic refractories. *Trans. Brit. Cer. Soc.* **57** [4], 187 (1958).   *356*, 357

G72. A. F. GREAVES-WALKER, C. W. OWENS, Jr., T. L. HURST and R. L. STONE, The development of pyrophyllite refractories and refractory cements. *State College Record (North Carolina State College of Agriculture and Engineering)* **36** [3], February (1937).   93

G73. A. F. GREAVES-WALKER, The origin, mineralogy and distribution of the refractory clays of the United States. *State College Record North Carolina State College of Agriculture and Engineering* **39** [4] (1939).   22, 28, 29, 93,

— A. T. GREEN, see G. R. RIGBY.

G74. R. L. GREENE, X-Ray diffraction and physical properties of potassium borate glasses. *J. A. Cer. S.* **25** [3], 83 (1942).   *209*, 210

G75. H. H. GREGER and M. BERG, Instrument for measuring workability of clay water systems. *J. A. Cer. S.* **39** [3], 98 (1956).   329

G76. J. W. GREIG, Immiscibility in silicate melts—I and II. *Amer. J. Sci.* **13** [73], 1 (1927); *Ibid.* **13** [74], 133 (1927).   *217*, 482

G77. J. W. GREIG and T. F. W. BARTH, *Amer. J. Sci.* **35A**, 93 (1938).   198

— J. W. GREIG, see N. L. BOWEN.

— GRIFFITHS, see POWELL.

G78. Grim, Bray and Bradley, The mica in argillaceous sedi-   21
ments. *Amer. Mineralogist* **22**, 813 (1937).
G79. Ralph E. Grim, Modern concepts of clay materials. *J.*   *16, 19*
Geol. L. No. 3, 225 (1942).
G80. R. E. Grim and R. A. Rowland, Differential and thermal   313
analysis of clays and shales—A control and prospecting
method. *J. A. Cer. S.* **27**, 5 (1944).
G81. Ralph E. Grim, Differential thermal curves of prepared   315
mixtures of clay minerals. *Amer. Mineralogist* **32**
[9–10], 493–501 (1947).
G82. R. E. Grim, Clay mineralogy in the U.S.A. *Ceramics* **9**   77, 78, 311
[109], March (1958).
G83. R. E. Grims and W. D. Johns, Jr., Reactions accompany-   857
ing firing of brick. *J. A. Cer. S.* **34** [3], 71 (1951).
G84. R. W. Grimshaw, E. Heaton and A. L. Roberts, The   29, 303, 313,
constitution of refractory clays. II—Thermal analysis   *314, 315*
methods. *Trans. Brit. Cer. Soc.* **44**, 76 (1945).
G85. R. W. Grimshaw, A. Westerman and A. L. Roberts, The   29
mineral constitution of some fireclays and its influence
on their properties. *International Ceramic Congress*,
1948.
G86. R. W. Grimshaw and A. L. Roberts, The constitution of   29, 303, 305
refractory clays. III—The determination of mineral
constitution. *Trans. Brit. Cer. Soc.* **51**, 327 (1952).
G87. R. W. Grimshaw, The Refractories Association of Great   29, 30
Britain. Fundamental aspects of fireclay research and
technology. *Refr. J.* **33** [12], 528 (1957).
G88. R. W. Grimshaw, Application of cation exchange in in-   55, 57, 63, 75,
dustrial practice. *Trans. Brit. Cer. Soc.* **57** [6], 340   78, 79
(1958).
— R. W. Grimshaw, see K. Carr.
— R. W. Grimshaw, see W. E. Worrall.
— R. O. Grisdale, see M. D. Rigterink.
— K. Grob, see W. Jander.
G89. P. Grodzinski and Felix Singer, *Report on the Hardness*   557
of Ceramic Glazes. Unpublished work.
G90. Vereinigte Grossalmeroder Thonwerke, (16) Gross-   39, 47, 48, 248
almerode, Bez. Kassel, Postschussfach 30, Germany.
G91. Grout, *J. A. Chem. S.* **27**, 1037 (1905).   328
G92. A. W. Groves, *Silicate Analysis*. Thomas Murby & Co.,   266
London (1937).
G93. J. W. Gruner, *Z. Krist.* **83**, 75 (1932); *Ibid.* **85**, 345 (1933).   *13*
— A. Gunzenhauser, see H. H. Hausner.

— Ch. Haeberle, see R. Rieke.
H1. Donald Hagar, *U.S. Pat.* 2213495.   446
H2. Donald Hagar, Personal communication from Donald   442–447, 472
Hagar, Zoenesville, Ohio.

H3. H. W. Hagen, *Ber. D. K. G.* [4] (1955); Schwinggitter zur Beseitigung der Strukturen im Massestrang. *Euro-Ceramic* [4] (1955). 731

H4. T. R. Haglund, Refractory composition for casting. *Brit. Pat.* 391662 (1931). 755

H5. Rudolf Hainbach, *Pottery Decorating* (English Trans.). Scott, Greenwood, London (1924). 638, 648, 653

H6. Hall China Co., East Liverpool, Ohio. *U.S. Pat.* 2296961. 767

H7. F. P. Hall, The plasticity of clays. *J. A. Cer. S.* **5**, 346 (1922). 326, 327

H8. F. P. Hall, The influence of chemical composition on the physical properties of glazes. *J. A. Cer. S.* **13**, 182 (1930). 548

H9. F. P. Hall and Herbert Insley, Phase diagrams for Ceramists—Third compilation. *J. A. Cer. S.* **30** [11, Pt. II] (1947). H. F. McMurdie and F. P. Hall, Supplement to above (1949). Ernest M. Levin, Howard F. McMurdie and F. P. Hall, Fifth compilation (1956). 215, 218

— F. P. Hall, see Edward Schramm.
— F. P. Hall, see R. E. Wilson.

H10. Louise Halm, Quelques aspects de la cristallisation mullitique. *Bull. Soc. Fran. Cér.* [1], October–December (1948). 192, *200*

H11. L. Halm and P. Lapoujade, Notes on the thermal conductivity of insulating blocks. *Bull. Soc. Fran. Cér.* [10], 20 (1951). 425, 897

H12. S. B. Hamilton, The history of hollow bricks. *Trans. Brit. Cer. Soc.* **58** [2], 41 (1959). 1069

H13. W. Hancock, The drying of tableware and other ceramic goods by the jet drying method. *Ceramics* **5, 6** [57–69], November (1953)–November (1954). 843, 844, 845

H14. J. H. Handwerk, L. L. Abernethy and R. A. Bach, Thoria and urania bodies. *Bull. A. Cer. S.* **36** [3], 99 (1957). 1144

— J. H. Handwerk, see C. A. Arenberg.

H15. C. J. Harbert, Less familiar elements in ceramic pigments. *Ind. Eng. Chem.* **30**, 770 (1938). 622–639

H16. C. J. Harbert, *F.I.A.T. Report* 794, pp. 6, 50 and 58 (1946). 610, 617

H17. Harbison-Walker Refractories Company, Barbosil. 577

H18. Harbison-Walker Refractories Company, *Modern Refractory Practice*. Pittsburgh, Pa. (1950). 420, 421, 423, 1213, 1273

H.18.1. R. S. Harding and A. N. Gilson, An automatic press for wall tile production. *Bull. A. Cer. S.* **37** [9], 405 (1958). 747, 1081

H19. Hardinge Company, Inc., York, Pa., U.S.A. *688*

H20. Hardinge Company, Inc., *Hardinge Conical Mills for Dry Grinding*. Bulletin No. 17B. 240, Arch St., York, Pa., U.S.A. 673, *674*, 689, 700

— C. W. Hardy, see H. R. Lahr.

H21. J. Hargreaves, The spalling of silica bricks. *Trans. Brit. Cer. Soc.* **57** [5], 242 (1958).   404

— J. Hargreaves, see K. Carr.

H22. H. Harkort, Die Schlämmanalyse mit dem verbessertem Schulze'schen Apparat als Betriebscontrolle. *Ber. D. K. G.* **8**, 6 (1927).   281, *282*

H23. H. Harkort, Untersuchungen über die Herstellung bleifester, gesundheitsunschädlicher Bleiglasuren. *Sprechs.* **67** [41], 621 (1934); *Ibid.* **67** [42], 637 (1934).   614

H24. H. Harkort, Untersuchungen über die Herstellung bleifester, gesundheitsunschädlicher Bleiglasuren *Sprechs.* **67**, 638 (1934).   570

— H. Harkort, see P. H. Dal.

H25. C. G. Harman and Felix Fraulini, Properties of kaolinite as a function of its particle size. *J. A. Cer. S.* **23**, 252 (1940).   298

H26. C. G. Harman and C. F. Schaefer, Study of factors involved in glaze slip control. I—Glaze slip specifications. *J. A. Cer. S.* **27**, 202 (1944).   370, 784

H27. C. G. Harman and M. K. Blanchard, Study of factors involved in glaze slip control. II—Correlation of glaze thickness with air permeability. *J. A. Cer. S.* **27**, 207 (1944).   370, 784

H28. C. G. Harman and H. C. Johnson, Study of factors involved in glaze slip control. III—Additional glazes and improved technique and apparatus. *J. A. Cer. S.* **27**, 209 (1944).   370, 784

H29. C. G. Harman and H. C. Johnson, Study of factors involved in glaze slip control. IV—Practical uses for coherence value and receptivity measurements. *J. A. Cer. S.* **27**, 214 (1944).   784

H30. C. G. Harman, Suggestions for solution of difficult glaze-fit problems. *J. A. Cer. S.* **27**, 231 (1944).   566

H31. C. G. Harman and C. W. Parmelee, Fundamental properties of raw clays influencing their use. *J. A. Cer. S.* **28** [4], 110 (1945).   52, 66, 67, 69, 75, *332*, 333

H32. C. G. Harman, Forming clay ware by hot pressing makes new products possible. *Brick & Clay Record* **94** [1], 15, 46 (1939).   752

— G. G. Harman, see C. W. Parmelee.

H33. Harrison and Son (Hanley) Ltd., Phoenix Chemical Works, Garth St., Hanley, Stoke-on-Trent.   *935*, 936, *937*

H34. W. N. Harrison and T. O. Hartshorn, A preliminary study of ceramic colors and their use in vitreous enamels. *J. A. Cer. S.* **10**, 747 (1927).   617

— R. Harrison, see A. W. Norris.

H35. W. N. Harrison, *High Temperature Ceramic Coatings for Molybdenum*. S. Automotive Engineers (1948).   513

H36. HARSHAW CHEMICAL CO., 1945, East 97th St., Cleveland, Ohio, U.S.A. Improvement in ceramic pigments. *Brit. Pat. Spec.* 625448, 28 June (1949).   639

— T. O. HARTSHORN, see W. N. HARRISON.

H37. F. A. HARVEY and R. E. BIRCH, Silica refractory. *U.S. Pat. Spec.* 2351204.   402, 1225

H38. FRED A. HARVEY and RAYMOND E. BIRCH, Olivine and forsterite refractories in America. *Ind. Eng. Chem.* **30** [1], 27, January (1938).   1273

H39. HASSIA HANDELSGESSELLSCHAFT, Kassel, Germany.   39

H40. HATHERNWARE LTD. Loughborough.   1108, 1111

H41. HAUCK MANUFACTURING CO., 124–136, Tenth St., Brooklyn 15, N.Y., U.S.A.   929

H42. E. A. HAUSER and C. E. REED, *J. A. Chem. Soc.* **58**, 1822 (1936).   60

H43. E. A. HAUSER, Colloid chemistry in ceramics. *J. A. Cer. S.* **24** [6], 179 (1941).   4, 11

H44. E. A. HAUSER and A. L. JOHNSON, Plasticity of clays. *J. A. Cer. S.* **25**, 223 (1942).   53

H45. HENRY H. HAUSNER, Influence of humidity on dielectric properties of high-frequency ceramics. *J. A. Cer. S.* **27** [6], 175 (1944).   365

H46. HENRY H. HAUSNER and ANGELA F. NAPORAN, Shrinkage of extruded steatite tubes. *Bull. A. Cer. S.* **24** [7], 246 (1945).   498

H47. H. H. HAUSNER, Metal ceramics. *Metal Industry* 405, 14 May (1948).   512

H48. H. H. HAUSNER and A. GUNZENHAUSER, Steatite becomes a major ceramic with a big future. *Cer. Ind.* July, August and September (1948).   89, 90

H49. W. E. HAUTH, Crystal chemistry in ceramics. *The Iowa State College Bulletin, Engineering Report* No. 7 (1951–1952); *Bull. A. Cer. S.* **30** [1, 2, 3, 4, 5, 6], 5, 47, 76, 137, 165, 203 (1951).   4, 6, 7, 8, 10, 169

H50. T. G. HAWKER, Magnetic separators in the clay industry. *Brit. Clywkr.* **61** [728], 277, December (1952).   690, 691

— W. G. HAWLEY, see H. BENNETT.

H51. R. F. HAYMAN, *Town's Gas For Drying By Radiant Heat.* Paper read at the Fuel Efficiency Committee Conference on 'Fuel and the Future', London, 8, 9 and 10, October 1946. Ministry of Fuel, London.   836

H52. A. HEATH, *A Handbook of Ceramic Calculations.* Webberley Ltd., Stoke-on-Trent (1937).   293

H53. F. R. C. HEATH, Modern developments in filter-pressing. *Trans. Brit. Cer. Soc.* **52** [4], 153 (1953).   703, 704

— E. HEATON, see R. W. GRIMSHAW.

H54. J. A. HEDVALL, Dissertation, Upsala (1915).   187, 189

H55. J. A. HEDVALL, *Z. anorg. Chem.* **98**, 57 (1916).   183, 185, 187

H56. J. A. Hedvall and J. Heuberger, *Z. anorg. Chem.* **116**, 137   181
(1921).
H57. J. A. Hedvall and J. Heuberger, *Z. anorg. Chem.* **128**, 1   183, 185
(1923)
H58. J. A. Hedvall and J. Heuberger, *Z. anorg. Chem.* **135**, 49   183, 185, 187
(1924).
H59. J. A. Hedvall, *Svensk. Kemisk Tidskr.* 166 (1925).   183, 185, 187
H60. J. A. Hedvall and E. Norström, *Z. anorg. Chem.* **151**, 1   183, 185, 187,
(1926).   587
H61. J. A. Hedvall and P. Sjöman, *Z. El. Chem.* **37**, 130 (1931).   189
H62. J. A. Hedvall, E. Garping, N. Lindekrantz and L.   185
Nelson, *Z. anorg. Chem.* **197**, 399 (1931).
H63. J. A. Hedvall and T. Nilsson, *Z. anorg. Chem.* **205**, 425   187
(1932).
H64. J. A. Hedvall and G. Schiller, *Z. anorg. Chem.* **221**, 97   189
(1934).
H65. J. A. Hedvall and A. Eldh, *Z. anorg. Chem.* **226**, 192   187
(1936).
H66. J. A. Hedvall and L. Leffler, *Z. anorg. Chem.* **234**, 129   189
(1937).
H67. J. A. Hedvall and V. Ny, *Z. anorg. Chem.* **235**, 149 (1937).   183
H68. J. A. Hedvall and S. O. Sandberg, *Z. anorg. Chem.* **240**,   183
15 (1938).
H69. J. A. Hedvall and Kaj Olsson, *Z. anorg. Chem.* **243**, 237   183
(1940).
H70 J. A. Hedvall, N. Isakson, G. Lander and S. Pålsson,   183
Z. anorg. Chem. **248**, 229 (1941).
H71. J. A. Hedvall and K. Andersson, *Scient. Papers, Inst.*   183, 185
Phys. Chem. Research, Tokyo **38**, 240 (1941).
H72. J. Arvid Hedvall, *Einführung in die Festkörperchemie.*   173, 180–189,
Friedr. Vieweg und Sohn, Braunschweig (1952).   194
H73. J. Arvid Hedvall, Alf Björkengren and Bengt Rähs,   485
Untersuchungen am System MgO–Al$_2$O$_3$–SiO$_2$ mit
BaO als Zusatzmittel. *Acta Chem. Scand.* **7**, 849
(1953).
H74. Fr. Hegemann and H. Zoellner, Über die quantitative   269, *271*
spektrochemische Gesamtanalyse von Silikaten. *Glas-*
Email-Keramo-Technik, **3** [8, 9, 10 and 11] (1952).
H75. Raymond A. Heindl and Lewis E. Mong, Length changes   79
and endothermic and exothermic effects during heating
of flint and aluminous clays. *J. Res. Nat. Bur. Stand.*
23, 427, September (1939).
H76. C. Heinel, Steinzeug als Werkstoff im Apparate- und   1110
Maschinenbau. *Chemische Apparatur* **10** [11], [12], 85,
93, June (1923).
— H. Hellbrugge, see H. Lehmann.
H77. P. W. Heller, 89 Windermere Ave., London, N.3.   671
— F. C. Henderson, see J. H. Koenig.

H78. S. B. HENDRICKS and L. T. ALEXANDER, *J. Soc. Agronom.* 455 (1940). — 307

H79. S. B. HENDRICKS, Base exchange of the clay mineral montmorillonite for organic cations and its dependence on adsorption due to van der Waal's forces. *J. Phys. Chem.* 45, 65 (1941). — 56

H80. D. B. HENDRYX, *Producer Gas as a Fuel for Burning Refractories.* Amer. Refr. Inst. Tech. Bull. No. 80, October (1944). — 919, 921

H81. DWIGHT B. HENDRYX, Control and segregation in dry grinding. *Brick & Clay Record* 124 [1], 45 (1954). — 684

H82. S. HERBST, Elektrisch beheizte Schmelzöfen für die Porzellanindustrie. *Euro-Ceramic* 6 [3], 65 (1956). — 1022

— EDWARD DAVID HERSEE, see LONDON BRICK CO. LTD.

H83. A. R. HEUBACH, Notes on the cause of matness in glazes. *Trans. A. Cer. S.* 15, 591, February (1913). — 592, 594–595

— J. HEUBERGER, see J. A. HEDVALL.

— E. HÉVÉZI, see G. J. BAJO.

H84. HAROLD HEYWOOD, The scope of particle size analysis and standardization. *Symposium on Particle Size Analysis*, p. 14. Suppl. to *Trans. Inst. Chem. Eng.* 25 (1947). — 278, 288–290

H85. J. C. HICKS and BEN DAVIES, High temperature properties of magnesia refractories. *Iron Age*, 11 August (1949). — 1256

H86. A. HIELSCHER, *Brennfehler, ihr Ursachen und Beseitigung.* Verlag Tonindustrie-Ztg., Berlin N.W.21. — 880, 997

H87. L. HIESINGER and H. KÖNIG, *Heraeus-Festschrift*, pp. 376–392 (1951). — 202

— R. HIGGINS, see N. F. EATON.

H88. HILGER AND WATTS, Hilger Division, 98, St. Pancras Way, Camden Rd., London, N.W.1. — *268, 269, 270, 272, 273, 312*

H89. CHARLES HILL, Continuous brick kilns. *Ceramics* 4 [47], 496 (1953). — 997–1001

H90. E. C. HILL, Chromium green stains. *J. A. Cer. S.* 15, 378 (1932). — 618

H91. W. F. HILLEBRAND, *The Analysis of Silicate and Carbonate Rocks.* Government Printing Office, Washington (1919). — 266

— A. HILLIARD, see W. S. NORMAN.

H92. F. R. HIMSWORTH, Determination of the thermal conductivity of insulating bricks. *Trans. Brit. Cer. Soc.* 56 [7], 345 (1957). — 354

H93. S. R. HIND, A visit to the osmosis plant at Carlsbad. *Trans. Cer. Soc.* 24, 73 (1924–1925). — 248

H94. S. R. HIND, Notes on fritting. *Trans. Cer. Soc.* 24, 329 (1924–1925). — 1057

H95. S. R. HIND, *Pottery Ovens, Fuels and Firing.* B.P.R.A. (1937). — 916, 917, 930

H96. STANLEY HIND, A new aspect of the chemical control of ceramic production. *Refr. J.* 25, 408 (1949). — 381

H97. S. R. HIND, Drying in the pottery industry. *J. Inst. Fuel* *845, 846,*
24, 116 (1951).   *848–849*
H98. P. E. HINES, Federation House, Stoke-on-Trent.   104
H99. TONWERK HINTERMEILINGEN GEBR. SCHMIDT, Limburg   49
a. d. Lahn, Germany.
H100. ARTHUR R. VON HIPPEL, *Dielectric Materials and Appli-*   518, 1170,
cations. Techn. Press. M.I.T. and John Wiley, New   1174, 1196,
York.   1204–1208
H101. SUKE HIRANO, SHINCHIRO OGAWA and SHIGEO SAWA-   440
MURA, Dolomitic earthenware. *Rep. Imp. Cer. Exp.*
Inst. Kyoto (Tojikishikensho) [14], 1–56, February (1936).
H102. HANS HIRSCH, Die wirkung verschiedener Formen von   466
Kieselsäure in Porzellanmassen. *Ker. Rund.* 34 [43]
(1926).
H103. H. HIRSCH and W. DAWIHL, The action of phosphoric acid   300
on ceramic raw and fired materials and a new method for
the rational analysis of clays. *Ber. D. K. G.* 13, 54
(1932).
— J. E. HOAGBIN, see R. W. SMITH.
— E. HOFFMANN, see W. JANDER.
H104. U. HOFMANN, K. ENDELL and D. WILM, *Z. Krist.* 86, 340   *20, 181*
(1933).
H105. U. HOFMANN and W. BILKE, *Koll. Z.* 77, 238 (1936).   *55*
H106. U. HOFMANN, Neue Erkenntnisse auf dem Gebiete der   59, *60*
Thixotropie, insbesondere bei tonhaltigen Gelen.
Koll.-Zt. 125 [2], 86 (1952). U. HOFMANN, ARMIN
WEISS and R. FAHN, Nachweis der Gerüststruktur in
thixotropen Gelen. *Naturwissenschaft* 39 [15], 351
(1952). U. HOFMANN, R. FAHN and ARMIN WEISS.
Thixotropie bei Kaolinit und innerkristalline Quellung
bei Montmorillonit. *Koll.-Z.* 151 [2], 97 (1957).
— U. HOFMANN, see K. ENDELL.
— U. HOFMANN, see R. FAHN.
H107. R. HOHLBAUM, Über keramische Farben. *Glashütte* 67   612
[41, 43, 44, 45], 633, 660, 677, 692 (1937).
H108. *Holdcroft and Co.'s Thermoscope.* Agents: Thomas Hulme   937
(Hanley) Ltd. and Harrison & Son (Hanley) Ltd.
— A. D. HOLDCROFT, see W. JACKSON.
H109. D. A. HOLDRIDGE, H. A. NANCARROW and MARCUS   97
FRANCIS, The effect of time and temperature on the
density and crushability of flint. *Trans. Brit. Cer. Soc.*
41, 149 (1941–1942).
H110. J. HOLLAND and W. J. GARDNER, Improvements in drying   847
refractories or other goods. *Trans. Cer. Soc.* 23, 16
(1923–1924).
H111. A. BERNARD HOLLOWOOD, The firing of pottery. *Ind.*   971, 974
Chem. 267, November (1941).
— M. E. HOLMES, see C. M. DODD.
— W. HORAK, see C. W. PARMELEE.

— Kazuhiko Hori, see Chiytoshi Kajisaki.
— A. M. Hossenlopp, see H. L. Crowley.
— Houden, see Cealglske.

H112. Ministry of Housing and Local Government, *Model Byelaws*, Series IV, *Buildings*. H.M.S.O. (1953).   605

H113. P. Howard and A. L. Roberts, A study of alumina suspensions. I—Effect of electrolytes on stability. II—Effect of ageing on the pH. *Trans. Brit. Cer. Soc.* **50** [8], 339 (1951); **52** [7], 386 (1953).   1138

H114. I. T. Howarth and W. E. S. Turner, *Soc. Glass Techn.* **14**, 402 (1937).   181

H115. J. Allen Howe, *A Handbook to the Collection of Kaolin, China-Clay and China-Stone in the Museum of Practical Geology*. Darling & Son, Bacon St. E., London (1914).   244

H116. J. M. Huber Corporation, *Kaolin Clays and Their Industrial Uses* (1949).   237, 244, 246

H117. G. F. Hüttig, M. Ehrenberg and H. Kittel, *Z. anorg. Chem.* **228**, 112 (1937).   187

H118. G. F. Hüttig, *Koll.-Z.* **99**, 266 (1942).   181, 183

H119. G. F. Hüttig, *Ber. deut. chem. Ges.* **75**, B, 1573 (1942).   181

— G. F. Hüttig, see H. Kittel.

H120. R. R. Hughan, Cordieritic saggars of increased durability from Australian talc and clays. *Aust. J. Appl. Sci.* **3** [2], 173 (1952).   412, 490

H121. F. A. Hummel, Thermal expansion properties of some synthetic lithia minerals. *J. A. Cer. S.* **34** [8], 235–240 (1951).   501

H122. F. A. Hummel, Ceramics for thermal shock resistance. *Cer. Ind.* (1955–1956).   1165–1167

H123. F. A. Hummel, Glazing composition for structural clay products and process for making same. *U.S. Pat.* 2871132; Abs. *Brit. Clywkr.* **67** [803], 336, March (1959).   582

— F. A. Hummel, see M. D. Karkhanavala.
— F. A. Hummel, see Krishma Murthy.
— F. A. Hummel, see T. I. Prokopowicz.

H124. Ercel B. Hunt and Kirkby K. Allen, The quality control in a refractories plant. *Refr. J.* **33** [10], 454 (1957).   1229

— K. Hunter, see G. W. Brindley.

H125. P. A. Huppert, *Cer. Age* **59** [4], 50 (1952).   610

H126. R. K. Hursh and W. R. Morgan, Determination of factors affecting size variation in clay products. *J. A. Cer. S.* **31** [11], 299 (1948).   83, 865

H127. R. K. Hursh, Development of a porcelain vacuum tube. *J. A. Cer. S.* **32** [3], 75 (1949).   472

H128. T. L. Hurst and E. B. Read, Survey of literature on slag tests for refractory materials. *J. A. Cer. S.* **25** [11], 283, July (1942).   363

— T. L. Hurst, see A. F. Greaves Walker.

H129. EDMOND P. HYATT, CARL J. CHRISTENSEN and IVAN B. CUTLER, Sintering of zircon and zirconia with the aid of certain additive oxides. *Bull. A. Cer. S.* **36** [8], 307 (1957). — 508, 509
— EDMOND P. HYATT, see IVAN B. CUTLER.
H130. COLLIN HYDE, J. F. QUIRK and W. H. DUCKWORTH, Preparation and properties of dense beryllium oxide. *Ceramics* **10** [120], 10 February (1959). — 505–506, 1139, 1140, 1141
H131. HYDE et al., *Nuclear Engineering*, Pt. IV *Chemical Engineering*. Progress Symposium Series No. 19, Vol. 52, p. 105. Amer. Inst. of Chem. Eng. — 1140
H132. J. F. HYSLOP and A. MCMURDO, The thermal expansion of some clay minerals. *Trans. Cer. Soc.* **37**, 180 (1938). — 79, *80*
H133. J. F. HYSLOP, J. STEWART and N. BURNS, Corrosion and the fluxing of refractory–glass mixtures. *Trans. Brit. Cer. Soc.* **46** [12], 377 (1947). — 364, 1059, 1060

I1. TOSHIO IKEGAKI, Chrome pink. *Bull. Govt. Res. Ins. Cers.* (*Kyoto*) **5** [4], 95 (1951). — 612, 622
— K. W. ILLNER, see C. KRÖGER.
I2. IMPERIAL INSTITUTE, *The Mineral Industry of the British Empire and Foreign Countries: China-Clay (Kaolin)*. H.M.S.O. (1931). — 244, 245, 248
I3. EARL INGERSON, G. W. MOREY and O. F. TUTTLE, The systems K_2O–ZnO–SiO_2, ZnO–B_2O_3–SiO_2, and Zn_2SiO_4 and Zn_2GeO_4. *Amer. J. Sci.* **31**, 246 (1948). — 133
I4. D. D. INNES, Spectrographic analysis of some minor constituents in refractories. *Trans. Brit. Cer. Soc.* **57** [1], 29 (1958). — 272
I5. H. INSLEY, The microstructure of earthenware. *J. A. Cer. S.* **19** [5], 317 (1927). — 543
I6. HERBERT INSLEY and RAYMOND H. EWELL, Thermal behaviour of kaolin minerals. *J. Res. Nat. Bur. Stand.* **14**, 615, May (1935). — 77, 192
— HERBERT INSLEY, see F. P. HALL.
I7. INTERNATIONAL COMBUSTION LTD., 19, Woburn Place, London, W.C.1. — 251, *252, 280, 283*
I8. INTERNATIONAL FURNACE EQUIPMENT CO. LTD., *Cascade Grates*. Imperial Works, Clement Street, Birmingham. — 909
I9. INTERNATIONAL NICKEL CO. — 166
— N. ISAKSON, see J. A. HEDVALL.
I10. HANS ITTNER, Salt glazing in continuous kilns. *Brit. Clywkr.* **68** [805], 58, May (1959). — *1003*
I11. I.V.O. ENGINEERING AND CONSTRUCTION CO. LTD., Wood Lane, London, W.12. — *840–841*

J1. C. E. JACKSON, Florida clay in bone china bodies. *Trans. Cer. Soc.* **27**, 151–155 (1927–1928). — 460

J2. Frederick G. Jackson, The oxidation of ceramic wares during firing. VI—The effect of varying the gas flow and of heating on the decomposition of pyrites in clay. *J. A. Cer. S.* **8**, 534 (1925). — 862

J3. Frederick G. Jackson, A chemical study of the absorption of sulphur dioxide from kiln gases by ceramic ware. *J. A. Cer. S.* **9** [3], 154 (1926). — 864

J4. Frederick G. Jackson, Oxidation of ceramic ware during firing. VII—Review. *Bull. A. Cer. S.* **14** [7], 225 (1935). — 862, 863

J5. W. Jackson, *Pottery Gazette* **25**, 1, 126 (1904); *Trans. Cer. Soc.* **3**, 16 (1904). *A Textbook on Ceramic Calculations*, p. 22. London (1904). — 292

J6. W. Jackson and A. D. Holdcroft, The fluxing effect of bone in English china. *Trans. Cer. Soc.* 6–23 (1904–1905). — 460

J7. William M. Jackson II, Low loss electrical bodies from from wollastonite. *Bull. A. Cer. S.* **32** [9], 306 (1953). — 500, 1193, 1206

— William M. Jackson II, see Cabot Minerals.

J8. K. Jacob, *Ker Rund.* **22**, 303 (1914). — 623

— E. Jäckel, see O. Krause.

J9. Carl Jäger, K. G., Höhr-Grenshausen I, Ringstrasse 4, Germany. — 38, 41, 43

J10. R. Jagitsch, Unpublished work. — 181

J11. R. Jagitsch, *Monatsh. f. Chem.* **68** [1], 1 (1936). — 185

J12. R. Jagitsch, *Monatsh. f. Chem.* **68** [2], 101 (1936). — 181

J13. R. Jagitsch, *K. Vet. Akad. (Stockholm) Ark. Kemi* **15A** [17] (1942). — 181, 183, 185

J14. R. Jagitsch, *K. Vet. Akad. (Stockholm) Ark. Kemi* **22A** [5] (1946). — 181

J15. R. Jagitsch and B. Bengston, *K. Vet. Akad. (Stockholm) Ark. Kemi.* **22A** [6] (1946). — 187

J16. J. Jakob, *Chemische Analyse der Gesteine und Silikatischen Mineralien*. Verlag Birkhäuser, Basle (1952). — 266

J17. W. James and A. W. Norris, Some physical properties of glazes. *Trans. Brit. Cer. Soc.* **55** [10], 601 (1956). — 370–372, 536–538, 547, 548, 553

J18. E. A. Jamison and W. H. Bateman, Propane and butane as industrial fuels. *Iron & Steel Engr.* **12** [4], 209 (1935). — 920

J19. W. Jander, *Z. anorg. Chem.* **163**, 1 (1927). — 185

J10. W. Jander, *Z. anorg. Chem.* **166**, 31 (1927). — 187

J21. W. Jander, *Z. anorg. Chem.* **168**, 113 (1928). — 181, 185

J22. W. Jander, *Z. anorg. Chem.* **174**, 11 (1928). — 185, 187, 189

J23. W. Jander, *Z. anorg. Chem.* **190**, 397 (1930). — 181, 187

J24. W. Jander and W. Stamm, *Z. anorg. Chem.* **190**, 65 (1930). — 187

J25. W. Jander and H. Frey, *Z. anorg. Chem.* **196**, 321 (1931). — 183, 185, 187, 189

J26. W. Jander and E. Hoffmann, *Z. anorg. Chem.* **200**, 245 (1931). — 187

J27. W. Jander and E. Hoffman, *Z. anorg. Chem.* **202**, 135 (1931).   183

J28. W. Jander and E. Hoffmann, Die Reaktionen zwischen Kalziumoxyd und Siliziumoxyd (Reaktionen in festen Zustande bei höheren Temperaturen. 11te. Mitteilung). *Z. anorg. Chem.* **218**, 211 (1934).   185, 191

J29. W. Jander and J. Wuhrer, *Z. anorg. Chem.* **226**, 225 (1936).   181, 191

J30. W. Jander and A. Krieger, *Z. anorg. Chem.* **235**, 89 (1937).   185

J31. W. Jander and K. Bunde, *Z. anorg. Chem.* **231**, 345 (1937).   187

J32. W. Jander and K. Bunde, *Z. anorg. Chem.* **239**, 418 (1938).   181

J33. W. Jander and M. Pfister, *Z. anorg. Chem.* **239**, 95 (1938).   181, 183

J34. W. Jander and G. Leuther, *Z. anorg. Chem.* **241**, 57 (1939).   181

J35. W. Jander and K. Grob, Reactions in the solid state at higher temperatures. XXV—Intermediate states occurring in the formation of nickel aluminate from nickel oxide and aluminium oxide in the solid state. *Z. anorg. Chem.* **245**, 67 (1940).   191

J36. W. Jander and W. Wenzel, *Z. anorg. Chem.* **246**, 67 (1941).   181

J37. W. Jander and N. Riel, *Z. anorg. Chem.* **246**, 81 (1941).   187

J38. W. Jander and G. Lorenz, *Z. anorg. Chem.* **248**, 105 (1941).   183

J39. H. Janert, Application of heat-of-wetting measurements to soil research problems. *Z. Pflanzenähr. Düngung u. Bodenk* **19A**, 281 (1931); *Ibid.* **34A**, 100 (1934).   332

J40. G. W. Jarman, Jr., Removal of 'Nonmagnetic' impurities from ceramic materials. *Bull. A. Cer. S.* [5], 126 (1934).   695

J41. K. Jasmund, *Die Silicatischen Tonminerale.* Monographien zu 'Angewandte Chemie' und 'Chemie-Ingenieur Technik.' Verlag Chemie, G.m.b.H., Weinheim (1951).   12, *13, 19, 20, 21, 22, 24*

J42. H. Jebsen-Marwedel, *J. Soc. Glass. Tech.* **21**, 436T (1937).   538

J43. Benjamin A. Jeffery, *U.S. Pat.* 2152738, 17 April (1936).   746

J44. H. W. Jewell, An engineer discusses merits of ceramic glazed clay pipe. *Brick & Clay Record* 59 March (1947).   604

— H. Johns, see G. H. Osborn.

— W. D. Johns, Jr., see R. E. Grims.

J45. Johnson, Matthey & Co. Ltd. and D. A. Davies, Metallization of ceramics. *Brit. Pat. Spec.* 594752, 3 December (1947).   657

J46. Johnson, Matthey & Co. Ltd., *Matthey On-Glaze Transfers for Pottery.* 73–83, Hatton Gardens, London, E.C.1.   830, *831*

J47. Johnson, Matthey & Co. Ltd., *Spectrographically Standardised Substances.* 73–83, Hatton Gardens, London, E.C.1.   269

— Johnson, Matthey & Co. Ltd., see L. Allen.

J48. Johnson and Thomas Machine Works, *The Davis Brown Laboratory Size Deairing-Extrusion Machine.* 123, East 23rd St., Los Angeles 11, California, U.S.A.   709

J49. A. L. JOHNSON and F. H. NORTON, Fundamental study of clay. II—Mechanism of deflocculation in the clay-water system. *J. A. Cer. S.* **24** [6], 189 (1941). 53, 54, 759

J50. A. L. JOHNSON and F. H. NORTON, Fundamental study of clay. III—Casting as a base-exchange phenomenon. *J. A. Cer. S.* **25** [12], 336 (1942). 762

— A. L. JOHNSON, see E. A. HAUSER.
— A. L. JOHNSON, see F. H. NORTON.
— H. C. JOHNSON, see C. G. HARMAN.

J51. J. A. JOHNSON and H. L. STEELE, Slip house technique in general earthenware production. *Trans. Brit. Cer. Soc.* **39** [6], 182 (1940). 764, 1094

J52. J. R. JOHNSON, S. D. FULKERSON and A. J. TAYLOR, *Technology of Uranium Dioxide, a Reactor Material.* 1143

J53. S. H. JOHNSON AND CO. LTD., *'Pyramid' Filter Presses.* Carpenters Rd, Stratford, E.15. 702–704

— W. JOHNSON, see A. G. COCKBAIN.

J54. WM. JOHNSON AND SONS (LEEDS) LTD., Castleton Foundry, Armley, Leeds 12. 672, 673, 698, 748

J55. THE JOINTLESS FIREBRICK CO. LTD., *Plibrico Suspended Arches and Roofs.* Egyptian House, 170, Piccadilly, W.1. 899

J56. J. S. JONES, *Notes on the Efficiencies of Ceramic Dryers.* Read at a Meeting of the Institute of Fuel, 22 April (1953). 851, 853

— P. R. JONES, see R. E. PYLE.

J57.—

— A. JOURDAIN, see H. CASSAN.

J58. DEANE B. JUDD, Methods of designating color. *Bull. A. Cer. S.* **20** [11], 375, November (1941). 228

J59. E. JUNKER, *Z. anorg. Chem.* **228**, 97 (1936). 170

J60. H. R. JUNKISON, Heat treatment of moulded carbon products. *Ceramics* **4** [38], 90, April (1952). 1118

K1. CHIYTOSHI KAJISAKI and KAZUHIKO HORI, Organic Defloculant. II—Effects of polyvinylamine and a new use of potassium polyacrylate. *Bull. Govt. Res. Inst. Cer. (Kyoto)* **6** [1], 25–28 (1952); *J. A. Cer. S.* Abs. 36, 10 (1953). 163, 761

K2. KALLAUNER and MATEJKA, *Sprechs.* **47**, 423 (1914). 299

K3. KALLE & Co. A.G., Wiesbaden-Biebrich, Germany. 156

— H. KALSING, see G. TAMMANN.
— E. KALTNER, see G. BOUVIER.

K4. M. KANTZER, Un nouvel appareil enregistreur d'analyse thermique différentielle. *Bull. Soc. Fran. Cér.* [15], 21 (1952). 313

K5. KAOLIN INC., (now Avery County Plant of Harris Clay Co.), *Carolina Kaolin. A Modern Industrial Romance.* North Carolina, U.S.A. (1937). 251

K6. M. D. KARKHANAVALA and F. A. HUMMEL, The polymorphism of cordierite. *J. A. Cer. S.* **36**, 389 (1953). 483

K7. M. KARSULIN, University of Zagreb. Personal communication. 759

K8. KARL M. KAUTZ, Enamels of standard colors for vitreous china sanitary ware at cone 6. *J. A. Cer. S.* **16**, 192 (1933). 619–643

K9. H. KEDESDY, Elektronenmikroskopische Untersuchungen über den Brennvorgang von Talk und Speckstein. (Electron microscope investigation of firing phenomena in talc and steatite.) *Ber. D. K. G.* **24** [6, 7], 201–232 (1943). 91, 192

K10. P. S. KEELING, Mineralogische Untersuchung von Tonen. *Sixth International Ceramic Cong., Wiesbaden.* Abs. *Euro-Ceram.* **8** [10], 260 (1958). 307

K11. S. V. KEELING & Co., Ltd., South St., Bradley, Stoke-on-Trent. 772

— J. J. KEILEN, see G. C. ROBINSON.

— W. B. KEITH, see G. W. MILES.

K12. K. K. KELLEY, *U.S. Bureau of Mines, Bulletin* No. 384 (1935). 158

K13. HERBERT H. KELLOGG, *Flotation of Kaolinite for removal of Quartz.* A.I.M.E. Tech. Publ. No. 1753; *Min. Tech.* January (1945). 251

K14. E. KEMPE, Gegen den Stückenkalk im Ton. *Tonind. Ztg.* **59** [6], 67 (1935). 694

— W. KENNEDY, see *Symposium on Ceramic Cutting Tools.* S219

K15. GEORGE KENT LTD., *The Kent Continuous pH Recorder.* Luton, Beds. *383*, 384

K16. GEORGE KENT LTD., *Kent Multelec Pyrometers; The Kent Multelec for Temperature Control.* Luton and London. 945, *946*, 949, 952

K17. KENYA KYANITE LTD., Incorporated in Kenya Colony. East Africa. Continental Representative: Kenya Kyanite (Sales), Ltd., 28, Eccleston St., London, S.W.1. 112

K18. GUSTAV KEPPELER, Unterscheidungsmerkmale der Keramisch wichtigen Tone. *Ber. D. K. G.* **10** [11], 50 (1929). 299

K19. GUSTAV KEPPELER and HANS GOTTHARDT, *Untersuchungen über Kaoline und Tone.* Verlag des Sprechsaal, Müller und Schmidt, Coburg (1930). 299

K20. G. KEPPELER, Neuere Untersuchungen über Tonmineralien. *Ber. D. K. G.* **19** [5], 159 (1938). 311

K21. KERABEDARF (Keramische Industrie-Bedarfs-Aktiengesellschaft), Berlin-Charlottenburg, Berliner Str. 23, Germany. 846, 916, 1042

K22. BRUNO KERL, *Handbuch der gesammten Thonwaarenindustrie*, p. 479. Vieweg und Sohn, Braunschweig (1907). 864

K23. W. KERSTAN, Beitrag zur Verdampfung keramischer Glasuren. *Ber. D. K. G.* **30** [11], 254 (1953). 540, 541

K24. Heinrich Kienberger, Der Spannungsriss und seine Ver- 442
hinderung bei sanitären Steingutspülwaren. *Ber. D. K. G.* 459, November (1940).

K24.1. Kilns and Furnaces Ltd., Keele St. Works, Tunstall, *981*
Stoke-on-Trent.

— B. W. King, see A. K. Smalley.
— R. M. King, see W. K. Carter.

K25. William David Kingery, Fundamental study of phosphate 522
bonding in refractories—I–IV. *J. A. Cer. S.* **33** [8],
239–250 (1950); *Ibid.* **35** [3], 61 (1952).

K26. W. D. Kingery, Note on thermal expansion and micro- 224
stresses in two-phase compositions. *J. A. Cer. S.* **40**
[10], 351 (1957).

K27. Percy C. Kingsbury, Stoneware in the electrochemical 1179
field. *Trans. Electrochem. Soc.* **75**, 131 (1939).

K28. D. Kirby, Pure oxide refractories. *Metallurgia.* June 1141
(1944).

— S. S. Kistler, see W. H. Davenport.

K29. H. Kittel and G. F. Hüttig, *Z. anorg. Chem.* **210** 26 187
(1933).

K30. H. Kittel and G. F. Hüttig, *Z. anorg. Chem.* **217**, 194 181, 183, 187
(1934).

K31. H. Kittel, *Z. phys. Chem.* **178**, 81 (1936). 181, 187

— H. Kittel, see G. F. Hüttig.

K32. A. A. Klein and G. H. Brown, Microscopic investigation 596
of some compounds noted in the systems soda–zinc oxide
–titanic oxide–silica. *Trans. A. Cer. S.* **17**, 745 (1915).

K33. Kleinpaul, Caolini Argille ed Affini S.A.R.L., Milan. 708
Massenaufbereitung ohne Filter pressen. *Ber. D. K. G.*
35 [9], 309 (1958); Erfahrung bei der Ausarbeitung einer
Arbeitsvorschrift für die 'Massenaufbereitung ohne
Filterpressen'. *Ber. D. K. G.* **36**, [4], 113 (1959).

K34. R. Kleinsteuber, Der Technische Fortschritt. *Euro-* 931
Ceramic **9** [6], 165 (1959).

K35. Theron A. Klinefelter, Robert G. O'Meara, Sidney 296
Gottlieb and Glenn C. Truesdell, *Syllabus of Clay
Testing*, Pt. I. Bureau of Mines Bulletin 451, United
States Department of the Interior (1943).

K36. Tonwerk Klingenberg am Main, Germany. 51

K37. J. Klug, Fehlerhafte Steingutmassen und Glasuren. 554
Sprechs. **66** [40, 41], 675, 693 (1933).

K38. Jacob Klug, Chemische oder rationelle Analyse als Grund- 301
lage für die Berechnung und Kontrolle keramische
Massen? *Sprechs.* **83**, 7, 25, 50, 65 (1950).

K39. Jacob Klug, Vorschlag für den Aufbau und die Berechnung 490
Steatitmassen. *Sprechs.* **91** [2], 22, 20 January (1958).

K40.—

K41. W. K. Knapp, Use of free energy data in the construction 215
of phase diagrams. *J. A. Cer.* **36** [2], 43 (1953).

BIBLIOGRAPHY AND AUTHORS INDEX

K42. R. W. Knauft, Bonded refractories for special purposes. 1230
The Glass Industry August and September (1949).
K43. R. W. Knauft, K. W. Smith, E. A. Thomas and W. C. 406, 408, 1247
Pittman, Bonded mullite and zircon refractories for
the glass industry. *Bull. A. Cer. S.* **36** [11], 412 (1957).
K44. Knett, *Tonind.-Ztg.* 352, 495, 531 (1896). 601
K45. Maurice A. Knight, Akron 9, Ohio, U.S.A. *1109*
K46. E. Koehler, L'analyse thermique complexe et son utilisa- 316
tion dans les recherches physicochimiques et techniques.
Bull. Soc. Fran. Cér. [38], 3, January–March (1958).
K47. C. J. Koenig and A. S. Watts, The durability of tableware 379, 609
decorations. *J. A. Cer. S.* **17** [9], 259 (1934).
K48. C. J. Koenig, Spit out. *J. A. Cer. S.* **19**, 287 (1936). 881
K49. C. J. Koenig, Cone 04 vitreous bodies with lepidolite and 440, 452
nepheline syenite. *Foote Prints* **122** [1], 19 (1950).
K50. E. W. Koenig, Calculation of mineralogical composition of 301
feldspar by chemical analysis. *J. A. Cer. S.* **25** [14], 420
(1942).
K51. E. W. Koenig, Froth flotation as applied to feldspar. *Cer.* 101
Age **47** [5], 192 (1946).
K52. F. J. Koenig, Coarser flint aids fitting of glaze. *Cer. Ind.* 783
40 [2], 62 (1943).
— H. König, see L. Hiesinger.
K53. J. H. Koenig, *Cer. Ind.* **27**, 10 (1936). 614
K54. J. H. Koenig, Lead frits and fritted glazes. *Cer. Ind.* **26**, 570–575
134 (1936); Lead frits and fritted glazes. *Ibid.* **27**, 108
(1936); *Lead Frits and Fritted Glazes.* Bulletin 95,
Ohio State University, Engineering Experiment Station,
Columbus, Ohio (1937).
K55. J. H. Koenig and F. C. Henderson, Particle size distribu- 778, *779–781*
tion of glazes. *J. A. Cer. S.* **24**, [9], 286 (1941).
K56. John H. Koenig, Ceramics. *Ind. Eng. Chem.* **41**, 2102, 1208
October (1949).
K57. John H. Koenig and Nicholas H. Snyder, Ceramics. 413, 511
Ind. Eng. Chem. **43**, 2208, October (1951).
K58. John H. Koenig and Edward J. Smoke, Adapting research 382, 384
tools for process controls in the whitewares industry.
Cer. Age **60** [2], 17 (1952).
K59 follows after **K93**.
— John H. Koenig, see Edward J. Smoke.
K60. Claus Koeppel, *Feuerfeste Baustoffe.* S. Hirzel, Leipzig 111, 195, 1249,
(1938). 1259
— N. Köppen, see A. Dietzel.
K61. Koerner, *Sprech.* **36**, 775 (1903). 299
K62. H. Kohl, Zur Trockenfestigkeit der Tone. *Ber. D. K. G.* 75, 76, 331,
11 [6], 325 (1930). 335.
K63. H. Kohl, Die Haltbarkeit der Gebräuchlichen Porzellan- 379
farben in der Geschirrspülmaschine und im Labora-
torium. *Sprechs.* **64** (1931).

K64 follows after **K65**.

K65. HANS KOHL, Die roten und gelben Farbkörper in der Keramik.
    (a) I—Die roten Farbkörper. *Ber. D. K. G.* **16** [4], 169 (1935).   617–637, 650
    (b) II—Die gelben Farbkörper. *Ber. D. K. G.* **17** [2], 597 (1936).   616–642

— N. E. KOLYUN, see N. K. ANTONEVICH.

— RENICHI KONDO, see TOSHIYOSHI YAMAUCHI.

K64. K. KONOPICKY, Chromite containing refractories. *Stahl. und Eisen* **61**, 53 (1941).   421

K66. HEINRICH KOPPERS AND EUROPAISCHE KOPPERS, Düsseldorf-Heerdt, Germany.   1243, 1251, 1258–1267

K67. M. KORACH, Theorie du four-tunnel et cuisson rapide 'Sandwich'. *Acta Technica* **11** [1–2], 161 (1955) (Budapest).   1011

— E. KORDES, see G. TAMMANN.

K68. H. V. KOUGH, Hardsurfacing for structural clay products machinery. *Bull. A. Cer. S.* **35** [6], 228 (1956).   165

K69. F. C. KRACEK, *J. Phys. Chem.* **34**, 2641 (1930).   198

K70. F. C. KRACEK, Cristobalite liquidus in alkali oxide–silica systems and heat of fusion of cristobalite. *J. A. Chem. Soc.* **52** [4], 1436 (1930); *Cer. Abs.* **9** [9], 783 (1930).   402

K71. F. C. KRACEK, Handbook of physical constants. *Geol. Soc. America Special Papers* [36], 140–174, 31 January (1942).   170

K72. F. C. KRACEK, *Phase Transformations in One-Component Silicate Systems*. Geophysical Laboratory, Carnegie Inst. of Washington, Bulletin No. 1150 (1951).   171, 194, 195

K73. HOBART M. KRANER, Use of eutectics as glazes. *J. A. Cer. S.* **9** [5], 319 (1926).   536

K74. HOBART M. KRANER and E. H. FRITZ, The rate of oxidation of porcelain and ball clays. *J. A. Cer. S.* **12** [1] 1 (1929).   859

K75. A. O. KRAUSE, Anwendung von Röntgenographie und Fluoreszenz-Strahlung in der Feinkeramik. *Ber. D. K. G.* **8** [5], 114 (1927).   78

K76. O. KRAUSE and E. JÄCKEL, Ueber specksteinhaltige Massen des Dreistoffsystems $MgO–Al_2O_3–SiO_2$. *Ber. D. K. G.* **15**, 485 (1934).   191

K77. F. KRAUSS and R. GRAMSS, Der Eisen-Titan-Spinell als Ursache beim Brennen von Porzellan auftretender Verfärbungen und deren Beseitigung. *Ker. Rund.* **43** [16–20] (1935).   465

— JUNIUS F. KREHBIEL, see ROSS C. PURDY.

K78. NORBERT J. KREIDL and WOLDEMAR A. WEYL, Phosphates in ceramic ware (phosphate glasses). *J. A. Cer. S.* **24**, 377 (1941).   588

K79–K80 follow **K81**.

K81. Norbert J. Kreidl and Woldemar A. Weyl, The development of low melting glasses on the basis of structural considerations. *The Glass Industry* **23** [9, 10, 11, 12], 335, 384, 426, 465 (1942).   533, 547, 559, 560

K79. Norbert J. Kreidl, Problems relating to glass structure. Crystal chemistry symposium. *Cer. Age* November (1948).   214

K80. A. Kreiling, Abbau und Förderung im Tonbergbau. *Euro-Ceramic* **8** [3], 52 (1958).   257

K81 precedes **K79**.

— A. Krieger, see W. Jander.

K82. H. J. S. Kriek, W. F. Ford and J. White, The effect of additions on the sintering and dead-burning of magnesia. *Trans. Brit. Cer. Soc.* **58** [1], 1 (1959).   418

— J. J. Krochmal, see W. H. Duckworth.

K83. C. Kröger and E. Fingas, *Z. anorg. Chem.* **213**, 12 (1933).   181
K84. C. Kröger and E. Fingas, *Z. anorg. Chem.* **223**, 257 (1935).   181
K85. C. Kröger and E. Fingas, *Z. anorg. Chem.* **224**, 289 (1935).   181
K86. C. Kröger and E. Fingas, *Z. anorg. Chem.* **225**, 1 (1935).   181
K87. C. Kröger and E. Fingas, *Festschr. d. T. H. Breslau* 298 (1935)   181

K88. C. Kröger and K. W. Illner, *Z. anorg. Chem.* **229**, 197 (1936).   183, 185

K89. C. Kröger and E. Fingas, *Glastechn. Ber.* **15** [9, 10, 11], 335, 371, 403 (1937).   181, 183, 185

K90. August Kruis and Hans Späth, Forschungen und Fortschritte auf dem Gipsgebiet seit 1939. *Tonind.-Ztg.* **75** [21–24], 341, 395 (1951).   163

— H. Krutter, see B. E. Warren.

K91.—

K92. J. Kwederawitsch, Development and possible uses of snakeskin glazes. *Ceramics* **8** [94], 304, December (1956), being an extract from *Silikat Technik.* **7** [4] (1956).   539, 581, 629

K94. Kyanite Mining Corporation, Cullen, Virgina, U.S.A.   112

K59. J. C. Kyonka and R. L. Cook, The effect of fluxes on the properties of semi-vitreous dinnerware. *Bull. A. Cer. S.* **32** [7], 233 (1953).   444

— Lachmann, see Appl.

L1. H. R. Lahr, Steel-ladle trials on fireclay bricks. *Trans. Brit. Cer. Soc.* **53**, 621 (1954).   1235–1237

L2. H. R. Lahr and C. W. Hardy, Influence of porosity on silica roof performance. *Trans. Brit. Cer. Soc.* **57** [5], 271 (1958).   404

L3. L. R. Lakin, Determination of the elastic constant of refractories by a dynamic method. *Trans. Brit. Cer. Soc.* **56**, [1], 1 (1957).   346

L4. L. R. LAKIN and C. S. WEST, Determination of the modulus of elasticity of refractories at high temperatures. *Trans. Brit. Cer. Soc.* **56** [1], 8 (1957). — 346

— G. LANDER, see J. A. HEDVALL.

L5. S. M. LANG, L. H. MAXWELL and R. F. GELLER, Some physical properties of porcelains in the systems magnesia–beryllia–zirconia and magnesia–beryllia–thoria and their phase relations. *J. Res. Nat. Bur. Stand.* **43** [5], 429 (1949). — 506, 1152

L6. S. M. LANG, L. H. MAXWELL and M. D. BURDICK, *J. Res. Nat. Bur. Stand.* **45** [5], 366 (1950). — 506, 1152

L7. S. M. LANG and C. L. FILLMORE, Refractory porcelains of beryllia–alumina–titania. *Bull. A. Cer. Soc.* **30** [3], 106 (1951). — 506, 1153

L8. BRUNO LANGE, *Kolorimetrische Analyse* (4th Ed.). Verlag Chemie (1952). — 267, 272

L9. RICHARD LAMAR, *Particle Shape and Differential Shrinkage of Steatite Talc Bodies.* Sierra Talc Co., April (1944). — 498

L10. R. S. LAMAR, Development of cordierite bodies with sierralite, a new ceramic material. *J. A. Cer. S.* **32** [2], 65 (1949). — 120, 484, 487

L11. R. S. LAMAR and M. F. WARNER, Reaction and fired property studies of cordierite compositions. *J. A. Cer. S.* **37** [12], 602 (1954). — 484, 486

L12. FRANK D. LAMB and JOHN RUPPERT, *Flotation of North Carolina Pyrophyllite Ore.* Bureau of Mines. Report of Investigation 4674 (1950). — 94, 306

L13. C. MAJOR LAMPMAN and H. G. SCHURECHT, Zinc-vapour glazing of clays. I—Effect of variable iron oxide and alkalies in clays on zinc-vapour glaze colours. *J. A. Cer. S.* **22**, 91 (1939). — 605

L14. C. MAJOR LAMPMAN, Flow of glazes on horizontal and inclined surfaces. *Bull. A. Cer. S.* **17**, 13 (1938). — 538, 587

L15. C. MAJOR LAMPMAN, The effect of different bodies on some wetting and flow characteristics of glazes. *J. A. Cer. S.* **21**, 253 (1938). — 538, 587

— P. LAPOUJADE, see L. HALM.

L16. *Lapp Chemical Porcelain*, Lapp Bulletin No. 196-4-43, 196-11-43. — 1126

L17. M. MARC LARCHEVÊQUE, *Industrie de la Porcelaine.* Revue des Materiaux de Construction et de Travaux Publics, Paris (1929). — 966

— M. L. LASSON, see E. HEIKEL VINTHER.

— JOHN S. LATHROP, see C. W. PARMELEE.

— W. G. LAWRENCE, see F. H. NORTON.

L18. F. M. LEA and R. W. NURSE, *J. Soc. Chem. Ind.* **58**, 277 (1939). — 294

L19. F. M. LEA and R. W. NURSE, Permeability methods of fineness measurement. *Symposium on Particle Size Analysis*, p. 47. Suppl. to *Trans. Inst. Chem. Engs.* **25** (1947).   294, *295*

L20. R. LECUIR, *L'Ind. Cér.* [391], 209 (1948).   157, 744

L21. THE LEEDS FIRECLAY CO. LTD., Wortley, Leeds 12.   *1086*, 1234–1237

L22. LEEDS AND NORTHRUP CO., *Speedomax Type G Pyrometers*. 4907, Stentor Ave., Philadelphia 44, Pa., U.S.A.   945

L23. LEEFRA (ENGLISH CLAYS, LOVERING, POCHIN & CO. LTD.), St. Austell, Cornwall.   1241

— L. LEFFLER, see J. A. HEDVALL.

L24. C. LEGRAND and J. NICOLAS, Contribution à l'étude du dosage du quartz dans les argiles à l'aide des rayons X. *Bull. Soc. Fran. Cér.* [38], 29, January–March (1958).   311

L25. R. G. LEGRIP, Considerations in the design of gas combustion systems on tunnel kilns affecting fuel economy. *Ceramics* **5** [58], 458 (1953).   1021

L26. HANS LEHMANN and URSULA WERTHER, Betriebskontrolle und Qualitätsprüfung in der Mosaikplattenindustrie. *Ber. D. K. G.* **13** [7], 281 (1932).   1078

L27. H. LEHMANN, K. ENDELL and H. HELLBRUGGE, Relation between chemical composition and fluidity of whiteware glazes and its technical significance. *Sprechs.* **73** [35], 307 (1940); *Ibid.* [36], 321 (1940).   372

L28. HANS LEHMANN and WALTER PRALOW, Über die Bestimmung von Natrium, Kalium und Calcium in den Rohstoffen und Fertigfabrikaten der Silikatindustrie mit dem Flammenphotometer. *Tonind.-Ztg.* **76** [3–4], 33 (1952).   272, 274

L29. H. LEHMANN and U. S. SINGH, Der Einfluss der Oberflächenspannung schmelzender Silikate auf den Angriff feuerfester Baustoffe. *Ber. D. K. G.* **34** [11], 353 (1957).   1214

L30. MARCEL LÉPINGLE, Les variations de volume des argiles a haute température. *Bull. Soc. Fran. Cér.* [16], 16, July–September (1952).   80, *81*

— I. G. LESOKLIM, see V. F. ZHURAVLEV.

L31. Y. LETORT, *Produits Réfractaires* (2nd Ed.). Dunod, Paris (1951).   1259–1269

— G. LEUTHER, see W. JANDER.

— ERNEST M. LEVIN, see F. P. HALL.

L32. EARL E. LIBMAN, Some properties of zinc oxide bodies. *J. A. Cer. S.* **5**, 488 (1922).   508

L33. IMFRIED LIEBSCHER and FRANZ WILLERT, *Technologie der Keramik*. Verlag der Kunst (1955).   327

— N. LINDEKRANTZ, see J. A. HEDVALL.

L34. MAX LINSEIS, Zusammenhänge zwischen mineralogischen Aufbau und keramischen Eigenschaften von Kaolinen und Tonen. *Sprechs.* **83** [18–22] (1950).   84–87, 298, 316–318

L35. Max Linseis, Eine kritische Betrachtung von Untersuchungsmethoden für tonige Rohstoffe. (A critical review of methods of investigation for argillaceous raw materials.) *Tonind.-Ztg.* **75**, 277 (1951). 316

L36. Max Linseis, Ein Beitrag zur Plastizitätsmessung. *Sprechs.* **84** [13] (1951). 303, 329, *330*

L37. Liptak Furnace Arches Ltd., *Suspended Walls and Roofs for all Kinds of Furnaces*. 59, Palace St., Victoria St., London, S.W.1. 899

L38. K. Litzow, Neue Wege in der Gipsformherstellung. *Sixth International Ceramic Congress, Wiesbaden*. Abs. *Euro-Ceramic* **8** [10], 262 (1958). 162, 163

L39. Lodge Plugs Ltd., Rugby. *1161*, 1179–1190

L40. W. Ja. Lokschin and M. S. Fedorowa, Rote Farbe zum Malen auf Glas (in Russian). *Stek. Pron* **14** [7], 31 (1938). 612

— Harold R. Lominac, see William C. Bell.

L41. London Brick Co. Ltd., Africa House, Kingsway, London, W.C.2. *238, 348, 668. 688, 749, 804, 957*

L42. London Brick Co. Ltd. and Edward David Hersee, Stewartby, Beds., Grinding mills of the edge runner type. *Brit. Pat.* 699848. 700

L43. George A. Loomis and A. R. Blackburn, Use of soda felspar in whiteware bodies. *J. A. Cer. S.* **29**, 48 (1946). 587

— G. Lorenz, see W. Jander.
— A. D. Loring, see B. E. Warren.

L44. J. Lougnet and H. Forestier, *Compt. rend. Acad. Sci.* **216**, 562 (1943). 183

— G. H. B. Lovell, see G. R. Rigby.
— F. Lowe, see K. Beck.

L45. Lübecker Maschinenbau-Gesellschaft, Karlstrasse 60–92, Lübeck, Germany. 263

— Josef Lucas, see Josef Robitschek.
— H. J. Lüngen, see H. Salmang.

L46. Alexis Luikov, Moisture content gradients in the drying of clay. *Trans. Brit. Cer. Soc.* **35**, 123 (1936). 74

L47. Daniel W. Luks, Beryl and ceramics. *Foote-Prints* **10** [1], 1, June (1937). 506

L48. O. Lutherer, Where do gas-oil burners fit in? *Ind. Heat.* **13**, 1959 (1946). 929

L49. Carolyn Banks Luttrell, Glazes for zircon porcelains. *J. A. Cer. S.* **32** [10], 327 (1949). 563

L50. J. F. Lutz, *Physiochemical Properties of Soils Affecting Soil Erosion*. Univ. of Mo. Agric. Expt. Station Research Bull. No. 212 (1934). 67

L51. K. C. Lyon, Calculations of surface tensions of glasses. *J. A. Cer. S.* **27**, 186 (1944). 539

— K. C. Lyon, see C. W. Parmelee.
L52. Clifford T. Lyons, Casting and slip transportation using compressed air. *Bull. A. Cer. S.* **29** [9], 321 (1950). 765
L53. S. C. Lyons, Practical aspects of particle-size control. *Bull. A. Cer. S.* **20**, 303 (1941). 304
L54. T. R. Lynam, A. Nicholson and P. F. Young, Low-alumina silica bricks. Manufacture. *Trans. Brit. Cer. Soc.* **51** [2], 113 (1952). 1220
L55. E. D. Lynch and A. W. Allen, Nepheline syenite-talc mixtures as a flux in low temperature vitrified bodies. *J. A. Cer. S.* **33** [4], 117 (1950). 454
L56. J. F. Lynch, J. F. Quirk and W. H. Duckworth, Investigation of ceramic materials in a laboratory rocket motor. *Bull. A. Cer. S.* **37** [10], 443 (1958). 1165
— Frobisher Lyttle, see Charles F. Binns.
— L. A. Lyutsareva, see G. N. Voldsevich.

M1. W. J. McCauhey, Mineralogy Department, Ohio State University, Columbus, Ohio. 220
M2. J. S. McDowell, A general view of the testing of refractories. *Amer. Refr. Inst. Tech. Bull.* [28], October (1928). 893, *896*
M3. J. Spotts McDowell, Refractories in the iron and steel industry. *Watkins Cyclopedia of The Steel Industry* (1955 Ed.). 417
— J. Spotts McDowell, see L. A McGill.
M4. H. M. Macey, Some considerations of the time factor in drying. *Brit. Cer. Soc.* **35**, 379 (1936). 834, 851
M5. H. H. Macey, The promotion of the drying of clay by the coagulating effect of acid. *Trans. Cer. Soc.* **34**, 396 (1935). 833
M6. H. H. Macey, Some observations on the safe drying of large fireclay blocks. *Brit. Cer. Soc.* **38**, 469 (1939). 835
M7. H. H. Macey, Clay-water relationships and the internal mechanism of drying. *Trans. Brit. Cer. Soc.* **41**, 73 (1942). 74
M8.—
M9. H. H. Macey, F. Moore and S. T. E. Davenport, *Brit. Cer. Res. Ass.* Private communication to **C41**. 338
M10. L. A. McGill and J. Spotts McDowell, Reaction temperatures between refractories. *Bull. A. Cer. Soc.* **30** [12], 425 (1951). 892, 894–895
— D. H. McGinnis, see W. C. Bell.
M11. G. Mackawa and B. Matsumura, *J. Soc. Chem. Ind. Japan* **46**, 1075 (1943). 181
— J. Mackensie, see E. I. Greaves.
M12. J. F. McMahon, *A Study of Clay Winning and Its Costs in the Provinces of Ontario and Quebec.* Canadian Department of Mines, Report No. 754 (1935). 257, 258

M13. J. C. McMullen, Transverse testing furnace. *J. A. Cer. S.* **25** [13], 389 (1942). — 984

— H. F. McMurdie, see F. P. Hall.
— A. McMurdo, see J. F. Hyslop.

M14. Malcolm McQuarrie, Structural behaviour in the system (Ba, Ca, Sr) TiO_3 and its relation to certain dielectric characteristics. *J. A. Cer. S.* **38** [12], 444 (1955). — 520

M15. Malcolm McQuarrie, Role of domain processes in polycrystalline barium titanate. *J. A. Cer. S.* **39** [2], 54 (1956). — 520

M16. Malcolm McQuarrie, Studies in the system (Ba, Ca, Pb) TiO_3. *J. A. Cer. S.* **40** [2], 35 (1957). — 520

M17. T. N. McVay and C. L. Thompson, X-ray investigation of the effect of heat on china clays. *J. A. Cer. S.* **11**, 829 (1928). — 77, 78, 192

M18. T. N. McVay, *The Effect of Iron and its Compounds on the Color and the Properties of Ceramic Materials and Engobes*. Thesis submitted to the University of Illinois (1936). — 800

M19. T. N. McVay, Mineralogical and rational analysis of ball clays. *Bull. A. Cer. S.* **21**, 267 (1942). — 300, 301

M20. J. M. McWilliam and J. McM. Gilmour, Experiments with firebrick batch for dry-pressing. *Refr. J.* **30** [1], 10 (1954). — 750, 751

M21. Mäckler, *Tonind.-Ztg.* 251, 440 (1905). — 598

— Richard Magner, see Hans Pulfrich.

M22. A. E. Maley, Some glost losses in the earthenware industry and their correction. *Trans. A. Cer. S.* **12**, 294 (1910). — 555

M23. F. Malkin & Co. Ltd., Campbell Road, Stoke-on-Trent. — *336*, 337, *352*, 353, *810*

— Manley W. Mallett, see John G. Thompson.

M24. Franz Mandt, Mineralmahlwerke. Werk Krohenhammer bei Wundsiedel. Bavaria, Germany. — 105

— Edward G. Manning, see William C. Bell.

M25. P. S. Manykin and S. G. Zlatkin, *J. Phys. Chem.* (*USSR*) **9**, 393 (1937). — 183

M26. Edward E. Marbaker, Improvements in the button test for determination of frit fluidity. *J. A. Cer. S.* **30** [12], 354 (1947). — 372

M27. Marconi Instruments Ltd., *Marconi Instruments for Industry* (1953). — 322

— Edward E. Marbaker, see Stuart M. Phelps.

M28. John Marquis, Personally contributed information. — 579

M29. J. Marquis. Ceramic leadless glazes for dinnerware. *Cer. Age* **45** [5], 192 (1945). — 578, 579

M30. John Marquis, Fritted glazes. *Ohio State Univ. Eng. Exp. Stn. News* **18**, 16 (1946). — 587

— John Marquis, see H. J. Orlowski.

— E. Marriott, see W. Gilbert.
M31. C. E. Marshall, Clays as minerals and as colloids. *Trans. Cer. Soc.* **30**, 10 (1931).   305
M32. G. Martin, Researches on the theory of fine grinding. VII—On the efficiency of grinding machines and grinding media with special reference to ball and tube mills. *Trans. Cer. Soc.* **26**, 34 (1927).   678
— D. W. Mashall, see O. J. Whittemore, Jr.
M33. R. Masson, Steinzeug und Temperaturwechselbeständigkeit. *Chimia* [8], 7 (1954).   1123, 1124
M34. R. Masson, Gefügespannungen und Zugfestigkeit von Hartporzellan. *Fifth International Ceramic Congress, Vienna*, 1956, p. 349.   468
— Matejka, see Kallauner.
M35. Edward B. Mathews, *The Fireclays of Maryland*. The Johns Hopkins Press (1922).   29
M36. Oscar E. Mathiasen, Effect of uranium in various types of glazes. *J. A. Cer. S.* **7**, 501 (1924).   587
M37. François Mathieu, U.S. Pat. 2328479, 31 August (1943).   982
M38. S. Matthews and R. A. Shonk, *The Use of Addition Agents in Conditioning a Dry-press Body*. Dept. of Mines and Technical Surveys, Ottawa, Canada.   744, 745
— B. Matsumura, see G. Mackawa.
M39. Paul H. Mautz, *The Function of Peptizing Protective Colloid in the Casting of Clay Slips*. Bulletin 55, *Ohio State Univ. Stud. Eng. Exp. Stn.* (1930).   761
— L. H. Maxwell, see S. M. Lang.
M40. Arthur E. Mayer, A peculiar property of some glazes. *Trans. A. Cer. S.* **11**, 369 (1909).   783
M41. E. T. Mayer, Ceramic colours. *Ceramics* **2** [24], February (1951).   607, 615
— O. P. Mchedlov, see A. I. Avgustinik.
M42. Edward Meeka, Drying and glazing of structural tile. *Bull. A. Cer. S.* **35** [6], 239 (1956).   795
M43. M. Mehmel, Die Bedeutung der opt. Untersuchungsmethoden für die feinkeram. Praxis. *Sprechs.* **91** [12], 283, June (1958); *Ibid.* **91** [13], 307, July (1958); *Ibid.* **91** [14], 329, July (1958).   341, 374, 384
M44. M. Mehmel, Die Bedeutung des Lithiums in Keramischen Massen und Glasuren. *Sprechs.* **90** [4, 5] (1957).   122/123
M45.—
M46. W. J. Meid, The 'Heart' of ceramic manufacturing processes. *Ceramics* **9** [105], 14, November (1957).   933
M47. G. E. Meir and J. W. Mellor, The ferric oxide colours. *Trans. Cer. Soc.* **36** [1], 31 (1937).   629, 630
M48. J. W. Mellor, Some chemical and physical changes in the firing of pottery, etc. *J. Soc. Chem. Ind.* **26** [8], 375, April (1907).   197, 216

M49. J. W. MELLOR, Studies on cylinder grinding. *Trans. Engl. Cer. Soc.* **9**, 50 (1910).   678

M50. J. W. MELLOR, Jackson's and Purdy's surface factors. *Trans. Engl. Cer. Soc.* **9**, 94 (1910).   292

M51. J. W. MELLOR, The behaviour of some glazes in the glost oven. *Trans. Engl. Cer. Soc.* **12** [1], 1 (1912–1913).   568

M52. J. W. MELLOR, Recent research on the bone china body. *Trans. Cer. Soc.* **18**, 497 (1919).   458–461

M53. J. W. MELLOR, On the plasticity of clays. *Trans. Cer. Soc.* **21**, 91 (1922).   66, 331

M54. J. W. MELLOR, A study of pan grinding. *Trans. Cer. Soc.* **29**, 271 (1929–1930); Note on study of pan grinding. *Ibid.* **37**, 126 (1937–1938).   667

M55. J. W. MELLOR, Crazing and peeling of glazes. *Trans. Cer. Soc.* **34**, 112 (1934–1935).   542, 543, 554, 555

M56. J. W. MELLOR, Durability of pottery frits, glazes, glasses and enamels in service. *Trans. Cer. Soc.* **34**, 113 (1934–1935).   560

M57. J. W. MELLOR, Spitting of glazes in the enamel kiln. *Trans. Cer. Soc.* **35** [1], 1 (1936); The spitting of glazes in the decorating furnace. *Cer. Ind.* March (1936).   881

M58. J. W. MELLOR, The chemistry of the Chinese copper red glazes. *Trans. Cer. Soc.* **35**, 364 (1936).   234

M59. J. W. MELLOR, Cobalt and nickel colours. *Trans. Cer. Soc.* **36** [1], 1 (1937).   626

M60. J. W. MELLOR, The chemistry of the chrome-tin colours. *Trans. Cer. Soc.* **36** [1], 16 (1937).   622

M61. J. W. MELLOR, The discolouration of chrome–green colours. *Trans. Cer. Soc.* **36** [1], 28 (1937).   618, 621

M62. J. W. MELLOR, The cobalt-blue colours. *Trans. Cer. Soc.* **36** [6], 264 (1937).   625

M63. J. W. MELLOR, A note on the molecular formulae of clays and glazes. *Trans. Cer. Soc.* **36** [8], 323 (1937).   527

— J. W. MELLOR, see G. W. MEIR.

— J. W. MELLOR, see BERNARD MOORE.

M64. W. D. MEREDITH, Slip house practice in glazed tile production. *Trans. Brit. Cer. Soc.* **39** [6], 163 (1940).   700, 1080

M65. C. W. MERRITT, A series of raw leadless glazes at low temperatures. *Bull. A. Cer. S.* **14**, 104 (1935).   578

M66. C. W. MERRITT, Raw leadless glazes for pottery and tile at cone 2. *J. A. Cer. S.* **19**, 24 (1936).   578

M67. BYRON W. MERWIN, Cerium oxide and rare earth oxide mixture in glazes. *J. A. Cer. S.* **20**, 96 (1937).   588

— H. E. MERWIN, see G. A. RANKIN.

M68. METAL AND THERMIT CORPORATION, *Ultrox Zirconium Glazes.* New York.   587, 588

M69. G. F. METZ, Grinding ceramic materials in ball, pebble, rod and tube mills. *Bull. A. Cer. S.* **16** [12], 461 (1937).   671–676

M70. GEORGE F. METZ, Grinding ceramic materials in pebble, tube, ball mills. *Cer. Ind.* **44** [5], 100 (1945). — 682, 1138
M71. W. MEYER, Über echte Goldfarben für die Glas- und Porzellan-malerei. *Farbe und Lack* [37], 437 (1937). — 652
— A. MICHEL, see E. POUILLARD.
M72. WILHELM MIEHR, Schamotte-Brennöfen. *Tonind.-Ztg.* **76** [5, 6] 65 (1952). — 1061
M73. WILHELM MIEHR, Die moderne Silika-Fabrik. *Tonind.-Ztg.* **75** [21–22], 365 (1951). — *1219*, 1221, 1223
M74. WILHELM MIEHR, Feuerfeste Mörtel, Stampf- und Anstrichmassen. *Sprechs.* **84** [14, 15], 277, 301 (1951). — 890, 892
M75. W. MIEHR, Kammer- und Kanal-bzw. Tunnelöfen. *Ullmanns Encyklopädie der technischen Chemie* (3rd Ed.), Vol. 1, p. 855. Urban & Schwarzenberg, Munich and Berlin (1951). — 973, 978
M76. G. W. MILES and W. B. KEITH, Steels for refractory and heavy clay industries. *Edgar Allen News* **33** [379, 380], 1, 28 (1954). — 164
M77. I. MILES, Pottery and potting, past, present and future. *Ceramics* 115–117, May (1955). — 444, 460
M78. MILLER POTTERY ENGINEERING CO., 2300, Palmer Street, Swissvale, Pittsburgh 18, Pa., U.S.A. — 725, 727, *728*
— B. K. MILLER, see R. S. BRADLEY.
— ESTHER MILLER, see EVERETT THOMAS.
M79. R. J. MILLER, Silk screen printing with bright gold on glass. *Cer. Ind.* **63** [4], 92, October (1954). — 818
M80. W. T. W. MILLER and R. J. SARJANT, The evolution of various types of crushers for stone and ore, and the characteristics of rocks as affecting abrasion in crushing machinery. *Trans. Cer. Soc.* **35**, 492 (1936). — 663, 670
— C. E. MILLIGAN, see M. H. BERNS.
— R. G. MILLS, see A. E. R. WESTMAN.
M81. ANGELA A. MILNE, Expansion of fired kaolin when autoclaved and the effect of additives. *Trans. Brit. Cer. Soc.* **57** [3], 148 (1958). — 88
M82. INSTITUTE OF MINING AND METALLURGY, Standard screens, weights and measures. *Eng. and Min. J.* **83**, 526 (1907). — 1301
M83. R. H. MINTON, The use of substitutes for tin oxide in glazes. *J. A. Cer. S.* **3**, 6 (1920). — 587
— V. D. MISHIN, see V. I. SMIRNOV.
M84. LANE MITCHELL, Ceramics. *Ind. Eng. Chem.* **47**, 1956, September (1955). — 500, 511, 513
— W. C. MOHR, see RALSTON RUSSELL.
M85. MOLER PRODUCTS LTD., Hythe Works, Colchester. — 900, 1286–1287
— M. MONAVAL, see R. PARIS.
— LEWIS E. MONG, see RAYMOND A. HEINDL.
M86. MONOMETER MANUFACTURING CO. LTD., *Oil or Gas Fired Semi-Rotary Frit-Kilns*. Savoy House, 115–116, Strand, London, W.C.2. — 1057

M87. BERNARD MOORE, The cause and prevention of the brown stain in China bodies in the enamel kiln. *Trans. Cer. Soc.* **5**, 37 (1905–1906). 460

M88. BERNARD MOORE and J. W. MELLOR, *Report on the Manufacture of Felspathic or Hard Porcelain from British Raw Materials.* British Pottery Manufactures Federation, 10 December (1925). 472, 583

— C. H. MOORE, JR., see L. R. BLAIR.

M89. F. MOORE, The rheology of ceramic slips and bodies. *Trans. Brit. Cer. Soc.* **58** [7–8], 470 (1959). 65, 329

— F. MOORE, see H. H. MACEY.

M90. H. MOORE, Structure and properties of glazes. *Trans. Brit. Cer. Soc.* **55** [10], 589 (1956). 205, 208, 211–215

— MOORE, see PURDY.

M91. MOREHOUSE INDUSTRIES, *Morehouse Speedline Equipment.* 1156, San Francisco Rd., Los Angeles 65, California, U.S.A. 683

— R. E. MORELAND, see M. D. BURDICK.

M92. G. W. MOREY and N. L. BOWEN, *Am. J. Sci.* **4**, 1 (1922). 198

M93. GEORGE W. MOREY, Glass: The bond in ceramics. *J. A. Cer. S.* **17** [6], 145 (1934). 199–201

M94. G. W. MOREY, *Properties of Glass.* Reinhold, New York, (1938). 538

— G. W. MOREY, see EARL INGERSON.

M95. MORGAN CRUCIBLE CO. LTD., Improvements in or relating to methods of and/or means for moulding articles from a 'Slip'. *Brit. Pat. Spec.* 313185. 767

M96. THE MORGAN CRUCIBLE CO. LTD. *792, 900, 931, 1130, 1131, 1143, 1199, 1238, 1276*

— M. O. MORGAN, see M. D. RIGTERINK.
— W. R. MORGAN, see R. K. HURSCH.
— O. MORNINGSTAR, see B. E. WARREN.

M97. JOHANNES MUELLER, The manufacture of graphite crucibles. *Ber. D. K. G.* **16** [8], 410 (1935). 416

M98. H. MÜLLER HESSE and E. PLANZ, Structure and properties of refractory bricks on a $BaO-Al_2O_3-SiO_2$ base. *Sixth International Ceramic Congress, Wiesbaden.* Abs. *Refr. J.* **35** [2], 80 (1959). 424

— W. MUHEILDON, see *Symposium on Ceramic Cutting Tools.* S219

M99. MUIRHEAD AND CO. LTD., *The Muirhead pH meter.* Beckenham, Kent. 321

M100. P. MUNIER, La prédétermination de la courbe température-temps pour la cuisson des produits céramiques. *Chaleur et Industrie.* July–August (1945). *861, 874, 875, 877, 878*

M101. P. Murray, I. Denton and E. Barnes, The preparation of dense thoria crucibles and tubes. *Trans. Brit. Cer. Soc.* **55** [3], 191 (1956). 1142
— P. Murray, see I. Denton.
M102. Krishma Murthy and F. A. Hummel, Phase equilibria in the system lithium metasilicate–eucryptite. *J. A. Cer. S.* **37**, 14–17 (1954). 503
M103. Krishma Murthy and F. A. Hummel, Phase equilibria in the system lithium metasilicate–forsterite–silica. *J. A. Cer. S.* **38** [2], 55–63 (1955). 503

N1. F. Nadachowski, Zur Frage der Abhängigkeit zwischen der Phasenzusammensetzung und Druckerweichungstemperatur feuerfester Magnesiamassen. *Silikattecknik* **6**, [11], 473 (1955). 418
N2. Paul G. Nahin, Infrared analysis of clays and some associated minerals. *Cer. Age* **60** [3], 32, September (1952). 313
N3. Walter Naish, Periodic kilns and their operation. *Clay Prod. News (Canada)* **30** [11], 10 (1957); *Ibid.* **30** [12], 12 (1957). 968, 980
N4. T. Nakai and Y. Fukami. Studies on the systems composed of silica, alumina and magnesia—I and II. *J. Soc. Chem. Ind. Japan* **39** [7], 230B, July (1936); *J. A. Cer. S. Abs.* **16** [1], 46 (1937). 191, 192, 195
— H. A. Nancarrow, see D. A. Holdridge.
— D. N. Nandi, see J. C. Banerjee.
— Angela F. Naporan, see Henry H. Hausner.
N5. National Carbon Co. Inc., Cleveland, Ohio, U.S.A. *1118*, 1128, 1129
N6. Carl Friedrich Naumann and Ferdinand Zirkel, *Elemente der Mineralogie* (14th Ed.). Wilhelm Engelmann, Leipzig (1901). 102
N7. J. F. Neilson, The drying of building bricks. *Clay. Prod. J. (Australia)* **21** [8], 7, June (1954). 840, 841, 847
— L. Nelson, see J. A. Hedvall.
N8. M. S. Nelson and Hewitt Wilson, Multiple-tunnel kilns. *Bull. A. Cer. S.* **20** [8], 270 (1941). 1008
N9. Robert T. Nelson, Destructive effect of alkali on carbon brick. *Bull. A. Cer. S.* **35** [5], 188 (1956). 1252
N10. Neuroder Schieferton, Gewerkschaft Neurode Kohlen- und Thonwerke. Neurode, bez. Breslau, Germany. 690
N11. Rexford Newcomb, Jr., *Ceramic Whitewares.* Pitman, New York (1947). 90, 95, 106, 107, 376, 395, 442, 452, 456, 478, 480, 481, 488, 494, 516, 518, 567, 974, [*See over.*

N12. D. M. NEWITT and M. COLEMAN, The mechanism of the drying of solids. III—The drying characteristics of china clay. *Trans. Inst. Chem. Engs.* **30**, 28 (1952). — 1096, 1102, 1178–1197, 1201, 1205, 1207; 69, 73, 74

— D. M. NEWITT, see J. F. PEARSE.

N13. H. W. NEWKIRK and H. H. SISLER, *J. A. Cer. S.* **41**, 93 (1958). — 224, 740

N14. NIAGARA SCREENS (GT. BRIT.) LTD., Straysfield Rd., Clay Hill, Enfield, Middx. — 685

— J. NICHOLAS, see C. LEGRAND.

N15. P. NICHOLLS, Determination of thermal conductivity of refractories. *Bull. A. Cer. S.* **15** [2], 37 (1936). — 353

— A. NICHOLSON, see T. R. LYNAM.

N16. C. M. NICHOLSON and G. A. BOLE, Cellulated ceramics for the structural clay products industry. *J. A. Cer. S.* **36** [4], 127 (1953). — 1290–1296

N17. R. NIEDERLEUTHER, *Unbildsame Rohstoffe keramischer Massen.* Julius Springer, Vienna (1928). — 94, 97, *100*, 109, 169

N18.—

— T. NILSSON, see J. A. HEDVALL.

N19. W. NOBLE, A. N. WILLIAMS and F. H. CLEWS, Influence of moisture content and forming-pressure on the properties of heavy clay products. *Trans. Brit. Cer. Soc.* **57** [7], 414 (1958). — 66

N20. NORTHERN MILL SUPPLY CO., 'Faradex' Thread Guides. East Didsbury Station, Manchester 20. — 1153

N21.—

N22. W. S. NORMAN, A. HILLIARD and C. H. V. SAWYER, Delanium carbon in chemical plant construction. *Materials of Construction in the Chemical Industry.* — 415, 1119

N23. A. W. NORRIS, The solubility of lead glazes. III—A standard specification for lead bisilicate. *Trans. Brit. Cer. Soc.* **50**, 246 (1950–1951); *Brit. Cer. Res. Ass. Paper* No. 66 December (1949). — 373–374

N24. A. W. NORRIS, Changes in a glass when used as a glaze. *Trans. Brit. Cer. Soc.* **55** [10], 674 (1956). — 376, 377, 542, 545

N25. A. W. NORRIS, F. VAUGHAN, R. HARRISON and K. C. J. SEABRIDGE, Size changes of porous bodies brought about by water and soluble salts. *Sixth International Ceramic Congress, Wiesbaden.* Abs. *Euro-Ceramic* **8** [10], 263 (1958). — 88

— A. W. NORRIS, see H. EDWARDS.

— A. W. NORRIS, see W. JAMES.

— E. NORSTRÖM, see J. A. HEDVALL.

N26. NORTON GRINDING WHEEL CO. LTD., *Norton Refractory Grain.* Welwyn Garden City, Herts. — *135*, 136

N27. Norton Company, *Norton Refractories*. Worcester 6, Mass, U.S.A.   118, 407, 511, 1137, 1149, 1153, 1217, 1223, 1236–1251, 1260–1269, 1292–1293

N28. C. L. Norton, Jr., Refractory. *Can. Pat.* 366784, 15 June 1937; *J. A. Cer. S.* Abs. **16** [9], 279 (1937).   195

N29. C. L. Norton, Jr., Apparatus for measuring thermal conductivity of refractories. *J. A. Cer. S.* **25** [15], 451 (1942).   *354*

N30. F. H. Norton, Control of crystalline glazes. *J. A. Cer. S.* **20** [7], 217 (1937).   593, 596, 597

N31. F. H. Norton, An instrument for measuring the workability of clays. *J. A. Cer. S.* **21** [1], 33 (1938).   329

N32. F. H. Norton and S. Speil, The measurement of particle sizes in clays. *J. A. Cer. S.* **21** [3], 89 (1938).   289

N33. F. H. Norton, Critical study of the differential thermal method for the identification of clay minerals. *J. A. Cer. S.* **22**, 54 (1939).   313, *314*

N34. F. H. Norton, A. L. Johnson and W. G. Lawrence, Fundamental study of clay. VI—Flow properties of kaolinite–water suspensions. *J. A. Cer. S.* **27** [5], 149 (1944).   59

N35. F. H. Norton, *Refractories* (3rd Ed.). McGraw-Hill, New York, Toronto and London (1949).   40, 50, 141, 170, 508, 511, *753*, 1140, 1142, 1146–1149, 1215, 1216

N36. F. H. Norton, *Elements of Ceramics*. Addison-Wesley, Cambridge 42, Mass., U.S.A. (1952).   12, 26, 100, 102, 161, 163, 452, 455, 456, 612, 763, 770, 1179–1194

N37. F. H. Norton, *Ceramics for the Artist Potter*. Addison-Wesley, Cambridge 42, Mass., U.S.A. (1956).   802, 803

— F. H. Norton, see *Symposium on Ceramic Cutting Tools.* **S219.**

— F. H. Norton, see A. L. Johnson.

— R. W. Nurse, see F. M. Lea.

— V. Ny, see J. A. Hedvall.

— C. R. Oberst, see H. S. Orth.

O1. A. S. W. Odelberg, Note on the durability of bone china hotel ware. *Trans. Cer. Soc.* **30** [6], 225 (1931).   464, 1099

— Francis Odell, see Pol Duwez.

O2. W. Oelsen and G. Taumann, *Z. anorg. Chem.* **193**, 17 (1930). 181
O3. S. Ogawa, Dolomitic faience I. *J. Jap. Cer. Assn.* **44** [520], 246 (1936). 594
— Shinchiro Ogawa, see Suke Hirano.
O4. Ellsworth P. Ogden, Problems in firing industrial kilns. *Brit. Clywkr.* 48, May (1937). 970
O5. H. Oliver, The development of insulating bricks for furnace construction. *Ceramics, A Symposium*, p. 592. British Ceramic Society (1953). 426, 427, 1284–1294
— T. R. Oliver, see J. F. Pearse.
— Kaj Olsson, see J. A. Hedvall.
— Robert G. O'Meara, see Theron A. Klinefelter.
O6. Ooms, Ittner & Cie, Cologne, Germany. *1003*, 1038, *1039*
O7. H. J. Orlowski and John Marquis, Lead replacements in dinnerware glazes. *J. A. Cer. S.* **28**, 344 (1945). 577, 578–579
O8. W. R. Ormandy, Electrical process for the purification of clays. *Trans. Engl. Cer. Soc.* **12**, 36 (1912–1913); British clays under the osmose purification process. *Ibid.* **13**, 35 (1913–1914). 248
O9. Gräfl. zu Ortenburg'sche Tongruben, Tambach, bei. Coburg, Germany. 37, 43
O10. H. S. Orth, Producer gas in connection with firing fire brick. *J. A. Cer. S.* **10** [9], 699 (1927). 1021
O11. H. S. Orth and C. R. Oberst, Better lubrication for kiln car bearings. *Bull. A. Cer. S.* **29** [6], 215 (1950). 1027
O12. Edward Orton, Jr., A study of a type of matte glaze maturing at cone 2–4. *Trans. A. Cer. S.* **10**, 547 (1908). 595
O13. The Edward Orton Jr. Ceramic Foundation, *Standard Pyrometric Cones.* Columbus, Ohio, U.S.A. 936
O14. G. H. Osborn and H. Johns, The rapid determination of sodium and potassium in rocks and minerals by flame photometry. *Analyst* **76**, 410 (1951). 274
— E. F. Osborn, see R. C. DeVries.
— E. F. Osborn, see R. Roy.
— C. W. Owens, Jr., see A. F. Greaves-Walker.

P1. Pacific Coast Clay Products Institute, *Handbook Vitrified Clay Pipe. Tables and Design Data for Southern California.* 605
— R. F. Paget, see W. H. Gamble.
— S. Pålsson, see J. A. Hedvall.
P2. A. V. Pamfilov and K. S. Fridmann, *Z. Gen. Chem.* (U.S.S.R.) **10**, 210 (1940). 187
P3. W. I. Panassjuk, Die Herstellung von Mahlfarben für Dekore von Glaserzeugnisse. (In Russian.) *Stekolm Prom.* **14** [10], 18 (1938). 610, 617–643

P4. R. Paris and M. Monaval, *Compt. rend. Acad. Sci.* **202**, 2075 (1936).   133

P5. T. W. Parker, Improvements in and relating to the treatment of clay. *Pat. Spec.* 243929 (1925).   249

P6. H. Parnham, The development of chrome-magnesite bricks. *Ceramics, A Symposium.* British Ceramic Society (1953).   1259–1272

P7. J. H. Partridge, *Refractory Blocks for Glass Tank Furnaces.* Society of Glass Techn., Sheffield (1935).   1231, 1232

P8. J. H. Partridge, *Refractory Materials.* Royal Soc. of Arts. Cantor Lectures (1939).   46, 47

P9. C. W. Parmelee and John S. Lathrop, Aventurine glazes. *J. A. Cer. S.* **7**, 567 (1924).   598

P10. C. W. Parmelee and D. T. H. Shaw, The effect of additions of CaO and MgO on enamel glasses. *J. A. Cer. S.* **13**, 498 (1930).   610

P11. C. W. Parmelee and C. G. Harman, Organic compounds as electrolytes and their effect on the properties of clay slip and on the life of plaster molds. *J. A. Cer. S.* **14**, 139 (1931).   761

P12. C. W. Parmelee and K. C. Lyon, A study of some frit compositions. *J. A. Cer. S.* **17** [3], 60 (1934).   576, 577

P13. C. W. Parmelee and W. Horak, The microstructure of some raw lead mat glazes. *J. A. Cer. S.* **17** [3], 67 (1934).   592

P14. C. W. Parmelee, *Clays and Some Other Ceramic Materials.* Edwards Brothers (1937).   11

P15. C. W. Parmelee and L. R. Barrett, Some pyrochemical properties of pyrophyllite. *J. A. Cer. S.* **21** [11], 388 (1938).   94, *95*

P16. C. W. Parmelee, Recent advances in grinding, control of particle size in glaze slips. *Cer. Ind.* **35**, 48 (1940).   782, 784

P17. C. W. Parmelee and Antonio R. Rodriquez, Catalytic mullitazation of kaolinite by metallic oxides. *J. A. Cer. S.* **25** [1], 1 (1942).   195, 196

P18. C. W. Parmelee and P. E. Buckles, Study of glaze and body interface. *J. A. Cer. S.* **25** [1], 11 (1942).   543

P19. C. W. Parmelee and V. D. Fréchette, Heat-of-wetting of unfired and fired clays. *J. A. Cer. S.* **25** [4], 108 (1942).   67, *68*, 332

— C. W. Parmelee, see J. H. Chesters.
— C. W. Parmelee, see C. G. Harman.
— C. W. Parmelee, see T. W. Talwalkar.

P20. Linus Pauling, *The Nature of the Chemical Bond.* Cornell University Press (1944).   4

P21. J. F. Pearse, T. R. Oliver and D. M. Newitt, The mechanism of the drying of solids. I—The forces giving rise to movement of water in granular beds, during drying. *Trans. Inst. Chem. Engs.* **27**, 1 (1949).   69, 70, 71, 73

P22. E. J. and J. Pearson Ltd., Stourbridge.   1234–1251

P23. C. J. Peddle, *J. Soc. Glass. Tech.* **5**, 72 (1921).   614

P24. G. N. PEEL, The ceramic applications of zircon and zirconia. *Ceramics* **7** [76, 77], 163, 207, June–July (1955). 1246–1248

— G. N. PEEL, see F. T. BOOTH.

P25. K. FORREST PENCE and E. B. BLOUNT, Vermicullite as raw material in ceramic manufacture. *J. A. Cer. S.* **27**, 51 (1944). 587

— A. C. PERRICONE, see D. J. WEINTRITT.

P26. F. T. PERRINS, Gas producers and their operation. *Claycraft* **18**, 238 (1945). 913

P27. EARL C. PETRIE, Notes on pyrometric cone equivalent determinations. *Bull. A. Cer. S.* **20** [9] (1941). 278

P28. PFÄLZISCHE CHAMOTTE- UND TONWERKE (SCHIFFER UND KIRCHER) A. G., Eisenberg/Pfalz, Germany. 49

P29. KARL F. PFEFFERKORN, Der Giessfleck. *Sprechs.* **55** [40–43], 441 (1922). 767

P30. K. PFEFFERKORN, Zur Kenntnis der Schlieren in keramischen Giessmassen. *Sprechs.* **67** [27], 401 (1934). 769

P31. K. PFEFFERKORN, Dispersant for ceramic cast mixes. *Sprechs.* **68** [19], 289 (1935). 761

— H. PFISTER, see W. JANDER.

P32. GIRARD W. PHELPS, Notes on casting. *Bull. A. Cer. S.* **20** [9], 313 (1941). 763, 1082

P33. G. W. PHELPS, Clays—deflocculation and casting control. *Cer. Age* April, May, June, August, September, October, November (1947); January, February (1948). 756–762

P34. S. M. PHELPS, Nomenclature for refractories. *Amer. Refr. Inst. Tech. Bull.* [6], December (1926). 893, *896*

P35. STUART M. PHELPS and EDWARD E. MARBAKER, New ceramic table tops. *Ind. Eng. Chem.* **29**, 541, May (1937). 1120, 1133

P36. STUART M. PHELPS and EDWARD E. MARBAKER, Karcite, a new ceramic material for laboratory equipment. *J. A. Cer. S.* **21** [3], 108 (1938). 1120, 1133

P37. PHILIPS LAMPS LTD., Shaftesbury Ave., London, W.C.2. 311, *312*, 341

P38. PHILIPS LAMPS LTD., Improvements in or relating to the manufacture of silver coated refractory non-metallic materials. *Pat. Spec.* 553441 (1943). 656

P39. J. G. PHILLIPS, *Improving the Properties of Clays and Shales.* Canada Dept. of Mines and Resources Bulletin No. 793, Ottawa (1938). 833

P40. J. ALLEN PIERCE and ARTHUR S. WATTS, A study of the $BaO-Al_2O_3-SiO_2$ deformation eutectic in mixtures with the deformation-eutectic of $CaO-Al_2O_3-SiO_2$, $MgO-Al_2O_3-SiO_2$, and KNa felspar in the raw and prefused state. *Bull. A. Cer. S.* **31** [11], 460 (1952). 220–222

P41. A. G. PINCUS, Glass from the atomic view. *Cer. Age* **39** [2], 38 (1942). *233*

P42. ALEXIS G. PINCUS, Invitation to glass technology. *Cer. Age* May–August (1949). 213, 214

- A. G. Pincus, see J. Biscoe.
- A. G. Pincus, see B. E. Warren.
- P43. A. A. Pirogov, *Ogneupory* **4**, 260 (1936); *Ibid.* **5**, 100 (1937); *Ibid.* **6**, 1286 (1938); **8**, 143 (1940). 1296
- W. C. Pittman, see R. W. Knauft.
- E. Planz, see H. Müller-Hesse.
- P44. C. W. Planje, Observation on crazing in terra cotta glazes. *J. A. Cer. S.* **14**, 747 (1931). 587
- P45. H. L. Podmore, The controlled milling of ceramic materials. *Ceramics* **36** [3], 621, February (1952). 680
- P46. W. Podmore and Sons, Ltd., Shelton, Stoke-on-Trent. 683
- P47. W. Podmore and Sons, Continuous production of glaze frits. *Pottery Gazette* **83** [973], 866, July (1958). 1059
- P48. G. I. Pokrovski, Plasticity mechanism of clay. *Pochvovedenie* [8], 38 (1940); *J. A. Cer. S. Abs.* **21**, 198 (1942). 65
- H. Polkinhorne, see G. C. Collins.
- E. A. Porai, see N. N. Valenkov.
- P49. Posey Iron Works Inc., Lancaster, Pa., U.S.A. 718
- P50. The Potteries Die Company, Knypersley Road, Norton, Stoke-on-Trent. 341
- P51. E. Pouillard and A. Michel, *Compt. rend. Acad. Sci.* **228**, 1232 (1949). 189
- P52. de la Poutroie et Sobole, Étude technique d'une etuve à radiations infra-rouge pour sechage d'objets creux en faience ou en porcelaine. *Bull. Soc. Fran. Cér.* [40], 29, July–September (1958). 850
- P53. Powell Duffryn Carbon Products Ltd., Springfield Rd., Hayes, Middx. *1119*, 1133
- P54. Powell and Griffiths, *Trans. Inst. Chem. Eng.* **13**, 175 (1935). 70
- P55. The Power-Gas Corporation Ltd., *Economic Utilisation of Coke Breeze.* Stockton-on-Tees. 918
- Walter Pralow, see Hans Lehmann.
- P56. P. R. Prior, Industrial instruments and the pottery industry. *Ceramics* **6** [71], 495, January (1955). 944, 945
- P57. Proctor and Schwartz Inc., *Proctor Dryers.* Seventh St. and Tabor Road, Philadelphia 20, Pa., U.S.A. 847
- P58. T. I. Prokopowicz and F. A. Hummel, Reactions in the system $Li_2O-MgO-Al_2O_3-SiO_2$. *J. A. Cer. S.* **39** [8], 266 (1956). 503
- P59. A. C. H. Pryce, Ceramics in the textile industry. *The Textile Manufacturer* 467, December (1940). 1113
- P60. A. C. H. Pryce, Chemical stoneware for pipe lines. *Ind. Chem.* **17** [201], 244 (1941). 1123
- A. C. H. Pryce, see R. Ward.
- P61. W. Pukall, Ueber Tonfilter, ihre Eigenschaften und ihre Verwendung in chemischen und bakteriologischen Laboratorien. *Ber. deut. chem. Ges.* **26**, 1159 (1893). 71

P62. W. Pukall, Beitrag zur Lösung der Bleifrage. *Sprechs.* 267
(1906).
P63. W. Pukall, My experience with crystal glazes. *Trans. A.* 596
Cer. S. **10**, 183 (1908).
P64. W. Pukall, Bunzlauer Feinsteinzeug. *Sprechs.* (1910). 535
P65. W. Pukall, Anorganische Synthesen. *Silikat-Z.* **2** [4, 5, 6] 173–179
(1914).
P66. W. Pukall, Ein neues Verfahren der Goldablösung mittelst 654
Königswasser. *Sprechs.* **49** [21], 150 (1916).
P67. W. Pukall, Ueber die Vorgänge beim Brennen keramischer 857
Waren. *Sprechs.* **52** [10, 11, 12] (1919).
P68. W. Pukall, *Grundzüge der Keramik*. Müller und Schmidt, 442
Coburg (1922).
P69. W. Pukall, Ueber die Vorgänge beim Trocknen kera- 71
mischer Rohwaren. *Sprechs.* [23] (1926).
P70. W. Pukall, Ueber die Vorgänge beim Trocknen kera- 71, *72*
mischer Rohwaren. *Sprechs.* [22–23], 429 (1928).
P71. W. Pukall, *Keramisches Rechnen* (4th Ed.), p. 81. Ferdi- 568
nand Hirt, Breslau (1927).
P72. Hans Pulfrich and Richard Magner, Manufacture of 1206
ceramic mass non shrinking during firing. *Ger. Pat.*
734385 (1943).
P73. Pulverising Machinery Company, *Mikro-Pulverizer and* 682
Mikro-Atomiser. Chatham Rd, Summit, N.J., U.S.A.
P74. R. C. Purdy, The calculation of the comparative fineness 292
of ground materials by means of a surface factor.
Trans. A. Cer. S. **7**, 441 (1905).
P75 and **P76.** R. C. Purdy and H. B. Fox, Fritted glazes: A 527, 543, 567
study of variations of the oxygen ratio and the silica–
boric acid molecular ratio. *Trans. A. Cer. S.* **9**, 95
(1907).
P77. Ross C. Purdy and Joseph K. Moore, Pyrochemical and 29
physical behavior of clays. *Trans. Amer. Cer. S.* **9**,
204, 239 (1907).
P78. Ross C. Purdy and Junius F. Krehbiel, Crystalline glazes. 593, 596
Trans. A. Cer. S. **9**, 319 (1907).
P79. R. C. Purdy, *Ill. Geol. Surv. Bulletin* No. 9, p. 270 (1908). 29
P80. Ross C. Purdy, Porcelain glazes. *Trans. A. Cer. S.* **13**, 588
550 (1911).
P81. Ross C. Purdy, Matte glazes. *Trans. A. Cer. S.* **14**, 671 594
(1912).
P82. R. E. Pyle and P. R. Jones, The effect of wetting agents 65
on the physical properties of clay bodies. *Bull. A.*
Cer. S. **31** [7], 233 (1952).

Q1. Quarzsand und Kies Verkaufs G.m.b.H., Dörentrup 98
(Lippe), Germany.

BIBLIOGRAPHY AND AUTHORS INDEX 1397

Q2. Norbert Quinn, Super refractories by the ethyl silicate casting process. *Industrial Chemist* October (1955).   771, 1230, 1246, 1247, 1254, 1257, 1274

— J. F. Quirk, see Collin Hyde.
— J. F. Quirk, see J. F. Lynch.
Q3. John F. Quirk, Factors affecting sinterability of oxide powders: BeO and MgO. *J. A. Cer. S.* **42** [4], 178 (1959).   1140

— Bengt. Rähs, see J. A. Hedvall.
R1. N. V. Raghunath, Electric tunnel kiln (at the Indian) Government porcelain factory, Bangalore. *J. Sci. Ind. Res. (India)* **2**, 40 (1943); *Brit. Chem. Physiol. Abs.* 1944 July, B.I. 236.   1025
— J. R. Rait, see G. R. Rigby.
— O. C. Ralston, see John Dasher.
R2. G. A. Rankin and H. E. Merwin, Ternary system MgO–Al$_2$O$_3$–SiO$_2$. *Am. J. Sci.* **45**, 301 (1918).   198, *217*, *482*
R3. D. E. Rase and Rustum Roy, Phase equilibria in the system BaO–TiO$_2$. *J. A. Cer. Soc.* **38** [3], 102 (1955).   520
R4. S. W. Ratcliffe and H. W. Webb, Rapid methods of grain-size measurement in pottery practice. *Trans. Brit. Cer. Soc.* **41**, 51 (1942).   288
R5. S. W. Ratcliffe and W. L. German, Fifty years of progress in ceramic whitewares. *Ceramics, A Symposium*. British Ceramic Society (1953).   790
— S. W. Ratcliffe, see W. L. German.
R6. Rawdon Ltd.,
  (a) *Horizontal and Vertical De-airing Machines*. Patent *Vertical De-airing Extrusion Machines*.   709
  (b) *Automatic Trickle Feed Kiln Stoker*.   910
  Moira, Nr. Burton-on-Trent.
— E. B. Read, see T. L. Hurst.
— C. E. Reed, see E. A. Hauser.
R7. The Refractory Brick Co. of England Ltd., *Stabilised Dolomite Bricks*.   1263
— H. P. Reinecker, see R. R. Danielson.
R8. M. Reitz, Über die Herstellung von Porzellanisolatoren. *Ker. Rund.* **36** (1928).   474
R9. V. H. Remington, Automatic machines and thermoplastic printing colors. *Bull. A. Cer. S.* **34** [11], 359 (1955).   816, 817
R10. Richard C. Remmey & Son Co., Hedley St. and Delaware River, Philadelphia 37, Pa., U.S.A.   1245
R11. F. H. Rhead, Mat glazes. *Trans. A. Cer. S.* **11**, 157 (1909).   594
R12. Rhône-Poulenc, Société des Usines Chimiques. 21, Rue Jean-Goujon, Paris-8°, France.   936

R13. Rhône-Poulenc, Société des Usines Chimiques, *Brit. Pat.* 578
594726.
— H. H. Rice, see C. A. Arenberg.
— R. Richards, see A. Dinnie.
R14. Lloyd D. Richardson, Ceramics for aircraft propulsion 1164
systems. *Bull. A. Cer. S.* **33** [5], 135 (1954).
R15. W. D. Richardson, The adaptability of the gas-fired com- 1001
partment kiln for burning clay products. *J. A. Cer. S.*
5 [5], 254 (1922).
R16. W. D. Richardson, Application of the downdraught 1015, 1036,
principle to tunnel kilns. *J. A. Cer. S.* **14** [8], 572 1037
(1931).
R17. W. D. Richardson, Should the downdraught periodic kiln 968
be round or rectangular? *Bull. A. Cer. S.* **17** [5],
197 (1938).
R18. H. Richter, Spektrometrische Gesamtanalyse silikatischer 267
Roh- und Werkstoffe. *Sixth International Ceramic
Congress, Wiesbaden*, 1958; Abs. *Euro-Ceramic* **8** [10],
264 (1958).
R19. Frank H. Riddle, Ceramic spark plug insulators. *J. A.* 1199
Cer. S. **32** [11], 333 (1949).
R20. R. R. Ridgeway and B. L. Bailey, Boron carbide articles 753
such as bearings, dies or sand blast nozzles. *U.S. Pat.*
2027786 (1934).
R21. Riedig, Einrichtungen zum Gewinnen von Ton. *Euro-* 263
Ceramic **8** [3], 55 (1958).
R22. Reinhold Rieke, Schwindungs und Porositätsbestim- 333, 334
mungen. *Fachbücher der Keramischen Rundschau* **10**
(1918).
R23. R. Rieke and Hans Thurnauer, Ueber den Einfluss eines 490
Specksteinzusatzes auf das Brennverhalten und einige
technischwichtige Eigenschaften von reinem Kaolin.
Ber. D. K. G. **13**, 245 (1932).
R24. R. Rieke and Pu-Yi Wen, Die Wirkung von Kaliglimmer 465
in Porzellanmassen. *Ber. D. K. G.* **20** [2], 43 (1939).
R25. R. Rieke and Ch. Haeberle, Ueber keramisch gebundene 479–481, 1156
Korund-Schleifscheiben mit leichtschmelzenden, ins-
besondere phosphathaltigen Bindungen. *Ber. D. K. G.*
24 [4–5], 117 (1943).
— N. Riel, see W. Jander.
R26. G. Riemann, Die Verwendung von Stampfmassen beim 891
Ofenbau. *Euro-Ceramic* **5** [9], 234 (1955).
R27.—
R28. Heinrich Ries, *Clays. Their Occurrence, Properties and* 25, 28, 64, 65,
Uses. Chapman & Hall, London (1927). 69, 76, 335,
337
R29. Hans B. Ries, Die Aufbereitung feuerfester und feinkera- 1080, 1202,
mischer Trockenpressmassen. *Euro-Ceramic* **7** [11], 1226
282 (1957).

R30. Rieterwerke, Konstanz-Peterhausen, Germany. *712*
R31. G. R. Rigby, G. H. B. Lovell and A. T. Green, An investigation of chrome ores. J. R. Rait, An investigation into the constitutions of chrome ore. *Third Report on Refractory Materials*, pp. 43 and 175. Joint Refr. Res. Comm. of Brit., Iron and Steel Res. Ass. and B.R.R.A. (1946). *421*
R32. G. R. Rigby, *The Thin-Section Mineralogy of Ceramic Materials.* B.R.R.A. (1948). *308, 309*
R33.—
— G. R. Rigby, see R. P. White.
R34. M. D. Rigterink, R. O. Grisdale and M. O. Morgan, *Relation between Chemical Composition and Dielectric Properties of Ceramic Materials with Low Dielectric Losses.* Bell Telephone Laboratories. *493, 497*
R35. A. Riley, Milling and materials. *Ceramics* **4** [44], 120, October (1952). *696*
R36. F. Rinne, *Z. Kryst.* **61**, 113 (1925). *78*
R37. P. von Rittinger, *Lehrbuch der Aufbereitungskunde*, p. 19. Berlin (1867). *680*
R38. A. Rivierè, Etude roentgenographique et identification minéralogique des argiles céramiques. *Bull. Soc. Fran. Cér.* **1**, 7 (1949). *311*
R39. A. L. Roberts, The constitution of refractory clays. *Trans. Brit. Cer. Soc.* **44**, 20 (1945). *29*
R40. A. L. Roberts, Séchage des produits céramiques par rayonnement infra-rouge. *Bull. Soc. Fran. Cér.* [40], 3, July–September (1958). *836*
— A. L. Roberts, see K. Carr.
— A. L. Roberts, see T. G. Carruthers.
— A. L. Roberts, see A. C. D. Chaklader.
— A. L. Roberts, see R. W. Grimshaw.
— A. L. Roberts, see P. Howard.
— A. L. Roberts, see W. E. Worrall.
R41. J. P. Roberts and W. Wyatt, The determination of the bend strength and bend creep of ceramic materials. *Trans. Brit. Cer. Soc.* **48**, 343 (1949). *506*
R42. Robert H. S. Robertson, *A Glossary of Clay Trade Names.* Clay Minerals Group of the Mineralogical Society of Great Britain and Ireland (1954). *31, 32–51*
— C. S. Robinson, Jr., see J. Biscoe.
R43. Gilbert C. Robinson, Limitations imposed by raw materials on firing schedules. *J. A. Cer. S.* **35** [1], 1 (1952). *857*
R44. G. C. Robinson and J. J. Keilen, The role of water in extrusion and its modification by a surface-active chemical. *Bull. A. Cer. S.* **36** [11], 422 (1957). *730*
R45. J. S. Robinson, The use of colliery waste for brickmaking. *Trans. Brit. Cer. Soc.* Preprint, 2 December (1958). *399*

R46. Josef Robitschek and Felix Singer, Chromite refractories. 137, 421
Brit. Clywkr. 310, January (1940).

R47. Josef Robitschek and Felix Singer, Chrome-magnesite 138
refractories. Brit. Clywkr. February (1940).

R48. Josef M. Robitschek, Heat shock resistance of chemical 358, 433
stoneware equipment. Cer. Age **43** [1], 8 (1944).

— Josef Robitschek is also known as Josef Lucas.

R49. J. T. Robson, *Operating the Tunnel Kiln*, pp. 28 and 114. 864, 865
Ind. Publ. Inc., Chicago (1954).

R50. W. F. Rochow and C. A. Brashares, Recent developments 1220
in refractories and their applications. Chem. Eng. Progress **44** [11] (1948).

— Antonio R. Rodriguez, see C. W. Parmelee.

— A. R. Rodriguez, see Hans Thurnauer.

R51. H. Rösler, *Beiträge zur Kenntniss einiger Kaolinlagerstätten.* 244
E. Schweizerbart'sche Verlagshandlung, (E. Nägele) Stuttgart (1902).

— Edwin J. Rogers, see John D. Sullivan.

R52. Helmut Rohn, Erfahrungen mit neuartigen Brennhilfs- 960
mitteln. Euro-Ceramic **8** [1], 17 (1958).

R53. Paul S. Roller, Size distribution of ceramic powders as 282
determined by a particle-size air analyser. *J. A. Cer. S.* **20**, 167 (1937).

R54. Rosenthal-Isolatoren G.M.B.H., Ger. Pat. 680205. 95
Werk Selb in Selb. Bayern, Germany.

R55. Ernst Rosenthal, *Porcelain and Other Ceramic Insulating* 1195–1197
Materials. Chapman & Hall, London (1944).

— W. Rosenthal, see G. Tammann.

R56.—

R57. Donald W. Ross, Nature and origin of refractory clays. 29
J. A. Cer. S. **10**, 704 (1927).

— M. A. S. Ross, see C. A. Beevers.

R58. Ernst Roth, Untersuchungen über die Änderungen einiger 126
physikalischer Eigenschaften von Hartporzellan durch wechseln den Gehalt an Kali- und Natron-feldspat und und durch unterschiedliche Brennweise. Bücher der Deutschen Keramischen Gesellschaft, Vol. 3. Berlin (1922).

R59. E. Rowden, Firing in the heavy clay and refractories in- *979*, 980, 984,
dustry, 1900–1950. *Ceramics, A Symposium*, p. 771. 1002, 1031,
British Ceramic Society (1953). 1035, 1036,
1037, 1043

R60. D. H. Rowland, The influence of glaze on insulator 549
strength. General Electric Review 136, March (1929).

R61. D. H. Rowland, Porcelain for high-voltage insulators. *549, 550,* 1177
Elect. Engineering **55** [6], 618, June (1936).

— R. A. Rowland, see R. E. Grim.

— Robert R. Rowlands, see Ralston Russell, Jr.

— Della M. Roy, see Rustum Roy.

— H. N. Roy, see Bhan Bushan.

R62. R. Roy and E. F. Osborn, *J. A. Chem. Soc.* **71**, 2087 (1949). 500

R63. Rustum Roy, Della M. Roy and E. F. Osborn, Compositional and stability relationships among the lithium aluminosilicates: eucryptite, spodumene and petalite. *J. A. Cer. S.* **33** [5], 152 (1950). 503

R64. Rustum Roy, Della M. Roy and E. E. Francis, New data on thermal decomposition of kaolinite and halloysite. *J. A. Cer. S.* **38** [6], 198 (1955). 77

— Rustum Roy, see D. E. Rase.

— R. Roy, see R. C. De Vries.

R65. R. Royer, Pâtes au talc vitrifiables à moyenne temperature. *L'Ind. Cér.* [493], 18, January (1958). 466

R66. R. Royer, Etudes de pâtes de faience au talc pour cuisson à 1000–1100°C. *L'Ind. Cér.* [496], 117, April (1958). 450

R67. W. C. Rueckel, Researches in dry press refractories. III— The effect of vacuum on the unfired properties of some dry press refractory batches. *J. A. Cer. S.* **14**, 764 (1931). 750

— Edwin Ruh, see N. H. Snyder.

— John Ruppert, see Frank D. Lamb.

R68. H. Russell and A. B. Searle, *The Making and Burning of Glazed Ware*, p. 112. London (1929). 578

R69. R. Russel and A. S. Watts, A study in earthenware body. *Cer. Age* **26**, 219–221, December (1935). 446

R70. Ralston Russell, Jr., Effect of thermal process on physical properties. I—Structural clay products. *Bull. A. Cer. S.* **19** [1], 1 (1940). 856

R71. Ralston Russell and W. C. Mohr, Characteristics of zircon porcelain. *J. A. Cer. S.* **30** [1], 32 (1947). 479, 1179–1191, 1205

R72. Ralston Russell, Jr. and Robert R. Rowlands, Glaze investigations. I—Effect of various silicas in typical glazes. *J. A. Cer. S.* **36** [1], 1 (1953). 568

— Ralston Russell, Jr., see John W. Donahey.

R73. Ruston-Bucyrus Ltd., Lincoln. *261*

R74. George E. Ryckman, Mechanised decoration of pottery and glass. *Bull. A. Cer. S.* **30** [12], 432 (1951). 810, 828

R75. E. Ryschkewitsch, The crushing and tensile strengths of some ceramic materials on the one-component basis. *Ber. D. K. G.* **22**, 54, 363 (1941). 1151–1153

— E. Ryshkewitch, see *Symposium on Ceramic Cutting Tools.* S219.

S1. H. E. Sachse, Artificial drying of bricks. *Tonind.-Ztg.* **67**, 193 (1943). 842

S2. P. D. S. St. Pierre, pH/Viscosity relationships in the system zirconia–water–polyvinyl alcohol–hydrochloric acid. *Trans. Brit. Cer. Soc.* **51** [4], 260 (1952). 760

S3. P. D. S. St. Pierre, The constitution of bone china. I— High temperature phase equilibrium studies in the system tricalciumphosphate–alumina–silica. *Technical Paper* No. 2, Canadian Dept. Mines and Technical Surveys, Mines Branch (1953). 459, 460

S4. P. D. S. St. Pierre, Bone china—II. *Chemistry in Canada* April, June and August (1954). 460, 1098

S5. P. D. S. St. Pierre, Bone china—III. *Technical Paper*, Canadian Dept. of Mines and Technical Surveys, Mines Branch (1955). 460

— K. V. Salazkina, see N. K. Antonevich.

S6. H. Salmang, Die Ursachen der Bildsamkeit der Tone. *Z. anorg. Chem.* **162**, 115 (1927). 65

S7. Hermann Salmang and Benno Wentz, Die Herstellung von Tridymitsteinen. I—Die Bedeutung des Tridymits im Silikastein. *Ber. D. K. G.* **12** [1], 1 (1931); Pt. II. *Ibid.* **14** [4], 141 (1933). Hermann Salmang and H. J. Lüngen, Technische Herstellung von Tridymitsteinen. *T.I.Z.* **57** [56] (1933). 402, 1218

S8. H. Salmang, *Die physikalischen und chemischen Grundlagen der Keramik*. 2nd Edition. Springer Verlag, Berlin, Göttingen and Heidelberg (1951). 1219

S9. H. Salmang, Das Gefüge des keramischen Scherbens. *Ber. D. K. G.* **30**, [11], 247 (1953). 397

— S. O. Sandberg, see J. A. Hedvall.

S10. E. B. Sandell, *Colorimetric Determination of Traces of Metals*. Interscience Publishers, New York and London. 267

S11. F. Sandford, Die Einwirkung der Ofenatmosphäre auf die Wasserbeständigkeit Keramischer Erzeugnisse. *Transactions of the International Ceramic Congress, Holland*, 1948, p. 167. 194

— C. H. V. Sawyer, see W. S. Norman.

S12. E. C. Sawyer, Refining of silica sand for glass manufacture. *Ohio State Univ. Eng. Exp. Stn. News* **19** [2], 57 (1947). 99

— R. J. Sarjant, see W. T. W. Miller.
— Shigeo Sawamura, see Suke Hirano.

S13. Scandinavisk Moler Industri. 1287

— C. F. Schaefer, see C. G. Harman.

S14. Léon Schätzer, *Keramik, Roh- und Werkstoffe, Prüfmethoden*. Verlag Technik, Berlin (1954). 412, 456

S15. J. F. Schairer and N. L. Bowen, *Bull. Soc. géol. Finlande*. **20**, 67 (1948). 198

S16. Alfred Scheidig, *Der Löss und seine geotechnischen Eigenschaften*. Theodor Steinkopff, Dresden and Leipzig (1934). 31

— G. Schiller, see J. A. Hedvall.

S17. P. Schleiss, Talkum und seine Verwendung in der Keramik. *Europäische Tonind.* **2** [9], 236–237 (1952). 442

S18. F. K. Schlünz, Mikroskopische und chemische Untersuchungen zweier Tone. *Diss. Rostock* (1933); *Chemie der Erde* **10**, 116 (1935). 304

S19. H. Schmellenmeier, Simple method for the manufacture of solid ceramic shapes for temperatures up to 1800°C. *Z. Tech. Physik* (German) **24** [9], 217 (1943). 1206

S20. A. Schmidt, *Die Brennöfen der Grob- und Feinkeramik und der Mörtelindustrie* (2nd Ed.). Carl Marhold, Halle (Saale) (1948). 977

S21. Boris E. Schmidt, Glanz- und Poliergold-Dekore im Siebdruckverfahren. *Sprechs.* **89** [14], 329, July (1956). 818

S22. Walther Schmied, Ein Beitrag zur Sedimentationsanalyse. *Euro-Ceramic* **8** [5] (1958). 285

S23. P. Schneider, Die Verwendung von Steinzeug in der Landwirtschaft. *Tonind.-Ztg.* **44** (1929). 1112

S24. R. Schober and E. Thilo, *Ber. deut. Chem. Ges.* **73B**, 1219 (1940). 187

S25. E. Schöne, *Bl. Soc. Moscoau*, 40 Pt. 1 (1867); *Z. Anal. Chem.* **7**, 29 (1868); *Über Schlämmanalysen*. Berlin (1867). 281

S26. H. Z. Schofield, A study of replacement of Cornwall stone by talc and felspar in a wall tile body. *Bull. A. Cer. S.* **16**, 203 (1937). 448

S28. H. Z. Schofield, Ball milling of pure ceramic bodies. *Bull. A. Cer. S.* **32** [2], 49 (1953). 682

— H. Z. Schofield, see C. A. Arenberg.

S27. R. K. Schofield, Clay mineral structures and their physical significance. *Trans. Brit. Cer. Soc.* **39**, 147 (1940). 12

S29. W. A. Scholes, Thermal conductivity of high BeO content bodies. *Brick & Clay Record* **114** [5], 60 (1949). 506, 1153

S30. E. Schramm and R. F. Sherwood, Some properties of glaze slips. *J. A. Cer. S.* **12**, 270 (1929). 782, 784

S31. Edward Schramm and F. P. Hall, Impact properties of china. *J. A. Cer. S.* **13** [12], 915 (1930). 350

S32. Edward Schramm, Dust elimination in the pottery industry. *J. A. Cer. S.* **16** [5], 205 (1933). 1323

— Edward Schramm, see P. L. Christensen.

S33. A. Schroeder, *Z. Kryst.* **66**, 493 (1928). 170

S34. W. Schröder, *Z. El. Chem.* **28**, 301 (1942); *Ibid.* **49**, 38 (1943). 187

S36. J. Schüffler, Der gasgefeuerte Hochtemperaturofen für Laboratorium und Betrieb. *Ber. D. K. G.* **15** [5], 261 (1934). 923–925, 1043, 1047

S35. H. G. Schurecht, The magnetic separation of iron-bearing minerals from clays. *J. A. Cer. S.* **6**, 615 (1923). 690

S37. H. G. Schurecht, *Brick & Clay Record* **94** [3], 18 (1939). 833

S38. H. G. SCHURECHT and KENNETH T. WOOD, *The Use of Borax and Boric Acid Together with Salt in Salt Glazing.* New York State Coll. of Cer., Cer. Exp. Stn., Bulletin No. 2, in co-operation with Pacific Coast Borax Co., February (1942).   600, 601, 1017

S39. H. G. SCHURECHT, Improving the quality of extruded mill brick. *J. A. Cer. S.* **31** [5], 122 (1948).   733

— H. G. SCHURECHT, see C. M. LAMPMAN.

S40. W. SCHWABE and Z. SYSKA, Studie über die Beeinflussung der Glasurspannung durch verschiedene Flussmittel. *Sprechs.* 487, 6 October (1938).   552

S41.—

S42. ARTHUR A. SCHWARTZ, *Carbides, the Royalty of Cutting Tools.* Bell Aircraft Corporation.   1155

S43. BERNARD SCHWARTZ, Fundamental study of clay. XII— A note on the effect of the surface tension of water on the plasticity of clay. *J. A. Cer. S.* **35**, 41 (1952).   65

— MURRAY A. SCHWARTZ, see BERTHOLD C. WEBER.

S44. SCHWEITZER EQUIPMENT CO., Cleveland, Ohio, U.S.A.   794

S45. H. E. SCHWEYER, *Ind. Eng. Chem. Anal. Ed.* **14**, 622–632 (1942).   290

S46. ALEX. SCOTT, *Ball Clays.* Special Reports on the Mineral Resources of Great Britain. D.S.I.R. Memoirs of the Geological Survey. H.M.S.O. (1929).   253, 255, 264

— K. C. J. SEABRIDGE, see A. W. NORRIS.

S47. J. G. SEANOR, The speed of your auger affects the quality of your ware. *Brick & Clay Record* **97** [1], 20, July (1940).   709, 730

S48. ALFRED B. SEARLE, *An Encyclopaedia of the Ceramic Industries.* Ernest Benn, London (1929).   460, 819, 861

S49. ALFRED B. SEARLE, *Modern Brickmaking.* Ernest Benn, London (1931).   969, 978, 997

S50. ALFRED B. SEARLE, *Refractory Materials* (3rd Ed.). Charles Griffin & Co., London (1940).   411, 416, 1218, 1277–1283

— A. B. SEARLE, see H. RUSSELL.

S51. H. SEGER, Über bleifreie Glasuren. *Tonind.-Ztg.* 523 (1889).   576

S52. H. SEGER, *Gesammelte Schriften* (2nd Ed.) (1908).
   (a) p. 38.   299
   (b) pp. 289–295.   864
   (c) p. 178; *Tonind.-Ztg.* 121 (1885); *Ibid.* 135, 229 (1886).   935
   (d) *Gesammelte Schriften*, pp. 450 and 471.   553
   (e) p. 461; *Tonind.-Ztg.* **6**, 227 (1882).   527

S53. SELAS CORPORATION OF AMERICA, *Gradiation.* Erie Avenue and D Street, Philadelphia 34, Pa., U.S.A.   922

S54. H. I. SEPHTON, A laboratory furnace with hydraulically positioned hot zone. *Bull. A. Cer. S.* **31** [9], 322 (1952).   983

S55. Service (Engineers) Ltd., Leek New Rd., Cobridge, Stoke-on-Trent. — 701

S56. G. Serwatzky, Über die Bestimmung der Mineralkomponenten in Tonen. (About the determination of the mineral components in clays.) *Sprechs.* **91** [17], 396, September (1958). — 311

S57. A. Humboldt Sexton, *Fuel and Refractory Materials.* Blackie & Son, London (1923). — 45, 46, 392, 907, 1277

S58. E. Sharratt and Marcus Francis, The durability of on-glaze decoration—II. *Trans. Brit. Cer. Soc.* **42**, 171 (1943). — 379, 380, 609, 615

S59. D. H. Sharp, Question box. Which is the cheaper fuel for firing a glost oven, electricity or town gas? *Trans. Brit. Cer. Soc.* **49**, 337 (1950). — 1013, 1014

S60. L. Shartsis and A. W. Smock, Surface tensions of some optical glasses. *J. A. Cer. S.* **30**, 130 (1947). — 538

S61. C. F. Shaw, Design and production of kiln furniture for fine china. *Bull. A. Cer. S.* **34** [3], 89 (1955). — 957, *959*, 964, 966

— D. T. H. Shaw, see C. W. Parmelee.

S62. M. Shayna, A. Canadian sewer pipe works. *Brit. Clywkr.* **66** [789], 290, January (1958). — 1075

— H. Sheard, see A. F. Dufton.

S62.1. Sheepbridge Equipment Ltd., Chesterfield. — 747

S63. Herbert D. Sheets, Jack J. Bulloff and Winston H. Duckworth, Phosphate bonding of refractory compositions. *Brick & Clay Record* July (1958). — 523

S64. H. R. Shell and W. P. Cartelyou, Soluble sulphate content of pottery bodies during preparation. *J. A. Cer. S.* **26** [6], 179 (1943). — 758

S64.1. Shelley Electric Furnaces Ltd., Longton, Stoke-on-Trent. — *981*

— G. R. Shelton, see E. N. Bunting.

— M. E. Shenina, see A. S. Berman.

S65. James W. Shenton, Great Bridge, Staffs. — *734*

S66. F. Sherwood Taylor, *Inorganic and Theoretical Chemistry*, p. 412. Heinemann, London (1946). — 913

— R. F. Sherwood, see E. Schramm.

— T. S. Shevling, see A. R. Blackburn.

S67. L. E. Shipley, Rational interpretation of the analyses of clays. *Claycraft* **23** [6], 268 (1950). — 300

S68. L. E. Shipley, Experimental cone series for kilns and furnaces. *Ceramics* **9** [113], 16, July (1958). — 936

S69. Yoichi Shiraki, Slip casting. III—Physical properties of sintered alumina porcelain. IV—Casting of amphoteric oxides, especially aluminium oxide. *J. Cer. Ass. Japan.* **58** [643], 9 [644], 41 (1950). — 760

— K. M. Shmukler, see P. P. Budnikow.

— R. A. Shonk, see S. Matthews.

S70. A. J. SHORTER, The measurement of the heat required in firing clays. *Trans. Brit. Cer. Soc.* **47** [1], 1 (1948).   869–872
S71. W. SIEGL, *Neues Jb. Mineral. Geol. Paläont.* 40 (1945).   307
S72. SIEG-LAHN BERGBAU. G.M.B.H., Bergverwaltung Weilburg/Lahn, Germany.   43
S73. SIEMENS-PLANIA, Chemische Fabrik, Griesheim, Germany.   1131, 1132
S74. SILBERGRUBE FELDSPATWERK, Kaolin und Tonbergbau Max Schmidt, Oberkotzau/Bayern, Germany.   47
S75. SILEX, Società per Azioni, Milan, Italy.   900
S76. SILEX, Società per Azioni, Milan, Italy. S. A. SILEX, Superisolants Thermiques. *Transactions of the International Ceramic Congress, Holland*, 1948, p. 423.   1288
S77. SILICA PRODUCTS COMPANY, *Bentonite Handbook*. Bulletin No. 107. 700, Baltimore Avenue, Kansas City, Missouri (1934).   258
S78. W. B. SILVERMAN, Surface tension of glass and its effects on chords. *J. A. Cer. S.* **25**, 168 (1942).   538
S79. H. E. SIMPSON, Classified review of refractory slag tests. *J. A. Cer. S.* **15** [10], 536 (1932).   363

FULL LIST OF PUBLICATIONS BY FELIX SINGER

This is a complete list of Dr. Singer's publications, including those referred to in this book numbered thus S80–S127; his personal numbering runs from (1)–(147a).

— (1) Modernes Luxussteinzeug. *Tonwarenindustrie* (1909).
— (2) Über Gläser und Glasuren. *Glasindustrie* (1909).
S80. (3) Concerning the position of boron in the glaze formula. *Trans. A. Cer. S.* **12**, 676 (1910).   527
— (4) Über künstliche Zeolithe und ihren konstitutionellen Zusammenhang mit anderen Silikaten. Dissertation, Berlin (1910).
— (5) Über die Stellung des Bor in der Glasurformel. *Sprechs.* (1911).
— (6) Ein Beitrag zur Theorie der Silikate. *Sprechs.* (1911).
— (7) Über Schmelzglasuren. *Tonwarenindustrie* (1911).
— (8) Zur Kenntnis der bleihaltigen Glasuren und deren Bleiabgabe an saure Flüssigkeiten. *Tonwarenindustrie* (1911).
— (9) Die Ausstellung der Bunzlauer Industrie-Erzeugnisse. *Deut. Töpfer Ziegler Zt.* (1911).
— (10) Die Bunzlauer Topfmesse. *Tonwarenindustrie* (1911).
— (11) Die Keramik auf der ostdeutschen Ausstellung. *Tonwarenindustrie* (1911).
— (12) Die Keramik auf der Gewerbe- und Industrieausstellung in Schweidnitz. *Tonwarenindustrie* (1911).
— (13) Bayrische Gewerbeschau 1912. Ausstellungsberichte für die Zeitschrift. *Tonwarenindustrie* (1912).

— (14) Über den Einfluss von Tonerde auf die Schmelzbarkeit von Glasuren. *Ker. Rund.* (1915), (1917).
— (15) Über den Glyzerinersatz in der Keramik. *Ker. Rund.* (1916).
— (16) Über die Zusammengehörigkeit keramischer Massen und Glasuren. *Ber. tech. wiss. Abt. Verbandes keram. Gewerke* (1917).
— (17) Keramische Farben. Beitrag zu ULLMANN's *Enzyklopädie der technischen Chemie* (1917).
— (18) Lüsterfarben. Beitrag zu ULLMANN's *Enzyklopädie der technischen Chemie* (1917).
— (19) Die Normalisierung der deutschen Industrie und ihre Bedeutung für die Keramik. Vortrag, Verband keramischer Gewerke (1918).
— (20) Die Porzellanindustrie im Kriege. (*Zeitschrift des deutschen Kriegswirtschaftsmuseums, Leipzig* (1918).
— (21) Glasartiges Erschmelzen und Verarbeiten von Porzellan. *Österreichische Glas- und Keram-Industrie* (1918).
— (22) Über Rosenthal-Porzellan für chemische und technische Zwecke. *Z. angew. Chem.* (1918).
— (23) Chemisches Porzellan. *Z. angew. Chem.* (1918).
— (24) Über glasartiges Erschmelzen von Porzellan. *Z. angew. Chem.* (1919).
— (25) Keramische Wirkungen des Elektro-Osmose-Verfahrens. *Ber. tech.-wiss. Abt. Verbandes keram. Gewerke* (1919).
— (26) Einige Probleme der Porzellanindustrie im Wechsel der Zeiten. *Dinglers Polytech. J.* (1920).
— (27) Die mechanischen Eigenschaften des Porzellans und exakte Prüfungsmethoden zu ihrer Bestimmung. *Elektrotechn. Z.* (1920). Gemeinsam mit DR. ERNST ROSENTHAL.
S81. (28) Porzellan als Werkstoff. Beitrag zum *Handwörterbuch der Werkstoffe von Paul Krais.* Joh. Ambrosius Barth, Leipzig (1921). 456, 460, 464, 470, 1126
—, (29) Die physikalischen Eigenschaften des Porzellans. *Ber. D. K. G.* (1920). Gemeinsam mit DR. ERNST ROSENTHAL.
— (30) Das Porzellan für Hochspannungsisolatoren. *Elektrotechn. Z.* (1920).
S82. (31) Über die Zähigkeit keramischer Massen. Dissertation Erlangen (1921). *Bücher der D. K. G.*, Vol. 2 (1921). 349
(32) Über die Zähigkeit keramischer Isolierstoffe der Elektrotechnik. *Elektro-J.* April (1921).
— (33) Die Keramik im Dienste der chemischen Industrie. *Z. angew. Chem.* (1921).

— (34) Steinzeugmaschinen. Wochenausgabe des *Berliner Tageblatts* (1922).
— (35) Die physikalischen Eigenschaften des Steinzeugs. *Z. angew. Chem.* (1923).
— (36) Keramische Isolierstoffe. *Z. elek. Betrieb* (1923).
— (37) Über die Temperaturwechselbeständigkeit keramischer Massen. *Ber. D. K. G.* (1923).
— (38) Der Einfluss der Keramik auf die Entwickelung der chemischen Industrie. *Chemiker-Z.* (1923), Keramikheft.

S83. (39) *Die Keramik im Dienste von Industrie und Volkswirtschaft.* Verlag von Friedr. Vieweg & Sohn, A.G. Braunschweig (1923). Herausgegeben mit 90 Mitarbeitern. 33, 37, 169, 434, 439, 455, 456, 460, 494, *868*, 976, 1084, 1094, 1102

— (40) Steatit. *Die Keramik* (1923).
— (41) Tabelle der physikalischen Eigenschaften keramischer Massen. *Die Keramik* (1923).
— (42) Neue physikalische Ziffern keramischer Massen. *Ber. D. K. G.* (1924).
— (43) Steinzeug als Konstruktionsmaterial für Hochspannungs-Isolatoren. *Elektro-J.* (1924).
— (44) Steinzeug als Werkstoff im chemischen Apparatebau. *Achema-Jahrbuch* (1925).
— (45) Füllkörper. *Z. Ver. deut. Ing.* (1925).
— (46) Fortschritte der Keramik und ihre Bedeutung für die chemische Industrie. *Z. angew. Chem.* (1926).
— (47) Die physikalischen Eigenschaften keramischer Massen. *Z. Elektrochem.* (1926).
— (48) Materiali di Diempimento. *Notiz. chim. Ind.* (1926).
— (49) Offener Brief an alle Keramiker. *Keramos* (1926).
— (50) Entwickelung und Ausblicke der Steinzeugindustrie. *Tonind.-Z.* (1926).

S84. — Ger. Pat. 605237 (18 December 1926). 488

— (51) Eine neue Säurehochleitungspumpe. *Achema-Jahrbuch* (1926–1927).

S85. (52) Steinzeug und Steatit als Isolierstoffe der Elektrotechnik. *Techn. Rund.* 25 October (1927). 1198

— (53) Steinzeug als Werkstoff. *Z. Ver. deut. Ing.* (1927).

S86. (54) Wege und Ziele der Steinzeugindustrie. *Chemiker-Z.* [7] (1927). 1122

— (55) Steinzeug als Werkstoff der Elektrotechnik. *Elektro-J.* (1927).
— (56) Kochbeständige keramische Massen. *Z. angew. Chem.*, Abt. 'Chemische Fabrik' (1927).
— (57) Keramische Massen als Werkstoffe. *Z. Ver. deut. Ing.* 67.

- (58) Chemiker in keramischen Betrieben. *Ker. Rund.* (1927).
- (59) Die Entwicklung des Steinzeugs als elektrischer Isolierstoff. *Elektrochem. Ztg.* (1927).
- (60) Die keramische Industrie in Deutschland (The German ceramic industry). *Ber. D. K. G.* (1928).
- (61) Stoneware. *Berliner Tageblatt* (1928).
- (62) Gres Ceramico. *Berliner Tageblatt* (1928).
- (63) Steinzeug als Werkstoff. Zeitschrift *Die Chemische Fabrik* (1929).
- (64) Die elektrische Leitfähigkeit keramischer Isoliermassen bei steigender Temperatur. *Ker. Rund.* (1929).

S87. (65) Über die Verwendung von Schamotte, Steinzeug und Porzellan als Werkstoffe der chemischen Industrie. Zeitschrift *Die Chemische Fabrik* 2 [48–49] (1929).   468

- (66) Salzglasuren. Zeitschrift *Die Chemische Fabrik* (1929).
- (67) The largest insulators in the world. *Helios, Electrical Export Trade Journal* (1929).
- (68) Weisses Steinzeug und andere weisse keramische Massen. *Chemiker-Z.* (1929).
- (69) Neue Fortschritte des DTS-Sillimanits. *Elektro-J.* (1929).
- (70) 10 Jahre Steinzeug. *Ker.-Rund.* (1929).

S88. (71) Keramische Farben. Beitrag zu ULLMANN's *Enzyklopädie der technischen Chemie* (2nd Ed.), Vol. 4, p. 815 (1929).   610, 620–642

S89. (72) Lüsterfarben. Beitrag zu ULLMANN's *Enzyklopädie der technischen Chemie* (2nd Ed.), Vol. 7, p. 409 (1929).   658, 660

- (73) Über neuartige Steinzeugmassen. *Ber. D. K. G.* (1929) Gemeinsam mit DR. WILLI M. COHN.
- (74) The uses of stoneware in machines and apparatus. *Engineering Progress* (1929).

S90. (75) *Das Steinzeug*. Monographie, herausgegeben anlässlich des 70. Geburtstags von DR. N. B. JUNGEBLUT. Friedr. Vieweg & Sohn, Braunschweig (1929).   44, 45, 66, 431–434, 438, 582, 716, 1105, 1106

- (76) 'Führer der Keramik: Nicolaus B. Jungeblut.' *Keramos* (1929).

S91. (77) Alternde und nicht alternde keramische Massen. *Ker. Rund.* 38, 167, 183, 216 (1930).   223

- (78) Geschmolzener Quarz. Beitrag zur *Elektrothermio* von DR. M. PIRANI. Julius Springer, Berlin (1930).
- (79) Neue Möglichkeiten für das Brennen von Steinzeug. *Keramos* (1930).
- (80) Keramik und Wissenschaft. *Keramos* (1930).
- (81) Chemical stoneware. *Cer. Age* (1931).

- (82) Ein amerikanischer Kleintunnelofen. *Tonind.-Ztg.* (1932).
- (83) Dachziegelbrand im Tunnelofen. *Tonind.-Ztg.* (1932).
- (84) Vergleichende Untersuchungen über die Wirtschaftlichkeit verschiedener Ofensysteme. *Tonind.-Ztg.* (1933).
- (85) Die Bedeutung des säurefesten Steinzeugs für die Kunstseiden-Industrie. *Deut. Kunstseiden-Ztg.* (1933).

S92. — Korrosionsbeständige Werkstoffe für die Textilveredlung. *Monatschr. Textil-Ind.* [2], 33, July (1933).   1113

S93. (86) *Der Tunnelofen.* Verlag 'Chemisches Laboratorium für Tonindustrie', Prof. Dr. H. Seger and E. Cramer, Berlin (1933).   407, 1234–1243

S94. (87) Über Vorgänge beim Mauken keramischer Massen. *Koll.-Z.* **64**, 234 (1933).   263, 711

- (88) Baumaterialien für den Tunnelofen. *Tonind.-Ztg.* [53] (1933).

S95. (89) Fuel economy in the ceramic industry—I. *Trans. Cer. Soc.* **33**, 215, June (1934).   903, 908

- (90) Fuel economy in the ceramic industry—II. *Trans. Cer. Soc.* **33**, 228, June (1934).

S96. (91) Alchemite, a new ceramic ware with a water absorption of nil. *Chem. Age.* June (1934).   434, 470, 1111

S97. (92) Chemical stoneware for special purposes. *Chem. and Ind.* November 23 (1934).   1123

- (93) Making blue bricks in Czechoslovakia. *Brit. Clywkr.* June 15 (1935).
- (94) Chemical stoneware old and new. *Ind. Chem.* October (1935).

S98. (95) White chemical stoneware. *Ind. Chem.* October (1935).   1112

S99. (96) Un des problèmes de l'industrie de la vaisselle de porcelaine. *La Céramique* February (1936).   472

- (97) Old and new stoneware. *Cer. Age* February (1936).

S100. (98) *Fuel Economy in the Ceramic Industry*, Pt. III. National Council of the Pottery Industry, March 30 (1936). See **S95**.

S101. (99) Barium aluminium silicates as refractories and their use for different technical purposes. *Trans. Cer. Soc.* **35** [9], 389, September (1936).   1016, 1249, 1252

- (100) Acid Resisting Industrial Filters. II—Porous Filters. *Proc. Chem. Engineering Group* **18** (1936).
- (101) Keramische Werkstoffe. *Handbuch der chemisch-technischen Apparate, maschinellen Hilfsmittel und Werkstoffe.* Herausgegeben von Dr. A. J. Kieser. Julius Springer, Berlin (1937).

—	(102)	Öfen. *Handbuch der chemisch-technischen Apparate, maschinellen Hilfsmittel und Werkstoffe.* Herausgegeben von DR. A. J. KIESER. Julius Springer, Berlin (1937).	
—	(103)	Modern methods of brick and tile colouring. *Brit. Clywkr.* February (1938).	
—	(104)	Refractory problems in the Iron and Steel Industry. *Iron and Steel Industry.* May, June and August (1938).	
S102.	(105)	Improvement of ceramic bodies—I. *Brit. Clywkr.* July (1938). Improvement of ceramic bodies—II. *Ibid.* August (1938).	695, 758
—	(106)	Isolation des Voutes de Fours Siemens-Martin. *Revue Universelle des Mines* **15**, January (1939). En collaboration avec le DR. JOSEPH M. LUCAS (*né* ROBITSCHEK), Akron, Ohio.	
S103.	(107)	Improvement of plaster moulds. *Brit. Clywkr.* March (1939).	163
—	(108)	Materials for sewerage pipes. *Trans. Cer. Soc.* **38**, April (1939).	
S104.	(109)	Carbon linings for blast furnaces. *Metal and Alloys* **10**, April (1939).	1250, 1252, 1253
S105.	(110)	Magnesite refractories. *Brit. Clywkr.* June (1939).	118, 1256
—	(111)	Aluminina from clay. *Brick & Clay Record.* June (1939).	
S106.	(112)	Water-resisting dolomite bricks, their uses in war time. *Brit. Clywkr.* July (1939).	1270
—	(113)	Chromite and chrome-magnesite refractories. *Brit. Clywkr.* January and February (1940). In collaboration with DR. JOSEPH M. LUCAS (*né* ROBITSCHEK), Akron, Ohio, **R46** and **R47**.	
S107.	(114)	The manufacture of refractory insulating bricks. *Brit. Clywkr.* **49**, 103, July and August (1940).	427, 428, 1296
S108.	(115)	*Ceramic Glazes.* Borax Consolidated Ltd., London, September (1940).	134, 223, 532, 535, 536, 554, 583–584
—	(116)	Stoneware for the food industry. *Food.* May and June (1941).	
S109.	(117)	Polishing of scratches on glazes. *Trans. Cer. Soc.* **39**, 513, September (1939); *Cer. Age* 101, October (1941).	558, 1061
S110.	(118)	Sanitary vitreous china. *Trans. Cer. Soc.* **40** [4], 119, April (1941).	35, 41, 94, 105, 106, 450, 452–455
—	(119)	*Ceramics Yesterday, To-day and To-morrow.* Address to the Association of Czechoslovak Scientists and Technicians on 23 February (1944).	
S111.	(120)	Slip-coated synthetic foundry sand. *Foundry Trade Journal.* March and April (1944).	54

- (121) Glazes and glasses. *Pottery and Glass* December (1944).
- S112. (122) *Coloured Bodies.* Read before the Pottery Managers' and Officials' Association 9 October (1945); *Pottery Gazette* January (1946).   231, 232, 592, 593, 624–636, 800
- (123) *Six Fundamental Principles of Modern Ceramics.* Address to the Association of Austrian Engineers, Chemists and Scientific Workers in Great Britain, 14 January (1946).
- (123a) Six Principes Fondamentaux des Céramiques modernes. *Verres et Réfractaires.* June (1947).
- (124) Ceramic cordierite bodies. *Trans. Canad. Cer. Soc.* **15** (1946).
- (125) Ceramics as engineering materials. *Engineering Materials.* December (1946). In collaboration with HANS THURNAUER, M.Sc.
- (126) Sinter alumina. *Metallurgia* **36** [215], [216] (1947). Together with HANS THURNAUER, M.Sc.
- S113. (127) Super-duty silica bricks. *Iron & Coal Trades Review* **157** [4201], September (1948).   1220, 1225
- S114. (128) *Low Solubility Glazes.* Borax Consolidated Ltd., London (1948).   133, 373
- (128a) *Les glaçures céramique. Les glaçures à faible solubilité.* French translation by J. BERNOT. Borax Français, Paris.
- (129) Sinter alumina as engineering material for cutting tools and turbine blades. *Ceramics* **1** [5], [6] (1949).
- (129a) French Translation. *Industrie Céramique*, November and December (1950).
- (129b) German Translation. Sintertonerde als Werkstoff für Drehwerkzeuge und Turbinenschaufeln. *Sprechs.* **84** [18–20] (1951).
- S115. (130) The nature of glasses and glazes and their hardness. *Glass Industry.* May, June and July (1950).   213
- (130a) German Translation. Die Konstitution der Gläser und Glasuren und Ihre Härte. *Europäische Tonind.*
- S116. (131) Weathering and maturing of clay for sewage pipes and other products. *Brit. Clywkr.* **70** [706], February (1951).   263, 711
- S117. (132) *Chemical Stoneware To-day.* Presented at the Twelfth International Congress of Pure and Applied Chemistry, New York, September, 1951; *Cer. Age* **58** [6], 33, December (1951).   432, 1107, 1122, 1123, 1128
- S118. (133) Die Schwierigkeiten des Entlüftens von Giessschlickern. *Europäische Tonind.* **2** [9], September (1952).   711, 764

S119. (134) Humid-pressing and dry-pressing of porcelain and   742, 746
steatite. *Brit. Clywkr.* **61** [727], November (1952).
— (134a) German Translation. Feucht- und Trockenpressen von Porzellan und Steatite. *Europäische Tonind.* **2** [11], [12], November, December (1952).
— (135) Zinc Glazes. *Zinc Bulletin* No. 10, October (1952).
S120. — Zinc Glazes. Duplicated Brochure by Zinc Pigment Development Association.   133, 214
— (136) 'Ceramics.' *Ind. Eng. Chem.* **44**, October (1952).
S121. — *Brit. Pat.* 669954; *Brit. Clywkr.* **61** [726], 27 October (1952).   157, 834
S122. (137) Olivine and Forsterite. *Brit. Clywkr.* **62** [735, 736, 737, 738], July–October (1953).   138, 422, 498, 1268–1274
— (137a) Olivin und Forsterit. *Sprechs.* **86** [14–16], (1953).
— (138) Vetrine Opache. (In Italian.) *La Ceramica* **8** (New Series) [11], 41 (1953).
— (139) Low temperature glazes. *Trans. Brit. Cer. Soc.* **53** [7], 398 (1954).
S123. (140) Progress in permeable porous ceramics. *Chem. Age* 9 January (1954).   345, 1121
S124. (141) Progress in porous and hollow ceramics for construction. *Brit. Clywkr.* **63** [745], 55, May (1954). Together with Sonja S. Singer.   426, 897, 972, 1218, 1284, 1285, 1295
— (142) Ceramics, yesterday, today and tomorrow. *Brit. Clywkr.* **63** [749], [750], September, October (1954).
— (143) Salt-glazed vitreous clay pipes. *Corrosion Technology* **3** [5] (1956).
— (144) Keramische Massen mit kleinem Ausdehnungskoeffizienten. *Sprechs.* **89** [17], 399, 5 September (1956). Together with Sonja S. Singer.
S125. (144a) Ceramic bodies with small expansion co-efficients. *Brit. Clywkr.* **66** [780–782], 19–23, 41–45, 88–92 (1957). Together with Sonja S. Singer.   484, 487, 503
— (145) Improving the plasticity of clay. *Brit. Clywkr.* **65** [774], 178, October (1956). Together with Sonja S. Singer.
— (146) Production of sinter alumina cutting tools. *Industrial Diamond Review* **17**, 126, July (1957). Together with Sonja S. Singer.
— (147) Sintertonerde als Schneidewerkzeuge. *Keram. Z.* **9** [10], 532 (1957). Together with Sonja S. Singer.
— (147a) Sinter alumina for cutting tools. *Interceram.* [6], 22 (1957); *Refr. J.* **34** [10], 456 (1958). Together with Sonja S. Singer.
— See P. Grodzinski.
S126. Sonja S. Singer, Using copper for coloring glazes. *Cer. Ind.* **57** [2, 3], 72, 69, August, September (1951).   234, 628

S127. Sonja S. Singer, The function and effect of boric oxide in ceramic glazes. *Brit. Clywkr.* **61** [730], 332, [731], 366; *Ibid.* **62** [733] 51, [734], 86 (1953).   131
— Sonja S. Singer, The flame photometer in ceramic laboratories. *Brit. Clywkr.* **64** [764], 262 (1955).
— Sonja S. Singer, Neue Fortschritte in verlustarmer Electrokeramik. *Keram. Z.* **13** [12], 664 (1961).
— Sonja S. Singer, Recent progress in low loss electroceramics. *Interceram.* [2], 93 (1961).
— Sonja S. Singer, Neuigkeiten in Grossbritannien. *Keram. Z.* **14** [3], 144 (1962).
— Sonja S. Singer, see Felix Singer, No. 141, 144, 144a, 145, 146, 147 and 147a.
— U. S. Singh, see H. Lehmann.
— H. H. Sisler, see H. W. Newkirk.
— P. Sjöman, see J. A. Hedvall.
S128. Skandinavisk Moler Industri A/S Sundby, *Moler Insulating Bricks.* Mors., Denmark.   900
S129. A. K. Smalley, B. W. King and W. H. Duckworth, V_2O_5 opacified white enamels. *J. A. Cer. S.* **40** [8], 253 (1957).   590
S130. Adolph G. Smekal, On the structure of glass. *J. Soc. Glass. Tech.* **35** [167], 411, December (1951).   201, 202
S131. V. I. Smirnov and V. D. Mishin. *Trudy. Ural. Ind. Inst.* [14], 58 (1940); *Khim Referat Zhur* [9], 73 (1940).   183
S132. V. I. Smirnov and V. D. Mishin. *Trudy. Ural. Ind. Inst.* [14], 69 (1940); *Khim Referat Zhur* [9], 75 (1940).   181, 189
S133. A. N. Smith, Some investigations on glaze body layers. *Transactions of the Third International Ceramic Congress, Paris*, 1952, p. 289.   378, 545
S134. A. N. Smith, Investigations on glaze-body layers. II— Influence of the layer on crazing and peeling. *Trans. Brit. Cer. Soc.* **53**, 220 (1954).   545, 546
— C. S. Smith, Jr., see J. Biscoe.
S135. K. E. Smith and A. Arbo, Further development of the underglaze colour crayon. *Bull. A. Cer. S.* **13** [2], 40 (1934).   806
— K. W. Smith, see R. W. Knauft.
S136. R. W. Smith and J. E. Hoagbin, Quantitative spectrographic analysis of ceramic materials. *J. A. Cer. S.* **29** [8], 222 (1946).   269
— A. W. Smock, see L. Shartsis.
S137. Edward J. Smoke, Steatites of higher thermal resistance. *Cer. Age* March (1948).   493, 494, 1184
S138. Edward J. Smoke, Thermal endurance of some vitrified industrial compositions. *Cer. Age* September (1949).   1166
S139. Edward J. Smoke, Ceramic compositions having negative linear expansion. *J. A. Cers.* **34** [3], 87 (1951).   501, *502*

S140. EDWARD J. SMOKE, Thermal shock behaviour of ceramic dielectrics. *Cer. Age* February (1951). — 1167

S141. EDWARD J. SMOKE and JOHN H. KOENIG, Recent developments in ceramic dielectrics. *Cer. Age* **60** [6], 20, December (1952). — *1169*

S142.—

S143. EDWARD J. SMOKE, Ceramic compositions having negative linear expansion—II. *Cer. Age* July 13 (1953). — 503

S144. EDWARD J. SMOKE, Spinel as dielectric insulation. *Cer. Age* May (1954). — 499

— EDWARD J. SMOKE, see JOHN H. KOENIG.

S145. W. J. SMOTHERS, YAO CHIANG and ALLAN WILSON, *Bibliography of Differential Thermal Analysis*. Univ. of Arkansas, Inst. of Sci. and Tech., Fayetteville, Research Series No. 21, November (1951). — 315

S146. J. L. SNOEK, *New Developments in Ferromagnetic Materials*. Elsevier, Amsterdam (1947). — 520

S147. N. H. SNYDER and EDWIN RUH, Properties of high magnesia whitewares. *Cer. Age* August (1954). — 503, 504, 1193

— NICHOLAS H. SNYDER, see JOHN H. KOENIG.

— SOBOLE, see POUTROIE.

S148. SOCIÉTÉ FRANÇAISE DE CÉRAMIQUE: Service de Recherches Techniques. Un four à haute température. *Ind. Cér.* [455], 171, July–Aug. (1954). — 983

S149. SOEST-FERRUM APPARATENBAU. G.M.B.H., Düsseldorf, 159 Hansa-Allee, Germany. — 665

S150. TONGRUBE SONNENBERG G.M.B.H., Eisenberg (Pfalz), Germany. — 43, 49

S151. H. SORTWELL, Earthenware bodies and glazes. *J. A. Cer. S.* **4**, 990 (1921). — 568

— HANS SPÄTH, see AUGUST KREUS.
— W. C. SPANGENBERG, see D. V. VAN GORDEN.
— S. SPEIL, see F. H. NORTON.
— G. SPENCE, see R. M. GILL.

S152. W. & C. SPICER, LTD., High Temperature Section, The Grange, Kingham, Oxon. — 1141

S153. H. M. SPIERS, *Technical Data on Fuel* (4th Ed.) British National Committee World Power Conference. London (1937). — 904–905

S154. W. D. SPORE, Water-spotting of semi-vitreous dinnerware in the decorating process. *Bull. A. Cer. S.* **25**, 82 (1946). — 830

S155. IRA E. SPROCAT, The use of pyrophyllite in wall tile bodies. *J. A. Cer. S.* **19** [5] (1936). — 440, 446

S156. H. SPURRIER, The use of ox gall to prevent crawling of glazes. *J. A. Cer. S.* **5**, 937 (1922). — 796

S157. H. SPURRIER, Servicing of kiln cars. *J. A. Cer. S.* **13** [12], 929 (1930). — 1027

S158. Staatliche Porzellan-Manufaktur Berlin (Alfred König), Gelbe keramische Unterglasurfarbe. D.B.P. 963138. — 618

S159. A. Staerker, Neue Erkenntnisse auf dem Gebiet des Oberflächenschutzes Feuerfester Baustoffe. *Tonind.-Ztg.* **75**, 33 (1951); Schutzschichten für die Anwendung in der Feuerfest- und Hüttenindustrie. *Ibid.* **76**, 93 (1952); Vanal, ein neuartiger Korrosionsschutz. *Ber. D. K. G.* **34** [10], 329 (1957). — 1213

S160. C. J. Stairmand, Some practical aspects of particle size analysis in industry. *Symposium on Particle Size Analysis*, p. 77. Suppl. *Trans. Inst. Chem. Eng.* **25** (1947). — *281, 284*, 290

S161. C. J. Stairmand, A new sedimentation apparatus for particle size analysis in the sub-sieve range. *Symposium on Particle Size Analysis*, p. 128. Suppl. *Trans. Inst. Chem. Eng.* **25** (1947). — 286

— C. J. Stairmand, see W. F. Carey.

S162. Homer F. Staley, The microscopic examination of twelve mat glazes. *Trans. A. Cer. S.* **14**, 691 (1912). — 594, 595

S163. Homer F. Staley, Antimony compounds as opacifiers in enamels. *Trans. A. Cer. S.* **17**, 173 (1915). — 589

— W. Stamm, see W. Jander.

S164. J. E. Stanier, *The Utilisation of Town Gas in the Ceramic Industry*. Paper read at the Conference on 'Fuel and the Future', London, 8, 9 and 10 October, 1946. Ministry of Fuel and Power, London. — 912, 914, 915

S165. J. E. Stanworth, *J. Soc. Glass. Tech.* **21**, 155 (1937). — 181

S166. J. E. Stanworth, The structure of glass. *Trans. S. Glass. Techn.* **30**, 54 (1946). — 212, 213

S167. J. E. Stanworth, On the structure of glass. *Trans. S. Glass. Techn.* **32**, 154 (1948). — 201, 212–215

S168. D. D. Starkey, Developing the de-airing process. *Cer. Age* **50**, 74 (1947). — 709

S169. Ernst Stauber, Herstellung von Seladonglasuren in reduzierendem Brand. *Euro-Ceramic* **9** [2], 40 (1959). — 632

— Julius Stawitz, see Friedrich Crössmann.

S170. Steatit Magnesia A.G. — 1187–1188

S171. Steatite & Porcelain Products Ltd., Stourport-on-Severn, Worcs. — 1179–1195

S172. M. E. C. Stedham, An automatic moisture indicator. *Trans. Brit. Cer. Soc.* **57** [7], 381 (1958). — 276

S173. Steele and Cowlishaw, Cooper St., Hanley, Stoke-on-Trent. — 672

S174. A. Preston Steele, Jr., Effect of additional pugging on clays and shales. *Bull. A. Cer. S.* **32** [6], 215 (1953). — 708

S175. G. Steele, *Enamel Mill Room Practice with Special Reference to Grinding Mills*. A paper read before The Institute of Vitreous Enamellers. Steel and Cowlishaw, Stoke-on-Trent. — 677, 679

— H. L. Steele, see J. A. Johnson.

S176. W. Steger, Neue Untersuchungen über Wärmeausdehnung und Entspannungstemperatur von Glasuren (mit besonderer Berücksichtigung der Anpassung der Glasur au den Scherben). *Ber. D. K. G.* **8**, 24 (1927). 535, 554

S177. W. Steger, *Ber. D. K. G.* **8**, 35 (1927); *Ibid.* **13**, 41 (1932). 555, 556

S178. W. Steger, *The Dependence of Stresses in Glazed Ware on the Distribution of the Glaze Constituency in the Frit and in the Mill Batch.* Ceramic Paper No. 15, The Euston Lead Co. (Extract from *Ber. D. K. G.* **12**, 43 (1931).) 568

S179. W. Steger, Über die Wärmeausdehnung von niedrig gebrannten kalkhaltigen keramischen Massen. *Ber. D. K. G.* **13**, 412 (1932). 554

S180. W. Steger, Das Auftreten von Glasurrissen an keramischen Waren im Muffelbrande. *Ber. D. K. G.* **13**, 42 (1932). 554

S181. W. Steger, Vergleichende Untersuchungen an Steingutglasuren. III.—Über die Einwirkung von Schwefeldioxyd auf Steingutglasuren mit Blei und Bor. *Ber. D. K. G.* **21**, 228 (1940). 573

S182. W. Steger, Fluor in Steingutglasuren ohne Blei und Bor. *Ber. D. K. G.* **22**, 73 (1941). 127

— P. Stegmuller, see K. Beck.
— B. L. S. Steierman, see R. F. Geller.

S183. John Stein & Co. Ltd., Bonnybridge, Scotland. 1235–1244, 1261–1267

S184. J. C. Stein and Co. Ltd. and H. C. Biggs, Casting and mixing ceramic material. *Brit. Pat.* 481217 (1936). 764

S185. Steinzeugfabrik Embrach A.G., Einteilung keramischer Erzeugnisse. Eigenschaftstafel keramischer Werkstoffe für die Technik. Embrach, Germany. 1088, 1125, 1127

S186. Theodor Stephan K.-G., 'Haiger', Haiger (Dill Kreis), Germany. 39

S187. J. M. Stevels, Some chemical aspects of opacification of vitreous enamels. *Sheet Metal Industries* November (1948). 585, 586

S188. J. M. Stevels, Comments by Dr. Stevels on the Paper of Mr. J. E. Stanworth (see **S166**). 212, 214

S189. D. K. Stevens and R. E. Birch, Shrinkage rates in firing fireclay refractories. *J. A. Cer. S.* **30** [4], 109 (1947). 868

S190. Frank J. Stevens, Notes on the use of pyrophyllite as an electrical insulator. *J. A. Cer. S.* **21** [9], 330 (1938). 500

— J. L. Stevens, see W. I. Garms.
— J. Stewart, see J. F. Hyslop.

S191. Stockdale Engineering Ltd., Royal London House, 196, Deansgate, Manchester 3. 706

S192. Stephen D. Stoddard and Adrian G. Allison, Casting of magnesium oxide in aqueous slips. *Bull. A. Cer. S.* **37** [9], 409 (1958). 760

S193.—
S194. R. O. STOKES AND CO. (Mechanical and Metallurgical 714
Engineers), *The Stokes Glandless Centrifugal Pump.*
815–817, Salisbury House, London Wall, E.C.2.
S195. R. O. STOKES AND Co., *The Stokes Diaphragm Pump.* 714
S196. R. O. STOKES AND CO. LTD. (Mechanical and Metallurgical 687
Engineers), *The Stokes Double-Acting Classifier.* 815–
817, Salisbury House, London Wall, E.C.2.
S197. R. L. STONE, *Bulletin* 25, Eng. Exp. Station, North 564
Carolina State College, The University of North
Carolina (1943).
S198. R. L. STONE, Method of controlling firing shrinkage of 494–497
dry-pressed steatite-bodies. *Cer. Age* 86, March
(1944).
S199. R. L. STONE, Binders and shrinkage control in dry- 743, 744
pressed steatite porcelains. *North Carolina State College Record* 44 [7], March (1945).
S200. ROBERT L. STONE, Temperature gradient method for 440, 984
determining firing range of ceramic bodies. *J. A.
Cer. S.* 36 [4], 140 (1953).
— R. L. STONE, see C. A. COWAN.
— R. L. STONE, see A. F. GREAVES WALKER.
S201. STOTHERT & PITT LTD., 38, Victoria St., London, S.W.1. *262*
S202. H. R. STRAIGHT, De-airing stiff mud bodies by other 709
means than the vacuum method. *J. A. Cer. S.* 16 [6],
251 (1933).
S203. H. R. STRAIGHT, *Why We Fire Ceramic Wares So Long.* 867
(Unpublished Work.)
S204. G. A. STRAKOV, *Jubileiny Sbornik Nauch. Trudov, Moskov.* 189
Inst. Tstvetnykh Metal i Zolota [9], 514 (1940); *Khim.
Referat. Zhur.* 4 [5], 80 (1941).
S205. HANS STRASSER, Improvements in and relating to the fixing *775*
of handles and other fittings to pottery articles. *Brit.
Pat. Spec.* 593520; described in *Ceramics* 3 [25], 41
(1951).
— HILDE STRUMM-BOLLENBACH, see WERNER BUSCH.
S206. L. STUKERT, Kadmiumgelb und Kadmiumrot als Farb- 616
körper in der Silikatindustrie. *Farbenzeitung* 39,
[1, 2, 3] 9, 36, 61 (1934).
S207. L. STUCKERT, Gelbfarbkörper in der Emailindustrie. 612
Glashütte 65 [39], 611 (1935).
S208. LUDWIG STUCKERT, Das Neapelgelb als feuerfester Unter- 634
glasurfarbkörper. *Sprechs.* [21–24] (1935).
S209. JASPER L. STUCKEY, Kaolins of North Carolina. *Trans.* 251
A.I.M.E. 173, 47 (1947).
S210. C. STUERMER, The pencil is also a ceramic product. *Euro-* 1144
päishe Tonind. 3 [6], 161 (1953).
S211. C. STÜRMER, Begüsse und Glasuren für Gefässkeramik. 798
Euro-Ceramic. 8 [11], 279 (1958).

BIBLIOGRAPHY AND AUTHORS INDEX 1419

S212. R. T. STULL, A cheap enamel for stoneware. *Trans. A. Cer. S.* **10**, 216 (1908); *Ibid.* **11**, 605 (1909).   588

S213. R. T. STULL and G. H. BALDWIN, Cobalt colors other than blue. *Trans. A. Cer. S.* **14** [3], 764 (1912).   626

S214. STUPAKOFF, Ceramic & Manufacturing Co. in Latholee, U.S.A.   1192

S215. STURTEVANT ENGINEERING CO. LTD., *Ring Roll Mills.* Publc. No. 1562. Sturtevant House, Macdonald Rd., Highgate Hill, London, N.19.   669

S216. JOHN D. SULLIVAN, Physico-chemical control of properties of clays. *Trans. Electrochem. Soc.* **75**, 71 (1939).   57, 63, 329

S217. JOHN D. SULLIVAN, CHESTER R. AUSTIN and EDWIN J. ROGERS, *Expanded Clay Products.* Amer. Inst. of Min. and Met. Eng., Techn. Publ. No. 1485, February (1942).   1290

— JOHN D. SULLIVAN, see ROBERT P. GRAHAM.

S218. S. W. SWAIN, Select refractories for the job they must do. *Brick & Clay Record* **124** [1], 94 (1954).   886–889, 893

S219. *Symposium on Ceramic Cutting Tools.*
    (a) F. H. NORTON, Processing ceramics.   1159
    (b) E. RYSHKEWITCH, Properties and use of ceramic tools.   1159
    (c) W. MUHEILDON, Notes on ceramic tools.   1159, 1160
    (d) E. DI CESARE, Molybdenum boride cutting tool.   513
    (e) B. BOVARNICK, Preparing ceramic tool material.
    (f) W. B. KENNEDY, Machining studies.   *1159*–1164
       U.S. Dept. Comm. O.T.S., PB 111757 9 February (1955).

— Z. SYSKA, see W. SCHWABE.

— HIDEO TAGAL, see TOSHIYOSHI YAMAUCHI.

T1. K. TAKAHASHI, *J. Soc. Glass. Tech.* **37**, 3N (1953).   548

T2. T. W. TALWALKAR and C. W. PARMELEE, Measurement of plasticity. *J. A. Cer. S.* **10**, 670 (1927).   327

T3. S. TAMARU and N. ANDÖ, *Z. anorg. Chem.* **184**, 385 (1929).   183

T4. S. TAMARU and N. ANDÖ, *Z. anorg. Chem.* **195**, 35 (1931).   187

T5. S. TAMURU and N. ANDÖ, *Z. anorg. Chem.* **195**, 309 (1931).   183

T6. G. TAMMANN and C. F. GREVEMEYER, *Z. anorg. Chem.* **136**, 114 (1924).   185

T7. G. TAMMANN, FR. WESTERHOLD, B. GARRE, E. KORDES and H. KALSING, *Z. anorg. Chem.* **149**, 21 (1925).   181, 183, 185, 187, 189

T8. G. TAMMANN and W. ROSENTHAL, *Z. anorg. Chem.* **156**, 20 (1926).   181, 185, 187, 189

T9. Y. TANAKA, *Bull. Chem. Soc. Japan* **16**, 455 (1941).   183

T10. Y. TANAKA, *Bull. Chem. Soc. Japan* **17**, 64 (1942).   183

T11. Y. TANAKA, *Bull. Chem. Soc. Japan* **17**, 186 (1942).   181, 183

T12. M. S. TARASENKO, The origin of S-shaped cracks. *Steklo i Keram.* **8** [6], 16 (1951); *Trans. Brit. Cer. Soc.* **51** [1], 5A (1952).   731

T13. CHARLES B. TAUBER and ARTHUR S. WATTS, A deformation study of mixtures of the deformation-eutectics of potash–soda felspar–CaO–Al_2O_3–SiO_2 and MgO–Al_2O_3–SiO_2, both in the raw and prefused state. *Bull. A. Cer. S.* **31** [11], 458 (1952). 220–222

— A. J. TAYLOR, see J. R. JOHNSON.

T14. THE CHAS. TAYLOR SONS Co., *Taylor Sillimanite*. Bulletin No. 318. Cincinnati, Ohio, U.S.A. 1241–1251, 1258–1269

— JACK L. TAYLOR, see POL. DUWEZ.

T15. NELSON W. TAYLOR, Reactions between solids in the absence of a liquid phase. *J. A. Cer. S.* **17**, 155 (1934). 195

T16. N. W. TAYLOR, *Bull. Geol. Soc. Am.* **46**, 1121 (1935). 181, 189

T17. JO MORGAN TEAGUE, JR., and ARTHUR S. WATTS, A study of deformation-eutectic development in mixtures of felspar and $CaCO_3$, $MgCO_3$, $BaCO_3$ and ZnO. *Bull. A. Cer. S.* **31** [11], 457 (1952). 220

— R. G. TEMPELMAN, see V. F. ZHURAVLEV.

T18. THE THERMAL SYNDICATE LTD., Wallsend, Northumberland. 1135, 1138, 1141

T19. L. E. THIESS, Influence of glaze composition on the mechanical strength of electrical porcelain. *J. A. Cer. S.* **19** [3], 70 (1936). 550

T20. L. E. THIESS, Some characteristics of steatite bodies. *J. A. Cer. S.* **20** [9], 311 (1937). 564

T21. L. E. THIESS, Steatite glazes. *J. A. Cer. S.* **29** [3], 84 (1946). 564

T22. L. E. THIESS, Vitrified cordierite bodies. *J. A. Cer. S.* **26** [3], 99 (1943). 484, 487–491, 563, 566

T23. EVERETT THOMAS, MILTON A. TUTTLE and ESTHER MILLER. Study of glaze penetration and its effect on glaze fit. *J. A. Cer. S.* **28** [2], 52 (1945). 543–545, 566

— E. A. THOMAS, see R. W. KNAUFT.

— M. THOMAS, see G. GEHLHOFF.

— R. R. THOMAS, JR., see HELEN BLAIR BONLETT.

T24. H. G. THOMASON, The effect of particle size of zinc oxide on the consistency of glaze slips. *J. A. Cer. S.* **12**, 581 (1929). 132

T25. C. L. THOMPSON and HERMAN G. WILCOX, High-temperature gas-fired test furnace. *Bull. A. Cer. S.* **19** [9], 336 (1940). 983

— C. L. THOMPSON, see T. N. MCVAY.

T26. JOHN G. THOMPSON and MANLEY W. MALLETT, Preparation of crucibles from special refractories by slip-casting. *J. Res. Nat. Bur. Stand.* **23**, RP 1236, August (1939). 1142

— WILLIAM M. THOMPSON, see BERTHOLD C. WEBER.

T27. P. THOR, Über die Länge des Ringofenbrennkanals. *Tonind.-Ztg.* **58** [17], 208 (1934). 997

T28. P. THOR, Künstliches Trocknen von Streichmaschinensteinen in Holland. *Der Ziegelmarkt* **6** [23], 5 (1953). 842

T29. M. W. THRING, *The Science of Flames and Furnaces.* Chapman & Hall, London (1952). — 899
T30. HANS THURNAUER and A. R. RODRIGUEZ, Notes on the constitution of steatite. *J. A. Cer. S.* **25** [15], 443–450, November (1942). — 494
T31. HANS THURNAUER, *Report of High Frequency Technical Ceramic Materials of Germany.* Report C-63, December (1945). — 518
T32. HANS THURNAUER, *High Frequency Insulation* (1949). — 1210
T33. HANS THURNAUER, Controls required and problems encountered in production dry pressing. *Ceramic Fabrication Processes.* W. D. KINGERY. Mass. Inst. of Tech. and John Wiley (1958). — *741, 742, 743,* 746
— HANS THURNAUER, see J. E. COMEFORO.
— HANS THURNAUER, see R. RIEKE.
T34.—
— KOJI TOMIURA, see GORO YAMAGUCHI.
T35. N. A. TOROPOW and F. JA. GALACHOW, Neue Ergebnisse über das System Al_2O_3–SiO_2. *Ber. Akad. d. Wiss. UdSSR* **78**, 299 (1951); Abs. *Ber. D. K. G.* **30** [11], R-408 (1953). — 198
T36. L. TOUTSCHEFF, Einige praktische Erfahrungen mit Salzglasuren. *Ker. Rund.* **44** [8], 84 (1936). — 601
T37. TRADE, BOARD OF, *Report of the Enquiry on the Ball-Clay Industry.* H.M.S.O., London (1946). — 256
T38. TRADE, BOARD OF, *Working Party Reports. China-Clay.* H.M.S.O., London (1948). — 241
— K. TRAUB, see C. A. BEST.
T39. W. E. TRAUFFER, Flint grinding pebble production. *Mine Quarry Eng.* **9**, 258 (1944). — 682
T40. R. S. TROOP, Some dryer considerations. *Sp. Bull. B.R.R.A.* [12], April (1926). — 838
T41. L. J. TROSTEL and D. J. WYNNE, Determination of quartz (free silica) in refractory clays. *J. A. Cer. S.* **23**, 18 (1940). — 306
— GLEN C. TRUESDELL, see THERON A. KLINEFELTER.
— EMIL TRUOG, see G. J. BARKER.
T42. L. G. TUBBS, The determination of soluble sulphates in clay. *Bull. A. Cer. S.* **32** [5], 181 (1953). — 322, 323
— W. E. S. TURNER, see I. T. HOWARTH.
T43. E. TUSCHHOFF, Ueber das Brennen und Kühlen gelber durch Eisenoxyd gefärbter Klinker aus kohlehaltigem Ton, die dabei auftretenden Porositäten und die unter gewissen Bedingungen sich zeigenden Farbveränderungen. *Ber. D. K. G.* **17** [7], 333 (1936). — 863
— MILTON A. TUTTLE, see EVERETT THOMAS.
— O. F. TUTTLE, see EARL INGERSON.
T44. ROBERT TWELLS, JR., Beryl as a constituent in high tension insulator porcelain. *J. A. Cer. S.* **5** [5], 228 (1922). — 506

— Robert Twells, see Eugene Fisher.
T45. W. S. Tyler Co., The profitable use of testing sieves. 1301
Catalogue 53. Cleveland, Ohio (1938).

U1.—
U2. A. Ungewiss, Untersuchungen über die Eignung von Press- 745
oelen für das Feuchtpressen elektrokeramischer Isolierteile. *Ber. D. K. G.* **25** [3–4], 42 (1944).
U3. Malcolm Upright, The planning and reconstruction of *1090–1092*
pottery factories. *Pottery Gazette.* May and June (1952); *Ceramics* **4** [41], 221, July (1952).
U4. U.S. Stoneware Co., Heat resisting ceramic. *Plater's* 1124
Guide **30** [10], 13 (1934); *Metal Cleaning and Finishing* **6** [10], 533 (1934); *Amer. Cer. Soc. Cer.* Abs. **14** [2], 42 (1935).
U5. U.S. Stoneware Co., Akron, Ohio, U.S.A. *1114*
U6. E. B. Uvarov, *A Dictionary of Science.* Penguin Books, 755
Harmondsworth, Middx. (1951).

— H. C. Vacher, see C. W. Davis.
— L. I. Vaganova, see O. E. Gorbunova.
V1. N. N. Valenkov, E. A. Porai and Koschits, *J. Phys. Chem.* 187
(*U.S.S.R.*) **6**, 757 (1935).
V2. Roland R. Van der Beck and J. O. Everhart, Firing and 867, *868*
cooling shrinkage behavior of structural clay bodies. *J. A. Cer. S.* **34** [12], 361 (1951).
V3. R. Vanderbilt (Suppliers), *Organic Clay Deflocculants.* 761
New York, U.S.A.
— D. V. Van Gordon, see R. R. Danielson.
V4. H. T. Van-Horn, Petroleum products in ceramic processing 155, 157
operations. *Bull. A. Cer. S.* **26** [8], 229 (1947).
V5. Albert Vasel, Die Fabrikation von Steinzeug-Bodenplatten. 436, 1078, 1079
Sprechs. **85** [2–4] (1952).
V6. B. E. Vassiliou and C. J. W. Baker, A simple laboratory 346, 347
method for determining the modulus of elasticity. *Trans. Brit. Cer. Soc.* **56** [10], 499 (1957).
— F. Vaughan, see A. W. Norris.
V7. R. J. Verba, Automatic spraying of glazes. *Bull. A. Cer.* 796
S. **33** [10], 307 (1954).
V8. Ralph G. Verdieck, A Foote note on lithium clay dispersions. 761
Foote Prints **22** [1], 17 (1950).
— K. Vet, see R. Jagitsch.
V9. Hans Vetter, Die Roh- und Hilfsstoffe der Keramik. 107
Europäische Tonind. **2** [11], 295 (1952).
V10. Hans Vetter, Lithium-Mineralien als Rohstoff der 122/123, 124
keramischen Industrie. *Euro-Ceramic* **8** [5], 119 (1958).

V11. Hans Vetter, Seltene Erden und seltene Elemente. 141, 142
Euro-Ceramic **8** [9], 219 (1958).

V12. Vickerys Ltd., 4 Lambeth Palace Rd., London S.E.1. *247*
Sole licensees of Bird Machine Co., South Walpole,
Mass., U.S.A.

V13. Fritz Viehweger, Die Herstellung und Verwendung von 811–813
Schablonen in der keramischen Industrie. *Sprechs.* **88**
[6], 109, March (1955).

V14. Fritz Viehweger, Feuerbeständige Farbkörper. *Sprechs.* 639, 640
89 [19], 449, October (1956).

V15. Fritz Viehweger, Farbige Lüster für SK 021–019. 658
Sprechs. **91** [17], 406, September (1958).

V16. Fritz Viehweger, Unterglasur-Farblösungen. *Sprechs.* 644–645
92 [3], 51, February (1959).

V17. L. Vielhaber, Wirkung einer Mischung verschiedener 590
Trübungsmittel. *Emailwaren-Industrie* **8** [12], 90 (1931).

V18. Hermann F. Vieweg, Some effects of grain size of china 84
clays. *J. A. Cer. S.* **16**, 77 (1933).

— V. S. Vigdergauz, see A. I. Avgustinik.

V19. E. Heikel Vinther and M. L. Lasson, Über Korngrössen- 285
Messungen von Kaolin- und Tonarten. *Ber. D. K. G.*
14, 259 (1933).

V20. Vogt, *Bull. Soc. d'Encouragement de l'Industrie nat.* **96**, 299
633 (1897); *Sprech.* **36**, 1483 (1903).

V21. Y. H. L. Vogt, The physical chemistry of the crystallisation 547
and magnetic differentiation of igneous rocks. *J. Geol.*
31 [3], 233 (1923).

— I. A. Voinovitch, see J. Debras.

V22. G. N. Voldsevich, V. D. Gerasimova and L. A. Lyut- 936
sareva, Ceramic pyroscopes for measuring temperatures
in reducing conditions. *Brit. Clywkr.* **67** [799], 200,
November (1958).

V23. Frank J. Vosburgh, Carbon, a Refractory Material. *Steel* 1117, 1252
5 and 12 April (1943).

— Julius Voss, see Friedrich Crössmann.

— R. C. De Vries, see J. A. Basmajian.

V24. R. W. P. de Vries, The gloss point of glazes. *Philips* 372
Technical Review **17** [5], 153, November (1955).

— R. B. Wagner, see J. E. Wiss.

W1. J. Walker, Development in kiln furniture. *Trans. Brit.* 409, 410, 955
Cer. Soc. **49**, 457 (1950).

W2. Raymond Francis Walker and Stefan George Bauer, 1142
Improvements in or relating to production of dense
oxide bodies. *Pat. Spec.* 665373, 23 January 1952.

W3. J. Wallich, Hartporzellan als Austauschstoff im Apparate- 1126
bau für die Mineralölindustrie. *Oel und Kohle* [15],
407, April (1942).

W4. LEO WALTER, High temperature measurement in ceramic manufacture. *Ceramics* 477, November (1950). 948

W5. LEO WALTER, Ceramic driers—III. *Ceramics* **3** [36], 630 (1952). 846, 847

W6. LEO WALTER, Programme control of gas-fired industrial furnaces. *Ceramics* **4** [37], 44, March (1952). 948

W7. W. H. WALTON, The application of electron microscopy to particle size analysis. *Symposium on Particle Size Analysis*, p. 64. Suppl. *Trans. Inst. Chem. Engs.* **25** (1947). 291

W8. R. WARD and A. C. H. PRYCE, Stoneware in industry. *Ind. Chem.* June and July (1945). 1110

— ROLAND WARD, see GEORGE GIBSON.

— M. F. WARNER, see R. S. LAMAR.

W9. B. WARREN and W. L. BRAGG, *Z. Krist.* **69**, 168 (1928). 5

W10. B. E. WARREN, X-ray determination of the structure of glass. *J. A. Cer. S.* **17** [8], 249 (1934). 203

W11. B. E. WARREN and A. D. LORING, X-ray diffraction study of the structure of soda-silica glass. *J. A. Cer. S.* **18** [9], 269 (1935). 205

W12. B. E. WARREN, H. KRUTTER and O. MORNINGSTAR, Fourier analysis of X-ray patterns of vitreous SiO_2 and B_2O_3 *J. A. Cer. S.* **19** [7], 202 (1936). 202, 203

W13. B. E. WARREN and J. BISCOE, Structure of silica glass by X-ray diffraction studies. *J. A. Cer. S.* **21** [2], 49 (1938). 203

W14. B. E. WARREN and J. BISCOE, Fourier Analysis of X-ray patterns of soda-silica glasses. *J. A. Cer. S.* **21** [7], 259 (1938). *205*

W15. B. E. WARREN and A. G. PINCUS, Atomic consideration of immiscibility in glass systems. *J. A. Cer. S.* **23** [10], 301 (1940). *206*, 207, 210

W16. B. E. WARREN, Summary of work on atomic arrangement in glass. *J. A. Cer. S.* **24** [8], 256 (1941). *211*

— B. E. WARREN, see J. BISCOE.

W17.—

— M. WARRINGTON, see H. C. CASTELL.

W18. TOM WATHEY, The mechanical edge lining of tableware. *Ceramics* **4** [47], 508, January (1953). 809

W19. T. WATHEY, A 'controlled flow' slip pump. *Ceramics* **5** [51], 129, May (1953). 714

W20. TOM WATHEY, An automatic pin loading machine. *Ceramics* **5** [57], 413, November (1953). 963

W21. T. WATHEY, Brushing machines. *Ceramics* **6** [51], 28, March (1954). 787

W22. HENRY WATKIN, *Watkins' Recorders.* Burslem, Stoke-on-Trent. 938

W23. WATTS, BLAKE AND BEARNE AND CO. LTD., Clay Mines, Newton Abbot, Devon. 240, 241, 242, *243, 254, 255*

W24. Arthur S. Watts, Bone china bodies. *Trans. A. Cer. S.* 460–463
7 [1], 204–231 (1905).

W25. A. S. Watts, *Mining and Treatment of Feldspar and Kaolin,* 244, 245
in the Southern Appalachian Region. Washington
Govt. Printing Office (1913).

W26. A. S. Watts, The practical application of Bristol glazes 547
compounded on the eutectic basis. *Trans. A. Cer. S.*
19, 301 (1917).

W27. Arthur S. Watts, Some whiteware bodies developed at 578
the Ohio State University. *J. A. Cer. S.* **10**, 148 (1927).

W28. A. S. Watts, Classification of ceramic dinnerware. *Bull.* 1089
A. Cer. S. **18**, 314 (1939).

W29. Arthur S. Watts, A Study of eutectic glasses as fluxes in 220, 450
whiteware bodies—Introduction. *Bull. A. Cer. S.* **31**
[11], 456 (1952).

— Arthur S. Watts, see J. Allen Pierce.

— A. S. Watts, see C. J. Koenig.

— A. S. Watts, see R. Russel.

— Arthur S. Watts, see Charles B. Tauber.

— Arthur S. Watts, see Jo. Morgan Teague, Jr.

W30. H. W. Webb, Alkaline casting slip—I. *Trans. Cer. S.* **33** 756–762
[4], 129 (1934).

— H. W. Webb, see C. Bettany.

— H. W. Webb, see W. A. Bloor.

— H. W. Webb, see S. W. Ratcliffe.

W31. Webcot Ltd., *Pottery Kilns.* 297, High Street, Bucknall, 983
Stoke-on-Trent.

W32. Berthold C. Weber, William M. Thompson, Hans O. 508
Bielstein, and Murray A. Schwartz. Ceramic
crucible for melting titanium. *J. A. Cer. S.* **40** [11], 363
(1957).

W33. B. C. Weber and M. A. Schwartz, Ueber Zirkonoxyd: 508
Seine Kristallpolymorphie und Eignung als Werkstoff
für hohe Temperaturen. *Ber. D. K. G.* **34** [12], 391
(1957).

W34. Arthur Wedgwood, *Ceramists Diary.* W. Podmore and 144/145
Sons Ltd., Shelton, Stoke-on-Trent.

W35. Josiah Wedgwood & Sons Ltd., 34 Wigmore St., London, *685, 701, 705,*
W.1. *710, 722, 739,*
768, 772, 774,
804, 807, 820,
822, 828, 961,
974, 1059,
1093, 1097,
1101

W36. C. Weelans, Inquiry into body dissolution and its causes. 555
Trans. A. Cer. S. **8**, 79 (1906).

W37. D. J. Weintritt and A. C. Perricone, New testing equipment for quality control. *Bull. A. Cer. S.* **36** [11], 401 (1957). — 326

— Armin Weiss, see R. Fahn.

W38. Wellman Smith Owen Engineering Corporation Ltd., (b) *The Wellman-Galusha Gas Producer for the Gasification of Anthracite and Coke.* Parnell House, Wilton Rd., S.W.1. — 916, 918

W39. Arthur A. Wells, Jr., Intaglio printing. *Cer. Age.* October (1948). — 821, 824

W40. A. F. Wells, *Structural and Inorganic Chemistry* (2nd Ed.). Clarendon Press, Oxford (1950). — 7, 101, 231

W41. H. Welte, Kadmiumfarben in der Glas-, Email- und Keramischen Industrie. *Emailwar.-Ind.* **14** [40], 277 (1937). — 616

— Pu-Yi Wen, see R. Rieke.
— B. Wentz, see H. Salmang.
— W. Wenzel, see W. Jander.

W42. L. C. Werking, Formed carbon and graphite in industry. *Bull. A. Cer. S.* **32** [2], 40 (1953). — 414, 1252

— Ursula Werther, see Hans Lehmann.
— C. S. West, see L. R. Lakin.
— J. West, see W. L. Bragg.

W43. J. H. Westbrook, Metal ceramic compositions—I and II. *Bull. A. Cer. S.* **31** [6, 7], 205, 248 (1952). — 512–514, 1142, 1145

W44. Westen & Co., Westerwälder Tongruben, Wirges/ Westerw., Germany. — 39

— Fr. Westerhold, see G. Tammann.
— A. Westerman, see R. W. Grimshaw.

W45. Westerwälder Thonindustrie-G.m.b.H., Breitscheid (Dillkreis), Germany. — 48

W46. A. E. R. Westman, The capillary suction of some ceramic materials. *J. A. Cer. S.* **12** [9], 585 (1929). — 73

W47. A. E. R. Westman and R. G. Mills, Some waste-heat drier calculations and charts. *J. A. Cer. S.* **12** [3], 162 (1929). — 851

W48. A. E. R. Westman, The drying and firing shrinkage of clays. *J. Can. Cer. S.* **2**, 24 (1933). — 80, *82*

W49. A. E. R. Westman, The effect of mechanical pressure on the drying and firing properties of typical ceramic bodies. *J. A. Cer. S.* **17**, 128 (1934). — 66

W50. Kurt Wetzel, Ueber den Einfluss verschiedener Zuzätze auf die physikalischen Eigenschaften des Porzellans. *Ber. D. K. G.* **6** [1], 23 (1925). — 465

W51. Woldemar A. Weyl (Chairman), Symposium of colour standards and measurements I–V. *Bull. A. Cer. S.* **20** [11], 375–411 (1941). — 341

W52. W. A. WEYL, Atomistic interpretation of the mechanism of solid state reactions and of sintering. *Cer. Age* **60** [5], 28, November (1952).   172
— WOLDEMAR A. WEYL, see NORBERT J. KREIDL.
W53. J. H. WEYMOUTH and W. O. WILLIAMSON, Some observations on the micro-structure of fired earthenware. *Trans. Brit. Cer. Soc.* **52** [6], 311 (1953).   216
— W. M. WHIELDON, see W. H. DAVENPORT.
W54. L. R. WHITAKER, Manufacture of brick and tile from extruded limestone. *Ceramics* **8** [91], 188, September (1956).   400, 587
— A. E. S. WHITE, see F. H. ALDRED.
W55. J. WHITE, Refractories in Great Britain. *Iron & Coal Trades Review* **170**, 1453, June (1955).   118, 403, 1223–1225
W56. J. WHITE, Some general considerations on thermal shock. *Trans. Brit. Cer. Soc.* **57** [10], 591 (1958).   223, *226*
— J. WHITE, see J. L. EVANS.
— J. WHITE, see H. J. S. KRIEK.
W57.—
W58. R. P. WHITE and G. R. RIGBY, The thermal expansion properties of compositions containing lithia, alumina and silica. *Trans. Brit. Cer. Soc.* **53**, 324 (1954).   503
W59. THEODORE WHITE, Effect of Quebracho extract on clays. *Cer. Age* October (1953).   761
W60. W. P. WHITE, *Amer. J. Sci.* **28** (1909).   870
W61. C. WHITTAKER AND Co. LTD., Dowdry St. Ironworks, Accrington, Lancs.   *748*
W62. D. G. WHITTAKER, Laboratory control of pottery processes. *Ceramics* **4** [42], 34, August (1952).   266
W63. H. WHITTAKER, Effect of particle size on plasticity of kaolinite. *J. A. Cer. S.* **22** [1], 16 (1939).   53
W64. JOHN W. WHITTEMORE, Industrial use of plasticizers, binders, and other auxiliary agents. *Bull. A. Cer. S.* **23** [11], 427 (1944).   154, 156
W65. O. J. WHITTEMORE, JR. and D. W. MASHALL, Fused stabilized zirconia and refractories. *J. A. Cer. S.* **35** [4], 85 (1952).   170, 1246
— O. J. WHITTEMORE, JR., see W. H. DAVENPORT.
— H. WICKERT, see A. DIETZEL.
W66. G. WIEGNER, Über eine neue Methode der Schlämmanalyse. *Landwirtsch. Versuchsstationen* **91**, 41 (1918).   287
— HERMAN G. WILCOX, see C. L. THOMPSON.
W67. FRANK A. WILDER, Gypsum: Its occurrence, origin, technology and uses. *Iowa Geological Survey* **28** (1917 and 1918).   160, 163
W68. WILHELM, Material- und Herstellungskostenersparnisse durch Spritzen der Emailmasse mit der Spritzpistole gegenüber dem Auftragen. *Sprechs.* **68** [18], 273, (1935).   791

W69. GORDON B. WILKES, The Specific heat of magnesium and aluminium oxides at high temperatures. *J. A. Cer. S.* **15**, 72 (1932). 870, 871

W70. WILKINSON RUBBER LINATEX LTD., *Linatex Ball Mill.* Frimley Rd., Camberley. 681

W71. J. WILKINSON AND SON, *'Metro' Ceramic Kiln* 1797 A/B. 42, Commercial Rd., Eastbourne, Sussex. 983

— FRANZ WILLERT, see IMFRIED LIEBSCHER.

— A. N. WILLIAMS, see W. NOBLE.

W72. WALTER E. WILLIAMS, A unique method for checking flow of clay-grog slip. *Bull. A. Cer. S.* **34** [1], 13 (1955). 325

— W. O. WILLIAMSON, see K. GRANT.

— W. O. WILLIAMSON, see J. H. WEYMOUTH.

— D. WILM, see U. HOFMANN.

W73. EARL O. WILSON, The plasticity of finely ground minerals with water. *J. A. Cer. S.* **19**, 115 (1936). 329

W74.—

— HEWITT WILSON, see M. S. NELSON.

W75. R. E. WILSON and F. P. HALL, The measurement of the plasticity of clay slips. *J. A. Cer. S.* **5**, 916 (1922). 327

W76. H. R. WISELEY and K. A. GEBLER, *Development of Cordierite Type Ceramic Bodies*. Progress Report No. 3, School of Ceramics, Rutgers University (1947). 485

W77. H. R. WISELY, Thermocouples for measurement of high temperatures. *Cer. Age* July (1955). 942, 944

— HARRIET R. WISELY, see KENNETH A. GEBLER.

W78 J. E. WISS, R. B. WAGNER and RICHARD BEAVER, Ceramic applications of flexible synthetic die and moldmaking equipment. *Bull. A. Cer. S.* **28** [2], 41 (1949). 163

W79.—

W80. JOSEF WOLF, Der Grundstoff Mangan als färbender Werkstoff in der Tonindustrie. Managanpinke, ihre Entstehung, Zusammensetzung und Verwendung. *Sprechs.* [43], [44], 659, 671 (1935). 635, 636

W81. JOSEF WOLF, Welche Grundstoffe und ihre Verbindungen werden als Färbemittel in der Tonwaren-, Glas- und Emailerzeugung verwendet? *Sprechs.* **70** [48, 49, 50], 601, 612, 625 (1937). 234, 235

W82. JOHANNES WOLLINGER, Manufacture of vitreous china sanitary ware. *Ker. Rund.* **47**, 549 (1939). 1086

W83. OTTO WOLPERT-WERKE G.M.B.H, Ludwigshafen am Rhein, Germany. *350*

— KENNETH T. WOOD, see H. G. SCHURECHT.

W84. WOODALL-DUCKAM, *Brit. Pat.* 353086. 1012

W85. J. G. WOODWARD, The vibrating-plate viscometer—An aid in slip-casting control. *Bull. A. Cer. S.* **31** [10], 389 (1952). 325

W86. WORCESTER ROYAL PORCELAIN CO. LTD., Royal Porcelain Works, Worcester. *1115*, 1127

BIBLIOGRAPHY AND AUTHORS INDEX 1429

W87. Wolsey G. Worcester, The function of alumina in a crystalline glaze. *Trans. A. Cer. S.* **10**, 450 (1908).   596

W88. W. E. Worrall, R. W. Grimshaw and A. L. Roberts, Structural significance of cation-exchange in fireclays. *Trans. Brit. Cer. Soc.* **57** [6] (1958).   17, 53, 57, 319

— J. Wuhrer, see W. Jander.
— W. Wyatt, see J. P. Roberts.
— D. J. Wynne, see L. J. Trostel.

Y1. Goro Yamaguchi and Koji Tomiura, Pink pigments of $MnO-P_2O_5-Al_2O_3$ and $Cr_2O_3-P_2O_5-Al_2O_3$ systems. *J. Ceram. Assoc. Japan* **62** [692], 111 (1954); *J. A. Cer. S.* Abs. **37** [8], 144 (1954).   625, 635

Y2. Toshiyoshi Yamauchi, Hideo Tagal and Renichi Kondo, Sintering of alumina porcelain. *J. Japan. Cer. Ass.* **57** [641], 137 (1949); *J. A. Cer. S.* Abs. **34** [1], 9 (1951).   195

— P. J. Yavorsky, see R. F. Geller.
— P. F. Young, see T. R. Lynam.

Z1. W. H. Zachariasen, *Z. Krist.* **74**, 139 (1930).   5

Z2. W. H. Zachariasen, The atomic arrangement in glass. *J. A. Chem. S.* **54** [10], 3841, October (1932).   *199*, 201, 203, 527

Z3. M. I. Zaïtseva and N. I. Chaïkovskaya, Zinc glazes for facing ceramic shapes. *Steklo i Keram.* **12** [3], 4 (1955).   589

Z4. P. A. Zemiatchensky, *Trans. Ceram. Research. Inst. Moscow* [7] (1927).   327

Z5. V. F. Zhuravlev, I. G. Lesoklim and R. G. Tempelman, *Zhur. Priklad. Khim.* **21**, 887 (1948).   183

Z6. K. Zimmermann, Zur Korngrössen bestimmung von Tonen. *Ziegelindustrie* **5**, 322 (1952).   285, 298

— Ferdinand Zirkel, see Carl Friedrich Naumann.
— S. G. Zlatkin, see P. S. Manykin.

Z7. Hans Zoellner, Die Rohbiegefestigkeit, ein betriebmässiges Mass für die Plastizität von Tonen, Kaolinen und keramischen Massen. *Sprechs.* **83** [14], 271 (1950).   335

Z8. Hans Zoellner, Über die quantitative spektrochemische Vollanalyse von keramischen Stoffen. (The quantitative spectrochemical analysis of ceramic materials.) *Ber. D. K. G.* **28**, 284 (1951).   269

Z9. H. Zoellner, Flammenphotometer zur Alkali-Bestimmung. *Glas.-Email-Keramo-Technik* **2** [9], 290 (1951).   272

Z10. H. Zoellner, Welche Faktoren beeinflussen die Genauigkeit der flammenphotometrischen Alkalibestimmung? *Glas-Email-Keramo-Technik* **2** [11], 378 (1951).   274

— H. Zoellner, see Fr. Hegemann.
— R. Zorn, see K. Endell.

Z11. B. ZSCHOKKE, *Tonind.-Ztg.* [120], 1658 (1905); *Baumaterial-* 329
kunde [24, 25–26] (1902); *Ibid.* [1–2, 3–4, 5–6] (1903).
Z12. A. ZWETSCH, Recherches relative aux felspaths Français. 277
Bull. Soc. Fran. Cér. [36], 3 (1957).

SUBJECT INDEX*

Abrasion resistance tests, 351
Abrasive grinding wheels, 1156
  bodies, 396, 479–481
Absorbers, porous, 1134
Absorption spectrophotometer, 272
Acid resistance tests, 363
  -resisting bricks and tiles, 1112
  -soluble iron determination, 363
Aerators, 1134
After-contraction measurement, 357
After-expansion measurement, 357
Agalmatolith, 93
Ageing, 711
Agricultural drains, 1074
  salt-glazed stoneware, 1112
Air elutriation, 282
  permeability apparatus, 294, *295*
  separators, 251, *252*, 689
Albite, 102
Alginates, 156
Alkaline earth–silica glasses, 206
Allen semi-automatic jigger, 726
Allied tunnel kilns, 1041
Allophane, 19, 22
Alloy cast irons, 165
α-β inversions, volume change of, and thermal stresses, 224
Alumina, polymorphism, 170
  raw materials, 114–115
  reactions, 191
  sintered ware, 1138, 1146, 1151
  bodies, 505
  -chromium cermet ware, 1145
  electrical insulators, 1189, 1199, 1204
  grinding wheels, 1157
  refractory bricks, 1230, 1245
  –silica eutectic, 405

Aluminates by solid state syntheses, 178
Aluminium in glass structure, 212
  in silicate structures, 9
Alumino-silicate refractory bodies, 404
Aluminous firebrick bodies, 405
  firebricks, 1229, 1235
Amblygonite, 122, 124
American fine china bodies, 454
  hotel china bodies, 395, 452
  hotel china ware, 1096
  household china bodies, 451
  household china ware, 1096
  Society for Testing Materials, Standards relating to ceramics, list of, 1317
Ammonia, dissociated, atmosphere in kilns, 933
Ammonium carbonates, 157
  dichromate, 146
  oxalate, 153
Amphiboles, 5
Analysis, chemical, 266–275
  complete or 'ultimate', 266
Anatase, 140, 170
Anauxite, 16, 22
Andalusite, 110–112
  reaction, 192
Andreasen pipette, 283, *285*
Andrews kinetic elutriator, *283*
Anhydrite, 158
Anorthite, 102
Anthracite, 136
Antimony oxide, 144, 146
Anti-sticking agents, 155
  in dry pressing, 743
Apatite, 108–109
Arc carbons, 1210

* Pages on which illustrations occur are given in italics.

Arsenic oxide, 144
Artware, 1090, 1097, *1101*
Asphalt, 136
Atmosphere, protective in kilns, 933
Atomic energy ceramics, 1165
   weights, 1300
Attapulgite, 19, *22, 23, 24*
Attenuator materials, 1210
Atterberg plasticity number, 326
Attritor, 682
Aventurine glazes, 598

Back stamping, 786
Baddeleyite, 135, 140, 170
Badilla, 158
Bagging of air-floated clay, *243*
Bag wall, 907
Ball clays, 27, 35–41, 253–256
Ball compression test, 349
Ball mills, 670, 671, 676
Banding, 808
Barite, 129
Barium aluminium silicate bodies, 424
      refractory bricks, 1247, 1249
  -anorthite bodies, 424
  -anorthite refractory bricks, 1247, 1249
  carbonate, 154, 695
  chromate, 146
  compounds, 129
  selenite, 148
  titanate bodies, 517
  zirconium silicate, 145
Barnett and Hadlington kiln, 1001
Barytes, 129
Basalt ware bodies, 455
Base-exchange capacity, determination, 318–319
Basic lead carbonate, 133
Basic refractory bodies, 415
Batch dryers, 839
Batching, 699
Battery carbons, 1211
Batting, 724
Bauxite, 30, 114, *115*
Bayerite, 114
Beehive kiln, 987
Beidellite, 17, 22, 24
Belgian kiln, 999, 1005
Belleek china, 454
    tableware, 1089
Benitoite, 5
Bentonite, 30, 51
  winning 258, *260*

Beryl, 5, 139
Beryllia bodies, 504, 505
  sintered, ware, 1139, 1146, 1152
Beryllium in glass structure, 212
  compounds, 139
Binders in dry pressing, 743
Bird continuous centrifuge, 246, *247*
Birefringence, 308
Biscuit firing, 854
Bismuth in glass structure, 214
  oxide, 134
Bitumen emulsions, 156
Black colourants, 642
Blending, 697–701
Block talc electrical insulators, 1181, 1201
Blue bricks, 1069
Blunging, 683, *685*
Bock kiln, 1001
Bodies, 393–524
  abrasives, 479
  beryllia, 504
  bone china, 457
  borides, 511
  brickware, 399
  carbides, 509
  cermet, 511
  composition, 397
  cordierite, 482
  definitions, 393–396
  earthenware, 438
  ferrite, 520
  fireclay, 438
  forsterite, 498
  hard porcelain, 464
  high-magnesia, 503
  lithia porcelain, 500
  nitrides, 510
  oxide bodies, 505
  permeable porous, 521
  porcelains, 451–479
  pyrophyllite, 500
  refractories, 393, 400–424
  rutile, 515
  semi-vitreous china, 439
  silicates, 509
  silicides, 511
  steatite, 487
  stoneware, 430
  soft porcelain, 451
  spinel, 499
  spinels, refractory, 508
  sulphides, 511
  thermal insulation, 425
  titanate, 516

SUBJECT INDEX 1433

vitreous china, 450
wollastonite, 500
zirconates, 509
Body decoration, 797, 799
  -glaze interaction, 542
  stains, 797, 799
Boehmite, 114
Bone ash, 108–110, 145
  china, 109
    bodies, 395, 457–464
    setting in the kiln, 964
    tableware, 1089, *1097*
Borax, 130
Boric acid, 130
  oxide, 130
    and network modifiers, 208
    anomaly, 208
    glass, 203
Boride bodies, 511
Borides, refractory, 1148
Borocalcite, 131
Boron carbide ware, 1136, 1147, 1150
  compounds, 130–132
  nitride, 1151
    bodies, 510
Bottle kiln, 973, *974*, 986
Breakdown strength, 1170
  test, 368
Bricesco-Harrop tunnel kilns, 1031
Bricesco tunnel kilns, 1048, 1049, 1052
Brick clays, 26, 30
  winning, 258
Bricks, 1065–1072
  analyses, 1070
  blue, 1069
  common, 1065
  engineering, 1069
  facing, 1066
  fletton, 1065
  glazed, 1072
  hollow, 1069
  labour requirements, 1067
  perforated, 1069
  properties, 1070
  stock, 1066
Brick shaping methods,
  semi-dry pressing, 746, *748*
  semi-stiff mud, 718
  soft-mud, 717
  stiff-plastic extrusion pressing, 740
  stiff-plastic making machine, 718
  wire cut, 733, *734*
Brickware bodies, 393, 399–400
Bright gold, 650

British ball clays, 253–256
British Standard Specifications relating to ceramics, List of, 1308
Brittle, 347
Brookite, 140, 170
Brown's Patent kiln, 1000
Brucite, 9, *10*, 118
Brush application of glazes, 791, *792*
Brushing, free-hand decoration, 805, *807*
Brushing off, 786
Bulk density, 342
Bullers rings, 938, *940*, *941*, 1304
Burnished gold, 653
Butane fuel, 920
Button test for glazes, 372

Cadmium compounds, 146
  in glass structure, 213
  -selenium orange and red colourants, 617
  sulphide, 146
  yellow and orange colourants, 616
Calcite, 116
Calcium carbonate, 116
    removal, 694
  chloride, 155
  compounds, 127–128
  oxide, raw materials, 115–119
  silicate inversion, 171
  sulphate, 158
  titanate bodies, 517
  zirconium silicate, 145
Calcspar, 116
Capacitors, 1208
Carbide bodies, 509
Carbides, boron, silicon, titanium, tungsten, zirconium, 1136–1137, 1147–1151
Carbocell bodies, 414
Carbon, 136
  bodies, 414
  electrical ware, 1210
  monoxide, refractories resistance to, measurement, 364
  oxidation in the firing, 857
  refractory blocks, 1249, 1252
  resistors, 931
  steel, 164
  ware, 1116, 1128
Carbonaceous bodies, 414
Carbonates, decomposition in firing, 860
Carolina Stone, 104
Casein, 156

Castability, 325
Casting, 716, 754–771
　control, 770
　defects, 767
　hollow or drain, 762
　slip, preparation and use, 764
　solid, 762
Cast irons, 165
Catapleite, 5
Cation exchange capacity, 56
　　determination, 318–319
Caustic soda, 153
C.E.C. determination, 318–319
Celestine, 128
Celestite, 128
Cellofas, 156
Celluloses, 156, 158
Celsian, 102
Cemented carbide ware, 1154
Centrifugal dewatering, 707
　sedimentation, 288
Centrifuge, continuous, 245
Centrifuging, 304
Ceramic bodies, see Bodies
　properties of clays, determination, 319–338
Cerium colourants, 618
　oxide, 144
Cermet bodies, 396, 511–515
　ware, 1145
Chalk, 116
Chamber dryers, 839
　kilns, 998
Charcoal, 136
　fuel, 902
Checker bricks, 1230
Chemical analysis, 266–275
　　for determining minerals in clay, 307
　　of clay, 299
　　porcelain bodies, 468
　　ware, 1113, 1126, 1135
　　properties of fired ceramic bodies, 361–364
　　resistance of glazes, 559
　　separation of minerals in clay, 306
　　stoneware, 1107, 1123, 1135
　　　for textile industries, 1112
　　　of zero absorption, 1111
　　　white, for pharmaceutical and foodstuffs industries, 1111, 1135
　　ware, 1103–1134
Chiastolite, 110
Chimney pots, 1073
China bodies, 451–464

clay, 27, 32–33
　winning, 237–242
　stone, 103
　preparation, 696
Chodrodites, 5
Chromates, 146
Chromatography, 275
Chrome-iron ore, 137
　-magnesite refractory bodies, 422
　　bricks, 1264, 1270
　pinks colourants, 625
　refractory bodies, 420
　　bricks, 1267, 1272
　-tin pink to crimson colourants, 622
Chromite, 137
Chromites by solid state syntheses, 179
Chromium-alumina cermet ware, 1145
Chromium brown, yellow, etc. colourants, 622
　green colourants, 618
　oxide, 146
　phosphate, 146
　steel, 164
Cimolite, 22
Clamp firing, 884
Classifiers, wet, 687
Claws, 962
Clay, colloids, 52
　constituents, 19, 23
　geology, 11
　minerals, 12–24, 301
　　and clay properties, 84–87
　　mineral chemical composition, 24
　　nomenclature, 22
　mineralogy, 12
　particles, 52–58
　plasticity of, 62–66
　shaving machine, 665, *666*
　substance, 12
　thixotropy of, 59–62
　varieties, 27–51
　winning machinery, 258–263
Clayite, 22
Clays, 9–88
　analyses of, 32–51
　brick, 26, 30
　classification, 25
　colluvial, 25
　drying, 67–74
　ferruginous, 43
　firing of, 76–84
　flow properties, 58
　high-alumina, 30, 47

SUBJECT INDEX 1435

investigation of, 296–338
preliminary tests, 296
properties of, 52–88
prospecting for, 236
red, 43
refractory, 26, 28, 45–51
residual, 11, 25
siliceous, 41–42, 47
slip, 26
soluble salts in, 249
stoneware, 26, 28, 44–45
transported, 25
weathering, 263
white burning, 26
winning and purification, 236–264
Clinoenstatite, 91
bodies, 487
electrical insulators, 1182
Closed circuit grinding, 662
Coal, 136
fuel, 903
gas fuel, 914
Cobalt compounds, 146
blue colourants, 625
oxide, 146
Coefficient of thermal expansion
measurement, *352*
of glazes, 547
Coherence value of glaze slip, 369, 784
Coke, 136
Cold crushing strength tests, 347
Colemanite, 131
Colorimetric analysis, 266
Colour, 227–235, 1307
by absorption, 229
by scattering, 234
by selective reflection and interference, 234
matching, 228, 340
solutions, 644
Colourants, 605–660
blacks, 642
cadmium selenium oranges and reds, 617
cadmium yellow and orange, 616
calcining of, 606
cerium colourants, 618
chrome pinks, 625
chrome-tin pinks to crimsons, 622
chromium browns, yellows, 622
chromium greens, 618
cobalt blues, 625
colour solutions, 644

copper greens, turquoise and blues, 628
copper reds, 628
effect of glaze composition on, 645
fading of underglaze colourants, 646
iridium, 629
iron-chrome browns, 632
iron greens, 632
iron-manganese browns, 632
iron yellows to reds, 629
ivories, 641
lead antimonate, 634
lead chromate coral red, 635
lustres, 657
manganese pinks, violets and browns, 635
metallic, 647–657
Naples yellow, 634
neodymium, 637
oils for, 615
on-glazes fluxes for, 608
recipes for, 615–643
underglaze fluxes for, 608
uranium red, 637
uranium browns and yellows, 638
vanadium yellows, blues and greens, 639
whites, 640
Coloured glazes, 560, 798, 803
Colouring agents, 143–151, 234
Colloidal silicic acid, 154
Collyrite, 22
Common bricks, 1065
Complex thermal analysis, 316
Compressive strength tests, 347, *348*
Conductance, thermal, 355
Conductivity, thermal, 1169
measurement, 353, *354*
Cones, pyrometric, 935, 1304
Conical mills, 671, *674*, 675
Constructional refractories, *see* Refractories, constructional
Continuous centrifuge, 245
dryers, 843
kilns, 884, 985–1006
Control testing, 265
Cooling after firing, 869
Copper compounds, 146–149
green, turquoise and blue colourants, 628
oxide, 146
red colourants, 628
Cordierite bodies, 396, 482–491
formation, 189, 191

1436   INDUSTRIAL CERAMICS

Cordierite (*contd.*)
   electrical insulators, 1186, 1203
   glazes, 563
   natural, 120
Core drilling, 236, *237*
Cornish china clay winning, 238–242
   stone, 103–105
Corridor dryers, 839, *840–841*
Corundum, 114
   bricks, 1230
   grinding wheels, 1157
Courlose, 156
Cover-coat transfers for decoration, 830, *831*
Crackle glazes, 581
Cranks, *962*
Crawling of glazes, 796
Crayon decoration, 806
Crazing, 551
Creep tests, *356*, 357
Cristobalite, 97, 168
   structure, 5
Cross-bending strength test, 349
Crucibles, 1275
Crucible shaping, 719
Crushers, 663–665, 684
   gyratory, 663
   hammer mills, 665, *666*
   jaw, *663*
   rolls, 663, *664*
Crushing, 661–665
   and grinding equipment, application, 684
Cryolite, 123, 145
Cryolithionite, 122
Crystal group, 308
Crystalline glazes, 593–598, 798, 803
Crystallisation, 198–201
Cup shaping, 721, *722*, 740, 774
Curie point, 1172
   temperature, 1172
Cutting tools, 1155, 1158–1164
Cyanite, 110
Cylindrical mills, 671, *673*, 675

Dannenberg's kiln, 1001
Davis tunnel kiln, 1050
De-airing of casting slip, 711
   pugmills, 708–711, *710*
Debye-Scherrer powder technique, 309
Decalomania, 'decal', 824
Decomposition period of firing, 857
Decorating firing, 854

Decoration, 797–831
   banding, 808
   body, 797, 799
   brushing, free-hand, 805, 807
   coloured bodies, 797, 799
   coloured glazes, 803
   cover coat transfers, 830, *831*
   crayons, 806
   crystalline glazes, 803
   decalomania, 'decal', 824
   durability of, 609
   durability tests, 378
   engobes, 798, 799
   engraving of printing plates, 821, *822*
   free-hand painting, 805, *807*
   glaze, 798, 803
   ground laying, *807*, 809
   in-glaze, 799, 805
   in-glaze colouring, 803
   inlaying, 798, 802
   intaglio printing of transfers, 821, *822*
   lining, hand, *807*, 808
   lining, machine, 808
   'litho', 824
   lithographed transfers, 825
   monochrome transfers, 821
   Murray curvex printing machine, 819, *820*
   offset printing, 818
   on-glaze, 799, 805
   opaque glazes, 803
   ornamenting, 803, *804*
   painting, free-hand, 805, *807*
   paper transfers, 819
   pâte sur pâte, 802
   photolithography, 825
   piercing, 802
   polychrome transfers, 824
   rejafix machine, 818
   relief work, 798, 802
   rubber stamping, 809, *810*
   Ryckman machine, 809
   sanding, 803, *804*
   screenprinted (paper backed) transfers, 826
   screen printing, 813
   sgraffito, 802
   silk-screen printing, 813
   slip-trailing, 802
   spraying, 806
   sprig work, 803
   stamping, 809, *810*

SUBJECT INDEX

stencilling, 811
transfers, application, 826, *828*, *829*
    paper, 819
under-glaze, 798, 805
underglaze spraying on hard porcelain, 806
waterslide, screenprinted, collodion backed transfers, 830
Deflocculants, 55, 152–154
Deflocculation, 55
    of casting slip, 755
Deformable, 347
Deformation eutectic mixtures, 219
Delanium bodies, 414
    carbon ware, 1117, 1133
    graphite ware, 1118, 1133
Demineralisation of water, 152
Density, 342
Dental porcelain bodies, 454
    ware, 1102
Deutsche Normen relating to ceramics, list of, 1313
Development research, 265
Devitrification and thermal stresses, 225
    of glazes, prevention, 546
Dextrin, 156
Diabon graphite ware, 1132
Diamond pyramid hardness number of glazes, 558
    test, 378
Diaspose, 30, 114
Diatomaceous earth, 97
    bodies, 426
Diatomite, 97
    insulating bricks, 1284, 1287
Dickite, 13, *15*, 16, 22, 29
Dielectric constant, 1170
    test, 368
    materials, 1170
    strength, 1170
    test, 368
Diesener kiln, 1001
Dies, extrusion and pressing, 165
Differential thermal analysis, 313
    for control tests, 382
    thermogravimetric analysis, 315
Diffusers, 1134
Diffusivity, 354
Dilithium sodium phosphate, 122
Diopside, 5
Dipping glazes, *787*
    machine, *790*
Dispersion of raw clay, 298
Disthene, 110

Ditcher, 259
Divers, for sedimentation apparatus, 288
Dobbins, 843, *845*
Doloma, 116
Dolomite, 116
    refractory bodies, 419
    bricks, 1257, 1263
    -serpentine mixtures, 419
Dorr Bowl classifier, 245, *246*
Dottling, *962*
Down-draught kilns, 969, 978, *979*, 987
D.P.H. number of glazes, 558
    test, 378
Dragline, *261*
Drags, Cornish clay pits, 239
Dragshovel, *261*
Drainage pipes, 1074
Draught measurement, 951
Drayton down-draught tunnel kiln, 1036
Dredges, 258
Dressler tunnel kilns, 1038–1041, 1051, 1052, 1053
Drum test, 351
Dryers for shaped ware, 834–853
    batch, 839
    chamber, 839
    continuous, 843
    corridor, 839, *840–841*
    dobbins, 843, *845*
    drying sheds, 835
    efficiencies, 850
    heated, 836
    high-frequency alternating current for drying, 838
    hot floor, 839
    infra-red heat for drying, 836
    jet, 842, 843
    mangles, 845, *846*
    potters' stoves, 843, *845*
    radiant heat for drying, 836
    tunnel, 846, *850*
    unheated, 835
Drying aids, 157
    kiln, Cornish clay pits, 239
    of clay, 67–74
    of shaped ware, 832–834
Drying of slip, 701–707
    filterpress, 702
    heated drum, 707
    spray, 707
    vacuum filter, 705
Drying sheds, 835
    shrinkage, 67
    measurement, 333

Dry pan grinding mills, 667
  pressing, 741–751
  purification methods for clays, 251
D.T.A., 313
  for control tests, 382
Dubosq colorimeter, 267
Ductile, 347
Dumortierite, 113
Dumps, *964*
Dunnachie kiln, 1002
Dunting, 551
Durability of on-glaze decoration, 609
  tests, 378
Durabon carbon ware, 1131
Dust control, 715
  extrusion, 738
  hazard, 1323
  pressing, 741–751
Dye absorption tests, 344

Earthenware bodies, 394, 438–451
  general methods of preparation
    English, 1093
    European, 1094
    Scandinavian, 1094
    U.S.A., 1094
  sanitary ware, 1084
  setting in the kiln, 962
  tableware, 1089, 1090, *1093*
Eddy current loss, 1172
Edge-runner grinding mill, 667, *668*
Efflorescence, 361
Elastic, 347
Electrical ceramics, 1170–1211
Electrical insulators, 1174–1210
  alumina, 1189, 1199, 1204
  barium titanate, 1208
  block talc, 1181, 1201
  calcium titanate, 1208
  clinoenstatite, 1182
  cordierite, 1186, 1203
  dry pressing of, 746
  forsterite, 1188, 1204
  high-frequency, 1200
  high permittivity capacitors, 1208
    rutile, 1207
  high temperature, 1198
  high-tension, low-frequency mullite
    porcelain, 1177
  high tension, low frequency
    porcelain, 1174
  lava, 1181, 1201
  lead zirconate-titanate, 1209
  lithia porcelain, 1192, 1205
  low-loss alumina, 1204
    cordierite, 1203
    forsterite, 1204
    steatite, 1201
    wollastonite, 1206
    zircon porcelain, 1205
  low-tension, low-frequency porcelain, 1174
  magnesia, 1193
  magnesium orthotitanate, 1196, 1207
  phosphate-bonded talc, 1181, 1201
  porcelain, 1174, 1178
  rutile, 1193, 1206
  shaping of, 719, 723, 732, 741, 746
  spacers for vacuum tubes, 1206
  sparking plug, 1180, 1199
  steatite, 1183, 1201
  stoneware, 1125, 1179, 1198
  strontium titanate, 1208
  titanates, 1196
  titania porcelain, 1194
  vacuum tube spacers, 1206
  wollastonite, 1192, 1206
  zircon porcelain, 1191, 1205
Electrical heating of kilns, 930
  porcelain bodies, 467
    British, 451
  glazes, 562
  properties of fired ceramic bodies, 364–369
  resistance thermometers, 943
  resistors, silicon carbide, 1210
  stoneware bodies, 437
Electrodialysis, 318
Electrolytes, 755, 1307
Electrolytic diaphragms, 1134
Electromagnets, *691, 692–693*
Electron diffraction, 311
  microscope, 291, 309
Electro-osmosis, 248
  filter press, 249
Electrostatic separation of minerals in
    clay, 306
Elutriation, 281
Emery, 114
  grinding wheels, 1157
Enamel, 1307
  firing, 854
Endothermic atmosphere in kilns, 933
Engineering bricks, 1069
  ware, 1135–1169
English kiln, 1000
English translucent china, 1096
Engobes, 798, 799

SUBJECT INDEX 1439

Engraving of printing plates, 821, *822*
Enslin's value, 326
Enstatite, 5, 91
   formation, 189, 191–192
   in steatite bodies, 492
Epoxy resin moulds, 163
Epsom salts, 154
Estimation of minerals in clay, 307
Ethyl silicate casting, 771
Eucryptite, 122
Eutectic points, 217
Exothermic atmosphere in kilns, 933
Expanded clay insulating blocks, 1285, 1290
Expansion, thermal, measurement, *352*
Extrusion, 729
   dies, 165

Facing bricks, 1066
Factory acts, 1323
Faience tiles, 1082
Fatigue, 223–226
Fatty oils, 155, 158
Faugeron tunnel kiln, 1031
Faults in fired ware, 880
Favas, 135
Fayalite, 138
Feeding, 697
Feldspar, 101–105
   preparation, 696
   structure, 5
Ferric oxide, reversible reactions, 190
Ferrite bodies, 396, 520–521
   ware, 1209
Ferrites by solid state syntheses, 176
Ferroelectric properties, 1172
   ware, 1208
Ferromagnetic materials, 1172
   ware, 1209
Fettling, *773*
Fibrolite, 110
Field drains, 1074
Filter cloth, 703
Filtering, 701–707
Filter press, 702–*705*
Filters, ceramic, 1121
Findlings quartzite, 96
Fine ceramic works control, 385–387
Fine china, U.S.A., setting in the kiln, 964
Firebrick bodies, 405
Fireclay bodies, 394, 404, 438–439
   refractory bricks, 1226, 1234
   sanitary ware, 1087

Fireclays, 28, 45–51
   winning of, 256
Fired properties, determination, 338
Firing, 854–884
   atmosphere in kiln, 863
   biscuit, 854
   body, 854
   carbonates, decomposition, 860
   carbon oxidation, 857
   cooling, 869
   decomposition period, 857
   decorating, 854
   enamel, 854
   faults, 880
   glaze, 855
   glost, 854
   hardening-on, 854
   heat requirements, 869
   iron oxides, decomposition, 863
   iron pyrites, decomposition, 862
   maturing temperatures, 866
   maturing the body, 865
   oxidation period, 857
   schedule, 874
   shrinkage, rate of, 867
   silver marking, 881
   spit out, 881
   sulphates, decomposition, 860
   sulphur dioxide, 864
   water smoking period, 855
Firing behaviour of clay determination, 337
   of clays, 76–84
   schedule of periodic kiln, 970
   shrinkage, 79
   under load, 337
Flame photometer, 272, *273*
   for control tests, 382
Flame-resistant ware, 1100
Flame temperatures, 928
Flash drying, 251
Fletton brick clays, 31
   pressing, *749*
   bricks, 1065
Flint, 97
   preparation, 696
Flocculants, 55, 154
Flocculation, 55
Floor tiles, 1075, 1078
Flower pots, 1073
Flowmeters, 951
Flow properties of clays, 58
Flue gas analysis, 952
Fluorine, 128

Fluorite, 127
Fluorspar, 127
Fluxes, 121–134
  auxiliary, 441
  for colourants, 608
Foamclay insulating blocks, 1290
Forsterite, 138
  bodies, 396, 498
  electrical insulators, 1188, 1204
  from olivine, 189, 191
  refractory bodies, 422
    bricks, 1268, 1273
Friable, 347
Frit kilns, 1047
  porcelain, 454
Fritting of glazes, 566–568
Front end loader, 259, *260*
Frostproof tiles, 1082
Froth flotation, *250*
  of minerals in clay, 305
Fuel for firing ceramic ware, 899–933
  butane, 920
  charcoal, 902
  coal, 903, 1019
  coal gas, 914
  consumption, 971, 1014, 1023, 1025
  electricity, 930, 1022
  flame temperatures, 928
  gas, 912, 1020
  gas burners, 920
  gas-oil burners, *929*
  grates, 907
  lignite, 902
  natural gas, 915
  oil, 925, 1021
  producer gas, 915, 1021
  propane, 920
  pulverised coal, 911
  radiant heat, 923, 1021
  resistors, 930
  semi-water gas, 915
  solid fuels, properties, 904–905
  town gas, 914
  water gas, 915
  wood, 902
Fuel testing, 391–392
Fused alumina grinding wheels, 1157
Fusibility of glazes, 533

G anister, 96
Garnet, 5
Gas fuel, 912
Gas-turbine ceramics, 1164
Gelatin, 156

German kiln, 976
Gibbons top-fired tunnel kiln, *1032–1034*
Gibbons tunnel kilns, 1051, 1053, 1054, 1055
Gibbsite, 30, 114
  structure, 9, *10*
Glass formation, 198–201
Glasshouse pots, 1281
  shaping, 718
Glass/refractory reactions, 1059
Glass, structure, *8*, 201–215
  tank blocks, 1231, 1236
Glassy frits as body fluxes, 441
Glaze/body interaction, 542
Glaze decoration, 798, 803
  fit tests, 375
  fit and thermal stresses, 225
  fluidity, tests, 372
  slip testing, 369
  viscosity tests, 370
Glazes, 201, 223, 369–380, 525–606
  annealing temperatures, 556
  aventurine, 598
  batch composition, 526
  chemical resistance, 559
  coefficient of expansion, 547
  coloured, 560
  composition and properties, 532–566
  cordierite, 563
  crackle, 581
  crazing, 551
  crystalline, 593–598
  delayed crazing, 555
  devitrification prevention, 546
  diamond pyramid hardness number, 558
  dunting, 551
  electrical ware, 561–566
  formulation, 526–532
  fritting, 566–568
  functions, 525
  fusibility, 533
  general list, 583–584
  hardness, 557
  high-titania body, 566
  influence on body strength, 549
  lead, 568–576
  leadless, 576–579
  legislation for use of lead, 569
  low-loss steatite, 563
  low-solubility lead, 569
  low temperature, 582
  matt, 592–595
  molecular formula, 526

SUBJECT INDEX 1441

opaque, 585–591
peeling, 551
percentage composition, 526
phase rule data, 223
properties and composition, 532–566
salt, 598–605
salting mixture, 600
semi-conducting, 562
snakeskin, *580*, 581
surface tension, 538
testing, 369–380
vapour glazing, 605
viscosity, 536
volatilisation of constituents, 540
Young's modulus of elasticity, 547
zinc salt glazing, 605
zircon porcelain, 563
Glazing, 777–797
    additives, 783
    back stamping, 786
    brush application, 791, *792*
    brushing off, 786
    consistency of glaze slip, 784
    crawling of glazes, 796
    dipping, *787*
    grinding of glazes, 777
    lead frit solubility, 779
    preparation of glazes, 777
    sponging, 796
    spraying, 791, *793*, *794*
Gloss point test, 372
Glost firing, 854
Glucose, 156
Glue, 156
Glut arch, 907
Glutrine, 156
Glycerin, 156
Gold, 148
    bright, 650
    burnished, 653
    chloride, 148
    colloidal, 647
    decoration, 647–654
    liquid, 650
    metallic, 650
Gottignies multi-passage kiln, 1043, *1045–1046*, 1056
Goulac, 156
Grain-size determination, 278–295
    clays, 297
Graphicell bodies, 414
Graphite, 136, 158
    bodies, 414
    refractories, 1250

ware, 1116, 1128
steel, 166
Grates, 907
Green strength, 75
    measurement, 335–337
Griffiths cracks, 227
Grinding, 661–683
    and crushing equipment, application, 684
    finished ware, *1060*
    in closed circuit, 662
    in open circuit, 662
Grinding mills, 665–683
    attritor, 682
    ball, 670, 671, 676
    conical, 671, *674*, 675
    cylindrical, 671, *673*, 675
    dry-pan, 667
    edge-runner, 667, *668*
    jar, 671, *672*
    mikroatomiser, 682
    mikropulveriser, 682
    mill linings, 681
    pan, 665
    pebble, 670, 671
    Podmore-Boulton vibro energy mill, 683
    pot, 671, *672*
    ring roll, 669
    rod, 670, *673*, 676
    running of ball and pebble mills, 676–680
    selection of, 673–676
    tube, 670, *672*, 675
    wet-pan, 667
Grinding wheels, 1156
Ground laying decoration, *807*, 809
Gulac, 156
Gums, various, 156
Gypsum, 158
Gyratory crushers, 663

Halloysite, *16*, *17*, 22, 24, 53
Hammer mills, 665
Handle shaping, 771
Handling, *774*
    equipment, 714
    machine, *775*
Hand modelling and moulding, 718
    throwing, 720, *722*
Hardening-on firing, 854
Hard metal ware, 1154
Hardness by Mohs scale, 351
    of glazes, 557

Hard porcelain bodies, 396, 464–475
  setting in the kiln, 964
  tableware, 1089, 1099
Harrop tunnel kilns, 1031
Health hazards, 1322
Heat, action on ceramic raw materials, 167–235
Heated drum drying, 707
Heat of wetting measurement, 332
  of clay, 67
Heat-resistant ware, 1100, 1124
Heat-work recorders, 934, 1304
Heavy clay bodies, 396
  liquid separation of minerals in clay, 305
  spar, 129
Heurty multi-passage kiln, 1056
Hewitt stiff-plastic brick machine, 718
Hiddenite, 102
High-alumina clays, 30, 47
  porcelain bodies, 476
  porcelain ware, 1127
High-frequency alternating current for drying, 838
High-magnesia bodies, 503
High-temperature insulating brick, 1294
High thermal conductivity ceramics, 1168
High-titania body glazes, 566
High-voltage insulator tests, 365
Hoffmann kilns, 884, 985, *994*, 1005
Holdcroft's bars, *937*, 1304
Hollow bricks, 1069
Horizontal draught kilns, 970
Hotel china bodies, 395, 452
  tableware, 1089, 1096
Hot-face insulating brick, 1294
Hot floor dryers, 839
Hot pressing, 752–754
Household china tableware, 1089, 1096
Humic acid, 154
Humidity drying, 832
Hyalophane, 102
Hydrargillite, 114
Hydration resistance tests, 362
Hydraulic mining, Cornish china clay, *240*
Hydraulic tests, 351
Hydrogen atmosphere in kilns, 933
Hydrous mica, 29
Hydrometer, 287

I.C.I. sedimentation apparatus, *286*
Ideal kiln, 999
Identification of minerals in clay, 307
Illite, 18, *21*, 24

Ilmenite, 140
Ilmenorutile, 140
Impact bending strength test, 349, *350*
  compressive strength test, 350
  strength test, 349
Impervite graphite ware, 1129
Impurity removal, 690–695
Infra-red drying, 836
  spectroscopy, 313
In-glaze colouring, 798, 803
  decoration, 799, 805
Inlaying, 798, 802
Inorganic deflocculants, 153
  syntheses by solid state reactions, 174
Insulating firebrick, 1294
Insulation of kilns, 894
Insulators, electrical, *see* Electrical insulators
  thermal, *see* Thermal insulators
Intaglio printing of transfers, 821, *822*
Intermittent kilns, 884, 967–994
Intrinsic solubility of lead glaze, 373
Inversion, high-low, *8*
Inversions, fast low-high, 167
Iridium colourants, 629
Iron, cast, 165
  -chrome brown colourants, 632
  chromite, 137
  dichromate, 148
  for crushing and grinding machinery, 670
  green colourants, 632
  impurity removal, 690–693
  -manganese brown colourants, 632
  orthosilicate, 138
  oxidation changes and thermal stresses, 225
  oxides, 148
    decomposition in firing, 863
    reversible reactions, 190
  pyrites, decomposition in firing, 862
  yellow to red colourants, 629
Isomorphous substitution, 9
Ivory colourants, 641

Jadeite, 5
Jar mills, 671, *672*
Jaspar ware bodies, 455
Jaw crushers, *663*
Jet drying, 842, 843
Jiggering, 721, *722*
Joining, 773
Jolleying, 721, *722*, *723*

SUBJECT INDEX 1443

Kaliophite–kalsilite inversion, 171
Kanthal resistors, 930
Kaolin, 27, 33–35
    deposits, 242
    purification, 245–253
    reactions, 191–192
    refractory bricks, 1241
    winning, 242–245
Kaolinite, *13*, *14*, 16, 22, 24, 27, 29. 53
Karbate bodies, 414
Karbate carbon and graphite ware, 1129
Karcite laboratory equipment, 1120, 1133
Kasseler kiln, 978
Kassler Braun, 154
Keene's cement, 159
Kerabedarf tunnel kilns, 1041
Kernite, 130
Kemite laboratory table tops, 1119, 1133
Kieselguhr, 97
Kiln atmosphere, 863
    furniture, 954, 1274
        bodies, 408
Kilns, 884–1061
    allied tunnels, 1041
    annular, 995, *996*
    archless, 997
    automatic stoking, 909
    bag wall, 907
    Barnett and Hadlington, 1001
    basket conveyors, 1030
    beehive, 987
    Belgian, 999, 1005
    Bock, 1001
    bottle, 973, *974*, 986
    Bricesco-Harrop tunnel, 1031
    Bricesco tunnel, 1048, 1049, 1052
    Brown's Patent, 1000
    burning zone of tunnel kiln, 1019
    cars, 1026
    cascade grates, 908
    chamber, 998
    clamp, 884
    coal-fired tunnel kilns, 1019
    construction materials, 885
    construction methods, 897
    continuous, 884, 985–1006
    conveyor belts, 1030
    cooling zone of tunnel kilns, 1025
    Dannenberg, 1001
    Davis tunnel, 1050
    Diesener, 1001
    down draught, 969, 978, *979*, 987
    draught establishment, 968

    draught measurement, 951
    Drayton down-draught tunnel, 1036
    Dressler tunnels, 1038–1041, 1051, 1052, 1053
    Dunnachie, 1002
    English, 1000
    electrically heated tunnel kilns, 1022
    electric kilns, 991
    Faugeron tunnel, 1031
    fire holes in kiln walls, 907
    firing holes in roof, 907
    firing schedule in intermittent kilns, 970
    flue gas analysis, 952
    frit, 1047
    fuel consumption of intermittent kilns, 971
    gas burners, 920
    gas-fired tunnel kilns, 1019
    gas-oil burners, *929*
    German, 976
    Gibbons top-fired tunnel, *1032–1034*
    Gibbons tunnels, 1051, 1053, 1054, 1055
    glut arch, 907
    Gottignies multi-passage, 1043, *1045–1046*, 1056
    grates for coal firing, 907
    Harrop tunnel, 1031
    heat-work recorders, 934
    Heurty multi-passage, 1056
    Hoffmann, 884, 985, *994*, 1005
    horizontal draught, 970
    Ideal, 999
    intermittent, 884, 967–994
    instruments, 934
    insulation, 894
    interconnecting intermittent, 984
    Kasseler, 978
    Kerabedarf tunnels, 1041
    kiln furniture, 954
    Koppers twin-track tunnel, 1037
    Lancashire, 1000
    main fire zone of tunnel kiln, 1019
    Manchester, 1000
    Mendheim, 1001, 1002
    Minton, 969
    Monnier top-fired tunnel, 1035
    Moore-Campbell electric tunnel, 1043, 1054, 1055
    muffle, 885, 970
    Newcastle, 978, 990
    oil-fired tunnel kilns, 1021

Kilns (contd.)
   Ooms-Ittner semi-muffle tunnel, 1038, *1039*
   open setting, 960
   periodic, 884, 967–994
   placing of ware, 885, 956
   preheating zone of tunnel kiln, 1018
   programme controllers, *949*
   propulsion for kiln cars, 1028
   pusher, 1026
   protective atmospheres in, 933
   refractories for kiln construction, 885
   refractory mortars, ramming mixtures and renderings, 890
   reverberatory frit, 1057
   Revergen tunnel, 1037, *1038*, 1051
   Robey, 969
   rotary, 1061
   rotary frit, 1057
   Rotolec electric enamel tunnel, 1043, *1044*, 1055
   saggars, 885, 954, 957
   Scotch, 976
   setting of ware, 885, 956
   shaft, *1058*, 1061
   Shaw, 1002, 1006
   Sherwin and Cotton, 990
   sparking plug kiln, 1043, *1047*
   Staffordshire, 1000, 1005
   step or cascade grates, 908, *909*
   suspended flat arches, 899
   temperature control, automatic, 948
   temperature measurement, 939
   test codes for performance and efficiency, 966
   testing furnaces, *982*
   thermoscopes, 934
   Thomas, 969
   top-fired tunnel, 1032–1035
   top hat, 980, *981*
   trolley hearth, 980, *981*
   truck kiln, 980, *981*
   trucks, 1026
   tunnel, 885, 1002–1056
   two-tier porcelain, 976, *977*
   up-draught, 969, 973, 986
   walking beam, 1029
   Webcot electric, 991
   Wilkinson, 969, 987
   Williamson down-draught tunnel, *1036*, 1949
   Williamson tunnel, 1035
   zigzag, 995, *997*
Kitchen ware, 1100
Kneading table, 708
Knopite, 140
Kochite, 22
Koppers twin-track tunnel kiln, 1037
Kunzite, 102
Kyanite, 110–112
   reaction, 192

Laboratory, ceramic, 265–392
   porcelain ware, *1115*, 1135
   ware, 1135–1169
Labradorite, 102
Lancashire kiln, 1000
Lava, 1181, 1201
Lead antimonate colourants, 634
   antimonates, 146
   chromate, 146
   chromate red colourant, 635
   compounds, 133
   frits and glazes, solubility determination, 373
   frit solubility with particle size, 779
   in glass structure, 214
   in glazes, 568–576
leadless glazes, 576–579
lead pencil, 1144
   silicates, 134
   titanate in bodies, 518
   zirconate-titanate, 1209
Lepidolite, 122, 124
Leverrierite, 22
Light-weight aggregate insulating blocks, 1285
Lignin, 154, 156
Lignite fuel, 902
Lime, 128
Lime-phosphoric acid glass, 208
Limestone, 116
Lining, hand decoration, *807*, 808
   machine decoration, 808
Liquid gold, 650
Litharge, 133
Lithia, 121
   porcelain electrical insulators, 1192, 1205
   porcelains and bodies, 500–503
Lithiophilite, 122
Lithium aluminosilicate bodies, 396, 500–503
   cobaltite, 146
   compounds, 121–125, 153
   in glass structure, 212
   silicate, inversion, 171
'Litho', 824

Lithographed transfers, 825
Livesite, 16, 29, 53
Loess, 11, 31
Loparite, 140
Loss angle, 1170
   factor, 1171
   test, 368
   on ignition for control tests, 382
Low-loss steatite glazes, 563
Low-solubility lead glazes, 569
Lubricants, 155, 158
   in dry pressing, 743
Lustres, 657–660

Machinery used in clay winning, 258–263
Magnesia alumina spinel refractory bricks, 1268
   electrical insulators, 1193
   reactions, 191
   sintered, ware, 1141, 1146, 1152
Magnesite, 117
   -chrome bodies, 422
   -chrome refractory bricks, 1266, 1272
   refractory bodies, 417
      bricks, 1254, 1258
Magnesium compounds, 126–127
   hydroxide, 118
   in glass structure, 212
   in silicate structures, 9
   metasilicate, 91
      in steatite bodies, 492
   orthosilicate, 138
   oxide, raw materials, 115–119
   phosphate bonds, 523
   silicate inversion, 171
   sulphate, 154
   titanate bodies, 517
      dielectric, 1196, 1207
   zirconium silicate, 145
Magnetic purification, 694
   separation of minerals in clay, 306
Magnets, 690, *691*, *692*–693
Majolica tableware, 1089
Malleable 347
   iron, 165
Manchester kiln, 1000
Manganese oxide, 148
   oxides, reversible reaction, 190
   pink, violet and brown colourants, 632, 635
   steel, 163
Manganic compounds, 148
Manganites by solid state syntheses, 179

Manganous compounds, 148
Mangles, 845, *846*
Marble, 116
Marls, 116
Matt glazes, 592–595
Maturing, 711
Maturing of the body during firing, 865
   temperatures, 866
   tower, *712*
Mechanical failure of brittle solids, 227
   plant, cost of, 259
   preparation of bodies and glazes, 661–683
   properties of fired ceramic bodies, 345–352
Melting, 198–201
   behaviour determination, 277
   point, congruent, 198
      incongruent, 198
Memory of clay, 65
Mendheim kiln, 1001, 1002
Metal bonded carbide ware, 1154
Metallic decoration, 647–657
   surface applications, non-decorative, 656
$MgO-Al_2O_3-SiO_2$ equilibrium diagram, *217*
Mica-clay, 239
Micas, Cornish clay pits, 239, *240*
Microchemical techniques, 266
Microcline, 102
Microcracks, 227
Microphotometer, *270*
Microscope in particle size analysis, 290
Microscopic examination of glazes, 374
   investigation of clays, 308
Microwave spectroscopy, 313
Mikroatomiser, 682
Mikropulveriser, 682
Miller automatic jigger, 727, *728*
Mineral content determination, 302–318
Mineralisers, 190
Mineralogical investigation of fired bodies, 338
Minerals, separation of, 303
Minton oven, 969
Mixing, 696, *698*–701
   ark, 701
Modelling, 718
Modulus of elasticity, 345
   of rupture (green) measurement, *336*
   of rupture test, 349
Mohs scale of hardness, 351

Moisture content determination, 275
　for determining minerals in clay, 307
　expansion of fired clay, 88
　movement tests, 344
Molasses, 156
Moler, 97
　thermal insulating bricks, 1284, 1286
Molochite refractory bricks, 1241
Molybdenum silicide resistors, 931
Monnier top-fired tunnel kiln, 1035
Montmorillonite, 17, *18–20*, 22, 24, 53
Moore-Campbell electric tunnel kiln, 1043, 1054, 1055
Mosaic, 1079
Mossite, 140
Motor brushes, 1211
Moulding, 718
Moulds, plaster of paris, 161–163
　synthetic plastics, 163
Muffle kilns, 885, 970
Mullite, 113, 405
　crystals, *200*
　formation, 191–192
　mineralisers for, 195–196
　porcelain bodies, 474
　　insulators, 1177, 1179
　　ware, 1135
　reaction, 192
　refractory bodies, 408
　　bricks, 1229, 1243
Multi-bucket, endless chain planer, 259, *262, 263*
Murray curvex printing machine, 819, *820*

Nacrite, *13*, 16, 22
Naphthalene for porous bodies, 428
Naples yellow, 634
National carbon and graphite ware, 1128
Natural gas fuel, 915
Neodymium colourants, 637
Nepheline, 106
　-carnegieite, inversion, 171
　syenite, 106
Network co-formers, 202, 212–215
　contractors, 212
　formers, 202
　modifiers, 202, 204–211
Neutral refractory bodies, 415
Newcastle kiln, 978, 990
Newtonite, 22
Nichrome resistors, 930
Nickel-chromium iron, 166
Nickel compounds, 148

Nitride bodies, 510
Nitride refractory ware, 1138, 1148
Nitrogen atmosphere in kilns, 934
Non-linear circuit elements, 1208
Nontronite, 17, 24
Normes Français relating to ceramics, list of, 1311

Ochres, 148
Offset printing decoration, 818
Oil fuel, 925
Oils, 155, 158
Oligoclase, 102
Olivine, 4, 5, 138
　reactions, 189, 191
　refractory bodies, 422
　　bricks, 1268, 1273
Once-fired ware, 854
On-glaze decoration, 799, 805
Oolitic limestone, 116
Ooms-Ittner semi-muffle tunnel kiln, 1038, *1039*
Opacifiers, 143–145
Opaque glazes, 585–591, 798, 803
Open-cast mining for kaolin, 244
　of ball clays, 253, *254*
Open-circuit grinding, 662
Optical radiation pyrometers, 947
Organic binders, 155–157
　deflocculants, 154
Original Pallman mill, 665
Ornamenting, 803, *804*
Orsat apparatus, 953
Orthoclase, 102
Orthoenstatite-clinoenstatite inversion, 171
Orthosilicates, 4, 5
Orton cones, 936, 1304
Over-feed stoking, 909
Ovenproof ware, 1089, 1100
Oxidation period of firing, 857

Painting, free-hand decoration, 805, *807*
Palladium, 148
Pandermite, 131
Pan grinding, 665
Paper transfers decoration, 819
Paraffin emulsion, 158
Parian ware, 455
Particle size analysis, 278–295
　distribution, clays, 298
　control tests, 384
　of clays, 297

SUBJECT INDEX 1447

Pâte sur pâte, 802
Paving brick pressing, 749
P.C.E., 277
Peat, 154
Pebble mills, 670, 671
Peeling, 551
Pegmatite dikes, 244
Pencil lead, 1144
Peptone, 156
Periclase bodies, 418
   refractory bricks, 1256, 1261
Periodic kilns, 884, 967–994
Periodic Table, 1300
Perlite, 108
   bodies, 426
Permanent linear change on reheating, 357
Permeability, 345
   (ease of magnetisation), 1172
   of clay, 66
   measurement, 332
Permeable porous bodies, 396, 521–522
Permittivity, 1170
Perowskite, 140
Petalite, 122, 124
Petrographic investigation in control tests, 384
Petroleum oils, 155, 158
pH and plasticity, 63–64
   measurement, 319
   of water, 151
   -recorder, continuous, *383*, 384
Phase diagrams, 215–223
   rule, 215
Phenacite, 5
Pholerite, 22
Phosphate bonding, 522–524
Phosphates, sodium, calcium, magnesium, 145
Phosphoric acid bonds, 523
Phosphorus pentoxide glass, 204
Photolithography for decoration, 825
Photosedimentation, 290
Physical tests, 275–295
Pickup of glaze slip, 369, 785
Piercing, 802
Piezoelectric coupling coefficient, 1171
   effect, 1171
   ware, 1208
Pin placing equipment, *963*
Pint weight, 700, 1302
Pipe machines, 736, *737*

Pipes
   drainage, field, agricultural, 1074
   saltglazed stoneware, 1074, *1076*
Pitches, 136
Placing of ware, 885, 956
Plagioclase, 102
Planers, 258
Plaskon thermalplastic, 156, 158
Plaster of Paris, 158–163
Plasticity, 62–66
   index, 327
   pH, 63–64
   measurement, 326–331
   surface tension, 65
   water of, 65
Plastic pressing, 719
   raw materials, 9–95
   shaping methods, 717–740
Plate shaping, 721, *722*, 726–728
Platinum, 148
   bright, 655
   grey, 654
Plumbago bodies, 416
Pneumatic spades, 253, *254*
Podmore-Boulton Vibro Energy Mill, 683
Polarograph, 275
Polarographic analysis for control tests, 382
Polishing glost fired ware, *1059*
Pollucite, 102
Polyester moulds, 163
Polymorphism, 167
Polyvinyl alcohol, 156
   chloride moulds, 163
Porcelain bodies, 395, 396, 451–479
   chemical ware, 1113
   insulators, 1174, 1178
   kiln, 976, *977*
   tableware, 1099
Pore distribution and surface texture, 345
   size, 344
Porosity, 342
Porous permeable bodies, 396, 521–522
Potash-silica glass, 206
Potassium antimonates, 146
   chromates, 146
   compounds, 126
   permanganate, 148
   silicates, inversions, 171
Pot mills, 671, *672*
Potters' stoves, 843, *845*
Potter's wheel, 720
Pottery, 1307
   bodies, 396

Pottery (*contd.*)
    (Health and Welfare) Special Regulations (1950), 373, 1324
    oils, 155
    works control, 385–387
Power factor, 1171
    test, 368
    shovels, 258, *261*
Praseodymium phosphate, 148
Press dust, preparation methods, 745, 1079
Pressing dies, 165
    dry and semi-dry, 741
    hot, 752
    plastic, 719
    stiff plastic, 740
Presses, 746, *747*, *748*, *749*
Process control, 380–392
Producer gas fuel, 915
Propane fuel, 920
Proportionating, *699*
Prospecting for clays, 236
Protective atmosphere in kilns, 933
    colloids, 154
Protoenstatite, 91
Pugging, 708–711
Pugmills, 708–711
Pumice, 107
Pumps, sliphouse, 713
Pure oxide bodies, 396, 505
    ware, 1138–1144, 1146, 1151
Purification of kaolin, 245–253
Purple of Cassius, 649
Pyrolusite, 148
Pyrometric cone equivalent, 277
    cones, *935*, 1304
Pyrophyllite, 22, 93–95
    bodies, 396, 500
Pyroxenes, 5

Quarry tiles, 1075
Quartz, 96, 168
    reactions, 191
    structure, 5
Quarzite, 96
Q value, 1171

Radiant heat for drying, 836
Radiation pyrometers, 946
Ram process, 719
Rare elements, 142
Rasorite, 130
Rational analysis of clay, 299
Raw materials, 3–166
    control tests, 380

Reactions, incomplete and complete, 197
Rearing, 962
Receptivity of ware, 785
    to glaze slip, 369
Recording of laboratory tests, 265
Red lead, 133
Refining tanks for Cornish china clay, 242
Refractive index, 308
Refractories, bodies, 393, 400–424
Refractories, constructional, 1212–1283
    alumina bricks, 1230, 1245
    aluminous firebrick, 1229, 1235
    barium aluminium silicate bricks, 1247, 1249
    barium-anorthite bricks, 1247, 1249
    carbon blocks, 1249, 1252
    checker bricks, 1230
    chrome bricks, 1267, 1272
    chrome magnesite bricks, 1264, 1270
    corundum bricks, 1230
    dolomite, semi-stable, bricks, 1257, 1263
    dolomite, stable, bricks, 1257, 1263
    fireclay bricks, 1226, 1234
    forsterite bricks, 1269, 1273
    glass tank blocks, 1231, 1236
    graphite, 1250
    heat confinement of, 1216
    interaction of, 1214
    kaolin bricks, 1241
    magnesia alumina spinel bricks, 1268
    magnesite bricks, 1254, 1258
    magnesite chrome bricks, 1266, 1272
    molochite bricks, 1241
    mullite bricks, 1229, 1243
    olivine bricks, 1268, 1273
    periclase brick, 1256, 1261
    resistance to destruction of, 1213
    semi-silica bricks, 1224, 1230, 1234
    silica brick, 1218, 1230
    silicon carbide bricks, 1251, 1254
    silicon nitride bonded silicon carbide, 1251
    silimanite bricks, 1241
    spalling of, 1215
    spinel bricks, 1268
    stabilised zirconia bricks, 1246, 1248
    strength of, 1212
    super-duty firebrick, 1229, 1236
    super-duty silica brick, 1220
    zircon bricks and hollow-ware, 1247, 1248
    zirconia, stabilised bricks, 1246, 1248

SUBJECT INDEX

Refractories for kiln construction, 885
    for very high temperatures, 890
    pressing, 750
    resistance to carbon monoxide measurement, 364
        to slag attack measurement, 363
    works control, 387–391
Refractoriness measurement, 356
    under load measurement, 356
Refractory/glass reactions, 1039
Refractory/refractory reactions, 1214
Refractory hollow-ware, 1274–1283
    crucibles, 1275
    glass-house pots, 1281
    kiln furniture, 1274
    retorts, 1282
    roasting dishes, 1275
    saggars, 1274
    scorifiers, 1275
Refractory mortars, ramming mixtures, renderings, 890
    oxides, 890
    raw materials, 134–141
Rejafix machine decoration, 818
Relative specific surface, 294
Relief work, 798, 802
Replacement leaching, 318
Research, 265
Residual losses, 1172
Resin, 136
Resistance thermometers, 943
Resistivity, 1171
    electrical, test, 369
Resistors for electrical heating, 930
    silicon carbide, 1210
Retorts, 1282
Revergen tunnel kiln, 1037, *1038*, 1051
Reversible reactions, 190
Richardson down-draught tunnel kiln, *1036*
Ring roll mills, 669
    test for glaze fit, 376
Robey kiln, 969
Rocket motor ceramic parts, 1165
Rod mills, 670, 673, 676
Roller bat-machine for roofing tiles, 730
    flat ware brushing machine, 786
    particle size air analyser, *284*
    plate-making machine, 727
Roofing tiles, 1072
    tile shaping, 717, 730, 736
Rosin emulsion, 156
Rotolec electric enamel tunnel kiln, 1043, *1044*, 1055

Rotary dryer for china clay, *242*
    vacuum filter, 249
Rubber stamping decoration, 809, *810*
Rubbing test, 352
Rutile, 140, 144, 148, 170
    bodies, 515
    body thread guides, 1113
    dielectrics, 1193, 1206
Ryckman edge-lining machine, 809

Saddles, *962*
Saggar bodies, 408
Saggars, 885, 954, 957, 1274
Saggar shaping, *719*
Salt glazes, 598–605
Salt-glazed stoneware pipes, 1074, *1076*
Salting mixture, 600
Sampling, clays, 297
Sand, 96
    blast test, 351
Sanding, 803, *804*
Sand pits, Cornish clay pits, 238
Sandstone, 96
Sanitary ware, 1084–1088
    earthenware, 1084
    fireclay, 1087
    stoneware, 1088
    vitreous china, 1085
Sapphire ware, 1139
Saponite, 17
Sauconite, 17
Scalloping, 740
Schöne elutriator, *281*
Schroetterite, 22
Schulze-Harkort elutriator, *282*
Scotch kiln, 976
Scrap, re-use, 713
Scrapers, 258
Screen-printed (paper-backed) transfers, 826
Screen-printing decoration, 813
Screens, *685*, 686, *688*
Sea-water/dolomite process, 118
    magnesia, 118
Sedimentary clays, winning, 253–258
Sedimentation, 283
Seger cones, 935, 1304
    porcelain, 451
Selenium, 148
Semi-conducting glazes, 562
Semi-dry pressing, 741–751
Semi-silica refractory bodies, 404
    bricks, 1224, 1230, 1234

Semi-stable dolomite bodies, 419
Semi-stiff mud shaping of bricks, 718
Semi-vitreous china bodies, 395, 439–450
   tableware, 1089, 1095
   porcelain bodies, 395
      tableware, 1089
Sericite, 95
   pyrophyllite, 95
Serpentine-dolomite mixtures, 419
Setting of ware, 885, 956
   time of plaster, 160
Settling pits, Cornish clay pits, 239, 240
Sgraffito, 802
Shaft kiln, *1058*, 1061
Shaft mining for ball clays, 253, *255*
Shaft mining for kaolin, 244
Shale planer, 259, *262*
Shales, winning, 258
Shaping of ware, 716–776
   Allen semi-automatic jigger, 726
   automatic jiggering, 726
   batting, 724
   casting, 754–771
   casting behaviour, 760
   casting control, 770
   casting defects, 767
   casting slip, preparation and use, 764
   deflocculation of casting slip, 755
   drain casting, 762
   dry and semi-dry pressing, 741–751
   dust extrusion, 738
   dust pressing, 741
   ethyl silicate casting, 771
   extrusion, 729
   fettling, *773*
   handles, 771
   handling, *774*
   handling machine, *775*
   hand modelling, 718
   hand moulding, 718
   hand throwing, 720, *722*
   hollow casting, 762
   hot pressing, 752–754
   jiggering, 721, *722*
   joining, 773
   jolleying, 721, *722*, 723
   Miller automatic jigger, 727, *728*
   plastic, 717–740
   plastic pressing, 719
   pipe machines, 736, *737*
   potter's wheel, 720
   ram process, 719
   roller bat-machine for roofing tiles, 730
   roller plate-making machine, 727
   semi-stiff mud, 718
   soft-mud, 717
   soft-plastic, 717
   solid casting, 762
   sponging and smoothing, *772*
   sticking up, 773
   stiff-plastic extrusion-pressing for bricks, 740
   Strasser handling machine, *775*
   tamping, 751
   throwing, 720, *722*
   turning, 738, *739*
   wire-cut bricks, 733, *734*
Sherwin and Cotton kiln, 990
Shrinkage, drying, 67
   measurement, 333
   equations for, 335
   firing, 79
   measurement, 337
   of porcelain, glaze influence on, 551
   rate of, firing, 867
Shaw kiln, 1002, 1006
Sierralite, 120
Sieves, 686
Sieving, 279, *280*
Silex bodies, 426
   thermal insulating bricks, 1284, 1288
Silica, 96–101
   and boric oxide glasses, 203, 211
   α-β inversion, *8*
   bricks, reactions, 189
   eutectics, 402, 405
   glass, 203
   modifications, 169
   polymorphism, 167–169
   refractory bodies, 401
      bricks, 1216, 1230
   structure, 5
Silicate bands, 5, *6*
   chains, 5, *6*
   chemistry, 3–9
   layers $(Si_2O_5)n$, 7
   rings, 5
   sheets $(Si_2O_5)n$, 5, 7
Silicates by solid state syntheses, 174
   refractory, 509
Siliceous firebrick bodies, 405
Silicides, refractory, 511

SUBJECT INDEX 1451

Silicon carbide, 136
  bodies, 410, 411
  electrical heating elements, 1210
  grinding wheels, 1157
  refractory bricks, 1251, 1254
  resistors, 930
  ware, 1137, 1148
  nitride bonded silicon carbide refractories, 1251
    ware, 1137
Silk-screen printed decoration, 813
Sillimanite, 110–112
  reactions, 192
  refractory bricks, 1241
Sillimanites, 5
Sillimanite ware, 1135
Silver, 148
  compounds, 148
  marking, 881
  metallic, 655
Sinter alumina ware, 1138, 1146, 1151
Sintered beryllia ware, 1139, 1146, 1152
  magnesia ware, 1141, 1146, 1152
  spinel ware, 1141, 1153
  thoria ware, 1142, 1146, 1153
  titania ware, 1141, 1147, 1153
  uranium dioxide ware, 1142
  zirconia ware, 1141, 1146, 1153
  zircon ware, 1144, 1153
Sinter-hearth machine, 429
Sintering, 171
$(SiO_2)n$ network, 5
SiO_4 tetrahedron, *4*, 5
Si_2O_7 unit, 5
Si_3O_9 unit, 5
Si_6O_{18} unit, 5
Size grading, 686–689
  of ceramic products, 341
Skimmers, 259, *261*
Slag attack, refractories resistance to, measurement, 363
Sliphouse pumps, 713
Slip-trailing, 802
Smoothing, 772
Snakeskin glazes, *580*, 581
Soda ash, 153
Soda-silica glass, *205*
Sodium antimonate, 144
  chromates, 146
  compounds, 122–126, 153
  diborate, 130
  phosphates, 145
  selenate, 148

  selenite, 148
  silicate, inversion, 171
  uranate, 150
Soft-mud shaping of bricks, 717
Soft-plastic shaping of bricks, 717
Soft porcelain bodies, 395, 451–457
Solid solutions, 173
  state reactions, 173–196
Solubility of lead frits and glazes, determination, 373
'Soluble' oils, 157
Soluble salts in clays, 249
  in fired ceramic bodies, determination, 361
  sulphates in clay, determination, 322
Sonic spalling test, *359*
Spacers for vacuum tubes, 1206
Spalling of refractories, 1215
  tests, 358
Sparking plug insulators, 1180, 1199
Specific gravity, 342
  determination, 276
  heat, 353
Specific heats, calcined alumina, 870
  calcined clays, 871–872
  fired silica brick, 873
  magnesia, 871
  quartz, 870
  silicon carbide, 871
Specific resistance, 1171
  surface, relative, 294
Spectrograph, *269*
Spectrographic analysis for control tests, 381
Spectrophotometric analysis, 267, *273*
Spekker, 267, *268*
Spinel, as colouring agents, 229
  bodies, 396, 499
  formation, 189, 191
  refractory bricks, 1268
  sintered ware, 1141, 1153
Spinels, refractory, 508
'Spit out' or 'Spitting', 881
Spodumene, 5, 102, 122
Sponging, 772
  glaze, 796
Spray drying, 707
Spraying decoration, 806
  glazes, 791, *793*, *794*
Sprig work, 803
Spurs, *962*
'Spurting', 881
Squeegee oil, 156

Stabilised zirconia refractory bricks, 1246, 1248
Stable dolomite bodies, 419
Staffordshire cones, 936, 1304
    kiln, 1000, 1005
    Seger cones, 936, *935*, 1304
Standard test sieves, 1301
Stamping decoration, 809, *810*
Stannates by solid state syntheses, 175
Stannic oxide, 144, 148
Starches, 156–158
Steatite, 89–93
    bodies, 396, 487–498
    electrical insulators, 1183, 1201
    ware, 1136
Stearates, 155–158
Steels, 163–166
    for crushing and grinding machinery, 669
Stencilling decoration, 811
Stevensite, 17
Sticking up, 773
Stiff-plastic brick machine, 718
    extrusion-pressing for bricks, 740
Stilts, *962*
Stock bricks, 1066
Stokes law, 279
Stoking, automatic, 909
    over-feed, 909
    under-feed, *910*
Storage, 689
Stone, *see* Cornish Stone
Stone preparation, 696
Stoneware bodies, 393, 430–438
    clays, 28, 44–45
    electrical insulators, 1179
    general methods of manufacture, 1103–1106
    in agriculture, 1112
    kitchen ware, 1100
    pipes, 1074, *1076*
    properties, 1122
    sanitary ware, 1088
    tableware, 1089
Stove tiles, 1082
Strasser handling machine, *775*
Strength, 347
Strontia, 129
Strontianite, 128
Strontium chromate, 146
    compounds, 128–129
    titanate bodies, 517
Sulphates, decomposition in firing, 860
    See also Soluble Salts
Sulphides, refractory, 511
Sulphite lye, 154, 157
Sulphur, 148
    dioxide in kiln atmosphere, 864
Super-duty firebricks, 1229, 1236
    silica refractory bodies, 402
    bricks, 1220
Surface factor, 292, 683
    resistivity, 1171
    tension, 65
    of glazes, 538
Swedish bone china hotel ware, 1098

Tableware bodies, 1089
Talc, 89–93
    natural or phosphate bonded, electrical insulators, 1181, 1201
    reactions, 191–192
Tamping, 751
Tannic acid, 154, 157
Tantalum in glass structure, 213
Tapiolith, 140
Tars, 136
Technical ware, 1103–1134
Temperature conductivity, 354
    control, automatic, 948
    measurement, 939
Tensile strength tests, *348*, 349
Terracotta bodies, 400
Test schedule for ceramic products, 366–367
Testing, ceramic, 265–392
    furnaces, 982
Textile industry chemical stoneware, 1112
Thallium in glass structure, 214
Thermal analysis, 313–316
    conductance, 355
    conductivity measurement, 353, 354
    of ceramic dielectrics, 1169
    expansion hysteresis, *226*
    measurement, *352*
    fatigue, 224
    gradients, 226
    insulation bodies, 393, 425–430
Thermal insulators, 1284–1297
    diatomite bricks, 1284, 1287
    expanded clay blocks, 1285, 1290
    foamclay, 1290
    high temperature insulating brick, 1294
    hot-face insulation, 1294
    insulating firebrick, 1294
    light-weight aggregate blocks, 1285
    moler bricks, 1284, 1286

SUBJECT INDEX

silex bricks, 1284, 1288
vermiculite bricks, 1285, 1286
Thermal properties measurement, 352, 361
  ratcheting, 225
  shock, 223–226
    resistance, 357
  shock-resistant ceramics, 1165
    porcelain bodies, 469
    stoneware bodies, 434
  stresses, 223–226
  resistance, 354
  transmittance, 354
Thermocouples, 943
Thermoelectric thermometers, 943
Thermometers, 939
Thermoscopes, 934, 1304
Thimbles, *962*
Thin sections, 308
Thixotropy, 59–62
Thomas kiln, 969
Thoria bodies, 508
  sintered ware, 1142, 1146, 1153
  -urania ware, 1143
Thorium oxide, 141
Thread guides, rutile body, 1113
Throwing, 720, *722*
Tiles
  decorated floor, 1078
  decorated wall, 1082
  exterior wall, 1082, *1083*
  faience, 1082
  floor, 1075
  frostproof, 1082
  hollow, 1069
  interlocking, 1072
  plain, 1072
  properties, 1073
  quarry, 1075
  rib and tile floor construction, 1069
  roofing, 1072
  stove (ofenkacheln), 1082
  wall, 1079
Tincal, 130
Tin oxide, 144, 148
  resistor, 932
Titanate bodies, 396, 516–520
Titanate dielectrics, 1196
Titanates by solid state syntheses, 176
Titania, 141, 148
  polymorphism, 170
  porcelain dielectrics, 1194
  ware, 1141, 1147, 1153
Titanite, 140

Titanium carbide ware, 1137, 1148, 1150
  compounds, 140, 144
  in glass structure, 213
  nitride bodies, 510
Topaz, 113
Top-fired tunnel kilns, 1032–1035
Tophat kiln, 980, *981*
Torsional strength test, 349
Total radiation pyrometers, 946
Tough, 347
Town gas fuel, 914
Transfers, application for decoration, 826, *828*, *829*
  monochrome, for decoration, 821
  paper, for decoration, 819
  polychrome, for decoration, 824
Translucent china, 1096
Transverse green strength measurement, 335
  strength test, 349
Tricalcium phosphate, 108–109
Tridymite, 97, 168
  structure, 5
Triphylite, 122
Trolley hearth kiln, 980, *981*
Truck kiln, 980, *981*
Tube mills, 670, *672*, 675
Tungsten carbide for cutting [tools, 515
  ware, 1137, 1148
  compounds, 150
Tunnel dryers, 846, *850*
Tunnel kilns, 885, 1002–1056
Turbidimeter, 289
Turbo-jet blades, 1164
Turning, 738, *739*
Two-tier porcelain kiln, 976, *977*
Tylose, 156

Ultramicroscope, 291
Ultraviolet light for microscopes, 309
Under-feed stoking, *910*
Under-glaze decoration, 798, 805
  spraying of hard porcelain decoration, 806
Union stone, 93
United States kaolin deposits, 244
Universal excavator, *261*
Up-draught kilns, 969, 973, 986
Uranium brown and yellow colourants, 638
  compounds, 150
  dioxide sintered ware, 1142
  red colourant, 637

Urea, 157
U value, 354

Vacuum filtration, 705, *706*
  tube spacers, 1206
Vanadium pentoxide, 150
  yellow, green and blue colourants, 639
Vapour glazing, 605
Vermiculite, 141
  bodies, 426
  insulating bricks, 1285, 1286
Viridine, 110
Viscometers, 324
Viscosity measurement, 323
  of glazes, 536
  tests, 370
Vitreous china bodies, 395, 450–455
  sanitary ware, 1085
  structure, *8*
Volatilisation of glaze constituents, 540
Volcanic ash, 107
Volume resistivity, 1171

Wall tile dust pressing, 746
  tiles, 1079
Washing soda, 153
Water, 150–152
  absorption tests, 343
  boiling point of, 151
  chemically held, in the firing, 856
  gas fuel, 915
  glass, 153
  hygroscopic, in the firing, 856
  jets, Cornish clay pits, *240*
  marking of decorated ware, 829
  mechanical, in the firing, 856
  re-use, 713
  -slide, transfers for decoration, 830
  smoking period of firing, 855
  -spotting of decorated ware, 829
Watkins recorders, 938, 1304
Wax emulsions, 157, 158
Weathering of clays, 263
Webcot electric kiln, 991
Wedging, 708–711
Wet classifiers, 687, *688*
Wet-pan grinding mill, 667
White colourants, 640
  lead, 133
  ware bodies, 396
Whiting, 116
Wiegner-Gessner sedimentation apparatus, *287*
Wiegner pipette, 287

Wilkinson kiln, 969, 987
Williamson tunnel kiln, 1035, 1049
Winning of clay, 236–264
Wire-cut bricks, 733, *734*
Witherite, 129
Wollastonite, 5, 119
  bodies, 396, 500
  electrical insulators, 1192, 1206
  -pseudowollastonite inversion, 171
Wonder stone, 93
Wood fuel, 902
Workability index, 327
Working range of plastic clay, 65
Works control in the fine ceramic industry, schedule, 385–387
  on refractory making, schedule, 387–391

X-ray diffraction, 309, *312*
  for control tests, 382
  goniometer, 311
  inspection of green and fired goods, 341
X-rays in particle size analysis, 291
X-ray spectrographic analysis for control tests, 382

Young's modulus of elasticity of glazes, 547

Zeolites, structure, 5
Zettlitz kaolin deposits, 242
Zigzag kilns, 995, *997*
Zinc chromate, 146
  compounds, 132–133, 144
  in glass structure, 213
  oxide bodies, 508
  spinel, 144
Zinnwaldite, 122
Zircon, 134, 144
Zirconates by solid state syntheses, 177
  refractory, 509
Zirconia, 135, 144
  bodies, 508
  cubic, 170
  polymorphism, 170
  resistors, 932
  sintered ware, 1141, 1146, 1153
  stabilised, 170
  stabilised, refractory bricks, 1246, 1248
Zirconium carbide ware, 1137, 1151
  compounds 134–136, 144–145

in glass structure, 213
spinel, 145
Zircon porcelain bodies, 478
  glazes, 563
  insulators, 1191, 1205
  ware, 1136
  refractory bricks and hollow-ware, 1247, 1248
  ware, 1144, 1153
Zirkite, 135

STOKE-ON-TRENT
PUBLIC
LIBRARIES

HORACE BARKS
Reference Library

STOKE-ON-TRENT
PUBLIC
LIBRARIES

HORACE BARKS
Reference Library